SYSTEM IDENTIFICATION
(SYSID'03)

A Proceedings volume from the 13th IFAC Symposium on System Identification, Rotterdam, The Netherlands, 27 – 29 August 2003

Edited by

P.M.J. Van den HOF
Delft Center for Systems and Control,
Delft University of Technology,
Delft, The Netherlands

B. WAHLBERG
Royal Institute of Technology,
Stockholm, Sweden

S. WEILAND
Department of Electrical Engineering
Eindhoven University of Technology,
Eindhoven, The Netherlands

(In four volumes)

Volume 1

Published for the

INTERNATIONAL FEDERATION OF AUTOMATIC CONTROL

by

ELSEVIER LTD

ELSEVIER Ltd
The Boulevard, Langford Lane
Kidlington, Oxford OX5 1GB, UK

Elsevier Internet Homepage
http://www.elsevier.com

Consult the Elsevier Homepage for full catalogue information on all books, journals and electronic products and services.

IFAC Publications Internet Homepage
http://www.elsevier.com/locate/ifac

Consult the IFAC Publications Homepage for full details on the preparation of IFAC meeting papers, published/forthcoming IFAC books, and information about the IFAC Journals and affiliated journals.

First edition 2004

Library of Congress Cataloging in Publication Data

A catalogue record for this book is available from the Library of Congress

British Library Cataloguing in Publication Data

A catalogue record for this book is available from the British Library

ISBN 0-08-043709 5
ISSN 1474-6670

Printed and bound in the United Kingdom
Transferred to digital print 2010

To Contact the Publisher

Elsevier welcomes enquiries concerning publishing proposals: books, journal special issues, conference proceedings, etc. All formats and media can be considered. Should you have a publishing proposal you wish to discuss, please contact, without obligation, the publisher responsible for Elsevier's industrial and control engineering publishing programme:

Christopher Greenwell
Publishing Editor
Elsevier Ltd
The Boulevard, Langford Lane
Kidlington, Oxford
OX5 1GB, UK

Phone: +44 1865 843230
Fax: +44 1865 843920
E.mail: c.greenwell@elsevier.com

General enquiries, including placing orders, should be directed to Elsevier's Regional Sales Offices – please access the Elsevier homepage for full contact details (homepage details at the top of this page).

13th IFAC SYMPOSIUM ON SYSTEM IDENTIFICATION (SYSID 2003)

Sponsored by
International Federation of Automatic Control (IFAC)
IFAC Technical Committees on:
- Modeling, Identification and Signal Processing (MISP)
- Adaptive Control and Tuning (ACT)

Co-sponsored by
IEEE Control Systems Society
Division of Automatic Control (MRBT) of the Royal Institution of Engineers in The Netherlands (KIVI)
The Netherlands Organisation of Scientific Research (NWO)
Royal Netherlands Academy of Arts and Sciences (KNAW)
Dutch Institute of Systems and Control (DISC)
Faculty of Applied Sciences, Delft University of Technology (TUD)
Delft Center for Systems and Control (TUD)
Department of Electrical Engineering, Eindhoven University of Technology, The Netherlands (TU/e)
Stichting Meten en Regelen ER-THE, The Netherlands

Organizing Committee
P.M.J. Van den Hof – Delft University of Technology, Delft, The Netherlands
B. Wahlberg – Royal Institute of Technology, Stockholm, Sweden
S. Weiland – Eindhoven University of Technology, Eindhoven, The Netherlands

IPC Task Force
M. Deistler
M. Gevers
L. Ljung
M. Morari
J. Schoukens
P.M.J. Van den Hof
M. Viberg
B. Wahlberg

International Programme Committee (IPC)
P.M.J. Van den Hof; The Netherlands (Co-Chair)
B. Wahlberg, Sweden (Co-Chair)

P. Albertos; Spain
B. Anderson; Australia
E. Bai; USA
M. Basseville; France
R. Bitmead; USA
S. Bittanti* **; Italy
M. Blanke; Denmark
J. Bokor; Hungary
M. Campi; Italy
H.F. Chen; P.R. China
J. Chen; USA
R. de Callafon; USA
M. Deistler; Austria
B. de Moor; Belgium
J.J. Fuchs; France
K. Godfrey; UK

G. Goodwin; Australia
M. Gevers; Belgium
P. Guillaume; Belgium
L. Guo; P.R. China
H. Hjalmarsson; Sweden
H. Kimura; Japan
R. Kosut; USA
V. Krishnamurty; Australia
K. Kumamaru; Japan
I. Landau; France
J.H. Lee; USA
L. Ljung; Sweden
P. Mäkilä; Finland
T. McKelvey; Sweden
M. Milanese; Italy
M. Morari; Switzerland
B. Ninness; Australia
R. Ortega**; France
G. Picci; Italy
R. Pintelon; Belgium
B. Polyak; Russia
P. Regalia; France
D. Rivera; USA
W. Scherrer; Austria
J. Schoukens; Belgium
R. Schumann; Germany
R. Smith*; USA
T. Söderström*; Sweden
T. Sugie; Japan
R. Tempo; Italy
J. van Schuppen; The Netherlands
M. Verhaegen; The Netherlands
S. Veres; UK
M. Viberg; Sweden
A. Vicino; Italy
E. Walter; France

* *Appointed by IFAC Technical Committee MISP*
** *Appointed by IFAC Technical Committee ACT*

National Organizing Committee (NOC)

P.M.J. Van den Hof (Finances, contacts NMO, Public Relations)
S. Weiland (PC Secretariat, Paper handling, Website)
A.C.P.M. Backx (Industrial participation, Sponsors)
M.H.G. Verhaegen (Publications)
Y. Zhu (Exhibitions)
T. Van der Weiden (Local arrangements, Technical and Social Events)

PREFACE

These Proceedings contain all the technical material presented at the 13th IFAC Symposium on System Identification (SYSID 2003), held in the Conference Center "De Doelen", Rotterdam, The Netherlands from 27 – 29 August 2003.

The SYSID symposium is organized every three years and is among the most successful symposia organized by IFAC. This has been the first SYSID symposium in the 3rd millennium and the second SYSID symposium to take place in The Netherlands, following The Hague symposium in 1973.

Being the only worldwide symposium that is fully directed towards system identification, it is the ideal opportunity for researchers and industrial engineers from very many disciplines to present and discuss the developments, the results and the future challenges in all aspects of modelling dynamical systems on the basis of experimental data.

The symposium covered all major aspects of system identification, experimental modelling, signal processing and adaptive control from theoretical and methodological developments to practical applications in a wide range of application areas. For the 13th edition of this symposium, the International Program Committee has taken steps to position SYSID 2003 as a meeting place where scientists and engineers from several research communities can meet to discuss issues related to these areas.

A total of 350 delegates from 40 different countries attended the conference. 100 of the participants were PhD students, showing that system identification is a very vital field of research. Out of a total of 422 papers that were submitted to SYSID 2003, the IPC selected 333 papers and these were incorporated in the final program. The selection was based on two referee reports per paper. The final program of the symposium was composed of 3 plenary papers, 6 semi-plenary papers, 232 papers in oral sessions, 82 posters and 10 software demonstrations. The Preprints of this Symposium appeared on CD-ROM and were distributed among the participants of the symposium. The Proceedings of SYSID-2003 contain 321 papers.

We hope that you, as reader or as researcher in the area of System Identification, will find the contents of these Proceedings useful and informative for your professional work.

We would like to thank all members of the International Program Committee (IPC), members of the IPC Taskforce and members of the National Organizing Committee for their work in the organization of this symposium and in the preparation of these Proceedings. We would also like to thank many friends and colleagues for their help and support in many practical matters related to SYSID 2003.

The editors,

Paul Van den Hof
Bo Wahlberg
Siep Weiland.

CONTENTS

VOLUME 1

PLENARY PAPER
FROM EXPERIMENTS TO CLOSED-LOOP CONTROL

From Experiments to Closed Loop Control 1
H. HJALMARSSON

IDENTIFICATION FOR CONTROL

Exploratory Modelling for Controller Optimization 15
S.M. VERES

Connecting PE Identification and Robust Control Theory: The Multiple-Input Single-Output Case.
Part I: Uncertainty Region Validation 21
X. BOMBOIS, P. DATE

Connecting PE Identification and Robust Control Theory: The Multiple-Input Single-Output Case.
Part II: Controller Validation 27
X. BOMBOIS, P. DATE

Relation Between Uncertainty Structures in Identification for Robust Control 33
S.G. DOUMA, P.M.J. van DEN HOF

Strong Robustness Measures for Sets of Linear SISO Systems 39
M. CADIC, S. WEILAND, J.W. POLDERMAN

Using a Sufficient Condition to Analyze the Interplay Between Identification and Control 45
H. HJALMARSSON, H. JANSSON

NONLINEAR IDENTIFICATION

Structure Selection with ANOVA: Local Linear Models 51
I. LIND, L. LJUNG

On Identification of Hammerstein Systems Using Excitation with a Finite Number of Levels 57
T. MCKELVEY, C. HANNER

Fast Approximate Identification of Nonlinear Systems 61
J. SCHOUKENS, J. NEMETH, P. CRAMA, Y. ROLAIN, R. PINTELON

Gaussian Processes Framework for Validation of Linear and Nonlinear Models 67
A. LUNDGREN, J. SJÖBERG

Functional Analytic Framework for Model Selection 73
M. SUGIYAMA

Robust Complexity Criteria for Nonlinear Regression in NARX Models 79
J. de BRABANTER, K. PELCKMANS, J.A.K. SUYKENS, B. de MOOR, J. VANDEWALLE

IDENTIFICATION OF MIMO COMMUNICATION CHANNELS

Analysis of MIMO Channel Measurements 85
G. Del GALDO, M. HENNHÖFER, M. HAARDT

Performance Evaluation of MIMO Channel Prediction Algorithms Using Measurements 91
T. SVANTESSON, J.W. WALLACE

High-Resolution Channel Parameter Estimation for Communication Systems Equipped with Antenna Arrays 97
B.H. FLEURY, X. YIN, P. JOURDAN, A. STUCKI

Analysis of Spectral-Based Localization of Spatially Distributed Sources 103
M. TAPIO, M. VIBERG

Ray Tracing Interpretation of Multiple-Input Multiple-Output Wireless Systems 109
P.F. DRIESSEN

Computationally Efficient Blind MMSE Receivers for Long Code WCMDA Using Time-Varying Systems Theory 115
A.-J. van der VEEN, L. TONG

ESTIMATION IN PHYSICAL AND MEDICAL SYSTEMS

Maximum Likelihood Identification of Quantum Systems for Control Design 121
R.L. KOSUT, H. RABITZ, I.A. WALMSLEY

Maximum Likelihood Estimation of Signal Amplitude and Noise Variance from Complex Valued Data 127
A.J. den DEKKER, J. SIJBERS

Reliable Nonlinear Identification in Medical Applications 133
L.Y. WANG, G.G. YIN, H. WANG

Pattern Recognition of EEG Signals During Right and Left Motor Imagery 139
K. INOUE, G. PFURTSCHELLER, C. NEUPER, K. KUMAMARU

From Dynamic Metabolic Modeling to Unstructured Model Identification of Complex Biosystems 145
J.E. HAAG, A.VANDE WOUWER, P. BOGAERTS

Flow Controlled Non-Invasive Ventilation Considering Mask Leakage and Spontaneous Breathing 151
F. DIETZ, A. SCHLOßER, D. ABEL

STOCHASTIC SYSTEMS

Estimation and Identification of Non-Stationary Functional Series TARMA Models 157
A.G. POULIMENOS, S.D. FASSOIS

Modelling Multivariate Pollutant Time Series with Wavelet Functions 163
G. NUNNARI, D. LONGO

Estimating the Lyapunov Exponents of Chaotic Time Series Based on Polynomial Modelling 169
M. ATAEI, A. KHAKI-SEDIGH, B. LOHMANN

Sampling Density Design for Particle Filters 175
M. ŠIMANDL, O. STRAKA

Diffusive Representation of *N*-Th Order Fractional Brownian Motion 181
J. SEMBIRING, K. SOEMINTAPOERA, T. KOBAYASHI, K. AKIZUKI

APPLICATIONS OF SYSTEM IDENTIFICATION

Multi-Channel Active Noise Control for Uncertain Secondary Channels 187
Y. OHTA, H. OHMORI, A. SANO

Channel Estimation and Coupling Wave Cancellation in OFDM Relay Station 193
L. SUN, A. SANO

Application of System Identification for the Prediction of Avalanche Hazard 199
J. MILEK, B. BRABEC

Models for Incoming Calls Forecasting in a Customer Attention Center 205
M.R. ARAHAL, P.P. FERNANDO, E.F. CAMACHO

Modeling the Relationships Between the Users DB and the Web-Log File of a Large Virtual Community 211
S.M. SAVARESI, S. GARATTI, S. BITTANTI

FINANCIAL ECONOMETRICS

A Short Introduction to Time-Varying Volatility in Financial Time Series 217
B. HANZON

Forecasting Emerging Equity Market Volatility Using Nonlinear GARCH Models 221
D. van DIJK

Stochastic Properties of Multivariate Time Series Equations with Emphasis on ARCH 227
A. RAHBEK

A Rational Probability Density Approach to Stochastic Volatility Estimation 233
B. HANZON

SEMI-PLENARY
SNIPPETS OF IDENTIFICATION THEORY IN COMPUTER VISION

Snippets of System Identification in Computer Vision 237
S. SOATTO, A. CHIUSO

SEMI-PLENARY
INTERVAL ANALYSIS FOR GUARANTEED NONLINEAR PARAMETER ESTIMATION

Interval Analysis for Guaranteed Nonlinear Parameter Estimation 249
E. WALTER, M. KIEFFER

IDENTIFICATION IN AUTOMOTIVE SYSTEMS

Online Detection of Tyre Pressure Deflation in Passenger Cars 261
J. SHAH, M. BÖRNER, R. ISERMANN, Y.G. SRINIVASA

A Subspace-Based Identification Approach for the Analysis of Road Vehicles Yaw Dynamics Around Steering-Pad Conditions 267
S.M. SAVARESI, E. SILANI, S. BITTANTI, F. FARACHI

Identification and Fault Detection of an Active Vehicle Suspension 273
D. FISCHER, M. ZIMMER, R. ISERMANN

Non-Adaptive Neural Automotive Sideslip Virtual Sensor 279
M. BATTIPEDE, D. DANESIN, P. KRIEF, G. SASSI, M. VELARDOCCHIA

Parametric Identification of the Car Dynamics 285
G. VENTURE, M. GAUTIER, W. KHALIL, P. BODSON

Simulating Energy Consumption of Auxiliary Units in Heavy Vehicles 291
N. PETTERSSON, K.H. JOHANSSON

SENSOR IDENTIFICATION AND MONITORING

Prior Characterization of the Performance of Software Sensors 297
I. BRAEMS, M. KIEFFER, E. WALTER

Model Based Source Localisation by Distributed Sensors for Point Sources and Diffusion 303
J. MATTHES, L. GRÖLL

Continuous-Time Model Identification by Using Adaptive Observer 309
K. IKEDA, Y. MOGAMI, T. SHIMOMURA

Optimal Filtering of Nonlinear Systems Based on Pseudo Gaussian Densities 315
U.D. HANEBECK

A Total Least Squares Approach to Sensor Characterisation 321
P.C.F. HUNG, S. MCLOONE, G. IRWIN, R. KEE

IDENTIFICATION OF NONLINEAR SYSTEMS I

Estimation and Validation of Semi-Parametric Dynamic Nonlinear Models 327
Y. ROLAIN, W. van MOER, J. SCHOUKENS

Nonlinear System Modelling Using the RBF Neural Network-Based Regressive Model 333
H. PENG, T. OZAKI, Y. TOYODA, K. NAKANO

Modeling and Linearization of Nonlinear Dynamic Systems 339
J.G. NÉMETH, J. SCHOUKENS

Linear Parameter Estimation and Predictive Constrained Control of Wiener/Hammerstein Systems 345
K.J. LATAWIEC, C. MARCIAK, R. ROJEK, G.H.C. OLIVEIRA

Identification of Wiener Systems Using Reduced Complexity Volterra Models 351
R. HACIOĞLU, G.A. WILLIAMSON

Structure Selection for Polynomial NARX Models Based on Simulation Error Minimization 357
L. PIRODDI, W. SPINELLI

MECHANICAL AND AEROSPACE APPLICATIONS

Nonlinear Identification of a Two Link Robotic System Using Dynamic Neural Networks 363
S. TORRES, V.M. BECERRA

Neural Network System Identification for a Low Pressure Non-Linear Dynamical Subsystem Onboard the Alicia II Climbing Robot 369
D. LONGO, G. MUSCATO, G. NUNNARI

Measurement of Young's Modulus Via Modal Analysis Experiments: A System Identification Approach 375
R. PINTELON, P. GUILLAUME, K. de BELDER, Y. ROLAIN

A Novel Algorithm for Fully Autonomous Star Identification 381
S. BITTANTI, E. de MARCHI, M. GIRANZANI, M. LOVERA, B. LÜBKE-OSSENBECK, E. SILANI

Fast Model Updates and Simulation for Efficient Flight Control Software Design 387
H. FRIEHMELT, D. ROHLF

CLOSED-LOOP IDENTIFICATION

Continuous-Time Identification of First-Order Plus Dead-Time Models from Step Response in Closed Loop 393
F.S. COELHO, P.R. BARROS

Identification of Simple Continuous-Time Models from Relay Feedback 399
G.H.M. de ARRUDA, P.R. BARROS

Continuous-Time Model Identification of Systems Operating in Closed-Loop 405
M. GILSON, H. GARNIER

Multivariable Closed-Loop System Identification of Plants Under Model Predictive Control 411
E. de KLERK, I.K. CRAIG

Dead Time Measurement of Closed Loop System by Wavelet 417
T. TABARU, S. SHIN

Closed Loop Identification Method Using a Subspace Approach 423
M. POULIQUEN, M. M'SAAD

INDUSTRIAL APPLICATION OF IDENTIFICATION

Model Identification of a Multivariable Industrial Furnace 429
M. BARRERAS, M. GARCÍA-SANZ

Extended Fuzzy GK Clustering with Application to Identification of an Automatic Voltage Regulation Loop Dynamics 435
L. REN, G.W. IRWIN

On Simplified Modelling Approaches to SMB Processes 441
V. GROSFILS, C. LEVRIE, M. KINNAERT, A.VANDE WOUWER

Optimal Filtering for Bilinear Systems and its Application to Terpolymerization Process State Identification 447
M. BASIN, M.A. ALCORTA-GARCIA

Neural Prediction of Cylinder Air Mass for AFR Control in SI Engine 453
G. BLOCH, Y. CHAMAILLARD, G. MILLERIOUX, P. HIGELIN

Contribution to Identification of Thermo-Mechanic Interaction at Vibrating Rubber-Like Materials 459
L. PEŠEK, L. PŮST, F. VANĚK

PROCESS CONTROL SYSTEMS

Identification of a High Efficiency Boiler by Support Vector Machines Without Bias Term 465
M. VOGT, K. SPREITZER, V. KECMAN

Implementing GA-Based Predictive Controller for on-line Control of a Process Mini-Plant 471
Y.Y. NAZARUDDIN, F. MAULANA

Long-Range Optimal Model and Multi-Step-Ahead Prediction Identification for Predictive Control 477
R. HABER, U. SCHMITZ, R. BARS

Predictive Control of Flow Quantity and Sloshing-Suppression During Back-Tilting of a Ladle for Batch-Type Casting Pouring Processes 483
K. TERASHIMA, K. YANO, M. KANEKO

CLOSED LOOP AND PERFORMANCE ISSUES

Optimal Prefiltering in Iterative Feedback Tuning 489
R. HILDEBRAND, A. LECCHINI, G. SOLARI, M. GEVERS

Identification of Performance Limitations in Control Using General SISO Models 495
J. MÅRTENSSON, H. HJALMARSSON

Control Loop Performance Monitoring by CUSUM Algorithms for Local Linear Hypotheses 501
M. KINNAERT, R. HANUS, C. PARLOIR

Model Approximation of Plant and Noise Dynamics on the Basis of Closed-Loop Data 507
J. ZENG, R.A. de CALLAFON

IV Methods for Closed-Loop System Identification 513
M. GILSON, P. van den HOF

Coprime Factor Perturbation Models for Closed-Loop Model Validation Techniques 519
M. CROWDER, R.A. de CALLAFON

REPRODUCING KERNELS I

An Introduction to Reproducing Kernel Hilbert Spaces and Why they are so Useful 525
G. WAHBA

An Introduction to Smoothing Spline ANOVA Models in RKHS, with Examples in Geographical Data, Medicine, Atmospheric Sciences and Machine Learning 531
G. WAHBA

Robust Design with Nonparametric Models: Prediction of Second-Order Characteristics of Process Variability by Kriging 537
L. PRONZATO, É. THIERRY

Geostatistical Models and Kriging 543
H. WACKERNAGEL

Hilbert Space Embeddings in Dynamical Systems 549
A.J. SMOLA, S.V.N. VISHWANATHAN

Bayesian Input Selection for Nonlinear Regression with LS-SVMs 555
T. van GESTEL, M. ESPINOZA, J.A.K. SUYKENS, C. BRASSEUR, B. de MOOR

VOLUME 2

BLIND ESTIMATION AND EQUALIZATION

Blind Turbo Equalization Using the Constant Modulus Algorithm 561
P.A. REGALIA

A New Method for Channel Estimation and Data Detection in the Context of Turbo Equalisation 567
S. PERREAU, G. GORLIER

On the Applicability to Correlated Sources of a Blind Channel Equalization Method Robust to Order Overestimation 573
R. LÓPEZ-VALCARCE

Blind Estimation with Signal Scrambling 579
H. XU, X. SONG, S. DASGUPTA

Blind Channel Shorteners 585
C.R. JOHNSON Jr., R.K. MARTIN, J.M. WALSH, A.G. KLEIN, C.E. ORLICKI, T. LIN

Multiple Antenna System Equalization Using Semi-Blind Subspace Identification Methods 591
C. ZHANG, R.R. BITMEAD

CONTINUOUS TIME IDENTIFICATION

The Identification of Continuous-Time Linear and Nonlinear Models: A Tutorial with Environmental Applications 597
P.C. YOUNG, H. GARNIER, A. JARVIS

Continuous-Time System Identification of A Food Extruder: Experiment Design and Data Analysis 609
L. WANG, P.J. GAWTHROP, C. CHESSARI, T. PODSIADLY

Identification of Continuous Time Models Using Discrete Time Data 615
N.R. KRISTENSEN, H. MADSEN, S.B. JØRGENSEN

On Possibilities for Estimating Continuous-Time ARMA Parameters 621
E.K. LARSSON, M. MOSSBERG

On the Interpretation of a Continuous–Time Model Identification Method in Terms of Regularization 627
S. MOUSSAOUI, D. BRIE, A. RICHARD

INPUT DESIGN

A Survey of Readily Accessible Perturbation Signals 633
K.R. GODFREY, A.H. TAN, H.A. BARKER

Multiple Input Design for Real-Time Parameter Estimation in the Frequency Domain 639
E.A. MORELLI

Minimizing the Worst-Case ν-Gap by Optimal Input Design 645
R. HILDEBRAND, M. GEVERS

Identification of Resonant Systems Using Periodic Multiplicative Reference Signals 651
W.J. DUNSTAN, R.R. BITMEAD

Aircraft Parameter Estimation by Using the Optimal Input Design and Linear Matrix Inequalities 657
C. JAUBERTHIE, L. DENIS-VIDAL, G. JOLY-BLANCHARD

The Performance of Multilevel Perturbation Signals for Nonlinear System Identification 663
H.A. BARKER, A.H. TAN, K.R. GODFREY

IDENTIFICATION FOR FLIGHT TEST EXPLORATION

Applying System Identification to Assess the Vibro-Acoustic Behaviour of Airplanes 669
B. PEETERS, R. RUOTOLO, A. VECCHIO, H. van der AUWERAER

Subspace Identification Combined with New Mode Selection Techniques for Modal Analysis of an Airplane 675
I. GOETHALS, B. de MOOR

Flight Flutter Analysis Using Frequency-Domain System Identification Techniques 681
P. GUILLAUME, P. VERBOVEN, B. CAUBERGHE

Real-Time Modal Analysis and its Application for Flutter Testing 687
T. UHL, M. BOGACZ

Statistical Approach to Flutter Monitoring 693
L. MEVEL, M. BASSEVILLE, A. BENVENISTE

Reliable System Identification for Large Flexible Space Structures 699
V. BABUŠKA, S.L. LACY, R.S. ERWIN, A.M. MELIN

IDENTIFIABILITY

Identifiability Analysis of a Class of Systems Described By Convolution Equations 705
L. BELKOURA

Identification of Fully Parameterized Linear and Nonlinear State-Space Systems by Projected Gradient Search 711
V. VERDULT, N. BERGBOER, M. VERHAEGEN

A Differential Geometric Viewpoint on Local Identifiability and Identification Part I: Theory 717
B. EITZINGER, K. SCHLACHER

A Differential Geometric Viewpoint on Local Identifiability and Identification Part II: Application 723
B. EITZINGER, K. SCHLACHER

Identifiability of Nonlinear Homogeneous Polynomial Systems 729
R. PEETERS, B. HANZON

PLENARY PAPER
SYSTEM IDENTIFICATION FOR STRUCTURAL DYNAMICS AND VIBROACOUSTICS DESIGN ENGINEERING

System Identification for Structural Dynamics and Vibroacoustics Design Engineering 735
H. van der AUWERAER

SELECTED TOPICS IN IDENTIFICATION

A Personal View on the Development of System Identification 747
M. GEVERS

System Identification Via a Computational Bayesian Approach 759
B. NINNESS, S. HENRIKSEN

A New Information Theoretic Approach to Order Estimation Problem 765
S. BEHESHTI, M.A. DAHLEH

Conditions for Local Convergence of Maximum Likelihood Estimation for Armax Models 771
G.C. GOODWIN, J.C. AGÜERO, R.E. SKELTON

A Nonparametric Approach to Model Selection 777
M. BEKARA, A.-K. SEGOUANE, F. GILLES

REPRODUCING KERNELS II

An Introduction to Learning with Reproducing Kernel Hilbert Spaces 783
M. PONTIL

Sparse Gaussian Processes: Inference, Subspace Identification and Model Selection 789
L. CSATÓ, M. OPPER

Sparse Kernel Methods 795
S.R. GUNN

A Generalised LS–SVM 801
J. VALYON, G. HORVÁTH

Adaptive Kernel Methods 807
A. KUH

Subspace Regression in Reproducing Kernel Hilbert Space 813
L. HOEGAERTS, J.A.K. SUYKENS, J. VANDEWALLE, B. de MOOR

IDENTIFICATION OF NONLINEAR BLOCK MODELS

Frequency Domain Identification of Wiener Models 819
E.-W. BAI

Non-Parametric Identification of Non-Linearity in Hammerstein Systems 825
W. GREBLICKI, P. ŚLIWIŃSKI

Generation of Enhanced Initial Estimates for Wiener Systems and Hammerstein Systems 831
P. CRAMA, J. SCHOUKENS, R. PINTELON

User Choices and Model Validation in System Identification Using Nonlinear Wiener Models 837
T. WIGREN

Approximation of Feasible Parameter Set in Worst Case Identification of Block-Oriented Nonlinear Models 843
L. GIARRÉ, G. ZAPPA

Parameters Set Evaluation of Wiener Models from Data with Bounded Output Errors 849
V. CERONE, M. MILANESE, D. REGRUTO

NEW RESULTS IN SUBSPACE IDENTIFICATION

Constructing the State of Random Processes with Feedback 855
A. CHIUSO, G. PICCI

Closed-Loop Subspace Identification with Innovation Estimation 861
S.J. QIN, L. LJUNG

A Frequency Domain Subspace Algorithm for Mixed Causal, Anti-Causal LTI Systems 867
R. FRAANJE, M. VERHAEGEN, V. VERDULT, R. PINTELON

A Stochastic Realization in a Hilbert Space Based on "LQ Decomposition" with Application to Subspace Identification 873
H. TANAKA, T. KATAYAMA

Subspace-Based Identification Methods Using Schur Complement Approach 879
Y. TAKEI, H. NANTO, S. KANAE, Z.-J. YANG, K. WADA

Recursive Subspace Identification for Continuous-/Discrete-Time Stochastic Systems 885
A. OHSUMI, Y. MATSUÜRA, K. KAMEYAMA

IDENTIFICATION FOR PROCESS CONTROL: INPUT DESIGN

"Plant-Friendly" System Identification: A Challenge for the Process Industries 891
D.E. RIVERA, H. LEE, M.W. BRAUN, H.D. MITTELMANN

Multi-Objective Input Signal Design for Plant-Friendly Identification 897
S. NARASIMHAN, R. SRINIVASAN, R. RENGASWAMY

Control-Relevant Design of Periodic Test Input Signals for Iterative Open-Loop Identification of Multivariable FIR Systems 903
J.H. LEE

Constrained Signal Design Using Approximate Prior Models with Application to the Tennessee Eastman Process 909
T. LI, C. GEORGAKIS

Constrained Minimum Crest Factor Multisine Signals for "Plant-Friendly" Identification of Highly Interactive Systems 915
H. LEE, D.E. RIVERA, H.D. MITTELMANN

IDENTIFICATION OF MECHANICAL SYSTEMS

Online Identification of a Robot Using Batch Adaptive Control 921
B. BUKKEMS, D. KOSTIĆ, B. de JAGER, M. STEINBUCH

Dynamic Identification of a Compactor Using Splines Data Processing 927
C.-E. LEMAIRE, P.-O. VANDANJON, M. GAUTIER

Non-Stationary Mechanical Vibration Modeling and Analysis Via Functional Series TARMA Models 933
A.G. POULIMENOS, S.D. FASSOIS

Globally Convergent Adaptive Tracking of Angular Velocity with Inertia Identification and Adaptive Linearization 939
A.K. SANYAL, M. CHELLAPPA, J.L. VALK, J. AHMED, J. SHEN, D.S. BERNSTEIN

On Vision-Based Kinematic Calibration of n-Leg Parallel Mechanisms 945
P. RENAUD, N. ANDREFF, G. GOGU, P. MARTINET

A Geometric Approach to Motion Tracking in Manifolds 951
J.G. SILVA, J.S. MARQUES, J.M. LEMOS

SOFTWARE SESSION I

Version 6 of the System Identification Toolbox 957
L. LJUNG

Process Identification, Controller Tuning and Control Circuit Simulation Using MS Excel 963
H.M. SCHAEDEL

Developments for the MATLAB CONTSID Toolbox 969
H. GARNIER, M. GILSON, E. HUSELSTEIN

detectNARMAX: A Graphical User Interface for Structure Detection of NARMAX Models Using The Bootstrap Method 975
E. SHAFAI, M. BIANCHI, H.P. GEERING

SEMI-PLENARY
DATA-BASED METHODS IN PROCESS CONTROL

Data-Based Methods for Process Analysis, Monitoring and Control 981
J.F. MacGREGOR

SEMI-PLENARY
SUBSPACE ALGORITHMS

Subspace Algorithms 993
D. BAUER

FILTERING AND ESTIMATION

Optimal Filtering for Linear Systems with Multiple Delays in Observations 1005
M. BASIN, R. MARTINEZ-ZUNIGA

The Information Analysis in Joint Problem of Continuous-Discrete Filtering and Generalized Extrapolation 1011
N.S. DYOMIN, I.E. SAFRONOVA, S.V. ROZHKOVA

Guaranteed Ellipsoidal State Estimation for Uncertain MIMO Models 1017
B.T. POLYAK, S.A. NAZIN, C. DURIEU, É. WALTER

Regularized Robust Estimators for Time Varying Uncertain Discrete-Time Systems 1023
A. SUBRAMANIAN, A.H. SAYED

Minimax L_2-E_2 FIR Filters for Deterministic Continuous-Time State Space Signal Models 1029
S.H. HAN, W.H. KWON

Numerically Reliable H_∞ – Synthesis of Estimators Based on J – Lossless Factorisations 1035
P. SUCHOMSKI

DIAGNOSIS, DETECTION AND TRACKING

Statistical Analysis of Subspace-Based Method for Direction Estimation Without Eigendecomposition 1041
J. XIN, A. SANO

Fault Detection of Non-Linear Systems Based on Multi-Form Quasi-Armax Modeling and its Application to the Ship Benchmark 1047
K. KUMAMARU, K. INOUE, Y. HOSOYAMADA, T. SÖDERSTRÖM

A Comparison of Two Methods for Stochastic Fault Detection: The Parity Space Approach and Principal Components Analysis 1053
A. HAGENBLAD, F. GUSTAFSSON, I. KLEIN

Identification of Object's Movement Models in a Radar Tracking Filter 1059
M. SANKOWSKI, Z. KOWALCZUK

Estimation and Tracking of Quasi-Periodically Varying Processes 1065
M. NIEDŹWIECKI, P. KACZMAREK

VOLUME 3

IDENTIFICATION OF NONLINEAR SYSTEMS II

A Pruning Method for the Identification of Polynomial NARMAX Models 1071
L. PIRODDI, W. SPINELLI

Generalized Orthonormal Basis Selection for Expanding Quadratic Volterra Filters 1077
A.Y. KIBANGOU, G. FAVIER, M.M. HASSANI

A Localised Forgetting Method for On-Line Adaptation of Gaussion RBFN Models 1083
D.L. YU, J.B. GOMM, D.W. YU, D. WILLIAMS

Subspace Identification of Switching Model 1089
K.M. PEKPE, K. GASSO, G. MOUROT, J. RAGOT

Application-Oriented Neural Modelling 1095
K. LI, G. IRWIN

IDENTIFICATION METHODS

Closed-Form Frequency Estimation Using Second-Order Notch Filters 1101
S.M. SAVARESI, S. BITTANTI, H.C. SO

L_1 Prediction Error System Identification: A Modified AIC Rule 1107
J.C. CARMONA, M. OULADSINE, M. EL ADEL

On Parameter Estimation of ARMAX Model Via BCLS Method 1113
L.-J. JIA, S. KANAE, Z.-J. YANG, K. WADA

Estimation in the Presence of Interferences 1119
J.J. FUCHS

Autoregressive Spectral Analysis with Randomly Missing Data 1125
P.M.T. BROERSEN, S. de WAELE, R. BOS

Estimating Unknown Probability Density Functions for Random Parameters of Stochastic ARMAX Systems 1131
H. WANG, Y. WANG

CONTROLLER TUNING AND IDENTIFICATION

Iterative Controller Tuning by Minimization of a Generalized Decorrelation Criterion 1137
L. MIŠKOVIĆ, A. KARIMI, D. BONVIN

Subspace Identification Based PID Control Tuning 1143
A. SANCHEZ, M.R. KATEBI, M.A. JOHNSON

Evolutionary Tuning of PID Parameters 1149
T. YAMAMOTO

Adaptive, Cautious, Predictive Control with Gaussian Process Priors 1155
R. MURRAY-SMITH, D. SBARBARO, C.E. RASMUSSEN, A. GIRARD

Controller Design for Systems Suffering Nonlinear Distortions 1161
M. SOLOMOU, D. REES, N. CHIRAS

How the Output Saturation of a Regulator Influences the Reachable Performance and Robustness Measures 1167
L. KEVICZKY, C. BÁNYÁSZ

APPLICATIONS OF IDENTIFICATION

Random Loading Identification of a Plastic Glass Cantilever Beam 1173
D. LI, X. GUO, H. LI

On Sequential Identification of a Diffusion Type Process with Memory 1179
U. KÜCHLER, V. VASIL'IEV

Incremental Identification of Transport Coefficients in Distributed Systems 1185
A. BARDOW, W. MARQUARDT

On the Structure of Static Balanced Flow Systems 1191
E. WEYER, A. GLEIß, M. DEISTLER, K. GRUBER, T. MATYUS

Endogeneity and Identification in Transportation Systems: Econometric Relationships to Partial Observability 1197
N.K. JUVVA, V.N. SHANKAR, S. CHAYANAN

Tool for Equal Opportunity Evaluation in Dynamic Organizations 1203
P. ALBERTOS, I. BENÍTEZ, J.L. DÍEZ, J.A. LACORT

BIOENGINEERING SYSTEMS

Linearization in the Parameters Via Differential Algebra Techniques 1209
M.P. SACCOMANI

A Penalty Function Approach to HIV/AIDS Model Parameter Estimation 1215
R. FILTER, X. XIA

Sensitivity Analysis and Parameter Identification of Wastewater Treatment System Based on Activated Sludge Models 1221
J. SATO, H. OHMORI

A Methodology for Nonlinear System Identification Using Volterra Series. Application to an Anaerobic Digestor 1227
G. BIBES, P. COIRAULT, R. OUVRARD, J.P. STEYER

Some Relations of Sensitivity Functions in Bio-Reactor Models 1233
J.A.R. PÉREZ, J.L.N. HERRERO

An Experimental Object-Oriented Modelling of an Hydraulic Valley 1239
T. BASTOGNE, A. LIBAUX

PARTICLE FILTERS

Particle Filters for System Identification with Application to Chaos Prediction 1245
F. GUSTAFSSON, P. HRILJAC

Particle Filters for System Identification of State-Space Models Linear in Either Parameters or States 1251
T. SCHÖN, F. GUSTAFSSON

Fault Detection, Isolation and Diagnosis with Particle Filters for Nonlinear Stochastic Systems 1257
V. KADIRKAMANATHAN, P. LI

Monte Carlo Mixture Kalman Filter and its Application to Space-Time Inversion 1263
T. HIGUCHI, J. FUKUDA

A Particle Implementation of the Recursive MLE for Partially Observed Diffusions 1269
A. GUYADER, F. LE GLAND, N. OUDJANE

Online Sampling for Parameter Estimation in General State Space Models 1275
C. ANDRIEU, A. DOUCET, V.B. TADIĆ

WIENER HAMMERSTEIN MODELS

Nonlinear Structure Identification with Application to Wiener-Hammerstein Systems 1281
D.J. LEITH, W.E. LEITHEAD, R. MURRAY-SMITH

Identification of a Wiener System with Some General Discontinuous Nonlinearities 1285
F. GUO, G. BRETTHAUER

Nonlinear Model Identification Using Working Point Variables 1291
Y. ZHU

Identification of Wiener-Hammerstein Models with Cubic Nonlinearity Using LIFRED 1297
A.H. TAN, K.R. GODFREY

Performance Investigation of SLICOT Wiener Systems Identification Toolbox 1303
V. SIMA

IDENTIFICATION USING BASIS FUNCTIONS

Rational Bases Generated by Blaschke Product Systems 1309
F. SCHIPP, J. BOKOR

More on Sparse Representations in Arbitrary Bases 1315
J.J. FUCHS

On Spectral Analysis Using Models with Pre-Specified Zeros 1321
B. WAHLBERG

Identification of Rational Spectral Densities Using Orthonormal Basis Functions 1327
A. BLOMQVIST, G. FANIZZA

Orthonormal Basis Functions for Modeling Continuous-Time Fractional Systems 1333
M. AOUN, R. MALTI, F. LEVRON, A. OUSTALOUP

Adaptive Laguerre Time Scaling Factor in Predictive Control 1339
M. EL ADEL, M. OULADSINE, J.C. CARMONA

SUBSPACE IDENTIFICATION AND APPLICATIONS

Identification of MIMO State Space Models for Helicopter Dynamics 1345
M. LOVERA

Estimation of Damped and Undamped Sinusoids with Application to Analysis of Electromagnetic FDTD Simulation Data 1351
T. McKELVEY, T. RYLANDER, M. VIBERG

Application of a Recursive Subspace Identification Algorithm to Change Detection 1357
H. OKU

Subspace-Based Modal Identification and Monitoring of Large Structures: A Scilab Toolbox 1363
L. MEVEL, M. GOURSAT, M. BASSEVILLE, A. BENVENISTE

Identifying Positive Real Models in Subspace Identification by Using Regularization 1369
I. GOETHALS, T. van GESTEL, J. SUYKENS, P. van DOOREN, B. de MOOR

Modeling Human Gaits with Subtleties 1375
A. BISSACCO, P. SAISAN, S. SOATTO

IDENTIFICATION IN LARGE SCALE SYSTEMS

Reduction of Large-Scale Groundwater Flow Models Via the Galerkin Projection 1381
P.T.M. VERMEULEN, A.W. HEEMINK, C.B.M TE STROET

Model Reduction for Large-Scale Linear Applications 1387
K. WILLCOX, A. MEGRETSKI

Reduced Order Modeling of an Industrial Feeder Model 1393
P. ASTRID, S. WEILAND, A. TWERDA

INDUSTRIAL APPLICATIONS OF IDENTIFICATION

Identification of the Topology of a Power System Network 1399
Y. HASSAINE, E. WALTER, M. DANCRE, B. DELOURME, P. PANCIATICI

LPV Identification of a Diesel Engine Torque Model 1405
X. WEI, L. DEL RE

Identification and Control of a PV-Supplied Separately Excited DC Motor Using Universal Learning Networks 1411
A. HUSSEIN, K. HIRASAWA, J. HU

Validation of Stability for an Induction Machine Drive Using Experiments 1417
H. MOSSKULL, B. WAHLBERG, J. GALIC

Automatic Steering Control System Design Utilizing a Visual Feedback Approach - System Identification and Control Experiments with a Radio-Controlled Car 1423
S. ADACHI, T. FUJIHIRA, Y. FUJIWARA

Application of RBF-Type ARX Modeling and Control to Gas Turbine Combined Cycle SCR Systems 1429
Y. TOYODA, H. PENG, T. OZAKI, K. NAKANO, H. SHIOYA

SOFTWARE SESSION II

Automatic Time Series Identification Spectral Analysis with MATLAB Toolbox ARMASA 1435
P.M.T. BROERSEN

MULTI-EDIP – An Interactive Software Package for Process Identification 1441
J. KASPRZYK

KALMTOOL for Use with MATLAB 1447
M. NØRGAARD, N.K. POULSEN, O. RAVN

The $ADAPT_X$ Software for Automated and Real-Time Multivariable System Identification 1453
W.E. LARIMORE

Frequency Domain System Identification Toolbox for MATLAB: Automatic Processing – from Data to Models 1459
I. KOLLÁR, R. PINTELON, Y. ROLAIN, J. SCHOUKENS, G. SIMON

PLENARY PAPER
PREDICTION ALGORITHMS: COMPLEXITY, CONCENTRATION AND CONVEXITY

Prediction Algorithms: Complexity, Concentration and Convexity 1465
P.L. BARTLETT

IDENTIFICATION AND PHYSICAL MODELING

Grey–Box Model Calibrator and Validator 1477
T. BOHLIN, A.J. ISAKSSON

Initialization of Physical Parameter Estimates 1483
P.A. PARRILO, L. LJUNG

Parameter Estimation in Linear Differential-Algebraic Equations 1489
M. GERDIN, T. GLAD, L. LJUNG

Model Validation in Non-Linear Continuous-Discrete Grey-Box Models 1495
J. HOLST, E. LINDSTRÖM, H. MADSEN, H.A. NIELSEN

Identification of Mechanical Parameters in Drive Train Systems 1501
A.J. ISAKSSON, R. LINDKVIST, X. ZHANG, M. NORDIN, M. TALLFORS

Identification and Model Predictive Control of a pH Neutralization Process Based on Linear and Wiener Models 1507
J.C. GÓMEZ, A. JUTAN

IDENTIFICATION OF NONLINEAR SYSTEMS

Local Modelling of Nonlinear Dynamic Systems Using Direct Weight Optimization 1513
J. ROLL, A. NAZIN, L. LJUNG

Optimality in SM Identification of Nonlinear Systems 1519
M. MILANESE, C. NOVARA

A Suboptimal Bootstrap Method for Structure Detection of Nonlinear Output-Error Models 1525
S.L. KUKREJA

Identification of Nonlinear Parametrically Varying Models Using Separable Least Squares 1531
F. PREVIDI, M. LOVERA

Modeling and Identification of Rate-Independent Hysteresis Using a Semilinear Duhem Model 1537
J. OH, D.S. BERNSTEIN

Least Squares Harmonic Signal Analysis Using Periodic Orbits of ODEs 1543
T. WIGREN, E. ABD-ELRADY, T. SÖDERSTRÖM

VOLUME 4

EDUCATION AND TRAINING

Educational Aspects of Identification Software user Interfaces 1549
L. LJUNG

An Identification Course on the Web: Rationale, Realization and Students' Evaluation 1555
R. GUIDORZI, I. PAGANI, R. DIVERSI

Control Related Topics in Identification - Closed Loop Experiments and Identification for Control 1561
R.R. BITMEAD, R.A. de CALLAFON

Teaching Semiphysical Modeling to Chemical Engineering Students Using a Brine-Water Mixing Tank Experiment 1567
D.E. RIVERA

Estimating Parameters in a Lumped Parameter System with First Principle Modeling and Dynamic Experiments 1573
R.A. de CALLAFON

RECURSIVE AND SUBSPACE IDENTIFICATION

Recursive Subspace Identification Based on Projector Tracking 1579
M. LOVERA

Subspace Identification and ARX Modeling 1585
M. JANSSON

Parallel QR Implementation of Subspace Identification with Parsimonious Models 1591
S.J. QIN, L. LJUNG

A New Recursive Method for Subspace Identification of Noisy Systems: EIVPM 1597
G. MERCÈRE, S. LECOEUCHE, C. VASSEUR

Canonical Correlation Partial Least Squares 1603
U. KRUGER, S.J. QIN

Frequency-Domain System Identification Techniques for Experimental and Operational Modal Analysis 1609
P. GUILLAUME, P. VERBOVEN, B. CAUBERGHE, S. VANLANDUIT, E. PARLOO, G. de SITTER

PROCESS CONTROL: THEORY

Data-Driven Modeling of Nonlinear and Time-Varying Processes 1615
D. BONNÉ, S.B. JØRGENSEN

PID Parameter Cycling to Tune Industrial Controllers: a New Model-Free Approach 1621
J. CROWE, M.A. JOHNSON, M.J. GRIMBLE

Stepwise Refinement of Sparse Grids in Data Mining Applications 1627
M. BRENDEL, W. MARQUARDT

Iterative Identification for Control and Robust Performance of Bioreactor 1633
K. BØJSTRUP, H.H. NIEMANN, N.K. POULSEN, S.B. JØRGENSEN

Modified Subspace Identification Method for Building a Long-Range Prediction Model for Inferential Control 1639
Y. PAN, J.H. LEE

Identification and Model Predictive Control of an Industrial Glass-Feeder 1645
L. HUISMAN, S. WEILAND

APPLICATION OF SYSTEM IDENTIFICATION

Computationally Efficient Estimation of Wave Propagation Functions of Viscoelastic Materials 1651
K. MAHATA, T. SÖDERSTRÖM, L. HILLSTRÖM

Identification of Underlying Intensity Processes of Interference Patterns 1657
L. NÁDAI, J. BOKOR, A EDELMAYER

Fractional Multimodels - Application to Heat Transfer Modeling 1663
R. MALTI, M. AOUN, J.-L. BATTAGLIA, A. OUSTALOUP, K. MADANI

A Recursive Algorithm for Estimating Parameters in a One Dimensional Diffusion System 1669
B. BHIKKAJI, T. SÖDERSTRÖM, K. MAHATA

Regularization Method in Infrared Image Processing 1675
S. DATCU, L. IBOS, Y. CANDAU, S. MATTEÏ, N. RAMDANI

Filtering of Stochastic Volatility Model 1681
S. AIHARA, A. BAGCHI

OPTIMAL FILTERING

State Estimation for Nonlinear Continuous Systems in a Bounded-Error Context 1687
T. RAÏSSI, N. RAMDANI, Y. CANDAU

Multigrid Design in Point-Mass Approach to Nonlinear State Estimation 1693
M. ŠIMANDL, J. KRÁLOVEC

An Efficient Nonlinear Adaptive Observer with Global Convergence 1699
Q. ZHANG, A. XU, G. BESANÇON

Adaptive Observer for Discrete Time Linear Time Varying Systems 1705
A. GUYADER, Q. ZHANG

Linear Dynamic Filtering with Noisy Input and Output 1711
I. MARKOVSKY, B. de MOOR

The *p*-Norm Generalization of the LMS Algorithm for Adaptive Filtering 1717
J. KIVINEN, M.K. WARMUTH, B. HASSIBI

SEMI-PLENARY
IDENTIFICATION OF LINEAR SYSTEMS WITH NONLINEAR DISTORTIONS

Identification of Linear Systems with Nonlinear Distortions 1723
J. SCHOUKENS, R. PINTELON, T. DOBROWIECKI, Y. ROLAIN

SEMI-PLENARY
SOME PROBLEMS IN STATISTICAL INFERENCE FOLLOWING MODEL SELECTION

Some Problems in Statistical Inference following Model Selection 1735
B.M. PÖTSCHER

USER CHOICES IN SUBSPACE IDENTIFICATION

Choosing Integer Parameters in Subspace Methods: A Survey on Asymptotic Results 1741
D. BAUER

Asymptotic Variances of Subspace Identification by Data Orthogonalization and Model Decoupling 1747
A. CHIUSO, G. PICCI

A Finite Sample Comparison of Automatic Model Selection Methods 1753
D. BAUER, S. de WAELE

On the Number of Rows and Columns in Subspace Identification Methods 1759
B.L.R. de MOOR

Aspects and Experiences of User Choices in Subspace Identification Methods 1765
L. LJUNG

Inferring Multivariable Delay and Seasonal Structure for Subspace Modeling 1771
W.E. LARIMORE

IDENTIFICATION OF STATIC AND DYNAMICAL NONLINEAR SYSTEMS

Mathematical Results Concerning Kernel Techniques 1777
R. SCHABACK

Multi-Output Suppport Vector Regression 1783
E. VAZQUEZ, E. WALTER

Set Membership Identification of Piecewise Affine Models 1789
A. BEMPORAD, A. GARULLI, S. PAOLETTI, A. VICINO

Piecewise-Linear Output-Error Models 1795
F. ROSENQVIST, A. KARLSTRÖM

CMAC with Linear Functional Weights 1801
Q. GAN, E. ROSALES

Optimal Expansions of Discrete-Time Volterra Models Using Laguerre Functions 1807
R.J.G.B. CAMPELLO, G. FAVIER, W.C. AMARAL

IDENTIFICATION AND MODEL VALIDATION

Quantification of the Variance of Estimated Transfer Functions in the Presence of Undermodeling 1813
R. HILDEBRAND, M. GEVERS

Reliable Parameter Estimation in Presence of Uncertain Variables that are not Estimated 1819
I. BRAEMS, L. JAULIN, M. KIEFFER, N. RAMDANI, E. WALTER

Validation Test Based Parameter Uncertainty Versus Analysis-Based Confidence Bounds 1825
S.G. DOUMA, X.J.A. BOMBOIS, P.M.J. van den HOF

Empirical Estimation of Parameter Distributions in System Identification 1831
W.J. DUNSTAN, R.R. BITMEAD

Uncertainty of Transfer Function Modeling Using Prior Estimated Noise Models 1837
R. PINTELON, J. SCHOUKENS, Y. ROLAIN

The Size of the Membership-Set in a Probabilistic Framework 1843
H. AKÇAY

MODEL APPROXIMATION

Connections Between L_2-Model Reduction and Balanced Truncation 1849
W. SCHERRER, F. TJÄRNSTRÖM

Recursive Exact H∞ Identification from Impulse Response Measurements 1855
O. KANEKO, P. RAPISARDA

Properties of Optimal Solutions in L_1 Identification Problem 1861
M. NAMVAR, A. BESANÇON-VODA

Optimal Approximation and Model Quality Estimation for Nonlinear Systems 1867
P.M. MÄKILÄ

Linear Models of Nonlinear FIR Systems with Gaussian Inputs 1873
M. ENQVIST, L. LJUNG

An Algebraic Method for System Reduction of Stationary Gaussian Systems 1879
D. JIBETEAN, J.H. van SCHUPPEN

PARAMETER ESTIMATION AND CONVERGENCE

Separable Least Squares Data Driven Local Coordinates 1885
T. RIBARITS, M. DEISTLER, B. HANZON

Optimal Yule Walker Method for Pole Estimation of ARMA Signals 1891
M. JANSSON, P. STOICA

Initializing Parameter Estimation Algorithms Under Scarce Measurements 1897
P. ALBERTOS, R. SANCHIS, I. PEÑARROCHA

Robust Parameter Estimation for Uncertain Gross-Error Models 1903
K. UOSAKI, K. SAITO, T. HATANAKA

Limit Covariance of Estimation Error for Quasistationary Functions 1909
A.E. BARABANOV

IDENTIFICATION OF HYDROLOGIC SYSTEMS

Structural Identification of Multivariate Neural Networks for Rainfall Runoff Modelling 1915
G. CORANI, G. GUARISO, S. CASTELLI

Parameter and State Regularization for Prediction of Distributed Hydrologic Systems 1921
E.E. van LOON, K.J. KEESMAN

Time–Delay Estimation of a Managed River Reach from Supervisory Data 1927
M. THOMASSIN, T. BASTOGNE, A. RICHARD, A. LIBAUX

Geohydrological Application of a Nonlinear Physically Based Time Series Model 1933
W.L. BERENDRECHT, A.W. HEEMINK, F.C. van GEER, J.C. GEHRELS

On Physical and Data Driven Modelling of Irrigation Channels 1939
S.K. OOI, M.P.M. KRUTZEN, E. WEYER

Identification and On-Line Estimation of the Unsaturated Hydraulic Conductivity in Presence of Forced Air Convection Based on a Distributed-Parameter Model 1945
O. SCHOEFS, D. DOCHAIN, R. CHAPUIS, R. SAMSON, M. PERRIER

ERRORS IN VARIABLE IDENTIFICATION

Confidence Regions for Non-Parametric Errors-In-Variables Estimates 1951
W.P. HEATH

A New Criterion in EIV Identification and Filtering Applications 1957
R. DIVERSI, R. GUIDORZI, U. SOVERINI

Strongly Consistent Parameter Estimate for Error-In-Variables Model 1963
H.-F. CHEN

Ellipsoid Set Refinement by Simultaneous Use of Multiple Hyperplane Cuts 1969
D. JOACHIM, J.R. DELLER Jr.

Identification Methods in a Unified Framework 1975
I. VAJK

Author Index 1981

IFAC
Publications
www.elsevier.com/locate/ifac

FROM EXPERIMENTS TO CLOSED LOOP CONTROL

Håkan Hjalmarsson *

* *Department of Signals, Sensors and Systems, Royal Institute of Technology, S-100 44 Stockholm, Sweden*

Abstract: In this paper we examine the links between identification and control. The main trends in this research area are summarized, with particular focus on design of low complexity controllers. It is argued that a guiding principle should be to model as well as possible before any model or controller simplifications are made, as this ensures the best statistical accuracy. Particular attention is given to the experiment design issue since well-designed experiments facilitates this task. Furthermore, the interaction between experimental constraints and performance specifications is discussed. *Copyright © 2003 IFAC*

Keywords: Identification for control; experiment design.

1. INTRODUCTION

Ever increasing productivity demands and environmental standards necessitates more and more advanced control methods to be employed in industry. However, such methods usually require a model of the process and modeling and system identification is expensive. Quoting (Ogunnaike, 1996):

"It is also widely recognized, however, that obtaining the process model is the single most time consuming task in the application of model-based control."

It has also been recognized that models for control pose special considerations. Again quoting (Ogunnaike, 1996):

"There is abundant evidence in industrial practice that when modeling for control is not based on criteria related to the actual end use, the results can sometimes be quite disappointing."

Hence, modeling and system identification techniques suited for industrial use and tailored for control design applications have become important enablers for industrial advances. The Panel for Future Directions in Control, (Murray *et al.*, 2003), has identified *automatic synthesis of control algorithms, with integrated validation and verification* as one of the major future challenges in control.

Spurred by this, identification for control has been one of the most active areas in system identification over the last decade. Since the joint identification and control problem shares the same elements as any engineering application where system identification is involved, much work under the "umbrella" of identification for control has general applicability.

So what are the issues? Well, to get a first hint consider the following (oversimplified) problem a control engineer might be faced with is:

We have this prior knowledge of the process. You are allowed to perform an identification experiment subject to these restrictions. Use this to design a controller with this performance and with robust stability guarantees. The complexity of the controller should be as low as possible.

Clearly, a useful theory should be capable of handling this type of questions. Below we will try to delineate the main issues involved.

The unforgiving nature of feedback In most applications, performance degrade gracefully as the accuracy of the model becomes worse. On the contrary, in feedback control *stability has to be maintained at all costs.* This issue has definitely put a finger on the approximative nature of system identification.

The forgiving nature of feedback With the numerous existing successful applications of PID-control to nonlinear processes, it is clear that simple, very approximative, models often suffice to give good or, perhaps more accurately, acceptable closed loop performance. Behind this is the rationale for feedback control: High loop gain makes the closed loop system insensitive to the quality of the model and the properties of the open loop system. This observation translates into the fact that one would want the model set produced by a system identification method to be shaped such that high performance can be obtained. This issue relates directly to the *design of the identification experiment.* Clearly, if the experiment ensures that all features of the system relevant for control design are present in the data, this objective can be achieved.

Compatibility requirements The uncertainty description obtained from system identification is dictated by the model structure and the prior information used; see Section 3. It may not be directly applicable to a particular control design method. Thus, it may be necessary to outer-bound the uncertainty description and this should be done so as to not introduce unnecessary conservatism, see (Van den Hof, 1998) for further discussion of this. Another aspect of this issue is that the order of a robust controller usually depends not only on the order of the nominal model but also on the orders of the weighting filters describing the robustness and performance requirements. Hence, the uncertainty description may also *influence the order of the controller.*

Summary The discussion above can be condensed as follows. The user has the experimental conditions and the performance specifications as design variables. To be able to select these in a systematic way such that stability and performance are guaranteed involves:

- ensuring that the 'true' system is accounted for in the set of delivered models.
- understanding which properties of the system have to be modeled accurately and which can be treated only superficially and how this relates to the performance specifications.
- designing experiments that reveal this information.
- representing this information mathematically in a way that is not overly complex.

The paper will unfold in the following way. Basic principles in prediction error identification are briefly reviewed in Section 2. In Sections 3–4 there is an effort to discuss general modeling principles and model validation. A statistical perspective on modeling is discussed in Section 5. We then zoom in on the interplay between identification and control in Section 6. Subsequent Sections 7–11 discuss identification for control and experiment design. The paper concludes with some comments.

2. THE PREDICTION ERROR MACHINERY

Sometimes we will make use of the assumption that the true system is LTI and given by

$$y(t) = G_\circ(q)u(t) + H_\circ(q)e_\circ(t) \tag{1}$$

where $e_\circ$ is white noise with variance λ_0 and $H_\circ$ stable, monic and minimum phase. We will denote the spectra of u and $H_\circ e_\circ$ by Φ_u and Φ_v, respectively.

2.1 Convergence and bias

Consider the LTI model structure

$$y(t) = G(q,\theta)u(t) + H(q,\theta)e(t)$$

where H is monic, stable and minimum phase and where $\theta \in \Theta \subset \mathbf{R}^n$. The signal e represents unmeasurable excitation signals. The prediction error is defined as

$$\varepsilon(t,\theta) = H^{-1}(q,\theta)(y(t) - G(q,\theta)u(t)). \tag{2}$$

The inverse of the noise model can also be interpreted as a prefilter that is used to filter the data y and u.

Given the data $Z^N = \{y^N, u^N\}$ ($x^N = [x(1),\ldots,x(N)]^T$), the parameter estimate is defined by

$$\hat{\theta}_N = \arg\min_{\theta\in\Theta} V_N(\theta),\; V_N(\theta) = \frac{1}{N}\sum_{t=1}^{N}\varepsilon^2(t,\theta).$$

Under weak assumptions (Ljung, 1999*b*), it holds that

$$\lim_{N\to\infty}\hat{\theta}_N = \theta^* \triangleq \arg\min_{\theta\in\Theta\subset\mathbf{R}^n}\lim_{N\to\infty}\mathbf{E}\{V_N(\theta)\} \text{ w.p.1.}$$

When the true system is given by (1) and operating in open loop, and when G and H are *independently parameterized*, the optimal G is characterized by

$$G(\theta^*) = \arg\min_{\theta\in\Theta} \int_{-\pi}^{\pi} |G_\circ(e^{j\omega}) - G(e^{j\omega},\theta)|^2 \frac{\Phi_u(\omega)}{|H(e^{j\omega},\theta^*)|^2}\,d\omega. \tag{3}$$

When (1) holds and the system is operating in closed loop,

$$\theta^* = \arg\min_{\theta\in\Theta} \int_{-\pi}^{\pi}\Big\{|G_\circ(e^{j\omega}) - G(e^{j\omega},\theta)|^2 |C\,S(G_\circ,C)|^2\Phi_r + \frac{|S(G_\circ,C)|^2}{|S(G(\theta),C)|^2}\Phi_v\Big\}\frac{1}{|H(e^{j\omega},\theta)|^2}\,d\omega, \tag{4}$$

where $S(G,C) = 1/(1+GC)$, holds. An alternative expression which characterizes the bias introduced by an erroneous noise model can be found in (Forssell and Ljung, 1999).

2.2 Variance

It holds also that the estimate $\hat{\theta}_N$ converges in law to a normally distributed random variable

$$\sqrt{N}(\hat{\theta}_N - \theta^*) \xrightarrow{\mathscr{D}} \mathscr{N}(0,P) \qquad \text{as } N\to\infty. \tag{5}$$

Furthermore,

$$\lim_{N\to\infty} N\cdot\mathrm{Cov}\{\hat{\theta}_N\} = P.$$

For our considerations, an interesting quantity is the variance of the frequency function estimate $\hat{G}_N(e^{j\omega}) \triangleq G(e^{j\omega},\hat{\theta}_N)$. Under the assumption of the true system in the model set, the following result (presented here for the case of open loop operation) holds (Ljung, 1999*b*)

$$\lim_{m\to\infty}\lim_{N\to\infty}\frac{N}{m}\mathrm{Var}(\hat{G}_N(e^{j\omega})) = \frac{\Phi_v(\omega)}{\Phi_u(\omega)}$$

where m is the model order. The result is based on a first order Taylor approximation, which for the case of fixed noise model is given by

$$N\cdot\mathrm{Var}(\hat{G}_N(e^{j\omega})) \approx \frac{dG^*(e^{j\omega},\theta)}{d\theta} P \frac{dG(e^{j\omega},\theta)}{d\theta}. \tag{6}$$

The above result suggests the intuitively appealing and well known approximation

$$\mathrm{Var}(\hat{G}_N(e^{j\omega})) \approx \frac{m}{N}\frac{\Phi_v(\omega)}{\Phi_u(\omega)} \tag{7}$$

which one would expect be valid for N and m large enough. An approximation with, in many cases, improved accuracy was proposed in (Ninness *et al.*, 1999). In (Xie and Ljung, 2001) an expression that is non-asymptotic in the model order m was derived for the case of a model with fixed denominator and fixed moving average noise model excited by an autoregressive (AR) input. In (Ninness and Hjalmarsson, 2002) this result was generalized and for the Box-Jenkins case of independently parameterized dynamics and noise models the result reads as follows.

Proposition 2.1. Suppose that the true system is operating in open loop and given by

$$y(t) = \frac{B_\circ(q)}{A_\circ(q)}u(t) + v(t)$$

where $v(t) = H_\circ(q)e_0(t)$ for some white noise sequence $e_0(t)$. Assume that the system is in the model set. Let $G(q,\theta) = q^{-k}B(q)/A(q)$ with m_b parameters in $B(q)$ and m_a parameters in $A(q)$.

Under the condition that

$$A_\dagger \triangleq A_\circ^2 H_\circ/\Phi_u^{1/2} \tag{8}$$

where $\Phi_u^{1/2}$ is the stable minimum-phase spectral factor of the input spectrum, is a polynomial in z^{-1} of degree at most $m_a + m_b$, it holds that

$$\lim_{N\to\infty} N\cdot\mathrm{Var}(\hat{G}_N(e^{j\omega})) = \kappa(\omega)\frac{\Phi_v(\omega)}{\Phi_u(\omega)}$$

where

$$\kappa(\omega) \triangleq \sum_{k=1}^{m_a+m_b}\frac{1-|\xi_k|^2}{|e^{j\omega}-\xi_k|^2} \tag{9}$$

where ξ_k, $k = 1,\ldots,m_a+m_b$, are the zeros of $z^{m_a+m_b}A_\dagger(z)$. ■

The preceding result suggests the following approximation for finite N

$$\mathrm{Var}(\hat{G}_N(e^{j\omega})) \approx \kappa(\omega)\,\frac{\Phi_v(\omega)}{N\Phi_u(\omega)}. \qquad (10)$$

Comparing with (7) we see that the factor m (the model order) is replaced by a frequency dependent factor $\kappa(\omega)$ which is a function of the poles of $A_{\dagger}$. Notice that $\int_{-\pi}^{\pi}\kappa(\omega)d\omega = m_a + m_b$ and hence that there is a "water-bed effect" in that a small variance in some frequency region has to be compensated for by high variance in another region.

Remark: Notice that the result holds, e.g., if the noise model is of MA-type and the input spectrum is of AR-type.

Since $\sqrt{N}(\hat{G}_N(e^{j\omega}) - G_\circ(e^{j\omega}))$ is also asymptotically normal distributed, we can compute the asymptotic distribution of $\sqrt{N}|\hat{G}_N(e^{j\omega}) - G_\circ(e^{j\omega})|$ and a $(1-\alpha)\cdot 100\%$ confidence region is given by

$$|\hat{G}_N(e^{j\omega}) - G_\circ(e^{j\omega})| \leq r_\alpha \cdot \sqrt{\kappa(\omega)\,\frac{\Phi_v(\omega)}{N\Phi_u(\omega)}} \qquad (11)$$

with r_α denoting the α-level for the distribution in question. We will use (11) as a generic description of confidence regions in the frequency domain. When Proposition 2.1 is not valid this means changing the definition of $\kappa(\omega)$.

3. INFORMATION CONTENT IN THE DATA

The question of what information the noisy measurement data contains regarding the system dynamics is really at the core of system identification. One can view system identification as the problem of cleaning up the data wrt noise as well as possible. In this section we will discuss the limitations of what can be achieved in this respect.

3.1 The concept of unfalsification

That a scientific theory may be falsified by contradicting evidence but never validated by corroborating evidence was elaborated on by the philosopher Karl Popper. In system identification, it is clear that all models (including unmeasurable external excitation signals) that are consistent with the observed input/output data Z^N cannot be discarded unless some prior information is available. We call this set of models *the set of unprejudiced unfalsified models*, which we denote by $\mathcal{G}(Z^N)$. This set represents the remaining uncertainty of the system dynamics given the observed data Z^N. The information contents in the observed data corresponds exactly to the set of models that are *falsified* by the observed data. The more "informative" data is, the larger is the set of candidate models that can be falsified.

3.2 Unfalsifying controllers

Suppose now that some apparatus is going to be constructed that when applied to the system performs a certain task and that, given input/output data Z^N from the system, we would like to verify if the designed system satisfies some performance specifications. To be specific, let us consider the problem of testing whether a certain controller C satisfies some given performance specifications. Well, if the controller satisfies the performance on at least one of the models in $\mathcal{G}(Z^N)$, then the controller C cannot be discarded since the model may correspond to the true system in which case the controller would satisfy the specifications. We would in this case say that the *controller is unfalsified.* However, since the models in $\mathcal{G}(Z^N)$ have completely arbitrary input/output behavior except for the specific trajectory defined by Z^N, which they all share, it is possible to find an unfalsified model such that the closed loop system consisting of this model and C satisfies any specifications, provided that these specifications are not violated when the closed loop system with C as controller exhibits the input/output behavior Z^N. Thus it is often very simple to check whether a given controller can be falsified or not by data only. The idea of controller unfalsification was introduced by Safonov and co-workers (Safonov and Tsao, 1997).

To illustrate the machinery suppose that a reference model T_d is given and that our performance specifications are

$$\|y - T_d r\|_{rms} \leq \bar{\rho}\,\|r\|_{rms} \quad \forall \|r\|_{rms} < \infty \qquad (12)$$

for a given constant $\bar{\rho}$ where $\|x\|_{rms}^2 = \lim_{N\to\infty} \frac{1}{N}\sum_{t=1}^{N}|x(t)|^2$. We now ask the following question. Given arbitrary input/output data Z^N, what can be said about which controllers that satisfy (12)? The key to resolving this problem is to note that for arbitrary input/output data u, y from the system, the signal

$$r_C(t) = \frac{1}{C}u(t) + y(t)$$

is the reference signal which with C in the loop would produce exactly the input/output data u, y. Hence given arbitrary input/output data $\{u(t), y(t)\}_1^N$, we can think of the corresponding $\{r_C(t),\, y(t)\}_1^N$ as a (fictitious) closed loop data set with the controller C in the loop. The controller is falsified precisely when the specifications (12) are not satisfied for the specific signals $\{r_C(t),\, y(t)\}_1^N$.

3.3 Introducing priors

The only possible way to reduce the size of the set of unprejudiced unfalsified models is to introduce prior information. For example, we might immediately be prepared to introduce the prior that the system is causal. We shall denote by $\bar{\mathcal{G}}(Z^N)$ the set of models consistent with data and the prior and we shall refer to this set as *the set of unfalsified models.*

A common approach is to introduce a parameterized model structure, e.g.

$$y = G(\theta)u + H(\theta)e + \Delta(u) \qquad (13)$$

where G and H are transfer functions parameterized by $\theta \in \Theta \subset \mathbf{R}^n$, where Δ is an unstructured dynamic term and where e is an unmeasurable excitation signal (noise). Notice that we can regard e^N as a vector of unknown parameters, just as θ. Since the unmeasurable excitation e may be taken such that any model in the above structure is unfalsified, assumptions on the unmeasurable excitation are also required. This issue is crucial to the system identification problem.

3.4 Set-membership identification

The set of unfalsified models becomes manageable for the model structure (13) by imposing that $\Delta \in S_\Delta$ and $e^N \in S_e$ for some suitably chosen sets S_Δ and S_e. Common choices for S_e are

$$S_e = \{e^N : |e(t)| \leq c,\, t = 1,\ldots,N\}, \qquad (14)$$

$$S_e = \{e^N : \sum_{t=1}^{N} e^2(t) \leq c\}. \qquad (15)$$

The unstructured uncertainty Δ is often taken as LTI and examples of S_Δ are

$$S_\Delta = \{\Delta = \sum_{k=1}^{\infty} \delta(k)q^{-k} : |\delta(k)| \leq C\lambda^k, 0 < \lambda < 1\},$$
$$S_\Delta = \{\Delta : \|\Delta\|_{\mathbf{H}_\infty} \leq \gamma\}. \quad (16)$$

The set of unfalsified models often becomes very complicated and outer- and inner-bounding techniques have to be used to provide simplified characterizations (Milanese and Vicino, 1991). We refer to (Milanese, 1998) and references therein for further details on set-membership identification.

3.5 *Uncertainty model unfalsification*

The combination (15) and (16) has been studied by Kosut and co-workers in a series of papers under the label *uncertainty model unfalsification.* As pointed out in (Kosut, 2001), one may compute $c_{\min}(\gamma)$, the smallest c for which there is some unfalsified model for a given bound γ on the unstructured uncertainty Δ. The graph $\gamma \to c_{\min}(\gamma)$ is referred to as the *uncertainty trade-off curve* and gives a hint on how dynamic versus unmeasurable excitation uncertainty may be traded-off.

3.6 *A likelihood approach to unfalsification*

One way to treat the unobservable excitation is to introduce a measure of how likely different excitation sequences are, i.e. a probability measure. Equipped with this measure, we can in principle order *all* models we can imagine according to how likely the corresponding external excitation is. The model corresponding to the most likely noise sequence is the maximum likelihood (ML) estimate.

To be specific, let us assume that the unobservable excitation is Gaussian white noise with variance λ and denote by $\varepsilon(t,M)$ the unobservable excitation signal corresponding to model M. Then the negative log-likelihood for this model is given by

$$V_N(M) = \frac{N}{2}\log(2\pi) + \frac{N}{2}\log(\lambda) + \frac{1}{2\lambda}\sum_{t=1}^{N}\varepsilon^2(t,M).$$

It is natural to take as the set of unfalsified models, the set of models that corresponds to noise sequences with a likelihood higher than a given level. When the noise variance is known, this corresponds to

$$\mathscr{G}(Z^N) = \left\{M : \frac{1}{2\lambda}\sum_{t=1}^{N}\varepsilon^2(t,M) \leq c\right\}$$

for some c. Hence we see that the stochastically motivated likelihood approach leads to *set-membership identification* with the set (15) used to characterize the unmeasurable excitation.

The principal difference between the likelihood approach and a deterministic approach lies in how c is chosen. In the likelihood approach one could argue that the constant c should be selected such that there is only a small probability that the true system is (erroneously) falsified.

In practice it is, of course, computationally infeasible to order all models and some parameterization has to be introduced. See (Ljung and Hjalmarsson, 1995) for an illustration.

Notice also that by instead assuming that the distribution has support $[-c,c]$ leads to a set of unfalsified models of the type (14) (Ninness and Goodwin, 1995).

3.7 *Stochastic embedding*

Just as e may be modeled in a stochastic framework, the unstructured uncertainty Δ in the model structure (13) may be modeled in a stochastic framework. This leads to what is known as the stochastic embedding approach (Goodwin *et al.*, 1992; Goodwin *et al.*, 2002).

3.8 *The prediction error approach*

When it is known that the true system is in the model set (1), and the asymptotic theory in Section 2 is valid, then (5) gives that the true parameter vector belongs to the set

$$\left\{\theta : (\theta - \hat{\theta}_N)^T P^{-1}(\theta - \hat{\theta}_N) \leq \frac{1}{N} \cdot \chi^2_\alpha(n)\right\}, \quad (17)$$

where $\chi^2(\alpha)_n$ is the α-level of the $\chi^2(n)$ distribution, with probability $1-\alpha$. Hence, it is natural to call this set, the set of unfalsified models $\mathscr{G}(Z^N)$. Furthermore, (11) gives, frequency by frequency, an approximate uncertainty set for the estimate $\hat{G}_N$ for a given confidence level α. Hence, a set that approximates the set of unfalsified models is given by

$$\left\{G : |\hat{G}_N(e^{j\omega}) - G(e^{j\omega})| \leq r_\alpha \sqrt{\kappa(\omega)\frac{\Phi_v(\omega)}{N\Phi_u(\omega)}}\right\}. \quad (18)$$

For a recent study on conditions for the asymptotic theory to be valid, see (Bittanti *et al.*, 2002).

3.9 *Maximum likelihood estimation*

When the true system is in the model set and the probability density function (pdf) for the unmeasurable excitation is known or can be parameterized, the ML-estimate is consistent and in addition *asymptotically efficient* under general conditions. This means that there is no other consistent estimator which will give a smaller set of unfalsified models when the data set is large enough.

3.10 *On the value of noise priors*

Let us return to the likelihood approach in Section 3.6. When the noise spectrum is known it is possible that the set of unfalsified models is empty, i.e. there is no single model in the model set which corresponds to a likely noise sequence and the whole model structure is falsified – a very powerful result. When the noise variance λ is known, but not the noise spectrum, it may still be possible to falsify an incorrect model structure. However, it will be considerably more difficult. The reason is that a flexible noise model can model some of the unmodeled dynamics so that the mean-square error of the prediction errors still is small. When the noise variance is unknown, the situation is much worse, since we then do not have a good way of selecting the threshold c. To conclude, we see that knowledge of the noise characteristics is extremely valuable. As pointed out, e.g. in (Ljung, 1999*a*), the noise sequence itself can be estimated if periodic inputs are used, c.f. the case when the input is zero.

3.11 *Summary*

In this section, we have presented identification as a way of producing sets of unfalsified models and illustrated that both deterministic and stochastic modeling paradigms fit into this framework. There is an on-going healthy cross-fertilization of ideas between deterministic and stochastic approaches, see e.g. (Tjärnström and Garulli, 2002; Hakvoort *et al.*, 1994). See also (Reinelt *et al.*, 2002) for a comparison of different approaches. For excellent overviews of different modeling frameworks we refer to (Mäkilä *et al.*, 1995) and (Ninness and Goodwin, 1995).

4. MODEL VALIDATION

When a model structure has been selected, the set of unfalsified models $\bar{\mathcal{G}}(Z^N)$ can by definition not be falsified by the data Z^N. One could say that the model builder is trapped inside the model structure. Hence, there is a need to "*look over the fence*" to ensure that there are no other model structures that can represent the data in a more plausible way, or alternatively test $\bar{\mathcal{G}}(Z^N)$ on *new* data. This is what model validation is all about!

4.1 Model error modeling

In (Ljung, 1999*a*) the concept of model error modeling is discussed. It is pointed out that standard model validation tests such as cross-correlation tests between residuals and inputs can be interpreted as first modeling the residuals, with the resulting model named *model error model*, and then testing whether the zero model is included in the set of unfalsified model error models. It is suggested that an intuitively appealing way of presenting these tests is by plotting the Bode-diagram of the model error model with uncertainty regions indicated. From this insight follows also that more complex models than finite impulse response (FIR) models (which correspond to standard cross-correlation tests between input and residuals) can be used and it is recommended that the model structure for the model error model should be considerably richer than the nominal model. The main message, however, is that if the nominal model is unfalsified, i.e. the uncertainty region for the model error model includes the zero model, then, even though the nominal model structure (with its own uncertainty description) is unfalsified, one should use the nominal model structure together with the *uncertainty region of the model error model*. Since the model structure for the model error model is more flexible than the model structure for the nominal model, this will give a larger, and hence "safer", set of unfalsified models. We are here at the crux of the modeling problem – the model builder wants to be sure that his model set includes the 'true' system. However, we stress that

- even though the set of unfalsified model based on the model error model structure is more conservative, there is still no guarantee that this set contains the true system.
- as we look wider and wider outside the selected model structure, our statements concerning the system will become weaker and weaker since less and less (assumed) prior information is used.

4.2 Validating with confidence

From the observations in the preceding section we conclude that it would boost the confidence of the model builder, if one could establish that a model (and its corresponding uncertainty set) would be falsified if the unmodeled dynamics is in a certain model structure $\mathcal{M}_{ME}$. Well, let us examine the outcome if a model error model is estimated using the structure $\mathcal{M}_{ME}$ to which the unmodeled dynamics belong. For simplicity, let us assume that the true system is LTI, c.f. (1), and that the asymptotic results in Section 2 are valid. In this case the residuals (2) are given by

$$\varepsilon = (G_\circ - \hat{G}_N)u_F + \frac{H_\circ}{\hat{H}_N}e_\circ$$

where $u_F = \hat{H}_N^{-1}u$, and the model $\hat{G}_N$ will be falsified if the uncertainty region for the model error model, which we denote by

$$\varepsilon = G_\varepsilon(\theta)u_F + H_\varepsilon(\theta)e,$$

does not include the zero model.

Using (11) and some simple algebra, this is guaranteed to happen if

$$|G_\circ(e^{j\omega}) - \hat{G}_N(e^{j\omega})| > 2r_\alpha\sqrt{\kappa(\omega)\cdot\frac{\Phi_v(\omega)}{N\Phi_u(\omega)}} \quad (19)$$

for some ω. Hence, if we want to ensure that model errors larger than some function $\delta(\omega)$ are detected, then the experiment should be carried out such that the right-hand side of (19) is less than $\delta(\omega)$. We emphasize, again, that this conclusion is predicated on the assumption that $\mathcal{M}_{ME}$ is flexible enough to capture the unmodeled dynamics.

Example 4.1. A third order system with a resonance is corrupted by white noise and excited with a low pass input, also with a resonance. The system is identified using a first order output error model. The model, together with its uncertainty region is shown in Figure 1. Clearly the model has missed the resonance peak and the uncertainty region is misleading. The model error is shown in Figure 2 together with the bound from (19) based on a 10th order FIR model error model. We see that we can expect to detect the resonance in our model error model but not any model error at other frequencies. In the same figure, the uncertainty bound for a 10th order FIR model error model is shown. As predicted, the resonance peak is detected since the uncertainty region for the model error model does not include zero around the resonance. ■

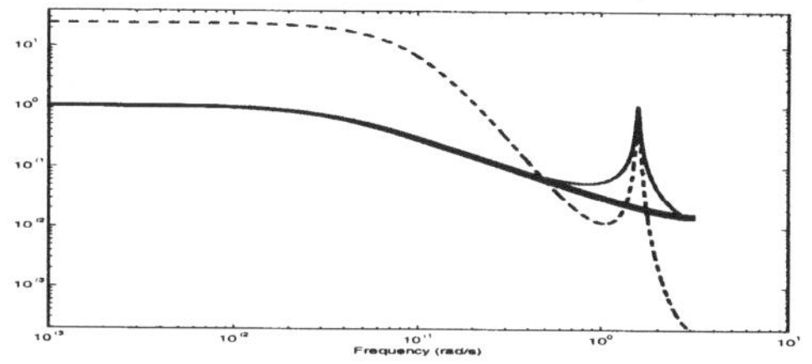

Fig. 1. Dashed line: Input spectrum. Solid line: True system. Thick line: Uncertainty region around estimated nominal first order OE-model.

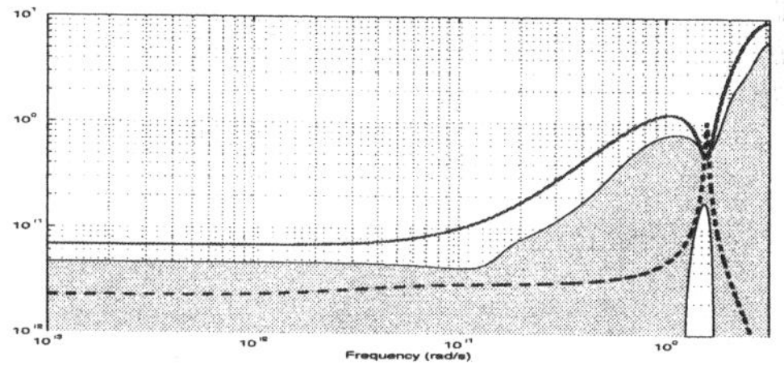

Fig. 2. Dashed line: Model error. Solid line: Lower bound for model errors that are guaranteed to be detected by a 10th order FIR model error model. Shaded area: Uncertainty region for estimated 10th order model error model.

Notice that the condition (19) depends on the input not only through the input spectrum but also through the factor $\kappa(\omega)$. This has a, perhaps unexpected, implication.

Example 4.2. (Example 4.1 continued). Suppose that the order of the model error model is increased from 10 to 100. One would then expect the lower bound (19) for detecting unmodeled dynamics to increase significantly. In view of (7), which is linear in the model order m, it should increase by a factor of $\sqrt{100/10} \approx 3.2$. The bound is shown in Figure 3 for the two cases. We see that there is actually an increase of approximately 3, except at low frequencies and especially around $\omega = \pi/2$, which happens to be where the peak of the input spectrum is located, where there is *only a minor increase*.

Notice also that the even though there is a peak in the input spectrum at $\omega = \pi/2$, it is a factor of 20 smaller than the input spectrum at low frequencies, c.f. Figure 1. Hence, the small increase around $\omega = \pi/2$ can not be explained by the magnitude of the input spectrum around this frequency.

The phenomenon is due to the factor $\kappa(\omega)$, defined in (9), which is present (19). A plot of $\kappa(\omega)$ is shown in Figure 4 for the two model orders. The poles in $\kappa(\omega)$ consists in this case of the poles 0.9, 0.9, $0.97e^{\pm j\pi/2}$ of the stable spectral factor of the input spectrum and the poles of the FIR-model. The double pole at 0.9 gives a large contribution to $\kappa(\omega)$ at low frequencies, whereas the complex poles give a large contribution around $\omega = \pi/2$. Since all the poles of an FIR-model are at the origin, $\kappa(\omega)$ gets a frequency independent contribution of m from an mth order FIR-model. Thus, $\kappa(\omega)$ increases linearly with the model order m and this is clearly seen in Figure 4 since the solid line (corresponding to $m = 100$) is offset by $100 - 10 = 90$ above the dashed line (corresponding to $m = 10$). However, the *relative* increase at different frequencies is vastly different. At frequencies where the poles of the input spectrum contribute very little, the increase is a factor 10 but at frequencies where the influence from the poles of the input spectrum is significant, the relative increase is much less. Hence, the relative increase in the uncertainty bound is much less at low frequencies and, especially, around $\omega = \pi/2$. ■

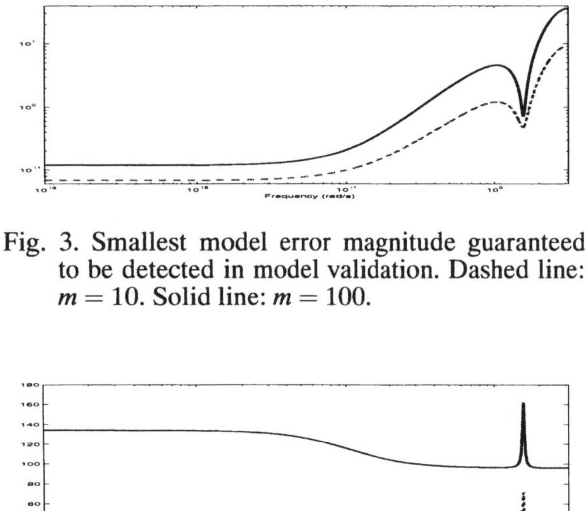

Fig. 3. Smallest model error magnitude guaranteed to be detected in model validation. Dashed line: $m = 10$. Solid line: $m = 100$.

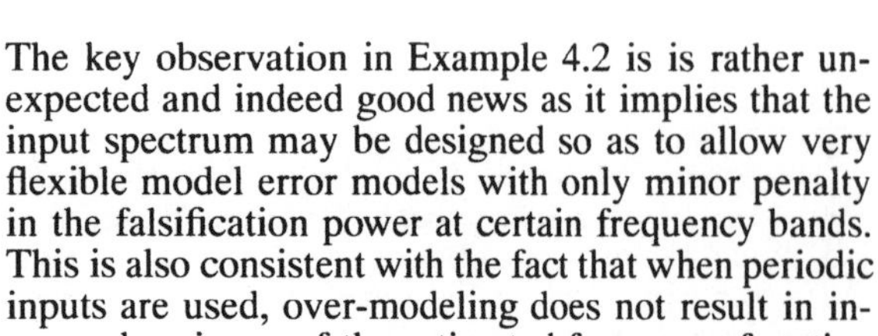

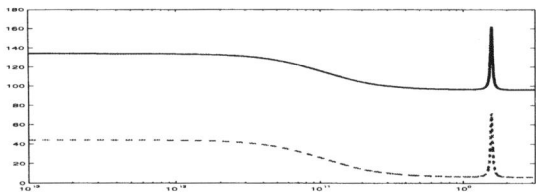

Fig. 4. $\kappa(\omega)$. Dashed line: $m = 10$. Solid line: $m = 100$.

The key observation in Example 4.2 is is rather unexpected and indeed good news as it implies that the input spectrum may be designed so as to allow very flexible model error models with only minor penalty in the falsification power at certain frequency bands. This is also consistent with the fact that when periodic inputs are used, over-modeling does not result in increased variance of the estimated frequency function for frequencies corresponding to the spectral lines of the input. However, notice also a large $\kappa(\omega)$ gives larger uncertainty bounds so, near peaks of the input spectrum, the bounds can be significantly worse than the noise to signal ratio.

5. A STATISTICAL VIEW ON APPROXIMATE MODELING FOR CONTROL

Figure 5 illustrates various ways of obtaining a restricted complexity controller via identification. The obvious question is of course if one of the paths is better than the others. Here we will discuss this problem from a statistical point of view.

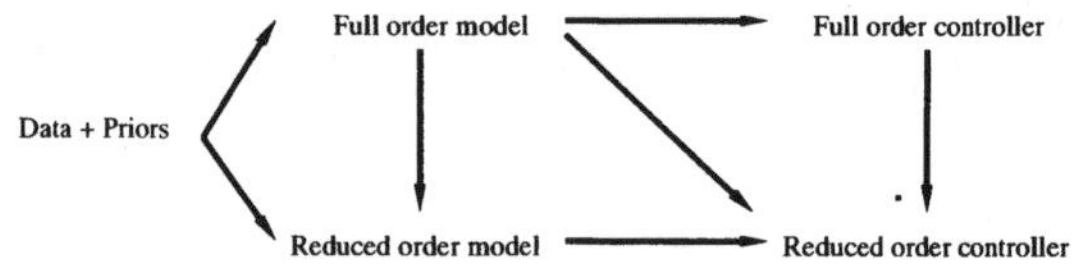

Fig. 5. Different possibilities of mapping data and prior information into a controller of reduced complexity.

5.1 Statistical advantages of biased models

From a statistical perspective, approximate modeling is usually motivated by examples such as the following.

Example 5.1. Consider the following high-order FIR system

$$y(t) = \sum_{k=1}^{n} g_k^{\circ}\, u(t-k) + e_{\circ}(t) \qquad (20)$$

where $e_{\circ}(t)$ is white Gaussian noise with variance σ_e^2, where the order n is *very* large and which is driven by white Gaussian noise u with variance σ_u^2.

Suppose that one is interested in estimating the static gain of the system. In the ML-approach one would then use a model structure of the same type as (20) and estimate $\theta = [g_1, \ldots, g_n]^T$ using least squares. The covariance matrix of $\hat{\theta}_N$ is approximately $\sigma_e^2/(N\sigma_u^2)\mathbf{I}$, and hence the variance of the estimated static gain $\hat{G}(e^{j0}) = \sum_{k=1}^{n} \hat{g}_k$ is approximately given by $n\sigma_e^2/(N\sigma_u^2)$ and we see that due to the high system order n, the uncertainty can be significant even if the input power is large.

This observation, naturally, prompts the idea that a (slightly) biased estimate of the transfer function may give an estimate at the zero frequency which is better. For example, using the model structure

$$y(t) = \eta u(t-1) + e(t) \qquad (21)$$

will give a mean-square error of approximately $(\sigma_e^2 + \sum_{k=2}^{n} |g_k^{\circ}|^2\, \sigma_u^2)/(\sigma_u^2 N) + |\sum_{k=2}^{\infty} g_k^{\circ}|^2$ which is significantly lower than for the ML estimate if only g_1° contributes significantly to the steady state gain! ■

The above example indicates that the ML-approach may be unsuitable when only approximate models are required for highly complex systems. However, the issue is a bit more subtle than at first glance.

5.2 A separation principle

As mentioned in Section 3.9, the ML-estimator is asymptotically efficient in the set of consistent estimators of the parameters θ describing the system. However, this property extends over to any quantity which is a smooth function $f(\theta)$ of the system parameters, (Lehmann, 1983). This provides us with a useful separation principle:

The estimator of some system dependent quantity that

i) first estimates a full order model using ML, and then

ii) uses the full order system estimate obtained in i), as if it were the true system, to estimate the desired quantity

is asymptotically efficient among all consistent estimators of this quantity.

Example 5.2. (Example 5.1 continued). Using the first impulse response only of the full-order ML estimate as a model, c.f. (21), will result in a biased estimate of the static gain with a mean square error $\sigma_e^2/(\sigma_u^2 N) + |\sum_{k=2}^{\infty} g_k^{\circ}|^2$ which is smaller than for the biased estimate suggested in Example 5.1. The reason is that the variance of the estimate obtained from (21) is inflated due to the unmodeled dynamics as compared to the ML-estimate of the first impulse response coefficient. ■

One application of this separation principle is to model reduction. Suppose that it is known that the true system $G_\circ$ belongs to some model structure parameterized by $\theta \in \Theta$ but that the desired quantity is a consistent estimate of the frequency function minimizing

$$\int_{-\pi}^{\pi} |G_\circ(e^{j\omega}) - G(e^{j\omega},\eta)|^2\, \Phi_u(\omega) d\omega$$

where $G(q,\eta)$ is a low order model parameterized by η. Then it is optimal, wrt to the variance of the estimated low order frequency function, to first estimate a full-order model $G(q,\hat{\theta}_N)$ using ML-estimation, and then to perform a model reduction step by minimizing

$$\int_{-\pi}^{\pi} |G(e^{j\omega},\hat{\theta}_N) - G(e^{j\omega},\eta)|^2\, \Phi_u(\omega) d\omega$$

wrt η. In (Tjärnström and Ljung, 2002), the increase in variance when a low order model is directly estimated is explicitly characterized when the prediction error method is used.

Another illustration of the separation principle can be found in (Zhu, 2000) where identification for simulation is considered. It is shown that modeling the spectrum of the noise is better than ignoring it, even though simulation does not require a noise model.

This principle applies also to control problems and indicates that, from an accuracy point of view, no matter what the ultimate objective is, be it modeling to tune a simple PID-controller or modeling suitable for high performance control, one should always *first try to model as well as possible*. After that, any simplifications can be performed without jeopardizing the statistical accuracy. Hence, returning to Figure 5, taking the lower path should be avoided if accuracy is a concern. We also conclude that going from a full order model directly to a low order controller or via a high order controller, will not significantly affect the statistical accuracy.

Finally we point out that a nice feature of first using a full-order model is that the derivation of the corresponding variance is simplified as the full order model has a parameter covariance matrix given by a simple expression (Ljung, 1999*b*).

5.3 Taking the experiment design into account

The issue of biased modeling vs full-order modeling has an additional dimension also, namely the consequences on the accuracy of experiment design. Let us return to Example 5.1 but let us now take the experiment design into consideration.

Example 5.3. (Example 5.1 continued). Suppose that the allowed input power is bounded by σ_u^2. Then, clearly a constant input with amplitude σ_u is optimal for estimating the static gain and even though the ML-estimate of the impulse response coefficients will be very poorly conditioned for finite data (singular asymptotically), it is easy to show that the estimate of the static gain will have variance approximately equal to $\sigma_e^2/(N\sigma_u^2)$. But now, the same accuracy is obtained with the static model (21) since the unmodeled dynamics do not influence the accuracy of the estimate $\hat{\eta}_N$ of η, in fact it is accounted for by $\hat{\eta}_N$ which is now an unbiased estimate of the static gain! ■

The example above suggests that when also the experiment design is taken into account, the difference between a biased method and ML decreases. Recall that a biased method has to make use of some prior information of the system to be useful. By making the comparison include the experiment design (which also requires system knowledge), the biased and the ML approaches are put on a more equal footing.

6. LINKS BETWEEN CONTROL AND IDENTIFICATION

In this section we will discuss how control and identification interact.

6.1 Robust control

Figure 6 illustrates a feedback configuration where the controller C and the true system $G_\circ$ are LTI. The closed loop system equations are

$$\begin{pmatrix} y \\ u \end{pmatrix} = \begin{pmatrix} \dfrac{G_0 C}{1+G_0 C} & \dfrac{G_0}{1+G_0 C} \\ \dfrac{C}{1+G_0 C} & \dfrac{1}{1+G_0 C} \end{pmatrix} \begin{pmatrix} r - v_\circ \\ w \end{pmatrix}. \quad (22)$$

Here r and w are known external excitations whereas $v_\circ$ represents an unmeasurable (disturbance/noise) term. We denote by $S(G_\circ,C) \triangleq 1/(1+G_\circ C)$, the achieved sensitivity function and by $T(G_\circ,C) \triangleq 1 - S(G_\circ,C)$, the achieved complementary sensitivity function. Let γ be a scalar performance measure. From a control point of view, the problem is to design the controller C such that the closed loop system in Figure 6 has as high performance γ as possible as well as guaranteed stability, for all systems in the set of unfalsified models $\mathscr{G}$.

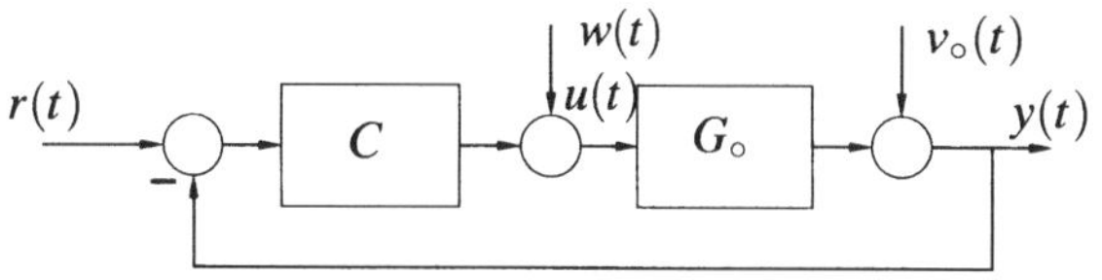

Fig. 6. Closed loop system.

One may, e.g., consider minimizing the mixed sensitivity $\|\gamma W_p S \;\; W_s T\|_{\mathbf{H}_\infty}$ where W_p is a weighting matrix representing the performance specifications on the sensitivity function and where W_s is a weighting matrix representing $\mathscr{G}(Z^N)$. We may here discern the relationship between our design variables. Increasing γ will require decreasing W_s which means increasing the input power spectrum. However, the relationship is implicit and hidden by the $\mathbf{H}_\infty$ machinery.

6.2 Linking identification and robust control

For future reference, let

$$\begin{aligned} J(G_\circ,C_\circ,G,C) &\triangleq \|T(G_\circ,C_\circ) - T(G,C)\|^2 \\ &= \|S(G_\circ,C_\circ) - S(G,C)\|^2 \\ &= \|((G_\circ - G)C_\circ + (C_\circ - C)G)S(G_\circ,C_\circ)S(G,C)\|^2 \end{aligned} \quad (23)$$

and let $C = C(G,\gamma)$ represent any model based control design method, e.g. internal model control or $\mathbf{H}_\infty$

loop-shaping, where, as above, γ denotes some performance index, e.g. designed bandwidth.

Robustness can also be ensured by choosing the model such that some performance relevant norm, e.g. $\mathbf{H}_2$ or $\mathbf{H}_\infty$, of the difference between the achieved and the designed complementary sensitivity functions, is small when both loops have $C(G,\gamma)$ as controller. This means minimizing $V(G,C(G,\gamma))$ wrt G where

$$V(G,C) \triangleq J(G_\circ,C,G,C). \quad (24)$$

We shall now analyze the cost function V in some more detail. We have

$$T(G_\circ,C) - T(G,C) = \underbrace{\frac{1}{1+\Delta}}_{\text{stability}} \times \underbrace{\Delta \times S(G,C)}_{\text{performance}} \quad (25)$$

where Δ is the weighted relative model error

$$\Delta = \Delta(G_\circ,G,C) \triangleq \frac{G_\circ - G}{G} T(G,C). \quad (26)$$

This interesting expression can be interpreted in the following way:

• The first factor, i.e. $1/(1+\Delta)$, is related to stability of the achieved closed loop system. If the designed closed loop is stable (which is a natural requirement), then stability of the achieved closed loop system is equivalent to the stability of the factor above. Using the small gain theorem, a sufficient stability condition is

$$|\Delta| < 1 \quad \forall \omega \quad (27)$$

which is valid if $G_\circ$ and G have the same number of unstable poles. We recognize this condition as a well known robust stability result.

Regarding performance, the following over-bound holds when $|\Delta| < 1$

$$|T(G_\circ,C) - T(G,C)| \leq \frac{1}{1-|\Delta|} |\Delta| \, |S(G,C)|.$$

This gives the following conditions on $|\Delta|$:

• At high frequencies $|S(G,C)| \approx 1$ so in this frequency band $|\Delta| \ll 1$ guarantees robust performance.

• Since the designed sensitivity $|S(G,C)|$ typically is small for low frequencies, we see that the condition on $|\Delta|$ is less severe in this frequency band compared to at high frequencies. This is a manifestation of the forgiving nature of feedback.

• $|S(G,C)|$ may be large around the designed bandwidth if the nominal design is not robust, e.g. when the designed bandwidth exceeds the limitations imposed by model non-minimum phase zeros. Hence, more severe restrictions on $|\Delta|$ may apply in this frequency band.

Notice that Δ is involved both in the performance condition and the stability condition. The magnitude of this factor depends on both the *quality of the model* G and the *control specifications* (represented by $T(G,C(G,\gamma))$). It is around Δ where the interplay between system identification and control is staged and below we will illustrate some of the trade-offs that it induces by considering the problem of making $|\Delta| \leq \alpha \ll 1$ for some α.

Trade-off 1: Performance specifications vs experimentation effort Suppose that the set of unfalsified models is given by (18). Then $\|\Delta\|_{\mathbf{H}_\infty}$ is with probability $1-\alpha$ bounded by

$$\sup_{\omega} r_\alpha \left| \frac{T(\hat{G}_N,C(\hat{G}_N,\gamma))}{\hat{G}_N} \right| \sqrt{\kappa \frac{\Phi_v}{N\Phi_u}}. \quad (28)$$

Here $T/\hat{G}_N$ is the designed transfer function from reference to input, which we could call the designed control effort. Hence, the above expression says that the square root of the signal energy density to noise power density ratio ($\sqrt{\text{SENPR}}$) $\sqrt{N\Phi_u/\Phi_v}$ has to be at least an order of magnitude larger than the control effort to guarantee robust performance.

In order to get some insight in what this implies, notice that in the passband of the designed complementary sensitivity function, $|T/\hat{G}_N| \approx |1/\hat{G}_N|$. Typically $G_\circ$, which $\hat{G}_N$ tries to approximate, is of low-pass character and hence this factor will have magnitude 1 or smaller for low frequencies up to when the system's gain drops below 1. Above this frequency the magnitude of $|T/\hat{G}_N|$ will increase until the bandwidth of T is reached where it will start to decrease if T rolls-off faster than $\hat{G}_N$. Thus for a system that rolls off like $1/\omega^n$, a $\sqrt{\text{SENPR}}$ proportional to ω^n is required in this frequency band. This indicates that increasing the closed loop bandwidth becomes increasingly expensive wrt experimentation beyond the open loop system's own bandwidth. We shall discuss experiment design further in Section 11.

Another important conclusion is that (28) involves quantities which a priori are unknown. Hence, it may be beneficial to update the experiment design as more information about the system is obtained. This leads to *adaptive experiment design* which we shall discuss in Section 11.3.

In summary, (28) captures the trade-off between the input spectrum and the performance specifications. The importance of adapting the performance specifications to the system uncertainty is the leading principle in the 'wind-surfer approach', see e.g. (Lee *et al.*, 1995). Here the performance is gradually increased along with improved confidence in the model. See also (Cadic *et al.*, 2003) for a recent discussion on the importance of this issue.

Trade-off 2: Performance specifications vs model complexity Suppose that in order to limit the complexity of the controller, a low order model G_{lo} is to be used. From the separation principle in Section 5.2 we have that the optimal model is obtained by first estimating a full order model $\hat{G}_{ML}$ using ML and then solving

$$\hat{G}_{lo} = \arg\min_{G_{lo}} \sup_{\omega} \left| \frac{\hat{G}_{ML} - G_{lo}}{G_{lo}} T(G_{lo},C(G_{lo},\gamma)) \right|.$$

The total error $\hat{G}_{lo} - G_\circ$ can be assessed by also computing the variance for $\hat{G}_{lo}$. However, already the minimum of the expression above gives useful information. For given bound α on $|\Delta|$, a lower bound on the complexity of G_{lo} is obtained. Conversely, given the complexity of G_{lo}, an upper bound on achievable performance γ is obtained.

6.3 Direct parameterization in C

Above we have considered choosing a model such that the performance difference $V(G,C(G))$ is small. When the control design $C(G)$ is injective, one may parameterize G directly in terms of the controller C. For model reference control

$$C = \frac{1}{G}\frac{T_d}{1-T_d}, \quad (29)$$

where T_d is the desired complementary sensitivity function, we have $G = \frac{1}{C}\frac{T_d}{1-T_d}$. In this case one may minimize $V(G(C),C)$ directly wrt the controller C instead of minimizing $V(G,C(G,\gamma))$ wrt G. The parameterization above is known as direct parameterization in adaptive control.

6.4 *Minimizing average performance degradation*

In Section 6.2, the same controller $C(\hat{G}_N,\gamma)$ was used on the true system as on the nominal design. It is possible to obtain an achieved performance closer to the nominal performance $T(\hat{G}_N,C(\hat{G}_N,\gamma))$ by replacing $C(\hat{G}_N,\gamma)$ with the controller C which minimizes the average performance deterioration

$$\mathbf{E}\{J(G_\circ,C,\hat{G}_N,C(\hat{G}_N,\gamma))\},$$

where the expectation is over $G_\circ$ in the set of unfalsified models (Goodwin *et al.*, 1999).

7. DIRECT IDENTIFICATION OF RESTRICTED COMPLEXITY MODELS FOR LTI SYSTEMS

Since the order of the controller typically depends on the order of the model, it is often desirable to restrict the complexity of the model. Thus there is a need to identify restricted complexity models that are suitable for control. This is the theme of this section. Formally, we consider minimizing $V(G,C(G))$ (24) wrt G.

7.1 *Asymptotic efficient identification*

From Section 5.2 it follows that it is optimal, wrt the asymptotic statistical accuracy, to first identify a full-order model using ML, and then reduce the complexity according to the following procedure:

Let $G(\theta)$ represent the full order estimate, i.e. $\exists\theta_\circ$ s.t. $G(\theta_\circ)=G_\circ$, and let $\hat{\theta}_{ML}$ denote the corresponding ML-estimate. Let $G(\eta)$ represent a restricted complexity model and define

$$\hat{J}(\theta,\eta,\gamma)\triangleq J(G(\theta),C(G(\eta),\gamma),G(\eta),C(G(\eta),\gamma)).$$

Then, take $\hat{\eta}$ as the minimizer of $\hat{J}(\hat{\theta}_{ML},\eta,\gamma)$ wrt η.

Here we note that the minimization of $\hat{J}(\hat{\theta}_{ML},\eta,\gamma)$ is nonlinear and, hence, it may be difficult to obtain the global optimum. The variance of $J(G_\circ,G(\hat{\eta}),\gamma)$ may be estimated numerically using Gauss' approximation formula and implicit differentiation of $\hat{J}(\theta,\eta,\gamma)$. Hence, for given data, the performance specification γ may be adjusted so that achieved and designed performance are guaranteed to be sufficiently close as well as to ensure stability. However, this procedure gives little insight in how design variables, such as input spectrum and γ, influence the accuracy. Note, though, that the considerations in Section 6.2 are of help in this respect.

Any consistent estimate of $G_\circ$ used in the first step will result in a consistent estimate of the optimal restricted complexity model, but the accuracy will be worse compared to the ML-estimate. In (Hjalmarsson and Lindqvist, 2001) a non-parametric estimate is used.

7.2 *Direct restricted complexity identification*

It may not be convenient to identify a full-order model and around 1990 several ideas of how to directly identify a model G which approximately minimize $V(G,C(G))$ (24) appeared. The idea is to choose the design variables in the identification method such that the asymptotic bias expression, e.g. (3), approximates (24).

The difference (25) can also be expressed as

$$\begin{aligned}T(G_\circ,C)-T(G,C)&=\Delta(G_\circ,G,C)S(G_\circ,C)\\&=S(G,C)\,(G_\circ-G)CS(G_\circ,C).\end{aligned}\tag{30}$$

Hence, (recall (24))

$$V(G,C)=\int_{-\pi}^{\pi}|G_\circ-G|^2|CS(G_\circ,C)|^2\,|S(G,C)|^2\,d\omega$$

if the $\mathbf{L}_2$-norm is used. Comparing this expression with the closed loop bias expression (4), we see that $V(G,C)$ corresponds to the first term in the closed-loop bias expression (4) if the noise model is taken as

$$H=\Phi_r^{1/2}S^{-1}(G,C).\tag{31}$$

Hence, the model that minimizes $V(G,C)$ wrt G will be obtained asymptotically if the identification is performed in closed loop with the controller C, the noise model (31) and the system is noise free. We will discuss this approach further in Section 7.3.

Before we proceed, observe that the above derivation can also be done in the time domain. Let u and y, denote the input and output, respectively, of the closed loop system in Figure 6 with $w=v_\circ\equiv 0$ and the controller C in the loop. The manipulations in (30) correspond in the time domain to

$$\begin{aligned}y-T(G,C)r&=(S(G,C)+T(G,C))y-T(G,C)r\\&=S(G,C)\,(y-GC\,(r-y))\\&=S(G,C)\,(y-Gu).\end{aligned}\tag{32}$$

The expression (32) is the prediction error (2) when data are collected in closed loop with C as the controller and $H=S^{-1}(G,C)$ as noise model. The difference compared to (31) is due to that the spectrum of (32) is weighted with the reference spectrum which is not the case in (30).

A simpler cost function is obtained by linearizing (23) wrt $G_\circ$

$$\begin{aligned}&J(G_\circ,C_\circ,G,C)\\&\approx\|((G_\circ-G)C_\circ+(C_\circ-C)G)S(G,C_\circ)S(G,C)\|^2\\&\triangleq\bar{J}(G_\circ,C_\circ,G,C).\end{aligned}\tag{33}$$

This gives the approximation

$$\begin{aligned}\bar{V}(G,C)&\triangleq\bar{J}(G_\circ,C,G,C)\\&=\|(G_\circ-G)CS^2(G,C)\|^2\end{aligned}\tag{34}$$

to $V(G,C)$ (24) and instead of minimizing $V(G,C(G))$, $\bar{V}(G,C(G))$ could be minimized (or $\bar{V}(G(C),C)$ if the direct parameterization in Section 6.3 is used).

Comparing (34), assuming the $\mathbf{L}_2$-norm, with the open loop bias expression (3), we see that open loop identification with the noise model

$$H=\Phi_u^{1/2}(S^2(G,C(G))C(G))^{-1}\tag{35}$$

under noise free conditions corresponds to minimizing $\bar{V}(G,C(G))$. We will discuss this approach in Section 7.4.

Remark 1: Comparing (34) with (24)–(25), we see that the stability guaranteeing term $1/(1+\Delta)$ is not present in the approximation $\bar{V}(G,C)$. Furthermore, comparing with the discussion on conditions on $|\Delta|$ in Section 6.2, we see that there is less emphasis on making $|\Delta|$ small at low frequencies. Hence, there is a possibility that $|\Delta|>1$ at low frequencies when (34) is used, with a potential risk for destabilization.

The expression (34) was, perhaps, first used for model reduction purposes (Rivera and Morari, 1987).

7.3 *Iterative methods*

In Section 7.2 we have seen that $V(G,C)$ can be minimized wrt G using closed loop identification with the controller C in the loop when the noise model (31) is used and data are noise free. Since the objective is to minimize $V(G,C(G))$ this suggests the following iterative procedure:

i) At iteration k: Identify a model G_k using the noise model (31) and using data collected with controller $C_k = C(G_{k-1})$ in the loop.
ii) Replace C_k with $C_{k+1} = C(G_k)$ in the closed loop. Let $k = k+1$ and GoTo i)

Surveys of this type of methods can be found in (Gevers, 1993; Van den Hof and Schrama, 1995). Despite the intuitive character, the above scheme will not converge to the minimum of $V(G,C(G))$ (Hjalmarsson *et al.*, 1995). However, for particular examples significant improvements have been reported.

The above derivation assumed noise free data. In the case of noisy data, the bias of the direct prediction error method can not be tuned at the user's will by the use of a prefilter/noise-model. This has spurred the development of a number of closed loop identification methods which have the ability to tune the bias as unifying feature. We refer to (Van den Hof, 1998; Forssell and Ljung, 1999; Landau *et al.*, 1997) for details on this.

7.4 *Prefiltering methods*

In Section 7.2 we introduced the approximation $\bar{V}(G,C)$ (34) to $V(G,C)$ (24) and showed that under noise free conditions $\bar{V}(G,C(G))$ can be minimized asymptotically if the noise model is chosen as (35). For noisy data, the bias expression (3) is no longer valid as the noise model (35) is not independently parameterized of G (Forssell and Ljung, 1999). However, this prefiltering method is simple to use and can be expected to work well if the signal-to-noise ratio is high and some care regarding the designed bandwidth is exercised, see Remark 1 in Section 7.2.

The above idea is the basis in (Rivera *et al.*, 1992) and also in the Virtual Feedback Reference Tuning method (VRFT) presented in (Campi *et al.*, 2002), where the direct parameterization presented in Section 6.3 is employed (Hjalmarsson and Lindqvist, 2001). In VRFT an instrumental variable approach is used to avoid noise induced bias.

Interestingly, also the idea of unfalsified controllers presented in Section 3.2 can, under LTI-assumptions, be interpreted as a prefiltering approach (Hjalmarsson and Lindqvist, 2001).

8. LTI MODELING OF NON-LINEAR SYSTEMS

There is abundant practical evidence that LTI models often are sufficient for control design for non-linear systems. In this section we will discuss some related issues.

8.1 *Performance aspects*

Let us assume that the system $G_\circ$ in Figure 6 is non-linear and noise-free ($v_\circ \equiv 0$) and also that $w \equiv 0$. Let us also assume that a linear model G is used to design a LTI model reference controller (29). We will now discuss how to find a suitable G such that the non-linear feedback system consisting of the non-linear system and the linear controller defined by (29) responds to r with the desired response $y_d = T_d r$. In Sections 7.1-7.2 we discussed how restricted complexity models for LTI systems could be identified in closed loop. When a non-linear full-order model is available, the method outlined in Section 7.1 can be adapted to the non-linear setting. When this is not the case, the ideas in Section 7.2 of closed loop identification can be used. Notice that no use of that the system is LTI was made in the derivation (32). Hence, for any system,

$$y - T_d r = y - y_d = (1 - T_d)(y - Gu) \qquad (36)$$

where u and y are the closed loop signals with the controller $C = C(G)$ in the loop. Thus, if there is a $C^* = C(G^*)$ such that $y = y_d$, then the right-hand side of (36) will be zero when $G = G^*$ and when data is collected with C^* as the controller in the loop. This implies that the model G^* corresponding to the desired controller will be obtained in closed loop identification when the desired $C(G^*)$ is operating in the loop and if the prefilter $1 - T_d$ is used (Henriksson *et al.*, 2001).

This observation supports the intuitively appealing idea that the identification experiments should be carried out under the *desired* operating conditions. A limitation of the argument above is that it is based on studying one single trajectory and, hence, does not give any information about the behavior for other reference signals, and in particular of closed loop stability. An interesting framework for characterizing how nonlinear a system is, can be found in (Eker and Nikolaou, 2002). It is based on a nonlinear generalization of (25). We will return to how to generate the desired operating conditions in Section 11.3.2.

8.2 *Stability aspects*

The robust stability condition (27) generalizes to the nonlinear setting in the following way (assuming both the model and the open loop system to be stable)

$$\|W_2\, T(G,C)\, W_1\|_{\mathbf{H}_\infty} \left\| W_1^{-1} \frac{G_\circ - G}{G} W_2^{-1} \right\| < 1,$$

where the second norm is the induced 2-norm, for arbitrary weighting filters W_1 and W_2. In (Ljung, 2000) the choice of weighting filters and how to estimate the non-linear gain $\left\| W_1^{-1} \frac{G_\circ - G}{G} W_2^{-1} \right\|$ are discussed. It is pointed out that only lower bounds can be obtained from data and that periodic inputs can be useful since they allow the noise to be averaged out. Another preliminary contribution in this important area is (Schoukens *et al.*, 2002).

This issue has also spurred activities in assessing how "small" nonlinearities may influence parameter estimates based on linear models. In (Enqvist and Ljung, 2002) it is illustrated that LTI-models may be extremely sensitive to non-linearities. In (Schoukens *et al.*, 1998) a general framework is developed for analyzing how linear estimates are affected by non-linearities of the system when the input excitation is periodic. Best LTI-approximants for non-linear systems are discussed in (Mäkilä and Partington, 2003).

9. EXAMPLES OF IDENTIFICATION FOR CONTROL METHODS

In this section we briefly outline a few methods for identification for control.

9.1 *Full order modeling*

When the true system is in the model set, and the asymptotic theory in Section 2 applies, we noted in Section 3.8 that the set of unfalsified models is defined by (17). The set (18) is an approximation due to the Taylor approximation in (6) and also usually conservative since the uncertainty usually is structured whereas (17) corresponds to unstructured uncertainty.

In a very interesting series of papers, summarized in (Gevers *et al.*, 2003), Gevers, Bombois and co-workers have developed a representation of the set of unfalsified models which avoids the Taylor approximation and which can be used to analyze robust performance and stability when the set of unfalsified models originate from prediction error identification. More specifically, robust stability for a controller designed for the nominal model and robust performance for a robustly stabilizing controller can be checked using convex optimization. Furthermore, it is shown that the worst case ν-gap, i.e. the largest ν-gap between the nominal model and any model in the set of unfalsified models, also can be computed by convex optimization. Based on these results, controller synthesis (Bombois *et al.*, 2002) and experiment design methods (Hildebrand and Gevers, 2003) have been developed.

9.2 High order modeling

The method ASYM, see (Zhu, 1998), is a fully integrated method for identification for control which is very close in spirit to the separation principle discussed in Section 5.2. The input design is based on high-order optimal input design; see Section 11.1. A high-order ARX-model is estimated from data, the motivation being its computational simplicity and that the high-order theory, c.f. Section 2.2 and (7), is applicable. Model reduction and model order selection are performed where the statistical properties of the high-order estimate are taken into account. To determine the unfalsified set of controllers, (18) is used with $\kappa(\omega) = m$ (=the model order) which follows from (7). The method was developed in the early 1990's and it has been successfully applied to numerous multivariable processes in process industry. High-order ARX-modeling is also advocated in (Rivera and Jun, 2000).

9.3 Set-membership identification

A method which integrates set-membership identification with robust control is presented in (Malan *et al.*, 2001). A prior exponential decaying bound on the impulse response and a variety of norm-bounded noise priors can be used. An $\mathbf{H}_\infty$ version of internal model control with robust stability constraint is used for the control.

In one of the iterative methods presented in (Veres, 2001) an interesting aspect is considered. Information from closed loop experiments when the controller is destabilizing is used, and found useful. It is shown that for the proposed method instability may only happen a finite number of times.

10. LOCAL MODELS

An idea that is close at hands when trying to model non-linear systems is to only try to model the behavior around operating points where the system has been operating. Below we discuss some of the methods that are based on this idea.

10.1 Iterative Feedback Tuning

Consider the problem of minimizing $V(G(C),C)$ (24) when model reference control (29) is used, i.e. we are trying to find a reduced order controller C that makes the complementary sensitivity function as close as possible to a fix reference model T_d. For simplicity, assume that the system is noise free and $w \equiv 0$. When the $\mathbf{L}_2$-norm is employed in (23), Parseval's formula gives

$$V(G(C),C) = \lim_{N\to\infty} \frac{1}{N} \sum_{t=1}^{N} ((T(G_\circ,C) - T_d)r(t))^2 \quad (37)$$

with r being white noise. One approach is to minimize (37) numerically using some descent algorithm such as Gauss-Newton. This was a popular approach in the 1950's and early 1960's.

For this, the sensitivity, i.e. gradient, of $(T(G_\circ,C) - T_d)r$ wrt C is required (or rather the parameters of C but we will omit this from the discussion). Straightforward differentiation gives

$$\begin{aligned} \frac{d}{dC}(T(G_\circ,C) - T_d)r &= \frac{1}{C}T(G_\circ,C)(1-T(G_\circ,C))r \\ &= \frac{1}{C}T(G_\circ,C)(r - y(C)) \quad (38) \end{aligned}$$

where $y(C)$ denotes the output of (22) with controller C in the loop, assuming $w \equiv v_\circ \equiv 0$. Thus we see that the above sensitivity can be obtained from the closed loop system (6) under noise-free conditions with controller C in the loop by 1) first performing an experiment with r as reference and collecting the output $y(C)$ and 2) using $r - y(C)$ as reference in a new experiment whose output is filtered through $1/C$.

Iterative Feedback Tuning (IFT) (Hjalmarsson *et al.*, 1998) is a generalization of the idea above. The sensitivities of the closed loop signals wrt the controller parameters are computed from two closed loop experiments as outlined above. It can be shown that, even in the presence of noise, the signal sensitivities are unbiased (modulo transient effects) and hence it is possible to guarantee that any convergence point of the algorithm corresponds to a *stationary point* of the desired objective function by the use of a stochastic approximation algorithm. The idea can be applied to any, differentiable, signal based objective function, i.e. not only (37). IFT has been applied by the chemical multinational Solvay SA for tuning of PID loops in distillation columns and evaporators (Hjalmarsson *et al.*, 1998).

From (38) we may also deduce other ways to approximate the sensitivity. The separation principle in Section 5.2 gives that if a good estimate $\hat{T}$ of $T(G_\circ,C)$ is available, then the best sensitivity estimate is obtained by replacing $T(G_\circ,C)$ by $\hat{T}$ everywhere in the middle expression of (38).

A middle-way is presented in (Kammer *et al.*, 2000) where it is suggested to avoid the second experiment by replacing $T(G_\circ,C)$ in the right-hand side expression of (38) by an estimate obtained using closed-loop data when C is operating in the loop.

The use of signal sensitivities can be seen as local modeling of how the closed loop signals depend on the controller. This has the important implication that IFT is able to cope with certain nonlinearities also, c.f. (Hjalmarsson, 1998; Sjöberg and Bruyne, 1999). We illustrate this on a DC-servo motor which exhibits backlash (Hjalmarsson *et al.*, 1998). The controller is a simple extension of a PID controller. The purpose is to obtain a controller which rejects the input load disturbance shown in Figure 7. When the initial controller is used, the system is in a limit cycle, see Figure 7. The output has clearly improved after 6 iterations as shown in Figure 8. From the Bode diagram (not shown) it can be deduced that a phase lead has been introduced exactly as suggested by describing function analysis.

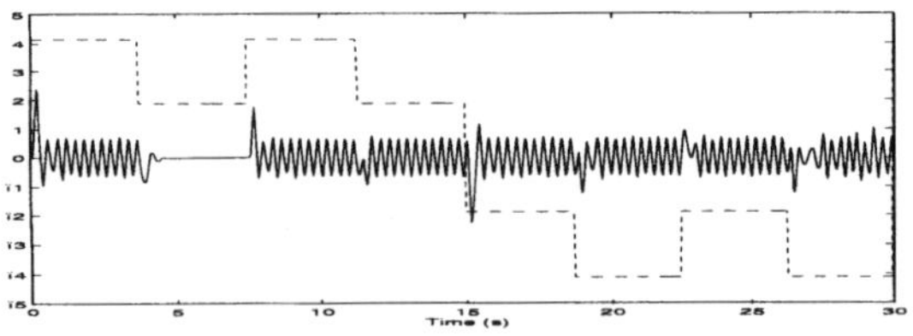

Fig. 7. Solid line: Output from DC-servo with the initial controller. Dashed line: The input load disturbance.

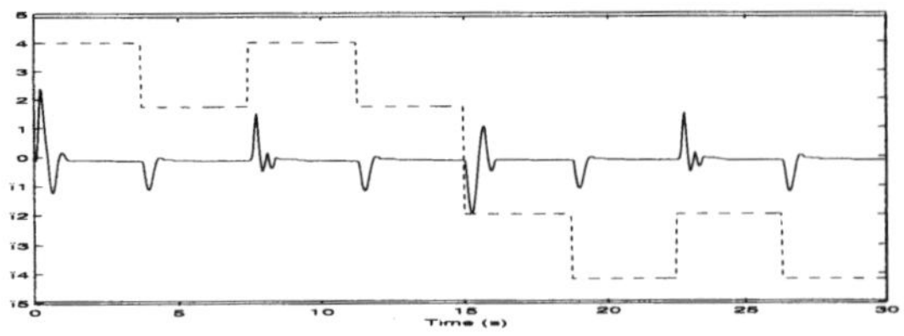

Fig. 8. Solid line: Output from DC-servo with the controller after six iterations. Dashed line: The input load disturbance.

10.2 De-correlation

An interesting approach is presented in (Karimi *et al.*, 2003) where the controller is tuned such that $y - y_d$, where y_d is the desired response, is un-correlated with an instrumental vector which is a function of lagged values of the reference r. It is shown that this can be done iteratively as in IFT but that only one experiment is required per iteration.

10.3 Model-on-demand

Another interesting method is Model-on-demand, a method with with close ties to non-parametric estimation in statistics. This method has been applied to model predictive control in (Braun *et al.*, 2000).

11. EXPERIMENT DESIGN

The reader may have noticed that input design has been a recurring theme up to now. Let us recapitulate:
• When the true system is in the model set, Eq. (18) gives the set of unfalsified models which depends on the input through the input spectrum as well as through $\kappa(\omega)$, c.f. (9).
• In Section 4.2 we saw that (18) also determines the strength of any validation statements that can be made.
• In Section 6.2 we saw explicitly in (28) how the important weighted model error (26) depends on the experimental conditions.
• In Section 8.1 it was indicated that using data from the desired operating conditions is very useful when trying to identify a restricted complexity model for a non-linear system. Going back to Section 7.3, we see that this is exactly the reason why closed-loop identification may help to obtain "control relevant" models for LTI systems as well.

In this section we shall further discuss how experiment design can be used to improve the closed loop performance.

11.1 Optimal input design

Starting with the now classical reference (Gevers and Ljung, 1986), there has been a series of contributions to optimal input design based on the high-order variance expression (7), and its closed loop counterpart, with specific applications to identification for control. The approach is to choose the experimental conditions such that the *average* performance degradation (recall (24))

$$\mathbf{E}\{V(\hat{G}_N, C(\hat{G}_N))\}$$

where the expectation is over $\hat{G}_N$, is minimized. An approximate solution can be derived using the first order approximation (34), i.e. by instead minimizing

$$\mathbf{E}\{\bar{V}(\hat{G}_N, C(\hat{G}_N))\} \tag{39}$$

When there is a constraint on the output variance, the optimal experiment is in closed loop with an LQG-controller determined solely by the constraints (Forssell and Ljung, 2000). Thus applications with different objective functions but with the same constraints share the same optimal experiment.

Alternatively, the parameter covariance matrix can be used directly for input design. When the input is generated by a FIR-filter, convex optimization can be used to compute the optimal FIR-coefficients (Lindqvist and Hjalmarsson, 2001).

11.2 Robust input design

The designs in the preceding section are all geared towards optimizing the average performance as they are based on (39). By instead using (28) as design criterion, an $\mathbf{H}_\infty$-type of design is obtained with guaranteed robust stability and robust performance (Hjalmarsson and Jansson, 2003).

11.3 Adaptive input design

As we have seen, in the early 1990's several schemes that iterated between identification and closed loop control were proposed. In fact, it was suggested, (Schrama, 1992), that high performance control based on restricted complexity models required such iterations. However, as we have indicated in Sections 6.2, 7.1 this is not necessarily so. It has also been shown, (Böling and Mäkilä, 1998), that the example in (Schrama, 1992) can be solved without iterations. So the question is rather *what* can be gained by iterating between identification and experimentation?

Clearly, when we are collecting data, we obtain new information about the system. Hence, iterating between identification and experiment is beneficial if we can improve our experiment design such that more *control relevant* information becomes available than if we would have stayed with our present design. We have seen that there are two quantities to consider:

- *Minimizing variance.* The input should be chosen such as to minimize the impact from the noise induced errors in the controller $C(\hat{G}_N)$ on the achieved closed loop performance.
- *Tuning the bias.* When the model set is restricted it is important that data contains the features relevant to control. This is embodied in that the desired operating conditions should be mimicked as closely as possible.

11.3.1. Minimizing variance: Adaptive input design for LTI systems For a sufficiently long experiment, the performance can be improved by iterating between experimentation and identification for certainty equivalence input design, i.e. the last estimated model is used instead of the true system in an optimal experiment design, (Hjalmarsson *et al.*, 1996). A potential problem with this approach is that instability may occur if the optimal experiment is in closed loop and the certainty equivalence design is based on a model of poor quality. An alternative which avoids this problem is to tune the input spectrum adaptively in open loop (Lindqvist and Hjalmarsson, 2001).

11.3.2. Tuning the bias: Adaptive input design for nonlinear systems For nonlinear systems, we have in Section 8.1 seen that it is advantageous to have the system operating under the desired conditions when data are collected. A systematic method to iteratively generate feed-forward controls that approach this objective is Iterative Learning Control (ILC), see (Moore, 1993) for details and references. One of the design variables in ILC corresponds to a model of the inverse system. By changing the model between iterations desired trajectories of nonlinear systems can be generated even though only simple LTI-models are used (Markusson *et al.*, 2002; Tao *et al.*, 1994).

11.3.3. Tuning the bias: Using non-linear feedback The relay experiment used in the relay auto-tuner method, originally proposed in (Åström and Hägglund, 1984) can be seen as a way of generating an experiment which provides information such that the bias in a model is "tuned" for control.

11.4 Input design for multivariable systems

Input design is perhaps most important for multivariable systems, especially for ill-conditioned processes such as high-purity distillation columns. To appreciate this, notice that the low gain directions of the system will be poorly identified unless precautions are taken to ensure that the signal to noise ratio is sufficient in these directions. The problem arises as the designed controller will use high gain in these poorly identified directions which may be disastrous for the closed loop behavior. The key to solving this is to use correlated inputs and both open loop and closed loop methods have been proposed to this end (Zhu, 2002; Jacobsen, 1994).

11.5 Pre-tests – Identifying performance limitations

The experiment design is eased significantly if the inherent limitations of the system are known. Often amplitude and slew-rate constraints on actuators, which limit the achievable bandwidth, are known and this knowledge should be incorporated in the experiment design, c.f. the discussion of the control effort in Section 6.2. Non-minimum phase zeros also limit the achievable bandwidth. An explicit variance expression has been derived which can be used for designing pretest experiments so that these zeros can be accurately identified (Mårtensson and Hjalmarsson, 2003). An interesting aspect is that the asymptotic accuracy of the identified zeros is basically independent of the model order when the prediction-error method is used. Hence, model order selection is not a critical issue here.

11.6 Open loop or closed loop experiments?

So is closed loop identification preferable? Well, while this certainly can be motivated by many reasons, e.g. safety, there is from an accuracy point of view no clear cut answer to this question. In Section 8.1 we saw that for bias reasons, the experimental conditions should reflect the *desired* operation characteristics. A closed loop experiment with a poorly tuned controller may in this respect give less useful data than a well designed open loop experiment, c.f. Section 11.3.2. In Section 11.1, we have seen that to counter-effect noise induced uncertainty (variance) effectively it is optimal with closed loop identification with a controller that depends on the *true system* and the experimental constraints. However, also here there is no guarantee in practice that closed loop data are more informative.

12. CONCLUDING REMARKS

Looking back at the wide variety of topics above, it is clear that much progress has been made in this research area since the SYSID plenary in Budapest in 1991 (Gevers, 1991), which in many respects can be seen as a trigger point for activities in this field.

During the course above two principles have emerged: 1) the benefits from a statistical perspective of modeling as well as possible regardless of the final objective. 2) the importance of a good experiment design – if you ask the right questions, you get the right answers and you don't have to worry.

While much of the work in the area has focused on LTI systems, c.f. Section 7, there has in recent years been a shift towards more realistic problem settings where the system is considered to exhibit different kinds of non-linear behaviors, c.f. Section 8. With the diversity of non-linear systems, one may expect this research area to proliferate in coming years; there are, e.g., results emerging for hybrid systems. This observation also accentuates the importance of close collaboration with application fields in order to address relevant problems.

A particularly important area is experiment design, especially for multivariable systems, and how it influences the accuracy of restricted complexity modeling, c.f. Section 5.3. There is also certainly more to learn from adaptive control and vice versa.

An observation is that it is often very difficult to assess merits of a method. For this, a set of generally acknowledged benchmarks would be most useful. Compare with numerical analysis and statistics where new algorithms are required to outperform existing ones to be found of value. It should be noted that there is always a great interest to participate whenever a benchmark problem appears, c.f., e.g., (Landau *et al.*, 2003).

To conclude, there is surely more to say and to be said on this fascinating subject. But as for now I rest my case.

Acknowledgments

During the last decade I have had the privilege to collaborate with

Franky De Bruyne, Laszlo Gerencsér, Michel Gevers, Svante Gunnarsson, Henrik Jansson, Kristian Lindqvist, Lennart Ljung, Ola Markusson, Brett Ninness, Johan Schoukens, Jonas Sjöberg, Bo Wahlberg and Sandor Veres,

on this topic; a process which has lead to many of the insights in this paper. Thanks guys for enlightening me, you're a great bunch! Special thanks goes to Brett and Bo for proof-reading this manuscript.

This research has been supported by The Swedish Research Council.

REFERENCES

Bittanti, S., M.C. Campi and S. Garatti (2002). New results on the asymptotic theory of system identification for the assessment of the quality of estimated models. In: *41th IEEE CDC*.

Böling, J.M. and P.M. Mäkilä (1998). On control relevant criteria in H_∞ identification. *IEEE Transactions on Automatic Control* **43**(5), 694–700.

Bombois, X., G. Scorletti, B.D.O. Anderson, M. Gevers and P. Van den Hof (2002). A new robust control design procedure based on a PE identification uncertainty set. In: *15th IFAC World Congress*.

Braun, M.W., B.A. McNamara, D.E. Rivera and A. Stenman (2000). Model-on-demand' identification for control: An experimental study and feasability analysis for mod-based predictive control. In: *SYSID 2000*.

Cadic, M., J.W. Polderman and S. Weiland (2003). Strong robustness measures for sets of linear SISO sysems. In: *SYSID 2003*.

Campi, M.C., A. Lecchini and S.M. Savaresi (2002). Virtual reference feedback tuning: a direct method for the design of feedback controllers. *Automatica* **38**(8), 1337–1346.

Eker, S.A. and M. Nikolaou (2002). Linear control of nonlinear systems: Interplay between nonlinearity and feedback. *AIChE Journal* **48**(9), 1957–1980.

Enqvist, M. and L. Ljung (2002). Estimating nonlinear systems in a neighorhood of LTI-approximants. In: *41st IEEE CDC*. pp. 1005–1010.

Forssell, U. and L. Ljung (1999). Closed-loop identification revisited. *Automatica* **35**, 1215–1241.

Forssell, U. and L. Ljung (2000). Some results on optimal experiment design. *Automatica* **36**(5), 749–756.

Gevers, M. (1991). Connecting identification and robust control: A new challenge. In: *Proc. IFAC/IFORS Symposium on Identification and System Parameter Estimation*.

Gevers, M. (1993). Towards a joint design of identification and control?. In: *Essays on Control: Perspectives in the Theory and its Applications* (H. L. Trentelman and J. C. Willems, Eds.). Birkhäuser.

Gevers, M. and L. Ljung (1986). Optimal experiment designs with respect to the intended model application. *Automatica* **22**, 543–554.

Gevers, M., X. Bombois, B. Cordrons, G. Scorletti and B.D.O. Anderson (2003). Model validation for control and controller validation in a prediction error identification framework – Part I:theory, Part II: applications. *Automatica* **39**, 403–427.

Goodwin, G.C., J.H. Braslavsky and M.M. Seron (2002). Non-stationary stochastic embedding for transfer function estimation. *Automatica* **38**(1), 47–62.

Goodwin, G.C., L. Wang and D. Miller (1999). Bias-variance trade-off issues in robust controller design using statistical confidence bounds. In: *14th IFAC World Congress*.

Goodwin, G.C., M. Gevers and B. Ninness (1992). Quantifying the error in estimated transfer functions with application to model order selection. *IEEE Trans. Automatic Control* **37**, 913–928.

Hakvoort, R.G., R.J.P. Schrama and P.M.J. Van den Hof (1994). Approximate identification with closed loop criterion and application to LQG feedback design. *Automatica* **30**, 679–690.

Henriksson, B., O. Markusson and H. Hjalmarsson (2001). Control relevant identification of nonlinear systems using linear models. In: *ACC 2001*.

Hildebrand, R. and M. Gevers (2003). Identification for control: Optimal input design with respect to a worst-case ν-gap cost function. *SIAM J. Control Optim* **41**, 1586–1608.

Hjalmarsson, H. (1998). Control of nonlinear systems using Iterative Feedback Tuning. In: *ACC 1998*. pp. 2083–2087.

Hjalmarsson, H. and H. Jansson (2003). Using a sufficient condition to analyze the interplay between identification and control. In: *SYSID 2003*.

Hjalmarsson, H. and K. Lindqvist (2001). Identification for control: L_2 and L_∞ methods. In: *40th IEEE CDC*.

Hjalmarsson, H., M. Gevers and F. De Bruyne (1996). For model based control design criteria, closed loop identification gives better performance. *Automatica* **32**, 1659–1673.

Hjalmarsson, H., M. Gevers, S. Gunnarsson and O. Lequin (1998). Iterative Feedback Tuning: theory and applications. *IEEE Control Systems Magazine* **18**(4), 26–41.

Hjalmarsson, H., S. Gunnarsson and M. Gevers (1995). Optimality and sub-optimality of iterative identification and control design schemes. In: *ACC 1995*. pp. 2559–2563.

Jacobsen, E.W. (1994). Identification for control of strongly interactive plants. In: *AIChE Annual Meeting*.

Kammer, L.C., R.R. Bitmead and P.L. Bartlett (2000). Direct iterative tuning via spectral analysis. *Automatica* **36**, 1301–1307.

Karimi, A., L. Miskovic and D. Bonvin (2003). Iterative controller tuning using the correlation approach. *Control Engineering Practice*. To appear.

Kosut, R.L. (2001). Uncertainty model unfalsification. *A. Rev. Control* **25**, 65–76.

Landau, I.D., A. Karimi and A. Constantinescu (1997). Recursive algorithms for identification in closed loop: A unified approach and evaluation. *Automatica* **33**(8), 1499–1523.

Landau, I.D., A. Karimi and H. Hjalmarsson (2003). Special issue on Design and optimization of restricted complexity controllers. *European Journal of Control* **9**(1), .

Lee, W.S., B.D.O. Anderson, I.M.Y. Mareels and R.L. Kosut (1995). On some key issues in the windsurfer approach to adaptive robust control. *Automatica* **31**(11), 1619–1636.

Lehmann, E.L. (1983). *Theory of Point Estimation*. John Wiley & Sons. New York.

Lindqvist, K. and H. Hjalmarsson (2001). Identification for control: Adaptive input design using convex optimization. In: *40th IEEE CDC*.

Ljung, L. (1999*a*). Model validation and model error modeling. In: *The Åström Symposiium on Control* (B. Wittenmark and A. Rantzer, Eds.). Studentlitteratur. pp. 15 –42.

Ljung, L. (1999*b*). *System Identification: Theory for the User*. 2nd edition ed.. Prentice-Hall. Englewood Cliffs, NJ.

Ljung, L. (2000). Model error modeling and control design. In: *SYSID 2000*. Santa Barbara, CA.

Ljung, L. and H. Hjalmarsson (1995). System identification through the eyes of model validation. In: *ECC*. pp. 949–954.

Malan, S., M. Milanese, D. Regruto and M. Taragna (2001). Robust control from data via uncertainty model sets identification. In: *40th IEEE CDC*.

Markusson, O., H. Hjalmarsson and M. Norrlöf (2002). A general framework for Iterative Learning Control. In: *15th IFAC World Congress*.

Milanese, M. (1998). Learning models from data: the set membership approach. In: *ACC 1998*. Vol. 1. pp. 178–182.

Milanese, M. and A. Vicino (1991). Optimal estimation theory for dynamic systems with set membership uncertainty: An overview. *Automatica* **27**(6), 997–1009.

Mäkilä, P. M. and J. R. Partington (2003). On linear models for nonlinear systems. *Automatica* **39**(1), 1–13.

Mäkilä, P.M, J.R. Partington and T.K. Gustafsson (1995). Worst-Case control-relevant identification. *Automatica* **31**(12), 1799–1819.

Moore, K.L. (1993). *Iterative Learning Control for Deterministic Systems*. Advances in Industrial Control. Springer Verlag.

Mårtensson, J. and H. Hjalmarsson (2003). Identification of performance limitations using general SISO structures. In: *SYSID 2003*.

Murray, R.M., K.J. Åström, S.P. Boyd, R.W. Brockett and G. Stein (2003). Future directions in control in an information-rich world. *IEEE Control Systems Magazine* **23**(2), 20–33.

Ninness, B. and G.C. Goodwin (1995). Estimation of model quality. *Automatica* **31**(12), 1771–1797.

Ninness, B. and H. Hjalmarsson (2002). Exact quantification of variance error. In: *15th World Congress on Automatic Control*.

Ninness, B., H. Hjalmarsson and F. Gustafsson (1999). The fundamental role of general orthonormal bases in system identification. *IEEE Transactions on Automatic Control* **44**, 1384–1406.

Ogunnaike, B.A. (1996). A contemporary industrial perspective on process control theory and practice. *A. Rev. Control* **20**, 1–8.

Reinelt, W., A. Garulli and L. Ljung (2002). Comparing different approaches to model error modeling in robust identification. *Automatica* **38**(5), 787–803.

Rivera, D.E. and K.S. Jun (2000). An integrated identification and control design methodology for multivariable process system applications. *IEEE Control Systems Magazine* **20**, 25–37.

Rivera, D.E. and M. Morari (1987). Control-relevant model reduction problems for SISO $\mathbf{H}_2$, $\mathbf{H}_\infty$ and μ-controller synthesis. *Int. J. Control* **46**(2), 505–527.

Rivera, D.E., J.F. Pollard and C.E. Garcia (1992). Control-relevant prefiltering: A systematic design approach and case study. *IEEE Transactions on Automatic Control* **37**(7), 964–974.

Safonov, M.G. and T-C Tsao (1997). The unfalsified control concept and learning. *IEEE Trans. on Automatic Control* **42**(6), 843–847.

Schoukens, J., R. Pintelon and T. Dobrowiecki (2002). Identification of the sability of feedback systems in the presence of nonlinear distortions. In: *15th IFAC World Congress*.

Schoukens, J., T. Dobrowiecki and R. Pintelon (1998). Parametric and nonparametric identification of linear systems in the presence of nonlinear distortions – A frequency domain approach. *IEEE Transactions on Automatic Control* **43**(2), 176–190.

Schrama, R. J. P. (1992). Accurate models for control design: the necessity of an iterative scheme. *IEEE Trans. Automatic Control* **37**, 991–994.

Sjöberg, J. and F. De Bruyne (1999). On a nonlinear controller tuning strategy. In: *14th IFAC World Congress*. Vol. I. pp. 343–348.

Åström, K.J. and T. Hägglund (1984). Automatic tuning of simple regulators with specifications on phase and amplitude margin. *Automatica* **20**, 645–651.

Tao, K.M., R.L. Kosut and M. Ekblad (1994). Feedforward learning – nonlinear processes and adaptation. In: *33rd IEEE CDC*. pp. 1060–1065.

Tjärnström, F. and A. Garulli (2002). A mixed probabilistic bounded-error approach to parameter estimation in the presence of amplitude bounded white noise. In: *41th IEEE CDC*.

Tjärnström, F. and L. Ljung (2002). L_2 model reduction and variance reduction. *Automatica* **38**(9), 1517–1530.

Van den Hof, P.M.J. (1998). Closed-loop issues in system identification. *A. Rev. Control* **22**, 173–186.

Van den Hof, P.M.J. and R.J.P. Schrama (1995). Identification and control – closed loop issues. *Automatica* **31**(12), 1751–1770.

Veres, S.M. (2001). Convergence of control performance by unfalsification of models – levels of confidence. *International Journal on Adaptive Control and Signal Processing* **15**, 471–502.

Xie, L.-L. and L. Ljung (2001). Asymptotic variance expressions for estimated frequency functions. *IEEE Trans. Automatic Control* **46**, 1887–1899.

Zhu, Y (1998). Multivariable process identification for MPC: the asymptotic method and its applications. *J. Proc. Cont.* **8**(2), 101–115.

Zhu, Y. (2000). Use of error criteria in identification for control. In: *SYSID 2000*.

Zhu, Y (2002). Case studies on closed-loop identification for MPC. *Control Engineering Practice* **10**, 403–417.

IFAC Publications
www.elsevier.com/locate/ifac

EXPLORATORY MODELLING FOR CONTROLLER OPTIMIZATION

Sandor M. Veres *,[1]

* *School of Engineering Sciences, University of Southampton, Highfield, SO17 1BJ, Email:sandy@mech.soton.ac.uk*

Abstract: Internal model control (IMC) is shown to have self-excitation for identification for control. Although IMC can be designed for tolerance against modelling errors for higher frequencies, tuned models can bring enhanced and more reliable performance. It is shown in this paper that the internal model control has synergistic interaction with system identification of the plant model. A new scheme called *exploratory modelling for controller optimisation* (EMCO) is introduced to obtain simple robust controllers. *Copyright © 2003 IFAC*

Keywords: Internal model control, system identification, adaptive control, robust control

1. INTRODUCTION

Internal model control (IMC) is an attractive control design method to practitioners for various reasons (Morari and Zafirou, 1989). It has a clear structure for studying the feedback mechanism and the effect of modelling errors can be taken into account to modify the controller transfer function for robust stability and robust performance. Figure 1 shows the block diagram of a two-degree of freedom IMC structure.

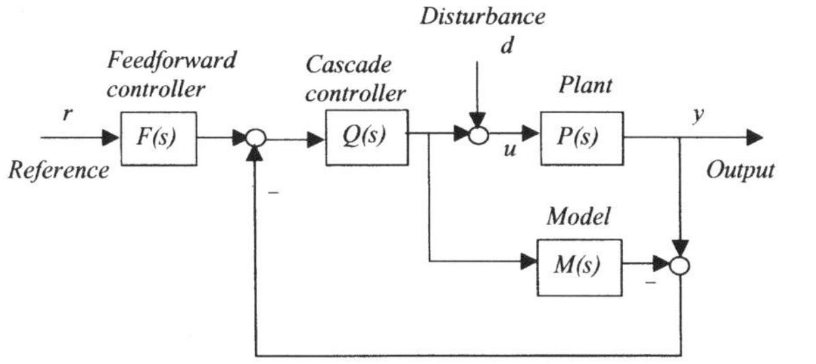

Fig. 1. Block diagram of an IMC control scheme.

[1] This work was partially supported by QinetiQ, Malvern, Enland.

The input signal to the plant can be expressed as

$$u = d + Q(Fr - \Delta M u - Md) \qquad (1)$$

where $\Delta M = P - M$ is the modelling error. The spectrum of input u to the plant in the closed-loop can be expressed as

$$\Phi_u = \frac{|QF|^2}{|1+Q\Delta M|^2}\Phi_r + \frac{|1-QM|^2}{|1+Q\Delta M|^2}\Phi_d \qquad (2)$$

This clearly shows that the input excitation to the plant is modified by modelling errors $\Delta M \neq 0$. If the $\Phi_r(\omega)$ and $\Phi_d(\omega)$ have constant spectra of two white noise signals, then the an increase in input excitation will occur at those frequencies where

$$\left|\frac{Q}{1+Q\Delta M}\right| = \frac{1}{|Q^{-1}+\Delta M|} \qquad (3)$$

increases, i.e. $|Q^{-1}+\Delta M|$ becomes smaller. The latter will be related to the generalized stability margin in Section 3. This suggests that IMC may support the more accurate identification of the model (i.e. reduction of ΔM) at frequencies where it is needed to reduce model uncertainty. Supportive interaction between modelling and IMC will be called synergy of identification and control (Anderson and Kosut, 1991; Van den Hof *et al.*, 1993; Veres and Wall, 2000; Hjalmarsson *et al.*, 1996; Veres, 2001).

Section 3 will show that $\Delta M + Q^{-1}$ varies in proportion with the "frequency dependent" stability margin. For small $|\Delta M(j\omega)|$ relative to $|Q(j\omega)^{-1}|$, it can happen that the main contributor to low generalized stability margin is the large control gain $Q(s)$ (intense control) within the closed loop bandwidth. This low margin remains however for any low ΔM, i.e. almost perfect identification of the plant dynamics, and therefore this is an issue for the nominal control performance design.

The problem can be addressed either in the stochastic framework or in the worst-case identification framework. Because of the problems of handling bias errors and estimating the variance of unmodelled dynamics in stochastic identification, the present paper addresses the problem using worst-case H_∞ control and frequency response bounding.

2. PERFORMANCE MEASURES

For clarity of ideas, only single-input single-output (SISO) plants will be considered, the scheme can be generalized to the multivariable case without much difficulty. The reformulated flow diagram of the setup of a plant P and controller C is indicated in Figure 2.

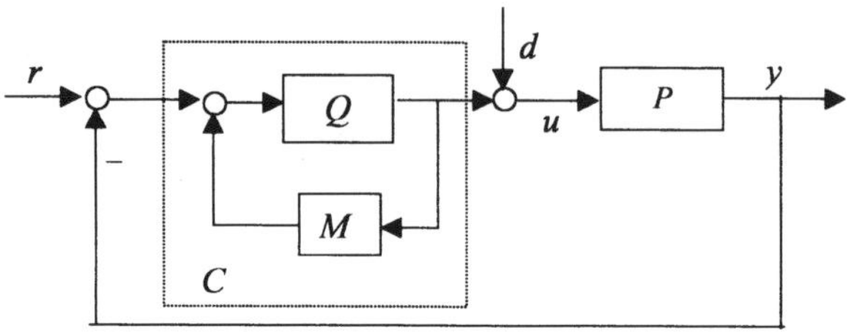

Fig. 2. Reformulation of the IMC structure.

The model-controller action within IMC is reorganized in Figure 2 into the form of a classical control loop where the controller is

$$C(s) = \frac{Q(s)}{1 - Q(s)M(s)} \tag{4}$$

It is assumed that the dynamics of the real plant can be well approximated by an unknown linear model of possibly very high complexity. Such plants occur, for instance, in active control of vibrations, flexible robots and active flow control (Veres and Wall, 2000). In these applications empirical modelling for control, based on limited amount of input-output data, becomes a non-trivial task.

Consider continuous time models and controllers with closed-loop equations

$$\begin{aligned} y(t) &= P(s)u(t) \\ u(t) &= -C(s)[r(t) - y(t)] + d(t) \end{aligned} \tag{5}$$

Introducing the signals

$$z = \begin{bmatrix} y \\ u \end{bmatrix}, \quad w = \begin{bmatrix} r \\ d \end{bmatrix} \tag{6}$$

the closed-loop system can be written as

$$z = \begin{bmatrix} PCS & PS \\ CS & S \end{bmatrix} w = T(P,C)w \tag{7}$$

where $S = (1 + PC)^{-1}$ is the sensitivity function.

The norm $b_{P,C} = \|T(P,C)\|_\infty^{-1}$ is the *generalised stability margin* and it measures the maximum allowed normalized H_∞-norm coprime factor dynamic perturbations of $P(s)$, which still preserve stability with the same controller $C(s)$. As $T(P,C)$ contains the sensitivity and complementary sensitivity functions, $\|T(P,C)\|_\infty$ also provides a rough measure of control performance, without frequency weighting.

Required *servo control performance* can be formulated in the general form $\|S(P,C)W\|_\infty \leq 1$ where W is a suitable weighting function (proper rational transfer function).

Robust performance for a nominal plant model M can be expressed by a worst-case performance criterion under frequency dependent plant uncertainty $|M(j\omega) - P(j\omega)| \leq \delta(\omega), \omega \geq 0$ (for short denoted by $|M - P| \leq \delta$) defined by

$$\begin{aligned} &M_{rp}(M,C,\delta) = \\ &= \sup\{\|S(P,C)W\|_\infty \mid |M - P| \leq \delta)\} \leq 1 \end{aligned} \tag{8}$$

M_{rp} is the measure of robust performance, dependent on the nominal plant M, controller C and uncertainty δ. The $\delta(\omega) > 0$, $\omega \geq 0$, is a frequency dependent bound of the plant uncertainty.

The stability margin $b_{M,C} = \|T(M,C)\|_\infty^{-1}$ has a geometric interpretation on the Riemann sphere (Vinnicombe, 1993). The Riemann sphere is a sphere of radius 0.5 placed over the origin of the complex plane as illustrated in Figure 3. The Nyquist curve of $M(s)$ is projected onto the Nyquist curve by connecting each point on the curve with the point $[0,0,1]$ at the top of the sphere, using straight lines. The chordal distance (Vinnicombe, 1993) between two projected points on the Riemann sphere is

$$\begin{aligned} &\kappa(M_1(j\omega), M_2(j\omega)) = \\ &= \frac{|M_1(j\omega) - M_2(j\omega)|}{(1 + |M_1(j\omega)|^2)^{1/2}(1 + |M_1(j\omega)|^2)^{1/2}} \end{aligned} \tag{9}$$

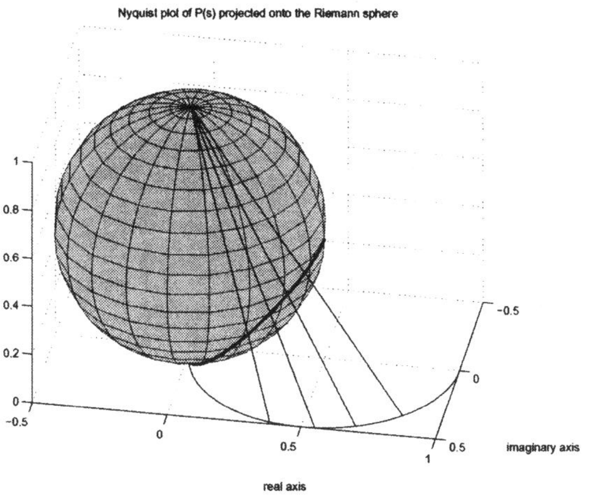

Fig. 3. The Riemann sphere and projection of a Nyquist curve onto the sphere

Later $b_{M,Q}$ is written for $b_{M,C}$ as Q is used in IMC. The following well known lemma is useful for the geometric interpretation of stability robustness.

Lemma 1. The generalised stability margin can be expressed as:

$$b_{M,Q} = \inf_{\omega \in [0,\Omega)} \kappa(M(j\omega), -C(j\omega)^{-1}) \quad (10)$$

This lemma means that the projected curve of $-C(j\omega)^{-1}$ has to be kept at a distance b from $M(j\omega)$ to achieve a stability margin $b > 0$.

3. SYNERGY OF IDENTIFICATION AND IMC

Lemma 1 in can also be used to quantify the frequency-dependent maximum allowed modelling error of a nominal plant model M for a given controller C to keep a given stability margin b and required servo performance given by W.

Lemma 2. For the IMC scheme the following statements hold.

(a) The generalized stability margin b is bounded as:

$$\inf_{\omega} \frac{|\Delta M + Q^{-1}|}{\sqrt{6}} \leq b_{M,Q} \leq$$

$$\leq \frac{|\Delta M + Q^{-1}|}{(1+|P|^2)^{1/2}(1+|Q^{-1}-M|^2)^{1/2}} \leq$$

$$\leq |\Delta M + Q^{-1}|, \ \forall \omega$$

if $\|P\|_\infty \leq 1$, $\|M\|_\infty \leq 1$, $\|Q^{-1}\|_\infty \leq 1$.

(b) The servo performance requirement can be expressed as

$$|SW| = \left| \frac{Q^{-1} - M}{\Delta M + Q^{-1}} \right| |W| \leq 1, \forall \omega \quad (11)$$

By gain adjustments of the plant one can normally achieve the conditions in (a). Q can normally take on larger ($>> 1$) values if the closed-loop bandwidth is increased with respect to the open loop bandwidth. IMC design allows that $|Q^{-1}(j\omega)| \leq 1$ can be ensured with the closed loop bandwidth. By (a) the stability will be lowered at frequencies ω where $(\Delta M + Q^{-1})(j\omega)$ is low due to $\Delta M(j\omega)$ approximately cancelling $Q^{-1}(j\omega)$. Exactly at these frequencies the plant input is large as outlined in the introduction and pointed out in (3). Large input amplitude at a frequency ω allows for more accurate identification of the response of the plant at that frequency. This proves that, under IMC, closed-loop plant input favours frequency response estimation at those frequencies which are responsible for lowering the stability margin. In turn more accurate frequency response estimation will reduce $\Delta M(\omega)$ relative to Q^{-1} and hence increases the stability margin. This explains the synergistic interaction between closed-loop identification and stability robustness of IMC.

Another nice thing about the above lemma is that the term $|\Delta M + Q^{-1}|$ occurs under (b) as well and this will allow the construction of a function to describe maximum frequency dependent plant perturbation to ensure a b stability margin and robust servo-performance at the same time.

4. ALLOWED PLANT GAIN ERRORS - AGE

Let $\bar{P}(\omega)$ be an estimated upper bound function of the plant gain, obtainable from frequency response testing of the plant.

Definition 3. The frequency dependent allowed gain error, i.e. AGE, is defined by

$$\beta_W^b(\omega|Q) = |Q^{-1}(j\omega)| - \max(\bar{B}_b(\omega|Q), B_W^M(\omega|Q))$$

where $B_W^M(\omega) = |Q^{-1}(j\omega) - M(j\omega)||W(j\omega)|$ and $\bar{B}_b(\omega|Q) = b(1+|\bar{P}(j\omega)|^2)^{1/2}(1+|Q^{-1}(j\omega) - M(j\omega)|^2)^{1/2}$ is an upper bound obtained from a plant-response upper bound $\bar{P}(\omega)$.

To achieve a robust stability margin b and the servo performance (as defined by $|W|$) for an uncertain plant $P = M + \Delta M$, it is a necessary condition that the AGE bound $\beta_W^b(\omega|Q)$ must bound the actual error ΔM of the plant model. In general this may not ensure stability of the closed loop under all plant deviations within the AGE. If, however, the nominal plant M is stable, then this is the case as the following result shows. Define the plant frequency response error function by $\Delta M(\omega) = |P(j\omega) - M(j\omega)|, \omega \geq 0$.

Theorem 4. Assume that the plant P is stable and the AGE $\beta^b_W(\omega|Q) > 0,\ \omega \in [0,\infty)$, has been computed for a model M and an associated controller Q. If

$$|\Delta M(\omega)| < \beta^b_W(\omega|Q),\ \omega \in [0,\infty) \quad (12)$$

then the closed-loop with controller Q and the plant P will be stable with generalized stability margin b and the servo performance will be achieved as required by W.

The relevance of this results is that it relates frequency domain identification with performance and stability robustness of the controller directly, and the test is the satisfaction of an inequality (12).

5. EXPLORATORY MODELLING FOR CONTROL

This section summarises the scheme of *exploratory modelling and controller optimisation* (EMCO).

Assume that frequency response error bounds are measurable on the plant, i.e. a nominal response $\hat{P}(j\omega),\ \omega \geq 0$ is obtained with an error bound function $0 < \delta(j\omega),\ \omega \geq 0$ so that

$$|\hat{P}(j\omega) - P(j\omega)| \leq \delta(j\omega),\ \omega \geq 0 \quad (13)$$

The essence of EMCO is to find a model structure and model parameters, i.e. model M, such that (12) is satisfied for the given stability robustness margin $b > 0$ and servo performance $W(j\omega), \omega > 0$ requirements. For a given model structure ν , the model parameter vector θ is to be determined by nonlinear optimisation of the cost function

$$L_\nu(\theta) = \sup_\omega\{|\hat{P}(j\omega) - M^\theta_\nu(j\omega)| + \delta(j\omega) - \beta^b_W(\omega|Q^\theta)\} \quad (14)$$

where Q^θ is an H_∞-controller associated with the model M^θ_ν and weighting functions W and W_2, the latter one being a proper rational upper approximation to the error bound $\gamma^\theta_\nu(\omega) = |\hat{P}(j\omega) - M^\theta_\nu(j\omega)| + \delta(j\omega)$ so that $\gamma^\theta_\nu(\omega) \leq |W_2(j\omega)|,\ \forall\omega$. The H_∞ design essentially requires the optimisation of $\||WS|^2 + |W_2(1-S)|^2\|_\infty$ by standard techniques (Doyle *et al.*, 1992).

The block diagram in Figure 4 shows the main steps if the EMCO scheme:

The EMCO scheme also easily lends itself to extension to the multivariable case as all the necessarily algebra is available in the multivariable case (Vinnicombe, 1993). To ensure that the simplest controller is found an exploration tree of models structures can be constructed. Based on such a

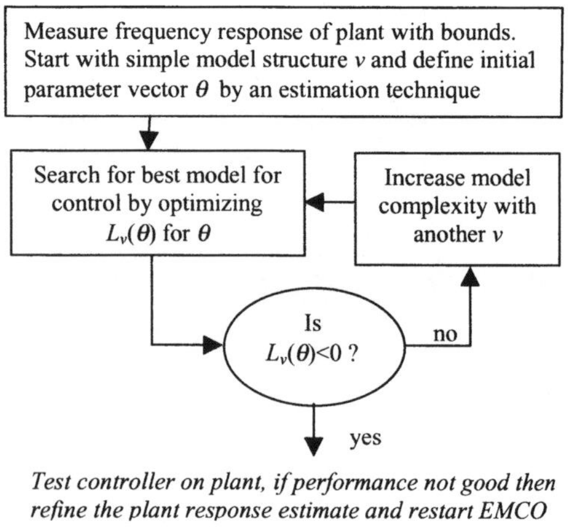

Fig. 4. Block diagram of EMCO.

tree the order of model structure switching policies can be defined, which will be the subject of future research. Figure 5 includes the block diagram

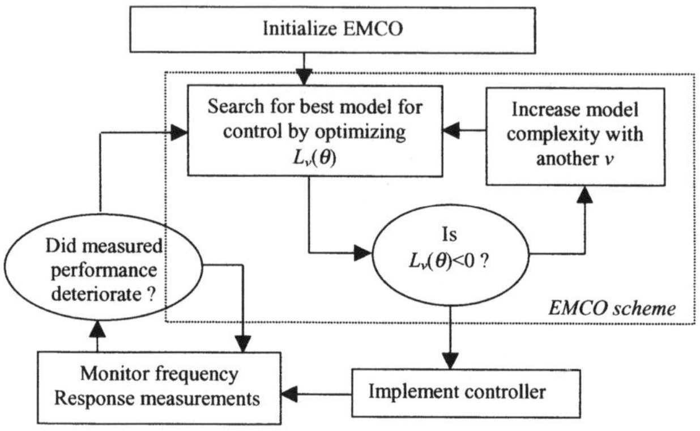

Fig. 5. Block diagram of an adaptive system using EMCO.

of EMCO within a continuously running adaptive control system. Such schemes can be applied in autonomous control systems.

The figures to follow illustrate the EMCO scheme starting from measured uncertainty bands on the open-loop plant response as shown in Figure 6 through the steps of obtaining control oriented models for three model structures. Because of lack of space the numerical details are omitted. The plant is a simulated 16-order resonant dynamics.

6. CONCLUSIONS

The paper first outlines the synergy between identification of plant dynamics performance of the IMC controller. This facilitates the identification and adaptation of the plant frequency response in

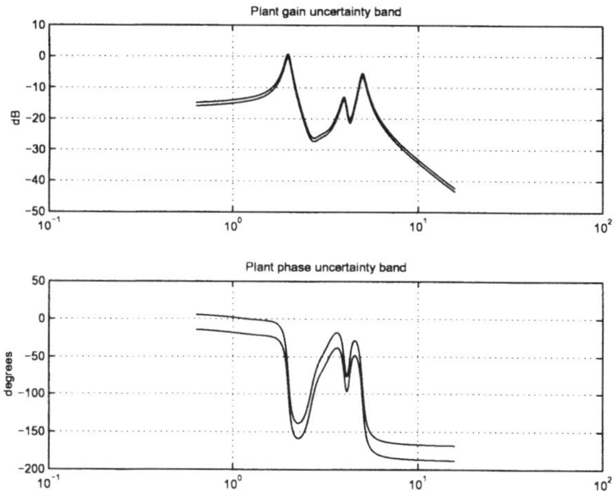

Fig. 6. Measured uncertainty bands of plant gain and phase.

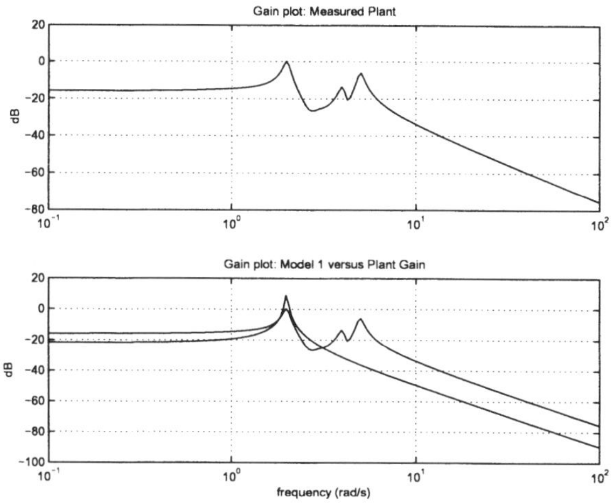

Fig. 7. Gain plot of the plant (top plot) and gain plot of the first control oriented model under model structure $c/(s^2 + as + b)$ superimposed (bottom plot).

closed loop to the degree the controller needs it. When the plant has a complex frequency response in areas such as active vibration control, flow control, flexible robots, etc. it is important to find the simplest controllers. The scheme presented solves this problem by a gradual exploration procedure of the parametric model space, starting from simple model structure and extending it only if it proves to be necessary. If the model structure tree is systematically searched, the scheme can guarantee to find the simplest robust controller up to a given robust performance specification. Future research will investigate model structure switching policies in detail.

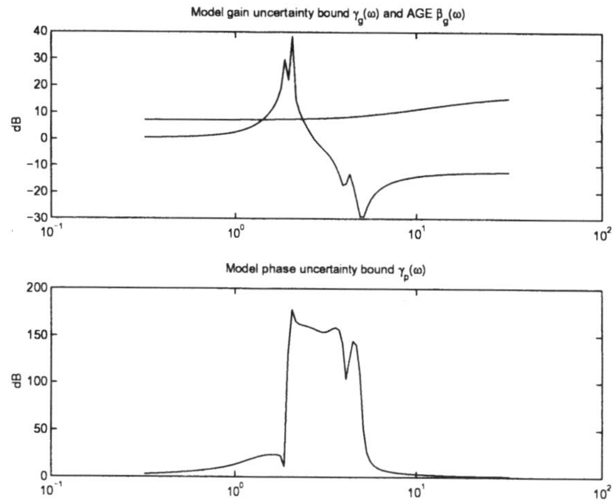

Fig. 8. Complex gain error bound $\gamma(\omega)$ with respect to the measured plant frequency response of Model 1. The gain error bound $\gamma(\omega)$ is superimposed with the AGE $\beta(\omega)$ (relatively flat curve on the top plot) and phase error bound $\gamma_p M(\omega)$ (bottom plot). The AGE $\beta(\omega)$ is violated by the model errors $\gamma(\omega)$.

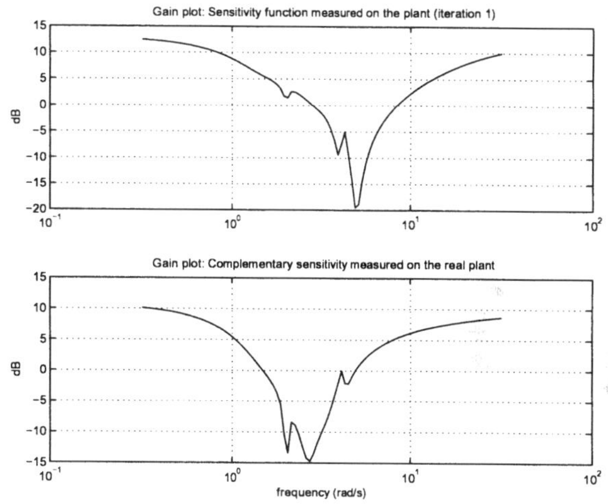

Fig. 9. Theoretical sensitivity and complementary sensitivity of the closed loop with Model 1 and the associated robust controller. The figure shows very bad performance. The closed loop is actually not stable with the real plant and the controller obtained from Model 1.

7. REFERENCES

Anderson, B. D. O. and R. L. Kosut (1991). Adaptive robust control: on-line learning. *Proc. Conference on Decision and Control, CDC'91, Brighton, England* pp. 297–298.

Van den Hof, P. M. J., R. J. P. Schrama, O. H. Bosgra and R. A. de Callafon (1993). Identification of normalized coprime plant factors for iterative model and controller enhancement. *Proc. Conference on Dicision and Control, CDC'93* pp. 2839–2844.

Doyle, J. C., B. A. Francis and A. R. Tannenbaum (1992). *Feedback Control Theory.* Mcmillan. New York.

Hjalmarsson, H., M. Gevers and F. de Bruyne (1996). For model-based control design, closed-loop identification gives better performance. *Automatica* **32**, 1659–1673.

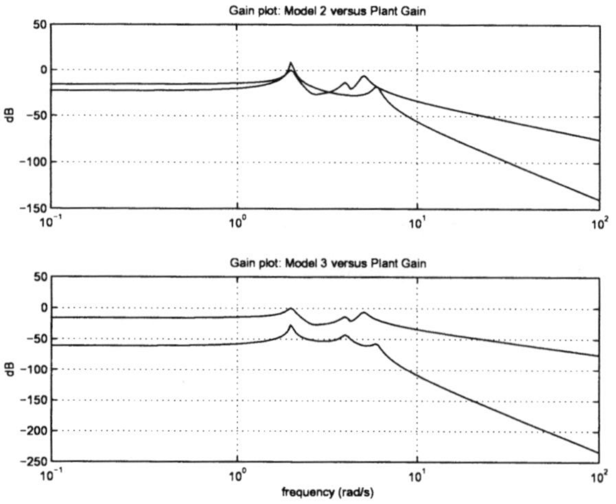

Fig. 10. Gain plots of Model 2 (fourth order transfer function, top plot) and Model 3 (sixth order transfer function, bottom plot) versus the average plant response. The large gain difference for Model 3 is due to difficulty to match the phase. Despite this mismatch, this is the model which provides and acceptable controller.

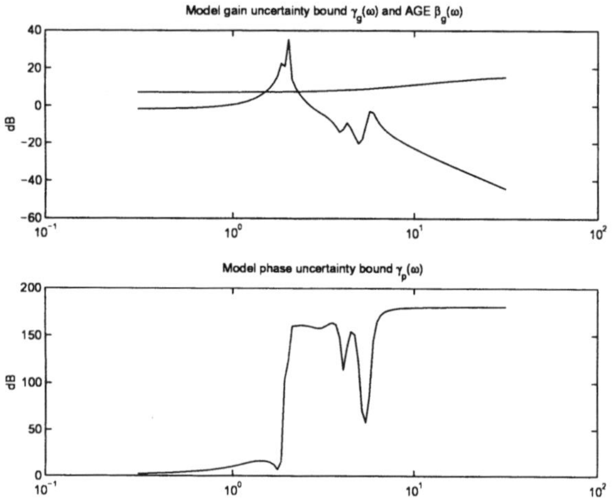

Fig. 11. Complex gain error bound $\gamma(\omega)$ with respect to the measured plant frequency response of Model 2. The real-valued-gain error bound of $\Delta(\omega)$ is superimposed with the AGE $\beta(\omega)$ (relatively flat curve on the top plot) and phase error bound of $\Delta M(\omega)$ (bottom plot). The AGE $\beta(\omega)$ is violated by the model errors $\gamma(\omega)$.

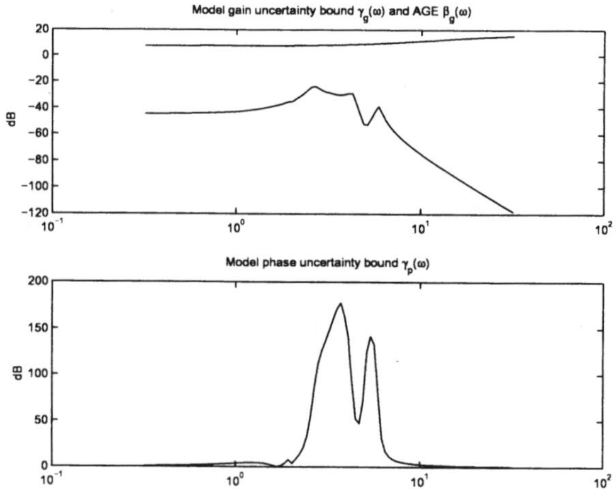

Fig. 12. Complex gain error bound $\delta(\omega)$ with respect to the measured plant frequency response of Model 2. The gain error bound of $\gamma(\omega)$ is superimposed with the AGE $\beta(\omega)$ (relatively flat curve on the top plot) and phase error bound of $\Delta M(\omega)$ (bottom plot). The AGE $\beta(\omega)$ bounds the model errors $\gamma(\omega)$ which indicates acceptable performance.

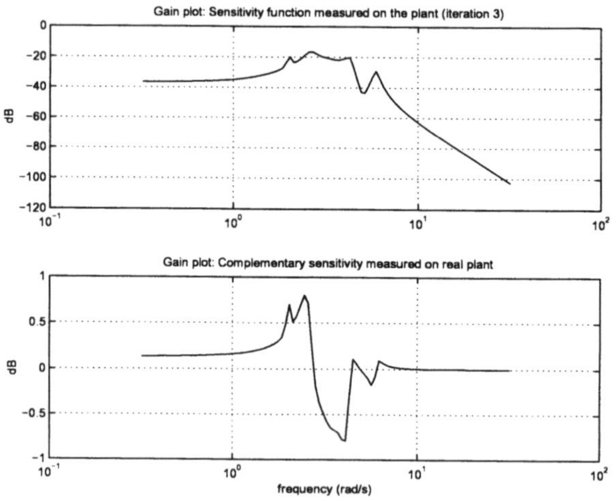

Fig. 13. Actual sensitivity and complementary sensitivity of the closed loop with Model 3 and the associated robust controller. The figure shows reasonable performance. The closed loop is stable with the real plant and the controller obtained from Model 3.

Morari, M. and E. Zafirou (1989). *Robust Process Control.* Prentice-Hall. New York.

Veres, S. M. (2001). Convergence of control performance by unfalsification of models - levels of confidence. *Int. J. Adaptive Control & Signal Processing* **15**, 471–502.

Veres, S. M. and D. S. Wall (2000). *Synergy and Duality of Identification and Control.* Taylor & Francis. London.

Vinnicombe, G. (1993). Frequency domain uncertainty and the graph topology. *IEEE Trans. on Automatic Control* **AC-38**, 1371–1383.

IFAC Publications
www.elsevier.com/locate/ifac

CONNECTING PE IDENTIFICATION AND ROBUST CONTROL THEORY: THE MULTIPLE-INPUT SINGLE-OUTPUT CASE. PART I: UNCERTAINTY REGION VALIDATION

Xavier Bombois * Paresh Date **

* *Department of Applied Physics, Delft University of Technology*
Lorentzweg 1, 2628 CJ Delft, The Netherlands
Email: X.J.A.Bombois@tnw.tudelft.nl
** *Department of Mathematical Sciences, Brunel University,*
London UB10 0BJ, United Kingdom
Email: Paresh.Date@brunel.ac.uk

Abstract: This paper and its companion paper extend previous results to connect PE identification and robust control theory in the SISO (single-input single output) case to the MISO (multiple-input single-output) case. *Copyright © 2003 IFAC*

Keywords: identification for robust control, PE identification, ν-gap , LMI, MISO systems

1. INTRODUCTION

This paper and its companion paper (Bombois and Date, 2002) are parts of the wide-spread effort to connect time-domain prediction error (PE) identification and robustness theory. This paper builds specifically on some of our earlier works (see e.g. (Bombois *et al.*, 2000; Bombois, 2000; Gevers *et al.*, 2002)). In these works, we have shown that PE identification with full-order model structures delivers both a model G_{mod} for control design and an uncertainty region $\mathcal{D}$ containing the true system at a certain probability level. This uncertainty region is a set of parametrized transfer functions whose (real) parameter vector is constrained to lie in an ellipsoid.

Since different PE identification experiments (under different experimental conditions) deliver different uncertainty regions $\mathcal{D}_i$ (containing all the true system G_0 at a certain probability level), a measure of robust stability was introduced, in (Bombois *et al.*, 2000), in order to assess the quality of an identified uncertainty region $\mathcal{D}$. This robust stability measure for $\mathcal{D}$ is defined as the worst case (i.e. the largest) ν-gap (Vinnicombe, 1993*a*; Vinnicombe, 1993*b*) between the model G_{mod} and the plants in $\mathcal{D}$. We proved that the smaller the worst case ν-gap between the model G_{mod} and all plants in some $\mathcal{D}$, the larger is the set of G_{mod}-based controllers [1] that are guaranteed by the ν-gap theory to robustly stabilize $\mathcal{D}$. The worst case ν-gap is thus an indicator of how well an uncertainty set $\mathcal{D}$ is tuned for robustly stable controller design based on the model G_{mod}. A too large indicator will therefore incite the designer to reject the uncertainty region and to perform a new identification experiment.

Once an uncertainty region $\mathcal{D}$ has been judged sufficiently tuned for robust control design with G_{mod}, a controller C for the true system G_0 can be designed from the identified model G_{mod}. Before being able to apply it to G_0, the controller C must be validated for stability and for performance. For this purpose, in (Bombois *et al.*, 2001), we have developed robustness analysis tools that are adapted to the uncertainty set $\mathcal{D}$ i.e. a method to verify whether a given controller stabilizes all systems in the uncertainty set $\mathcal{D}$ (and therefore also the true system), and a method to compute the worst case performance achieved by a given controller over all plants in $\mathcal{D}$.

[1] The G_{mod}-based controllers are the controllers designed from G_{mod}

In all previous works, the attention was restricted to single-input single-output (SISO) systems. For this particular type of systems, we showed that the worst case ν-gap can be exactly computed using an optimization problem involving LMI constraints (Bombois *et al.*, 2000), that the condition for the stabilization of all plants in $\mathcal{D}$ by a given controller is necessary and sufficient (Bombois *et al.*, 2001) and that the worst case performance can be also exactly computed using an optimization problem involving LMI constraints (Bombois *et al.*, 2001).

Most often, the real-life systems G_0 have multiple inputs and outputs. It was therefore crucial to extend our computational tools (developed in the SISO case) to multiple-inputs, multiple-outputs (MIMO) systems. Indeed, while the concepts that allowed us to connect PE identification and robust control theory in the SISO case carry over to the MIMO case without too many problems, the extension of the computational tools is by no means trivial. In this paper and in the companion paper (Bombois and Date, 2002), first steps in this direction are achieved. Indeed, we extend our computational tools to the case of multiple-inputs, single-output (MISO) systems.

For this purpose, we have first expressed the uncertainty region $\mathcal{D}$ delivered by the PE identification of a MISO true system in an appropriate way for the development of these computational tools. Using this appropriate expression of the uncertainty region $\mathcal{D}$, we give, in this paper, a new LMI-based optimization problem that exactly computes the robust stability measure of the uncertainty region $\mathcal{D}$, the worst case ν-gap . In the companion paper (Bombois and Date, 2002), we give both a necessary and sufficient condition for the stabilization of all plants in $\mathcal{D}$ by a given controller and a new LMI-based optimization problem that exactly computes the worst case performance achieved by a controller over all plants in $\mathcal{D}$.

Paper Outline. In Section 2, we give the general expression of the uncertainty region $\mathcal{D}$ delivered by PE identification in the case of a MISO true system. In Section 3, the concept of worst case ν-gap is defined and the new method to compute it is given. We finish by a numerical illustration and some conclusions in Sections 4 and 5.

2. UNCERTAINTY REGION FOR MISO SYSTEMS

In this section, we give an expression of the uncertainty regions $\mathcal{D}$ delivered by PE identification of MISO true systems in a full order model structure (Ljung, 1999). In the ensuing discussion, we will concentrate on MISO systems with two inputs for notational simplicity. However all the results are applicable for any (finite) number of inputs. Assume thus that the true system is linear and time-invariant, with a rational MISO transfer function G_0 such that

$$y = \overbrace{\begin{pmatrix} G_{0,1} & G_{0,2} \end{pmatrix}}^{G_0} \overbrace{\begin{pmatrix} u_1 \\ u_2 \end{pmatrix}}^{u} + v$$

where v is additive noise.

It is of great importance for the sequel of this paper that the identification is performed using a parametrization of G_0 with an identical auto-regressive part for $G_{0,1}$ and $G_{0,2}$ i.e.

$$G_0 = G(z, \delta_0) = \frac{\begin{pmatrix} e_1 + Z_{N_1}\delta_0 & e_2 + Z_{N_2}\delta_0 \end{pmatrix}}{1 + Z_D\delta_0} \quad (1)$$

where $\delta_0 \in \mathbf{R}^{k\times 1}$ is the *unknown* true parameter vector, $Z_{N_1}(z)$, $Z_{N_2}(z)$ and $Z_D(z)$ are row vectors of size k of known transfer functions. $e_1(z)$ and $e_2(z)$ are known transfer functions. This is the case when G_0 (and the noise model) has an ARX structure or an ARMAX structure. This is also the case when $G_{0,1}$ (resp. $G_{0,2}$) is parametrized by the sum of a given model $e_1(z)$ (resp. $e_2(z)$) and an expansion of linear basis functions (such as Laguerre filters). In this last case, $Z_D = 0$.

When $G_{0,1} = n_{0,1}/d_{0,1}$ and $G_{0,2} = n_{0,2}/d_{0,2}$ [2] have different auto-regressive parts, a structure such as in (1) can be obtained by considering a parametrization $G(z, \delta_0)$ such that $Z_{N_1}\delta_0 = n_{0,1}\, d_{0,2}$, $Z_{N_2}\delta_0 = n_{0,2}\, d_{0,1}$ and $1 + Z_D\delta_0 = d_{0,1}\, d_{0,2}$. Here $e_1(z)$ and $e_2(z)$ are equal to 0.

Proposition 1. Consider $G_0 = G(z, \delta_0)$, the true MISO system with two inputs parametrized as in (1). A PE identification experiment (with a full order model structure) performed on G_0 delivers an identified model $G_{mod} = G(z, \hat{\delta})$ and an uncertainty region $\mathcal{D}$ containing the true system G_0 at a prescribed probability level α. This uncertainty region is centered at $G(z, \hat{\delta})$ and can be described by the following generic form:

$$\mathcal{D} = \left\{ G(z,\delta) \mid G(z,\delta) = \frac{\begin{pmatrix} e_1 + Z_{N_1}\delta & e_2 + Z_{N_2}\delta \end{pmatrix}}{1 + Z_D\delta} \ \ and\ \delta \in U \right\} (2)$$

where $U = \{\delta \mid (\delta - \hat{\delta})^T R(\delta - \hat{\delta}) < \chi^2\}$, $\delta \in \mathbf{R}^{k\times 1}$ is a real parameter vector, $\hat{\delta}$ is the estimated parameter vector defining the identified model, R is a symmetric positive definite matrix $\in \mathbf{R}^{k\times k}$ that is equal to the inverse of the covariance matrix of $\hat{\delta}$, χ^2 is determined by the desired probability level α, $Z_{N_1}(z)$, $Z_{N_2}(z)$ and $Z_D(z)$ are row vectors of size k of known transfer functions. $e_1(z)$ and $e_2(z)$ are known transfer functions.

PROOF. Trivial extension of the results in (Bombois, 2000; Bombois *et al.*, 1999). □

---

[2] Here, $n_{0,i}(z)$ and $d_{0,i}(z)$ are the numerator and denominator of $G_{0,i}$, respectively.

Note that different identification experiments (i.e. open-loop or closed-loop identification, different measured data, ...) lead to different identified parameter vectors, different covariance matrices, and therefore also different uncertainty sets $\mathcal{D}$.

Due to the multivariable structure in (2), the numerical method that we had developed in our previous work (Bombois *et al.*, 2000) for a SISO uncertainty region in order to compute the robust stability measure of $\mathcal{D}$ is no more valid. In this paper, a new numerical methodology adapted to the multivariable structure of $\mathcal{D}$ will be presented in order to compute this measure.

3. A ROBUST STABILITY MEASURE FOR $\mathcal{D}$

As said in the introduction, the robust stability measure for the uncertainty region $\mathcal{D}$ is based on the concept of *worst-case* ν*-gap* between the identified model $G_{mod} = G(z, \hat{\delta})$ and the uncertainty set $\mathcal{D}$ that has been introduced in (Bombois *et al.*, 2000). The worst-case ν-gap is an extension of the ν-gap, introduced by Vinnicombe (Vinnicombe, 1993*a*), which is a measure of distance between two transfer functions.

Definition 2. Consider the uncertainty region $\mathcal{D}$ given in Proposition 1 and centered at the identified model $G_{mod} = G(z, \hat{\delta})$. The worst case ν-gap $\delta_{WC}(G_{mod}, \mathcal{D})$ is given by:

$$\delta_{WC}(G_{mod}, \mathcal{D}) = \sup_{G_{in} \in \mathcal{D}} \delta_{\nu}(G_{mod}, G_{in}) \quad (3)$$

where $\delta_{\nu}(G_{mod}, G_{in})$ is the ν-gap between the plants G_{mod} and G_{in} (see (Vinnicombe, 1993*a*)).

The definition of another important quantity is also recalled: the worst case chordal distance $\kappa_{WC}(G_{mod}(e^{j\omega}), \mathcal{D})$. This quantity, whose computation is the result of a convex optimization problem involving LMI constraints in the MISO case as will be shown in Section 3.1, will allow us to give an alternative expression for $\delta_{WC}(G_{mod}, \mathcal{D})$.

Definition 3. At a particular frequency ω, we define $\kappa_{WC}(G_{mod}(e^{j\omega}), \mathcal{D})$ as the maximum chordal distance between $G_{mod}(e^{j\omega})$ and the frequency responses of all plants in $\mathcal{D}$ at this frequency:

$$\kappa_{WC}(G_{mod}(e^{j\omega}), \mathcal{D}) = \sup_{G_{in} \in \mathcal{D}} \kappa(G_{mod}(e^{j\omega}), G_{in}(e^{j\omega})) \quad (4)$$

where the chordal distance κ is defined as follows:

$$\kappa(G_{mod}(e^{j\omega}), G_{in}(e^{j\omega})) \triangleq \overline{\sigma}\left(\alpha_{in}^{-\frac{1}{2}} \left(G_{mod}(e^{j\omega}) - G_{in}(e^{j\omega})\right) \beta_{mod}^{-\frac{1}{2}}\right) \quad (5)$$

with $\alpha_{in} = 1 + G_{in}(e^{j\omega})G_{in}^{*}(e^{j\omega})$, $\beta_{mod} = I_2 + G_{mod}^{*}(e^{j\omega})G_{mod}(e^{j\omega})$ and $\overline{\sigma}(A)$, the largest singular value of A.

This last quantity can now be used to give an alternative expression of the worst case ν-gap . This is done in the following lemma, which is an extension of a property presented in (Vinnicombe, 1993*b*, page 66).

Lemma 4. The worst case Vinnicombe distance $\delta_{WC}(G_{mod}, \mathcal{D})$ defined in (3) can also be expressed in the following way using the worst case chordal distance:

$$\delta_{WC}(G_{mod}, \mathcal{D}) = \sup_{\omega} \kappa_{WC}(G_{mod}(e^{j\omega}), \mathcal{D}) \quad (6)$$

where $\kappa_{WC}(G_{mod}(e^{j\omega}), \mathcal{D})$ is defined in (4).

PROOF. See (Bombois *et al.*, 2000). □

3.1 *Computation of the worst case chordal distance and worst case* ν*-gap*

We have recalled the definition of the worst case ν-gap between the model G_{mod} and all plants in an uncertainty region $\mathcal{D}$. In this subsection, we give a procedure to compute this worst case ν-gap $\delta_{WC}(G_{mod}, \mathcal{D})$ in the considered case of an uncertainty region delivered by the PE identification of a MISO true system G_0. This LMI-procedure is the extension to the MISO case of the procedure developed in (Bombois *et al.*, 2000) for the SISO systems.

Theorem 5. Consider the uncertainty region $\mathcal{D}$ given in Proposition 1 and centered at the identified model $G_{mod} = G(z, \hat{\delta})$. Then $\kappa_{WC}(G_{mod}(e^{j\omega}), \mathcal{D}) = \sqrt{\gamma_{opt}}$, where γ_{opt} is the optimal value of γ in the following standard convex optimization problem involving LMI constraints (Boyd *et al.*, 1994) evaluated at ω:

$$\begin{aligned} &\textit{minimize } \gamma \\ &\textit{over} \quad \gamma,\ \tau \\ &\textit{subject to } \tau \geq 0 \textit{ and} \end{aligned}$$

$$\begin{pmatrix} Re(a_{11}) & Re(a_{12}) \\ Re(a_{12}^{*}) & Re(a_{22}) \end{pmatrix} - \tau \begin{pmatrix} R & -R\hat{\delta} \\ (-R\hat{\delta})^{T} & \hat{\delta}^{T} R \hat{\delta} - \chi^{2} \end{pmatrix} < 0 \quad (7)$$

where

- $a_{11} = \left((k_1 Z_1^{*} + k_2 Z_2^{*}) Z_1 + (k_2^{*} Z_1^{*} + k_3 Z_2^{*}) Z_2 \right) - \gamma \left(Z_{N_1}^{*} Z_{N_1} + Z_{N_2}^{*} Z_{N_2} + Z_D^{*} Z_D \right)$,
- $a_{12} = \left(f_1 (k_1 Z_1^{*} + k_2 Z_2^{*}) + f_2 (k_2^{*} Z_1^{*} + k_3 Z_2^{*}) \right) - \gamma \left(e_1 Z_{N_1}^{*} + e_2 Z_{N_2}^{*} + Z_D^{*} \right)$,
- $a_{22} = \left(f_1 (k_1 f_1^{*} + k_2 f_2^{*}) + f_2 (k_2^{*} f_1^{*} + k_3 f_2^{*}) \right) - \gamma \left(1 + e_1 e_1^{*} + e_2 e_2^{*} \right)$,
- $f_1 = x_1 - e_1$ and $f_2 = x_2 - e_2$
- $Z_1 = x_1 Z_d - Z_{N_1}$ and $Z_2 = x_2 Z_d - Z_{N_2}$
- $\begin{pmatrix} k_1 & k_2 \\ k_2^{*} & k_3 \end{pmatrix} = BB^{*}$ with $B = \left(I_2 + X^{*} X \right)^{-\frac{1}{2}}$
- $X = \left(x_1 \ x_2 \right) = G_{mod}(e^{j\omega})$.

The worst case ν-gap is then obtained as

$$\delta_{WC}(G_{mod}, \mathcal{D}) = \sup_{\omega} \kappa_{WC}(G_{mod}(e^{j\omega}), \mathcal{D})$$

PROOF. We prove that the square root of the solution of the LMI optimization problem gives the worst case chordal distance $\kappa_{WC}(G_{mod}(e^{j\omega}), \mathcal{D})$ at some frequency ω. The derivation of the worst case ν-gap is a direct consequence of Lemma 4.

If we denote the frequency response of the model $G_{mod}(e^{j\omega})$ by X, and that of any plant $G(e^{j\omega}, \delta) \in \mathcal{D}$ by $Y(\delta)$. Then a convenient way to state the problem of computing the worst case chordal distance at some frequency ω is as follows:

$$\begin{gathered} minimize\ \gamma\ such\ that \\ \kappa(X, Y(\delta))^2 < \gamma\ for\ all\ \delta \in U \end{gathered}$$

Let us rewrite $\kappa(X, Y(\delta))^2$ in an appropriate way for the LMI formulation. Using the fact that X and $Y(\delta)$ are in $\mathcal{C}^{1\times 2}$, the fact that $\overline{\sigma}^2(A) = \overline{\lambda}(A^*A)$ ($\overline{\lambda}(A^*A)$ is the largest eigenvalue of A^*A) and the definition (5) of $\kappa(X, Y(\delta))$:

$$\kappa(X, Y(\delta)) =$$

$$\overline{\sigma}\left((1 + Y(\delta)Y(\delta)^*)^{-\frac{1}{2}} (X - Y(\delta)) \overbrace{(I_2 + X^*X)^{-\frac{1}{2}}}^{B} \right),$$

we obtain successively that

$$\kappa(X, Y(\delta))^2 < \gamma \Longleftrightarrow$$

$$(1 + Y(\delta)Y(\delta)^*)^{-1}\ \overline{\lambda}\left(B^*(X - Y(\delta))^*(X - Y(\delta))B \right) < \gamma \Longleftrightarrow$$

$$(X - Y(\delta))BB^*(X - Y(\delta))^* < \gamma(1 + Y(\delta)Y(\delta)^*) \qquad (8)$$

where the last equivalence is a consequence of the fact that, when ϕ is a row vector, $\overline{\lambda}(\phi^*\phi) = \phi\phi^*$. By pre-multiplying (8) by $d_Y \triangleq (1 + Z_D\delta)$ and post-multiplying the same expression by d_Y^*, we obtain:

$$(d_Y X - N_Y)BB^*(d_Y^* X^* - N_Y^*) < \gamma(d_Y d_Y^* + N_Y N_Y^*) (9)$$

where $N_Y = \begin{pmatrix} e_1 + Z_{N_1}\delta & e_2 + Z_{N_2}\delta \end{pmatrix}$. Using the notations used in the statement of Theorem 5, Expression (9) is equivalent with:

$$\begin{gathered} \begin{pmatrix} f_1 + Z_1\delta & f_2 + Z_2\delta \end{pmatrix} \begin{pmatrix} k_1 & k_2 \\ k_2^* & k_3 \end{pmatrix} \begin{pmatrix} f_1 + Z_1\delta & f_2 + Z_2\delta \end{pmatrix}^* \\ < \gamma(d_Y d_Y^* + N_Y N_Y^*) \end{gathered}$$

Some trivial manipulations show then that this last expression is equivalent with the following constraint on δ with variable γ:

$$\begin{pmatrix} \delta \\ 1 \end{pmatrix}^* \begin{pmatrix} a_{11} & a_{12} \\ a_{12}^* & a_{22} \end{pmatrix} \begin{pmatrix} \delta \\ 1 \end{pmatrix} < 0 \qquad (10)$$

with a_{11}, a_{12} and a_{22} as defined in the statement of Theorem 5. Since δ is real, it can be shown that (10) is equivalent with

$$\overbrace{\begin{pmatrix} \delta \\ 1 \end{pmatrix}^T \begin{pmatrix} Re(a_{11}) & Re(a_{12}) \\ Re(a_{12}^*) & Re(a_{22}) \end{pmatrix} \begin{pmatrix} \delta \\ 1 \end{pmatrix}}^{\psi(\delta)} < 0 \quad (11)$$

This last expression is equivalent to stating that $\kappa(G_{mod}(e^{j\omega}), G(e^{j\omega}, \delta))^2 < \gamma$ for a particular $\delta \in U$. However, this must be true for all $\delta \in U$. Therefore, (11) must be true for all δ such that:

$$\overbrace{\begin{pmatrix} \delta \\ 1 \end{pmatrix}^T \begin{pmatrix} R & -R\hat{\delta} \\ (-R\hat{\delta})^T & \hat{\delta}^T R\hat{\delta} - \chi^2 \end{pmatrix} \begin{pmatrix} \delta \\ 1 \end{pmatrix}}^{\rho(\delta)} < 0 (12)$$

which is equivalent to the statement "$\delta \in U$". Let us now recapitulate. Computing $\kappa_{WC}(G_{mod}(e^{j\omega}), \mathcal{D})^2$ is equivalent to finding the smallest γ such that $\psi(\delta) < 0$ for all δ for which $\rho(\delta) < 0$. By the $\mathcal{S}$ procedure (Boyd *et al.*, 1994), this problem is equivalent to finding the smallest γ and a positive scalar τ such that $\psi(\delta) - \tau\rho(\delta) < 0$, for all $\delta \in \mathbf{R}^{k\times 1}$ which is precisely (7). To complete this proof, note that the worst case chordal distance at ω is thus equal to $\sqrt{\gamma_{opt}}$ where γ_{opt} is the optimal value of γ. □

Remarks. The LMI-procedure given in Theorem 5 for a MISO system with two inputs can be easily extended to all MISO systems. Theorem 5 can also be easily adapted to SIMO systems i.e. single-input multiple-outputs systems. For two SIMO systems G_1 and G_2, $G_1^T(e^{j\omega})$ and $G_2^T(e^{j\omega})$ are indeed row vectors and $\kappa(G_1(e^{j\omega}), G_2(e^{j\omega})) = \kappa(G_1^T(e^{j\omega}), G_2^T(e^{j\omega}))$. Finally, Theorem 5 allows us also to compute frequency weighted worst case ν-gap : $\delta_{WC}^{\alpha}(G_{mod}, \mathcal{D}) = \sup_{G_{in}\in\mathcal{D}} \delta_\nu(\alpha G_{mod}, \alpha G_{in})$ where α is a (scalar) transfer function. This especially makes sense in the $\mathcal{H}_\infty$ loopshaping design paradigm, where $\alpha(z)$ would be a weighting function (see (Glover and McFarlane, 1989) for details). The weighting α in this case would only modify e_1, e_2, Z_{N_1}, Z_{N_2} and as such would not have an effect on the suggested algorithm.

3.2 *The worst case ν-gap as a robust stability measure for $\mathcal{D}$*

In the previous subsections, the notion of worst case ν-gap between a model G_{mod} and an uncertainty region $\mathcal{D}$ has been recalled and a procedure has been given to compute this distance for a MISO uncertainty region $\mathcal{D}$. This worst case ν-gap can be considered as a robustness measure of $\mathcal{D}$ with respect to robustly stable controller design based on the model G_{mod}. We have indeed the following result.

Proposition 6. Consider the uncertainty region $\mathcal{D}$ given in Proposition 1 and centered at the identified model $G_{mod} = G(z, \hat{\delta})$. Consider also the worst case ν-gap $\delta_{WC}(G_{mod}, \mathcal{D})$ defined in (3). All controllers C that stabilize G_{mod} and that lie in the set

$$\mathcal{C}(G_{mod}, \mathcal{D}) = \{C \mid b_{G_{mod},C} > \delta_{WC}(G_{mod}, \mathcal{D})\} (13)$$

are guaranteed to stabilize all plants in the uncertainty region $\mathcal{D}$. The quantity $b_{G_{mod}C}$ is the generalized stability margin of the closed loop system $[G_{mod}\ C]$ and is defined as follows, if $[G_{mod}\ C]$ is stable,

$$b_{G_{mod}C} \triangleq \left\| \begin{pmatrix} G_{mod} \\ I_2 \end{pmatrix} \left(I_2 + CG_{mod} \right)^{-1} \left(C\ I_2 \right) \right\|_{\infty}^{-1} \quad (14)$$

and $b_{G_{mod}C} \triangleq 0$, if $[G_{mod}\ C]$ is unstable.

PROOF. Using the definition of the worst case ν-gap given in (3), we see that $\mathcal{D}$ is embedded in the uncertainty region $\{G|\delta_\nu(G_{mod}, G) \leq \delta_{WC}(G_{mod}, \mathcal{D})\}$. This theorem is thus a direct consequence of the main robust stability result in (Vinnicombe, 1993*a*). □

Note that the generalized stability margin $b_{G_{mod}C}$ defined in (14) has a maximum $b_{opt}(G_{mod})$ (see e.g. (Zhou and Doyle, 1998)) given by

$$b_{opt}(G_{mod}) = \max_C b_{G_{mod}C} = \sqrt{1 - \|\ (\,N\ \ M\,)\ \|_H^2}, \quad (15)$$

where $\|\ A\ \|_H$ is the Hankel norm of the operator A and $\{N, M\}$ is the normalized left coprime factorization of G_{mod} i.e. the coprime factorization $G_{mod} = M^{-1}N$ (N and $M \in \mathcal{R}H_\infty$) such that $NN^* + MM^* = 1$.

Proposition 6 tells us that the worst case ν-gap between the model G_{mod} and an uncertainty set $\mathcal{D}$ is a measure of size of the set $\mathcal{D}$ that is directly connected to the size of the set $\mathcal{C}(G_{mod}, \mathcal{D})$ of G_{mod}-based controllers that are guaranteed to stabilize all plants in $\mathcal{D}$. Proposition 6 shows also that *the smaller $\delta_{WC}(G_{mod}, \mathcal{D})$, the larger is this robustly stabilizing controller set.* Thus, the worst case ν-gap is a nice and compact measure of the ability of an uncertainty set $\mathcal{D}$ to be robustly stabilized by a large set of controllers designed from G_{mod} and therefore of how well the uncertainty region $\mathcal{D}$ is tuned for robustly stable controller design based on G_{mod}.

It is to be noted that there may be additional controllers outside the set $\mathcal{C}(G_{mod}, \mathcal{D})$ that stabilize all models in $\mathcal{D}$. Indeed, according to (Vinnicombe, 1993*a*), the set $\mathcal{C}(G_{mod}, \mathcal{D})$ contains all controllers that stabilize all systems in the uncertainty set $\{G|\delta_\nu(G_{mod}, G) \leq \delta_{WC}(G_{mod}, \mathcal{D})\}$ that embeds $\mathcal{D}$. However, the advantage of this description is that the size of the set $\mathcal{C}(G_{mod}, \mathcal{D})$ (i.e. $\delta_{WC}(G_{mod}, \mathcal{D})$) is only a function of G_{mod} and $\mathcal{D}$. In the companion paper (Bombois and Date, 2002), a necessary and sufficient condition will be given for the stabilization of all plants in $\mathcal{D}$ by a given controller. However, this necessary and sufficient condition may not be used as a measure of robust stability for $\mathcal{D}$, as shown in (Bombois, 2000).

Quality assessment of $\mathcal{D}$. As said above, the worst case ν-gap is a nice and compact measure of how well the uncertainty region $\mathcal{D}$ is tuned for robust control design with respect to G_{mod}. In order to present the practical use of this measure, let us consider that we have performed one PE identification experiment leading to an uncertainty region $\mathcal{D}$ centered at the model $G_{mod} = G(z, \hat{\delta})$. The worst case ν-gap $\delta_{WC}(G(z, \hat{\delta}), \mathcal{D})$ can then be used as a tool to validate the uncertainty region $\mathcal{D}$. Indeed, if the worst case ν-gap is small with respect to the optimal stability margin $b_{opt}(G(z, \hat{\delta}))$ (see (15)), then the set $\mathcal{C}(G(z, \hat{\delta}), \mathcal{D})$ of $G(z, \hat{\delta})$-based controllers that are guaranteed to robustly stabilize $\mathcal{D}$ is large and the designer will be therefore generally incited to keep the uncertainty region $\mathcal{D}$: a controller can be designed from G_{mod} and the results of the companion paper (Bombois and Date, 2002) can be used to validate the controller for stability and for performance with respect to the "validated" uncertainty region $\mathcal{D}$. On the contrary, if the worst case ν-gap is large with respect to the optimal stability margin $b_{opt}(G(z, \hat{\delta}))$, then the set $\mathcal{C}(G(z, \hat{\delta}), \mathcal{D})$ of $G(z, \hat{\delta})$-based controllers that are guaranteed to robustly stabilize $\mathcal{D}$ is small. Therefore, even though the set $\mathcal{C}(G(z, \hat{\delta}), \mathcal{D})$ is not guaranteed to contain all robustly stabilizing controllers, the designer will be nevertheless generally incited to reject the uncertainty region $\mathcal{D}$ and to perform a new validation experiment in order to obtain a new uncertainty region $\mathcal{D}_{bis}$ with a larger set of stabilizing controllers. For this purpose, the designer could e.g. use the guidelines that are presented in (Bombois, 2000).

Remark. In the $\mathcal{H}_\infty$ loopshaping design paradigm, the performance of a closed-loop system $[C\ G]$ is measured by its generalized stability margin b_{GC} (defined in (14)). In that sense, the worst case ν-gap $\delta_{WC}(G_{mod}, \mathcal{D})$ can be seen as a measure of the worst case performance degradation in the set $\mathcal{D}$. Indeed, given any controller C in $\mathcal{C}(G_{mod}, \mathcal{D})$ (see (13)), we have the following result: $\inf_{G \in \mathcal{D}} b_{GC} \geq b_{G_{mod}C} - \delta_{WC}(G_{mod}, \mathcal{D})$. This result is a direct consequence of the properties of the ν-gap metric (Vinnicombe, 1993*b*) and of the fact that $\mathcal{D}$ is embedded in the ν-gap uncertainty region $\{G|\delta_\nu(G_{mod}, G) \leq \delta_{WC}(G_{mod}, \mathcal{D})\}$. Note that, in practice, in $\mathcal{H}_\infty$ loopshaping design, we have to consider a shaped nominal plant and thus this result has to be rewritten with this shaped plant and with a *frequency weighted* worst case ν-gap (see the remark at the end of Section 3.1).

4. NUMERICAL ILLUSTRATION

Let us now illustrate our results by an example. Consider thus the following MISO true system in the ARX structure:

$$y(t) = \frac{\overbrace{(\,n_{0,1}(z)\ \ n_{0,2}(z)\,)}^{G_0}}{d_0(z)} u(t) + \frac{e(t)}{d_0(z)}$$

where $e(t)$ is a white noise of variance 0.1, $n_{0,1} = 0.1047z^{-1} + 0.0872z^{-2}$, $n_{0,2} = 0.2z^{-1} + 0.05z^{-2}$

and $d_0 = 1 - 1.5578z^{-1} + 0.5769z^{-2}$. A PE identification experiment is performed on this true system using an input signal whose two components are a white noise of variance 1. 5000 data are collected and the following model is identified:

$$G_{mod} = G(z, \hat{\delta}) =$$

$$\frac{\left(0.099z^{-1} + 0.086z^{-2} \quad 0.2013z^{-1} + 0.048z^{-2} \right)}{1 - 1.5554z^{-1} + 0.578z^{-2}}$$

Along with this model, the uncertainty region $\mathcal{D}$ centered at G_{mod} and containing the true system G_0 at a probability 0.95 is built. This uncertainty region has the general structure (2) with e1 and $e_2 = 0$, $\chi^2 = 12.6$, $Z_D = (z^{-1}\ z^{-2}\ 0\ 0\ 0\ 0)$, $Z_{N_1} = (0\ 0\ z^{-1}\ z^{-2}\ 0\ 0)$, $Z_{N_2} = (0\ 0\ 0\ 0\ z^{-1}\ z^{-2})$ and $\hat{\delta} = (-1.554, 0.578, 0.099, 0.086, 0.2013, 0.048)^T$.

Quality assessment of $\mathcal{D}$. The results of Section 3 are now used to verify if $\mathcal{D}$ is stabilized by a large set of controllers stabilizing G_{mod}. This can be achieved by computing the worst case ν-gap $\delta_{WC}(G_{mod}, \mathcal{D})$ between G_{mod} and the plants in $\mathcal{D}$. For this purpose, we first compute the worst case chordal distances $\kappa_{WC}(G_{mod}(e^{j\omega}), \mathcal{D})$ at each frequency using the LMI tools developed in Section 3.1. The worst case chordal distances are represented in Figure 1

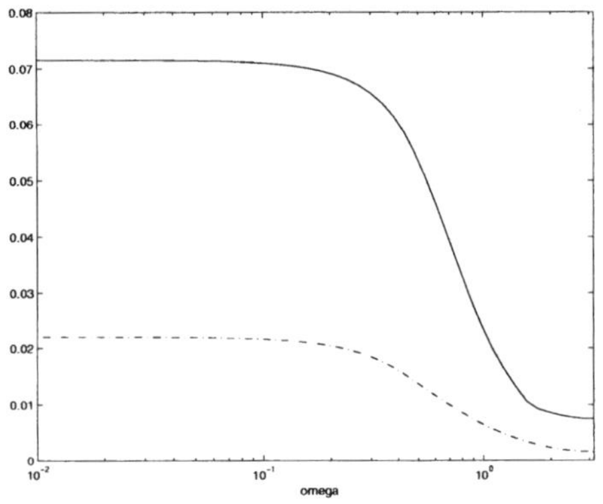

Fig. 1. $\kappa_{WC}(G_{mod}(e^{j\omega}), \mathcal{D})$ (solid) and $\kappa(G_{mod}(e^{j\omega}), G_0(e^{j\omega}))$ (dashdot) at each frequency

According to Lemma 4, we can derive the worst case ν-gap $\delta_{WC}(G_{mod}, \mathcal{D})$ from the worst chordal distances: $\delta_{WC}(G_{mod}, \mathcal{D}) = \sup_\omega \kappa_{WC}(G_{mod}(e^{j\omega}), \mathcal{D}) = 0.0716$. The optimal stability margin $b_{opt}(G_{mod})$ can also be computed using (15) and is here equal to 0.5325. We may therefore conclude that the set $\mathcal{C}(G_{mod}, \mathcal{D})$ of G_{mod}-based controllers that are guaranteed by the ν-gap theory to robustly stabilize $\mathcal{D}$, is relatively large. We are therefore incited to keep the pair $\{G_{mod}\ \mathcal{D}\}$ in order to make the design of a G_{mod}-based controller C for G_0 and to validate this controller for stability and for performance over the systems in $\mathcal{D}$. This analysis will be performed in the companion paper (Bombois and Date, 2002).

5. CONCLUSIONS

In this paper and in the companion paper, we have extended to the more general case of MISO systems previous results to connect PE identification and robust control theory that were developed in the SISO case. To simplify the notations, the results were here presented for the case of MISO systems with two inputs, but can be easily extended to all MISO systems and also to the dual case of SIMO (single-input output-output) systems. Research is now pursued to extend these results to the case of systems with multiple inputs and multiple outputs.

6. REFERENCES

Bombois, X. (2000). Connecting Prediction Error Identification and Robust Control Analysis: a new framework. PhD thesis. Université Catholique de Louvain.

Bombois, X. and P. Date (2002). Connecting PE identification and robust control theory: the multiple-input single-output case. part II: Controller validation. Submitted to the 13th IFAC Symposium on System Identification.

Bombois, X., M. Gevers and G. Scorletti (2000). A measure of robust stability for a set of parametrized transfer functions. *IEEE Transactions on Automatic Control* **45**(11), 2141–2145.

Bombois, X., M. Gevers and G. Scorletti (1999). Controller validation based on an identified model. In: *Proc. 38th IEEE Conference on Decision and Control*, Phoenix, Arizona.

Bombois, X., M. Gevers, G. Scorletti and B.D.O. Anderson (2001). Robustness analysis tools for an uncertainty set obtained by prediction error identification. *Automatica* **37**(10), 1629–1636.

Boyd, S., L. El Ghaoui, E. Feron and V. Balakrishnan (1994). *Linear Matrix Inequalities in Systems and Control Theory*. Vol. 15 of *Studies in Appl. Math.*. SIAM. Philadelphia.

Gevers, M., X. Bombois, B. Codrons, G. Scorletti and B.D.O. Anderson (2002). Model validation for control and controller validation in a prediction error identification framework- part I : Theory. Accepted for publication in Automatica.

Glover, K. and D. McFarlane (1989). Robust stabilisation of normalised coprime factor plant descriptions with H_∞-bounded uncertainty. *IEEE Trans. Automat. Contr.* **34**, 821–830.

Ljung, L. (1999). *System Identification: Theory for the User, 2nd Edition*. Prentice-Hall. Englewood Cliffs, NJ.

Vinnicombe, G. (1993*a*). Frequency domain uncertainty and the graph topology. *IEEE Trans Automatic Control* **AC-38**, 1371–1383.

Vinnicombe, G. (1993*b*). Measuring the Robustness of Feedback Systems. PhD thesis. Cambridge University.

Zhou, K. and J. Doyle (1998). *Essentials of Robust Control*. Prentice Hall, Upper Saddle River, New Jersey.

IFAC Publications
www.elsevier.com/locate/ifac

CONNECTING PE IDENTIFICATION AND ROBUST CONTROL THEORY: THE MULTIPLE-INPUT SINGLE-OUTPUT CASE. PART II: CONTROLLER VALIDATION

Xavier Bombois * **Paresh Date** **

* *Department of Applied Physics, Delft University of Technology*
Lorentzweg 1, 2628 CJ Delft, The Netherlands
Email: X.J.A.Bombois@tnw.tudelft.nl
** *Department of Mathematical Sciences, Brunel University,*
London UB10 0BJ, United Kingdom
Email: Paresh.Date@brunel.ac.uk

Abstract: This paper and its companion paper extend previous results to connect PE identification and robust control theory in the SISO (single-input single output) case to the MISO (multiple-input single-output) case. *Copyright © 2003 IFAC*

Keywords: identification for robust control, PE identification, LMI, MISO systems

1. INTRODUCTION

This paper and its companion paper (Bombois and Date, 2002) are parts of the wide-spread effort to connect time-domain prediction error (PE) identification and robustness theory (see e.g. (Goodwin *et al.*, 1992; Hakvoort and Van den Hof, 1997; Ljung, 2000)). This paper builds specifically on some of our earlier works (see e.g. (Bombois *et al.*, 2000; Bombois, 2000; Gevers *et al.*, 2002)). In these works, we have shown that PE identification with full-order model structures delivers both a model G_{mod} for control design and an uncertainty region $\mathcal{D}$ containing the true system at a certain probability level. This uncertainty region is a set of parametrized transfer functions whose (real) parameter vector is constrained to lie in an ellipsoid.

Since different PE identification experiments (under different experimental conditions) deliver different uncertainty regions $\mathcal{D}_i$ (containing all the true system G_0 at a certain probability level), a measure of robust stability was introduced, in (Bombois *et al.*, 2000), in order to assess the quality of an identified uncertainty region $\mathcal{D}$. This robust stability measure for $\mathcal{D}$ is defined as the worst case (i.e. the largest) ν-gap (Vinnicombe, 1993*a*; Vinnicombe, 1993*b*) between the model G_{mod} and the plants in $\mathcal{D}$. We proved that the smaller the worst case ν-gap between the model G_{mod} and all plants in some $\mathcal{D}$, the larger is the set of G_{mod}-based controllers [1] that are guaranteed by the ν-gap theory to robustly stabilize $\mathcal{D}$. The worst case ν-gap is thus an indicator of how well an uncertainty set $\mathcal{D}$ is tuned for robustly stable controller design based on the model G_{mod}. A too large indicator will therefore incite the designer to reject the uncertainty region and to perform a new identification experiment.

Once an uncertainty region $\mathcal{D}$ has been judged sufficiently tuned for robust control design with G_{mod}, a controller C for the true system G_0 can be designed from the identified model G_{mod}. Before being able to apply it to G_0, the controller C must be validated for stability and for performance. For this purpose, in (Bombois *et al.*, 2001), we have developed robustness analysis tools that are adapted to the uncertainty set $\mathcal{D}$ i.e.

- a method to verify whether a given controller stabilizes all systems in the uncertainty set $\mathcal{D}$ (and therefore also the true system),

[1] The G_{mod}-based controllers are the controllers designed from G_{mod}

- a method to compute the worst case performance achieved by a given controller over all plants in $\mathcal{D}$.

In all previous works, the attention was restricted to single-input single-output (SISO) systems. For this particular type of systems, we showed that the worst case ν-gap can be exactly computed using an optimization problem involving LMI constraints (Bombois *et al.*, 2000), that the condition for the stabilization of all plants in $\mathcal{D}$ by a given controller is necessary and sufficient (Bombois *et al.*, 2001) and that the worst case performance can be also exactly computed using an optimization problem involving LMI constraints (Bombois *et al.*, 2001).

Most often, the real-life systems G_0 have multiple inputs and outputs. It was therefore crucial to extend our computational tools (developed in the SISO case) to multiple-inputs, multiple-outputs (MIMO) systems. Indeed, while the concepts that allowed us to connect PE identification and robust control theory in the SISO case carry over to the MIMO case without too many problems, the extension of the computational tools is by no means trivial. In this paper and in the companion paper (Bombois and Date, 2002), first steps in this direction are achieved. Indeed, we extend our computational tools to the case of multiple-inputs, single-output (MISO) systems.

For this purpose, we have first expressed the uncertainty region $\mathcal{D}$ delivered by the PE identification of a MISO true system in an appropriate way for the development of these computational tools. Using this appropriate expression of the uncertainty region $\mathcal{D}$, we give, in the companion paper (Bombois and Date, 2002), a new LMI-based optimization problem that exactly computes the robust stability measure of the uncertainty region $\mathcal{D}$, the worst case ν-gap . In this paper, we give both a necessary and sufficient condition for the stabilization of all plants in $\mathcal{D}$ by a given controller and a new LMI-based optimization problem that exactly computes the worst case performance achieved by a controller over all plants in $\mathcal{D}$. Note that the performance of a closed-loop system is here defined as the largest singular value of four closed-loop transfer functions and vectors describing the closed-loop system.

Paper Outline. In Section 2, we give the general expression of the uncertainty region $\mathcal{D}$ delivered by PE identification in the case of a MISO true system. In Sections 3 and 4, the controller validation procedures for stability and performance are presented. We finish by a numerical illustration and some conclusions in Sections 5 and 6.

2. UNCERTAINTY REGION FOR MISO SYSTEMS

In this section, we give an expression of the uncertainty regions $\mathcal{D}$ delivered by PE identification of MISO true systems in a full order model structure (Ljung, 1999). In the ensuing discussion, we will concentrate on MISO systems with two inputs for notational simplicity. However all the results are applicable for any (finite) number of inputs. Assume thus that the true system is linear and time-invariant, with a rational MISO transfer function G_0 such that

$$y = \overbrace{\begin{pmatrix} G_{0,1} & G_{0,2} \end{pmatrix}}^{G_0} \overbrace{\begin{pmatrix} u_1 \\ u_2 \end{pmatrix}}^{u} + v$$

where v is additive noise.

It is of great importance for the sequel of this paper that the identification is performed using a parametrization of G_0 with an identical auto-regressive part for $G_{0,1}$ and $G_{0,2}$ i.e.

$$G_0 = G(z, \delta_0) = \frac{\begin{pmatrix} e_1 + Z_{N_1}\delta_0 & e_2 + Z_{N_2}\delta_0 \end{pmatrix}}{1 + Z_D\delta_0} \quad (1)$$

where $\delta_0 \in \mathbf{R}^{k\times 1}$ is the *unknown* true parameter vector, $Z_{N_1}(z)$, $Z_{N_2}(z)$ and $Z_D(z)$ are row vectors of size k of known transfer functions. $e_1(z)$ and $e_2(z)$ are known transfer functions. This is the case when G_0 (and the noise model) has an ARX structure or an ARMAX structure. This is also the case when $G_{0,1}$ (resp. $G_{0,2}$) is parametrized by the sum of a given model $e_1(z)$ (resp. $e_2(z)$) and an expansion of linear basis functions (such as Laguerre filters). In this last case, $Z_D = 0$.

When $G_{0,1} = n_{0,1}/d_{0,1}$ and $G_{0,2} = n_{0,2}/d_{0,2}$ [2] have different auto-regressive parts, a structure such as in (1) can be obtained by considering a parametrization $G(z, \delta_0)$ such that

$$Z_{N_1}\delta_0 = n_{0,1}\, d_{0,2}, \quad Z_{N_2}\delta_0 = n_{0,2}\, d_{0,1},$$
$$1 + Z_D\delta_0 = d_{0,1}\, d_{0,2}.$$

Here $e_1(z)$ and $e_2(z)$ are equal to 0.

Proposition 1. Consider $G_0 = G(z, \delta_0)$, the true MISO system with two inputs parametrized as in (1). A PE identification experiment (with a full order model structure) performed on G_0 delivers an identified model $G_{mod} = G(z, \hat{\delta})$ and an uncertainty region $\mathcal{D}$ containing the true system G_0 at a prescribed probability level α. This uncertainty region is centered at $G(z, \hat{\delta})$ and can be described by the following generic form:

[2] Here, $n_{0,i}(z)$ and $d_{0,i}(z)$ are the numerator and denominator of $G_{0,i}$, respectively.

$$\mathcal{D} = \left\{ G(z,\delta) \mid G(z,\delta) = \frac{\begin{pmatrix} e_1 + Z_{N_1}\delta & e_2 + Z_{N_2}\delta \end{pmatrix}}{1 + Z_D\delta} \text{ and } \delta \in U \right\} \quad (2)$$

where $U = \{\delta \mid (\delta - \hat{\delta})^T R(\delta - \hat{\delta}) < \chi^2\}$, $\delta \in \mathbf{R}^{k\times 1}$ is a real parameter vector, $\hat{\delta}$ is the estimated parameter vector defining the identified model, R is a symmetric positive definite matrix $\in \mathbf{R}^{k\times k}$ that is equal to the inverse of the covariance matrix of $\hat{\delta}$, χ^2 is determined by the desired probability level α, $Z_{N_1}(z)$, $Z_{N_2}(z)$ and $Z_D(z)$ are row vectors of size k of known transfer functions. $e_1(z)$ and $e_2(z)$ are known transfer functions.

PROOF. Trivial extension of the results in (Bombois, 2000; Bombois *et al.*, 1999). □

Due to the multivariable structure in (2), the numerical methods that we had developed in our previous work (Bombois *et al.*, 2001) for a SISO uncertainty region in order to perform the validation of a controller for stability and for performance are no more valid. In this paper, new numerical methodologies adapted to the multivariable structure of $\mathcal{D}$ will be presented.

3. CONTROLLER VALIDATION FOR STABILITY

Once an uncertainty region $\mathcal{D}$ has been identified and its quality assessed using the results of (Bombois and Date, 2002), a controller can be designed from the model G_{mod}. This controller has then to be validated for stability. In other words, we have to verify whether the controller designed from the model G_{mod} stabilizes all systems in the uncertainty region $\mathcal{D}$ (and therefore also the true system G_0). This can be achieved using the following theorem. Indeed this gives a necessary and sufficient condition for the stabilization of all plants in $\mathcal{D}$ by a given controller. Theorem 2 is thus the extension to the MISO case of the results presented in (Bombois *et al.*, 2001) in the SISO case.

Theorem 2. Consider an uncertainty set $\mathcal{D}$ of the form (2) and a controller $C(z)$ that stabilizes the center of that set, $G(z, \hat{\delta})$. Then all models in $\mathcal{D}$ are stabilized by $C(z)$ if and only if

$$\sup_{\omega} \mu(M_{\mathcal{D}}(e^{j\omega})) \leq 1, \quad (3)$$

where

- $M_{\mathcal{D}}(z)$ is defined as

$$M_{\mathcal{D}}(z) = \frac{-(Z_D + \frac{C_1(Z_{N_1} - e_1Z_D) + C_2(Z_{N_2} - e_2Z_D)}{1 + e_1C_1 + e_2C_2})T^{-1}}{1 + (Z_D + \frac{C_1(Z_{N_1} - e_1Z_D) + C_2(Z_{N_2} - e_2Z_D)}{1 + e_1C_1 + e_2C_2})\hat{\delta}}, \quad (4)$$

- $C(z) = \begin{pmatrix} C_1(z) \\ C_2(z) \end{pmatrix}$
- T is a square root of the matrix R/χ^2 defining U: $R/\chi^2 = T^T T$.
- $\phi = T(\delta - \hat{\delta})$, whereby $\delta \in U \Leftrightarrow |\phi|_2 < 1$
- $\mu(M_{\mathcal{D}}(e^{j\omega}))$ is called the (real) *stability radius* of the loop $[M_{\mathcal{D}}(z)\ \ \phi]$. For a real vector ϕ it is computed as follows, if $Im(M) \neq 0$,

$$\mu(M(e^{j\omega})) = \sqrt{|Re(M)|_2^2 - \frac{(Re(M)Im(M)^T)^2}{|Im(M)|_2^2}}$$

and $\mu(M(e^{j\omega})) = |M|_2$, if $Im(M) = 0$.

PROOF. The proof consists of showing that the set of feedback loops $[C\ \ G(z,\delta)]$ can be recast in a framework to which the results of (Rantzer, 1992) can be applied. We first prove that the closed-loop connection of a plant $G(z,\delta)$ in $\mathcal{D}$ with the controller C can be restated in the general LFT framework of robust stability analysis. Let us for this purpose introduce the notations $Z_1 = Z_{N_1} - e_1Z_D$ and $Z_2 = Z_{N_2} - e_2Z_D$, and introduce the signals p_1 and q_1 such that $p_1 = \delta^T q_1$:

$$\begin{cases} y = G(z,\delta)u \\ u = -Cy \end{cases} \iff$$

$$\begin{cases} y = \begin{pmatrix} e_1 + \frac{\delta^T Z_1^T}{1 + \delta^T Z_D^T} & e_2 + \frac{\delta^T Z_2^T}{1 + \delta^T Z_D^T} \end{pmatrix} u \\ u = -Cy \end{cases} \iff$$

$$\begin{cases} p_1 = \delta^T q_1 \\ \begin{pmatrix} q_1 \\ y \end{pmatrix} = \begin{pmatrix} -Z_D^T & \begin{pmatrix} Z_1^T & Z_2^T \end{pmatrix} \\ 1 & \begin{pmatrix} e_1 & e_2 \end{pmatrix} \end{pmatrix} \begin{pmatrix} p_1 \\ u \end{pmatrix} \\ u = -Cy \end{cases} \iff$$

$$\begin{cases} p_1 = \delta^T q_1 \\ q_1 = \overbrace{(-Z_D^T - \frac{C_1Z_1^T + C_2Z_2^T}{1 + e_1C_1 + e_2C_2})}^{M_1^T(z)} p_1 \end{cases}$$

The loop $[C\ \ G(z,\delta)]$ is equivalent (from the stability point of view) with the loop $[M_1^T(z)\ \ \delta^T]$. This loop $[C\ \ G(z,\delta)]$ is thus also equivalent (from the stability point of view) with the loop $[M_1(z)\ \ \delta]$. The latter loop can be represented as follows by introducing two new signals p_2 and q

$$\begin{cases} p_2 = \delta q \\ q = M_1(z)p_2 \end{cases} \quad (5)$$

We now show the equivalence (from the stability point of view) between the set of loops $[C\ \ G(z,\delta)]$ for all $\delta \in U$ and the set of loops $[M_{\mathcal{D}}(z)\ \ \phi]$ for all ϕ such that $|\phi|_2 < 1$, by replacing δ and its uncertainty domain U by the real vector $\phi \triangleq T(\delta - \hat{\delta})$ and its uncertainty domain $|\phi|_2 < 1$. With $p \triangleq \phi q$ and $\delta = \hat{\delta} + T^{-1}\phi$, we have

$$\begin{cases} p_2 = \delta q \\ q = M_1(z)p_2 \end{cases} \Leftrightarrow \begin{cases} p = \phi q \\ q = \frac{M_1T^{-1}}{1 - M_1\hat{\delta}} p = M_{\mathcal{D}}(z)p \end{cases} \quad (6)$$

The necessary and sufficient condition then follows from the fact that $M_{\mathcal{D}}(z) \in \mathcal{R}H_{\infty}$ and from a result in (Rantzer, 1992). This result states that the stability radius of a set of loops $[M_{\mathcal{D}}(z)\ \phi]$ whose uncertainty part is a real vector constrained to lie in a two-norm unit ball can be computed as shown in the statement of Theorem 2. □

4. CONTROLLER VALIDATION FOR PERFORMANCE

In Section 3, we have presented a procedure to check whether a controller C stabilizes all plants in the uncertainty region $\mathcal{D}$ whose expression is given in (2). However, stabilization does not imply good performance with all plants in $\mathcal{D}$. In this section, we show that we can evaluate the worst case performance in the uncertainty region $\mathcal{D}$, i.e. the worst level of performance of a closed loop made up of the connection of the considered controller and any plant in $\mathcal{D}$. Modulo the probability that $G_0 \in \mathcal{D}$, the worst case performance in $\mathcal{D}$ is of course a lower bound for the closed-loop performance achieved with the true system.

4.1 *The worst case performance*

There is no unique way of defining the performance of a closed-loop system. However, most commonly used performance criteria can be derived from the largest singular value of the transfer operators representing the closed-loop system $[C\ G]$ made up of the system G in feedback with the controller C. In this paper, we will restrict attention to the following closed-loop transfer functions and vectors:

$$T_1(G,C) \triangleq C(1+GC)^{-1} \tag{7}$$

$$T_2(G,C) \triangleq (1+GC)^{-1}G \tag{8}$$

$$T_3(G,C) \triangleq (1+GC)^{-1} \tag{9}$$

$$T_4(G,C) \triangleq (1+GC)^{-1}GC. \tag{10}$$

The choice of these four transfer functions and vectors may be justified as follows. T_3 and T_1 are the transfer operators from the disturbance at the output of the plant to the output and the input of the plant respectively. Similarly, T_2 and $T_{4,bis} = C(I+GC)^{-1}G$ represent the transfer operators from the disturbance at the plant input to the output and the input of the plant respectively. The choice of T_4 instead of $T_{4,bis}$ is driven by mathematical tractability, and by the fact that $\overline{\sigma}(T_{4,bis}) = \overline{\sigma}(T_4)$.

The following proposition gives an elegant way to represent the performance of a closed-loop system $[C\ G]$ in the case where G is one system $G(z,\delta)$ in $\mathcal{D}$.

Proposition 3. Consider one of the closed-loop transfer functions or vectors $T_i(G,C)$ defined in (7)-(10). Consider also a MISO plant $G(z,\delta)$ having a structure as in (2) and a SIMO controller:

$$C(z) = \begin{pmatrix} C_1 \\ C_2 \end{pmatrix}$$

which stabilizes $G(z,\delta)$. At the frequency ω, define the performance $J(G(z,\delta),C(z),\omega,T_i)$ of the closed-loop system $[C\ G(z,\delta)]$ with respect to T_i as the largest singular value of $T_i(G(e^{j\omega},\delta),C(e^{j\omega}))$:

$$J(G(z,\delta),C(z),\omega,T_i) \triangleq \overline{\sigma}(T_i(G(e^{j\omega},\delta),C(e^{j\omega}))) \tag{11}$$

Then there exist some known transfer functions $h_1(z)$, $h_2(z)$ and $f(z)$, and some known row transfer vectors $Z_{num,1}(z)$, $Z_{num,2}(z)$ and $Z_{den}(z)$ of size k such that

$$J(G(z,\delta),C(z),\omega,T_i) = \overline{\sigma}\Bigg(\overbrace{\frac{h_1+Z_{num,1}\delta}{f+Z_{den}\delta}\ \ \frac{h_2+Z_{num,2}\delta}{f+Z_{den}\delta}}^{T_{gen}(e^{j\omega},\delta)}\Bigg). \tag{12}$$

PROOF. We must prove that $\overline{\sigma}(T_i(G(e^{j\omega},\delta),C(e^{j\omega})))$ $(i = 1...4)$ can be expressed as in (12). For this purpose, denote $f(z) \triangleq (1+e_1C_1+e_2C_2)$ and $Z_{den}(z) \triangleq (Z_D + C_1 Z_{N_1} + C_2 Z_{N_2})$ and observe that:

$$T_1 = \begin{pmatrix} \dfrac{C_1}{f+Z_{den}\delta} \\ \dfrac{C_2}{f+Z_{den}\delta} \end{pmatrix} \qquad T_2 = \begin{pmatrix} \dfrac{e_1+Z_{N_1}\delta}{f+Z_{den}\delta} & \dfrac{e_2+Z_{N_2}\delta}{f+Z_{den}\delta} \end{pmatrix}$$
$$T_3 = \frac{1+Z_D\delta}{f+Z_{den}\delta} \qquad T_4 = \frac{(e_1C_1+e_2C_2)+(C_1Z_{N_1}+C_2Z_{N_2})\delta}{f+Z_{den}\delta} \tag{13}$$

From (13), we see that $\overline{\sigma}(T_2)$ is of the form (12). Since $\overline{\sigma}(T_1) = \overline{\sigma}(T_1^T)$, this is also the case for T_1. For the two scalar transfer functions T_3 and T_4, $\overline{\sigma}(T_i)$ $(i = 3$ and $4)$ is also of the form (12) when choosing $h_2 = 0$ and $Z_{num,2} = 0$ since $\overline{\sigma}((T_i\ 0)) = \overline{\sigma}(T_i)$ $(i = 3$ and $4)$. □

We can now define the worst case performance (with respect to T_i) over all plants in an uncertainty region $\mathcal{D}$ as follows.

Definition 4. Consider an uncertainty region $\mathcal{D}$ of systems $G(z,\delta)$ with $\delta \in U$. Consider also a controller $C(z)$ and one of the closed-loop transfer functions (or vectors) $T_i(G,C)$ defined in (7)-(10). The worst case performance (with respect to T_i) achieved by this controller at a frequency ω over all systems in $\mathcal{D}$ is defined as:

$$J_{WC}(\mathcal{D},C,\omega,T_i) = \sup_{G(z,\delta)\in\mathcal{D}} J(G(z,\delta),C(z),\omega,T_i), \tag{14}$$

where $J(G(z,\delta),C(z),\omega,T_i)$ is defined in (11).

4.2 *Computation of the worst case performance*

We now present a procedure for the computation of $J_{WC}(\mathcal{D}, C, \omega, T_i)$ at a given frequency ω. This procedure must be repeated at each frequency in order to obtain the shape of the frequency function J_{WC}.

Theorem 5. Consider an uncertainty region $\mathcal{D}$ defined in (2) and a controller $C(z)$. The worst case performance at ω (i.e. $J_{WC}(\mathcal{D}, C, \omega, T_i)$) defined in (14) is equal to $\sqrt{\gamma_{opt}}$, where γ_{opt} is the optimal value of γ for the following standard convex optimization problem involving LMI constraints evaluated at the frequency ω:

$$\begin{array}{ll} \textit{minimize} \ \gamma & \\ \textit{over} \quad \gamma, \tau & \\ \textit{subject to} \ \tau \geq 0 \ \textit{and} & \end{array} \qquad (15)$$

$$\begin{pmatrix} Re(a_{11}) & Re(a_{12}) \\ Re(a_{12}^*) & Re(a_{22}) \end{pmatrix} - \tau \begin{pmatrix} R & -R\hat{\delta} \\ (-R\hat{\delta})^T & \hat{\delta}^T R \hat{\delta} - \chi^2 \end{pmatrix} < 0$$

where

- $a_{11} = (Z_{num,1}^* Z_{num,1} + Z_{num,2}^* Z_{num,2}) - \gamma(Z_{den}^* Z_{den})$,
- $a_{12} = Z_{num,1}^* h_1 + Z_{num,2}^* h_2 - \gamma(Z_{den}^* f)$,
- $a_{22} = h_1^* h_1 + h_2^* h_2 - \gamma f^* f$

and where $h_1(z)$, $h_2(z)$, $f(z)$, $Z_{num,1}(z)$, $Z_{num,2}(z)$ and $Z_{den}(z)$ are the transfer functions and vectors such that (12) holds.

PROOF. Using (12) and (14), we see that proving Theorem 5 is equivalent to proving that the solution γ_{opt} of the LMI problem (15), evaluated at ω, is such that:

$$\sqrt{\gamma_{opt}} = \sup_{\delta \in U} \overline{\sigma}(T_{gen}(e^{j\omega}, \delta)) \Longleftrightarrow$$
$$\gamma_{opt} = \sup_{\delta \in U} T_{gen}(e^{j\omega}, \delta) T_{gen}(e^{j\omega}, \delta)^*.$$

An equivalent and convenient way of restating this problem is as follows:

$$\textit{minimize} \ \gamma \ \textit{such that}$$
$$T_{gen}(e^{j\omega}, \delta) T_{gen}(e^{j\omega}, \delta)^* - \gamma < 0 \quad \forall \delta \in U.$$

Using now the expression of $T_{gen}(z, \delta)$ given in (12), we can rewrite $T_{gen}(e^{j\omega}, \delta) T_{gen}(e^{j\omega}, \delta)^* - \gamma < 0$ into the following constraint on δ with variable γ:

$$\begin{pmatrix} \delta \\ 1 \end{pmatrix}^* \begin{pmatrix} a_{11} & a_{12} \\ a_{12}^* & a_{22} \end{pmatrix} \begin{pmatrix} \delta \\ 1 \end{pmatrix} < 0 \qquad (16)$$

with a_{11}, a_{12} and a_{22} as defined in the statement of the theorem (see (15)). Since δ is real, it can be shown that (16) is equivalent with

$$\overbrace{\begin{pmatrix} \delta \\ 1 \end{pmatrix}^T \begin{pmatrix} Re(a_{11}) & Re(a_{12}) \\ Re(a_{12}^*) & Re(a_{22}) \end{pmatrix} \begin{pmatrix} \delta \\ 1 \end{pmatrix}}^{\psi(\delta)} < 0 \quad (17)$$

This last expression is equivalent to stating that $T_{gen}(e^{j\omega}, \delta) T_{gen}(e^{j\omega}, \delta)^* - \gamma < 0$ for a particular $\delta \in U$. However, this must be true for all $\delta \in U$. Therefore, (17) must be true for all δ such that:

$$\overbrace{\begin{pmatrix} \delta \\ 1 \end{pmatrix}^T \begin{pmatrix} R & -R\hat{\delta} \\ (-R\hat{\delta})^T & \hat{\delta}^T R \hat{\delta} - \chi^2 \end{pmatrix} \begin{pmatrix} \delta \\ 1 \end{pmatrix}}^{\rho(\delta)} < 0 (18)$$

which is equivalent to the statement "$\delta \in U$".

Let us now recapitulate. Computing $\sup_{\delta \in U} T_{gen}(e^{j\omega}, \delta) T_{gen}(e^{j\omega}, \delta)^*$ is equivalent to finding the smallest γ such that $\psi(\delta) < 0$ for all δ for which $\rho(\delta) < 0$. By the $\mathcal{S}$ procedure (Boyd *et al.*, 1994), this problem is equivalent to finding the smallest γ and a positive scalar τ such that $\psi(\delta) - \tau\rho(\delta) < 0$, for all $\delta \in \mathbf{R}^{k \times 1}$ which is precisely (15). □

5. NUMERICAL ILLUSTRATION

Let us now illustrate our results by the same example as in the companion paper (Bombois and Date, 2002). Consider thus the following MISO true system in the ARX structure:

$$y(t) = \frac{\overbrace{\left(n_{0,1}(z) \ n_{0,2}(z) \right)}^{G_0}}{d_0(z)} u(t) + \frac{e(t)}{d_0(z)}$$

where $e(t)$ is a white noise of variance 0.1, $n_{0,1} = 0.1047z^{-1} + 0.0872z^{-2}$, $n_{0,2} = 0.2z^{-1} + 0.05z^{-2}$ and $d_0 = 1 - 1.5578z^{-1} + 0.5769z^{-2}$. A PE identification experiment has been performed on this true system using an input signal whose two components are a white noise of variance 1. 5000 data have been collected and the following model has been identified:

$$G_{mod} = G(z, \hat{\delta}) =$$

$$\frac{\left(0.099z^{-1} + 0.086z^{-2} \quad 0.2013z^{-1} + 0.048z^{-2} \right)}{1 - 1.5554z^{-1} + 0.578z^{-2}}$$

Along with this model, the uncertainty region $\mathcal{D}$ centered at G_{mod} and containing the true system G_0 at a probability 0.95 has been built. This uncertainty region has the general structure (2) with e1 and $e_2 = 0$, $\chi^2 = 12.6$, $Z_D = (z^{-1} \ z^{-2} \ 0 \ 0 \ 0 \ 0)$, $Z_{N_1} = (0 \ 0 \ z^{-1} \ z^{-2} \ 0 \ 0)$, $Z_{N_2} = (0 \ 0 \ 0 \ 0 \ z^{-1} \ z^{-2})$ and $\hat{\delta} = (-1.554, 0.578, 0.099, 0.086, 0.2013, 0.048)^T$.

Since this uncertainty region $\mathcal{D}$ has been assessed sufficiently tuned for robust control design in the companion paper (Bombois and Date, 2002), the model G_{mod} can be used to deduce the following proportional controller: $C = \left(0.1 \ 0.1 \right)^T$. We now verify if this controller stabilizes and achieves sufficient performance with all plants in $\mathcal{D}$ and therefore also with the true system G_0.

Controller validation for stability. Following the procedure of Theorem 2, we build the dynamic vector $M_{\mathcal{D}}\left(e^{j\omega}\right)$ corresponding to the candidate controller C, and we compute its stability radius at each frequency. The supremum of the stability radii is

$$\sup_{\omega} \mu(M_{\mathcal{D}}(e^{j\omega})) = 0.08$$

Since this supremum is smaller than one, we may conclude that the controller C stabilizes all plants in the uncertainty set $\mathcal{D}$ and thus also the system G_0. The controller C is thus validated for stability.

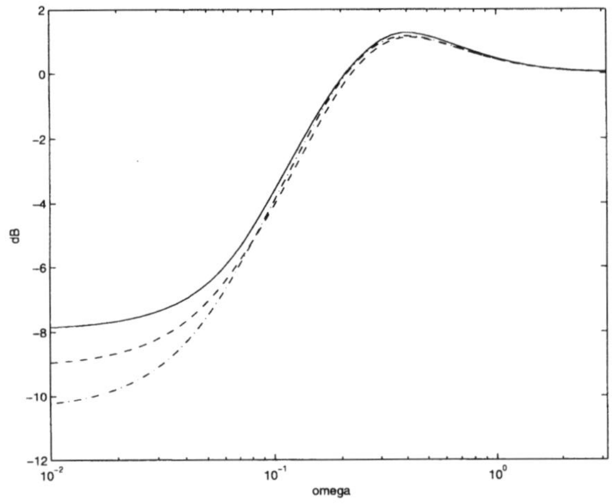

Fig. 1. $J_{WC}(\mathcal{D}, C, \omega, T_3)$ (solid), $\mid T_3(G(\hat{\delta}), C) \mid$ (dashed) and $\mid T_3(G_0, C) \mid$ (dashdot)

Controller validation for performance. In order to verify that C gives satisfactory performance with all plants in $\mathcal{D}$, we compute at each frequency the worst case modulus $J_{WC}(\mathcal{D}, C, \omega, T_3)$ of the scalar sensitivity function T_3 achieved by C over all plants in $\mathcal{D}$. This can be done by using the LMI procedure of Theorem 5. The worst case performance $J_{WC}(\mathcal{D}, C, \omega, T_3)$ for T_3 over all models in $\mathcal{D}$ is represented in Figure 1. In this figure, the worst case performance level is compared with the sensitivity functions of the designed closed loop $[C \; G(z, \hat{\delta})]$ and of the achieved closed loop $[C \; G_0]$. We see that the worst case performance for T_3 remains acceptable at every frequency and so the controller C can be stated validated for performance.

6. CONCLUSIONS

In this paper and in the companion paper (Bombois and Date, 2002), we have extended to the more general case of MISO (multiple-input single-output) systems previous results to connect PE identification and robust control theory that were developed in the SISO case. To simplify the notations, the results were here presented for the case of MISO systems with two inputs, but can be easily extended to all MISO systems and also to the dual case of SIMO (single-input output-output) systems. Research is now pursued to extend these results to the case of systems with multiple inputs and multiple outputs.

7. REFERENCES

Bombois, X. (2000). Connecting Prediction Error Identification and Robust Control Analysis: a new framework. PhD thesis. Université Catholique de Louvain.

Bombois, X. and P. Date (2002). Connecting pe identification and robust control theory: the multiple-input single-output case. part I: Uncertainty region validation. Submitted to the 13th IFAC Symposium on System Identification.

Bombois, X., M. Gevers and G. Scorletti (2000). A measure of robust stability for a set of parametrized transfer functions. *IEEE Transactions on Automatic Control* **45**(11), 2141–2145.

Bombois, X., M. Gevers and G. Scorletti (1999). Controller validation based on an identified model. In: *Proc. 38th IEEE Conference on Decision and Control*, Phoenix, Arizona.

Bombois, X., M. Gevers, G. Scorletti and B.D.O. Anderson (2001). Robustness analysis tools for an uncertainty set obtained by prediction error identification. *Automatica* **37**(10), 1629–1636.

Boyd, S., L. El Ghaoui, E. Feron and V. Balakrishnan (1994). *Linear Matrix Inequalities in Systems and Control Theory*. Vol. 15 of *Studies in Appl. Math.*. SIAM. Philadelphia.

Gevers, M., X. Bombois, B. Codrons, G. Scorletti and B.D.O. Anderson (2002). Model validation for control and controller validation in a prediction error identification framework- part I : Theory. Accepted for publication in Automatica.

Goodwin, G.C., M. Gevers and B. Ninness (1992). Quantifying the error in estimated transfer functions with application to model order selection. *IEEE Trans. Automatic Control* **37**, 913–928.

Hakvoort, R.G. and P.M.J. Van den Hof (1997). Identification of probabilistic system uncertainty regions by explicit evaluation of bias and variance errors. *IEEE Trans. Automatic Control* **42**(11), 1516–1528.

Ljung, L. (1999). *System Identification: Theory for the User, 2nd Edition*. Prentice-Hall. Englewood Cliffs, NJ.

Ljung, L. (2000). Model error modeling and control design. In: *CD-ROM Proc. IFAC Symposium on system identification*, Santa Barbara, California.

Rantzer, A. (1992). Convex robustness specifications for real parametric uncertainty in linear systems. In: *Proc. American Control Conference*. pp. 583–585.

Vinnicombe, G. (1993*a*). Frequency domain uncertainty and the graph topology. *IEEE Trans Automatic Control* **AC-38**, 1371–1383.

Vinnicombe, G. (1993*b*). Measuring the Robustness of Feedback Systems. PhD thesis. Cambridge University.

IFAC Publications
www.elsevier.com/locate/ifac

RELATION BETWEEN UNCERTAINTY STRUCTURES IN IDENTIFICATION FOR ROBUST CONTROL

Sippe G. Douma * **Paul M.J. Van den Hof** *

* *Delft Center for Systems and Control,*
Delft University of Technology, Mekelweg 2, 2628 CD Delft,
The Netherlands,
s.g.douma@dcsc.tudelft.nl; p.m.j.vandenhof@dcsc.tudelft.nl

Abstract: Various techniques of system identification exist providing for a nominal model and uncertainty bound. An important question is what the implications are for the particular choice of the structure in which the uncertainty is described when dealing with robust stability/performance analysis of a given controller and when dealing with robust synthesis. An amplitude-bounded (circular) uncertainty set can equivalently be described in terms of an additive, Youla parameter and ν-gap uncertainty. Closed-loop performance functions based on these sets are again bounded by circles in the frequency domain, allowing for exact worst-case performance calculation and for the evaluation of the consequences of uncertainty for robust design. *Copyright © 2003 IFAC*

Keywords: Uncertainty, Identification for Control, Robust Control, (ν) gap, dual Youla

1. INTRODUCTION

In identification for robust control an identified model has to be accompanied by a bound on its uncertainty, while the representation of this uncertainty should allow for robustness analysis and robust controller synthesis. A large number of such uncertainty descriptions is available from robust control theory, as e.g. a (H_∞)-norm-bounded additive or multiplicative uncertainty on the plant model, a norm-bounded uncertainty on a closed-loop plant representation (e.g. its dual Youla parameter), uncertainties bounded in the gap or ν-gap metric, and real parametric uncertainties, see e.g. (Vinnicombe, 2001; Zhou *et al.*, 1996; Tay *et al.*, 1989).

In the past the necessity to deliver model uncertainty bounds has generated a new class of identification techniques directed towards the construction of worst-case (H_∞) error bounds (Chen and Gu, 2000). However in many situations a worst-case bounded approach has been shown to lead to unnecessarily conservative results. On the other hand, a range of identification techniques exists providing for uncertainty structures identified from the data, where also account is given of the stochastic nature of disturbances in the data. The resulting probabilistic or combined probabilistic/worst-case approaches to the problem deliver a variety of uncertainty sets as e.g. parametrically structured (ellipsoidal) uncertainty, norm-bounded additive, non-parametric (boxed, ellipsoidal) additive in the frequency domain (Ljung, 1987; Goodwin *et al.*, 1992; Hakvoort and den Hof, 1997; de Vries and Van den Hof, 1995).

Amongst such a variety of possible uncertainty structures a relevant question is what the implications are of a particular choice of structure for the identification for robust control problem. An ultimate question to be answered would be what is, for a given purpose (robust stability/performance analyis or synthesis), the best model uncertainty structure in which to identify the model set (nominal model and uncertainty bound). And consequently, what would be the best experiment allowing for minimisation of the uncertainty.

In this paper the former problem will be taken at hand. While extensive literature exists dealing with characteristics of each uncertainty structure, answering the posed question requires a thorough comparison and a bridging of the gap between identification and ro-

bust control that goes beyond the present state of the art. A first attempt with limited scope only directed towards robust stability issues was made in (Douma *et al.*, 2003). This paper is intended to highlight aspects in which the various uncertainty structures differ in their consequences for robust analysis and design and in their potentials to be determined on the basis of realistic experimental data.
In the next section the uncertainty structures and the performance measure is specified. The third section explores the link between the uncertainty structures and their behaviour under a linear fractional transformation. In the last three sections the uncertainty (structure) is analyzed with respect to, respectively, robust stability and robust performance analysis.

2. FRAMEWORK

We consider single-input-single-output linear time-invariant finite dimensional systems $G(s)$ and controllers $C(s)$. Coprime factorizations of plants and controllers are defined as $G(s) = N(s)D^{-1}(s)$ and $C(s) = N_c(s)D_c^{-1}(s)$ where $N(s),D(s),N_c,D_c \in \mathbb{R}H_\infty$ satisfy the usual conditions (Vidyasagar, 1985). The factorizations are normalized, denoted by $\bar{(\cdot)}$, if they additionally satisfy $\bar{N}(s)^*\bar{N}(s)+\bar{D}(s)^*\bar{D}(s)=1$, where $(\cdot)^*$ denotes complex conjugate transpose. This paper considers three model sets based on a specific uncertainty structure:

Additive uncertainty

$$\mathcal{G}_a(G_x,W_a) := \{G_\Delta(s) \mid G_\Delta(s) = G_x(s)+\Delta_a(s)\ , \\ |\Delta_a(i\omega)| \le |W_a(i\omega)| \quad \forall\omega\in\mathbb{R}\}, \quad (1)$$

with $G_x(s)$ a nominal model and $W_a(s)$ a weighting function.

Youla-uncertainty

$$\mathcal{G}_Y(G_x,C,Q,Q_c,W_Y) := \\ \left\{G_\Delta(s) \mid G_\Delta(s) = \frac{\bar{N}_x(s)+\bar{D}_c(s)\Delta_G(s)}{\bar{D}_x(s)-\bar{N}_c(s)\Delta_G(s)}, \right. \quad (2) \\ \left.\left|Q_c^{-1}(i\omega)\Delta_G(i\omega)Q(i\omega)\right| \le |W_Y(i\omega)| \quad \forall\omega\in\mathbb{R}\right\}.$$

with $G_x(s) = \bar{N}_x(s)\bar{D}_x^{-1}(s)$ a nominal model, $C(s) = \bar{N}_c(s)\bar{D}_c^{-1}(s)$ a present controller and $Q(s),Q_c(s)$ stable and stably invertible weighting functions reflecting the freedom in choosing the coprime factorizations of $G_x(s)$ and $C(s)$ (Vidyasagar, 1985). An additional weighting can be provided by $W_Y(s)$. The Youla parameter $\Delta_G(s)$ is uniquely determined by $\Delta_G(s) = \bar{D}_c^{-1}(s)(1+G_\Delta(s)C(s))^{-1}\,(G_\Delta(s)-G_x(s))\bar{D}_x(s)$.

ν-gap uncertainty (Vinnicombe, 2001)

$$\mathcal{G}_\nu(G_x,W_\nu) := \quad (3) \\ \{G_\Delta(s) \mid \kappa(G_\Delta(i\omega),G_x(i\omega)) \le |W_\nu(i\omega)| \quad \forall\omega\in\mathbb{R}\},$$

with $\kappa(G_\Delta(\omega),G_x(\omega))$ denoting the chordal distance defined, for a plant $G_\Delta(s) = \bar{N}_\Delta(s)\bar{D}_\Delta^{-1}(s)$ with respect to the nominal model $G_x(s) = \bar{N}_x(s)\bar{D}_x^{-1}(s)$, by

$$\kappa(G_\Delta(i\omega),G_x(i\omega)) := |\bar{N}_x(i\omega)\bar{D}_\Delta(i\omega)-\bar{D}_x(i\omega)\bar{N}_\Delta(i\omega)| \\ = \frac{|G_x(i\omega)-G_\Delta(i\omega)|}{\sqrt{\left(1+|G_\Delta(i\omega)|^2\right)\left(1+|G_x(i\omega)|^2\right)}}.$$

Note that at this point the usual stability conditions are not yet imposed on either the $G_\Delta(s),G_x(s),\Delta(s),P(s)$ or $W(s)$. The focus here lies on the frequency domain conditions. In section 4 the stability conditions will be discussed.

We consider a performance measure formulated in the frequency domain:

$$J(G_\Delta,C,V,W) = \bar{\sigma}(V(i\omega)T(G_\Delta(i\omega),C(i\omega))W(i\omega)), \quad (4)$$

with $\bar{\sigma}$ the maximum singular value and

$$T(G_\Delta,C) := \begin{bmatrix} G_\Delta \\ 1 \end{bmatrix}(1+G_\Delta C)^{-1}\begin{bmatrix} C & 1 \end{bmatrix}. \quad (5)$$

The weighting matrices V and W are diagonal. These diagonal weighting functions allow for a large range of performance specifications (e.g. bounds on the (complementary) sensitivity function).

Remark 1. From here onwards arguments are omitted to include both an evaluation in terms of transfer functions (e.g. $G(s)$) with frequency responses over the whole frequency axis, and an evaluation over a frequency grid with, e.g., $G(i\omega_k) \in \mathbb{C}$.

In all three of the above uncertainty sets a bounding condition is imposed on the frequency response of their members $G_\Delta(s)$. At each frequency $G_\Delta(i\omega)$ is constrained to within a circle in the complex plane. That this is the case for $\mathcal{G}_\nu(G_x,W_\nu)$ as well will be made clear in the next section. Further, note that the mapping of the open-loop transfer functions G_Δ to the closed-loop performance functions in (5) can be described with a linear fractional transformation (LFT), $F(P,C) := P_{22}+P_{21}C(1+P_{22}C)^{-1}P_{12}$. The behaviour of uncertainty sets under a linear fractional transformation is the topic of the next section.

3. EFFECT OF LINEAR FRACTIONAL TRANSFORMATIONS

3.1 *Mapping of circles*

As is well known, a linear fractional transformation (Möbius transformation) maps circles into circles. The explicit formulation of the mapping allows for great insight when comparing uncertainty structures and their influence on performance.

Proposition 1. A set of frequency responses $F(P,\Delta)$ described by the (SISO) LFT

$$F(P,\Delta) = P_{22}+P_{21}\Delta(1+P_{11}\Delta)^{-1}P_{12}\ \text{, with } \left|W^{-1}\Delta\right| \le 1$$

and a one-dimensional uncertainty block Δ can at each frequency be described in an additive structure as

$$F(P,\Delta) = F_{centre} + \Delta_a \quad , \quad \left|W_a^{-1}\Delta_a\right| \leq 1,$$
$$F_{centre} = P_{22} + \frac{-P_{21}P_{12}P_{11}^{*}\,|W|^2}{1-|P_{11}W|^2}$$
$$W_a = \frac{|P_{21}P_{12}|}{\left(1-|P_{11}W|^2\right)}\,|W|,$$

provided that $|P_{11}W| < 1$. Whenever $|P_{11}W| > 1$, the frequency responses of the set $F(P,\Delta)$ lie in the area outside the circle $F_{centre} + \Delta_a$, $\left|W_a^{-1}\Delta_a\right| \leq 1$.

The proposition is formulated in terms of frequency responses as this will be our main interest. Subtitution of $|W|^2$ by $W(s)W^*(s)$ would allow for a formulation in terms of (rational) systems. The same applies to the following sections. The proposition shows that any circular region in the frequency domain is again mapped into a circle. However, the original centre (P_{22}) is not the new centre unless the linear fractional transformation happens to be affine ($P_{11} = 0$).

3.2 *Performance functions*

When using proposition 1 to study the effect of an uncertainty (structure) on the closed-loop performance of a controller, the entry P contains both the controller C and G_x. For example the set of sensitivity functions S_Δ induced by an additive uncertainty set $G_\Delta = G_x + \Delta_a$, $\left|W_a^{-1}\Delta_a\right| \leq 1$ and a controller C is given, with $P_{22} = 0$, $P_{12}P_{21} = \frac{1}{1+G_xC}$ and $P_{11} = \frac{C}{1+G_xC}$, by

$$S_\Delta = \frac{1}{1+(G_x+\Delta_a)C}\ , \quad \left|W_a^{-1}\Delta_a\right| \leq 1$$
$$= \frac{(1+CG_x)^{-1}}{1-\left|(1+CG_x)^{-1}CW_a\right|^2} + \Delta_S,$$
$$|\Delta_S| \leq |W_S| = \frac{\left|(1+CG_x)^{-2}C\right|}{1-\left|(1+CG_x)^{-1}CW_a\right|^2}\,|W_a|.$$

The circular representation allows for an analytical expression of minimal and maximal (worst-case) performance when evaluated in the frequency domain. E.g. $\max_{\Delta_a,\left|W_a^{-1}\Delta_a\right|\leq 1} |S_\Delta| = |S_{centre}| + |W_S|$.
The example shows that the robust stability condition (cf. section 4) appears naturally in the denominators and how the nominal sentivity $(1+CG_x)^{-1}$ is not the centre of the set of sensitivity functions. This property becomes critically important when considering non-circular boundaries and/or probability density functions over the uncertainty set.

3.3 *Probability density and non-circular bounds*

From an identification point of view the probabilistic uncertainty regions usually follow from a complex probability density function (de Vries and Van den Hof, 1995)(Hakvoort and den Hof, 1997)(Goodwin *et al.*, 1992)(Ljung, 1987). The uncertainty region per frequency is bounded, with respect to a certain probability, either only in terms of the amplitude (circular) or the real and imaginary part separately (ellipsoidal or boxed). A (SISO) LFT, being a conformal mapping, will map closed contours into closed contours and will leave angles locally intact. However, the mapping will in general not preserve shape, as depicted in Figure 1. Worst-case performance analysis and robust stability evaluation on such non-circular sets will require special, adapted procedures.
As indicated in (Heath, 2001) the transformation will change the structure of the probability distribution. For example, when a closed-loop identified object is used to obtain the open-loop plant by recalculation with the present controller the statistical properties change drastically. An unbiased estimate of the closed-loop object does not imply an unbiased estimate of the recalculated open-loop plant. An important exception here is formed by all affine transformations ($P_{11} = 0$) such as closed-loop functions of a Youla uncertainty set $\mathcal{G}_Y(G_x,C,Q,Q_c,W_Y)$ based on the present controller C (cf. section 5).

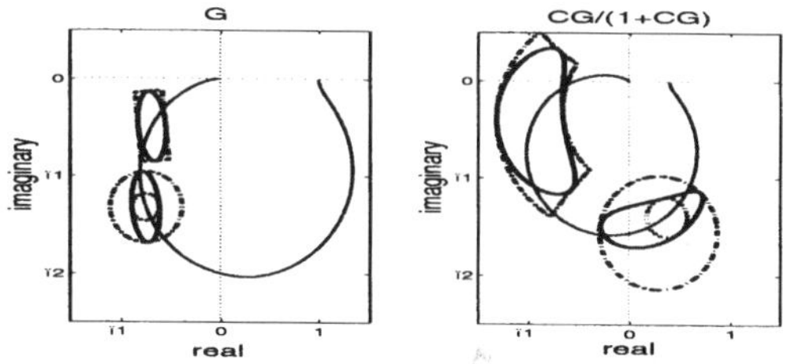

Fig. 1. *Transformation of circular, ellipsoidal and boxed uncertainty bounds from plant model G (left) to closed-loop transfer CG/(1+CG) (right).*

3.4 *Circular uncertainty structures*

All the uncertainty structures of section 2 can equivalently be described by an additive structure in terms of their frequency domain properties. While this follows directly from Proposition 1 for e.g. the Youla parameter uncertainty structure and (inverse) multiplicative structure, the ν-gap structure requires a separate proposition to show this fact. First the use of Proposition 1 for the Youla parameter uncertainty structure is made explicit.

Corollary 1. The set of frequency responses of all plants $G_\Delta \in \mathcal{G}_Y(G_x,C,Q,Q_c,W_Y)$ (see (2)) is equivalently described as an additive uncertainty set $\mathcal{G}_a(G_{centre},W_a)$ (see (1)) with

$$G_{centre} =$$
$$G_x\left(\frac{|D_x|^2}{|D_x|^2-|N_cW_Y|^2}\right) - C^{-1}\left(\frac{-|N_cW_Y|^2}{|D_x|^2-|N_cW_Y|^2}\right)$$

$$W_a = \frac{|\Lambda|}{|D_x|^2 - |N_c W_Y|^2} |W_Y|,$$

where $\Lambda = N_x N_c + D_c D_x$ and $N_x = \bar{N}_x Q$, $D_x = \bar{D}_x Q$, $N_c = \bar{N}_c Q_c$, $D_c = \bar{D}_c Q_c$.

Note that the centre of the Youla uncertainty set is given by a convex combination of the nominal model G_x and the negative inverse of the controller C used in the Youla parametrization. The transformation of the ν-gap uncertainty set to an additive structure is indicated in the following proposition.

Proposition 2. The set of frequency responses of all plants $G_\Delta \in \mathcal{G}_\nu(G_x, W_\nu)$ (as defined in (3)) is equivalently described as an additive uncertainty set $\mathcal{G}_a(G_{centre}, W_a)$ (see (1)) with

$$G_{centre} = \frac{G_x}{1 - \left(1 + |G_x|^2\right)|W_\nu|^2}$$

$$W_a = \frac{\sqrt{\left(1 - |W_\nu|^2\right)\left(|G_x|^2 + 1\right)}\,|W_\nu|}{1 - \left(1 + |G_x|^2\right)|W_\nu|^2}.$$

The fact that both the Youla uncertainty set and the ν-gap uncertainty set allow for an additive description shows that the two sets can be transformed into one another. The explicit formulations above of the uncertainty sets in terms of an additive structure allows for a thorough comparison, as is further explored in the coming section.

3.5 *Interpretation*

Consider an uncertainty bound in the frequency domain of any shape following from an identification experiment, as e.g. an ellipsoidal region following from (de Vries and Van den Hof, 1995)(Hakvoort and den Hof, 1997). It is clear that the smallest (unique) circle embedding the uncertainty can equivalently be described in all structures of section 2 (with different nominal models and weighting functions). Alternatively, this also pleads for an identification criterium minimizing over both nominal model and uncertainty, rather than first identifying a nominal model (with any method) and subsequently bounding the model uncertainty.
In general, however, embedding is sought while maintaining a particular nominal model, in which case all structures will provide different embedding regions. Propositions 1, 2 and Corollary 1 show in which directions the centres move and the circles expand. For example, it is immediate that when embedding an additive set $\mathcal{G}_a(G_x, W_a)$ with a ν-gap set $\mathcal{G}_\nu(G_x, W_\nu)$, the size W_ν (chordal distance) is determined by the element $G_x\,(|G_x - W_a| / |G_x|)$. That is, the direction away from the nominal model G_x towards the origin is most costly, in terms of increase of the uncertainty region (see figure 2). However, the increase of the uncertainty region should be judged against the effect on the attainable performance (cf. section 5).

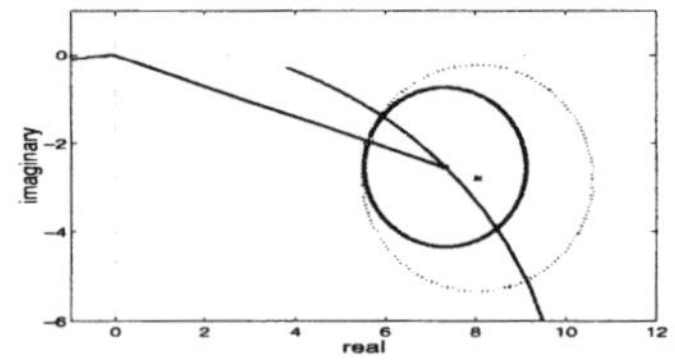

Fig. 2. *Embedding an additive uncertainty (solid) with a ν-gap uncertainty (dashed-dot). The point in line with the nominal model G_x ('x') will cause the largest chordal distance.*

4. ROBUST STABILITY

4.1 *Robust stability and the frequency domain*

Checking for internal stability for all elements $G_\Delta(s) \in \mathcal{G}$ is feasible when certain conditions are imposed on $G_\Delta(s)$. For example, all controllers $C(s)$, satisfying the condition $\left|C(i\omega)\left(1 + C(i\omega)G_x(i\omega)\right)^{-1}\right| < \left|W_a^{-1}(i\omega)\right|$, stabilize the additive set $\mathcal{G}_a(G_x, W_a)$ if and only if $G_x(s)$ is stabilized by $C(s)$ and the number of unstable poles of all $G_\Delta(s)$ is equal to the number of unstable poles of $G_x(s)$. In such conditions for robust stability three parts can be discerned:

i. the condition that $C(i\omega) \neq -G_\Delta^{-1}(i\omega)$ for all $G_\Delta(i\omega) \in \mathcal{G}$ and for all $\omega \in \mathbb{R}$.
ii. stability of C with a nominal model $G_x \in \mathcal{G}$.
iii. conditions on all $G_\Delta \in \mathcal{G}$ with respect to the nominal model G_x.

The first condition on the frequency responses seems most characteristic for different uncertainty structures: $\left|C(i\omega)\left(1 + C(i\omega)G_x(i\omega)\right)^{-1}\right| < \left|W_a^{-1}(i\omega)\right|$ for additive, $\left|Q^{-1}(i\omega)\Delta_C(i\omega)Q_c(i\omega)\right| \leq \left|W_Y^{-1}(i\omega)\right|$ for the Youla parameter uncertainty and $\bar{\sigma}\left(T(C(i\omega), G_x(\omega)\right) < \left|W_\nu^{-1}(i\omega)\right|$ for the ν-gap uncertainty. However, they are simply ensuring the condition that
$C(i\omega) \neq -G_\Delta^{-1}(i\omega)$ for all $G_\Delta(i\omega) \in \mathcal{G}$. And from Proposition 1 it was clear that all three uncertainty sets can be transformed into one another with respect to the frequency responses of the members. Naturally the 'nominal' model and weighting function will change. For example, part *i.* of the robust stability condition for the ν-gap set $\bar{\sigma}\left(T(C(i\omega), G_x(i\omega)\right) < \left|W_\nu^{-1}(i\omega)\right|$, is equivalently described by the (additive) condition $\left|C\left(1 + CG_{centre}\right)^{-1}\right| < \left|W_a^{-1}\right|$, with G_{centre} and W_a from Proposition 2. The uncertainty sets do differ in terms of parts *ii.* and *iii.*. That is, they differ in terms of a winding number condition or a condition on unstable poles and zeros.

4.2 *Identification and robust stability*

Identification techniques as (de Vries and Van den Hof, 1995)(Goodwin *et al.*, 1992)(Hakvoort and den Hof, 1997), which take bias effects into consideration, characterize the plant identification uncertainty in terms of bounds on the frequency response. However, we have seen that for robust stability the distinguishing factor between the uncertainty sets is a winding-number/pole/zero condition on the members of the set. It is important to note that this information does not come directly from the identification procedure, but has to be provided based on an assumption. A realistic assumption would be that the identified object is stable. That is, an open-loop identification could lead to an additive uncertainty with the condition that all elements $G_\Delta \in \mathcal{G}_a(G_x, W_a)$ are stable. Or, a closed-loop identification could lead to a Youla parameter uncertainty where a stability condition on Δ_G can automatically be satisfied. The Youla parameter can directly be identified from closed-loop data as described in (Van den Hof, 1998).

5. ROBUST PERFORMANCE ANALYIS

A controller is said to perform robustly for a set of plants if a certain performance level is reached for all plants in the uncertainty set. A robust performance analysis comes down to a worst-case performance evaluation over the uncertainty set. To this end, μ-analysis or LMI based procedures are available for general (LFT based) uncertainty structures (Zhou *et al.*, 1996). For SISO systems with a one dimensional uncertainty block, bounded in amplitude, both methods provide for an unconservative answer. However, in the following we will derive analytical expressions since this will allow for much more insight. In case non-circular uncertainty regions are considered, in some cases an adapted μ-analysis could be employed, as for ellipsiodal regions (Bombois *et al.*, 2001), but in general the procedure would become complex.

5.1 *Analytical worst-case calculation*

Section 3.2 has shown how Proposition 1 allows for an analytical expression of the worst-case performance for all SISO closed-loop performance functions. Moreover Proposition 1 reveals the fact that every increase in uncertainty (W) at a particular frequency results in a decrease of the worst-case performance cost at that frequency (due to W_a). A new experiment could be designed to decrease the worst-case performance at a particular frequency by inducing that amount of power there that could reduce the uncertainty with a factor explicitly given by Proposition 1. Note, however that the new experiment would also yield a new nominal controller.

The Youla parameter uncertainty structure plays a special rôle when considering robust performance for the auxiliary controller C used in the parametrization. For example,

$$\frac{G_\Delta C}{1+CG_\Delta} = \frac{N_c(N_x + D_c\Delta_G)}{D_c(D_x - N_c\Delta_G) + N_c(N_x + D_c\Delta_G)}$$
$$= \frac{G_x C}{1+CG_x} + \frac{N_c D_c}{D_c D_x + N_c N_x}\Delta_G.$$

For the auxiliary controller the closed-loop functions of the set $\mathcal{G}_Y(G_x, C, Q, Q_c, W_Y)$ are affine in the uncertainty Δ_G. This implies that the nominal performance will be the centre of the set of performance associated with $\mathcal{G}_Y(G_x, C, Q, Q_c, W_Y)$. Moreover, probability density functions and non-circular uncertainty structures (e.g. boxed/ellipsoidal/irregular) will maintain their shape under the mapping to the performance functions.

In case the weights V and W in (4) are chosen such that the performance function is not SISO, Proposition 1 cannot be used. For this case the following Lemma is presented.

Lemma 1. Consider the following weighted $T(G_\Delta, C)$ matrix for SISO systems G_Δ and C:

$$\begin{pmatrix} V_1 & 0 \\ 0 & V_2 \end{pmatrix} \begin{pmatrix} G_\Delta \\ 1 \end{pmatrix} (1+CG_\Delta)^{-1} \begin{pmatrix} C & 1 \end{pmatrix} \begin{pmatrix} W_1 & 0 \\ 0 & W_2 \end{pmatrix}.$$

The maximum singular value $\bar{\sigma}(VT(G_\Delta, C)W)$ is given by

$$\bar{\sigma}(VT(G_\Delta,C)W) = \left(\left(S_\Delta - \frac{|V_1|^2}{(|V_1|^2+|V_2|^2|C|^2)}\right)\left(S_\Delta - \frac{|V_1|^2}{(|V_1|^2+|V_2|^2|C|^2)}\right)^* + \frac{|V_1|^2|V_2|^2|C|^2}{(|V_1|^2+|V_2|^2|C|^2)^2}\right)\frac{(|V_1|^2+|V_2|^2|C|^2)(|W_2|^2+|W_1|^2|C|^2)}{|C|^2},$$

where $S_\Delta = (1+CG_\Delta)^{-1}$. A similar result is derived in (Vinnicombe, 2001) in connection with loop-shaping.

With Lemma 1 the worst-case performance over a set $\mathcal{G}$ can at each frequency be made explicit using Proposition 1. Any (circular) uncertainty set $\mathcal{G}$ can be transformed with a controller C into an associated set of sensitivity functions $\mathcal{S}_\mathcal{G}(G_x, C, W_a) := \{S_\Delta \mid S_\Delta = S_{centre} + \Delta_a, |W_a^{-1}\Delta_a| < 1\}$. Lemma 1 then shows that the worst-case performance $\bar{\sigma}(VT(G_\Delta, C)W)$ is achieved for that particular $S_{WC} \in \mathcal{S}_\mathcal{G}(G_x, C, W_a)$ most removed from the 'centre' $|V_1|^2/\left(|V_1|^2 + |V_2|^2|C|^2\right)$.

5.2 *Performance cost and uncertainty sets*

Lemma 1 can also be read as to give a description of all sensitivity functions achieving $\bar{\sigma}(VT(G_\Delta, C)W) < 1$. Using the fact that $G_\Delta = \frac{1-S_\Delta}{CS_\Delta}$, all plants achieving $\bar{\sigma}(VT(G_\Delta, C)W) < 1$ with C must have their frequency responses in the region described by the following corollary,

Corollary 2. All plants G_Δ achieving $\bar{\sigma}(VT(G_\Delta,C)W) < 1$ are characterized by

$$G_\Delta = \frac{D_c|V_2|^2|C|^2 + D_c\Delta}{N_c|V_1|^2 - N_c\Delta}, \text{ with } |\Delta| \le W_Y \text{ and}$$

$$W_Y = |C|\sqrt{\left(\frac{\left(|V_1|^2 + |V_2|^2|C|^2\right)}{\left(|W_2|^2 + |W_1|^2|C|^2\right)} - |V_1|^2|V_2|^2\right)}$$

Due to the LFT structure, this set can also be described in an additive structure with Proposition 1 ,

$$G_\Delta = G_{centre} + \Delta_a \;,\; \left|W_a^{-1}\Delta_a\right| \le 1$$

$$G_{centre} = C^{-1}\frac{|W_Y|^2 + |V_2|^2|C|^2|V_1|^2}{|V_1|^4 - |W_Y|^2}$$

$$W_a = \left|C^{-1}\right|\frac{\left||V_2|^2|C|^2 + |V_1|^2\right|}{|V_1|^4 - |W_Y|^2}W_Y.$$

Note again that the set equals the exterior of the circle in case $|V_1|^4 < |W_Y|^2$.

What can Corollary 2 show us with regards to robust performance analysis? It provides us with a clear indication of the 'cost' of an uncertainty region. The worst-case performance over an uncertainty set $\mathcal{G}$ is determined by the measure in which the set of Corollary 2 has to expand in order to contain all the members G_Δ of the set $\mathcal{G}$. As extreme example, a certain nominal performance with G_x is achieved by many a plant G_Δ according to Corollary 2. The set of all these plants (with G_x on the boundary) could be taken as an uncertainty around G_x for free, i.e. without changing the worst-case performance.
The centre of the set of plants performing with the controller C lies in the direction of C^{-1}. This indicates a direction in which the performance cost is most sensitive to an increase in uncertainty. The Youla parametrization was shown to expand in the direction of a convex combination of the nominal model G_x and the negative inverse of the auxiliary controller C_{aux} (cf. Corr. 1). In case the auxiliary controller C_{aux} is close to the controller C to be evaluated, the Youla parameter uncertainty is 'pulled' in the right direction, i.e. in the direction being least sensitive for an increase in the worst-case performance. However, at this point this is still a matter of research. It is further interesting to research the connection between this corollary and the results of (Glover and Doyle, 1988) based on computational techniques which do include robust stability.

6. CONCLUSIONS

A first attempt is made to identify and discuss implications of the choice of an uncertainty structure to the robust control problem. An amplitude-bounded (circular) uncertainty set following from system identification can equivalently be described in terms of an additive, Youla parameter and ν-gap uncertainty. Closed-loop performance functions based on these sets are again bounded by circles in the frequency domain, allowing for exact worst-case performance calculation and for the evaluation of the consequences of uncertainty for robust design. Uncertainty sets with noncircular bounds and their underlying probability density functions do not retain their properties when transformed to closed-loop functions.

7. REFERENCES

Bombois, X., M. Gevers, G. Scorletti and B.D.O. Anderson (2001). Robustness analysis tools for an uncertainty set obtained by prediction error identification. *Automatica* **37**, 1629–1636.

Chen, J. and G. Gu (2000). *Control Oriented System Identification*. Wiley Interscience.

de Vries, D.K. and P.M.J. Van den Hof (1995). Quantification of uncertainty in transfer function estimation: a mixed probabilistic – worst-case approach. *Automatica* **31**(4), 543–557.

Douma, S.G., P.M.J. Van den Hof and O.H. Bosgra (2003). Controller tuning freedom under plant identification uncertainty: double youla beats gap in robust stability. *Automatica* **39**, 325–333.

Glover, K. and J.C. Doyle (1988). State space formulae for all stabilizing controllers that satisfy an H_∞-norm bound and relations to risk sensitivity .. *Systems and Control Letters* **11**, 167–172.

Goodwin, G.C., M. Gevers and B.M. Ninness (1992). Quantifying the error in estimated transfer functions with application to model order selection. *IEEE Trans. Autom. Control* **37**, 913–928.

Hakvoort, R.G. and P.M.J. Van den Hof (1997). Identification of probabilistic uncertainty regions by explicit evaluation of bias and variance errors. *IEEE Trans. Autom. Control.*

Heath, W.P. (2001). Bias of indirect non-parametric transfer function estimates for plants in closed loop.. *Automatica* **37**, 1529–1540.

Ljung, L. (1987). *System Identification: Theory for the User*. Prentice-Hall, Englewood Cliffs, New Jersey, USA.

Tay, T.T., J.B. Moore and R. Horowitz (1989). Indirect adaptive techniques for fixed controller performance enhancement. *International Journal on Control* **50**(5), 1941–1959.

Van den Hof, P.M.J. (1998). Closed-loop issues in system identification. *Annual reviews in control* **22**, 173–186.

Vidyasagar, M. (1985). *Control System Synthesis: A Factorization Approach*. MIT Press, Cambridge, Massachusetts, USA.

Vinnicombe, G. (2001). *Uncertainty and feedback : H_∞ loop-shaping and the ν-gap metric*. Imperial College Press, London, UK.

Zhou, K., J.C. Doyle and K. Glover (1996). *Robust and Optimal Control*. Prentice-Hall, New Jersey, USA.

IFAC Publications
www.elsevier.com/locate/ifac

STRONG ROBUSTNESS MEASURES FOR SETS OF LINEAR SISO SYSTEMS

Maria Cadic * Siep Weiland ** Jan Willem Polderman *

* *Fac. of Electrical Eng., Math. and Computer Sciences, University of Twente, P.O. Box 217, 7500 AE Enschede, The Netherlands*
** *Dept. of Electrical Engineering, Eindhoven University of Technology, P.O. Box 513, 5600 MB Eindhoven, The Netherlands*

Abstract: In this paper the issue of strong robustness of sets of systems is introduced and motivated in a scenario of adaptive control. A set of systems is said to be time-invariant strongly robust if the optimal control law designed on the basis of any element in this set stabilizes any other fixed member of the set. A stronger notion amounts to requiring that time-varying stabilizing control laws designed on the basis of any sequence of elements in the set asymptotically or quadratically stabilizes any other fixed member of the set. In this case, the set of systems is said to be time-varying strongly robust or time-varying strongly quadratically robust. The framework is exhibited for a class of linear and time-invariant SISO systems in discrete-time with fixed order n for a specified class of control objectives. It is established that if a given set of systems satisfies some robustness measures involving the classical structured real or complex radius, then the set is strongly robust. Then, for a more specific class of control objectives, attention is paid to the case of polyhedral sets of systems. A characterization of time-varying strongly quadratically robust sets of systems is given in the form of a feasibility test on a finite set of linear matrix inequalities (LMI's). *Copyright © 2003 IFAC*

Keywords: strong robustness, time-varying systems, adaptive control, structured stability radii, linear matrix inequalities

1. INTRODUCTION

Most of the classical adaptive control systems are based on the *certainty equivalence principle*. The parameters describing the unknown plant are estimated via some reliable estimation process and the model is used to compute the control law as if there was no mismatch between the true parameters values and their estimates. However, due to the unavoidable uncertainty, there is a priori little reason to expect the model-based controller to stabilize the true plant, what can be manifested in the adaptive closed-loop behavior through bad transient performance. Hence, to guarantee closed-loop stability of the control system, it is not sufficient that some robustness condition is satisfied by the model obtained at a certain time. Instead, such a robustness condition should hold for *any model candidate*, i.e., for the whole uncertainty set.
When, indeed, the identified uncertainty set would be such that the controller based on any model taken from this set asymptotically stabilizes any other model in this set, even when switched at each instant of time (as it is the case under adaptation), no bad transient behavior could occur. Uncertainty sets that possess this property, we call *time-varying strongly robust*. In the case where the stability notion would be quadratic stability rather than asymptotic stability, a similar notion can be defined, leading to *time-varying strongly quadratically robust* sets of systems. Alternatively, if the time-variation condition on the controller is re-

[1] The first and the third authors acknowledge the support of the Department of Electrical and Electronic Engineering of the University of Melbourne for their visit in November 2002, during which period this paper was written.

leased, the set is said to be *time-invariant strongly robust*.

The main purpose of this paper is to establish conditions to guarantee that a given set shows one of these strong robustness notions. Necessary and sufficient conditions are stated using classical robustness measures and give insight in the structure of strongly robust sets of systems. However, to the best of our knowledge, these tests are not computationally tractable. To obtain numerically efficient conditions, an algebraic criterion is further presented for a refined class of systems in the case of pole placement.

The paper is organized as follows. Next, in Section 2, the issue of strong robustness is introduced and motivated in the context of adaptive control. The general framework is then presented. Further, in Section 3, the notion of strong robustness is revisited exploiting the classical notions of real and complex structured stability radii. Then, in Section 4, attention is paid to a computable test for strong quadratical robustness. Finally Section 5 concludes the paper.

2. STRONG ROBUSTNESS AND ADAPTIVE CONTROL

We first show how adaptive control motivates the notion of strong robustness.

2.1 *Strong robustness: motivation*

The key issue in the adaptive control philosophy is the certainty equivalence principle (Mareels and Polderman, 1996), (Goodwin *et al.*, 1981): at each time, based on data measurements and an update law, a new model is chosen within a class of admissible models to describe the system to be controlled. The controller is then computed on the basis of this model to be applied to the real plant. Hence, at each time instant it is as if the present model was exactly matching the plant. During the adaptive process, some information is collected on the system behavior, hence the uncertainty level is reduced. In turn, since the uncertainty level decreases, the controller can be tuned so that the control result of the adaptive system improves asymptotically. However, when little information is available on the real system, as it is common in the initial phase of an adaptive control algorithm, it is probable that the model is poor and therefore this model cannot be expected *a priori* to lead to a controller stabilizing the true system. The well-documented possibility of this event is shown in the adaptive closed loop behavior through bad transient behavior. From a practical point of view, a good or bad transient behavior of a control system might be the criterion deciding on the quality of the controller, hence it is crucial to prevent bad transients to occur. In addition, since the model is kept updated at each iteration, the model-based controller itself is time-varying. Even if at each time instant the controller stabilizes the real plant, it is well-known that the time-varying closed-loop system might not be asymptotically stable in the case where the time variations are too fast. For this reason, the second requirement is that the time-varying controller asymptotically stabilizes the real system. Hence, classical adaptive control suffers from two drawbacks: an initial insufficient knowledge on the system to control might induce some undesired transients in the closed-loop system behavior, and the time-variations of the model-based controller might destroy the asymptotic stability of the control system.

To this effect, our strategy is to split the adaptive control design in two phases (Cadic and Polderman, 2002). In the first phase there is no guarantee that the time-varying controller stabilizes the true unknown system, therefore no model-based controller is applied to the system to be controlled, but effort is put on collecting information on the real system so as to reduce uncertainty about the plant. The second phase starts when closed-loop stability can be guaranteed for the class of uncertain plants that are identified at that time when controlled by a controller that meets the control objective for one identified system. When the process shifts to the second phase, effort is put on control according to the certainty equivalence principle. Moreover, it is a requirement that if the adaptive system is allowed to step in the second phase, a switch back to the first phase is not permitted, so that no bad transient can ever occur. Clearly, such a switching criterion must depend on the true system to be controlled, or, since this system is not known, on the *best* information available on the system, i.e., the uncertainty set. We define the switching criterion in the following way: the adaptive system stays in the first learning phase until the uncertainty set is such that the time-varying controller based on any sequence of models taken in this set stabilizes any other system in the set. Since the real system to be controlled belongs to the uncertainty set at any time, it is guaranteed to be stabilized by the time-varying model-based controller, wherever the model is chosen within the uncertainty set. If the uncertainty set has this property, it is said to be *time-varying strongly robust*.

2.2 *Strong robustness: definitions*

We now give the definition of time-invariant and time-varying strong robustness notions.

2.1 DEFINITION
Let Ω denote a class of systems and σ a given control objective such that the closed-loop control system is asymptotically stable. For any system $\theta \in \Omega$, we denote by $f(\theta)$ the controller based on the model θ achieving σ.

- A set $S \subset \Omega$ of systems is *time-varying strongly robust* with respect to σ if $\forall\theta \in S$ and $\forall\{\theta_i\}_{i>0} \in$

S, the time-varying closed-loop system $(\theta, f(\theta_i))$ is asymptotically stable.
- A set $S \subset \Omega$ of systems is *time-invariant strongly robust* with respect to σ if $\forall \theta, \theta' \in S$, the time-invariant closed-loop system $(\theta, f(\theta'))$ is asymptotically stable.

To simplify terminology, 'strongly robust' will, by default, refer to time-varying strong robustness, whereas we clearly specify when refering to strong robustness in the time-invariant case.

2.2 REMARK
- Any subset $S \subset \Omega$ which is strongly robust is time-invariant strongly robust.
- For any controllable system $\theta \in \Omega$, the point-set $\{\theta\}$ is strongly robust. Existence of non trivial strongly robust sets of systems depends on the nature of the class of systems Ω and of the control objective σ. In the case of linear stable discrete time invariant SISO systems with a known order n, and for pole placement in some given poles, existence of non-trivial strongly robust sets of systems is shown (Cadic and Polderman, 2003).
- Any subset of a strongly robust set of systems is strongly robust. In the adaptive control framework, this ensures that once the uncertainty set is strongly robust, it keeps this property at any further time, provided that it cannot grow with time.
- Whether or not a given set is strongly robust depends on the control objective. This is desirable, as robustness indicates a link between performance and uncertainty. If an uncertainty set is not strongly robust with respect to a particular control objective, it may be strongly robust with respect to another control objective. Information about the uncertainty set may be used to find the most suitable control objective. This is an important property in control design, normally lacking from classical adaptive control discussions.

Definition 2.1 shows that the notion of strong robustness involves the notion of asymptotic stability. To further refine this requirement, we now define the notion of *strongly quadratically robust* sets of systems.

2.3 DEFINITION (QUADRATICALLY ROBUST SETS)
Let Ω denote a class of systems and σ a given control objective. For any system $\theta \in \Omega$, we denote by $f(\theta)$ the controller based on θ achieving σ. A set $S \subset \Omega$ of systems is *strongly quadratically robust* with respect to σ if for any system $\theta \in S$ and any sequence of systems $\{\theta_i\} \in S$, there exists a quadratic Lyapunov function for the time-varying system defined by $(\theta, f(\theta_i))$.

Hence, for a specified control objective, if a set $S \subset \Omega$ is strongly quadratically robust, then S is strongly robust.

2.3 *Problem statement*

We denote by Ω the class of linear time-invariant SISO systems in discrete-time with fixed order n of the form:

$$y(k+n) + \sum_{i=0}^{n-1} [a_i y(k+i) - b_i u(k+i)] = 0, \ \forall k \qquad (1)$$

where u and y denote the input and output respectively, and where the parameters $\{(a_i, b_i)\}_{0 \le i \le n-1} \in \mathbb{R}^2$ are unknown. Moreover, any system in Ω is supposed to be controllable, i.e., the polynomials $a(s) := s^n + \sum_{i=0}^{n-1} a_i s^i$ and $b(s) := \sum_{j=0}^{n-1} b_j s^j$ are co-prime for any system in the class Ω. Any system in Ω can be also be described in terms of a state space representation:

$$\begin{aligned} x(k+1) &= Ax(k) + bu(k) \\ y(k) &= Cx(k) \end{aligned} \qquad (2)$$

where $x \in \mathbb{R}^{2n-1}$ denotes the non-minimal state vector:

$$x(k) = \mathrm{col}(y(k) \cdots y(k-n+1)\ u(k-1) \cdots u(k-n+1))$$

and where the matrices $A \in \mathbb{R}^{(2n-1)\times(2n-1)}$, $b \in \mathbb{R}^{2n-1}$ and $C \in \mathbb{R}^{1\times(2n-1)}$ are given by

$$A = \begin{bmatrix} -a_{n-1} & \cdots & \cdots & -a_1 & -a_0 & b_{n-2} & \cdots & \cdots & b_1 & b_0 \\ 1 & 0 & \cdots & 0 & 0 & 0 & \cdots & \cdots & 0 & 0 \\ 0 & \ddots & & \vdots & \vdots & \vdots & & & \vdots & \vdots \\ \vdots & & \ddots & 0 & \vdots & \vdots & & & \vdots & \vdots \\ \vdots & & & 1 & \vdots & \vdots & & & \vdots & \vdots \\ 0 & \cdots & \cdots & 0 & 0 & 0 & \cdots & \cdots & 0 & 0 \\ \vdots & & & 0 & \vdots & 1 & & & \vdots & \vdots \\ \vdots & & & \vdots & \vdots & 0 & \ddots & & \vdots & \vdots \\ \vdots & & & \vdots & \cdots & 0 & & \ddots & 0 & \vdots \\ 0 & \cdots & \cdots & 0 & 0 & 0 & \cdots & \cdots & 1 & 0 \end{bmatrix} \qquad (3)$$

$$b = \mathrm{col}(b_{n-1}, 0, \cdots, 0, 1, 0, \cdots, 0) \qquad (4)$$

$$C = \begin{bmatrix} 1 & 0 & \cdots & 0 & 0 \end{bmatrix} \qquad (5)$$

and such that the controllability matrix associated with A, b is of full row rank. Any system in Ω is identified with its matrix description (A, b) as given by (3). Therefore, Ω is a subset in $\mathbb{R}^{(2n-1)\times(2n-1)} \times \mathbb{R}^{2n-1}$.

2.4 *Control objective*

The control objective σ is left unspecified. However we make the following assumption.

2.4 ASSUMPTION
- For any system $(A, b) \in \Omega$, there exists a *unique* controller described by

$$u(k) = f(A, B)x(k), \ f(A, b) \in \mathbb{R}^{1\times(2n-1)} \qquad (6)$$

that achieves the control objective σ. Moreover, σ is such that the closed-loop system given by

$$\begin{aligned} x(k+1) &= (A + bf(A, b))x(k) & (7) \\ y(k) &= Cx(k) & (8) \end{aligned}$$

is asymptotically stable.

- The map $f(\cdot)$ assigning to each system in Ω the controller

$$f : (A,b) \in \Omega \longmapsto f(A,b) \in \mathbb{R}^{1\times(2n-1)} \qquad (9)$$

according to (6) is continuous with respect to the parameters $\{a_i, b_i\}_{i\leq n-1}$. This is to guarantee existence of nontrivial strongly robust sets of systems in Ω (Cadic and Polderman, 2003).

2.5 REMARK
Pole placement in some given desired stable poles $\{\alpha_i\}_{i=0\cdots 2n-1}$ satisfies Assumption 2.4. In this case, the control law based on any system (A,b) is computed according Ackermann's Formula (Mareels and Polderman, 1996):

$$f(A,b) := -\begin{pmatrix}0 & \cdots 0 & 1\end{pmatrix}[\mathscr{C}(A,b)]^{-1}\Pi(A) \qquad (10)$$

where $\Pi(\xi) = \prod_{i=0}^{2n-1}(\xi - \alpha_i)$ and where $\mathscr{C}(A,b)$ denotes the controllability matrix associated with A,b defined by $\mathscr{C}(A,b) = \begin{pmatrix}b & Ab & \cdots & A^{2n-1}b\end{pmatrix}$.

3. STRONG ROBUSTNESS MEASURES IN TERMS OF STRUCTURED STABILITY RADII

Often in control, the plant to be controlled is unknown and even a good mathematical model cannot exactly describe the dynamics of this plant. Hence any controller designed on the basis of a nominal model must satisfy some robustness criterion to perform well when applied to the real system. An extensive literature deals with robustness measures for linear state space systems, under unstructured perturbation (Qiu and Davidson, 1986) or structured perturbation (Hinrichsen and Pritchard, 1988). We now use some of these results to characterize time-invariant strong robustness and strong robustness.

3.1 *Time-invariant strong robustness and real structured stability radius*

We now recall the definition of real stability radius of a Schur matrix under structured perturbation.

3.1 DEFINITION (REAL STRUCTURED STABILITY RADIUS)
Let $M \in \mathbb{R}^{(2n-1)\times(2n-1)}$ denote a strictly Schur-stable matrix. The real stability radius of M with respect to the perturbation structure (b, I_{2n-1}), for $b \in \mathbb{R}^{(2n-1)\times 1}$, is defined by (Hinrichsen and Pritchard, 1988):

$$r_{\mathbb{R}}(M,b,I_{2n-1}) = \inf\{||D||_{\mathbb{R}} : D \in \mathbb{R}^{1\times(2n-1)},\ M+bD \text{ is not Schur stable}\}, \qquad (11)$$

where $||.||_{\mathbb{R}}$ denotes the matrix norm in $\mathbb{R}^{1\times(2n-1)}$.

We now introduce the set of stabilizing controllers.

3.2 DEFINITION (SET OF STABILIZING CONTROLLERS)
Given a system $(A,b) \in \Omega$, we define the set of all stabilizing controllers for (A,b) by $\mathscr{S}_{(A,b)}$:

$$\mathscr{S}_{(A,b)} = \{f \in f(\Omega) : A + bf \text{ is Schur stable.}\} \qquad (12)$$

We then have the following result:

3.3 THEOREM

$$S \subset \Omega \textit{ is time-invariant strongly robust} \qquad (13)$$
$$\Leftrightarrow f(S) \subset \bigcap_{(A,b)\in S} \mathscr{S}_{(A,b)}$$

Proof: from Definition 2.1, $S \subset \Omega$ is time-invariant strongly robust if and only if $\forall (A,b),(A',b') \in \Omega$, $f(A,b)$ stabilizes (A',b'). This is equivalent to say that $\forall (A,b),(A',b') \in \Omega$, $f(A',b') \subset \mathscr{S}_{(A,b)}$, equivalently $\forall (A,b) \in \Omega$, $f(S) \subset \mathscr{S}_{(A,b)}$. □

$\forall (A,b) \in \Omega$, let $r_{(A,b)}$ denote the radius of the largest open ball centered about $f(A,b)$ contained in $\mathscr{S}_{(A,b)}$:

$$r_{(A,b)} = \sup\{\varepsilon \geq 0 : \forall \varphi \in f(\Omega), \qquad (14)$$
$$||\varphi - f(A,b)||_{\mathbb{R}} \leq \varepsilon \Rightarrow \varphi \in \mathscr{S}_{(A,b)}\}.$$

3.4 LEMMA
$\forall (A,b) \in \Omega$, $r_{(A,b)} \geq r_{\mathbb{R}}(A + bf(A,b), b, I_{2n-1})$ where $r_{\mathbb{R}}(A + bf(A,b), b, I_{2n})$ is given in Definition 3.1.

Proof: $r_{(A,b)}$ can be expressed as follows

$$r_{(A,b)} = \sup_{\varphi\in f(\Omega)} \{||\varphi - f(A,b)||_{\mathbb{R}} : A + b\varphi \text{ is schur}\}, \text{ or:}$$

$$r_{(A,b)} \geq \inf_{\varphi\in f(\Omega)} \{||\varphi - f(A,b)||_{\mathbb{R}} : A + b\varphi \text{ is not schur}\},$$

i.e., $r_{(A,b)} \geq \inf\{||\varphi - f(A,b)||_{\mathbb{R}} : \varphi \in \mathbb{R}^{2n-1},$
$A + bf(A,b) + b(\varphi - f(A,b))$ is not Schur stable $\}$.

Therefore, $r_{(A,b)} \geq r_{\mathbb{R}}(A + bf(A,b), b, I_{2n})$. □

From Lemma 3.4 we obtain the following result:

3.5 COROLLARY
For any set $S \subset \Omega$, if $\forall (A,b),(A',b') \in S$,

$$||f(A,b) - f(A',b')||_{\mathbb{R}} \leq r_{\mathbb{R}}(A + bf(A,b), b, I), \quad (15)$$

then S is time-invariant strongly robust.

Proof: for a given set $S \in \Omega$, suppose that (15) holds. Therefore, using Lemma 3.4, $\forall (A,b),(A',b') \in S$,

$$||f(A',b') - f(A,b)||_{\mathbb{R}} \leq r_{\mathbb{R}}(A + bf(A,b), b, I) \leq r_{(A,b)}$$

hence $\forall (A,b),(A',b') \in S, f(A',b') \in \mathscr{S}_{(A,b)}$. According to Theorem 3.3, it implies that S is time-invariant strongly robust. □

3.2 *Strong robustness and complex structured stability radius*

We now recall the definition of complex stability radius of a Schur matrix under structured perturbation.

3.6 DEFINITION (COMPLEX STRUCTURED STABILITY RADIUS)
Let $M \in \mathbb{R}^{(2n-1)\times(2n-1)}$ denote a strictly Schur stable matrix. The complex stability radius of M with respect to the perturbation structure $(b, I_{(2n-1)}, b \in \mathbb{R}^{(2n-1)\times 1}$ is defined by (Hinrichsen and Pritchard, 1988):

$$r_{\mathbb{C}}(M, b, I_{(2n-1)}) = \inf\{||D||_{\mathbb{C}} : D \in \mathbb{C}^{1\times(2n-1)}, \\ M + bD \text{ is not Schur stable}\}, \quad (16)$$

where $||.||_{\mathbb{C}}$ denotes the matrix norm in $\mathbb{C}^{1\times(2n-1)}$.

3.7 REMARK
It is shown in (Hinrichsen and Pritchard, 1988) that Definition 3.6 is the same for complex time-invariant perturbations or complex time-varying perturbations.

We now introduce the following definition.

3.8 DEFINITION
For any system $(A, b) \in \Omega$, we denote by $\mathscr{B}(A, b)$ the matrix ball of systems in Ω centered in $A + bf(A, b)$ with radius the complex stability radius $r_{\mathbb{C}}(A + bf(A, b), b, I_{2n-1})$. It is defined by:

$$\mathscr{B}(A, b) = \{A + b\varphi : \varphi \in f(\Omega) \text{ and} \quad (17) \\ ||f(A, b)) - \varphi||_{\mathbb{R}} < r_{\mathbb{C}}(A + bf(A, b), b, I_{2n-1})\}$$

We then obtain the following result:

3.9 THEOREM
For a given set $S \subset \Omega$, if $f(S) \subset \bigcap_{(A,b)\in S} \mathscr{B}(A, b)$, then S is strongly robust.

Proof: suppose $S \subset \Omega$ to be such that $f(S) \subset \bigcap_{(A,b)\in S} \mathscr{B}(A, b)$. Then $\forall (A, b) \in S$, and for all sequence $\{(A_k, b_k)\}_{k\geq 0} \subset S$, $||f(A, b) - f(A_k, b_k)||_{\mathbb{R}} \leq r_{\mathbb{C}}(A + bf(A, b), b, I_{2n-1})$. Therefore, using Remark 3.7, $\forall (A, b) \in S$, and for all sequence $\{(A_k, b_k)\}_{k\geq 0} \subset S$, the closed-loop time-varying system with system matrix $A + bf(A, b) + b(f(A, b) - f(A_k, b_k))$ is asymptotically stable. Hence S is strongly robust. □

Theorem 3.9 yields the following result.

3.10 COROLLARY
For any open set $S \subset \Omega$, if $\forall (A, b), (A', b') \in S$,

$$||f(A, b) - f(A', b')||_{\mathbb{R}} \leq r_{\mathbb{C}}(A + bf(A, b), b, I_{2n}),$$

then S is strongly robust.

Proof: suppose (18) holds for a given set $S \subset \Omega$. Then $\forall (A, b), (A', b') \in S$, $f(A', b') \in \mathscr{B}(A, b)$. Hence, from Theorem 3.9, S is strongly robust. □

3.11 REMARK
Corollary 3.10 reduces the problem of checking if any time-varying controllers in $f(S)$ do stabilize any system in S into a test about the maximum distance between controllers taken in $f(S)$. Therefore, a conclusion about the stability of the closed-loop varying system can be drawn from a test on time-invariant controllers.

4. STRONG ROBUSTNESS AND POLYHEDRAL SETS OF SYSTEMS

In this section we consider systems of the form (1) together with their control-canonical state space representations

$$x(k+1) = Ax(k) + Bu(k) \\ y(k) = Cx(k), \quad (18)$$

where $A \in \mathbb{R}^{n\times n}$, $B \in \mathbb{R}^{n}$ and $C \in \mathbb{R}^{1\times n}$ are defined by:

$$A = \begin{bmatrix} 0 & 1 & 0 & \cdots & 0 \\ \vdots & \ddots & \ddots & \ddots & \vdots \\ \vdots & & \ddots & \ddots & 0 \\ 0 & \cdots & \cdots & 0 & 1 \\ -a_0 & -a_1 & \cdots & \cdots & -a_{n-1} \end{bmatrix} \quad B = \begin{bmatrix} 0 \\ 0 \\ \vdots \\ 0 \\ 1 \end{bmatrix} \quad (19)$$

$$C = \begin{bmatrix} b_0 & b_1 & \cdots & b_{n-1} \end{bmatrix} \quad (20)$$

For fixed $n > 0$, the class of all nth order systems obtained in this way is denoted by Ω, i.e, Ω consists of all triples (A, B, C) of real matrices of dimension $n \times n$, $n \times 1$ and $1 \times n$ that assume this control canonical form. Let $S \subset \Omega$ be those systems whose coefficients a_i and b_i satisfy bounds according to

$$\underline{a}_i \leq a_i \leq \bar{a}_i \text{ and } \underline{b}_i \leq b_i \leq \bar{b}_i, \ \forall i \leq n-1. \quad (21)$$

for known values of the bounds $\underline{a}_i$, $\bar{a}_i$, $\underline{b}_i$ and $\bar{b}_i$.
Suppose that the control objective σ amounts to locating the closed-loop poles in the roots of a given characteristic polynomial $p(s) = s^n + \sum_{i=0}^{n-1} p_i s^i$ with known real coefficients p_i. Under the controllability assumption of Ω this means that for any $(A, B, C) \in \Omega$ we have that the controller is uniquely defined by the feedback law $u(k) = f(A, b, C)x(k)$ where $f : \Omega \to \mathbb{R}^{1\times n}$ is defined by

$$f(A, B, C) = \begin{bmatrix} a_0 - p_0 & \ldots & a_{n-1} - p_{n-1} \end{bmatrix}. \quad (22)$$

Note that the use of control canonical forms implies that $f(A, B, C)$ does not depend on C. Given any pair of systems $(A, B, C), (A', B', C') \in \Omega$, the closed loop system $(A, B, C), f(A', B', C'))$ is defined by $x(k+1) = ((A + Bf(A', B', C'))x(k)$ where the closed loop state evolution matrix $A + Bf(A', B', C')$ takes the form:

$$A + Bf(A', B', C') = \\ \begin{bmatrix} 0 & 1 & 0 & \cdots & 0 \\ \vdots & \ddots & \ddots & \ddots & \vdots \\ \vdots & & \ddots & \ddots & 0 \\ 0 & \cdots & \cdots & 0 & 1 \\ (a'_0 - a_0) - p_0 & \cdots & & \cdots & (a'_{n-1} - a_{n-1}) - p_{n-1} \end{bmatrix} \quad (23)$$

For $i = 0, \ldots, n-1$, let M_i denote the $n \times n$ matrix which is zero except at its (n, i) entry where it is 1. Let $M_n = A + Bf(A, B, C)$ denote the nominal closed loop desired matrix. Then, when (A, B, C) and (A', B', C')

range over S, the matrix $A+Bf(A',B',C')$, given in (23), is of polyhedral form:

$$M(\delta) := A+Bf(A',b',C') = M_n + \sum_{i=0}^{n-1} \delta_i M_i, \quad (24)$$

where, for $i = 0,\ldots,n-1$, the parameter δ_i assumes its values in the interval:

$$\underline{\delta}_i := \underline{a}_i - \overline{a}_i \leq \delta_i \leq \overline{a}_i - \underline{a}_i := \overline{\delta}_i \quad (25)$$

Let $\delta = \mathrm{col}(\delta_0,\cdots,\delta_{n-1})$ be the uncertainty vector, and define

$$\Delta_0 := \{\delta = \mathrm{col}(\delta_0,\cdots,\delta_{n-1}) \mid \delta_i = \pm\overline{\delta}_i\}, \quad (26)$$

$$\Delta := \{\delta = \mathrm{col}(\delta_0,\cdots,\delta_{n-1}) : |\delta_i| \leq \overline{\delta}_i\}. \quad (27)$$

Δ_0 is the finite set consisting of all 'corner points' of the uncertainty region (25) and Δ is the convex hull of Δ_0. The set of all possible closed-loop state-evolution matrices is defined by the affine set $M(\delta)$ where $\delta \in \Delta$. We have the following result:

4.1 THEOREM (MAIN RESULT)
With Ω as defined in this section, the subset $S \subset \Omega$ is strongly quadratically robust if and only if there exists $K = K^T > 0$ such that

$$[M(\delta)]^T K M(\delta) - K + I < 0, \quad \forall \delta \in \Delta_0 \quad (28)$$

where Δ_0 is defined by (27).

Proof: to prove necessity, we go along the following lines. Suppose S is as specified. Define Δ_0 according to (27). If there exists $K = K^T > 0$ such that (28) holds for any $\delta \in \Delta_0$, then convexity of the function $h_x(\delta) := x^T([M(\delta)]^T K M(\delta) - K)x$ for any $x \in \mathbb{R}^n$ implies that (Weiland, 2002)

$$[M(\delta)]^T K M(\delta) - K + I < 0, \quad \forall \delta \in \Delta \quad (29)$$

Now, define $V : \mathbb{R}^n \to \mathbb{R}$ according to $V(x) = x^T K x$. We then claim that V defines a Lyapunov function for any of the interconnected systems. Indeed, $V(\cdot)$ is non-negative and for any system $(A,B,C),(A_i,B_i,C_i) \in S$, the interconnection $((A,B,C),f(A_i,B_i,C_i))$ takes the state-space form $x(k+1) = M(\delta)x(k)$ with $M(\delta)$ defined by (24), where $\delta \in \Delta$. Hence we have that

$$V(x(k+1)) = x^T(k)M(\delta)^T K M(\delta)x(k) < x(k)^T K x(k) = V(x(k))$$

for any $\delta \in \Delta$. Therefore, it follows from Definition 2.3 that S is strongly quadratically robust. Conversely, if no such positive definite matrix K exists, from Definition 2.3, there is no quadratic stability of the parameterized closed-loop system and hence S is not strongly quadratically robust. □

We would like to emphasize the following consequences of Theorem 4.1.

(1) Theorem 4.1 involves a *finite number* of LMI's (at most 2^n LMI's). Therefore, the strong quadratic robustness characterization, placing a priori an infinite number of constraints on the set of systems to be tested, has been converted into a numerically tractable and efficient test.
(2) In an adaptive control framework, the parameters δ_i range over the diameter of the uncertainty region $|\delta_i| \leq \overline{\delta}_i$, for $i = 0,\cdots,n-1$. Hence, it is an interesting problem to guarantee that the intervals $[\underline{\delta}_i, \overline{\delta}_i]$ will be uniformly decreasing as function of the iteration time of an adaptive algorithm. This amounts to reducing the uncertainty diameters $\overline{a}_i - \underline{a}_i$, $i = 0,\ldots,n-1$.
(3) For given uncertainty intervals $\overline{\delta}_i := \overline{a}_i - \underline{a}_i$, the feasibility test of Theorem 4.1 depends on the desired pole locations defined by the characteristic polynomial p. This is in accordance with Remark 2.2 and shows that some pole locations might be better suited to obtain strong robustness than others.
(4) Because of the assumption that the controller defined by (22) does not depend on the matrix C, the results presented in the present section only hold for a restricted class of systems in Ω.

5. CONCLUSION

The notion of strongly robust sets of systems has been introduced and strong robustness measures have been exhibited using some classical structured stability measures. Besides the fact that the notion of strong robustness is a further extension of the notion of classical robustness, it is motivated by applications in adaptive control. Adaptive control systems based on strong robustness would indeed have the ability to cope with lack of knowledge on the system or time-variations of the controller.

6. REFERENCES

Cadic, M. and J.W. Polderman (2002). Strong robustness in multi-phase adaptive control: the basic scheme. *Proc. 15th MTNS Symposium, Notre Dame, USA*.

Cadic, M. and J.W. Polderman (2003). Strong robustness in adaptive control. *LNCIS 281* pp. 45–54.

Goodwin, G.C., P.J. Ramadge and P.E. Caines (1981). Discrete-time stochastic adaptive control. *SIAM J. Control and Optim.* **19**, 829–853.

Hinrichsen, D. and A.J. Pritchard (1988). Robustness measures for linear state space systems under complex and real perturbations. *Perspectives in Control Theory, Proc. Sielpa Conference*.

Mareels, I.M.Y. and J.W. Polderman (1996). *Adaptive Systems: An Introduction*. Birkhäuser. Boston.

Qiu, L. and E.J. Davidson (1986). New perturbation bounds for the robust stability of linear state-space models. *Proc. 25th Conference on Decision and Control* pp. 751–755.

Weiland, S. (2002). *Lecture notes, Disc Course Linear Matrix Inequalities in Control.*

IFAC Publications
www.elsevier.com/locate/ifac

USING A SUFFICIENT CONDITION TO ANALYZE THE INTERPLAY BETWEEN IDENTIFICATION AND CONTROL

Håkan Hjalmarsson and Henrik Jansson *,[1]

* *Department of Signals, Sensors and Systems, Royal Institute of Technology, S-100 44 Stockholm, Sweden*

Abstract: In this contribution we use a standard condition for robust stability and performance to discuss the interplay between identification and control. This leads to a criterion which can be used in the control design to analyze the connections between closed loop bandwidth, model complexity, noise characteristics and input design. Based on this criterion experiment design of H_∞-type is derived. *Copyright © 2003 IFAC*

Keywords: Identification for control; robust performance; robust stability.

1. INTRODUCTION

Identification and control has received extensive interest during the last years (Goodwin *et al.*, 1999; Forssell and Ljung, 2000; Henriksson *et al.*, 2001; Ljung, 2000; Zhu, 1998). The general problem of control design based on an identified model can be divided into the following three steps

1) Performing an identification experiment, whereby specific experimental details, such as whether performing an open or closed loop experiment, and which input excitation to employ, have to be considered.
2) Identifying a suitable model, whereby specific identification details, such as which model structure to use, which prefilters to employ, have to be considered
3) Designing a controller, whereby control specific details, such as which bandwidth to design for, have to be considered.

Inarguably these steps should be considered together. For example, the designed closed loop bandwidth should depend on the quality of the model which in turn depends on the experimental conditions. A number of contributions discuss this interplay between identification and control, e.g. (Gevers, 1993; Lee *et al.*, 1995; Van den Hof and Schrama, 1995), in a very insightful way. Still, relatively little in terms of simple rule's of thumb have emerged. Our objective here is, through the use of a simple criterion, to discuss quantities that can be useful to roughly solve questions such as:

1) Here is a data set and some prior information we have on the system. Design a controller with as high performance as possible and with a certain robustness margin. We can only use PID controllers.
2) Here is some prior information about the system. Which experiment should be carried out in order that enough information is available to allow the design of a controller that satisfies our prespecified performance and robustness specifications or to determine that the specifications cannot be met due to fundamental limitations of the system?
3) Here is some prior information about the system. You are allowed to do an experiment for this long and with these restrictions on the input/output behaviour of the system. Which performance and robustness specifications for the closed loop system will be possible with this information?

We will limit the discussion to the single input single output (SISO) case.

[1] This research has been supported by the Swedish Research Council.

2. SISO SYSTEMS

Let the true system be

$$y = G_0 u + v \qquad (1)$$

with $v = H_0 e$ where e is white noise – we omit the time dependence of the signals in the notation.

With the feedback given by

$$u = C(r - y) + w \qquad (2)$$

where r and w are external excitation signals, the closed loop system is described by

$$\begin{aligned}\begin{pmatrix} y \\ u \end{pmatrix} &= \begin{pmatrix} \frac{G_0 C}{1+G_0 C} & \frac{G_0}{1+G_0 C} \\ \frac{C}{1+G_0 C} & \frac{1}{1+G_0 C} \end{pmatrix} \begin{pmatrix} r \\ w \end{pmatrix} \\ &= \frac{1}{1+G_0 C} \begin{bmatrix} G_0 \\ 1 \end{bmatrix} [C \ 1] \\ &\triangleq P(G_0, C) \begin{pmatrix} r \\ w \end{pmatrix} \end{aligned} \qquad (3)$$

if we, for the moment, neglect the disturbance v.

2.1 Comparing the achieved and designed closed loops

We will assume that C is a controller designed with some arbitrary model based method such as LQG, IMC or H_∞ loop-shaping. Let G and H denote the model of G_0 and H_0, respectively, used in the control design. Then the difference between the achieved and designed complementary sensitivity functions is given by

$$T_0 - T = \frac{G_0 C}{1+G_0 C} - \frac{GC}{1+GC} = (G_0 - G) C S_0 S, \qquad (4)$$

where $S_0 = 1/(1+G_0 C)$ and $S = 1/(1+GC)$ respectively are the true and the designed sensitivity functions. This expression has been the starting point for many methods that tries to make the achieved closed and the designed closed loop similar. In particular, it has lead to a number of iterative control and identification methods (Zang *et al.*, 1995; Åström, 1993).

The difference $T_0 - T$ can also be massaged into the following form

$$T_0 - T = \underbrace{\frac{1}{1+\Delta(G_0, G, T)}}_{\text{stability}} \underbrace{\Delta(G_0, G, T)}_{\text{performance}} (1 - T) \qquad (5)$$

where $\Delta(G_0, G, T)$ is the weighted relative model error

$$\Delta(G_0, G, T) = \frac{G_0 - G}{G} T \qquad (6)$$

This interesting expression can be interpreted in the following way:

- The first factor, i.e.
$$\frac{1}{1+\Delta(G_0, G, T)}$$
is related to stability of the achieved closed loop system. If the designed closed loop is stable (which is a natural requirement), then stability of the achieved closed loop system is equivalent to the stability of the factor above. Using the small gain theorem, a sufficient stability condition is that (6) should have gain less than one at all frequencies. We recognize this condition as a well known robust stability result.
- The second factor in (5), i.e. $\Delta(G_0, G, T)$, is related to performance. By making this factor small, *performance* close to the designed performance is guaranteed.
  That $\Delta(G_0, G, T)$ should be small in order for the true performance to be close to the designed performance can be given an intuitive interpretation. First, recall that the objective of the controller is to generate a suitable input to the system given a desired output response, i.e. to invert the system (approximately). Suppose that $w \equiv v \equiv 0$. Then the designed output is Tr. In order to obtain the designed output, the signal
$$u_d = G_0^{-1} Tr$$
should be applied to the system. However, it is easily verified that when $y = Tr$, the input is in reality given by
$$u = G^{-1} Tr$$
For a controller which provides similar achieved performance as designed performance, u should be close to u_d, i.e. the difference
$$u - u_d = (G^{-1} - G_0^{-1}) Tr$$
should be small. Thus one could argue that identification for control should focus on models that are good for input prediction rather than output prediction. While there is a certain truth in this, the issue is slightly more complex. One is often more interested in how the output of the achieved closed loop behaves as compared to the designed output. When the two input signals above are fed to the true system, the resulting difference in the outputs become
$$(G_0 G^{-1} - 1) Tr$$
Thus, for frequencies where $\Delta(G_0, G, T)$ is small, the achieved and the designed output will be similar. We have thus been able to interpret the weighted relative model error (6) in terms of that a good model for control should be able to provide a good approximation to the inverse of the system in the frequency range represented by the bandwidth of the control design. This is a very useful insight that we will explore in the following.
- The third factor in (5) can be viewed just as weighting function and we postpone the discussion of this factor until later.

Notice that the same quantity, i.e. (6), is involved both in the performance condition and the stability condition. The size of this factor depends on both the quality of the model G and the control specifications (represented by T).

It is of course not sufficient that T_0 is close to T in order for the achieved closed loop to be close to the designed closed loop. More generally, one has to consider the difference $P(G_0,C)-P(G,C)$ which after some manipulations can be written

$$P(G_0,C)-P(G,C)=\frac{\Delta(G_0,G,T)}{1+\Delta(G_0,G,T)}(1-T) \times \begin{bmatrix} 1 \\ -C \end{bmatrix} \begin{bmatrix} 1 & \frac{1}{C} \end{bmatrix}. \quad (7)$$

It is straightforward to establish that the largest singular value of this matrix is given by

$$\bar{\sigma}=\left|\frac{1}{1+\Delta(G_0,G,T)}\right| |\Delta(G_0,G,T)|\,|1-T| \times \sqrt{1+\frac{1}{|C|^2}}\sqrt{1+|C|^2} \quad (8)$$

which we can re-write as

$$\bar{\sigma}=\left|\frac{1}{1+\Delta(G_0,G,T)}\right| |\Delta(G_0,G,T)| \times \sqrt{1+\frac{|1-T|^2\,|G|^2}{|T|^2}}\sqrt{|1-T|^2+\left|\frac{T}{G}\right|^2}$$

The first square-root factor becomes large in the rare situation that the designed bandwidth is smaller than the bandwidth of the open loop model.

The second square-root factor becomes large when the designed bandwidth is larger than the bandwidth of the open loop model.

2.2 *A sufficient condition for stability and performance*

We will now use the insight acquired in the previous subsection to design a method which can guarantee that P_0-P is small and which allows us to capture some intuitive insight into the problem.

From the previous subsection it should be clear that $|\Delta|<1$ guarantees stability and $|\Delta|<<1$ guarantees achieved performance close to the designed performance.[2]

Let $\hat{G}$ be a known arbitrary model of G_0 and let $G=G(\theta)$ be a parametrization of the model that is to be used for the control design. Then by the triangle inequality

$$|\Delta| \le \left|\frac{T}{G}\right| |G_0-\hat{G}| + \left|\frac{T}{G}\right| |\hat{G}-G(\theta)| \quad (9)$$

[2] Notice though, that the larger T/G is, the smaller Δ has to be for this to be true.

If the error $|G_0-\hat{G}|$ was known, the upper bound on the right-hand side would include only known quantities and it would, at least in principle, be possible to choose θ so as to minimize this over bound.

Here we shall use the following result to bound $|G_0-\hat{G}|$.

Proposition 2.1. Suppose that the true system is operating in open loop and given by

$$y(t)=\frac{B_\circ(q)}{A_\circ(q)}u(t)+v(t) \quad (10)$$

where $v(t)=H_\circ(q)e_0(t)$ for some white noise sequence $e_0(t)$. Assume that the system is in the model set. Let $G(q,\theta)=q^{-k}B(q)/A(q)$ with m_b parameters in $B(q)$ and m_a parameters in $A(q)$.

Under the condition that

$$A_\dagger \triangleq A_\circ^2 H_\circ/\Phi_u^{1/2} \quad (11)$$

is a polynomial in z^{-1} of degree at most m_a+m_b, it holds that

$$\lim_{N\to\infty} N \cdot \operatorname{Var}(\hat{G}_N(e^{j\omega}))=\kappa(\omega)\frac{\Phi_v(\omega)}{\Phi_u(\omega)} \quad (12)$$

where

$$\kappa(\omega) \triangleq \sum_{k=1}^{m_a+m_b} \frac{1-|\xi_k|^2}{|e^{j\omega}-\xi_k|^2}. \quad (13)$$

where $\xi_k, k=1,\dots,m_a+m_b$ are the zeros of $z^{m_a+m_b}A_\dagger(z)$.
■

Under the assumptions in the proposition we have, asymptotically in N, that

$$|\hat{G}-G_0| \le \frac{c_\alpha}{\sqrt{N}}\sqrt{\kappa\frac{\Phi_v}{\Phi_u}} \quad (14)$$

with $(1-\alpha)\cdot 100$ % probability (e.g. 99%) for a suitable choice of c_α.

Using the high order estimate $\hat{G}$ in (9) gives that

$$|\Delta| \le \frac{c_\alpha}{\sqrt{N}}\left|\frac{T}{G}\right|\sqrt{\kappa\frac{\Phi_v}{\Phi_u}}+\left|\frac{T}{G}\right||\hat{G}-G(\theta)| \quad (15)$$

with $(1-\alpha)\cdot 100$ % probability.

We now take the L_∞ norm of the upper bound in (15) as design criterion

$$J(\theta,\gamma)=\sup_\omega\left\{c_\alpha\left|\frac{T(\theta,\gamma)}{G(\theta)}\right|\sqrt{\kappa\frac{\Phi_v}{N\cdot\Phi_u}} + \left|\frac{T(\theta,\gamma)}{G(\theta)}\right||\hat{G}-G(\theta)|\right\} \quad (16)$$

Above we have introduced the scalar parameter γ in T to denote a parameter that controls the bandwidth of the designed closed loop.

The following is immediately evident:

- The factor $|T(\theta,\gamma)/G(\theta)|$ is the designed gain from the reference to the input.

- Poor model accuracy, either through poor statistical accuracy (term 1) or bad approximation accuracy (term 2), can be counteracted by the control design through the factor $T(\theta,\gamma)/G(\theta)$.
- In the passband of the design, $|T/G| \approx 1/G$, hence poor quality data and modeling is allowed where G is large. However, if it is desired to extend the designed bandwidth outside the bandwidth of the model, more accurate data and modeling is required.

The last item can be used to see how much the SNR has to be increased when the bandwidth is increased in order to keep the statistical accuracy constant. A rough estimate is obtained as follows. Let the signal energy to noise power ratio $SENPR(\omega,\gamma)$ be defined as the value of $\sqrt{\frac{N \cdot \Phi_u}{\Phi_v}}$ required to keep J at a certain level. Suppose that the system is of order n and given by $G_0 = a^n/(j\omega + a)^n$ (we will work in continuous time for easier interpretation of the results), and that the designed complementary sensitivity function is $T(\gamma) = \gamma^n/(j\omega+\gamma)^n$, i.e. the $-3n$ dB bandwidth is γ. Straightforward calculations give that

$$\frac{\partial \, \text{SENPR}(\gamma,\gamma)}{\partial \gamma} \sim \gamma^{n-1}$$

i.e. the change in SENPR at $\omega = \gamma$, the designed bandwidth, required for a small change in the bandwidth is proportional to the bandwidth to the power of $n-1$. Hence, the higher the bandwidth is, the higher the required change of signal-to-noise ratio is to an increase in the bandwidth γ.

The criterion (16) can be used in several ways.

- *Joint identification and control.* Given data and a model based control design method, one can
  - for a given complexity of the model, find the largest bandwidth such that the designed and achieved performance are guaranteed to be close.
  - for a given bandwidth, find the necessary complexity of the model (and hence controller) such that the designed and achieved performance are guaranteed to be close.
- *Input design.* Given a certain desired bandwidth and an upper bound on the noise level, one can determine the experiment length N and input spectrum such that a given bandwidth can be obtained for a
  - full order controller.
  - low order controller if an upper bound on the approximation is available.

We will take a closer look at experiment design in the next section.

3. $\mathbf{H}_\infty$ INPUT DESIGN

For given input spectrum, a good model is as we have seen in the previous section obtained by minimizing

$$J(\theta,\gamma,\Phi_u) = \sup_{\omega} \frac{c_\alpha}{\sqrt{N}} \left|\frac{T(\theta,\gamma)}{G(\theta)}\right| \sqrt{\kappa \frac{\Phi_v}{\Phi_u}} + \left|\frac{T(\theta,\gamma)}{G(\theta)}\right| |\hat{G} - G(\theta)|$$

over θ. Denote the corresponding θ by θ^*. Notice that θ^* is a function of Φ_u. A good input spectrum should be such that $J(\theta^*,\gamma,\Phi_u)$ is small.

Unfortunately we cannot perform input design based on J since most of the factors in J are unknown. To cope with this let $U(\omega,f)$ denote a frequency by frequency upper bound on f and let us instead consider the following upper bound on J (with $\bar{c} = c_\alpha/\sqrt{N}$)

$$\bar{J}(\theta,\gamma,\Phi_u) = \sup_{\omega} V(\omega,\gamma,\Phi_u)$$

$$V(\omega,\gamma,\Phi_u) = \bar{c}\, U\left(\left|\frac{T(\theta,\gamma)}{G(\theta)}\right|\right) \sqrt{U(\kappa)\frac{U(\Phi_v)}{\Phi_u}} \tag{17}$$

$$+ U\left(\left|\frac{T(\theta,\gamma)}{G(\theta)}\right| |\hat{G} - G(\theta)|\right). \tag{18}$$

If we can manage to make this small, J is guaranteed to be small. Now we pose the input design problem

$$\begin{aligned} &\min_{\Phi_u} \max_{\omega} V(\omega,\gamma,\Phi_u) \\ &\text{s.t.} \int_{-\pi}^{\pi} \Phi_u(\omega) d\omega \le \alpha \end{aligned} \tag{19}$$

Theorem 1. The optimal input spectrum to (19) is given by

$$\Phi_u = \frac{\bar{c}^2\, U(\kappa)\, U(\Phi_v)\, U\left(\left|\frac{T(\theta^*,\gamma)}{G(\theta^*)}\right|\right)^2}{\left(\lambda - U\left(\left|\frac{T(\theta^*,\gamma)}{G(\theta^*)}\right| |\hat{G} - G(\theta^*)|\right)\right)^2} \tag{20}$$

where λ is defined by

$$\int_{-\pi}^{\pi} \frac{\bar{c}^2\, U(\kappa)\, U(\Phi_v)\, U\left(\left|\frac{T(\theta^*,\gamma)}{G(\theta^*)}\right|\right)^2}{\left(\lambda - U\left(\left|\frac{T(\theta^*,\gamma)}{G(\theta^*)}\right| |\hat{G} - G(\theta^*)|\right)\right)^2} d\omega = \alpha \tag{21}$$

and where the minimum cost λ satisfies

$$\lambda \ge \max_{\omega} U\left(\left|\frac{T(\theta^*,\gamma)}{G(\theta^*)}\right| |\hat{G} - G(\theta^*)|\right). \tag{22}$$

Proof: Introduce

$$\gamma(\Phi_u) = \max_{\omega} V(\omega,\gamma,\Phi_u) \tag{23}$$

and let $U_N = \bar{c}^2\, U(\kappa)\, U(\Phi_v)\, U\left(\left|\frac{T(\theta^*,\gamma)}{G(\theta^*)}\right|\right)^2$ and $U_D = U\left(\left|\frac{T(\theta^*,\gamma)}{G(\theta^*)}\right| |\hat{G} - G(\theta^*)|\right)$. Notice that $\gamma(\Phi_u) \ge \|U_D\|_\infty$. The spectrum Φ_u can be extracted from (17) as

$$\Phi_u = \frac{U_N}{(V(\cdot) - U_D)^2}.$$

This is a feasible solution to the input design problem if

$$\int_{-\pi}^{\pi} \frac{U_N}{(V(\cdot) - U_D)^2} d\omega \le \alpha.$$

Using (21) and (23) the left hand side of this expression is bounded by

$$\int_{-\pi}^{\pi} \frac{U_N}{(\gamma(\Phi_u) - U_D)^2} d\omega \leq \int_{-\pi}^{\pi} \frac{U_N}{(V(\cdot) - U_D)^2} d\omega$$
$$\leq \int_{-\pi}^{\pi} \frac{U_N}{(\lambda - U_D)^2} d\omega = \alpha.$$

Since

$$\frac{d}{d\gamma} \int_{-\pi}^{\pi} \frac{U_N}{(\gamma(\Phi_u) - U_D)^2} d\omega \leq 0$$

for $\gamma \geq ||U_D||_\infty$ it follows that $\lambda \leq \gamma$. The optimal input spectrum corresponds to the smallest value of γ which fulfills the constraint. Hence the smallest feasible value is $\gamma = \lambda$, which gives

$$\alpha = \int_{-\pi}^{\pi} \frac{U_N}{(\lambda - U_D)^2} d\omega \leq \int_{-\pi}^{\pi} \frac{U_N}{(V(\cdot) - U_D)^2} d\omega$$
$$\leq \int_{-\pi}^{\pi} \frac{U_N}{(\lambda - U_D)^2} d\omega = \alpha$$

and thus $V(\omega, \gamma, \Phi_u) = \lambda$. Solving for Φ_u gives (20).
□

Remark: Were it not for the fact that $\kappa(\omega)$ depends on the input, the solution would be exact if the upper bounds were replaced by the true quantities. This is a major complicating factor. To circumvent this, one may use a high-order variance approximation, see (Ljung and Yuan, 1985), instead of (12). This amounts to replacing κ with m, the model order. An approximate solution will then be obtained.

4. NUMERICAL EXAMPLES

We will illustrate some of the ideas on a discrete-time model of a flexible transmission system which was proposed in (Landau *et al.*, 1995) as a benchmark problem for digital control design. The transfer function of the system is

$$\begin{aligned} G_0(z^{-1}) &= \frac{z^{-2}B(z^{-1})}{A(z^{-1})} \\ B(z^{-1}) &= 0.28261z^{-1} + 0.50666z^{-2} \\ A(z^{-1}) &= 1 - 1.41833z^{-1} + 1.58939z^{-2} \\ &\quad - 1.31608z^{-3}0.88642z^{-4} \end{aligned} \quad (24)$$

The designed complementary sensitivity function, T, is chosen as the zero-order-hold discretization of

$$T_c(s)) = \frac{\omega_0(\gamma)}{(s + \omega_0(\gamma))^2} \quad (25)$$

where $\omega_0(\gamma) = \gamma/(\sqrt{\sqrt{2}-1})$ and γ is the desired bandwidth. We use IMC for the control design, i.e. if the model is minimum phase the controller becomes $C = G^{-1}T/(1-T)$.

If we introduce $\bar{\Delta} = \min_\theta \frac{c_\alpha}{\sqrt{N}} \left|\frac{T}{G}\right| \sqrt{\kappa \frac{\Phi_v}{\Phi_u}} + \left|\frac{T}{G}\right| |\hat{G} - G(\theta)|$, an upper bound on the difference $T_0 - T$ can be derived from (7) as

$$U_{\tilde{T}}(\omega) = \frac{\bar{\Delta}}{1 - \bar{\Delta}} |1 - Q| \quad (26)$$

γ	0.5	0.75	1.0	1.25	1.5
$J(\theta, \gamma) \geq$	0.0496	0.1046	0.1707	0.2414	0.3115

Table 1.

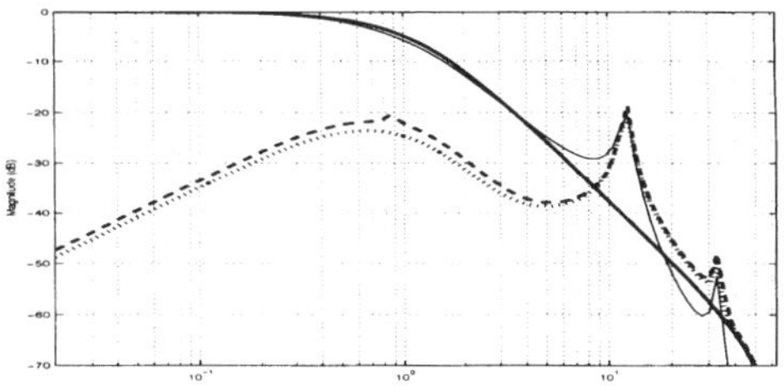

Fig. 1. Magnitude plots. The achieved closed loop T_0 (thin line), designed closed loop T (thick line), $T_0 - T$ (dotted) and $U_{\tilde{T}}$ (dashed).

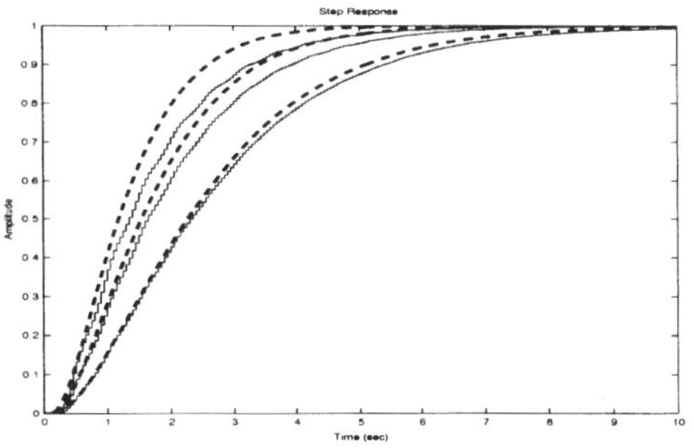

Fig. 2. Step responses. Achieved responses(solid) and desired responses(dashed) for the bandwidths $\gamma = [0.5\ 0.75\ 1.0]$.

The input used for identification is $u(t) = 1/(1 - 0.9q^{-1})e(t)$, where $e(t)$ is white noise ($\Phi_e = 1$). The output is corrupted with white noise with variance 0.2.

4.1 Experiments

- A set of 1000 samples of data and a prescribed order of the model is given. We want to determine the largest bandwidth such that the designed and the achieved performance is guaranteed to be close for a fixed complexity of the model/controller. Here we consider a low order model of the form $G(\theta) = bq^{-3}/(1 - aq^{-1})$. The minimal values of (16), i.e. the upper bound of Δ, as a function of the bandwidth are shown in Table. 1. The obtained result indicates that a bandwidth of 0.75 yields a good performance close to the designed, given the complexity restrictions on the model/controller.

  Figure. 1 shows the magnitude plots of T_0, T and the difference $T_0 - T$ together with the estimated upper bound (26). Figure. 2 shows the achieved and the desired step responses for the bandwidths $\gamma = [0.5\ 0.75\ 1.0]$.
- A set of 1000 samples of data is given and the desired bandwidth is $\gamma = 3$. The task is to determine a suitable order of G, where G belongs to the set $G = bq^{-1}/(1 - \sum_i a_i q^{-i})$.

  The minimal values of (16) as a function of the number of poles if G is given in Table. 2. The obtained result indicates that a model order

order	1	2	3	4
$J(\theta,\gamma) \geq$	0.6203	0.2052	0.1699	0.0563

Table 2.

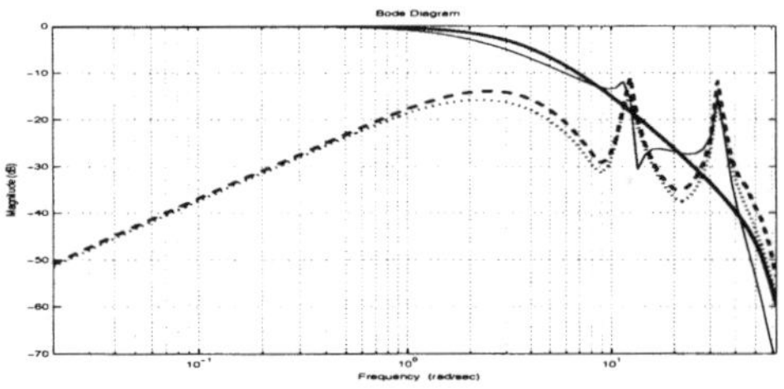

Fig. 3. Magnitude plots. The achieved closed loop T_0 (thin line), designed closed loop T (thick line), $T_0 - T$ (dotted) and $U_{\hat{T}}$ (dashed).

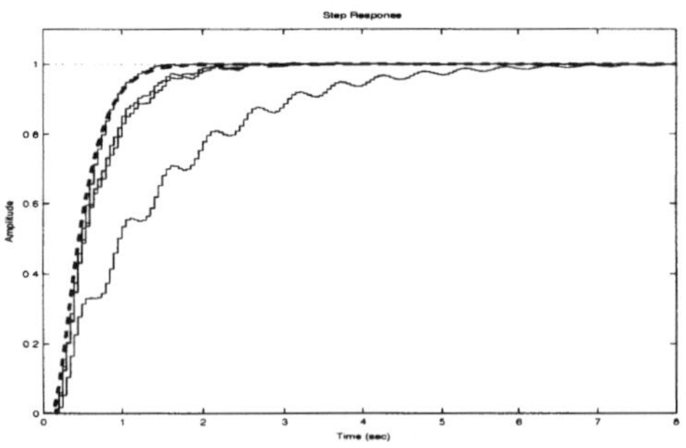

Fig. 4. Step responses. Achieved responses(solid) and desired response(dashed) for the obtained models of order 1-4(from right to left).

of at least 2 is required to obtain reasonable performance.

Figure. 3 shows the magnitude plots of T_0, T and the difference $T_0 - T$ together with $U_{\hat{T}}$ for the obtained second order model. Figure. 4 shows the achieved and the desired step responses for the orders 1-4.

5. EXTENSIONS

The same type of derivations as in Section 2 can be applied to the robust performance problem with unstructured uncertainty, c.f. (Skogestad and Postlethwaite, 1996). The problem is

$$\max_C \gamma \text{ subject to. } \gamma W_p \, |S(G,C)| < 1, \ \forall\omega$$
$$\forall G = \hat{G}(1+\Delta), \ |\Delta| \leq W_s \, \forall\omega \quad (27)$$

Here $\hat{G}$ is a nominal model, W_s upper bounds the relative model uncertainty frequency by frequency and γW_p is the performance specification for the sensitivity function. The problem can be expressed as

$$\max_C \gamma \quad \text{subject to}$$
$$\gamma W_p \, |S(\hat{G},C)| + W_s \, |T(\hat{G},C)| < 1, \ \forall\omega \quad (28)$$

Now by replacing $W_s \, |T(\hat{G},C)|$ with the right-hand side of (15) a criterion is obtained which apart from the terms in (16) includes the additional term $\gamma W_p \, |S(\hat{G},C)|$.

6. CONCLUSIONS

We have shown that using a standard robust stability criterion, extended to robust performance, leads to one criterion which encapsulates how control performance requirements, the approximation ability of the low order model structure, the noise characteristics and the input design interact in a fairly simple way.

REFERENCES

Åström, K.J. (1993). Matching criteria for control and identification. In: *Proc. ECC*. Groningen, The Netherlands. pp. 248–251.

Forssell, U. and L. Ljung (2000). Some results on optimal experiment design. *Automatica* **36**(5), 749–756.

Gevers, M. (1993). Towards a joint design of identification and control?. In: *Essays on Control: Perspectives in the Theory and its Applications* (H. L. Trentelman and J. C. Willems, Eds.). Birkhäuser.

Goodwin, G.C., L. Wang and D. Miller (1999). Bias-variance trade-off issues in robust controller design using statistical confidence bounds. In: *14th IFAC World Congress*. Beijing, P.R. China.

Henriksson, B., O. Markusson and H. Hjalmarsson (2001). Control relevant identification of nonlinear systems using linear models. In: *American Control Conference (ACC 2001)*. Arlington, USA.

Landau, I. D., D. Rey, A. Karimi, A. Voda and A. Franco (1995). "A flexible transmission system as a benchmark for robust digital control". *European Journal of Control* **1**(2), 77–96.

Lee, W.S., B.D.O. Anderson, I.M.Y. Mareels and R.L. Kosut (1995). On some key issues in the windsurfer approach to adaptive robust control. *Automatica* **31**(11), 1619–1636.

Ljung, L. (2000). Model error modeling and control design. In: *Proc IFAC Symposium SYSID 2000.*. Santa Barbara, CA.

Ljung, L. and Z.D. Yuan (1985). Asymptotic properties of black-box identification of transfer functions. *IEEE Trans. Automatic Control* **30**, 514–530.

Skogestad, S. and I. Postlethwaite (1996). *Multivariable Feedback Control, Analysis and Design.* John Wiley and Sons.

Van den Hof, P.M.J. and R.J.P. Schrama (1995). Identification and control – closed loop issues. *Automatica* **31**(12), 1751–1770.

Zang, Z., R.R. Bitmead and M. Gevers (1995). Iterative weighted least-squares identification and weighted LQG control design. *Automatica* **31**, 1577–1594.

Zhu, Y (1998). Multivariable process identification for MPC: the asymptotic method and its applications. *J. Proc. Cont.* **8**(2), 101–115.

IFAC Publications
www.elsevier.com/locate/ifac

STRUCTURE SELECTION WITH ANOVA: LOCAL LINEAR MODELS

Ingela Lind and Lennart Ljung

Division of Automatic Control, Department of Electrical Engineering, Linköpings Universitet, SE-581 83 Linköping, Sweden.
E-mail: {ingela, ljung}@isy.liu.se

Abstract: The structure identification problem when estimating local linear models can be eased by using Analysis of Variance (ANOVA) as a prior step in the estimation procedure. The information gained from using ANOVA on the input/output data is what regressors that should be used to partition the input space and what regressors are needed only for the linear models in each part. Also the complexity of the partitioning can be restricted due to the extra information. *Copyright © 2003 IFAC*

Keywords: Identification, Nonlinear models, Analysis of variance

1. PROBLEM DESCRIPTION

The problem that will be discussed in this paper is the one of easing the process of building good, parsimonious models of nonlinear systems. The tool used for obtaining this goal is the statistical method analysis of variance (ANOVA). (Miller, 1997) is a comprehensive reference.

In earlier work, it has been shown that ANOVA can be used to extract structure information from input/output data, see (Lind, 2000; Lind, 2001; Lind, 2002). This can be done without estimating any complex model, which is a great benefit due to the complexity of the estimation task. In this contribution it will be shown how the structure information from ANOVA can be applied to the local linear model structure.

2. THE KEY TO STRUCTURE INFORMATION

The key to extracting structure information from the data lies in utilising a point model with a special parameterisation, called a linear statistical model (here for a two-dimensional function):

$$y_{ijk} = \mu + \tau_i + \beta_j + (\tau\beta)_{ij} + \varepsilon_{ijk}, \qquad (1)$$

The overall mean effect is denoted by μ. τ_i is the mean effect of the ith level of the factor A, where $i = 1,\dots,a$, while β_j is the mean effect of the jth level of the factor B, where $j = 1,\dots,b$. The means are taken over all the levels of the other factor. The interactions between the factors are denoted by $(\tau\beta)_{ij}$, which is indexed by the levels of both factors. The random error component ε_{ijk} is assumed to come from a Gaussian distribution with constant variance σ^2 and the index k denotes the sample number from the ij:th level combination. The number of free parameters in the model is restricted by the following equations, that define all the effects as deviations from lower order interactions (with the overall mean as interaction of order zero): $\sum_{i=1}^{a} \tau_i = 0$, $\sum_{j=1}^{b} \beta_j = 0$, $\sum_{i=1}^{a} (\tau\beta)_{ij} = 0, \forall\, j$ and $\sum_{j=1}^{b} (\tau\beta)_{ij} = 0, \forall\, i$.

It is possible to divide the residual quadratic sum from this point model into orthogonal parts related to the τ_i:s, β_j:s, $(\tau\beta)_{ij}$:s and the random error component respectively. These then form F-distributed test variables that can be used to determine whether factor A and/or B affect the output at all, additively or with interaction.

For a higher-dimensional point model, the used parameterisation has many more than a sufficient number of parameters. The reason to use this parameterisation is that the high-order interactions are very seldom present in the data, so many of the interaction effect parameters can be discarded with the F-tests, thereby simplifying the model while giving the wanted structural information.

The linear statistical model structure also benefit the transparency of the model. People tend to think about functional relationships in two- and three-dimensional views and have a hard time understanding complex interactions between many variables. If the model should be useful for physical interpretations the order of the interactions should be kept low. This type of model thinking has been incorporated by (Harris *et al.*, 2002) in their SUPANOVA algorithm which is a support vector machine built on the ideas to separate low-order interaction effects from high-order interaction effects by a special function expansion.

3. LOCAL MODELS

A local model, as described in (Nelles, 2001; Töpfer *et al.*, 2002), is a nonlinear model of the following form:

$$\hat{y} = \sum_{i=1}^{M} \Phi_i(\varphi) g_i(\varphi), \tag{2}$$

where a set of local models, the $g_i(\varphi)$:s, are weighted by the weighting, activation or membership functions $\Phi_i(\varphi)$. The regressor vector φ can consist of, e. g., different input time lags, old output lags and/or transformations of these. This model is an alternative function expansion to the one used in neural networks. The idea is that the input space is divided into parts in which a simple model g_i, most often a linear one, describes the output satisfactory. The Φ_i:s take care of to which part of the input space the input belongs and the possible smoothing of the transitions between the parts. The partitioning of the input space can be done in the following ways; a grid structure, recursive partitioning or a partitioning of arbitrary form. The choice of partitioning is called structure identification. The model can also be seen as a linear model with operating-point dependent parameters;

$$\hat{y} = \sum_{i=1}^{M} (w_{i0} + w_{i1}\varphi_1 + \dots w_{ip}\varphi_p)\Phi_i(\varphi) =$$
$$= w_0(\varphi) + w_1(\varphi)\varphi_1 + \dots + w_p(\varphi)\varphi_p,$$

with

$$w_q = \sum_{i=1}^{M} w_{qi}\Phi_i(\varphi).$$

It is not necessarily the case that all regressors in φ are included in both $g_i(\varphi)$ and $\Phi_i(\varphi)$. In the following, the part of φ that occurs in g_i will be called $\mathbf{x}$ and the part that occurs in Φ_i will be called $\mathbf{z}$. $\mathbf{x}$ and $\mathbf{z}$ can be partly constituted by the same regressors. As already mentioned the $g_i(\mathbf{x})$ are often linear(or affine) functions. Also the notion of cells will be used. Each part of the input space defined in an axis-orthogonal grid is called a cell, see Figure 1.

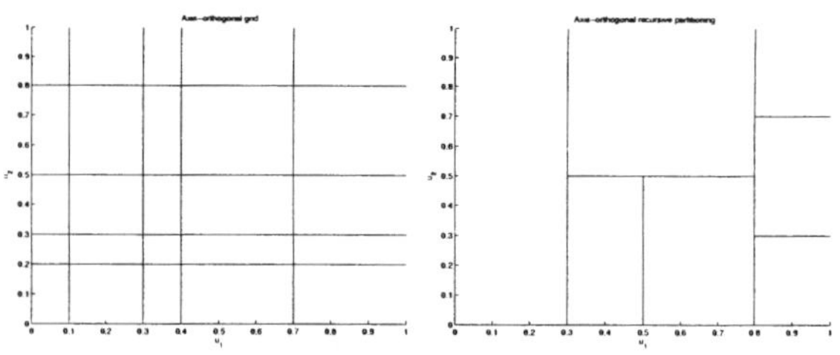

Fig. 1. These are examples of an axis-orthogonal grid to the left and an axis-orthogonal recursive partitioning to the right.

Sometimes the $\Phi_i(\mathbf{z})$ are decomposed into functions of the form $f_{i1}(z_1)f_{i2}(z_2)\cdot \ldots \cdot f_{ik}(z_k)$, where each $f_{ij}(z_j)$ is a bell shaped function. The spread of the bell depends on whether a smooth transition between different parts of the input space is wanted or not. This decomposition gives an axis-orthogonal partitioning of the input space, see Figure 1.

Here the questions are how many and which $f_{ij}(z_j)$ are needed for successful modelling and which parts of φ that should be included in $\mathbf{x}$ and $\mathbf{z}$ respectively. This is exactly the kind of information ANOVA can contribute with. Also process knowledge will be helpful.

4. REDUCING STRUCTURAL COMPLEXITY

In this section, data from the system

$$y = \mathrm{sgn}(u_1)\cdot u_2 + u_3 + e$$

will be used as example. If u_1 is the derivative of the position, this system is a very simple example in which different dynamics apply when braking or accelerating. Here, e comes from a zero mean normal distribution with variance 1. The data are collected such that u_1, u_2 and u_3 each assume the levels -1, 1, 3 and 5 and all level combinations of these factors are present four times. This gives 256 input/output data. If the factors are different lags of the same signal, the level combinations can be obtained by using a pseudo-random multi-level signal (Godfrey, 1993).

The reduction of the structural complexity can be seen as a three-step procedure:

(1) Apply ANOVA to your data (Lind, 2001). The result is an ANOVA table, see Table 1. From this table, the significant regressors and their interaction pattern can be concluded. In this case, the three-factor interaction is not significant, i. e., all the factors contribute non-additively to the output, and the two-factor interactions between u_1 and u_3 and between u_2 and u_3 are not significant. The main effect from u_3 is significant, but it is not interesting to consider the main effects from u_1 or u_2 since the interaction between these factors is significant.
(2) The significant effects can be investigated further by looking at plots of the cell means, i. e., the

Effect	Degrees of Freedom	Mean Square	F	p-level
u_1	3	248	224	0.0000
u_2	3	111	100	0.0000
u_3	3	427	386	0.0000
$u_1 * u_2$	9	111	100	0.0000
$u_1 * u_3$	9	0.9	0.8	0.64
$u_2 * u_3$	9	1.4	1.2	0.28
$u_1 * u_2 * u_3$	27	1.0	1.0	0.53
Error	192	1.1		

Table 1. Analysis of Variance Table. The columns are from the left; the degrees of freedom associated with each sum of squares, the sum of squares divided by its degrees of freedom, the value of the F-distributed test variable associated with the corresponding interaction effect and, finally, the probability that the data could originate from a system where the interaction is not present. A p-level less than, e. g., 0.01 is interpreted as a significant effect.

mean over all the data in each cell, together with their confidence intervals. If a straight line can be drawn through the confidence intervals, in a way that will be explained further, the effect of the factor on the x-axis can be considered to be linear, given the other factor(s) involved in the plot.

(3) Use the information when deciding what possible partitionings to use for estimating the local linear model. See section 4.2.

4.1 *Investigating interaction effects*

To see if an interaction effect is linear in one of the factors given the others, e. g., $y = \text{sgn}(u_1) \cdot u_2$ which is linear in u_2 given u_1, it is possible to check the plots of the cell means, see Figure 2. These plots are computationally "free" and they also give an indication of how reliable the linear assumption is. No extra computations are needed for the plots of cell means, since all the cell means are computed in order to make the ANOVA table. A good estimate of the error variance is also obtained, so confidence intervals for the cell means are easily computed. The linearity tests done by looking at these plots can be done formally by defining linear contrasts, which is a linear combination of the cell means nominally adding to zero. A confidence interval for the linear contrast will tell if the difference from zero is significant (no linearity) or not.

It is enough to look at the cell mean plots for the highest order interaction of significance (for each factor combination). Assume that the cell means (in a two-dimensional grid) are linear functions of factor B for each fixed level i of factor A, that is,

$$\mu_{ij} = \mu + \tau_i + \beta_j + (\tau\beta)_{ij} = v_i + \rho_i \delta_j,$$

where μ_{ij} is the cell mean in cell (i, j), v_i is a constant offset, ρ_i is the slope and δ_j is the distance between level 1 and level j of factor B. The row means μ_j are computed as

$$\mu_j = \frac{1}{a}\sum_{i=1}^{a} \mu_{ij} = \mu + \beta_j =$$
$$= \frac{1}{a}\sum_{i=1}^{a} v_i + (\frac{1}{a}\sum_{i=1}^{a} \rho_i)\delta_j = v + \rho\delta_j,$$

where the second equality is due to the parameter restrictions and $v = \frac{1}{a}\sum_{i=1}^{a} v_i$ and $\rho = \frac{1}{a}\sum_{i=1}^{a} \rho_i$. Hence, it is obvious that if the two-dimensional cell means show a linear relation for factor B, also the one-dimensional cell means (the row means) will show a linear relation.

If high-order interaction effects are present, as shown by the ANOVA table, there will be more plots to look at. The axes of the high-dimensional cell means plot have to be permuted such that each regressor in the significant interaction effect gets to be at the x-axis once. Then the regressors that only need to be included in **x**, the regressor vector for the local linear models, will be detected. The number of plots for each permutation will vary depending on how many levels each factor has and the order of the significant interaction.

4.2 *Corresponding local linear model structure*

Listed below are the possible outcomes of steps 1 and 2 and the corresponding choices of **x**, **z** and $\Phi_i(\mathbf{z})$, for an analysis with three factors. How many $\Phi_i(\mathbf{z})$:s that are needed cannot be concluded from these tests. The idea is quite simple, but there are many slightly different cases, hence the length of the list.

1a; Three-factor interaction between u_1, u_2 and u_3. No linearities detected in the cell means plots. Let $\mathbf{x} = [u_1, u_2, u_3]$ and $\mathbf{z} = [u_1, u_2, u_3]$. Possible $\Phi_i(\mathbf{z})$:s are $f_{i1}(u_1) \cdot f_{i2}(u_2) \cdot f_{i3}(u_3)$.

1b; Three-factor interaction between u_1, u_2 and u_3. Linear in u_1 given the levels of u_2 and u_3. Let $\mathbf{x} = [u_1, u_2, u_3]$ and $\mathbf{z} = [u_2, u_3]$. Possible $\Phi_i(\mathbf{z})$:s are $f_{i2}(u_2) \cdot f_{i3}(u_3)$.

1c; Analogue to case 1b, but with u_2 linear given the others.

1d; Analogue to case 1b, but with u_3 linear given the others.

1e; Three-factor interaction between u_1, u_2 and u_3. Linear in u_1 given the levels of u_2 and u_3, and in u_2 given the levels of u_1 and u_3. Choose between using the setup in case 1b (u_2 as regime variable) or the setup in case 1c (u_1 as regime variable).

1f; Analogue to case 1e, but with u_1 or u_3 linear given the others.

1g; Analogue to case 1e, but with u_2 or u_3 linear given the others.

1h; Three-factor interaction between u_1, u_2 and u_3. Linear in u_1 given the levels of u_2 and u_3, in u_2 given the levels of u_1 and u_3, and in u_3

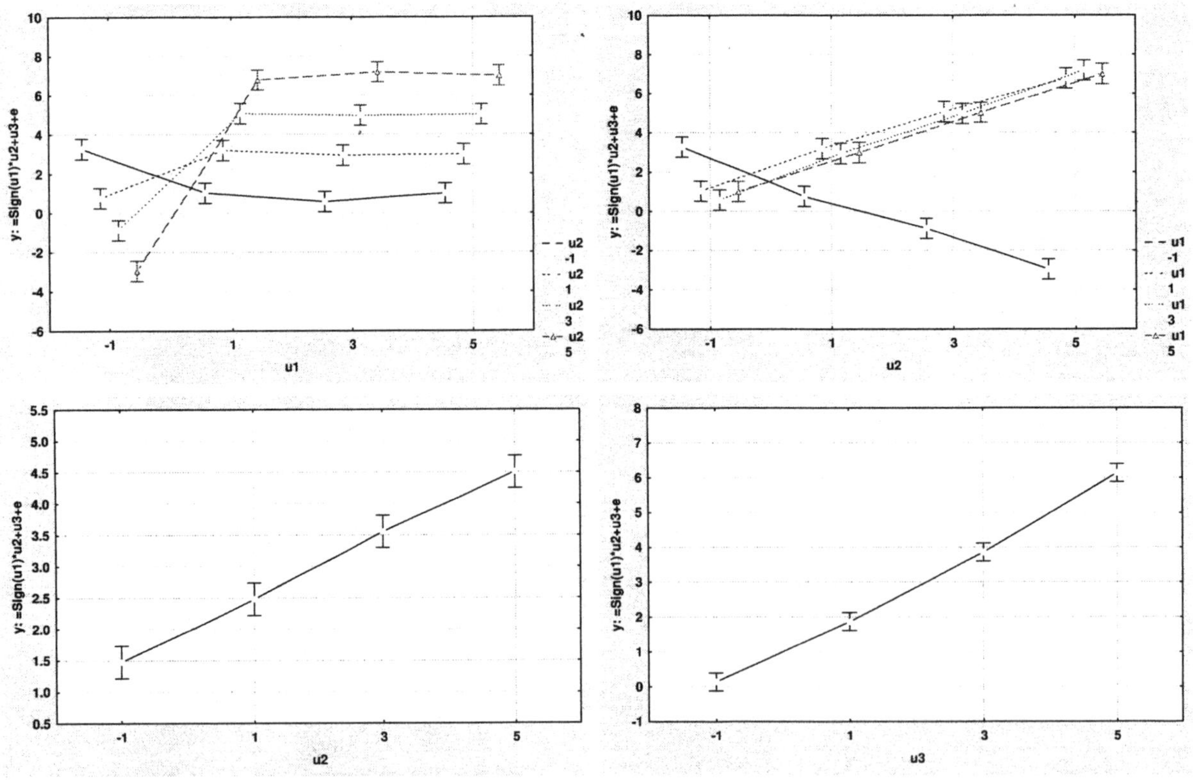

Fig. 2. Cell mean plots with confidence intervals. The upper two plots show the two-dimensional cell means, corresponding to the interaction between u_1 and u_2. The plot to the left have u_1 on the x-axis and one line for each value of u_2, while the plot to the right have u_2 on the x-axis. From the plots, the following conclusions can be drawn: u_1 and u_2 affect y with interaction, since the curves within each plot do not have the same shape. u_2 affects y linearly if the value of u_1 is fixed, since it is possible to draw a straight line through the confidence intervals for μ_{ij} for each value of i. (The lines in the plot only connects the cell means, they are not straight lines). The lower plots show the one-dimensional cell means with u_2 and u_3 on the x-axis respectively. The left plot confirms that the important information is present in the two-dimensional cell means plots. The right plot shows that the effect from u_3 could be linear.

given the levels of u_1 and u_2. Choose between using the setup in case 1b (u_2 and u_3 as regime variables), the setup in case 1c (u_1 and u_2 as regime variables), or the setup in case 1d (u_1 and u_3 as regime variables).

2; Two-factor interactions between u_1 and u_2, between u_1 and u_3 and between u_2 and u_3 but no three-factor interaction. The model can be decomposed into three additive sub-models. See further cases 5.1, 5.2 and 5.3. When there are possible choices of which variables should act as regime variables it could be wise to consider the possible choices in the other sub-models too.

3.1; Two-factor interactions between u_1 and u_2, and between u_1 and u_3. The model can be decomposed into two additive sub-models. See further cases 5.1 and 5.3.

3.2; Two-factor interactions between u_1 and u_3, and between u_2 and u_3. The model can be decomposed into two additive sub-models. See further cases 5.1 and 5.2.

3.3; Two-factor interactions between u_1 and u_2, and between u_2 and u_3. The model can be decomposed into two additive sub-models. See further cases 5.2 and 5.3.

4.1; Two-factor interaction between u_1 and u_2 and main effect from u_3. The model can be decomposed into two additive sub-models. See further cases 5.1 and 7.3.

4.2; Two-factor interaction between u_1 and u_3 and main effect from u_2. The model can be decomposed into two additive sub-models. See further cases 5.2 and 7.2.

4.3; Two-factor interaction between u_2 and u_3 and main effect from u_1. The model can be decomposed into two additive sub-models. See further cases 5.3 and 7.1.

5.1a; Two-factor interaction between u_1 and u_2. No linearities. Let $\mathbf{x} = [u_1, u_2]$ and $\mathbf{z} = [u_1, u_2]$. Possible $\Phi_i(\mathbf{z})$:s are $f_{i1}(u_1) \cdot f_{i2}(u_2)$.

5.1b; Two-factor interaction between u_1 and u_2. Linear in u_1 given the levels of u_2. Let $\mathbf{x} = [u_1, u_2]$ and $\mathbf{z} = [u_2]$. Possible $\Phi_i(\mathbf{z})$:s are $f_{i2}(u_2)$.

5.1c; Two-factor interaction between u_1 and u_2. Linear in u_2 given the levels of u_1. Let $\mathbf{x} = [u_1, u_2]$ and $\mathbf{z} = [u_1]$. Possible $\Phi_i(\mathbf{z})$:s are $f_{i1}(u_1)$.

5.1d; Two-factor interaction between u_1 and u_2. Linear in u_1 given the levels of u_2 and linear in u_2 given the levels of u_1. Choose between the setup in case 5.1b (u_2 as regime variable), and the setup in case 5.1c (u_1 as regime variable).

5.2; Two-factor interaction between u_1 and u_3. Analogue to case 5.1.

5.3; Two-factor interaction between u_2 and u_3. Analogue to case 5.1.

6.1; Main effects from u_1 and u_2. No interactions. The model can be decomposed into two additive sub-models. See further cases 7.1 and 7.2.

6.2; Main effects from u_1 and u_3. The model can be decomposed into two additive sub-models. See further cases 7.1 and 7.3.

6.3; Main effects from u_2 and u_3. The model can be decomposed into two additive sub-models. See further cases 7.1 and 7.2.

7.1a; Main effect from u_1. u_1 is linear. Let $\mathbf{x} = [u_1]$ and $\mathbf{z}$ empty. No partitioning is needed.

7.1b; Main effect from u_1. u_1 is nonlinear. Let $\mathbf{x} = [u_1]$ and $\mathbf{z} = [u_1]$. Possible $\Phi_i(\mathbf{z})$:s are $f_{i1}(u_1)$.

7.2; Main effect from u_2. Analogue to case 7.1.

7.3; Main effect from u_3. Analogue to case 7.1.

8; No significant effects. None of the tested factors show systematic effects in the data.

From the example data set, we have gained the information that all the factors u_1, u_2 and u_3 are present in the model. The model structure according to the ANOVA table is

$$y = h_1(u_1, u_2) + h_2(u_3) + e,$$

which is case 4.1 above. The interaction between u_1 and u_2 was then investigated further (Figure 2), giving the result that $h_1(u_1, u_2) = f(u_1) \cdot u_2$, that is, u_1 belongs to the regime variables $\mathbf{z}$ while u_2 belongs to $\mathbf{x}$. If $h_2(u_3)$ is a nonlinear function, u_3 need to be in both $\mathbf{z}$ and $\mathbf{x}$. Since u_1 and u_3 do not interact the weighting functions Φ_i will be either $f_{i1}(u_1)$ or $f_{i3}(u_3)$, but not $f_{i1}(u_1) \cdot f_{i3}(u_3)$. The linearity can be tested by plotting the cell means against the levels of factor C (u_3), see Figure 2, which shows that $h_2(u_3)$ could be linear and u_3 does not need to be included in $\mathbf{z}$. The total model consist of two additive sub-models, one corresponding to case 5.1c and the other to case 7.3a.

4.3 *Another example*

The second example is a bit more complicated than the first one;

$$y = \mathrm{sgn}(u_1) \cdot u_2 \cdot u_3 + u_3^2 + e.$$

The inputs and the noise are chosen exactly as in the first example and the amount of data is the same. The ANOVA table is given in Table 2 and shows that we have case 1 in the list in Section 4.2. The linearity is investigated in the cell mean plots, see Figures 3, 4 and 5. These show that we have case 1c. That means that we should choose $\mathbf{x}$ as $[u_1, u_2, u_3]$ and $\mathbf{z}$ as $[u_1, u_3]$. Possible $\Phi_i(\mathbf{z})$:s are $f_{i1}(u_1) \cdot f_{i3}(u_3)$, but exactly where the partitionings should go is an identification issue.

Effect	Degrees of Freedom	Mean Square	F	p-level
u_1	3	1008	910	0.0000
u_2	3	435	392	0.0000
u_3	3	12122	10941	0.0000
$u_1 * u_2$	9	435	392	0.0000
$u_1 * u_3$	9	423	382	0.0000
$u_2 * u_3$	9	183	165	0.0000
$u_1 * u_2 * u_3$	127	170	154	0.0000
Error	192	1.1		

Table 2. Analysis of Variance Table for the second example.

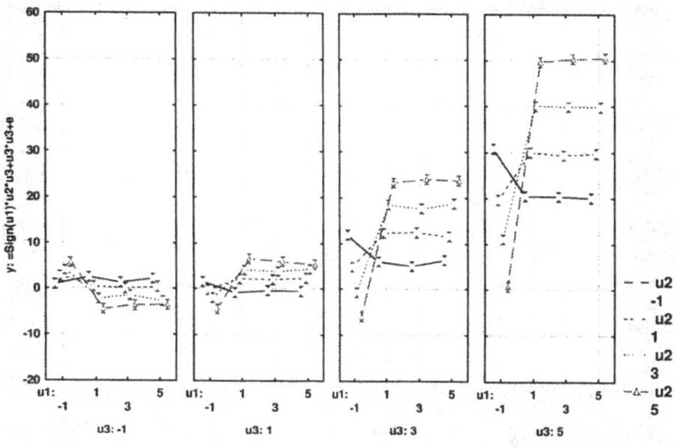

Fig. 3. Cell mean plots with confidence intervals. u_1 on the x-axis.

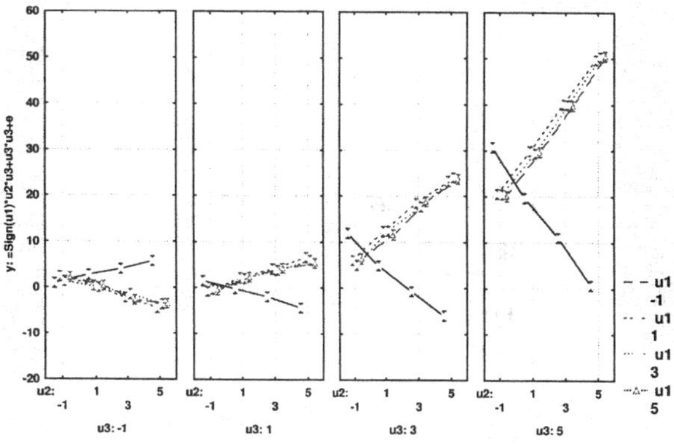

Fig. 4. Cell mean plots with confidence intervals. u_2 on the x-axis. The effects seem linear.

5. CONCLUSIONS

It is possible to gain enough structure information from the input/output data to assist the structure identification task in local linear modelling by using ANOVA. The information that can be extracted from the ANOVA table and the cell means plots are what regressors that should be used for the linear models in each part of the input space, what regressors that should determine the partitioning of the input space and how complex the partitioning should be (that is, how many regressors interact in each weighting func-

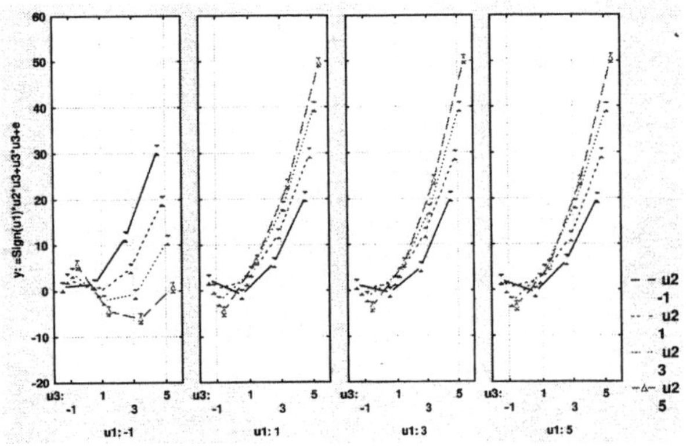

Fig. 5. Cell mean plots with confidence intervals. u_3 on the x-axis.

tion). The cell means plots with confidence intervals appeal also to intuition. They are very similar to scatter plots, but with the benefit that the statistical properties of the input/output data are made visible. Left to consider in the structure identification task is how many partitionings to use and where the limits of each part should be.

6. ACKNOWLEDGEMENTS

This work has been supported by the Swedish Research Council (VR), which is gratefully acknowledged.

7. REFERENCES

Godfrey, Keith (1993). *Perturbation Signals for System Identification*. Prentice Hall. New York.

Harris, C., X. Hong and Q. Gan (2002). *Adaptive Modelling, Estimation and Fusion from Data: a neurofuzzy approach*. Springer-Verlag. Berlin Heidelberg.

Lind, Ingela (2000). Model order selection of N-FIR models by the analysis of variance method. In: *Proc IFAC Symposium SYSID 2000*. Santa Barbara. pp. 367–372.

Lind, Ingela (2001). Regressor selection in system identification using ANOVA. Technical Report Licentiate Thesis no. 921. Department of Electrical Engineering, Linköping University. SE-581 83 Linköping, Sweden.

Lind, Ingela (2002). Regressor selection with the analysis of variance method. In: *Proc 15th IFAC World Congress, 21-26 July, 2002*. Barcelona, Spain. pp. T–Th–E 01 2.

Miller, Jr., Rupert G. (1997). *Beyond ANOVA*. Chapman and Hall. London.

Nelles, O. (2001). *Nonlinear System Identification*. Springer-Verlag. Berlin Heidelberg.

Töpfer, S., A. Wolfram and R. Isermann (2002). Semiphysical modelling of nonlinear processes by means of local approaches. In: *Proc 15th IFAC World Congress, 21-26 July, 2002*. Barcelona, Spain. pp. T–Th–M 01 5.

IFAC
Publications
www.elsevier.com/locate/ifac

ON IDENTIFICATION OF HAMMERSTEIN SYSTEMS USING EXCITATION WITH A FINITE NUMBER OF LEVELS

Tomas McKelvey and **Christian Hanner**

* *Dept. of Signals and Systems, Chalmers University of Technology, SE-412 96 Gothenburg, Sweden, mckelvey@s2.chalmers.se*

Abstract: Identification of nonlinear systems of Hammerstein type is considered in a scenario where the input signal is constrained to attain only a finite number of amplitude levels. This leads to the possibility to obtain asymptotically unbiased point-wise estimates of the nonlinear function without the need for a priori assumptions about the shape of the function. A three step identification procedure is presented which starts with a sub-space based method to estimate the poles of the linear system. Two alternative techniques are then outlined for the estimation of the zeros of the linear system and the non-linearity. The simpler procedure can be used if the initial condition of the system is known while an iterative procedure is employed when the initial condition is unknown. In a third step the preliminary estimate given by the first two steps are refined by the prediction error method. *Copyright © 2003 IFAC*

Keywords: system identification, non-linear modeling, Hammerstein system, state-space model, sub-space methods, prediction error method, bilinear parametrization

1. INTRODUCTION

The class of Hammerstein systems is one of the simplest nonlinear extensions to linear dynamical systems and is composed of two parts; a static nonlinear function in series with a linear dynamical system. In this paper we particularly discuss the case when the linear system is sampled and has a finite state dimension. We adopt a discrete time formulation to describe the dynamics. This (sub-)class of Hammerstein system can conveniently be described in a state-space form as

$$\begin{aligned} x(t+1) &= Ax(t) + bf(u(t)) + Ke(t), \quad x(0) = x_0 \\ y(t) &= cx(t) + df(u(t)) + e(t) \end{aligned} \tag{1}$$

where $u(t) \in \mathbb{R}$ is the input, $y(t) \in \mathbb{R}$ the output, $x(t) \in \mathbb{R}^n$ the state vector, $f(\cdot)$ the scalar real valued static nonlinear function, $e(t)$ is a white noise signal, x_0 is the initial condition and A, b, c, d, K, x_0 are real valued matrices and vectors of compatible dimensions.

Several methods for estimating the linear and nonlinear parts of a Hammerstein system exists. One way of characterizing the existing approaches is by how the nonlinearity is estimated. The most common approach is to use a parametric way of representing a limited set of nonlinearities, e.g. (Narendra and Gallman, 1966; Billings, 1980; Rangan *et al.*, 1995; Verhaegen and Westwick, 1996; Bai, 2002). The main advantage is that the estimation problem becomes parametric and a wide range of techniques can be employed to estimate the parameters. The disadvantage is that the nonlinear model structure has to be selected properly otherwise the best parametric approximation might be far away from the true nonlinear function. The other branch of techniques takes a nonparametric approach to the modeling of the nonlinearity. In these approaches the input to the linear part of the system is first estimated and then subsequently used to estimate the non-linearity. Some approaches are based on kernel estimation techniques, e.g. (Greblicki, 2002) and references therein. Other approaches uses fast sampling to blindly estimate the linear part e.g. (Bai and Fu, 2002). An obvious drawback with any nonparametric technique is that the nonlinearity never can be recovered without error even in the limit.

This paper suggests a mix between the unprejudiced non-parametric technique and the parametric one by aiming to estimate the non-linear function only at a discrete set of input levels. Such an approach has two main good properties:

- Asymptotically (in data) the estimates of the non-linearity at a finite set of points will be error-free, i.e. under mild assumptions the convergence will be with probability one (w.p. 1).
- No model class needs to be chosen *a priori* to the identification. The non-linearity of the Hammerstein system can be any function as long as the input signal to the linear system part is sufficiently exciting.

An obvious drawback with the suggested technique is that the non-linearity will be completely unknown outside the finite set of discrete points used in the excitation. This leads to the question of the intended use of the identified model. If the model is going to be used in a subsequent control design, two alternatives arises. If an *a posteriori* assumption can be enforced on the smoothness of the nonlinear function $f(\cdot)$, an interpolation approach can be used to obtain an estimate of the function between the discrete points. This function estimate can (if invertible) then be used to invert the non-linearity and subsequently apply linear robust control design tools. A second possible approach is to constrain the control problem to employ only the same input levels as used in the identification. As will be demonstrated later, the non-linear Hammerstein control problem can then be recast as multi-input linear control problem with constraints on the inputs. An advantage with this formulation is that optimal controllers can be designed using model predictive control techniques (Maciejowski, 2002; Goodwin, 2002; Quevedo *et al.*, 2002).

1.1 *Reformulation as a linear system*

As mentioned above the Hammerstein system can be recast as a MISO linear system if only a finite number of excitation levels are used.

Assume the input $u(t)$ of the system are drawn from the finite set

$$\mathscr{U}_m \triangleq \{u_k\}_{k=1}^{m} \tag{2}$$

where m denotes the cardinality of the set. Under the finite level input assumption the Hammerstein system (1) is observationally identical to the following multi-input linear system

$$\begin{aligned} x(t+1) &= Ax(t) + Bz(t) + Ke(t), \quad x(0) = x_0 \\ y(t) &= cx(t) + Dz(t) + e(t) \end{aligned} \tag{3}$$

where

$$B = b\alpha^T \in \mathbb{R}^{n\times m}, \quad D = d\alpha^T \in \mathbb{R}^{1\times m} \tag{4}$$

$$\alpha = \left[f(u_1) \; f(u_2) \; \dots \; f(u_m) \right]^T \in \mathbb{R}^{m\times 1}, \tag{5}$$

and finally

$$z(t) = e_i \quad \text{when} \quad u(t) = u_i \tag{6}$$

where e_i is the $i-$th unit vector, i.e. column i of the identity matrix. Note that the considered state-space formulation is not necessary for applying the proposed technique. A transfer function approach could just as well have been used.

Based on the presentation above, the Hammerstein identification problem has been reformulated to the following a linear identification problem:

> Given the data $Z^N = \{y(t), z(t)\}_{t=1}^{N}$ the objective is to estimate the parameters of the linear state-space model given by (3) taking into account that the D and B matrices has the particular structure given by (4).

2. SOME NOTES ON IDENTIFIABILITY

Before proposing an estimation algorithm for the stated problem we will briefly address some issues regarding the identifiability of the linear model (3). It is well known (Ninness and Gibson, 2002) that the linear part represented by A, b, c, d and the non-linear part represented by α are identifiable up to a scalar gain. Hence to make the Hammerstein model identifiable from an input/output point of view a restriction on the linear system model set is introduced by the constraint $\| \left[\begin{smallmatrix} b \\ d \end{smallmatrix}\right] \| = 1$. Another possibility, which can be more tractable from an implementation point of view, is to set one of elements in b to a fixed value.

A necessary condition for identifiability is that the signal $f(u(t))$ which drives the linear part of the Hammerstein system is a signal which is *persistently exciting* of an order larger than or equal to the McMillan degree of the linear model (Ljung, 1999).

3. ESTIMATION PROCEDURE

The model parameters will be estimated using a three step procedure presented below.

Sub-space based estimate In the first step the available data in Z^N is used in a sub-space based procedure to get an initial estimate of the A and c parameters of the state-space model. We use the first steps in the MOESP procedure (Verhaegen, 1993).

Estimate of b,d and α We now consider A and c to be known and focus on estimating the remaining parameters b, d, α and possibly x_0. The easiest route is to use a linear regression approach to estimate B, D and x_0 as they all appear linearly in $y(t)$ if A,c and $z(t)$ are known. However, it can be shown (Hanner, 2003) that a simultaneous linear regression estimate of B, D and x_0 is not possible due to the properties of the used input signal $z(t)$. If on the other hand x_0 can be assumed known, B and D can be obtained via linear regression. Since the matrix

$$\begin{bmatrix} B \\ D \end{bmatrix} = \begin{bmatrix} b \\ d \end{bmatrix} \alpha^T \tag{7}$$

has rank one the singular value decomposition (Golub and Van Loan, 1989) can be employed to obtain an estimate of b, d and α based on the estimate from the linear regression. If σ_1 is the largest singular value of $\left[\begin{smallmatrix} B \\ D \end{smallmatrix}\right]$ and u_1 and v_1 are the left and right singular vectors respectively, then u_1 is the estimate of $\left[\begin{smallmatrix} b \\ d \end{smallmatrix}\right]$ and $\sigma_1 v_1$ is the estimate of α.

If x_0 is unknown it must be estimated together with the other parameters. Without taking into account the structure of B and D a joint estimation of x_0, B and D is not possible. In order to make the identification unique the internal structure of the B and D matrices must be taken into account. Thus instead we view x_0, b and d and α as the unknown parameters. To simplify the notation in what follows we denote by $\bar{b}$ the matrix $\left[\begin{smallmatrix} b \\ d \end{smallmatrix}\right]$. Note that x_0 appears linearly in the output $y(t)$ while α together with $\bar{b}$ appear bilinearly in the output. Hence, there exists a matrix Ψ which only depends on A,c and $z(t)$ such that the following relation holds:

$$\Psi\theta = \mathbf{y} \tag{8}$$

Here the parameter vector θ contains the initial state vector x_0 together with α and $\bar{b}$ where the latter two occur in a bilinear pattern.

$$\theta = \begin{bmatrix} x_0 \\ \alpha_1\bar{b} \\ \alpha_2\bar{b} \\ \vdots \\ \alpha_m\bar{b} \end{bmatrix}$$

Let θ_B denote the undermost part of θ so that

$$\theta = \begin{bmatrix} x_0 \\ \theta_B \end{bmatrix}.$$

With use of the Kronecker product we observe that θ_B can be rewritten in two alternative ways:

$$\theta_B = \begin{bmatrix} \alpha_1\bar{b} \\ \vdots \\ \alpha_m\bar{b} \end{bmatrix} = (\alpha \otimes I_n)\bar{b} = (I_m \otimes \bar{b})\alpha. \tag{9}$$

Suppose now that α is known. Then combining (8) and (9) results in

$$\begin{bmatrix} \Psi_{x_0} & \Psi_B(\alpha \otimes I_n) \end{bmatrix} \begin{bmatrix} x_0 \\ \bar{b} \end{bmatrix} = \mathbf{y},$$

and $\begin{bmatrix} x_0 \\ \bar{b} \end{bmatrix}$ can be solved for analytically as

$$\begin{bmatrix} x_0 \\ \bar{b} \end{bmatrix} = \begin{bmatrix} \Psi_{x_0} & \Psi_B(\alpha \otimes I_n) \end{bmatrix}^{\dagger} \mathbf{y}, \tag{10}$$

where † denotes the *pseudo-inverse*. If instead $\bar{b}$ is known it follows analogously that

$$\begin{bmatrix} x_0 \\ \alpha \end{bmatrix} = \begin{bmatrix} \Psi_{x_0} & \Psi_B(I_m \otimes \bar{b}) \end{bmatrix}^{\dagger} \mathbf{y}. \tag{11}$$

The expressions (10) and (11) can be solved alternately in an iterative process. It can be proved (Ljung, 1999) that this process will converge to a solution that is a local minimizer of the least-squares quality function

$$V(x_0, \bar{b}, \alpha) = ||\Psi_{x_0}x_0 + \Psi_B(\alpha \otimes I_n)\bar{b} - \mathbf{y}||^2.$$

Structured PEM In the third and final step we employ the prediction error method (PEM) (Ljung, 1999) using the special parametrization of B and D given in equation (4). The optimization is started from the initial estimate given by the previous two steps. The final estimate delivered by the prediction error method will, under some general assumptions, be asymptotically efficient.

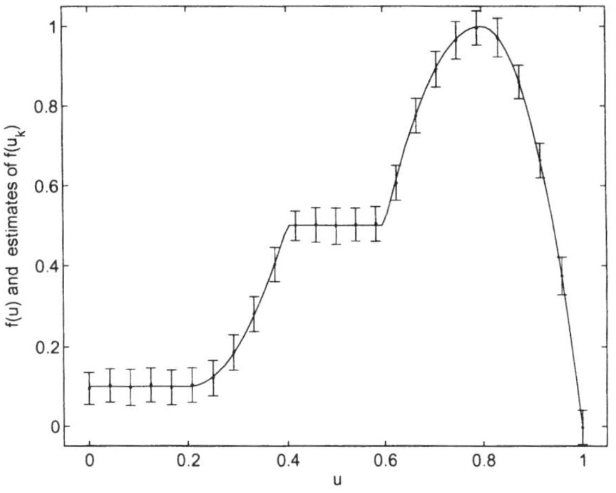

Fig. 1. Nonlinear function $f(u)$ together with estimated values of $f(u_k)$. The error bars display $\pm$ one standard deviation.

4. EXAMPLE

To illustrate the performance of the identification technique outlined above, two small illustrating examples are here presented. In both examples the same system is used:

LTI system:

$$\left[\begin{array}{c|c|c} A & b & x_0 \\ \hline C & d & \end{array}\right] = \left[\begin{array}{cc|c|c} 0 & 1 & 1 & 0.6 \\ -0.022 & 0.42 & 0.29 & 0.3 \\ \hline 1 & 0 & 0.077 & \end{array}\right]$$

Here $K = 0$ referring to (1) and the shape of the nonlinear function is illustrated in Figure 1.

Example 1. The first example illustrates how the output value of a nonlinear function can be estimated in a number of points. The input $u(t)$ was selected uniformly random from the input set $\mathscr{U}_{25}$ which consists of 25 values uniformly distributed over the input range of the nonlinear function. The noise $e(t)$ was selected as an independent normal distributed zero mean random variable with a variance of $2.5 \cdot 10^{-4}$ times the square of the maximum amplitude of the noise free output signal. The model parameters were estimated from data consisting of 200 input/output pairs and 100 realizations were generated and the sample statistics for the parameters were calculated. In each of the realizations the noise and input sequences were regenerated. In Figure 1 the estimates of the output values, $f(u_k)$, are plotted together with the true nonlinear function.

Example 2. This example shows how the standard deviation of the parameter estimates decrease as the length of the data set is increased. This time the input $u(t)$ is drawn from the input set $\mathscr{U}_{10}$. Just like above we make 100 realizations and calculate sample

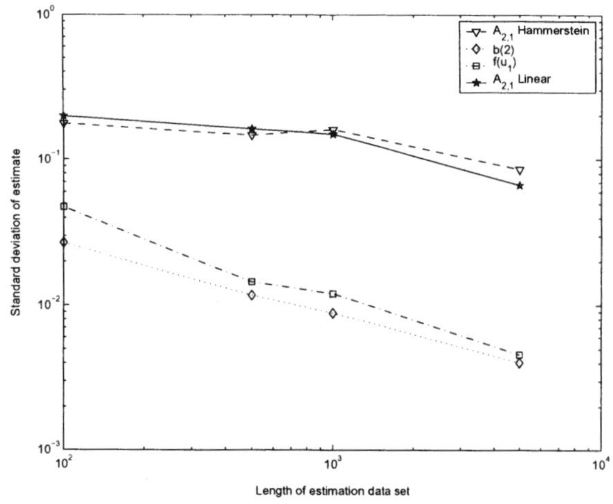

Fig. 2. Estimator performance based on 100 Monte Carlo simulations.

statistics from them. The results of the Monte Carlo simulation is shown in Figure 2.

5. CONCLUSIONS

A method for identifying Hammerstein systems has been presented. The method probes the system with an input with only a finite number of amplitudes. The use of this particular input enables a simultaneous estimation of the linear dynamical system as well as the gain of the non-linear function at the used input amplitudes. The method is based on a reformulation of the Hammerstein identification problem into a more standard multi-input linear problem. A three step identification procedure is presented which starts with a sub-space based method to estimate the poles of the linear system. Two alternative techniques are then outlined for the estimation of the zeros of the linear system and the non-linearity. The simpler procedure can be used if the initial condition of the system is known while an iterative procedure is employed when the initial condition is unknown. In a third step the preliminary estimate given by the first two steps are refined by the prediction error method. A small example gives a numerical illustration of the proposed method.

6. REFERENCES

Bai, Er-Wei (2002). Identification of linear systems with hard input nonlinearities of known structure. *Automatica* **38**(6), 853–860.

Bai, Er-Wei and Minyue Fu (2002). A blind approach to hammerstein model identification. *IEEE trans. on Signal Processing* **50**(7), 1610–1619.

Billings, S.A. (1980). Identification of nonlinear systems: A servey. *Proc. Inst. Elect. Eng.* **127**, 272–285.

Golub, G. H. and C. F. Van Loan (1989). *Matrix Computations*. second ed.. The Johns Hopkins University Press. Baltimore, Maryland.

Goodwin, G. C. (2002). Inverse problems with constraints. In: *Proc. 15th IFAC World Congress, Barcelona*. Plenary paper.

Greblicki, W. (2002). Stochastic approximation in nonparametric identification of hammerstein systems. *IEEE Trans. on Automatic Control* **AC-47**(11), 1800–1810.

Hanner, C. (2003). Identification of hammerstein systems using finite number of excitation levels. Master's thesis. Electrical Engineering, Chalmers University of Technology.

Ljung, L. (1999). *System Identification: Theory for the User*. second ed.. Prentice-Hall. Englewood Cliffs, New Jersey.

Maciejowski, J. M. (2002). *Predictive control with constraints*. Prentice Hall.

Narendra, K. S. and P. G. Gallman (1966). An iterative method for the identification of nonlinear systems using hammerstein model. *IEEE Trans. on Automatic Control* **AC-11**(31), 546–550.

Ninness, B. and S. Gibson (2002). Quantifying the accuracy of hammerstein model estimation. *Autiomatica* **38**(12), 2037–2051.

Quevedo, D. E., J. A. De Dona and G. C. Goodwin (2002). Receding horizon linear quadratic control with finite constraint sets. In: *Proc. 15th IFAC World Congress, Barcelona*.

Rangan, S., G. Wolodkin and Poolla K. (1995). New results for hammerstein system identification. In: *Proc. of the 34th Conference on Decision & Control*. New Orleans, LA, USA. pp. 697–702.

Verhaegen, M. (1993). Subspace model identification, Part III: Analysis of the ordinary output-error state space model identification algorithm. *Int. J. Control* **58**, 555–586.

Verhaegen, M. and D. Westwick (1996). Identifying mimo hammerstein systems in the context of subspace model identification methods. *Int. J. Control* **63**(2), 331–349.

IFAC
Publications
www.elsevier.com/locate/ifac

Fast approximate identification of nonlinear systems

J. Schoukens (*), J. Nemeth (**), P. Crama (*), Y. Rolain(*), R. Pintelon (*)

(*): Vrije Universiteit Brussel, dep. ELEC, Pleinlaan 2, B1050 Brussels, Belgium
(**): Budapest University of Technology & Economics, Dep. MIS, H-1521 Budapest
email: Johan.Schoukens@vub.ac.be

Abstract:A method is presented to extend the classical linear system identification methods towards nonlinear modelling. A well chosen nonlinear model structure is proposed that is identified in a 2-step procedure. First, a best linear approximation is identified using the classical linear identification methods. Next, the nonlinear extensions are identified with a linear least squares method. The proposed model not only includes Wiener and Hammerstein systems, it is also suitable to model nonlinear feedback systems. The method is illustrated on experimental data. *Copyright © 2003 IFAC*

Keywords: system identification, linear systems, nonlinear systems

1. Introduction

Identification of linear systems became a routine task. A number of well understood methods are available to solve the problem in the time- or in the frequency domain, using iterative and non iterative identification schemes (Ljung, 1999; Söderström and Stoica, 1989; Pintelon and Schoukens, 2001; Van Overschee and De Moor, 1994, Verhaegen, 1994). Although these techniques are very popular, they offer in most cases only an approximation to reality, because most phenomena are nonlinear in nature. There are a number of possibilities to deal with these unmodelled phenomena:

- The most simple solution is just to neglect the presence of nonlinear distortions. They are considered as an acceptable distortion to the modelled output. This option is often made, even if the distortions are quite important, because the available alternatives are not attractive.
- The nonlinear distortions are explicitly identified using a dedicated model. Although this seems to be the most powerful approach, it suffers from a severe drawback. Dedicated models are mostly very time consuming to build and to identify. Moreover, it is very difficult to develop general purpose packages to identify such models, because each problem is different from the other. Model selection, identifiability issues, and nonlinear optimization issues have to be solved.
- In order to avoid the above mentioned problems, approximate nonlinear modelling like neural nets popped up. Although this is a nice solution to many problems, the structure is not really suited to be used for the identification of dynamic systems, that have a dominant linear behaviour.

In this paper block oriented nonlinear models are used as an alternative for the neural net approach, combining the power of the linear identification machinery with the needs to capture at least a significant part of the nonlinear distortions. Due to a well chosen model structure, the second step, that extends the linear model to include the nonlinear distortions, is almost for free because it consists of a linear least squares problem and no new experiments are needed. For that reason there is almost no cost to make a trial to identify the extension. If it turns out that we can reduce the remaining errors by a factor 2 or more, a significant increase in model quality is obtained. If the trial fails, only a few seconds of additional computer time are lost.

The use of block-oriented models is not new (Bendat, 1998; Billings and Fakhouri, 1980 and 1982; Billings and Tsang, 1990; Billings 1980). Compared with these methods, the major contribution of the approach in this paper is its simplicity. The identification of the nonlinear structure pops up as a natural extension of the linear system identification approach. While it becomes possible to include in a wide range of applications a significant fraction of the nonlinear effects, the simplicity of the linear identification methodology is maintained.

First we introduce and motivate the extended model and next an identification procedure is proposed. After studying the properties of the method, it is illustrated on a number of experiments, ranging from a nonlinear feedback structure, over a chemical sensor with direction dependent dynamics, to a Hammerstein system.

2. Model structure

Consider the single-input, single-output, continuous or discrete time nonlinear dynamic system

$$y_0(t) = g_{\mathrm{NL}}(q, u(t)), \tag{1}$$

and the discrete observations (for a normalized sampling period).

$$y(t) = y_0(t) + n_y(t), \; t = 0, 1, \ldots, N \tag{2}$$

where $n_y(t)$ is zero mean noise.

Assumption 1: *Noise model*
The input is assumed to be known exactly. $n_y(t)$ is filtered white noise with finite moments up to order 4 and with absolutely summable cumulants up to order 4 (the 'correlation' length of the noise is not too large).

The model structure that will be used to describe the input-output relation for this system is given in Fig. 1

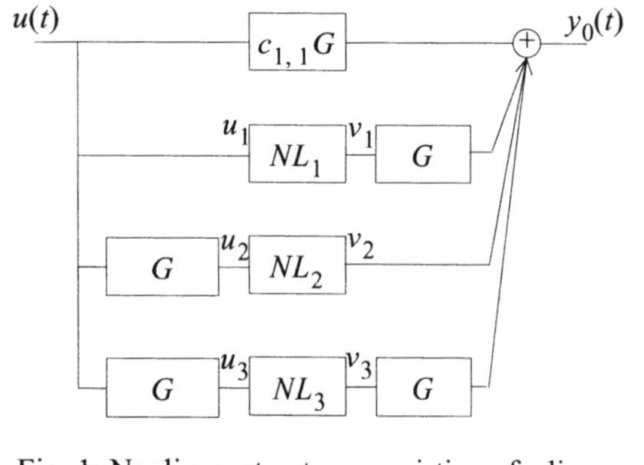

Fig. 1: Nonlinear structure consisting of a linear, a Hammerstein, a Wiener, and a feedback branch.

As it can be seen, it consists of a number of parallel branches, each taking care for a typical nonlinear behaviour. It can also be noticed that only one linear model block is allowed. The major reason for this choice is the need for fast and simple identification methods (see Section 3.).

- The first branch (called linear branch) captures all linear dynamics. G is a rational form in s (continuous time systems), or in z^{-1} (discrete time systems), and $c_{1,1}$ a real number.

- The second branch (called Hammerstein branch) is a Hammerstein system. NL_1 is a static nonlinear system, and can be described using a simple polynomial model $v_1 = \sum_{k=2}^{N_1} c_{1,k} u_1^k$. u_1^k can be replaced by a set of basis functions $B_k(u_1)$ in order to get better numerical properties during the identification step. Note that the sum starts from 2 in order to avoid identifiability problems: $k = 1$ would result in the same contribution as the first branch.

- The third branch (called Wiener branch) is a Wiener system, where NL_2 is again a static nonlinearity $v_2 = \sum_{k=2}^{N_2} c_{2,k} u_2^k$.

- The last branch has a special structure. It also consists of a static nonlinearity, but this time it is pre- and post filtered by the linear system G. It is based on the p-th order inverse structure of a nonlinear system, and it turns out that it is very well suited to take care for nonlinear feedback dynamics. The static nonlinearity is described by $v_3 = \sum_{k=1}^{N_3} c_{3,k} u_3^k$. Note that this time the linear term is included. This can be tolerated because no identifiability problem with the first branch will appear as long as G is a frequency dependent system. This branch is called the feedback branch.

Since we identify an approximate model, it is necessary to specify the class of input signals for which the approximation is valid. Moreover, also the approximation criterion should be specified.

Assumption 2: *Class of excitations*
$u(t)$ is assumed to be a random excitation with a user specified amplitude distribution (e.g. normal, binary, uniform) and power spectrum.

Note that more than one specification can be put into the set of excitations (e.g. uniformly and normally distributed signals). The wider the set, the more generally valid the model will be, but the lower the quality of the approximation.

Approximation criterion
The model parameters will be tuned in order to minimize the least squares distance between the measured and modelled output:

$$V(\theta, y) = \|y(t) - y(t, \theta)\|_2^2. \tag{3}$$

For simplicity and without loss of generality, we do not add a frequency weighting to the approximation criterion. The reader should be aware that it is not obvious to make an optimal choice in the presence of nonlinear distortions. Only if it can be guaranteed that the noise errors will dominate the model errors, it makes sense to consider the noise variance as a weighting function. Otherwise, alternative weightings (e.g. the relative error) may even be preferred.

3. Identification procedure

Although the structure in Fig. 1 looks quite complicated, it can be easily identified using a 2 step procedure.

Step 1: Identify the best linear approximation $\hat{G}$ of the nonlinear system using the classical linear identification procedures (Ljung, 1999; S der-str m, T. and P. Stoica, 1989, Pintelon and Schoukens, 2001).

Step 2: Fix the linear model G everywhere in Fig. 1 to $\hat{G}$. Next identify the parameters in the nonlinear models. Note that this problem is linear-in-the-parameters.

Step 3: The 2 step procedure is not optimal, as it is known that it does not necessarily reach the global minimum. A nonlinear search, leaving all model parameters free, should be added. We do not consider this step in the paper since our goal is to setup a fast method.

3.1 Identification of the best linear approximation

The aim of the first step is to identify the linear block $G(j\omega)$ (or $G(e^{j\omega})$ for discrete time systems). In Billings and Fakhouri (1980, 1982) and Schoukens *et al.* (1998), it was shown that the best linear approximation of a Hammerstein or Wiener system (branch 2 and 3), is asymptotically given within a constant by G, if the excitation is normally distributed. This result is generalized to a wider class of excitations by Billings and Fakhouri (1982). The 4th branch results in $G(j\omega)^2 = G(j\omega)G(j\omega)$, the squared transfer function corresponding to the cascaded linear systems in the feedback branch. Hence the best linear approximation of the overall structure in Fig. 1 is

$$\hat{G}(j\omega) = d_1 G(j\omega) + d_2 G^2(j\omega), \tag{4}$$

which means that an error of order $d_2 G^2$ appears with respect to the desired result (which is $d_1 G$). However, by adding a linear term to the third nonlinear branch, this error can be eliminated in the linear least squares approximation of the 2nd step. Consider the linear contribution of the feedback branch $c_{3,1}\hat{G}^2$:

$$c_{3,1}\left\{d_1^2 G^2 + 2d_1 d_2 G^3 + d_2 G^4\right\}. \tag{5}$$

Choosing $c_{3,1} = -d_2/d_1^2$ cancels the error term in the first branch, and leaves an error of $O(d_2 G^2)$ in the 2nd and 3th branch, and an error of $O(G^3 d_2^2/d_1)$ in the 4th one. Since these branches describe only the nonlinear distortions, we get an error on a correction term which is only a 2nd order effect (The assumptions that are needed to give a formal proof of these results are given in the appendix).

Remarks

- In practice the best linear approximation $\hat{G}$ is obtained by identifying a parametric model $\hat{G}(j\omega, \theta)$ using the classical identification methods on the measured data $u(t), y(t)$. This is a kind of smoothing step, that eliminates significantly the impact of the nonlinear distortions on the best linear approximation.
- In practice we apply this approach also to non Gaussian excitation signals. For such excitations, there is in general no proof available that the best linear approximation of the Wiener- and Hammerstein-branch is still proportional to G. However, this is the best we can do within this framework.

3.2 Identifying the nonlinear model parameters

Once the best linear approximation is identified, the parameters $c_{i,k}$ in the nonlinear static models are estimated. Because G is fixed to $\hat{G}$, this is a simple least squares problem that is linear-in-the-parameters. The basic idea is to calculate for each parameter $c_{i,k}$ the contribution to the output, order these as the columns in the matrix J, and solve the least squares problem:

$$[y] \approx J_t \theta_{NL}, \text{ (time domain) or}$$
$$[Y] \approx J_f \theta_{NL}, \text{ (freq. domain)} \tag{6}$$

with
$\theta_{NL} = \left[c_{1,1}\ c_{1,2}, \ldots, c_{1,N_1} \ldots c_{3,1}, \ldots, c_{3,N_3}\right]^T$.

The frequency domain formulation allows easy selection of the frequency bands of interest. In the time domain this is not that simple, because prefiltering is not allowed in nonlinear modelling. Care should be taken to use robust numerical procedures to maintain good numerical conditioning in the second step.

3.3 Model selection

Although the model selection problem looks at first glance very complicated, it is quite easily solved, due to the fact that a two step procedure is followed. In the first step, the order of the linear model $\hat{G}$ is selected using the classical order selection schemes. Next the active nonlinear branches should be selected. In all the examples below, we followed the same strategy. In a first run, all branches are switched on, using a nonlinearity of degree 5. Next the contribution of each branch to the output is verified (e.g. checking the output power of each branch, or by comparing the output power spectra). Next the dominating branches are selected, and the order of the static nonlinearity is tuned. It turned out that this procedure is very effective.

4. Impact of the disturbing noise

The impact of disturbing noise on the estimation of the best linear approximation is explicitly studied in (Schoukens *et al.*, 1998). There it is shown that under Assumption 1, the estimate $\hat{G}(j\omega, \theta)$ of $G(j\omega)$ is consistent, even in the presence of nonlinear distortions, if the model $G(j\omega, \theta)$ is flexible enough to capture the dynamics of $G(j\omega)$. The stochastic convergence of the 2nd step is straight forwards (under the same noise conditions as the first step) since it is linear-in-the-parameters. The estimated parameters will converge to the true parameters, if the nonlinear structure is rich enough to capture the nonlinear part of the system, otherwise an approximation will be obtained.

5. Experimental verification

In this section the results from 3 experiments are reported, illustrating the flexibility of the method with respect to different types of nonlinearity, and the robustness with respect to the distribution of the excitation signal.

5.1 Nonlinear feedback system

The first experiment is done on a nonlinear electrical circuit that is ideally described by a 2nd order differential equation with a cubic feedback. A detailed study of this nonlinear system can be found in Pintelon and Schoukens (2001). In a first step, the system is identified using an odd random multisine (almost normally distributed periodic signal with only odd harmonics excited, see Pintelon and Schoukens, 2001), next the identified model is validated using a normally distributed noise excitation with slowly increasing standard deviation. In the first step, a continuous time linear model G of order 2 was selected and identified. Next, we checked the output power for each branch as explained in Section 3.3. For simplicity we give the results in dB, normalizing the linear output to 0 dB. The following results were found: P_{Lin}=0 dB, P_{Ham}=-55 dB, P_{Wien}=-41 dB, P_{FdBck}=-24 dB. Only the feedback branch was retained in parallel with the linear model branch. A static nonlinearity of degree 5 is used.

In Fig. 2, the identification result is shown. The measured output spectrum is compared to the modelled one, and this for the best linear approximation and the nonlinear model. The validation result is shown in Fig. 3 The measured output is again compared to the linear and nonlinear modelled output. From Fig. 2 and 3, it is clearly seen that the errors drop with about a factor 10 when the linear model is replaced by the nonlinear one. Note that both models completely fail once the output is larger than 0.1 V, which was exactly the range that was used during the identification step. This shows that these approximate models, that are not built on physical principles, may not be extrapolated.

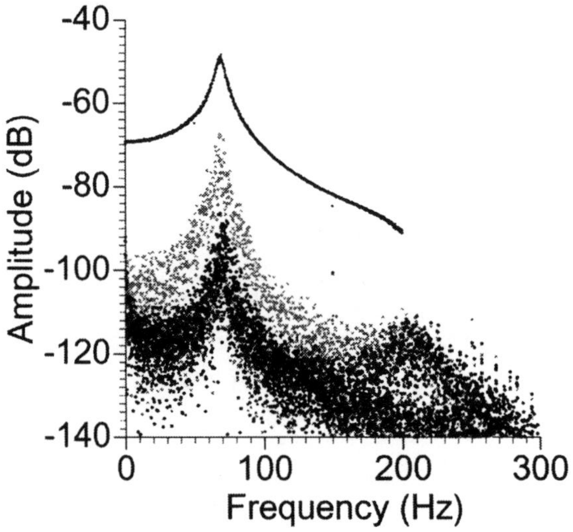

Fig. 2: Comparing the identified output to the measured one. gray dots: $|Y - Y_{\text{Lin}}|$, black dots: $|Y - Y_{\text{NLin}}|$, full line: measured output.

5.2 System with direction dependent dynamics

In the second experiment, an 'electronic nose' is modelled. The sensor is based on an adsorption/desorption process. Since both processes have different dynamics, the behaviour of the system depends on the slope of the output. The system is excited with a binary signal (an inverse repeat periodic signal). Two data sets with different excitations were available. These data were obtained by Tan and Godfrey (2002). In order to show the flexibility of the modelling approach, we selected this time a discrete time dynamic model for G of order 5 (similar results were obtained with a continuous time model). The scan of the nonlinear branches resulted in: P_{Lin}=0 dB, P_{Ham} = not present (binary input), P_{Wien}=-27 dB, P_{FdBck}=-21 dB. From this, the Wiener branch and also the feedback branch were selected. The degree of the static nonlinearity was tuned to 4 for both branches. The first experiment was used to identify the model. The measured output is compared to the identified one in Fig. 4 From this figure it is seen that especially in the lower frequency band a

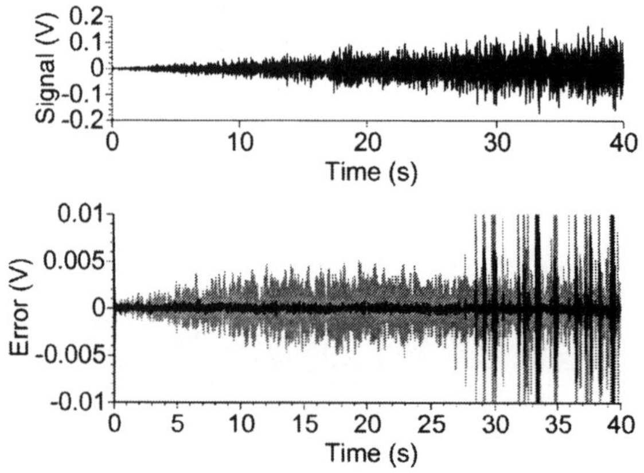

Fig. 3:Validation results. Top: the measured output; Bottom: gray: the linear simulated output error; black: the nonlinear simulated output error.

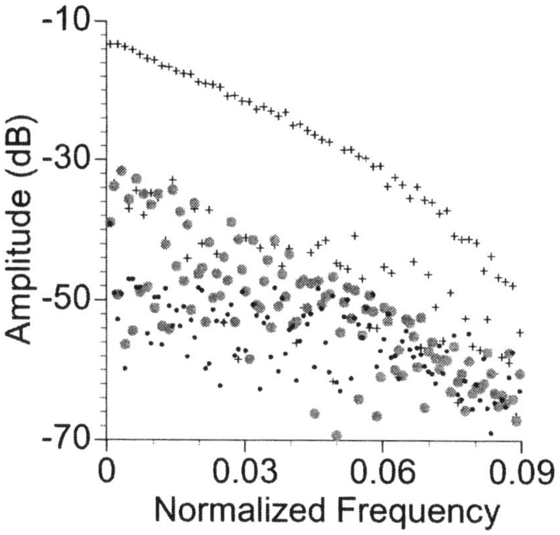

Fig. 4: Comparing the identified output to the measured one. gray dots: $|Y - Y_{\text{Lin}}|$, black dots: $|Y - Y_{\text{NLin}}|$, +: measured output.

significant gain is made by switching from a linear model to a nonlinear one. The validation results are shown in Fig. 5. Also in this case it can be clearly seen that the nonlinear model performs significantly better than the linear one.

5.3 Hammerstein system

In a last experiment, we identify a Hammerstein device. A detailed description of the setup is given in Crama and Schoukens (2001). The first scan of the branches resulted in: P_{Lin}=0 dB, P_{Ham}=-14dB, P_{Wien}=-47 dB, P_{FdBck}=-48 dB. Only the Hammerstein branch was retained in parallel with the linear model branch. The final degree of the static nonlinearity was chosen equal to 10. The identification results are shown in Fig. 6 and the validation results in Fig. 7 Again a significant improvement is found by using the nonlinear model. It can also be observed that in this case, the model was also able to give a reasonable prediction of the output outside the excitation band (0-10KHz). It should be noted that the identified linear

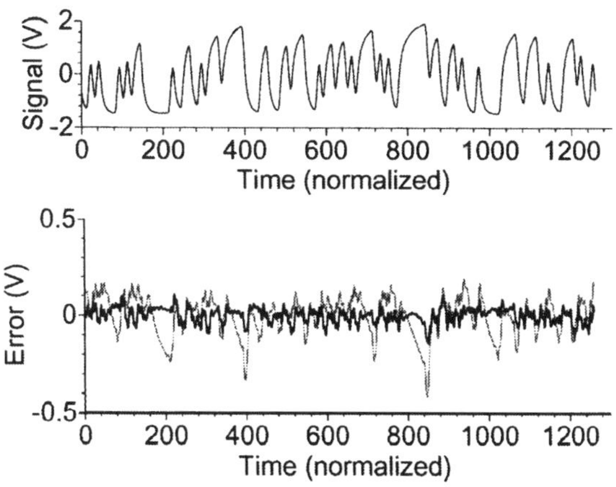

Fig. 5:Validation results. Top: the measured output; Bottom: gray: the linear simulated output error; black: the nonlinear simulated output error.

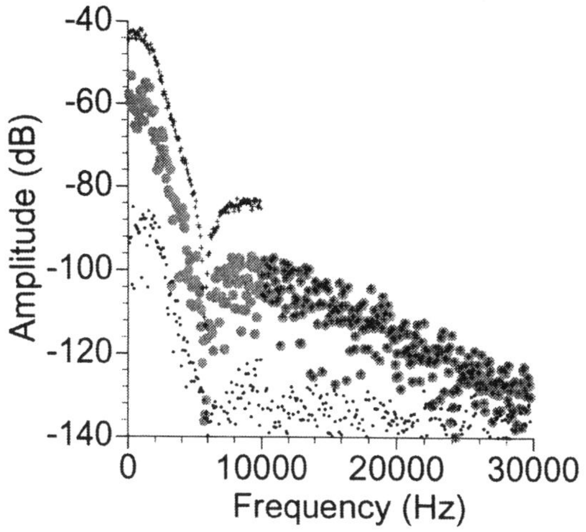

Fig. 6: Comparing the identified output to the measured one. gray dots: $|Y - Y_{\text{Lin}}|$, black dots: $|Y - Y_{\text{NLin}}|$, +: measured output.

model (continuous time, order 3) is extrapolated in that case, since no direct information about the linear contribution in the band 10-30 KHz is available.

6. Conclusions

In this paper, an approximate nonlinear modelling technique is proposed. The major goals of the method are: simple in use, fast to identify at a low experimental cost. Instead of going for the best possible nonlinear model, we explicitly preferred a reasonable approximation that reduces the nonlinearity errors significantly, using a model that can be easily used during design, simulation, etc. To reach this goal, a nonlinear model structure that can be easily identified using a two step procedure is proposed. In the first step, the linear dynamics are estimated. In the next step the nonlinear branches in the model are identified. A simple model selection procedure allows to select the active nonlinear branches. The method allows to deal with Hammerstein, Wiener, and nonlinear feedback systems. Illustrations on 3 different experiments showed the flexibility and the power of the method.

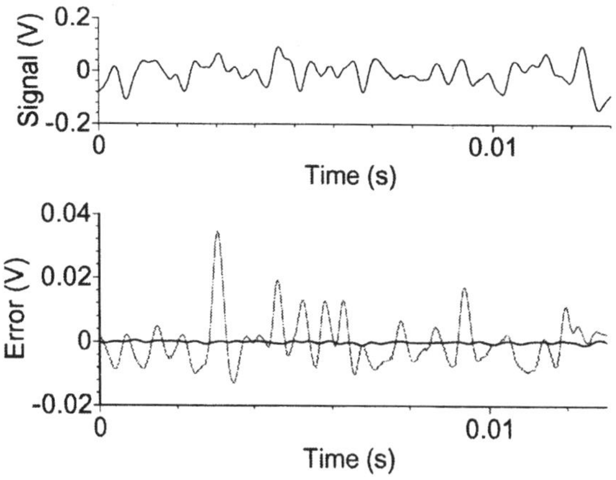

Fig. 7:Validation results. Top: the measured output; Bottom: gray: the linear simulated output error; black: the nonlinear simulated output error.

7. Acknowledgement

This work was supported by the FWO-Vlaanderen, the Flemish Community (Concerted actions IMMI and ILiNoS, bilateral agreement 99/18), and the Belgian Government (IUAP-5).

8. Appendix

Assumption 3: The nonlinear output contributions are small compared to the linear output:

Define $\max(\|y_2\|_2, \|y_3\|_2, \|y_4\|_2) = M_{NL}$, with y_k the output of the kth branch, and assume that $M_{NL} \ll \|y_1\|_2$ (small distortions). This can be rephrased by noticing that the nonlinear distortions are a first order distortion:

$$\frac{\max(\|y_2\|_2, \|y_3\|_2, \|y_4\|_2)}{\|y_1\|_2} = O\left(\frac{M_{NL}}{\|y_1\|_2}\right), \ M_{NL}/\|y_1\|_2 \ll 1 . \tag{7}$$

Assumption 4 : $\|\hat{G}(u) - G(u)\|_2 = O(M_{NL}/\|y_1\|_2)$.

From this assumption the following results follow:

$$\|\hat{G}(NL_1(u)) - G(NL_1(u))\|_2/\|y_2\|_2 = O(M_{NL}/\|y_1\|_2)$$

$$\|NL_2(\hat{G}(u)) - NL_2(G(u))\|_2/\|y_3\|_2 = O(M_{NL}/\|y_1\|_2)$$

$$\|\hat{G}(NL_3(\hat{G}(u))) - G(NL_3(G(u)))\|_2/\|y_4\|_2 = O(M_{NL}/\|y_1\|_2)$$

Remarks:

1) This assumption is not too hard if the considered frequency band is well excited during the experiments, for example using white noise. However, if in some excited frequency bands, the nonlinear power would dominate the linear power due to a poor excitation (a frequency band is almost not excited), large uncontrolled errors can appear on a nonparametric estimate of G. Parametric estimation methods that identify $\hat{G}(j\omega, \theta_G)$ from the input u and output y, like for example the output error method in Ljung (1999) or the frequency domain estimator in Pintelon and Schoukens (2001), have an intrinsic weighting in their cost function that is proportional to the power spectrum of u, and consequently the critical frequency bands (that are almost not excited) are already suppressed by these methods.

2) Combining Assumption 4 with Assumption 3 results in

$$\|\hat{G}(NL_1(u)) - G(NL_1(u))\|_2/\|y_1\|_2 = O(M_{NL}/\|y_1\|_2)^2$$

$$\|NL_2(\hat{G}(u)) - NL_2(G(u))\|_2/\|y_1\|_2 = O(M_{NL}/\|y_1\|_2)^2$$

$$\|\hat{G}(NL_3(\hat{G}(u))) - G(NL_3(G(u)))\|_2/\|y_1\|_2 = O(M_{NL}/\|y_1\|_2)^2$$

9. References

Bendat, J. S. (1998). *Nonlinear Systems Techniques and Applications.* Wiley, New York.

Billings, S.A. (1980). Identification of nonlinear systems. A survey. *IEE proc. D*, vol. **127**, no. 6, pp. 272-285.

Billings, S.A. and S.Y. Fakhouri (1980). Identification of non-linear systems using correlation analysis and pseudorandom inputs. *Int. J. Systems SCI.*, vol. **11**, no. 3, pp. 261-279.

Billings, S. A. and S.Y. Fakhouri (1982). Identification of systems containing linear dynamic and static nonlinear elements. *Automatica*, vol. **18**, no. 1, pp. 15-26.

Billings, S.A. and K.M. Tsang (1990). Spectral analysis of block structured non-linear systems. *Mechanical Systems and Signal Processing*, vol. **4**, no. 2, pp. 117-130.

Crama, P. and J. Schoukens (2001). First Estimates of Wiener and Hammerstein Systems Using Multisine Excitation. *Proceedings of the 18th IEEE Instrumentation and Measurement Technology Conference*, Budapest, Hungary, May 21–23, vol. **2**, pp. 1365–1369

Haber, R. and H. Unbehauen (1990). Structure identification of nonlinear dynamic systems - A survey on input/output approaches. Automatica, vol. **26**, no. 4, pp. 651-677.

L. Ljung (1999). *System Identification - Theory for the user.* Prentice-Hall, Upper Saddle River, N.J., 2nd edition.

Pintelon R. and J. Schoukens (2001). *System Identification: A frequency domain approach.* IEEE Press, Picataway.

Schetzen, M. (1980). *The Volterra and Wiener Theories of Nonlinear Systems.* Wiley and Sons, New York.

Schoukens, J., T. Dobrowiecki, and R. Pintelon (1998). Identification of linear systems in the presence of non-linear distortions. A frequency domain approach. *IEEE Trans. on Automatic Control*, vol. **43**, no. 2, pp. 176-190.

Söderström, T. and P. Stoica (1989). *System Identification.* Prentice-Hall, Englewood Cliffs,

Tan, A.H. and K.R. Godfrey (2002). Modelling of Direction-Dependent Dynamic Processes: A Comparison of Wiener Models and Neural Networks. *IEEE Instrumentation and Measurement Technology Conference, Anchorage.*

Van Overschee, P. and B. DeMoor (1994). N4SID: subspace algorithms for the identification of combined deterministic-stochastic systems. *Automatica,* vol. **30**, no. 1, pp. 75-93.

Verhaegen, M. (1994). Identification of the deterministic part of MIMO state space models given in innovations form from input-output data. *Automatica,* vol. **30**, no. 1, pp. 61-74.

IFAC Publications
www.elsevier.com/locate/ifac

GAUSSIAN PROCESSES FRAMEWORK FOR VALIDATION OF LINEAR AND NONLINEAR MODELS [1]

Astrid Lundgren and Jonas Sjöberg

Department of Machine and Vehicle Systems, Chalmers University of Technology, 412 96 Göteborg, Sweden, Email: astrid.lundgren@me.chalmers.se, jonas.sjoberg@me.chalmers.se

Abstract: A statistical nonlinear model validation method is suggested based on the *Gaussian processes* framework. Instead of testing for correlation between the residuals and certain test variables, as in traditional statistical tests, a parameterized model of the correlation is proposed and the significance for this model is tested. The test makes it possible to validate against nonlinear models without making detailed assumptions on the structure of the nonlinearities. *Copyright © 2003 IFAC*

Keywords: system identification, validation, nonlinear systems, Gaussian processes, nonlinear models

1. INTRODUCTION

Model selection and model validation are important steps in the system identification process and effective tools are needed. Strictly, the model validation test should indicate if the model is good enough for its purposes, however validation tests are more often used to indicate the answer to the question "Is it possible to obtain a better model?". This paper concerns this second formulation of the validation question.

Classic statistical model validation methods, as described in eg (Ljung, 1999; Söderström and Stoica, 1989) are based on correlation tests between the residuals and various types of test variables, for example lagged values of the input signal. Such tests give a measure of the model mismatch which is a projection of the mismatch onto the space of the test variables. This means that an auto-correlation test of the residuals and a correlation test between the input signal and the residuals can only reject the model if better linear models exist.

The classical linear validation methods can easily be extended to nonlinear validation by incorporating nonlinear elements among the test variables. Many papers on such ideas have been published, eg (Billings and Tao, 1991; Leontaritis and Billings, 1987; Stoica *et al.*, 1986; Haber, 1985). In (Sjöberg, 1995) these ideas are exploited not only to validate a model, but also to indicate if a better model is possible. However, also with nonlinear test variables, only projections of the model deficiencies are indicated. Especially in nonlinear system identification, where the possible choices of test variables are huge, this might be a problem, since it is not possible to test them all.

Here an alternative approach to statistical validation based on the *Gaussian Processes* framework is investigated. This is a modelling technique where a model is proposed for the *covariance* of a signal instead of for the signal itself. Since regression with Gaussian processes is a nonlinear black-box identification method, the validation test works for both linear and nonlinear models. A main drawback with the framework is that only moderate sized data sets can be handled since a matrix of dimension equal to the size of the data set has to be inverted inside a loop in the estimation

[1] Support by the Swedish Research Council for Engineering Science (VR) is gratefully acknowledged.

algorithm. However, for cases with small data sets this technique might be useful, see for example (Rasmussen, 1996; Gibbs, 1997) for more information.

The new contributions in this work is the proposed use of the Gaussian processes framework, and the use of a hypothesis test to decide upon the relevance of the covariance model.

The paper is organized as follows: in section 2 the problem is defined. A brief introduction to the framework of regression with Gaussian processes and to different covariance model structures is given in section 3. Section 4 defines the null hypothesis and the statistic test variable.

2. PROBLEM DEFINITION

Consider the situation where a batch of N input-output data from an unknown process is available $\{u_t, y_t\}_{t=1}^N$. A model $g(\varphi_t)$ is proposed, giving predictions of the output

$$\hat{y}_t = g(\varphi_t) \tag{1}$$

The (pseudo)-regressor vector φ_t contains components available at time t. It might be lagged input values, u_{t-k}, lagged output values y_{t-k}, past model outputs, $\hat{y}_{t-k}$, states, or combinations of any of these. Also assume that the data can be described by a true process described by

$$y_t = f(\tilde{\varphi}_t) + e_t \tag{2}$$

where $\tilde{\varphi}_t$ is the true regressor, which not necessarily equals φ_t. The data are disturbed by the white noise sequence e_t.

If the model is perfect, then the residuals

$$\varepsilon_t = y_t - \hat{y}_t \tag{3}$$

are a realization of the white noise sequence e_t. Hence, with the validation test we try to indicate non-whiteness of the residuals.

For simplicity it is assumed that e_t is stationary and the variance of e_t is estimated as

$$\hat{\lambda} = \frac{1}{N}\sum_{t=1}^{N} \varepsilon_t^2 \tag{4}$$

Hence,

$$R_e(\tau) = E[e_t e_{t+\tau}] = \begin{cases} \lambda & \text{if } \tau = 0 \\ 0 & \text{otherwise} \end{cases} \tag{5}$$

Traditional statistical validation techniques use the residuals to form an estimate of $R_e(\tau)$

$$\hat{R}_\varepsilon^N(\tau) = \frac{1}{N}\sum_{t=1}^{N} \varepsilon(t)\varepsilon(t-\tau) \tag{6}$$

This is actually an estimate of the auto-correlation as a function of time. Then the significance of the difference between $R_e(\tau)$ and $\hat{R}_\varepsilon(\tau)$ is tested. If the difference is not significant ε_t is a relisation of e_t.

With a model mismatch the residuals will not only be a function of time but also a function of the regressors $\tilde{\varphi}_t$ and φ_t

$$\varepsilon_t = f(\tilde{\varphi}_t) - g(\varphi_t) + e_t \tag{7}$$

The proposed method estimates correlation of the residuals as a function of φ (note that $\tilde{\varphi}$ is unknown). Introduce the parameterized covariance model

$$C(\varphi_t, \varphi_k, t-k, \theta) \tag{8}$$

where θ is a vector of adjustable parameters, as an estimate of

$$E[\varepsilon_k(\varphi_k)\varepsilon_t(\varphi_t)] \tag{9}$$

where a change of notation has been made to indicate the dependence on the regressor.

For a perfect model g the best covariance model would be

$$C(\varphi_t, \varphi_k, t-k, \theta) = \begin{cases} \lambda & \text{if } \tau = 0 \\ 0 & \text{otherwise} \end{cases} \tag{10}$$

which only depends on the time lag $t-k$. The proposed method can now be described as *testing for significance for a covariance model which also depends on the regressors*. The test is constructed as a traditional hypothesis test, where the H_1 hypothesis that a larger model than (10) is needed. This hypothisis is then tested against the H_0 hypothesis described by (4) and (10).

3. BRIEF INTRODUCTION TO GAUSSIAN PROCESSES FOR REGRESSION

Here follows a brief introduction to the system identification technique regression with Gaussian processes. The aim is to describe the technique from a practical user point of view and most of the underlying theory is left out, see for example (Rasmussen, 1996; Gibbs, 1997) for further details.

As mentioned earlier regression with Gaussian processes uses a covariance model to describe the covariance between the output data as a function of a regressor rather than constructing a model of the output data as a function of a regressor. The covariance model gives an estimate of the elements of the $N \times N$ covariance matrix $C_N(\theta)$. One example of a covariance model, making use of the assumption that the system is continuous, is

$$C(\varphi_t, \varphi_k, t-k, \theta) =$$
$$= v_0 \exp\left[-\frac{1}{2}\sum_{l=1}^{p} \alpha_l(\varphi_{t,l} - \varphi_{k,l})^2\right] + \lambda\delta(t,k) \tag{11}$$

where $\theta = [\alpha_1 \ldots \alpha_p \; v_0 \; \lambda]^T$ are adjustable parameters, p the number of regressor dimensions and

$$\delta(t,k) = \begin{cases} 1 & \text{if } t = k \\ 0 & \text{otherwise} \end{cases} \tag{12}$$

This model structure will be referred to as $\mathbf{M}_L$ and it has $p_L = \dim(\varphi) + 2$ number of parameters. To use the model structure $\mathbf{M}_L$ is equivalent with making the assumption that close regressor-values, φ_t and φ_k, should give highly correlated residuals, ε_t and ε_k. The parameter α is a bandwidth parameter which states how close two regressor-values have to be to one another to give highly correlated residuals. Each dimension of φ has its own α-parameter. The parameter λ is the estimated variance of the noise, e, and v_0 is a scaling parameter.

The covariance model (11) is the most natural to use when no prior knowledge about the system suggests that some other choice of covariance model might be better.

If the residuals ε_t and ε_k are not correlated, for $t \neq k$, the parameter $v_0 = 0$ and the covariance model (11) equals covariance model (10). The smaller model structure (10) will be called $\mathbf{M}_{L_D}$ and has number of parameters $p_{L_D} = 1$. For other examples on covariance models see for example (Rasmussen, 1996).

The parameters, θ, are chosen to maximize the logarithmic likelihood function

$$L = -\frac{1}{2}\log\det C_N(\theta) - \frac{1}{2}\varepsilon^T C_N^{-1}(\theta)\varepsilon - \frac{N}{2}\log 2\pi \tag{13}$$

which is a function of the covariance matrix with elements $C(\varphi_t, \varphi_k, t-k, \theta)$. To maximize (13) its derivative

$$\frac{dL}{d\theta} = -\frac{1}{2}\text{trace}\left(C_N^{-1}(\theta)\frac{dC_N(\theta)}{d\theta}\right) + \\ + \frac{1}{2}\varepsilon^T C_N^{-1}(\theta)\frac{dC_N(\theta)}{d\theta}C_N^{-1}(\theta)\varepsilon \tag{14}$$

is needed.

4. NULL HYPOTHESIS AND TEST VARIABLE

The hypothesis test compares the two model structures $\mathbf{M}_L$ and $\mathbf{M}_{L_D}$. For the covariance model structure $\mathbf{M}_{L_D}$ the maximum of the logarithmic likelihood (13) can be calculated analytically. If the residuals have zero mean the result is

$$L_D = -\frac{N}{2}\left(\log\left(\frac{1}{N}\varepsilon^T\varepsilon\right) + 1 + \log(2\pi)\right) \tag{15}$$

See appendix A for details.

The null hypothesis, that the two model structures $\mathbf{M}_{L_D}$ and $\mathbf{M}_L$ performs equally well, i.e. $v_0 = 0$, can be formulated:
H_0: L, calculated from equation (13), is approximatly equal to L_D, equation (15)
and the alternative hypothesis, that the larger model structure $\mathbf{M}_L$ is better than the smaller $\mathbf{M}_{L_D}$, can be formulated:
H_1: $L > L_D$
To tell if where is statistic significance to reject H_0 a test variable, κ_D, which is χ^2-distributed if H_0 can not be rejected, is necessary.

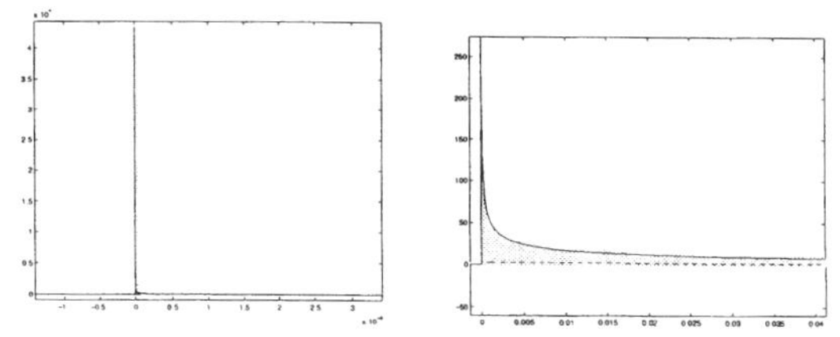

(a) The χ^2-distribution compared with simulated data

(b) Close-up of 1(a)

Fig. 1. Simulation: L calculated with ε_t as random numbers, $N = 40$, which means that H_0 is true. The histograms shows the distribution of 52145 different κ_D values. The solid line is the $\chi^2(1)$-distribution, with one degree of freedom, $p_L - p_{L_D} - 1 = 1$. The histogram coincide with the true χ^2 distribution indicating that κ_D is χ^2 distributed.

4.1 *Test variable*

To say if the difference between $\mathbf{M}_L$ and $\mathbf{M}_{L_D}$ is significant or not, an *F-test* is used, see e.g. (Söderström and Stoica, 1989). If H_0 can not be rejected and ε_t has unit variance the test quantity

$$\kappa_D = \frac{4}{N}(L - L_D) \tag{16}$$

is asymptotically χ^2-distributed with $p_L - p_{L_D} - 1$ degree of freedom, for proof see appendix B. Figure 1 shows that when ε_t is a vector of random numbers, H_0 can not be rejected, the distribution of κ_D coincide with the χ^2-distribution. Thus if $\kappa_D \leq \chi^2_\beta(p_L - p_{L_D} - 1)$ there is not statistic significance to reject H_0 with the significance level β. And $\mathbf{M}_{L_D}$ should be chosen over $\mathbf{M}_L$ which means that ε_t is assumed to not be a function of φ_t.

If ε_t do not have unit variance it has to be normalized

$$\varepsilon_{t,norm} = \frac{\varepsilon_t}{\sqrt{\hat{\lambda}}} \tag{17}$$

before applying the validation test. The unit variance means that

$$\hat{\lambda} = \frac{1}{N}\varepsilon^T\varepsilon = 1 \tag{18}$$

and what the expression for the logarithmic likelihood function (15) can be simplified to

$$L_D = -\frac{N}{2}(1 + \log(2\pi)) \tag{19}$$

The test procedure of a model sums up like this:

- compute the residuals, using (5)
- estimate λ and normalize the residuals, using (2) and (17)
- choose a covariance model, use (11) as default
- estimate the parameters which maximize L, using (13)
- calculate L_D, using (15)

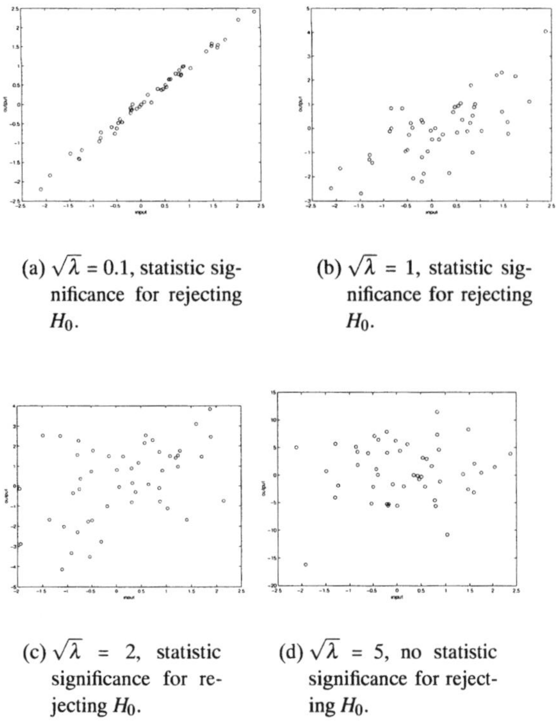

(a) $\sqrt{\lambda} = 0.1$, statistic significance for rejecting H_0.

(b) $\sqrt{\lambda} = 1$, statistic significance for rejecting H_0.

(c) $\sqrt{\lambda} = 2$, statistic significance for rejecting H_0.

(d) $\sqrt{\lambda} = 5$, no statistic significance for rejecting H_0.

Fig. 2. Four data sets computed from a straight line with different amounts of noise added to them. Applying the validation test gives that for $\sqrt{\lambda} = 5$ there is no statistic significance for rejecting H_0. The significance level β is chosen to 0.025.

- calculate κ_D, using (16)
- choose a significance level β
- test if H_0 holds

If there is no significance for H_0 to be rejected, one can be satisfied with the model. If not a change of the model structure should be considered.

4.2 Illustrating examples

In this subsection the validation test will be illustrated with two examples. In both cases a MATLAB-program by Rasmussen (2001) has been used to maximize L .

Assume that the residuals can be described by a straight line

$$\varepsilon_t = u_t + e_t \tag{20}$$

with white noise e_t added to it. This can illustrate the case where the model does not reproduce a linear trend in the system. Four data sets, $N = 50$, with $\sqrt{\lambda}$ equal to 0.1, 1, 2 and 5 respectively are created and plotted in figure 2. The residuals are normalized, the covariance model (11) is chosen and the maximum L is computed and the significance level, β, is chosen to 0.025.

For the first three data sets there is statistic significance for rejecting H_0 in favor of H_1, but for the data set with $\sqrt{\lambda} = 5$ the test shows no significance for rejecting H_0. The result is consistent with a visual inspection of

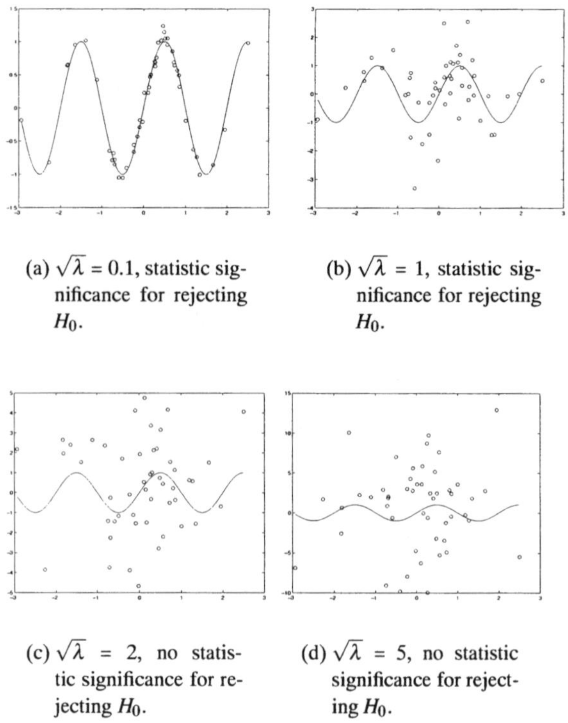

(a) $\sqrt{\lambda} = 0.1$, statistic significance for rejecting H_0.

(b) $\sqrt{\lambda} = 1$, statistic significance for rejecting H_0.

(c) $\sqrt{\lambda} = 2$, no statistic significance for rejecting H_0.

(d) $\sqrt{\lambda} = 5$, no statistic significance for rejecting H_0.

Fig. 3. Four data sets computed from a sinusoid with different amounts of noise added to it. Applying the validation test gives that for $\sqrt{\lambda} = 2$ and $\sqrt{\lambda} = 5$ there is no statistic significance for rejecting H_0. The significance level β is chosen to 0.025. The solid line is the noise free sinusoid.

figure 2. For noise with standard deviation 0.1, 1 and 2 the linear trend is visible, but not when the standard deviation of the noise is equal to 5.

The validation test is also applied on four data sets from a sinusoid with white noise added to it

$$\varepsilon_t = \sin(\pi u_t) + e_t \tag{21}$$

where the standard deviations are the same as in the case with the straight line. The data sets are plotted in figure 3. Here the test shows no statistic significance to reject H_0 both for $\sqrt{\lambda} = 2$ and $\sqrt{\lambda} = 5$.

Of course these examples do not say anything about the test performance on a more complex problem but it is just meant to illustrate the basic idea.

5. DISCUSSION

A validation method based on with Gaussian processes has been presented. The parameterized covariance model (11) gives a statistical test variable without any other assumptions on the structure of the system than that it is continuous. More traditional extensions of linear methods require some prior knowledge about the system, if the validation process will not be too time consuming. The main drawback with the validation test is that the $N \times N$ covariance matrix has to be inverted in the maximizing algorithm of the logarithmic likelihood function (13) this limits the size of the data set which can be tested.

Appendix A. DERIVATION OF L_D

In this appendix equation (15) will be derived.

If the covariance model (10) is used the covariance matrix, C_N, is a diagonal matrix

$$C_N = C_D = \lambda I \tag{A.1}$$

where the diagonal element, λ, is the noise variance.

The value of the parameter λ which maximizes the logarithmic likelihood function (13) can be calculated analytically by setting the derivative of (13) with respect to λ to zero, as defined by equation (14)

$$\frac{dL_D}{d\lambda} = -\frac{1}{2}\text{trace}\left(C_D^{-1}\frac{dC_D}{d\lambda}\right) + \\ + \frac{1}{2}\varepsilon^T C_D^{-1}\frac{dC_D}{d\lambda}C_D^{-1}\varepsilon = 0 \tag{A.2}$$

The derivative of the covariance matrix with respect to the parameter λ:

$$\frac{dC_D}{d\lambda} = I \tag{A.3}$$

The inverse of the covariance matrix:

$$C_D^{-1} = \frac{1}{\lambda}I \tag{A.4}$$

Inserted in equation (A.2) gives:

$$-\frac{1}{2}\text{trace}\left(\frac{1}{\lambda}I\right) + \frac{1}{2\lambda^2}\varepsilon^T I \varepsilon = 0 \tag{A.5}$$

Further simplified:

$$-\frac{N}{\lambda} + \frac{1}{\lambda^2}\varepsilon^T\varepsilon = 0 \tag{A.6}$$

and solved with respect to λ:

$$\hat{\lambda} = \frac{1}{N}\varepsilon^T\varepsilon \tag{A.7}$$

which equals equation (4). Inserting (A.7) in the expression for the logarithmic likelihood function (13) gives:

$$L_D = -\frac{1}{2}\log\left(\left(\frac{1}{N}\varepsilon^T\varepsilon\right)^N\right) - \frac{N}{2} - \frac{N}{2}\log(2\pi) = \\ = -\frac{N}{2}(\log(\lambda) + 1 + \log(2\pi)) \tag{A.8}$$

Which is equal to (15).

Appendix B. PROOF F-TEST

Here it will be proved that the test variable, κ_D, is χ^2-distributed with $p_L - p_{L_D} - 1$ degrees of freedom, if $\lambda = 1$. This proof is based on a proof by Söderström and Stoica (1989), in appendix A11.2.

First two lemmas from (Söderström and Stoica, 1989):

- Lemma B.13 If ψ is an n-dimensional Gaussian vector, $\psi \sim \mathbf{N}(0,I)$, and M is a $n \times n$ idempotent matrix [2], of rank r. When is $\psi^T M \psi$ $\chi^2(r)$ distributed.
- Lemma A.28 (ii) For an idempotent matrix M, rank(M) = trace(M).

Assume that H_0 holds

$$L_D(\lambda) = L(\theta) \tag{B.1}$$

Taylor expansion of L for the true parameter vector θ around the estimated parameter $\hat{\theta}$:

$$L(\theta) \approx L(\hat{\theta}) + \frac{dL(\hat{\theta})}{d\theta}(\theta - \hat{\theta}) + \\ \frac{1}{2}(\theta - \hat{\theta})^T\frac{d^2L}{d\theta^2}(\theta - \hat{\theta}) \\ \approx \left\{\frac{dL(\hat{\theta})}{d\hat{\theta}} \approx 0\right\} \approx L(\hat{\theta}) + \frac{1}{2}(\theta - \hat{\theta})^T\frac{d^2L}{d\theta^2}(\theta - \hat{\theta}) \tag{B.2}$$

Taylor expansion of $\frac{dL}{d\theta}$ for the estimated parameter around the true parameter:

$$0 = \frac{dL(\hat{\theta})}{d\hat{\theta}}^T \approx \frac{dL(\theta)}{d\theta} + \frac{d^2L(\theta)}{d\theta^2}(\hat{\theta} - \theta) \tag{B.3}$$

$$\Rightarrow \hat{\theta} - \theta \approx -\frac{d^2L}{d\theta^2}^{-1}\frac{dL}{d\theta}^T \tag{B.4}$$

Insert (B.4) in (B.2):

$$L(\theta) \approx \\ L(\hat{\theta}) + \frac{1}{2}\left(\frac{d^2L}{d\theta^2}^{-1}\frac{dL}{d\theta}^T\right)^T\frac{d^2L}{d\theta^2}\frac{d^2L}{d\theta^2}^{-1}\frac{dL}{d\theta}^T \approx \\ \approx L(\hat{\theta}) + \frac{1}{2}\frac{dL}{d\theta}\frac{d^2L}{d\theta^2}^{-1}\frac{dL}{d\theta}^T \tag{B.5}$$

Take $L(\theta) - L_D(\lambda)$ as defined by (B.5):

$$L(\theta) - L_D(\lambda) \approx L(\hat{\theta}) + \frac{1}{2}\frac{dL}{d\theta}\frac{d^2L}{d\theta^2}^{-1}\frac{dL}{d\theta}^T \\ - L_D(\hat{\lambda}) - \frac{1}{2}\frac{dL_D}{d\lambda}\frac{d^2L_D}{d\lambda^2}^{-1}\frac{dL_D}{d\lambda}^T = \\ = \frac{1}{2}\frac{dL}{d\theta}\frac{d^2L}{d\theta^2}^{-1}\frac{dL}{d\theta}^T - \frac{1}{2}\frac{dL_D}{d\lambda}\frac{d^2L_D}{d\lambda^2}^{-1}\frac{dL_D}{d\lambda}^T \tag{B.6}$$

The model structure $\mathbf{M}_{L_D}$ is a subset of the model structure $\mathbf{M}_L$, therefore is the parameters in $\mathbf{M}_L$, θ, a function, f, of the parameter in $\mathbf{M}_{L_D}$, λ. Define:

$$S(\lambda) = \frac{df(\lambda)}{d\lambda} \tag{B.7}$$

[2] idempotent matrix $M = M^2$

$$\theta = f(\lambda) \quad (B.8)$$

$$L_D(\lambda) = L(f(\lambda)) \quad (B.9)$$

$$\frac{dL_D}{d\lambda} = S(\lambda)L'(f(\lambda)) \quad (B.10)$$

$$\frac{d^2L_D}{d\lambda^2} = S^T(\lambda)L''(f(\lambda))S(\lambda) + L'(f(\lambda))\frac{dS(\lambda)}{d\lambda} \quad (B.11)$$

$$P = \frac{d^2L}{d\theta^2}^{-1} \quad (B.12)$$

$$S = S(\lambda) \quad (B.13)$$

S is a matrix of size $p_L \times p_{L_D}$ Definitions (B.7) to (B.13) and equation (B.6)

$$L - L_D \approx \frac{1}{2}\frac{dL}{d\theta}P\frac{dL}{d\theta}^T - \frac{1}{2}\frac{dL}{d\theta}S(S^TP^{-1}S)^{-1}S^T\frac{dL}{d\theta}^T \quad (B.14)$$

The distribution of $L - L_D$ depends on the distribution of the random variable $\frac{dL}{d\theta}$:
If $L \approx L_D$ then $C_N \approx I_N$ which leads to

$$\frac{dL}{d\theta} = -\frac{1}{2}\text{trace}\left(C_N^{-1}\frac{dC_N}{d\theta}\right) + \frac{1}{2}\varepsilon^T C_N^{-1}\frac{dC_N}{d\theta}C_N^{-1}\varepsilon$$

$$\approx -\frac{1}{2}\text{trace}(I_N) + \frac{1}{2}\varepsilon^T I_N \varepsilon = -\frac{N}{2} + \frac{1}{2}\varepsilon^T I_N \varepsilon \quad (B.15)$$

The residuals ε are normal distributed with zero mean and unit variance $\mathbf{N}(0,I)$. This means what $\varepsilon^T I_N \varepsilon$ according to lemma B.13 in (Söderström and Stoica, 1989) is $\chi^2(N)$-distributed.

$$\chi^2(N) \to \mathbf{N}(N, 2N) \text{ when } N \to \infty \quad (B.16)$$

(Chatfield, 1983). The distribution of $\frac{dL}{d\theta}^T$ is when $\mathbf{N}\left(0, \frac{N}{2}\right)$ and (B.4) gives

$$(\hat{\theta} - \theta) \text{ distributed } \mathbf{N}(0, \frac{N}{2}P^{-1}) \quad (B.17)$$

Introduce:

$$z = \sqrt{\frac{2}{N}}P^{1/2}\left(\frac{dL}{d\theta}\right)^T \quad (B.18)$$

which is distributed $\mathbf{N}(0,I)$ because of the distribution of $\frac{dL}{d\theta}^T$, where $P^{1/2}$ is defined by $P = P^{T/2}P^{1/2}$ and $P^{T/2} = (P^{1/2})^T$.

Combiening equations (B.14) and (B.18) leads to

$$L - L_D \approx$$

$$\approx \frac{N}{4}z^T P^{-T/2}[P - S(S^TP^{-1}S)^{-1}S^T]P^{-1/2}z =$$

$$= \frac{N}{4}z^T[I - A(A^TA)^{-1}A^T]z \quad (B.19)$$

where

$$A = P^{-T/2}S \quad (B.20)$$

A is a $p_L \times p_{L_D}$ matrix

$$z^T[I - A(A^TA)^{-1}A^T]z \approx \frac{4}{N}(L - L_D) = \kappa_D \quad (B.21)$$

According to lemma B.13 in (Söderström and Stoica, 1989), κ_D is χ^2-distributed with rank$[I - A(A^TA)^{-1}A^T]$ degrees of freedom, if $[I - A(A^TA)^{-1}A^T]$ is a idempotent matrix, since z is a random vector with distribution $\mathbf{N}(0,I)$.
Show that $[I - A(A^TA)^{-1}A^T]$ is idempotent:

$$[I - A(A^TA)^{-1}A^T][I - A(A^TA)^{-1}A^T] =$$

$$I - 2A(A^TA)^{-1}A^T + A(A^TA)^{-1}A^TA(A^TA)^{-1}A^T =$$

$$= [I - A(A^TA)^{-1}A^T] \quad (B.22)$$

The matrix is idempotent. According to lemma A.28 in (Söderström and Stoica, 1989) rank(M) = trace(M) if M is an idempotent matrix.

$$\text{rank}[I - A(A^TA)^{-1}A^T] =$$

$$= \text{trace}[I - A(A^TA)^{-1}A^T] =$$

$$= \text{trace}I - \text{trace}(A^TA)(A^TA)^{-1} =$$

$$= p_L - p_{L_D} \quad (B.23)$$

This shows that κ_D is χ^2 distributed with $p_L - p_{L_D}$ degrees of freedom but since ε_t has to be normalized to unit standard deviation with the estimated standard deviation, see equation (17), one degree of freedom will be lost (Chatfield, 1983) so the degrees of freedom will be $p_L - p_{L_D} - 1$.

REFERENCES

Billings, S.A. and Q.H. Tao (1991). Model validity tests for non-linear signal processing applications. *Int. J. Control* **54**(1), 157–194.

Chatfield, C. (1983). *Statistics for technology*. third ed.. Chapman and Hall.

Gibbs, M.N. (1997). Bayesian Gaussian Processes for Regression and Classification. PhD thesis. University of Cambridge, United Kingdom.

Haber, R. (1985). Nonlinearity tests for dynamic processes. In: *IFAC identification and system parameter estimation.*

Leontaritis, I.J. and S.A. Billings (1987). Model selection and validation methods for non-linear systems. *Int. J. Control* **45**(1), 311–341.

Ljung, L. (1999). *System identification*. second ed.. Prentice hall.

Rasmussen, C. E. (1996). Evaluation of Gaussian processes and other methods for non-linear regression. PhD thesis. University of Toronto, Canada.

Rasmussen, C.E. (2001). Matlab program. www.kyb.tuebingen.mpg.de/bs/people/carl/code/.

Sjöberg, J. (1995). Non-linear System Identification with Neural Networks. PhD thesis. Linköping University, Sweden.

Söderström, T. and P. Stoica (1989). *System identification*. Prentice hall.

Stoica, P., P. Eykhoff, P. Janssen and T. Söderström (1986). Model-structure by cross-validation. *Int. J. Control* **43**(6), 1841–1878.

IFAC Publications
www.elsevier.com/locate/ifac

FUNCTIONAL ANALYTIC FRAMEWORK FOR MODEL SELECTION

Masashi Sugiyama *,[1]

* *Tokyo Institute of Technology, Tokyo, Japan*

Abstract: Model selection is one of the most important tasks in the identification of black-box systems. In this paper, we give a novel model selection method from the viewpoint of functional analysis. We formulate the system identification problem as a function approximation problem in a reproducing kernel Hilbert space (RKHS), where the approximation error is measured by the RKHS norm. Within this framework, we derive an estimator of the approximation error called the subspace information criterion (SIC) and show its properties.

Keywords: functional analysis, model selection, reproducing kernel Hilbert space, subspace information criterion.

1. INTRODUCTION

Model selection is one of the most important tasks in the identification of black-box systems. The goal of model selection is to determine the model such that the approximation error between an estimated system and the true system is minimized. However, the approximation error usually depends on the unknown system so it can not be directly calculated. One of the general approaches to model selection is to derive an estimator of the approximation error and then to determine the model such that the estimator is minimized. So far, a number of methods for estimating the approximation error have been proposed from various different standpoints, e.g., methods based on the asymptotic statistics (Akaike, 1974; Murata *et al.*, 1994), the Vapnik-Chervonenkis (VC) theory (Vapnik, 1995), resampling techniques (Efron, 1979; Wahba, 1990; Efron and Tibshirani, 1993), and the Bayesian statistics (Schwarz, 1978; Akaike, 1980).

In this paper, we give a novel model selection method from the viewpoint of functional analysis and show its properties.

[1] The author would like to thank Prof. Hidemitsu Ogawa for his valuable comments and discussions. This research is partially supported by MEXT, Grants-in-Aid for Scientific Research, 14380158 and 14780262.

2. PROBLEM FORMULATION

Let us regard a black-box system as a real-valued function $f(\boldsymbol{x})$ of d variables defined on a subset $\mathcal{D}$ of the d-dimensional Euclidean space $\mathbb{R}^d$. We would like to identify the function $f(\boldsymbol{x})$ from a set of n input-output samples. A sample consists of an input value $\boldsymbol{x}_i$ in $\mathcal{D}$ and a corresponding output value y_i in $\mathbb{R}$. We assume that the output value y_i is degraded by the additive noise ϵ_i with mean zero. That is, the set of samples are expressed as

$$\{(\boldsymbol{x}_i, y_i) \mid y_i = f(\boldsymbol{x}_i) + \epsilon_i\}_{i=1}^{n}. \quad (1)$$

We consider the case where the unknown function $f(\boldsymbol{x})$ belongs to a specified *reproducing kernel Hilbert space* (RKHS) $\mathcal{H}$. The *reproducing kernel* of a functional Hilbert space $\mathcal{H}$ is a bivariate function defined on $\mathcal{D} \times \mathcal{D}$. Let us denote the reproducing kernel of $\mathcal{H}$ by $K(\boldsymbol{x}, \boldsymbol{x}')$. The reproducing kernel of $\mathcal{H}$ satisfies the following conditions (Aronszajn, 1950):

- For any fixed $\boldsymbol{x}'$ in $\mathcal{D}$, $K(\boldsymbol{x}, \boldsymbol{x}')$ is a function of $\boldsymbol{x}$ in $\mathcal{H}$.
- For any function f in $\mathcal{H}$ and for any $\boldsymbol{x}'$ in $\mathcal{D}$, it holds that

$$\langle f(\cdot), K(\cdot, \boldsymbol{x}')\rangle_{\mathcal{H}} = f(\boldsymbol{x}'), \quad (2)$$

where $\langle \cdot, \cdot \rangle_{\mathcal{H}}$ stands for the inner product in $\mathcal{H}$.

Let $\hat{f}(\boldsymbol{x})$ be an estimate of the function $f(\boldsymbol{x})$. The goal of system identification is to find $\hat{f}(\boldsymbol{x})$ such that it is as 'close' to $f(\boldsymbol{x})$ as possible. We shall measure the closeness between $\hat{f}(\boldsymbol{x})$ and $f(\boldsymbol{x})$ by the expected squared norm in the RKHS $\mathcal{H}$:

$$\mathrm{E}_{\boldsymbol{\epsilon}}\|\hat{f} - f\|_{\mathcal{H}}^2, \tag{3}$$

where $\mathrm{E}_{\boldsymbol{\epsilon}}$ denotes the expectation over the noise $\{\epsilon_i\}_{i=1}^n$ and $\|\cdot\|_{\mathcal{H}}$ denotes the norm in the RKHS $\mathcal{H}$. That is, the goal is to find $\hat{f}$ from $\mathcal{H}$ such that

$$\min_{\hat{f}\in\mathcal{H}} \mathrm{E}_{\boldsymbol{\epsilon}}\|\hat{f} - f\|_{\mathcal{H}}^2. \tag{4}$$

Note that we do not take the expectation over input points $\{\boldsymbol{x}_i\}_{i=1}^n$, as is done in some statistical model selection theories (Akaike, 1974; Murata *et al.*, 1994). Therefore, our approach may be more data-dependent. Since $\|f\|_{\mathcal{H}}^2$ does not depend on $\hat{f}$, we subtract it and use the following measure as the approximation error.

$$\begin{aligned} J &= \mathrm{E}_{\boldsymbol{\epsilon}}\|\hat{f} - f\|_{\mathcal{H}}^2 - \|f\|_{\mathcal{H}}^2 \\ &= \mathrm{E}_{\boldsymbol{\epsilon}}\|\hat{f}\|_{\mathcal{H}}^2 - 2\mathrm{E}_{\boldsymbol{\epsilon}}\langle \hat{f}, f\rangle_{\mathcal{H}}, \end{aligned} \tag{5}$$

where $\langle\cdot,\cdot\rangle_{\mathcal{H}}$ denotes the inner product in the RKHS $\mathcal{H}$. The approximation error J defined by Eq.(5) can not be directly calculated since it includes the unknown function $f(\boldsymbol{x})$. The aim of this paper is to derive an estimator of the approximation error J.

3. ESTIMATING APPROXIMATION ERROR J

In this section, we derive an estimator of the approximation error J called the subspace information criterion (SIC) [2].

3.1 *Preliminary*

Our key idea for estimating the approximation error J is to use a linear unbiased estimate $\hat{f}_u$ of the unknown function f, instead of f itself. Here let us assume that we have a linear operator X_u such that

$$\hat{f}_u = X_u\boldsymbol{y}, \quad \mathrm{E}_{\boldsymbol{\epsilon}}\hat{f}_u = f, \tag{6}$$

where $\boldsymbol{y} = (y_1, y_2, \ldots, y_n)^\top$. We will discuss how to obtain the linear operator X_u in the following sections.

Letting $\boldsymbol{\epsilon} = (\epsilon_1, \epsilon_2, \ldots, \epsilon_n)^\top$, we have the following lemma.

Lemma 1. The approximation error J is expressed as

$$J = \mathrm{E}_{\boldsymbol{\epsilon}}\left(\|\hat{f}\|_{\mathcal{H}}^2 - 2\langle\hat{f}, \hat{f}_u\rangle_{\mathcal{H}} + 2\langle\hat{f}, X_u\boldsymbol{\epsilon}\rangle_{\mathcal{H}}\right). \tag{7}$$

Based on the above lemma, let us define 'preSIC' by

$$\mathrm{preSIC} = \|\hat{f}\|_{\mathcal{H}}^2 - 2\langle\hat{f}, \hat{f}_u\rangle_{\mathcal{H}} + 2\mathrm{E}_{\boldsymbol{\epsilon}}\langle\hat{f}, X_u\boldsymbol{\epsilon}\rangle_{\mathcal{H}}. \tag{8}$$

The above quantity is named preSIC because SIC will be derived based on this quantity. It is clear from Lemma 1 that preSIC satisfies

$$\mathrm{E}_{\boldsymbol{\epsilon}}\mathrm{preSIC} = J. \tag{9}$$

The third term in preSIC is expected over the noise, and it can not be directly calculated. In the following, we shall give methods of calculating or approximating the third term in preSIC under some conditions.

3.2 *SIC for Linear Estimates*

Let us consider the case where $\hat{f}$ is a linear estimate, i.e., with a linear operator X, $\hat{f}$ is given by

$$\hat{f} = X\boldsymbol{y}. \tag{10}$$

Eq.(10) includes, for example, least-squares or ridge estimation (Hoerl and Kennard, 1970) for linear or kernel regression models. A particular form of the Gaussian process regression (Williams and Rasmussen, 1996) and the least-squares support vector machines (Suykens *et al.*, 2002) are also included.

Let $\boldsymbol{Q}$ be the noise covariance matrix, $\mathrm{tr}(\cdot)$ be the trace of an operator, and X_u^* be the adjoint of X_u. Then we have the following lemma.

Lemma 2. When $\hat{f}$ is a linear estimate, it holds that

$$\mathrm{E}_{\boldsymbol{\epsilon}}\langle\hat{f}, X_u\boldsymbol{\epsilon}\rangle_{\mathcal{H}} = \mathrm{tr}\left(X\boldsymbol{Q}X_u^*\right). \tag{11}$$

Based on the above lemma, we define SIC for linear estimates as follows [3].

$$\mathrm{SIC} = \|\hat{f}\|_{\mathcal{H}}^2 - 2\langle\hat{f}, \hat{f}_u\rangle_{\mathcal{H}} + 2\mathrm{tr}\left(X\boldsymbol{Q}X_u^*\right). \tag{12}$$

It is clear that the above SIC is an unbiased estimator of the approximation error J.

$$\mathrm{E}_{\boldsymbol{\epsilon}}\mathrm{SIC} = J. \tag{13}$$

---

[2] The name 'subspace information criterion' came from the fact that in our early work (Sugiyama and Ogawa, 2001), the criterion was derived for the purpose of selecting subspace models in linear regression.

[3] In our early work (Sugiyama and Ogawa, 2001), SIC is defined as

$$\|\hat{f} - \hat{f}_u\|_{\mathcal{H}}^2 - \mathrm{tr}\left((X - X_u)\boldsymbol{Q}(X - X_u)^*\right) + \mathrm{tr}\left(X\boldsymbol{Q}X^*\right),$$

which is an unbiased estimator of Eq.(3). In the current paper, we ignored some constant terms that correspond to an estimate of $\|f\|_{\mathcal{H}}^2$ thus do not depend on X.

3.3 *SIC for Non-Linear Differentiable Estimates*

Here let us consider the case where $\hat{f}$ is a smooth non-linear estimate, i.e., with a twice almost differentiable (Stein, 1981) non-linear operator X, $\hat{f}$ is given by

$$\hat{f} = X(\boldsymbol{y}). \tag{14}$$

For example, some of the M-estimators such as Huber's robust estimation (Huber, 1981) for linear or kernel regression models are expressed by Eq.(14).

When $X(\boldsymbol{y})$ is almost differentiable, $[X_u^* X](\boldsymbol{y})$ is also almost differentiable since X_u^* is linear. Note that $[X_u^* X](\boldsymbol{y})$ is a vector-valued function from $\mathbb{R}^n$ to $\mathbb{R}^n$. Then we have the following lemma.

Lemma 3. When $\hat{f}$ is a smooth non-linear estimate and the noise $\{\epsilon_i\}_{i=1}^n$ is independently and identically drawn from the normal distribution with mean 0 and variance σ^2, it holds that

$$\mathrm{E}_{\boldsymbol{\epsilon}}\langle \hat{f}, X_u \boldsymbol{\epsilon}\rangle_{\mathcal{H}} = \sigma^2 \mathrm{E}_{\boldsymbol{\epsilon}} \sum_{i=1}^{n} \frac{\partial [X_u^* X]_i(\boldsymbol{y})}{\partial y_i}, \tag{15}$$

where $[X_u^* X]_i(\boldsymbol{y})$ is the i-th output of the vector-valued function $[X_u^* X](\boldsymbol{y})$.

Based on the above lemma, we define SIC for smooth non-linear estimates as follows.

$$\mathrm{SIC} = \|\hat{f}\|_{\mathcal{H}}^2 - 2\langle \hat{f}, \hat{f}_u\rangle_{\mathcal{H}} + 2\sigma^2 \sum_{i=1}^{n} \frac{\partial [X_u^* X]_i(\boldsymbol{y})}{\partial y_i}. \tag{16}$$

Note that even for smooth non-linear estimates, SIC is an unbiased estimator of J, i.e., Eq.(13) holds. It is easy to confirm that Eq.(16) agrees with Eq.(12) when $\hat{f}$ is a linear estimate. Therefore, Eq.(16) can be regarded as a natural extension of Eq.(12).

3.4 *Bootstrap Approximation of SIC for Non-Linear Estimates*

Finally, let us consider the case where $\hat{f}$ is a general non-linear estimate, i.e., with a general non-linear operator X, $\hat{f}$ is given by

$$\hat{f} = X(\boldsymbol{y}). \tag{17}$$

This includes, for example, ℓ_1-norm regularized estimation for linear or kernel regression models (Williams, 1995; Tibshirani, 1996; Chen *et al.*, 1998) or the support vector regression (Vapnik, 1995; Schölkopf and Smola, 2002). For general non-linear estimates, we shall approximate the third term in preSIC by the bootstrap method (Efron, 1979; Efron and Tibshirani, 1993). We define the bootstrap approximation of SIC (BASIC) as follows.

$$\mathrm{BASIC} = \|\hat{f}\|_{\mathcal{H}}^2 - 2\langle \hat{f}, \hat{f}_u\rangle_{\mathcal{H}} + 2\mathrm{E}_{\boldsymbol{\epsilon}}^b \langle \hat{f}^b, X_u \hat{\boldsymbol{\epsilon}}^b\rangle_{\mathcal{H}}. \tag{18}$$

More specifically, we calculate the third term $\mathrm{E}_{\boldsymbol{\epsilon}}^b \langle \hat{f}^b, X_u \hat{\boldsymbol{\epsilon}}^b\rangle_{\mathcal{H}}$ by bootstrapping residuals as follows.

1. Obtain an approximation $\hat{f}$ with samples $\{(\boldsymbol{x}_i, y_i)\}_{i=1}^n$ as usual.
2. Estimate the noise by $\hat{\epsilon}_i = y_i - \hat{f}(\boldsymbol{x}_i)$.
3. Create bootstrap noise samples $\{\hat{\epsilon}_i^b\}_{i=1}^n$ by sampling with replacement from $\{\hat{\epsilon}_i\}_{i=1}^n$.
4. Obtain an approximation $\hat{f}^b$ with the bootstrap samples $\{(\boldsymbol{x}_i, y_i^b) \mid y_i^b = \hat{f}(\boldsymbol{x}_i) + \epsilon_i^b\}_{i=1}^n$.
5. Calculate $\langle \hat{f}^b, X_u \hat{\boldsymbol{\epsilon}}^b\rangle_{\mathcal{H}}$.
6. Repeat 3. to 5. for a number of times and output the mean of $\langle \hat{f}^b, X_u \hat{\boldsymbol{\epsilon}}^b\rangle_{\mathcal{H}}$.

4. WHEN UNBIASED ESTIMATE OF f IS AVAILABLE

In the previous section, we derived SIC and its approximation for linear, smooth non-linear, and general non-linear estimates. In their derivations, we assumed that a linear unbiased estimate $\hat{f}_u$ of the learning target function f is available. In this section, we show how to obtain $\hat{f}_u$.

4.1 *Existence Condition for Unbiased Estimate of f*

The following theorem shows the existence condition for $\hat{f}_u$.

Theorem 4. A linear unbiased estimate $\hat{f}_u$ of the learning target function f exists if and only if the functions $\{K(\boldsymbol{x}, \boldsymbol{x}_i)\}_{i=1}^n$ span the whole RKHS $\mathcal{H}$.

Now we shall show how to obtain $\hat{f}_u$ under the situation where the condition in the above theorem is fulfilled. To this end, let us introduce the notion of the *Neumann-Schatten product* (Schatten, 1970). For any fixed g in a Hilbert space $\mathcal{H}_1$ and any fixed f in a Hilbert space $\mathcal{H}_2$, the Neumann-Schatten product of f and g, denoted by $(f \otimes \overline{g})$, is an operator from $\mathcal{H}_1$ to $\mathcal{H}_2$ that satisfies for any h in $\mathcal{H}_1$

$$(f \otimes \overline{g})\, h = \langle h, g\rangle f. \tag{19}$$

When both $\mathcal{H}_1$ and $\mathcal{H}_2$ are the Euclidean spaces, $(f \otimes \overline{g})$ is simply expressed as $(f \otimes \overline{g}) = f g^\top$. Using the Neumann-Schatten product, let us define the linear operator A by

$$A = \sum_{i=1}^{n} \left(\boldsymbol{e}_i \otimes \overline{K(\cdot, \boldsymbol{x}_i)}\right), \tag{20}$$

where $\boldsymbol{e}_i$ is the i-th standard basis in $\mathbb{R}^n$, i.e., it is the n-dimensional vector with the i-th element 1 and others 0. The property of the reproducing kernel implies that

$$Af = (f(\boldsymbol{x}_1), f(\boldsymbol{x}_2), \ldots, f(\boldsymbol{x}_n))^\top. \quad (21)$$

For this reason, A is called the sampling operator.

Let $A^\dagger$ be the Moore-Penrose generalized inverse of A (Albert, 1972). Then we have the following theorem.

Theorem 5. If the functions $\{K(\boldsymbol{x}, \boldsymbol{x}_i)\}_{i=1}^n$ span the whole RKHS $\mathcal{H}$, a linear operator X_u that provides an unbiased estimate $\hat{f}_u$ is given by

$$X_u = A^\dagger. \quad (22)$$

It can be confirmed that $A^\dagger$ provides the best linear unbiased estimate of f (Albert, 1972).

4.2 *SIC for Linear Regression Models*

Here we consider a standard linear regression problem, and show how SIC can be applied.

Let us consider the case where the unknown function f is of the form

$$f(\boldsymbol{x}) = \sum_{i=1}^{p} \alpha_i^* \varphi_i(\boldsymbol{x}), \quad (23)$$

where $\{\varphi_i(\boldsymbol{x})\}_{i=1}^p$ are the specified basis functions and $\{\alpha_i^*\}_{i=1}^p$ are unknown. We estimate $f(\boldsymbol{x})$ by the following linear regression model.

$$\hat{f}(\boldsymbol{x}) = \sum_{i=1}^{p} \alpha_i \varphi_i(\boldsymbol{x}). \quad (24)$$

$\{\alpha_i\}_{i=1}^p$ are parameters estimated by

$$\hat{\boldsymbol{\alpha}} = (\hat{\alpha}_1, \hat{\alpha}_2, \ldots, \hat{\alpha}_p)^\top = \boldsymbol{X}(\boldsymbol{y}), \quad (25)$$

where $\boldsymbol{X}$ is a vector-valued function from $\mathbb{R}^n$ to $\mathbb{R}^p$. The approximation error is measured by the expected weighted distance in the input domain $\mathcal{D}$.

$$\mathrm{E}_{\boldsymbol{\epsilon}} \int_{\mathcal{D}} \left(\hat{f}(\boldsymbol{x}) - f(\boldsymbol{x})\right)^2 w(\boldsymbol{x}) d\boldsymbol{x}, \quad (26)$$

where $w(\boldsymbol{x})$ is a specified weight function. Let $\boldsymbol{A}$ be the so-called *design matrix* whose (i, j)-th element is given by $\varphi_j(\boldsymbol{x}_i)$.

This setting corresponds to the case where the RKHS $\mathcal{H}$ is spanned by $\{\varphi_i(\boldsymbol{x})\}_{i=1}^p$ and the inner product is defined by

$$\langle f, g \rangle_{\mathcal{H}} = \int_{\mathcal{D}} f(\boldsymbol{x}) g(\boldsymbol{x}) w(\boldsymbol{x}) d\boldsymbol{x}. \quad (27)$$

Indeed, the reproducing kernel is given by

$$K(\boldsymbol{x}, \boldsymbol{x}') = \sum_{i=1}^{p} \varphi_i(\boldsymbol{x}) \tilde{\varphi}_i(\boldsymbol{x}'), \quad (28)$$

where $\tilde{\varphi}_i(\boldsymbol{x})$ is the dual of $\varphi_i(\boldsymbol{x})$ (Ogawa, 1998). When $\{\varphi_i(\boldsymbol{x})\}_{i=1}^p$ is the orthonormal basis in the RKHS $\mathcal{H}$, $\tilde{\varphi}_i(\boldsymbol{x})$ simply agrees with $\varphi_i(\boldsymbol{x})$.

Then we have the following theorem.

Theorem 6. When the rank of $\boldsymbol{A}$ is p, the functions $\{K(\boldsymbol{x}, \boldsymbol{x}_i)\}_{i=1}^n$ always span the whole RKHS $\mathcal{H}$.

Therefore, when the rank of $\boldsymbol{A}$ is p, an unbiased estimate $\hat{f}_u$ of the learning target function f exists. In this case, SIC can be calculated as follows.

- When $\boldsymbol{X}$ is linear,

$$\mathrm{SIC} = \hat{\boldsymbol{\alpha}}^\top \boldsymbol{U} \hat{\boldsymbol{\alpha}} - 2\hat{\boldsymbol{\alpha}}^\top \boldsymbol{U} \boldsymbol{A}^\dagger \boldsymbol{y} + 2\mathrm{tr}\left(\boldsymbol{U} \boldsymbol{X} \boldsymbol{Q} (\boldsymbol{A}^\top)^\dagger\right), \quad (29)$$

where $\boldsymbol{U}$ is the p-dimensional matrix whose (i, j)-th element is given by

$$\boldsymbol{U}_{ij} = \int_{\mathcal{D}} \varphi_i(\boldsymbol{x}) \varphi_j(\boldsymbol{x}) w(\boldsymbol{x}) d\boldsymbol{x}. \quad (30)$$

- When $\boldsymbol{X}$ is smooth non-linear,

$$\mathrm{SIC} = \hat{\boldsymbol{\alpha}}^\top \boldsymbol{U} \hat{\boldsymbol{\alpha}} - 2\hat{\boldsymbol{\alpha}}^\top \boldsymbol{U} \boldsymbol{A}^\dagger \boldsymbol{y} + 2\sigma^2 \sum_{i=1}^{n} \frac{\partial [(\boldsymbol{A}^\top)^\dagger \boldsymbol{U} \boldsymbol{X}]_i(\boldsymbol{y})}{\partial y_i}. \quad (31)$$

- When $\boldsymbol{X}$ is general non-linear,

$$\mathrm{BASIC} = \hat{\boldsymbol{\alpha}}^\top \boldsymbol{U} \hat{\boldsymbol{\alpha}} - 2\hat{\boldsymbol{\alpha}}^\top \boldsymbol{U} \boldsymbol{A}^\dagger \boldsymbol{y} + 2\mathrm{E}_{\boldsymbol{\epsilon}}^{b} \hat{\boldsymbol{\epsilon}}^{b\top} (\boldsymbol{A}^\top)^\dagger \boldsymbol{U} \hat{\boldsymbol{\alpha}}^b. \quad (32)$$

4.3 *Estimating Prediction Error and Test Error*

One of the common approximation error measures may be the prediction error (or the expected test error) defined by

$$\int_{\mathcal{D}} \left(\hat{f}(\boldsymbol{x}) - f(\boldsymbol{x})\right)^2 p(\boldsymbol{x}) d\boldsymbol{x}, \quad (33)$$

where $p(\boldsymbol{x})$ is the probability density function from which the (future) test input points are drawn. Letting $w(\boldsymbol{x}) = p(\boldsymbol{x})$, we can use SIC for estimating the prediction error.

However, $p(\boldsymbol{x})$ is often unknown so the matrix $\boldsymbol{U}$ can not be calculated. One of the options is to use the empirical distribution of $\{\boldsymbol{x}_i\}_{i=1}^n$ instead of $p(\boldsymbol{x})$ under the assumption that $\{\boldsymbol{x}_i\}_{i=1}^n$ are independently and identically drawn from $p(\boldsymbol{x})$. That is, $\boldsymbol{U}$ is estimated by

$$\boldsymbol{U}_{ij} \approx \frac{1}{n}\sum_{k=1}^{n} \varphi_i(\boldsymbol{x}_k)\varphi_j(\boldsymbol{x}_k). \tag{34}$$

In this case, it can be confirmed that SIC for linear estimates essentially agrees with Mallows's C_L (Mallows, 1973) and SIC for smooth non-linear estimates essentially agrees with Stein's unbiased risk estimator (Stein, 1981).

When input points without output values (which is often referred to as unlabeled samples) are available, another option comes in handy. That is, we estimate $p(\boldsymbol{x})$ by the empirical distribution of the unlabeled samples. Then $\boldsymbol{U}$ is estimated by

$$\boldsymbol{U}_{ij} \approx \frac{1}{n'}\sum_{k=1}^{n'} \varphi_i(\boldsymbol{x}'_k)\varphi_j(\boldsymbol{x}'_k), \tag{35}$$

where $\{\boldsymbol{x}'_i\}_{i=1}^{n'}$ are the unlabeled samples. Note that ordinary samples $\{\boldsymbol{x}_i\}_{i=1}^{n}$ can also be included in the set of unlabeled samples.

In some cases, test input points $\{\boldsymbol{x}''_i\}_{i=1}^{n''}$ are known in advance, and the goal is to estimate the output values $\{f(\boldsymbol{x}''_i)\}_{i=1}^{n''}$ corresponding to the test input points (which is often referred to as the transductive inference). In such cases, SIC can be used for estimating the error at the test points $\{\boldsymbol{x}''_i\}_{i=1}^{n''}$ by defining $\boldsymbol{U}$ as

$$\boldsymbol{U}_{ij} = \frac{1}{n''}\sum_{k=1}^{n''} \varphi_i(\boldsymbol{x}''_k)\varphi_j(\boldsymbol{x}''_k). \tag{36}$$

5. WHEN UNBIASED ESTIMATE OF f IS NOT AVAILABLE

We showed in Section 4 that a linear unbiased estimate $\hat{f}_u$ of the unknown function f exists if and only if the functions $\{K(\boldsymbol{x},\boldsymbol{x}_i)\}_{i=1}^{n}$ span the whole RKHS $\mathcal{H}$. In this section, we consider the case where the functions $\{K(\boldsymbol{x},\boldsymbol{x}_i)\}_{i=1}^{n}$ do not span the whole RKHS $\mathcal{H}$.

5.1 *Existence Condition for Unbiased Estimate of Projection of f*

So far, we searched the approximation $\hat{f}$ in the whole RKHS $\mathcal{H}$. Here we restrict ourselves to searching the approximation $\hat{f}$ within a subspace $\mathcal{S}$ of the RKHS $\mathcal{H}$.

Let $f_{\mathcal{S}}$ be the orthogonal projection of f onto the subspace $\mathcal{S}$. Recalling that $\langle \hat{f}, f\rangle_{\mathcal{H}} = \langle \hat{f}, f_{\mathcal{S}}\rangle_{\mathcal{H}}$ for $\hat{f} \in \mathcal{S}$, the approximation error J defined by Eq.(5) is expressed as

$$J = \mathrm{E}_{\boldsymbol{\epsilon}}\|\hat{f}\|_{\mathcal{H}}^2 - 2\mathrm{E}_{\boldsymbol{\epsilon}}\langle \hat{f}, f_{\mathcal{S}}\rangle_{\mathcal{H}}, \tag{37}$$

where f in Eq.(5) is simply replaced by $f_{\mathcal{S}}$. Therefore, if a linear unbiased estimate of the projection $f_{\mathcal{S}}$ is available, we can make the same discussion as Section 3. Indeed, the following theorem shows the existence condition for a linear unbiased estimate of the projection $f_{\mathcal{S}}$.

Theorem 7. (Sugiyama and Müller, 2002) A linear unbiased estimate of the projection $f_{\mathcal{S}}$ exists if and only if the subspace $\mathcal{S}$ is included in the span of the functions $\{K(\boldsymbol{x},\boldsymbol{x}_i)\}_{i=1}^{n}$.

This theorem means that a linear unbiased estimate of the projection $f_{\mathcal{S}}$ exists if $\hat{f}$ is searched in the span of the functions $\{K(\boldsymbol{x},\boldsymbol{x}_i)\}_{i=1}^{n}$, i.e., $\hat{f}$ is searched of the form

$$\hat{f}(\boldsymbol{x}) = \sum_{i=1}^{n} \alpha_i K(\boldsymbol{x},\boldsymbol{x}_i), \tag{38}$$

where $\{\alpha_i\}_{i=1}^{n}$ are parameters to be estimated.

The following theorem shows how to obtain a linear unbiased estimate of the projection $f_{\mathcal{S}}$.

Theorem 8. If the subspace $\mathcal{S}$ is included in the span of the functions $\{K(\boldsymbol{x},\boldsymbol{x}_i)\}_{i=1}^{n}$, a linear operator X_u that provides an unbiased estimate of the projection f is given by Eq.(22).

Theorems 5 and 8 show that the same $A^{\dagger}$ gives an unbiased estimate of f or the projection of f.

When parameters $\{\alpha_i\}_{i=1}^{n}$ in the kernel regression model (38) are estimated by

$$\hat{\boldsymbol{\alpha}} = (\hat{\alpha}_1, \hat{\alpha}_2, \ldots, \hat{\alpha}_n)^{\top} = \boldsymbol{X}(\boldsymbol{y}), \tag{39}$$

where $\boldsymbol{X}$ is a vector-valued function from $\mathbb{R}^n$ to $\mathbb{R}^n$, SIC can be calculated as follows.

- When $\boldsymbol{X}$ is linear,

$$\mathrm{SIC} = \hat{\boldsymbol{\alpha}}^{\top}\boldsymbol{K}\hat{\boldsymbol{\alpha}} - 2\hat{\boldsymbol{\alpha}}^{\top}\boldsymbol{y} + 2\mathrm{tr}\left(\boldsymbol{X}\boldsymbol{Q}\right). \tag{40}$$

- When $\boldsymbol{X}$ is smooth non-linear,

$$\mathrm{SIC} = \hat{\boldsymbol{\alpha}}^{\top}\boldsymbol{K}\hat{\boldsymbol{\alpha}} - 2\hat{\boldsymbol{\alpha}}^{\top}\boldsymbol{y} + 2\sigma^2\sum_{i=1}^{n}\frac{\partial[\boldsymbol{X}]_i(\boldsymbol{y})}{\partial y_i}. \tag{41}$$

- When $\boldsymbol{X}$ is general non-linear,

$$\mathrm{BASIC} = \hat{\boldsymbol{\alpha}}^{\top}\boldsymbol{K}\hat{\boldsymbol{\alpha}} - 2\hat{\boldsymbol{\alpha}}^{\top}\boldsymbol{y} + 2\mathrm{E}_{\boldsymbol{\epsilon}}^{b}\hat{\boldsymbol{\epsilon}}^{b\top}\hat{\boldsymbol{\alpha}}^{b}. \tag{42}$$

5.2 *Restriction on Generalization Measure*

As shown above, even when the functions $\{K(\boldsymbol{x},\boldsymbol{x}_i)\}_{i=1}^{n}$ do not span the whole RKHS $\mathcal{H}$, SIC can be applied if the kernel regression model (38) is used.

However, in this case, we should care about the fact that the shape of the kernel function $K(\boldsymbol{x}, \boldsymbol{x}')$ and the definition of the norm in the RKHS $\mathcal{H}$ relate each other. That is, if we use a desired kernel function, then we can no longer define the approximation error measure as desired. Conversely, if we use the desired approximation error measure, then the shape of the kernel function can no longer be chosen as desired.

For example, if the following Gaussian kernel is used

$$K(\boldsymbol{x}, \boldsymbol{x}') = g(\boldsymbol{x} - \boldsymbol{x}') = \exp\left(-\frac{\|\boldsymbol{x} - \boldsymbol{x}'\|^2}{2}\right), \quad (43)$$

then the norm in the Gaussian RKHS is given by

$$\|\hat{f} - f\|_{\mathcal{H}}^2 = \int \frac{\left(F[\hat{f}](\boldsymbol{\omega}) - F[f](\boldsymbol{\omega})\right)^2}{F[g](\boldsymbol{\omega})} d\boldsymbol{\omega}, \quad (44)$$

where $F[\cdot]$ denotes the Fourier transform. This norm has a property that high frequency components are strongly penalized.

6. CONCLUSIONS

In this paper, we formulated the system identification problem as a function approximation problem in a reproducing kernel Hilbert space (RKHS), and derived an estimator of the approximation error defined by the RKHS norm called the subspace information criterion (SIC). When the approximation function is estimated in a linear or smooth non-linear fashion, SIC has an analytic form and is an unbiased estimator of the true approximation error. When the approximation function is estimated in a general non-linear fashion, we proposed approximating SIC by the bootstrap method. SIC can be applied when a linear unbiased estimate of the unknown target function is available. We provided the necessary and sufficient condition for the existence of the linear unbiased estimate. For the cases where such a linear unbiased estimate does not exist, we further showed that SIC can be still applied if the approximation function is a kernel regression model.

7. REFERENCES

Akaike, H. (1974). A new look at the statistical model identification. *IEEE Transactions on Automatic Control* **AC-19**(6), 716–723.

Akaike, H. (1980). Likelihood and the Bayes procedure. In: *Bayesian Statistics* (N. J. Bernardo et al. Eds.). University Press. Valencia. pp. 141–166.

Albert, A. (1972). *Regression and the Moore-Penrose Pseudoinverse*. Academic Press. New York and London.

Aronszajn, N. (1950). Theory of reproducing kernels. *Transactions of the American Mathematical Society* **68**, 337–404.

Chen, S. S., D. L. Donoho and M. A. Saunders (1998). Atomic decomposition by basis pursuit. *SIAM Journal on Scientific Computing* **20**(1), 33–61.

Efron, B. (1979). Bootstrap methods: Another look at the jackknife. *The Annals of Statistics* **7**(1), 1–26.

Efron, B. and R. J. Tibshirani (1993). *An Introduction to the Bootstrap*. Chapman & Hall. New York.

Hoerl, A.E. and R.W. Kennard (1970). Ridge regression: Biased estimation for nonorthogonal problems. *Technometrics* **12**(3), 55–67.

Huber, P. J. (1981). *Robust Statistics*. John Wiley. New York.

Mallows, C. L. (1973). Some comments on C_P. *Technometrics* **15**(4), 661–675.

Murata, N., S. Yoshizawa and S. Amari (1994). Network information criterion — Determining the number of hidden units for an artificial neural network model. *IEEE Transactions on Neural Networks* **5**(6), 865–872.

Ogawa, H. (1998). Theory of pseudo biorthogonal bases and its application. In: *Research Institute for Mathematical Science, RIMS Kokyuroku, 1067, Reproducing Kernels and their Applications*. number 1067. pp. 24–38.

Schatten, R. (1970). *Norm Ideals of Completely Continuous Operators*. Springer-Verlag. Berlin.

Schölkopf, B. and A. J. Smola (2002). *Learning with Kernels*. MIT Press. Cambridge, MA.

Schwarz, G. (1978). Estimating the dimension of a model. *The Annals of Statistics* **6**, 461–464.

Stein, C. M. (1981). Estimation of the mean of a multivariate normal distribution. *The Annals of Statistics* **9**(6), 1135–1151.

Sugiyama, M. and H. Ogawa (2001). Subspace information criterion for model selection. *Neural Computation* **13**(8), 1863–1889.

Sugiyama, M. and K.-R. Müller (2002). The subspace information criterion for infinite dimensional hypothesis spaces. *Journal of Machine Learning Research* **3**(Nov), 323–359.

Suykens, J. A. K., T. Van Gestel, J. De Brabanter, B. De Moor and J. Vandewalle (2002). *Least Squares Support Vector Machines*. World Scientific Pub. Co.. Singapore.

Tibshirani, R. (1996). Regression shrinkage and selection via the lasso. *Journal of the Royal Statistical Society, Series B* **58**(1), 267–288.

Vapnik, V. N. (1995). *The Nature of Statistical Learning Theory*. Springer-Verlag. Berlin.

Wahba, H. (1990). *Spline Model for Observational Data*. Society for Industrial and Applied Mathematics. Philadelphia and Pennsylvania.

Williams, C. K. I. and C. E. Rasmussen (1996). Gaussian processes for regression. In: *Advances in Neural Information Processing Systems* (D. S. Touretzky, M. C. Mozer and M. E. Hasselmo, Eds.). Vol. 8. The MIT Press. pp. 514–520.

Williams, P. M. (1995). Bayesian regularization and pruning using a Laplace prior. *Neural Computation* **7**(1), 117–143.

IFAC Publications
www.elsevier.com/locate/ifac

ROBUST COMPLEXITY CRITERIA FOR NONLINEAR REGRESSION IN NARX MODELS

Jos De Brabanter * Kristiaan Pelckmans * Johan A.K. Suykens * Bart De Moor * Joos Vandewalle *

* *K.U.Leuven, ESAT-SCD-SISTA*
Kasteelpark Arenberg 10, B-3001 Leuven - Heverlee (Belgium)
{Jos.Debrabanter,Johan.Suykens}@esat.kuleuven.ac.be

Abstract: Many different methods have been proposed to construct a smooth regression function, including local polynomial estimators, kernel estimators, smoothing splines and LS-SVM estimators. Each of these estimators use hyperparameters. In this paper a robust version for general cost functions based on the Akaike information criterion is proposed.

Keywords: Akaike information criterion (AIC), smoother matrix, weighted LS-SVM, nonparametric variance estimator, influence function, breakdown point

1. INTRODUCTION

Most efficient learning algorithms in neural networks, support vector machines and kernel based methods (Vapnik, 1998) require the tuning of some extra learning parameters, or hyperparameters, denoted here by θ. In statistics, various hyperparameter selection criteria have been developed. One class of selection criteria is based on resampling. Cross-validation (Stone, 1974) and bootstrap (Efron, 1979) are methods in problems with i.i.d. data. Applications to time series and other dependent data (Markov chains and stochastic processes that are not indexed by time) are not straightforward. Another class, denoted by complexity criteria, (e.g. Akaike Information Criterion (Akaike, 1973), Baysian Information Criterion (Schwartz, 1979) and C_p statistic (Mallows, 1973) take the general form of a prediction error which consists of the sum of the training set error and a complexity term. The complexity term represents a penalty which grows as the number of free parameters in the model grows.

In the context of dependent data, it is often preferable to have a complexity criterion to select θ. These methods can handle both dependent and independent data. Complexity criteria have been developed often in the context of linear models, for assessing the generalization performance of trained models without the use of validation data. (Moody, 1992) has generalized such criteria to deal with nonlinear models and to allow for the presence of a regularization term. One advantage of cross-validation and bootstrap over complexity criteria is that they do not require estimates of the error variance. This means that complexity criteria require roughly correct working model to obtain the estimate of the error variance. In order to overcome such a problem, a nonparametric variance estimator (independent of the model) is used and the complexity criteria will work well even if the models being assessed are far from correct. (Akaike, 1973) found a relationship between the Kullback-Leibler distance and Fisher's maximized log-likelihood function. This relationship leads to an effective and very general methodology for selecting a parsimonious model for the analysis of empirical data. But when there are outliers in the observations (or if the distribution of the random errors has a heavy tail so that $E[|e_k|] = \infty$), it becomes very difficult to obtain good asymptotic results for the complexity criteria. In order to overcome such problems, robust complexity criteria are proposed in this paper.

In Section 2 we discuss the classical model selection criteria. Section 3 explains the robustness problem (extreme sensitivity) of a complexity criterion. In Section 4 we propose a new robust criterion for prediction. Section 5 illustrates the applications.

2. CLASSICAL MODEL SELECTION CRITERIA

Consider the nonparametric regression model

$$y_k = f(x_k) + e_k \tag{1}$$

where $x_1, ..., x_N$ are points in the Euclidean space $\mathbb{R}^d$, $f : \mathbb{R}^d \to \mathbb{R}$ is an unknown real-valued smooth function that we wish to estimate and the e_k are random errors with $E[e_k] = 0$ and $Var[e_k] < \infty$. Let $\mathcal{P}$ be a finite set of parameters. For $p \in \mathcal{P}$, let $\mathcal{F}_p$ be a set of functions f , let $Q_N(p) \in \mathbb{R}^+$ be a complexity term for $\mathcal{F}_p$ and let $\hat{f}$ be an estimator of f in $\mathcal{F}_p$. A crucial step in estimating $f(x)$ is choosing the hyperparameters. The hyperparameters are chosen to be the minimizer of a cost function defined as

$$J(\theta) = \frac{1}{N}\sum_{k=1}^{N} L\left(y_k, \hat{f}(x_k; \theta)\right) + \lambda\left(Q_N(p)\right)\hat{\sigma}_e^2 \tag{2}$$

where $\sum_{k=1}^{N} L(y_k, \hat{f}(x_k; \theta))$ is the residual sum of squares (RSS). The general cost function depends ony on $\hat{f}$ and the data. If $\hat{f}$ is defined by minimizing the empirical L_2 risk over some linear vector space $\mathcal{F}_p$ of functions with dimension K_p, then $J(\theta)$ will be of the form:

- Let $\lambda = 2$ and $Q_N(p) = N^{-1}K_p$
$$AIC(\theta) = \frac{1}{N}RSS + 2\left(\frac{K_p}{N}\right)\hat{\sigma}_e^2$$
- Let $\lambda = \log N$ and $Q_N(p) = N^{-1}K_p$
$$BIC(\theta) = \frac{1}{N}RSS + (\log N)\left(\frac{K_p}{N}\right)\hat{\sigma}_e^2$$
- Let $\lambda = \log\log N$ and $Q_N(p) = N^{-1}K_p$
$$J_1(\theta) = \frac{1}{N}RSS + (\log\log N)\left(\frac{K_p}{N}\right)\hat{\sigma}_e^2,$$
which has been proposed by (Hannon and Quinn, 1979) in the context of autoregressive model order determination.

The AIC was originally designed for parametric models as an approximately unbiased estimate of the expected Kullback-Leibler information. For linear regression and time series models, Hurvich and Tsai (Hurvich and Tsai, 1989) demonstrated that in small samples the bias of the AIC can be quite large, especially as the dimension of the candidate model approaches the sample size (leading to overfitting of the model), and they proposed a corrected version, $AICC$, which was found to be less biased than the AIC. The $AICC$ for hyperparameter selection is given by

$$AICC(\theta) = \frac{1}{N}RSS + \left(1 + \frac{2(K_p + 1)}{N - K_p - 2}\right)\hat{\sigma}_e^2 \tag{3}$$

where $\lambda = 1$ and $Q_N(p) = 1 + \frac{2(K_p+1)}{N-K_p-2}$.

An important property is that all the estimators are linear, in the sense that $\hat{f}(x; \hat{\theta}) = S(x; \hat{\theta})y$, where the matrix $S(x; \hat{\theta})$ is called the smoother matrix and depends on x but not on y (e.g. regression spline estimators, wavelet and LS-SVM estimators are linear estimators). Let $\mathcal{Q}_N$ be a finite set of effective number of parameters. For $q \in \mathcal{Q}_N$, let $\mathcal{F}_{N,q}$ be a set of functions f, let $Q_N(q) \in \mathbb{R}^+$ be a complexity term for $\mathcal{F}_{N,q}$ and let $\hat{f}$ be an estimator of f in $\mathcal{F}_{N,q}$. Based on (Moody, 1992) and by analogy, the hyperparameters are chosen to be the minimizer of a more generalized cost function defined as

$$J_C(\theta) = \frac{1}{N}RSS + \left(1 + \frac{2\mathrm{tr}(S(x; \hat{\theta})) + 2}{N - \mathrm{tr}(S(x; \hat{\theta})) - 2}\right)\hat{\sigma}_e^2. \tag{4}$$

Each of these selectors depends on $S(x; \hat{\theta})$ through its trace ($\mathrm{tr}(S(x; \hat{\theta})) < N - 2$), which can be interpreted as the effective number of parameters used in the fit. Because $J_C(\theta)$ is based on least squares estimation (via $L\left(y_k, \hat{f}(x_k; \theta)\right)$, $\hat{\sigma}_e^2$), it is very sensitive to outliers and other deviations from the normality assumption on the error distribution.

3. ANALYSIS OF ROBUSTNESS

3.1 *Contamination scheme*

In order to understand why certain estimators behave the way they do, it is necessary to look at the various measures of robustness. The effect of one outlier, on the expected Kullback-Leibler information estimator, in the sample can be described by the influence function (Hampel, 1974) which (roughly speaking) formalizes the bias caused by one outlier. Another measure of robustness is which amount of contaminated data an expected Kullback-Leibler information estimator can stand before it becomes useless. This aspect is covered by the breakdown point of the estimator. Given a random sample $\{(x_1, y_1), ..., (x_N, y_N)\}$ with common distribution F. To allow for the occurrence of outliers and other departures from the classical model we will assume that the actual distribution F of the data belongs to the contamination neighborhood

$$F_\epsilon = \{F : F = (1-\epsilon)F_0 + \epsilon H, \;\; H \text{ arbitrary}\} \tag{5}$$

where $0 \leq \epsilon < 0.5$. The interpretation is that we have a model F_0 which is not thought to hold exactly and to allow for this one considers an amount of contamination ϵ which is represented by the distribution H. If samples are generated from a F then on average a proportion $(1 - \varepsilon)$ of the observations will come from the model F_0 and a proportion ϵ of observations, the so called contaminants, will come from the distribution H. Next we give definitions of influence function, breakdown point and show that the $J_C(\theta)$ as described above is not robust.

3.2 *Measure of robustness*

Let F be a fixed distribution and $T(F)$ a statistical functional defined on a set $\mathcal{F}$ of distributions satisfying some regularity conditions (Hampel, 1974). Statistics which are representable as functionals $T(F_N)$ of the sample distribution F_N are called statistical functionals. Let $\mathcal{X}$ be a random variable, the relevant functional for the variance parameter σ^2 is $T(F) = \int (x - \int x dF(x))^2 dF(x)$. Let the estimator $T_N = T(F_N)$ of $T(F)$ be the functional of the sample distribution F_N. Then the influence function $IF(x; T, F)$ is defined as

$$IF(x, T; F) = \lim_{\epsilon \downarrow 0} \frac{T[(1-\epsilon)F + \epsilon\Delta_x] - T(F)}{\epsilon}. \tag{6}$$

Here Δ_x is the pointmass distribution at the point x. The IF reflects the bias caused by adding a few outliers at the point x, standardized by the amount ϵ of contamination.

3.3 *Influence function of the expected Kullback-Leibler information estimator*

Let $\mathcal{U}$ be a random variable with realizations $\xi_k = (y_k - \hat{f}(x_k; \theta))$. The corresponding statistical functional for the expected Kullback-Leibler information estimator is given by

$$T(F) = T_1(F) + T_2(F) = \int L(\xi)\, dF(\xi) + A \int \left(\xi - \int \xi dF(\xi)\right)^2 dF(\xi), \tag{7}$$

where $A = 1 + \frac{2[\mathrm{tr}(S(x;\hat{\theta}))+1]}{N - \mathrm{tr}(S(x;\hat{\theta})) - 2}$ and by model assumption $\int \xi dF(\xi) = 0$. From (6), (7) and $IF(x; T, F) = IF(x; T_1, F) + IF(x; T_2, F)$, it follows that $IF(\xi; T, F) = L(\xi) - T_1(F) + A\xi^2 - T_2(F) = A\xi^2 + \xi^2 - T(F_N)$. We see that the influence function is unbounded in $\mathbb{R}$. This means that an added observation at a large distance from $T(F_N)$ gives a large value in absolute sense for the influence function.

3.4 *L-estimators*

Consider the class of L-estimators in the location model, $x_k = \vartheta + e_k$, $k = 1, ..., N$. Let $x_{N(1)} \leq ... \leq x_{N(N)}$ be the order statistics corresponding to $x_1, ..., x_N$ with distribution F. The L-estimator of ϑ in the location model is of the form

$$T_N = \sum_{k=1}^{N} C_k x_{N(k)}, \tag{8}$$

where $\sum_{k=1}^{N} C_k = 1$ and the weights are usually chosen to be nonnegative and symmetric (i.e., $C_k \geq 0$ and $C_k = C_{N-k+1}$, $k = 1, ..., N$). This class covers a big scale of estimators from the sample mean to the sample median. To get a robust L-estimator of ϑ, we should put $C_k = 0$ for $k \leq g$ and for $k \geq N - g + 1$ with a proper $g > 0$. A typical example of such an estimator is the β-trimmed mean $(0 < \beta < 0.5)$

$$T_N(\beta) = \frac{1}{N - 2g} \sum_{k=g+1}^{N-g} x_{N(k)}, \tag{9}$$

where the trimming proportion β is selected so that $g = \lfloor N\beta \rfloor$. When F is no longer symmetric, it may sometimes be preferable to trim asymmetrically if the tail is expected to be heavier in one direction than the other. If the trimming proportions are β_1 on the left and β_2 on the right, the (β_1, β_2)-trimmed mean is defined as

$$T_N(\beta_1, \beta_2) = \frac{1}{N - (g_1 + g_2)} \sum_{k=g_1+1}^{N-g_2} x_{N(k)}, \tag{10}$$

where β_1 and β_2 are selected so that $g_1 = \lfloor N\beta_1 \rfloor$ and $g_2 = \lfloor N\beta_2 \rfloor$.

The trimmed mean has a simple structure and it is easy to use for practitioners. The corresponding statistical functional for the $(0, \beta_2)$-trimmed mean is given in terms of a quantile function and is defined as

$$T(0, \beta_2; F) = \frac{1}{1 - \beta_2} \int_0^{F^-(1-\beta_2)} x dF(x). \tag{11}$$

The quantile function of a cumulative distribution function F is the generalized inverse $F^- : (0,1) \to \mathbb{R}$ given by $F^-(q) = \inf\{x : F(x) \geq q\}$. In the absence of information concerning the underlying distribution function F of the sample, the empirical distribution function F_N and the empirical quantile function F_N^- are reasonable estimates for F and F^-, respectively. The empirical quantile function is related to the order statistics $x_{N(1)} \leq, \ldots, \leq x_{N(N)}$ of the sample and $F^-(q) = x_{N(k)}$, for $q \in \left(\frac{k-1}{N}, \frac{k}{N}\right)$.

4. ROBUST MODEL SELECTION CRITERIA

4.1 *A robust cost function*

A natural approach to robustify the expected Kullback-Leibler information estimator $J_C(\theta)$ is to replace: the $RSS = \sum_{k=1}^{N} (y_k - \hat{f}(x_k; \theta))^2$ by the corresponding robust counterpart RSS_{robust}:

$$RSS_{robust} = \sum_{k=1}^{N} v_k^2 \left(y_k - \hat{f}_{robust}(x_k; v_k, \theta)\right)^2 \tag{12}$$

where $\hat{f}_{robust}(x_k; v_k, \theta)$ is a weighted representation of the function estimate based on (e.g. M-estimator (Huber, 1964) or weighted LS-SVM (Suykens *et al.*, 2002*a*)). The weighting of $\hat{f}_{robust}(x_k; v_k, \theta)$ corresponding with $\{x_k, y_k\}$ is denoted by v_k. This means that data point k will be retained in the procedure if its residual is small to moderate, but disregarded if it is an outlier. An alternative is to replace the variance estimator by a robust estimator of variance.

4.2 *Estimation of variance*

The class of estimators of σ^2 that covers all approaches is that of quadratic forms in the output variable. The first subclass contains variance estimators based on the residuals from an estimator of the regression itself (Neumann, 1994). Difference-based variance estimators lead to the second subclass (Gasser *et al.*, 1986). These estimators are independent of a model fit, but the order of difference (the number of related outputs involved to calculate a local residual) has some influence.

Consider the NARX model (Ljung, 1987)

$$\hat{y}(t) = f\left(y(t-1), ..., y(t-l), u(t-1), ..., u(t-l)\right). \quad (13)$$

In practice, it is usually the case that only the ordered observed data $y(t)$ according to the discrete time index t, are known. The variance estimator suggested by (Gasser *et al.*, 1986) is used

$$\hat{\sigma}_e^2(y(t)) = \frac{1}{N-2}\sum_{k=2}^{N-1}\frac{(y(t-1)a + y(t+1)b - y(t))^2}{a^2+b^2+1} \quad (14)$$

where $a = \frac{(t+1)-t}{(t+1)-(t-1)}$ and $b = \frac{t-(t-1)}{(t+1)-(t-1)}$.

Let $\mathcal{Z} = \mathcal{U}^2$ be a function of a random variable with distribution $F(\mathcal{Z})$. Let $F_N(\mathcal{Z})$ be the empirical estimate of $F(\mathcal{Z})$. In (14), a realization of the random variable $\mathcal{U}$ is given by

$$\xi_k = \frac{(y(t-1)a + y(t+1)b - y(t))}{\sqrt{a^2+b^2+1}}. \quad (15)$$

The variance estimator (14) can now be written as an average of the random sample $\xi_1^2, ..., \xi_N^2$:

$$\hat{\sigma}_e^2 = \frac{1}{n-2}\sum_{k=2}^{n-1}\xi_k^2.$$

This can be viewed as an indication of the central value of the density function $g(\mathcal{Z})$. The mean is therefore sometimes referred to as a location parameter. The median, a robust counterpart of the mean, is also a location parameter. But in the asymmetric case the median is not necessarily equal to the mean. Other robust location estimators are: M-estimators (obtained as solutions of equations), L-estimators (linear functions of order statistics) and R-estimators (rank scores). Based on the knowledge of the distribution $F_N(\mathcal{Z})$, in practice $F_N(\mathcal{Z})$ is always concentrated on the positive axis with an asymmetric distribution, we can select the candidate location estimators. First, the Huber-type M-estimators have been found to be quite robust.For asymmetric distributions on the other hand, the computation of the Huber type M-estimators requires rather complex iterative algorithms and its convergence cannot be guaranteed in some important cases. Secondly, the trimmed mean (L-estimator) can be used when the distribution is symmetric and even when the distribution is asymmetric. In this paper we use the trimmed mean.

4.3 *Robust Kullback-Leibler information estimator*

The final robust expected Kullback-Leibler information estimator is given by

$$J_C(\theta)_{robust} = \frac{1}{\sum_{k=1}^N v_k^2} RSS_{robust} + \left(1 + \frac{2[\text{tr}(S^*(x;v_k,\hat{\theta})) + 1]}{n - \text{tr}(S^*(x;v_k,\hat{\theta})) - 2}\right)(\hat{\sigma}_e^2)_{robust} \quad (16)$$

where the smoother matrix $S^*(x;v_k,\hat{\theta})$ is now based on the weighting elements v_k and the robust variance estimator is given by

$$\left(\hat{\sigma}_e^2\right)_{robust} = \frac{1}{M - \lfloor M\beta_2 \rfloor}\sum_{m=\lfloor M\beta_2\rfloor+1}^{M-\lfloor M\beta_2\rfloor}\xi_{M(m)}^2 \quad (17)$$

where $m = 1, ..., M,\ \ M = N-1,\ \ m = k-2.$

The corresponding statistical functional for the robust expected expected Kullback-Leibler information estimator is given by

$$T(F) = \int v^2(\xi)L(\xi)dF(\xi) + A^*\frac{1}{(1-\beta_2)}\int_0^{F^-(1-\beta_2)}\xi^2 dF(\xi) \quad (18)$$

where $A^* = [1 + \frac{2tr(S^*(x,\theta))+2}{N - tr(S^*(x,\theta))-2}]$. ¿From (6) and (18) it follows that $IF(\xi, T_1; F) = v^2(\xi)\xi^2 - T_1(F_N)$ and

$$IF(\xi, T_2; F) = \begin{cases} \frac{\xi^2 - \beta_2 A^*(F^-(1-\beta_2))^2}{(1-\beta_2)} - T_2(F_N) \\ \quad \text{if } |\xi| \le F^{-1}(1-\beta_2) \\ A^*(F^-(1-\beta_2))^2 - T_2(F_N) \\ \quad \text{if } |\xi| > F^-(1-\beta_2). \end{cases}$$

We see that the influence functions are bounded in $\mathbb{R}$. This means that an added observation at a large distance from $T(F_N)$ gives a bounded value in absolute sense for the influence functions.

5. APPLICATION

5.1 *LS-SVM for nonlinear function estimation*

Given a training data set of N points $\{(x_k, y_k)\}_{k=1}^N$ with output data $y_k \in \mathbb{R}$, at input data $x_k \in \mathbb{R}^n$ according to equation (1). where the e_k are i.i.d. random errors with $E[e_k] = 0$, $Var[e_k] = \sigma^2 I_N$ and $f(x)$ is an unknown real-valued regression function that we wish to estimate. One considers the following optimization problem in primal weight space:

$$\min_{w,b,e} \mathcal{J}(w,e) = \frac{1}{2}w^T w + \frac{1}{2}\gamma\sum_{k=1}^N e_k^2 \quad (19)$$

such that $y_k = w^T\varphi(x_k) + b + e_k,\ \ k = 1, \ldots, N$ with $\varphi(.) : \mathbb{R}^n \rightarrow \mathbb{R}^{n_h}$ a function which maps the input space into a so-called higher dimensional (possibly infinite dimensional) feature space, weight vector $w \in \mathbb{R}^{n_h}$ in primal weight space, random errors $e_k \in \mathbb{R}$ and bias term b. Note that the cost function $\mathcal{J}$ consists of a RSS fitting error and a regularization term, which is also a standard procedure for the training of MLPs and is related to ridge regression.The relative importance of these terms is determined by the positive real constant γ. In the case of noisy data one avoids overfitting by taking a smaller γ value. Formulations of this form have been investigated in (Suykens *et al.*, 2002*b*). Without the bias term the mathematical solution is equivalent to regularization networks (Poggio and Girosi, 1990).

In primal weight space one has the model $y(x) = w^T\varphi(x) + b$ The weight vector w can be infinite dimensional, which makes a calculation in w from (19) impossible in general. Therefore, one computes the model in the dual space instead of the primal space. One defines the Lagrangian $\mathcal{L}(w, b, e; \alpha) = \mathcal{J}(w, e) - \sum_{k=1}^{N} \alpha_k \left(w^T\varphi(x_k) + b + e_k - y_k\right)$ with Lagrangian multipliers $\alpha_k \in \mathbb{R}$ (called support values). The conditions for optimality are given by $\frac{\partial \mathcal{L}}{\partial w} = \frac{\partial \mathcal{L}}{\partial b} = \frac{\partial \mathcal{L}}{\partial e_k} = \frac{\partial \mathcal{L}}{\partial \alpha_k} = 0$. After elimination of w and e one obtains the solution

$$\left[\begin{array}{c|c} 0 & 1_N^T \\ \hline 1_N & \Omega + \frac{1}{\gamma}I_N \end{array}\right] \left[\begin{array}{c} b \\ \hline \alpha \end{array}\right] = \left[\begin{array}{c} 0 \\ \hline y \end{array}\right] \tag{20}$$

with $y = [y_1; ...; y_N]$, $1_N = [1; ...; 1]$, $\alpha = [\alpha_1; ...; \alpha_N]$ and $\Omega_{kl} = \varphi(x_k)^T \varphi(x_k)$ for $k, l = 1, ..., N$. According to Mercer's theorem, there exists a mapping φ and an expansion $K(t, z) = \sum_k \varphi_k(t)\varphi_k(z)$, $t, z \in \mathbb{R}^n$ As a result, one can choose a kernel $K(.,.)$ such that $K(x_k, x_l) = \varphi(x_k)^T \varphi(x_l)$, $k, l = 1, ..., N$. The resulting LS-SVM model for function estimation becomes

$$\hat{f}(x; \theta) = \sum_{k=1}^{N} \hat{\alpha}_k K(x, x_k) + \hat{b} \tag{21}$$

where $\hat{\alpha}$, $\hat{b}$ are the solution to (20). We focus on the choice of an RBF kernel $K(x_k, x_l; h) = \exp\left\{-\|x_k - x_l\|_2^2/h^2\right\}$. One obtains $\hat{f}(x;\theta) = \Omega\alpha + 1_N b = S(x;\theta)y$ where $\theta = (h, \gamma)^T$. The LS-SVM for regression corresponds to

$$S(x;\theta) = \Omega(Z^{-1} - Z^{-1}\frac{J_N}{c}Z^{-1}) + \frac{J_N}{c}Z^{-1} \tag{22}$$

where $c = 1_N^T(\Omega + \frac{1}{\gamma}I_N)^{-1}1_N$, J_N is a square matrix of 1's and $Z = (\Omega + \frac{1}{\gamma}I_N)$. Therefore, the LS-SVM for regression is an example of a linear smoother. The linear operator $S(x; \lambda, \gamma)$ is known as the smoother matrix. An important property of the smoother matrix based on RBF kernels is that the $\mathrm{tr}[S(x;\theta)] < N$, except in the case $(h \to 0, \gamma \to \infty)$ where $\mathrm{tr}[S(x;\theta)] \to N$.

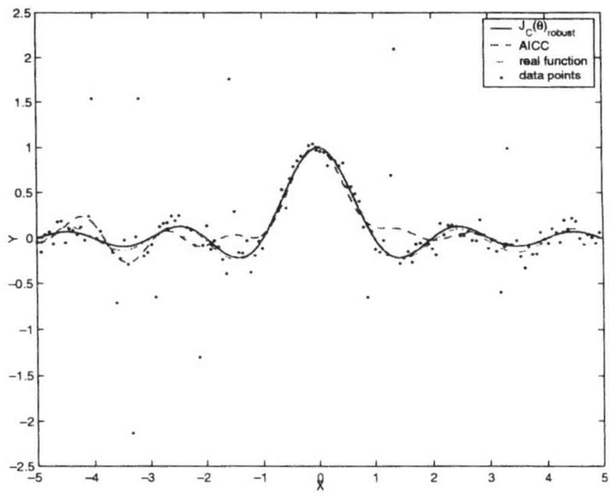

Fig. 1. *Experimental results comparing the LS-SVM tuned by AICC and the weighted LS-SVM tuned by $J_C(\theta)_{robust}$ on a noisy sinc dataset with heavy contamination.*

5.2 *Weighted LS-SVM for nonlinear function estimation*

In order to obtain a robust estimate based upon the previous LS-SVM solution, in a subsequent step, one can weight the error variables $e_k = \alpha_k/\gamma$ by weighting factors v_k (Suykens *et al.*, 2002*a*). This leads to the optimization problem:

$$\min_{w^*, b^*, e^*} \mathcal{J}(w^*, e^*) = \frac{1}{2}w^{*T}w^* + \frac{1}{2}\gamma\sum_{k=1}^{N} v_k e_k^{*2} \tag{23}$$

such that $y_k = w^{*T}\varphi(x_k) + b^* + e_k^*$, $k = 1, \ldots, N$ The Lagrangian is constructed in a similar way as before. The unknown variables for this weighted LS-SVM problem are denoted by the $*$ symbol. From the conditions for optimality and elimination of w^*, e^* one obtains the Karush-Kuhn-Tucker system:

$$\left[\begin{array}{c|c} 0 & 1_N^T \\ \hline 1_N & \Omega + V_\gamma \end{array}\right] \left[\begin{array}{c} b^* \\ \hline \alpha^* \end{array}\right] = \left[\begin{array}{c} 0 \\ \hline y \end{array}\right] \tag{24}$$

where the diagonal matrix V_γ is given by $V_\gamma = \mathrm{diag}\left\{\frac{1}{\gamma v_1}, ..., \frac{1}{\gamma v_N}\right\}$ The choice of the weights v_k is determined based upon the error variables $e_k = \alpha_k/\gamma$ from the (unweighted) LS-SVM case (20). Robust estimates are obtained then (see (Rousseeuw and Leroy, 1986)) e.g. by taking

$$v_k = \begin{cases} 1 & \text{if } |e_k/\hat{s}| \le c_1 \\ \frac{c_2 - |e_k/\hat{s}|}{c_2 - c_1} & \text{if } c_1 \le |e_k/\hat{s}| \le c_2 \\ 10^{-4} & \text{otherwise} \end{cases} \tag{25}$$

where $\hat{s} = 1.483\ \mathrm{MAD}(e_k)$ is a robust estimate of the standard deviation of the LS-SVM error variables e_k and MAD stands for the median absolute deviation.

5.3 *Illustrative examples*

Two examples illustrate the advantage of the proposed criteria. The first example is a static regression problem for a noisy sinc with heavy contamination (see Fig. 1). The second example considers a stochastic version of the logistic map $x_{t+1} = cx_t(x_t -$

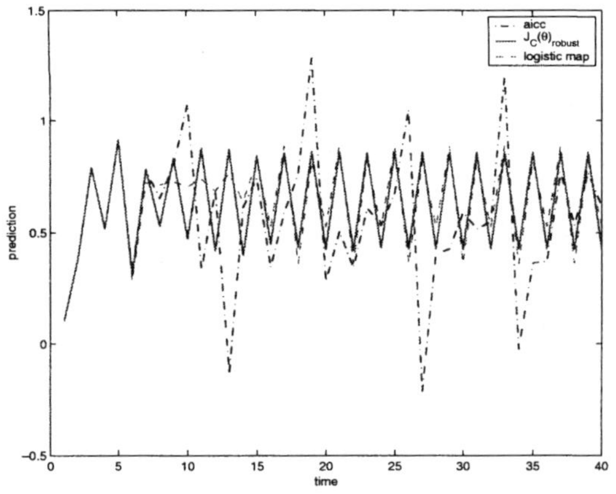

Fig. 2. *Recurrent predictions based on the logistic map, the LS-SVM tuned by AICC and the weighted LS-SVM tuned by $J_C(\theta)_{robust}$.*

Method	L_2	L_1	L_∞
Static Regression			
LS-SVM & AICC:	0.0133	0.0816	0.3241
WLS-SVM & $J_C(\theta)_{robust}$	0.0009	0.0247	0.1012
Stochastic Logistic Map, 40 points predicted recurrently			
LS-SVM & AICC:	0.1088	0.2251	1.0338
WLS-SVM & $J_C(\theta)_{robust}$	0.0091	0.0708	0.2659

Table 1. *Numerical comparison in different norms*

$1) + e_t$ with $c = 3.55$ and contaminating process noise e_t. The recurrent prediction illustrates the difference of the NAR model based on the LS-SVM tuned by AICC and the weighted LS-SVM tuned by $J_C(\theta)_{robust}$ (see Fig 2). The numerical test set performances of both are shown in Table 1 with improved results (in L_2, L_1, L_∞ norm) by applying the robust model selection criteria.

6. CONCLUSION

In this paper, robust complexity criteria based on AIC have been introduced. This can be used to tune the hyperparameters of weighted LS-SVMs in the case of non-Gaussian noise and outliers on the data for function approximation and NARX models.

Acknowledgements. This research work was carried out at the ESAT laboratory of the Katholieke Universiteit Leuven. It is supported by grants from several funding agencies and sources: Research Council KU Leuven: Concerted Research Action GOA-Mefisto 666 (Mathematical Engineering), IDO (IOTA Oncology, Genetic networks), several PhD/postdoc & fellow grants; Flemish Government: Fund for Scientific Research Flanders (several PhD/postdoc grants, projects G.0407.02 (support vector machines), G.0256.97 (subspace), G.0115.01 (bio-i and microarrays), G.0240.99 (multilinear algebra), G.0197.02 (power islands), research communities ICCoS, ANMMM), AWI (Bil. Int. Collaboration Hungary/ Poland), IWT (Soft4s (softsensors), STWW-Genprom (gene promotor prediction), GBOU-McKnow (Knowledge management algorithms), Eureka-Impact (MPC-control), Eureka-FLiTE (flutter modeling), several PhD grants); Belgian Federal Government: DWTC (IUAP IV-02 (1996-2001) and IUAP V-10-29 (2002-2006) (2002-2006): Dynamical Systems and Control: Computation, Identification & Modelling), Program Sustainable Development PODO-II (CP/40: Sustainibility effects of Traffic Management Systems); Direct contract research: Verhaert, Electrabel, Elia, Data4s, IPCOS. JS is a postdoctoral researcher with the National Fund for Scientific Research FWO - Flanders and a professor with KU Leuven. BDM and JVDW are full professors at KU Leuven, Belgium.

7. REFERENCES

Akaike, H. (1973). Statistical predictor identification. *Ann. Inst. Statist. Math.* **22**, 203–217.

Efron, B. (1979). Bootstrap methods: another look at the jackknife. *Ann. of Statist.* **7**(1), 1–26.

Gasser, T., L. Sroka and C. Jennen-Steinmetz (1986). Residual variance and residual pattern in nonlinear regression. *Biometrika* **73**, 625–633.

Hampel, F.R. (1974). The influence curve and its role in robust estimation. *J. Am. Statist. Ass.* **69**, 383–393.

Hannon, E.J and B.G. Quinn (1979). The determination of the order of autoregression. *J. Roy. Statist. Soc. Ser.* **B**(41), 190–195.

Huber, P.J. (1964). Robust estimation of a location parameter. *Ann. Math. Statist.* **35**, 73–101.

Hurvich, C.M. and C.L. Tsai (1989). Regression and time series model selection in small samples. *Biometrika* **76**, 297–307.

Ljung, L. (1987). *System Identification, Theory for the User*. Prentice Hall.

Mallows, C.L. (1973). Some comments on C_p. *Technometrics* **15**, 661–675.

Moody, J.E. (1992). The effective number of parameters: An analysis of generalization and regularization in nonlinear learning systems. In: *Neural Information Processing Systems*. Vol. 4. Morgan Kaufmann. San Mateo CA. pp. 847–854.

Neumann, M.H. (1994). Fully data-driven nonparametric variance estimators. *Statistics* **25**, 189–212.

Poggio, T. and F. Girosi (1990). Networks for approximation and learning. *Proceedings of the IEEE* **78**(9), 1481–1497.

Rousseeuw, P.J. and A.M. Leroy (1986). *Robust Regression and Outlier Detection*. Wiley & sons.

Schwartz, G. (1979). Estimating the dimension of a model. *Ann. of Statist.* **6**, 461–464.

Stone, M. (1974). Cross-validatory choice and assessment of statistical predictions. *J. Roy. Statist. Soc. Ser.* **B**(36), 111–147.

Suykens, J.A.K., J. De Brabanter, l. Lukas and J. Vandewalle (2002*a*). Weighted least squares support vector machines: robustness and sparse approximation. *Neurocomputing* **48**(1-4), 85–105.

Suykens, J.A.K., T. Van Gestel, De Brabanter J., B. De Moor and J. Vandewalle (2002*b*). *Least Squares Support Vector Machines*. World Scientific, Singapore.

Vapnik, V.N. (1998). *Statistical Learning Theory*. Wiley and Sons.

IFAC Publications
www.elsevier.com/locate/ifac

ANALYSIS OF MIMO CHANNEL MEASUREMENTS

Giovanni Del Galdo, Marko Hennhöfer, and Martin Haardt

Ilmenau University of Technology
Communications Research Laboratory
P.O. Box 100565
D-98684 Ilmenau, Germany

{Giovanni.Delgaldo, Marko.Hennhoefer, Martin.Haardt}@tu-ilmenau.de

Abstract: In this paper, we introduce a new method to estimate the Rice factor in MIMO (Multiple Input - Multiple Output) systems with a line of sight path (LOS) and omnidirectional scattering. This method is based on eigenvalue decompositions of the spatial correlation matrices at the transmitter and the receiver. To improve the performance, the correlation between subsequent temporal snapshots is used to get a more reliable estimate of the subspaces. The performance of the new technique is illustrated with MIMO channel sounder measurements obtained from a measurement campaign in Ilmenau. *Copyright © 2003 IFAC*

Keywords: Ricean channel, subspace projection, MIMO measurements, Rice factor estimation

1. INTRODUCTION

The trend towards MIMO systems in mobile communications promises huge capacity gains for the users. In many publications the optimistic Rayleigh channel model is used for MIMO simulations. When looking at measurement data it can be observed that in most cases a Ricean channel model is better suited.

In this paper we analyze MIMO channel measurements obtained with the multidimensional MIMO RUSK channel sounder (Thomä *et al.*, 2001) in Ilmenau, Germany. In Section 2 we define a four-dimensional measurement channel transfer array in space, frequency, and time. Section 3 reviews the data model and basic properties of the Ricean K-factor. Section 4 describes how the subspaces are determined and tracked. This information is used in Section 5 to remove the LOS (line of sight) component from the measured channel transfer functions via the spatial subtraction. In Sections 6 and 7 the method is verified and tested with MIMO channel sounder measurements. The paper concludes with Section 8.

2. STRUCTURE OF THE CHANNEL

The multidimensional MIMO channel sounder uses multiple antennas at the transmitter as well as at the receiver (Thomä *et al.*, 2001). By measuring the channel transfer function between any pair of the antennas on both sides, a four-dimensional channel transfer array $\mathcal{H} \in \mathbb{C}^{M_R \times M_T \times W \times N}$ can be constructed as depicted in Fig. 1, where M_R denotes the number of receive antennas, M_T is the number of transmit antennas, W is the number of samples in the frequency domain, and N denotes the number of temporal snapshots. The

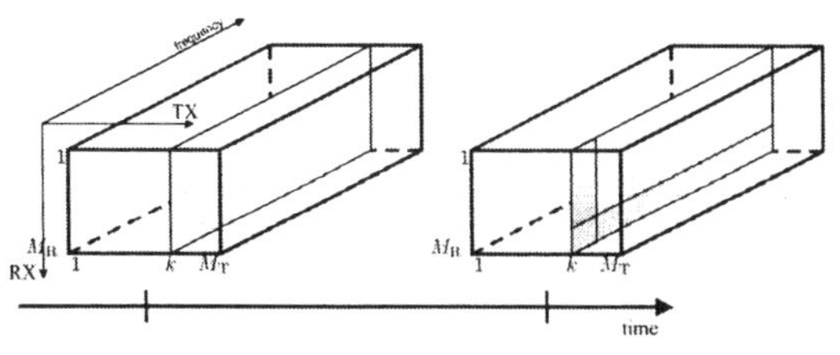

Fig. 1. Illustration of the channel array $\mathcal{H}$

high measurement repetition rate of the sounder hardware and its long-term recording capability enable the resolution of fast fading and the assessment of the long-term variations of the channel as well.

3. DATA MODEL AND BASIC PRINCIPLES

The Ricean K-factor is defined as the power of the line of sight (LOS) component divided by the total power of the scattered components.

$$K = \frac{P_{\text{LOS}}}{P_{\text{scatt}}} \tag{1}$$

Therefore it is convenient to model the channel as a sum of two different processes, one characterized by a strong LOS component and the other by diffuse scattering.

If we consider a single snapshot in time and flat-fading, the channel can be completely described by a single matrix $\boldsymbol{H}$ of size $M_{\text{R}} \times M_{\text{T}}$, where M_{R} and M_{T} represent the number of antennas. The two processes are additive and the final channel matrix can thus be expressed as a sum of two channel matrices, representing the LOS component and the scattering, respectively.

$$\boldsymbol{H} = \boldsymbol{H}_{\text{LOS}} + \boldsymbol{H}_{\text{scatt}} \tag{2}$$

The K-factor, being a ratio of two powers, suggests that the absolute power levels are not significant and we will thus take only the normalized channel matrix $\tilde{\boldsymbol{H}}$

$$\tilde{\boldsymbol{H}} = \frac{\boldsymbol{H}}{\sqrt{P_{\text{LOS}} + P_{\text{scatt}}}} \tag{3}$$

A similar normalization of the matrices $\boldsymbol{H}_{\text{LOS}}$ and $\boldsymbol{H}_{\text{scatt}}$ leads to

$$\tilde{\boldsymbol{H}} = \frac{\sqrt{P_{\text{LOS}}} \cdot \tilde{\boldsymbol{H}}_{\textbf{LOS}} + \sqrt{P_{\text{scatt}}} \cdot \tilde{\boldsymbol{H}}_{\textbf{scatt}}}{\sqrt{P_{\text{LOS}} + P_{\text{scatt}}}}, \tag{4}$$

where $\tilde{\boldsymbol{H}}_{\textbf{LOS}}$ and $\tilde{\boldsymbol{H}}_{\textbf{scatt}}$ are the normalized matrices representing a fully correlated channel and a fully uncorrelated channel, respectively. Equation (4) suggests that if we could remove from $\tilde{\boldsymbol{H}}$ the first term, the one corresponding to the LOS component, we would be able to extract the normalized scattering power $\tilde{P}_{\text{scatt}}$.

$$\tilde{P}_{\text{scatt}} = \frac{P_{\text{scatt}}}{P_{\text{LOS}} + P_{\text{scatt}}} \tag{5}$$

It is possible to rewrite equation (4) in terms of K as

$$\tilde{\boldsymbol{H}} = \sqrt{\frac{K}{1+K}} \cdot \tilde{\boldsymbol{H}}_{\textbf{LOS}} + \sqrt{\frac{1}{1+K}} \cdot \tilde{\boldsymbol{H}}_{\textbf{scatt}} \tag{6}$$

revealing that

$$\sqrt{\frac{1}{1+K}} = \frac{\sqrt{P_{\text{scatt}}}}{\sqrt{P_{\text{LOS}} + P_{\text{scatt}}}} = \sqrt{\tilde{P}_{\text{scatt}}} \tag{7}$$

equation (7) leads directly to the calculation of K as

$$K = \frac{1 - \tilde{P}_{\text{scatt}}}{\tilde{P}_{\text{scatt}}} \tag{8}$$

The basic idea of the proposed K-factor estimation scheme (presented in Sections 4 and 5) is to subtract from $\tilde{\boldsymbol{H}}$ the best estimate of $\tilde{P}_{\text{LOS}} \cdot \tilde{\boldsymbol{H}}_{\textbf{LOS}}$ in order to extract from the remaining matrix an estimate of $\tilde{P}_{\text{scatt}}$ which leads directly to the Ricean K-factor. This can be achieved via a subspace analysis of the available data in two separate steps.

First, the LOS subspace has to be identified and has to be tracked in time. This procedure determines good estimates of $\tilde{\boldsymbol{H}}_{\text{LOS}}$ for every time snapshot. The second step is the actual subtraction. The main problem is the determinination of $\tilde{P}_{\text{LOS}}$ in order to subtract all the power contained in the LOS component. This scaling problem can be solved by a subtraction of the subspaces via projection matrices. This spatial subtraction method (Del Galdo, 2002) is described in Section 5.

4. SUBSPACE IDENTIFICATION AND TRACKING

In this section we show how to estimate the power of the LOS path and the scattering components, respectively. There are several ways to estimate the LOS component. One way is to estimate the angle of arrival (AoA) and the angle of departure (AoD) via a parameter estimation technique such as MUSIC (Schmidt, 1981), ESPRIT (Roy *et al.*, 1986) or a multidimensional extension (Haardt and Nossek, 1998). However to perform this estimation, the array manifold is needed and in some cases special array geometries are required. A higher order statistics-based solution for the estimation of the K-factor is presented in (Abdi *et al.*, 2001)

In the approach presented in this paper, we identify the strongest path in the correlation matrices at the receiver and the transmitter side. To assess the structure of the channel, eigenvalue decompositions (EVD) of the correlation matrices can be computed. As the Ricean fading channel models a LOS component and some scattering components impinging from uniformly distributed angles, there will be one large eigenvalue representing the LOS component. The corresponding eigenvectors span a rank one subspace.

In MIMO measurements we can benefit from the temporal correlations between successive snapshots (Haardt *et al.*, 2001). In case of frequency selective channels, the channel transfer functions can be averaged over the frequency bins, where W is the number of samples in the frequency domain. By applying an exponential window, the effect of the noise is reduced and a more stable estimate of the subspaces is obtained. As the channel array $\mathcal{H}$ is four dimensional, the following Matlab notation is used: indexing with the colon operator (:) returns all elements of the corresponding dimension. Furthermore, let any singleton dimension be squeezed out, forming a new array of a smaller dimensionality. Then the long-term spatial correlation matrices at the transmitter and the receiver can be estimated as

$$\begin{aligned}\boldsymbol{R}_{\mathrm{R}}(n) &= \rho\, \boldsymbol{R}_{\mathrm{R}}(n-1) \\ &\quad + \frac{1-\rho}{W} \sum_{w=1}^{W} \mathcal{H}(:,:,w,n) \cdot \mathcal{H}(:,:,w,n)^H \\ \boldsymbol{R}_{\mathrm{T}}(n) &= \rho\, \boldsymbol{R}_{\mathrm{T}}(n-1) \\ &\quad + \frac{1-\rho}{W} \sum_{w=1}^{W} \mathcal{H}(:,:,w,n)^H \cdot \mathcal{H}(:,:,w,n).\end{aligned}$$

The forgetting factor ρ effects the window length of the averaging process. Its choice depends on the time variance of the channel and the resolution of the measurements in the time domain. To successfully use the strongest eigenbeam as the weight vector pointing only at the LOS path, it is crucial to assume that the scatterers are uniformly distributed around the arrays. In fact, if the scatterers were distributed in clusters, the first eigenbeam would map also these directions yielding to an inaccurate estimate of the LOS power. In this case, multidimensional high resolution parameter estimation schemes, e.g., (Haardt and Nossek, 1998), should be used to extract the LOS component.

5. SUBTRACTION OF THE SUBSPACE

Once the rank one subspace that corresponds to the LOS component has been identified for every time snapshot, it is possible to apply the spatial subtraction method outlined in this section. The spatial subtraction scheme removes all the power that is present in the estimated one-dimensional subspace. If $\boldsymbol{R}_{\mathrm{R}}(n)$ and $\boldsymbol{R}_{\mathrm{T}}(n)$ have been estimated as the spatial correlation matrices for a given time snapshot n, the eigenvectors corresponding to their largest eigenvalues are denoted as $\boldsymbol{u}_{\mathrm{R}}(n)$ and $\boldsymbol{u}_{\mathrm{T}}(n)$, respectively.[1] The corresponding projection matrices are given by

$$\begin{aligned}\boldsymbol{P}_{\text{left}} &= \boldsymbol{u}_{\mathrm{R}}(n) \cdot \boldsymbol{u}_{\mathrm{R}}^H(n) \\ \boldsymbol{P}_{\text{right}} &= \boldsymbol{u}_{\mathrm{T}}(n) \cdot \boldsymbol{u}_{\mathrm{T}}^H(n)\end{aligned}$$

Note that every matrix depends on the current time index n. The index has been omitted for a more convenient notation. Now we can remove from $\tilde{\boldsymbol{H}}$ the part of it which lies in the space spanned by $\boldsymbol{P}_{\text{left}}$ and $\boldsymbol{P}_{\text{right}}$, i.e., the space which contains all the power attributed to the LOS component.

$$\boldsymbol{H}_{\text{sub}} = \tilde{\boldsymbol{H}} - \boldsymbol{P}_{\text{left}} \cdot \tilde{\boldsymbol{H}} \cdot \boldsymbol{P}_{\text{right}}. \tag{9}$$

[1] Note that if multidimensional high resolution parameter estimation scheme is used (as indicated at the end of Section 4), the vectors $\boldsymbol{u}_{\mathrm{T}}(n)$ and $\boldsymbol{u}_{\mathrm{R}}(n)$ should be replaced by the array steering vectors corresponding to the AoD and the AoA of the LOS component, respectively.

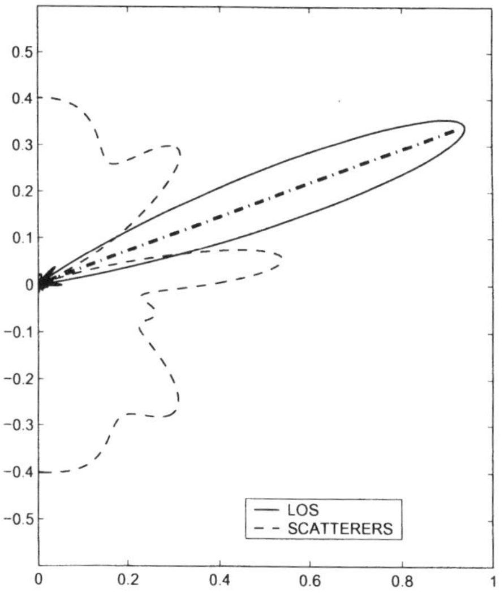

Fig. 2. Example of the spatial subtraction scheme. The dashed line shows the remaining normalized beampattern

The effect of this spatial subtraction procedure can be illustrated by analyzing the beampatterns. Fig. 2 shows a SIMO example. A strong line of sight component impinges on an eight-element uniform linear array (ULA) from an AoA of 20° (dash-dotted line). The one-dimensional subspace that is subtracted from the $\tilde{\boldsymbol{H}}$ matrix corresponds to the beampattern plotted with the solid line. With the spatial subtraction all the power collected by this beampattern will be subtracted. The power counted as power of the scatterers corresponds to the dashed beampattern. As Fig. 2 clearly shows, the power accredited to the LOS component is acquired mostly from the direction of the impinging wavefront while the scattering component is collected from all other directions. However, the estimate calculated in such a way is affected by a systematic bias. In fact, the solid beampattern will collect also the power of the scatterers within its main beam that will be added to the LOS component. In other words, the same amount of power will be subtracted from the scatterers and added to the power attributed to the LOS component leading to the following biased estimate

$$\hat{K} = \frac{P_{\text{LOS}} + \alpha \cdot P_{\text{scatt}}}{P_{\text{scatt}} - \alpha \cdot P_{\text{scatt}}}, \tag{10}$$

where $\alpha \cdot P_{\text{scatt}}$ is the fraction of the power incorrectly attributed to the LOS component. With an increasing number of antenna elements, the LOS beam gets narrower such that the bias of the estimate gets smaller. Under the assumption that the scatterers are uniformly distributed it is possible to assume that α does not depend on K. The normalized power that remains after the subtraction is

$$\tilde{P}_{\text{sub}} = \tilde{P}_{\text{scatt}} \cdot (1-\alpha) = \frac{P_{\text{scatt}}}{P_{\text{scatt}} + P_{\text{LOS}}}(1-\alpha)$$

and thus the correct estimate for the Ricean K-factor is equal to

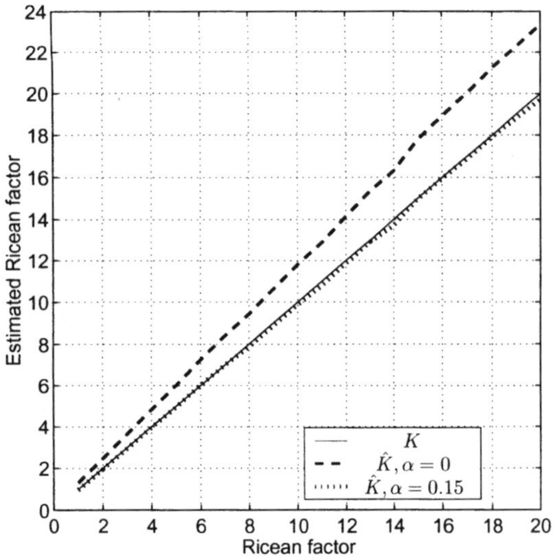

Fig. 3. Estimated Rice factors and the effect of α as a function of the K-factor used to generate the synthetic data, cf. Section 6

$$K = \frac{1 - \frac{\tilde{P}_{\mathrm{sub}}}{1-\alpha}}{\frac{\tilde{P}_{\mathrm{sub}}}{1-\alpha}}. \tag{11}$$

The factor α depends entirely on the beampattern and is independent of the actual powers involved, as seen in Fig. 3. A more detailed explanation of this figure is given in the next section. Under the assumption that the scatterers remain uniformly distributed around the array, that there is a LOS component, and that the array has a circular geometry, we can assume that any beampattern generated will have a main lobe with the same constant beamwidth. Such an assumption is realistic for UCAs (Uniform Circular Arrays) and CUBAs (Circular Uniform Beam Array). In other words, if the beamwidth of the main lobe remains constant for any angle, then α can be estimated and applied in any case. For other geometries, like ULAs (Uniform Linear Array) we observe that the beamwidth of the main lobe changes significantly with the AoA of the LOS path. Therefore, we have to estimate α for every eigenbeam obtained. If the array steering matrix is known, or in other words, if the antenna response is known for every angle θ, then it is possible to evaluate the fraction of power collected by the mainlobe of the array caused by the uniformly distributed scatterers around the antenna.

6. SYNTHETIC DATA ANALYSIS

To verify the accuracy of the estimation method it is appropriate to test the algorithm in a controlled scenario in which the Rice factor K is known. This can be easily be achieved by developing a synthetic channel model. The channel matrix $\tilde{\boldsymbol{H}}$ must match the initial assumptions. This means that it must be generated from two independent processes corresponding to a perfect LOS component and uniformly distributed scatterers around the receiver, respectively, i.e.,

$$\boldsymbol{H} = \boldsymbol{H}_{\mathrm{LOS}} + \boldsymbol{H}_{\mathrm{scatt}} \tag{12}$$

The matrix $\boldsymbol{H}_{\mathrm{LOS}}$ can be built conveniently using a geometric model. Assuming θ_{R} and θ_{T} for the angle of arrival (AoA) and angle of departure (AoD), respectively, and ULAs (Uniform Linear Arrays) at both sides of the link (Rx and Tx) we can write the channel matrix as

$$\boldsymbol{H}_{\mathrm{LOS}} = \boldsymbol{a}_{\mathrm{R}} \cdot \sqrt{\boldsymbol{P}_{\mathrm{LOS}}} \cdot \boldsymbol{a}_{\mathrm{T}}^{H} \tag{13}$$

where $\boldsymbol{a}_{\mathrm{R}}$ is the normalized array steering vector for the receiver

$$\boldsymbol{a}_{\mathrm{R}} = \left[1 \; e^{j\mu} \; e^{j2\mu} \ldots \; e^{j(M_{\mathrm{R}}-1)\mu}\right] \tag{14}$$

$$\mu = -j\frac{2\pi}{\lambda}\Delta \sin(\theta_{\mathrm{R}}) \tag{15}$$

and $\boldsymbol{a}_{\mathrm{T}}$ is constructed in the same way for the transmitter. The spacing between the antennas is Δ and the wavelength is equal to λ. To generate the matrix representing the scattering, a statistical model which better suits this scenario was used. In fact, the channel matrix for a uniform distribution of the scatterers corresponds to a full rank matrix in which the amplitude of every element follows the same Rayleigh distribution. To obtain such a distribution we ensure that every element has its real and imaginary parts generated from the same Gaussian random process. A proper normalization has to be applied in order to obtain $\tilde{\boldsymbol{H}}_{\mathrm{scatt}}$. At this point the estimation algorithm proposed can be applied and the results can be seen in Fig. 3. The channel matrix was generated for different values of K and for 8 antennas both at the receiver and at the transmitter.

7. ESTIMATE WITH MIMO MEASUREMENTS

The data for the following simulation resulted from a measurement campaign acquired at Ilmenau University of Technology. The location was the courtyard of a big building as seen in Fig. 4.

Looking at Fig. 4 it can be seen that at the starting position the receiver is in a corner of the courtyard (Rx2). Therefore a lot of scattered power is present at the receiver, resulting in a low K-factor. Then the transmitter and the receiver are moved simultaneously along the two lines (Tx7 ... Tx4 ... Tx3 and Rx2 ... Rx8 ... Rx9). At the end of the paths, the receiver gets less reflected components such that the K-factor increases. Fig. 5 shows the estimated values of K as a function of the measurement time. The standard deviation is plotted as well. It is calculated across the different frequencies and, therefore, it should not be interpreted as an indication of the quality of the estimate.

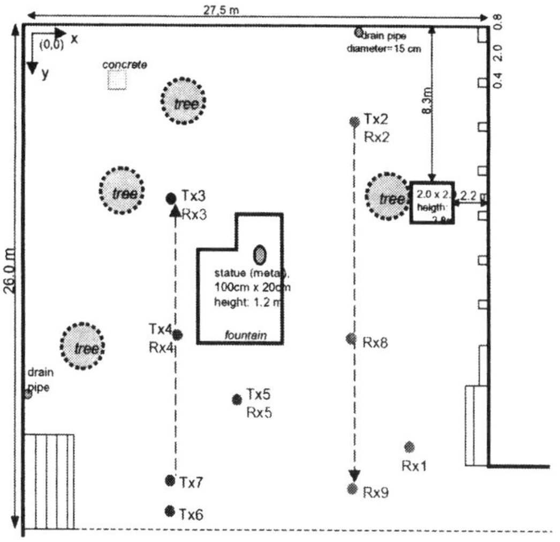

Fig. 4. Map of the measured scenario

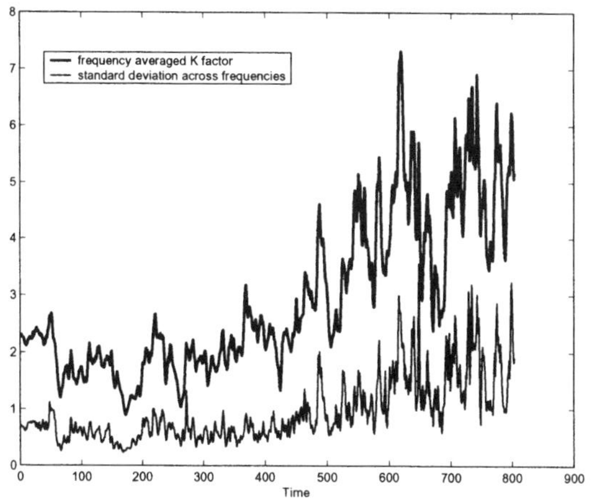

Fig. 5. The estimate for the K factor has been calculated for every frequency bin and time snapshot. The plot shows the mean and standard deviation across frequency.

8. CONCLUSIONS

The proposed algorithm yields to a good estimate of the Rice factor if the channel has a relatively strong LOS path compared to the scattered paths. If high resolution parameter estimation schemes are not used to determine the LOS component, the estimate of the K-factor still has a bias. This bias can be eliminated if the array responses are known or it might be neglected as the number of antennas increases, e.g., $M_{\mathrm{R/T}} \geq 16$.

9. REFERENCES

Abdi, A., C. Tepedelenlioglu, M. Kaveh and G. B. Giannakis (2001). On the estimation of the k parameter for the rice fading distribution. *IEEE Communication Letters* **5**(3), 92–94.

Del Galdo, G. (2002). *Efficient Beamforming in MIMO Systems via Subspace Analysis of Multidimensional Channel-Sounding Measurements*. Master Thesis, Ilmenau University of Technology. Ilmenau, Germany.

Haardt, M. and J. A. Nossek (1998). Simultaneous schur decomposition of several non-symmetric matrices to achieve automatic pairing in multidimensional harmonic retrieval problems. *IEEE Transactions on Signal Processing* **46**, 161–169.

Haardt, M., C. F. Mecklenbräuker, M. Vollmer and P. Slanina (2001). Smart antennas for UTRA TDD. *European Transactions on Telecommunications, Special Issue on Smart Antennas* **12**(5), 393–406.

Roy, R., A. Paulraj and T. Kailath (1986). Estimation of signal parameters via rotational invariance techniques ESPRIT. In: *Proc. IEEE ICASSP*. Tokyo, Japan. pp. 2495–2498.

Schmidt, R.O. (1981). *A signal subspace approach to multiple emitter location and spectral estimation*. Ph.D. Dissertation, Stanford University. Palo Alto, CA.

Thomä, R.S., D. Hampicke, A. Richter, G. Sommerkorn and U. Trautwein (2001). Mimo vector channel sounder measurement for smart antenna system evaluation. *European Transaction on Telecommunications, Special issue on smart antennas*.

IFAC Publications
www.elsevier.com/locate/ifac

PERFORMANCE EVALUATION OF MIMO CHANNEL PREDICTION ALGORITHMS USING MEASUREMENTS

Thomas Svantesson and Jon W. Wallace [1]

Department of Electrical and Computer Engineering
Brigham Young University, Provo, UT 84602-4099
E-mail: tomaso@ee.byu.edu, wall@ieee.org

Abstract: Knowledge of future channel conditions can increase the performance of many types of wireless systems. This is especially true for radio channels with multiple transmit and receive antennas, i.e. Multiple-Input Multiple-Output (MIMO) systems. This paper investigates the performance of several prediction schemes for MIMO channels using both simulations and measurements. It is found that in scenarios with few scattering clusters corresponding to outdoor scenarios, a significant performance gain is possible. Using a MIMO AR predictor, the prediction horizon almost doubles compared to the corresponding SISO AR predictor. However, for indoor scenarios, prediction algorithms that exploit the special structure of the MIMO radio channel are needed. *Copyright © 2003 IFAC*

Keywords: multivariable, predictor theory, telecommunication, measurement analysis

1. INTRODUCTION

In many wireless communications applications, knowledge of the future mobile radio channel is important (Duel-Hallen *et al.*, 2000). For instance, systems employing adaptive coding and modulation benefit of knowledge of future channel conditions. Power control and radio-resource allocation are other applications where channel knowledge can improve performance. Thus, there are many advantages to have knowledge of future channel conditions. Several different methods of predicting the future channel, i.e. channel prediction, based on past samples of the channel have been suggested (Andersen *et al.*, 1999; Duel-Hallen *et al.*, 2000; Ekman, 2002; Teal and Vaughan, 2001) for the case of single transmit and receive antennas. This channel will be referred to as the Single-Input Single-Output (SISO) channel.

Systems with multiple antennas at both the transmitter and receiver, so called Multiple-Input Multiple-Output (MIMO) systems, have recently been proposed for wireless communications (Paulraj *et al.*, 2003). These systems achieve high data rates and high spectral efficiency by exploiting the spatial domain to a larger extent by employing arrays at both the transmitter and receiver. In many applications, the performance of the system can be significantly increased if the transmitter knows the channel and can transmit the data using antenna weights that optimize performance. For example, large performance benefits are possible in communication scenarios where several users share the same frequency and each user has several antennas. Such a scenario is usually referred to as a multi-user MIMO (MU-MIMO) scenario which currently is an area of intense research (Paulraj *et al.*, 2003). Estimating the channel at the receiver and feeding the information back to the transmitter introduces a time delay which potentially can reduce the performance since outdated channel information is used in the transmit algorithm. Here, channel prediction could be used to provide current channel information.

Another example where channel prediction potentially could increase performance is in antenna selection schemes which currently are being investigated for MIMO systems (Paulraj *et al.*, 2003). It has been

[1] This work was supported in part by the National Science Foundation under Wireless Initiative Grant CCR-9979452 and Information Technology Research Grant CCR-0081476.

found that with channel knowledge, the performance achieved by adaptively selecting a subset of antennas is close to the performance when all antennas are utilized. In this case, channel prediction can be used to provide current channel knowledge to the algorithms for selecting the antenna subset. Hence, knowledge of the future MIMO channel via prediction is an interesting problem that potentially can increase the performance of wireless communication schemes.

Prediction of MIMO channels has an interesting benefit over traditional channel prediction using a single transmit and receive antenna since the signal is sampled at several points in space. Hence, more of the spatial structure of the incoming wave field is revealed and a better reconstruction of the wave field may be possible that would allow for better prediction of future channel samples. This paper will investigate the performance of channel prediction algorithms for MIMO channels using both simulated and measured data. It is found that in certain channel scenarios, the denser spatial sampling of MIMO systems indeed can increase prediction performance.

The performance of several standard prediction algorithms, applied to the MIMO channel, will be examined in this paper. To understand the behaviour of these algorithms when applied to real data, several simulation scenarios will be studied and the impact of important physical channel properties investigated. before the algorithms are applied to measurement data.

2. PROBLEM FORMULATION

The higher spectral efficiency of MIMO systems is achieved by employing multiple transmit and receive antennas that exploits the spatial properties of the radio channel. For narrowband systems, a single complex valued coefficient h_{ij} can be used to describe the path gain from transmit antenna j to receive antenna i at baseband. If the system has N_T transmit antennas and N_R receive antennas, these coefficients may be stacked into an $N_R \times N_T$ channel matrix $\mathbf{H}$. The received complex baseband signal $\mathbf{x}(t)$ of size $N_R \times 1$ at time t is then related to the transmitted $N_T \times 1$ signal $\mathbf{s}(t)$ as

$$\mathbf{x}(t) = \mathbf{H}(t)\mathbf{s}(t) + \mathbf{e}(t), \tag{1}$$

where $\mathbf{e}(t)$ denotes additive noise. In the following it will be assumed that the channel matrix $\mathbf{H}(t)$ is estimated at the receiver. Hence, the observations of the channel $\tilde{\mathbf{H}}(t)$ will usually contain some noise which can be modeled as

$$\tilde{\mathbf{H}}(t) = \mathbf{H}(t) + \mathbf{e}_H(t). \tag{2}$$

The problem considered in this paper is then to estimate or predict future samples of the channel matrix $\hat{\mathbf{H}}(t+L)$ given past samples of the channel $\{\tilde{\mathbf{H}}(t), \tilde{\mathbf{H}}(t-1), \ldots, \tilde{\mathbf{H}}(t-N)\}$. In the development of prediction schemes it will prove useful to stack the channel coefficients into a vector instead of a matrix $\mathbf{h}(t) = \text{vec}[\mathbf{H}(\text{t})]$.

A common way of modeling the wireless channel is to represent an electromagnetic wave as a ray. In a typical radio channel there are many paths through which electromagnetic waves or rays can reach the receiver from the transmitter. The total received field at the receiver is the sum of all these paths which sometimes add constructively and sometimes destructively contributing to a time-varying amplitude of the channel. This phenomenon is usually referred to as multipath fading and an accepted way of modeling this is to simply regard the channel as the sum of many individual rays (Jakes, 1974). Here, a simplified version of a MIMO channel model based on this concept (Svantesson, 2002) will be used to study the impact of different channel properties on prediction performance. Using this model, the channel matrix $\mathbf{H}(t)$ can be expressed as

$$\mathbf{H}(t) = \sum_{n=1}^{N_p} \alpha_n(t)\mathbf{a}_R(\phi_{R,n}(t))\mathbf{a}_T(\phi_{T,n}(t))^T, \tag{3}$$

where $\alpha_n(t)$ is the complex amplitude of ray n, the array response vectors at the transmitter and receiver are denoted $\mathbf{a}_R(x)$ and $\mathbf{a}_T(x)$, and the number of rays is denoted N_p. The angles $\phi_{R,n}(t)$ and $\phi_{T,n}(t)$ represents the Direction Of Arrival (DOA) and Direction Of Departure (DOD) of the nth ray at time t. For well-calibrated MIMO systems, the array response vectors can be considered known. However, maintaining an accurate calibration can be difficult so for some systems it may be better to consider them unknown. The time-varying complex amplitude $\alpha_n(t)$ is modeled as a complex exponential corresponding to the total time-delay due to the path length multiplied with a complex scalar representing the strength of the path. Hence, the channel can be viewed as a sum of complex exponentials which is known to capture the basic properties of a fading channel (Jakes, 1974).

In general both the amplitudes of the rays and the angles are time-varying but previous investigations of outdoor scenarios have found that the angles can vary much slower than the amplitudes (Arredondo *et al.*, 2002). Making that assumption, the channel can be re-written as

$$\mathbf{H}(t) = \sum_{n=1}^{N_p} \alpha_n(t)\mathbf{a}_R(\phi_{R,n})\mathbf{a}_T(\phi_{T,n})^T. \tag{4}$$

Here, it is interesting to observe that the introduction of antenna arrays at both ends results in more samples of the time-varying amplitudes $\alpha_n(t)$. At each time instant $N_T N_R$ samples are obtained compared to a single sample for a SISO system. Intuitively, this should allow for a more accurate prediction of the channel especially if well-calibrated antennas are used so that the array response vectors are known and only the angles are estimated. However, even with unknown array response vectors it seems likely that better prediction performance should be possible for a MIMO system than for a SISO system. Of course, if the angles instead are changing rapidly and equation (4) no longer is valid, the benefit of using multiple antennas diminishes.

Two channel scenarios will be simulated using the model in (3). An outdoor scenario where the rays arrive in two clusters each having an angular spread of $5°$. Note that there are clusters surrounding both the transmitter and the receiver as suggested in (Svantesson, 2002). Each cluster consists of ten rays with a corresponding complex amplitude for each ray. Sequences of channel data is then easily generated by moving the transmitter through the channel geometry defined by the transmit and receive angles as explained in (Svantesson, 2002). By creating channel data this way, many physical channel properties are incorporated which allows for a realistic study of the prediction performance. A second channel scenario will also be studied where the number of clusters is increased to ten to better represent a channel with dense multipath which may be encountered in indoor scenarios. For more details regarding the channel model, see (Svantesson, 2002).

3. PREDICTION ALGORITHMS

The purpose of this introductory study is to investigate the main differences between MIMO and SISO channel prediction and not to find the optimal predictor for the fading MIMO channel. Therefore, several prediction models based on standard data models will be used for prediction and the performance of SISO and MIMO channel prediction will be compared. For a more detailed description on the various predictors and their performance, see for instance (Ljung, 1999).

3.1 *AutoRegressive*

A common approach for channel prediction is to assume that the channel obeys the following AutoRegressive (AR) model

$$\mathbf{A}(q)\mathbf{h}(t) = \mathbf{e}(t), \qquad (5)$$

where $\mathbf{A}$ is a filter of order n_A and q is the forward shift operator. The prediction is easily calculated by extrapolating an AR model that is fitted to the training data using for example least-squares. The AR approach has been found to be a reasonable model for SISO channels with single clusters of scatterers (Ekman, 2002).

3.2 *AutoRegressive Moving Average*

A more general approach is to assume the following AutoRegressive Moving Average (ARMA) model

$$\mathbf{A}(q)\mathbf{h}(t) = \mathbf{C}(q)\mathbf{e}(t), \qquad (6)$$

where $\mathbf{C}$ is a filter of order n_C. The prediction values are calculated in the same manner as for the AR approach but the more general structure in (6) can often provide better predictions. It has also been found that the ARMA approach is a good model for SISO channels with several clusters of scatterers (Ekman, 2002).

3.3 *State-Space*

Another approach for channel prediction is to base the predictor on a State-Space (SS) model of the channel

$$\mathbf{z}(t) = \mathbf{A}\mathbf{z}(t-1) + \mathbf{B}\mathbf{u}(t) \qquad (7)$$

$$\mathbf{h}(t) = \mathbf{C}\mathbf{z}(t) + \mathbf{w}(t), \qquad (8)$$

where the channel states at time t are denoted $\mathbf{z}(t)$. The system matrices $\mathbf{A}, \mathbf{B}$, and $\mathbf{C}$ are estimated by minimizing the prediction error (Ljung, 1999). This approach has the advantage of being easy to extended to also estimating the channel at the receiver (Liu *et al.*, 2002). Once, the data model has been estimated, the prediction values are calculated using a Kalman predictor.

3.4 *Beamspace Approach*

The previous prediction algorithms are general approaches that can be applied to any signals. However, the signal model in (3) contains a lot of structure which can be exploited in the predictor algorithm. It is outside the scope of this paper to derive an optimal predictor given the signal model in (3) since the main interest here is to examine the main differences between MIMO and SISO channel prediction. However, an interesting low complexity approach is to apply the above predictors to a transformed version of the channel matrix $\mathbf{H}$. The important observation is that if the channel matrix is projected onto the subspace spanned by the outer product of the array responses, it is possible to separate out the different rays. Thus, the predictors are applied to the signal $\mathbf{Y}$

$$\mathbf{Y}(t) = \mathbf{W}^H\mathbf{H}(t)\mathbf{W} \qquad (9)$$

instead of $\mathbf{H}$ directly. Once the predictions of $\mathbf{Y}$ are calculated, the inverse transform of (9) is applied to obtain the predictions of $\mathbf{H}$. The transformation matrix $\mathbf{W}$ is easily obtained by calculating the antenna weights that form a set of beams which span the entire angular region from which signals are transmitted/arriving. Essentially, the received data is just projected into a number of angular regions. In general, it will not be possible to resolve each ray due to insufficient number of antennas or too closely spaced signals, but the projection reduces the number of rays present in each signal that is to be predicted. Therefore, it is expected that the prediction performance should improve since the signal experiences less variation when there are less rays present. For details in selecting the transformation (or beamformer), see for instance (Trees, 2002).

4. MEASUREMENT EQUIPMENT AND SETUP

A narrowband custom made MIMO communications system designed and built at Brigham Young University (BYU) in Utah was used to collect measurements. The system was equipped with ten monopoles forming a uniform circular array at each end. However, since the elements were mounted over a ground plane the monopoles behave as dipoles and essentially have the same radiation patterns as dipoles. Furthermore, the elements were positioned in a circle with radius 0.86 wavelengths (λ) that approximately gives an element separation of a half wavelength. The operating frequency was 2.43GHz. For a detailed description of the measurement equipment, see (Wallace and Jensen, 2001).

Although the most important application of channel prediction may be in a mobile outdoor scenario, many of the properties of outdoor channels are also present indoors. In fact, the indoor channel is much harder to predict since the indoor environment in general has significantly more multipath than outdoor environments. However, since several indoor measurement campaigns already have been conducted at BYU, these measurements will be used in this paper to examine the potential of MIMO channel prediction. Measurements were collected within the Clyde building at the BYU campus in both Line Of Sight (LOS) and Non Line Of Sight (NLOS) in order to obtain both slowly and rapidly varying channels.

5. PREDICTION RESULTS

In this section, the performance of the different prediction approaches of Section 3 will be studied and the potential benefit of MIMO prediction over SISO prediction examined. Several performance measures have been proposed in the literature for SISO channel prediction (Andersen *et al.*, 1999; Teal and Vaughan, 2001). However, the introduction of multiple subchannels in the MIMO case requires a new performance measure. A natural criterion in the MIMO case is the following Root Mean Square Error (RMSE) measure

$$\varepsilon(L) = \sqrt{\frac{E\left[||\hat{\mathbf{H}}(t+L) - \mathbf{H}(t+L)||_F^2\right]}{E\left[||\mathbf{H}(t)||_F^2\right]}} \tag{10}$$

where L denotes the prediction horizon and $||\cdot||_F$ denotes the Frobenius norm. This error measure essentially represents the average error of all the elements of the channel matrix $\mathbf{H}$. Hence, the error levels for MIMO and SISO can be compared using this measure.

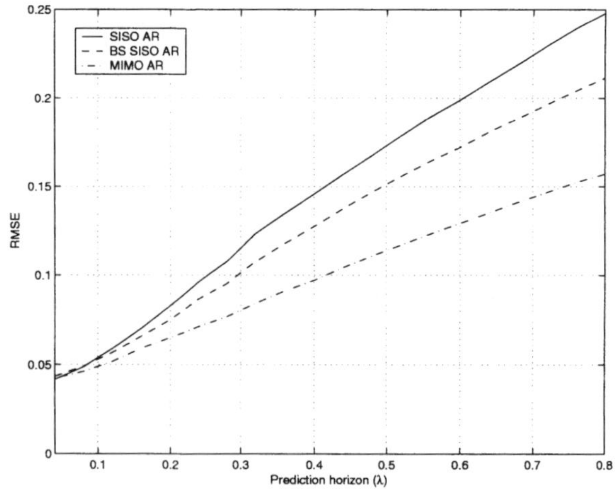

Fig. 1. The RMSE error ε for a MIMO system with N_T=N_R=3 versus prediction horizon for a two cluster scenario. The SISO AR and its beamspace version are of model order N_o=20 while the MIMO-AR predictor is N_o=4 and the SNR=30dB.

First the performance of the prediction schemes is examined for a simulated scenario with two clusters as described in Section 2. This scenario can be seen as an outdoor scenario with two distant clusters of scatterers that contribute to the multipath. Since few rays arrive at the receiver, the multipath fading is not severe and the correlation between signals received by spatially separated antennas decays slowly with separation distance. Hence, the elements of the channel matrix are correlated and using a MIMO approach should improve prediction performance.

The RMSE error ε for a MIMO system with a uniform linear array of size N_T=N_R=3 when $\mathbf{H}$ is observed in an SNR of 30dB is shown in Figure 1. A MIMO version of the AR approach is used as well as running nine SISO AR predictors corresponding to all elements of the $\mathbf{H}$ matrix. Finally, a beamspace approach is applied where the channel matrix first is transformed according to (9) and then using the nine SISO AR predictors. For the SISO predictors a model order N_o=20 was used while the MIMO AR only used N_o=4. A training length corresponding to about 8λ was used and the predictors were evaluated over 2λ.

It is clear that the MIMO version of the AR predictor manages to exploit the correlation between the elements of the $\mathbf{H}$ matrix to improve the prediction performance. In fact, at an error of 0.15 the prediction horizon for the MIMO approach is 0.75λ while the corresponding horizon for the SISO approach only is 0.42λ. Using the transformation approach yields a small gain. However, increasing the number of antennas increases the angular discrimination of the array which should result in better prediction performance. This is confirmed in Figure 2 where the same scenario is used but for a MIMO system with N_T=N_R=10. A significant performance gain for the beamspace ap-

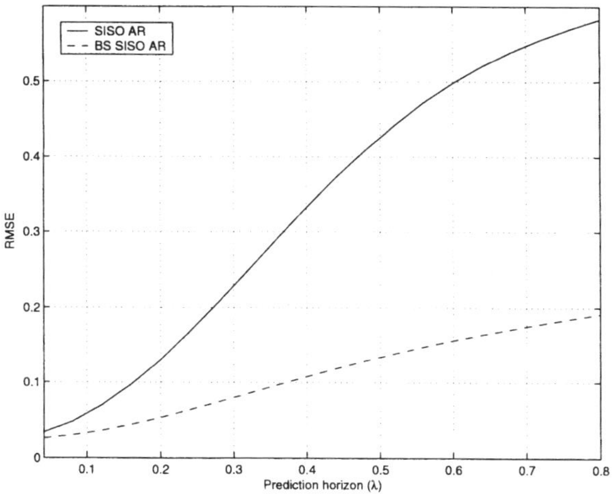

Fig. 2. The RMSE error ε for a MIMO system with N_T=N_R=10 versus prediction horizon for a two cluster scenario. The SISO AR and its beamspace version are of model order N_o=10 and the SNR=30dB.

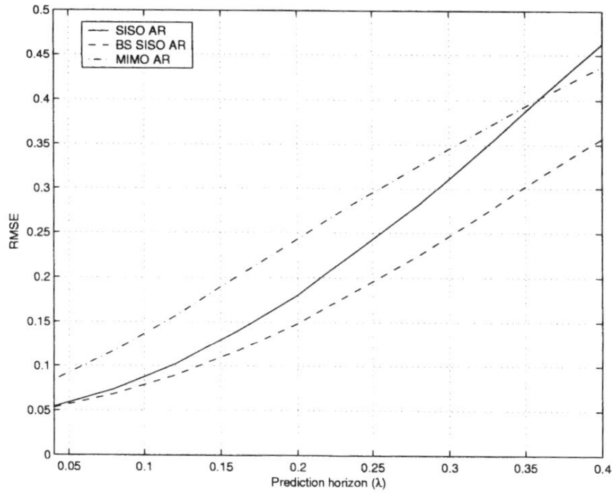

Fig. 3. The RMSE error ε for a MIMO system with N_T=N_R=3 versus prediction horizon for a ten cluster scenario. The SISO AR and its beamspace version are of model order N_o=10, MIMO AR is of order N_o=4 and the SNR=30dB.

proach is clearly visible. In fact, at the 0.15 error level the prediction horizon is more than doubled using the beamspace approach. Thus, exploiting the angular structure in the problem can improve prediction performance considerably. The MIMO AR approach could not be evaluated for this number of antennas due to the computational complexity of calculating the multivariable predictor for 100 variables. A very long training length is needed to estimate all the parameters for such a predictor. In practical scenarios, the time-variant nature of the radio channel limits the measurement length over which the channel can be considered stationary making the MIMO AR approach impractical for larger number of antennas. Trials with a MIMO version of the state-space approach resulted in about the same performance as the MIMO AR approach. However, since the SS approach is more general the performance suffers more from the limited training length than the MIMO AR since more parameters are required for the MIMO SS. In general it is only practical to use MIMO versions for small number of antennas.

The prediction in denser multipath is investigated in Figure 3 where all channel parameters are the same as in Figure 1 except for using ten clusters instead of two. With denser multipath, the correlation between signals received by spatially separated antennas decreases rapidly with separation distance. Hence, there is less correlation between the elements of the **H** matrix to exploit. This is clearly indicated in Figure 3 where the performance of the MIMO AR scheme is actually worse than the SISO AR. The reduced performance is due to the fact that there is less correlation to exploit and increased estimation error since the MIMO AR has to estimate more parameters than the SISO approach. Again, the beamspace version of SISO AR manages to exploit some of the angular structure to reach a small performance gain even for this three

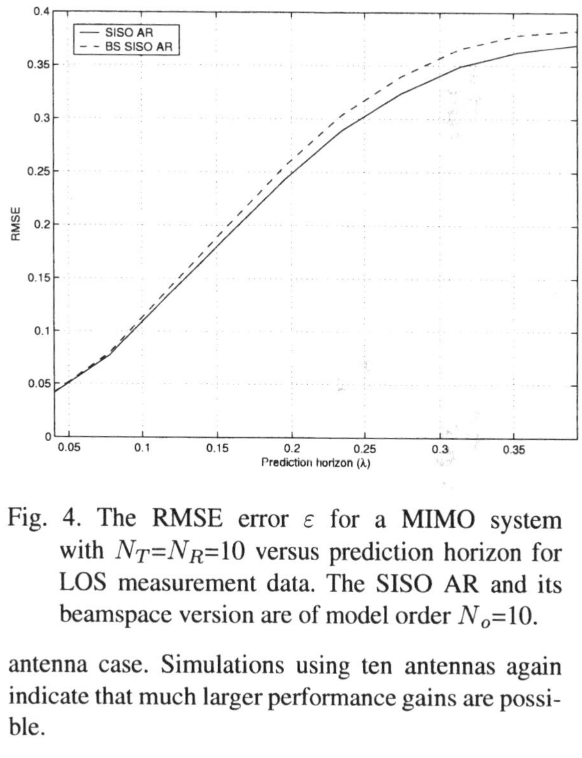

Fig. 4. The RMSE error ε for a MIMO system with N_T=N_R=10 versus prediction horizon for LOS measurement data. The SISO AR and its beamspace version are of model order N_o=10.

antenna case. Simulations using ten antennas again indicate that much larger performance gains are possible.

The antenna arrangement used for the measurements was a ten element (N_T=N_R=10) uniform circular antenna which requires a different beamforming matrix **W** than for the linear array used in Figure 1-3. Unfortunately, it is not straightforward to find a suitable beamforming matrix since the standard phase steering approach (Trees, 2002) yields high sidelobes which reduce performance. Here, a least-squares approach was used to get low sidelobes (Trees, 2002). The results from a LOS measurement is shown in Figure 4 where this beamspace approach is compared to a standard SISO AR. No noise was added to the measured data but it can be assumed that the data was collected under high SNR conditions. A training length corresponding to about 2λ was used and the predictors

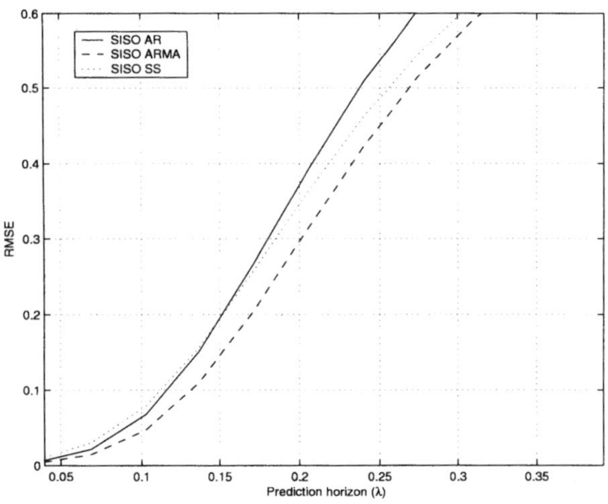

Fig. 5. The RMSE error ε for a MIMO system with N_T=N_R=10 versus prediction horizon for NLOS measurement data. The SISO AR, SISO ARMA, and SISO SS are of model order N_o=10.

were evaluated over 2λ. It is clear that the beamspace version does not provide any performance gain in this case. However, by careful design of the beamformer **W**, it may be possible to improve performance if there is angular structure exploit. Furthermore, the prediction horizon for the measured data is shorter than for the simulated data but not by several magnitudes. A possible explanation for the shorter prediction length in LOS is that even in LOS there are many rays arriving from many different directions contributing to a rapidly decaying correlation.

The prediction results using the same measurement setup but in a NLOS scenario are shown in Figure 5. The prediction horizon for the SISO AR method compared to the LOS scenario is slightly shorter. This indicates that the NLOS scenario is slightly harder to predict, possibly due to more rays. Furthermore, the more general approach of ARMA modeling yields slightly better performance. However, the general SS approach does not perform better indicating that too many parameters are estimated given the short training length.

6. CONCLUSIONS

The performance of several prediction schemes for MIMO channels were investigated using both simulations and measurements. It was found using simulations that in scenarios with relatively few clusters corresponding to outdoor scenarios, a significant performance gain is possible. Using a MIMO AR predictor, the prediction horizon almost doubled. Furthermore, a beamspace approach that exploited the angular structure of the channel was found to work well for MIMO systems with many antennas. Unfortunately, the MIMO AR predictor becomes impractical for larger number of antennas, excluding its use for many MIMO systems.

The measurement data was found to be harder to predict possibly indicating the presence of significant multipath. It is suspected that outdoor measurements would yield longer prediction horizons. The SISO ARMA predictor was found to be a good trade-off between complexity and performance for the measurement data.

The investigation in this paper indicates that in certain scenarios it is possible predict the MIMO radio channel further into the future than for a SISO channel. However, this study was performed for standard prediction algorithms. Significant improvements should be possible with predictors specifically designed for the MIMO radio channel which is an important topic of future research.

7. REFERENCES

Andersen, J.B., J. Jensen, S.H. Jensen and F. Fredriksen (1999). “Prediction of Future Fading Based on Past Measurements”. In: *Proc. IEEE VTC 1999 Fall*. Amsterdam. pp. 151–155.

Arredondo, A., K.R. Dandekar and G. Xu (2002). “Vector Channel Modeling and Prediction for the Improvement of Downlink Received Power”. *IEEE Trans. on Comm.* **50**(7), 1121–1129.

Duel-Hallen, A., S. Hu and H. Hallen (2000). “Long-Range Prediction of Fading Signals”. *IEEE Signal Processing Magazine* **17**(3), 62–75.

Ekman, T. (2002). Prediction of Mobile Radio Channels: Modeling and Design. PhD thesis. Signals and Systems, Uppsala Univeristy, Sweden.

Jakes, W. (1974). *Microwave Mobile Communications*. Wiley-Interscience. New York.

Liu, Z., X. Ma and G.B. Giannakis (2002). “Space-Time Coding and Kalman Filtering for Time-Selective Fading Channels”. *IEEE Trans. on Comm.* **50**(2), 183–186.

Ljung, L. (1999). *System Identification: Theory for the User*. second ed.. Prentice-Hall. Englewood Cliffs, NJ.

Paulraj, A., R. Nabar and D. Gore (2003). *Introduction to Space-Time Wireless Communications*. Cambridge University Press.

Svantesson, T. (2002). “A Double-Bounce Channel Model for Multi-Polarized MIMO Systems”. In: *Proc. IEEE VTC 2002 Fall*. Vol. 2. Vancouver BC, Canada. pp. 691–695.

Teal, P.D. and R.G. Vaughan (2001). “Simulation and Performance Bounds for Real-Time Prediction of the Mobile Multipath Channel”. In: *Proc. IEEE SSP 01*. Singapore. pp. 548–551.

Trees, H.L. Van (2002). *Optimum Array Processing*. Vol. IV. John Wiley and sons, Inc.. New York.

Wallace, J.W. and M.A. Jensen (2001). “Measured Characteristics of the MIMO Wireless Channel”. In: *Proc. IEEE VTC 01 Fall*. Atlantic City, NJ, USA. pp. 2038–2042.

IFAC Publications
www.elsevier.com/locate/ifac

HIGH-RESOLUTION CHANNEL PARAMETER ESTIMATION FOR COMMUNICATION SYSTEMS EQUIPPED WITH ANTENNA ARRAYS

Bernard H. Fleury * Xuefeng Yin * Patrik Jourdan ** Andreas Stucki **

* *Digital Communications Division, Department of Communication Technology, Aalborg University, DK-9220 Aalborg, Denmark*
** *Elektrobit AG, Rosswiesstrasse 29, CH-8608 Bubikon, Switzerland*

Abstract: This contribution describes an extension of the ISI-SAGE (initialization-and-search-improved space-alternating generalized expectation maximization) algorithm originally published in (Fleury *et al.*, 2002*b*) and (Fleury *et al.*, 2002*c*) to include polarization estimation. The proposed scheme allows for joint estimation of the relative delay, the direction (i.e. azimuth and co-elevation) of departure, the direction of incidence, the Doppler frequency and the polarization matrix of propagation paths between the transmitter and receiver sites in mobile radio environments.
Experimental investigations in a non-line-of-sight pico-/micro-cellular environment show that the polarization characteristics of individual propagation paths can be related directly to the types of interaction that the waves experience along their paths, such as reflection, diffraction, scattering. This detailed insight into the propagation mechanisms is of paramount importance for the design of accurate propagation models, i.e. stochastic models for optimization and performance simulation of communication systems equipped with multiple transmit and receive antennae as well as deterministic models for field prediction.

Keywords: High-resolution channel parameter estimation, polarization, maximum-likelihood estimation, multiple-input multiple-output (MIMO) channel.

1. INTRODUCTION

Recent works have demonstrated that the deployment of multiple-element arrays at both transmitter (Tx) and receiver (Rx) combined with appropriate coding can substantially increase the capacity of mobile radio communication systems (Foschini and Gans, 1998), (Telatar, 1999), and (Lozano *et al.*, 2001). Optimization and performance assessment of these communication systems require realistic models of the propagation channel that incorporate all kinds of dispersion effects, i.e. in delay, direction (azimuth and co-elevation) of departure, direction of arrival, Doppler frequency, and polarization. These models are designed based on extensive experimental investigations. The design process embraces selection of the appropriate models, estimation of the model parameters, and finally validation of these models. All three steps heavily rely on advanced signal processing techniques, e.g. in order to extract the critical channel parameters from measurement data.

The recently proposed high-resolution algorithm ISI-SAGE (Fleury *et al.*, 2002*a*), (Fleury *et al.*, 2002*c*) performs joint estimation of the complex weight, the relative delay, the Doppler frequency, the direction of departure, and the direction of arrival of propagation paths between the Tx and the Rx sites. The acronym ISI stresses the fact that the initialization procedure and the search procedures required for finding maxima are optimized in order to enhance the convergence and the path detection probability. Experimental investigations (Fleury *et al.*, 2002*b*) demonstrate the high potential of the ISI-SAGE algorithm for detailed propagation investigations as it makes it possible to infer the main propagation mechanisms between the Tx and Rx sites. This detailed insight into the propagation mechanisms is of paramount importance for the design of realistic channel models for the multiple-input multiple-output (MIMO) channel comprising the multiple Tx antennae, the propagation channel and the multiple Rx antennae.

The paper describes an extension of the ISI-SAGE algorithm that incorporates polarization estimation. The polarization of each propagation path is characterized by a 2×2 dim. polarization matrix of which the elements describe the co- and cross-polarization characteristics of the path.

The organization of the paper is as follows. Section 2 describes the signal model including polarization. Section 3 presents the channel sounding technique. The expectation and maximization steps of the SAGE algorithm are derived in Section 4. Sections 5 reports results from the experimental investigation in a Non-Line-Of-Sight (NLOS) pico-/micro-cellular environment. Finally some concluding remarks are addressed in Section 6.

2. SIGNAL MODEL

Fig. 1 illustrates the underlying model of the propagation environment. A certain number L of waves propagate along different paths from the M_1 Tx antennae forming Array 1 to the M_2 Rx antennae forming Array 2. Along its path a wave interacts with a certain number of obstacles/objects, so-called scatterers. At each interaction, the amplitudes and phases of the two polarization components of the incidence wave are altered depending on the type of interaction, the geometrical and electrical properties of the scatterer, as well as the incidence and re-radiation directions.

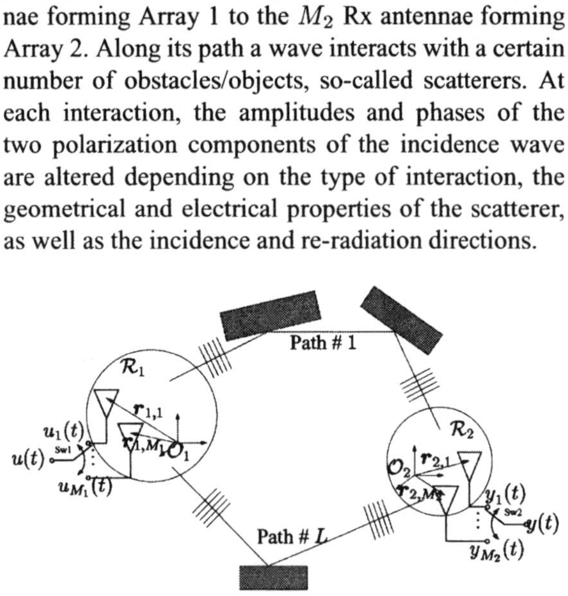

Fig. 1. Propagation model for channel sounding with multiple Tx and Rx antennae.

We assume that the far-field condition holds and that the elements of the arrays are confined in regions denoted as $\mathcal{R}_1$ and $\mathcal{R}_2$ respectively where the small-scale characterization applies (Fleury *et al.*, 1999). As a consequence of these assumptions, when any array is transmitting, the waves impinging in the region surrounding the other (receiving) array can be assumed to be plane waves. In the sequel Array 1 is transmitting and Array 2 is receiving.

Each antenna element in both arrays is characterized by its (complex electric) field patterns for the ϕ- and the θ-polarization (Jakes, 1974). For notational convenience we use the index $p \in \{1, 2\}$ to distinguish between the two types of polarization: $p = 1$ and $p = 2$ denote respectively the θ-polarization and the ϕ-polarization. Using this convention, $f_{k,m,p}(\boldsymbol{\Omega})$, $p = 1, 2$, denote the field patterns of the mth element ($m = 1, ..., M_k$) of Array k ($k = 1, 2$). Employing a coordinate system specified at an arbitrary origin O_k in $\mathcal{R}_k$, the location of this element is determined by a vector $\boldsymbol{r}_{k,m} \in \mathbb{R}^3$. Here $\mathbb{R}$ denotes the real line.

We define the input signal vector to be $\boldsymbol{u}(t) \doteq [u_1(t), ..., u_{M_1}(t)]^{\mathrm{T}}$, where $[\cdot]^{\mathrm{T}}$ denotes the transpose operator and the entry $u_m(t)$ ($m = 1, ..., M_1$) represents the (baseband representation of the) signal at the input of the mth element of Array 1. Then under the above assumptions, the contribution of the wave propagating along the ℓth path to the signal at the output of Array 2 can be written in vector notation as

$$\begin{aligned} \boldsymbol{s}(t; \boldsymbol{\theta}_\ell) &\doteq [s_1(t; \boldsymbol{\theta}_\ell), \ldots, s_{M_2}(t; \boldsymbol{\theta}_\ell)]^{\mathrm{T}} \\ &= \exp\{j2\pi\nu_\ell t\} \boldsymbol{C}_2(\boldsymbol{\Omega}_{2,\ell}) \boldsymbol{A}_\ell \boldsymbol{C}_1(\boldsymbol{\Omega}_{1,\ell}) \\ &\cdot \boldsymbol{u}(t - \tau_\ell). \end{aligned} \quad (1)$$

In this expression $\boldsymbol{\theta}_\ell \doteq [\boldsymbol{\Omega}_{1,\ell}, \boldsymbol{\Omega}_{2,\ell}, \tau_\ell, \nu_\ell, \boldsymbol{A}_\ell]$ is a vector of which the entries characterize the ℓth propagation path: $\boldsymbol{\Omega}_{1,\ell}, \boldsymbol{\Omega}_{2,\ell}, \tau_\ell, \nu_\ell$ and $\boldsymbol{A}_\ell$ denote the direction of departure, the direction of arrival, the propagation delay, the Doppler frequency, and the polarization matrix respectively.

We describe a direction as a unit vector $\boldsymbol{\Omega}$ with its initial point anchored at the reference location or equivalently as the tip of this vector, i.e. a point located on a sphere of unit radius centered at the reference point (Jakes, 1974). This unit vector $\boldsymbol{\Omega}$ is uniquely determined by its spherical coordinates $(\phi, \theta) \in [-\pi, \pi) \times [0, \pi]$ according to $\boldsymbol{\Omega} = [\cos(\phi)\sin(\theta), \sin(\phi)\sin(\theta), \cos(\theta)]^{\mathrm{T}}$. The angles ϕ and θ are referred to as, respectively, the azimuth and the co-elevation of $\boldsymbol{\Omega}$.

The matrix $\boldsymbol{C}_k(\boldsymbol{\Omega})$ ($k = 1, 2$) arising in (1) is defined to be $\boldsymbol{C}_k(\boldsymbol{\Omega}) \doteq [\boldsymbol{c}_{k,1}(\boldsymbol{\Omega}), \boldsymbol{c}_{k,2}(\boldsymbol{\Omega})]$, where the M_k-dim. vector

$$\begin{aligned} \boldsymbol{c}_{k,p}(\boldsymbol{\Omega}) \doteq [&f_{k,m,p}(\boldsymbol{\Omega}) \exp\{j2\pi\lambda_0^{-1}(\boldsymbol{\Omega} \cdot \boldsymbol{r}_{k,m})\}; \\ &m = 1, \ldots, M_k,]^{\mathrm{T}} \ (k = 1, 2, \ p = 1, 2) \end{aligned}$$

represents the response of Array k in direction $\boldsymbol{\Omega}$ for the pth polarization. In this expression λ_0 is the wavelength and $(\,\cdot\,)$ denotes the scalar product.

The diagonal (off-diagonal) elements of the polarization matrix

$$\boldsymbol{A}_\ell \doteq \begin{bmatrix} \alpha_{\ell,1,1} & \alpha_{\ell,1,2} \\ \alpha_{\ell,2,1} & \alpha_{\ell,2,2} \end{bmatrix}$$

represent the weights or transmission coefficients for the co- (cross-) polarization components. The Frobenius norm of $\boldsymbol{A}_\ell$, $\|\boldsymbol{A}_\ell\| \doteq \left(\sum_{p_1,p_2} |\alpha_{\ell,p_2,p_1}|^2\right)^{1/2}$ is the power weight of Path ℓ.

The signal at the output of Array 2 can be written in vector notation according to

$$\begin{aligned}\boldsymbol{Y}(t) &\doteq [Y_1(t), ..., Y_{M_2}(t)]^{\mathrm{T}} \\ &= \sum_{\ell=1}^{L} \boldsymbol{s}(t;\boldsymbol{\theta}_\ell) + \sqrt{\frac{N_0}{2}}\boldsymbol{W}(t). \qquad (2)\end{aligned}$$

In this expression $\boldsymbol{W}(t) \doteq [W_1(t), \ldots, W_{M_2}(t)]^{\mathrm{T}}$ denotes standard M_2-dim. complex temporally and spatially white noise and N_0 is a positive constant.

3. CHANNEL SOUNDING TECHNIQUE

The details of the adopted time-division multiplexed (TDM) measurement mode as well as of the resulting signal model can be found in (Fleury *et al.*, 2002*a*). Following the notation used in this reference the signal at the output of the sensing switch (SW2 in Fig. 1) reads

$$Y(t) = \sum_{\ell=1}^{L} s(t;\boldsymbol{\theta}_\ell) + \sqrt{\frac{N_0}{2}} q_2(t) W(t). \qquad (3)$$

Here $W(t)$ denotes complex standard white Gaussian noise and $q_2(t)$ is an indicator function, i.e. a function with range $\{0, 1\}$, which takes value one if, and only if, some element of the Rx array is connected to the output of SW2. Moreover,

$$\begin{aligned} s(t;\boldsymbol{\theta}_\ell) &= \exp(j2\pi\nu_\ell t) \sum_{p_2=1}^{2}\sum_{p_1=1}^{2} \alpha_{\ell,p_2,p_1} \\ &\cdot \boldsymbol{c}_{2,p_2}^T(\boldsymbol{\Omega}_{2,\ell}) \boldsymbol{U}(t;\tau_\ell) \boldsymbol{c}_{1,p_1}(\boldsymbol{\Omega}_{1,\ell}). \qquad (4)\end{aligned}$$

The $M_2 \times M_1$ sounding signal matrix $\boldsymbol{U}(t;\tau_\ell)$ is defined to be $\boldsymbol{U}(t;\tau_\ell) \doteq \boldsymbol{q}_2(t)\boldsymbol{q}_1(t)^T u(t-\tau_\ell)$, where $u(t)$ is the signal at the input of the sounding switch (SW1 in Fig.1) at the Tx. The M_k-dim. vector-valued functions $\boldsymbol{q}_k(t)$ $(k = 1, 2)$ characterizes the switch-timing of SWk. More specifically, the mth entry of $\boldsymbol{q}_k(t)$ is an indicator function which takes value one if, and only if, SWk switches the mth element of Array k (see (Fleury *et al.*, 2002*a*)).

4. SAGE ALGORITHM

Referring to (3) the problem at hand is the joint estimation of the parameter vectors $\boldsymbol{\theta}_\ell, \ell = 1, \ldots, L$, specifying the propagation paths as well as the number L of these paths from an observation $Y(t) = y(t)$. The estimation of L is not addressed in the paper. In practice, L is set to a value large enough to capture all the significant propagation paths.

Maximum Likelihood (ML) estimation of $\boldsymbol{\theta} \doteq [\boldsymbol{\theta}_1, \ldots, \boldsymbol{\theta}_L]$ based on an observation $Y(t) = y(t)$ requires solving a $6 \times L$ dim. non-linear optimization problem (Fleury *et al.*, 2002*c*) of which the high complexity prevents any implementation. The expectation-maximization (EM) algorithm (McLachlan and Krishnan, 1997), (Moon, 1997) and a generalization of it referred to as the space-alternate generalized EM (SAGE) algorithm (Fessler and Hero, 1994) can be employed to derive schemes of feasible complexity that approximate the ML solution.

In (Fleury *et al.*, 2002*a*) and (Fleury *et al.*, 2002*c*) the SAGE algorithm is used to estimate the direction of departure, the direction of incidence, the propagation delay, the Doppler frequency and the complex weight of propagation paths. Moreover, the algorithm is augmented with an initialization and search-improved (ISI) procedure that improves its convergence and increases the probability of detecting paths with low amplitude. In this contribution, the ISI-SAGE algorithm is extended to include estimation of the polarization matrix of paths. In the sequel we keep the same nomenclature and definitions as in (Fessler and Hero, 1994), (Fleury *et al.*, 1999), (Fleury *et al.*, 2002*a*), and (Fleury *et al.*, 2002*c*).

4.1 *Admissible data and loglikelihood function*

Following (Fleury *et al.*, 1999), $Y(t)$ in (3) is the incomplete data and

$$X_\ell(t) \doteq s(t;\boldsymbol{\theta}_\ell) + \sqrt{\beta_\ell}\sqrt{\frac{N_o}{2}} q_2(t) W_\ell(t), \qquad (5)$$

where $W_\ell(t)$ is standard complex white Gaussian noise and β_ℓ is a real coefficient satisfying $\beta_\ell \in [0, 1]$, is an admissible data for $\boldsymbol{\theta}_\ell$. An appropriate selection for β_ℓ is $\beta_\ell = 1$. From (5) the loglikelihood function of $\boldsymbol{\theta}_\ell$ given the observation $X_\ell(t) = x_\ell(t)$ reads

$$\begin{aligned}&\Lambda(\boldsymbol{\theta}_\ell; x_\ell) \propto \\ &2\Re\underbrace{\left\{\int s(t;\boldsymbol{\theta}_\ell)^* x_\ell(t)\mathrm{d}t\right\}}_{G_1} - \underbrace{\int |s(t;\boldsymbol{\theta}_\ell)|^2 \mathrm{d}t}_{G_2}. \qquad (6)\end{aligned}$$

Inserting (4) in the first integral and rearranging, G_1 can be expressed according to

$$G_1 \doteq \boldsymbol{\alpha}_\ell^{\mathrm{H}} \boldsymbol{f}(\bar{\boldsymbol{\theta}}_\ell), \qquad (7)$$

where $(\cdot)^{\mathrm{H}}$ denotes the Hermitian operator and

$$\boldsymbol{\alpha}_\ell \doteq [\alpha_{\ell,1,1}, \alpha_{\ell,1,2}, \alpha_{\ell,2,1}, \alpha_{\ell,2,2}]^{\mathrm{T}},$$

$$\boldsymbol{f}(\bar{\boldsymbol{\theta}}_\ell) \doteq \begin{bmatrix} \boldsymbol{c}_{2,1}^H(\boldsymbol{\Omega}_{2,\ell})\boldsymbol{X}_\ell(\tau_\ell,\nu_\ell)\boldsymbol{c}_{1,1}(\boldsymbol{\Omega}_{1,\ell})^* \\ \boldsymbol{c}_{2,1}^H(\boldsymbol{\Omega}_{2,\ell})\boldsymbol{X}_\ell(\tau_\ell,\nu_\ell)\boldsymbol{c}_{1,2}(\boldsymbol{\Omega}_{1,\ell})^* \\ \boldsymbol{c}_{2,2}^H(\boldsymbol{\Omega}_{2,\ell})\boldsymbol{X}_\ell(\tau_\ell,\nu_\ell)\boldsymbol{c}_{1,1}(\boldsymbol{\Omega}_{1,\ell})^* \\ \boldsymbol{c}_{2,2}^H(\boldsymbol{\Omega}_{2,\ell})\boldsymbol{X}_\ell(\tau_\ell,\nu_\ell)\boldsymbol{c}_{1,2}(\boldsymbol{\Omega}_{1,\ell})^* \end{bmatrix}.$$

In the above expression $\boldsymbol{X}_\ell(\tau_\ell, \nu_\ell)$ is a $M_2 \times M_1$ dim. matrix with entries

$$\begin{aligned} X_{\ell,m_2,m_1}(\tau_\ell,\nu_\ell) &= \sum_{i=1}^{I} \exp(-j2\pi\nu_\ell t_{i,m_2,m_1}) \\ &\cdot \int_0^{T_{\mathrm{sc}}} u^*(t-\tau_\ell) x_\ell(t + t_{i,m_2,m_1}) \mathrm{d}t, \qquad (8)\end{aligned}$$

where T_{sc} is the sensing interval for an individual Rx antenna element, $i \in [1, ..., I]$ denotes the index of the measurement cycles with a total number of I and t_{i,m_2,m_1} represents the beginning of the sensing interval for the m_2th Rx element when the m_1th Tx element is transmitting. In the derivation of (8) the phase change over the sensing interval $[t_{i,m_2,m_1}, t_{i,m_2,m_1}+T_{sc}]$ due to the Doppler frequency is assumed to be negligible. Inserting (4) in the second integral in (6) and solving while considering that the cross-correlation between any two distinct elements of $\boldsymbol{U}(t;\tau_\ell)$ vanishes we obtain for G_2

$$G_2 \doteq IPT_{sc} \cdot \boldsymbol{\alpha}_\ell^H \boldsymbol{D}(\boldsymbol{\Omega}_{2,\ell}, \boldsymbol{\Omega}_{1,\ell}) \boldsymbol{\alpha}_\ell. \qquad (9)$$

Here P represents the power of $u(t)$ and the matrix $\boldsymbol{D}(\boldsymbol{\Omega}_{2,\ell}, \boldsymbol{\Omega}_{1,\ell})$ is given by

$$\boldsymbol{D}(\boldsymbol{\Omega}_{2,\ell}, \boldsymbol{\Omega}_{1,\ell}) \doteq \left[\boldsymbol{C}_2(\boldsymbol{\Omega}_{2,\ell})^{\mathrm{H}} \; \boldsymbol{C}_2(\boldsymbol{\Omega}_{2,\ell})\right] \otimes \left[\boldsymbol{C}_1(\boldsymbol{\Omega}_{1,\ell})^{\mathrm{H}} \; \boldsymbol{C}_1(\boldsymbol{\Omega}_{1,\ell})\right].$$

The symbol $\otimes$ denotes the Kronecker product. Provided $\boldsymbol{D}(\boldsymbol{\Omega}_{2,\ell}, \boldsymbol{\Omega}_{1,\ell})$ is invertible, the ML estimate of $\boldsymbol{\theta}_\ell$, $(\widehat{\boldsymbol{\theta}}_\ell)_{\mathrm{ML}}(x_\ell) \doteq \arg\max_{\boldsymbol{\theta}_\ell}\{\Lambda(\boldsymbol{\theta}_\ell; x_\ell)\}$ is obtained by performing the next two steps:

$$(\widehat{\bar{\boldsymbol{\theta}}}_\ell)_{\mathrm{ML}}(x_\ell) = \arg\max_{\bar{\boldsymbol{\theta}}_\ell}\{z(\bar{\boldsymbol{\theta}}_\ell; x_\ell)\}, \qquad (10)$$

$$(\widehat{\boldsymbol{\alpha}})_{\mathrm{ML}}(x_\ell) = (IPT_{sc})^{-1} \boldsymbol{D}(\boldsymbol{\Omega}_{2,\ell}, \boldsymbol{\Omega}_{1,\ell})^{-1} \cdot \boldsymbol{f}(\bar{\boldsymbol{\theta}}_\ell)\big|_{\bar{\boldsymbol{\theta}}_\ell = (\widehat{\bar{\boldsymbol{\theta}}}_\ell)_{\mathrm{ML}}(x_\ell)}. \qquad (11)$$

In these expressions $\bar{\boldsymbol{\theta}}_\ell \doteq [\boldsymbol{\Omega}_{1,\ell}, \boldsymbol{\Omega}_{2,\ell}, \tau_\ell, \nu_\ell]$ and

$$z(\bar{\boldsymbol{\theta}}_\ell; x_\ell) \doteq \boldsymbol{f}(\bar{\boldsymbol{\theta}}_\ell)^{\mathrm{H}} \boldsymbol{D}(\boldsymbol{\Omega}_{2,\ell}, \boldsymbol{\Omega}_{1,\ell})^{-1} \boldsymbol{f}(\bar{\boldsymbol{\theta}}_\ell).$$

Due to space limitation, the necessary and sufficient condition for $\boldsymbol{D}(\boldsymbol{\Omega}_2, \boldsymbol{\Omega}_1)$ to be invertible will be addressed in a forthcoming paper.

4.2 *Expectation (E) step*

Because of the particular form of $\Lambda(\boldsymbol{\theta}_\ell; x_\ell)$ (see (6)), the E-step reduces to calculation of (Fleury *et al.*, 1999)

$$\hat{x}_\ell(t) \doteq y(t) - \sum_{\ell'=1, \ell' \neq \ell}^{L} s(t; \widehat{\boldsymbol{\theta}}'_{\ell'}), \qquad (12)$$

with $\widehat{\boldsymbol{\theta}}'_\ell$ denoting the latest estimate of $\boldsymbol{\theta}_\ell$.

4.3 *Maximization (M) step*

The M-step for updating the estimate of $\boldsymbol{\theta}_\ell$, reads $\widehat{\boldsymbol{\theta}}''_\ell = (\widehat{\boldsymbol{\theta}}_\ell)_{\mathrm{ML}}(\hat{x}_\ell)$ (Fleury *et al.*, 1999). From (10) this operation requires an optimization procedure in the 6-dim. range of $\bar{\boldsymbol{\theta}}_\ell$. A further complexity reduction can be achieved by splitting the joint optimization (10) in separate optimizations in one-dim. spaces (Fleury *et al.*, 2002*a*) and (Fleury *et al.*, 2002*b*):

$$\hat{\tau}''_\ell = \arg\max_{\tau_\ell} z(\hat{\phi}'_{1,\ell}, \hat{\theta}'_{1,\ell}, \hat{\phi}'_{2,\ell}, \hat{\theta}'_{2,\ell}, \tau_\ell, \hat{\nu}'_\ell; \hat{x}_\ell)$$

$$\hat{\nu}''_\ell = \arg\max_{\nu_\ell} z(\hat{\phi}'_{1,\ell}, \hat{\theta}'_{1,\ell}, \hat{\phi}'_{2,\ell}, \hat{\theta}'_{2,\ell}, \hat{\tau}''_\ell, \nu_\ell; \hat{x}_\ell)$$

$$\hat{\theta}''_{2,\ell} = \arg\max_{\theta_{2,\ell}} z(\hat{\phi}'_{1,\ell}, \hat{\theta}'_{1,\ell}, \hat{\phi}'_{2,\ell}, \theta_{2,\ell}, \hat{\tau}''_\ell, \hat{\nu}''_\ell; \hat{x}_\ell)$$

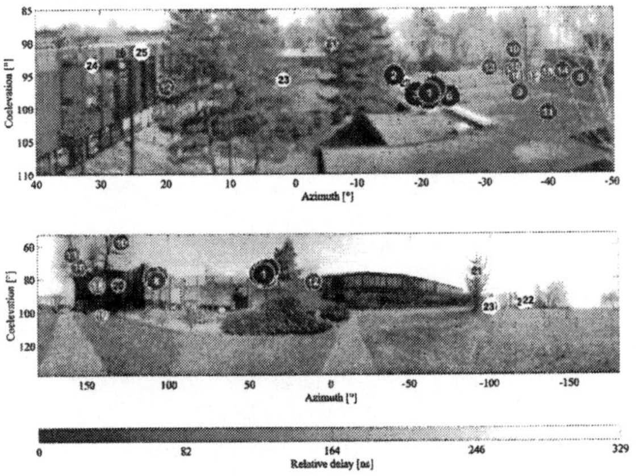

Fig. 2. Panoramic views from the Rx (top) and from the Tx (bottom), on which the estimated directions of incidence and directions of departure, respectively, are superimposed.

$$\hat{\phi}''_{2,\ell} = \arg\max_{\phi_{2,\ell}} z(\hat{\phi}'_{1,\ell}, \hat{\theta}'_{1,\ell}, \phi_{2,\ell}, \hat{\theta}''_{2,\ell}, \hat{\tau}''_\ell, \hat{\nu}''_\ell; \hat{x}_\ell)$$

$$\hat{\theta}''_{1,\ell} = \arg\max_{\theta_{1,\ell}} z(\hat{\phi}'_{1,\ell}, \theta_{1,\ell}, \hat{\phi}''_{2,\ell}, \hat{\theta}''_{2,\ell}, \hat{\tau}''_\ell, \hat{\nu}''_\ell; \hat{x}_\ell)$$

$$\hat{\phi}''_{1,\ell} = \arg\max_{\phi_{1,\ell}} z(\phi_{1,\ell}, \hat{\theta}''_{1,\ell}, \hat{\phi}''_{2,\ell}, \hat{\theta}''_{2,\ell}, \hat{\tau}''_\ell, \hat{\nu}''_\ell; \hat{x}_\ell).$$

From (11) $\hat{\boldsymbol{\alpha}}_\ell$ is updated according to

$$(\hat{\boldsymbol{\alpha}}_\ell)'' = (IPT_{sc})^{-1} \boldsymbol{D}(\widehat{\boldsymbol{\Omega}}''_{2,\ell}, \widehat{\boldsymbol{\Omega}}''_{1,\ell})^{-1} \boldsymbol{f}(\widehat{\bar{\boldsymbol{\theta}}}''_\ell). \qquad (13)$$

We define a SAGE iteration step for updating $\hat{\boldsymbol{\theta}}_\ell$ to be the combination of the E-step (12) with the six above maximization procedures and (13) in the M-step.

The initialization procedure of the ISI-SAGE algorithm will be addressed in a forthcoming paper.

5. EXPERIMENTAL INVESTIGATIONS

Measurements have been performed using the channel sounder PROPSound (Stucki *et. al.*, 2001). The Tx array consists of three conformal sub-arrays of 8 dual-polarized patches uniformly spaced on a cylinder together with a uniform rectangular 2×2 sub-array of 4 similar dual-polarized patches. At the Rx a planar antenna array consisting of 4×4 dual-polarized patches is used. The spacing between the Rx array elements and the elements of the Tx sub-arrays is half a wavelength. The selected carrier frequency is 2.45 GHz. The sounding signal of power $P = 100$mW is a pseudo-noise (PN) sequence of length $K = 255$ chips with chip duration $T_c = 10$ns. The sensing interval coincides with one period of the PN-sequence, i.e. $T_{sc} = KT_c = 2.55\mu$s.

The measurements were performed in the same propagation environment as described in (Fleury *et al.*, 2002*b*). The reader is referred to this reference for a detailed description of the surroundings. The NLOS scenario is considered here. The Rx is located on the second floor of Building B1 (Position Rx on the map in Fig. 3), while the Tx is positioned behind Building B3 (Position Tx). The extended ISI-SAGE algorithm

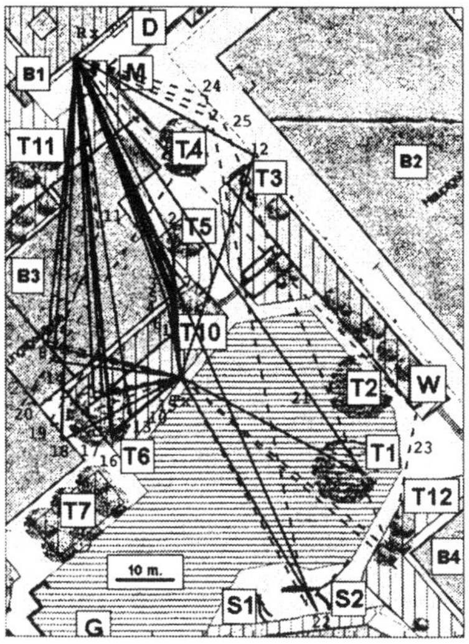

Fig. 3. Reconstructed one-bounce (solid) and two-bounce (dashed) propagation paths.

was applied to extract the parameters of 30 paths using $I = 4$ sounding cycles. The selected quantization steps in delay, azimuth, elevation and Doppler frequency are 0.5ns, 0.1^o, 0.1^o and 0.1Hz respectively. Twenty iteration cycles were performed to calculate the above estimates. The estimated paths are ranked in reverse order of their estimated delays, i.e. the path with the shortest delay has rank 1. Among the estimated paths, those for which the product of the norms of the array responses evaluated at the corresponding estimated path directions is below a predefined level are discarded. This selection is reasonable as the estimate accuracy is decreased in the direction ranges where the norms of the array responses are low. According to this criterion, totally twenty-five estimated paths are retained.

Fig. 2 depicts the estimated directions of incidence and directions of departure of the twenty-five paths in form of dots superimposed to panoramic photographs of the surroundings taken from the Rx location and the Tx location respectively. The radius of a dot increases linearly with the power weight expressed in dB with respect to the noise floor from a minimum value (Path 21: 25.60dB) to a maximum value (Path 4: 48.54dB). The darkness of a dot codes the estimated relative delay according to the bar reported in the figure.

By using the ray-tracing technique described in (Fleury *et al.*, 2002*b*), eighteen paths were identified to result from one-bounce scattering and seven were categorized to originate from two-bounce scattering. The uncertainty range in delay for the one-bounce scenario is set equal to $\pm 0.5T_c = \pm 5$ns. The uncertainty in azimuth and coelevation is set to be $\pm 5^o$. The two-bounce paths are reconstructed while considering an uncertainty range only in delay of $\pm 0.1T_c = \pm 1$ns. Fig. 3 depicts the twenty five reconstructed paths. It can be seen that one-bounce propagation occurs via diffraction around roof corner (B3: Paths 1 to 7 but Path 2), scattering/reflection on facade (B3: Paths 8

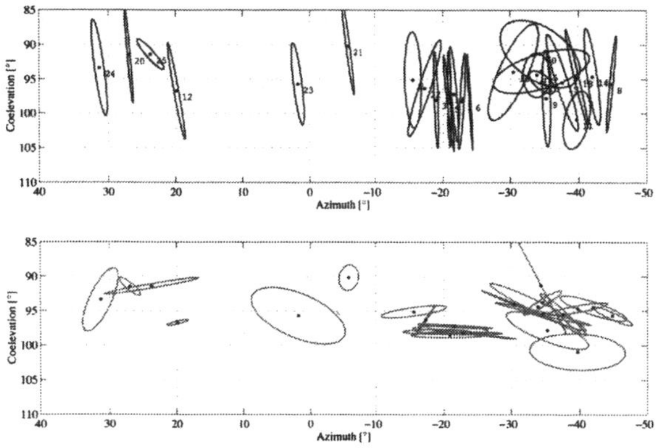

Fig. 4. Normalized polarization ellipses for vertical (top) and horizontal (bottom) transmitted polarization of the paths. The centers of the ellipses coincide with the estimated incidence directions.

and 14; B2: Path 12), and trees (T5: Path 2; T6: Paths 10, 13, 15, 16, 17 and 18; T1: Path 21), and metallic sculptures (S2: Path 22). Two-bounce propagation occurs via facade - roof (B3-B3: Paths 9, 11 and 19), facade - facade (B3-B2: Path 20), tree - wall (T12-W: Path 23), sculpture - facade (S2-B2: Path 24) and tree - facade (T12-B2: Path 25). Notice that the one-bounce path No. 18 and the two-bounce paths No. 19 and No. 20 have interaction points not exactly on the facade of B3 but slightly inside the building. This is probably due to multi-bounce scattering inside the tree T6 (see Fig. 2 and Fig. 4), which accounts for additional propagation delay that is not considered in the current ray-tracing tool.

5.1 *Normalized polarization ellipses*

Fig. 4 depicts the normalized polarization ellipses calculated for each estimated propagation path from the first column (vertical transmitted polarization) and the second column (horizontal transmitted polarization) of the estimated polarization matrix normalized by the estimated root power weight. The major axis of most ellipses are observed to be approximately vertical in the top plot of Fig. 4 and horizontal in the bottom plot. This indicates that the co-polarization components are dominant. However, some paths exhibit large minor axes, which indicates significant cross-polarization components.

5.2 *Cross-polarization discrimination (XPD)*

The co- and cross-polarization characteristics of each path can be assessed quantitatively by calculating the XPD $\hat{r}_{\ell,p} \doteq |\hat{\alpha}_{\ell,p,p}/\hat{\alpha}_{\ell,\bar{p},p}|^2$ for vertical ($p = 1$) and for horizontal ($p = 2$) transmitted polarization. In the above definition $\bar{p} = 3 - p$ represents the complementary of p. Fig. 5 depicts the scatter plot of the estimated XPDs. It is observed that the estimated paths can be categorized into two groups: Group 1 consisting of Paths 1 to 7 but Path 2, which are identified to originate from diffraction around the south roof edge of B3; Group 2 that comprises the remaining paths. We have further subdivided Group 2 into two categories: Category 2a that comprises paths resulting

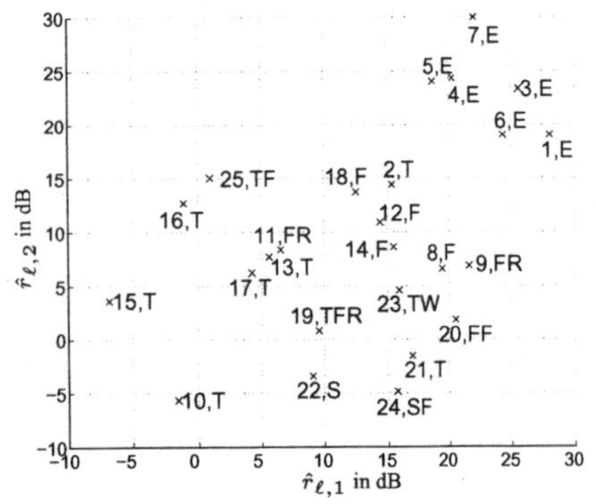

Fig. 5. Scatter plot of the estimated XPDs of the individual paths. The symbols denote the types of scatterers identified along the paths: facade (F), roof (R), edge (E) of buildings as well as tree (T), sculpture (S) and wall (W).

from an interaction with at least one tree (Paths 2, 10, 13, 15, 16, 17, 19, 21, 23 and 25) and Category 2b that embodies paths interacting with only man-made structures (Paths 8, 9, 11, 12, 14, 18, 20, 22, and 24). Table 1 reports the ranges of the XPDs for the three groups.

Table 1. XPDs versus the interaction type.

Group	Interaction type/scatterers along the propagation path	XPDs in dB	
		$\hat{r}_{\ell,1}$	$\hat{r}_{\ell,2}$
1	Diffraction around the roof edge of B3	$[15, 28]$	$[16, 30]$
2a	Reflection/scattering by at least one tree	$[-10, 17]$	$[-6, 16]$
2b	Reflection/scattering only by man-made structures	$[5, 22]$	$[-6, 16]$

6. CONCLUSIONS

In this contribution an extension of the ISI-SAGE algorithm originally described in (Fleury *et al.*, 2002*b*) and (Fleury *et al.*, 2002*c*) to include estimation of the polarization matrix of propagation paths is reported. Hence, the new scheme allows for estimation of the following parameters of path propagating between the Tx and Rx sites: directions of departure, directions of arrival, propagation delay, Doppler frequency, and polarization matrix. The additional complexity resulting from the extension is moderate: the estimation of the coefficients of the polarization matrix of an individual path is a linear estimation problem that requires inversion of a 4×4 dim. matrix.

Experimental results are reported on the polarization characteristics of individual propagation paths estimated in a NLOS pico-/micro-cellular environment. Joint estimation of the path parameters makes it possible to gain a detailed insight into the propagation mechanisms, e.g. to identify the type of scatterers with which the waves interact along their propagation paths and to relate the polarization characteristics of the paths to the interaction types.

This insight is of paramount importance for the design of realistic models of the propagation channel. Based on this experimental knowledge accurate stochastic channel models can be developed for the design and performance assessment of communication systems equipped with multiple transmit and receive antennae. Furthermore, this knowledge can also be exploited to refine deterministic models for field prediction in order to enhance their prediction accuracy. The latter models are extensively used in planning and optimization of wireless networks.

7. REFERENCES

Fessler, J. A. and A. O. Hero (1994). Space-alternating generalized expectation-maximization algorithm. *IEEE Trans. on Signal Processing* **42**(10), 2664–2677.

Fleury, B. H., M. Tschudin, R. Heddergott, D. Dahlhaus and K. L. Pedersen (1999). Channel parameter estimation in mobile radio environments using the SAGE algorithm. *IEEE Journal on Selected Areas in Communications* **17**(3), 434–450.

Fleury, B. H., P. Jourdan and A. Stucki (2002*a*). High-resolution channel parameter estimation for MIMO applications using the SAGE algorithm. *2002 Int. Zurich Seminar on Broadband Communications* **30**, 1–9.

Fleury, B. H., X. Yin, K. G. Rohbrandt, P. Jourdan and A. Stucki (2002*b*). High-resolution bidirection estimation based on the SAGE algorithm: Experience gathered from field experiments. *Proc. XXVIIth General Assembly of the Int. Union of Radio Scientists (URSI)*. # 2127.

Fleury, B. H., X. Yin, K. G. Rohbrandt, P. Jourdan and A. Stucki (2002*c*). Performance of a high-resolution scheme for joint estimation of delay and bidirection dispersion in the radio channel. *Proc. IEEE Vehicular Technology Conference, VTC 2002 Spring* pp. 522–526.

Foschini, G. J. and M. J. Gans (1998). On limits of wireless communications in a fading environment when using multiple antennas. *Wireless Personal Communications* **6**, 311–335.

Jakes, W. C. (1974). *Microwave Mobile Communications*. IEEE Press.

Lozano, A., F. R. Farrokhi and R. A. Valenzuela (2001). Lifting the limits on high-speed wireless data access using antenna arrays. *IEEE Communications Magazine* **Sept.**, 156–162.

McLachlan, G. J. and T. Krishnan (1997). *The EM Algorithm and Extensions*. Wiley Series in Probability and Statistics.

Moon, T. (1997). The expectation-maximization algorithm. *IEEE Signal Processing Mag.* pp. 47–60.

Stucki *et. al.*, A. (2001). PropSound System Specifications Document: Concept and Specifications. Internal Report. Elektrobit AG, Switzerland.

Telatar, I. E. (1999). Capacity of multi-antenna Gaussian channels. *European Transactions on Telecommunication* **10**, 585–595.

IFAC Publications
www.elsevier.com/locate/ifac

ANALYSIS OF SPECTRAL-BASED LOCALIZATION OF SPATIALLY DISTRIBUTED SOURCES

Mikael Tapio [1] Mats Viberg

Department of Signals & Systems
Chalmers University of Technology, Sweden
Email: {mito,viberg}@s2.chalmers.se

Abstract: Antennas that are able to adaptively direct the transmitted (and received) energy are of great interest in future wireless communication systems. The focussing implies reduced transmit power and interference, and also a potential for increased capacity. Due to e.g. local scattering around the transmitter, the source as seen from the receiver appears spatially distributed. A characterization of the spatial channel, in particular mean direction of arrival and spatial spread, is of great interest for system optimization and performance prediction. Despite not exploiting the model, the variance of the proposed non-parametric estimators is found to be close to the Cramer-Rao lower bound for Gaussian spreads. *Copyright © 2003 IFAC*

Keywords: Robust estimation, Performance analysis, Distributed models, Communication channels, Spectral estimation

1. INTRODUCTION

Many classical Direction Of Arrival (DOA) estimation methods are based on point source models. However in some cases, for example, in indoor radio communications the point source assumption is violated, and a distributed source model would be a better approximation (Tholl and Fattouche, 1993). Another example is fast fading in mobile communications, where the elevated base-station antenna, due to local scattering around the mobile, experiences the received signal as distributed in space rather than being emitted from a point source (Zetterberg, 1997). In conventional high-resolution DOA estimation methods this leads to deterioration in performance, and therefore a number of high-resolution estimators for distributed sources have recently been proposed, e.g. (Bengtsson and Ottersten, 2000; Bengtsson and Ottersten, 2001; Meng *et al.*, 1996; Trump and Ottersten, 1996; Valaee *et al.*, 1995). Many of these new estimators make assumptions on the shape of the signal distribution, assume narrow spatial spreads, and eigen-decompose the full-rank covariance matrix into a pseudo-signal subspace and a pseudo-noise subspace. Most often they render a multi-dimensional optimization problem, implying high computational loads.

In many applications, e.g. fast fading mobile communications, wireless indoor communications etc., we know little or nothing about the spatial distribution of the signals. Therefore, it would be attractive to make use of a robust and simple beamformer that does not make any assumptions on signal distribution nor uses any hard-to-choose design parameters.

In this paper, the possibility of using low-complexity spectral-based techniques for localization of distributed signals is addressed and analyzed. It is noted that the signal power emanating from a small spatial band is approximately proportional to the spatial distribution, or more correctly the *spatial power density function (SPDF)*, of the distributed signal. As a consequence, we can estimate the nominal (or mean) DOA and angular spread from the Power Azimuth Spectrum (PAS). We derive simple spectral-based estimators and

[1] This work was supported in part by the Personal Computing and Communication Program (PCC/PCC++). PCC/PCC++ is funded by the Swedish Foundation for Strategic Research.

analyze their performance. The advantage of these non-parametric beamforming-based techniques is that they do not rely on any underlying models, and therefore are more robust. Finally, numerical examples are presented to support the theoretical results and the estimators are applied to indoor measurement data.

2. DISTRIBUTED SOURCE MODEL

We use the distributed signal model suggested in (Bengtsson, 1999), in which the distributed signal impinging upon a Uniform Linear Array (ULA) is modeled as being emitted from a tight cluster of L spatially separated point sources (or scatterers), each with random complex gains. Thus, the received signal can then be written as

$$\mathbf{x}(t) = s(t)\sum_{\ell=1}^{L}\gamma_\ell(t)\mathbf{a}(\theta_0+\tilde{\theta}_\ell(t)) + \mathbf{n}(t) , \qquad (1)$$

where $\gamma_\ell(t)$, θ_0, and $\tilde{\theta}_\ell(t)$, are, respectively, the complex gain of ray ℓ, nominal (or mean) DOA, and random spatial (angular) deviation of ray ℓ. The K-sensor array steering vector is denoted by

$$\mathbf{a}(\theta) = [1, e^{-jkd\sin\theta}, \ldots, e^{-j(K-1)kd\sin\theta}]^T ,$$

where $k = 2\pi/\lambda$ (wavelength – λ) denotes the circular wave number, and d is the element separation. Finally, $s(t)$ denotes the transmitted source signal, and $\mathbf{n}(t)$ the noise, which is modeled as a zero-mean complex circularly symmetric, and spatio-temporally white Gaussian process with variance σ^2.

As in (Bengtsson, 1999; Trump and Ottersten, 1996), it is also assumed that the scattering environment changes rapidly compared to the mean DOA and spread parameters, i.e. the random complex gains $\gamma_\ell(t)$ are assumed to be temporally white. Further, they are also assumed to be independent from ray to ray, zero-mean, and circularly symmetric

$$E[\gamma_\ell(t)] = 0 , \qquad E[\gamma_\ell(t)\gamma_k(s)] = 0 ,$$
$$E[\gamma_\ell(t)\gamma_k^*(s)] = \frac{1}{L}\sigma_\gamma^2\delta_{\ell k}\delta_{ts} , \quad \forall\, \ell,k,t,s , \qquad (2)$$

where $\delta_{\ell k}$ is the Kronecker delta function.

The random spatial deviation $\tilde{\theta}_\ell(t)$ is assumed to be a zero-mean random variable described by a Probability Density Function (PDF) $p(\tilde{\theta};\sigma_\theta)$. The PDF is assumed to be symmetric in $\tilde{\theta}$ and parameterized by a spread (or standard deviation) parameter σ_θ. Note that the resulting estimators do not require a symmetric PDF.

By assuming large L, and using the central limit theorem, it is argued in (Bengtsson, 1999; Trump and Ottersten, 1996) that $\mathbf{x}(t)$ is a zero-mean complex Gaussian vector with covariance matrix

$$\begin{aligned}\mathbf{R}_x &= E\left[\mathbf{x}(t)\mathbf{x}^H(t)\right] \\ &= E_{\tilde{\theta}}\left[E\left[\mathbf{x}(t)\mathbf{x}^H(t)\,|\,\tilde{\theta}\right]\right] \\ &= \sigma_s^2\sigma_\gamma^2\int_{\tilde{\theta}} p(\tilde{\theta};\sigma_\theta)\mathbf{a}(\theta_0+\tilde{\theta})\mathbf{a}^H(\theta_0+\tilde{\theta})d\tilde{\theta} + \sigma^2\mathbf{I} \\ &= S\mathbf{R}_v(\theta_0,\sigma_\theta) + \sigma^2\mathbf{I} , \end{aligned} \qquad (3)$$

where $S = \sigma_s^2\sigma_\gamma^2 = E[|s(t)|^2]\sigma_\gamma^2$ is the source signal power including the path gain factor, and $\mathbf{R}_v(\theta_0,\sigma_\theta)$ is the channel covariance matrix (excluding the path gain) of the zero-mean Gaussian channel vectors. In the limit, the cluster of scatterers can be viewed upon as a continuum of scatterers, i.e. the spatio-temporal complex gain can be described by a temporal stochastic process $\gamma(\tilde{\theta};t)$, which has a continuous spatial distribution with covariance kernel

$$E[\gamma(\tilde{\theta}_1;t)\gamma^*(\tilde{\theta}_2;t)] = \sigma_\gamma^2 p(\tilde{\theta}_1;\sigma_\theta)\delta(\tilde{\theta}_1-\tilde{\theta}_2) .$$

In this context, $p(\tilde{\theta};\sigma_\theta)$ can be viewed upon as a continuous *Spatial Power Density Function (SPDF)*.

The spatial deviation is commonly modeled by a Gaussian distribution, i.e. $\tilde{\theta} \sim \mathcal{N}(0,\sigma_\theta)$. For small Gaussian spatial spreads, the $(m,n)^{\text{th}}$ entry of the covariance matrix can then be written as (Bengtsson, 1999; Bengtsson and Ottersten, 2000)

$$\begin{aligned}[\mathbf{R}_v(\omega_0,\sigma_\omega)]_{m,n} &= \exp(j(m-n)\omega_0) \\ &\quad\times\exp(-0.5((m-n)\sigma_\omega)^2), \end{aligned} \qquad (4)$$

where $\omega_0 = kd\sin\theta_0$ denotes the *spatial frequency*, and $\sigma_\omega = kd\sigma_\theta\cos\theta_0$ is the corresponding standard deviation.

3. SPECTRAL-BASED LOCALIZATION

The DOA of a single deterministic point source is usually taken as the peak of the Power Azimuth Spectrum (PAS)

$$\hat{\theta} = \arg\max_\theta \hat{P}(\theta) , \qquad (5)$$

where $\hat{P}(\theta)$ denotes the estimated PAS, from the sample covariance matrix. Using the true covariance matrix, the PAS is given by

$$P(\theta) = \begin{cases} \dfrac{\mathbf{a}^H(\theta)\mathbf{R}_x\mathbf{a}(\theta)}{\mathbf{a}^H(\theta)\mathbf{a}(\theta)} & \text{- CBF} \\ \dfrac{1}{\mathbf{a}^H(\theta)\mathbf{R}_x^{-1}\mathbf{a}(\theta)} & \text{- Capon.} \end{cases} \qquad (6)$$

Capon's beamformer (Capon, 1969) has higher resolution than the Conventional Beamformer (CBF). This is achieved at the expense of reduced white noise suppression, and therefore Capon's beamformer is more suited for high Signal-to-Noise Ratio (SNR). The performance of CBF for distributed sources was analyzed in (Raich *et al.*, 1998), where the nominal DOA estimate was taken from (5). In general, the performance of the peak-finding algorithm is poor for distributed sources and it is found that the Center of Mass of the PAS is a better estimate (Tapio, 2003).

Assume that the steering vectors corresponding to distinct directions are orthogonal for large K, i.e.

$$\lim_{K\to\infty}\frac{1}{K}\mathbf{a}^H(\theta)\mathbf{a}(\eta)=\begin{cases}1\,, & \theta=\eta\\ 0\,, & \theta\neq\eta.\end{cases} \tag{7}$$

Hence, for large K we have $\frac{1}{K}\mathbf{A}^H(\theta)\mathbf{A}(\theta)\approx\mathbf{I}$, where $\mathbf{A}(\theta)$ denotes the $K\times L$ steering matrix of the vector $\theta=[\theta_1,\ldots,\theta_L]^T$.

The discrete version of the channel covariance matrix in (3) can be written

$$\mathbf{R}_v=\sum_{\ell=1}^{L}p(\theta_\ell)\mathbf{a}(\theta_\ell)\mathbf{a}^H(\theta_\ell)=\mathbf{A}(\theta)\mathbf{P}(\theta)\mathbf{A}^H(\theta)\,, \tag{8}$$

where $\mathbf{P}(\theta)$ denotes a diagonal matrix with the power mass entries $[\mathbf{P}(\theta)]_{\ell,\ell}=p(\theta_\ell)$. Since it is only non-zero values of $p(\theta_\ell)$ that contribute to the channel covariance matrix, we assume $p(\theta_\ell)>0$ to make $\mathbf{P}$ full rank. Note that in this context, θ_ℓ is equal to the nominal DOA θ_0 plus the previously used random deviation $\tilde{\theta}_\ell$ of ray ℓ.

Using the matrix inversion lemma and assuming large K, the inverse covariance matrix becomes

$$\begin{aligned}\mathbf{R}_x^{-1}&=[S\mathbf{R}_v+\sigma^2\mathbf{I}]^{-1}=[S\mathbf{A}\mathbf{P}\mathbf{A}^H+\sigma^2\mathbf{I}]^{-1}\\ &\approx\mathbf{A}\check{\mathbf{P}}\mathbf{A}^H+\sigma^{-2}\mathbf{I},\end{aligned} \tag{9}$$

where $\check{\mathbf{P}}$ is the diagonal matrix with entries $[\check{\mathbf{P}}]_{\ell,\ell}=-\frac{Sp(\theta_\ell)}{\sigma^4+\sigma^2SKp(\theta_\ell)}$. It is now straightforward to insert $\mathbf{R}_x$ and $\mathbf{R}_x^{-1}$ into the expressions for the PAS given by (6). Doing so, we find the following relationship

$$P(\theta_\ell)\approx\beta p(\theta_\ell;\theta_0,\sigma_\theta)+\varepsilon\,,\quad\beta\geq 0,\ \varepsilon\geq 0\,, \tag{10}$$

where $\beta=SK$, $\varepsilon=\sigma^2$ for the CBF spectrum, and $\beta=S$, $\varepsilon=\frac{\sigma^2}{K}$ for Capon's spectrum.

Now, to find our parameters $\{\theta_0,\sigma_\theta\}$ we have to decide the support Ω of the underlying PDF. Also, since the underlying PDF does not necessarily have a finite support (e.g. Gaussian PDF), we have to restrict the support to be finite. For example, the support of a Gaussian PDF with mean θ_0 and standard deviation σ_θ can with negligible error be restricted to $[\theta_0-3\sigma_\theta,\theta_0+3\sigma_\theta]\subset(-90°,90°)$. The obvious problem that arises here is how to choose the support. One criterion for choosing it is to find the region where the signal part of the PAS is distinct from the noise power level. For a single distributed source, for example, taking the median value of the PAS may give a good enough estimate of the noise level.

When we have decided the support Ω, the scaling factor β is easily found. We simply subtract the minimum value of $P(\theta)$ on Ω, which corresponds to ε, from $P(\theta)$. Then we use the fact that $p(\theta_\ell;\theta_0,\sigma_\theta)$ is a PDF (or rather a probability mass function) and should therefore sum to unity, i.e. β can be found as $\beta=\sum_{\theta_\ell\in\Omega}\tilde{P}(\theta_\ell)$, where $\tilde{P}(\theta)=P(\theta)-\min_{\theta\in\Omega}P(\theta)$ and is non-negative.

Putting it all together and using the standard definitions of expected value and standard deviation, the nominal DOA actually becomes the Center of Mass (CM) of $\tilde{P}(\theta)$ on Ω

$$\theta_0=\frac{\sum_{\theta_\ell\in\Omega}\theta_\ell\tilde{P}(\theta_\ell)}{\sum_{\theta_\ell\in\Omega}\tilde{P}(\theta_\ell)}\,, \tag{11}$$

and the spatial spread parameter or standard deviation is given by

$$\sigma_\theta=\left(\frac{\sum_{\theta_\ell\in\Omega}(\theta_\ell-\hat{\theta}_0)^2\tilde{P}(\theta_\ell)}{\sum_{\theta_\ell\in\Omega}\tilde{P}(\theta_\ell)}\right)^{1/2}\,. \tag{12}$$

As mentioned earlier, when $L\to\infty$, the distributed source can be described by a continuous SPDF. Consequently, if we compute the PAS densely enough, the SPDF can be acquired from the PAS. It should also be mentioned that there is no requirement for the SPDF to be symmetric.

Finally, since the true covariance matrix $\mathbf{R}_x$ is unknown, we have to use the sample covariance matrix which is defined as $\hat{\mathbf{R}}_x=\frac{1}{N}\sum_{t=1}^{N}\mathbf{x}(t)\mathbf{x}^H(t)$ to estimate the PAS. It is well known (Wu and Fuhrmann, 1991) that $\hat{\mathbf{R}}_x$ converges (with probability one) to $\mathbf{R}_x$ as $N\to\infty$. Hence, the estimated PAS is a consistent estimate of the true PAS, and our estimate of $p(\theta;\theta_0,\sigma_\theta)$ becomes consistent, thus leading to consistent estimates of the nominal DOA and spread parameters. Note that we have used the assumption of orthogonal steering vectors. If the number of sensors is too small, the estimate of σ_θ will deteriorate while the estimate of θ_0 (being the CM) is less sensitive to the number of sensors.

The proposed algorithm is summarized below

(1) Form $\hat{\mathbf{R}}_x=\frac{1}{N}\sum_{t=1}^{N}\mathbf{x}(t)\mathbf{x}^H(t)$
(2) Compute the PAS $\hat{P}(\theta)$ using (6), and choose the support Ω
(3) Form $\hat{\tilde{P}}(\theta)=\hat{P}(\theta)-\min_{\theta\in\Omega}\hat{P}(\theta)$
(4) $\hat{\theta}_0=\frac{\sum_{\theta_\ell\in\Omega}\theta_\ell\hat{\tilde{P}}(\theta_\ell)}{\sum_{\theta_\ell\in\Omega}\hat{\tilde{P}}(\theta_\ell)}$
(5) $\hat{\sigma}_\theta=\left(\frac{\sum_{\theta_\ell\in\Omega}(\theta_\ell-\hat{\theta}_0)^2\hat{\tilde{P}}(\theta_\ell)}{\sum_{\theta_\ell\in\Omega}\hat{\tilde{P}}(\theta_\ell)}\right)^{1/2}$

Fig. 1 illustrates $\tilde{P}(\theta)/\beta$ for the CBF and Capon spectra when the theoretical angular distribution is Gaussian. Also included are Non-Linear Least Squares (NLLS) curve fits of $\beta p(\theta;\theta_0,\sigma_\theta)+\varepsilon$ (where $p(\theta)$ is a Gaussian PDF) to the PAS. The NLLS problem is optimized over all four parameters but since it is linear in β and ε it can be separated and therefore a two-dimensional search is needed to find θ_0 and σ_θ. Both of the spectral-based PDF estimates $\tilde{P}(\theta)/\beta$ show good fit to the underlying Gaussian PDF and motivate the relationship given by (10). Note that the NLLS fits are slightly worse than their corresponding PDF estimates. The support for the CBF and Capon estimates is limited to $\Omega = [\theta_0 - 3\sigma_\theta, \theta_0 + 3\sigma_\theta]$, which causes a better fit than their corresponding NLLS fits.

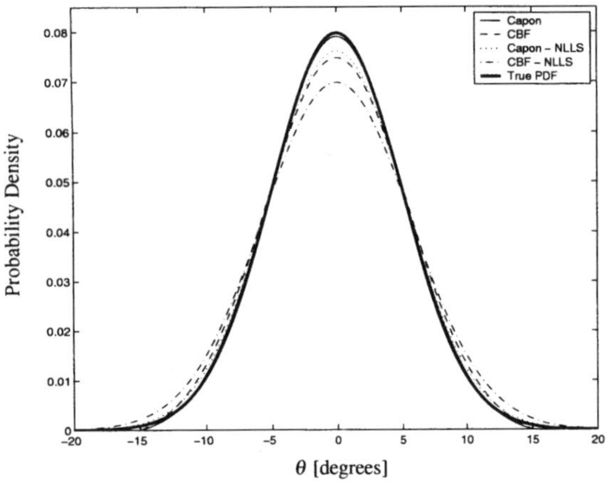

Fig. 1. Estimated PDFs. Gaussian spread, $\theta_0 = 0°$, $\sigma_\theta = 5°$, 18 sensors, theoretical covariance matrix with $S/\sigma^2 = 10$ dB.

4. PERFORMANCE ANALYSIS

The Mean Square Error (MSE) performances of the estimators are analytically calculated and approximated by using the first terms of a geometric series expansion and known expectations of complex Wishart forms (Tague and Caldwell, 1994). The derivations are excluded here and only the results are presented. All results are approximations and are only valid for large numbers of snapshots, i.e. we assume that $N \gg 1$ and for the Capon beamformer we also assume that $N \gg K$.

The MSE of the CBF-based DOA estimate is found to be approximated by

$$\mathrm{mse}(\hat{\theta}_0) \approx \frac{\sum_{k,l}(\theta_k - \theta_0)(\theta_\ell - \theta_0)\mathbf{a}_k^H \mathbf{R}_x \mathbf{a}_\ell \cdots}{N[\sum_k(\mathbf{a}_k^H \mathbf{R}_x \mathbf{a}_k - K\varepsilon)]^2} \cdots \mathrm{Tr}\{\mathbf{a}_k \mathbf{a}_\ell^H \mathbf{R}_x\}, \quad (13)$$

where the summation is over $\{k, \ell : \theta_k, \theta_\ell \in \Omega\}$ and $\mathbf{a}_k = \mathbf{a}(\theta_k)$. For the Capon-based DOA estimate, the MSE is approximated by

$$\mathrm{mse}(\hat{\theta}_0) \approx \frac{\sum_{k,\ell}(\theta_k - \theta_0)(\theta_\ell - \theta_0)\bar{\mathbf{a}}_k^H \mathbf{R}_x^{-1} \bar{\mathbf{a}}_\ell \cdots}{(N-K)[\sum_k((\mathbf{a}_k^H \mathbf{R}_x^{-1} \mathbf{a}_k)^{-1} - \varepsilon)]^2} \cdots \mathrm{Tr}\{\bar{\mathbf{a}}_k \bar{\mathbf{a}}_\ell^H \mathbf{R}_x^{-1}\}, \quad (14)$$

where $\bar{\mathbf{a}}_k = \mathbf{a}(\theta_k)/(\mathbf{a}^H(\theta_k)\mathbf{R}_x^{-1}\mathbf{a}(\theta_k))$. In both expressions, the nominal DOA estimate is assumed to be unbiased, an assumption that is supported by the numerical examples.

To simplify the calculations, the true nominal DOA value is used instead of its estimate in the calculations of the squared angular spread MSE. The expressions for the MSE of the squared angular spread are, respectively,

$$\mathrm{mse}(\hat{\sigma}_\theta^2) \approx (\bar{\sigma}_\theta^2 - \sigma_\theta^2)^2 + \frac{\sum_{k,\ell}[(\theta_k - \theta_0)^2 - \bar{\sigma}_\theta^2][(\theta_\ell - \theta_0)^2 - \bar{\sigma}_\theta^2]\mathbf{a}_k^H \mathbf{R}_x \mathbf{a}_\ell \cdots}{N[\sum_k(\mathbf{a}_k^H \mathbf{R}_x \mathbf{a}_k - K\varepsilon)]^2} \cdots \mathrm{Tr}\{\mathbf{a}_k \mathbf{a}_\ell^H \mathbf{R}_x\} \quad (15)$$

for the CBF-based estimate and

$$\mathrm{mse}(\hat{\sigma}_\theta^2) \approx (\bar{\sigma}_\theta^2 - \sigma_\theta^2)^2 + \frac{\sum_{k,\ell}[(\theta_k - \theta_0)^2 - \bar{\sigma}_\theta^2][(\theta_\ell - \theta_0)^2 - \bar{\sigma}_\theta^2]\bar{\mathbf{a}}_k^H \mathbf{R}_x^{-1} \bar{\mathbf{a}}_\ell \cdots}{(N-K)[\sum_k((\mathbf{a}_k^H \mathbf{R}_x^{-1} \mathbf{a}_k)^{-1} - \varepsilon)]^2} \cdots \mathrm{Tr}\{\bar{\mathbf{a}}_k \bar{\mathbf{a}}_\ell^H \mathbf{R}_x^{-1}\} \quad (16)$$

for the Capon-based estimate. The first term in each expression corresponds to the asymptotic squared bias (i.e. $\bar{\sigma}_\theta^2 \approx \mathrm{E}\{\hat{\sigma}_\theta^2\}$) and the second term to the variance of the squared angular spread estimate. Moreover, the squared asymptotic biases in (15) and (16) are acquired by evaluating the squared difference between (12) (in where the true covariance matrix is used) and the true angular spread value. The bias is very sensitive to the choice of support – too a small support typically leads to under-estimating the spread and too a large support leads to over-estimating the spread. With knowledge of the true support together with antenna beamwidth (resolution), the support can be chosen in such a way that the bias almost vanishes.

To evaluate the analytical expressions, a number of numerical examples are studied. In all the simulations a ULA consisting of twenty elements with half-wavelength separation is used to compute the PAS. The source is Gaussian distributed with a nominal DOA of $\theta_0 = 0°$ and an angular spread of $\sigma_\theta = 5°$. The SNR is 10 dB. To choose the support, a constant, c, times the median value of the PAS is used (denote the resulting product as α). The signal part of the spectrum which is above this value α is used to define the support. The constant c is adjusted so that the support "on average" becomes $\Omega = [\theta_0 - 3\sigma_\theta, \theta_0 + 3\sigma_\theta]$.

Fig. 2 shows the theoretical and simulated MSE for the nominal DOA estimate versus different numbers of snapshots. The simulated values show good fit to the theoretical curves and are close to the Cramer-Rao lower bound (CRLB).

Since the bias is large compared to the variance of the squared spread estimate, the variance is plotted separately. Fig. 3 shows the simulated and theoretical variances of the squared spread. Finally, the squared difference between the asymptotic expected value, $\bar{\sigma}_\theta^2$, and the simulated mean value, $\bar{m}(\hat{\sigma}_\theta^2)$, of the squared angular spread (not the squared bias) is shown in Fig. 4. In this numerical example, the theoretical bias is $|\bar{\sigma}_\theta^2 - \sigma_\theta^2| = 1.12°$ for the CBF-based estimate and $|\bar{\sigma}_\theta^2 - \sigma_\theta^2| = 0.87°$ for the Capon-based estimate. The true value is $\sigma_\theta^2 = 25°$. As seen in Fig. 4, the simulated mean of the CBF-based estimate of the squared angular spread converges faster to its asymptotical expected value than the simulated mean of the Capon-based estimate does.

All simulated results show good fit to their analytical counterparts.

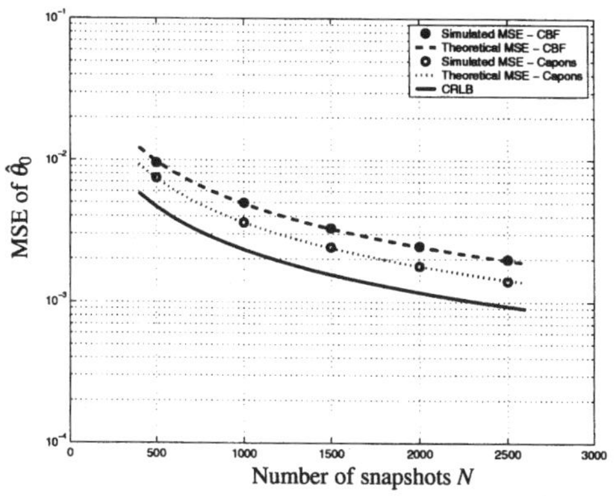

Fig. 2. MSE of $\hat{\theta}_0$ for different numbers of snapshots N.

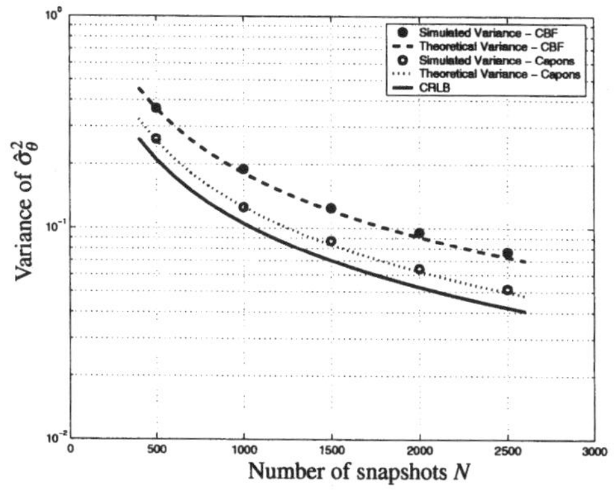

Fig. 3. Variance of $\hat{\sigma}_\theta^2$ for different numbers of snapshots N.

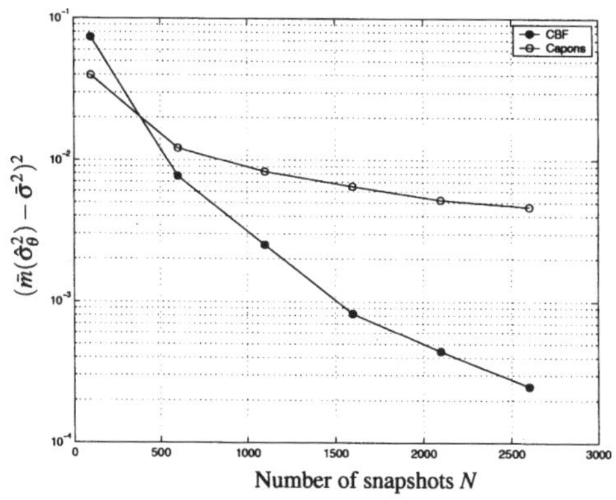

Fig. 4. Squared difference between the theoretical expected value, $\bar{\sigma}_\theta^2$, and the simulated mean value, $\bar{m}(\hat{\sigma}_\theta^2)$, of the squared spread estimate vs. snapshots N.

5. MEASUREMENT DATA EXAMPLES

Indoor multiple-input multiple-output (MIMO) channel measurements have been conducted and reported in (Svantesson and Wallace, 2002). Fig. 5 shows the conventional beamforming spectrum evaluated for three different receiver positions. These spectra have been acquired using circular antenna arrays, each with ten half-wavelength separated monopoles. The transmitter is mainly focussing its energy in an angular band along the horizontal corridor. First we identify the distributed sources and then we choose their corresponding supports in an *ad-hoc* manual fashion. The results acquired from applying our estimators to the three spectra are summarized in Table 1, where the DOA is measured counter-clockwise relative the left-right horizontal axis. For detailed information on the measurement setup and channel estimation procedure, we refer the reader to (Svantesson and Wallace, 2002).

RX position	$\hat{\theta}_0$	$\hat{\sigma}_\theta$
1	80.3°	14.2°
1	265.1°	25.0°
2	1.3°	8.9°
2	178.3°	7.1°
3	49.5°	7.4°
3	129.0°	14.6°
3	203.5°	6.6°
3	275.9°	6.1°
3	333.8°	7.6°

Table 1. Estimated nominal DOAs and angular spreads from the conventional beamforming spectra in Fig. 5.

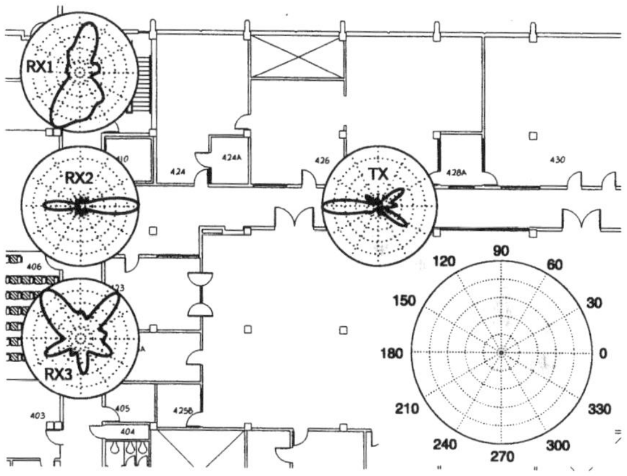

Fig. 5. Conventional beamforming spectra in three receiver positions.

Of course, using real data there are no true values available. However, the obtained values agree reasonably well with the appearance of the "local power maxima" that are visible in the spectra. It is also clear that the local spatial distributions are not Gaussian.

6. CONCLUSIONS

This paper has presented a statistical analysis of a previously proposed method for estimating nominal direction and spread for spatially distributed sources. The method is based on non-parametric estimates of the spatial spectrum, using either the conventional or the adaptive (Capon's/MVDR) beamformer. Closed-form expressions of the asymptotic mean-square errors of the parameter estimates were provided. The theoretical expressions were found to agree well with results from simulated data. Additionally the MSE of the nominal DOA and the variance of the squared spread parameter estimates are close to the Cramer-Rao lower bound for a Gaussian distributed source. Further, the non-parametric approach was used to determine nominal DOA and spatial spread for real indoor data. Although no quantitative evaluation is available for the real data, the obtained results agree well with a visual inspection of the spatial spectra.

ACKNOWLEDGEMENT

The authors wish to acknowledge Dr. Thomas Svantesson for providing with indoor MIMO channel measurements from Brigham Young University, USA.

7. REFERENCES

Bengtsson, M. (1999). "Antenna Array Signal Processing for High Rank Data Models". PhD thesis. Royal Institute of Technology. Stockholm, Sweden.

Bengtsson, M. and B. Ottersten (2000). "Low-Complexity Estimators for Distributed Sources". *IEEE Trans. on Signal Processing* **48**(8), 2185–2194.

Bengtsson, M. and B. Ottersten (2001). "A Generalization of Weighted Subspace Fitting to Full-Rank Models". *IEEE Trans. on Signal Processing* **49**(5), 1002–1012.

Capon, J. (1969). "High Resolution Frequency Wave Number Spectrum Analysis". *In Proc. IEEE* **57**, 1408–1418.

Meng, Y., P. Stoica and K.M. Wong (1996). "Estimation of the Directions of Arrival of Spatially Dispersed Signals in Array Processing". *Proc. IEE Radar, Sonar and Nav.* **143**(1), 1–9.

Raich, R., J. Goldberg and H. Messer (1998). "Bearing Estimation for a Distributed Source via the Conventional Beamformer". *Proc. Stat. Signal Array Process. Workshop* pp. 5–8.

Svantesson, T. and Jon Wallace (2002). "Statistical Characterization of the Indoor MIMO Channel Based on LOS/NLOS Measurements". In: *Proc. 36th Asilomar Conf. Sig., Syst., Comput.*. Pacific Grove, CA.

Tague, J.A. and C.I. Caldwell (1994). "Expectations of Useful Complex Wishart Forms". *Multidimensional Systems and Signal Processing* **5**, 263–279.

Tapio, M. (2003). "On the use of beamforming for estimation of spatially distributed signals". In: *Proc. IEEE ICASSP 03*. Hong Kong, China.

Tholl, D. and M. Fattouche (1993). "Angle of Arrival Analysis of Indoor Radio Propagation Channel". *Proc. Int. Conf. on Univ. Pers. Comm.* **1**, 79–83.

Trump, T. and B. Ottersten (1996). "Estimation of Nominal Direction of Arrival and Angular Spread using an Array of Sensors". *Signal Processing* **50**(1-2), 57–69.

Valaee, S., B. Champagne and P. Kabal (1995). "Parametric Localization of Distributed Sources". *IEEE Trans. on Signal Processing* **43**(9), 2144–2153.

Wu, Q. and D.R. Fuhrmann (1991). "A Parametric Method for Determining the Number of Signals in Narrow-Band Direction Finding". *IEEE Trans. on Signal Processing* **39**(8), 1848–1857.

Zetterberg, P. (1997). "Mobile Cellular Communications with Base Station Antenna Arrays: Spectrum Efficiency, Algorithms and Propagation Models". PhD thesis. Royal Institute of Technology. Stockholm, Sweden.

IFAC Publications
www.elsevier.com/locate/ifac

RAY TRACING INTERPRETATION OF MULTIPLE-INPUT MULTIPLE OUTPUT WIRELESS SYSTEMS

Peter F. Driessen

Electrical and Computer Engineering
University of Victoria
peter@ece.uvic.ca, www.driessen.ca

Abstract: A wireless system with multiple antennas at both transmitter and receiver has a channel capacity which grows linearly (rather than logarithmically) with the number of antennas (assuming fixed bandwidth and total radiated power and many scatterers). In this paper, we show that such high capacities can be achieved on line-of-sight (LOS) paths (no scatterers) by explicitly spreading out the antennas well beyond a wavelength. Roughly speaking, the wide spacing, replicates the effect of scatterers which create images and thus serves to spread out the apparent source of the signals over a wider angular range. In this way capacities on the order of $C_{lin} = nlog_2(1+\rho)$ bps/Hz can be obtained for LOS channels with n transmit and n receive antennas. In contrast, when the transmit antennas are closely spaced, the number of degrees of freedom on a LOS channel degenerate, resulting in capacities of only $C_{log} = log_2(1+n\rho)$.

1. INTRODUCTION

Wireless communications systems with multiple antennas at both transmitter and receiver (MIMO systems) split (demultiplex) the transmitted data into multiple substreams. These substreams are separately independently coded and modulated *on the same carrier frequency* and connected to separate antennas. The receiver demodulates and decodes each substream despite the apparent interference from other substreams on the same carrier frequency. In this paper, a geometric interpretation based on ray tracing helps to explain how such MIMO systems can work.

As previously reported (Foschini and Gans, 1998), wireless communications channels with multiple antennas at both transmitter and receiver have an information-theoretic capacity which can grow linearly (rather than logarithmically) with the number of antennas, for fixed power and bandwidth. Such linear growth is found when the matrix channel transfer function between the transmitter and receiver antennas is modelled statistically, as a matrix with independent complex gaussian (scalar) random variable entries (to represent Rayleigh fading), so that then capacity is expressed as a random variable. At a specified outage level, the capacity is increased in direct proportion to the number of degrees of freedom (Foschini and Gans, 1998).

In a Rayleigh fading environment, with scattered signals arriving from many directions, and n antennas at each site, n degrees of freedom are possible (Foschini and Gans, 1998), assuming an antenna spacing of about half-wavelength for which mutual coupling effects are negligible. However, in certain propagation environments, particularly when there is a deterministic or line-of-sight (LOS). component, the number of degrees of freedom can degenerate (Foschini and Gans, 1998). For example, with n-level receive diversity of the maximum ratio combining type,which is a well known form of diversity, the capacity increases only logarithmically with n. The increase is due only to the SNR gain from the array of the n receive antennas.

The question arises as to whether it is possible to retain the linear capacity growth with n for channels with a significant deterministic component. Thus this paper explores the capacities which may be achieved in propagation environments where there is a strong deterministic component adding to the statistical character of the matrix channel.

1.1 Ray tracing for the deterministic component of capacity

Our investigations use ray tracing to construct an $n \times n$ matrix channel response explicitly for a specified environment with obstructions and scatterers, including smooth walls as well as rough surfaces (P.Beckmann and A.Spizzichino, 1963). Such a channel response will change as the receiver is moved, so that a capacity distribution is obtained from the ensemble of sample matrix elements at different receiver locations. Alternatively, a deterministic matrix may be added to the Rayleigh matrix to form a matrix of Ricean scalars, from which a capacity distribution is obtained. As part of this study, we will find the capacities which may be achieved for particular realizations of the channel matrix (i.e. one sample of the ensemble of sample matrix elements), and show explicitly how the capacity depends on the rank of this matrix. We will show how spreading out the antennas can result in near-full rank channel matrices (i.e. with close to n significant (non-negligible) eigenvalues), with corresponding high capacity, even in a purely LOS scenario without any scatterers. This work is based on (Driessen and Foschini, 1999).

This paper is organized as follows: In section 2, we review the capacity expressions, and show explicitly the relationship between capacity and the rank of the channel matrix. In section 3, we consider geometric ray tracing in a purely LOS environment, and show how the capacity increases as the spacing or angular spread between the n antenna elements is increased. In section 4, we consider a specified environment (LOS down an urban street with building and ground reflections), with an n-element array with half-wavelength spacing, and show how the channel matrix approaches near-full rank (and the capacity increases) as the number of reflections considered is increased from zero to five. Section 5 contains a summary and conclusions.

2. CAPACITY EXPRESSIONS

Next we will present some capacity expressions. We stress that the expressions are for a single communication link in a single cell with no adjacent channel interference. While capacity studies in a multi-user context, are of great importance (Catreux *et al.*, 2000), they are beyond the scope of this paper. We will be using T and R as convenient abbreviations for transmit and receive respectively, with subscript n to enumerate the number of antenna elements at a T site and R site respectively.

2.1 Basic Capacity expression

The fundamental result for the capacity in bps/Hz of a wireless system with n_T transmit antennas and n_R receive antennas with an average received SNR ρ (independent of n_T) at each receive antenna was obtained in (Foschini and Gans, 1998) as

$$C = log_2(det[I_{n_R} + (\rho/n_T)HH^\dagger]) \qquad (1)$$

where the normalized channel matrix H contains complex scalars with unity average power loss, and $H^\dagger$ is the complex conjugate transpose of H. The capacity is expressed in bps/Hz in the narrowband limit with no frequency dependence. Normalization is achieved by dividing out the free space power loss and setting the parameter ρ to the desired SNR [1]. This result assumes that H is unknown to the receiver but n_R and ρ are known (Foschini and Gans, 1998)(Foschini, 1996). The signal from each antenna is different. The transmitted data has been demultiplexed and the demultiplexed substreams are separately independently coded and modulated. Moreover, instead of just committing the resulting subsignals one-to-one to the n_T transmit antennas, the association between substreams and antenna elements is cycled to encourage that over time each substream experiences a similar propagation environment.

If the matrix elements H_{ij} are random variables (e.g. Rayleigh, Ricean), then C is also a random variable. In this case, we define an outage threshold x (say 0.01), and define C_x to be that capacity for which $Prob\{C > C_x\} = 1 - x$.

2.2 Line-of-sight systems with closely spaced antenna elements

We consider an environment with only LOS propagation and T and R arrays of $n_T = n_R = n$

[1] This avoids the need to compute absolute propagation loss and then set the transmitted power to obtain the desired SNR.

closely spaced antennas. Here we designate the base and subscriber ends of the link as T and R, respectively, but reciprocity applies, and all subsequent results apply to both the downlink (base-to-subscriber) and the uplink (subscriber-to-base). The T and R arrays are far apart relative to the array size. For this case, all the matrix elements have essentially the same amplitude and phases such that that $H = H_1$ (rank 1).

An evaluation of $HH\dagger$ reveals all elements essentially equal to n, so that using the normalized H_1, the matrix

$$A = [I_n + (\rho/n)HH^\dagger] \quad (2)$$

has all diagonal elements equal to $1+\rho$, and all off-diagonal elements equal to ρ, so that $detA = 1 + n\rho$. Thus for this case

$$C = log_2(1 + n\rho) = C_{log} \quad (3)$$

i.e. C_{log} increases logarithmically with n. For n closely spaced antennas, the capacity gain is essentially due to the n-fold gain in ρ.

2.3 Line-of-sight systems with widely spaced antenna elements

In general, with arrays of n more widely spaced antennas at T and R for which the path lengths between any pair of transmit and receive antennas are approximately the same to within the array size, the complex scalars H_{ik} all have magnitude 1 but different phases θ_{ik}. For such H, is it possible to obtain capacity exhibiting linear growth with n? In this subsection we show, by mathematical example, that the answer is yes.

By choosing the θ_{ik} so that $HH^\dagger = nI_n$ in (2), all diagonal elements of A are $1 + \rho$, resulting in $detA = (1 + \rho)^n$ and therefore

$$C = nlog_2(1 + \rho) = C_{lin} \quad (4)$$

i.e. C_{lin} increases linearly with n.

For a mathematical example of this particular type of H (denoted H_n) for which $HH^\dagger = nI_n$, take the generic ik entry of H_n to be given by

$$H_{ik} = exp(j\theta_{ik}) \quad \text{where} \quad (5)$$

$$\theta_{ik} = \frac{\pi}{n}\left((i - i_0) - (k - k_0)\right)^2. \quad (6)$$

For $n = 2$ and $i_0 = k_0 = 0$,

$$H_n = \begin{pmatrix} 1 & j \\ j & 1 \end{pmatrix}. \quad (7)$$

A second example for which $HH^\dagger = nI_n$ is

$$H_{ik} = exp(j\theta_{ik}) \quad \text{where} \quad (8)$$

$$\theta_{ik} = \frac{2\pi}{n}(i - i_0)(k - k_0). \quad (9)$$

For $n = 2$ and $i_0 = k_0 = 0$,

$$H_n = \begin{pmatrix} -1 & 1 \\ 1 & 1 \end{pmatrix}. \quad (10)$$

Both types of $H \simeq H_n$ can be attained in practice by an appropriate geometrical arrangement of the T and R arrays. The first example above is two 2-element broadside arrays of appropriate spacing and separation to yield a path length difference $\delta = d(T_1, R_1) - d(T_1, R_2) = \lambda/4$. The second example is a 2-element array with $\lambda/2$ spacing at T with R comprising 2 elements equidistant from T, one broadside to T and one aligned (endfire) with T. The question of how widely spaced the antennas need to be to promote H tending to H_n is considered in section IV along with an investigation of robustness.

3. MAXIMUM LOS CAPACITY

To find the maximum capacity which may be realized in practice in a LOS environment, we seek to construct geometric arrangements of the T and R arrays such that $H_{LOS} \simeq H_n$. Specifically, we ask what antenna spacing at T and R is required to obtain the θ_{ik} in H_n, and what spacing at T is required if the spacing at R is constrained to a half-wavelength. For these spacings, $H_{LOS} \simeq H_n$ as a direct result of ray-tracing calculations for the chosen geometry. We stress that the transmitter does not use detailed channel knowledge in any way. We are maintaining our assumption that each of the n antennas carries a separate component of the information and that the transmitter does not know the channel and does not adjust the power or phase of the n information components to fit the channel.

To achieve $C \simeq C_{lin}$, the H_{ik} need not be exactly according to (6)(9). Numerical results show that $C \simeq C_{lin}$ even if the off-diagonal elements in $HH^\dagger$ are non-zero with values up to $n/3$. Thus the capacity is robust in the presence of small variations in the geometric arrangements, as will be observed in Section V.

3.1 R array constrained to practical size

We consider a 3-element transmit and receive array (Figure 3), where the R array is constrained to have a fixed small interelement spacing y_r (e.g. $y_r = \lambda/2$), so that the size of R is practical. We seek to find the interelement spacing of T for which capacity is maximum. To preserve the

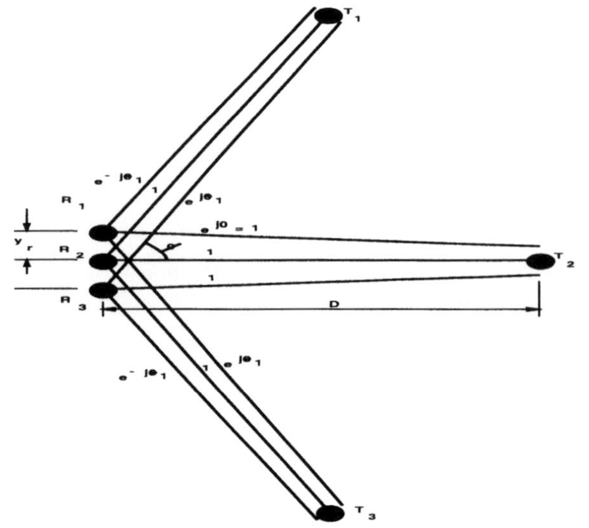

Fig. 1. Transmit and receive arrays, small spacing at R

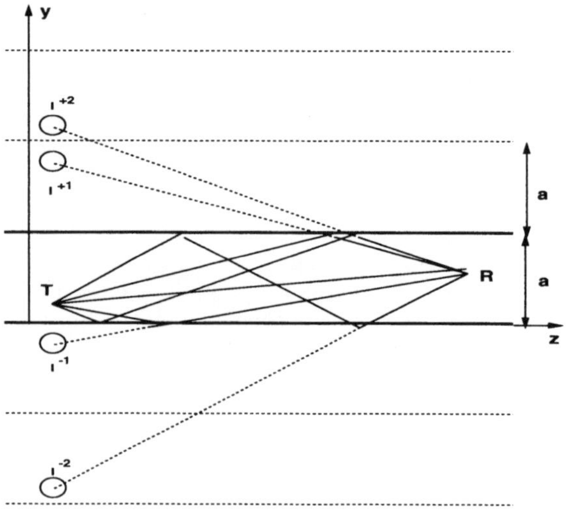

Fig. 2. Images in street canyon - top view

property that all elements of H have unit magnitude, all rays from T to R must be approximately the same length, and thus the T array is in the shape of an arc. We will assume that the T-R distance $D \gg y_a = y_r(n-1)$, where y_a is the size of the linear R array, so that we can assume plane wave arrivals from each element of T across the entire array R. Thus the T interelement spacing is defined by the angle ϕ between T array elements as seen at R, and ϕ will be independent of D.

The geometrical symmetry allows us to write the normalized matrix channel

$$H_{LOS} = \begin{pmatrix} e^{-j\theta_1} & 1 & e^{j\theta_1} \\ 1 & 1 & 1 \\ e^{j\theta_1} & 1 & e^{-j\theta_1} \end{pmatrix} \quad (11)$$

For this particular case, H_{LOS} is in the form (9) with $i_0 = k_0 = 2$, and solving for θ_1 such that $H_{LOS} = H_n$ (i.e. $HH^\dagger = nI_n$) yields $\theta_1 = 120$ degrees, with correponding path length difference

$$\delta = d(T_1, R_1) - d(T_1, R_2) = \frac{120}{360}\lambda = \frac{\lambda}{3} \quad (12)$$

(Figure 3). The geometrical interpretation is that the radiation pattern of R with all elements in phase has nulls in the direction of all but the center element of T.

3.2 Discussion

In practice, the path lengths between different antennas pairs are not all the same, so that H_{LOS} is not precisely equal to H_n, and $C = C_{lin}$ is not quite attained. The ray tracing calculations in Section V are needed to obtain precise results for the capacity as a function of spacing y_r or the angular size of the arc 2ϕ.

We will find that for the R array constrained to half-wavelength spacing,the angular range of the arc of the T array increases slowly with n, reaching $2\phi = 108$ degrees for $n = 8$. The large n result is consistent with the angular spread of signals $2\triangle \simeq \frac{30}{y_r/\lambda}$ (J.Salz and J.H.Winters, 1994) required to obtain near zero correlation between array elements for $y_r = \lambda/2$, and thus is intuitively satisfying. This result remains consistent with other values of y_r.

4. CAPACITY IN SPECIFIED ENVIRONMENT USING RAY TRACING

Capacities available in a specific environment were obtained in (G.J.Foschini and Valenzuela, 1996) using ray tracing for the case of a single transmit antenna ($n_T = 1$) and multiple receive antennas. In this section, we seek to extend this work for the case $n_T > 1$. A simple ray-tracing program which can accommodate some simple illustrative geometries was used. We will find H_{LOS} for various LOS geometries with no scatterers, as well as an urban street with 3 reflecting (scattering) surfaces, i.e. two walls (buildings) and the ground.

4.1 Geometry and coordinates

The urban street geometry contains two parallel reflectors representing the building walls, separated by the street width a. A top view is shown in Figure 4, where the ground is the y-z plane. The n array elements at each of T and R and the images I^k have coordinates $R_j : (x_{r_j}, y_{r_j}, z_{r_j})$, $T_i : (x_{t_i}, y_{t_i}, z_{t_i})$, $I_i^0 : (-x_{t_i}, y_{t_i}, z_{t_i})$, $I_i^{+1} : (x_{t_i}, 2a - y_{t_i}, z_{t_i})$, $I_i^{-1} : (x_{t_i}, -y_{t_i}, z_{t_i})$, with $i, j = 1, ..., n$, where we define the reference location to be the first element $T_1 : (x_t, y_t, z_t) = (x_{t_1}, y_{t_1}, z_{t_1})$ and similarly for R. $I^{\pm k}$ represents an image due to k specular reflections from the walls, and I_i^0 is the 'ground reflection' image not visible in the figure. A general expression for the coordinates of I^k is given in (Driessen *et al.*, 1992). In this first example, the T and R arrays are oriented perpendicular to the street direction, with half-wavelength antenna spacing, so that $x_{t_i} = x_t, y_{t_i} = y_t + i\lambda/2, z_{t_i} = z_t$, and $x_{r_j} = x_r, y_{r_j} = y_r + j\lambda/2, z_{r_j} = z_r$,

In the absence of any scattering surfaces, we have only the LOS path and $C \simeq C_{log}$ [2] However, with scattering surfaces present, we anticipate that the images may be sufficiently spread apart for C to approach C_{lin}. We make explicit calculations of C in the sequel.

For the street geometry outlined above, the elements of H may be written

$$\frac{H_{ij}}{|T_1 - R_1|} = \frac{exp(-j2\pi|T_i - R_j|/\lambda)}{|T_i - R_j|} + \sum_{k=-m}^{m} \Gamma^k \frac{exp(-j2\pi|I_i^k - R_j|/\lambda)}{|I_i^k - R_j|} \quad (13)$$

where Γ is the amplitude reflection coefficient [3], m is the maximum number of reflections considered, and we have assumed isotropic antennas. H_{ij} is normalized by the distance between the reference locations T_1, R_1, so that the absolute attenuation need not be calculated. Here we assume a fixed frequency $f = c/\lambda$.

4.2 Capacity results

We use the expressions for H_{ij} in the preceding sections to obtain numerical results for capacity in the specified geometries. For all these results, $\rho = 100$ (20 dB) at the reference distance D.

4.2.1. Street scenario We consider the street scenario with the images added, and the two arrays oriented broadside to each other, each with interelement spacing fixed at a half-wavelength. The parameters used are: wavelength 1/3 meter, street width a=25 meters with walls at $y = 0$ and 25, $(x_t, y_t, z_t) = (10, 15, 0)$, $(x_r, y_r, z_r) = (1, 5, 30)$, all dimensions in meters. Γ is set to 0.6, so that the average total power received from reflections only, with 5 reflections, is equal to the average LOS power. The ground reflection coefficient is set to -1. Figure 5 shows the capacity as a function of (x_r, y_r, z_r) ranging in 200 steps from (1,5,30) to (1,6,40) meters (about 30 wavelengths). The capacity increases as more images are added, and matches the 90 percent Rayleigh capacity obtained in Figure 2 for $n = 8$ with 7 images ($k = 0, ..., \pm 3$ in eqn. (13)) plus T. Furthermore, the received signal envelope looks increasingly Rayleigh-like as more images are added. The power increase resulting from adding images accounts for some, but not all of the capacity increase.

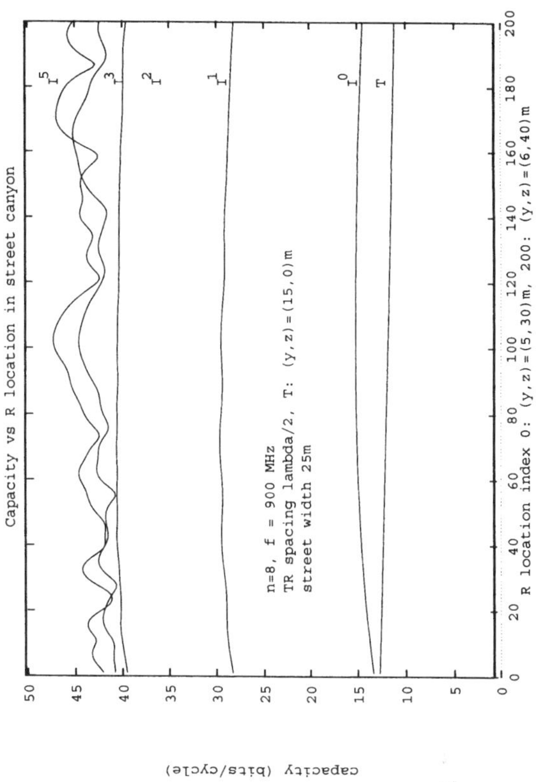

Fig. 3. Capacity versus R location in street canyon

These results show that capacity can be increased by spreading out the elements of the T array (implicitly) by creating images of the (closely-spaced) T array. Thus we effectively have an $m \times n$ element T array, where $m - 1$ is the number of images, and n distinct signals are each simulcast m times. The implicit spreading of the T array may be achieved in an urban scenario by lowering the array below the rooftops [4].

5. SUMMARY AND CONCLUSIONS

Wireless communications systems with multiple antennas at both transmitter and receiver have an information-theoretic channel capacity which can grow linearly (rather than logarithmically) with

---

[2] In fact, $C = C_{log}$ for zero spacing between elements, and C is slightly greater than C_{log} for half-wavelength spacing.
[3] In general, the Γ are different for each image, since they depend on the angles of incidence and reflection. Here we assume Γ has the same constant value for all reflections, except the ground reflection Γ^0 which is set to -1. This approximation suffices to illustrate the capacity gain.

[4] This idea of lowering the transmit antennas below the rooftops was first suggested by M. J. Gans.

the number of antennas (Foschini and Gans, 1998) for fixed total radiated power and fixed bandwidth. This result assumes that the transmitter does not know the channel, and that separate information is sent out of each transmitter antenna. Here we show that such capacities can be achieved not only on Rayleigh channels with many scatterers, but also on deterministic channels with direct LOS paths only and no scatterers, by explicitly spreading out the antennas well beyond a wavelength. This has significant implications for application areas such as indoor wireless LANs.

Capacities approaching $C_{lin} = nlog_2(1+\rho)$ can be achieved in a deterministic LOS (non-Rayleigh) environment. This is achieved by spreading out the elements of T either explicitly (by placing one element of T at each of n sites), or implicitly (by adding reflectors which create images of T). This result suggests that in the absence of reflectors, the T array can be explictly spread apart, but using n antennas instead of one at each of n sites, thus duplicating the effect of images (Figure 7). The models for $H = H_{LOS} + H_{Rayleigh}$ described here may be useful for the testing of wireless systems which exploit the available capacity (e.g.(Foschini, 1996)).

The results for a linear array at R may be interpreted physically in two ways. First, capacities approaching C_{lin} are attained in a deterministic environment by spreading the n array elements of T evenly around an arc with the same angular range required to obtain zero correlation between R elements in a scattering enviroment (J.Salz and J.H.Winters, 1994). Thus the continuum of sources (scatterers) along the arc assumed in (J.Salz and J.H.Winters, 1994) is replaced by n discrete elements of T spread out over the same angular range. Second, C_{lin} is precisely attained when the locations of the T array elements along the arc corresponds precisely with the nulls in the radiation pattern of R, i.e. they are spaced almost, but not quite, evenly around the arc, assuming equal phase at all T elements. However, the spacing need not be so precise to attain capacities close to C_{lin}. For given spacing at R, increasing n results in more closely spaced nulls, and the angular range of the n array elements of T does not change. In both interpretations, the angular range of the arc depends only on the spacing of the R elements, and is independent of n.

We conclude that high capacities may be achieved in several ways: 1) The transmit antenna array can be spread out. The interelement spacing or angular spread may be such that n separate base station sites are needed (Driessen, 2000). However, capacity may be lost if some of the base station sites are shadowed from the mobile. 2a) The n antennas at a single base site are implicitly duplicated by placing them in a scattering environment where images are created. This may be viewed as a kind of implicit spreading of the transmit array. 2b) The n antennas are duplicated at each of the n base sites, to minimize shadowing loss (n-conductor leaky feeder with n leaks, thus using n^2 radiators for n-times simulcasting of the n signals). This scheme may be viewed as explicitly creating images (duplicates) where there are no scatterers to create them implicitly.

6. ACKNOWLEDGMENTS

The author thanks G.J. Foschini for our delightful collaboration on (Driessen and Foschini, 1999), and L.J. Greenstein for supporting this work during my visit to AT&T.

REFERENCES

Catreux, S., P.F. Driessen and L.J. Greenstein (2000). Simulation results for an interference-limited multiple-input multiple-output cellular system. *IEEE Communications Letters.*

Driessen, P.F. (2000). Cellular communication system with multiple same frequency broadcasts in a cell. US Patent 6,052,599.

Driessen, P.F. and G.J. Foschini (1999). On the capacity of multiple-input multiple-output wireless channels: a geometric interpretation. *IEEE Trans. Commun.* **47**(2), 173–176.

Driessen, P.F. et al. (1992). Ray model of indoor propagation. In: *Personal Communications Networks* (T. Rappaport and M. Feuerstein, Eds.). Kluwer Academic Press.

Foschini, G.J. (1996). Layered space-time architecture for wireless communication in a fading environment when using multi-element antennas. *Bell Labs Technical Journal* **2**(2), 41–59.

Foschini, G.J. and M.J. Gans (1998). On limits of wireless communication in a fading environment when using multiple antennas. *Wireless Personal Communications* **6**(3), 331–335.

G.J.Foschini and R.A. Valenzuela (1996). Initial estimation of communication efficiency of indoor wireless channels. *International Journal of Wireless Information Networks.*

J.Salz and J.H.Winters (1994). Effect of fading correlation on adaptive arrays in digital mobile radio'. *IEEE Trans. Vehic. Tech.* **43**(4), 1049–1057.

P.Beckmann and A.Spizzichino (1963). *The scattering of electromagnetic waves from rough surfaces.* Permagon.

IFAC Publications
www.elsevier.com/locate/ifac

COMPUTATIONALLY EFFICIENT BLIND MMSE RECEIVERS FOR LONG CODE WCMDA USING TIME-VARYING SYSTEMS THEORY

Alle-Jan van der Veen [*,1] Lang Tong [**]

[*] *Delft Univ. of Technology, Dept. Electrical Eng./DIMES, Mekelweg 4, 2628 CD Delft, The Netherlands*
[**] *Cornell University, Dept. Electrical Eng., 326 Frank H.T. Rhodes Hall, Ithaca, NY 14853, USA*

Abstract: UMTS systems will employ long-code wideband CDMA modulation schemes. Receivers for this system are for computational reasons usually based on simple matched-filter techniques, and hence suffer from multiaccess interference. Decorrelating RAKE and MMSE receivers do not have this problem but, until now, were considered as too complex, due to the inversion of a large code matrix. As is shown in this paper, the code matrix can be interpreted as a time-varying system. Efficient implementations are then possible by carrying out the inversion using time-varying state space theory, yielding a complexity comparable to that of the conventional RAKE receiver. *Copyright © 2003 IFAC*

Keywords: Long-code W-CDMA, decorrelating RAKE receiver, MMSE receiver, time-varying system theory, computational structures

1. INTRODUCTION

Current receivers for long-code (or aperiodic spreading code) wideband CDMA are typically based on RAKE receivers, i.e. banks of matched filters which correlate the received data with the desired user's code, followed by a combining of the outputs (RAKE fingers). Since multiuser interference is not completely cancelled, the performance degrades, especially when the network is heavily loaded and power control imperfect.

In this paper, we consider the uplink (mobiles to base station) and assume that the base station knows all codes. We model multiuser interference explicitly and propose a blind decorrelating RAKE and MMSE receiver to estimate the channel and user symbols, based on all samples in a frame. The decorrelating RAKE was presented earlier by us in (Tong *et al.*, 2002*b*; Tong *et al.*, 2002*a*) with an emphasis on identifiability and performance; the MMSE receiver is similar. Here, we also take the noise covariance into account and focus in particular on the efficient implementation of these receivers.

The decorrelating matched filter asks for the inversion of a code matrix whose long dimension is equal to the number of chips over the complete frame. This is a formidable task, but fortunately, the sparse structure of this matrix admits computationally efficient techniques. As an application of the work in (Dewilde and van der Veen, 1998) on the inversion of infinite-size matrices, we derive efficient time-varying state-space implementations of the various steps in the algorithm.

Blind channel estimation and multiuser detection for long code CDMA has been considered by a number of other authors. In particular, second order moment techniques (Zoltowski *et al.*, 1996; Liu and Zoltowski, 1997; Sidiropoulos and Bro, 1999; Xu and Tsatanis, 2000; Escudero *et al.*, 2001) rely on the convergence of time averages, which often requires hundreds to thousands of symbols. Although related, Weiss and Friedlander (Weiss and Friedlander, 1999) focus on the down link where users can be considered synchronous.

[1] The research of A.J. van der Veen was supported in part by the Dutch Min. Econ. Affairs under TSIT 1025 "Beyond-3G".

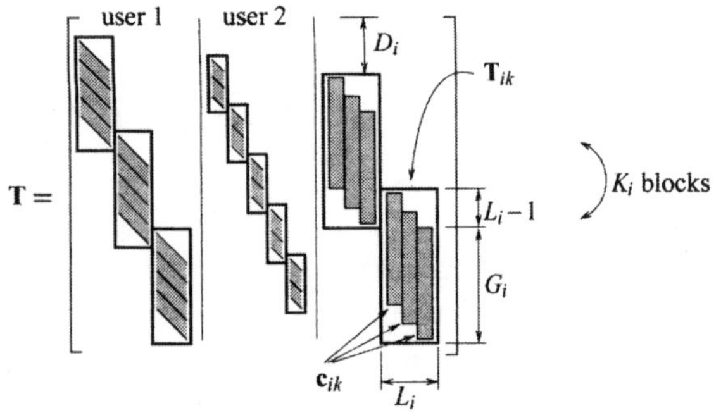

Fig. 1. Structure of the code matrix **T**.

2. DATA MODEL

We consider the uplink of a slotted system with I asynchronous users. In a frame, the i-th user transmits a vector $\mathbf{s}_i$ consisting of K_i symbols s_{ik}. Each symbol is spread by an aperiodic code $\mathbf{c}_{ik}$ of length G_i. After multipath propagation over a channel with length L_i chips and relative delay D_i, pulse shape matched filtering and sampling at the chiprate, the receiver stacks the received samples in a frame in a vector $\mathbf{y}$. (Oversampling is equivalent to a system with multiple receive antennas.) The contribution of s_{ik} is a linear combination of the transmitted signal $\mathbf{c}_{ik}s_{ik}$, plus delays of it, properly scaled by the L_i channel coefficients collected in a vector $\mathbf{h}_i$, or

$$\mathbf{y}_{ik} = \mathbf{T}_{ik}\mathbf{h}_i\mathbf{s}_{ik}, \quad k = 1, \cdots, K_i.$$

$\mathbf{T}_{ik}$ is a Toeplitz matrix whose L_i columns consist of shifts of the code $\mathbf{c}_{ik}$. Including all users and the noise, we have

$$\begin{aligned} \mathbf{y} &= \mathbf{THs} + \mathbf{w} \qquad (1)\\ \mathbf{T} &:= [\mathbf{T}_{11} \cdots \mathbf{T}_{1,G_1}, \cdots, \mathbf{T}_{I1} \cdots \mathbf{T}_{I,G_I}]\\ \mathbf{H} &:= \mathrm{diag}(\mathbf{I}_{K_1} \otimes \mathbf{h}_1, \cdots, \mathbf{I}_{K_I} \otimes \mathbf{h}_I). \end{aligned}$$

where matrix $\mathbf{H}$ is block diagonal with $\mathbf{I} \otimes \mathbf{h}_i$ as the ith block, vector $\mathbf{s}$ is a stacking of all symbol vectors, and $\mathbf{w}$ is a vector representing the additive Gaussian noise. The structure of the code matrix $\mathbf{T}$ is illustrated in figure 1. Note that different spreading gains G_i are part of the model. Multiple antennas are a simple extension.

We will assume that the code matrix $\mathbf{T}$ is known, "tall" and has full column rank. This implies that the receiver knows the codes, the delay offsets D_i, and the number of paths L_i of all users.

3. BLIND RECEIVER ALGORITHMS

3.1 *Conventional RAKE*

The conventional RAKE receiver consists of a bank of matched filters and projects the received signal into the code domains of the individual users, by correlating with several shifts of the code vectors, or $\mathbf{r} = \mathbf{T}^H\mathbf{y}$. Since the codes are not exactly orthogonal (let alone shift-orthogonal), $\mathbf{T}^H\mathbf{T} \neq \mathbf{I}$, and contributions of each user enter into the projections of any other user. This makes the performance interference-limited.

3.2 *Decorrelating RAKE*

The proposed decorrelating RAKE uses a decorrelating matched filter, or $\mathbf{T}^\dagger = (\mathbf{T}^H\mathbf{T})^{-1}\mathbf{T}^H$. This removes all multi-user interference. The output of the decorrelating matched filter is given by

$$\mathbf{u} = \mathbf{T}^\dagger\mathbf{y} = \mathrm{diag}(\mathbf{I} \otimes \mathbf{h}_1, \cdots, \mathbf{I} \otimes \mathbf{h}_I)\,\mathbf{s} + \mathbf{n}, \qquad (2)$$

where $\mathbf{n} = \mathbf{T}^\dagger\mathbf{w}$ is now a colored noise vector. After computing $\mathbf{u}$, we estimate the channel and the data symbols, blindly and independently for each user. Partition $\mathbf{u}$ into segments $\mathbf{u}_{ik}$ of length L_i. The structure of $\mathbf{u}$ implies that $\mathbf{u}_{ik}$ corresponds to symbol k of user i,

$$\mathbf{u}_{ik} = \mathbf{h}_i s_{ik} + \mathbf{n}_{ik}, \quad k = 1, \cdots, K_i, \qquad (3)$$

and is free from multiuser interference. Collecting all data for user i gives $\mathbf{U}_i = [\mathbf{u}_{i1}, \cdots, \mathbf{u}_{iK_i}] = \mathbf{h}_i\mathbf{s}_i^T + \mathbf{N}_i$. This is a rank-1 data model, and estimates of $\mathbf{h}_i$ and $\mathbf{s}_i$ (with an unknown scaling factor) are found from a rank-one factorization of $\mathbf{U}_i$. In other words, denoting

$$\mathbf{\Psi}_i := \frac{1}{K_i}\sum_{k=1}^{K_i} \mathbf{u}_{ik}\mathbf{u}_{ik}^H, \qquad (4)$$

we obtain the least squares estimates

$$\hat{\mathbf{h}}_i = \arg\max_{\|\mathbf{g}\|=1} \mathbf{g}^H\mathbf{\Psi}_i\mathbf{g}, \quad \hat{s}_{ik} = \hat{\mathbf{h}}_i^H\mathbf{u}_{ik}. \qquad (5)$$

The solution $\hat{\mathbf{h}}_i$ is given as the dominant eigenvector of $\mathbf{\Psi}_i$. The scaling ambiguity is resolved by a single pilot symbol. See (Tong *et al.*, 2002*b*; Tong *et al.*, 2002*a*) for further results and performance simulations.

3.3 *Whitened Estimator*

The channel and symbol estimator given in (5) did not take into account that the noise process $\mathbf{n}_{ik}$ is colored, both in k and in its components. If we ignore the coloring in k, then a simple whitening approach can be applied. Specifically, since $\mathbf{n} = \mathbf{T}^\dagger\mathbf{w}$, we have that $\mathbf{n}_{ik} \sim \mathcal{N}(\mathbf{0}, \sigma^2\mathbf{\Sigma}_{ik})$ where $\mathbf{\Sigma}_{ik}$ is an $L_i \times L_i$ submatrix on the diagonal of $\mathbf{T}^\dagger(\mathbf{T}^\dagger)^H$. We have

$$\begin{aligned} \mathrm{E}(\mathbf{\Psi}_i) &= \frac{\|\mathbf{s}_i\|^2}{K_i}\mathbf{h}_i\mathbf{h}_i^H + \sigma^2\mathbf{\Delta}_i,\\ \mathbf{\Delta}_i &:= \frac{1}{K_i}\sum_{k=1}^{K_i}\mathbf{\Sigma}_{ik} \end{aligned}$$

where $\mathbf{\Delta}_i$ is a known matrix. The channel can then be estimated from the following modification which whitens the noise on $\mathbf{\Psi}_i$:

$$\mathbf{g}_* = \arg\max_{\|\mathbf{g}\|=1} \mathbf{g}^H(\mathbf{\Delta}_i^{-1/2}\mathbf{\Psi}_i\mathbf{\Delta}_i^{-H/2})\mathbf{g}$$
$$\hat{\mathbf{h}}_i = \mathbf{\Delta}^{1/2}\mathbf{g}_* .$$

The symbol estimator given in (5) is replaced by $\hat{s}_{ik} = \hat{\mathbf{h}}_i^H \mathbf{\Sigma}_{ik}^{-1}\mathbf{u}_{ik}$.

3.4 *MMSE Receiver*

Based on the data model (1), the estimated data sequence by a linear minimum mean square error (MMSE) receiver is known to be

$$\hat{\mathbf{s}} = (\mathbf{H}^H\mathbf{T}^H\mathbf{T}\mathbf{H} + \sigma^2\mathbf{I})^{-1}\mathbf{H}^H\mathbf{T}^H\mathbf{y}. \qquad (6)$$

This receiver can be implemented using the previously estimated channel matrix $\mathbf{H}$, and assuming that the noise power σ^2 is known. Compared to the decorrelating RAKE, the MMSE can have a significantly improved performance. It is also one of two similar steps in an iterative LS estimator (Hieu and van der Veen, 2003).

4. EFFICIENT IMPLEMENTATIONS

The code matrix $\mathbf{T}$ can be very large. Without an efficient technique to compute and apply the left inverse $\mathbf{T}^\dagger = (\mathbf{T}^H\mathbf{T})^{-1}\mathbf{T}^H$, the proposed receiver structures would not be feasible. Fortunately, $\mathbf{T}$ is sparse. Using the Matlab sparse toolbox, $\mathbf{u} = \mathbf{T}^\dagger\mathbf{y}$ can be computed efficiently via a sparse QR factorization $\mathbf{T} = \mathbf{Q}\mathbf{R}$, and $\mathbf{u} = \mathbf{R}^{-1}\mathbf{Q}^H\mathbf{y}$, or, avoiding the storage of $\mathbf{Q}$, as

$$\begin{bmatrix} \mathbf{R} & \mathbf{v} \\ \mathbf{0} & \varepsilon \end{bmatrix} := \text{qr([sparse(T) y])}$$
$$\mathbf{u} := \mathbf{R}\backslash\mathbf{v}$$

$\mathbf{v} = \mathbf{Q}^H\mathbf{y}$, and $\mathbf{R}\backslash\mathbf{v}$ denotes $\mathbf{R}^{-1}\mathbf{v}$, implemented efficiently via backsubstitution. This does not reveal how the sparse computations can actually be implemented in a practical system. It is also unclear how the noise whitening (computation of $\mathbf{\Sigma}_{ik}$) can be implemented efficiently. Explicit computation of $\mathbf{\Sigma} = \mathbf{R}^{-1}\mathbf{R}^{-H}$ is to be avoided because $\mathbf{R}^{-1}$ is not sparse even if $\mathbf{R}$ is. In this section, we show how time-varying state space representations can be used for this purpose. The theory behind it is available in (Dewilde and van der Veen, 1998).

4.1 *State Space Representation of a Matrix*

Consider an input signal $\mathbf{u}$ and output signal $\mathbf{y}$, with arbitrary block-partitioning $\mathbf{u} = [\mathbf{u}_1^T, \cdots, \mathbf{u}_N^T]^T$, $\mathbf{y} = [\mathbf{y}_1^T, \cdots, \mathbf{y}_N^T]^T$. The partitioning introduces the notion of "time", or a stage in a computational procedure. The blocks do not need to be of equal size, and some dimensions can even be zero (such a block is denoted by "$\cdot$").

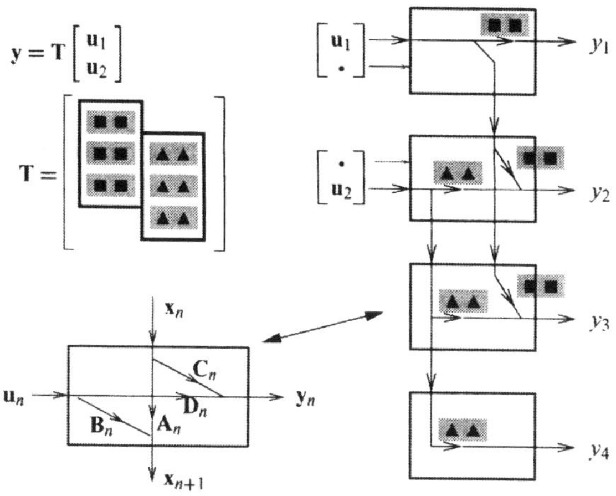

Fig. 2. Computational network for $\mathbf{T} = [\mathbf{T}^{(1)}\ \mathbf{T}^{(2)}]$.

A time-varying state space realization has the form

$$\begin{cases} \mathbf{x}_{n+1} = \mathbf{A}_n\mathbf{x}_n + \mathbf{B}_n\mathbf{u}_n \\ \mathbf{y}_n = \mathbf{C}_n\mathbf{x}_n + \mathbf{D}_n\mathbf{u}_n \end{cases}$$
$$\Leftrightarrow$$
$$\begin{bmatrix} \mathbf{x}_{n+1} \\ \mathbf{y}_n \end{bmatrix} = \mathbf{T}_n \begin{bmatrix} \mathbf{x}_n \\ \mathbf{u}_n \end{bmatrix}, \quad \mathbf{T}_n = \begin{bmatrix} \mathbf{A}_n & \mathbf{B}_n \\ \mathbf{C}_n & \mathbf{D}_n \end{bmatrix}$$

The realization starts at time 1 with $\mathbf{x}_1 = \cdot$ (or: no state), and ends with $\mathbf{x}_{N+1} = \cdot$. Hence $\mathbf{A}_1 = \cdot$, $\mathbf{A}_N = \cdot$, $\mathbf{C}_1 = \cdot$, $\mathbf{B}_N = \cdot$.

A time-varying state-space realization specifies a linear mapping of $\mathbf{u}$ to $\mathbf{y}$, hence a matrix $\mathbf{T}$ such that $\mathbf{y} = \mathbf{T}\mathbf{u}$. In particular, it defines a factorization of $\mathbf{T}$ into factors $\mathbf{T}_n$.

Lemma 1. Let be given a time-varying realization $\mathbf{T}_n = \{\mathbf{A}_n, \mathbf{B}_n, \mathbf{C}_n, \mathbf{D}_n\}$ of $\mathbf{T}$. Then $\mathbf{T} = \tilde{\mathbf{T}}_N \cdots \tilde{\mathbf{T}}_2\tilde{\mathbf{T}}_1$ where $\tilde{\mathbf{T}}_n$ is an embedding of $\mathbf{T}_n$,

$$\tilde{\mathbf{T}}_n := \left[\begin{array}{c|cccccc} \mathbf{A}_n & & & \mathbf{B}_n & & & \\ \hline & \mathbf{I} & & & & & \\ & & \ddots & & & & \\ & & & \mathbf{I} & & & \\ \mathbf{C}_n & & & \mathbf{D}_n & & & \\ & & & & \mathbf{I} & & \\ & & & & & \ddots & \\ & & & & & & \mathbf{I} \end{array}\right]$$

(there are $n-1$ and $N-n$ identity matrices in the diagonal sequences, respectively.) Moreover, matrix $\mathbf{T}$ is block-lower triangular and has the form

$$\mathbf{T} = \begin{bmatrix} \mathbf{D}_1 & & \\ \mathbf{C}_2\mathbf{B}_1 & \mathbf{D}_2 & \\ \vdots & \ddots & \ddots \\ \mathbf{C}_N\mathbf{A}_{N-1}\cdots\mathbf{A}_2\mathbf{B}_1 & \cdots\ \mathbf{C}_N\mathbf{B}_{N-1} & \mathbf{D}_N \end{bmatrix}.$$

Conversely, if a matrix $\mathbf{T}$ has this form, then it has a state space realization $\mathbf{T}_n = \{\mathbf{A}_n, \mathbf{B}_n, \mathbf{C}_n, \mathbf{D}_n\}$.

The inherent causality translates to $\mathbf{T}$ being block-lower triangular. However, by playing with dimensions, *any* matrix can fit this model, as the next examples illustrate. Consider first an an arbitrary $N \times L$ matrix $\mathbf{T}$, with rows $\mathbf{t}_n^H$. A (trivial) realization that models $\mathbf{y} = \mathbf{T}\mathbf{u}$ is obtained by setting $\mathbf{u}_1 = \mathbf{u}$, $\mathbf{u}_2 = \cdots = \mathbf{u}_N = \bullet$ (i.e., the complete input vector is entered at time 1), and

$$\begin{bmatrix} \mathbf{A}_1 & \mathbf{B}_1 \\ \mathbf{C}_1 & \mathbf{D}_1 \end{bmatrix} = \begin{bmatrix} \bullet & \mathbf{I} \\ \bullet & \mathbf{t}_1^H \end{bmatrix}$$

$$\begin{bmatrix} \mathbf{A}_n & \mathbf{B}_n \\ \mathbf{C}_n & \mathbf{D}_n \end{bmatrix} = \begin{bmatrix} \mathbf{I} & \bullet \\ \mathbf{t}_n^H & \bullet \end{bmatrix}, \quad n = 2, \cdots, N-1$$

$$\begin{bmatrix} \mathbf{A}_N & \mathbf{B}_N \\ \mathbf{C}_N & \mathbf{D}_N \end{bmatrix} = \begin{bmatrix} \bullet & \bullet \\ \mathbf{t}_N^H & \bullet \end{bmatrix}$$

As a second example, let $\mathbf{T} = [\mathbf{T}^{(1)} \ \mathbf{T}^{(2)}]$ be an arbitrary block-partitioned matrix, where $\mathbf{T}^{(1)}$ has realization $\{\mathbf{A}_n^{(1)}, \mathbf{B}_n^{(1)}, \mathbf{C}_n^{(1)}, \mathbf{D}_n^{(1)}\}$ and $\mathbf{T}^{(2)}$ has realization $\{\mathbf{A}_n^{(2)}, \mathbf{B}_n^{(2)}, \mathbf{C}_n^{(2)}, \mathbf{D}_n^{(2)}\}$. Then

$$\mathbf{T}_n = \left[\begin{array}{cc|cc} \mathbf{A}_n^{(1)} & 0 & \mathbf{B}_n^{(1)} & 0 \\ 0 & \mathbf{A}_n^{(2)} & 0 & \mathbf{B}_n^{(2)} \\ \hline \mathbf{C}_n^{(1)} & \mathbf{C}_n^{(2)} & \mathbf{D}_n^{(1)} & \mathbf{D}_n^{(2)} \end{array}\right]$$

is a realization of $\mathbf{T}$. Its structure is shown in Fig. 2.

The code matrix $\mathbf{T}$ in our case has a block structure as shown in Fig. 1. By combining the two examples, we can represent any code matrix $\mathbf{T}$. The number of state space time points equals the number of rows of $\mathbf{T}$. The input vector is partitioned in blocks of L_i entries which enter the system at appropriate time points, determined by the starting points of the individual code blocks. The state dimension at each time point is the number of nonzero entries in the corresponding row of $\mathbf{T}$.

4.2 *QR Factorization and Inversion in State Space*

To compute the left inverse $\mathbf{T}^\dagger$, our aim is to first compute a QR factorization $\mathbf{T} = \mathbf{Q}\mathbf{R}$ where $\mathbf{Q}^H\mathbf{Q} = \mathbf{I}$ and $\mathbf{R}$ is square and lower triangular, and then to invert each of the factors: $\mathbf{T}^\dagger = \mathbf{R}^{-1}\mathbf{Q}^H$. The computation of the QR factorization can be done in state space, as is demonstrated by the following theorem (cf. (Dewilde and van der Veen, 1998), p.156]).

For $\mathbf{T}$ with realization $\{\mathbf{A}_n, \mathbf{B}_n, \mathbf{C}_n, \mathbf{D}_n\}_{n=1,\cdots,N}$, consider the recursion (economy-size QR factorizations)

$$\mathbf{Y}_{N+1} = \bullet$$

$$\begin{bmatrix} \mathbf{Y}_{n+1}\mathbf{A}_n & \mathbf{Y}_{n+1}\mathbf{B}_n \\ \mathbf{C}_n & \mathbf{D}_n \end{bmatrix} =: \underbrace{\begin{bmatrix} \mathbf{A}_n^Q & \mathbf{B}_n^Q \\ \mathbf{C}_n^Q & \mathbf{D}_n^Q \end{bmatrix}}_{\mathbf{Q}_n} \begin{bmatrix} \mathbf{Y}_n & 0 \\ \mathbf{C}_n^R & \mathbf{D}_n^R \end{bmatrix}, \quad (7)$$

$$n = N, N-1, \cdots, 1$$

where $\mathbf{Q}_n$ is isometric ($\mathbf{Q}_n^H\mathbf{Q}_n = \mathbf{I}$), and the right factor is lower triangular (possibly staircase) and partitioned such that $\mathbf{Y}_n$ has the same number of columns as $\mathbf{A}_n$, $\mathbf{D}_n^R$ has the same number of columns as $\mathbf{D}_n$, and both $\mathbf{Y}_n$ and $\mathbf{D}_n^R$ are full row rank.

Theorem 2. If $\mathbf{T}$ is full column rank, then all $\mathbf{D}_n^R$ are square, lower triangular and invertible. Define the realizations

$$\mathbf{Q}_n = \begin{bmatrix} \mathbf{A}_n^Q & \mathbf{B}_n^Q \\ \mathbf{C}_n^Q & \mathbf{D}_n^Q \end{bmatrix}, \qquad \mathbf{R}_n = \begin{bmatrix} \mathbf{A}_n & \mathbf{B}_n \\ \mathbf{C}_n^R & \mathbf{D}_n^R \end{bmatrix}.$$

Then $\mathbf{T} = \mathbf{Q}\mathbf{R}$, where $\mathbf{Q}$ is specified by $\mathbf{Q}_n$ and is isometric ($\mathbf{Q}^H\mathbf{Q} = \mathbf{I}$), and $\mathbf{R}$ is specified by $\mathbf{R}_n$ and is lower triangular and invertible.

PROOF Recall the factorization $\mathbf{T} = \tilde{\mathbf{T}}_N\tilde{\mathbf{T}}_{N-1}\cdots\tilde{\mathbf{T}}_1$ and consider the first factor, $\mathbf{T}_N$. Since $\mathbf{A}_N = \bullet$, $\mathbf{B}_N = \bullet$, and $\mathbf{Y}_{N+1} = \bullet$,

$$\mathbf{T}_N = \begin{bmatrix} \mathbf{A}_N & \mathbf{B}_N \\ \mathbf{C}_N & \mathbf{D}_N \end{bmatrix} = \begin{bmatrix} \mathbf{Y}_{N+1}\mathbf{A}_N & \mathbf{Y}_{N+1}\mathbf{B}_N \\ \mathbf{C}_N & \mathbf{D}_N \end{bmatrix}.$$

The first step in the recursion is the QR factorization

$$\mathbf{Q}_N^H\mathbf{T}_N = \begin{bmatrix} \mathbf{A}_N^Q & \mathbf{B}_N^Q \\ \mathbf{C}_N^Q & \mathbf{D}_N^Q \end{bmatrix}^H \begin{bmatrix} \mathbf{Y}_{N+1}\mathbf{A}_N & \mathbf{Y}_{N+1}\mathbf{B}_N \\ \mathbf{C}_N & \mathbf{D}_N \end{bmatrix} = \begin{bmatrix} \mathbf{Y}_N & 0 \\ \mathbf{C}_N^R & \mathbf{D}_N^R \end{bmatrix}$$

Premultiplying $\mathbf{T}$ by $\tilde{\mathbf{Q}}_N^H$ gives

$$\tilde{\mathbf{Q}}_N^H\mathbf{T} =$$

$$= \left[\begin{array}{c|c|c} \mathbf{A}_N^Q & & \mathbf{B}_N^Q \\ & \mathbf{I} & \\ \mathbf{C}_N^Q & & \mathbf{D}_N^Q \end{array}\right]^H \left[\begin{array}{c|c} \mathbf{Y}_{N+1}\mathbf{A}_N & \mathbf{Y}_{N+1}\mathbf{B}_N \\ \hline & \mathbf{I} \\ \mathbf{C}_N & \mathbf{D}_N \end{array}\right] \tilde{\mathbf{T}}_{N-1}\cdots\tilde{\mathbf{T}}_1$$

$$= \left[\begin{array}{c|c|c} \mathbf{Y}_N & & 0 \\ \hline & \mathbf{I} & \\ & \ddots & \\ & & \mathbf{I} \\ \mathbf{C}_N^R & & \mathbf{D}_N^R \end{array}\right] \left[\begin{array}{c|c} \mathbf{A}_{N-1} & \mathbf{B}_{N-1} \\ \hline & \mathbf{I} \\ & \ddots \\ \mathbf{C}_{N-1} & \mathbf{D}_{N-1} \\ & \mathbf{I} \end{array}\right] \tilde{\mathbf{T}}_{N-2}\cdots\tilde{\mathbf{T}}_1$$

$$= \left[\begin{array}{c|cc} \mathbf{Y}_N\mathbf{A}_{N-1} & \mathbf{Y}_N\mathbf{B}_{N-1} & \\ \hline & \mathbf{I} & \\ & \ddots & \\ \mathbf{C}_{N-1} & \mathbf{D}_{N-1} & \\ \mathbf{C}_N^R\mathbf{A}_{N-1} & \mathbf{C}_N^R\mathbf{B}_{N-1} & \mathbf{D}_N^R \end{array}\right] \tilde{\mathbf{T}}_{N-2}\cdots\tilde{\mathbf{T}}_1$$

We subsequently obtain

$$\tilde{\mathbf{Q}}_{N-1}^H\tilde{\mathbf{Q}}_N^H\mathbf{T} =$$

$$\left[\begin{array}{c|cc} \mathbf{Y}_{N-1} & 0 & \\ \hline & \mathbf{I} & \\ & \ddots & \\ \mathbf{C}_{N-1}^R & \mathbf{D}_{N-1}^R & \\ \mathbf{C}_N^R\mathbf{A}_{N-1} & \mathbf{C}_N^R\mathbf{B}_{N-1} & \mathbf{D}_N^R \end{array}\right] \left[\begin{array}{c|c} \mathbf{A}_{N-2} & \mathbf{B}_{N-2} \\ \hline & \mathbf{I} \\ \mathbf{C}_{N-2} & \mathbf{D}_{N-2} \\ & \mathbf{I} \\ & \mathbf{I} \end{array}\right] \cdot\tilde{\mathbf{T}}_{N-3}\cdots\tilde{\mathbf{T}}_1 =$$

$$\left[\begin{array}{c|ccc} \mathbf{Y}_{N-1}\mathbf{A}_{N-2} & \mathbf{Y}_{N-1}\mathbf{B}_{N-2} & & \\ \hline & \mathbf{I} & & \\ \mathbf{C}_{N-2} & \mathbf{D}_{N-2} & & \\ \mathbf{C}_{N-1}^R\mathbf{A}_{N-2} & \mathbf{C}_{N-1}^R\mathbf{B}_{N-2} & \mathbf{D}_{N-1}^R & \\ \mathbf{C}_N^R\mathbf{A}_{N-1}\mathbf{A}_{N-2} & \mathbf{C}_N^R\mathbf{A}_{N-1}\mathbf{B}_{N-2} & \mathbf{C}_N^R\mathbf{B}_{N-1} & \mathbf{D}_N^R \end{array}\right] \cdot\tilde{\mathbf{T}}_{N-3}\cdots\tilde{\mathbf{T}}_1$$

Following the recursion this way, we finally obtain

$$\tilde{\mathbf{Q}}_1^H \cdots \tilde{\mathbf{Q}}_N^H \mathbf{T} = \left[\begin{array}{c|c} \mathbf{Y}_1 & \\ \hline \mathbf{C}_1^R & \mathbf{D}_1^R \\ \mathbf{C}_2^R \mathbf{A}_1 & \mathbf{C}_2^R \mathbf{B}_1 \quad\quad \mathbf{D}_2^R \\ \vdots & \vdots \quad\quad \ddots \quad \ddots \\ \mathbf{C}_N^R \mathbf{A}_{N-1} \cdots \mathbf{A}_1 & \mathbf{C}_N^R \mathbf{A}_{N-1} \cdots \mathbf{A}_2 \mathbf{B}_1 \cdots \cdots \mathbf{D}_N^R \end{array}\right].$$

Note that $\mathbf{A}_1 = \bullet$ so that the first column has zero width. Hence $\mathbf{Y}_1 = \bullet$ (since the $\mathbf{Y}_k$ are wide) and also the first row has empty dimensions. It follows that

$$\tilde{\mathbf{Q}}_1^H \cdots \tilde{\mathbf{Q}}_N^H \mathbf{T} = \begin{bmatrix} \mathbf{D}_1^R & & \\ \mathbf{C}_2^R \mathbf{B}_1 & \mathbf{D}_2^R & \\ \vdots & \ddots & \ddots \\ \mathbf{C}_N^R \mathbf{A}_{N-1} \cdots \mathbf{A}_2 \mathbf{B}_1 & \cdots \cdots & \mathbf{D}_N^R \end{bmatrix} = \mathbf{R}$$

This is equal to $\mathbf{Q}^H \mathbf{T} = \mathbf{R}$, where $\mathbf{R}$ is lower triangular. Lemma 1 shows that $\mathbf{R} = \tilde{\mathbf{R}}_N \cdots \tilde{\mathbf{R}}_1$, so that $\mathbf{R}$ has the advertised state space realization. Since $\mathbf{T}$ is full column rank, all $\mathbf{D}_N^R$ are square and invertible, so that $\mathbf{R}$ is square and invertible. $\mathbf{Q}$ is isometric since each of its factors $\mathbf{Q}_n$ is isometric. □

The structure of the factorization is shown in Fig. 3(a). Note that in our application, $\mathbf{A}_n$ and $\mathbf{B}_n$ are trivial: embeddings of identity matrices of appropriate sizes. Hence the multiplication by $\mathbf{Y}_{n+1}$ is trivial and the only actual work in (7) is the QR factorization.

Theorem 3. Suppose that $\mathbf{R}$ is a square invertible lower triangular matrix. Then its inverse is lower triangular too. If $\mathbf{R}$ has state space realization

$$\mathbf{R}_n = \begin{bmatrix} \mathbf{A}_n^R & \mathbf{B}_n^R \\ \mathbf{C}_n^R & \mathbf{D}_n^R \end{bmatrix}, \qquad n = 1, \cdots, N$$

then $\mathbf{S} := \mathbf{R}^{-1}$ has state space realization

$$\mathbf{S}_n = \begin{bmatrix} \mathbf{A}_n^R - \mathbf{B}_n^R \mathbf{D}_n^{R^{-1}} \mathbf{C}_n^R & \mathbf{B}_n^R \mathbf{D}_n^{R^{-1}} \\ -\mathbf{D}_n^{R^{-1}} \mathbf{C}_n^R & \mathbf{D}_n^{R^{-1}} \end{bmatrix}, \quad n = 1, \cdots, N$$

PROOF Note that $\mathbf{Ru} = \mathbf{y} \Leftrightarrow \mathbf{Sy} = \mathbf{u}$, hence $\mathbf{S}$ maps $\mathbf{y}$ to $\mathbf{u}$. Since $\mathbf{S}$ is lower triangular (causal),

$$\begin{aligned} \mathbf{y}_n &= \mathbf{C}_n^R \mathbf{x}_n + \mathbf{D}_n^R \mathbf{u}_n \\ \Leftrightarrow \mathbf{u}_n &= -\mathbf{D}_n^{R^{-1}} \mathbf{C}_n^R \mathbf{x}_n + \mathbf{D}_n^{R^{-1}} \mathbf{y}_n \end{aligned}$$

Backsubstitution in $\mathbf{x}_{n+1} = \mathbf{A}_n^R \mathbf{x}_n + \mathbf{B}_n^R \mathbf{u}_n$ gives the result.

□

The left-inverse of the isometric factor $\mathbf{Q}$ is $\mathbf{Q}^H$, with *anticausal* state space realization (backward recursion)

$$\begin{cases} \mathbf{x}_n = \mathbf{A}_n^{QH} \mathbf{x}_{n+1} + \mathbf{C}_n^{QH} \mathbf{u}_n \\ \mathbf{y}_n = \mathbf{B}_n^{QH} \mathbf{x}_{n+1} + \mathbf{D}_n^{QH} \mathbf{u}_n \end{cases}$$
$$n = N, N-1, \cdots, 1.$$

The preceding theorems can be used to invert more general matrices, in particular the code matrix $\mathbf{T}$. We obtain an implementation of $\mathbf{T}^\dagger = \mathbf{SQ}^H$ in factored form, where $\mathbf{T}^\dagger$, $\mathbf{R}$ and $\mathbf{Q}$ are never explicitly evaluated. The structure of the computational network is shown in Fig. 3(b). As is seen from this structure, the "complexity" of $\mathbf{T}$ and $\mathbf{T}^\dagger$ is the same, even if $\mathbf{T}^\dagger$ is a full matrix without visible sparse structure.

4.3 *Computation of* $\mathbf{\Sigma}_{ik}$

In the computation of the noise covariance, expressions for $\mathbf{\Sigma}_{ik}$ are needed. We can apply the following theorem:

Theorem 4. Let $\mathbf{T}$ have state space realization $\{\mathbf{A}_n, \mathbf{B}_n, \mathbf{C}_n, \mathbf{D}_n\}$. A realization for the lower triangular part of $\mathbf{N} := \mathbf{TT}^H$ is given by

$$\mathbf{N}_n = \begin{bmatrix} \mathbf{A}_n & \mathbf{A}_n \mathbf{\Lambda}_n \mathbf{C}_n^H + \mathbf{B}_n \mathbf{D}_n^H \\ \mathbf{C}_n & \mathbf{C}_n \mathbf{\Lambda}_n \mathbf{C}_n^H + \mathbf{D}_n \mathbf{D}_n^H \end{bmatrix}$$

where $\mathbf{\Lambda}_n$ is specified by the forward recursion

$$\mathbf{\Lambda}_1 = \bullet; \quad \mathbf{\Lambda}_{n+1} = \mathbf{A}_n \mathbf{\Lambda}_n \mathbf{A}_n^H + \mathbf{B}_n \mathbf{B}_n^H$$
$$n = 1, 2, \cdots, N,$$

PROOF By inspection of Fig. 3(c) and following the mapping of $\mathbf{x}_n, \mathbf{u}_n$ to $\mathbf{x}_{n+1}, \mathbf{y}_n$. The causal part of the state is $\mathbf{x}_n$, the non-causal part is $\mathbf{x}'_n$, and $\mathbf{\Lambda}_n$ represents the transfer of $\mathbf{x}'_n$ to $\mathbf{x}_n$. (A formal proof appears in (Dewilde and van der Veen, 1998, p.366).) □

The preceding recursions are useful in the computation of the noise covariance after the decorrelating matched filter. If $\mathbf{w}$ is a white noise vector with power normalized to $\sigma^2 = 1$, and $\mathbf{n} = \mathbf{T}^\dagger \mathbf{w} = (\mathbf{T}^H \mathbf{T})^{-1} \mathbf{T}^H \mathbf{w}$, then the covariance of $\mathbf{n}$ is given by

$$\mathbf{\Sigma} := \mathrm{E}(\mathbf{nn}^H) = (\mathbf{T}^H \mathbf{T})^{-1} = \mathbf{SS}^H$$

where $\mathbf{T} = \mathbf{QR}$ and $\mathbf{S} = \mathbf{R}^{-1}$. A state space realization of $\mathbf{S}$ was derived before. Thus, theorem 4 (applied to $\mathbf{S}$) gives a recursion to compute a realization for the lower part of $\mathbf{SS}^H$. The upper part is simply the transpose.

In the identification algorithm in section 3.3, we are only interested in the main (block)-diagonal of $\mathrm{E}(\mathbf{nn}^H)$ (the auto-covariances of size $L_i \times L_i$). In this case, it suffices to compute

$$\begin{aligned} \mathrm{E}(\mathbf{n}_n \mathbf{n}_n^H) &= \mathbf{C}_n^S \mathbf{\Lambda}_n \mathbf{C}_n^{SH} + \mathbf{D}_n^S \mathbf{D}_n^{SH} \\ \mathbf{\Lambda}_{n+1} &= \mathbf{A}_n^S \mathbf{\Lambda}_n \mathbf{A}_n^{SH} + \mathbf{B}_n^S \mathbf{B}_n^{SH} \end{aligned}$$

4.4 *Computation of the MMSE Receiver in State Space*

Recall the MMSE receiver (6). It is known that equations of this form can be efficiently computed via a QR factorization. Indeed, note that

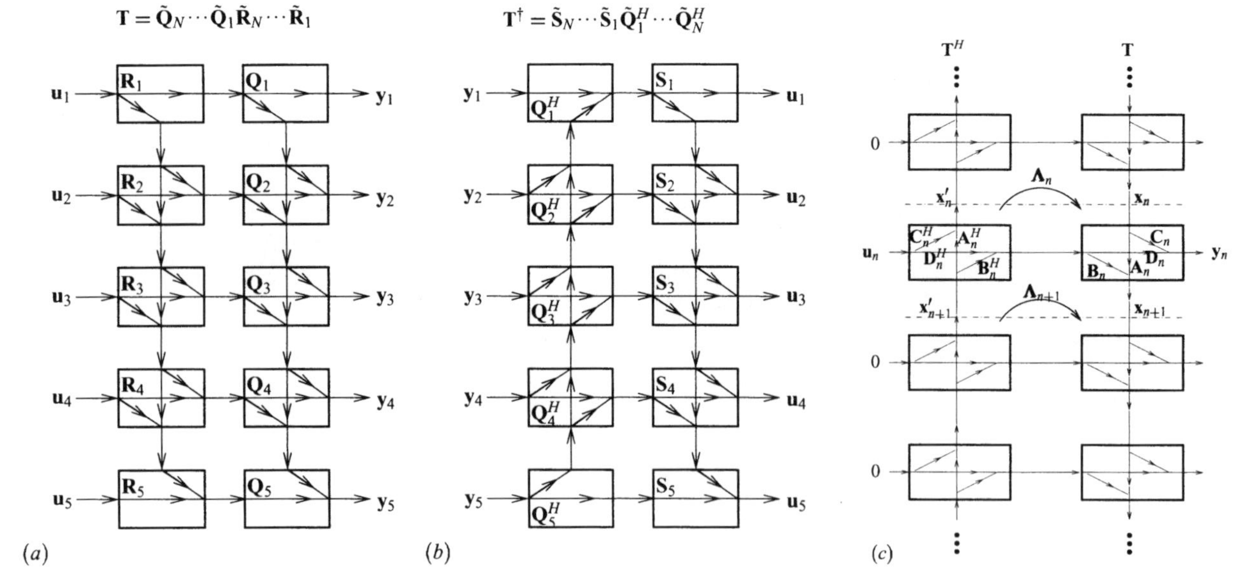

Fig. 3. (*a*) Structure of the QR factorization, (*b*) structure of the inverse, (*c*) structure of $\mathbf{TT}^H$.

$$\hat{\mathbf{s}} = (\mathbf{H}^H\mathbf{T}^H\mathbf{T}\mathbf{H} + \sigma^2\mathbf{I})^{-1}\mathbf{H}^H\mathbf{T}^H\mathbf{y} \qquad (8)$$
$$= (\mathbf{H}^H\mathbf{T}^H\mathbf{T}\mathbf{H} + \sigma^2\mathbf{I})^{-1}\begin{bmatrix}\mathbf{H}^H\mathbf{T}^H & \sigma\mathbf{I}\end{bmatrix}\begin{bmatrix}\mathbf{y}\\ \mathbf{0}\end{bmatrix}$$
$$= \underbrace{\begin{bmatrix}\mathbf{TH}\\ \sigma\mathbf{I}\end{bmatrix}^\dagger}_{\mathbf{M}}\begin{bmatrix}\mathbf{y}\\ \mathbf{0}\end{bmatrix}$$

Thus, if $\mathbf{M} =: \mathbf{Q}^M\mathbf{R}^M$ is an economy-size QR factorization for $\mathbf{M}$ (where $\mathbf{R}^M$ is square triangular, and $\mathbf{Q}^M$ is tall and isometric), then

$$\hat{\mathbf{s}} = (\mathbf{R}^M)^{-1}(\mathbf{Q}^M)^H\begin{bmatrix}\mathbf{y}\\ \mathbf{0}\end{bmatrix}.$$

The QR factorization and factor inversion can be done in state space as before. Thus, $\hat{\mathbf{s}}$ is the output of a computational structure similar to the one in Fig. 3(*b*). The only new aspect is the derivation of a realization for **M**.

A realization $\{\mathbf{A}_n, \mathbf{B}_n, \mathbf{C}_n, \mathbf{D}_n\}$ for **T** is already known. **H** is block-diagonal, with blocks $\mathbf{h}_i$ matching the inputs of **T**. Define

$$\mathbf{H}_n := \begin{bmatrix}\boldsymbol{\beta}_{1,n} & & \\ & \ddots & \\ & & \boldsymbol{\beta}_{I,n}\end{bmatrix}$$
$$\boldsymbol{\beta}_{i,n} := \begin{cases}\mathbf{h}_i, & \mathbf{T} \text{ has an input for user } i \text{ at } n\\ \cdot, & \text{otherwise.}\end{cases}$$

A realization for **TH** is then given by

$$(\mathbf{TH})_n = \begin{bmatrix}\mathbf{A}_n & \mathbf{B}_n\mathbf{H}_n\\ \mathbf{C}_n & \mathbf{D}_n\mathbf{H}_n\end{bmatrix}, \quad n = 1,\cdots,N.$$

Finally, a realization for **M** is simply obtained by extending the *D*-matrix by $\sigma\mathbf{I}$:

$$\mathbf{M}_n = \left[\begin{array}{c|c}\mathbf{A}_n & \mathbf{B}_n\mathbf{H}_n\\ \hline \mathbf{C}_n & \mathbf{D}_n\mathbf{H}_n\\ \mathbf{0} & \sigma\mathbf{I}\end{array}\right], \quad n = 1,\cdots,N.$$

5. REFERENCES

Dewilde, P. and A.J. van der Veen (1998). *Time-Varying Systems and Computations*. Kluwer Academic Publishers. Dordrecht, The Netherlands.

Escudero, C., U. Mitra and D. Slock (2001). A Toeplitz displacement method for blind multipath estimation for Long Code DS/CDMA signals. *IEEE Trans. Signal Processing* **SP-48**(3), 654–665.

Hieu, D.Q. and A.J. van der Veen (2003). Single- and multi-user blind receivers for long code WCDMA. In: *IEEE workshop on Signal Proc. Adv. in Wireless Comm. (SPAWC)*. Rome (Italy).

Liu, H. and M. Zoltowski (1997). Blind equalization in antenna array CDMA systems. *IEEE Trans. Signal Processing* **45**(1), 161172.

Sidiropoulos, N. and R. Bro (1999). User separation in DS-CDMA Systems with unknown Long PN Spreading Codes. In: *Proc. SPAWC*. Annapolis, MD.. pp. 194–197.

Tong, Lang, A.J. van der Veen and P. Dewilde (2002*a*). Channel estimation for long-code WCDMA. In: *Proc. IEEE ISCAS*. IEEE. Scotsdale (AZ).

Tong, Lang, A.J. van der Veen and P. Dewilde (2002*b*). A new decorrelating RAKE receiver for long-code WCDMA. In: *36th Ann. Conf. Information Sciences and Systems (CISS)*. Princeton (NJ).

Weiss, A. and B. Friedlander (1999). Channel estimation for DS-CDMA downlink with aperiodic spreading codes. *IEEE Tr. Comm.* **47**(10), 1561–1569.

Xu, Z. and M. Tsatanis (2000). Blind channel estimation for long code multiuser CDMA systems. *IEEE Trans.Signal Processing* **SP-48**(4), 988–1001.

Zoltowski, M., Y. Chen and J. Ramos (1996). Blind 2D RAKE receivers based on space-time adaptive MVDR processing for IS-95 CDMA system. In: *Proceedings of the 15th IEEE MILCOM*. Atlanta, GA. pp. 618–622.

IFAC Publications
www.elsevier.com/locate/ifac

MAXIMUM LIKELIHOOD IDENTIFICATION OF QUANTUM SYSTEMS FOR CONTROL DESIGN

Robert L. Kosut *,[1] **Herschel Rabitz** **,[1] **Ian A. Walmsley** ***,[1]

* *SC Solutions, Sunnyvale, CA, USA*
** *Princeton University, Princeton, NJ, USA*
*** *Oxford University, Oxford, UK*

Abstract:
Maximum likelihood estimation is explored for use in the identification of quantum systems for control. *Copyright © 2003 IFAC*

Keywords: system identification, quantum systems, adaptive control

1. INTRODUCTION

The construction of quantum devices and computers will probably not be possible without a combination of control and error correcting codes [2]. And it is more than likely that the control will have to be found using data from the actual system rather than solely from a theoretical model. Thus we are led to developing strategies for *adaptive control* of quantum systems. The necessity for adaptive control of quantum systems prompted the learning algorithm proposed in [6] for using the results of the experiment to adapt a laser pulse-shaper. This is a *direct* adaptive method: no model is used and no reference trajectory is defined. Only a measurable target is posed. To date over 40 experiments have been successfully controlled in this way. In this paper we begin to explore how far we can go with a model based approach, specifically, an *indirect* adaptive scheme making use of maximum likelihood estimation (MLE) as the identification step. This continues [7] where an observer structure was presented which can be used to reconstruct the state and/or parameters of a quantum system using recorded data together with a model Hamiltonian which is assumed to have the same structure as the true Hamiltonian (except for the unknown parameters). In that approach the parameters were estimated using a gradient algorithm.

MLE has been previously proposed for quantum system identification. In [9], which includes the references to the authors earlier work, MLE is applied to quantum systems with *non-continuing measurements*, *i.e.*, data is taken from repeated identical experiments at a finite number of (discrete) sample times. For continuing measurements a general (Bayesian) formulation is presented in [10].

In this paper we further explore the use of MLE with non-continuing discrete-time measurements, specifically for estimating Hamiltonian parameters and/or the system state with the ultimate purpose being control design. A distinction is drawn between estimating Hamiltonian parameters and the initial density matrix. We focus especially on estimating the Hamiltonian parameters for control, assuming the initial state is known, *e.g.*, prepared in a calibration mode for tuning the control. We also explore the use of the Cramér-Rao Inequality to assess the limit of performance of estimating the Hamiltonian parameters or the state. In the case when a good model of the Hamiltonian is available, the MLE of the density matrix can be cast as a convex optimization problem. Optimal experiment design for state estimation is briefly discussed. Due to space limitations proofs are omitted and are available in an expanded article upon request.

---

[1] Supported by the DARPA QUIST Program
[2] See [8] and the references therein.

2. NON-CONTINUING MEASUREMENTS

Consider a quantum system with m distinct measurement outcomes, labeled $\{ \alpha \mid \alpha = 1, \ldots, m \}$. Data is collected from the system using the following procedure which is a standard way to circumvent the "observe is to disturb" dictum of quantum measurement. At each of $n_{\rm sa}$ sample times, measurements are recorded from identical experiments repeated $n_{\rm expt}$ times. Let $\{ t_k \mid k = 1, \ldots, n_{\rm sa} \}$ denote the sample times relative to the start of each experiment. Let $y_i(t_k)$ denote the measurement recorded at the (relative) sample time t_k in the i-th experiment. Each $y_i(t_k)$ is, of course, one of the m possible outcomes. All the $n_{\rm sa} n_{\rm expt}$ recorded measurements are collected into the data set,

$$Y = \left\{ y_i(t_k) \,\middle|\, \begin{array}{l} i = 1, \ldots, n_{\rm expt} \\ k = 1, \ldots, n_{\rm sa} \end{array} \right\} \tag{1}$$

The problem addressed here is to use the recorded data Y to estimate unknown parameters in a model of the quantum system.

3. MODEL OF THE SYSTEM

The process of modeling a quantum system usually begins with the construction of a Hamiltonian *operator* on an infinite dimensional Hilbert space [3]. *Eventually*, a finite dimensional approximation is invoked in order to calculate anything. In some cases a finite dimensional model is immediately appropriate, *e.g.*, spin systems. The finite dimensional model is the starting point here.

Evolution model The quantum system generating the data (1) is modeled by a finite dimensional Hamiltonian matrix $H(t, \theta) \in \mathbf{C}^{n \times n}$, having a known dependence on the time t and on an unknown parameter vector $\theta \in \mathbf{R}^p$. The model density matrix will depend on θ and the initial state of the model $r \in \mathbf{C}^{n \times n}$. Thus, the density, $\rho(t, \theta, r) \in \mathbf{C}^{n \times n}$ evolves according to,

$$\rho(t, \theta, r) = U(t, \theta) r U(t, \theta)^* \tag{2}$$

where $U(t, \theta) \in \mathbf{C}^{n \times n}$ is the unitary propagator associated with $H(t, \theta)$ which satisfies,

$$i\hbar \dot{U} = H(t, \theta) U,\ U(0, \theta) = I_n \tag{3}$$

To insure that $\rho(t, \theta, r)$ is a density matrix it is necessary that r is a density matrix, *i.e.*, r must be in the set,

$$\left\{ r \in \mathbf{C}^{n \times n} \,\middle|\, r = r^* \geq 0,\ {\rm Tr}\, r = 1 \right\} \tag{4}$$

[3] The notation and definitions used here is standard for quantum dynamics, *e.g.*, [4],[2], and quantum computation/information systems, *e.g.*, [8].

This set includes, of course, both pure and mixed states. Constraints on the set of possible θ-parameters are not as generic; they will depend on the specific physics involved, *e.g.*, some parameters may be strictly positive, some with known ranges, and so on.

Measurement model The probability that the outcome from the i-th experiment, $y_i(t_k)$, is α is given by,

$$p_\alpha(t_k, \theta, r) = {\rm Tr}\ M_\alpha \rho(t_k, \theta, r) \tag{5}$$

where the $M_\alpha \in \mathbf{C}^{n \times n}$ are non-negative Hermitian matrices, referred to as *observables*, which satisfy the *completion relation*, $\sum_{\alpha=1}^{m} M_\alpha = I_n$. Using (2), the outcome probabilities are given equivalently by,

$$\begin{aligned} p_\alpha(t_k, \theta, r) &= {\rm Tr}\ O_\alpha(t_k, \theta) r \\ O_\alpha(t, \theta) &= U(t, \theta)^* M_\alpha U(t, \theta) \end{aligned} \tag{6}$$

The properties of the M_α transfer to $O_\alpha(t, \theta)$, that is, each is non-negative Hermitian and satisfies the completion relation.

4. MAXIMUM LIKELIHOOD ESTIMATION

Let $n_\alpha(t_k)$ be the number of times the outcome α is recorded from the $n_{\rm expt}$ experiments at the sample time t_k; thus $\sum_{\alpha=1}^{m} n_\alpha(t_k) = n_{\rm expt}$. If the experiments are independent, then the probability predicted by the model (2)-(6) of obtaining the data set (1) for a particular (θ, r) is a product of the individual model probabilities (6). Consequently, for a particular value of (θ, r), the probability of obtaining the complete data set (1) is given by,

$$\mathbf{Prob}\,\{Y, \theta, r\} = \prod_{k,\ \alpha} p_\alpha(t_k, \theta, r)^{n_\alpha(t_k)} \tag{7}$$

The data is thus captured in the outcome counts $\{n_\alpha(t_k)\}$ whereas the model terms have a (θ, r)-dependence. The function $\mathbf{Prob}\,\{Y, \theta, r\}$ is called the *likelihood* function and since it is positive, the *maximum likelihood estimate* (MLE) of (θ, r) is obtained by maximizing the *log-likelihood function* as follows:

$$\begin{array}{ll} \text{maximize} & L(Y, \theta, r) \\ \text{subject to} & (2) - (6) \end{array} \tag{8}$$

Using (6) yields the equivalent expressions,

$$\begin{aligned} L(Y, \theta, r) &= \sum_{k,\ \alpha} n_\alpha(t_k) \log\, p_\alpha(t_k, \theta, r) \\ &= \sum_{k,\ \alpha} n_\alpha(t_k) \log {\rm Tr}\ O_\alpha(t_k, \theta) r \end{aligned} \tag{9}$$

$L(Y, \theta, r)$ is a log-concave function of the $\{p_\alpha(t_k, \theta, r)\}$, and since $p_\alpha(t_k, \theta, r)$ is linear in $\rho(t_k, \theta, r)$, $L(Y, \theta, r)$

is log-concave in the $\{\rho(t_k,\theta,r)\}$. If these were convex in θ, r, then (8) would fall into the category of a class of well studied convex optimization problems and could be solved to within any desired accuracy, *e.g.*, [1]. Unfortunately, however, the dependence of $\rho(t_k,\theta,r)$ on θ, r is not generically convex, and hence, neither is $L(Y,\theta,r)$. In an example to follow we will see that $L(Y,\theta,r)$ can have many local maxima. Nonetheless, there are also many instances where $L(Y,\theta,r)$ is convex for θ, r in a bounded convex set. A special situation where convex optimization applies is in estimation of the density matrix under the assumption that the Hamiltonian is known. This is discussed in section 8. We next present some well known properties of the likelihood function and apply these to quantum system estimation.

5. ESTIMATOR PERFORMANCE

In this section we examine the fundamental limit of estimator performance of *any* estimator for the case when the initial state, r, is known. This would arise, for example, in a calibration mode in a quantum computer where known input states are prepared for testing quantum logic gates. In section 8 we examine the case when the Hamiltonian parameters, θ, is known.

The basis for the performance analysis is the following classical theorem.

Theorem 1. **Cramér-Rao Inequality**[3]
Suppose there is a parameter (θ_0, r_0) such that the distribution of the data, Y, is given by $\mathbf{Prob}\,\{Y,\theta_0,r_0\}$, *i.e.*, $\mathbf{Prob}\,\{Y\} = \mathbf{Prob}\,\{Y,\theta_0,r_0\}$. Suppose r_0 is known and $\widehat{\theta}(Y|r_0)$ is any unbiased estimate of θ_0, *i.e.*, $\mathbf{E}\;\widehat{\theta}(Y|r_0) = \theta_0$. Then, the covariance of the estimate satisfies,

$$\mathbf{cov}\;\widehat{\theta}(Y|r_0) \geq F^{-1} \tag{10}$$

where F is the *Fisher information matrix*,

$$F = -\mathbf{E}\,\nabla_{\theta\theta} L(Y,\theta,r_0)\Big|_{\theta=\theta_0} \tag{11}$$

The result above requires that r_0 is known and there is a "true" parameter value θ_0 in the model set which is consistent with the statistics of the actual data, *i.e.*, the system and model distributions are identical. In other words, the system generating the data is a member of the model set used for identification. This is always a questionable assumption and in most engineering practice is never true. Although an important issue, particularly when identification is to be used for control design, the case when the system is not in the model set will not be explored any further here.

Fisher information matrix For the log-likelihood function (9), the elements of the Fisher information matrix requires evaluating $\mathbf{E}\,n_\alpha(t_k)$. Since the counts, $n_\alpha(t_k)$, come from the distribution $p_\alpha(t_k,\theta_0,r_0)$, it follows by definition that,

$$\mathbf{E}\,n_\alpha(t_k) = n_{\text{expt}}\,p_\alpha(t_k,\theta_0,r_0) \tag{12}$$

Hence, the elements of the Fisher information matrix are,

$$F_{ij} = n_{\text{expt}} \sum_{k,\,\alpha} \left(\frac{\nabla_{\theta_i} p_\alpha(t_k,\theta_0,r_0)\;\nabla_{\theta_j} p_\alpha(t_k,\theta_0,r_0)}{p_\alpha(t_k,\theta_0,r_0)} - \nabla_{\theta_i\theta_j} p_\alpha(t_k,\theta_0,r_0) \right)$$

Shape of parameter space Another aspect of estimator performance is the shape of the parameter space, *i.e.*, $L(Y,\theta,r_0)$ for θ in some region. It is convenient to examine the average likelihood function,

$$\begin{aligned} \mathcal{L}(\theta,r_0) &= \mathbf{E}\,\frac{1}{n_{\text{sa}} n_{\text{expt}}} L(Y,\theta,r_0) \\ &= \frac{1}{n_{\text{sa}}} \sum_{k,\,\alpha} p_\alpha(t_k,\theta_0,r_0) \log\, p_\alpha(t_k,\theta,r_0) \end{aligned} \tag{13}$$

The last line follows from (12). As n_{expt} becomes large, it can be shown that $\mathcal{L}(\theta,r_0)$ expresses the (asymptotic) shape of the parameter space. The maximum of $\mathcal{L}(\theta,r_0)$ occurs at $\theta = \theta_0$, and hence, the negative average likelihood function, $-\mathcal{L}(\theta_0,r_0)$, is (to within a scale factor) the *entropy* of the system generating the data. Thus when $\mathcal{L}(\theta,r_0)$ is maximized, the entropy is as small as possible, *i.e.*, the system and the model have the same entropy.

The following example illustrates some of these maximum likelihood properties.

6. EXAMPLE: ONE HAMILTONIAN PARAMETER

The system generating the data is a single spin system. [4]

Hamiltonian

$$\begin{aligned} H(t) &= \frac{1}{\sqrt{2}}\left(\varepsilon_z(t)\theta_0 Z + \varepsilon_x(t) X\right) \\ &= \frac{1}{\sqrt{2}} \begin{bmatrix} \varepsilon_z(t)\theta_0 & \varepsilon_x(t) \\ \varepsilon_x(t) & -\varepsilon_z(t)\theta_0 \end{bmatrix} \end{aligned}$$

Observables

$$M_1 = \begin{bmatrix} 1 \\ 0 \end{bmatrix} [1\;0],\; M_2 = \begin{bmatrix} 0 \\ 1 \end{bmatrix} [0\;1]$$

[4] (X, Z) are two of the three 2×2 Pauli spin matrices: $X = \begin{bmatrix} 0 & 1 \\ 1 & 0 \end{bmatrix}$, $Y = \begin{bmatrix} 0 & -i \\ i & 0 \end{bmatrix}$, $Z = \begin{bmatrix} 1 & 0 \\ 0 & -1 \end{bmatrix}$ See [4, III,Ch.12-9],[2] on models of spin systems.

Initial state

$$r_0 = \rho(0) = \psi_0\psi_0^*, \; \psi_0 = \begin{bmatrix} 1 \\ 0 \end{bmatrix}$$

$(\varepsilon_z, \varepsilon_x)(t)$ are known inputs to the system. Also, for this example, the initial state of the model, r_0, is assumed to be known, thus only θ_0 is to estimated. The model and the system are identical except that in the model the unknown parameter θ replaces θ_0.

Simulation results Simulations were performed with

$$\theta_0 = 1 \tag{14}$$

and for two input cases. Constant inputs:

$$\begin{array}{l} 0 \leq t \leq \pi/2 \\ \varepsilon_z(t) = 1 \\ \varepsilon_x(t) = 1 \end{array} \tag{15}$$

Random time-varying inputs:

$$\begin{array}{l} 0 \leq t \leq \pi/2, \; \theta_0 = 1 \\ \varepsilon_z(t) \in \text{unif}(.75, 1.25) \\ \varepsilon_x(t) \in \text{unif}(.75, 1.25) \end{array} \tag{16}$$

In this case the inputs are drawn independently at every time instant from a uniform distribution whose amplitude ranges from .75 to 1.25, *i.e.*, the mean is 1 and the deviation is $\pm.25$.

Using (10), for the constant input case the table below shows the minimum number of experiments, $n_{\text{sa}}n_{\text{expt}}$, required to achieve an RMS relative error of no more than 0.1. The number of sample times, n_{sa}, are uniformly spaced in the time interval $0 \leq t \leq \pi/2$.

n_{sa}	n_{expt}	$n_{\text{sa}}n_{\text{expt}}$	$\|\|(\widehat{\theta}(Y\|r_0) - \theta_0)/\theta_0\|\|_{\text{rms}}$
2	100	200	0.1
10	52	520	0.1
20	30	600	0.1
50	14	700	0.097
100	7	700	0.095

The table values increase slightly (not shown) for the random input case. The table suggest that for an error of 0.1, taking 2 time samples with 100 experiments each is sufficient. This is true but not the full story. We need also to examine the shape of the parameter space which gives an indication of the difficulty in obtaining the maximum likelihood estimate.

Figure 1 shows a comparison of the average likelihood function, $\mathcal{L}(\theta, r_0)$, for the two input cases: the constant input case is shown in black, and a particular time-varying realization drawn from the above distribution, is shown in red. Much is revealed from these plots. First we see that there are numerous local maxima. The density of these increases with the number of sample times n_{sa}, but the relative amplitude between local maximum and minimum decreases; it is very large for $n_{\text{sa}} = 2$ and very small for $n_{\text{sa}} = 100$. In all cases, the maximum occurs for $\theta = \theta_0 = 1$ and $\mathcal{L}(\theta, r_0)$ is concave for θ approximately in the interval $-2.5 \leq \theta \leq 2.5$ (only positive θ is shown as the function is symmetric). The implication is that it is easy to get stuck at a local maximum using hill climbing methods (*e.g.*, Newton's method) particularly if the algorithm "step-size" is small. In consequence, although $n_{\text{sa}} = 2, n_{\text{expt}} = 100$ can produce a small error, it may be difficult to achieve unless a good starting value is known, *e.g.*, in this example any initial θ in the interval $-2.5 \leq \theta \leq 2.5$, which is a large interval compared to the nominal $\theta_0 = 1$.

The time-varying input shown was selected because it significantly increases the concavity range of $\mathcal{L}(Y, \theta)$ as n_{sa} increases. For $n_{\text{sa}} = 2, 10, 20$, the local variations are greatly reduced, and for $n_{\text{sa}} \geq 50$ no variations appear on the scale shown; almost the entire range of θ shown exhibits concavity. Hence, it is unlikely that even a sensitive hill climbing method would falter. Of course we arrived at this input choice by simply running simulations with randomly drawn inputs until we got what we liked. Although it worked here this is not a generic method, *e.g.*, we would never draw all signals, such as sinusoids, which could be optimal.

7. ITERATIVE ADAPTIVE CONTROL

In this section we explore the effect of controlling the quantum system using the identified Hamiltonian parameters in a calibration mode where the initial state, r_0, is prepared (known). A typical iterative adaptation scheme has an identification step followed by a control design step.

(1) **Maximum Likelihood Estimation**

$$\widehat{\theta}^{(i)} = \arg\max_{\theta} \; L(Y(\varepsilon^{(i-1)}), \theta, r_0) \tag{17}$$

(2) **Control design**

$$\varepsilon^{(i)} = \overline{\varepsilon}(\widehat{\theta}^{(i)}) \tag{18}$$

The data in the log likelihood function, $Y(\varepsilon^{(i-1)})$, depends on the control action, $\varepsilon^{(i-1)}$, from the previous iteration which is obtained from a design method dependent on the previous estimate, symbolically denoted by the function $\overline{\varepsilon}(\cdot)$.

Convergence analysis of this two-step iteration is very difficult because the data at iteration i depends on *all* past control sequences, which in turn depend on all past plant parameter estimates, and so on. However, for a large number of experiments per sample, the likelihood function can be replaced by it's average. Hence, in this limiting case, the parameter estimation step at iteration i is,

$$\widehat{\theta}^{(i+1)} = \arg\max_{\theta} \; \mathcal{L}(\theta, r_0)\big|_{\varepsilon = \overline{\varepsilon}(\widehat{\theta}^{(i)})} \tag{19}$$

As observed in [5], convergent solutions of (19) solve,

$$\widehat{\theta} = \arg\max_{\theta} \mathcal{L}(\theta, r_0)|_{\varepsilon=\overline{\varepsilon}(\widehat{\theta})} \tag{20}$$

Equivalently, $\widehat{\theta}$ is a *fixed-point* of the function defined by the above right hand side.

Consider the single-spin example again, with Hamiltonian model written explicitly as,

$$H(\theta, \varepsilon(t)) = \frac{1}{\sqrt{2}} (\varepsilon(t)\theta Z + X) \tag{21}$$

To simplify the problem further, we will restrict $\varepsilon(t)$ to a constant value. Suppose the control goal is to make the Hadamard transformation [8, Ch.1]. If the true value of θ is θ_0, and the final time is $t_f = \pi/2$, then the obvious "optimal" solution is, $\varepsilon^{\text{opt}} = 1/\theta_0$. Therefore, the control design step in our iterative scheme is simply, $\overline{\varepsilon}(\theta) = 1/\theta$. Figure 2 shows the iteration results with $\theta_0 = 1$, r_0 as in the previous example, initial control $\varepsilon = 1.2$, number of time samples $n_{\text{sa}} = 3$, and the initial parameter estimate found by the local optimization,

$$\widehat{\theta}^0 = \arg \max_{2.5 \leq \theta \leq 4} \mathcal{L}(\theta, r_0)|_{\varepsilon=1.2} = 3.06$$

It takes only two identification steps (locally hill climbing) to converge to the true parameter. This, of course, is the ideal case with the system in the model set. With finite data the estimates will be random variables with the plots in figure 2 being the mean.

We performed more simulations with different starting values. Not all cases showed convergence to the true value. A full understanding of the conditions for convergence is not yet available, but in general we know that for large data length, stability of the fixed-point defined by (20) is required, *e.g.*, [5].

8. ESTIMATING THE DENSITY MATRIX

In this section we examine the problem of estimating the density matrix using prior knowledge of the Hamiltonian. This problem arises, for example, in quantum tomography [11]. It is thus assumed that the only unknown in the model (2)-(6) is the initial state $r \in \mathbf{C}^{n \times n}$ constrained to be a member of the set (4). The Hamiltonian parameter θ is known to be θ_0. Thus, the log-likelihood function (9) becomes,

$$L(Y, \theta_0, r) = \sum_{k,\ \alpha} n_\alpha(t_k) \log \text{Tr}\ O_\alpha(t_k, \theta_0) r \tag{22}$$

The *maximum likelihood* estimate of r is obtained by solving:

$$\begin{array}{ll} \text{maximize} & L(Y, \theta_0, r) \\ \text{subject to} & r \geq 0,\ \text{Tr}\ r = 1 \end{array} \tag{23}$$

If $\widehat{r}(Y|\theta_0)$ is a solution to (23), then an estimate of the density follows from (2), *i.e.*,

$$\widehat{\rho}(t) = U(t\theta_0)\widehat{r}(Y|\theta_0)U(t, \theta_0)^* \tag{24}$$

Since the constraint set (4) is convex in r, and $L(Y, \theta_0, r)$ is log-concave in r, (23) is a convex optimization problem, and hence, the optimal solution can be obtained to within any desired accuracy [1]. In general, the most efficient algorithm for solving this type of problem is an *interior point method* or *primal-dual interior point method*. For not-so-large problems, a *primal interior point* method will suffice, such as a *barrier methods*. This approach will still work for even very large problems but tends to become slow.

The limit of estimator performance can be evaluated using the Cramér-Rao inequality. As before, the assumption is that the system generating the data is in the model set used for estimation.

Theorem 2. **Cramér-Rao Inequality**
Suppose there exist parameters (θ_0, r_0) such that the distribution of the data, Y, is given by $\mathbf{Prob}\,\{Y, \theta_0, r_0\}$. Let the estimate $\widehat{r}(Y|\theta_0)$ be a density matrix and also an unbiased estimate of r_0. Then, the RMS estimation error satisfies,

$$\begin{aligned} &\mathbf{E}\, \|\widehat{r}(Y|\theta_0) - r_0\|_{\text{frob}}^2 \geq \text{Tr}\ G^{-1} - \frac{b^T G^{-2} b}{b^T G^{-1} b} \\ &G = n_{\text{expt}} \sum_{k=1}^{n_{\text{sa}}} \sum_{\alpha=1}^{m} \frac{z_\alpha(t_k, \theta_0) z_\alpha(t_k, \theta_0)^*}{p_\alpha(t_k, r_0)} \\ &z_\alpha(t_k, \theta_0) = \text{vec}\ O_\alpha(t_k, \theta_0),\ b^T = [1_n^T\ 0_{n^2-n}^T] \end{aligned} \tag{25}$$

provided the indicated inverse exists.

The vec$(\cdot)$ operator takes the columns of a matrix and stacks them into a vector, *e.g.*, the matrix $O_\alpha(t_k, \theta_0) \in \mathbf{C}^{n \times n} \Rightarrow \text{vec}\ O_\alpha(t_k, \theta_0) \in \mathbf{C}^{n^2}$. The lower bound corresponds to the minimum RMS error for the *constrained* estimate, *i.e.*, the theoretical unbiased estimator, $\widehat{r}(Y|\theta_0)$, is forced to satisfy the equality constraint, $\text{Tr}\ \widehat{r}(Y|\theta_0) = 1$. The RMS error of the unconstrained estimate is $\text{Tr}\ G^{-1}$ – the constrained lower bound is smaller since there is one fewer parameter to estimate. In most of the examples we have examined the difference between the two terms is small but that should not be taken as a general rule.

The lower bound (25) can also be used for optimal experiment design. In more general terms, let the index k in t_k refer to possible configurations of the system. The problem is to determine the minimum number of experiments per configuration, ℓ_k, in order to obtain an estimate of a specified quality., *i.e.*, the size of (25). The experiment design problem we would like to solve is the following:

$$\begin{array}{ll} \text{minimize} & V(\ell) \\ \text{subject to} & \sum_k \ell_k = \ell_{\text{expt}} \\ & \ell_k \text{ non-negative integers} \end{array} \quad (26)$$

where ℓ_{expt} is the desired number of total experiments and $V(\ell)$ is the lower bound in (25) with G replaced by $G(\ell) = \sum_k \ell_k \sum_\alpha z_{\alpha k} z_{\alpha k}^* / p_{\alpha k}$. The good news is that $V(\ell)$ is convex in ℓ. Unfortunately, restricting ℓ to a vector of integers makes the problem combinatorial. As shown in [1], a good approximate solution can be found by introducing the variables $\lambda_k = \ell_k/\ell_{\text{expt}}$, each of which is the fraction of the total number of experiments performed in configuration k. Clearly, $\sum_k \lambda_k = 1$. If λ_k is only otherwise constrained to the non-negative reals, then this has the effect of relaxing the constraint that the ℓ_k are integers. The *relaxed* problem is a convex optimization in λ. This solution is useful for quantum tomography applications where the number of configurations (phase shifter settings) can be very large and is somewhat time consuming to set them all [11].

9. REFERENCES

[1] S. Boyd and L. Vandenberghe, *Convex Optimization with Engineering Applications*, Course Reader for EE364, Dec. 2001.

[2] C. Cohen-Tannoudji, B. Diu, and F. Laloë, *Quantum Mechanics*, Wiley, 1977.

[3] H. Cramér, *Mathematical Methods of Statistics*, Princeton Press, 1946.

[4] R. P. Feynman, R. B. Leighton, and M. Sands, *The Feynman Lectures on Physics*, Addison-Wesley, 1963-1965.

[5] H. Hjalmarsson, S. Gunnarsson, and M. Gevers, "A convergent iterative restricted complexity control design scheme," *Proc. 1994 CDC*,pp. 1735-1740, Orlando, Florida, Dec. 1994.

[6] R.S. Judson and H. Rabitz, "Teaching lasers to control molecules," Phys. Rev. Lett., 68, 1500 (1992).

[7] R.L. Kosut and H. Rabitz, "Identification of quantum systems," *15th IFAC World Congress*, Barcelona, Spain, 21-26 July 2002.

[8] M. A. Nielsen and I. L. Chuang, *Quantum Computation and Quantum Information*, Cambridge, 2000.

[9] M. G. A. Paris, G. M. D'Ariano, and M. F. Sacchi, "Maximum likelihood method in quantum estimation," arXiv: quant-ph/ 0101071 v1, 16 Jan 2001.

[10] F. Verstraete, A. C. Doherty and H. Mabuchi, "Sensitivity optimization in quantum parameter estimation," *Phys. Rev. A* 64, 032111 (2001)

[11] I. A. Walmsley and L. Waxer, "Emission tomography for quantum state measurement in matter," *J. Phys. B: At. Mol. Opt. Phys.*, **31** (1998)

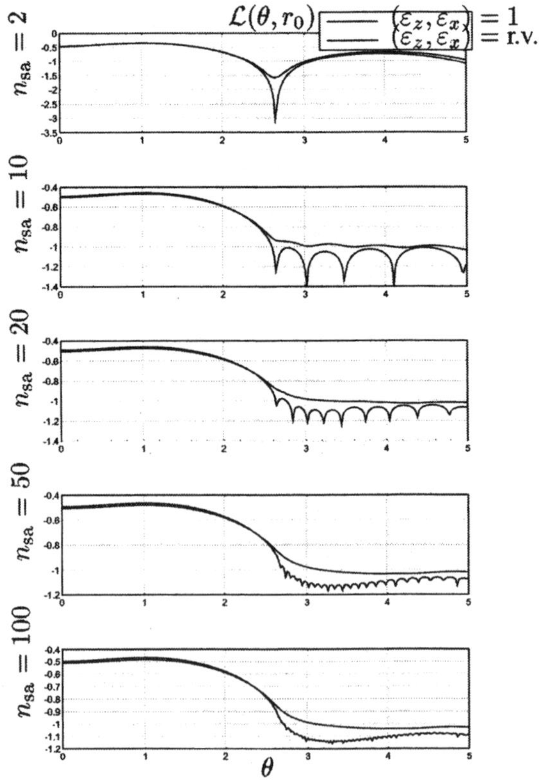

Fig. 1. $\mathcal{L}(\theta, r_0)$ for $n_{sa} = [2, 10, 20, 50, 100]$. Black: $(\varepsilon_z, \varepsilon_x) = 1$. Red: $(\varepsilon_z, \varepsilon_x) \in$ unif(.75, 1.25).

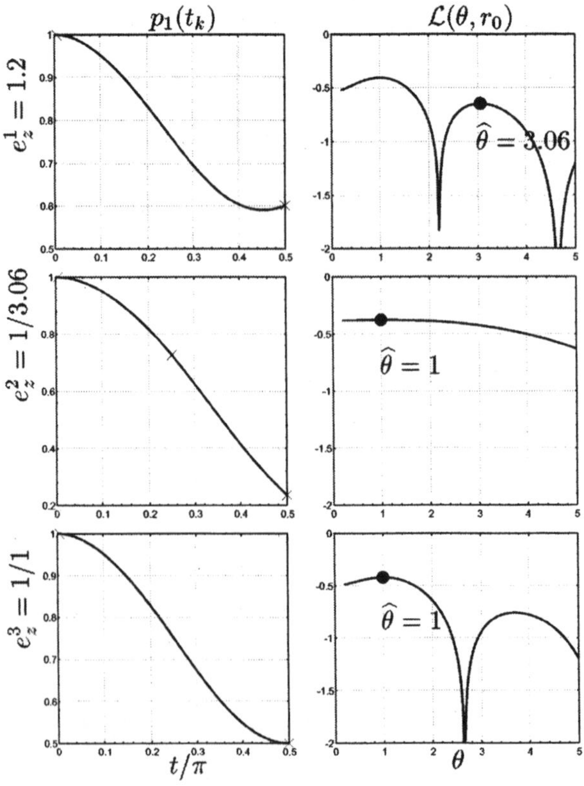

Fig. 2. iterative adaptive control: $n_{sa} = 3$ at X's in column 1. ● in column 2 marks the identified parameter.

IFAC
Publications
www.elsevier.com/locate/ifac

MAXIMUM LIKELIHOOD ESTIMATION OF SIGNAL AMPLITUDE AND NOISE VARIANCE FROM COMPLEX VALUED DATA

A. J. den Dekker * **J. Sijbers** **,[1]

* *Delft Center for Systems and Control, Delft University of Technology, The Netherlands.*
** *Vision Lab, University of Antwerp, Belgium*

Abstract: Complex valued data may be obtained by means of acquisition systems such as magnetic resonance imaging (MRI) or digital communication systems. If the signal amplitude is to be estimated from these (inevitably noise corrupted) complex data, one has two options. Either the signal amplitude is directly estimated from the complex valued data set, or, the complex data is first transformed into a magnitude data set after which the signal amplitude is estimated. Similarly, the noise variance can be estimated from both data sets.
This paper addresses the question whether it is better to use complex valued data or magnitude data for the estimation of these parameters using the maximum likelihood method. As a performance criterion, the mean-squared error (MSE) is used. *Copyright © 2003 IFAC*

Keywords: parameter estimation, mean-square error, accuracy, precision, complex variables, maximum likelihood estimators, magnetic resonance microscopy, amplitude, probability density function

1. INTRODUCTION

Data received from an acquisition system may be complex valued. For example, data received from a Magnetic Resonance Imaging (MRI) system, are intrinsically complex valued. In general, these data are corrupted by zero-mean Gaussian distributed noise, of which the variance is assumed to be the same for each data point.

Now, consider a data processing application (e.g., noise filtering) that requires estimation of the underlying signal amplitude from a number of noise corrupted, complex valued data points. Thereby, for each complex data point, belonging to the data set under concern, the underlying signal amplitude is assumed to be the same. Then, this amplitude can be estimated either by first transforming the complex data points into a set of magnitude data points and afterwards estimating the signal amplitude from the thus obtained data set, or, by directly estimating the signal amplitude from the original complex valued data points. Indeed, both data sets contain the signal amplitude to be estimated.

In general, if N complex points are available, this data set consists of $2N$ observations (N real and N imaginary data points) and $N + 2$ unknowns (N true phase values, 1 true signal amplitude, and the noise variance). On the other hand, if the N complex data points are transformed into a set of N magnitude data points, such a data set has only 2 unknown parameters (the true signal amplitude and the noise variance). Hence, questions may rise like "Should we use the complex data set or the magnitude data set when estimating the unknown signal amplitude?" and "Does it matter whether or not the true phase values of the complex data points, from which the signal amplitude is estimated, are the same?". In this paper, these questions are addressed.

[1] Jan Sijbers is a Postdoctoral Fellow of the F.W.O. (Fund for Scientific Research - Flanders, Belgium)

For both types of data sets (complex and magnitude), the Maximum Likelihood (ML) estimators for the signal amplitude will be derived. Several estimators of the signal amplitude have been proposed in the literature, based on complex (Pintelon *et al.*, 1988) as well as on magnitude data (McGibney and Smith, 1993; Miller and Joseph, 1993; Sijbers *et al.*, 1998). In the present paper, we will only consider the ML estimator, which will be derived for both complex and magnitude data.

The use of the ML estimator is justified by the fact that it has a number of favorable statistical properties (Van den Bos, 1982). First, it is asymptotically most precise, i.e., it achieves the so called Cramér-Rao Lower Bound (CRLB) for an infinite number of observations. The CRLB defines a lower bound on the variance of any unbiased estimator of a parameter. Second, the ML estimator is consistent, which means that it converges to the true value of the parameter(s) in a statistically well defined way if the number of observations increases. Third, the ML estimator is asymptotically normally (i.e., Gaussian) distributed, with a mean equal to the true value of the parameter(s) and a (co)variance (matrix) equal to the CRLB. If these asymptotic properties also apply if the number of observations is finite, depends on the particular estimation problem under concern. In the present case, this can be found out analytically (for complex data points) or by means of simulations (for magnitude data points). Anyhow, it is known that if there exists an estimator that attains the CRLB, it is given by the Maximum Likelihood estimator (Van den Bos, 1982). For both types of data sets, the performance of the corresponding ML estimators of the signal amplitude will be evaluated in terms of the mean-squared error, being a measure of both accuracy (bias) and precision (variance). Moreover, for the complex valued data set, the variance of the ML estimator will be compared with the CRLB, which can been computed analytically. In addition, for both types of data sets, the ML estimators of the variance of the noise will be derived and their performance will be evaluated in terms of both accuracy and precision.

2. SIGNAL ESTIMATION

In this section, the problem of signal estimation from complex as well as magnitude data is addressed. Both options are studied in terms of accuracy and precision.

2.1 *Signal estimation from complex data*

We start by considering complex, Gaussian distributed data. The CRLB for unbiased estimation of the underlying amplitude signal as well as the ML estimator of this signal will be derived. This will be done for the case of identical underlying phase values, as well as for the case of different underlying phase values.

2.1.1. *Identical phase values* Assume that we have N independent, Gaussian distributed complex data points $\{(R_n, I_n)\}$ with underlying true amplitude and phase values, A and φ, respectively. This means that $A\cos\varphi$ and $A\sin\varphi$ represent the true real and imaginary values, respectively. As the real and imaginary data are independent, the joint probability density function (PDF) of the complex data, p_c, is simply the product of the real and imaginary PDF's:

$$p_c = \prod_{n=1}^{N} \frac{e^{-\frac{(r_n - A\cos\varphi)^2}{2\sigma^2}}}{\sqrt{2\pi\sigma^2}} \frac{e^{-\frac{(i_n - A\sin\varphi)^2}{2\sigma^2}}}{\sqrt{2\pi\sigma^2}} \tag{1}$$

where σ denotes the standard deviation of the noise, and $\{(r_n, i_n)\}$ are the real and imaginary variables corresponding with the complex data points $\{(R_n, I_n)\}$.

2.1.1.1. *CRLB* The CRLB for unbiased estimation of A and φ can be computed from the Fisher information matrix $\boldsymbol{I}$ (Van den Bos, 1982):

$$\boldsymbol{I} = -E\left[\begin{pmatrix} \frac{\partial^2 \log p_c}{\partial A^2} & \frac{\partial^2 \log p_c}{\partial A \partial\varphi} \\ \frac{\partial^2 \log p_c}{\partial\varphi\partial A} & \frac{\partial^2 \log p_c}{\partial\varphi^2} \end{pmatrix}\right] \tag{2}$$

with the joint PDF p_c given by (1). Evaluating (2) and applying the inverse operator yields for the CRLB:

$$\text{CRLB} = \boldsymbol{I}^{-1} = \begin{pmatrix} \sigma^2/N & 0 \\ 0 & \sigma^2/NA^2 \end{pmatrix} \tag{3}$$

2.1.1.2. *ML estimation* The Likelihood function L is obtained by substituting the observations $\{(R_n, I_n)\}$ for $\{(r_n, i_n)\}$ into the joint PDF (1). Then, the ML estimates of A and φ are found by maximizing this function with respect to A and φ. Notice that maximizing L is equivalent to maximizing the log-likelihood function, $\log(L)$. Taking the logarithm yields:

$$\log L = -N\log(2\pi\sigma^2) + \frac{1}{2\sigma^2}\sum_{n=1}^{N}\left[(R_n - A\cos\varphi)^2 + (I_n - A\sin\varphi)^2\right]$$

For A or φ to be a maximum, the first derivative of $\log L$ with respect to A and φ should be zero. From the resulting equations, the ML estimators of A and φ are found to be:

$$\widehat{A}_{\text{ML}} = \sqrt{\left(\frac{1}{N}\sum_{n=1}^{N} R_n\right)^2 + \left(\frac{1}{N}\sum_{n=1}^{N} I_n\right)^2} \tag{4}$$

$$\widehat{\varphi}_{\text{ML}} = \arctan\left(\frac{\sum_n I_n}{\sum_n R_n}\right) \tag{5}$$

Notice that the estimator $\widehat{A}_{\text{ML}}$ is obtained by taking the square root of the quadratic sum of two normally distributed variables. Hence, $\widehat{A}_{\text{ML}}$ is Rician distributed.

2.1.1.3. *MSE* As $\widehat{A}_{\mathrm{ML}}$ is Rician distributed, we find for its Mean Squared Error (MSE), which is the sum of its bias (b) squared and its variance:

$$\mathrm{MSE}(\widehat{A}_{\mathrm{ML}}) = \left[b(\widehat{A}_{\mathrm{ML}})\right]^2 + \mathrm{Var}(\widehat{A}_{\mathrm{ML}}) \quad (6)$$

in which

$$b(\widehat{A}_{\mathrm{ML}}) = A - E\left[\widehat{A}_{\mathrm{ML}}\right] \quad (7)$$

$$\mathrm{Var}(\widehat{A}_{\mathrm{ML}}) = E\left[\widehat{A}^2_{\mathrm{ML}}\right] - E\left[\widehat{A}_{\mathrm{ML}}\right]^2 \quad (8)$$

with

$$E\left[\widehat{A}_{\mathrm{ML}}\right] = \sqrt{\frac{2}{N}}\sigma\Gamma\left(\frac{3}{2}\right)\,{}_1F_1\left[-\frac{1}{2};1;-\frac{NA^2}{2\sigma^2}\right] \quad (9)$$

$$E\left[\widehat{A}^2_{\mathrm{ML}}\right] = A^2 + \frac{2\sigma^2}{N} \quad (10)$$

where ${}_1F_1$ is the confluent hypergeometric function of the first kind, and Γ denotes the Gamma function.

2.1.2. *Different phase values* Now assume that the complex observations $\{(R_n, I_n)\}$ have an underlying noiseless amplitude value A and arbitrary phase values $\varphi_1, \cdots, \varphi_N$.

2.1.2.1. *CRLB* It can easily be shown that the CRLB for unbiased estimation of $A, \varphi_1, ..., \varphi_N$ is given by:

$$\mathrm{CRLB} = \begin{pmatrix} \sigma^2/N & 0 & \cdots & 0 \\ 0 & \sigma^2/A^2 & \cdots & 0 \\ \cdots & \cdots & \cdots & \cdots \\ 0 & 0 & \cdots & \sigma^2/A^2 \end{pmatrix} \quad (11)$$

2.1.2.2. *ML estimation* The likelihood function for multiple independent Gaussian distributed complex observations $\{(R_r, I_n)\}$ with underlying noiseless signal A and arbitrary phase values $\varphi_1, \cdots, \varphi_N$ is given by:

$$L = \prod_{n=1}^{N} \frac{e^{-\frac{(R_n - A\cos\varphi_n)^2}{2\sigma^2}}}{\sqrt{2\pi\sigma^2}} \frac{e^{-\frac{(I_n - A\sin\varphi_n)^2}{2\sigma^2}}}{\sqrt{2\pi\sigma^2}} \quad (12)$$

Taking the logarithm and maximizing the result with respect to A and φ_n yields the ML estimator of A and φ_n, respectively:

$$\widehat{A}_{\mathrm{ML}} = \frac{1}{N}\sum_{n=1}^{N}\sqrt{R_n^2 + I_n^2} \quad (13)$$

$$\widehat{\varphi}_{n,\mathrm{ML}} = \arctan\left(\frac{I_n}{R_n}\right) \quad (14)$$

2.1.2.3. *MSE* The ML estimator, given by Eq. (13), is distributed as the average of N independent, Rician distributed variables. Therefore, its mean value is simply given by the average of the mean values of the individual Rician distributed variables, whereas its variance is given by the sum of their variances, divided by N^2. Hence, we have for the MSE:

$$\begin{aligned}\mathrm{MSE}(\widehat{A}_{\mathrm{ML}}) &= \left[b(\widehat{A}_{\mathrm{ML}})\right]^2 + \mathrm{Var}(\widehat{A}_{\mathrm{ML}}) \quad (15)\\ &= \left(A - E[\widehat{A}_{\mathrm{ML}}]\right)^2 \\ &\quad + \left(A^2 + 2\sigma^2 - E[\widehat{A}_{\mathrm{ML}}]^2\right)/N \quad (16)\end{aligned}$$

where the first moment $E[\widehat{A}_{\mathrm{ML}}]$ is now given by

$$E\left[\widehat{A}_{\mathrm{ML}}\right] = \sqrt{2}\sigma\Gamma\left(\frac{3}{2}\right)\,{}_1F_1\left[-\frac{1}{2};1;-\frac{A^2}{2\sigma^2}\right] \quad (17)$$

2.2 *Signal estimation from magnitude data*

In this subsection, we consider the estimation of the underlying noiseless amplitude signal A from N independent magnitude data points $\{M_n\}$, where A is assumed to be constant and the noise variance is assumed to be known. It is well known that magnitude data are Rician distributed, where the Rice PDF is given by:

$$p_M(m) = \frac{m}{\sigma^2}e^{-\frac{m^2+A^2}{2\sigma^2}} I_0\left(\frac{Am}{\sigma^2}\right) \quad , \quad (18)$$

with I_0 the 0th order modified Bessel function of the first kind.

2.2.1. *CRLB* The CRLB for unbiased estimation of A is given by (Karlsen *et al.*, 1999):

$$\mathrm{CRLB} = \frac{\sigma^2}{N}\left(Z - \frac{A^2}{\sigma^2}\right)^{-1} \quad . \quad (19)$$

where

$$Z = E\left[\frac{m^2}{\sigma^2}\frac{I_1^2\left(\frac{Am}{\sigma^2}\right)}{I_0^2\left(\frac{Am}{\sigma^2}\right)}\right] \quad , \quad (20)$$

with $\underline{m}$ a Rician distributed random variable with true parameters (A, σ).

2.2.2. *ML estimator* The joint PDF of the magnitude observations $\{M_n\}$ is given by:

$$p_M(\{m_n\}) = \prod_{n=1}^{N}\frac{m_n}{\sigma^2}e^{-\frac{m_n^2+A^2}{2\sigma^2}} I_0\left(\frac{Am_n}{\sigma^2}\right) \quad (21)$$

Then, the ML estimator of A can be constructed by substituting the set of N available observations $\{M_n\}$ for the $\{m_n\}$ in the expression for the joint PDF (21), considering A as variable, and maximizing the resulting function $L(A)$ with respect to A. The log-Likelihood function, only as a function of A, is given by:

$$\log L \sim \sum_{n=1}^{N} \log I_0\left(\frac{AM_n}{\sigma^2}\right) - \frac{NA^2}{2\sigma^2} \tag{22}$$

Note that $\log L$ is symmetric about $A = 0$. The ML estimate is the global maximum of $\log L$:

$$\widehat{A}_{\mathrm{ML}} = \arg\left\{\max_A (\log L)\right\} \quad . \tag{23}$$

Notice, that Eq. (23) cannot be solved analytically. Finding the maximum of the (log-)likelihood is therefore a numerical optimization problem.

3. NOISE ESTIMATION

In this section, we describe how the noise variance or the noise standard deviation can be estimated from complex as well as magnitude data using ML estimation.

3.1 *Estimation of the noise variance from complex data*

Suppose the noise variance needs to be estimated from N complex valued points $\{(R_n, I_n)\}$. We will consider the case of identical underlying phase values, as well as the case of different underlying phase values.

3.1.1. *Estimation for identical phases*

3.1.1.1. *CRLB* It can easily be shown that the CRLB for unbiased estimation of (A, φ, σ^2) is given by:

$$\mathrm{CRLB} = \begin{pmatrix} \sigma^2/N & 0 & 0 \\ 0 & \sigma^2/NA^2 & 0 \\ 0 & 0 & \sigma^4/N \end{pmatrix} \tag{24}$$

3.1.1.2. *ML estimation* For identical true phase values, the ML estimator of σ^2 can be shown to be:

$$\widehat{\sigma^2}_{\mathrm{ML}} = \frac{1}{2N}\sum_n \left[\left(\widehat{A}_{\mathrm{ML}}\cos\widehat{\varphi}_{\mathrm{ML}} - R_n\right)^2 + \left(\widehat{A}_{\mathrm{ML}}\sin\widehat{\varphi}_{\mathrm{ML}} - I_n\right)^2\right] \tag{25}$$

with $\widehat{A}_{\mathrm{ML}}$ and $\widehat{\varphi}_{\mathrm{ML}}$ given by Eq. (4) and (5), respectively. Notice, that for $N = 1$ the numerator, and thus the estimator $\widehat{\sigma^2}_{\mathrm{ML}}$, will be equal to zero.

3.1.1.3. *MSE* It can be shown that, for large N, the quantity $2N\widehat{\sigma^2}_{\mathrm{ML}}/\sigma^2$ is approximately distributed as χ^2_{2N-2} (i.e., chi-square distributed with $2N - 2$ degrees of freedom). This means, that

$$E\left[\widehat{\sigma^2}_{\mathrm{ML}}\right] \simeq \frac{\sigma^2}{2N}(2N-2) = \sigma^2\left(1 - \frac{1}{N}\right) \tag{26}$$

and

$$\mathrm{Var}\left(\widehat{\sigma^2}_{\mathrm{ML}}\right) \simeq \frac{\sigma^4}{N}\left(1 - \frac{1}{N}\right) \tag{27}$$

It follows from Eq. (26) that $\widehat{\sigma^2}_{\mathrm{ML}}$ has a bias equal to $\frac{\sigma^2}{N}$. Notice, that one can simply correct for this bias, but only at the expense of a higher variance. The MSE of $\widehat{\sigma^2}_{\mathrm{ML}}$ is given by:

$$\mathrm{MSE}\left(\widehat{\sigma^2}_{\mathrm{ML}}\right) \simeq \frac{\sigma^4}{N} \tag{28}$$

3.1.2. *Estimation for different phases*

3.1.2.1. *CRLB* It can easily be shown that the CRLB for unbiased estimation of $(A, \varphi_1, ..., \varphi_N, \sigma^2)$ is given by

$$\mathrm{CRLB} = \begin{pmatrix} \sigma^2/N & 0 & ... & 0 & 0 \\ 0 & \sigma^2/A^2 & ... & 0 & 0 \\ ... & ... & ... & ... & ... \\ 0 & 0 & ... & \sigma^2/A^2 & 0 \\ 0 & 0 & ... & 0 & \sigma^4/N \end{pmatrix} \tag{29}$$

3.1.2.2. *ML estimation* For different phase values, we have:

$$\widehat{\sigma^2}_{\mathrm{ML}} = \frac{1}{2N}\sum_n \left[\left(\widehat{A}_{\mathrm{ML}}\cos\widehat{\varphi}_{n,\mathrm{ML}} - R_n\right)^2 + \left(\widehat{A}_{\mathrm{ML}}\sin\widehat{\varphi}_{n,\mathrm{ML}} - I_n\right)^2\right] \tag{30}$$

with $\widehat{A}_{\mathrm{ML}}$ and $\widehat{\varphi}_{n,\mathrm{ML}}$ given by Eq. (13) and (14), respectively.

3.1.2.3. *MSE* It can be shown that, for large N, the quantity $2N\widehat{\sigma^2}_{\mathrm{ML}}/\sigma^2$ is approximately distributed as $\chi^2_{2N-(N+1)}$. This means, that

$$E\left[\widehat{\sigma^2}_{\mathrm{ML}}\right] \simeq \sigma^2\left(1 - \frac{N+1}{2N}\right) \tag{31}$$

and

$$\mathrm{Var}\left(\widehat{\sigma^2}_{\mathrm{ML}}\right) \simeq \frac{\sigma^4}{N}\left(1 - \frac{N+1}{2N}\right) \tag{32}$$

It follows from Eq. (31) that $\widehat{\sigma^2}_{\mathrm{ML}}$ has a bias equal to $\frac{N+1}{2N}\sigma^2$. Notice, that one can simply correct for this bias, but only at the expense of a higher variance. The MSE of $\widehat{\sigma^2}_{\mathrm{ML}}$ is given by:

$$\mathrm{MSE}\left(\widehat{\sigma^2}_{\mathrm{ML}}\right) \simeq \left(\frac{N+1}{2N}\sigma^2\right)^2 + \frac{\sigma^4}{N}\left(1 - \frac{N+1}{2N}\right) \tag{33}$$

3.2 *Estimation of the noise variance and standard deviation from magnitude data*

We will now describe ML estimation of the noise variance and standard deviation from background magnitude data (i.e., a region in which the underlying signal is zero). In subsection 3.2.1, we will consider ML estimation of the noise variance from a so-called background region, that is, . The CRLB for unbiased estimation of both noise variance and standard deviation will be computed and the ML estimators will be derived.

3.2.1. *Noise Variance* Suppose that a set of N independent magnitude data points $\{M_n\}$ is available from a region where the true signal value A is zero for each data point (background region). Hence, these data points are governed by a Rayleigh distribution and their joint PDF, p_M, is given by:

$$p_M(\{m_n\}) = \prod_{n=1}^{N} \frac{m_n}{\sigma^2} e^{-\frac{m_n^2}{2\sigma^2}} \tag{34}$$

Notice, that the Rayleigh distribution is a special case of the Rice distribution (21) with $A = 0$.

3.2.1.1. *CRLB (variance)* The Fisher information matrix $\boldsymbol{I}$ with respect to σ^2 is simply given by:

$$\boldsymbol{I} = -E\left[\frac{\partial^2 \log p_M}{(\partial\sigma^2)^2}\right] \tag{35}$$

from which the CRLB for unbiased estimation of σ^2 is easily found:

$$\text{CRLB} = \sigma^4/N \tag{36}$$

3.2.1.2. *CRLB (standard deviation)* From the knowledge of the CRLB for unbiased estimation of σ^2, the CRLB for unbiased estimation of σ can be derived (Sijbers *et al.*, 1999):

$$\text{CRLB} = \sigma^2/4N \tag{37}$$

3.2.1.3. *ML estimation* We will now describe the ML method for the estimation of the noise standard deviation and the noise variance from background magnitude data.

The likelihood function is obtained by substituting the available background data points $\{M_n\}$ for the m_n in Eq. (34). Then the log-likelihood function, only as a function of σ^2, is given by:

$$\log L \sim -N \log \sigma^2 - \frac{1}{2\sigma^2}\sum_{n=1}^{N} M_n^2 \tag{38}$$

Maximizing with respect to σ^2, yields the ML estimate of σ^2:

$$\widehat{\sigma^2}_{\text{ML}} = \frac{1}{2N}\sum_{n=1}^{N} M_n^2 \quad , \tag{39}$$

It can be shown that (39) is an unbiased estimator, that is, its mean is equal to σ^2. Furthermore, the variance of the ML estimator (39) is equal to σ^4/N, which equals the CRLB given by Eq. (36) for all values of N.

One might as well be interested in the value of the standard deviation σ, e.g., to estimate the signal-to-noise ratio (SNR): A/σ. Simply taking the square root of the ML estimator of σ^2 (39) yields an estimator of σ:

$$\widehat{\sigma}_{\text{ML}} = \sqrt{\frac{1}{2N}\sum_{n=1}^{N} M_n^2} \quad . \tag{40}$$

This estimator is identical to the ML estimator of σ, as the square root operation has a single valued inverse (cfr. Invariance property of ML estimators (Mood *et al.*, 1974)). Its variance is approximately equal to:

$$\text{Var}(\widehat{\sigma}_{\text{ML}}) \simeq \frac{\sigma^2}{4N} \quad , \tag{41}$$

which equals the CRLB (cfr. 37). The estimator (40) is, however, biased because of the square root operation. Its expectation value is approximately equal to:

$$E[\widehat{\sigma}_{\text{ML}}] \simeq \sigma\left(1 - \frac{1}{8N}\right) \quad . \tag{42}$$

4. NUMERICAL RESULTS

The MSE of the ML estimator of A from complex data with identical and different phase values has been calculated analytically using Eqs. (6-10) and (16-17), respectively. The MSE of the ML estimator from magnitude data has been obtained from a Monte-Carlo simulation experiment with sample size 10000. Thereby, the number of data points was set to $N = 25$ and the variance was set to $\sigma^2 = 1$. In addition, the CRLB for unbiased estimation of the signal amplitude from complex data with identical and different phase values was computed using Eqs. (3) and (11), respectively, whereas the CRLB for unbiased estimation of the signal amplitude from magnitude observations was computed using Eqs. (19) and (20).

In Figs. 1 and 2, the MSE of $\widehat{A}_{\text{ML}}$ has been plotted as a function of the SNR for magnitude data and complex data with identical and different phases, respectively. In addition, in both figures, the CRLB for unbiased estimation of the signal amplitude from complex as well as from magnitude data is shown.

5. DISCUSSION

5.1 *CRLB*

From Eq. (3) and (11) it is clear that the CRLB for unbiased estimation of A from complex observations

- does not depend on the phase values. On the other hand, the CRLB for unbiased estimation of

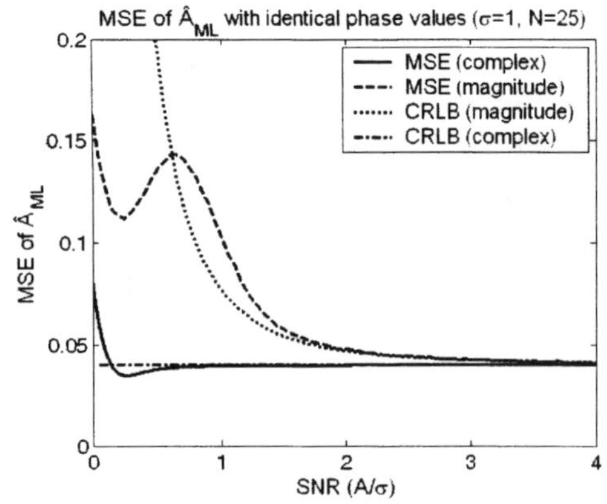

Fig. 1. MSE of $\widehat{A}_{\mathrm{ML}}$ from magnitude data and complex data with identical phase values. The CRLB's are also shown.

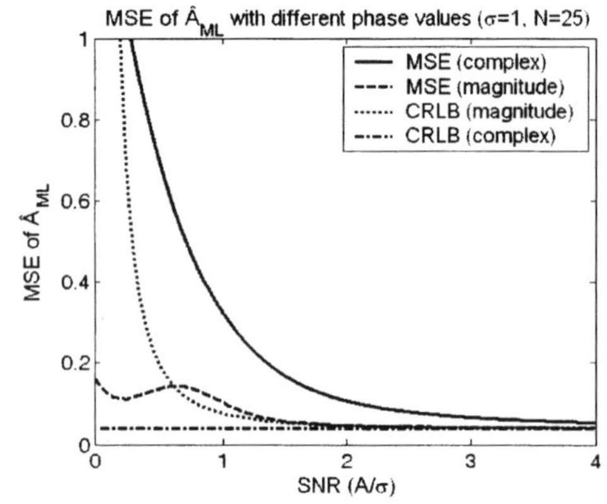

Fig. 2. MSE of $\widehat{A}_{\mathrm{ML}}$ from magnitude data and complex data with different phase values. The CRLB's are also shown.

the true phase values is inversely proportional to the squared signal amplitude A.

- does not depend on whether or not the true phase values are identical.
- is inversely proportional to the number of data points used for the estimation.

The CRLB for unbiased estimation of A from magnitude observations is given by (19). Note that for high SNR, this CRLB equals the CRLB from complex observations.

5.2 *MSE*

From the simulation results of the MSE from magnitude data, one can observe that, in general, $\widehat{A}_{\mathrm{ML}}$ for complex data performs better than the one for magnitude data in case the true phase values are equal. If the true phase values are different, the ML estimator based on magnitude data performs significantly better in terms of the MSE.

Recall that the MSE of an estimator is the sum of its bias squared and its variance.

For complex valued data, the variance turns out to be inversely proportional to the number of data points, for identical as well as for different phase values. The contribution of the bias of the ML estimator is less straightforward. In general, however, the bias rapidly decreases with increasing SNR. For high SNR (i.e., SNR > 5), the bias can be neglected. With respect to the dependence of the bias on the number of data points: if the phase values are different, the bias of $\widehat{A}_{\mathrm{ML}}$ does not decrease with the number of data points. On the other hand, if the phase values are identical, the bias decreases generally with the number of data points used for the estimation.

6. CONCLUSIONS

It has been shown that ML (amplitude) signal estimation from complex data points with equal true phase values is generally better in terms of the MSE compared to ML estimation from magnitude data points. However, in practice, phase values usually vary within the data set from which the amplitude signal is estimated. In that case, estimation from magnitude data is significantly better in terms of the MSE.

7. REFERENCES

Karlsen, O. T., R. Verhagen and W. M. Bovée (1999). Parameter estimation from Rician-distributed data sets using maximum likelihood estimator: application to T_1 and perfusion measurements. *Magnetic Resonance in Medicine* **41**, 614–623.

McGibney, G. and M. R. Smith (1993). An unbiased signal-to-noise ratio measure for magnetic resonance images. *Medical Physics* **20**(4), 1077–1078.

Miller, A. J. and P. M. Joseph (1993). The use of power images to perform quantitative analysis on low SNR MR images. *Magnetic Resonance Imaging* **11**, 1051–1056.

Mood, A. M., F. A. Graybill and D. C. Boes (1974). *Introduction to the Theory of Statistics*. 3rd ed.. McGraw-Hill. Tokyo.

Pintelon, R., J. Schoukens and J. Renneboog (1988). The geometric mean of power (amplitude) spectra has a much smaller bias than the classical arithmetic (RMS) averaging. *IEEE Transactions on Instrumentation and Measurement* **37**(2), 213–218.

Sijbers, J., A. J. den Dekker, E. Raman and D. Van Dyck (1999). Parameter estimation from magnitude MR images. *International Journal of Imaging Systems and Technology* **10**(2), 109–114.

Sijbers, J., A. J. den Dekker, P. Scheunders and D. Van Dyck (1998). Maximum likelihood estimation of Rician distribution parameters. *IEEE Transactions on Medical Imaging* **17**(3), 357–361.

Van den Bos, A. (1982). *Handbook of Measurement Science*. Chap. 8: Parameter Estimation, pp. 331–377. Vol. 1. Edited by P.H. Sydenham. Wiley, Chichester, England.

IFAC
Publications
www.elsevier.com/locate/ifac

RELIABLE NONLINEAR IDENTIFICATION IN MEDICAL APPLICATIONS

Le Yi Wang* G. George Yin Hong Wang*****

* *ECE Dept., Wayne State University, Detroit, USA*
** *Math Dept., Wayne State University*
*** *Anesthesiology Dept., Wayne State University*

1. INTRODUCTION

In this paper, we discuss several fundamental challenges in real-time nonlinear system identification for medical applications and offer some ideas in evaluating the reliability of identification algorithms and methods in face of these challenges by using a benchmark example to study real-time identification of patient models for anesthesia decisions in operating rooms.

Real-time medical decisions are exemplified by general anesthesia for attaining an adequate anesthetic depth (consciousness level of a patient), analgesia (pain management), sedation control in ICU (intensive care units), fluid resuscitation in trauma cases, circulation control (ventilation and mechanical circulation), etc. In all these medical decision processes, one of the most critical requirements is to predict the impact of the inputs (drug infusion rates, fluid flow rates, airway pressure and flow rates, etc.) on the outcomes (consciousness levels, pain scores, blood pressures, heart rates, oxygen saturation, etc.) This prediction capability is not necessarily "control-oriented," although its control implication is obvious. It can be used for display, warning, predictive diagnosis, decision analysis, outcome comparison, etc.

The core function of this prediction capability is embedded in establishing, in real-time, a reliable model that relates the (multiple) inputs to the (multiple) outcomes. Apparently, this is a real-time identification problem. This problem offers a great challenge and opportunity for advancing system identification. The major challenges that must be addressed include: (1) Reliability and safety are mandatory. Theoretical analysis and limited simulation must be further enhanced by a more comprehensive evaluation process. Eventually reliability must be established statistically via clinical trials. (2) The characteristics of patient responses demonstrate significant nonlinearity and time variation. (3) Unlike industry applications, patient responses depend critically on patient medical conditions, surgical procedures, and drug interactions; and hence they are not repeatable. Models must be established individually and in real-time. (4) This is not a data rich environment. At the starting time interval before substantial data become available one must rely on expert assessment to initiate the model and decision process. This implies that the modeling process must allow a seamless combination of assessment and data-generated estimation. (5) Since data points are limited, low-complexity models and fast identification are always preferred. (6) Due to medical constraints, the inputs are not allowed to be arbitrarily selected for best identification experiments or probing excitation.

These challenges, together with traditional focus of system identification on linear, well defined, and repeatable environment, have resulted in a gap between theoretical results in identification and applications to medical applications.

2. PATIENT MODEL STRUCTURES

In an attempt to understand the potential of system identification in this important application area, we have identified anesthesia decisions as a typical setting for studying the system identification problem. Some preliminary findings on the utility of Wiener model structures (Hammerstein models can be similarly applied), combination of assessment and real-time estimation, and identification algorithms are summarized below.

The goals of general anesthesia are to achieve hypnosis and analgesia simultaneously throughout a surgical operation while maintaining the vital functions of the body. One of the most critical tasks of anesthesia is to attain an adequate anesthetic depth. The newly marketed BIS monitor provides a viable direct measurement of anesthesia depth. To facilitate control strategy development for improved anesthesia performance, it is highly desirable to develop representative models of BIS responses to drug infusion. Since patients differ dramatically in metabolism, preexisting medical conditions, and surgical procedures, individualized models must be established for each patient with limited patient information and data points. A knowledge-based and control-oriented modeling approach was introduced recently and applied to develop a feedback and predictive control strategy for anesthesia infusion [6]. The models retain only the key characteristics of patient responses that are essential for control strategy development and can be determined by either data or expert knowledge. Furthermore, an identification algorithm was introduced in [7] for updating the patient model in real-time.

Due to complications in human metabolism and nerve systems, we do not perceive that the system dictates any specific model structure as especially superior. Simple model structures that can capture the key characteristics and maintain physiological meanings of model parameters will be preferable since they will allow easy interface with physicians and fast model updates.

Here we have tested the validity of Wiener structures in representing the patient dynamics. The patient dynamics is defined by three basic elements: (1) Initial time delay τ_p after drug infusion; (2) Time constant T_p representing the speed of response; (3) A nonlinear static function f_p representing sensitivity of the patient to a drug dosage at steady state. The meanings of these three components are depicted in Figure 1.

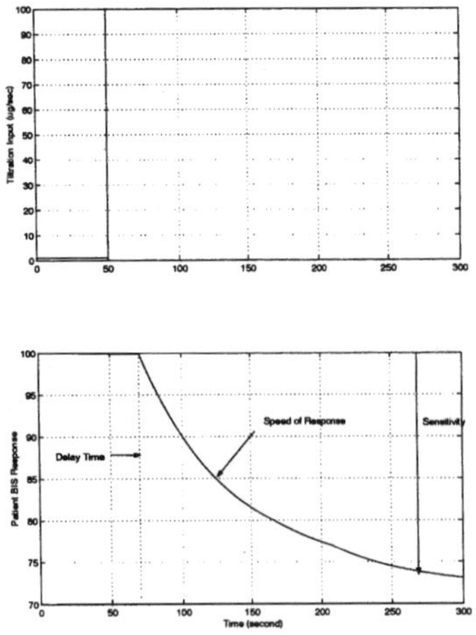

Fig. 1. Titration Model Characterization

This model structure can be mathematically represented by a Wiener model with input u and output y, $x = Gu + \widetilde{w}, y = f(x) + \widetilde{v}$ where G is a dynamic linear system, f is a memoryless nonlinear function, and $\widetilde{w}$ and $\widetilde{v}$ are noises. x is an intermediate variable that cannot be directly measured. In our applications here, G and f represent the transient and steady-state behavior of a patient's response to a drug infusion input, respectively. G can be discretized and represented by an ARMA model

$$x_t = \phi_t^T \theta + \widetilde{w}_t \qquad (1)$$

where $\phi_t^T = [-x_{t-1}, \ldots, -x_{t-n}, u_t, \ldots, u_{t-n}]$, $\theta^T = [a_1, \ldots, a_n, b_0, \ldots, b_n]$ or approximated by a MA model with sufficiently high orders. The sensitivity function f can be modeled by either a linear combination of some pre-selected base functions or a parameterized nonlinear function. The delay time τ_p, time constant T_p, and the sensitivity function vary greatly among patients and must be established individually for each patient.

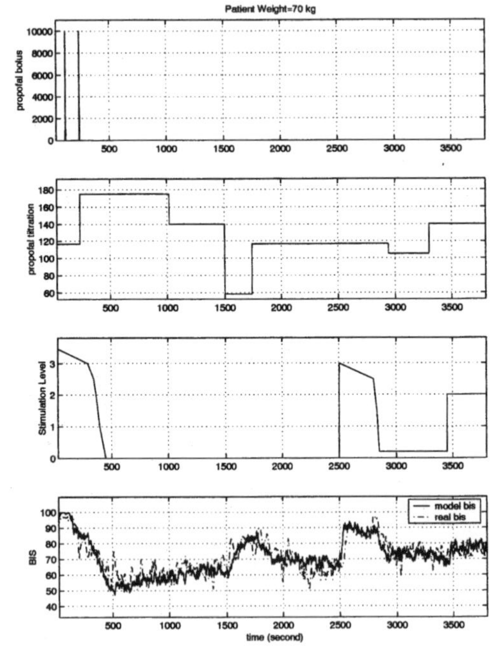

Fig. 2. Patient Model Responses

To verify the utility of the model structure and develop individual patient models, clinical data are collected. One of these data sets is described below to illustrate the process and results. The anesthesia process was administered manually by an anesthesiologist and lasted about 76 minutes, starting from the initial drug administration and continuing until the last dose of administration. To verify the capability of the model structure, the data were used to determine model parameters and function forms, by an off-line estimation method (an optimal nonlinear LS estimator). The actual response is compared with the model response over the entire surgical procedure. Comparison results are shown in Figure 2. Here, the inputs of drug titration (continuous drug flow) and bolus (drug injection of a short duration) are the recorded real-time data. The model captures the key trends and magnitudes of the output variations in the surgical procedure. This indicates that the model structure, though very simple, contains sufficient freedom in representing the main features of the patient response. Also, the impact of surgical stimulation is captured by the model.

3. RELIABILITY EVALUATION OF IDENTIFICATION ALGORITHMS

Despite well established theoretical results on many identification algorithms, our studies indicate that to enhance reliability of system identification in medical applications, some well developed algorithms must be significantly re-evaluated and modified beyond routine convergence analysis. Our criteria for evaluating an algorithm include (1) Accuracy and convergence (estimation errors); (2) Robustness (error excursion frequencies); (3) Model complexity (number of parameters); (4) Time complexity (data points required); (5) Usage convenience (requirements of delicate tuning or sensitive dependence on initial values are not desirable).

To understand the reliability of various identification methods on Wiener models and anesthesia patient models, the following platforms are used as a benchmark.

Platform 1: A Simulated Wiener System

The example plant is expressed by the Wiener system $x_t = a_1 u_t + a_2 u_{t-1} + w_t$,

$$y_t = 2x_t + \frac{C_1 x_t(x_t - \xi_1)}{\xi_1(\xi_1 - \xi_2)} + \frac{C_2 x_t(x_t - \xi_2)}{\xi_2(\xi_2 - \xi_1)} + v_t (2)$$

where the true parameters are $a_1 = 2$; $a_2 = 1$; $C_1 = 4.1$; $C_2 = 3.5$. The testing points for the expert information on the nonlinear function are $\xi_1 = 5$; $\xi_2 = 10$. For anesthesia applications, we always have $y_t = 0$ when $x_t = 0$. The nominal function $y_t = 2x_t$ is known and contained in the expression. Prior information on unknown parameters are given by: range of $a_1 = [0, 3]$; range of $a_2 = [0, 2]$; range of $C_1 = [2, 6]$; range of $C_2 = [1, 5]$. w_t and v_t are i.i.d. uniformly distributed random processes, in the range $[-1.5, 1.5]$.

Platform 2: An Anesthesia Patient Model

From the patient data illustrated in Figure 2, the titration model is first extracted by an off-line method. The method eliminates the impact of surgical stimulation and drug impact from bolus injection, and produces a BIS response of the drug propofol titration on the patient as shown in Figure 3.

3.1 Recursive Algorithms

Model parameter updating algorithms must be used to estimate the parameters of the Wiener model, especially the linear dynamic part. There

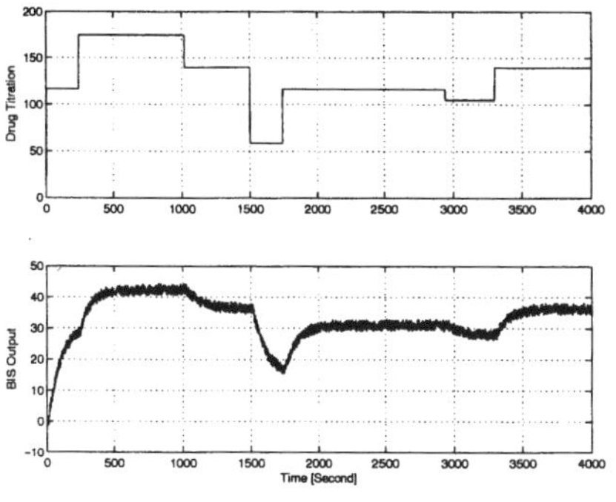

Fig. 3. Patient BIS Response to Propofol Titration

have been established many valid recursive algorithms in a wide variety of applications. Our objective here is to evaluate the accuracy and convergence of different recursive algorithms under this application. Here a linear system is used $y_t = 1.5y_{t-1} - 0.7y_{t-2} + 0.9u_{t-1} + 0.5u_{t-2} + e_t = \phi_t^T \theta + e_t$, where e is an i.i.d. Gaussian disturbance with zero mean and unit variance.

The evaluated algorithms include the following four types. For the first three algorithms, since the tunable parameters have dramatic impact on their performance, their values are optimized in each run.

(1) Adaptive Filtering

$$\theta_{t+1} = \theta_t + \frac{c}{t^r}\phi_t(y_t - \phi_t^T \theta_t), \quad 0 < r \le 1.$$

(2) Adaptive Filtering with Averaging

$$\widehat{\theta}_{t+1} = \theta_t + \frac{c}{t^r}\phi_t(y_t - \phi_t^T \widehat{\theta}_t), \quad 0 < r < 1;$$

$$\theta_{t+1} = \theta_t - \tfrac{1}{t+1}\theta_t + \tfrac{1}{t+1}\widehat{\theta}_{t+1}.$$

(3) One-step Modified Optimal Projection

$$\theta_{t+1} = \theta_t + r\frac{\phi_t}{a + \phi_t^T \phi_t}(y_t - \phi_t^T \theta_t).$$

(4) Recursive Least Squares

$$K_t = \frac{P_t \phi_t}{1 + \phi_t^T P_t \phi_t}; P_{t+1} = (I - K_t \phi_t^T)P_t;$$

$$\theta_{t+1} = \theta_t + K_t(y_t - \phi_t^T \theta_t).$$

The algorithms are started with the same initial estimate, which is obtained by using an LS estimate with 10 data points. Then the recursive algorithms are performed for 1000 data points. For each algorithm, the estimation errors during the last 20 time points are averaged. The averaged error is used as a measure of identification accuracy. The simulation is repeated 50 times. The 50 estimation errors are plotted in the right column of Figure 4 for algorithm comparison. Also, the 50th error trajectory for each algorithm is illustrated in the left column.

It seems clear that the recursive least squares algorithm outperforms the rest in two aspects: (1) It provides more accurate estimates; (2) It does not require any variable tuning. It is noted that we have optimized variables for each run in other algorithms. In real applications, these

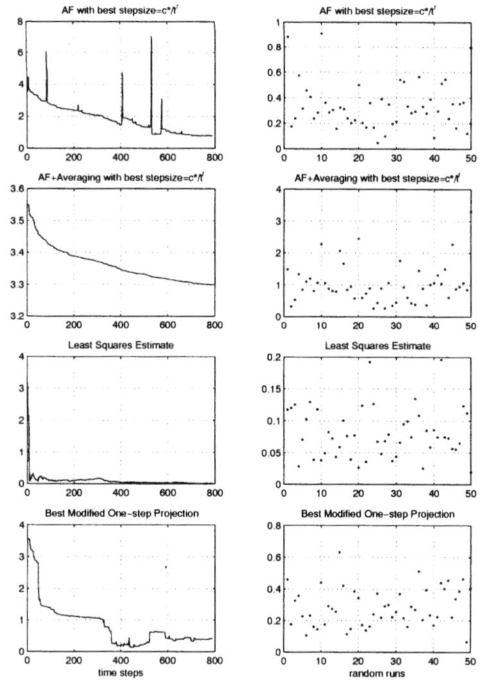

Fig. 4. Comparison of Recursive Algorithms

variables cannot be optimized over each sample path. Consequently, the actual performance of these algorithms will be worse than what we have demonstrated in this simulation. In fact, when the variables are fixed, we encountered occasionally parameter divergence in adaptive filtering or projection algorithms. In general, the RLS may require slightly more computational time. However, this is not an issue in this case since the model order is small in patient models. Consequently, we will use LS algorithms in our identification of patient models.

3.2 Block Recursion

In [7], a two-step recursive estimation algorithm was proposed for identifying the patient model in real-time. The procedure can be briefly summarized as follows.

Two-step Procedure. We start at $k = 0$ with $\widehat{\theta}_0$ (the parameters for the linear part) and $\widehat{\Theta}_0$ (the parameters for the nonlinear part).

Step 1: At time k, the available $\widehat{\Theta}_k$ is used to map the output y_k backwards to obtain the "intermediate" output $\widehat{x}_k$. Then u_k and $\widehat{x}_k$ are used to update $\widehat{\theta}_{k+1}$.

Step 2: From the obtained $\widehat{\theta}_{k+1}$, we map u_{k+1} forward to obtain $\widehat{x}_{k+1}$. Then $\widehat{x}_{k+1}$ and y_{k+1} are used to update $\widehat{\Theta}_{k+1}$ from $\widehat{\Theta}_k$.

Selection of recursive algorithms, averaging, projection into a bounded set, step size selection, and convergence analysis were presented in [7]. This algorithm is computationally very efficient.

To evaluate its reliability, the system (2) is used. It works reasonably well when initial estimates are close to true values. However, when initial estimates are further away from true values the algorithm demonstrated irregular behavior that includes parameter divergence, slow parameter drifting, and large output prediction errors. In other words, it is not a reliable algorithm.

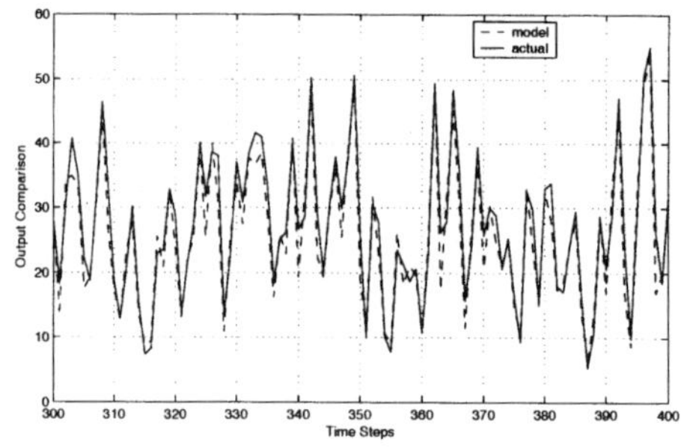

Fig. 5. Block Recursion

Consequently, the algorithm is modified into a block recursive algorithm. In this modified algorithm, the parameters are not updated upon receiving each observation. But rather, a block of m observations is used collectively to update the model parameters. Starting with the most recent estimate as an initial condition, the algorithm searches for the optimal parameters that minimize the output errors between the model output and measured output in the data block. This modified algorithm have shown a greatly improved reliability. For the system (2), we used the block size 10 to run a simulation of 800 data points with a random input. The simulation was repeated 50 times and demonstrated good prediction capability in each run. A typical result is shown in Figure 5. On the other hand, parameter convergence is not always demonstrated.

3.3 Optimization Efficiency

For a nonlinear system, optimization carries potentially large computational burdens. As shown in the previous section, overly simplified algorithms may suffer loss of reliability. In search of efficient optimization methods during a block recursion, an embedded optimization method was employed. In this method, for given parameters of the nonlinear part of the model, the parameters of the linear part are optimized by the least squares algorithm that is computationally very simple. Consequently, only the parameters of the nonlinear function must be searched. In our evaluation, global grid search, local grid search (around the initial estimate), simplex optimization methods were tested. The simplex method finds similar parameter values but requires on average only about 1/20 of computational time than the grid search methods.

3.4 *Parameterization of Nonlinear Functions*

The memoryless function f in the Wiener model represents steady-state sensitivity: The impact of the input (drug amount) on the outcome. Theoretically, a memoryless and continuous nonlinear function can be approximated by a polynomial function to any degree of accuracy when the order of the polynomial is allowed to increase. However, the parameters of the polynomial usually do not carry any physical meanings. In [6,7], it was proposed that the following functions are more suitable to connect with expert assessment.

A physician's knowledge can be expressed as follows: Given m typical x amounts $\xi_1, \ldots, \xi_m$, the approximate ranges of the predicted steady-state outcomes are $[\underline{y}_1, \overline{y}_1], \ldots, [\underline{y}_m, \overline{y}_m]$. We parameterize f by

$$y = f(x) = \sum_{i=1}^{m} C_i \Phi_i(x) \tag{3}$$

where $\Phi_i(x)$ is the mth order polynomial $\Phi_i(x) = \frac{\prod_{j=1,j\neq i}^{m}(x-\xi_j)}{\prod_{j=1,j\neq i}^{m}(\xi_i-\xi_j)}$. $\Phi_i(x)$ satisfies $\Phi_i(\xi_j) = 0$, $j \neq i$; $\Phi_i(\xi_i) = 1$. Consequently, the prior expert information becomes precisely the prior interval information on the parameters $\Theta = [C_1, \ldots, C_m]'$: $C_i \in [\underline{y}_i, \overline{y}_i]$, $i = 1, \ldots, m$. This knowledge will serve as the prior information on the unknown parameters θ and Θ. Consequently, the parameters will have clear physical meanings and expert assessment can be integrated seamlessly into this parameterization.

Interestingly, for computational purposes this parameterization implics further complications. Understanding that the nonlinear function in this application represents a drug's steady-state impact on the patient outcome. Hence it must obey the basic monotone principle: the more the drug infusion rate, the more the impact on the patient. Unfortunately, this requirement is not naturally embedded in the polynomial parameterization. In fact, in a box of parameter values for the polynomials in (3) the resulting nonlinear functions are not always monotone. Also, for the same parameters, the corresponding function may become non-monotone when the range of the drug rates is enlarged. Consequently, in search of optimal parameters during system identification, monotonicity must be laboriously checked in each evaluation and sometimes leads to the failure of a local search.

A remedy of this situation is found by using a different parameterization of the nonlinear function. The following two-parameter function is employed: Suppose that x takes values in $[0, b]$.

$$y = r\left(x \pm \left(\frac{\text{erf}(\alpha x)}{\text{erf}(\alpha b)} - x\right)\right)$$

where $\text{erf}(\cdot)$ is the standard error function, and the parameters r and α define variation in y at $x = b$ and curvatures of the erf function in $x \in [0, b]$. Since it is very simple to find a box in $r - \alpha$ space under which the resulting nonlinear functions are always monotone, it greatly improves time complexity in search algorithms. In Figures 8,9,10, the ranges of searched nonlinear functions are included in the bottom plots.

3.5 *Model Complexity and Time Complexity*

Model complexity and time complexity are especially important in medical applications due to limited data points, input probing richness, requirements of prompt decisions, and time varying natures of patient models. In this evaluation the patient model in Figure 3 is used. Since actual model order and structures are unknown to the designer, we simply use either MA or ARMA models of various orders and evaluate their prediction capability. The input data are the real drug infusion rates which do not provide much richness in excitation. The total surgery lasted about 4000 seconds. We used the first 400 seconds of the data for system identification for ARMA models, and 1000 data points for MA models since their orders are much higher. Then the prediction capability of the identified model is evaluated by applying it to the entire data. It is noted that during the first 400 seconds, the input changes its value only once.

Conceptually, it is arguable that for a linear stable system, it is always possible to approximate it by a FIR model (moving average). This model structure can significantly simplify theoretical analysis. However, a comparison of Figures 6 and 7 shows clearly that to capture the dynamics of the actual system, the order of the MA model must be very high. This is highly undesirable in this application since 200 parameters in this model is a very high model complexity. It also increases identification time complexity. Since an overall patient model contains many model blocks of similar types, the MA structure will imply a very large parameter set for system identification when the overall system is considered.

Figures 8, 9 and 10 show the prediction capability of ARMA models of orders 1, 4 and 10 respectively. It is seen that ARMA models can capture the dynamics of the system with much lower orders than MA models.

4. CONCLUDING REMARKS

This paper highlights three key points: (1) Wiener or Hammerstein models have great potential utility in modeling patient dynamics; (2) Reliable identification is a stringent requirement that mandates a modified evaluation criterion on identification algorithms; (3) Within the framework of reliable identification, all components of a system

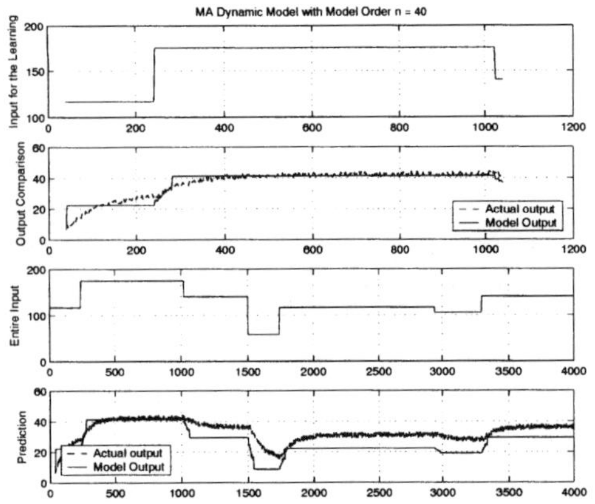

Fig. 6. Moving Average Model of Order 40

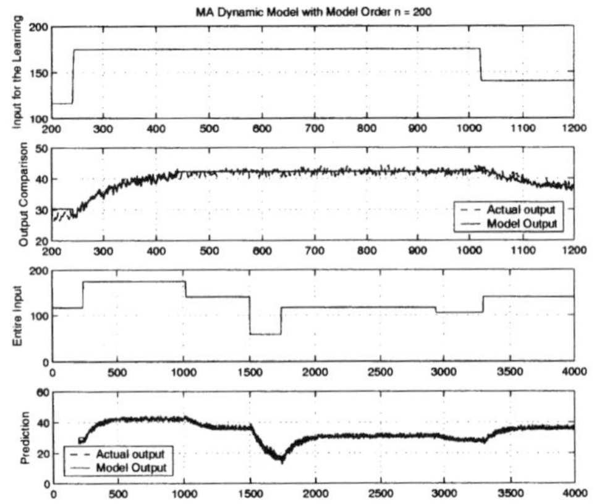

Fig. 7. Moving Average Model of Order 200

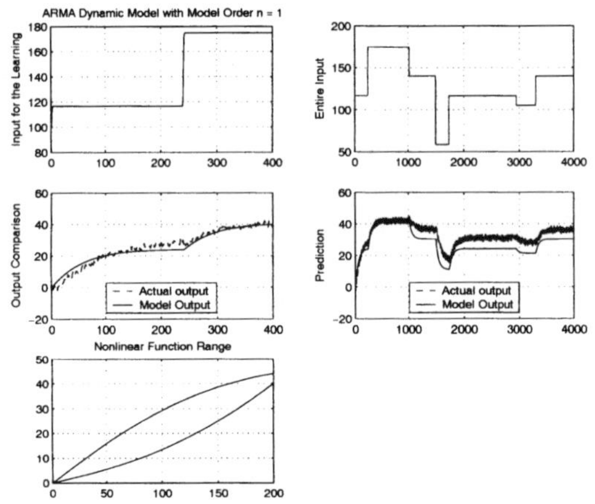

Fig. 8. ARMA Model of Order 1

identification must be carefully examined in terms of robustness, reliability, model complexity, time complexity, accuracy, and tracking capability for time varying parameters. A study using anesthesia patient modeling is presented to illuminate these aspects of reliable system identification.

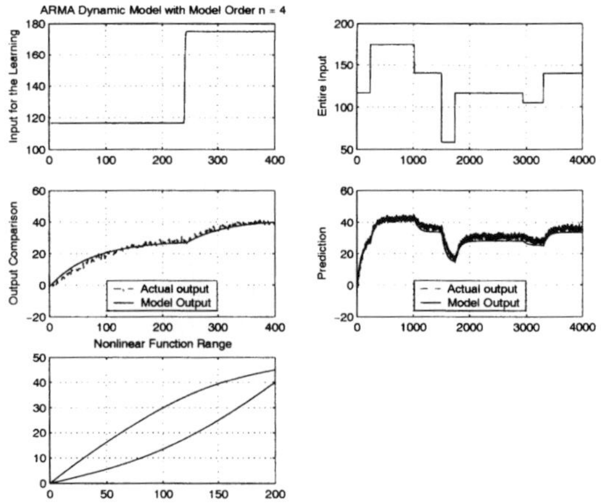

Fig. 9. ARMA Model of Order 4

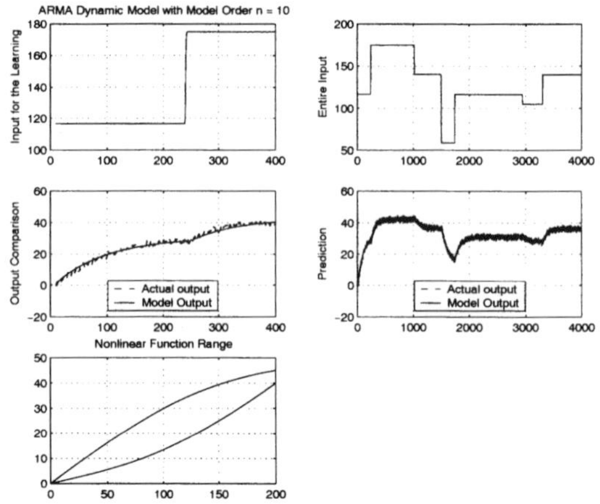

Fig. 10. ARMA Model of Order 10

REFERENCES

[1] H.J. Kushner and G. Yin, *Stochastic Approximation Algorithms and Applications*, Springer-Verlag, New York, 1997.

[2] Linkens, D.A., Adaptive and intelligent control in anesthesia, IEEE Control Syst. Magazine, 6-11, Dec., 1992.

[3] L. Ljung and T. Söderström, *Theory and Practice of Recursive Identification*, MIT Press, Cambridge, MA,1983.

[4] M. Milanese and A. Vicino, Optimal estimation theory for dynamic systems with set membership uncertainty: An overview, *Automatica*, **27** (1991), 997-1009.

[5] Rosow, C., and P.J. Manberg, Bispectral index monitoring, annual of anesthetic pharmacology, 2:1084-2098, 1998.

[6] L.Y. Wang and H. Wang, Part I: Control-oriented modeling of BIS-based patient response to anesthesia infusion, Part II: Feedback and predictive control of anesthesia infusion using control-oriented patient models, *2002 Intern. Conf. Math. and Eng. Tech. in Medicine and Biological Sci.*, Las Vegas, June 24-27, 2002.

[7] L.Y. Wang, H. Wang, G. Yin, Anesthesia infusion models: Knowledge-based real-time identification via stochastic approximation, *41st IEEE Cont. and Dec. Conf.*, Las Vegas, 2002.

IFAC Publications
www.elsevier.com/locate/ifac

PATTERN RECOGNITION OF EEG SIGNALS DURING RIGHT AND LEFT MOTOR IMAGERY

Katsuhiro Inoue*, Gert Pfurtscheller, Christa Neuper** and Kousuke Kumamaru***

** Department of Control Engineering and Science, Faculty of Computer Science and Systems Engineering, Kyushu Institute of Technology, 680-4 Kawazu, Iizuka, Fukuoka, 820-8502, Japan*
*** Department of Medical Informatics, Institute of Biomedical Engineering, Graz University of Technology, Inffeldgasse 16a 8010 Graz, Austria*

Abstract: Electroencephalograph (EEG) recordings during right and left motor imagery can be used to move a cursor to a target on a computer screen. Such an EEG-based brain-computer interface (BCI) can provide a new communication channel to replace an impaired motor function. It can be used by e.g., handicap users with amyotrophic lateral sclerosis (ALS). In this study, statistical pattern recognition method based on AR model was introduced to discriminate the EEG signals recorded during right and left motor imagery. And learning methods (processing period for parameter estimation, AR order, etc.) were investigated. Finally, the effectiveness of our method was confirmed through the experimental studies. *Copyright © 2003 IFAC*

Keywords: EEG, AR-model, statistical pattern recognition, brain computer interface, motor imagery.

1 .INTRODUCTION

Classification of EEG signals is a difficult task, especially when the derived classification result is to be used to control an electronic device, because in this case the classification has to be performed on a single-trial basis (i.e. not averaged). Such a system which transforms signals from the brain into control signals is known as a brain-computer interface (BCI) (Vidal, 1973, Wolpaw *et al.*, 1994). For the single-trial signal recognition method, there are many methods were proposed. For examples, artificial neural network (ANN) with frequency components as a feature vectors (Björn O, 1997), and linear discrimination method based on adaptive auto-regressive (AR) parameter (Pfurtscheller, G. 1998). In the case of ANN and linear discriminant analysis, it is difficult to normalize feature parameters.
In this paper, statistical pattern recognition method based on AR model which does not need normalization of the feature vector, is introduced and is applied to discriminate the EEG signals.

2. METHOD AND MATERIALS

Three subjects (20-28 years old) participated at this study. All were right-handed and free of medication and central nervous system abnormality.

2.1 Experimental paradigm, training and testing sessions

During the experiment, the subject fixated a computer monitor 100 cm in front of her. Each trial was 8 sec-long and started with the presentation of a fixation cross at the centre of the monitor, followed by a short warning tone ('beep') at 2000 ms (see Fig.1). At 3000 ms, the fixation cross was overlaid with an arrow at the centre of the monitor for 1250 ms, pointing either to the left or to the right. Depending on the direction of the arrow, the subject was instructed to imagine a movement of the left or the right hand. Prior to the experiment, each subject was given the opportunity to practice and perform actual movements of the left and right hand according to the arrow direction displayed on the monitor.

Feedback consisted of a symbol presented in the centre of the monitor at 6000 ms; the type of symbol (large or small '+' or '-', or '0') depending on how well a subject-specific classifier could recognize movement-dependent EEG characteristics in a fixed time window from 3250 to 4250 ms .

There were two types of sessions: in the training sessions, data were collected for the creation of a subject-specific classifier and, therefore, no feedback was provided. In the following test sessions, the classifier was then used to classify the subject's EEG on-line while she imagined the requested kind of hand movement, and feedback was given to the subject as described above. Each of the initial subjects participated in 3-4 training sessions, all on different days; the number of feedback (testing) sessions varied between 7 and 8. Each session consisted of 4 experimental runs of 40 trials (20 'left' and 20 'right' trials) and lasted for about 1 h. The sequence of 'left' and 'right' trials, as well as the duration of the breaks between consecutive trials (ranging between 500 and 2500 ms), were randomized throughout each experimental run.

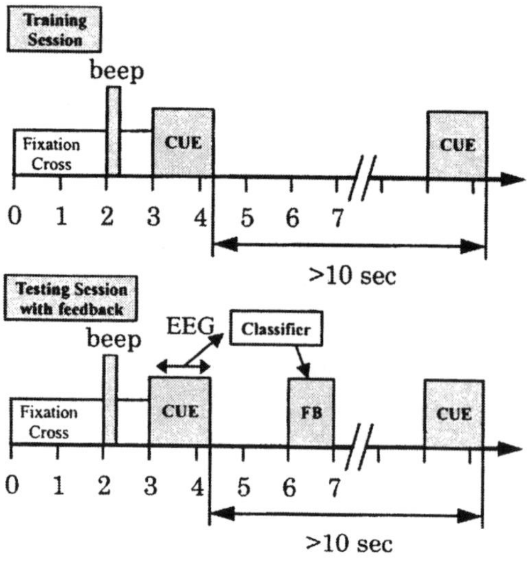

Fig.1 Timing of training and test session.

2.2 *EEG recording and data acquisition*

Two types of EEG recording method were adopted as shown in Fig.2. One is a small Laplacian filtered signals(SL). Channel C3 was derived from following 5 electrodes.

1. an electrode placed C3 (y_0)
2. an electrode placed 2.5cm anterior to C3 (y_A)
3. an electrode placed 2.5cm posterior to C3 (y_P)
4. an electrode placed 2.5cm left to C3 (y_L)
5. an electrode placed 2.5cm right to C3 (y_R)

Channel C3 is derived by following equation.

$$y = y_0 - (y_A + y_P + y_L + y_R)/4 \qquad (1)$$

Another is bipolar signals (BP). EEG was recorded using two bipolar leads over left and right central area. Channel C3 was derived from an electrode placed 2.5cm anterior to C3 and an electrode placed 2.5cm posterior to C3. Channel C4 was derived similarly. The EEG signals were amplified and band-pass filtered between 0.5 and 33Hz by a Nihon Khoden amplifier and then sampled at 128Hz. EOG was derived from two electrodes, one placed medially just above the right eye and the other laterally just below the right eye, in order to detect vertical as well as horizontal eye movements. These signals were used to screen the EEG recordings for eye movement artefacts. In addition, the surface EMG of the extensor muscles of both hands was monitored during each feedback session.

In this study, only feed back session's data were used. Available data (artefact free data) were follows.

Subject f3:	Left 74 trials	Right 75 trials
Subject f5:	Left 75 trials	Right 67 trials
Subject f7:	Left 59 trials	Right 56 trials

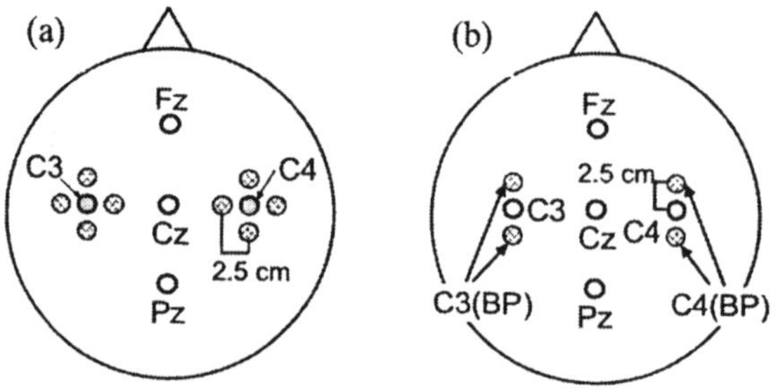

Fig.2 Position of the electrodes
(a) Small Laplacian filtered signal
(b) Bipolar signal

3 PATTERN RECOGNITION METHOD BASED ON AR MODEL

The EEG signals are assumed to be generated from an autoregressive (AR) model,

$$y_t = \sum_{j=1}^{m} \phi_j y_{t-j} + v_t = \Phi^T z_{t-1} + v_t \qquad (2)$$

$$\Phi = [\phi_1, \phi_2, \cdots, \phi_m]^T$$

$$z_{t-1} = [y_{t-1}, y_{t-2}, \cdots, y_{t-m}]^T .$$

Where, y_t is the observed EEG signal at time t, and v_t is the independent random variable with normal distribution.

$$E[v_t v_s] = \rho\delta_{t,s}, \qquad v_t = N[v_t; 0, \rho]$$

A feature vector θ is defined as follows

$$\theta = [\Phi^T, \rho]^T .$$

The first m observations $y_1, y_2, \cdots, y_m$ serve as initial conditions for the equation (2), following equation is assumed.

$$p(y_1, y_2, \cdots, y_m, \theta_k) = p(y_1, y_2, \cdots, y_m) \cdot p(\theta_k) \quad (3)$$

And it is assumed that the feature parameter θ in the class ω_k is deterministic. This assumption is expressed as follows

$$p(Z_N / \omega_k) = p(Z_N / \theta_k) \qquad (4)$$

$$Z_N = \{y_1, y_2, \cdots, y_N\}.$$

These assumptions yield the following explicit expression for the conditional probability density functions.

$$p(Z_N/\omega_k) = P(y_1, y_2, \cdots, y_m)\left(\frac{1}{2\pi\rho_k}\right)^{\frac{N-m}{2}} \cdot \exp\left[-\frac{1}{2\rho_k}\sum_{t=m+1}^{N}\left(y_t - \Phi_k^T \cdot z_{t-1}\right)^2\right] \tag{5}$$

The purpose of this paper is to identify the human's will (right and left motor imagery) by using 2 channels EEG signals (C3 and C4). Therefore, following Byes decision rule is adopted.

$$k^* = Arg \underset{k}{Max} \ \Pr(\omega_k / Z_{DN}) \tag{6}$$

Where,

$Z_{DN} = \{ Y_{1N}, Y_{2N} \}$

$Y_{1N} = \{y_{11}, y_{12}, \cdots, y_{1N}\}$: Signals (C3)

$Y_{2N} = \{y_{21}, y_{22}, \cdots, y_{2N}\}$: Signals (C4)

ω_k : Class

1: Left Movement Class

2: Right Movement Class

In this study, EEG signals (electrode C3 and C4) are pre-processed by spatial filter. Then, it is conceivable that these signals are mutually independent. Therefore, following assumption is adopted.

$$\begin{aligned} p(Z_{DN}, \omega_k) &= p(Y_{1N}, Y_{2N} / \omega_k) \\ &= p(Y_{1N}/\omega_k) \cdot p(Y_{2N}/\omega_k) \end{aligned} \tag{7}$$

$$p(Y_{iN}/\omega_k) = P(y_{i,1}, y_{i,2}, \cdots, y_{i,m})\left(\frac{1}{2\pi\rho_{ik}}\right)^{\frac{N-m}{2}} \cdot \exp\left[-\frac{1}{2\rho_{ik}}\sum_{t=m+1}^{N}\left(y_{i,t} - \Phi_{ik}^T \cdot z_{i,t-1}\right)^2\right] \tag{8}$$

[Parameter Estimation]

For the estimation of parameters, following likelihood function is considered.

$$L_i = \sum_{j=1}^{N_p} \ln p(Y_{iN}^j / \omega_k) \tag{9}$$

where, N_P is the number of samples of given class k, and Y_{iN}^j is *j*-th sample of channel *i* (C3 or C4). To maximize the criterion function L_i, partial differentiation of L_i with respect to each parameter is done and then equated to zero. Then the obtained simultaneous equations are solved to give rise the below equations.

$$\hat{\Phi}_{ik} = \left[\sum_{j=1}^{N_p}\sum_{t=m+1}^{N} z_{i,t-1}^j z_{i,t-1}^{jT}\right]^{-1} \cdot \left[\sum_{j=1}^{N_p}\sum_{t=m+1}^{N} y_{i,t}^j z_{i,t-1}^{jT}\right] \tag{10}$$

$$\hat{\rho}_{ik} = \frac{1}{N_p(N-m)}\sum_{j=1}^{N_p}\sum_{t=m+1}^{N}\left(y_{i,t}^j - \hat{\Phi}_k^T z_{i,t-1}^j\right)^2 \tag{11}$$

Now, applying Bayes rule to the equation (6) and substituting equation (8), the concrete decision rule is obtained.

[Decision Rule]

$$\begin{aligned} k^* = Arg \underset{k}{Max} \ [&\ln(\Pr(\omega_k)) \\ &- \frac{N-m}{2}\ln(2\pi\rho_{1k}) \\ &- \frac{1}{2\rho_{1k}}\sum_{t=m+1}^{N}\left(y_{1,t} - \Phi_{1k}^T \cdot z_{1,t-1}\right)^2 \\ &- \frac{N-m}{2}\ln(2\pi\rho_{2k}) \\ &- \frac{1}{2\rho_{2k}}\sum_{t=m+1}^{N}\left(y_{2,t} - \Phi_{2k}^T \cdot z_{2,t-1}\right)^2 \ ] \end{aligned} \tag{12}$$

In this study, $\Pr(\omega_1) = \Pr(\omega_2) = 0.5$ is assumed.

4 PATTERN RECOGNITION OF THE EEG

In order to realize the discernment machine of the EEG signal at the time of imagination of a right hand and a left hand of operation, it is necessary to clarify the optimal feature extraction method. That is, the following problems have to be solved.

1) What is the best AR order?
2) Where is the bet processing period?
3) How to treat the signals measured from electrode C3 and C4.

For the question 2), 112 kinds of period shown in Fig.3 are considered.

Decision Point [sec]	Data Length [sec]						
	0.50	0.75	1.00	1.25	1.50	1.75	2.00
3.50							
3.75							
4.00							
4.25							
4.50							
4.75							
5.00					O		
5.25							
5.50							
5.75							
6.00							
6.25							
6.50							
6.75							
7.00							
7.25							
7.50							
7.75							
8.00							

Fig.3 Processing period

Data Length: 7 kinds 0.5～2.0 [sec]

Decision Time: 13～19 kinds, 3.5～8.0 [sec]

For example, "O" on 1.5sec (data length) and 5.00 sec (decision point) means the period(3.5-5.0sec).

For the question 3), recognition results based on SL signals and BP signals will be discussed. And recognition results based on C3, C4 and C3&C4 will also be discussed.

To obtain a more general view of the ability of classification, a 10 times 10 fold cross validation is performed. The 10 times 10 fold cross validation mixes the data set randomly and divides it into 10

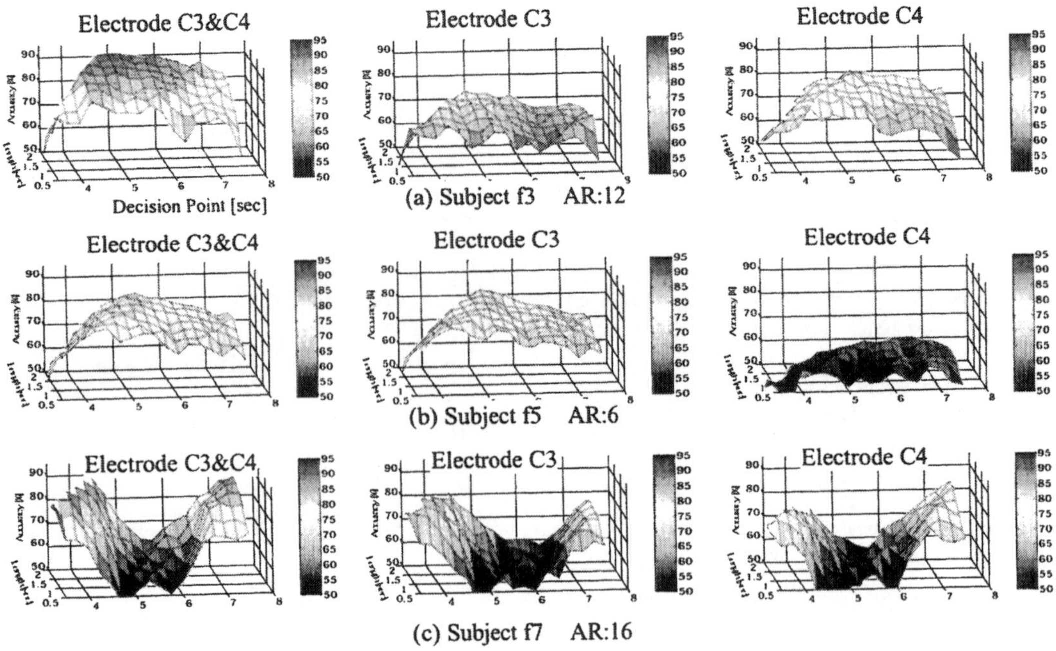

Fig.4 Pattern recognition results by using SL signals
(a) Subject f3 (b) Subject f5 (c) Subject f7
The figures of a left column show the results based on both of C3 and C4 signals, the figures in a middle column show the results based on only C3 signals. And the figures in a right column show the results based on only C4 signals

equally sized disjunct partitions. Each partition is then used once for testing, the other partitions are used for training. This results in 10 different accuracies (the percent ratios of the number of the samples classified correctly and the total number of the testing samples), which are averaged. This is the accuracies of a 10 fold cross validation. To further improve the estimate the procedure is repeated 10 time and again all accuracies are averaged.

The examples of the recognition result are shown in Fig.4. These results are obtained based on 10 times 10 fold cross validation. Testing data are classified based on the parameter which is estimated by using the same period of training data.

4.1 AR order

The following criteria are introduced in order to determine the optimal order of the AR model.

Criterion 1 : The order which takes the greatest value of the average of the accuracy over the best 20 period in each AR order. (shown in Fig.5)

Criterion 2 : The order which takes the greatest value in the histogram of the optimal AR order in each period (shown in Fig.6)

Criterion 3 : The order which takes the greatest value in the histogram of the optimal AR order in each data length (average over the start points). (shown in Fig.7).

Criterion 4 : The order which takes the greatest value in the histogram of the optimal AR order in each start point (average over the data length). (shown in Fig.8).

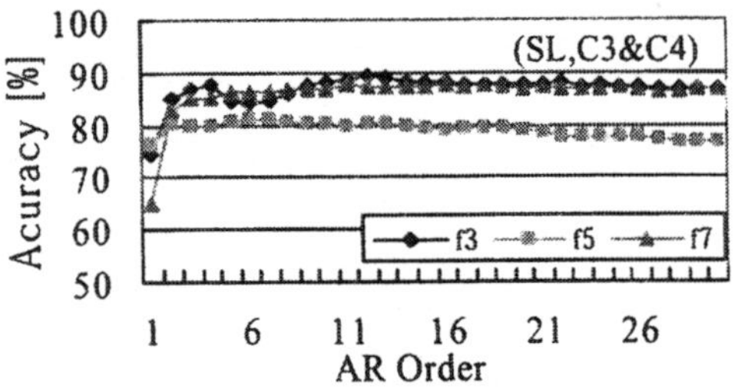

Fig.5 The average of the accuracy over the best 20 period in each AR order (Criterion 1)

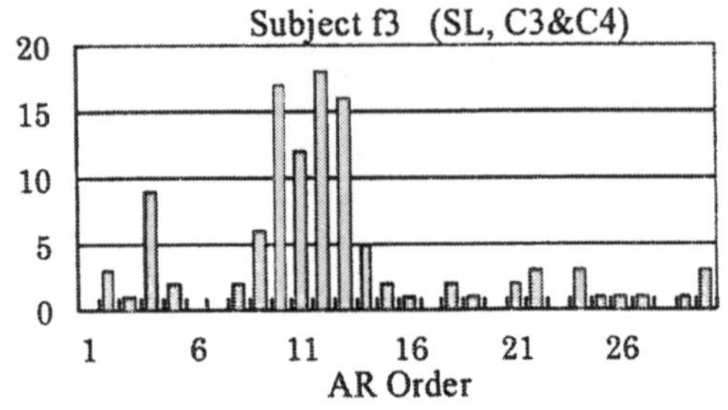

Fig.6 Histogram for AR order (Criterion 2)
Subject f3

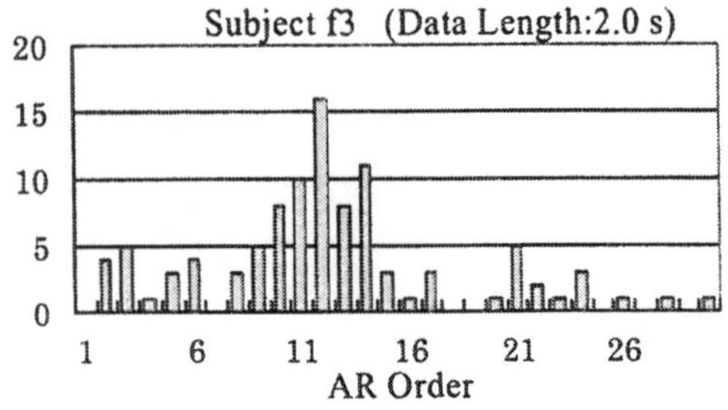

Fig.7 Histogram for AR order (Criterion 3)
(SL signals, C3&C4)

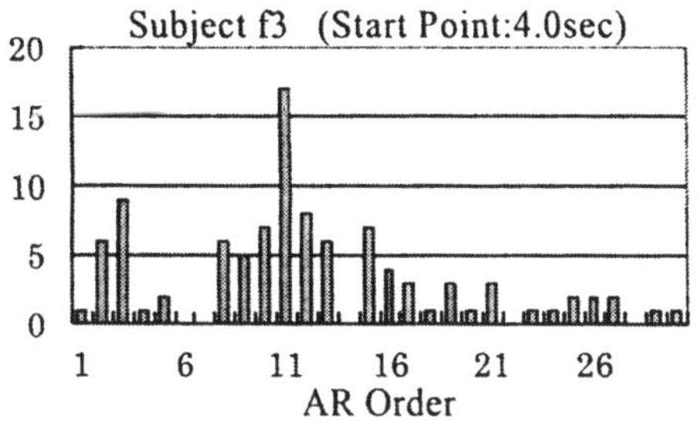

Fig.8 Histogram for AR order (Criterion 4)
(SL signals, C3&C4)

Table.1 The Optimal AR order in Each Criterion

		Criterion			
		1	2	3	4
f3	SL	12	12	10-13	12
	BP	11	11	11-13	11
f5	SL	7	5	5,6	5
	BP	5	5	3,5	5
f7	SL	16	7	10	16
	BP	12	12	8-11	17

The Table.1 shows the summary of the optimal AR order based on each criterion. From these results, it fixes the AR order (subject f3:12, subject f5:5, subject f7:16) and subsequent discussion will be performed.

4.2 Comparison of SL and BP

Fig.9 shows the recognition result based on SL(Small Laplacian filtered) signal and BP(bipolar) signal. These results suggest that SL signal is better than BP signals.

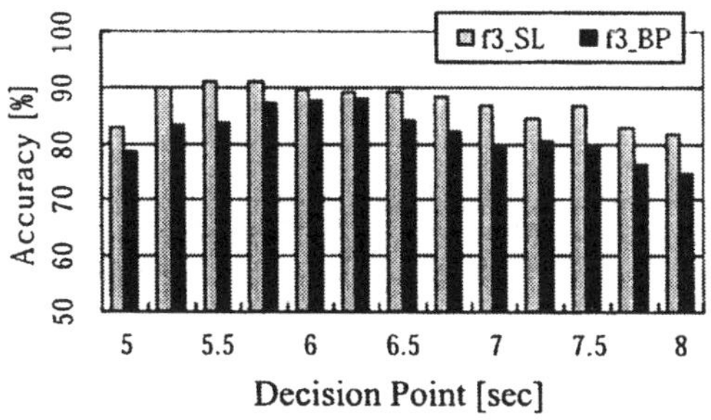

(a) Subject f3 (AR order :12)

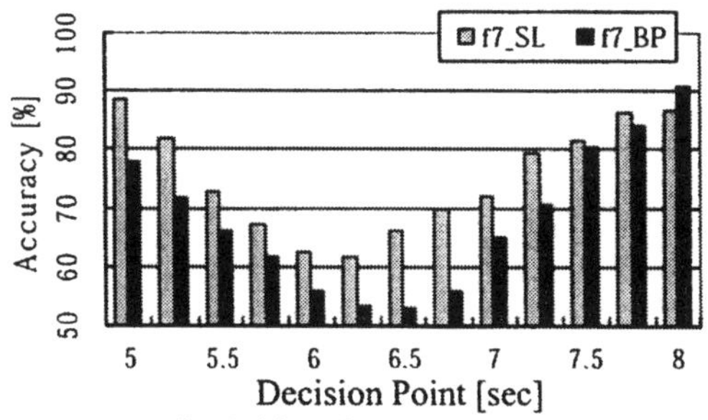

(b) Subject f7 (AR order :16)

Fig.9 Pattern recognition results in each signals (SL and BP) Data Length:2.0sec
(a):Subject f3, (b) Subject f7

4.3 Data length

Fig.10 shows the recognition results in each data length. Generally speaking, it is better that data length is long. But, for the subject f7, the shorter data length is better than the longer data length at 7.25~8.0 desicion points.

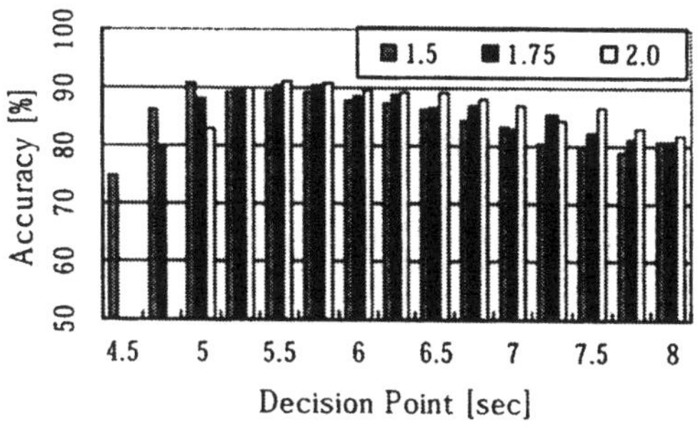

(a) Subject f3 (AR order :12)

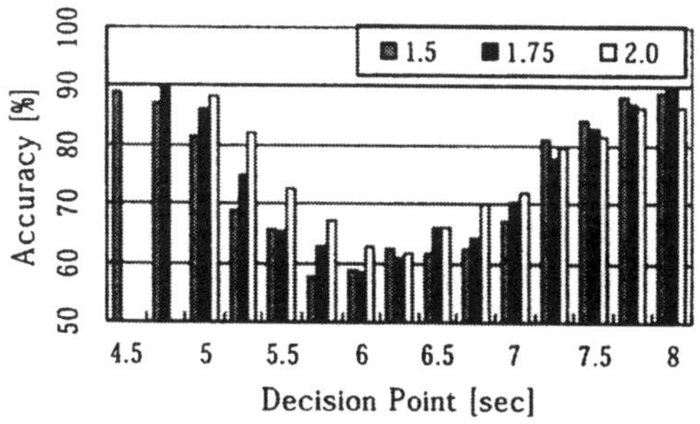

(b) Subject f7 (AR order :16)

Fig.10 Pattern recognition results in each data length. (a):Subject f3, (b) Subject f7

4.4 Comparison of C3 and C4

Fig. 11 shows the recognition results by using both of C3 and C4 signal, only C3 signal and only C4 signal, respectively. It is clear that it is the best to use both of C3 and C4 signals. But, It depends on a subject for the contribution of the C3 and C4 signal in discrimination. For the subject f3, C3 and C4 signals equivalently contribute the discrimination. For the subject f5, C4 signal do not contribute the discrimination. And for the subject f7, C3 signals have more information than C4 on a first half period, but C4 signals have more information than C3 on a last half period.

4.5 Synchronized period recognition and fixed period recognition

Two kinds of recognition are considered in this subsection.

Synchronized Period Recognition: Testing data are classified based on the parameter which is estimated by using the same period of training data

Fixed Period Recognition: Testing data are classified based on the parameter which is estimated by using the period [3.0, 5.0] of training data.

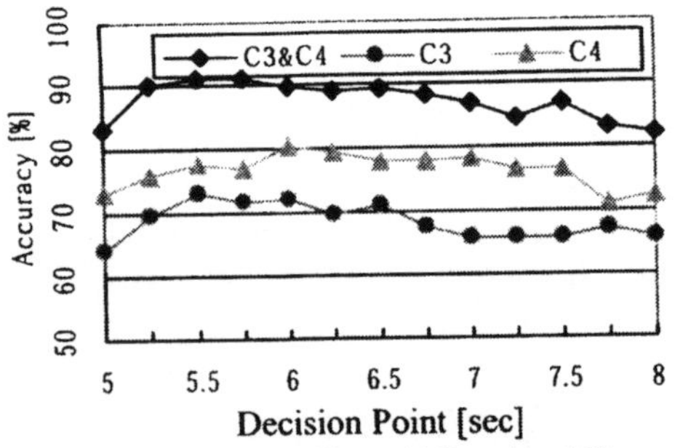

(a) Subject f3 (AR order :12)

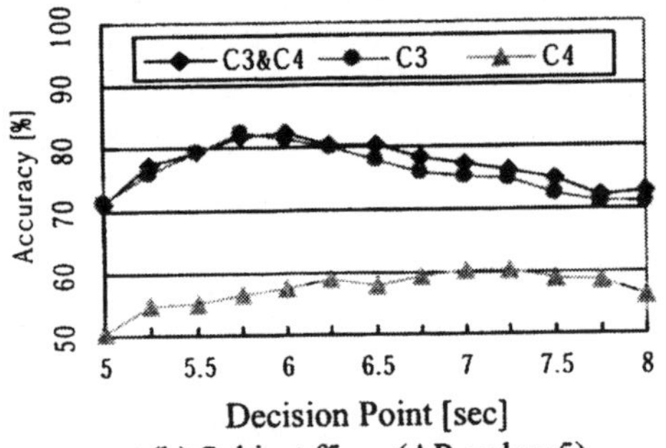

(b) Subject f5 (AR order :5)

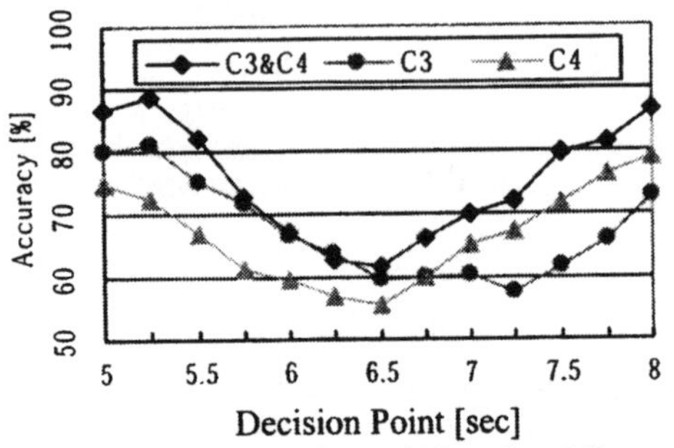

(c) Subject f7 (AR order :16)

Fig.11 Pattern recognition results in each signals (C3, C4, C3&C4) Data Length:2.0sec (a):Subject f3, (b) Subject f7 (c) Subject f7

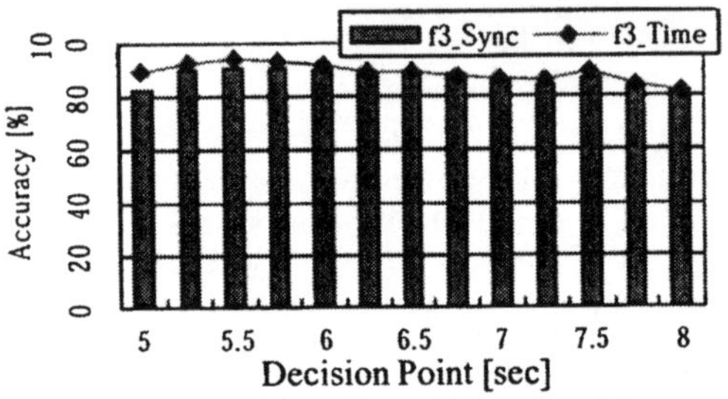

(a) Subject f3 (AR order :12)

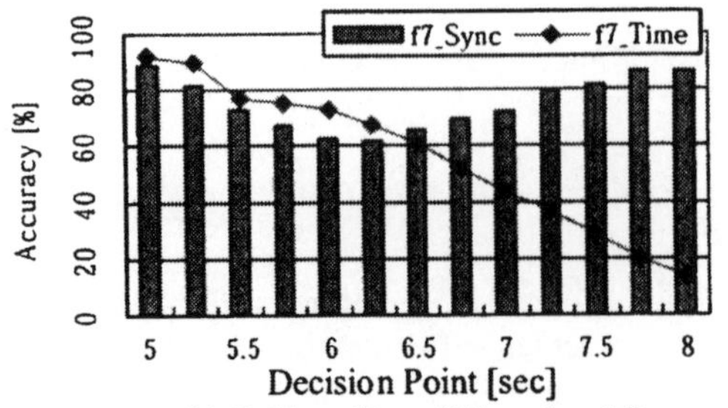

(b) Subject f7 (AR order :16)

Fig.12 Pattern recognition results by *Synchronized Period* recognition and *Fixed Period* recognition Data Length:2.0sec (a):Subject f3, (b) Subject f7

Fig.12 shows the results. *Synchronized Period Recognition* result and *Fixed Period Recognition* result have the same tendency for subject f3 and f5. But for the subject f7, these are quite different. High accuracy is obtained at the just after the appearance of CUE (part 1) and the end part of trial (part 2) on the *Synchronized Period* recognition result, but *Fixed Period* recognition result is quite low at the part 2. This means that EEG wave patterns are reverse patterns between the part 1 and part 2.

5 CONCLUSION

In this paper, statistical pattern recognition method based on AR model is proposed in order to discriminate EEG signals recorded during right and left motor imagery. Following characteristics have been confirmed through the experiments..

- The decision accuracy is improved by using both of C3 and C4 signals.
- The SL signal is more suitable than the BP signal.
- The optimal period of the parameter estimation depends on each subject.
- For some subjects, the response EEG wave patterns change with time.

Although our method needed to tune up according to the subject, 80%~90% accuracies were obtained. And by modifying the decision rule (equation (12)) to recursive form, on-line recognition can be executed. These facts suggest that our method will be applicable to real system.

REFERENCES

Björn O. Peters, G. Pfurtscheller and H. Flyvbjerg (1997). Prompt Recognition of Brain States by their EEG Signals. *Theory Bioscience,* **116,** pp.290-301.

Inoue K., G. N. Rao, A. Mizutori, Y. Morita and K. Sekoguchi (1991). Classification of Gas-Liquid Two-Phase Flow Regime by an AR-Model. *Proc. of IFAC Symposium on Identification and System Parameter Estimation*, pp.597-602.

Pfurtscheller G., Ch. Neuper, D. Flotzinger and M. Pregenzer (1997). EEG-based discrimination between imagination of right and left hand movement. *Electroencephalography and clinical Neurophyiology,* **103**, pp.642-651.

Pfurthscheller G., C. Neuper, A. Schlögl and K. Lugger (1998). Separability of EEG Signals Recorded During Right and Left Motor Imagery Using Adaptive Autoregressive Parameters. IEEE Trans. on Rehabilitation Engineering, 6-3, pp.316-325.

Vidal, J. (1973). Toward direct brain-computer communication, *Annual Rev. Biophysics Bio-engineering.* pp.157-180

Wolpaw. J. and D. McFarland (1997). Multichannel EEG-based brain-computer communications, *Electroencephalography and clinical Neuro-phyiology.* **90**, pp.444-449

IFAC Publications
www.elsevier.com/locate/ifac

FROM DYNAMIC METABOLIC MODELING TO UNSTRUCTURED MODEL IDENTIFICATION OF COMPLEX BIOSYSTEMS

Jens E. Haag [*,1] **Alain Vande Wouwer** [*] **Philippe Bogaerts** [**]

[*] *Laboratoire d'Automatique, Faculté Polytechnique de Mons, 31 Boulevard Dolez, 7000 Mons, Belgium, tel. +32 65 37 4141, fax +32 65 37 4136*
[**] *Service d'Automatique et d'Analyse des Systèmes, Université Libre de Bruxelles, 50 Avenue F. D. Roosevelt, CP 165/55, 1050 Bruxelles, Belgium, tel. +32 2 650 2676, fax +32 2 650 2677*

Abstract: In this study, a class of dynamic models based on metabolic reaction pathways is analyzed, showing that systems with complex intracellular reaction networks can be represented by macroscopic reactions relating extracellular components only. Based on rigorous assumptions, the model reduction procedure is systematic and allows an equivalent 'input–output' representation of the system to be derived. The procedure is illustrated with a few examples. *Copyright © 2003 IFAC*

Keywords: Biotechnology, mathematical modeling, model reduction, parameter estimation, nonlinear systems.

1. INTRODUCTION

For bioprocess control and optimization purposes, it is necessary to deal with mathematical models of rather low complexity. Macroscopic and unstructured modeling plays therefore an important role in this field. However, with the increasing knowledge of the cell metabolism, more detailed models have gained attention in recent years describing certain metabolic features. The objective of these structured modeling approaches are mostly pathway analysis for metabolic engineering. Metabolic Flux Analysis is applied to chemostat cultures in order to quantitatively determine the fluxes under different steady-state culture conditions based on molar balances (Zupke and Stephanopoulos, 1995; Bonarius *et al.*, 1996).

Until recently, both branches of research – macroscopic modeling for process engineering and microscopic modeling for metabolic engineering – have lacked a physical link. Haag *et al.* (2002) provided the justification for the use of macroscopic dynamic models in biotechnology, which is derived from the ideas of Metabolic Flux Analysis and a few rigorous assumptions.

In the model reduction approach of (Haag *et al.*, 2002), some degrees of freedom are left for the definition of the macroscopic reaction scheme based on the detailed metabolic pathway network. A systematic search procedure for these missing constraint equations is presented in this contribution.

The legitimation of macroscopic models motivates the use of modeling tools, like the procedure for systematic generation and evaluation of C-identifiable macroscopic reaction schemes proposed by Bogaerts and Vande Wouwer (2001), which does not take the underlying metabolic pathways into account, because it is of little interest for the control community. Parallels and

[1] Corresponding author, e-mail: jens.haag@fpms.ac.be

differences between both approaches are also shown in this study.

The paper is organized as follows. In Section 2, dynamic metabolic modeling of bioprocesses is introduced. A model reduction procedure is developed in Section 4, which allows complex intracellular reaction networks to be represented by macroscopic reactions. In Section 4, a systematic search for the constraint Matrix defining the macroscopic reaction scheme is presented. Based on this legitimation of the use of macroscopic representation of the cell metabolism, the systematic procedure for identification of C-identifiable reaction scheme proposed by Bogaerts and Vande Wouwer (2001) is introduced for comparison purposes in Section 5. Both modeling approaches are illustrated with a simple example throughout this contribution. Section 6 summarizes and concludes the theoretical aspects and the properties of both systematic modeling procedures.

2. DYNAMIC METABOLIC MODELING

Cell cultures in bioprocess engineering are described by a differential equation for the extracellular concentration vector $c \in \mathbb{R}^{n_c}$ in the following state-space form:

$$\frac{\mathrm{d}c(t)}{\mathrm{d}t} = K_1 q(c(t), x(t)) c_{X_v}(t) + F(c(t), t), \quad (1)$$

featuring n_r metabolic reactions. The first RHS term is the reaction term and consists of the specific reaction rate vector $q \in \mathbb{R}^{n_r}$ and the matrix $K_1 \in \mathbb{R}^{n_c \times n_r}$ containing the respective stoichiometric coefficients, multiplied with the viable cell concentration $c_{X_v} \in \mathbb{R}$. The second term $F \in \mathbb{R}^{n_c}$ accounts for mass transfer in and out of the bioreactor The balance equation for the intracellular concentrations $x \in \mathbb{R}^{n_x}$ is written analogously

$$\frac{\mathrm{d}x(t)}{\mathrm{d}t} = K_2 q(c(t), x(t)) - x(t)\mu(c(t), x(t)), \quad (2)$$

with the respective stoichiometric matrix $K_2 \in \mathbb{R}^{n_x \times n_r}$. The last term in (2) describes the dilution due to biomass growth with the specific cell growth rate $\mu \in \mathbb{R}$ defined by the balance for the total biomass concentration c_{X_t}, i. e. the sum of viable and dead cells, in equation (1):

$$\frac{\mathrm{d}c_{X_t}(t)}{\mathrm{d}t} = \frac{\mathrm{d}c_{X_v}(t)}{\mathrm{d}t} + \frac{\mathrm{d}c_{X_d}(t)}{\mathrm{d}t} = \mu(c(t), x(t)) c_{X_v}(t) + F_{X_v}(c(t), t). \quad (3)$$

System $\{(1),(2)\}$ defines completely the dynamics of a bioreaction system and takes intracellular metabolites and the corresponding metabolic pathways into account. In practice, the major drawback of this approach is the difficulty to get measurements of the intracellular concentrations. It is therefore delicate to set up and use such a system representation for parameter and/or state estimation due to the lack of information.

The following section deals with the systematic simplification of the metabolic reaction scheme.

3. MODEL REDUCTION

If the assumption holds that the cell growth term in equation (2) is negligibly small and the intracellular concentrations are in a quasi-steady state (Zupke and Stephanopoulos, 1995),

$$\frac{\mathrm{d}x(t)}{\mathrm{d}t} \approx 0, \quad (4)$$

the intracellular balance (2) turns into the set of algebraic equations:

$$K_2 q = 0. \quad (5)$$

Since the number of reactions n_r (and in turn, the number of unknown reaction rates) is necessarily greater than the number of resulting algebraic equations n_x resulting from (5), $n_f = n_r - n_x$ conditions have to be formulated for the complete determination of the metabolic flux vector q. Assume that these constraints are in the form:

$$K_3 q = f(c) \quad (6)$$

with $K_3 \in \mathbb{R}^{n_f \times n_r}$ and $f(c) \in \mathbb{R}^{n_f}$, a vector-function of the extracellular concentrations. In the simplest case, the constraint matrix K_3 contains only one nonzero element in each row, which is equivalent to the definition of single reaction rates in the pathway, e.g. by Michaelis-Menten laws.

If the combined matrix $\begin{bmatrix} K_2^{\mathrm{T}} & K_3^{\mathrm{T}} \end{bmatrix}^{\mathrm{T}}$ is of full rank, the flux vector q becomes a linear function of the vector-function $f(c)$

$$q(c) = \begin{bmatrix} K_2 \\ K_3 \end{bmatrix}^{-1} \begin{bmatrix} 0 \\ I \end{bmatrix} f(c) \quad (7)$$

with the $(n_x \times n_f)$ zero matrix 0 and the n_f-dimensional identity matrix I. The stoichiometric matrix $K \in \mathbb{R}^{n_c \times n_f}$ (instead of $\mathbb{R}^{n_c \times n_r}$) of the new reduced-order reaction system is therefore defined by

$$K = K_1 \begin{bmatrix} K_2 \\ K_3 \end{bmatrix}^{-1} \begin{bmatrix} 0 \\ I \end{bmatrix}, \quad (8)$$

and the extracellular mass balance (1) becomes

$$\frac{\mathrm{d}c(t)}{\mathrm{d}t} = K f(c(t)) c_{X_v}(t) + F(c(t), t). \quad (9)$$

Thus, the transformation (7) maps the metabolic pathway network into a reduced-order global reaction scheme, which gives a black-box 'input–output' representation of the biomass.

Example 1. Consider the simple metabolic reaction scheme of $n_r = 4$ reactions with two substrates (S), one product (P), $n_x = 2$ metabolites (M) and the biomass (X):

(1) $S_1 \longrightarrow 2M_1$
(2) $S_2 \longrightarrow M_2$
(3) $M_1 \longrightarrow 2M_2 + P$
(4) $M_2 \longrightarrow 0.1X$

With the state vectors $c^T = [c_{S_1}\ c_{S_2}\ c_P\ c_X]$ and $x^T = [x_{M_1}\ x_{M_2}]$, the respective stoichiometric matrices are defined as

$$K_1 = \begin{bmatrix} -1 & 0 & 0 & 0 \\ 0 & -1 & 0 & 0 \\ 0 & 0 & 1 & 0 \\ 0 & 0 & 0 & 0.1 \end{bmatrix} \quad (10)$$

$$K_2 = \begin{bmatrix} 2 & 0 & -1 & 0 \\ 0 & 1 & 2 & -1 \end{bmatrix}. \quad (11)$$

Since $n_r - n_x = 2$, two additional equations are required to complete the definition of the reaction system, which are chosen as follows in this example:

$$q_1 = f_1(c) \quad (12)$$
$$q_2 = f_2(c). \quad (13)$$

The constraint matrix is then written:

$$K_3 = \begin{bmatrix} 1 & 0 & 0 & 0 \\ 0 & 1 & 0 & 0 \end{bmatrix}. \quad (14)$$

From equation (8), the following global stoichiometric matrix results:

$$K = \begin{bmatrix} -1 & 0 \\ 0 & -1 \\ 2 & 0 \\ 0.4 & 0.1 \end{bmatrix}, \quad (15)$$

which is equivalent to the global reaction scheme

(1) $S_1 \longrightarrow 2P + 0.4X$
(2) $S_2 \longrightarrow 0.1X$.

Metabolic stoichiometry is subject of intense research led by biologists, and can be considered as relatively well known a priori. In contrast, there is usually no information about the choice of the constraint matrix K_3. The next section deals with the systematic search for the best choice of the constraint matrix.

4. SYSTEMATIC CONSTRAINT SEARCH

Since the metabolic stoichiometry is supposed to be known, the search for a global, macroscopic reaction scheme reduces to the search for an appropriate constraint matrix K_3. A full matrix K_3 would lead to the problem of estimating $n_f(n_r - 1)$ uncorrelated unknown parameters related to the stoichiometry. In order to avoid additional stoichiometric parameters, a special form of the constraint matrix is assumed in the following, which contains only one non-zero element in each row. This element, which can be taken equal to one, corresponds to the definition of a specific reaction rate. In this case, the maximum number of possible global reaction schemes is the number of permutations

$$N = \binom{n_r}{n_f} \quad (16)$$

In order to evaluate the quality of each individual choice for K_3, the kinetics functions f have to be structurally defined and parametrized. This must be done in a general way in order to treat each possible scheme equally. In this study, the following general structure is proposed:

$$f_i(c) = q_{\max,i} \prod_{j=1}^{n_c} \frac{c_j^{\operatorname{sgn}\{K_{ij}\}}}{c_j^{\operatorname{sgn}\{K_{ij}\}} + K_{ij}^2} \qquad i = 1 \ldots n_f \quad (17)$$

featuring limitation, according to the Michaelis-Menten law (f_i monotonically increasing with c_j) for $K_{ij} > 0$ with the Michaelis constant $K_{M,ij} = K_{ij}^2$, and inhibition (f_i monotonically decreasing with c_j) for $K_{ij} < 0$. At $K_{ij} = 0$, the influence of c_j vanishes, i. e. component j has no impact on the reaction i. Note that the unit of K_{ij} changes at this point. The influence of each component on the kinetics is multiplicative. Squaring K_{ij} renders the transition at $K_{ij} = 0\ (c_j > 0)$ differentiable, which facilitates the parameter estimation.

Hence, the unknown model parameters are

- n_f maximum reaction rates $q_{\max,i}$ and
- $n_f n_c$ modulation constants K_{ij}.,

i.e. the model parameter vector p contains $n_f(n_c + 1)$ elements. The structural parameters have to be estimated for each permutation k, with $1 \le k \le N$, and the cost functions $j(p^*)$ at the optimum have to be evaluated and compared. The best choice for the constraint matrix is then the permutation corresponding to the lowest cost:

$$k^* = \arg\min_k \{ j(K_{3,k}, p_k^*) \} \quad (18)$$

with

$$p_k^* = \arg\min_p \{ j(K_{3,k}, p) \} \quad (19)$$

Due to the structure of the kinetics functions f_i, the parameter estimation problem is smooth and structurally identifiable. Of course, the quality of the estimates depends also on the experimental conditions chosen to identify all unknown parameters. It is intuitively clear that all concentrations must undergo a dynamic variation in order to allow a good estimation of their corresponding K-values. Consequently, steady state experiments would be inappropriate.

Example 2. To illustrate the effectiveness of the constraint search procedure, consider again the example of Section 3. The reference, i. e. the "true" system is characterized by the following kinetics:

$$f_1 = k_1 \frac{c_{S_1}}{c_{S_1} + K_1} \quad (20)$$

$$f_2 = k_2 \frac{c_{S_2}}{c_{S_2} + K_2} \frac{K_I}{c_P + K_I} \quad (21)$$

with the parameter values given in Table 1. Note that equations (20) and (21) match the general kinetics structure (17).

Table 1. Parameter values for "true" reference system.

parameter	value	
k_1	1	$= q_{max,1}$
k_2	2	$= q_{max,2}$
K_1	1	$= K_{11}^2$
K_2	2	$= K_{22}^2$
K_I	5	$= K_{23}^{-2}$

The culture is operated in perfusion mode, i.e. the biomass is retained in the bioreactor. The experimental function in the extracellular balance (1) then takes the following form:

$$F(c(t),t) = [c_{in} - c_{out}(c(t))] D(t) \quad (22)$$

with $c_{out}^T(c) = \begin{bmatrix} c_{S_1} & c_{S_2} & c_P & 0 \end{bmatrix}$.

The objective is to assess the correct constraint matrix K_3 defined in (14) among the $N = 6$ candidate schemes from the experiment with the reference scheme. To this end, the reference system is simulated numerically with the initial value solver routine ode15s in MATLAB® with outputs every time step Δt. The experimental parameters and functions are listed in Table 2. Measurement noise with a relative standard deviation σ_{rel} is added to the results. This data set is used to compute a maximum-likelihood estimate of the unknown model parameters. Note that the initial conditions c_0^T of the differential equation system (1) are also unknown because of the measurement error and have to be estimated. Thus, 14 unknowns are determined from 124 experimental data values, i.e. the redundancy factor of the estimation problem is 8.9.

Table 2. Experimental conditions for model identification.

experimental parameter	value
experiment duration	$t_{end} = 30$
sampling interval	$\Delta t = 1$
perfusion rate	$D(t) = 1 - \left(\frac{2t}{t_{end}} - 1\right)^2$
initial state	$c_0^T = \begin{bmatrix} 1 & 10 & 0.1 & 0.1 \end{bmatrix}$
feed concentrations	$c_{in}^T = \begin{bmatrix} 10 & 1 & 0 & 0 \end{bmatrix}$
relative measurement error	$\sigma_{rel} = 1\%$

The initial estimates for the parameter estimation are chosen as follows:

- $\hat{c}_0 = y_0$
- $\hat{k}_i = 1$
- $\hat{K}_{ij} = \begin{cases} 1 & \text{if } \{K\}_{ij} < 0 \\ 0.1 & \text{otherwise} \end{cases}$

The evaluation of the minimum cost for all candidate schemes leads to the ranking given in Table 3. Moreover, Table 4 computes the estimated kinetic parameter values corresponding to the best scheme with those of the reference model. From these tables and Figure 1, it is apparent that the systematic search procedure provides good estimates of the exact pseudo-stoichiometry and kinetics. An important underlying assumption in this illustrative example is the knowledge of the kinetic structure (17). A careful analysis of the kinetic structure is required in real case studies.

Table 3. Ranking of candidate schemes .

rank	flux combination	cost
"true"	{1,2}	259.7
1	{1,2}	264.1
2	{2,3}	578.6
3	{2,4}	1276.4
4	{3,4}	83865.4
5	{1,4}	126889.5
6	{1,3}	$\begin{bmatrix} K_2^T & K_3^T \end{bmatrix}$ singular

Table 4. Kinetic parameter values for the "best" reaction scheme (flux combination {1,2}).

parameter	identified value	"true" value
$q_{max,1}$	1.00	1.00
$q_{max,2}$	1.92	2.00
K_{11}	0.96	1.00
K_{12}	−0.05	0.00
K_{13}	−0.04	0.00
K_{14}	−0.03	0.00
K_{21}	0.00	0.00
K_{22}	1.36	1.41
K_{23}	−0.45	−0.45
K_{24}	0.00	0.00

5. SYSTEMATIC IDENTIFICATION OF GLOBAL REACTION SCHEMES

Bogaerts and Vande Wouwer (2001) have developed a systematic procedure to generate and evaluate all the macroscopic reaction schemes, which possess the C-identifiability property (Chen and Bastin, 1996), i.e. the property that the pseudo-stoichiometry can be identified univoquely independently of the kinetics. A necessary condition of C-identifiability is that the number of global reactions n_f must be smaller than the number of extracellular components n_c. In addition, C-identifiable schemes involve at least $n_f < n_c$ components, which are either occurring as reactant only or as product only (in some sense, these components are the evidence or testimony of each reaction).

Consider the following special case of the general state-space model (1) in partitioned form (Bastin and Dochain, 1990):

$$\frac{d}{dt}\begin{bmatrix} c_a(t) \\ c_b(t) \end{bmatrix} = \begin{bmatrix} K_a \\ K_b \end{bmatrix} r(c_a(t), c_b(t)) - \begin{bmatrix} c_a(t) \\ c_b(t) \end{bmatrix} D(t) + \begin{bmatrix} u_a(t) \\ u_b(t) \end{bmatrix}, \quad (23)$$

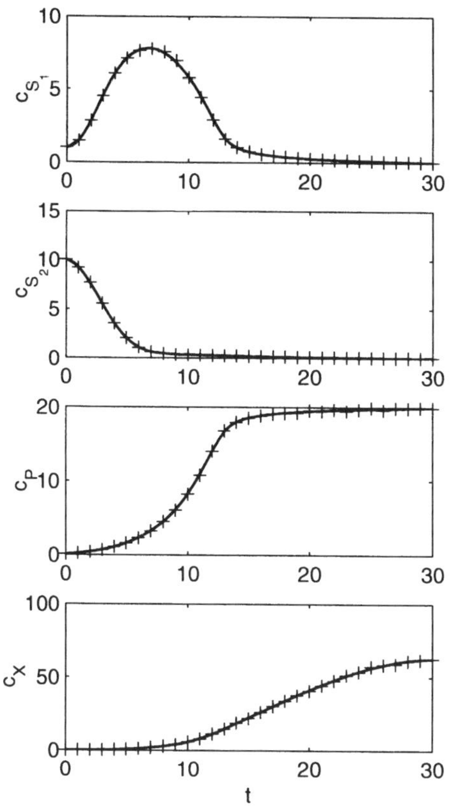

Fig. 1. "Experimental" data (+) from reference model in comparison with identified general model (–).

where $c_a \in \mathbb{R}^{n_K}$, $c_b \in \mathbb{R}^{n_c - n_K}$ and $r \in \mathbb{R}^{n_f}$ with $n_K = \text{rank}\{K\} \leq n_r$ and $K^T = [K_a^T \; K_b^T]$.

If rank$\{K_a\} = n_K > 0$, then there exists a transformation (Bastin and Dochain, 1990)

$$z = Cc_a + c_b \tag{24}$$

with the transformation matrix C uniquely defined by

$$CK_a + K_b = 0, \tag{25}$$

such that the reduced transformed system in $z \in \mathbb{R}^{n_c - n_K}$ is written:

$$\frac{dz(t)}{dt} = -z(t)D(t) + Cu_a(t) + u_b(t), \tag{26}$$

which is independent of the reaction rate vector $r(c)$.

Let κ_i be a vector containing the unknown elements in the ith column of the matrix K. Then κ_i is C-identifiable, if and only if there exists at least one partition $K_a \in \mathbb{R}^{n_K \times n_f}$ with rank$\{K_a\} = n_K$, which does not contain any element of κ_i (Chen and Bastin, 1996).

One partition fulfilling this condition is a diagonal matrix K_a containing the elements ± 1 at the positions corresponding to components known as being only produced or only consumed, respectively. Then K_b might contain all unknown stoichiometric coefficients $\kappa_i = \{K_b\}_{*,i}$. With (25) and a known K_a, the transformation matrix C thus contains the full information on K_b (Bogaerts and Vande Wouwer, 2001).

It is easily shown that, through another transformation (Bogaerts, 1999), the analytical solution of the transformed system (26) can be written in the linear form

$$Y(c_b, u_b, t) = \Theta^T(z(t_0), C)\, U(c_a, u_a, t) \tag{27}$$

with the matrix Θ^T containing the unknown parameters of the system: the uncertain initial condition $z(t_0)$ and the unknown matrix C defining the unknown stoichiometric coefficients of K_b through equation (25). Assuming known Gaussian white noise on the output with zero mean, the appropriate maximum-likelihood cost function for the determination of the unknown pseudo-stoichiometry is written:

$$j(\Theta) = \sum_{k=0}^{n_t} \left[Y_k - \Theta^T U_k\right]^T \cdot \left[Q_{Y,k} + \Theta^T Q_{U,k} \Theta\right]^{-1} \left[Y_k - \Theta^T U_k\right] \tag{28}$$

with the maximum index of the output sampling instants, n_t, and Q_Y, Q_U being the covariance matrices of the measurement errors on Y and U, respectively. The most likely pseudo-stoichiometry is characterized by the stoichiometric coefficients given by the estimated matrix C:

$$\hat{K}_b = -\hat{C}(\hat{\Theta})\, K_a \tag{29}$$

with

$$\hat{\Theta} = \arg\min_{\Theta} \{j(\Theta)\}. \tag{30}$$

In order to find the most likely among all C-identifiable reaction schemes, all permutations have to be evaluated and compared with respect to their cost function values. The number of candidate schemes is given by

$$N = \binom{n_c}{n_K} \tag{31}$$

with the number of components $n_K < n_c$, which are certain to be consumed or produced in each reaction.

Example 3. Consider again the example introduced in the previous sections with the "true" global reference reaction system given by (15). Assume that the substrates S_1 and S_2 are known to be always consumed and the biomass X to be always produced. Furthermore, assume that there is no a priori knowledge about the product P with this respect. Consequently, there are $N = 3$ possible partitions K_a with $n_K = 3$ and therefore three candidate schemes to be evaluated and compared. In contrast with example 2, no assumption on the structure of the kinetics has to be made since the transformed system (26) is decoupled from the reaction kinetics r.

Table 5 shows all the C-identifiable candidate reaction schemes and their ranking according to the (quasi-) maximum-likelihood cost function.

It is apparent that the reaction scheme with the two substrates $\{S_1, S_2\}$ as reference components is most likely. Their consumption rates are therefore equivalent with the respective reaction rates. The other two combinations with the pairs $\{S_1, X\}$ and $\{S_2, X\}$ of

Table 5. Ranking of C-identifiable reaction scheme for the example problem.

rank	reaction scheme	cost
1	$S_1 \longrightarrow 2.00P + 0.39X$	
	$S_2 + 0.08P \longrightarrow 0.08X$	20.02
2	$S_2 + 0.40P \longrightarrow -$	
	$2.43S_1 \longrightarrow X + 4.94P$	39.99
3	$S_1 \longrightarrow 2.00P$	
	$1.07S_2 + 0.09P \longrightarrow X$	1821.23

reference components as representatives of the considered two reactions have much higher costs.

The stoichiometric coefficients are very close to those given in Section 3. The observed differences are due to the measurement errors and the "experimental" conditions resulting in non-ideal sensitivities of the cost function with respect to the unknown parameters.

6. CONCLUSIONS

The results in Section 3 legitimate the use of unstructured models for the description of complex metabolic biosystems, like mammalian cell cultures, under the conditions that

- the biomass growth is slow,
- the intracellular kinetics are so fast that the dynamics of the intracellular concentrations can be neglected compared to extracellular dynamics.

The metabolic reaction system is equivalent to an unstructured representation of the cell behavior. The reduced reaction system dimension depends on the degree of freedom left to the system and is equal to the difference of the metabolic reaction system dimension and the number of balancing metabolites assumed in quasi-steady-state.

One possible method for systematically using these degrees of freedom in the search for the reaction kinetics is proposed in this contribution. At this stage, a general formulation of the specific reaction rates featuring modulation effects (from limitation to inhibition) is introduced,which keeps the number of structural model parameters low and offers good identifiability properties.

The fact that metabolic reaction schemes can be reduced to equivalent macroscopic reaction relating extracellular components only legitimates the use of unstructured models. It is therefore recommended to use the systematic identification of the most likely C-identifiable reaction scheme (Bogaerts and Vande Wouwer, 2001) described in Section 5. C-identifiable schemes can be represented in a transformed reduced-order system and have the properties of good practical identifiability.

Note that a global scheme as obtained with the metabolic model reduction technique is not necessarily C-identifiable, because the number of considered reactions could be equal to or higher than the number of external components.

Note also that from the knowledge of the global stoichiometry (systematically obtained as described in Section 5) the constraint matrix K_3 (see Section 3) cannot be determined uniquely, because this matrix contains more elements than the known elements in K provided by the identification of the C-matrix. Some additional assumptions have therefore to be made in order to deduce uniquely the unknown conditions K_3 from K with an underlying metabolic pathway network. Current research entails theoretical investigations of such a deduction and will be applied to a real biotechnological process.

ACKNOWLEDGEMENTS

This work is partly supported by the company *4C Biotech s.a.*, Seneffe (Belgium).

REFERENCES

Bastin, G. and D. Dochain (1990). *On-Line Estimation and Adaptive Control of Bioreactors*. Vol. 1 of *Process Measurement and Control*. Elsevier. Amsterdam.

Bogaerts, P. (1999). Contribution À la Modélisation Mathématique Pour la Simulation et l'Observation D'États Des Bioprocédés. PhD thesis. Université Libre de Bruxelles. Brussels.

Bogaerts, P. and A. Vande Wouwer (2001). Systematic generation of identifiable macroscopic reaction schemes. In: *8th International Conference on Computer Applications in Biotechnolgy* (D. Dochain and M. Perrier, Eds.). IFAC. Québec City. pp. 13–18.

Bonarius, H. P. J., V. Hatzimanikatis, K. P. H. Meesters, C. D. de Gooijer, G. Schmid and J. Tramper (1996). Metabolic flux analysis of hybridoma cells in different culture media using mass balances. *Biotechnol. Bioeng.* **50**(3), 299–318.

Chen, L. and G. Bastin (1996). Structural identifiability of the yield coefficients in bioprocess models when the reaction rates are unknown. *Math. Biosci.* **132**, 35–67.

Haag, J. E., A. Vande Wouwer and M. Remy (2002). A systematic model reduction procedure for complex bioprocesses described by metabolic pathway networks. In: *15th IFAC World Congress b'02*. IFAC. Barcelona.

Zupke, C. and G. Stephanopoulos (1995). Intracellular flux analysis in hybridomas using mass balances and in vitro ^{13}C NMR. *Biotechnol. Bioeng.* **45**(4), 292–303.

IFAC Publications
www.elsevier.com/locate/ifac

FLOW CONTROLLED NON-INVASIVE VENTILATION CONSIDERING MASK LEAKAGE AND SPONTANEOUS BREATHING

Florian Dietz * Axel Schloßer * Dirk Abel *

* *Institute of Automatic Control, Aachen University, Aachen, Germany*

Abstract: This contribution introduces a procedure for volume-controlled non-invasive mechanical ventilation under consideration of leakages and spontaneous breathing. The identification and flow feedback control implemented were tested with a nonlinear simulation model of a single-hose home-care ventilator with sensors inside the device. Measured values of clinical tests were used to examine the influence of spontaneous breathing on the identification. The suggested method permits a compensation of the leakage and at the same time a classification of the breathing effort of the mechanically ventilated patients for the first time. Since only data of the inspiration interval is needed, the suggested procedure applies even to simple and economical ventilators common in home care. As a consequence sophisticated monitoring, adaptation of the mechanical ventilation level according to the patient's needs and better acceptance by the patient are possible, since the procedure is accomplished without interrupting the routine mechanical ventilation.

Keywords: nonlinear, parameter estimation, bioengineering and medical systems, flow control, leakage, unobservable

1. INTRODUCTION

For many years non-invasive positive pressure mechanical ventilation (NIV) used in home care is successfully applied by nose or face mask. Also in intensive-care medicine the NIV is more and more applicable in order to prevent e.g. a higher morbidity rate at pneumonia due to the invasive employment of an endotracheal tube or tracheal cannula (Dittmann, M., 1993).

Usually the inspiration under non-invasive ventilation is pressure controlled, i.e. no defined tidal volume (V_t) is specified. Volume-controlled or volume-targeted non-invasive ventilation with adjustable tidal volume though has yielded better results in some studies in comparison to pressure-controlled NIV. However, the advantages are not proven yet (Schönhofer, B.; Sortor-Leger, S., 2002). But the volume-controlled NIV is so far hardly applicable, since some basic problems prevent its reliable implementation. Losses of tidal volume up to 50% and more caused by leakages between ventilation mask and patient's face lead to an unpredictable tidal volume (Lofaso, F.; Fodil, R.; Lorino, H.; Leroux, K.; Quintel, A.; Leroy, A.; Harf, A., 2000). Fig. 1 shows schematically a home-care ventilator with single hose and nose mask as interface for mechanical ventilation. Between ventilation mask and patient's face time-variant leakages ($\dot{V}_{leakage}$) can appear in relevant magnitude, even with carefully selected masks. With sleeping patients a leakage can also result from an opened mouth when using nose masks.

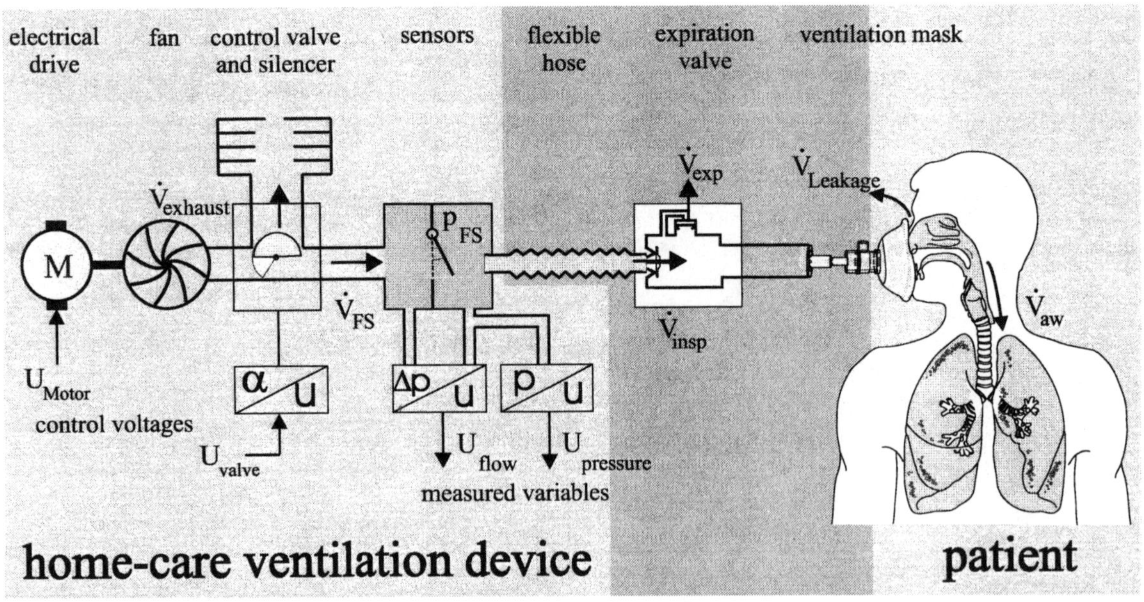

Fig. 1. Principle sketch of a home-care ventilation device and a patient's lung

The identification of lung parameters and leakage is possible in case of completely passive patients (no spontaneous breathing, e.g. due to sedation and relaxation) even for nonlinear interrelation (Dietz, F.; Rake, H., 2001; Dietz, F.; Schloßer, A., 2002; Schüttler, F., 1998). A dynamic compensation of the volume loss can be accomplished by means of the identified parameters. This can be done even with very simple ventilation devices. The arrangement of the sensors for pressure (p_{FS}) and flow ($\dot{V}_{FS}$) near or inside the ventilation device and the active principle of the expiration valve lead to a shortened identification window. Only data of the inspiration interval ($\dot{V}_{insp}$) can be acquired. During the expiration the patient breathes directly to the environment through a blow valve ($\dot{V}_{exp}$), a check valve secludes the ventilation device from the patient. So, the sensor technology is sort of blind during expiration.

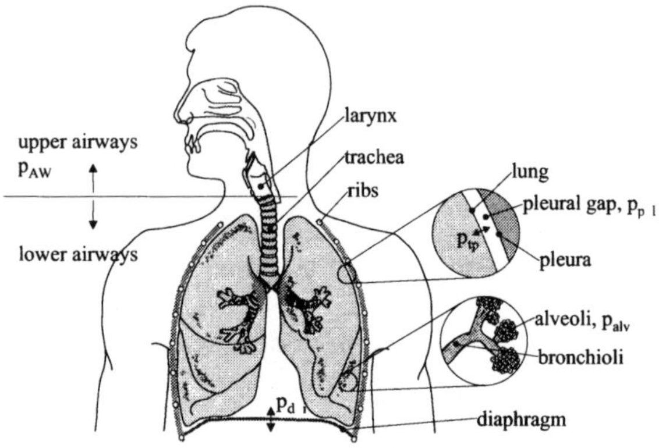

Fig. 2. Respiratory system of a human being with the pressure of the upper airways p_{aw}, the alveolar pressure p_{alv} and the pleural pressure p_{pl} as well as the pressure differences p_{tp} and p_{di}

In practice the breathing effort of the patient must be considered, since the patients usually are conscious and spontaneously breath cooperatively or contrary to the ventilation targets. A valid value for the breathing effort is the transpulmonary pressure p_{tp} that corresponds to the pressure difference between the alveolary and pleural space ($p_{tp} = p_{alv} - p_{pl}$, see fig. 2). The natural drive of the ventilation mainly takes place via the diaphragm.

Since the alveolar and the pleural pressure are not measurable at justifiable expenses, the medicine uses the transdiaphragmatic pressure p_{tp} as equivalent parameter to p_{di}. The transdiaphragmatic pressure represents the difference between gastric pressure p_{ga} inside the stomach and the esophageal pressure p_{es} inside the esophagus ($p_{di} = p_{ga} - p_{es}$) (Haberthür, C.; Guttmann, J.; Osswald, P.; Schweitzer, M., 2001). The measurement of p_{ga} and p_{es} is possible using a catheter system with two pressure sensors fixed on. The first sensor is attached to the tip of the catheter and will be positioned inside the stomach, the second is mounted about 0.2 m proximal to the tip and is positioned inside the esophagus. Still the utilisation of the catheter is an invasive engagement, which interferes with food intake and the non-invasive idea of NIV. Therefore this technique is not available for home-care ventilation. Thus a method to take consideration of the spontaneous breathing into account has to get along without direct measurement of physiological values. The verification of implemented methods and the acquisition of test data however profit meaningfully by the measurement of the transdiaphragmatic pressure p_{di}.

2. IDENTIFICATION

An equivalent circuit diagram of the respiratory system including leakages (fig. 3) shows that the patient's diaphragm activity p_{mus} adds to the airways pressure p_{aw} and thus to the air flow to the lung ($\dot{V}_{aw}$). Hence a leakage identification has also to take the diaphragm activity p_{mus} into account to compensate the leakage flow $\dot{V}_{leakage}$ sufficiently. In such a way the desired tidal volume is guaranteed during non-invasive ventilation. Additionally the knowledge of the breathing effort permits an evaluation of the mechanical ventilation quality. Namely one can conclude e.g. from a synchronous cooperative spontaneous breathing on a good acceptance of the mechanical ventilation by the patient.

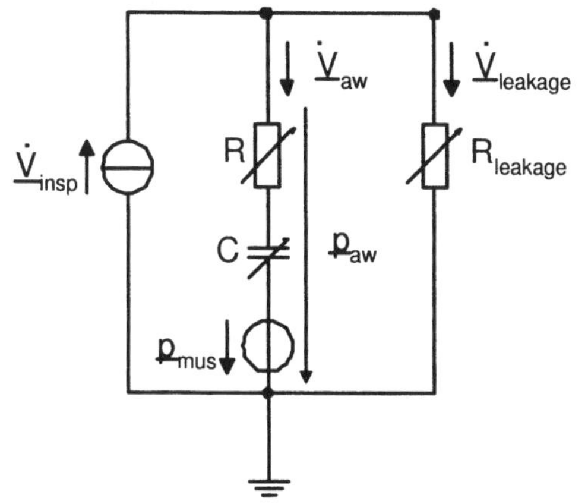

Fig. 3. Equivalent circuit diagram of lung (R, C), leakages ($R_{leakage}$) and spontaneous ventilation (p_{mus}) with excitation flow $\dot{V}_{insp}$

Usual procedures for parametric identification e.g. regression analysis by least squares fitting lead to wrong estimated values of the lung parameters resistance R and compliance C (lung elasticity) and the leakage resistance $R_{leakage}$ due to the influence of the spontaneous breathing. If the patient breathes synchronously respectively cooperative with the ventilation device, the compliance C is overrated by up to two orders of magnitude. Lung and leakage resistances however are underestimated or even become negative. If the patient breathes against the equipment's excitation then C and $R_{leakage}$ are underestimated while R is quite well estimated.

Following the linear and nonlinear identification of the lung parameters R, C and the leakage resistance $R_{leakage}$ an approach to consider the spontaneous breathing p_{mus} is sought after. Because of the arbitrary time-variant dynamic behaviour of the diaphragm activity p_{mus} its contribution to the ventilation cannot be calculated in terms of constant parameters as it is done with the lung parameters.

A basic approach for the determination of lung parameters under spontaneous breathing (without consideration of leakages) has been presented in detail for the usage of pressure-controlled mechanical ventilation. The suggested procedure is realised by alternately changing pressure levels in successive breaths (Navajas, D.; Alcaraz, J.; Peslin, R.; Roca, J.; Farré, R., 2000). On the assumption that the breathing effort of the patient between the successive inspirations changes only insignificantly, the patient's fraction of the total ventilation work can be eliminated by calculating the difference of the consecutive time series. This approach was modified for the volume-controlled non-invasive ventilation and extended to an identification of the leakages. The following interrelation between the airways pressure p_{aw} and the inspiratory flow $\dot{V}_{insp}$ is valid in laplacian domain:

$$p_{aw}(s) = \left[p_{mus}(s) + \dot{V}_{insp}(s)\frac{1+sRC}{sC}\right] \cdot \frac{sR_{leakage}C}{1+s(R+R_{leakage})C} \quad (1)$$

The differences of airways pressure Δp_{aw} and inspiratory flow $\Delta\dot{V}_{insp}$ are calculated by

$$\Delta p_{aw}(s) = p_{aw,k}(s) - p_{aw,k-1}(s), \quad (2)$$

$$\Delta\dot{V}_{insp}(s) = \dot{V}_{insp,k}(s) - \dot{V}_{insp,k-1}(s), \quad (3)$$

with

$k \equiv$ *serial number of breath.*

Assuming that the breathing effort has an equal dynamic course in both consecutive breaths according to

$$p_{mus,k}(s) = p_{mus,k-1}(s) \quad (4)$$

one gets a more simple equation for the interrelation between pressure difference Δp_{aw} and flow difference $\Delta\dot{V}_{insp}$ in laplacian domain by inserting eq. (1) in eq. (2) and using eqs. (3) and (4):

$$\Delta p_{aw}(s) = \Delta\dot{V}_{insp}(s)\frac{(1+sRC)R_{leakage}}{1+s(R+R_{leakage})C} \quad (5)$$

3. IDENTIFICATION RESULTS

Different levels of flow differences $\Delta\dot{V}_{insp}$ (10%, 20% and 30%) were applied to a nonlinear computer simulation of a respiratory system. The identification results with the new procedure were compared with identification results without considering spontaneous breathing. First only linear behaviour for the lung parameters and the leakage are considered. In tab. 1 and tab. 2 results

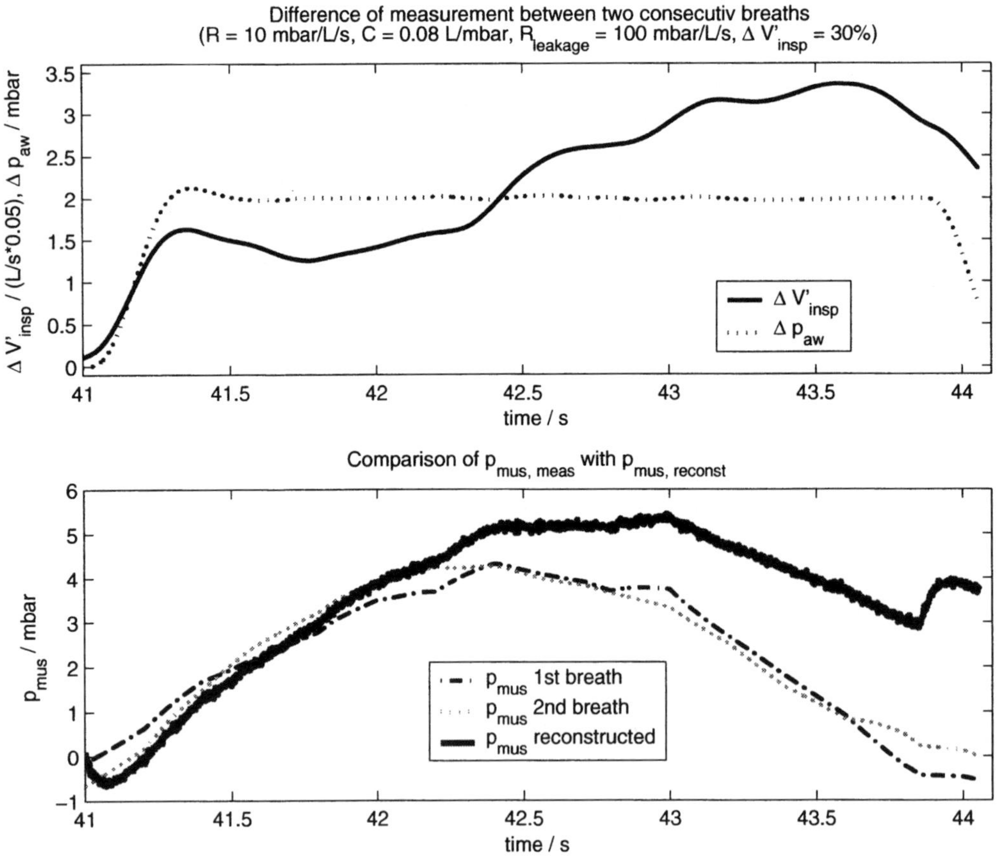

Fig. 4. Parametric identification of R, C, $R_{leakage}$ and p_{mus} as well as reconstruction of the time behaviour of the spontaneous breathing effort with the help of pressure and flow differences (Δp_{aw} and $\Delta \dot{V}_{insp}$) between two consecutive breaths with an inspiration flow difference ΔQ_{insp} of 30%

of identifications for flow differences of 30% are indicated.

reference values	cooperation	counterbreath
R (10 mbar/L/s)	0.64 (9.5)	11.5 (0.6)
C (0.08 L/mbar)	2.6 (2.7)	0.02 (0.01)
$R_{leakage}$ (100 mbar/L/s)	-1.9 (14.1)	36 (0.0)
variance of identification	9.2e-6 (4e-6)	1.1e-5 (3e-6)

Table 1. Identification **without** consideration of the spontaneous respiration. Average values of several identifications (standard deviations in parentheses) are indicated

reference values	cooperation	counterbreath
R (10 mbar/L/s)	9.2 (2.5)	9.1 (2.7)
C (0.08 L/mbar)	0.048 (0.11)	0.049 (0.11)
$R_{leakage}$ (100 mbar/L/s)	43.6 (3.8)	43.1 (4.0)
variance of identification	1.9e-5 (7e-6)	1.9e-5 (7e-6)

Table 2. Identification **with** consideration of the spontaneous respiration. Average values of several identifications (standard deviations in parentheses) are indicated

Looking at the average values and standard deviations in table 1 and table 2 one can see that with patient's cooperation clearly better results can be obtained particularly by considering spontaneous breathing. With breathing against the ventilator's excitation advantages show up with consideration of the spontaneous breathing in particular for parameter C.

The lower part of fig. 4 shows an example of the time behaviour of the spontaneous breathing in the case of a patient's breath against the device's excitation. Two different time behaviours of p_{mus} for consecutive breaths are given. In the upper part of fig. 4 the time behaviour of the computed differences for pressure Δp_{aw} and flow $\Delta \dot{V}_{insp}$ (twentyfold increased) can be seen. The regression analysis of the discrete time series of Δp_{aw} and $\Delta \dot{V}_{insp}$ delivers the appropriate estimated parameters $\tilde{R}$, $\tilde{C}$ and $\tilde{R}_{leakage}$. As a result of the identification the parameters are used for the compensation of the leakage flow and to reconstruct of the spontaneous breathing time behaviour. Using the absolute data of pressure and flow of the last inspiration and the estimated parameters the time behaviour for the diaphragm activity (p_{mus}) can be reconstructed ($\tilde{p}_{mus}$). The lower part of fig. 4 shows the result of the reconstruction, too.

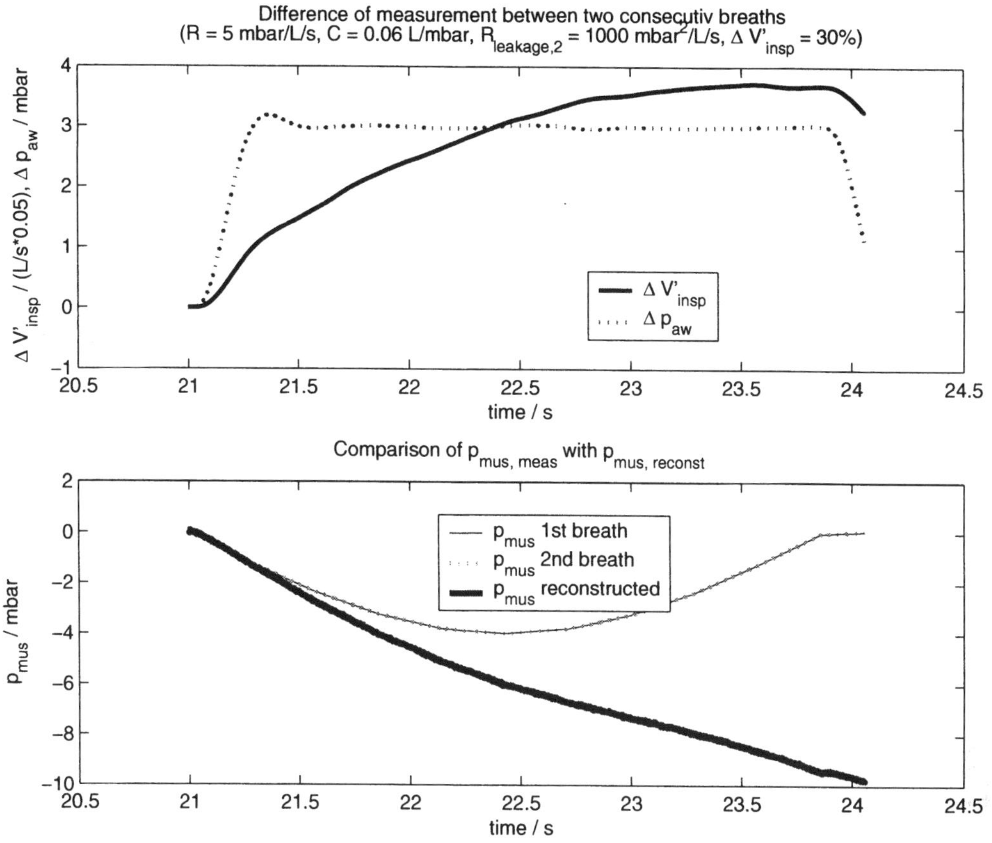

Fig. 5. Parametric identification of R, C, $R_{leakage,2}$ and p_{mus} as well as reconstruction of the time behaviour of the spontaneous breathing effort with the help of pressure and flow differences (Δp_{aw} and $\Delta \dot{V}_{insp}$) between two consecutive breaths with an inspiration flow difference ΔQ_{insp} of 30%

This reconstruction permits a judgement of the spontaneous breathing. In the simplest case it can be stated whether the patient cooperates with the mechanical ventilation or not by integrating the reconstructed inspiratory values of $\tilde{p}_{mus}$. More complex procedures for a classification of the spontaneous breathing can derive a more precise evaluation.

As a result of the investigations it can be noted that an identification of the lung parameters and leakage with good synchronisation between patient and ventilator functions well and permits a classification of the spontaneous breathing. With similar behaviour of the spontaneous breathing, but poor synchronisation at least the identification of the leakage is very robust. If the identification results with and without consideration of the spontaneous breathing are compared individually, the variance of the identification results is a reliable reference to the quality of the identification results with and without consideration of p_{mus}.

With nonlinear mask leakages e.g. quadratic resistance $R_{leakage,2}$ the results of an identification are kind of ambiguous. For regressive analysis additionally the signal $p_{aw,k}^2 - p_{aw,k-1}^2$ is taken into account. The results are shown in fig. 5 with ideally equal breathing effort. As can be seen in the reconstructed time behaviour of p_{mus} the nonlinear resistance interfere with the breathing effort.

With regard to the implementation each breath cycle is partitioned in three sections (fig. 6). During the inspiration phase an identification without consideration of the spontaneous breathing is accomplished by means of recursive regression analysis. During the expiration an identification with consideration of the spontaneous breathing is calculated with stored data of the current and last inspiration. If the second identification result is better, i.e. the result has a lower variance, then additionally a reconstruction of the spontaneous breathing time behaviour of the preceding inspiration is calculated and a classification of the patient's breathing effort is made.

4. FLOW CONTROL

After an examination of the identified parameters on plausibility an additional reference variable input is calculated using $R_{leakage}$ and then added to $\dot{V}_{insp}$ to realise a leakage compensation for the next inspiration (see fig. 7). Yet another additional reference variable input is made of $\pm\Delta\dot{V}_{insp}/2$, with sign change after each breath.

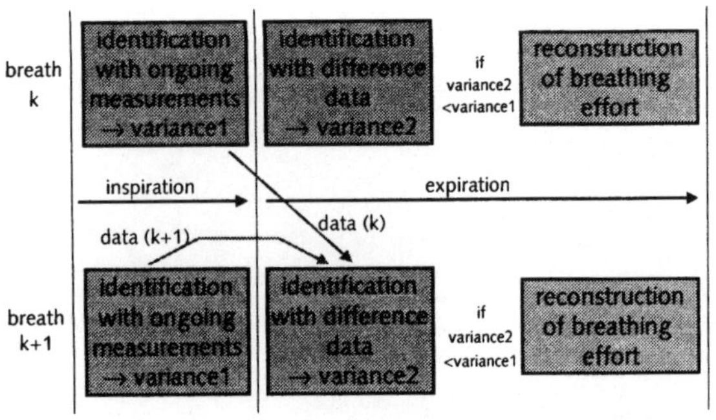

Fig. 6. Three stage implementation of the online identification of respiratory parameters and leakage

So the desired and physiologically important tidal minute volume can be kept at the average value. Lung parameters R and C are used to optimise an adaptive or model-based predictive control. During flow control the valid pressure limits are constantly supervised for security reasons. By reaching or exceeding these limits a pressure control at the pressure limit is activated instantly in order to avoid any physical harm for patients.

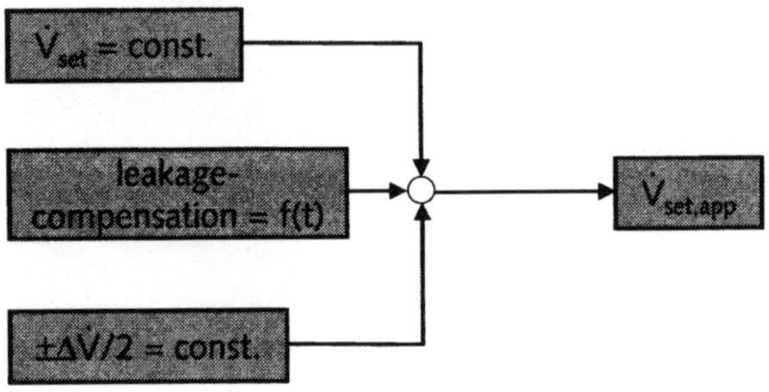

Fig. 7. Additional reference variable input for leakage compensation and alternating excitation

5. CONCLUSION

For the non-invasive mechanical ventilation (NIV) a procedure was presented which provides an online leakage estimation and compensation in spite of non-negligible spontaneous breathing effort. Thus a reliable volume-controlled or volume-targeted NIV can be established. The procedure is suitable also for common home-care ventilation devices, which exhibit a simple sensor technology within the equipment and permit data acquisition during the inspiration only because of the usage of single hose systems and expiration valves. Beside leakage compensation a classification of the spontaneous breathing effort is additionally possible under certain conditions. Thus the procedure offers large progress and new possibilities for NIV regarding volume-controlled or volume-targeted mechanical ventilation and patient monitoring also for the intensive care medicine.

REFERENCES

Dietz, F.; Rake, H. (2001). Leckagekompensation und Flowregelung bei der Heimbeatmung. In: *Automatisierungstechnik im Spannungsfeld neuer Technologien* (G. Gerlach, U. Jumar and K.-H. Lachmann, Eds.). VDI Berichte 1608. Baden-Baden, Germany. pp. 159–166. ISBN 3-18-091608-7.

Dietz, F.; Schloßer, A. (2002). Heimbeatmung: Herausforderung volumenkontrollierte Beatmung. In: *Simulationstechnik ASIM 2002* (D. Tavangarian and R. Grützner, Eds.). SCS–The Society for Modeling and Simulation International. Rostock, Germany. pp. 215–220. ISBN 3-936150-19-2.

Dittmann, M. (1993). *Respiratoren in der klinischen Praxis.* Springer. Berlin. ISBN 3-540-55929-9.

Haberthür, C.; Guttmann, J.; Osswald, P.; Schweitzer, M. (2001). *Beatmungskurven - Kursbuch und Atlas.* Springer. Berlin. ISBN 3-540-67830-1.

Lofaso, F.; Fodil, R.; Lorino, H.; Leroux, K.; Quintel, A.; Leroy, A.; Harf, A. (2000). Inaccuracy of tidal volume delivered by home mechanical ventilators. *European Respiratory Journal* **15**, 338–341. ISSN 0903-1936.

Navajas, D.; Alcaraz, J.; Peslin, R.; Roca, J.; Farré, R. (2000). Evaluation of a method for assessing respiratory mechanics during noninvasive ventilation. *European Respiratory Journal* **16**, 704–709. ISSN 0903-1936.

Schönhofer, B.; Sortor-Leger, S. (2002). Equipment needs for noninvasive mechanical ventilation. *European Respiratory Journal, 20* pp. 1029–1036. ISSN 0903-1936.

Schüttler, F. (1998). Parameterschätzung atmungsmechanischer Modelle zur Vorhersage pharyngealer Obstruktionen bei Patienten mit schlafbezogenen Atmungsstörungen. Ph. d. dissertation. Philipps-Universität Marburg.

IFAC Publications
www.elsevier.com/locate/ifac

ESTIMATION AND IDENTIFICATION OF NON-STATIONARY FUNCTIONAL SERIES TARMA MODELS ⋆

Aggelos G. Poulimenos *and* **Spilios D. Fassois**

Stochastic Mechanical Systems (SMS) Group
Department of Mechanical & Aeronautical Engineering
University of Patras, GR 265 00 Patras, Greece
E-mail: {poulimen, fassois}@mech.upatras.gr
Internet: http://www.mech.upatras.gr/~sms

Abstract: A complete estimation and identification method for the modeling of non-stationary signals via Functional Series Time-dependent ARMA (FS-TARMA) models is developed. Model estimation is based upon the generalization of an isomorphic matrix algebra type method to the case of "incomplete" functional subspaces. Model identification is based upon a two-phase optimization scheme that utilizes genetic algorithms and backward regression. The effectiveness of the complete method is demonstrated via a Monte Carlo study.

Keywords: Non-stationary signals, time-varying systems, ARMA parameter estimation, identification.

1. INTRODUCTION

Non-stationary stochastic signals, that is signals with time-dependent characteristics, are frequently encountered in engineering, and have been receiving increasing attention in recent years (Niedzwiecki 2000, Kitagawa and Gersh 1996). Typical examples include signals relating to seismic motion, types of structural vibration, speech, automotive and aircraft systems, rotating machinery, and so on. From a physical standpoint non-stationarity stems from either time-dependent dynamics and transient phenomena and/or inherently non-linear behavior.

Parametric methods for the modeling and analysis of non-stationary stochastic signals complement their non-parametric counterparts (Cohen 1995) and offer advantages such as representation parsimony, improved accuracy, resolution, and tracking (Petsounis and Fassois 2000). A notable class of such methods includes *Functional Series Time-dependent ARMA (FS-TARMA) methods*, which postulate deterministic parameter evolution within specific functional subspaces (Ben Mrad *et al.* 1998*a*, Petsounis and Fassois 2000). These are particularly attractive over alternative *adaptive methods* [unstructured parameter evolution (Ljung 1999)] or *stochastic parameter evolution methods* (Kitagawa and Gersh 1996), as: (a) The imposition of deterministic (maximum) structure on parameter evolution is often motivated by the physics of the problem; (b) leads to high parsimony; (c) capability of tracking "fast" or "slow" variations; and, (d) high accuracy and resolution.

Functional Series methods may be classified as either pure TAR or full TARMA; the latter being more general and allowing for the accurate representation of both resonances and antiresonances in the signal's time-dependent power spectral density (Petsounis and Fassois 2000). Yet, the estimation of FS-TARMA models is more difficult and has attracted limited attention (Grenier 1989). A promising Polynomial-Algebraic (P-A) estimation method was recently introduced by Ben Mrad *et al.* (1998*a*), and further developed (within an isomorphic matrix algebra context) by Fouskitakis and Fassois (2001). Based upon it, non-stationary Functional Series TARMA modeling has

⋆ Research sponsored by the VolkswagenStiftung.

been, thus far, successfully applied to automotive systems (Ben Mrad *et al.* 1998*b*), mechanical vibration (Petsounis and Fassois 2000, Poulimenos and Fassois 2003*b*), mechanical reliability analysis (Stavropoulos and Fassois 2000), and earthquake ground motion modeling (Fouskitakis and Fassois 2002).

The goal of the present study is two-fold: *(i)* The generalization of the model (parameter) estimation method to the case of *"incomplete"* (not including all consecutive functions up to the subspace dimensionality) polynomial functional subspaces, and *(ii)* the formulation of an effective model identification (model structure estimation) method. The importance of both objectives is obvious: Functional subspaces may be often "incomplete" in practice, while the rich structure of FS-TARMA models renders model identification critical. The resulting, complete, model estimation and identification method is expected to significantly facilitate FS-TARMA based non-stationary signal modeling.

2. FUNCTIONAL SERIES TARMA MODELS

Functional Series TARMA (FS-TARMA) models constitute conceptual extensions of their conventional, stationary, ARMA counterparts (Ljung 1999), with parameters belonging to functional subspaces spanned by specific time functions *(basis functions)*. A TARMA$(n_a, n_c)_{[p_a,p_c]}$ model, with n_a, n_c designating its AR, MA orders and p_a, p_c its AR, MA subspace dimensionalities, thus is of the form:

$$\underbrace{\left[1+\sum_{i=1}^{n_a} a_i[t]\cdot B^i\right]}_{A[B,t]}\cdot x[t] = \underbrace{\left[1+\sum_{i=1}^{n_c} c_i[t]\cdot B^i\right]}_{C[B,t]}\cdot w[t] \quad (1)$$

$$a_i[t] = \sum_{j=1}^{p_a} a_{i,j}\cdot G_{b_a(j)}[t], \quad c_i[t] = \sum_{j=1}^{p_c} c_{i,j}\cdot G_{b_c(j)}[t] \quad (2)$$

with t designating normalized discrete time, B the backshift operator ($B^i \cdot x[t] \triangleq x[t-i]$), $x[t]$ the non-stationary signal modelled, $w[t]$ an innovations (uncorrelated) sequence with zero mean and variance σ_w^2, and $A[B,t]$, $C[B,t]$ the time-varying AR and MA polynomial operators respectively.

$G_{b_a(j)}[t]$, $G_{b_c(j)}[t]$ designate the j-th function in the AR, MA functional subspaces, respectively, and $a_{i,j}$, $c_{i,j}$ the corresponding *coefficients of projection*. The vectors [1] $\boldsymbol{b}_a$ and $\boldsymbol{b}_c$:

$$\boldsymbol{b}_a \triangleq [b_a(1) \ldots b_a(p_a)]^T, \quad \boldsymbol{b}_c \triangleq [b_c(1) \ldots b_c(p_c)]^T \quad (3)$$

contain the AR and MA, respectively, basis function *indices*, which indicate the specific functions (from a selected ordered set) that span each subspace. The AR and MA functional subspaces thus are:

$$\mathcal{F}_{AR} \triangleq \mathcal{F}_A \triangleq \{G_{b_a(1)}[t], \ldots, G_{b_a(p_a)}[t]\} \quad (4)$$

[1] Bold face lower/upper case symbols designate vector/matrix quantities, respectively.

$$\mathcal{F}_{MA} \triangleq \mathcal{F}_C \triangleq \{G_{b_c(1)}[t], \ldots, G_{b_c(p_c)}[t]\} \quad (5)$$

For convenience, the following maximum indices and order are defined:

$$\bar{b}_a \triangleq \max_i \{b_a(i)\} \qquad \bar{b}_c \triangleq \max_i \{b_c(i)\} \quad (6)$$

$$\bar{b} \triangleq \max(\bar{b}_a, \bar{b}_c) \qquad \bar{n} = \max(n_a, n_c) \quad (7)$$

3. THE INVERSE FUNCTION REPRESENTATION

The TARMA model of Equation (1) may be re-written in the *inverse function form*:

$$I[B,t]\cdot x[t] = w[t] \quad (8)$$

where the inverse function operator $I[B,t]$ (Ben Mrad *et al.* 1998*a*) is defined as:

$$I[B,t] \triangleq 1 + \sum_{q=1}^{\infty} i_q[t] B^q \triangleq C^{-1}[B,t] \circ A[B,t] \quad (9)$$

with $i_q[t]$ ($q = 1, 2, \ldots$) designating the model's time-varying inverse function and "$\circ$" the non-commutative multiplication operation of the time-varying polynomial operator algebra (Fouskitakis and Fassois 2001). The non-commutative multiplication operation satisfies the properties:

$$B^i \circ B^j = B^{i+j} \qquad B^i \circ d[t] = d[t-i]\cdot B^i \quad (10)$$

The functional subspace of the inverse function polynomial operator is now examined.

Definition 1. The functional subspace $\mathcal{F}_C$ of a time varying polynomial operator $C[B,t]$ is defined as the union of the functional subspaces of its parameters, that is:

$$\mathcal{F}_C = \bigcup_{0\le i\le n_c} \mathcal{F}_{c_i[t]} \quad (11)$$

∇

The following propositions then hold (Poulimenos and Fassois 2003*a*):

Proposition 1. Given a polynomial operator $C[B,t]$, the functional subspace of the r-th parameter $\kappa_r[t]$ of the inverse operator $C^{-1}[B,t] = K[B,t] = 1 + \sum_{r=1}^{\infty} \kappa_r[t]\cdot B^r$ is given as:

$$\mathcal{F}_{\kappa_r[t]} = \mathcal{F}_C \cup \left\{ \bigcup_{\substack{1\le \ell\le r-1 \\ 1\le i\le p_{\kappa_\ell} \\ 1\le j\le p_c}} \mathcal{F}_{G_{b_{\kappa_\ell}(i)}[t]\cdot G_{b_c(j)}[t-\ell]} \right\} \quad (12)$$

$$(r \le n_c)$$

$$\mathcal{F}_{\kappa_r[t]} = \bigcup_{\substack{\max(1,r-n_c)\le \ell\le r-1 \\ 1\le i\le p_{\kappa_\ell} \\ 1\le j\le p_c}} \mathcal{F}_{G_{b_{\kappa_\ell}(i)}[t]\cdot G_{b_c(j)}[t-\ell]} \quad (13)$$

$$(r > n_c)$$

with $b_{\kappa_r}(i)$ and p_{κ_r} designating the i-th basis function index and the functional subspace dimensionality, respectively, for $\kappa_r[t]$. ∇

Proposition 2. The functional subspace of the r-th inverse function parameter $i_r[t]$ is given as:

$$\mathcal{F}_{i_r[t]} = \\ = \mathcal{F}_A \cup \mathcal{F}_{\kappa_r[t]} \cup \left\{ \bigcup_{\substack{1\le \ell\le r-1 \\ 1\le i\le p_{\kappa_\ell} \\ 1\le j\le p_a}} \mathcal{F}_{G_{b_{\kappa_\ell}(i)}[t]\cdot G_{b_a(j)}[t-\ell]} \right\} \quad (14)$$

$$(r \le n_a)$$

$$= \mathcal{F}_{\kappa_r[t]} \cup \left\{ \bigcup_{\substack{\max(1, r-n_a) \leq \ell \leq r-1 \\ 1 \leq i \leq p_{\kappa_\ell} \\ 1 \leq j \leq p_a}} \mathcal{F}_{G_{b_{\kappa_\ell}(i)}[t] \cdot G_{b_a(j)}[t-\ell]} \right\} \quad (r > n_\alpha) \tag{15}$$

where $\mathcal{F}_{\kappa_r[t]}$ is given by Equations (12) and (13). ∇

In the case of *polynomial* functional subspaces, the previous results may be simplified. Toward this end the functional subspace $\mathcal{F}_{G_p[t] \cdot G_n[t-\ell]}$ of the product $G_p[t] \cdot G_n[t-\ell]$ is first examined:

Lemma 1. The functional subspace $\mathcal{F}_{G_p[t] \cdot G_n[t-\ell]}$ $(p, n \geq 0,\ \ell \geq 1)$ in the case of polynomial subspaces is given as:

$$\mathcal{F}_{G_p[t] \cdot G_n[t-\ell]} = \begin{cases} \{G_p[t]\} & \text{n} = 0 \\ \{G_0[t], ..., G_{p+n}[t]\} & \text{n} \geq 1 \end{cases} \tag{16}$$

∇

Proof: See Appendix. ∇

Using Equation (16) and Propositions 1 and 2, the following results are obtained:

Proposition 3. Given a polynomial operator $C[B,t]$ characterized by polynomial functional subspace, the functional subspace of the r-th parameter $\kappa_r[t]$ of its inverse is given as:

$$\mathcal{F}_{\kappa_r[t]} = \begin{cases} \mathcal{F}_C & \text{r} = 1 \\ \{G_0[t], ..., G_{r \cdot \bar{b}c}[t]\} & \text{r} \geq 2 \end{cases} \tag{17}$$

∇

Proof: See Poulimenos and Fassois (2003*a*). ∇

Proposition 4. The functional subspace of the inverse function polynomial operator $I[B,t]$ of a model with polynomial functional subspaces is given as:

$$\mathcal{F}_{i_r[t]} = \begin{cases} \mathcal{F}_A \cup \mathcal{F}_C & \text{r} = 1 \\ \{G_0[t], ..., G_{(r-1) \cdot \bar{b}c + \bar{b}}[t]\} & \text{r} \geq 2 \end{cases} \tag{18}$$

∇

Proof: See Poulimenos and Fassois (2003*a*). ∇

4. FUNCTIONAL SERIES TARMA MODEL ESTIMATION

Consider the general case of an FS-TARMA model characterized by *distinct* and *"incomplete"* (that is not including all consecutive functions up to the subspace dimensionality) functional subspaces. Assuming, for the moment, the model structure $\mathcal{M}$ (defined by the orders n_a, n_c, a selected family of functions, and the vectors $\boldsymbol{b}_a$, $\boldsymbol{b}_c$) a-priori known, estimation of its parameter (that is coefficient of projection) vector:

$$\boldsymbol{\theta} \triangleq [\, \boldsymbol{a}^T \vdots \boldsymbol{c}^T \,]^T \tag{19}$$

$$\boldsymbol{a}^T \triangleq [\, a_{1,1} \ldots a_{1,p_a} \vdots \ldots \vdots a_{n_a,1} \ldots a_{n_a,p_a} \,] \tag{20}$$

$$\boldsymbol{c}^T \triangleq [\, c_{1,1} \ldots c_{1,p_c} \vdots \ldots \vdots c_{n_c,1} \ldots c_{n_c,p_c} \,] \tag{21}$$

as well as its innovations variance σ_w^2, may be achieved [based upon available signal samples $x[t]$ $(t = 1, 2, \ldots, N)$] via proper generalization of the isomorphic matrix algebra method of Fouskitakis and Fassois (2001).

The formulation of the method is based upon the generalization (capable of accounting for "incomplete" functional subspaces) of a matrix algebra that is isomorphic to the algebra of time-varying polynomial operators. Let $\mathbb{M}$ designate the set of all finite-dimensional real matrices. Designate as f a mapping from the set $\mathbb{A}[B]$ of finite order time-varying polynomial operators to the set $\mathbb{M}$, defined such that the operator $A[B,t]$ is mapped to a $(\bar{b}_a + 1) \times (n_a + 1)$ matrix with elements the operator's coefficients of projection (with each column corresponding to the coefficients of projection of each particular time-varying parameter):

$$f(A[B,t]) \triangleq \boldsymbol{A} = \begin{bmatrix} a_{0,0} & \cdots & a_{n_a,0} \\ \vdots & \ddots & \vdots \\ a_{0,\bar{b}_a} & \cdots & a_{n_a,\bar{b}_a} \end{bmatrix}_{(\bar{b}_a+1) \times (n_a+1)} \tag{22}$$

$$a_{i,j} \triangleq 0, \quad \forall\, i,j : G_j[t] \notin \mathcal{F}_{a_i[t]} \tag{23}$$

The set $\mathbb{A}[B]$ of finite order time-varying polynomial operators, equipped with the addition "$+$" and skew multiplication operations "$\circ$", defines an algebraic ring $(\mathbb{A}[B], +, \circ)$. Defining a proper addition "$\oplus$" and multiplication "$\star$" operation within the set $\mathbb{M}$, in such a way as to correspond to their counterparts within $\mathbb{A}[B]$, establishes an algebraic ring $(\mathbb{M}, \oplus, \star)$ that is isomorphic to the original $(\mathbb{A}[B], +, \circ)$. This allows for the formulation of the estimation method within the new ring.

The addition "$\oplus$" and multiplication "$\star$" operations within the set $\mathbb{M}$ are now defined as follows:

$$\boldsymbol{A} \oplus \boldsymbol{C} \triangleq \left[\begin{array}{c|c} \boldsymbol{A} & \boldsymbol{0} \\ \hline \boldsymbol{0} & \boldsymbol{0} \end{array} \right]_{\bar{b} \times (\bar{n}+1)} + \left[\begin{array}{c|c} \boldsymbol{C} & \boldsymbol{0} \\ \hline \boldsymbol{0} & \boldsymbol{0} \end{array} \right]_{\bar{b} \times (\bar{n}+1)} \tag{24}$$

$$\boldsymbol{A} \star \boldsymbol{C} = \boldsymbol{K} \qquad (\bar{b}_k + 1) \times (n_k + 1) \tag{25}$$

with:

$$\boldsymbol{k}^i \triangleq \sum_{j=0}^{i} \left(\boldsymbol{\Sigma}_{\bar{b}_a, \bar{b}_c}^{j, \bar{b}_k} \right) \cdot \left[\boldsymbol{a}^j \otimes \boldsymbol{c}^{i-j} \right], \quad i \in [0, n_k] \tag{26}$$

where $\otimes$ designates Kronecker product (Magnus and Neudecker 1988), and $\boldsymbol{k}^i$, $\boldsymbol{a}^i$, $\boldsymbol{c}^i$ the $(i+1)$-th column of the matrices $\boldsymbol{K}$, $\boldsymbol{A}$, $\boldsymbol{C}$, respectively. $\boldsymbol{\Sigma}$ is a matrix containing coefficients of projection of time-shifted basis function products on the functional subspace. Its elements are of the form $S_{a,b}^{j,m}$, which designates the projection of the basis function product $G_a[t] \cdot G_b[t-j]$ on $G_m[t]$, and are computed via a procedure outlined in the Appendix.

The estimation method consists of five steps as follows:

Step 1. Inverse Function Estimation.
In this step a truncated (n_i-th order) inverse function model of the form [compare with Equation (8)]:

$$I[B, t, \boldsymbol{i}] \cdot x[t] = e[t, \boldsymbol{i}] \tag{27}$$

with functional subspace defined via Proposition 4 [Equation (18)] is estimated based upon minimization of the residual $(e[t, \boldsymbol{i}])$ sum of squares via linear regression. In this expression $\boldsymbol{i}$ designates the projection coefficient vector of the truncated inverse function, and $b_{i_r}(j)$ the index designating the j-th function in the subspace of $i_r[t]$.

Step 2. Initial MA/AR Polynomial Estimation.
Initial estimates of the AR and MA projection coefficients (vectors $\boldsymbol{a}$ and $\boldsymbol{c}$) are obtained via deconvolution, using the identity:

$$A[B,t,\boldsymbol{a}] = C[B,t,\boldsymbol{c}] \circ I[B,t] \iff \boldsymbol{A} = \boldsymbol{C} \star \boldsymbol{I} \quad (28)$$

in which the theoretical inverse function is replaced by its truncated estimate. This yields the following expressions (within the set $\mathbb{M}$) for $i > \bar{n}$:

$$-\boldsymbol{i}^i = \sum_{j=1}^{i-1} \begin{bmatrix} \left(\boldsymbol{\Sigma}_{\bar{b}_c, \bar{b}_{i_{i-j}}}^{j, \bar{b}_c + \bar{b}_{i_{i-j}}}\right) \cdot \left[\boldsymbol{c}^j \otimes \boldsymbol{i}^{i-j}\right] \\ \boldsymbol{0}_{(\bar{b}_{i_i} - \bar{b}_c - \bar{b}_{i_{i-j}})} \end{bmatrix} \quad (29)$$

with $\bar{b}_{i_r} \triangleq \max_j\{b_{i_r}(j)\}$ and $\boldsymbol{c}^i \equiv \boldsymbol{0}$ for $i > n_c$. Based upon these, a linear system of equations may be set up and solved in terms of the MA coefficient of projection vector $\boldsymbol{c}$.

The AR coefficient of projection vector $\boldsymbol{a}$ may be obtained from the following expression [also obtained from Equation (28)]:

$$\boldsymbol{a}^i = \boldsymbol{c}^i \oplus \boldsymbol{i}^i \oplus \sum_{j=1}^{i-1} \left\{ \left(\boldsymbol{\Sigma}_{\bar{b}_c, \bar{b}_{i_{i-j}}}^{j, \bar{b}_a}\right) \cdot \left[\boldsymbol{c}^j \otimes \boldsymbol{i}^{i-j}\right] \right\} \quad (30)$$

with $\boldsymbol{a}^i \equiv \boldsymbol{0}$ for $i > n_a$.

Step 3. Signal Filtering.
Once initial estimates of the AR and MA polynomial operators are available, the filtering operations [2]:

$$C[B,t,\hat{\boldsymbol{c}}] \cdot z[t] = A[B,t,\hat{\boldsymbol{a}}] \cdot x[t] \quad (31)$$

$$A[B,t,\hat{\boldsymbol{a}}] \cdot \bar{x}[t] = z[t] \quad (32)$$

are performed, and the signal $\bar{x}[t]$ is obtained.

Step 4. Final AR/MA Polynomial Estimation.
The filtered signal $\bar{x}[t]$ theoretically obeys the FS-TAR part of the original FS-TARMA model, that is:

$$A[B,t] \cdot \bar{x}[t] = w[t] \quad (33)$$

Hence the AR projection coefficient estimates are obtained via linear regression by minimizing the residual sum of squares of the corresponding model:

$$A[B,t,\boldsymbol{a}] \cdot \bar{x}[t] = e[t,\boldsymbol{a}] \quad (34)$$

The final MA coefficient of projection estimates (vector $\boldsymbol{c}$) are subsequently obtained based upon Equation (29), using the updated AR estimates (deconvolution).

Step 5. Residual Variance Estimation.
The variance of the residual series [computed via the FS-TARMA expression (1) with the estimated AR/MA parameters] is estimated as:

$$\hat{\sigma}_w^2 = \hat{\sigma}_e^2 = \frac{1}{N-d} \cdot \sum_{t=d+1}^{N} e^2[t, \hat{\boldsymbol{\theta}}] \quad (35)$$

where $d \triangleq \dim \boldsymbol{\theta}$ and $e[t, \hat{\boldsymbol{\theta}}]$ designates the obtained model residuals.

[2] The hat designates estimator/estimate.

5. FUNCTIONAL SERIES TARMA MODEL IDENTIFICATION

Given a basis function family (such as Chebyshev, Laguerre, Jacobi, and so on), model identification (structure selection) refers to the estimation of the set of integers:

$$\mathcal{M} \triangleq \{n_a,\ n_c,\ p_a,\ p_c,\ b_a(j),\ b_c(j)\} \quad (36)$$

where n_a and n_c designate the AR and MA orders, p_a and p_c the AR and MA subspace dimensionalities, and the indices $b_a(j)$, $b_c(j)$ the specific functions used in the AR and MA subspaces, respectively.

The method postulated for model identification is a direct, iterative, optimization scheme based upon minimization of the Bayesian Information Criterion (BIC) (Schwarz 1978) [which may be confirmed to hold in the present case (Poulimenos and Fassois 2003*a*)]:

$$\hat{\mathcal{M}}(\hat{\boldsymbol{\theta}}) = \arg\min_{\mathcal{M}, \boldsymbol{\theta}} BIC(\mathcal{M}, \boldsymbol{\theta}) \quad (37)$$

The direct nature of the method, in which the search is guided only by the criterion value, renders it effective for the discrete variable optimization problem at hand. The method is *hybrid* (Syrjakow and Szczerbicka 1999), consisting of two distinct phases:

Phase I. Coarse ("global") optimization.
This phase aims at exploring the complete search space with the objective of locating promising regions within which optimal model structures (either in the local or global sense) might be located. This is achieved via genetic algorithms (King 1999) which maximize the negative BIC ("fitness" function).

Phase II. Fine ("local") optimization.
This phase aims at refining the results of phase I and selecting the global optimum point along with its exact coordinates. It operates in a (suitably defined) neighborhood of each initial solution (as provided by phase I and presently used as starting point), and is based upon the concept of *backward regression* (Haber and Keviczky 1999). It thus starts with maximum values of the arguments (within the selected neighborhood) and subsequently reduces either the AR/MA orders (by neglecting a set of p_a or p_c regressors, respectively) or the AR/MA functional subspace dimensionalities (by neglecting a set of n_a or n_c regressors, respectively), until no further reduction in the BIC is achieved. The procedure is repeated for all initial solutions (phase I results), and the model structure corresponding to the globally optimum BIC is selected.

6. MONTE CARLO STUDY

The effectiveness of the FS-TARMA estimation and identification method is presently examined via a Monte Carlo study. A non-stationary signal generated by an FS-TARMA$(2,1)_{[3,3]}$ model with AR and

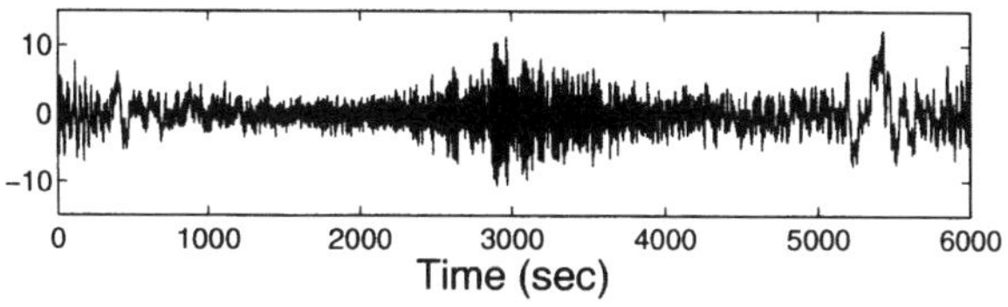

Fig. 1. Chebyshev TARMA(2,1)$_{[3,3]}$ signal realization.

Table 1. Parameter estimation for the selected model (30 Monte Carlo runs).

Coefficient	True Value	Estimate (Mean ± Std Deviation)
$a_{1,1}$	-0.68	-0.6821 ± 0.0173
$a_{1,2}$	0.16	0.1576 ± 0.0096
$a_{1,3}$	-0.26	0.2589 ± 0.0078
$a_{2,1}$	0.16	0.1573 ± 0.0144
$a_{2,2}$	-0.42	0.4164 ± 0.0115
$a_{2,3}$	0.38	0.3784 ± 0.0088
$c_{1,1}$	0.20	0.1972 ± 0.0215
$c_{1,2}$	0.25	0.2496 ± 0.0109
$c_{1,3}$	-0.15	-0.1491 ± 0.0100
σ_w^2	1.00	0.9959 ± 0.0142

MA subspaces belonging to the Chebyshev II family (William 1997) and specified by the index vectors:

$$\boldsymbol{b}_a = [0 \quad 2 \quad 4]^T, \quad \boldsymbol{b}_c = [0 \quad 1 \quad 3]^T$$

is utilized. The true model coefficients are indicated in Table 1. A typical signal realization is depicted in Figure 1.

The study consists of 30 runs, each one being based upon a $N = 6000$ sample-long signal realization. Model structure estimation in this case includes selection of the proper functional family among four candidate families: The (true) Chebyshev II family, and three Jacobi families (William 1997) characterized by the parameters $(\alpha, \beta) = (-0.3, 0.8)$, $(0.8, -0.3)$, and $(1.5, 0.3)$. Coarse optimization is based upon a genetic algorithm with population size 300, number of generations equal to 3, crossover probability 0.85, and mutation probability 0.001.

The method's ability to select the proper functional family is demonstrated in Table 2, in which the correct Chebyshev II family is shown to attain the minimum mean (over the 30 runs) BIC, as well as RSS (Residual Sum of Squares). It is worth noting that the relatively small differences in the attained mean RSS values suggest that models based upon different functional families may approximate the given signal with sufficient accuracy. Yet, the relatively larger differences in the mean BIC values suggest that such approximations are characterized by increased parametric complexity.

The model identification (model structure estimation) results within the Chebyshev II family are also very accurate, as the true model orders (n_a, n_b), functional subspace dimensionalities (p_a, p_c), as well as indices ($\boldsymbol{b}_a$, $\boldsymbol{b}_c$), are all correctly estimated in all 30 runs.

Parameter estimation results for the identified Chebyshev II TARMA$(2,1)_{[3,3]}$ model are presented in Table 1, and are characterized by high accuracy. Indeed, the mean estimates are very close to their true counterparts, and the standard deviations are small.

Table 2. Functional family selection: Globally optimal BIC and RSS values for four families (means of 30 Monte Carlo runs.)

Family	BIC	RSS
Chebyshev II	0.0085	5964.3
Jacobi $(-0.3, 0.8)$	0.0123	5981.3
Jacobi $(0.8, -0.3)$	0.0136	5983.9
Jacobi $(1.5, 0.3)$	0.0221	6023.8

7. CONCLUDING REMARKS

The problem of modeling non-stationary stochastic signals via FS-TARMA models was considered, and a complete model (parameter) estimation and identification (structure estimation) method was formulated. Model estimation was based upon the generalization of an isomorphic matrix algebra type method to the case of "incomplete" functional subspaces. Model identification was based upon a direct two-phase optimization method which utilizes genetic algorithms and backward regression. The effectiveness of the complete method was demonstrated via a Monte Carlo study, in which both the model structure and parameters were shown to be accurately estimated.

8. REFERENCES

Ben Mrad, R., S.D. Fassois and J.A. Levitt (1998*a*). A polynomial-algebraic method for non-stationary TARMA signal analysis. Part I: The method. *Signal Processing* **65**, 1–19.

Ben Mrad, R., S.D. Fassois and J.A. Levitt (1998*b*). A polynomial-algebraic method for non-stationary TARMA signal analysis. Part II: Application to modelling and prediction of power consumption in automobile active suspension systems. *Signal Processing* **65**, 21–38.

Cohen, L. (1995). *Time-Frequency Analysis*. Prentice Hall PTR.

Fouskitakis, G.N. and S.D. Fassois (2001). On the estimation of non-stationary functional series TARMA models: An isomorphic matrix algebra based method. *ASME Journal of Dynamic Systems, Measurement, and Control* **123**, 601–610.

Fouskitakis, G.N. and S.D. Fassois (2002). Functional series TARMA modelling and simulation of earthquake ground motion. *Earthquake Engineering and Structural Dynamics* **31**, 399–420.

Grenier, Y. (1989). Parametric time-frequency representations. In: *Time and Frequency Representations of Signals and Systems* (G. Longo and B. Picinbono, Eds.). Springer-Verlag.

Haber, R. and L. Keviczky (1999). *Nonlinear System Identification: Input-Output Modelling Approach*. Kluwer Academic Publishers. Dordrecht.

King, R.E. (1999). *Computational Intelligence in Control Engineering*. Marcel Dekker. New York.

Kitagawa, G. and W. Gersh (1996). *Smoothness Priors Analysis of Time Series*. Springer-Verlag. New York.

Ljung, L. (1999). *System Identification: Theory for the User (second edition)*. Prentice-Hall PTR.
Magnus, J.R. and H. Neudecker (1988). *Matrix Differential Calculus*. Wiley. New York.
Niedzwiecki, M. (2000). *Identification of Time-Varying Processes*. John Wiley.
Petsounis, K.A. and S.D. Fassois (2000). Non-stationary functional series TARMA vibration modelling and analysis in a planar manipulator. *Journal of Sound and Vibration* **231**, 1355–1376.
Poulimenos, A.P. and S.D. Fassois (2003*a*). Estimation and identification of non-stationary signals using functional series TARMA models. Under preparation for publication.
Poulimenos, A.P. and S.D. Fassois (2003*b*). Non-stationary mechanical vibration modelling and analysis via functional series TARMA models. *These conference proceedings*.
Schwarz, G. (1978). Estimating the dimension of a model. *Annals of Statistics* **6**(2), 461–464.
Stavropoulos, Ch.N. and S.D. Fassois (2000). Non-stationary functional series modelling and analysis of hardware reliability series: A comparative study using rail vehicle interfailure times. *Reliability Engineering and System Safety* **68**, 169–183.
Syrjakow, M. and H. Szczerbicka (1999). Efficient parameter optimization based on combination of direct global and local search methods. In: *Evolutionary Algorithms* (L.D. Davis, K.D. Jong, M.D. Vose and L.D. Whitley, Eds.). Spinger. New York.
William, T.J. (1997). *Atlas for Computing Mathematical Functions*. John Wiley. New York.

Appendix A. PROOF OF LEMMA 1 AND COMPUTATION OF THE ELEMENTS OF Σ [EQUATION 26].

Each family of orthogonal polynomials satisfies a relationship of the form (William 1997):

$$G_n[t] = [K_{1,n} + K_{2,n} \cdot t] \cdot G_{n-1}[t] + K_{3,n} \cdot G_{n-2}[t] \quad (n \geq 1) \tag{A.1}$$

$$G_n[t] \equiv 0 \;\; (n \leq -1), \qquad G_0[t] = 1 \tag{A.2}$$

where the quantities $K_{1,n}$, $K_{2,n}$ and $K_{3,n}$ are functions of the integer variable n. It may be shown (Poulimenos and Fassois 2003*a*), using Equations (A.1) and (A.2), that the product $G_p[t] \cdot G_n[t-\ell]$ $(p, n \geq 0,\ \ell \geq 1)$ is given as:

a. Case $n = 0$:

$$G_p[t] \cdot G_n[t-\ell] = G_p[t] \tag{A.3}$$

b. Case $n \geq 1, p = 0$:

$$\begin{aligned} G_p[t] \cdot G_n[t-\ell] &= \\ &= R_{1,n} \cdot S^{c,0}_{0,n-1} \cdot G_1[t] \\ &+ \sum_{j=2}^{n} \frac{R_{1,n}}{R_{1,j}} \cdot S^{c,j-1}_{0,n-1} \cdot G_j[t] \\ &+ \sum_{j=1}^{n-1} \frac{-R_{1,n} \cdot R_{2,j+1}}{R_{1,j+1}} \cdot S^{c,j}_{0,n-1} \cdot G_j(t) \\ &+ \sum_{j=0}^{n-2} \frac{-R_{1,n} \cdot K_{3,j+2}}{R_{1,j+2}} \cdot S^{c,j+1}_{0,n-1} \cdot G_j[t] \\ &+ \sum_{j=0}^{n-1} [R_{2,n} - \ell \cdot K_{2,1} \cdot R_{1,n}] \cdot S^{c,j}_{0,n-1} \cdot G_j[t] \\ &+ \sum_{j=0}^{n-2} K_{3,n} \cdot S^{c,j}_{0,n-2} \cdot G_j[t] \end{aligned} \tag{A.4}$$

c. Case $n \geq 1, p = 1$:

$$\begin{aligned} G_p[t] \cdot G_n[t-\ell] &= \\ &= \sum_{j=0}^{n+1} \frac{1}{R_{1,n+1}} \cdot S^{c,j}_{0,n+1} \cdot G_j[t] \\ &+ \sum_{j=0}^{n} \left[\ell \cdot K_{2,1} - \frac{R_{2,n+1}}{R_{1,n+1}}\right] \cdot S^{c,j}_{0,n} \cdot G_j[t] \\ &+ \sum_{j=0}^{n-1} \frac{-K_{3,n+1}}{R_{1,n+1}} \cdot S^{c,j}_{0,n-1} \cdot G_j[t] \end{aligned} \tag{A.5}$$

d. Case $n \geq 1, p \geq 2$:

$$\begin{aligned} G_p[t] \cdot G_n[t-\ell] &= \\ &= R_{2,p} \cdot S^{c,0}_{p-1,n} \cdot G_0[t] + R_{1,p} \cdot S^{c,0}_{p-1,n} \cdot G_1[t] \\ &+ \sum_{j=2}^{p+n} \frac{R_{1,p}}{R_{1,j}} \cdot S^{c,j-1}_{p-1,n} \cdot G_j[t] \\ &+ \sum_{j=1}^{p+n-1} \left[R_{2,p} - \frac{R_{1,p} \cdot R_{2,j+1}}{R_{1,j+1}}\right] \cdot S^{c,j}_{p-1,n} \cdot G_j(t) \\ &+ \sum_{j=0}^{p+n-2} \frac{-R_{1,p} \cdot K_{3,j+2}}{R_{1,j+2}} \cdot S^{c,j+1}_{p-1,n} \cdot G_j[t] \\ &+ \sum_{j=0}^{p+n-2} K_{3,p} \cdot S^{c,j}_{p-2,n} \cdot G_j[t] \end{aligned} \tag{A.6}$$

where $R_{1,n} \triangleq \frac{K_{2,n}}{K_{2,1}}$, $R_{2,n} \triangleq K_{1,n} - \frac{K_{1,1}}{K_{2,1}} \cdot K_{2,n}$, and the quantity $S^{c,j}_{p,n}$ indicates the coefficient of projection of the product $G_p[t] \cdot G_n[t-c]$ on the basis function $G_j[t]$. From Equations (A.3) - (A.6) it follows that:

$$G_p[t] \cdot G_n[t-\ell] = \sum_{j=0}^{p+n} S^{c,j}_{p,n} \cdot G_j[t], \;\; (n \geq 1) \tag{A.7}$$

which proves Equation (16). The same equations may be used for the computation of the projection coefficients $S^{c,j}_{p,n}$ utilized in Steps 2 and 4 of the model estimation method.

IFAC Publications
www.elsevier.com/locate/ifac

MODELLING MULTIVARIATE POLLUTANT TIME SERIES WITH WAVELET FUNCTIONS

Giuseppe Nunnari and Domenico Longo

Dipartimento di Ingegneria Elettrica, Elettronica e dei Sistemi, Università di Catania
Viale A. Doria, 6, 95122, Catania (Italy)

Abstract: This paper deals with a new approach based on wavelet functions to model multivariate time series. Time series are formalised in terms of NARX (Non Linear Auto Regressive with eXogenous inputs) models and the vector of unknown parameters is determined by using a genetic algorithms (GAs) optimisation approach, since GAs allow finding the global minimum of a function with many variables, overcoming the limitation of typical gradient based techniques. A case study, referring to the modelling of daily averages of SO_2 time series recorded in the industrial area of Syracuse (Italy) is reported. The performance of the proposed approach is compared with other traditional approaches such as ARX and Multi-layer neural networks. The results obtained show that while there are no significant differences between the neural and the wavelet approach in terms of model performance and computational effort, there is an appreciable advantage in using the proposed technique in terms of internal model complexity. *Copyright © 2003 IFAC*

Keywords: 5–10 time series, wavelets, neural networks, genetic algorithms, air pollution

1. INTRODUCTION

Modelling air pollution is a complex task which has drawn attention of many scientists all over the world since the early 1960's (Pasquill, 1961). To address this topic, which is relevant to the problem of controlling the levels air pollution, several modelling approaches have been proposed ether of deterministic and statistical type. Deterministic models are based on fundamental mathematical descriptions of atmospheric processes in which the output is represented by the air pollution concentration field and the inputs by emissions. The main drawback using deterministic models is that they involve the solution of PDEs equation requiring accurate information about emissions, wind speed and direction and turbulent dispersion coefficients in the overall integration volume. Statistical methods can be preferred to deterministic approaches for short horizon (e.g. one day) predictions at specific points. This is, for instance, the case of predicting the level of pollution at the recording points of an air quality monitoring network. Indeed statistical modelling approaches do not necessarily need data about emissions (which are sometimes unavailable, especially not in real-time) since they are based on the use of past air quality, meteorological measurements, traffic data and any kind of information that can help to model the concentration of a pollutant time series recorded at a point. The main drawback using statistical models is that of predicting rare but critical pollution episodes (i.e. the so-called exceedances of a prefixed threshold). Wavelets have been used for signal processing and prediction for a long time (see for instance Mallat, (1999)). They are known to be very useful for time series with a short time horizon, since one of the peculiarity of wavelets is their ability to represent the transitory features of time series. Time series of daily average concentration of pollutants, such as SO_2, recorded in areas characterised by marine breezes, have a short memory due to the daily fluctuations of wind directions. For this reason a wavelet based approach may be appropriate. In this paper a new approach for modelling multivariate pollutant time series based on wavelet is proposed. The paper is organised as follows. The basic features of wavelet functions are first reviewed. Then a formalisation of multivariate time series in terms of a NARX (Non-linear Auro-Regressive with eXogenous inputs) model is introduced together with an approximating scheme based on a finite sum of scalar wavelet functions. A procedure to obtain the vector of unknown parameters of the approximating model is than obtained using a GAs optimisation algorithms. The performance of the proposed approach is finally reported in comparison with other modelling techniques proposed in literature both of linear and non linear type.

2. BACKGROUND ABOUT WAVELETS

Wavelets are families of functions obtained by taking the dilatations and translations of a particular function with sufficient decay in both the time and the frequency domain. The use of wavelets for system identification was introduced by Benveniste et. al., (1994) and successively considered by others authors such as Zhang (1997), Mallat, (1999) and Rying et al. (2002). The main advantage of modelling with wavelets lies in the fact that is possible to represent the transitory characteristics of the time series in more efficient way. This advantage derives from the fact that wavelets are limited duration functions; moreover, the shape of the wavelets used for the modelling can be chosen according to the peculiarities of the time series to be modelled.

Let $\psi(t)$ be a basic wavelet and let s ($s \neq 0$) and u be real numbers; the family of wavelets corresponding to $\psi(t)$ is

$$\Psi_{s,u}(t) = \frac{1}{\sqrt{s}}\Psi\left(\frac{t-u}{s}\right) \qquad (1)$$

where s and u gives the dilatation and the translation respectively. A set of wavelets can be obtained by giving different values to s and u The term $1/\sqrt{s}$ normalises $|\psi(t)|$. Wavelet families which are constructed by allowing s and u to vary continuously are called continuous wavelet families. Discrete wavelets are constructed by constraining the values *s* and *u* to be a discrete lattice of points. It should to be observed that the discrete wavelet does not imply that the wavelets are discrete-valued, but that the dilates and translates are discretized.

Some of the most popular wavelet functions are the Mexican hat (2) and the Morlet (3) represented in Fig. 1 a) and b) respectively

$$\Psi(t) = \left(\frac{2}{\sqrt{3}}\pi^{-\frac{1}{4}}\right)\left(1-t^2\right)e^{-\frac{t^2}{2}} \qquad (2)$$

$$\Psi(t) = Ce^{-\frac{t^2}{2}}\cos(5t) \qquad (3)$$

3. MODELLING MULTIVARIATE TIME SERIES

Let y(t) be an air pollution time series. It can be considered a random distributed variable whose

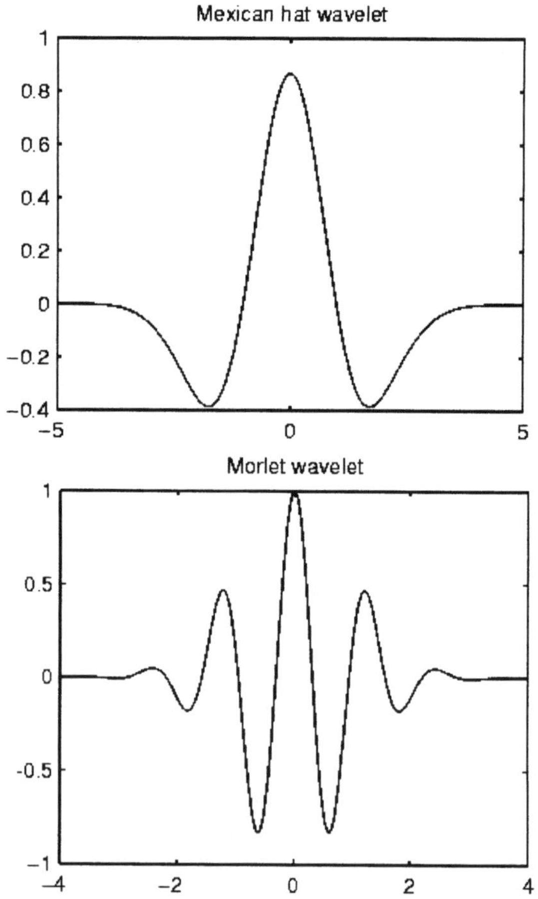

Fig. 1 a) Mexican hat, b) Morlet wavelet functions

behaviour is influenced by other correlated time series, namely meteorological time series (e.g. wind direction, wind speed, temperature, pressure etc) and others air pollutants. Hence the modelling of y(t) usually starts with a correlation analysis to determine what kind of variables (i.e. the exogenous model inputs) the time series mainly depends on and to what extent (number of regressions). Since models of pollutant time series are usually non-linear, they can be represented as NARX (Non-linear Auto-Regressive with eXogenous inputs) models which in form of one-step ahead predictors can be represented as:

$$\begin{aligned} y(t+1) = f(&y(t), y(t-1), \ldots y(t-n_y+1), \\ &u_1(t),\ u_1(t-1), \ldots, u_1(t-n_1+1), \\ &\ldots, u_q(t), u_q(t-1), \ldots, u_q(t-n_q+1)) + e(t) \end{aligned} \qquad (4)$$

Here *f* is an unknown non-linear function, $u_1, \ldots, u_q$ are the exogenous model inputs and n_y, n_1, ... n_q are integer numbers related to the model order. NARX models can be considered a generalisation of the well-known ARX models. The decision to use exogenous inputs, which can always be done using a trial and error procedure, should be made not only on the availability of such data but also on the basis of their degree of correlation with the time series being modelled. Formally, the modelling problem is that of finding a suitable approximation of the unknown function *f*. In this paper we show how the approximation problem can be formulated, by operating an appropriate choice of the variables involved in a NARX model. To this end let us indicate by:

$$X^T(t) = [y(t), y(t-1), \cdots y(t-n_y+1), \\ u_1(t), \cdots, u_1(t-n_1+1), \cdots, \\ u_q(t), \cdots u_q(t-n_1+1)] \in R^d \qquad (5)$$

the argument of function f (see expression (4)), i.e. the pattern of input data at the time t. Moreover let

A_j, *and* $U_j \in R^d$, $j = 1, \cdots, M$ (6) be appropriate vectors of unknown parameters. Furthermore, let us introduce the following expression

$$t_j = \frac{[A_j \otimes (X(t) - U_j)] \cdot [A_j \otimes (X(t) - U_j)]^T}{s_j} \qquad (6)$$

where t_j and s_j are scalar quantities, as a map between the vectorial argument of expression (4) and the scalar argument t of the generic wavelet function (1). On the basis of on these positions we propose approximating $y(t+1)$ in expression (4) as:

$$y(t+1) = \sum_{j=1}^{M} wj \cdot \Psi_j(t_j) \qquad (7)$$

In our procedure the number of approximating wavelet functions M is obtained by a *trial and error* iterative procedure.

4. FINDING THE MODEL PARAMETERS

One of the problems that immediately arise in using the introduced approach is that of finding the unknown parameters, namely A_j,, U_j, s_j, w_j $(j=1, M)$. To this end, a genetic algorithms (GAs) optimisation approach has been considered in this paper to searching for optimal prediction model. The reason for using GAs is that they are capable of finding the global minimum of a function with many variables overcoming the limitation of typical gradient based optimisation techniques (Goldberg, 1989). A description of GAs is beyond the scope of this paper and the interested reader is referred to the numerous text books on this subject, such as Holland, (1955) and Fortuna et. al., (2001). The cost function was defined by appropriately modifying the traditional root mean square error as indicated in (8),

$$J = \sum_{i=1}^{N} \sqrt{\left(y_i - \hat{y}_i \right)^2} \bullet a \qquad (8)$$

$$\text{where} \begin{cases} a = a^{\bullet} > 1 & \text{if } y_i > Th \\ a \quad 1 & \text{if } y_i \leq Th \end{cases}$$

Here, y and $\hat{y}_i$ indicate the true and simulated time series values respectively. The structure of expression (8) can be explained bearing in mind that one of the most important capabilities for an air pollution prediction model is that of predicting concentration exceedances above a prefixed threshold (e.g. 125 μg/m^3 for the daily average level of SO_2). The coefficient a represents an appropriate *weighting* constant depending on the prefixed threshold value *Th*. Due to the presence of this weighting coefficient the prediction errors corresponding to high concentrations are considered *more important* than prediction errors corresponding to lower concentrations and hence *weighted more.*

5. THE PROPOSED IDENTIFICATION ALGORITHM

The proposed identification algorithm can be synthesised as follows:

1) let us assign $i = 1$ and $R_i = y$;
2) find the best parameters of the chosen wavelet function ψ_i by using genetic algorithms, after defining the object function (or cost function) as indicated above,
3) compute the residual function as: $R_{i+1} = R_i - \psi_i$
4) increment i by 1 and repeat steps 1) through 3) until $i \leq k$

where k is a predefined integer value.

6. CHOOSING THE MODEL STRUCTURE

The proposed procedure was applied to modelling the daily mean value of SO_2 recorded in the industrial area of Syracuse (Italy). The monitoring network covers a wide area comprising several towns in the Syracuse neighbourhood. The atmospheric pollution monitored in the area is essentially caused by industrial emissions, as there are large petrochemical plants nearby. A number of exceedances of the attention level threshold (125 μg/m^3, daily mean

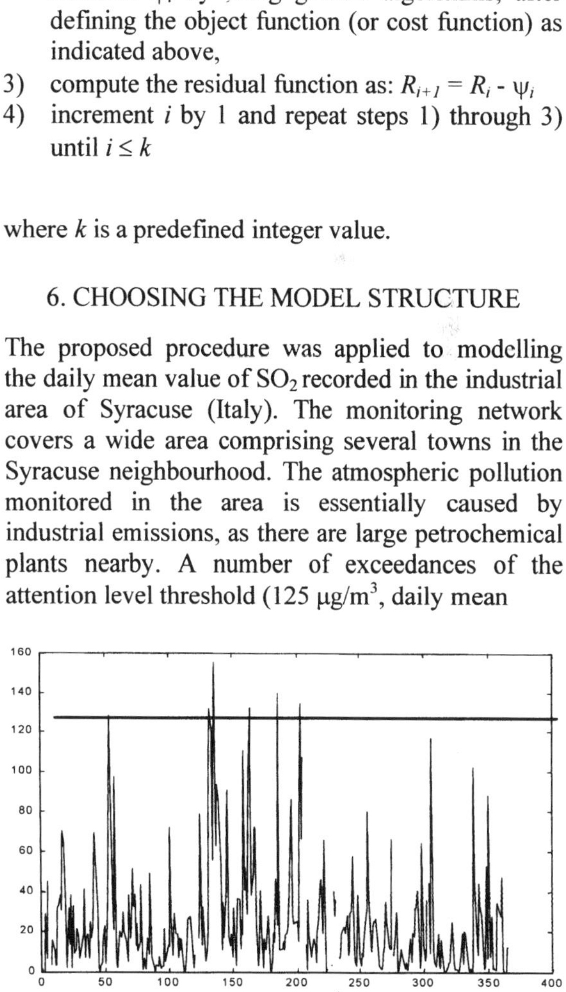

Fig. 2. SO_2 daily mean (DMEA) concentration in μg/m^3. recorded during 1997 The horizontal line indicate the threshold attention level (125 μg/m^3).

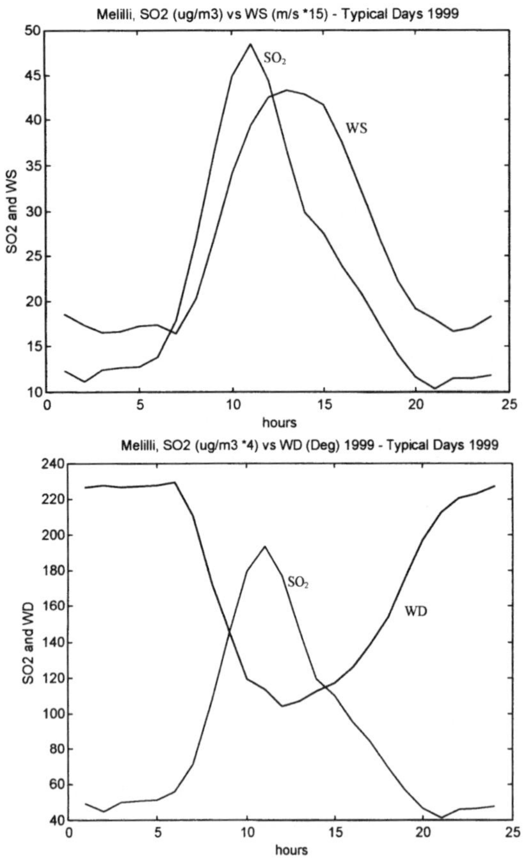

Fig. 3. a) (upper). Typical daily SO_2 concentration and wind speed (multiplied by a factor 15) at the Melilli station b) (lower). Typical daily SO_2 (multiplied by 4) concentration and wind direction.

value) were observed in this area during 1995-1999. The results reported below refer to the recording station called Melilli. Daily mean SO_2 values recorded at this station during 1997 are given in Fig. 2. while typical daily concentration of SO_2 and typical wind speed and direction are reported in Fig. 3. By typical daily concentrations we mean the curve obtained by averaging the values of a given variable (e.g. SO_2) recorded at the same time of day throughout the year. From Fig. 3 it is possible to see that peaks of SO_2 emissions are recorded between 10-12 a.m. However, the presence of these peaks cannot be attributed to particular emission policies, since from data emissions provided by local authorities, not reported here for the sake of simplicity, it is possible to see that the SO_2 emission rate is almost constant during the day. From Fig. 3 it is also possible to observe that the SO_2 peaks are correlated with wind direction (WD) and wind speed (WS). The topography of the area where the considered recording station is located, which features a chain of mountains to the west and the sea to the east helps to interpret the daily distribution of SO_2 levels previously shown. Peaks in the SO_2 concentration are recorded during the daytime when the wind blows from the sea towards the land (typical behaviour of marine breezes) where SO_2 probably accumulates due to the presence of mountains on the west side. Lower SO_2 concentrations are recorded when the wind blows from the land to the sea. Hence it seems reasonable to attribute the peaks to the local atmospheric conditions and in particular to the action of wind.

On the basis of these consideration it was decided to consider the wind speed and direction as exogenous inputs for the prediction model. In more detail, it was experimentally found that the average WD values between 7 a.m. and 7 p.m. and the average values of WS from 6 a.m. to 6 p.m. are a suitable choice for the exogenous model inputs. The target for the prediction model was to perform one-day ahead forecasting of the DMEA concentration of SO_2. When meteo forecasts are allowed, prediction of WS and WD on day (t+1) were also considered to predict the level of SO_2 on day (t+1).

7. PERFORMANCE INDEXES

In order to evaluate the performance of prediction models, an appropriate number of performance indexes can be evaluated evaluated. Generally speaking these indexes were organised into two different sets. The first set, referred to here as *global fit indexes*, evaluate the fitting capabilities of the overall time series. This set includes the Bias (see expression (9), the *RMSE* (Root Square Mean Error) (10), the *MAE* (Mean Absolute Error) (11) that give estimates of the average error, and the index of agreement d (12) which is a bounded relative measure capable of measuring the degree to which predictions are error-free.

$$Bias: \quad \frac{1}{N}\sum_{i=1}^{N}(P_i - O_i) \tag{9}$$

$$MAE: \quad \frac{1}{N}\sum_{i=1}^{N}\left|P_i - O_i\right| \tag{10}$$

$$RMSE: \quad \sqrt{\frac{1}{N}\sum_{i=1}^{N}(P_i - O_i)^2} \tag{11}$$

$$d: \quad 1-\frac{\sum_{i=1}^{N}(P_i - O_i)^2}{\sum_{i=1}^{N}\left(\left|P_i - \bar{O}\right| + \left|O_i - \bar{O}\right|\right)^2} \tag{12}$$

In expressions above, O and P indicate the observed and predicted time series respectively, the overbar indicates the mean value and the suffix i the generic element of the time series.

The second set of indexes was studied to specifically evaluate model capabilities in predicting critical SO_2 episodes. The most important index of this set is the success index *SI* which indicates how well the exceedances were predicted. It is not affected by a large number of correctly forecasted non-exceedances and therefore it is useful for evaluating rare events. Others indexes in this second set are: the probability of detection index *SP* which assess the ability to predict SO_2 exceedances; the false alarm rate index *FA* which indicates the percentage frequency of instances when a forecast of an SO_2 concentration exceedance did not actually occur.

These indexes are recommended by the ETC-AQ (Van Aalst and De Leeuw, 1997)) and the EPA, (1986). Expressions for SP, SR, FA and SI are given below:

$$SP\% = 100\frac{N_P}{N_O}, \quad SR\% = 100\frac{N_P}{N_F},$$

$$FA\% = 100 - SR\%, \qquad (13)$$

$$SI\% = 100(\frac{N_P}{N_O} + \frac{N + N_P - N_O - N_F}{N - N_O} - 1)$$

where N_o is the total number of observed exceedance, N_P is the number of correctly predicted exceedances, N_F is the total number of forecasted exceedances and N the total number of data points.

8. INTERCOMPARED TECHNIQUES

In order to compare the wavelet based approach introduced in this paper, referred as WAG (Wavelets with Genetic Algorithms) below, two of the most popular time series identification approaches considered in literature were also implemented: the Multi-Layer Perceptron (MLP) neural network and the ARX (Auto Regressive with eXogenous inputs) approaches. Both these techniques have been widely considered in literature (see for instance, Finzi et. al. 1998, Boznar et al. 1993), Arena et al. (1996), Nunnari et al. (1998). The available data set, which consisted of hourly data recorded from 1995 to 1999 at the Melilli station, was arranged into two subsets: data from 1996 to 1998 was considered as the identification set while the 1999 data was used as the testing set. Two type of prediction models were implemented, referred here as NFOR and FOR. NFOR models do not make use of WS and WD forecasts, while FOR type models consider meteo forecasts of these meteorological variables as model exogenous inputs.

9. NUMERICAL RESULTS

For the NFOR type models, the global performance obtained are reported In Fig 4, while the exceedance performance indexes corresponding to two different thresholds, namely 80 and 125 μg/m^3 are reported in Fig. 5 and Fig. 6 respectively.

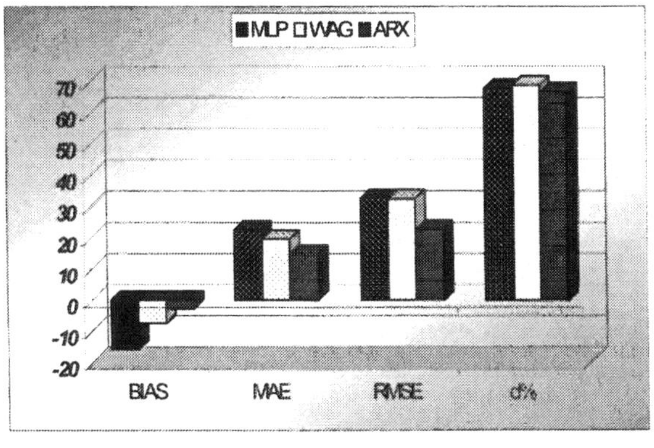

Fig. 4 - Global performance indexes for the NFOR type models

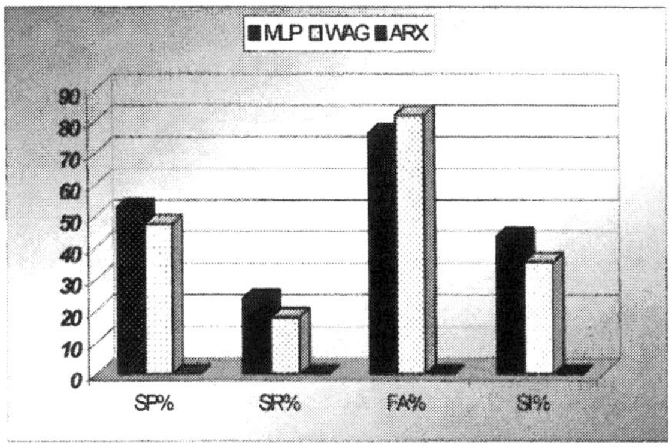

Fig. 5 - Exceedance indexes for NFOR models assuming a threshold of 80 μg/m^3.

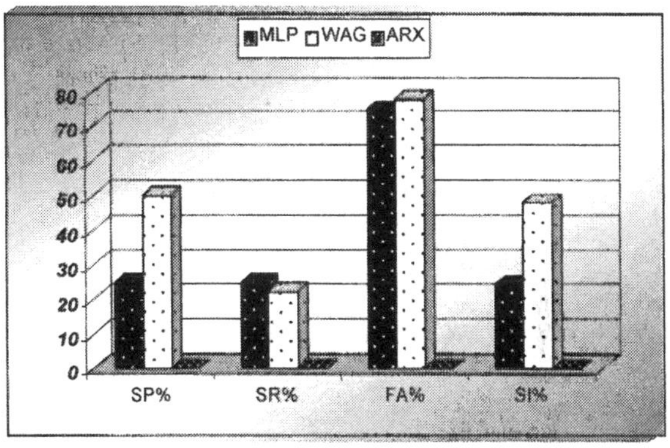

Fig. 6 - Exceedance indexes for NFOR models assuming a threshold of 125 ug/m3.

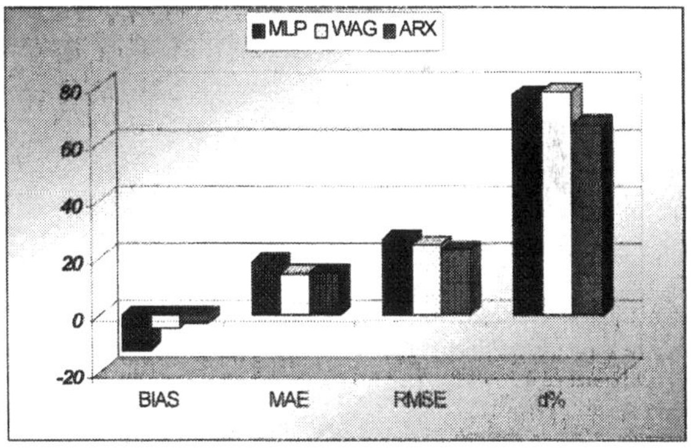

Fig. 7 - Global performance indexes for the FOR type models

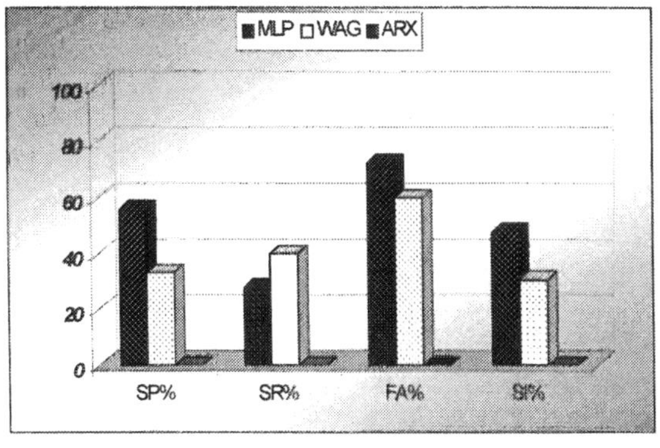

Fig. 8- Exceedance indexes for FOR models assuming a threshold of 80 μg/m^3.

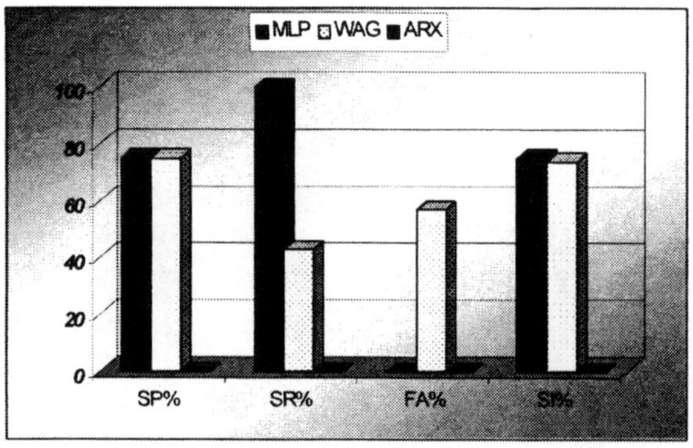

Fig. 9 - Exceedance indexes for FOR models assuming a threshold of 125 μg/m^3.

For the FOR type models, the global performance obtained are reported In Fig 7, while the exceedance performance indexes corresponding to the considered thresholds, are reported in Fig. 8 and Fig. 9 respectively. From the figures it is possible to observe that all the compared techniques perform quite similar in terms of global performance indexes. In particular, ARX models perform slightly better than MLP and WAG in terms of BIAS, MAE and RMSE, but these latter models perform better in terms of index of agreement (d%). The main difference between linear models (e.g. ARX) and non-linear models such as MLP and WAG becomes evident looking at the indexes of exceedance. ARX models exhibits SP%=SR%=0 at both the considered thresholds, i.e. they were not able to predict any exceedances of the considered threshold. MLP and WAG models were instead able to predict exceedances. Comments on the performances of MLP and WAG models in terms of SP% or FA% is not so straightforward For instance, if we consider the SP% index, i.e. the capability of models to predict exceedances of a given threshold, then it seems that when meteo forecasts are not considered (NFOR models), at lower thresholds (80) MLP performs better than WAG; instead at higher thresholds (125) WAG performs better than MLP. For this reason it is probably more appropriate to discuss the performances in terms of the SI% (i.e. the success index), which give an average measure of how well the exceedances were predicted. Generally speaking it seems that the use of meteo forecasts allows both MLP and WAG to improve the level of SI% at higher thresholds (see Fig. 9). Moreover it seems that at the higher threshold (125) MLP and WAG performs quite similar.

CONCLUSIONS

In this paper a novel approach to identify air pollution prediction models is introduced. The results obtained so far show that the proposed approach performs as well as the MLP neural approach, which nowadays is considered one of the best techniques available for the identification of NARX models. However, even if there are no significant differences between the MLP and the proposed WAG approach in terms of performance, there is a significant advantage in using WAG in terms of model complexity. Experience gained in identifying air pollutant time series shows that usually a few wavelet functions (from one to three) allow to implement a suitable prediction model, while the MLP neural model normally require from 10 to 15 neurons in the hidden layer, thus indicating a higher complexity. There are no significant differences between the two considered approaches in terms of the computational effort need to identify the models.

REFERENCES

Arena P, Baglio S., Castorina C., Fortuna L., Nunnari G. (1996) A Neural Architecture to Predict Pollution in Industrial Areas. Proc. ICNN, 4, 2107-2112, Washington.

Benveniste A., Juditsky A., Delion B., Zhang Q., Glorennec P. Y. (1994) Wavelets in identification. Proc. SYSIS '94, 2, 27-48 .

Boznar M., Lesjak M., Mlakar P. (1993) A Neural Network-Based Method for Short-Term Predictions of Ambient SO_2 Concentrations in Highly Polluted Industrial Areas of Complex Terrain. Atmosph. Environment, 27B, 221-230.

Fortuna L., Rizzotto G., Lavorgna M., Nunnari G., Xibilia M.G., Caponetto R., (2001) Soft Computing – New Trends and Applications, Springer Verlag.

Finzi G., Volta M., Nucifora A., Nunnari G., (1998), Real-Time Ozone Episode Forecast: a Comparison between Neural Network and Grey-Box Models, Proc. of the Intern. ICSC/IFAC Symposium on Neural Computation - NC'98, Vienna.

Goldberg D.E., (1989), Genetic Algorithm in Search, Optimization and Machine Learning, Addison Wesley.

Holland J.H. (1975) Adaptation in Natural and Artificial System, University of Michigan Press.

Mallat S. (1999) A Wavelet Tour of Signal Processing, 2nd ed., New York , Academic 1999.

Nunnari G., Nucifora A., Randieri C. (1998) The Application of Neural Techniques to the Modelling of Time Series of Atmospheric Pollution Data. Ecological Modelling, 111, 187-205.

Pasquill F., (1961), The estimation of the dispersion windborne material, Meteorological Magazine, N. 90, pp 33-49.

Rying E.A. Bilbro G.L., J. Lu, (2002), Focused Local Learning with Wavelt Neural Networks, 13, 2 304-316.

Vose M.D. (1999), The simple genetic algorithm. Foundation and Theory MIT Press 1999.

Van Aalst R. M., De Leeuw F. A. (1997), National Ozone Forecasting System and International Data Exchange in Northwest Europe, European Topic Centre on Air Quality.

Zhang Q., (1997) Using wavelet network in non parametric estimation, IEEE trans on Neural Networks, 8, 227-236.

IFAC Publications
www.elsevier.com/locate/ifac

ESTIMATING THE LYAPUNOV EXPONENTS OF CHAOTIC TIME SERIES BASED ON POLYNOMIAL MODELLING

M. Ataei [1], A. Khaki-Sedigh [2], and B. Lohmann [3]

[1] *Institute of Automation, University of Bremen*
Otto-Hahn-Allee./ NW1, D-28359, Bremen, Germany
E-mail: ataei@iat.uni-bremen.de

[2] *Department of Electrical Engineering, K. N. Toosi University of Technology*
Sayyed Khandan Bridge, P.O. Box 16315-1355, Tehran, Iran
E-mail: sedigh@eetd.kntu.ac.ir

[3] *Institute of Automation, University of Bremen*
Otto-Hahn-Allee./ NW1, D-28359, Bremen, Germany
E-mail: BL@iat.uni-bremen.de

Abstract: The problem of Lyapunov Exponents (LEs) estimation from chaotic data based on Jacobian approach by polynomial models is considered. The optimum embedding dimension of reconstructed attractor is interpreted as suitable order of model. Therefore, based on global polynomial modeling of system, a novel criterion for selecting the embedding dimension is presented. By considering this dimension as the model order, the best nonlinearity degree of polynomial is estimated. The selected structure is used for local estimating of Jacobians to calculate the LEs. This suitable structure of polynomial model leads to better results and avoids of sporious LEs. Simulation results show the effectiveness of proposed methodology. *Copyright © 2003 IFAC*

Keywords: Chaos, time series, polynomial models, Lyapunov exponents, factorization, Jacobian matrices.

1. INTRODUCTION

The evolutionary motion of the processes can be identified by applying appropriate techniques of modelling. The pattern in output measurements reflects different behaviours, which can be classified as linear, non-linear, chaotic non-linear or pure random processes. Obviously, different types of models and analysis are needed for each case such that improper selection of required tool may lead to lose the deterministic structure. For example, output data from some deterministic low dimensional non-linear chaotic systems seem stochastic when linear techniques are used for analysis. However, uncovering this deterministic structure is important because it leads to more realistic and better modelling (Lillekjendlie, *et al.*, 1994).

Deterministic chaos appears in variety of fields of science like biology, phisiology and engineering. Therefore, recognizing the chaotic behaviour of dynamical systems when only output data are available, is an important field of research. To achieve this, there are some quantitative measures included fractal dimension, entropy and Lyapunov exponents.

The Lyapunov exponents (LEs) which are related to the exponentially divergence or convergence of nearby orbits in phase space, is conceptually the most basic indicator of deterministic chaos. A system with one or more positive LEs is defined to be chaotic. These exponents provide not only a qualitative characterization of dynamical behaviour but also the exponent itself determines the measure of predictability. Hence, the estimation of the LEs as the useful dynamical classifier for deterministic chaotic systems is an important issue in nonlinear time series analysis. Two general approaches for computing the LEs from output time series are geometrical and Jacobian approaches. In geometrical approach, LEs are calculated based on the long term evolution of an infinitesimal sphere of initial conditions (Wolf, *et al.*, 1985; Kantz, 1994). In the Jacobian approach, local Jacobian matrices are estimated and the long term product of matrices is used for LEs computation

(Eckmann, *et al.*, 1986; Brown, *et al.*, 1991; Oiwa and Fiedler-Ferrara, 1998). The important step in later approach is to estimate the Jacobian matrices. Since the LEs are derived from the eigenvalues of the Jacobians, any small error in the computation of Jacobians can cause major error in the LE computation. Some general perturbation results and error analysis in QR algorithms for computing LEs can be found in (Diect, *et al.*, 1997).

In the previous works on the subject, it is usually tried to find the Jacobians by locally linear mapping of neighbourhoods near the reference trajectory to neighbourhood at a subsequent time. In (Brown, *et al.*, 1991), it is shown that using local neighbourhood-to-neighbourhood mappings with higher order Taylor series, rather than just local linear maps is advantageous. In (Eckmann, *et al.*, 1986), the linearized map from the neighbour data set into m step ahead of this set is considered as an approximation for the tangent map. In addition, in (Briggs, 1990) it is suggested that the estimation of Jacobian matrices is best achieved in the case of noisy data by least squares polynomial fitting. An adaptive method for the computation of the Jacobian matrices has been presented in (Khaki-Sedigh, *et al.*, *2002*). Four other methods for estimating the Jacobian has been referred to in (McCaffrey, *et al.*, 1992), including the local thin-plate splines, radial basis functions, projection pursuit and neural nets.

In this paper, the problem of LE computation from chaotic time series based on Jacobian approach by using polynomial modeling is considered. The main idea is similar to (Briggs, 1990), however, the model fitting process in (Briggs, 1990) is done on the vectors in reconstructed state space without any transparent decision criteria for choosing the proper order of model and also degree of nonlinearity. On the other hand, different model orders lead to different Jacobians which result in various LEs. In addition, large value of orders obtains the sporious LEs which confuses the selection of true LEs. In the first part of this paper, it is discussed that, the order of model plays the role of embedding dimension in the state space reconstruction. Therefore, by using global polynomial modeling of underlying dynamical system and considering the first step ahead prediction error, a criterion for estimating the suitable model order is presented. Then, the best degree of nonlinearity of polynomial model is estimated based on prediction error of different models. The final structure is used in each point of reconstructed space to model the dynamics of system locally and estimate the Jacobian matrices in each point. Finally, LEs are estimated by using QR factorization method.

The background materials are given in sec. 2. In sec. 3.1, by considering the global polynomial modeling, a new idea for selecting the suitable order of model is presented. The estimation of LEs based on local polynomail modeling is presented in sec. 3.2. Finally, simulation results are provided to show the effectiveness of the proposed methodology for the well-known chaotic dynamical systems in section 4.

2. BACKGROUND MATERIALS

In order to use the Jacobian approach to calculate the LEs, related mathematical background is explained in this section. In addition, the general form of polynomial models which used in the next section is presented.

2.1. Computation of LEs by Jacobian Approach

Consider the discrete dynamical system described in the following form:

$$X_{k+1} = F(X_k) \quad k = 0,1,\cdots \tag{1}$$

where X_k is the state vector in the R^m space and $F(\cdot)$ is a continuously differentiable nonlinear function. The linearized system for a small range around the operational trajectory in the phase space can be written as:

$$X_{k+1} = J_k . X_k \, , \, J_k = \left.\frac{\partial F}{\partial X}\right|_{X_k} \in R^{m\times m} \quad k = 0,1,\cdots \tag{2}$$

where J_k is the Jacobian matrix in point k. The LE is defined as follows (Diect, *et al.*, 1997):

Definition 1- Let $Y^k = J_{k-1} \cdot J_{k-2} \cdots J_0$, then the following symmetric positive definite matrix exists:

$$\Lambda = \lim_{k\to\infty}\left((Y^k)^T \cdot Y^k\right)^{\frac{1}{2k}} \tag{3}$$

and the logarithms of its eigenvalues are called the *Lyapunov Exponents.*

However, the computation of LEs by definition 1 has some problems in practice that is treated in sec. 2.2.

2.2. QR Algorithm

Computation of the LEs by using definition 1 has some problems. The first problem is that for large value of k, the fundamental solution Y^k may go to very large values and actually, the calculation of Λ is not possible. Further, the computation of Y^k should be such that the linear independency of the columns is maintained. Otherwise, this computation leads only to the largest LEs. To deal with these problems, the QR factorisation algorithm is used for approximation of LEs (Eckmann, *et al.,* 1986; Brown, *et al.,* 1991; Oiwa and Fiedler-Ferrara, 1998; Diect, *et al.,* 1997). The steps involved in this method can be summarized as follows:

I. Given orthogonal Q_0 such that $Q_0^T \cdot Q_0 = I$.

II. Solve $Z_{k+1} = J_k \cdot Q_k$, $k = 0, 1, \cdots$ and obtain the decomposition: $Z_{k+1} = Q_{k+1} \cdot R_{k+1}$ where Q_k is an orthogonal matrix and R_{k+1} is upper triangular matrix with positive diagonal elements.

III. The LEs can be calculated as follows:

$$
\begin{aligned}
\lambda_j &= \lim_{k \to \infty} \frac{1}{k} \log\left((R_k)_{jj} \cdots (R_1)_{jj}\right) \\
&= \lim_{k \to \infty} \frac{1}{k} \sum_{i=1}^{k} \log\left((R_i)_{jj}\right) \quad j = 1, \cdots, m
\end{aligned}
\tag{4}
$$

2.3. Polynomial Model

The dynamical behavior of system is considered with the following nonlinear difference equation:

$$
y(k) = f(X(k-1)) \tag{5}
$$

where $f(.)$ is a continuously differentiable function and $X(k)$ is state vector. It is supposed that the output data of the dynamical system is available as the following univariate time series:

$$
y(t + t_s),\ y(t + 2t_s),\ \ldots.,\ y(t + Nt_s) \tag{6}
$$

where t_s is the sampling time and N is total number of measurements.

In many practical situations the structure of the underlying dynamical system which generates the data is unknown. Depending on the objectives, there are different theories, such as the functional theory, which are suitable for special analysis of nonlinear systems. In this paper, an arbitrary degree polynomial as follows is selected to fit the output data:

$$
\begin{aligned}
y(k) = \theta_0 + \cdots + \sum_{i=1}^{d}\sum_{j=i}^{d} \theta_{2ij}\, y(k-i) y(k-j) + \\
\sum_{i=1}^{d}\sum_{j=i}^{d}\sum_{p=j}^{d} \theta_{3ijp}\, y(k-i) y(k-j) y(k-p) + \cdots
\end{aligned}
\tag{7}
$$

where d is the order of model and degree of nonlinearity is determined by number of summation terms. For the model order d and degree of nonlinearity n the number of parameters in vector Θ that should be estimated to identify the underlying model is:

$$
no = \frac{(d+n)!}{d!.n!} \tag{8}
$$

This identification can be done by using Least Squares method.

3. ESTIMATING THE LES BASED ON POLYNOMIAL MODELLING

In order to estimate the LEs based on polynomial modelling, at first the idea for selecting the most suitable structure of polynomial model is presented. Then by using this structure, the LE estimation procedure is explained.

3.1. Estimation of theEmbedding Dimension

In practical situations the dynamical equations (5) and state variables of chaotic system are not available. However, the original phase space geometery can be reconstructed by applying the method of delays by Takens' theorem (Takens, 1981) and the invariant measures of chaotic system, like LEs, can be calculated from this reconstrucred space. In method of delays, the delay coordinates as follows are used to form a *d* dimensional vector space:

$$
X(t) = [y(t-(d-1)\tau), \ldots, y(t-\tau), y(t)] \tag{9}
$$

where τ is the lag time and the dimension of this reconstructed space, *d*, is called the embedding dimension which its minimum value is looked for. There are many methods that concern this dimension including *False Neighbour* method (Kennel, *et al., 1992*) and *Singular Value Decomposition* method (Broomhead and King, 1986) and related papers for their modifications and using the extension for multivariate time series case (Ataei, *et al.,* 2003a, b; Mees, *et al., 1987*).

In this paper, in order to model the reconstructed state space, the vector (9) is considered as the state vector. The lag time can be selcted by using well known methods based on correlation coefficient or mutual information (Fraser and Swinney, 1986). Here, the lag time is fixed and the vector (9) with normalized step is considered. Then for deriving the state equations, the function $f(\cdot)$ is estimated by polynomial model such that:

$$
y(t+1) = f(X(t)) \tag{10}
$$

By this, the number of required state variables which, is the order of autorgressive polynomial model, will also be equal *d*. Therefore, the minimum embedding dimension is chosen as the most suitable order of the polynomial model. The under-estimation of order causes the loss of dynamics of data generator, and

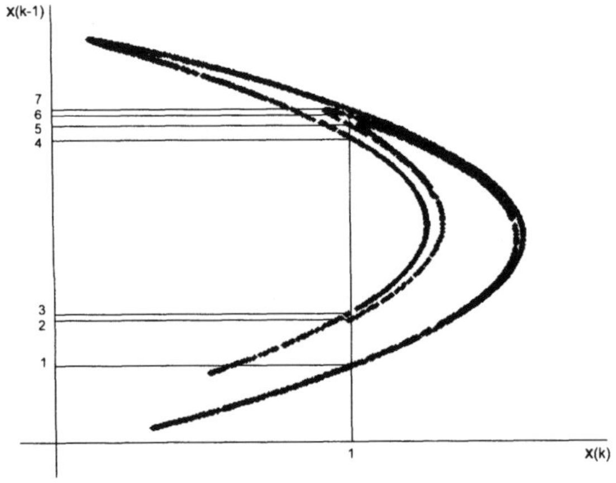

Fig. 1. Attractor of a two dimensional chaotic system, if order is under-estimated to d=1 all the points 1, ..., 7 on X(k-1) axis are projected on point 1 in X(k) axis.

the Lyapunov spectrum can not be obtained neither complete nor accurate from this model. On the other hand, the over-estimation of model order leads to sporious LEs which perturbs the decision on true LEs.
In this subsection, by using global polynmial model of system, a criterion for choosing the suitable model order or embedding dimension is presented. To show the main idea, consider a two dimensional chaotic system with the state trajectory as shown in figure 1. The objective is to find the model as (5) by using the structure of (7). If the order of model is under-estimated to d=1, the obtained model will project the points $(i,1)$ $i=1,\cdots,7$ to the same one step ahead value, say $\hat{x}_{k+1}$. Therefore, the first step ahead prediction error for each transition of this point is:

$$e(i,1)=\hat{x}_{k+1}-x_{k+1}(i,1) \quad i=1,\cdots,7 \qquad (11)$$

where $x_{k+1}(i,1)$ denotes to the true first step ahead value. These errors will be large since only one fixed projection has been considered for all points. If the order of model is selected to d=2, then for each points of $x_{k+1}(i,1)$ $i=1,\cdots,7$ different one step ahead value is estimated. The prediction error in this case is:

$$e(i,1)=\hat{x}_{k+1}(i,1)-x_{k+1}(i,1) \quad i=1,\cdots,7 \qquad (12)$$

The errors in this case are much smaller than the previous case since the error is only due to the capability of selected model to predict the next value. The Least Mean Squares (LMS) of these errors for all the points of attractor as follows is also different value in these two case:

$$\sigma=\left(\frac{1}{N}\left(\sum_{k=1}^{N}e_k^2\right)\right)^{\frac{1}{2}} \qquad (13)$$

where N is the total number of points. Therefore, for selecting the optimum model order, the value of σ is considered for different model order. The dimension that σ is decreased to a lower level and after that takes approximately the same value is the suitable model order and is denoted by d_{opt}. It should be noted this idea can be used independent of the type of model if the selected function for modeling satisfies the continous differentiability property.

After finding the d_{opt}, for this fixed order, values of σ for different degree of nonlinearity is calculated and the degree which σ takes approximately the same values is selected as the nonlinearity degree of polynomial and denoted by n_p.

3.2. LEs estimation

The objective of this subsection is the estimation of LEs for time series (6) by Jacobian approach. Therefore, the accurate estimation of Jacobian matrices are required. For this, by considering the polynomial model as (7) and using the procedure of subsection (3.1), d_{opt} and n_p is estimated. Then, by defining the state vector as the following delay vector

$$X(t)=\left(y(t-(d_{opt}-1)\tau),\cdots,y(t-\tau),y(t)\right) \qquad (14)$$

where $\tau=n.t_s$ is the lag time, a canonical state space representation of the system is obtained as:

$$X(t+1)=\begin{bmatrix} y(t-(d_{opt}-2)\tau) \\ y(t-(d_{opt}-3)\tau) \\ \vdots \\ f(X(t)) \end{bmatrix}=\begin{bmatrix} x_2(t) \\ x_3(t) \\ \vdots \\ f(X(t)) \end{bmatrix} \qquad (15)$$

Then, the Jacobian matrix J_k in each point k of the typical trajectory for this canonical representation is as:

$$J_k=\begin{bmatrix} 0 & 1 & \cdots & 0 & 0 \\ 0 & 0 & 1 & \cdots & 0 \\ \vdots & \vdots & \vdots & \ddots & \vdots \\ Df_1 & Df_2 & \cdots & Df_{m-1} & Df_m \end{bmatrix} \qquad 16)$$

where $Df_i=\frac{\partial f}{\partial x_i}$ and the Df_i $i=1,...,d$ are computed in terms of the parameters of the model. As it is seen, the accuracy of Jacobians depend on estimated parameter of the model. In order to have accurate parameters estimation in each step, the local polynomial modeling is used. For this, a general polynomial with order d_{opt} and degree of nonlinearity n_p which have been selected in the last part is considered. To estimate the parameters of the model, for each point of the reconstructed state space with dimension d_{opt} like (14), some nearer neighbours are selected. The number of neighbours should be more than number of parameters which should be estimated. Then the LMS method is used to estimate the model parameters in each point. The obtained local model is used to calculate the Jacobian matrix as (16). This is accomplished for all the points on the reconstructed attractor and the calculated Jacobians are used in the QR algorithm for calculating the LEs.

4. SIMULATION RESULTS

To show the effectiveness of the proposed procedure in sec. 3, the algorithms are applied to some well-known chaotic systems. These chaotic systems and their characteristics are defined in table 1 (Briggs, 1990; Schuster, 1995; Wolf, *et al.,* 1985). In the first part, the presented method for estimating the suitable order of model based on global polynomial modeling is implemented. For obtained time series of each system, the developed general program of polynomial modeling is applied for various d and n and σ is computed for all the cases in a look up table. Based on the theoretic discussions on sec. 3.1, then the optimum order of model is selected in each case. The mean squared of error, σ, for defined chaotic systems have been shown in table 2. According to these results, the optimum order of model and nonlinearity degree of each system are estimated as the table 3.

By using the values in table 3 for model order and nonlinearity degree, an structure of polynomial model is selected for each chaotic system. The LEs, then can be estimated by the method explained in subsection 3.2.

Table 1. The characteristics of chaotic systems for simulations

Systems	Dynamical Equation	Coeffs.	LEs
Logistic	$x_{k+1} = r \cdot x_k(1-x_k)$	$r=4$	0.693
Triangular	$x_{k+1} = r \cdot (1-2\cdot\|0.5-x_k\|)$	$r=0.91$	0.5988
Henon	$\begin{cases} x_{k+1} = 1-ax_k^2+y_k \\ y_{k+1} = bx_k \end{cases}$	$\begin{cases} a=1.4 \\ b=0.3 \end{cases}$	$\begin{cases} 0.4188 \\ -1.62 \end{cases}$
Lorenz	$\begin{cases} \dot{x}=\sigma(y-x) \\ \dot{y}=x(R-z)-y \\ \dot{z}=xy-bz \end{cases}$	$\begin{cases} \sigma=16 \\ R=45.92 \\ b=4 \end{cases}$	$\begin{cases} 1.497 \\ 0 \\ -22.45 \end{cases}$

Table 2. Mean squared of first step ahead prediction error of defined chaotic systems for different values of model order (d) and degree of nonlinearity (n)

n \ d	1	2	3	4	5
Logistic map, N=500					
1	0.9992	1.0016	1.0032	1.0027	1.0065
2	5.43e-6	5.69e-6	5.95e-6	5.99e-6	6.0e-6
3	2.30e-5	2.50e-5	2.72e-5	2.86e-5	3.09e-5
4	6.83e-5	7.92e-5	9.04e-5	9.72e-5	1.07e-4
Triangular map, N=500					
1	0.9018	0.8570	0.8627	0.8607	0.8550
2	0.2587	0.2597	0.2599	0.2596	0.2618
3	0.2233	0.2043	0.2045	0.2049	0.2069
4	0.1337	0.1228	0.1224	0.1231	0.1245
Henon map, N=500, time series of x variable					
1	0.9394	0.9283	0.8763	0.8649	0.8553
2	0.2787	1.75e-7	1.52e-7	1.96e-7	3.07e-7
3	0.2748	4.13e-7	2.76e-7	4.80e-7	1.76e-6
4	0.2742	3.05e-6	9.71e-6	2.69e-6	2.62e-5
Lorenz system, N=1000, time series of x variable, lag time= 0.05					
1	0.4115	0.2784	0.2284	0.2062	0.1988
2	0.4115	0.2786	0.2292	0.2074	0.2004
3	0.3830	0.0781	0.0167	0.0096	0.0092
4	0.3830	0.0802	0.0105	0.0100	0.0096

Table 3. The estimated optimum order of model and nonlinearity degree of polynomial model

	d_{opt}	n_p
Logistic map	1	2
Triangular map	1	4
Henon map	2	2
Lorenz system	3	3

The estimated LEs are shown in the table 4. The results for the Logistic and Henon maps that polynomial models exactly fit the dynamics, are almost the same as results obtained by computation from dynamical equations. For triangular map and Lorenz flow that the equations are not in polynomial form the results are also well acceptable. This accuracy is due to optimum selection of structure for

Table 4. The estimated LEs by using local polynomial modeling

Systems	Data	Estimated LEs
Logistic	N=1000	0.6926
Triangular	N=1000	0.5871
Henon	N=1500	0.4134, -1.6172
Lorenz	N=1500	1.5254, 0.0128, -18.4338

Table 5. The estimated LEs of Lorenz system with improper choice of model structure

d	n_p	Lyapunov Exponents
4	3	2.3044, 0.7538, -1.6386, -21.5328
3	2	1.5551, 1.1270, -12.8595

the model which was suggested in sect.3. To show the disadvantage of improper choice of model structure, the estimated LEs of Lorenz system for non-optimum structure of model are shown in table 5. The improper structure of model leads not only to inaccurate LEs but also obtains the sporious LEs. At the end, it should be noted that the obtained polynomial models can also be used for prediction of chaotic time series as is done in (Casdagli, 1989).

5. CONCLUSIONS

In this paper, an improved method for the estimation of LEs based on polynomial modelling is proposed. Since in practical situations the structure of the system is unknown, a criteria for the selection of optimum structure of polynomial model is presented. This is done by using the global polynomial modeling of underlying system and considering the first step ahead prediction error. Then, this structure is used in every point of reconstructed space to model the dynamic of system locally and estimate the Jacobian matrices in each point. Finally, LEs are estimated by using QR factorization method. This procedure ensures the effective polynomial based calculation of LEs in the face of unknown chaotic dynamical systems and avoids calculation of sporious LEs. To show the effectiveness of the proposed methodology, it is applied to some well-known chaotic systems and simulation results are provided.

ACKNOWLEDGEMENTS

We would like to thank the DAAD (German Academic Exchange Service) for providing financial assistance to make this research possible.

REFERENCES

Ataei, M., A. Khaki-Sedigh, B. Lohmann, and C. Lucas (2003a). Determining embedding dimension from output time series of dynamical systems- Scalar and multiple output cases. Proceedings of the 2nd *International Conference on System Identification and Control Problems*, Moscow, Russia, pp. 1004-1013.

Ataei, M., A. Khaki-Sedigh, B. Lohmann, and C. Lucas (2003b). Determination of embedding dimension using multiple time series based on singular value decomposition. Proceedings 4th *International Symposium on mathematical Modeling*, Vienna, Austria, pp. 190-196.

Briggs, K. (1990). An improved method for estimating Liapunov exponents of chaotic time series. *Phys. Lett. A*, **151**, pp. 117-121.

Broomhead, D. S., and G. P. King (1986). Extracting qualitative dynamics from experimental data. *Physica D*, **20**, pp. 217-236.

Brown, R., P. Bryant, and H. D. I. Abarbanel (1991) Computing the Lyapunov spectrum of a dynamical system from an observed time series. *Phys. Rev. A*, **43**, no. 6, pp. 2787-2806.

Casdagli, M. (1989). Non-linear prediction of chaotic time series. *Physica D*, **35**, pp. 335-356.

Diect, L., R. D. Russell, and E. S. Van Vleck (1997). On the computation of Lyapunov exponents from continuous dynamical systems. *SIAM J. Numer. Anal.*, **34**, no.1, pp. 402-423.

Eckmann, J. P., S. O. Kamphorst, D. Ruelle, and S. Ciliberto (1986). Liapunov exponents from time series. *Phys. Rev. A*, **34**, no. 6, pp. 4971-4979.

Fraser, A. M., and H. L. Swinney (1986). Independent coordinates for strange attractors from mutual information. *Phys. Rev. A*, **33**, no. 2, pp. 1134-1140.

Kantz, H. (1994). A robust method to estimate the maximum Lyapunov exponent of a time series. *Phys. Lett. A*, **185**, pp. 77-87.

Kennel, M. B., R. Brown, and H. D. I Abarbanel (1992). Determining embedding dimension for phase space reconstruction using a geometrical construction. *Phys. Rev. A*, **45**, no 6., pp.3403-3411.

Khaki-Sedigh, A., M. Ataei, B. Lohmann, and C. Lucas (2002). Adaptive calculation of Lyapunov exponents from time series observations of chaotic time varying dynamical systems. submitted for publication in *International Journal of Bifurcation and Chaos*.

Lillekjendlie, B., D. Kugiumtzis, and N. Christophersen (1994). Chaotic time series, Part II: System identification and prediction. *Modelling, Identification and Control*, **15**, no. 4, pp. 225-243.

McCaffrey, D. F., S. Ellner, A. R. Gallant, and D. W. Nychka (1992). Estimating the Lyapunov exponent of a chaotic system with nonparametric regression. *Journal of the Amer. Sta.l Assoc.*, **87**, no. 419, pp. 682-695.

Mees, A. I., P. E. Rapp, and L. S. Jenning (1987). Singular value decomposition and embedding dimension. *Phys. Rev. A*, **36**, no. 1, pp. 340-346.

Oiwa, N. N., and N. Fiedler-Ferrara (1998). A fast algorithm for estimating Lyapunov exponents from time series. *Phys. Let. A* **246**, pp. 117-121.

Schuster, H. G. (1995). *Deterministic Chaos: an introduction*, VCH Verlasgesellschaft, Germany

Takens, F. (1981). Detecting strange attractors in turbulence. In: *Lecture Notes in Mathematics*. (D.A. Rand, L.S. Young. (Ed)) **898**, pp. 366-381, Springer, Berlin.

Wolf, A., J. B. Swift, H. L. Swinney, J. A. Vastano (1985). Determining Lyapunov exponents from a time series. *Physica D*, **16**, pp. 285-317.

IFAC Publications
www.elsevier.com/locate/ifac

SAMPLING DENSITY DESIGN FOR PARTICLE FILTERS

Miroslav Šimandl Ondřej Straka

Department of Cybernetics, University of West Bohemia in Pilsen
Univerzitni 8, Pilsen, Czech Republic
simandl@kky.zcu.cz, straka30@kky.zcu.cz

Abstract: Particle filters for state estimation of discrete time dynamic stochastic systems are treated. The stress is laid on design of sampling pdf which is significant for quality of the particle filters. A new functional sampling density design based on utilization of transient and measurement pdf's is proposed. The functional approach compares two pdf's of a reference variable which are obtained by transformation of the transient and measurement pdf's, using Kullback J-divergence. The functional approach to sampling density function synthesis using transient and measurement density functions can be understood as improvement of the sampling density design of the auxiliary particle filter which uses only point estimate of the transient pdf. The higher quality of the functional particle filter with respect to the bootstrap and auxiliary particle filter is illustrated in a numerical example.

Keywords: state estimation, nonlinear systems, particle filters, sampling density

1. INTRODUCTION

Recursive state estimation of discrete-time nonlinear dynamic systems from noisy measurement data has been a subject of considerable research interest for the last three decades. Bayesian approach can be used for design of general solution of state estimation problem. The closed form solution of the Bayesian recursive relations (BRR) is available only for a few special cases. There are three main ways of BRR solution: analytical approach based on model linearization and Gaussian sum approximation of probability density functions (Sorenson and Alspach, 1970), numerical approach based on numerical solution of the integral in BRR and leading to grid-based filters (Šimandl *et al.*, 2002) and simulation approach using Monte Carlo (MC) approximation (Gordon *et al.*, 1993) which has been dominating in nonlinear estimation since nineties. Cheap and formidable computational power, easy implementation and applicability in very general settings are the main reason for such growing interest in nonlinear estimation field.

Sequential Monte Carlo methods, also known as particle, bootstrap or condensation filters, are a set of simulation-based methods which provide a convenient approach to computing the posterior probability density function of (pdf) the state. The fundamental paper dealing with Monte Carlo solution of BRR was given by Gordon *et al.* (1993) where the first effective filtering method in MC technique framework was proposed. The principal of the paper was introducing the resampling step to sequential importance sampling method which disposed convergence problem. On the other hand the resampling step causes some degeneracy issues which were treated in and Berzuini and Gilks (2001) using Markov Chain Monte Carlo simulation methods.

Another two important issues for particle filter design are sampling density design and efficient sample size setting which affect estimation quality significantly. Sampling density design was treated in e.g. (Liu and Chen, 1998) but the important contribution to this topic was made in Pitt and Shephard (2001) where the sampling probability density function included information from current measurement and moreover this sampling pdf was suitable for large class of systems. Efficient sample size setting has been disregarded for a long time although it represents the key parameter of the particle filter design, Some advances in effective sample size setting were done in (Šimandl and Straka, 2002*b*) where the Cramér Rao bound was used as a gauge for quality evaluation of the particle filter and in (Fox, 2001) where the number of samples is based on the likelihood of observations.

During nineties the combined state and parameter estimation and state estimation with heavy-tailed pdf were treated as well because the standard approaches fail in these cases.The correction of the basic algorithm for combined state and parameter estimation was done in e.g. Liu and West (2001) using artificial parameter evolution. The particle filter design for heavy-tailed pdf was shown in e.g. van der Merwe *et al.* (2000) using the unscented particle filter.

Goal of the paper is an attempt on quality improvement of particle filters using a new sampling density design. The intention is to utilize transient pdf for state prediction instead of its point estimates, see (Pitt and Shephard, 2001), that is standardly used in sampling density design. The motivation for the step is the belief that the taking in account the complete available information about the state space model enables to obtain more appropriate sampling density and consequently to improve the particle filter quality.

The paper is organized as follows: The usage of the particle filter in state estimation is introduced in Section 2. Problem formulation is accomplished in Section 3 and the functional sampling density design is dealt with in Section 4. Properties of the density design are discussed in Section 5. Further, the numerical illustration of particle filter with the functional sampling density design is provided in Section 6 and finally, main results are summarized in Section 7.

2. STATE ESTIMATION BY PARTICLE FILTER

This section provides usage of the particle filter in the state estimation of a discrete-time nonlinear stochastic system. Consider the discrete time stochastic system:

$$\mathbf{x}_{k+1} = \mathbf{f}_k(\mathbf{x}_k, \mathbf{w}_k), \quad k = 0, 1, 2, \dots \tag{1}$$

$$\mathbf{z}_k = \mathbf{h}_k(\mathbf{x}_k) + \mathbf{v}_k, \quad k = 0, 1, 2, \dots \tag{2}$$

where the vectors $\mathbf{x}_k \in \mathbb{R}^n$, $\mathbf{z}_k \in \mathbb{R}^m$ represent the state of the system and the measurement at time k, respectively, $\mathbf{w}_k \in \mathbb{R}^n$ and $\mathbf{v}_k \in \mathbb{R}^m$ are state and measurement zero mean white noise sequences, mutually independent and independent of $\mathbf{x}_0$, with known pdf's $p(\mathbf{w}_k)$ and $p(\mathbf{v}_k)$ respectively, $\mathbf{f}_k : \mathbb{R}^n \times \mathbb{R}^n \to \mathbb{R}^n$, $\mathbf{h}_k : \mathbb{R}^n \to \mathbb{R}^m$ are known vector functions and the pdf $p(\mathbf{x}_0)$ of the initial state $\mathbf{x}_0$ is known.

The general solution of state estimation problem is provided by the BRR which produce filtering pdf $p(\mathbf{x}_k|\mathbf{z}^k)$ and predictive pdf $p(\mathbf{x}_{k+1}|\mathbf{z}^k)$, where $\mathbf{z}^k = [\mathbf{z}_0^T, \mathbf{z}_1^T, \dots, \mathbf{z}_k^T]^T$

The idea of the particle filter in nonlinear state estimation is to approximate the filtering pdf $p(\mathbf{x}_k|\mathbf{z}^k)$, $k = 0, 1, 2 \dots$, by the empirical filtering pdf $r_N(\mathbf{x}_k|\mathbf{z}^k)$ given by ν random samples of state at time instant k denoted as $\mathbf{x}_k^{(i)}$, $i = 1, \dots, \nu$ and corresponding weights $w_k^{(i)}$, $i = 1, \dots, \nu$.

Initialization: The filter initializes by generating samples $\{\mathbf{x}_0^{(i)}\}$ from the prior pdf $p(\mathbf{x}_0|\mathbf{z}^{-1})$

$$\mathbf{x}_0^{(i)} \sim p(\mathbf{x}_0|\mathbf{z}^{-1}); \quad \text{i=1,2,}\dots,\nu\,.$$

The weights $\{\mathfrak{w}_0^{(i)}\}$ are associated with the samples $\{\mathbf{x}_0^{(i)}\}$, where $\mathfrak{w}_0^{(i)} \propto p(\mathbf{z}_0|\mathbf{x}_0^i)$; i=1,2,...,$\nu$.

The empirical pdf $r_N(\mathbf{x}_0|\mathbf{z}^0)$ given by the samples $\{\mathbf{x}_0^{(i)}\}$ and the weights $\{\mathfrak{w}_0^{(i)}\}$

$$r_N(\mathbf{x}_0|\mathbf{z}^0) = \sum_{i=1}^{\nu} \mathfrak{w}_0^{(i)} \delta(\mathbf{x}_0 - \mathbf{x}_0^{(i)})$$

approximates the filtering pdf $p(\mathbf{x}_0|\mathbf{z}^0)$. The function $\delta(\cdot)$ is Dirac function defined as $\delta(\mathbf{x}) = 0$ for $\mathbf{x} \neq 0$ and $\int \delta(\mathbf{x})d\mathbf{x} = 1$. It holds that $\sum_{i=1}^{\nu} \mathfrak{w}_0^{(i)} = 1$. Let the time step k be one $k = 1$.

Resampling: This step is dedicated to resampling which avoids samples degeneration. Simply it multiplies samples with corresponding high weights and suppresses samples of low weights.

Filtering: Samples for the next time instant k are generated from the global sampling pdf $\pi(\mathbf{x}_k|\mathbf{x}_{k-1}, \mathbf{z}_k)$

$$\mathbf{x}_k^{(i)} \sim \pi(\mathbf{x}_k|\mathbf{z}^{k-1}, \mathbf{z}_k); \quad \text{i=1,2,}\dots,\nu\,,$$

where

$$\pi(\mathbf{x}_k|\mathbf{x}_{k-1}, \mathbf{z}_k) = \sum_{i=1}^{\nu} \mathfrak{v}^{(i)} \pi(\mathbf{x}_k|\mathbf{x}_{k-1}^{(i)}, \mathbf{z}_k). \tag{3}$$

To generate samples $\mathbf{x}_k^{(i)}$; i=1,2,...,ν one has to first simulate ν indices $j(i)$; i=1,2,...,ν from multinomial distribution with parameters given by primary weights $\mathfrak{v}^{(i)}$. Then each sample $\mathbf{x}_k^{(i)}$ is generated from the local sampling pdf $\pi(\mathbf{x}_k|\mathbf{x}_{k-1}^{(j(i))}, \mathbf{z}_k)$. The weights $\{\mathfrak{w}_k^{(i)};\ \text{i=1,2,}\dots,\nu\}$ associated to the samples $\{\mathbf{x}_k^{(i)};\ \text{i=1,2,}\dots,\nu\}$ are calculated using the following form

$$\mathfrak{w}_k^{(i)} \propto \frac{p(\mathbf{z}_k|\mathbf{x}_k^{(i)})\, p(\mathbf{x}_k^{(i)}|\mathbf{x}_{k-1}^{(j(i))})}{\pi(\mathbf{x}_k^{(i)}|\mathbf{x}_{k-1}^{(j(i))}, \mathbf{z}_k)} \mathfrak{w}_{k-1}^{(j(i))}. \tag{4}$$

The empirical pdf $r_N(\mathbf{x}_k|\mathbf{z}^k)$ given by the samples $\{\mathbf{x}_k^{(i)}\}$ and the weights $\{\mathfrak{w}_k^{(i)}\}$

$$r_N(\mathbf{x}_k|\mathbf{z}^k) = \sum_{i=1}^{\nu} \mathfrak{w}_k^{(i)} \delta(\mathbf{x}_k - \mathbf{x}_k^{(i)})$$

approximates the filtering pdf $p(\mathbf{x}_k|\mathbf{z}^k)$.

Let $k \leftarrow k + 1$ and continue with **Resampling**.

3. PROBLEM FORMULATION

Standard particle filter algorithm presented in Section 2 consists of three basic steps whereas quality of the particle filter is strongly affected especially by the step 4. Object of this step is to generate samples $\{\mathbf{x}_k^{(i)};\ \ i=1,2,\ldots,\nu\}$ for the next time instant k. In order to be able to draw the samples, it is necessary to define the global sampling probability density function $\pi(\mathbf{x}_k|\mathbf{x}_{k-1},\mathbf{z}_k)$

$$\pi(\mathbf{x}_k|\mathbf{x}_{k-1},\mathbf{z}_k) = \sum_{i=1}^{\nu} \mathfrak{v}(\mathbf{x}_{k-1}^{(i)},\mathbf{z}_k)\pi(\mathbf{x}_k|\mathbf{x}_{k-1}^{(i)},\mathbf{z}_k), \tag{5}$$

where $\mathfrak{v}(\mathbf{x}_{k-1}^{(i)},\mathbf{z}_k) > 0\ i = 1,\ldots\nu$ are primary weights, $\sum_{i=1}^{\nu}\mathfrak{v}(\mathbf{x}_{k-1}^{(i)},\mathbf{z}_k) = 1$ and the pdf $\pi(\mathbf{x}_k|\mathbf{x}_{k-1}^{(i)},\mathbf{z}_k)$ will be called local sampling density. It is probability density function of the state $\mathbf{x}_k$ at time instant t_k conditioned by two quantities, $\mathbf{x}_{k-1}^{(i)}$ and the current measurement $\mathbf{z}_k$. Note that the quantity $\mathbf{x}_{k-1}^{(i)}$ contains information given by measurements $\mathbf{z}_0,\ldots,\mathbf{z}_{k-1}$ and thus the pdf $\pi(\mathbf{x}_k|\mathbf{x}_{k-1}^{(i)},\mathbf{z}_k)$ of state $\mathbf{x}_k$ is conditioned by measurements $\mathbf{z}_0,\ldots,\mathbf{z}_k$. The best choice of the local sampling pdf $\pi(\mathbf{x}_k|\mathbf{x}_{k-1}^{(i)},\mathbf{z}_k)$ is $p(\mathbf{x}_k|\mathbf{x}_{k-1}^{(i)},\mathbf{z}_k)$ and then the global sampling pdf $\pi(\mathbf{x}_k|\mathbf{x}_{k-1},\mathbf{z}_k)$ is given by:

$$\pi(\mathbf{x}_k|\mathbf{x}_{k-1},\mathbf{z}_k) = \sum_{i=1}^{\nu} \frac{1}{\nu} p(\mathbf{x}_k|\mathbf{x}_{k-1}^{(i)},\mathbf{z}_k). \tag{6}$$

The choice (6) is the best because the samples generated from this global sampling pdf do not increase variance of weights $\mathfrak{w}_k^{(i)}$ in (4) which is directly related to estimation quality (Pitt and Shephard, 2001). Except a few special cases of system, e.g. measurement function $\mathbf{h}_k : \mathbb{R}^n \to \mathbb{R}^m$ in (2) is linear, the pdf $p(\mathbf{v}_k)$ of measurement noise $\mathbf{v}_k$ is Gaussian and the transitional pdf $p(\mathbf{x}_k|\mathbf{x}_{k-1})$ is Gaussian as well, it is not possible to find closed form of thc pdf $p(\mathbf{x}_k|\mathbf{x}_{k-1}^{(i)},\mathbf{z}_k)$.

The simplest choice of the local sampling density $\pi(\mathbf{x}_k|\mathbf{x}_{k-1}^{(i)},\mathbf{z}_k)$ is used in so called bootstrap filter (BF) (Gordon *et al.*, 1993) where the pdf is chosen as the transient density $p(\mathbf{x}_k|\mathbf{x}_{k-1}^{(i)})$. Then the global sampling pdf $\pi(\mathbf{x}_k|\mathbf{x}_{k-1},\mathbf{z}_k)$ is given by

$$\pi(\mathbf{x}_k|\mathbf{x}_{k-1},\mathbf{z}_k) = \sum_{i=1}^{\nu} \frac{1}{\nu} p(\mathbf{x}_k|\mathbf{x}_{k-1}^{(i)}). \tag{7}$$

That means the measurement $\mathbf{z}_k$ together with some correspondent system properties are absent in sampling density design.

A better choice of the sampling density $\pi(\mathbf{x}_k|\mathbf{x}_{k-1}^{(i)},\mathbf{z}_k)$ contrary to the BF choice is introduced in Auxiliary particle filter (APF) by Pitt and Shephard (2001). Each local sampling pdf $p(\mathbf{x}_k|\mathbf{x}_{k-1}^{(i)})$ is attached by primary weight $\mathfrak{v}(\mathbf{x}_{k-1}^{(i)},\mathbf{z}_k)$. Contrary to the choice of the sampling pdf (7), the primary weight $\mathfrak{v}(\mathbf{x}_{k-1}^{(i)},\mathbf{z}_k)$ does depend on the sample $\mathbf{x}_{k-1}^{(i)}$ and the measurement $\mathbf{z}_k$. The global sampling pdf $\pi(\mathbf{x}_k|\mathbf{x}_{k-1},\mathbf{z}_k)$ has the form

$$\pi(\mathbf{x}_k|\mathbf{x}_{k-1},\mathbf{z}_k) = \sum_{i=1}^{\nu} \mathfrak{v}(\mathbf{x}_{k-1}^{(i)},\mathbf{z}_k)p(\mathbf{x}_k|\mathbf{x}_{k-1}^{(i)}). \tag{8}$$

It was shown in Pitt and Shephard (2001) that the particle filter using (8) has weights $\mathfrak{w}_k^{(i)}$ in (4) with lower variance than the weights $\mathfrak{w}_k^{(i)}$ of the BF.

4. FUNCTIONAL SAMPLING DENSITY DESIGN

The aim of the section is to introduce a technique of sampling pdf design which is based on new approach to primary weights setting which uses more prior information concerning the system then the proposal (8). The key point of the new technique is setting of primary weights using probability density functions and thus it will be called functional sampling pdf design. Prior to the new technique introduction the primary weights proposal in (8) will be treated.

The primary weight $\mathfrak{v}(\mathbf{x}_{k-1}^{(i)},\mathbf{z}_k)$ proposed in APF has the following form

$$\mathfrak{v}(\mathbf{x}_{k-1}^{(i)},\mathbf{z}_k) \propto p(\mathbf{z}_k|\mu_k^{(i)}), \tag{9}$$

where $\mu_k^{(i)}$ is mean, mode, median or another likely value of the state $\mathbf{x}_k$ which is described by the transient pdf $p(\mathbf{x}_k|\mathbf{x}_{k-1}^{(i)})$. The implication is simulation from particles associated with large predictive likelihoods in (9). However the sampling pdf (8) does not respect complete information which is given in system description. Both sources of information should be considered in primary weight setting: the information given by the sample $\mathbf{x}_{k-1}^{(i)}$ described by the transient pdf $p(\mathbf{x}_k|\mathbf{x}_{k-1}^{(i)})$ which expresses the way of particles $\mathbf{x}_k^i$ generation and the information from the measurement $\mathbf{z}_k$ given by the measurement pdf $p(\mathbf{z}_k|\mathbf{x}_k)$ which characterize the way of particles weighting. Both sources of information are present in (9) as well. The second source given by the pdf $p(\mathbf{z}_k|\mathbf{x}_k)$ is used nevertheless the first source given by the pdf $p(\mathbf{x}_k|\mathbf{x}_{k-1}^{(i)})$ is considered only partially because only corresponding point estimate is employed. Utilization of both transient and measurement pdf is fundamental idea for the new technique of sampling pdf design which will be described in the following part of the section.

To find quality of the samples $\{\mathbf{x}_{k-1}^{(i)};\ \ i=1,2,\ldots,\nu\}$ reflected in weights, it is necessary to compare both sources of information. However, the straightforward comparison procedure is not obvious because both pdf's are functions of different variables. Nevertheless the comparison can be realized after a suitable transformation of the variables.

Considering deterministic part $\mathbf{h}_k(\mathbf{x}_k)$ of the measurement $\mathbf{z}_k$ in (2) as the reference variable $\mathbf{y}_k$, i.e.

$$\mathbf{y}_k = \mathbf{h}_k(\mathbf{x}_k), \tag{10}$$

then the comparison procedure can be achieved using the pdf $p(\mathbf{y}_k|\mathbf{z}_k)$ and the pdf $p(\mathbf{y}_k|\mathbf{x}_{k-1}^{(i)})$.

The pdf $p(\mathbf{y}_k|\mathbf{z}_k)$ can be obtained from measurement equation (2) and has the form $p(\mathbf{y}_k|\mathbf{z}_k) = p_{\mathbf{v}_k}(\mathbf{z}_k - \mathbf{y}_k)$. The pdf $p(\mathbf{y}_k|\mathbf{x}_{k-1}^{(i)})$ is pdf of the random variable $\mathbf{y}_k$ obtained by transformation $\mathbf{h}_k : \mathbb{R}^n \to \mathbb{R}^m$ of random variable $\mathbf{x}_k$ which is described by the pdf $p(\mathbf{x}_k|\mathbf{x}_{k-1}^{(i)})$.

Different measures can be applied for comparison of the pdf's. The Kullback J-divergence (KJD) (Kullback and Leibler, 1951)

$$J(p(\mathbf{y}_k|\mathbf{z}_k)\|p(\mathbf{y}_k|\mathbf{x}_{k-1}^{(i)})) \triangleq \int [p(\mathbf{y}_k|\mathbf{z}_k) - p(\mathbf{y}_k|\mathbf{x}_{k-1}^{(i)})]$$
$$\log p(\mathbf{y}_k|\mathbf{z}_k) - \log p(\mathbf{y}_k|\mathbf{x}_{k-1}^{(i)})d\mathbf{y}_k \quad (11)$$

can be applied here as a standard tool. Note that the measure is symmetric.

The primary weight $\mathfrak{v}(\mathbf{z}_k|\mathbf{x}_{k-1}^{(i)})$ corresponding to the particle $\mathbf{x}_{k-1}^{(i)}$ should the higher the KJD is lower. A sound relation between the primary weight and corresponding KJD is given by the following form

$$\mathfrak{v}(\mathbf{z}_k|\mathbf{x}_{k-1}^{(i)}) = e^{-c\cdot J(p(\mathbf{y}_k|\mathbf{z}_k)\|p(\mathbf{y}_k|\mathbf{x}_{k-1}^{(i)}))}, \quad (12)$$

where the $c > 0$ is an arbitrary constant.

Now, the procedure of functional sampling pdf design can be summarized.

Algorithm of functional sampling density design

(1) **Prerequisites:** The samples $\{\mathbf{x}_{k-1}^{(i)},\ i = 1, \dots \nu\}$ are given.
(2) **Specification of reference variable pdf's:**
  (a) *Find* $p(\mathbf{y}_k|\mathbf{z}_k)$ by
  $$p(\mathbf{y}_k|\mathbf{z}_k) = p_{\mathbf{v}_k}(\mathbf{z}_k - \mathbf{y}_k)$$
  (b) *Find* $p(\mathbf{y}_k|\mathbf{x}_{k-1}^{(i)})$ *by*
  $$p(\mathbf{y}_k|\mathbf{x}_{k-1}^{(i)}) = p_{\mathbf{x}_k|\mathbf{x}_{k-1}}(\mathbf{h}_k^{-1}(\mathbf{y}_k)|\mathbf{x}_{k-1}^{(i)})\left|\frac{d\mathbf{h}_k^{-1}(\mathbf{y}_k)}{d\mathbf{y}_k}\right|,$$
(3) **Primary weights evaluation:**
  (a) *Enumerate the KJD measure*
  $$J(p(\mathbf{y}_k|\mathbf{z}_k)\|p(\mathbf{y}_k|\mathbf{x}_{k-1}^{(i)}))$$
  according to (11).
  (b) *Enumerate primary weights* $\mathfrak{v}^{(i)}$ *by*
  $$\mathfrak{v}^{(i)} = e^{-c\cdot J(p(\mathbf{y}_k|\mathbf{z}_k)\|p(\mathbf{y}_k|\mathbf{x}_{k-1}^{(i)}))},$$
  where c is a constant and $\mathfrak{v}^{(i)}$ is short notation for $\mathfrak{v}(\mathbf{x}_{k-1}^{(i)}, \mathbf{z}_k)$, $i = 1, 2, \dots, \nu$.
(4) **Functional sampling pdf generation:** Using the primary weights $\{\mathfrak{v}^{(i)};\ i=1,2,\dots,\nu\}$, transient pdf $p(\mathbf{x}_k|\mathbf{x}_{k-1})$ and samples $\{\mathbf{x}_{k-1}^{(i)};\ i=1,2,\dots,\nu\}$, the functional sampling pdf can be specified as

$$\pi(\mathbf{x}_k|\mathbf{x}_{k-1}, \mathbf{z}_k) = \sum_{i=1}^{\nu} \mathfrak{v}^{(i)} p(\mathbf{x}_k|\mathbf{x}_{k-1}^{(i)}). \quad (13)$$

5. PROPERTIES OF THE SAMPLING DENSITY

The functional sampling density design described in the previous section embodies two severe operations. The matter is to specify the pdf $p(\mathbf{y}_k|\mathbf{x}_{k-1}^{(i)})$ in step (2b) and to enumerate the KJD measure $J(p(\mathbf{y}_k|\mathbf{z}_k)\|p(\mathbf{y}_k|\mathbf{x}_{k-1}^{(i)}))$ in step (3a) of the algorithm.

To obtain the pdf $p(\mathbf{y}_k|\mathbf{x}_{k-1}^{(i)})$, in case of transformation $\mathbf{h}_k : \mathbb{R}^n \to \mathbb{R}^m$ where $m = n$, the standard rule of random variable transformation may be mostly used. In case that $m \neq n$, an alternative approach has to be applied, e.g. the Unscented transformation (Julier and Uhlmann, 1996) which is based on specification of set of points according to shape of the original pdf, transformation of the set by $\mathbf{h}_k : \mathbb{R}^n \to \mathbb{R}^m$ and specification of the consequent pdf using transformed set of points. Special cases, when usage of standard rule of transformation can be used, will be treated in the section.

The KJD measure $J(p(\mathbf{y}_k|\mathbf{z}_k)\|p(\mathbf{y}_k|\mathbf{x}_{k-1}^{(i)}))$ has to be calculated numerically except a few special cases. (e.g. both pdf's $p(\mathbf{y}_k|\mathbf{z}_k)$ and $p(\mathbf{y}_k|\mathbf{x}_{k-1}^{(i)})$ are Gaussian). Both cases of analytical and numerical calculation of KJD will be applied.

Closed form of functional sampling pdf

Consider system (1), (2) where $p(\mathbf{x}_k|\mathbf{x}_{k-1}^{(i)})$ is Gaussian $p(\mathbf{x}_k|\mathbf{x}_{k-1}^{(i)}) = \mathcal{N}\{\mathbf{x}_k; \hat{\mathbf{x}}_k^{(i)}, \mathbf{Q}_k^{(i)}\}$, the pdf $p(\mathbf{v}_k)$ is Gaussian as well, i.e. $p(\mathbf{v}_k) = \mathcal{N}\{\mathbf{v}_k; 0, \mathbf{R}_k\}$. the transformation $\mathbf{h}_k : \mathbb{R}^n \to \mathbb{R}^m$ is linear, thus $\mathbf{y}_k = \mathbf{H}_k\mathbf{x}_k$, where $\mathbf{H}_k$ is a regular matrix, and $n = m$.

The pdf $p(\mathbf{y}_k|\mathbf{z}_k)$ of the reference variable $\mathbf{y}_k$ is Gaussian $p(\mathbf{y}_k|\mathbf{z}_k) = \mathcal{N}\{\mathbf{y}_k; \mathbf{z}_k, \mathbf{R}_k\}$. Linearity of the transformation $\mathbf{h}_k : \mathbb{R}^n \to \mathbb{R}^m$ and equality of the dimensions n of state and m of measurement admits application of standard random variable transformation rule. The transformed pdf $p(\mathbf{y}_k|\mathbf{x}_{k-1}^{(i)})$ of the reference variable $\mathbf{y}_k$ is therefore Gaussian

$$p(\mathbf{y}_k|\mathbf{x}_{k-1}^{(i)}) = \mathcal{N}\{\mathbf{y}_k; \mathbf{H}_k\hat{\mathbf{x}}_k^{(i)}, \mathbf{H}_k\mathbf{Q}_k^{(i)}\mathbf{H}_k^T\}.$$

for each sample $\{\mathbf{x}_{k-1}^{(i)};\ i=1,2,\dots,\nu\}$.

The KJD measure between two Gaussian pdf's $p(\mathbf{y}_k|\mathbf{z}_k)$ and $p(\mathbf{y}_k|\mathbf{x}_{k-1}^{(i)})$ can be evaluated analytically in the form

$$J(p(\mathbf{y}_k|\mathbf{z}_k)\|p(\mathbf{y}_k|\mathbf{x}_{k-1}^{(i)})) = \frac{|\mathbf{R}_k|}{2|\mathbf{F}_k|} + \frac{|\mathbf{F}_k|}{2|\mathbf{R}_k|} - 1 +$$
$$\frac{1}{2}\operatorname{tr}[(\mathbf{R}_k^{-1} + \mathbf{F}_k^{-1})(\mathbf{z}_k - \mathbf{H}_k\hat{\mathbf{x}}_k^{(i)})(\mathbf{z}_k - \mathbf{H}_k\hat{\mathbf{x}}_k^{(i)})^T], \quad (14)$$

where $\operatorname{tr}(\cdot)$ is trace of the matrix and $\mathbf{F}_k = \mathbf{H}_k\mathbf{Q}_k^{(i)}\mathbf{H}_k^T$. Using notation $\mathbf{S}_k^{-1} \triangleq \mathbf{R}_k^{-1} + \mathbf{F}_k^{-1}$ and introducing $\tilde{\mathbf{z}}_k^{(i)} \triangleq \mathbf{z}_k - \mathbf{H}_k\hat{\mathbf{x}}_k^{(i)}$ as local measurement estimation error (LMEE), the primary weight $\mathfrak{v}^{(i)}$ is given by

$$\mathfrak{v}^{(i)} \propto e^{-c\left[\frac{|\mathbf{R}_k|}{2|\mathbf{F}_k|} + \frac{|\mathbf{F}_k|}{2|\mathbf{R}_k|} + \frac{1}{2}\operatorname{tr}[\mathbf{S}_k^{-1}\tilde{\mathbf{z}}_k^{(i)}\tilde{\mathbf{z}}_k^{(i)T}] - 1\right]}. \quad (15)$$

The relation (15) can be rewritten in the following way

$$\mathfrak{v}^{(i)} \propto e^{-c\left[\frac{|\mathbf{R}_k|}{2|\mathbf{F}_k|} + \frac{|\mathbf{F}_k|}{2|\mathbf{R}_k|} - 1\right]} e^{-c\frac{1}{2}\tilde{\mathbf{z}}_k^{(i)T}\mathbf{S}_k^{-1}\tilde{\mathbf{z}}_k^{(i)}}. \quad (16)$$

The term $e^{-c\left[\frac{1}{2}\left(\frac{|R_k|}{|Q_k|}+\frac{|Q_k|}{|R_k|}\right)-1\right]}$ in (16) does not depend on sample $\mathbf{x}_{k-1}^{(i)}$, it represents a multiplicative constant only and thus can be omitted.

Hence the primary weight $\mathfrak{v}^{(i)}$ can be considered in the form

$$\mathfrak{v}^{(i)} \propto e^{-c\frac{1}{2}\tilde{\mathbf{z}}_k^{(i)T}\mathbf{S}_k^{-1}\tilde{\mathbf{z}}_k^{(i)}]}. \tag{17}$$

The primary weights $\mathfrak{v}^{(i)}$ used in APF given by $\mathfrak{v}^{(i)} \propto p(\mathbf{z}_k|\mu_k(i))$ for choice μ_k as mean have for the considered system the following form

$$\mathfrak{v}^{(i)} \propto e^{-\frac{1}{2}\tilde{\mathbf{z}}_k^{(i)T}\mathbf{R}_k^{-1}\tilde{\mathbf{z}}_k^{(i)}]}. \tag{18}$$

Comparing (17) and (18) it is obvious that for $c = 1$ the matrices of quadratic form $\mathbf{R}_k^{-1}$ and $\mathbf{S}_k^{-1}$ fulfill the relation $\mathbf{R}_k^{-1} < \mathbf{S}_k^{-1}$ because $\mathbf{S}_k^{-1} \stackrel{\triangle}{=} \mathbf{R}_k^{-1} + (\mathbf{H}_k\mathbf{Q}_k^{(i)}\mathbf{H}_k^{\mathrm{T}})^{-1}$. The consequence of the functional approach is the fact the good samples have higher primary weights at the expense of poor samples as far as considered system representation with respect to primary weights obtained from Pitt's approach (Pitt and Shephard, 2001). The design parameter c affects area of state space $\mathbb{R}^n$ covered by empirical filtering pdf $r_N(\mathbf{x}_k|\mathbf{z}_k)$. Influence of choice of the design parameter c on primary weights of samples for a system with $m = n = 1$ is illustrated in Fig 1. Alternatively, the

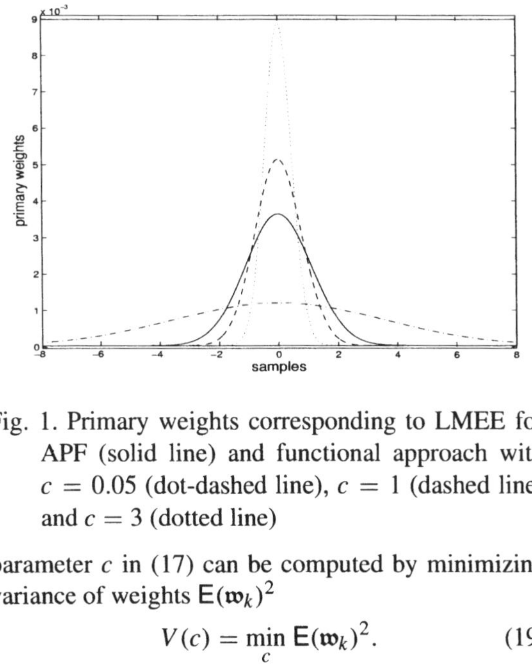

Fig. 1. Primary weights corresponding to LMEE for APF (solid line) and functional approach with $c = 0.05$ (dot-dashed line), $c = 1$ (dashed line) and $c = 3$ (dotted line)

parameter c in (17) can be computed by minimizing variance of weights $\mathsf{E}(\mathfrak{w}_k)^2$

$$V(c) = \min_{c} \mathsf{E}(\mathfrak{w}_k)^2. \tag{19}$$

The minimization of (19) can be realized analytically (Šimandl and Straka, 2002*a*). For systems with $m = n = 1$ and $\mathbf{Q}_k^{(i)} = \mathbf{Q}_k$; i=1,2,...,$\nu$ the optimal c^* is given by

$$c^* = \frac{Q_k R_k}{(Q_k + R_k)(2H_k^2 Q_k + R_k)}, \tag{20}$$

Closed form of reference variable pdf

Consider system (1), (2) where $p(\mathbf{x}_k|\mathbf{x}_{k-1}^{(i)})$ is mixture of Gaussians $p(\mathbf{x}_k|\mathbf{x}_{k-1}^{(i)}) = \sum_{j=1}^{M}\alpha_j\mathcal{N}\{\mathbf{x}_k; \hat{\mathbf{x}}_{k,j}^{(i)}, \mathbf{Q}_{k,j}\}$, the pdf $p(\mathbf{v}_k)$ is Gaussian , i.e. $p(\mathbf{v}_k) = \mathcal{N}\{\mathbf{v}_k; 0, \mathbf{R}_k\}$. the transformation $\mathbf{h}_k : \mathbb{R}^n \to \mathbb{R}^m$ is linear, thus $\mathbf{y}_k = \mathbf{H}_k\mathbf{x}_k$, where $\mathbf{H}_k$ is a regular matrix, and $n = m$.

The pdf $p(\mathbf{y}_k|\mathbf{z}_k)$ of the reference variable $\mathbf{y}_k$ is Gaussian $p(\mathbf{y}_k|\mathbf{z}_k) = \mathcal{N}\{\mathbf{y}_k; \mathbf{z}_k, \mathbf{R}_k\}$. Linearity of the transformation $\mathbf{h}_k : \mathbb{R}^n \to \mathbb{R}^m$ and equality of the dimensions n of state and m of measurement admits application of standard random variable transformation rule. The transformed pdf $p(\mathbf{y}_k|\mathbf{x}_{k-1}^{(i)})$ of the reference variable $\mathbf{y}_k$ is therefore Gaussian mixture

$$p(\mathbf{y}_k|\mathbf{x}_{k-1}^{(i)}) = \sum_{j=1}^{M}\alpha_j\mathcal{N}\{\mathbf{y}_k; \mathbf{H}_k\hat{\mathbf{x}}_{k,j}^{(i)}, \mathbf{H}_k\mathbf{Q}_{k,j}\mathbf{H}_k^{\mathrm{T}}\}.$$

for each sample $\{\mathbf{x}_{k-1}^{(i)};\quad$ i=1,2,...,ν }.

The KJD measure between the Gaussian pdf $p(\mathbf{y}_k|\mathbf{z}_k)$ and Gaussian mixture pdf $p(\mathbf{y}_k|\mathbf{x}_{k-1}^{(i)})$ can not be evaluated analytically so it is necessary to compute the J-divergence numerically.

The Fig 2 illustrates primary weights evaluated by APF algorithm and functional algorithm for the case when the pdf $p(\mathbf{y}_k|\mathbf{x}_{k-1}^{(i)})$ is given by

$$\begin{aligned} p(\mathbf{y}_k|\mathbf{x}_{k-1}^{(i)}) =& 0.5\mathcal{N}\{\mathbf{y}_k; x_{k-1}^{(i)} - 2, P_k\} + \\ & 0.5\mathcal{N}\{\mathbf{y}_k; x_{k-1}^{(i)} + 2, P_k\} \end{aligned} \tag{21}$$

and $p(\mathbf{y}_k|\mathbf{z}_k)$ is given by

$$p(\mathbf{y}_k|\mathbf{z}_k) = \mathcal{N}\{\mathbf{y}_k; 0, R_k\}. \tag{22}$$

Both choices of $\mu_k^{(i)}$ representing the corresponding likely value of the sample $x_{k-1}^{(i)}$ (mean and mode) in APF are considered however none of them respect multi-modality of the transient density. On the other hand primary weights evaluated by functional approach appears from both transient and measurement pdf's and consequently samples are rated by primary weights more efficiently. Then variance of weights $\mathfrak{w}_k^{(i)}$ and variance of estimates of this approach will be lower with respect to the APF approach.

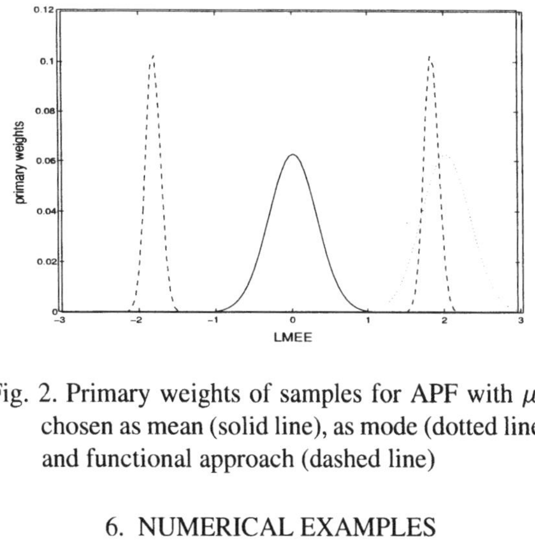

Fig. 2. Primary weights of samples for APF with μ_k chosen as mean (solid line), as mode (dotted line) and functional approach (dashed line)

6. NUMERICAL EXAMPLES

To show different properties of bootstrap filter (BF), APF and functional particle filter (FPF), the system with multi-modal pdf of state noise is considered:

$$x_k = 0.9\, x_{k-1} + w_k, \quad z_k = x_k + v_k, \qquad (23)$$

$p(w_k) = 0.5\, \mathcal{N}(w_k; -3, 0.1) + 0.5\, \mathcal{N}(w_k; 3, 0.1)$, $p(v_k) = \mathcal{N}(v_k; 0, 0.01)$ a $p(x_0) = \mathcal{N}(v_k; 10, 0.01)$. The system was simulated for $k = 1, 2$. The BF, APF and FPF were used for estimation of state x_k. The sample size $\nu = 100$ and initial samples $x_0^{(i)}$; i=1,2,...,ν were common for the three filters. MC estimates of weights variance $\mathsf{E}[(\mathfrak{w}_2)^2]$ at time instant $k = 2$ and estimates variance $\mathsf{var}[\hat{x}_2]$ for the three filters were compared. The $\hat{x}_2$ is point estimate of state x_2 described by the empirical filtering pdf $r_N(x_2|z^2)$, $\hat{x}_2 = \mathsf{E}[x_2|z^2]$ and

$$\mathsf{E}[(\mathfrak{w}_2)^2] = \frac{1}{100} \sum_{s=1}^{100} \mathsf{E}[(\mathfrak{w}_2(s))^2] \qquad (24)$$

$$\mathsf{var}[\hat{x}_2] = \frac{1}{100} \sum_{s=1}^{100} (\hat{x}_2(s) - x_2(s))^2, \qquad (25)$$

where $\mathsf{E}[(\mathfrak{w}_2(s))^2]$ is variance of weights for simulation s, $\hat{x}_2(s)$ is state estimate for simulation s and $x_2(s)$ is real value of estimated state for simulation s.

The comparison of the filters is presented in Table 1 The results in Table (1) demonstrates that the functional particle filter gives the best results and the APF filter with choice μ_k as mean provides poor results comparing to the bootstrap filter for the considered system. The results correspond with discussion provided in Section 5.

Table 1. Variance of weights and point estimates for filters BF, APF and FPF

	BF	*APF*	*FPF*
$\mathsf{E}[(\mathfrak{w}_2)^2]$	$5.47\ 10^{-4}$	$7.72\ 10^{-4}$	$3.73\ 10^{-4}$
$\mathsf{var}[\hat{x}_2]$	$42.46\ 10^{-4}$	$52.38\ 10^{-4}$	$37.47\ 10^{-4}$

7. CONCLUSION

The paper dealt with particle filters for nonlinear state estimation. Crucial aspect affecting estimation quality of the particle filter is sampling probability density function.A new functional approach for design of sampling density was developed using both the transient and measurement density pdf's. Contrary to the well known auxiliary particle filter that uses sampling density design based on comparing measurement pdf with point estimate of the transient pdf, the functional sampling density design comes from comparison of the both pdf's through pdf's a of reference variable. The analysis of the functional particle filter demonstrated profitable properties resulting to better estimates quality with respect to the bootstrap filter and auxiliary particle filter which is illustrated in a numerical example as well.

8. ACKOWLEDGEMENTS

The work was supported by the Grant Agency of the Czech Republic, project GA ČR 102/01/0021 and the Ministry of Education, Youth and Sports of the Czech Republic, project MSM 2352 00004.

9. REFERENCES

Berzuini, C. and Gilks, W., Eds.) (2001). *Sequential Monte Carlo Mathods in Practise*. Chap. RESAMPLE MOVE Filtering with Cross-Model Jumps. Springer. (Ed. A. Doucet, N. de Freitas and N. Gordon).

Fox, D. (2001). Kld-sampling: Adaptive particle filters and mobile robot localization. In: *Advances in Neural Information Processing Systems (NIPS)*.

Gordon, N., D. Salmond and A.F.M. Smith (1993). Novel approach to nonlinear/non-gaussian Bayesian state estimation. *IEE proceedings-F* **140**, 107–113.

Julier, S.J. and J.K. Uhlmann (1996). A general method for approximating nonlinear transformations of probability destributions. Technical report. Robotics Research Group, Department of Engineering Science, University of Exford.

Kullback, S. and R. Leibler (1951). On Information and Sufficiency. *Annals of Math. Statist.* (22), 79–86.

Liu, J. and M. West (2001). *Sequential Monte Carlo Mathods in Practise*. Chap. Combined Parameter and State Estimation in Simulation-Based Filtering, pp. 197–223. Springer. (Ed. A. Doucet, N. de Freitas and N. Gordon).

Liu, J.S. and R. Chen (1998). Sequential Monte Carlo Methods for Dynamic Systems. *J. Amer. Statist. Assoc.* **93**(443), 1032–1044.

Pitt, M.K. and Shephard, N., Eds.) (2001). *Monte Carlo Methods in Practise*. Chap. Auxiliary Variable Based Particle Filters. Springer. (Ed. A. Doucet, N. de Freitas and N. Gordon).

Šimandl, M. and O. Straka (2002*a*). Functional sampling density design for particle filters. Technical report. University of West Bohemia in Pilsen.

Šimandl, M. and O. Straka (2002*b*). Nonlinear estimation by particle filters and cramér-rao bound. In: *Preprints of the 15th IFAC World Congress on Automatic control*.

Šimandl, M., J. Královec and T. Söderström (2002). Anticipative Grid Design in Point-Mass Approach to Nonlinear State Estimation. *IEEE Transactions on Automatic Control*.

Sorenson, H.W. and D.L. Alspach (1970). Approximation of density function by a sum of gaussians for nonlinear bayesian estimation. In: *Proceedings of the 1st Symposium for Nonlinear Estimation Theory and Its Applications*. San Dieago. pp. 88–90.

Van der Merwe, R., A. Doucet, N. de Freitas and E. Wan (2000). The unscented particle filter. Technical Report CUED/F-INFENG/TR380. Cambridge University Engineering Department.

IFAC Publications
www.elsevier.com/locate/ifac

DIFFUSIVE REPRESENTATION OF N-TH ORDER FRACTIONAL BROWNIAN MOTION

Jaka Sembiring *,[1] **Kudrat Soemintapoera** **
Tetsunori Kobayashi * **Kageo Akizuki** *

* *Waseda University, Japan*
** *Insitut Teknologi Bandung, Indonesia*

Abstract: This paper describes an effort to give a different representation of a newly introduced n-th order fractional Brownian motion (n-fBm). The new representation is called diffusive representation which has been successfully applied to the $1/f^{\alpha}$ fractional noise. Thus this paper generalizes such representation to cover also n-fBm which is an extension to the ordinary fBm, due to the fact that the spectral properties of n-fBm cover larger range of parameter α. Different from $1/f^{\alpha}$ case, the solution involves finite part concept of theory of distribution. The advantage of the proposed method on synthesizing sample of n-fBm is presented. *Copyright © 2003 IFAC*

Keywords: Brownian motion, Gaussian processes, Signal processing, Stochastic systems

1. INTRODUCTION

This paper generalizes the work of (Levernhc *et al.*, 2001) to cover a newly introduced n-fBm processes (Perrin *et al.*, 2001). The n-fBm process itself is an expansion of a long established theory of fBm (Mandelbrot and Ness, 1968). In fBm one is restricted to the value of Hurst parameter H in the range of $0 < H < 1$. Observations have revealed that the value of H could be above one. In their paper, (Perrin *et al.*, 2001) tried to overcome this limitation by introducing what the so called n-fBm. More precisely this extension leads to the $n-$th order of fBm with H parameter lies in $]n-1, n[$, where n is positive integer. Surely it includes ordinary fBm when $n = 1$. A note on the fBm class processes is that they are Gaussian, non-stationary and continuous stochastic processes as well as posses a statistically self-similar property (Mandelbrot and Ness, 1968). This family of processes has long been used to model signal or texture with multi-scale pattern. Due to their long memory property, such processes have been widely become a subject of interest in the area of hydrology and communications, to name a few.

In view of such diverse applications and impacts of such processes, it is only natural if one has a suitable and simple tool for analysis and calculation. But in fact, the mathematical complexity in dealing with fBm processes class can not be regarded lightly. Time domain representation is always the norm, but recalling the long memory behavior of the processes it is clear that classical representation of fBm class will leads to stochastic integro-differential equation which is usually complex and time consuming to compute. These facts lead (Levernhe *et al.*, 2001) to come up with idea of diffusive representation which has been successfully applied to the deterministic context (Montseny *et al.*, 1993). Their paper focused mainly on $1/f^{\alpha}$ fractional noises. This paper expands their method to cover general form of fBm processes. With the proposed method integro-differential problem is converted to the classical iterated stochastic differential equation, since the representation possesses Markovian property. Numerical simulation on synthesizing a particular sample of n-fBm is shown to demonstrated the advantage of the proposed method.

[1] Partly supported by The Hitachi Scholarship Foundation. He is now with the Dept. of Electrical Engineering, Insitut Teknologi Bandung, Indonesia.

2. N-FBM AND DIFFUSIVE REPRESENTATION PROPERTIES

2.1 *n-fBm Processes*

The starting point of discussion is by recalling the ordinary fBm. The mostly known form is the one constructed by Mandelbrot and Van Ness in (Mandelbrot and Ness, 1968). The fBm is represented through the following formula,

$$B_H(t) - B_H(0) = \frac{1}{\Gamma(H+1/2)} \left\{ \int_{-\infty}^{0} \left[(t-u)^{H-1/2} - (-u)^{H-1/2} \right] dB(u) + \int_0^t (t-u)^{H-1/2} dB(u) \right\}. \tag{1}$$

Then Perrin *et. al.* obtained a new model they called n-fBm through limit expansion of the kernel of ordinary fBm, see (Perrin *et al.*, 2001). To suit the subsequent derivation, this paper will introduce a slightly different notation than the one described in their paper. Here the n-fBm is written as:

$$B^n_{H'}(t) = \frac{1}{\Gamma(H'+1/2)} \left\{ \int_{-\infty}^{0} \left[(t-u)^{H'-1/2} - (-u)^{H'-1/2} - \cdots - \left(H' - \frac{1}{2}\right) \cdots \left(H' - \frac{(2n-3)}{2}\right) \cdot (-u)^{H'-n+1/2} \frac{t^{n-1}}{(n-1)!} \right] dB(u) + \int_0^t (t-u)^{H'-1/2} dB(u) \right\}, \tag{2}$$

where $n = \{1, 2, \cdots\}, n \in \mathbb{B}$ and $H' = n - 1 + H$, with H denotes Hurst parameter on the range of $0 < H < 1$. It is clear that when $n = 1$, Eq. (2) becomes ordinary fBm. It has been proved in (Perrin *et al.*, 2001) that the n-fBm processes possess several properties such as zero mean, self-similar and non-stationary just as the ordinary fBm. One important thing to be covered is the concept of fractional Gaussian noise (fGn) and its general form n-fGn. This is due to the well known fact that the fBm class processes are better analyzed using their increments. Following the derivation in (Perrin *et al.*, 2001), here the n-fGn is defined as

$$\triangle^{(n)}_{\tau} y(t) = \triangle^{(n-1)}_{\tau} y(t+\tau) - \triangle^{(n-1)}_{\tau} y(t) \tag{3a}$$

$$= \sum_{j=0}^{n} (-1)^{n-j} \binom{n}{j} y(t + j\tau), \tag{3b}$$

where n denotes n-th order increments of n-fBm which provides stationarity of the increments.

The spectral shape of n-fBm is roughly equal to $1/\omega^{2H+1}$, thus with $n = \{1, 2, \cdots\}$ one can get the spectral of the form $1/f^{\alpha}$ with α in the range of $]2n-1, 2n+1[$. In general this form of spectrum is called a fractional noise.

2.2 *Diffusive Representation Notations*

The diffusive representation originated from the so called diffusive representation of fractional integrators, see (Montseny *et al.*, 1993). It is an input-output diffusive system defined by

$$\frac{\partial Y(r,t)}{\partial t} = \frac{\partial^2 Y(r,t)}{\partial r^2} + \delta(r) u(t)$$
$$Y(r,0) = 0, r \in \mathbb{R}, \tag{4a}$$

$$z(t) = \int_{-\infty}^{+\infty} m_{\alpha}(r) Y(r,t) dr, \tag{4b}$$

where m_{α} is a suitable distribution that should be determined later. Using Fourier transform Eq. (4) can be written equivalently as

$$\frac{\partial \hat{Y}(\rho,t)}{\partial t} = -4\pi^2 \rho^2 \hat{Y}(\rho,t) + u(t), \ \rho \in \mathbb{R}, \tag{5a}$$

$$z(t) = \int_{-\infty}^{+\infty} \hat{m}_{\alpha}(\rho) \hat{Y}(\rho,t) d\rho, \tag{5b}$$

Following (Levernhe *et al.*, 2001), introducing variable change $\xi = 4\pi^2\rho^2$ and considering Wiener process input $B(t)$, one will have

$$dy_{\xi}(t) = -\xi y_{\xi}(t) dt + dB(t)$$
$$y_{\xi}(0) = y_{\xi_o}, \xi > 0, \tag{6a}$$

$$y(t) = \int_0^{+\infty} \mu_{\alpha}(\xi) [y_{\xi}(t) - y_{\xi_o}] d\xi. \tag{6b}$$

Thus theoretically $y(t)$ can be obtained from Eq. (6).The y_{ξ_o} in Eq. (6b) is included to be consistent with the definition of n-fBm.

3. DIFFUSIVE REPRESENTATION OF N-FBM

Following the definition of fractional integral of order $n + \alpha, n \in \mathbb{B}, 0 < \alpha < 1$, see (Oldham and Spanier, 1974)

$$I^{(n+\alpha)} f(t) = \frac{1}{\Gamma(n+\alpha)} \int_0^t \frac{f(t)}{(t-s)^{(1-n-\alpha)}} ds, \tag{7}$$

and recalling the spectral shape of n-fBm as discussed before, the process $B^n_{H'}(t)$ can be written as a fractional noise

$$y(t) \triangleq B^n_{H'}(t) = I^{(n+\alpha)} + \varsigma(t), \tag{8}$$

with $\alpha = H - 1/2$, and the $\varsigma(t)$ is the correction factor. The proposed representation $y(t)$ is an n-fBm process provided that it satisfies two properties below.

$$y(t) - I^{(n+\alpha)} w(t) \xrightarrow[t\to+\infty]{\text{q.m}} 0, \tag{9a}$$

$$\triangle^{(n)}_{\tau} (y(t) - I^{(n+\alpha)} w(t)) \xrightarrow[t\to+\infty]{\text{q.m}} 0, \tag{9b}$$

where n is the order of increment, $w(t)$ denotes white noise. Eq (9a) is a stationarized process of n-fBm, and Eq. (9b) states the incremental property of n-fGn process.

Several properties of the representation in Eq. (4) have been proved in (Levernhe *et al.*, 2001), and the one which is relevant to the derivation in this paper is re-written as follows. For exact representation, the process $\{y_\xi\}$ is Gaussian and stationary with covariance

$$\mathbb{E}[(y_{\xi_0} y_{\eta_0})] = \frac{1}{\xi + \eta}, \tag{10a}$$

$$\mathbb{E}[(y_{\xi_0})^2] = \frac{1}{2\xi}. \tag{10b}$$

In addition to the property in Eq (10) above, this paper introduces the following Lemma to support the subsequent derivation.

Lemma 1. The process Eq. (6) is a fractional noise process of order $(n + \alpha)$ only if

$$\mu_\alpha(\xi) = (-1)^n \frac{\sin(\alpha\pi)}{\pi} \text{fp}\left(\frac{1}{\xi^{n+\alpha}}\right), \tag{11}$$

with fp denotes finite part in Schwartz distribution sense (Schwartz, 1966).

PROOF. Begin with the stochastic integration of Eq. (6a) which can be written as

$$y_\xi(t) = e^{-\xi t} y_{\xi_0} + \int_0^{+\infty} e^{-\xi(t-s)} dB(s). \tag{12}$$

Then recalling the Fubini theorem, the fractional process $y(t)$ of Eq. (6b) can be expressed as

$$y(t) = \int_0^t \left(\int_0^{+\infty} e^{-\xi(t-s)} \mu_\alpha(\xi) d\xi\right) dB(s) + \int_0^{+\infty} (e^{-\xi t} - 1)\, y_{\xi_0} \mu_\alpha(\xi) d\xi. \tag{13}$$

Here the introduction of finite part concept becomes clear by noting the boundary of the integration. With μ_α as Eq. (11) above the first term of the right hand side of Eq. (13) will be

$$\begin{aligned}
\text{1st r.h.s.} &= \int_0^t \left[\int_0^{+\infty} e^{-\xi(t-s)} (-1)^n \frac{\sin(\pi\alpha)}{\pi} \text{fp}\left(\frac{1}{\xi^{n+\alpha}}\right) d\xi\right] dB(s) \\
&= \int_0^t \left[\int_0^{\infty} e^{-\xi(t-s)} \frac{\sin(\pi(n+\alpha))}{\pi} \text{fp}\left(\frac{1}{\xi^{n+\alpha}}\right) d\xi\right] dB(s).
\end{aligned}$$

Using well known relations

$$\frac{\sin(\pi(n+\alpha))}{\pi} = \Gamma(n+\alpha)\Gamma(1-(n+\alpha)),$$

and

$$\int_0^{+\infty} \xi^{-(n+\alpha)} e^{-\xi t} d\xi = \frac{\Gamma(1-(n+\alpha))}{t^{(1-(n+\alpha))}},$$

the 1st r.h.s. can be recomposed into

$$\begin{aligned}
\text{1st r.h.s.} &= \int_0^t \left[\int_0^{+\infty} e^{-\xi(t-s)} \frac{1}{\Gamma(n+\alpha)\Gamma(1-(n+\alpha))} \text{fp}\left(\frac{1}{\xi^{n+\alpha}}\right) d\xi\right] dB(s) \\
&= \frac{1}{\Gamma(n+\alpha)} \int_0^t \frac{1}{(t-s)^{(1-(n+\alpha))}} dB(s).
\end{aligned} \tag{14}$$

Now recall Eq. (7), one can conclude that for μ_α defined in Eq. (11) the first term in Eq. (13) is equal to $I^{(n+\alpha)} dB(t)/dt$ where $dB(t)/dt = w(t)$ is white noise.

Theorem 1. The diffusive representation process defined in Eq (6) is an n-fBm if and only if Eq. (10), Eq. (11) and Eq. (9) are satisfied.

PROOF. The consequences of Eq. (10) and Eq. (11) are clear. The remaining parts are the validity of both properties expressed in Eq. (9). First the statement that $y(t)$ is a stationary process. Since $y(t)$ is indeed a stationary process, see discussion before, then the property of Eq. (9a), $\mathbb{E}[(y(t) - \tilde{y}(t)^2] \xrightarrow[t\to+\infty]{\text{q.m}} 0$, can be simplified to $\mathbb{E}[(y(t) - \tilde{y}(t)] \xrightarrow[t\to+\infty]{\text{q.m}} 0$. From $\mathbb{E}[(y_{\xi 0})^2] = 1/(2\xi)$ one can have $\mathbb{E}[|y_0|] = c_1/(\sqrt{\xi})$ so that using Eq. (13) and ordinary definition of expectation,

$$\begin{aligned}
\mathbb{E}[y(t) - \tilde{y}(t)] &= \mathbb{E}\left[\left|\int_0^{+\infty} (e^{-\xi t} - 1) y_{\xi_0} \mu_\alpha(\xi) d\xi\right|\right] \\
&\le \int_0^{+\infty} (e^{-\xi t} - 1)\mathbb{E}[|y_{\xi_0}|]|\mu_\alpha(\xi)| d\xi \\
&\le (e^{-\xi t} - 1)\frac{c_1}{\xi^{1/2}} \left|(-1)^n \frac{\sin(\pi\alpha)}{\pi} \text{fp}\left(\frac{1}{\xi^{n+\alpha}}\right)\right| d\xi \\
&\le c_2 \int_0^{+\infty} \frac{(e^{-\xi t} - 1)}{\xi^{1/2}} \left|\text{fp}\left(\frac{1}{\xi^{n+\alpha}}\right)\right| d\xi \\
&\le c_2 \int_0^{+\infty} \frac{e^{-\xi t}}{\xi^{1/2}} \left|\text{fp}\left(\frac{1}{\xi^{n+\alpha}}\right)\right| d\xi.
\end{aligned}$$

Using Taylor expansion around $\xi = 0$ and taking absolute value of the fp expression as in (Hadamard, 1952), the above expression becomes

$$\begin{aligned}
\mathbb{E}[y(t) - \tilde{y}(t)] &\le c_2 \int_0^{+\infty} \frac{e^{-\xi t}}{\xi^{1/2}} \left[\frac{1}{(n-\alpha)\xi^{n-\alpha}} + \cdots + \frac{(-1)^{n-1}}{(n-1)!\,\alpha\xi^\alpha} + \frac{2}{n!}\xi^{1/2}\right] d\xi, \\
&\le \infty
\end{aligned} \tag{15}$$

for $\forall t > t_0 > 0$ and $0 < \alpha < 1$. From Lebesgue dominated convergence theorem

$$\lim_{t\to\infty} \mathbb{E}[y(t) - \tilde{y}(t)] \le c_2 \int_0^{+\infty} \lim_{t\to\infty} \frac{e^{-\xi t}}{\xi^{1/2}} \left[\frac{1}{(n-\alpha)\xi^{n-\alpha}} + \cdots + \frac{(-1)^{n-1}}{(n-1)!\,\alpha\xi^{\alpha}} + \frac{2}{n!}\xi^{1/2}\right] d\xi = 0. \tag{16}$$

Now consider the n-fGn case on Eq. (9b)

$$\begin{aligned}
&\mathbb{E}[\triangle_\tau^{(n)} y(t) - \triangle_\tau^{(n)} \tilde{y}(t)] \\
&= \mathbb{E}\left[\left|\sum_{j=0}^{\infty}(-1)^{n-j}\binom{n}{j} \int_0^{+\infty} (e^{-\xi(t+j\tau)} - 1) y_{\xi_0} \mu_\alpha(\xi) d\xi\right|\right] \\
&\le \sum_{j=0}^{\infty}(-1)^{n-j}\binom{n}{j} \int_0^{+\infty} \frac{(e^{-\xi(t+j\tau)} - 1)}{\xi^{1/2}} (-1)^n \frac{\sin(\pi\alpha)}{\pi} \operatorname{fp}\left(\frac{1}{\xi^{n+\alpha}}\right) d\xi \\
&\le c_1 \sum_{j=0}^{\infty}(-1)^{n-j}\binom{n}{j} \int_0^{+\infty} \frac{(e^{-\xi(t+j\tau)} - 1)}{\xi^{1/2}} \operatorname{fp}\left(\frac{1}{\xi^{n+\alpha}}\right) d\xi.
\end{aligned}$$

Using the same method as before

$$\begin{aligned}
&\mathbb{E}[\triangle_\tau^{(n)} y(t) - \triangle_\tau^{(n)} \tilde{y}(t)] \\
&\le c_1 \sum_{j=0}^{\infty}(-1)^{n-j}\binom{n}{j} \int_0^{+\infty} \frac{e^{-\xi(t+j\tau)}}{\xi^{1/2}} \left[\frac{1}{(n-\alpha)\xi^{n-\alpha}} + \cdots + \frac{(-1)^{n-1}}{(n-\alpha)!\,\alpha\xi^{\alpha}} + \frac{2}{n!}\xi^{1/2}\right] d\xi \\
&\le c_1 \left\{\binom{n}{0} \int_0^{+\infty} \frac{e^{-\xi(t)}}{\xi^{1/2}} \left[\frac{1}{(n-\alpha)\xi^{n-\alpha}} + \cdots + \frac{(-1)^{n-1}}{(n-\alpha)!\,\alpha\xi^{\alpha}} + \frac{2}{n!}\xi^{1/2}\right] d\xi \right. \\
&\quad - \binom{n}{1} \int_0^{+\infty} \frac{e^{-\xi(t+\tau)}}{\xi^{1/2}} \left[\frac{1}{(n-\alpha)\xi^{n-\alpha}} + \cdots + \frac{(-1)^{n-1}}{(n-\alpha)!\,\alpha\xi^{\alpha}} + \frac{2}{n!}\xi^{1/2}\right] d\xi \\
&\quad \left. + \cdots + \binom{n}{n} \int_0^{+\infty} \frac{e^{-\xi(t+n\tau)}}{\xi^{1/2}} \left[\frac{1}{(n-\alpha)\xi^{n-\alpha}} + \cdots + \frac{(-1)^{n-1}}{(n-\alpha)!\,\alpha\xi^{\alpha}} + \frac{2}{n!}\xi^{1/2}\right] d\xi \right\},
\end{aligned} \tag{17}$$

for $\forall t > t_0 > 0$ and $0 < \alpha < 1$. Again using Lebesgue dominated convergence one can obtain

$$\begin{aligned}
&\lim_{t\to\infty} \mathbb{E}[\triangle_\tau^{(n)} y(t) - \triangle_\tau^{(n)} \tilde{y}(t)] \\
&= c_1 \left\{\binom{n}{0} \int_0^{+\infty} \lim_{t\to+\infty} \frac{e^{-\xi(t)}}{\xi^{1/2}} \left[\frac{1}{(n-\alpha)\xi^{n-\alpha}} + \cdots + \frac{(-1)^{n-1}}{(n-\alpha)!\,\alpha\xi^{\alpha}} + \frac{2}{n!}\xi^{1/2}\right] d\xi \right. \\
&\quad - \binom{n}{1} \int_0^{+\infty} \lim_{t\to+\infty} \frac{e^{-\xi(t+\tau)}}{\xi^{1/2}} \left[\frac{1}{(n-\alpha)\xi^{n-\alpha}} + \cdots + \frac{(-1)^{n-1}}{(n-\alpha)!\,\alpha\xi^{\alpha}} + \frac{2}{n!}\xi^{1/2}\right] d\xi \\
&\quad \left. + \cdots + \binom{n}{n} \int_0^{+\infty} \lim_{t\to+\infty} \frac{e^{-\xi(t+n\tau)}}{\xi^{1/2}} \left[\frac{1}{(n-\alpha)\xi^{n-\alpha}} + \cdots + \frac{(-1)^{n-1}}{(n-\alpha)!\,\alpha\xi^{\alpha}} + \frac{2}{n!}\xi^{1/2}\right] d\xi \right) \\
&= 0.
\end{aligned} \tag{18}$$

The above derivation combined with the properties expressed in Eq. (10) complete the proof.

4. NUMERICAL SIMULATION

It is commonly understood that synthesis of fBm, not to mention n-fBm, is difficult and cumbersome. But using diffusive representation this effort is easier and computationally manageable. This is clear from Eq. (6) that the formulation takes the form of classical stochastic differential equation and possesses Markovian property. Thus the sample of n-fBm signal can be thought as a filtering with N bank of low-pass filter whose inputs are white noise, and followed with the integration over the whole range of ξ. Approximation representation can be written as

$$dy_{\xi_i}(t) = -\xi y_{\xi_i}(t)dt + dB(t)$$
$$y_\xi(0) = y_{\xi_0}, \xi > 0, i = 1 \cdots N, \tag{19a}$$
$$\hat{y}(t) = \sum_{i=1}^{N} \mu_{\alpha_i}(\xi)[y_{\xi_i}(t) - y_{\xi_0}] d\xi_i. \tag{19b}$$

In the simulation, a realization of a sample of n-fBm with $\alpha = 0.25$ is generated through Eq. (19) with $0.001 \le \xi_i \le 10{,}000$, and $N = 7$. The parameter ξ is determined from the range of frequency encountered in fBm classes signal, and N is the trade-off between computation and accuracy of the model. The Eq. (19a) is generated through classical method such as (Kloeden *et al.*, 1991). A particular sample of classical fBm as well as 2-fBm can be seen on Fig. 1. Note that the Fig. 1 (Bottom) illustrates a low frequency evolution of 2-fBm.

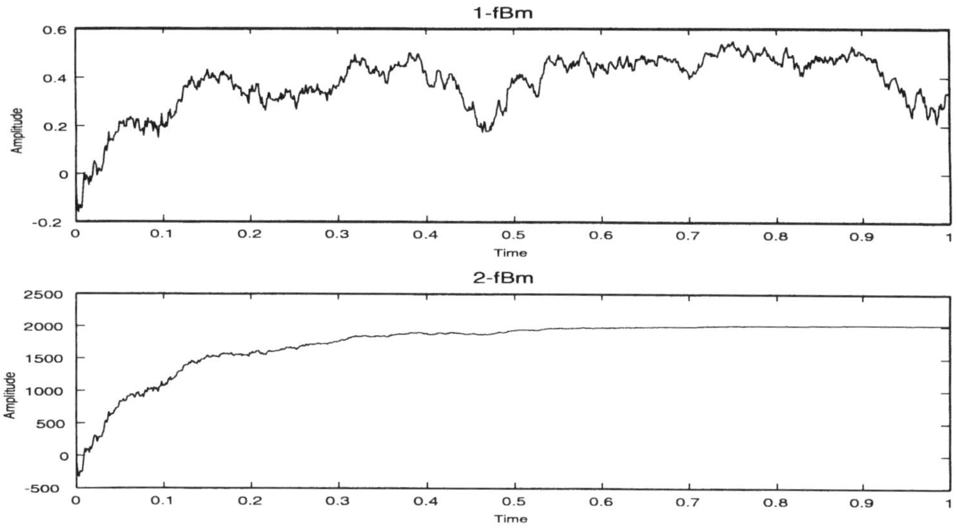

Fig. 1. Sample of n-fBm with $\alpha = 0.25$ ($H = 0.75$) using the proposed method. (Top) The sample path of ordinary fBm or 1-fBm. (Bottom) The sample path of 2-fBm. The 2-fBm is smoother than 1-fBm, it contains a low frequency evolution. Note that these sample paths are directly constructed from Eq. (19).

5. CONCLUSION

A new representation of n-fBm process is discussed through what the so called diffusive representation. This paper shows that this new representation can simplify the complex and tedious effort of standard time domain representation. The discussion began with the introduction of an even function as the key solution the problem, and followed with the proof of derived properties. The main feature of the solution is the finite part term adapted from the theory of distribution. The subsequent derivation is centered from this notations. The numerical simulation provides a simple procedure to synthesize the process, instead of long and often cumbersome process in time domain.

6. REFERENCES

Hadamard, Jaques (1952). *Lectures on Cauchy's Problem in Linear Partial Differential Equations.* Dover Publications. New York.

Kloeden, Peter E., Eckhard Platen and Henri Schurz (1991). *Numerical Solution of SDE Through Computer Experiments.* Springer-Verlag. New York.

Levernhe, Francis, Gérard Montseney and Jaques Audonet (2001). Markovian diffusive representation of $1/f^{\alpha}$ noises and application to fractional stochastic differential models. *IEEE Trans. on Signal Processing* **49**, 414–423.

Mandelbrot, Benoit B. and John W. Van Ness (1968). Fractional Brownian motion, fractional noises and applications. *SIAM Review* **10**, 422–438.

Montseny, G., J. Audonet and B. Mbodje (1993). Optimal models of fractional integrators and application to systems with fading memory. In: *Proc. IEEE SMC Conf.* Le Touquet, France. pp. 65–70.

Oldham, Keith B. and Jerome Spanier (1974). *The Fractional Calculus.* Academic Press. New York.

Perrin, Emmanuel, Rachid Harba, Corrine Berzin-Joseph, Ileana Iribarren and Aline Bonami (2001). nth-Order fractional Brownian motion and fractional gaussian noises. *IEEE Trans. on Signal Processing* **49**, 1049–1059.

Schwartz, Laurent (1966). *Théorie Des Distributions.* Hermann. Paris.

IFAC Publications
www.elsevier.com/locate/ifac

MULTI-CHANNEL ACTIVE NOISE CONTROL FOR UNCERTAIN SECONDARY CHANNELS

Yuhsuke Ohta *, **Hiromitsu Ohmori** * **and Akira Sano** *

* *Department of System Design Engineering, Keio University, Yokohama 223-8522, Japan.*
e-mail: sano@sd.keio.ac.jp

Abstract: Fully adaptive feedforward control algorithm is proposed for general multi-channel active noise control (ANC) when all the noise transmission channels are uncertain. To reduce the actual canceling error, two kinds of virtual errors are introduced and are forced into zero by adjusting three adaptive FIR filter matrices in an on-line manner, which can result in the canceling at the objective points. Unlike other conventional approaches, the proposed algorithm does not need exact identification of the secondary paths, and so requires neither any dither signals nor the PE property of the source noises, which is a great advantage of the proposed adaptive approach. *Copyright © 2003 IFAC*

Keywords: daptive control, active noise control, adaptive filter, adaptive identification

1. INTRODUCTION

Active noise control (ANC) is efficiently used to suppress unwanted low frequency noises generated from primary sources by emitting artificial secondary sounds to objective points (Elliot *et al.*, 1993; Kuo *et al.*, 1996). Adaptive feedforward control schemes using the primary noises measured by reference microphones are effective to ANC, since the noise transmission channels cannot be precisely modeled and may be uncertainly changeable. A variety of filtered-x algorithms have been adopted to attain the feedforward adaptation (Kuo *et al.*, 1996; Jiang *et al.*, 1997), on the assumption that all the secondary path channels are known a priori. The stability and convergence of the adaptive algorithm using the extended error has also been investigated in (Bjarnason, 1995; Jiang *et al.*, 1997; Shimizu *et al.*, 2002).

To deal with a general case when the secondary path channels are also unknown, almost previous works were based on indirect adaptive approaches which employ on-line identification of the secondary channels. Therefore, dither noise should be added to artificial control sound to assure the PE condition. By taking an extension of the filtered-x type of approaches, the identified models of the secondary channels are used correspondingly in the filtered-x algorithms (Eriksson, 1994; Saito *et al.*, 1996; Kim *et al.*, 1997; Zhang *et al.*, 2001). However, the stability of the adaptive approaches is hardly assured. Recently stability analysis has been done by (Meurers *et al.*, 2002) in a limited case. By utilizing the identified models of the secondary channels, the feedforward controller can also be updated and redesigned in a real-time manner (Jiang *et al.*, 1997; Shimizu *et al.*, 2002). The apporach will suffer from computational burden especially in multi-channel cases. Thus the indirect adaptive approaches need the exact on-line identification of the secondary path channels. However, because the persistently exciting (PE) property of the primary source noises is insufficient and the control sounds are strongly correlated with the sensed source noises, additional dither sounds are needed to secure the identifiability for the secodary channels.

The aim of this paper is to propose a fully direct adaptive control approach which does not need explicit identification of the secondary channels. To reduce the actual canceling errors, two kinds of virtual errors are introduced and are forced into zero by adjusting three adaptive FIR filter matrices in an on-line manner, which enables the noise cancellation at the objective points. The idea has been given by the authors in a single channel case (Kohno *et al.*, 2002), and its effectiveness is also shown in active noise control of an air duct. In a multi-channel case, some modifications are needed since the order exchange of two matrices does not give the same product. Unlike the previous indirect approaches based on the explicit on-line identification, neither dither sounds nor the PE property of the source noises are required in the proposed scheme. Finally its effectiveness is examined in numerical simulations and and experiments in active noise suppression in a room.

2. FEEDFORWARD ADAPTIVE ACTIVE NOISE CONTROL

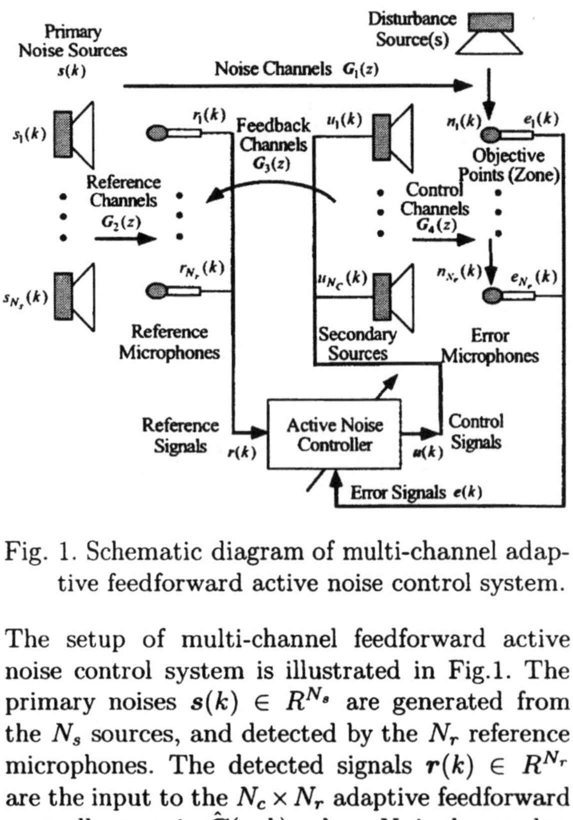

Fig. 1. Schematic diagram of multi-channel adaptive feedforward active noise control system.

The setup of multi-channel feedforward active noise control system is illustrated in Fig.1. The primary noises $\boldsymbol{s}(k) \in R^{N_s}$ are generated from the N_s sources, and detected by the N_r reference microphones. The detected signals $\boldsymbol{r}(k) \in R^{N_r}$ are the input to the $N_c \times N_r$ adaptive feedforward controller matrix $\hat{\boldsymbol{C}}(z,k)$, where N_c is the number of the secondary loudspeakers which produce artificial sounds to cancel the primary noises at the N_c objective points. The canceling errors are detected as $\boldsymbol{e}_c(k) \in R^{N_c}$ by the N_c error microphones. $\boldsymbol{G}_1(z) \in \mathcal{Z}^{N_c \times N_s}$ and $\boldsymbol{G}_2(z) \in \mathcal{Z}^{N_r \times N_s}$ represent the primary channel matrices from the primary noise $\boldsymbol{s}(k)$ to the reference microphones and error microphones, respectively. $\boldsymbol{G}_3(z) \in \mathcal{Z}^{N_r \times N_c}$ and $\boldsymbol{G}_4(z) \in \mathcal{Z}^{N_c \times N_c}$ are the secondary channel matrices from the secondary control sounds $\boldsymbol{u}(k)$ to the reference and error microphone sets, respectively, where $\boldsymbol{G}_3(z)$ is referred to as the coupling channel matrix. Because these channels may contain model uncertainty and parameter changeability, adaptive control approaches are significantly needed to deal with the problems.

It follows from Fig.1 that

$$\boldsymbol{e}_c(k) = \boldsymbol{G}_1(z)\boldsymbol{s}(k) - \boldsymbol{G}_4(z)\boldsymbol{u}(k) \qquad \text{(1a)}$$

$$\boldsymbol{r}(k) = \boldsymbol{G}_2(z)\boldsymbol{s}(k) + \boldsymbol{G}_3(z)\boldsymbol{u}(k) \qquad \text{(1b)}$$

$$\boldsymbol{u}(k) = \boldsymbol{C}(z)\boldsymbol{r}(k) \qquad \text{(1c)}$$

where $\boldsymbol{C}(z)$ is an FIR type of feedforward control matrix, and let $\hat{\boldsymbol{C}}(z,k)$ be an adaptive implement of $\boldsymbol{C}(z)$, then the secondary control sounds are generated in an adaptive case by

$$\boldsymbol{u}(k) = \hat{\boldsymbol{C}}(z,k)\boldsymbol{r}(k) = \boldsymbol{\Phi}^T(k)\hat{\boldsymbol{\theta}}(k) \qquad (2)$$

where $\boldsymbol{\Phi}^T(k) \equiv \mathrm{Diag}[\boldsymbol{\phi}_1^T(k), \boldsymbol{\phi}_2^T(k), \cdots, \boldsymbol{\phi}_{N_c}^T(k)]$, $\boldsymbol{\phi}_i(k) = [\boldsymbol{\phi}_{i1}^T(k), \boldsymbol{\phi}_{i2}^T(k), \cdots, \boldsymbol{\phi}_{iN_r}^T(k)]^T$ and $\boldsymbol{\phi}_{ij}(k) = [r_j(k-1), r_j(k-2), \cdots, r_j(k-m_{ij})]^T$. Let the parameter vector be defined by $\hat{\boldsymbol{\theta}}(k) \equiv [\hat{\boldsymbol{\theta}}_1^T(k), \hat{\boldsymbol{\theta}}_2^T(k), \cdots, \hat{\boldsymbol{\theta}}_{N_c}^T(k)]^T$ where $\hat{\boldsymbol{\theta}}_i(k) = [\hat{\boldsymbol{c}}_{i1}^T(k), \hat{\boldsymbol{c}}_{i2}^T(k), \cdots, \hat{\boldsymbol{c}}_{iN_r}^T(k)]^T$ and $\hat{\boldsymbol{c}}_{ij}(k) = [\hat{c}_{ij}^{(1)}(k), \hat{c}_{ij}^{(2)}(k), \cdots, \hat{c}_{ij}^{(m_{ij})}(k)]^T$. Thus, the FIR adaptive controllers in $\hat{\boldsymbol{C}}(z,k)$ are expressed by

$$\hat{C}_{ij}(z,k) = \hat{c}_{ij}^{(1)}(k)z^{-1} + \cdots + \hat{c}_{ij}^{(m_{ij})}(k)z^{-m_{ij}}$$

By eliminating $\boldsymbol{s}(k)$ from (1) and rewriting the canceling error $\boldsymbol{e}(k)$ in terms of the accessible signals $\boldsymbol{r}(k)$ and $\boldsymbol{u}(k)$, the error system can be obtained as

$$\begin{aligned} e_c(k) &= \bar{\boldsymbol{G}}_1(z)\boldsymbol{r}(k) - \bar{\boldsymbol{G}}_1(z)\boldsymbol{u}(k) \\ &= \bar{\boldsymbol{G}}_4(z)[\bar{\boldsymbol{G}}_4^{-1}(z)\bar{\boldsymbol{G}}_1(z) - \hat{\boldsymbol{C}}(z,k)]\boldsymbol{r}(k) \\ &= \bar{\boldsymbol{G}}_4(z)[\bar{\boldsymbol{C}}^*(z) - \hat{\boldsymbol{C}}(z,k)]\boldsymbol{r}(k) \end{aligned} \qquad (3)$$

where $\bar{\boldsymbol{G}}_1(z)$ and $\bar{\boldsymbol{G}}_4(z)$ are referred to as the equivalent primary and secondary channel matricies respectively, which are expressed by

$$\bar{\boldsymbol{G}}_1(z) = \boldsymbol{G}_1(z)[\boldsymbol{G}_2^T(z)\boldsymbol{G}_2(z)]^{-1}\boldsymbol{G}_2^T(z) \qquad \text{(4a)}$$

$$\begin{aligned} \bar{\boldsymbol{G}}_4(z) &= \boldsymbol{G}_4(z) + \boldsymbol{G}_1(z)[\boldsymbol{G}_2^T(z)\boldsymbol{G}_2(z)]^{-1} \\ &\quad \cdot \boldsymbol{G}_2^T(z)\boldsymbol{G}_3(z) \end{aligned} \qquad \text{(4b)}$$

If the secondary channels $\bar{\boldsymbol{G}}_4(z)$ and $\bar{\boldsymbol{G}}_3(z)$ are known a priori and not changeable, the following stability-assured adaptive algorithm proposed by one of the authors can be applied to update the FIR parameters of the feedforward controller matrix $\hat{\boldsymbol{C}}(z,k)$ (Shimizu *et al.*, 2002):

$$\hat{\boldsymbol{\theta}}(k+1) = \hat{\boldsymbol{\theta}}(k) + \gamma(k)\boldsymbol{\Psi}(k)\boldsymbol{\eta}(k) \qquad \text{(5a)}$$

$$\gamma(k) = \frac{2\alpha\boldsymbol{\eta}^T(k)\boldsymbol{\eta}(k)}{d + \boldsymbol{\eta}^T(k)\boldsymbol{\Psi}^T(k)\boldsymbol{\Psi}(k)\boldsymbol{\eta}(k)} \qquad \text{(5b)}$$

$$\boldsymbol{\eta}(k) = \boldsymbol{e}_c(k) + \boldsymbol{G}_4(z)\boldsymbol{u}(k) - \boldsymbol{\Psi}^T(k)\hat{\boldsymbol{\theta}}(k) \qquad \text{(5c)}$$

$$\boldsymbol{\Psi}^T(k) = \boldsymbol{G}_4(z)\boldsymbol{\Phi}^T(k) \qquad \text{(5d)}$$

where $0 < \alpha < 1$, and $d > 0$ is a sufficient small number to prevent division by zero.

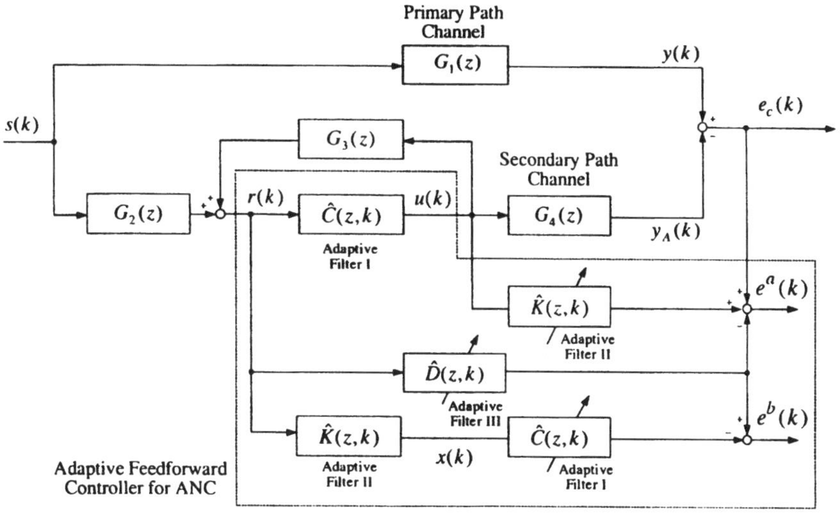

Fig. 2. Direct fully adaptive algorithm for single channel ANC

3. NEW ADAPTIVE ALGORITHM FOR UNCERTAIN SECONDARY CHANNELS

3.1 *Key idea for new adaptation algorithm in single channel case*

We give a new direct adaptive algorithm which does not need explicit identification of the secondary path dynamics, unlike the ordinary filtered-x algorithms using the identified model of $\overline{G}_4(z)$. The basic structure of the proposed adaptive feedforward control algorithm is illustrated in Fig.2, where $e_c(k)$, $e^a(k)$, $e^b(k)$ can be expressed as:

$$e_c(k) = \overline{G}_1(z)r(k) - \overline{G}_4(z)u(k) \qquad (6a)$$

$$e^a(k) = e_c(k) + \hat{K}(z,k)u(k) - \hat{D}(z,k)r(k) \qquad (6b)$$

$$e^b(k) = \hat{D}(z,k)r(k) - \hat{C}(z,k)x(k) \qquad (6c)$$

where $\overline{G}_1(z)$ and $\overline{G}_4(z)$ are defined in the previous section, and the control input $u(k)$ and the auxiliary signal $x(k)$ are also defined as

$$u(k) = \hat{C}(z,k)r(k) \qquad (7)$$

$$x(k) = \hat{K}(z,k)r(k) \qquad (8)$$

It follows from Fig.2 and the definitions that

$$\begin{aligned} e^a(k) + e^b(k) = & \\ & [e_c(k) + \hat{K}(z,k)u(k) - \hat{D}(z,k)r(k)] \\ & +[\hat{D}(z,k)r(k) - \hat{C}(z,k)x(k)] \\ = & \; e_c(k) + \hat{K}(z,k)\hat{C}(z,k)r(k) \\ & -\hat{C}(z,k)\hat{K}(z,k)r(k) \end{aligned} \qquad (9)$$

Thus, if $e^a(k)$ and $e^b(k) \to 0$ for $k \to \infty$ is satisfied and the FIR parameters of $\hat{C}(z,k)$ and $\hat{K}(z,k)$ converge to any constants, then the second and third terms in the right hand side of (9) can be cancelled, then by the relation $e^a(k) + e^b(k) = e_c(k)$, thus it can also be attained that $e_c(k) \to 0$ (Kohno *et al.*, 2002).

It seems that $\hat{D}(z,k)$ and $\hat{K}(z,k)$ are the identified models for $\overline{G}_1(z)$ and $\overline{G}_4(z)$ respectively and the adaptive controller $\hat{C}(z,k)$ is adjusted according to the identified models of the secondary path dynamics. However, even when the source noise does not satisfy the PE property, the proposed algorithm does not require the convergence of the adjusted parameters to their true values, but only the convergence of their parameters to any constants such that the errors $e^a(k)$ and $e^b(k)$ can converge to zero. Therefore, any dither sounds are not needed unlike the conventional indirect adaptive algorithm. The degradation and complexity caused by the dither signals can also be overcome by the proposed direct adaptive algorithm.

3.2 *New adaptation algorithm in two-channel case*

In a multi-channel case, the exchange of two matrices gives a different result, so the algorithm should be modified. To mitigate the problem, we employ a diagonal matrix as $\hat{\boldsymbol{K}}(z,k)$ to derive a fully adaptive algorithm for adjusting the controller parameters. Here for the simplicity of notation, we show the adaptive algorithm in two-channel case.

Similar to the single-channel case, we define two kinds of virtual error vectors $\mathbf{e}^a(k)$ and $\mathbf{e}^b(k)$ which can be described by

$$\begin{bmatrix} e_1^a(k) \\ e_2^a(k) \end{bmatrix} = \begin{bmatrix} e_{c1}(k) \\ e_{c2}(k) \end{bmatrix} + \begin{bmatrix} \hat{K}_{11}(z) & 0 \\ 0 & \hat{K}_{22}(z) \end{bmatrix} \begin{bmatrix} u_1(k) \\ u_2(k) \end{bmatrix} - \begin{bmatrix} \hat{D}_{11}(z) & \hat{D}_{12}(z) \\ \hat{D}_{21}(z) & \hat{D}_{22}(z) \end{bmatrix} \begin{bmatrix} r_1(k) \\ r_2(k) \end{bmatrix} \qquad (10)$$

$$\begin{bmatrix} e_1^b(k) \\ e_2^b(k) \end{bmatrix} = \begin{bmatrix} \hat{D}_{11}(z) & \hat{D}_{12}(z) \\ \hat{D}_{21}(z) & \hat{D}_{22}(z) \end{bmatrix} \begin{bmatrix} r_1(k) \\ r_2(k) \end{bmatrix} - \begin{bmatrix} \hat{C}_{11}(z) & \hat{C}_{12}(z) & 0 & 0 \\ 0 & 0 & \hat{C}_{21}(z) & \hat{C}_{22}(z) \end{bmatrix} \begin{bmatrix} x_{11}(k) \\ x_{12}(k) \\ x_{21}(k) \\ x_{22}(k) \end{bmatrix} \qquad (11)$$

where

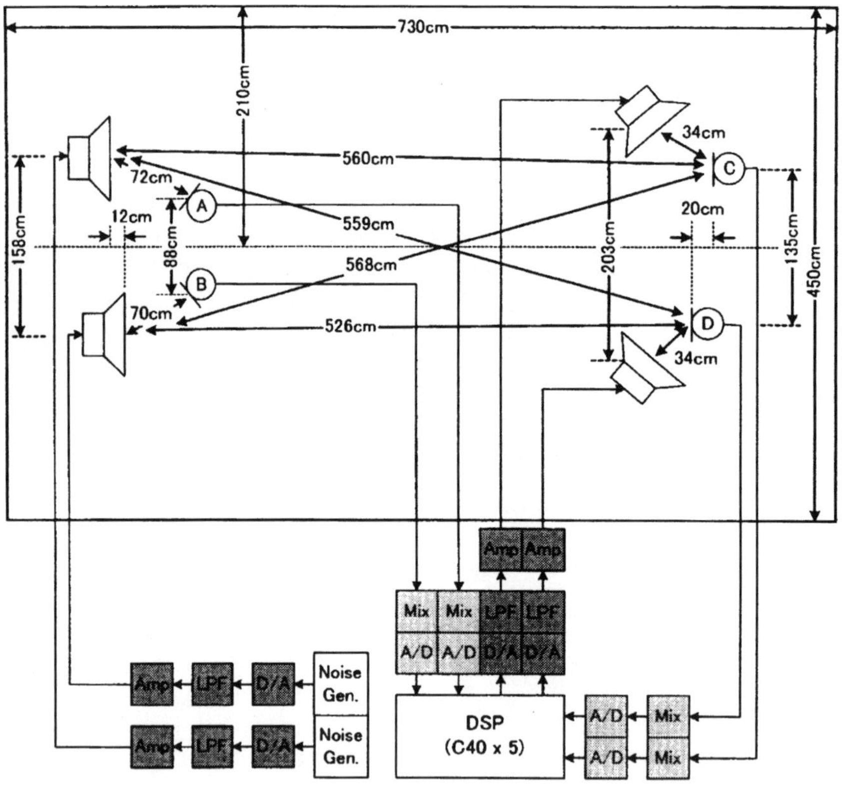

Fig. 3. Experimental setup of ANC in room

$$\begin{bmatrix} x_{11}(k) \\ x_{12}(k) \\ x_{21}(k) \\ x_{22}(k) \end{bmatrix} = \begin{bmatrix} \hat{K}_{11}(z) & 0 \\ 0 & \hat{K}_{11}(z) \\ \hat{K}_{22}(z) & 0 \\ 0 & \hat{K}_{22}(z) \end{bmatrix} \begin{bmatrix} r_1(k) \\ r_2(k) \end{bmatrix} \tag{12}$$

$$\begin{bmatrix} u_1(k) \\ u_2(k) \end{bmatrix} = \begin{bmatrix} \hat{C}_{11}(z) & \hat{C}_{12}(z) \\ \hat{C}_{21}(z) & \hat{C}_{22}(z) \end{bmatrix} \begin{bmatrix} r_1(k) \\ r_2(k) \end{bmatrix} \tag{13}$$

In the proposed adaptive algorithm, the coefficient parameters of $\hat{K}_{ii}(z)$ and $\hat{D}_{ij}(z)$ are updated so that the first virtual errors $e_1^a(k)$ and $e_2^a(k)$ can be eliminated, while the parameters of $\hat{C}_{ij}(z)$ are updated so that the second virtual errors $e_1^b(k)$ and $e_2^b(k)$ can be eliminated. If these parameters converge to constant and the two virtual errors can be reduced almost to zero, then it follows from (10) $\sim$ (13) that

$$\begin{aligned} e_i^a(k) + e_i^b(k) &= e_{ci}(k) + \hat{K}_{ii}(z)u_i(k) \\ &\quad -\hat{D}_{i1}(z)r_1(k) - \hat{D}_{i2}(z)r_2(k) \\ &\quad +\hat{D}_{i1}(z)r_1(k) + \hat{D}_{i2}(z)r_2(k) \\ &\quad -\hat{C}_{i1}x_{i1}(k) - \hat{C}_{i2}(z)x_{i2}(k) \\ &= e_{ci}(k) + \hat{K}_{ii}(z)\hat{C}_{i1}(z)r_1(k) \\ &\quad +\hat{K}_{ii}(z)\hat{C}_{i2}(z)r_2(k) - \hat{C}_{i1}(z)\hat{K}_{ii}(z)r_1(k) \\ &\quad -\hat{C}_{i2}(z)\hat{K}_{ii}(z)r_2(k) \quad \text{for } i = 1 \text{ and } 2 \end{aligned} \tag{14}$$

Thus, if the virtual errors $e_i^a(k)$ and $e_i^b(k)$ can be reduced to zero and all the adjustable parameters converge to constant, then other four terms in the right hand side can be cancelled and finally it follows that the actual errors $e_{ci}(k)$ can also eliminated.

Let the adjustable parameters be the coefficients of the FIR filters defined as:

$$\hat{C}_{ij}(z,k) = \hat{c}_{ij}^{(1)}(k)z^{-1} + \cdot + \hat{c}_{ij}^{(m_{ij}^c)}(k)z^{-m_{ij}^c}$$

$$\hat{K}_{ii}(z,k) = \hat{k}_{ii}^{(1)}(k)z^{-1} + \cdot + \hat{k}_{ii}^{(m_{ii}^k)}(k)z^{-m_{ii}^k}$$

$$\hat{D}_{ij}(z,k) = \hat{d}_{ij}^{(1)}(k)z^{-1} + \cdot + \hat{d}_{ij}^{(m_{ij}^d)}(k)z^{-m_{ij}^d}$$

where $i,\ j = 1$ and 2. The adaptive parameters $\{\hat{k}_{ii}^{(m)}(k)\}$ and $\{\hat{d}_{ij}^{(m)}(k)\}$ are updated so that the instantaneous error norm $\|e^a(k)\|^2$ may be minimized, while the parameters $\{\hat{d}_{ij}^{(m)}(k)\}$ are updated so that the error norm $\|e^a(k)\|^2$ may be minimized.

Let the parameter vectors be defined as

$$\hat{\boldsymbol{\theta}}_{Kii}(k) = (\hat{k}_{ii}^{(1)}(k), \cdots, \hat{k}_{ii}^{(m_{ii}^k)}(k))^T$$

$$\hat{\boldsymbol{\theta}}_{Dij}(k) = (\hat{d}_{ij}^{(1)}(k), \cdots, \hat{d}_{ij}^{(m_{ij}^d)}(k))^T$$

$$\hat{\boldsymbol{\theta}}_{Cij}(k) = (\hat{c}_{ij}^{(1)}(k), \cdots, \hat{c}_{ij}^{(m_{ij}^c)}(k))^T$$

and let the regressor vectors be denoted as

$$\zeta_i(k) = [u_i(k-1), \cdots, u_i(k-m_{ii}^k)]^T$$

$$\xi_j(k) = [r_j(k-1), \cdots, r_j(k-m_{ij}^d)]^T$$

$$\varphi_{ij}(k) = [x_{ij}(k-1), \cdots, x_{ij}(k-m_{ij}^c)]^T$$

where $i = 1$ and 2, and $x_{ij}(k) = \hat{K}_{ii}(z)r_j(k)$.

Thus the adaptive algorithm for updating these parameters is summarized as follows:

$$\hat{\boldsymbol{\theta}}_{Kii}(k+1) = \hat{\boldsymbol{\theta}}_{Kii}(k) - \gamma_K \boldsymbol{\zeta}_i(k)\varepsilon_i^a(k) \quad (15)$$

$$\hat{\boldsymbol{\theta}}_{Dij}(k+1) = \hat{\boldsymbol{\theta}}_{Dij}(k) + \gamma_D \boldsymbol{\xi}_j(k)\varepsilon_i^a(k) \quad (16)$$

$$\varepsilon_i^a(k) = e_i^a(k)/[1 + \gamma_D(\|\boldsymbol{\xi}_1(k)\|^2 + \|\boldsymbol{\xi}_2(k)\|^2) + \gamma_K\|\boldsymbol{\zeta}_i(k)\|^2] \quad (17)$$

$$\hat{\boldsymbol{\theta}}_{Cij}(k+1) = \hat{\boldsymbol{\theta}}_{Cij}(k) + \gamma_C \boldsymbol{\varphi}_{ij}(k)\varepsilon_i^b(k) \quad (18)$$

$$\varepsilon_i^b(k) = \frac{e_i^b(k)}{1 + \gamma_C(\|\boldsymbol{\varphi}_{i1}(k)\|^2 + \|\boldsymbol{\varphi}_{i2}(k)\|^2)} \quad (19)$$

where $i = 1$ and 2.

Other adaptive algorithms can also be derived, for instance, the norm $\|\boldsymbol{e}(k)\|^2$ of the augmented error $\boldsymbol{e}(k) = (\boldsymbol{e}^{aT}(k),\ \boldsymbol{e}^{bT}(k))^T$ can also minimized, which gives an althernative algorithm which requries more computation. Therefore, we adopted the proposed algorithm given by (15) to (19) in the experiment.

4. COMPARATIVE RESULTS

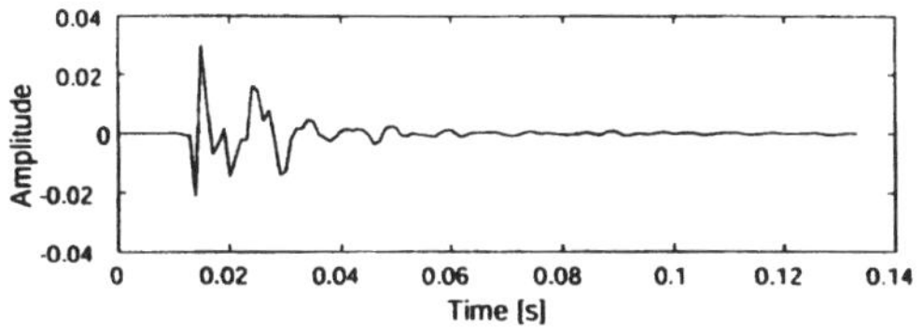

(a) $G_{111}(z)$

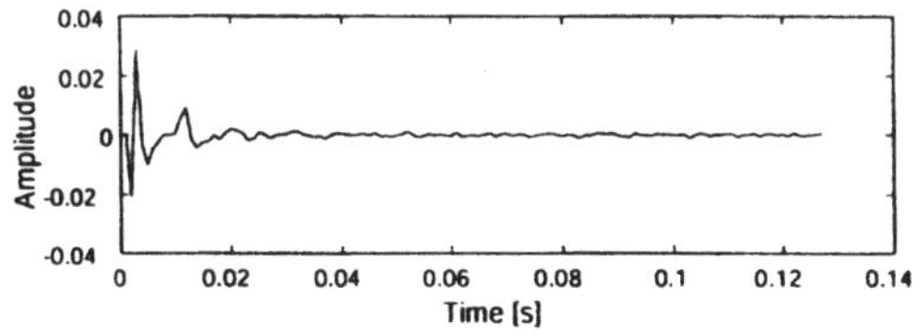

(b) $G_{411}(z)$

Fig. 4. Example of identified FIR channels

Fig.3 depicts an experimental setup for the active noise control in a room. Sound reflections on the room walls are suppressed by passive means. Two primary loudspeakers, two reference microphones A and B, two secondary loudspeakers, and two error microphones C and D are placed initially at the indicated locations, which configures a two-channel active noise control system ($N_s = N_r = N_c = N_e = 2$). To implement the two-channel adaptive ANC algorithm, five DSPs (TMS320-C40) are used for the parallel calculation, where the sampling period is 1 ms. The power spectra of the primary source noises $s_1(k)$ and $s_2(k)$ are limited in low frequency range from 50 to 400 Hz, or sometimes are sinusoids with unknown frequencies which do not satisfy the PE condition. Therefore in the conventional indirect approaches need dither sounds which are additively generated from the secondary loudspeakers to identify the secondary channels under the PE condition. On the other hand, the proposed approach does not need any dither sounds and the PE condition, but it can attain the noise attenuation even in the presence of uncertainties of the secondary channels. Identified FIR model of the primary and secondary channels are illustrated in Fig.4(a) and 4(b), where $G_{111}(z)$ and $G_{411}(z)$ are plotted for instance.

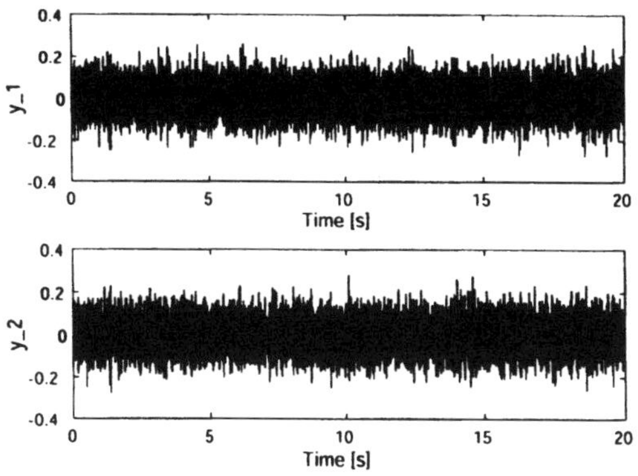

Fig. 5. $e_{c1}(k)$ and $e_{c2}(k)$ in case without control.

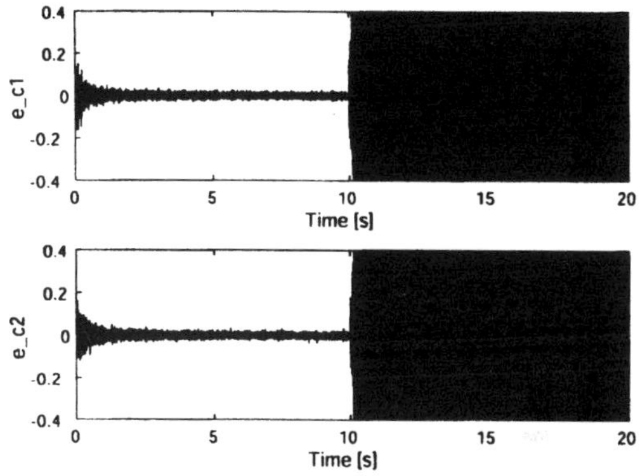

Fig. 6. $e_{c1}(k)$ and $e_{c2}(k)$ obtained by the extended error based filtered-x algorithm (Shimizu *et al.*, 2002).

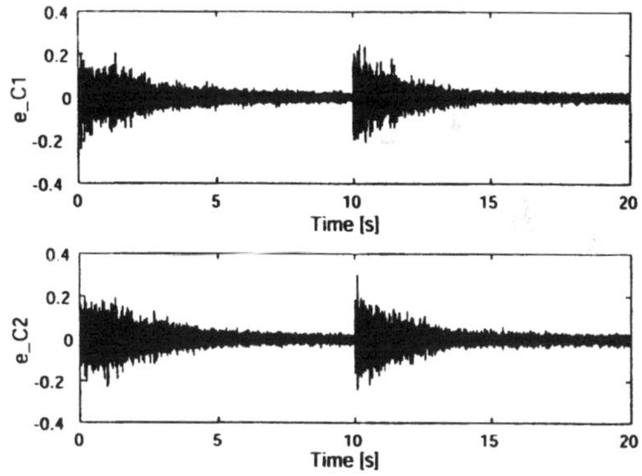

Fig. 7. $e_{c1}(k)$ and $e_{c2}(k)$ obtained by the direct fully adaptive algorithm.

Figs.5 to 7 show the actual canceling errors $e_{c1}(k)$ and $e_{c2}(k)$. In the numerical simulations, the locations of the two error microphones are changed by 34[cm] instantaneously far from the original positions by using the switches at 10[s] after the start of control. This causes the uncertain changes of the secondary path channels, and the ex-

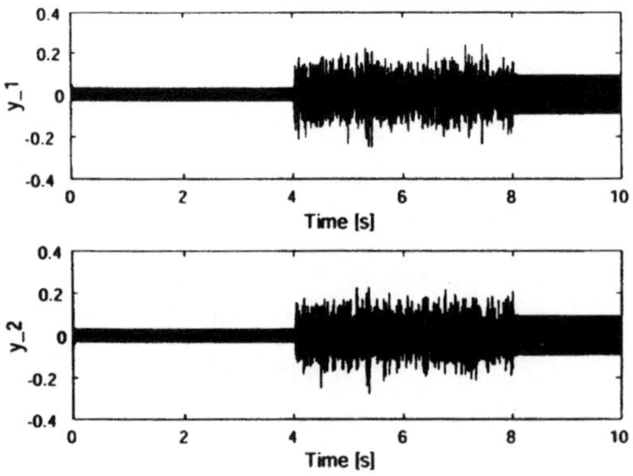

Fig. 8. Canceling error $e_{c1}(k)$ and $e_{c1}(k)$ in case without control

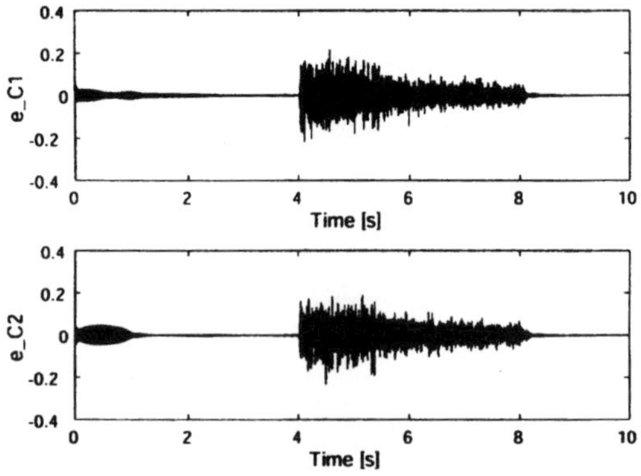

Fig. 9. Canceling error $e_{c1}(k)$ and $e_{c1}(k)$ obtained by the proposed fully adaptive algorithm

tended error based filtered-x algorithm (Shimizu *et al.*, 2002) could not keep the stable attenuation performance as shown in Fig.6, since it needs precise knowledge on the secondary channels. On the other hand, the proposed method can attain the stable control performance even if the secondary channels change very rapidly. Thus, the proposed algorithm does not need on-line identification of the secondary channels.

Figs.8 and 9 show the control results when the primary source noises consist of sinusoids with unknown frequencies (actually about 150 Hz and 250 Hz in the interval (0s, 4s), and 100 Hz in the interval (8s, 10s). Even when the primary source noises have an insufficient PE condition of the primary source noises like sinusoids, the proposed algorithm can give very nice attenuation performance still in the interval (0s, 4s). During the interval, the adaptive algorithm updates only small number of parameters required for reducing the canceling errors. During the interval (4s, 8s), the primary source noises have rather much PE property and then the many parameters of the adaptive filters should be updated, so the convergence time is need to achieve nice attenuation performance. During the interval (8s, 10s) the primary source noises are sinusoids again, however, since the adjustment of almost all adaptive parameters has been completed , then no parameters are required to update. Thus, the proposed control scheme is very robust to the insufficiency of the primary source noises unlike the conventional approaches.

5. CONCLUSION

The proposed fully direct adaptive control approach can work effectively even when all of the primary and secondary channel dynamics are uncertainly changeable. To reduce the canceling error, two virtual errors are introduced and are forced into zero by adjusting the three adaptive FIR filter matrices in an on-line manner, which enables the noise cancellation at the objective points. Unlike the previous methods, neither dither noises nor the PE property of the source noises are required.

6. REFERENCES

Bjarnason E. (1995). Analysis of the filtered-x LMS algorithm. *IEEE Trans. Speech and Audio Processing* **3-3** , 504–514.

Elliot S. J. and P. A. Nelson (1993). *Active Noise Control*, Academic Press.

Eriksson L. J. (1991). Development of the filtered-U algorithm for active noise control. *J. Acoust. Soc. Am.* **89** , 257–265.

Jiang F., H. Tsuji, H. Ohmori and A. Sano (1997). AAdaptation for active noise control. *IEEE Control Systems* **17-6** , 275–283.

Kim S. and Y. J. Park (1997). Unified-error filtered-x LMS algorithm for on-line active control of noise in time-varying environment. *Proc. Active 97* , 801–810, Budapest, Hungary.

Kuo M. and D. R. Morgan (1996). *Active Noise Control Systems* John Wiley and Sons.

Meurers T.and S.M. Veres (2002). Stability analysis of secondary path estimation during FSF-based feedback control. *Proc. IEEE Conf. Control Applications*, pp.483-488, Glasgow, UK.

Kohno, T., Y. Ohta, H. Ohmori and A. Sano (2002). New direct adaptive active noise control algorithms in case of uncertain path dynamics. *Proc. Amer. Contr. Conf.*, pp.1767-1772, Anchorage, USA.

Saito N., T. Sone, T. Ise and M. Akiho (1996). Optimal on-line modeling of primary and secondary paths in active noise control systems. *J. Acoust. Soc. Jpn (E)* **17-6** , 275–283.

Shimizu T., T. Kohno, H. Ohmori and A. Sano (2002). Multichannel active noise control, *Active Sound and Vibration Control,* Eds. by M.O. Tohki and S.M. Veres IEE.

Zhang M., H. Lan and W. Ser (2001). Cross-updated active noise control system with on-line secondary path modeling. *IEEE Trans. Speech and Audio Processing* **9-5** , 598–602.

IFAC Publications
www.elsevier.com/locate/ifac

CHANNEL ESTIMATION AND COUPLING WAVE CANCELLATION IN OFDM RELAY STATION

Lianming Sun * Akira Sano **

* *Department of Information & Media Sciences, The University of Kitakyushu*
1-1 Hibikino, Wakamatsu-ku, Kitakyushu 808-0135, Japan
** *Department of System Design Engineering, Keio University*
3-14-1 Hiyoshi, Kohoku-ku, Yokohama 223-8522, Japan

Abstract: The paper is concerned with adaptive cancellation of coupling effect in SFN relay station of an OFDM system. Due to the band limit of the source signal, the precise estimation of coupling wave channel outside the signal bandwidth is almost infeasible and that makes the performance of coupling effect cancellation very poor, even deteriorates the system stability. A new approach to estimate the channel model and to evaluate its estimation error is proposed for stable canceller updating. *Copyright © 2003 IFAC*

Keywords: Identification, frequency estimation, interpolation approximation, stability, uncertainty

1. INTRODUCTION

Orthogonal Frequency Division Multiplexing (OFDM) systems have been considered to be a reliable choice for high rate transmission and are widely adopted in communication applications such as digital audio or video broadcasting, broadband wireless LAN, ADSL communication networks.

The OFDM signal is transmitted through Single Frequency Networks (SFN), where the transmission efficiency is higher than that in multi-frequency networks. Nevertheless, a serious problem occurs in the SFN relay station. Due to the antennas of transmitter and receiver use the same carrier frequency, the radio wave from the transmitter antenna couples into the receiver antenna in the relay station, as illustrated in Fig.1. The coupling wave deteriorates the quality of the signal transmission greatly, even causes the serious oscillation problem. Hence an adaptive cancellation technique, which requires dynamic property of the coupling wave channel, should be utilized to reduce the coupling effect.

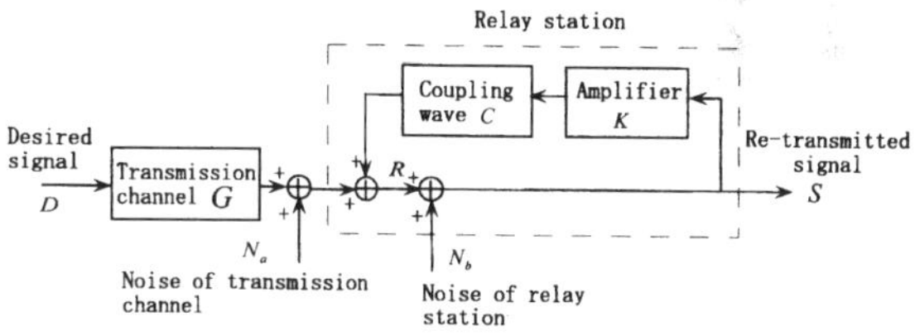

Fig. 1. Block diagram of SFN relay station

Several methods have been proposed to deal with the coupling wave problem. In the time domain algorithm, the dynamic property C of the coupling wave path is estimated then the coupling wave canceller W, which is often a FIR filter as illustrated in Fig.2, is designed to cancel the coupling wave (Hamazumi *et al.*, 1999). In the space processing technique based approaches, the antenna arrays are used for beam-forming and direction-of-arrival estimations (Godara, 1997). And some space-time joint algorithms are also proposed (Yang *et al.*, 2001; Tomitsuka *et al.*, 2001).

In both the time domain and space domain algorithms, the estimation of the coupling wave path C and the transmission multi-path channel G are necessary for canceller design or source symbol

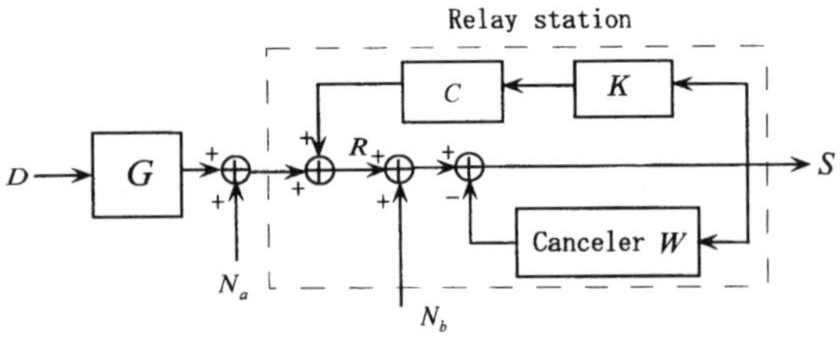

Fig. 2. Coupling wave cancellation in time-domain

recovery, and many estimation algorithms have been developed. For example, the subspace-based blind or semi blind OFDM channel estimation algorithm (Muquet *et al.*, 2002), cyclostationarity based algorithm (Heath and Giannakis, 1999), etc. However, the computational complexity increases greatly with the channel order increases. Moreover, in order not to interfere the adjacent channels, the desired signal in the OFDM systems is generally a band limited signal, consequently it is very difficult to estimate all the frequency property for the whole frequency range by most of the existing methods. Then it implies that the stability of the system by using the insufficient information is not always guaranteed in the relay station, especially outside the signal bandwidth.

In this paper, a new canceller design approach in time domain is considered in presence of both coupling effect in SFN relay station and multi-path in the transmission channel from the main station to relay station. In the new scheme, the algorithms of the system model estimation, estimation error evaluation and adaptive coupling wave cancellation with consideration of stability are presented. In contrast to the existing methods, the proposed algorithm makes use of the distinct property of OFDM signals to reveal the dynamics of channel model, even outside of the signal bandwidth.

2. PROBLEM STATEMENT

The desired base band signal $d(t)$ in one symbol period T is given by

$$d(t) = \sum\nolimits_{n=-(N-1)/2}^{(N-1)/2} d_n e^{jn\omega_0 t} \tag{1}$$

where ω_0 is the frequency interval and d_n is the source symbol. Here t is a time variable. Usually $d(t)$ is generated by inverse Fourier transform with FFT size $M > N$. Then the spectrum of $d(t)$ is concentrated inside the range $|n| \leq (N-1)/2$, as shown in Fig.3. It implies that the direct estimation of dynamic property of the unknown plant model is almost impossible for $|n| > (N-1)/2$ (Ljung, 1999).

Though the source symbol is generally unknown to the receivers, the pilot symbols used for detection or synchronization are inserted in the OFDM signals, as shown in Fig.4. When the receiver of the relay station is well synchronized with the

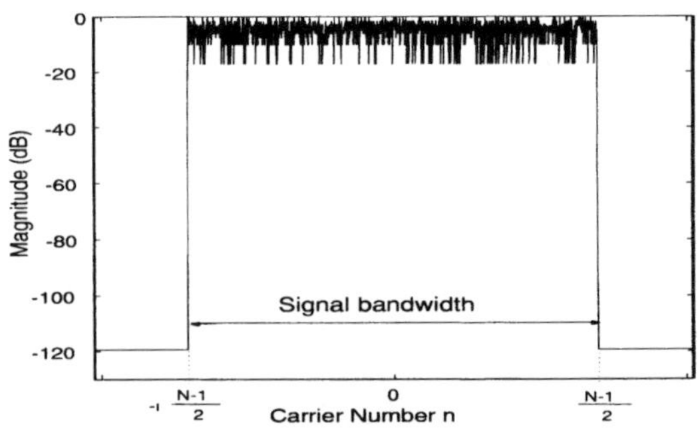

Fig. 3. Normalized spectrum of OFDM signal

source signal, the pilot symbols can be regarded as known information in the desired signal spectrum $\mathcal{D}(j\omega_P)$ at the corresponding frequency points ω_P.

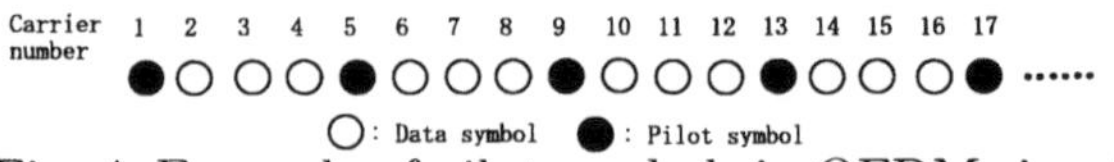

Fig. 4. Example of pilot symbols in OFDM signal

Another distinctive feature of the OFDM system is that, a small amount of guard interval, i.e., a cyclic prefix (CP), whose length is commonly larger than the channel memory at the transmitter, is inserted to each symbol, enables OFDM to avoid the multiple-access interference and helps the model estimation greatly.

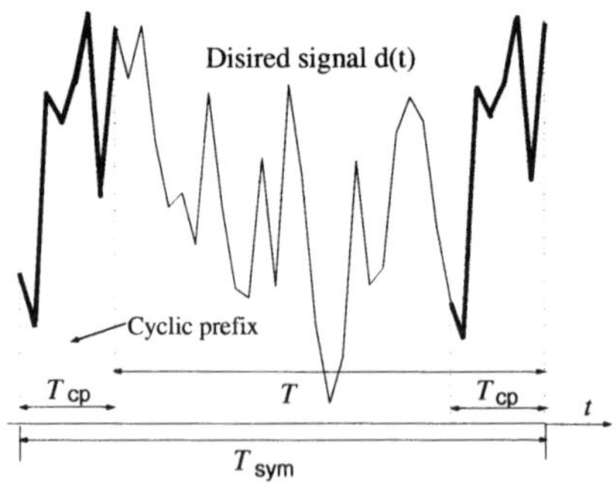

Fig. 5. CP in OFDM signal

As illustrated in Fig.5, the period of the original real symbol is denoted as T. The cyclic prefix is inserted before the original symbol by copying the last part of the original signal within time T_{cp}. Then the total period of transmitted symbol becomes to $T + T_{\text{cp}}$, which is denoted as T_{sym}.

Following the block diagram of coupling wave cancellation in Fig.2, the transfer function from D to S in frequency domain is expressed by

$$\mathcal{S}(j\omega) = \frac{\mathcal{G}(j\omega)\mathcal{D}(j\omega)}{1 + \mathcal{W}(j\omega) - \mathcal{K}(j\omega)\mathcal{C}(j\omega)} + \frac{\mathcal{N}_a(j\omega) + \mathcal{N}_b(j\omega)}{1 + \mathcal{W}(j\omega) - \mathcal{K}(j\omega)\mathcal{C}(j\omega)} \tag{2}$$

Denote H as the transfer function of the system, and Γ the sensitivity function, then

$$\mathcal{H}(j\omega) = \frac{\mathcal{G}(j\omega)}{1 + \mathcal{W}(j\omega) - \mathcal{K}(j\omega)\mathcal{C}(j\omega)}$$

$$\Gamma(j\omega) = \frac{1}{1 + \mathcal{W}(j\omega) - \mathcal{K}(j\omega)\mathcal{C}(j\omega)}$$

Here K is the dynamics of the power amplifier part which is assumed to work in its linear range. Then the objects of the algorithm are

(1) To estimate the channel model property.
(2) To evaluate the estimation error.
(3) To design the filter W such that the system is stable and

$$\mathcal{H}(j\omega) \to 1, \text{ for } |\omega| \leq \frac{N-1}{2}\omega_0 \qquad (3)$$

while all the poles of sensitivity function $\Gamma(q^{-1})$ are inside the unit circle in complex q-plane.

3. CHANNEL ESTIMATION AND EVALUATION

Let the canceller W be a FIR filter with length $L \geq \max(L_g, L_c)$, where L_g and L_c are the FIR model lengths of multi-path $\mathcal{G}(j\omega)$ and coupling effect $\mathcal{C}(j\omega)$. Channel estimation is performed by using the re-transmitted signal $s(t)$, pilot symbols collected for one symbol period T_{sym}, and the CP property of the OFDM system.

3.1 *Channel Estimation inside Signal Bandwidth*

Assume that the receiver of the relay station is well synchronized with the transmitter of main station, then the pilot symbols at the frequency points ω_P, i.e., $\mathcal{D}(j\omega_P)$, are considered to be known to the relay station. Furthermore, the re-transmitted signal $s(t)$ as well as its Fourier transform $\mathcal{S}(j\omega)$ using the data within one symbol period T can also be obtained. It yields that the estimate $\hat{\mathcal{H}}(j\omega)$ corresponding to the pilot symbol frequency ω_P can be calculated as follows.

$$\hat{\mathcal{H}}(j\omega_P) = \frac{\mathcal{S}(j\omega_P)}{\mathcal{D}(j\omega_P)} \qquad (4)$$

Then using the interpolation methods, the estimates of $\hat{\mathcal{H}}(j\omega)$ inside the signal band can be obtained. For example, the linear interpolation method is used in (Hamazumi *et al.*, 1999), which is summarized as follows.

$$\hat{\mathcal{H}}(j\omega) = \hat{\mathcal{H}}(j\omega_{P_1}) + \frac{\omega - \omega_{P_1}}{\omega_{P_2} - \omega_{P_1}}\left(\hat{\mathcal{H}}(j\omega_{P_2}) - \hat{\mathcal{H}}(j\omega_{P_1})\right) \qquad (5)$$

where ω_{P_1} and ω_{P_2} are the adjacent pilot frequencies, and ω is the frequency between ω_{P_1} and ω_{P_2}. Since the pilot symbols are distributed between the frequency $-(N-1)\omega_0/2$ and $(N-1)\omega_0/2$, $\hat{\mathcal{H}}(j\omega)$ can be obtained for $-(N-1)\omega_0/2 \leq \omega \leq (N-1)\omega_0/2$.

3.2 *Estimation of Sensitivity Function*

To guarantee the system stability, the poles of sensitivity function $\Gamma(q^{-1})$ must be assigned inside and bounded away from the unit circle.

Recall the cyclic prefix inserted in the OFDM signals. During one symbol time T_{sym}, let the new signals for $0 < t \leq T_{\text{cp}}$ be defined by

$$\begin{aligned} d_1(t) &= d(t+T) - d(t) \\ s_1(t) &= s(t+T) - s(t) \end{aligned} \qquad (6)$$

Then $d_1(t) = 0$ following Fig.5 so $s_1(t)$ can be described by an AR model

$$\frac{1}{\Gamma}s_1(t) = v(t), \quad \text{for } 0 < t \leq T_{\text{cp}} \qquad (7)$$

where

$$\begin{aligned} \frac{1}{\Gamma(j\omega)} &= 1 + \mathcal{W}(j\omega) - \mathcal{K}(j\omega)\mathcal{C}(j\omega) \\ v(t) &= N_a(t) + N_b(t) \end{aligned}$$

When $v(t) \neq 0$, (7) can further be approximated by an ARMA model with low complexity

$$\frac{\Gamma_2}{\Gamma_1}s_1(t) = v(t), \quad \text{for } 0 < t \leq T_{\text{cp}} \qquad (8)$$

where $\Gamma_2(j\omega)/\Gamma_1(j\omega) \approx 1/\Gamma(j\omega)$.

Then the parameters of $\Gamma(j\omega)$ can be estimated by LS algorithm for fast convergence or LMS algorithm for low computational complexity from (7) or (8).

3.3 *Channel Estimation of Multi-path Channel*

Filtering signal $s_1(t)$ by $s_2(t) = s_1(t)/\Gamma$ yields that

$$s_2(t) = Gd_1(t), \quad \text{for } -T_{\text{cp}} \leq t \leq T_{\text{cp}} \qquad (9)$$

To estimate G, we recover the desired source signal $d(t)$ by using $\hat{\mathcal{H}}(j\omega)$ and $\mathcal{S}(j\omega)$ within one symbol period.

$$\hat{d}(t) = \text{IFFT}\left(\frac{\mathcal{S}(j\omega)}{\hat{\mathcal{H}}(j\omega)}\right), \quad \text{for } 0 \leq t < T \quad (10)$$

Then $\hat{d}_1(t)$ can be calculated following (6), so $\hat{\mathcal{G}}(j\omega)$ is obtained by solving the MA model estimation problem.

3.4 *Evaluation of Estimation Errors*

The estimation of $\mathcal{H}(j\omega)$, $\mathcal{G}(j\omega)$ or $\Gamma(j\omega)$ themselves only is not adequate for canceller design. The corresponding estimation error, especially the estimation error of the sensitivity function $\Gamma(j\omega)$, is also very important. The estimation error is mainly caused by the limited signal bandwidth, the corrupted noise in observed data and the model uncertainty when using low complexity model for computational simplicity.

Recall the model in (7). If the AR model is converted to a model such that the initial conditions are 0 for $t < 0$, then the impulse response of Γ can be obtained from the transient response directly.

Define the signal $\bar{d}_1(t)$ as follows.

$$\bar{d}_1(t)=\begin{cases}-\frac{1}{\hat{\Gamma}}s_1(t), & \text{for } L_g-L+1<t<L_g \\ 0, & \text{others}\end{cases} \tag{11}$$

Then the signal $\bar{s}_1(t)$ given by

$$\bar{s}_1(t) = s_1(t) + \hat{\Gamma}\bar{d}_1(t) \tag{12}$$

is a signal such that

$$\frac{1}{\hat{\Gamma}}\bar{s}_1(t) = \bar{v}(t) \tag{13}$$

with initial conditions $\bar{s}_1(t) = 0$ for $L_g - L + 1 < t < L_g$, which is the impulse response of $\hat{\Gamma}$ to $d_1(0) + \bar{d}_1(L_g)$ with noise $\bar{v}(t)$. Denote the new estimate as $\hat{\bar{\Gamma}}$ whose impulse response is $\hat{\bar{\gamma}}(t)$ and its frequency response is $\hat{\bar{\Gamma}}(j\omega)$. Next we evaluate the error of $\hat{\Gamma}(j\omega)$ by using $\hat{\Gamma}(j\omega)$ and $\hat{\bar{\gamma}}(t)$.

Assume that $\hat{\bar{\Gamma}}(j\omega)$ can be expressed by

$$\hat{\bar{\Gamma}}(j\omega) = \Gamma(j\omega) + \bar{\mathcal{V}}(j\omega) \tag{14}$$

Now introduce a window function $w_m = \{w_{m,k}\}_{k=0}^{m-1}$ where m is the window length. The window function is chosen as a rectangular window or a triangular window. $\bar{\mathcal{V}}(j\omega)$ can be evaluated by some nonlinear algorithms (Chen and Gu, 2000), here a linear algorithm is adopted where the model with low complexity is given for the simplicity of computation

$$\hat{\bar{\Gamma}}_{\text{low-c}}(z) = \sum_{k=0}^{m-1} w_{m,k}\hat{\bar{\gamma}}(k)z^k \tag{15}$$

then the following approximation

$$\left\|\hat{\bar{\Gamma}}_{\text{low-c}}(j\omega) - \Gamma(j\omega)\right\| \approx Q_w\|\bar{\mathcal{V}}(j\omega)\| - \Lambda(\omega) \geq 0 \tag{16}$$

holds, where $Q_w > 1$ is a constant given by

$$Q_w = \frac{1}{M}\sum\left|\text{FFT}\{w_{m,k}\}\right| \tag{17}$$

Here M is the length of FFT. Generally $\Lambda(\omega)$ is difficult to evaluated. Here we just assume that $\Lambda(\omega)$ can be approximated by

$$\Lambda(\omega) \approx Q_w\left\|\hat{\bar{\Gamma}}_{\text{low-c}}(j\omega) - \hat{\bar{\Gamma}}(j\omega)\right\| + \left\|\hat{\bar{\Gamma}}_{\text{low-c}}(j\omega) - \hat{\Gamma}(j\omega)\right\| \tag{18}$$

Moreover, the following inequality holds

$$\left\|\hat{\bar{\Gamma}}_{\text{low-c}}(j\omega) - \Gamma(j\omega)\right\| \leq \left\|\hat{\bar{\Gamma}}_{\text{low-c}}(j\omega) - \hat{\bar{\Gamma}}(j\omega)\right\| + \|\bar{\mathcal{V}}(j\omega)\| \tag{19}$$

Then substituting (16) into (19) leads to

$$\|\bar{\mathcal{V}}(j\omega)\| \leq \frac{1}{Q_w - 1}\left(\left\|\hat{\bar{\Gamma}}_{\text{low-c}}(j\omega) - \hat{\bar{\Gamma}}(j\omega)\right\| + \Lambda(\omega)\right)$$

$$\approx \frac{Q_w + 1}{Q_w - 1}\left\|\hat{\bar{\Gamma}}_{\text{low-c}}(j\omega) - \hat{\bar{\Gamma}}(j\omega)\right\| + \frac{1}{Q_w - 1}\left\|\hat{\bar{\Gamma}}_{\text{low-c}}(j\omega) - \hat{\Gamma}(j\omega)\right\| \tag{20}$$

Therefore the error evaluation for $\hat{\Gamma}(j\omega)$ given in Section 3.2 is given by

$$\begin{aligned}\|\Delta\Gamma(j\omega)\| &= \left\|\hat{\Gamma}(j\omega) - \Gamma(j\omega)\right\| \\ &\leq \left\|\hat{\Gamma}(j\omega) - \hat{\bar{\Gamma}}(j\omega)\right\| + \left\|\hat{\bar{\Gamma}}(j\omega) - \Gamma(j\omega)\right\| \\ &\leq \left\|\hat{\Gamma}(j\omega) - \hat{\bar{\Gamma}}(j\omega)\right\| + \|\bar{\mathcal{V}}(j\omega)\|\end{aligned} \tag{21}$$

4. CANCELLER UPDATING

The estimation and canceller design are performed iteratively at every symbol period. Let the superscript (l) denote the estimation and error evaluation in the l-th iteration.

Let the frequency response of W at ω_k be

$$\mathcal{W}(j\omega_k) = \alpha_k + j\beta_k \tag{22}$$

Define the updating criterion for W as follows.

$$Q = \frac{1}{N_P + N_H}\left(\sum_{k=1}^{N_P}\mathcal{E}^*(j\omega_{P,k})\mathcal{E}(j\omega_{P,k}) + \sum_{k=1}^{N_H}\mathcal{E}^*(j\omega_{H,k})\mathcal{E}(j\omega_{H,k})\right) \tag{23}$$

where $\omega_{P,k}$ is the frequency of pilot symbol, $\omega_{H,k}$ is the frequency outside of the signal bandwidth, $\mathcal{E}(j\omega)$ is the error function given by

$$\mathcal{E}(j\omega)=\begin{cases}\hat{\mathcal{G}}(j\omega)\left(1 - \frac{\mathcal{D}(j\omega)}{\mathcal{S}(j\omega)}\right), & |\omega| \leq \frac{N-1}{2}\omega_0 \\ \hat{\mathcal{G}}(j\omega) - \frac{1}{\hat{\Gamma}(j\omega)}, & |\omega| > \frac{N-1}{2}\omega_0\end{cases} \tag{24}$$

Then the updating weights of W in frequency domain are given by LMS algorithm in the $(l+1)$-th iteration.

$$\Delta\mathcal{W}^{(l+1)}(j\omega) = \boldsymbol{Q}'(j\omega) = -\mathcal{E}(j\omega) \tag{25}$$

The corresponding implementation of updated weights vector $\Delta\boldsymbol{W}$ in time domain satisfies that

$$\Delta\mathcal{W}^{(l+1)}(j\omega)=\left[1, e^{-j\omega}, \cdots, e^{-j\omega(L-1)}\right]\Delta\boldsymbol{W}^{(l+1)} \tag{26}$$

Then for $\omega = \omega_{P,1}, \cdots, \omega_{P,N_P}$ and $\omega_{H,1}, \cdots, \omega_{H,N_H}$ we have that

$$\Delta\boldsymbol{\mathcal{W}}^{(l+1)}(j\omega) = \boldsymbol{\Phi}(e^{-j\omega})\Delta\boldsymbol{W}^{(l+1)} \tag{27}$$

Then $\Delta\boldsymbol{W}$ can be calculated by weighted least squares algorithm as follows.

$$\Delta\boldsymbol{W}^{(l+1)} = \boldsymbol{\Omega}\boldsymbol{\Phi}^*(e^{-j\omega})\boldsymbol{A}(\omega)\Delta\boldsymbol{\mathcal{W}}^{(l+1)}(j\omega) \tag{28}$$

where the superscript $*$ denotes complex conjugate transpose. $\boldsymbol{\Omega}$ is the matrix given by

$$\boldsymbol{\Omega} = \left(\boldsymbol{\Phi}^*(e^{-j\omega})\boldsymbol{A}(\omega)\boldsymbol{\Phi}(e^{-j\omega})\right)^{-1}$$

And $\boldsymbol{A}(\omega)$ is a diagonal weighting matrix. So the new canceller is updated by

$$\boldsymbol{W}^{(l+1)} = \boldsymbol{W}^{(l)} + \mu^{(l+1)} \Delta \boldsymbol{W}^{(l+1)} \quad (29)$$

where the stable step size $\mu^{(l+1)}$ in (25) is determined by the following Theorem.

Theorem 1. The system stability is guaranteed if the updating step size $\mu^{(l+1)}$ is given by

$$\min \left\{ \frac{\lambda}{\left(\left\| \hat{\Gamma}^{(l+1)}(j\omega) \right\| + \left\| \Delta \Gamma^{(l+1)}(j\omega) \right\| \right)} \times \frac{1}{\left\| \Delta \bar{\mathcal{W}}^{(l+1)}(j\omega) \right\|}, 1 \right\} \quad (30)$$

where $0 < \lambda < 1$ and $\Delta \bar{\mathcal{W}}^{(l+1)}(j\omega) = \text{FFT}(\Delta \boldsymbol{W}^{(l+1)})$.

The proof is given by using the well-known small gain theorem and the structure of the relay station model. Then in the next symbol period, the estimation and canceller design will be performed in the same way after collecting new data.

5. NUMERICAL EXAMPLES

In the simulation examples, the source symbols $d_k = a_k + jb_k$ are the 64 QAM signals, where a_k, $b_k \in \{\pm 1, \pm 3, \pm 5, \pm 7\}$. The carrier number is $N = 1405$, among which there are 118 pilot symbols. The carrier frequency interval is $\omega_0 = 2\pi f_0$, where $f_0 = 4$ kHz, and FFT/IFFT size is $M = 2048$. The guard interval of cyclic prefix $T_{\text{cp}} = T/4$. Let the ratio of the desired signal to coupling waves, i.e., the DU ratio be defined by

$$\text{DU} = 10 \log_{10} \left(\frac{D^2}{C_i^2} \right) \text{dB}$$

Here D^2 and C_i^2 are the signal power of the desired signal, and i-th coupling wave, respectively. Whereas the signal to noise ratio is 40dB in the transmission channel. Without loss of the generality, $\mathcal{K}(j\omega)$ is set to a constant.

The taps length of canceller W is 100, and its initial value is $W = 0$. Two simulation examples are considered in the paper.

5.1 *Simple Case: 1 Coupling Wave and 1 Multi-path*

Consider the simulation conditions in (Hamazumi *et al.*, 1999). Assume that there is only one coupling wave in the relay station and one multi-path in the channel from the main station to the relay station. Their arrival angles are 45° and 180°, the delay times are 17 taps and 8 taps, respectively. The DU ratios are 2dB and 10dB.

For comparison with the algorithm in (Hamazumi *et al.*, 1999), where the updating step size is chosen as $\mu = 0.2$, $\mu = 0.5$ and $\mu = 1.0$ respectively, and the proposed algorithm in this paper are used to the problem of coupling effects cancellation. The simulations are performed for 50 iterations and the DU ratios for the re-transmitted signals are depicted in Fig.6, from which the convergence of the proposed algorithm is almost the same as that of the method in (Hamazumi *et al.*, 1999). However, as illustrated in Fig.7, the frequency property outside the signal bandwidth is almost not compensated due to the insufficient information in the existing method. When more coupling waves entering into the receiver antenna, the stability of the canceller cannot be guaranteed.

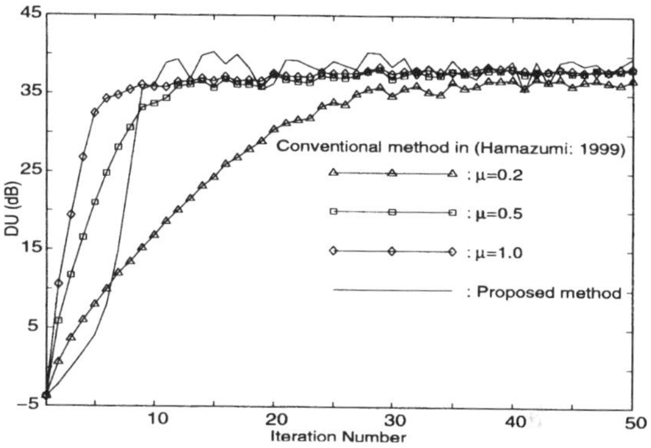

Fig. 6. Comparison of the convergence of proposed algorithm and existing method

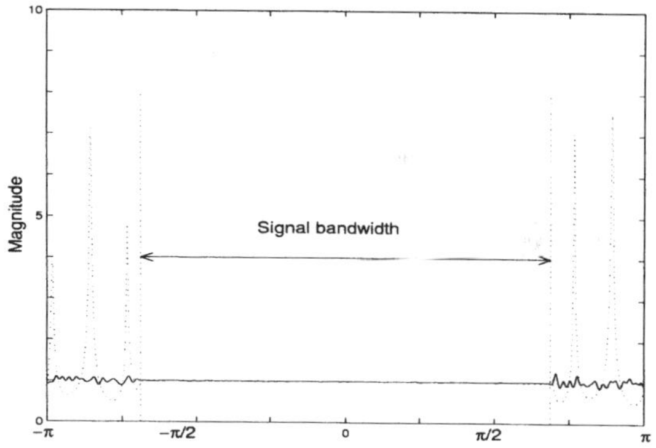

Fig. 7. Comparison of the convergence of proposed algorithm and existing method. Solid line: proposed algorithm; Dot line: conventional algorithm in (Hamazumi *et al.*, 1999) where $\mu = 0.2$

5.2 *General Case: Many Coupling Waves Case*

Assume that the coupling effects have five waves entering the receiver antenna at angles 58°, 30°, −165°, −92°, 45° in SFN station. The delay time of the coupling waves are 4, 6, 7, 9, 12 taps later than the desired signal $d(t)$. The DU ratios of coupling waves are 7.7dB, 3.25dB, 61dB, 10.9dB, 65dB respectively. In the transmission channel from main station to the relay station, there is a multiple path whose direction of arrival angle is

-60° with DU ratio 3.87dB, and its time delay is 9 taps.

The frequency response of the sensitivity function in the first iteration is given in Fig.8. It shows that the sensitivity function is well estimated inside the signal bandwidth. Though the power of desired signal $d(t)$ outside the bandwidth is near to zero, the frequency response of the sensitivity function outside signal bandwidth can also be evaluated by using the information of cyclic prefix.

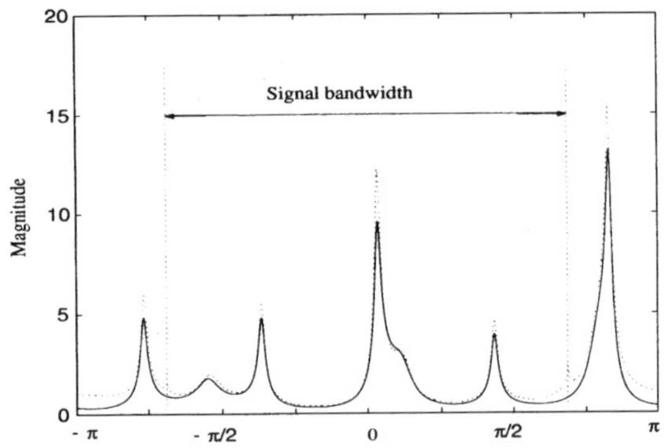

Fig. 8. The estimated sensitivity function of relay station model. The solid line: true value of $\|\Gamma(j\omega)\|$; The dash line: estimate

In Fig.9, the frequency response of the channel model after 100 iterations is given. It shows that $\mathcal{H}(j\omega)$ converges to 1 inside as well as outside the signal bandwidth. Fig.10 illustrates the convergence of the proposed algorithm by plotting DU ratios with respect to the iteration number. In the conventional method in (Hamazumi *et al.*, 1999), the cancellation system is only stabilized for small updating step size. When the step size is large, for example, the system fails to work in the second iteration for $\mu = 0.5$, and the system is fall into oscillation for $\mu = 1.0$.

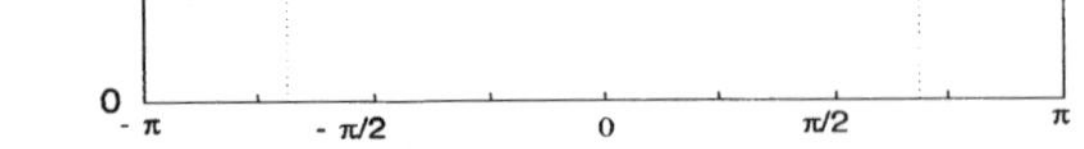

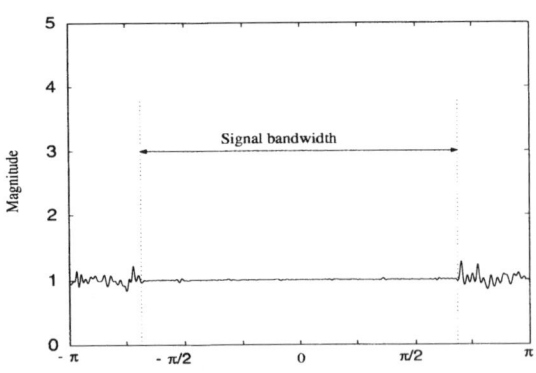

Fig. 9. The transmission channel frequency property after 100 iterations

6. CONCLUSIONS

The problem for estimation and cancellation of the coupling wave effect in SFN relay station is

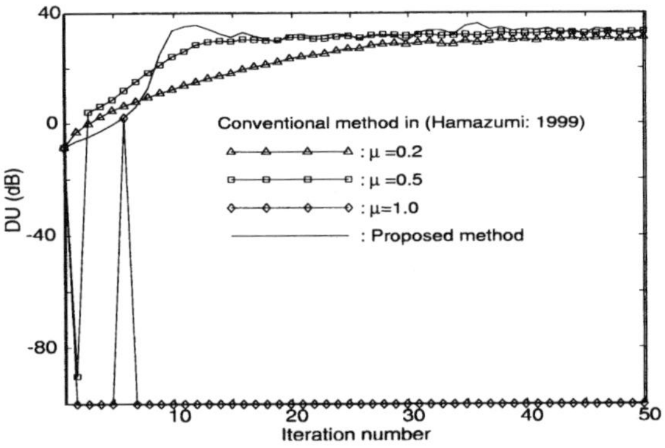

Fig. 10. Convergence of DU ratios

discussed for the OFDM communication system. The new approach to deal with the estimation both inside and outside the signal band using the property of OFDM signal is proposed, and the adaptive algorithm for coupling wave cancellation is also given. It is illustrated that the estimation of the unknown channel and the estimation of sensitivity function should be considered for the stable canceller design. The numerical examples demonstrate the effectiveness of the proposed algorithms. The new approach can also be extended into the space-time joint algorithms.

7. REFERENCES

Chen, J. and G. Gu (2000). *Control-Oriented System Identification–An H_∞ Approach.* John Wiley & Sons.

Godara, L.C. (1997). Applications of antenna arrays to mobile communications, Part II: Beam-forming and direction-of-arrival considerations. *Proc. of IEEE* **85**(8), 1195–1245.

Hamazumi, H., K. Imamura, N. Iai and K. Shibuya (1999). A study on coupling wave canceler for relay station in digital terrestrial broadcasting SFN. *Technical Report of IEICE* **EMCJ98-111**, 49–56.

Heath, R. and G.B. Giannakis (1999). Exploiting input cyclostationarity for blind channel identification in OFDM systems. *IEEE Trans. Signal Processing* **47**(3), 848–856.

Ljung, L. (1999). *System Identification – Theory for the User.* Prentice-Hall.

Muquet, B., M. Courville and P. Duhamel (2002). Subspace-based blind and semi-blind channel estimation for OFDM systems. *IEEE Trans. Signal processing* **50**(7), 1699–1712.

Tomitsuka, K., Y. Ichikawa, S. Obote and K. Kagoshima (2001). Realization by dsp of adaptive array antenna with space-temporal joint equalization. *Technical Report of IEICE* **CS2001-39**, 33–40.

Yang, K., Y. Zhang and Y. Mizuguchi (2001). A single subspace-based subband approach to space-time adaptive processing for mobile communications. **49**(2), 401–413.

IFAC Publications
www.elsevier.com/locate/ifac

APPLICATION OF SYSTEM IDENTIFICATION FOR THE PREDICTION OF AVALANCHE HAZARD

Janusz Milek and Bernhard Brabec

Insightful AG,
Christoph Merian-Ring 11, CH-4153 Reinach, Switzerland; http://www.insightful.com

Abstract: System identification methods are applied to predict avalanche hazard. A ten-year record of avalanche hazard and snow cover from Davos and Weissfluhjoch in the Swiss Alps is analyzed. Several models are estimated and validated. Dynamic models present a substantial improvement in the quality of hazard predictions compared to predictions derived from a commonly used static nearest neighbor model structure. Further possible modeling improvements are briefly discussed. *Copyright © 2003 IFAC*

Keywords: avalanche forecasting, modeling, discrete-time, linear models, neural networks, physical measurement systems.

1. INTRODUCTION

During winter, avalanches are the prominent natural hazard endangering inhabitants of alpine villages and participants in recreational outdoor activities. Therefore, several avalanche research centers have been established in Austria, Switzerland, Italy, and France. These countries also have active avalanche warning programs (see http://www.avalanches.org for an overview). Nevertheless, about 25 people are killed on average each year in the Swiss Alps (Tschirky, *et al.*, 2000) by avalanches, most while participating in recreational activities. Up-to-date avalanche forecasts are an essential means of managing the risks on a recreational tour.

During the last ten years the process of constructing avalanche forecasts has been improved by the Swiss Federal Institute for Snow and Avalanche Research (SLF) by the introduction of computer-assisted avalanche forecasting (Brabec, 2001; Brabec *et al.*, 2001). Two complementary approaches are used to support avalanche forecasters in determining the hazard level: (1) physical models simulate the snow cover at the location of automatic weather station (Lehning, *et al.*, 1999) and (2) statistical models are used to calculate the avalanche hazard for whole area of the Swiss Alps (Brabec and Meister, 2001). The statistical models use the nearest-neighbor approach for daily predictions of avalanche hazard.

In Switzerland, the first attempts to develop models to forecast avalanche releases were undertaken by Bois *et al.* (1974). Buser (1983; 1989) further developed the nearest-neighbor approach for local avalanche forecasting. Bolognesi (1998), Kristensen *et al.* (1994), and McClung *et al.* (1994) have also adapted the nearest-neighbor approach. Schweizer *et al.* (1996) have developed a series of expert system-based models that simulate the reasoning of a pragmatic avalanche forecaster. In France the model-chain Safran-Crocus-Mepra (Durand, *et al.*, 1999) calculates the stability of a mean snow pack for each of 35 massifs of the French Alps for different elevations, aspects, and slope angles.

2. PROBLEM FORMULATION

This paper examines how regional avalanche forecasting as defined in Brabec and Meister (2001) can be improved by system identification methods. Salway (1979) takes a time series approach to a related problem, specifically, the modeling of observed avalanche activity.

Regional avalanche forecasts are given using the five-level European avalanche hazard scale; see (Meister, 1998) for details. Unit areas are defined by homogeneous snow pack stability and avalanche activity. One of the five levels of the avalanche hazard-scale (with corresponding slope aspect and altitude range) is assigned to each unit area each day. Here, predicting the avalanche hazard is considered as defined by the European avalanche hazard scale for the unit area around Davos from the manual observation stations based at Weissfluhjoch peak (2834 m above sea level) and Davos. Table 1 gives descriptions of the measurements at these stations.

Table 1 Input data of avalanche forecast model.

Variable	Description
Hn	new snow depth [cm]
Hs	snow depth [cm]
Wi	weather and intensity [code]
Ee	east component of wind [knts]
Nn	north component of wind [knts]
Ta	air temperature [°C]
Ts	snow temperature [°C]
Sf	snow surface [code]
Ps	penetration depth [cm]
Hnd	density of new snow [kgm^{-3}]

Figure 1 shows the general schema of all the models that are described in this paper and the models used for comparison.

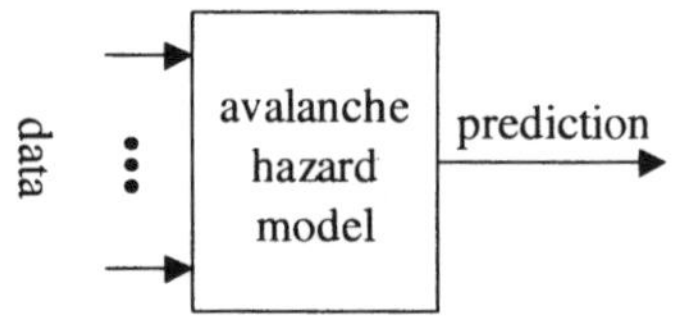

Fig. 1. Black-box avalanche forecast model.

The models developed in the paper are compared to the nearest-neighbor model presented in (Brabec and Meister, 2001) using the same evaluation method they applied: cross-validation. Each winter is excluded from the historic database and then forecasted by the model applied to remaining data. These predictions are then summed in contingency tables and the percentage of equally predicted days is calculated (*q* performance index). Another applied performance index (denoted *Q*) takes into account the misclassification degree by using, correspondingly, weight 1 for difference of 1 degree, 4 for 2, 9 for 3 and 16 for 4. Hence, the index corresponds to a mean square error measure that is minimized during the estimation of dynamical models.

2.1 Nearest Neighbor Model

Brabec and Meister (2001) have proposed a nearest-neighbor model (Hastie, *et al.*, 2001) for regional avalanche forecasting. This model uses the data described in Table 1 as input augmented by two derived "dynamic" variables: *Hn_sum3d* is the 3-day sum of *Hn* and *ta_dif1d* is the 1-day difference of *ta*. Weights are applied to each variable to integrate them into a complete distance function that is used to determine the 10 nearest neighbors (*e.g.*, days with similar conditions). The average hazard level for these 10 most similar days is the predictor of the current day. This model resulted in a cross-validated prediction rate of 52%.

The nearest-neighbor model averages regressands, which correspond to "similar" regressors. To make averaging of the model admissible, the avalanche hazard is treated as real rather than ordinal. This approach is applied here also for dynamic models.

3. RATIONALE FOR DYNAMIC MODELING

The avalanche hazard depends on factors such as terrain, snow cover properties, and weather. (The terrain properties remain constant for a given region and are not considered here.) Physical phenomena such as snow accumulation, thermal dynamics, change of the snow crystals, melting, and sublimation constantly change the snow cover. Hence, the snow cover is a dynamical system, driven by stochastic weather conditions (Salway, 1979). Experimental snow cover model is described by Lehning *et al.*, 1999. Similarly, also the avalanche hazard itself is a dynamic process. Its dynamic nature is evidenced by the general observation that changes in average daily avalanche hazard are slow, or, using statistical terminology, the avalanche hazard is highly auto-correlated.

3.1 Dynamic Snow Cover Model

Lehning *et al.* (1999) describe the snow-cover model *SNOWPACK*. This model simulates the evolution of a snow pack for a given location based on measurements from automatic weather stations.

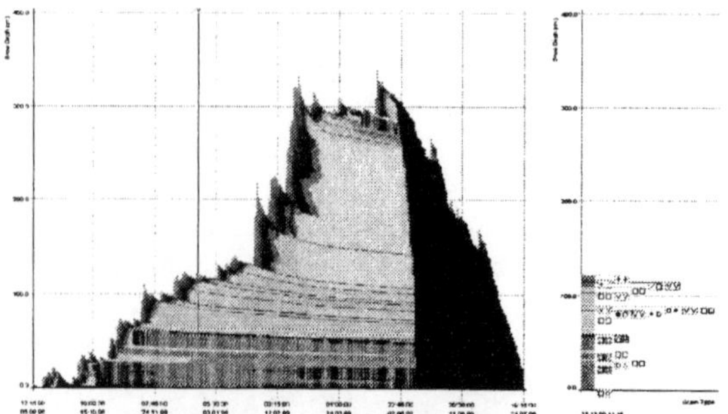

Fig. 2. Simulated time evolution of the snow cover in the Davos area. (By courtesy of SLF, www.slf.ch.)

Many important processes and features of the layering of the snow cover can be simulated. Figure 2 shows an example of such a simulation. Beginning with the first snow measurements the snow cover is built up and continuously updated according to new measurements. Unfortunately, the stability of the snow cover cannot yet be derived and furthermore the operational version of the model does not take terrain into account. Therefore, a hazard level cannot be derived from this model.

3.2 *Dynamic Effects Observed in the Data*

The autocorrelation function (ACF) of the daily sampled avalanche hazard time series $y(k)$, shown in Fig. 3, reveals presence of memory effects.

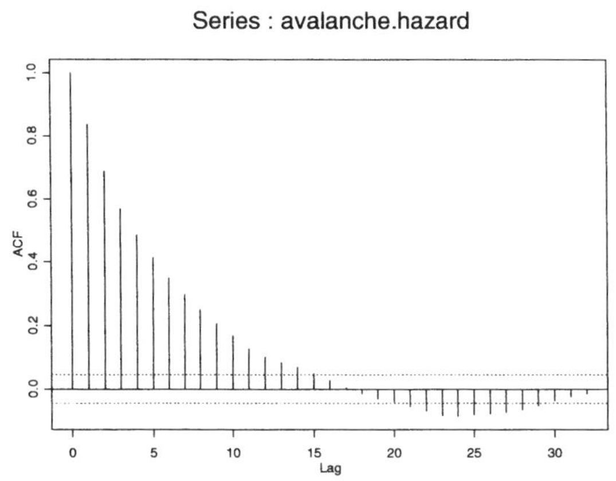

Fig. 3. Autocorrelation function of the avalanche hazard gives $R_{yy}(1)= 0.84$.

The ACF can be seen as approximately generated by a first-order auto-regressive process. (The physical meaning of the negative ACF values for the lags 18–32, anti-correlation, remains un-explained.) Fitting an AR(1) model of the form

$$y(k) = -ay(k-1) + s + e(k) \tag{1}$$

into the hazard time series produces reasonably white residuals $e(k)$ and $a = -0.84$, a coefficient value that coincides with $R_{yy}(1)$. From here a "time constant" of the avalanche hazard can be derived, giving

$$T = \frac{-1}{\ln(-a)} = 5.74 \text{ days.} \tag{2}$$

4. FAMILY OF DYNAMIC MODELS

A priori determination of the avalanche hazard model structure is not possible, since the complex underlying physical relationships are unknown. The approach applied here is to use both physical insight and some experimentation, and trying first simple model structures. The results obtained are then evaluated using the previously described cross-validation procedure. The models use the regressors new snow depth in Davos (*Hn.D*) as well as new snow depth (*Hn.W*), total snow depth (*Hs.W*), penetration depth (*Ps.W*), and new snow density (*Hnd.W*) on Weissfluhjoch. Hydrological day of the year is also used as a regressor for some models. It is interesting to note that all other input variables used in the nearest neighbor model do not increase the quality of the dynamic models. Hence, these inputs are not used. The regressor variables are normalized to zero-mean unit-variance.

4.1 *ARX Model*

The ARX (Auto-Regressive with eXogenous input) model (Ljung, 1987) is estimated using linear regression and a one-step-ahead predictor equation:

$$\hat{y}(k) = -ay(k-1) + \sum_{i=1}^{5} b_i u_i(k) + s\,, \tag{3}$$

where s denotes the shift parameter.

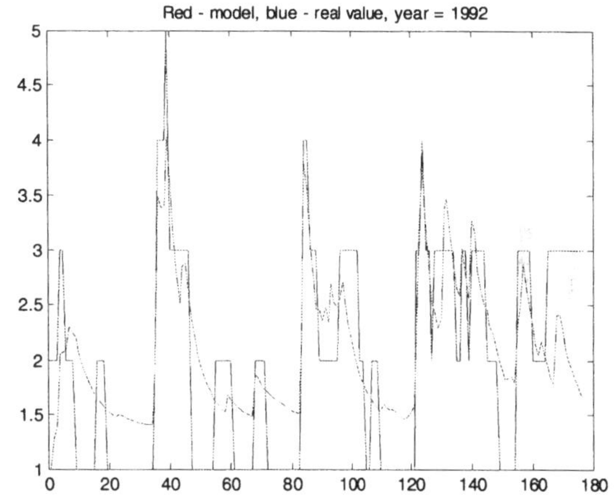

Fig. 4. True and ARX-predicted (not quantized) avalanche hazard, validated for the year 1992.

The performance index q achieved by this model is 56.4%, the contingency table (after rescaling, discussed in Section 5) is shown in Table 2:

Table 2 Example contingency table for ARX model.

	true avalanche hazard				
model	1	2	3	4	5
1	**411**	177	48	0	0
2	177	**441**	128	3	0
3	9	160	**207**	18	0
4	0	1	16	**18**	1
5	0	0	2	0	**2**

The model has only 7 parameters and outperforms the nearest neighbor model. The values of the model parameters $[a \; b_1 \dots b_5]$ are

$$[-.76 \quad .087 \quad .12 \quad .025 \quad .034 \quad -0.023].$$

The most important inputs are new snow heights in Davos and on Weissfluhjoch, penetration depth, total snow depth, and new snow density (with negative gain). The time constant of the equivalent low-pass filter equals $T = -1/\ln(-a) = 3.64$ days. This value is close to the 3-day averaging horizon of *Hn_sum3d* variable used in the nearest–neighbor model.

4.2 OE Models

The ARX model is estimated by minimizing one-step-ahead prediction errors. However, for the purpose of avalanche hazard prediction, the infinite-step-ahead predictor has to be used. Hence, the ARX model does not deliver the optimal predictor. The optimal predictor can be estimated using an Output Error (OE) model structure (Ljung, 1987). To avoid over-parameterization, the parsimonious model, given by (4), uses a common denominator for all regressors.

$$\hat{y}(k) = \frac{\sum_{i=1}^{5} b_i u_i(k)}{1 + fq^{-1}} + s \qquad (4)$$

Due to the above-mentioned optimality, the model performs better than ARX (performance index q = 57.2%), but the improvement is not substantial. Similarly, increasing the model order does not lead to a better quality model. Instead, it is advantageous to use hydrological day of year as an additional regressor. The quality of the model, denoted OE+, is summarized in Table 3.

4.3 NNARX Model

It can be expected for the considered application that non-linear dynamical models are more appropriate than the linear ones. Dynamic neural networks constitute a particularly flexible class of models (Nørgaard *et al.*, 2000). An NNARX (Neural Network ARX) model with one neuron in the hidden layer, defined by equation (5), is estimated.

$$\hat{y}(k) = \beta \tanh(-ay(k-1) + \sum_{i=1}^{5} b_i u_i(k) + s_1) + s_2 \quad (5)$$

The performance index q achieves 59.6 % for this model. The model results are improved indeed, while the number of parameters is increased only by one. However, extending the number of nodes and layers in the network structure does not lead to a better model.

4.4 NNOE Models

Further improvement of the model quality is achieved by applying NNOE (Neural Network OE) model structure (Nørgaard, *et al.*, 2000). The predictor equation applied in the model estimation has a recurrent form and can be directly applied for infinite-step prediction. The model structure with one neuron in the hidden layer is given by

$$\hat{y}(k) = \beta \tanh(-a\hat{y}(k-1) + \sum_{i=1}^{5} b_i u_i(k) + s_1) + s_2 \,. \quad (6)$$

The model (denoted NNOE I) achieves a high performance index q, namely 63.8%. To make a comparison to the nearest-neighbor model fair, the NNOE I model without *hn.D* is investigated (NNOE Ia). It performs slightly worse: Q = 62.7% and q = 56.9%, but is still superior to the nearest-neighbor model. Increased model flexibility can be obtained by adding a second neuron to the hidden layer, see Fig. 5. The model has 17 parameters and is denoted NNOE II. Model results are summarized in Table 3.

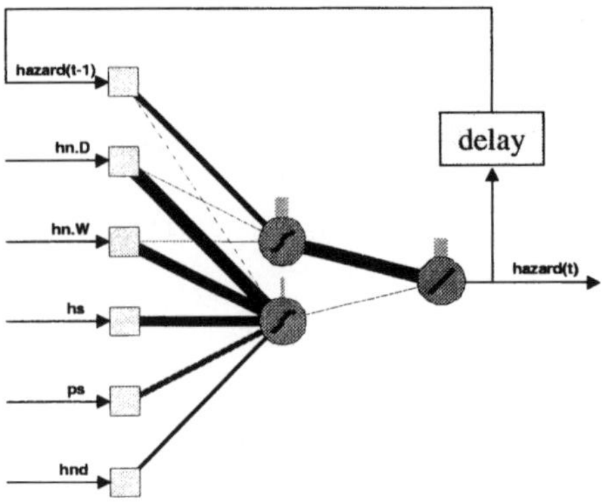

Fig. 5. Structure and weights of the NNOE II model.

The NNOE I model can also be improved by incorporating hydrological day of year as an additional regressor. The values of the performance indices for the model (denoted NNOE+) are the highest: Q = 65.6% and q = 63.8%.

5. COMPARISON OF MODELING RESULTS

Modeling results for all the considered models are summarized in the following paragraphs. Since the regression-based models have a tendency not to preserve the *a priori* estimated distribution of hazard levels, an additional rescaling (calibration) of the model output quantization thresholds has been conducted, as described in (Brabec and Meister, 2001) for the nearest neighbor model. The results before and after rescaling are, correspondingly, shown on the left and right side of Table 3:

Table 3 Summary of the modeling quality.

Model	q[%] without rescaling	Q[%] without rescaling	q[%] with rescaling	Q[%] with rescaling	# of params
H=2	43.3	35.5	–	–	1
10-NN	–	–	52.0	39.5	–
LR	50	46.6	48.5	30.2	6
NN	52.5	48.7	49.6	32.2	8
ARX	56.4	53.6	59.3	48.9	7
OE	57.2	54.1	59.6	50.1	7
OE+	61.2	59.7	59.4	51.3	8
NNARX	59.6	57.8	60.6	52.4	9
NNOE I	63.8	59.4	62.0	54.4	9
NNOE Ia	62.7	56.9	61.0	52.3	8
NNOE II	64.0	58.9	63.1	55.5	17
NNOE +	65.6	63.8	60.6	55.5	10

LR denotes a static linear regression model, while NN, denotes a static neural network model, which performs better. Both models work well without rescaling, particularly for Q index. Linear dynamic ARX and OE models bring improvement, especially in the Q index after rescaling, and clearly outperform the nearest-neighbor model. Including hydrological day of year as a regressor in model OE+ increases model quality. Further improvement is obtained for dynamic neural network models. As can be seen in Fig. 6, the quality increase is twice the difference between the trivial predictor (avalanche hazard = 2) and the nearest neighbor (10-NN) model.

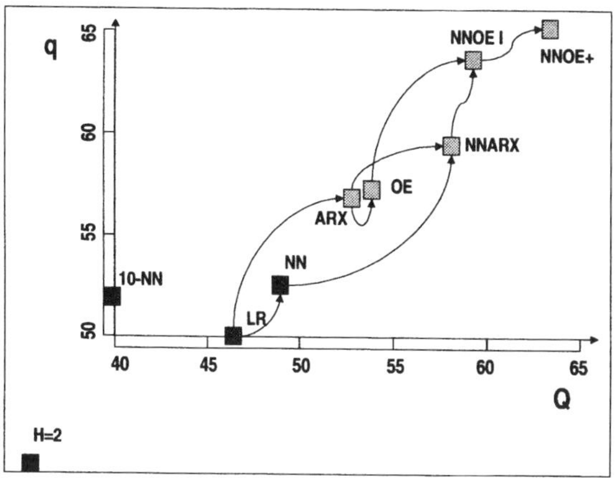

Fig. 6. Model results without rescaling, shown in (Q, q) coordinates for selected models.

The achieved results suggest an upper limit that is less than 100% for model quality. There are three main reasons for the existence of such a limit: (1) incomplete information in the regressors, (2) inconsistent, unpredictable data in the regressand (which is generated by human experts), and, (3) quantization noise, due to treating categorical hazard variable as a continuous variable. Consequently, the limit can be pushed closer towards 100% by (1) use of additional regressor data, generated, *e.g.*, by a snow cover model, (2) application of additional, more reliable regressands (*e.g.*, obtained by processing satellite imagery data), and (3) use of a non-quantized regressand. These issues are considered in the next section.

6. OUTLOOK

The presented modeling approach could be extended by applying existing multivariate modeling techniques to construct joint models for whole regions. Other extensions are discussed below.

6.1 Use of the Snow Cover Model to Generate Additional Regressor Data

The *SNOWPACK* model is based on physical modeling and tuned using real data and statistical approaches. Since snow cover properties are highly correlated with the avalanche hazard, the model could be used to generate relevant regressor data as in Fig. 7. Note that both models in Fig. 7 are dynamical and can be tuned via system identification methods.

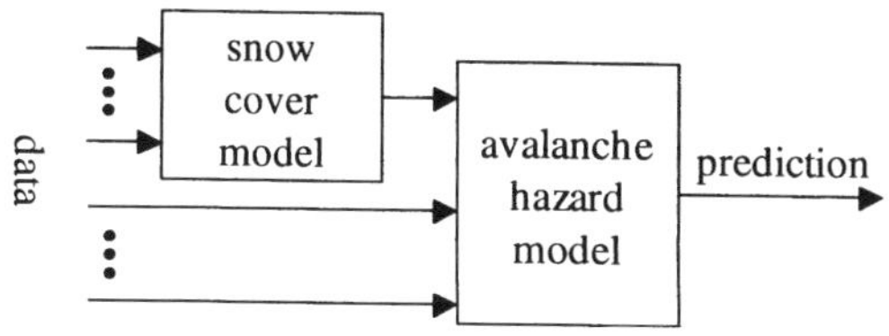

Fig. 7. Potential use of the snow cover model for avalanche hazard prediction.

6.2 Use of Satellite Imagery to Generate Additional Regressands

The estimate of avalanche hazard data is subjective and may be inconsistent. Consistency could possibly be improved by utilizing avalanche-related data generated from satellite imagery via multi-spectral methods (Koperski and Marchisio, 2000). Already existing software tools such as *VisiMine*™ by Insightful are capable of extraction of texture-related data from satellite images. However, this application of satellite imagery has not yet been explored.

6.3 Data Validation with Dynamic Models

The models presented here require numerous measurements and high data quality. The latter can be ensured by application of data validation and cleansing approaches (Milek, *et al.*, 2000).

Consider a simple example of data validation at the peak of Weissfluhjoch. Let $y(k)$ denote the daily measurement of new snow depth and $u(k)$ the total snow depth. The following first-order ARX model

$$y(k) = -ay(k-1) + bu(k) + e(k) \qquad (7)$$

is capable of modeling snow accumulation and compression phenomena.

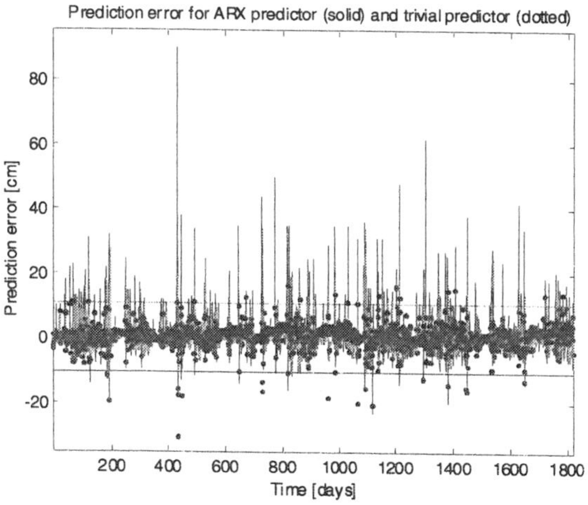

Fig. 8. Data validation by a simple ARX model.

The model is identified from the measured data. The estimated coefficients are a = –0.973 (average old snow compression) and b = 0.78 (average new snow compression). The resulting one-step ahead predictor is suitable for data validation:

$$\varepsilon_1(k) = y(k) + ay(k-1) - bu(k). \qquad (8)$$

As shown in Fig. 8, the magnitudes of the residuals of the model (dots) are usually smaller than 10 [cm]. Hence, they are much smaller than the residuals of the trivial model, $\varepsilon_0(k) = y(k) - y(k-1)$ (solid line) which could be used for the most simple validation.

7. CONCLUSIONS

The paper demonstrates that avalanche hazard can be thought as of a dynamic phenomenon. System identification methods can be used to improve the quality of experimental models, particularly with respect to static nearest-neighbor models. Additionally, simple model-based mechanisms for data validation are devised.

Further improvements could be made by including new relevant regressors to the model and improving the quality of the regressand.

ACKNOWLEDGEMENTS

We gratefully acknowledge the following sources of data and support. The Swiss Federal Institute of Snow and Avalanche Research SLF in Davos delivered the avalanche hazard and snow data. The project was partially supported by Automatic Control Laboratory of the Swiss Federal Institute of Technology in Zurich. The neural network models have been computed using freeware *NNSYSID Toolbox* (Nørgaard, 1997), see also http://www.iau.dtu.dk/research/control/nnsysid.htm . Maria Silkey substantially improved the readability of this paper.

REFERENCES

Bois, P., C. Obled and W. Good. (1974). Multivariate data analysis as a tool for day-by-day avalanche forecast. In: *Snow Mechanics - Symposium - Mechanique de la Neige*, Vol. 114. Grindelwald, IAHS, pp. 391–403.

Bolognesi, R. (1998). NivoLogTM: An avalanche forecast support system. *Inter. Snow Science Workshop*. Sunriver, Oregon, pp. 412–418.

Brabec, B. (2001). *Computergestützte Regionale Lawinenprognose*. PhD Thesis, Swiss Fed. Inst. of Snow and Avalanche Research (SLF), Davos.

Brabec, B. and R. Meister. (2001). A nearest-neighbor model for regional avalanche forecasting. *Annals of Glac.*, **32**, pp. 130–134.

Brabec, B., R. Meister, U. Stöckli, A. Stoffel and T. Stucki. (2001). RAIFoS: Regional Avalanche Inf. and Forecasting System. *Cold Regions Science and Technology*, **33**, pp. 303–311.

Buser, O. (1983). Avalanche forecast with the method of nearest neighbours: an interactive approach. *Cold Regions Science and Technology*, **8**, pp. 155–163.

Buser, O. (1989). Two years experience of operational avalanche forecasting using the nearest neighbors method. In: *Annals of Glaciology*, Vol. 13. Lom, Norway, International Glaciological Soc., pp. 31–34.

Durand, Y., G. Giraud, E. Brun, L. Mérindol and E. Martin. (1999). A computer-based system simulating snowpack structures as a tool for regional avalanche forecasting. *Journal of Glaciology*, **45** (151), pp. 469–484.

Hastie, T., R. Tibshirani and J. Friedman. (2001). The Elements of Statistical Learning. *Data Mining, Inference, and Prediction*. Springer-Verlag, New York.

Koperski, K. and G. Marchisio. (2000). Multilevel Indexing and GIS Enhanced Learning for Satellite Imageries. In: *International Workshop on Multimedia Data Mining, MDM/KDD '2000*, pp. 8–13.

Kristensen, K. and Ch. Larsson. (1994). An avalanche forecasting program based on a modified nearest neighbor method. In: *International Snow Science Workshop*. Snowbird, Utah, pp. 22–30.

Lehning, M., P. Bartelt, B. Brown, T. Russi, U. Stöckli and M. Zimmerli. (1999). SNOWPACK model calculations for avalanche warning based upon a new network of weather and snow stations. *Cold Regions Science and Technology*, **30**, pp. 145–157.

Ljung, L. (1987). *System Identification: Theory for the User*. Prentice Hall, Englewood Cliffs.

McClung, D. M. and J. Tweedy. (1994). Numerical avalanche prediction: Kootenay Pass, British Columbia, Canada. *Journal of Glaciology*, **40** (135), pp. 350–358.

Meister, R. (1998). Interpretationshilfe zum Lawinenbulletin des Eidg. Inst. für Schnee- und Lawinenforschung. *Mitteilungen 50. SLF.*

Milek, J. and F. Kraus. (2000). Use of Analytic Redundancy in Fault-tolerant Sensor Systems. In: *Proc. of XVI IMEKO World Congress*, Vol. V, pp. 121–127.

Nørgaard, M. (1997). Neural Network based System Identification Toolbox. *Techn. Report 97-E-851*, Department of Automation, DTU, Lyngby.

Nørgaard, M., O. Ravn, N. K. Poulsen and L. K. Hansen (2000). *Neural Networks for Modelling and Control of Dynamic Systems*. Springer-Verlag, London.

Salway, A. A. (1979). Time Series Modeling of Avalanche Activity from Meteorological Data. *Journal of Glaciology*, **22**, (88), pp. 513–528.

Schweizer, J. and P. M. B. Föhn. (1996). Avalanche forecasting - an expert system approach. *Journal of Glaciology*, **42** (141), pp. 318–332.

Tschirky, F., B. Brabec and M. Kern. (2000). Avalanche rescue systems in Switzerland: experience and limitations. In: *International Snow Science Workshop*, Big Sky, Montana.

IFAC
Publications
www.elsevier.com/locate/ifac

MODELS FOR INCOMING CALLS FORECASTING IN A CUSTOMER ATTENTION CENTER

Manuel R. Arahal, Fernando Pavón P., Eduardo F. Camacho

Departamento de Ingeniería de Sistemas y Automática.
Escuela Superior de Ingenieros.
Camino de Los Descubrimientos s/n. 41092 Sevilla

Abstract: Telephone customers attention centers (CAC) are complex systems. In order to provide the best service to clients with minimum costs a careful scheduling of human resources (agents) is needed.
Call centers often receive thousands of incoming calls. A large data base of services is in many cases available for modelling. Such data has been used in different ways to improve the quality of service. In this particular case, the schedule of attention staff a week in advance.
In this paper the number of incoming calls in the hour is modelled using autoregressive models, both linear and nonlinear (neural networks). As it turns out, the most important part of the modelling procedure is the selection of appropriate input variables. *Copyright © 2003 IFAC*

Keywords: Call center, forecasting, linear models, nonlinear models, neural networks

1. INTRODUCTION

Call centers are nowadays ubiquitous in companies that have to deal with large number of customers, some (like airlines, hotels, and credit card companies) relay almost exclusively in call centers for service providing and customer feedback.

The capacity of a call center is mostly determined by the human resources employed. Since these are expensive, the quality of service is often balanced with capacity so that the call center can provide the best service with minimum costs (Pinedo *et al.*, 1999).

Forecasting techniques can be useful to schedule the number of agents needed to provide a desired quality of service at any time.

Simple time-series methods have been used to forecast call center load, such as in (Sze, 1984). Andrews and Cunningham (Andrews and Cunningham, 1995) develop ARIMA models that estimate the number of daily calls at L.L. Bean. A statistical method for predicting arrival rates can be found in (Jongbloed and Koole, 2001).

Apart from that we have found very few papers describing the application of time series analysis to call centers load forecasting.

The model in (Andrews and Cunningham, 1995) is used to predict daily call volumes for the next three weeks in order to produce efficient agent scheduling. The ARIMA model is a linear combination of lagged values of the independent variable (daily call volumes) and past prediction errors. The number of regressors and their lag were chosen using a 5-year data base. The resulting models yield a mean absolute percent error of about ten in a year of use after construction (ex-post forecasts).

In this paper we present different autoregressive models for forecasting the hourly call load for an information call center of a Spanish phone company. Such forecast is then used to schedule the number of agents needed to attain a prescribed level of efficiency. The paper also discuses different modelling techniques and the results obtained. The work extends the results obtained previously in the same context ((Andrews and Cunningham, 1995)) by using neural models and a systematic procedure for selecting input variables. The paper also shows that for this particular problem, the key issue is an appropriate selection of input variables and not the complexity of the model chosen. Furthermore, the best results from a forecasting point of view are obtained for simple models with few input variables.

Once a phone call is received an agent is designed depending of some characteristics of the call: code area, customer, kind of information requested, language of the caller. A call can be transferred from an agent to another. Each time an agent manages a call an attention counter is increased. Thus a single call can produce many attentions. Thousands of calls and many more attentions are managed each day in the call center.

Lost calls are a figure of merit used by the company to assess the quality of the service. Three are the reasons for a lost call:

- Calls lost while in queue. These are due to lack of agents to manage all incoming calls. The client hangs because the waiting is to long.
- Calls lost by saturation. These are due to lack of space to enqueue more calls. The call is not answered in any way.
- Calls lost not in saturation. Most are due to technical failure of communication. These calls are considered correctly attended because is not a fault in the part of the CAC.

A large data base is available consisting of hourly values of variables such as number of incoming calls, number of internal calls, average length of calls, etc. In order to produce a schedule for the agents the most important variable is the number of incoming calls. In the next section the forecasting problem is posed in a more precise way. Section 3 shows the linear and neural models developed followed by the results obtained in the application. The paper ends with some conclusions.

2. DATA

Let us denote by $v(d,h)$ the number of incoming calls received at the CAC at day d and hour h. Days are numbered consecutively, thus d represents a unique day in the data base. Hours are numbered from 1 to 24. The number of incoming calls will also be referred to as load.

For scheduling reasons, it is paramount to forecast the load with a week in advance. That means that if k is a Monday, the forecast for next-week Monday $k+7$, next-week Tuesday $k+8$, and so on until next-week Sunday $k+13$ are needed. Since the forecast has to be made the very day k that means that information about k is not available yet. This amounts to a lead-time in the predictions ranging from 8 to 14 days.

In order to simplify the problem it is better to consider just the maximum lead time and make predictions with a 14-day horizon. To be more precise: we are to produce $\hat{v}(d+14,h)$ for hours $h = 1, ..., 24$ using information about past loads $v(k,h)$ for days $k = 1, ..., d$.

Data is composed of historical information from the call centers of a Spanish phone company. It consists of hourly load (number of incoming calls every hour) for a number of months. The data base consists of values of load $v(d,h)$ for each hour h and day d. It can also be viewed as a time series $x(t)$ being $t = 24d + h$.

Figure 1 shows the number calls for 31 days. The vertical axis has been scaled (to zero mean and variance unity) for confidentiality reasons.

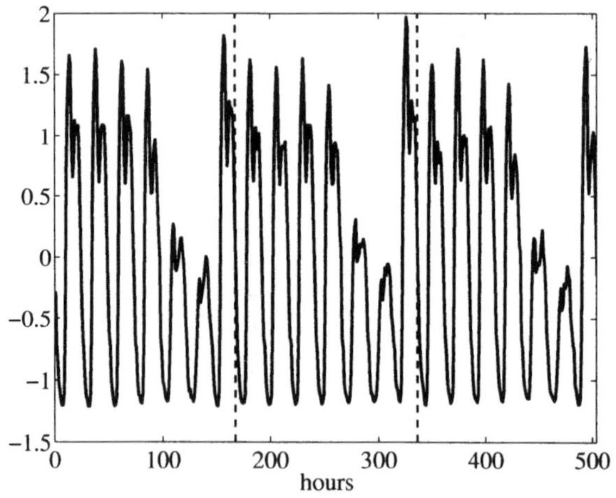

Fig. 1. Scaled number of incoming calls during 21 days at the CAC.

Periodicity in the data is clearly visible. The power spectrum (Figure 2) shows a number of peaks at 8, 12, 24, 28, 84 and 178 hours per cycle. The most prominent is the 24 hour cycle. This observation opens the way for autoregressive models.

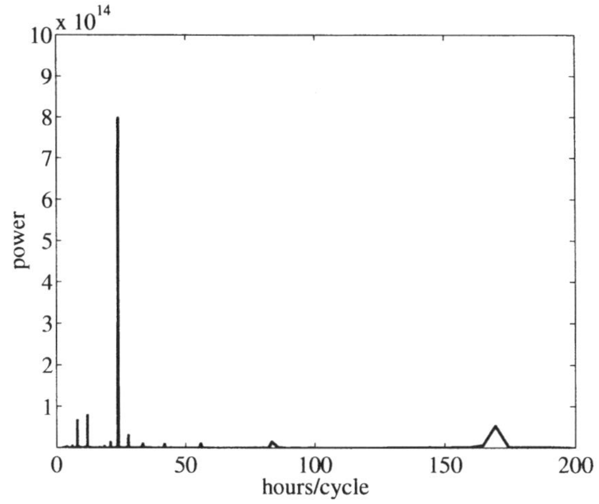

Fig. 2. Power Spectrum of $x(t)$.

3. MODELS

Autoregressive linear models are the first choice for the problem considered. Different orders and time lags can be easily tested using computer programs such as ident for MATLAB.

The general structure for a linear autoregressive (AR) model is

$$\hat{x}(t) = \sum_{k=1}^{na} a_k x(t - \tau_k) \qquad (1)$$

being na the order of the model, $a_k, k = 1, ..., na$ the coefficients and $\tau_k, k = 1, ..., na$ the time lags. The past value $x(t - \tau_k)$ is referred to as regressor. An AR models produces predictions as a linear combination of past values of the observed variable. In order to construct a model, appropriate values for the order and time lags have to be searched for. Then, the coefficients can be obtained by least squares.

In many cases time lags are consecutive multiples of a basic lag-unit T plus a constant dead-time d, that is: $\tau_k = d + (k - 1)T$. In this case, the only choices are na, d and T. Normally d is the lead-time of the prediction, fourteen days in the present case.

Akaike's criterion and others may be used to choose the value of na that achieves a balance between accuracy and complexity of the model.

Regressors can also be selected using a forward inclusion technique. In forward inclusion lagged variables are included in the model one by one, selected according to how much they help reducing the model error (see (Arahal *et al.*, 2002)). The problem arises on when to stop since any new variable will introduce new degrees of freedom that will allow to reduce the error in a controlled set of observations. Validation techniques must be carefully used to perform model selection.

In the application we are tackling the order must be kept as low as possible since we do not have large amounts of data needed to develop higher order models.

The adjustable parameters of the models are obtained minimizing a quadratic criterion of the prediction error. A set of observations is set aside to test the validity of the model. The root mean squared error over this set is defined as:

$$J = \sqrt{\frac{1}{N}\sum_{t=1}^{N}(x(t) - \hat{x}(t))^2} \qquad (2)$$

being N the number of observations in the validation set. J will be used through the paper as a figure of merit to compare different models.

3.1 Linear models

Models of the form:

$$\hat{x}(t) = \sum_{k=1}^{na} a_k x(t - d - (k-1)T) \qquad (3)$$

will be considered first, being $d = 14 \cdot 24 = 336$ the lead time of the prediction. Table 1 shows the root mean squared prediction error for different values of na and T. It can be seen that variables lagged 24 hours ($T = 24$) are the most informative but even in this case a large model order is needed to reduce the error significantly.

$T = 1$		$T = 8$		$T = 24$	
na	J	na	J	na	J
1	0.306	1	0.306	1	0.306
2	0.306	2	0.306	2	0.304
3	0.306	3	0.305	3	0.304
10	0.304	10	0.303	10	0.278
20	0.303	20	0.303	15	0.268
50	0.301	25	0.276	30	0.247

Table 1. Root mean squared error for different AR linear models.

Forward inclusion of variables can be easily tested. Models now have the form

$$\hat{x}(t) = \sum_{k=1}^{na} a_k x(t - d - l_k) \qquad (4)$$

being l_k selected lags not necessarily consecutive. The first lag l_1 is selected comparing the correlation coefficients between the corresponding regressor $x(t - d - l_1)$ and the observed variable $x(t)$. The variable with the highest correlation is chosen because it is the best suited to approximate $x(t)$.

The model with just one regressor yields predictions $\hat{x}_1(t) = a_1 x(t - d - l_1)$, being a_1 a coefficient estimated via least squares. The error of this model is $e_1(t) = x(t) - \hat{x}_1(t)$. The selection of a new input variable is done as previously but looking for high correlations with $e_1(t)$.

Table 2 shows the root mean squared prediction error for different models. The first column na is the number of input variables used by the model. The second column shows the values of lags l_k (see 4). It can be seen that the more variables the better results are obtained. We must keep in mind that these results are not forecasting but estimation results since data has been used up in building the models.

na	lags l_k	J
1	0	0.306
2	0, 504	0.274
3	0, 504, 215	0.271
4	0, 504, 215, 599	0.270
5	0, 504, 215, 599, 583	0.269

Table 2. Root mean squared for linear AR models with selected non consecutive lags.

3.2 Neural models

The latter strategy of forward inclusion can be of use for nonlinear models such as neural networks. One-hidden layer neural networks with linear output node have been selected for the sake of simplicity. A neural net computes a nonlinear function of its input vector $\mathbf{z}$ as:

$$\mathrm{NN}(\mathbf{z}) = \sum_{n=1}^{nn} w_n^o s_n(\mathbf{z}) \tag{5}$$

being nn the number of nodes in the hidden layer and $s_n(\mathbf{z})$ the output of the n-th node, calculated as

$$s_n(\mathbf{z}) = \tanh(\sum_{k=1}^{na} w_k^i z_k) \tag{6}$$

being na the dimension of the input vector. Coefficients w_n^o and w_k^i are the adjustable parameters of the network and are referred to as weights. Training is the procedure (gradient based in most cases) for assigning a value to weights so that the approximation error is made small.

In order to use the net as a AR model, the input vector must contain lagged values of $x(t)$ such that $z_k = x(t - d - l_k)$.

Forward inclusion now poses a problem. Since the expected nature of relationship between input variables and observed output is nonlinear it does not make sense to look for linear correlations among them. The procedure to select variables should be done in a different manner. A possible approach is to use brute force and compute models for all possible combinations of two input variables, then of three and so on. Although time consuming this approach should be taken into account. Another possible way of proceeding is to use a nonlinear measure of correlation such as the one proposed in (Yuan and Fine, 1998).

Table 3 shows the results obtained at different stages of the forward inclusion procedure.

na	lags l_k	J
1	0	0.308
2	0, 336	0.249
3	0, 336, 48	0.245
4	0, 336, 48, 288	0.242
5	0, 336, 48, 288, 480	0.237

Table 3. Root mean squared error for neural AR models with lags selected using a nonlinear correlation measure among variables.

The neural network training has been done using the MATLAB neural toolbox. The data base of input/output pairs has been split in two random separate sets of equal cardinality. One set (training set) is used for adjusting the weights and the other (validation set) to avoid overtraining. The value of J presented in table 3 is calculated after training using both sets.

3.3 Ad hoc models

Results obtained so far are somehow poor. The estimated error in the prediction is about 20 % of the mean volume of incoming calls. In order to obtain sharper scheduling operations it is necessary to improve the forecasting ability of the models.

A study of the data reveals some facts:

- Week-days and week-ends have very different loads.
- Load in any particular day is more similar to load of previous days of the same kind.
- Holidays in the middle of a week have a load similar to that of a Sunday.

According to this it seems sensible to modify the way regressors are obtained. Instead of using fixed values for lags, ones should always try to use past values from similar kind of days. That is, to forecast a Monday data from past Mondays should be used and the same holds for Tuesdays to Sundays and Holidays.

The new model structure can be mathematically described as:

$$\hat{x}(t) = \sum_{k=1}^{na} a_k x(t - d - 24r_k(t)) \tag{7}$$

where $r_k(t)$ represents a variable number of days. The correct value for $r_k(t)$ is obtained searching the database for days similar to the forecasted day.

This can be clarified with an example. Suppose that today is a Monday and we are to forecast next week. In order to forecast the load for next Monday we can use load from previous Monday $r_1 = 0$, and the one before $r_2 = 7$, etc. Unless one of these Mondays happened to be a holiday. Then this day is skipped and the previous one is considered.

With this in mind the r_1 can be described as the minimum "distance" (in days) that one has to regress to obtain a day of the same type as the forecasted day (discounting the lead time of fourteen days). Similarly r_2 is the distance to the second closest day of the same kind, and so on.

Note that in this way holidays are never used as regressors to forecast a day that is not a holiday. In the same way data from a Monday is not used to forecast load of a day that is not a Monday.

Model order can be selected just trying out increasing values for na . Table 4 shows the results obtained.

na	J
1	0.210
2	0.197
3	0.188
4	0.186

Table 4. Root mean squared error for linear AR models with variable lags.

Neural networks can also be used with this new way of selecting regressors. The new model structure can be mathematically described as $\hat{x}(t) = \text{NN}(\mathbf{z})$, being NN() a nonlinear function implemented by a neural net with input vector $\mathbf{z}$. The na components of this vector are selected past values of $x(t)$ obtained as $z_k = x(t - d - 24r_k(t))$ as explained above. Table 5 shows the results obtained.

na	J
1	0.195
2	0.189
3	0.180
4	0.176

Table 5. Root mean squared error for neural AR models with variable lags.

It can be seen that results have improved, being now possible to achieve root mean squared errors of about 15% of the mean load with low order models.

4. APPLICATION RESULTS

The models generated in the previous section have been tested during a period of 41 days. Table 6 shows the results in terms of root mean squared error of normalized variables J^n as in the previous sections. Also, for the sake of comparison with other papers, other measures of forecasting error are given. In the third column the root mean squared error as a percentage of the mean load is shown (J^p). This figure of merit is nothing but the root mean error of the non-normalized variable divided by the mean load and multiplied by one hundred. In some applications the arithmetic mean of the absolute value of the error is used instead. This is shown in the fourth column as MAE again as a percentage of the mean load.

A) linear models with lags $d + 24(k-1), k = 1, ..., na$

na	J^n	J^p	MAE
1	0.095	7.73 %	5.41 %
2	0.093	7.54 %	5.17 %
3	0.094	7.63 %	5.36 %
4	0.095	7.72 %	5.45 %
5	0.095	7.66 %	5.42 %
6	0.094	7.58 %	5.34 %
10	0.076	6.15 %	**4.31** %

B) nonlinear models with lags $d + l_k$

na	l_k	J^n	J^p	MAE
1	0	0.110	8.89 %	6.42 %
2	0, 336	0.069	5.62 %	**4.00** %
3	0, 336, 48	0.087	7.05 %	4.23 %
4	0, 336, 48, 288	0.080	6.51 %	4.37 %
5	0, 336, 48, 288, 480	0.088	7.12 %	4.95 %

C) linear models with variable lags

na	J^n	J^p	MAE
1	0.087	7.07 %	4.86 %
2	0.078	6.29 %	**4.45** %
3	0.101	8.16 %	5.59 %
4	0.116	9.41 %	6.40 %

D) nonlinear models with variable lags

na	J^n	J^p	MAE
1	0.090	7.28 %	5.06 %
2	0.085	6.88 %	**4.72** %
3	0.109	8.79 %	5.50 %
4	0.194	15.73 %	8.02 %

Table 6. Summary of application results of different models.

As can be seen in Table 6 the best results (typed in boldface) are achieved by very low order models. The performance is in some cases deteriorated by the use of new input variables. This was not previously detected although a large validation set was used to avoid overtraining.

Figure 3 shows the normalized number of incoming calls and the forecast made by the best model of each group during a typical week. A portion of the curve has been amplified to show the differences among models.

To make the comparison more apparent the error has been also drawn for each model in Figure 4.

It is also noticeable that simple but intuitive models, can perform as well as other derived by more sophisticate procedures.

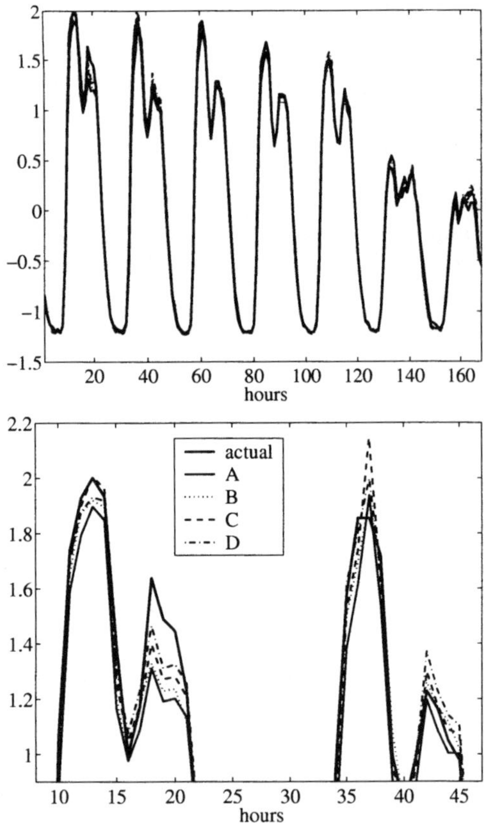

Fig. 3. Normalized load and forecast for the bests models of each group (see table 6 and text for details).

Going back to the scheduling of agents, any of the proposed models produce predictions with root mean squared error under 10% of the mean load, allowing to attain a balance between lost calls and staffing.

5. CONCLUSIONS

It has been shown that techniques of system identification and neural networks can be used to obtain models to forecast the number of incoming calls to a CAC.

The results obtained in this application shows the importance of selecting input variables and model order for forecasting. In this particular case it has been shown that overestimation of model order has a negative effect on model forecasting performance.

REFERENCES

Andrews, B. and S.M. Cunningham (1995). L.L. bean improves call-center forecasting. *Interfaces* **25**, 1–13.

Arahal, M.R., Alfonso Cepeda and E.F. Camacho (2002). Input variable selection for forecasting models. *Preprints of the 15th triennial world congress of the IFAC.*

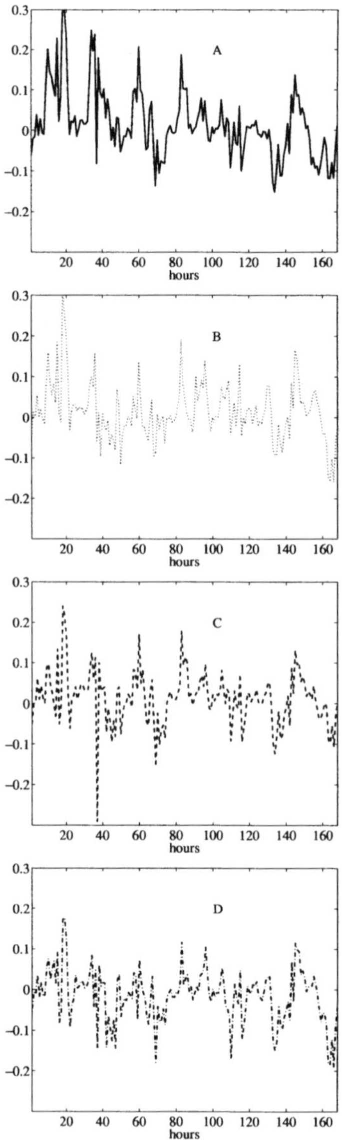

Fig. 4. Normalized forecasting error for the bests models of each group (see table 6 and text for details).

Jongbloed, G. and G.M. Koole (2001). Managing uncertainty in call centers using poisson mixtures. *Applied Stochastic Models in Business and Industry* **17**, 307–318.

Pinedo, M., S. Seshadri and J.G. Shanthikumar (1999). Call centers in financial services: strategies, technologies, and operations. In: *Creating Value in Financial Services: Strategies, Operations and Technologies* (E.L.Melnick, P. Nayyar, M.L. Pinedo and S. Seshadri, Eds.). Chap. 18, pp. 357–388. Kluwer.

Sze, D.Y. (1984). A queuing model for telephone operator staffing. *Operations Research* **32**, 229–249.

Yuan, J.-L. and T.L. Fine (1998). Neural-network design for small training sets of high dimension. *IEEE Transactions on neural networks* **9**, 266–280.

IFAC Publications
www.elsevier.com/locate/ifac

MODELING THE RELATIONSHIPS BETWEEN THE USERS DB AND THE WEB-LOG FILE OF A LARGE VIRTUAL COMMUNITY

Sergio M. Savaresi, Simone Garatti, Sergio Bittanti
Dipartimento di Elettronica e Informazione, Politecnico di Milano, Piazza L. da Vinci, 32, 20133 Milano, ITALY.

Abstract: In this paper the analysis and modeling of a large data-set related to a very popular Italian virtual community is presented. The community is constituted by more than half-million registered users, characterized by a unique nickname. Each user has its own "profile", which is filled during the registration procedure, on a voluntary basis. Two data-sets are used: the database of the users (nickname and profile), and the database of their web navigation sessions. The latter has been obtained from the log-file of the servers hosting the community web-site. This work is constituted by three main parts: 1) analysis and clustering of the users DB; 2) analysis and clustering of the navigation sessions; 3) correlation of users clusters and navigation sessions clusters. This analysis provides a complete and full-rounded picture of the virtual community users.

Keywords: Data Modeling; Data Mining; Virtual Community; profiled users; web-log files; sessions; unsupervised clustering; PDDP; heterogeneous data;

1. INTRODUCTION AND PROBLEM STATEMENT

This paper deals with the analysis and modeling of a large data-set related to a very popular Italian virtual community which is constituted by more than 550.000 registered users. Each user is characterized by a unique nickname and has an own "profile" which is filled (choosing among a list of items) during the registration, on a voluntary basis (the profile can be left completely blank or can be only partially filled).

The analysis is made using two different data-sets:

- the database (DB) of the users (nicknames and profiles);
- 1-week log-file of the servers hosting the community web site.

Needless to say, these two data sets are extremely different: they deliver complementary pieces of information, and they must be processed and analyzed using completely different techniques.

The main goal of this work (which is also its main original contribution, from a methodological point of view) is to establish relationships between these two heterogeneous data sets. This is inherently a very challenging task, and – to the best of our knowledge – this is one of the first attempts documented in the Data-Modeling literature to "merge" and to find relationships between Users DB and web-navigation behaviors of a very large Virtual Community ([7,8]).

The search for relationships between half-million Users and millions of page-views cannot be faced directly from the raw data sets. The basic idea and methodological approach proposed in this work is the following:

- the Users DB has been analyzed and clustered into a small number (12) of clusters; each class represents a "prototype" of User (Section 2);
- the log-file of the web server has been first sessionized and then analyzed and clustered (using an unsupervised bisecting divisive clustering approach) into 8 clusters; each cluster represents a "navigation behavior" (Section 3);
- thanks to the huge dimensional reduction of the two data-sets (the Users DB has been reduced to 12 items; the 1-week log-file has been reduced to 8 items), it is possible to find the association map between Users and navigation sessions (Section 4). Note that this can be done since the page-views registered in the log-file contain the nickname of the User, stored in a *cookie*. This allows the linking between the Users DB and the log-file.

This not-trivial analysis and modeling – although

preliminary – provides a very general and full-rounded picture of the virtual community users.

2. ANALYSIS AND CLUSTERING OF THE USERS DB

The bulk of the Users database has a simple structure, which is condensed into a single table, where each row is given by:

- the nickname (*primary-key* of the table)
- 12 fields, describing the "profile" of the user. For each field the user can select among a finite (well-defined) set of items. In the data-base only the numeric code of the selected item is stored.

Each user is, thus, characterized by a sequence of 12 values (plus the nickname) which are:

- *categorical*, since each field value represents a category, not a quantity (e.g. *job*=2 means that the user is a student)
- *non-ordinal* (e.g. saying that a student is greater than a clerk has non sense).

According to the indications expressed by the management of the virtual community, the analysis of this Data-Base has been done by focusing on the willingness of a User to fill a specific field during the registration procedure. This is a very interesting piece of information since the profiling is made on a voluntary basis (i.e. the user can leave undefined one or more fields in his profile).

As first step, the entire data-set of the Users DB has been transformed into a real-valued matrix *M*, of size 550.000×12. The element M_{ij} of *M* represents the item of the *j*-th field selected by the *i*-th user.

Using this data-set, preliminarily, the amount of users leaving undefined a specific field has been computed for each field. The result is displayed in Fig.1.

It is interesting to observe that:

- The gender is – by far – the most "filled" field. This confirms that this kind of Virtual Community is mainly seen as a mean for meeting (dating...) people.
- Age, language, sexual orientation, and "alone with" are the less voted fields.

The second step of the analysis made on the matrix *M* was the search of hidden relationships (also known as *Association Rules* – [8]) between the 12 fields of the profile. This analysis has been done by computing a sort of normalized "correlation" (or "dependence") index $\Gamma(h|k)$ $(h,k=1,2,...,12)$. $\Gamma(h|k)$ has been computed as follows.

First, the *average mutual information* $I(h,k)$ between fields *h* and *k* ([8]) has been computed. It is defined as:

$$I(h,k)=\sum_{ij}\ln\left(\frac{p(h=i,k=j)}{p(h=i)p(k=j)}\right)\cdot p(h=i,k=j),$$

where *i* and *j* take all the possible values for the fields *h* and *k*, respectively. *p(E)* is the sample probability

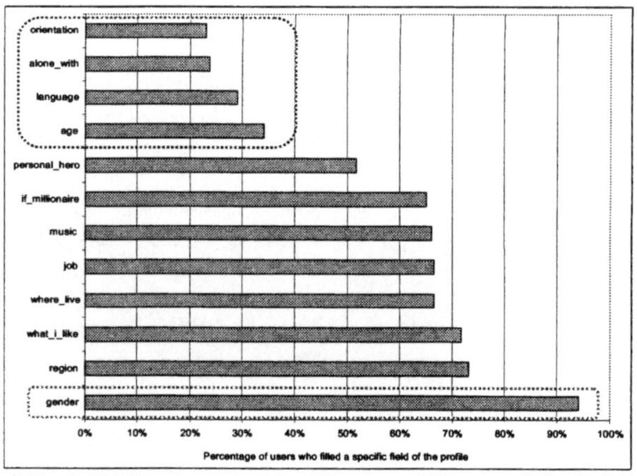

Fig.1. Willingness of the Users to fill a specific field of the profile

of the event *E*; it has been computed by exhaustive search in *M*. Using $I(h,k)$, the correlation index $\Gamma(h|k)$ hence can be computed as:

$$\Gamma(h|k)=\frac{I(h,k)}{I(h,h)}.$$

Note that $\Gamma(h|k)\in[0,1]$. $\Gamma(h|k)$ measures the dependence between *h* and *k*. More precisely, $\Gamma(h|k)$ measures the information level on *h* which can be obtained from the knowledge of *k*. For example, if *h* and *k* are independent, then $\Gamma(h|k)=0$ since the knowledge of *k* gives no information on *h*; on the contrary, if *k*=*h* then $\Gamma(h|k)=1$ since *k* describes completely *h*.

Note that $\Gamma(h|k)$ is not symmetric (in general $\Gamma(h|k)\neq\Gamma(k|h)$). As a matter of fact, the information on *h* given by *k* may be different from the information on *k* given by *h*.

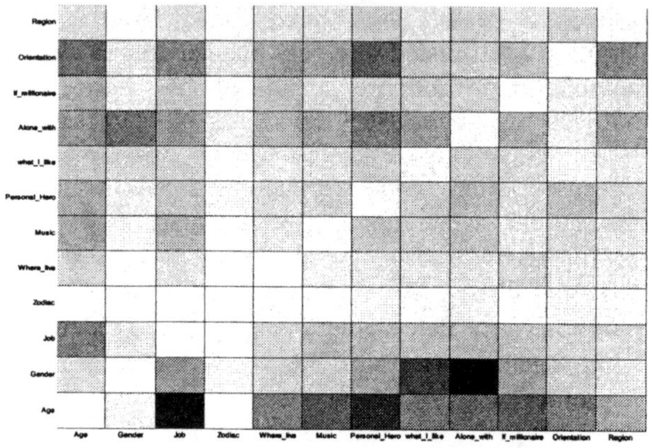

Fig.2. Association rules between the 12 fields of the profile (dark = strong correlation).

The results of this field-correlation analysis are condensed in Fig.2, where $\Gamma(h|k)$ is plotted as follows:

- each cell has been coloured proportionally to the value of $\Gamma(h|k)$, where *h* is the field on the row

and k is the field on the column; the darker the cell is, the more $\Gamma(h|k)$ is close to 1 (hence a dark cell means strong correlation);
- the values on the diagonal (all equal to 1 by definition) have been set to 0, in order to enhance the colour contrast of the plot.

The analysis of the correlation plot in Fig.2 reveals many interesting things. Among others:
- Age depends on most of the other fields (e.g. from the choice of the "Personal Hero", the age of the users can be easily predicted), but, at the same time, many fields can be predicted from the age of the user. This is somehow expected and suggests that the age is a good field for clustering. Note that the strongest correlation is between age and job.
- The gender is strongly dependent on fields: "Job", "Personal Hero", "What I like" and "Alone With".
- "Region" seems to be independent of other fields, showing that it represents a piece of information which cannot be predicted from other fields.

The map in Fig.2 delivers many interesting association rules between the 12 fields. It represents, per-se, an interesting result.

The third step of the Users DB analysis was the clustering of the data-set into a limited number of clusters. According to the marketing goals of the web site provider and to the results of the correlation analysis, the clustering has been done using 3 fields only: gender, age, and geographic region. These fields are the most filled by users and are significant from a socio-demographic point of view.

Since data were categorical and non-ordinal, the most appealing clustering procedure, in this case, was a hierarchy of univariate decision on the 3 chosen fields ([8]). Thus, it has been built a *classification tree* such that each internal node specifies a subset member-ship test on a singular field (for example a test on Age could be: "verify if age is greater than 35, less than 35 or undefined"). Then, each row-vector **u** in M (representing a user) descends a unique path from the root node to a leaf node depending on how the values of individual components of **u** match the tests of internal nodes. The set of users reaching the same leaf node is a cluster of the data-set and the characteristics of each cluster can be obtained by following the path connecting the root-node of the classification tree to the corresponding leaf node.

Fig.3 displays the size and the characteristics of the 12 clusters obtained by applying the algorithm above to M. This partition is a simple socio-demographic partition, actionable for target marketing.

3. SESSIONIZATION, ANALYSIS AND CLUSTERING OF A 1-WEEK WEB-LOG FILE

The second data set analyzed in this work is the log-file of the servers hosting the Virtual Community web-site, collected during a week of January 2002.

It is a standard log-file delivered by an Apache 1.3

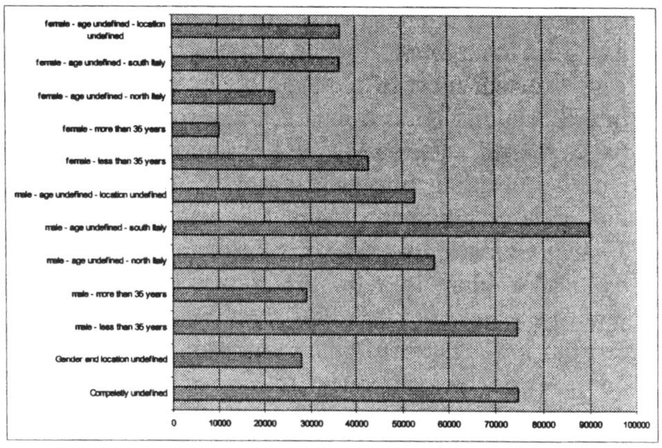

Fig.3. Partition of the whole Users DB (550.000 registered Users) into 12 clusters.

web server ([1]). In Fig.4 a small sample of this huge file is shown.

Each item (row) of the log-file represents a single "page-view" of a navigation session. The log-file contains, among other, the following data (see Fig.4):
- IP address of the remote host (User) retrieving the web page;
- complete time-stamp;
- URL of the web page requested by the remote host;
- a *cookie*, containing the indication of the nickname of the user.

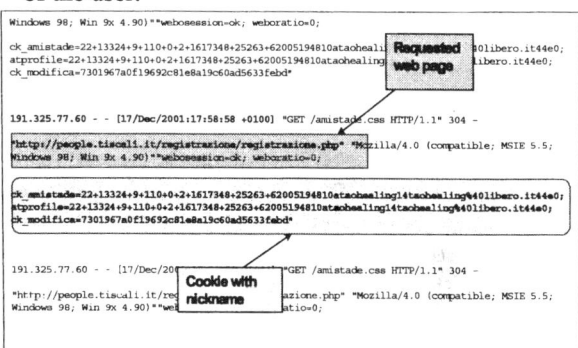

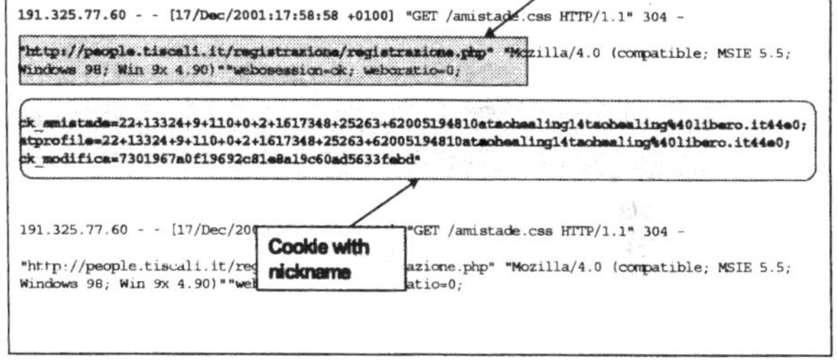

Fig.4. Sample of the raw log-file delivered by an Apache web server.

The log-file analyzed in this work is quite huge (about 2.7 GBytes). It is referred to one week (from 00:00 of Monday, to 24:00 of Sunday) of January 2002. The treatment of such a log-file has required some non-trivial pre-processing. After pre-processing, the log-file has been stored into a single table of a Data-Base. Each record of the table is a "page-view" registered by the web-server (see Fig.8). The table has 10 fields : 4 fields are used for the IP address; 4 fields are used for the complete time-stamp; 1 field for the URL of the requested web-page; 1 field for the nickname (if any).

All the (thousands of) different URLs registered by the web-server have been manually grouped into 30 sets, in order to be easily managed and interpreted.

Using this single-table Data-Base, a preliminary analysis has been done by computing the sample distribution of the page views on the 30 sets of URLs has been computed. This distribution results to be very skewed: most of the page views are condensed into 8 sets of URLs. It is interesting to note that the pages related to *messenger* and *chat* are very popular. Another peculiar thing which is worth to be noted is that a large number of hits are pages out of the Virtual Community web site. This can be explained by the fact that, in a fashion similar to the famous www.geocities.com Virtual Community, the User can put in his/her profile the link to his/her personal home page, which often is hosted on different domains. This confirms that personal web-pages are intensely visited during web navigation.

Starting from the raw data-base extracted from the 1-week log-file, the next step has been the "sessionization" of the page views. Sessionizing a log-file is known to be a tricky and subtle task, which requires some heuristics and a-priori assumptions (see e.g. [2]).

The navigation sessions have been stored into a real-valued matrix *S*. It is a 30×460.000 matrix, where the element $S_{i,j}$ is the number of seconds spent on the the *i*-th URL in the *j*-th session of the week. The matrix *S* has been built as follows:

- a session is constituted by a set of time-contiguous URLs requested by the same host (namely by the same IP address);
- a timeout of 15 minutes has been used (two URLs requested by the same IP address, separated by more than 15 minutes are assumed to belong to different sessions);
- the last page of a session has been assigned a nominal visiting time of 30 seconds (all other visiting times can be computed as the time difference between two subsequent page-views made by the same host).

In contrast with the matrix *M* of user profiles, each elements $S_{i,j}$ of *S* assumes quantitative and ordinal values (in fact, $S_{i,j}$ is the time spent by a user on a web page). Therefore, each row of *S* (i.e. each session) can be seen as a vector in a Euclidean space of dimensionality equals to 30 and the Euclidean metric can be used as distance between two sessions.

In this particular setting, using iteratively bisecting divisive partitioning algorithm (i.e. the data-set is divided in 2 clusters maximizing the distance between clusters itself and minimizing the distance between items in each cluster, see [9]) proved to be particularly suitable for the clusterization of matrix *S*. To be more precise, the bisection of the clusters was done, according to the analysis developed in [11,12], using the cascade of the Principal Direction Divisive Partitioning (PDDP) algorithm and the bisecting K-means algorithm. For the sake of self-consistency of this paper, these two algorithms are here briefly recalled in Tables 1 and 2.

K-means is probably the most celebrated and widely used clustering technique; hence it is the best representative of the class of iterative centroid-based divisive algorithms ([10,13]). PDDP is a recently proposed technique ([4,5]). It is representative of the non-iterative techniques based upon the Singular Value Decomposition (SVD) of a matrix built from the data-set ([3,6]).

Table 1: PDDP clustering algorithm.

Step 1.	Compute the centroid *w* of *S*.
Step 2.	Compute the auxiliary matrix $\tilde{S}$ as: $\tilde{S} = S - we$, where *e* is a *N*-dimensional row vector of ones, namely $e = [1,1,1,1,1,\ldots 1]$.
Step 3.	Compute the Singular Value Decompositions (SVD) of $\tilde{S}$: $\tilde{S} = U\Sigma V^T$, where Σ is a diagonal $p\times N$ matrix, and *U* and *V* are ortonormal unitary square matrices having dimension $p\times p$ and $N\times N$, respectively (see [7] for an exhaustive description of SVD).
Step 4.	Take the first column vector of *U*, say $u = U_1$, and divide $S = [x_1, x_2, \ldots, x_N]$ into two sub-clusters S_L and S_R, according to the following rule: $\begin{cases} x_i \in S_L & if \quad u^T(x_i - w) \le 0 \\ x_i \in S_R & if \quad u^T(x_i - w) > 0 \end{cases}$.

Table 2: Bisecting K-means.

Step 1.	(Initialization). Randomly select a point, say $c_L \in \Re^p$; then compute the centroid *w* of *S*, and compute $c_R \in \Re^p$ as $c_R = w - (c_L - w)$.
Step 2.	Divide $S = [x_1, x_2, \ldots, x_N]$ into two sub-clusters S_L and S_R, according to the following rule: $\begin{cases} x_i \in S_L & if \quad \lVert x_i - c_L \rVert \le \lVert x_i - c_R \rVert \\ x_i \in S_R & if \quad \lVert x_i - c_L \rVert > \lVert x_i - c_R \rVert \end{cases}$
Step 3.	Compute the centroids of S_L and S_R, w_L and w_R.
Step 4.	If $w_L = c_L$ and $w_R = c_R$, stop. Otherwise, let $c_L := w_L$, $c_R := w_R$ and go back to Step 2.

The main difference between K-means and PDDP is that K-means is based upon an **iterative** procedure, which, in general, provides different results for different initializations, whereas PDDP is a **one-shot** algorithm, which provides a unique solution. Is has been proven ([11,12]) that the best performance (in terms of quality of partition and of computational effort) can be obtained by applying PDDP, followed by K-means initialized with the PDDP result.

The PDDP+K-means algorithms has been first

applied to S; after the first bi-sectioning step, the algorithm has been iterated, each step bisecting one the clusters obtained in the step before. The decision on the cluster to split has been made heuristically, by direct inspection of the actual clusters.

The final result is a 8-cluster partition. The details on the whole partition (taxonomy of the all clustering steps) of S are displayed in Fig.5. The final clusters along with a brief characterization of them are displayed in Fig.6.

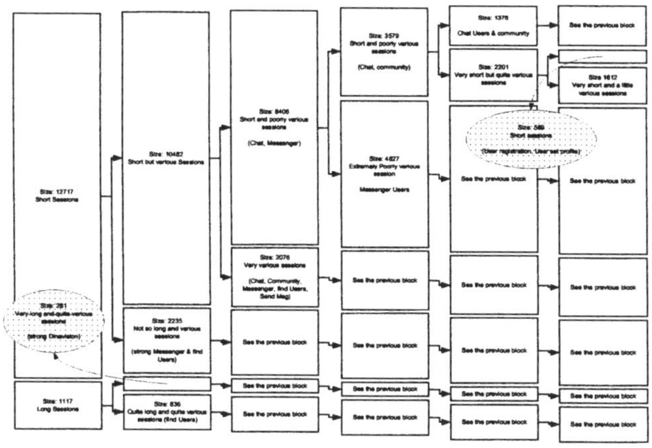

Fig.5. Complete partition-tree of the session matrix S.

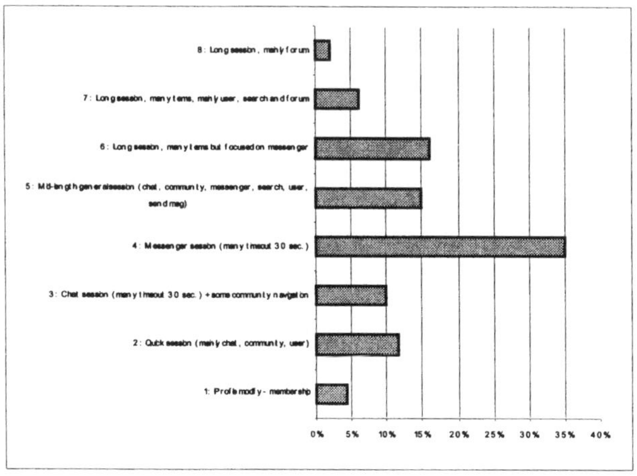

Fig.6. Leaves of the partition of the session matrix S.

As expected, the partition made by PDDP+K-means shows the most typical navigation sessions. Note, in particular, the relevance of messenger-based or chat-based sessions, and the navigations spent out of the Virtual Community domains.

4. ESTABLISHING RELATIONSHIPS BETWEEN USERS AND SESSIONS

The last step of the analysis presented in this work is the search of the main relationships between the Users DB and the navigation log-file. As already said in the Introduction, this task cannot be faced directly from the raw data-sets. The basic idea was to pre-process and reduce the Users DB to 12 clusters, and the log-file to 8 "prototype" clusters of sessions. The correlation then is searched between clusters. In this way the complexity of the problem is enormously reduced, and the results can be more easily interpreted.

To perform the correlation analysis between Users and sessions, a matrix C' of dimension 12×8 has been built as follows:

- According to his/her profile, each user has been classified into one of the 12 user clusters (hence it has been associated to one row of the matrix C'); the whole set of sessions made by such User has been extracted from the sessions-matrix S. Each session then has been classified into one of the 8 session clusters.
- A row vector of size 8 is built for each user; this vector represents the sample probability distribution (its sum is normalized to 1) of the sessions made by that user during the week.
- All the rows of the Users belonging to the i-th cluster have been summed. The result has been normalized and represents the i-th row of the matrix C'.

Thus, $C'(i,j)$ represents the average frequency of performing a session in the j-th session cluster by a user in the i-th user cluster. The plot of C' is in Fig.7. The colour (darkness) of each cell of Fig.7 is proportional to the value of $C'(i,j)$. The coding of i and j is that displayed in Fig. 4 and Fig.13.

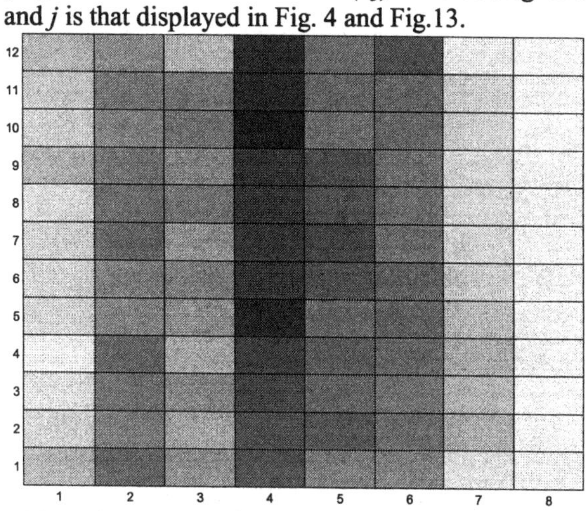

Fig.7. Correlation between the 12 clusters of Users and the 8 clusters of sessions.

As it appears, each cluster of users performs Messenger-based sessions in mainly (column 4). This is an interesting result since it highlights that all the prototypes of users behave similarly in average. However, this predominance of Messenger session hides the difference between user clusters.

To avoid this, a new correlation matrix C has been computed from C' by dividing (scaling) each column of C' by the average value of the column. The plot of C is in Fig.8. Again, the colour (darkness) of each cell of Fig.8 is proportional to the value of $C(i,j)$, where i is the User cluster (row) and j is the session cluster (column).

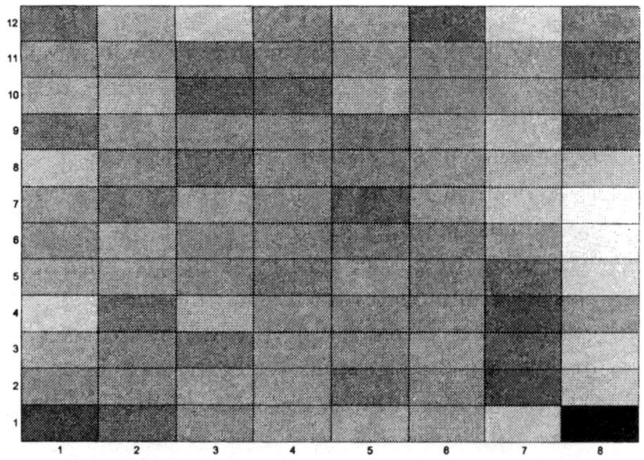

Fig.8. Scaled correlation between the 12 clusters of Users and the 8 clusters of sessions.

By analysing the results displayed in Fig.8, many interesting pieces of information can be drawn. For example the map of the main associations between clusters of Users and clusters of sessions can be built. This map is displayed in Fig.9 and, among others things, it can be seen that:

- males seems to be very related to long and various sessions;
- females seems to be primarily interested to sessions with forum or chat;
- long sessions focused on the messenger seem very correlated with Users who left the gender blank.

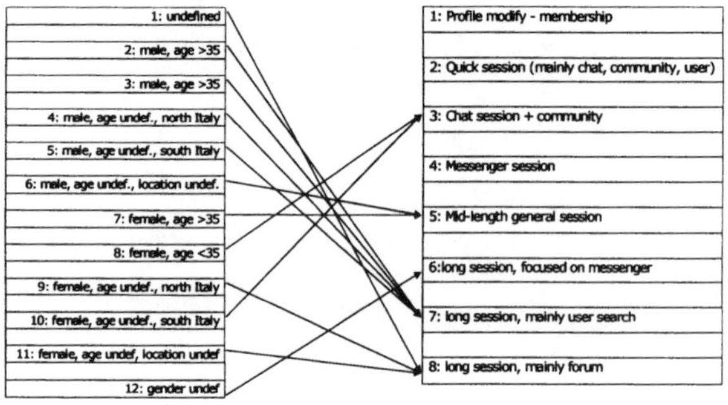

Fig.9. Main association rules between Users and sessions

5. CONCLUSIONS

In this paper a case study of Data-Modeling is presented: two heterogeneous and very large Data-Bases of a Virtual Community have been analyzed and correlated. The approach used for this analysis has been the preliminary pre-processing and independent clustering of the two Data-Bases, and then the correlation of the clusters only. This approach revealed well-suited to manage this kind of data, and a complete and easy to interpret picture of the Virtual Community Users has been built.

ACKNOWLEDGMENTS

This work has been supported by *Tiscali S.p.A.* and by MIUR project *"New Methods for Identification and Adaptive Control for Industrial Systems"*. Thanks are also due to Luca La Brocca of *Tiscali S.p.A.* and Davide Romieri and Paolo Prestinari of *n-Machines s.r.l* for enlightening discussions on Virtual Communities.

REFERENCES

[1] Aulds C. (2000). *Linux Apache Web Server Administration*. Sybex.

[2] Berent B., Mobasher B., Spiliopoulou M., and Wiltshire J. (2001). Measuring the Accuracy of Sessionizers for Web Usage Analysis. *Web Mining Workshop, at 1st SIAM International Conference on Data Mining.*

[3] Berry, M.W., Z. Drmac, E.R. Jessup (1999). "Matrices, Vector spaces, and Information Retrieval". *SIAM Review*, vol.41, pp.335-362.

[4] Boley, D.L. (1998). "Principal Direction Divisive Partitioning". *Data Mining and Knowledge Discovery*, vol.2, n.4, pp. 325-344.

[5] Boley, D.L., M. Gini, R. Gross, S. Han, K. Hastings, G. Karypis, V. Kumar, B. Mobasher, J. Moore (2000). "Document Categorization and Query Generation on the World Wide Web Using WebACE". *AI Review*, vol.11, pp 365-391.

[6] Golub, G.H, C.F. van Loan (1996). *Matrix Computations (3rd edition)*. The Johns Hopkins University Press.

[7] Hagel J.III, Armstrong A.G. (1999). *Net Gain: Expanding Markets Through Virtual Communities*. Harvard Business School Press.

[8] Hand D., Mannila H., Smyh P. (2001). *Principles of Data Mining*. MIT Press.

[9] Jain, A.K, M.N. Murty, P.J. Flynn (1999). "Data Clustering: a Review". *ACM Computing Surveys*, Vol.31, n.3, pp.264-323.

[10] Selim, S.Z., M.A. Ismail (1984). "K-means-type algorithms: a generalized convergence theorem and characterization of local optimality". *IEEE Trans. on Pattern Analysis and Machine Intelligence*, vol.6, n.1, pp.81-86.

[11] Savaresi S.M., D.L. Boley (2001). On the performance of bisecting K-means and PDDP. *1st SIAM Conference on Data Mining*, Chicago, IL, USA, paper n.5, pp.1-14.

[12] Savaresi S.M., D.L. Boley, S. Bittanti, G. Gazzaniga (2002). "Cluster selection in divisive clustering algorithms". *2nd SIAM International Conference on Data Mining*, Arlington, VI, USA, pp.299-314.

[13] Steinbach, M., G. Karipis, V. Kumar (2000). "A comparison of Document Clustering Techniques". *Proceedings of World Text Mining Conference, KDD2000*, Boston.

IFAC
Publications
www.elsevier.com/locate/ifac

A SHORT INTRODUCTION TO TIME-VARYING VOLATILITY IN FINANCIAL TIME SERIES

B. Hanzon *

* *Mathematical Institute, Leiden University, P.O. Box 9512, 2300 RA Leiden, The Netherlands, E-mail:* `bhanzon@math.leidenuniv.nl`

Abstract: This is a tutorial introduction to the invited session entitled *Links between financial econometrics and system identification.*

Keywords: Financial econometrics, stochastic volatility, ARCH-models, GARCH-models, non-linear filtering, option pricing, implied volatility.

1. INTRODUCTION

This presentation should serve as a tutorial introduction to the SYSID2003 invited session entitled *Links between Financial Time Series and System Identification.* In the area of financial time series the Black-Scholes model is often used. However, it is well-known that although the volatility is assumed to be constant in the Black-Scholes model, estimates of the volatility based for example on observed prices of call options and other financial derivatives are known to be time-varying. This phenomenon lies at the basis of the so-called VIX volatility index. In financial time series this has led to investigation of more general models in which the volatility is allowed to vary. Two important sub-classes of this set are

(i) (G)ARCH (Generalized Auto-Regressive Conditional Heteroskedasticity) models.
(ii) Stochastic volatility models.

In this presentation we want to compare these two classes of models and in this way explain the basic features of these models to a wider system identification-oriented audience.

It is also possible to approach the problem of observed time-varying volatility in another way. One can also relax other basic assumptions in the Black-Scholes model. One possibility is to work with non-Gaussian probability density functions. For example one can work with probability density functions with heavy tails, or one can try to work with non-linear volatility models. Several of these approaches are treated in the invited session.

2. THE ROLE OF VOLATILITY IN FINANCIAL TIME SERIES

Time series of prices of stocks, bonds, and financial derivatives like options are the topic of study in the area of financial time series. An important ingredient of the theory underlying financial time series is the Efficient Market Hypothesis (EMH). A market will be called *perfectly efficient* if the prices fully reflect available information, so that prices adjust fully and instantaneously when new information becomes available. One of the effects of the EMH is that prices cannot be predicted in a systematic fashion. The basic argument is that if they could be predicted in a systematic fashion then this could be used in trading to ones profit. However this would imply that not all available information is yet reflected in the prices, which contradicts the EMH. The result of the EMH is that the attention in the analysis of financial time series has shifted over time from prediction of price levels, to estimation and prediction of other aspects of the time series. The first main focus has become the standard deviation of the series or in the vocabulary of financial time series the *volatility*. But also higher order moments, like the third order moment (related to skewness) and the fourth order moment (related to the kurtosis) are investigated (and often found to be *not* corresponding to Gaussian probability density functions). Another important part of quantitative financial theory nowadays is the theory of *option pricing*. In the seminal paper of Black and Scholes ((Black and Scholes, 1973)) it was recognized that options and other financial contracts that depend on the future prices of some assets can be mimicked by a dynamic "feedback" trading rule that describes the continuous adaptation of a portfolio depending on the price movements of the stock. The effect of this so-called "dynamic hedge" is the same as the contract: it has the same pay-off at expiry. Therefore the price has to be the same. The present price of the dynamic hedge is the price of what is in the dynamic portfolio at present. This is the basis for the modern theory of option pricing. A surprising fact is that the price of an option is in general *not* equal to its expected value (the expected value of a stochastic "game" is also called the Huygens-Bernoulli value). Instead the option price is a function of the *volatility* of the returns of the asset price on which it is based. This is therefore a second reason for the interest in the behaviour of the volatility of a financial time series.

3. VOLATILITY PARAMETER ESTIMATION

In classical models for financial time series, including linear stochastic systems (ARMA models) and the Black-Scholes model, the volatility (the volatility of the returns of the asset price to be precise) is constant, it is a parameter of the model. As is discussed in the paper by Martens and Zein ((Martens and Zein, 2003)) in this session there tends to be a certain skepticism in the financial time series literature concerning the possibility to estimate the volatility parameter with standard time series methods, which are solely based on the past values of the time series at hand. In this context one speaks of *volatility estimation based on historical data*. Instead what is often used is so-called *implied volatility*. The idea is to use the fact that the theory describes a precise relation between an option price and the volatility of the return on the underlying asset. By observing the value of an option on the options market one can therefore deduce the corresponding implicit (or "implied") volatility. The argument is that the value of the volatility that the traders that buy and sell options use is this implied volatility rather than the volatility estimate that is based on historical data. For a further discussion we refer to (Martens and Zein, 2003). What is remarkable in this respect is that the estimates that are obtained by both methods, turn out to be *time-varying* in practice. This is however in conflict with the assumption of a constant volatility parameter that is underlying the classical ARMA model and the Black-Scholes model.

4. MODELING TIME-VARYING VOLATILITY

Broadly speaking there are two classes of models with time-varying volatility for financial time series. Recent investigations show that there are in fact a lot of connections between these two classes and in some sense they differ mostly in the way they are usually represented. The first class will be called the stochastic volatility models and the second class is formed by the so-called ARCH and GARCH models and their generalizations. A simple example of the first class of models is given by the following stochastic state-space model (cf. e.g (Taylor, 1986), (Andersen, 1994))

$$\begin{aligned} X_t &= aX_t + W_t \\ Y_t &= V(X_t).U_t, \; t = 0, 1, 2, \ldots, \end{aligned} \tag{1}$$

where $V(x)$ is a non-negative function of x and $\{U_t\}$ and $\{V_t\}$ all jointly stochastically independent random variables and a a real number. In financial time series usually $Y_t = \log(S_{t+1}) - \log(S_t),\; t = 0, 1, 2, \ldots$ are the returns on the asset prices $S_t,\; t = -1, 0, 1, 2, \ldots$ Let us assume further that all U_t have the same distribution, $t = 0, 1, 2, \ldots$ and all W_t have the same distribution, $t = 0, 1, 2, \ldots$. One well-known further specification is $V(x) = \exp\left((x + \gamma)/2\right)$ and U_t zero-mean Gaussian with variance σ^2 and V_t zero-mean Gaussian with unit variance. The problem of estimating the state X_t given the observations up till time t is in fact a *non-linear filtering problem*. No closed-form solution is known for this filtering problem. Several approaches to approximate solutions to this filtering problem have been proposed in the literature. See e.g. (A. Harvey, 1994), (Mahieu and Schotman, 1998), (Brigo and Hanzon, 1998) and the references given there. In the paper (Hanzon, 2003) in this session an exact solution of the filtering problem will be presented for an analogous stochastic volatility model in which however the Gaussian distributions are replaced by rational distributions and the function V is taken to be a non-zero, non-negative polynomial. From the available literature one can conclude that the non-linear filtering problem of dynamically estimating the volatility based on a stochastic volatility model is a hard problem. The non-linear filtering problem can be formulated as the problem of finding the *conditional probability density* of the state variable X_t given the observations $Y_t,\; Y_{t-1},\; Y_{t-2},\; \ldots$. The corresponding conditional mean, assuming it exists, can be used as an estimate for the state X_t, while the corresponding conditional expectation of $V(X_t)$, assuming it exists, is the conditional variance of the observation Y_t. Note that the conditional variance will in general depend on the history of the observations $Y_t,\; Y_{t-1},\; Y_{t-2},\; \ldots$ hence the conditional variance will be time-varying in general. This is in sharp contrast with the conditional variance of the observations in a finite dimensional linear Gaussian dynamical system, in which case the conditional variance does *not* depend on the history of the observations and can in fact be calculated ahead of time. This follows from the well-known theory of the Kalman filter (see e.g. (Anderson and Moore, 1979)). In the econometrics literature, the property that the conditional variance is time-varying is called conditional heteroskedasticity. An alternative approach to specify a stochastic system, pioneered by (Engle, 1982), is to start with specifying a functional form for the conditional variance of the output, as a function of past outputs. Once it has been shown that the specification of the conditional variance is admissible in the sense that there is a stochastic system for which this is the conditional variance, the parameters appearing in the conditional variance can be estimated directly from the data. In this way the non-linear filtering problem is in some sense circumvented by specifying the functional form of the conditional variance and estimating the unknown parameters in this functional form directly from the data. The first class of models of this kind that was introduced by (Engle, 1982) was the class of ARCH (autoregressive conditional heteroskedastic) models. An ARCH(q) model for an error process $\{\epsilon_t\}$ is given by $\epsilon = \delta\sqrt{\epsilon^2}$, where δ is the *sign* of ϵ which is assumed to have the symmetric probability distribution on the set $\{-1,\ 1\}$ and ϵ_t^2 is given by

$$\epsilon_t^2 = c + \sum_{i=1}^{q} a_i \epsilon_{t-i}^2 + u_t, \qquad (2)$$

where $E(\epsilon_t|\epsilon_{t-1}, \epsilon_{t-1}, \ldots) = 0$ and $\{U_t\}$ is (also) a martingale difference sequence. The conditional variance of ϵ_t is

$$V(\epsilon_t|\epsilon_{t-1}, \epsilon_{t-2}, \ldots) = c + \sum_{i=1}^{q} a_i \epsilon_{t-i}^2 \qquad (3)$$

and therefore depends cleary on the past through the q most recent values of ϵ_t^2. The ARCH model is based on an autoregressive representation of the conditional variance. Just as AR models can be generalized to ARMA models, ARCH models can be generalized by adding a moving average (MA) part. The resulting models are called *generalized ARCH*, or GARCH models. They were introduced in (Bollerslev, 1986). In GARCH models the conditional mean is zero and the conditional variance is given by

$$V(\epsilon_t|\epsilon_{t-1}, \epsilon_{t-2}, \ldots) =$$
$$h_t = c + \sum_{i=1}^{q} \alpha_i \epsilon_{t-i}^2 + \sum_{j=1}^{p} \beta_j h_{t-j} \qquad (4)$$

A number of further generalizations have appeared in the literature, see e.g. (Gouriéroux, 1997). For an application of nonlinear GARCH models to financial data see the paper (van Dijk, 2003) in this session. Note that in ARCH and GARCH models the specification of the conditional variance is usually easier than in models that are not specified directly in terms of their conditional variance, but instead in terms of their unconditional stochastic evolution. However for ARCH and GARCH models and their generalizations it is often more involved to determine the stochastic and asymptotic properties of the model. See e.g. the paper (Rahbek, 2003) in this session for results on the stochastic properties of a multivariate extension of the ARCH model.

5. SOME REMARKS ON OPTION PRICING IN CASE OF TIME-VARYING VOLATILITY

In the context of this short introduction to time-varying volatility we want to make only a few remarks concerning option pricing based on models with time-varying volatility.

As was noted above, in the Black-Scholes theory of option pricing the price of the option is obtained by constructing a continuous-time dynamic hedge which mimicks the option perfectly. This was extended to continuous-time stochastic volatility models in several papers, including (Hull and White, 1987), (Wiggins, 1987), (Heston, 1993), (Scott, 1987), (Stein and Stein, 1991). It should be noted however, that for these models due to what is called *incompleteness* of the market of stock and bond prices, the option price can in general not be based on no-arbitrage arguments only. Therefore additional hypotheses are introduced in the literature needed to arrive at a unique option price for these models.

Option pricing based on ARCH and GARCH models was studied by (Duan, 1995) and by (Kallsen and Taqqu, 1998) among others. As was noted in (Garcia and Renault, 1997) in these two papers one arrives at the same option pricing formula, but at a different hedging strategy! That this is possible has to do with market incompleteness which is rather the rule than the exception for discrete-time models.

A link between ARCH models and continuous-time stochastic volatility models was first noted by (Nelson, 1990). Concerning the relation of these two classes of models let us give two quotes from (Garcia and Renault, 1997): *To summarize, the endogenous or exogenous characterization of the source of randomness introduced in the asset price volatility to explain the smile is not as clear-cut as it appears. Apart from the fact that both the stochastic volatility (SV) models and the GARCH-type models can reproduce the stylized facts associated with the smile according to the simulations performed by (Hull and White, 1987) and (Duan, 1995), we argued that these models are not as far apart as originally believed.*

and

In conclusion we stress that ARCH-type and stochastic volatility option pricing models should not be seen as competitors (as it is commonly believed) but rather as complements, since the ARCH model offers a useful discrete-time filter for SV models.

6. CONCLUSION

We have presented a short introduction to mathematical financial models with time-varying volatility. Broadly speaking there are two classes of such models, namely SV models and GARCH-type models. For SV models the problem of dynamically estimating the state (the filtering problem) is usually hard. For GARCH type models the conditional variance is specified explicitly, however to establish stochastic and asymptotic properties for these models is more involved. Especially due to the theory of option pricing and due to the availability of high-frequency data from financial markets continuous-time models are very important in this area. The real-time aspects including rules for continuous trading to replicate an option make that this area is much closer in spirit to systems engineering than other more traditional areas of quantitative economics. Parameter estimation for nonlinear stochastic systems, nonlinear filtering, feedback aspects in dynamic hedging all make this area very interesting from the point of view of systems identification. It is to be expected that input from systems identifiation and control theory to this area can bring in new and surprising approaches to the problem formulations in this field.

7. REFERENCES

A. Harvey, E. Ruiz, N. Shephard (1994). Multivariate stochastic variance models. *Review of Economic Studies* **61**, 247–264.

Andersen, T. G. (1994). Stochastic autoregressive volatility: a framework for volatility modelling. *Mathematical Finance* **4**, 75–102.

Anderson, B.D.O. and J.B. Moore (1979). *Optimal Filtering*. Prentice Hall. Englewood Cliffs.

Black, F. and M. Scholes (1973). The pricing of options and corporate liabilities. *Journal of Political Economy* **81**, 637–659.

Bollerslev, T. (1986). Generalized autoregressive conditional heteroskedasticity. *Journal of Econometrics* **31**, 301–327.

Brigo, D. and B. Hanzon (1998). On some filtering problems arising in mathematical finance. *Insurance: Mathematics and Economics* **22**, 53–64.

Duan, J.-C. (1995). The garch option pricing model. *Mathematical Finance* **5**, 13–32.

Engle, R. (1982). Autoregressive conditional heteroskedasticity with estimates of the variance of uk inflation. *Econometrica* **50**, 987–1008.

Garcia, R. and E. Renault (1997). A note on hedging in arch and stochastic volatility option pricing models. *Report 97s-13, Centre Universitaire de Recherche et Analyse des Organisations CIRANO, Montréal.*

Gouriéroux, C. (1997). *ARCH Models and Financial Applications*. Springer. New York.

Hanzon, B. (2003). A rational probability density approach to stochastic volatility estimation. *Proceedings SYSID 2003 Rotterdam, this volume, paper ISC-384.*

Heston, S. (1993). A closed form solution for options with stochastic volatility with application to bond and currency options. *Review of Financial Studies* **6**, 327–343.

Hull, J. and A. White (1987). The pricing of options on assets with stochastic volatilities. *Journal of Finance* **42**, 281–299.

Kallsen, J. and M. Taqqu (1998). Option pricing in arch-type models. *Mathematical Finance* **8**, 13–26.

Mahieu, R. and P. Schotman (1998). An empirical application of stochastic volatility models. *Journal of Applied Econometrics* **13**, 333=360.

Martens, M. P. E. and J. Zein (2003). Predicting financial volatility: high frequency time-series forecasts vis-a-vis implied volatility. *Proceedings SYSID 2003, this volume, paper ISC-140.*

Nelson, D. B. (1990). Arch models as diffusion approximations. *Journal of Econometrics* **45**, 7–39.

Rahbek, A. (2003). Stochastic properties of multivariate time series–with an emphasis on arch. *Proceedings SYSID 2003 Rotterdam, this volume, paper ISC-225.*

Scott, L. (1987). Option pricing when the variance changes randomly: theory, estimation and an application. *Journal of Financial and Quantitative Analysis* **22**, 419–438.

Stein and Stein (1991). Stock price distributions with stochastic volatility: an analytic approach. *Review of Financial Studies* **4**, 727–752.

Taylor, S. (1986). *Modelling Financial Time Series.* John Wiley and Sons. London.

van Dijk, D.J.C. (2003). Forecasting emerging equity market volatility using nonlinear garch models. *Proceedings SYSID 2003 Rotterdam, this volume, paper ISC-408.*

Wiggins, J. (1987). Option values under stochastic volatility: theory and empirical estimates. *Journal of Financial Economics* **19**, 351–372.

IFAC Publications
www.elsevier.com/locate/ifac

FORECASTING EMERGING EQUITY MARKET VOLATILITY USING NONLINEAR GARCH MODELS

Dick van Dijk *

* *Econometric Insitute, Erasmus University Rotterdam*

Abstract: In this paper we examine the usefulness of nonlinear Generalized Autoregressive Conditionally Heteroskedastic (GARCH) models for forecasting daily volatility in a number of Asian and Latin American emerging equity markets. Two of the most popular nonlinear GARCH specifications, the GJR model and the Exponential GARCH model, are found to outperform a linear GARCH model in terms of one-day ahead out-of-sample volatility forecasts. This conclusion holds both when volatility forecasts are evaluated by means of traditional criteria that rely upon a proxy for unobserved volatility or by means of indirect probability forecasts. *Copyright © 2003 IFAC*

Keywords: Conditional heteroskedasticity, Leverage effect, Volatility forecasts, Probability forecasts, Forecast evaluation

1. INTRODUCTION

Over the last two decades, since the introduction of the (Generalized) Autoregressive Conditionally Heteroskedastic ((G)ARCH) model by (Engle, 1982) and (Bollerslev, 1986), the literature on modeling and forecasting volatility of high-frequency asset returns has grown enormously. On the one hand, much effort has been devoted to developing more advanced models to describe the salient features of volatility, which has led to a plethora of nonlinear GARCH models and stochastic volatility models, see (Franses and van Dijk, 2000) for an overview. This also spurred interest in techniques for evaluating different volatility models by means of misspecification tests and comparison of out-of-sample forecasting performance, see (Lopez, 2001) for a recent example. On the other hand, empirical applications of these models to describe volatility of stock returns, exchange rate returns and interest rates have become abundant, see the survey in (Bollerslev *et al.*, 1992). The empirical applications of GARCH-type models have been largely limited, however, to data from developed financial markets. In this paper, we consider volatility in a number of Asian and Latin American emerging equity markets. While emerging markets finance has been a topic of growing interest among academics and practitioners alike,[1] only few studies have considered the application of GARCH-type models for describing and forecasting volatility in these markets, see (Santis and Imrohoroğlu, 1997) for an example. Among the possible reasons for this neglect is the fact that emerging markets data have certain characteristics not observed in developed markets and which generally are thought difficult to model. For example, equity prices in emerging markets are subject to extremely large shifts, as documented in (Bekaert *et al.*, 1998), among others. Also, the effect of market liberalizations, which in many countries occurred in the early 1990s, on volatility is uncertain. Some studies find an increase in volatility following the opening of an emerging equity market, where others document a decline, see (Richards, 1996), (Bekaert and

[1] See (Bekaert and Harvey, 2003) for a recent survey.

Harvey, 1997), (Santis and Imrohoroğlu, 1997), and (Aggarwal *et al.*, 1999). In this paper, we consider daily stock index data from a number of Latin American and Asian countries. Our purpose is to examine the potential for successful application of nonlinear GARCH-type models to describe and forecast volatility for these equity markets.

2. DATA

We use daily returns measured in US dollars on the MSCI index for the following Latin American and Asian emerging equity markets: Argentina, Brazil, Chile, Mexico, Korea, Malaysia, Philippines, Taiwan, and Thailand. The sample period runs from January 1, 1988 to December 31, 2002. The summary statistics in Table 1 show that all returns series are skewed and highly leptokurtic, such that normality of the series is convincingly rejected. Moreover, for many countries the daily returns exhibit significantly positive first-order autocorrelation. The empirical autocorrelation functions for the squared and absolute returns have the typical pattern for high-frequency financial returns of a small but significant positive first-order autocorrelation followed by a slow decay towards zero, as shown in Figure 1 for Brazil.

Table 1. Summary statistics for daily emerging market equity returns

Country	Mean	Min	Max	SD	Skew	Kurt	ρ_1
Argentina	0.12	−60.4	57.8	4.04	0.88	46.8	0.07
Brazil	0.08	−23.2	23.7	2.96	−0.01	9.5	0.15
Chile	0.05	−15.0	9.1	1.30	−0.26	11.9	0.24
Mexico	0.09	−19.6	19.5	2.01	0.35	15.0	0.12
Korea	0.04	−19.5	30.8	2.45	0.90	17.7	0.09
Malaysia	0.03	−30.9	29.5	2.00	0.78	49.3	0.10
Philippines	0.01	−10.4	24.6	1.79	1.10	19.2	0.18
Taiwan	0.04	−10.5	13.5	2.19	0.16	5.2	0.07
Thailand	0.02	−13.4	19.8	2.23	1.05	13.7	0.19

Summary statistics for daily emerging market equity returns. The sample period is January 1, 1988 until December 31, 2002, which equals 3913 daily observations. ρ_1 denotes the first-order autocorrelation coefficient.

3. MODELS

We employ the following set-up to describe the daily emerging market equity returns r_t

$$r_t = \mu + \delta'\mathbf{D}_t + \theta\varepsilon_{t-1} + \varepsilon_t, \tag{1}$$

$$\varepsilon_t = z_t\sqrt{h_t} \quad \text{with } z_t \sim \text{i.i.d.}F(0,1), \tag{2}$$

where $\mathbf{D}_t = (D_{\mathrm{MON},t} - D_{\mathrm{WED},t}, D_{\mathrm{TUE},t} - D_{\mathrm{WED},t}, D_{\mathrm{THU},t} - D_{\mathrm{WED},t}, D_{\mathrm{FRI},t} - D_{\mathrm{WED},t}, D_{\mathrm{HOL},t})'$ is a vector consisting of demeaned daily dummies[2]

[2] $D_{\mathrm{MON},t}$ is a dummy which takes the value 1 if day t is a Monday and 0 otherwise, and $D_{\mathrm{TUE},t}$, $D_{\mathrm{WED},t}$, $D_{\mathrm{THU},t}$ and $D_{\mathrm{FRI},t}$ are similarly defined for the other days of the week.

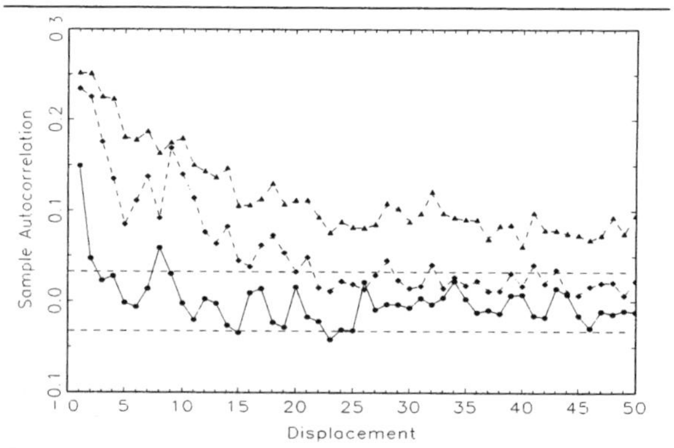

Fig. 1. Empirical autocorrelation functions for daily returns (circles), absolute returns (triangles) and squared returns (diamonds) for the MSCI index of Brazil, over the period January 1988-December 2002. The dashed lines are Bartlett 95% confidence bounds, computed as $\pm 2/\sqrt{T}$, where T denotes the number of observations.

and a "holiday" dummy $D_{\mathrm{HOL},t}$ which is equal to 1 if day t follows a weekday on which the exchange was closed. The MA(1) term $\theta\varepsilon_{t-1}$ is included in the specification for the conditional mean of r_t to capture the first-order autocorrelation in the returns. The decomposition of the shock ε_t in (2) implies that ε_t is conditionally heteroskedastic, that is $\varepsilon_t|\Omega_{t-1} \sim F(0, h_t)$, where Ω_{t-1} is the information set available at time $t-1$. Most (nonlinear) GARCH and stochastic volatility models assume that

$$h_t = g(\varepsilon_{t-1}, h_{t-1}, \mathbf{D}_t; \xi) \tag{3}$$

where $g(\varepsilon_{t-1}, h_{t-1}, \mathbf{D}_t; \xi)$ is a possibly nonlinear function describing the evolution of the conditional variance h_t. To keep the number of volatility models limited, we restrict ourselves to three of the most popular GARCH models. Specifically, we consider the GARCH(1,1) model as introduced by (Bollerslev, 1986), where

$$h_t = \omega + \alpha_1\varepsilon_{t-1}^2 + \beta_1 h_{t-1} + \zeta'\mathbf{D}_t. \tag{4}$$

For identification and to ensure that h_t is positive almost surely, the parameters in (4) need to satisfy the restrictions $\omega > 0$, $\alpha > 0$ and $\beta_1 \geq 0$. Stationarity of the GARCH process requires $\alpha_1 + \beta_1 < 1$. In empirical applications of this model, it is typically found that α_1 is small, while the sum $\alpha + \beta_1$ is close to 1. For such parameter combinations, the autocorrelations of the squared returns are such that $\rho_1 \approx \alpha_1$ and $\rho_k = (\alpha_1 + \beta_1)^{(k-1)}\rho_1$, demonstrating that the GARCH model can capture the typical shape of the empirical autocorrelation function of r_t^2 which is characterized by a small positive value of ρ_1 followed by a slow decline towards 0. The term $\zeta'\mathbf{D}_t$ is included in (4) to allow for day-of-the-week effects in volatility as well, and for different volatility following holidays.

Note that the GARCH(1,1) model in (4) is "linear" in the sense that the effect of the lagged shock ε_{t-1} on the conditional variance h_t is symmetric and does not depend on its sign. However, a stylized fact of equity returns is the presence of a so-called *leverage effect*, which means to say that negative shocks have a larger effect on volatility than positive shocks of the same magnitude. Many nonlinear GARCH models in fact are designed especially to capture this effect. Here we employ the intuitively attractive GJR model of (Glosten *et al.*, 1993). In this case, the conditional variance h_t is specified as

$$h_t = \omega + \alpha_1 \varepsilon_{t-1}^2 (1 - \mathsf{I}[\varepsilon_{t-1} > 0]) + \gamma_1 \varepsilon_{t-1}^2 \mathsf{I}[\varepsilon_{t-1} > 0] + \beta_1 h_{t-1} + \zeta' \mathbf{D}_t, \quad (5)$$

where $\mathsf{I}[A] = 1$ if the event A occurs, and 0 otherwise. Note that in the presence of a leverage effect, we would expect the estimate of α_1 to be larger than the estimate of γ_1. In this case, $\omega > 0$, $\alpha_1 > 0$, $\gamma_1 > 0$ and $\beta_1 \geq 0$ are required for identification of the parameters and to guarantee that h_t is non-negative for all t, while $(\alpha_1 + \gamma_1)/2 + \beta_1 < 1$ is a necessary condition for covariance stationarity.

As an alternative nonlinear specification which allows for asymmetric effects of positive and negative shocks on h_t, we use the exponential GARCH [EGARCH] model of Nelson (1991), where

$$\ln(h_t) = \omega + \alpha_1 z_{t-1} + \gamma_1(|z_{t-1}| - \mathsf{E}[|z_{t-1}|]) + \beta_1 \ln(h_{t-1}) + \zeta' \mathbf{D}_t. \quad (6)$$

The advantage of the EGARCH model is that no restrictions on the parameters have to be imposed to ensure that h_t is positive.

Finally, we assume that the standardized shocks z_t follow either a standard normal distribution, or a Student t distribution with ν degrees of freedom, where ν is estimated along with the other parameters in the model. The reason for not restricting ourselves to the normal distribution is that the resulting GARCH models cannot adequately capture the excess kurtosis of the daily emerging market returns due to the occasionally large jumps in equity prices.

4. METHODOLOGY

We recursively estimate the different volatility models by means of maximum likelihood, using a moving window of five years of data, starting with January 1, 1988 - December 31, 1992. Using a moving window instead of an expanding window is motivated by the fact that the volatility process in these emerging markets is likely to have experienced changes due to, for example, market liberalizations. The estimated models are used to construct out-of-sample one-day ahead forecasts or returns and volatility, denoted $\hat{r}_{t+1|t}$ and $\hat{h}_{t+1|t}$, respectively.

A number of out-of-sample performance measures are used to evaluate and compare the volatility forecasts from different models. First, we employ several popular error metrics, namely the Mean Squared Prediction Error (MSPE; MSPE $= \frac{1}{R}\sum_{i=1}^{R}(h_{t+1} - \hat{h}_{t+1|t})^2$ where R denotes the number of forecasts), the Mean Absolute Error (MAE; MAE $= \frac{1}{R}\sum_{i=1}^{R}|h_{t+1} - \hat{h}_{t+1|t}|$), and the Heteroskedasticity-adjusted MSPE (HMSPE; HMSPE $= \frac{1}{R}\sum_{i=1}^{R}(\frac{h_{t+1}}{\hat{h}_{t+1|t}} - 1)^2$). As the true conditional variance h_{t+1} is not directly observable, we use the squared innovation $\hat{\varepsilon}_{t+1}^2$ as a proxy when computing these measures. We employ the tests of equal forecast accuracy developed by Diebold and Mariano (1995) to assess whether differences in the error metrics of two competing models are significant; see the discussion below.

Furthermore, the quality of individual forecasts is assessed by regressing "true" volatility on the corresponding forecast,

$$h_{t+1} = b_0 + b_1 \hat{h}_{t+1|t} + \nu_t, \quad (7)$$

where b_0 and b_1 should be equal to 0 and 1, respectively, for the forecast to be unbiased and efficient. Again, we use $\hat{\varepsilon}_{t+1}^2$ as a proxy for h_{t+1} to implement this regression.

As discussed at length in (Andersen and Bollerslev, 1998), the usefulness of the above evaluation measures is limited because of the use of the proxy $\hat{\varepsilon}_{t+1}^2$ for the unobserved true conditional volatility at day $t+1$. Even though the squared innovation is an unbiased estimator of h_{t+1}, it is also a very imprecise or "noisy" estimator, which can lead to misleading conclusions regarding the forecasting performance of volatility models. For example, (Andersen and Bollerslev, 1998) show that in case the standardized shocks z_t are normally distributed and h_t follows a standard GARCH(1,1) process (4), the regression R^2 from (7) is bounded from above by 1/3 for volatility forecasts constructed from a GARCH(1,1) model. Hence, it is not surprising that in empirical applications of (7), the R^2 is invariably found to be rather low, usually in the range from 0.05 to 0.20. Also, the estimates of b_0 and b_1 often differ significantly from their target values of 0 and 1, respectively.

An alternative evaluation framework which circumvents these problems has been suggested by (Lopez, 2001). This involves first converting the volatility forecast into probability forecasts of certain observable events. For example, given a volatility forecast $\hat{h}_{t+1|t}$, the forecast of the probability that the return at day $t+1$ is between $L_{t+1|t}$ and $U_{t+1|t}$ is given by

$$
\begin{aligned}
\widehat{P}_{t+1|t} &= \mathsf{P}[L_{t+1|t} \leq r_{t+1} \leq U_{t+1|t}|\Omega_t] \\
&= \mathsf{P}[L_{t+1|t} - \hat{r}_{t+1|t} \leq \varepsilon_{t+1} \leq U_{t+1|t} - \hat{r}_{t+1|t}|\Omega_t] \\
&= \mathsf{P}\left[\frac{L_{t+1|t} - \hat{r}_{t+1|t}}{\sqrt{\hat{h}_{t+1|t}}} \leq z_{t+1} \leq \frac{U_{t+1|t} - \hat{r}_{t+1|t}}{\sqrt{\hat{h}_{t+1|t}}}\right] \\
&= \int_{l_{t+1|t}}^{u_{t+1|t}} f(z_{t+1})dz_{t+1}, \qquad (8)
\end{aligned}
$$

where $\hat{r}_{t+1|t}$ is the 1-day ahead return forecast obtained from (1), $l_{t+1|t} = (L_{t+1|t} - \hat{r}_{t+1|t})/\sqrt{\hat{h}_{t+1|t}}$ and $u_{t+1|t}$ similarly defined, and $f(\cdot)$ is the density function of the standardized shocks z_t. The probability forecasts $\widehat{P}_{t+1|t}$ can be evaluated by comparing them with the corresponding indicator function R_{t+1}, which is equal to one if the event of interest occurs and zero otherwise. Here we employ the quadratic probability score, which is the analogue of the MSPE for probability forecasts computed as

$$
\text{QPS} = \frac{1}{R}\sum_{j=1}^{R} 2(\widehat{P}_{t+1|t} - R_{t+1})^2. \qquad (9)
$$

Hence, the QPS $\in [0, 2]$ with smaller values indicating more accurate forecasts. As discussed in (Lopez, 2001), probability forecasts from two competing models can be compared by means of Diebold-Mariano tests of equal forecast accuracy. Specifically, define the "loss differential" d_t as

$$
d_t = 2(\widehat{P}^{(1)}_{t+1|t} - R_{t+1})^2 - 2(\widehat{P}^{(2)}_{t+1|t} - R_{t+1})^2,
$$

where $\widehat{P}^{(1)}_{t+1|t}$ and $\widehat{P}^{(2)}_{t+1|t}$ are the probability forecasts from the two models being compared. The null hypothesis of equal probability forecast accuracy then implies that $\mathsf{E}[d_t] = 0$, which can be tested by means of the statistic

$$
S = \frac{\overline{d}}{\sqrt{\widehat{V(\overline{d})}}} \xrightarrow{asy} N(0,1), \qquad (10)
$$

where $\overline{d}$ is the sample mean loss differential and $\widehat{V(\overline{d})}$ is a consistent estimate of the asymptotic variance of $\overline{d}$.

5. EMPIRICAL RESULTS

Table 2 reports the results from "traditional" out-of-sample forecast evaluation metrics, including MSPE, MAE and HMSPE and parameter estimates and R^2 values from the regression (7). Only results for the volatility models using a normal distribution for the standardized shocks z_t are shown to conserve space. A number of conclusions emerge from this table. The most interesting finding perhaps is that allowing for the leverage effect clearly improves forecast accuracy, in the sense that with very few exceptions the GARCH model (4) is outperformed by either the GJR model or the EGARCH model, or both, in terms of MSPE, MAE and HMSPE. In many cases, the differences are statistically significant

Table 2. Out-of-Sample Volatility Forecast Evaluation, January 1993-December 2002

	b_0	b_1	R^2	MSPE	MAE	HMSPE
Argentina						
GARCH	1.90	0.59	0.059	5.94	7.36	5.50
	(0.72)	(0.15)				
GJR	2.25	0.52	0.059	6.06	7.42	5.48
	(0.75)	(0.14)		(−1.34)	(−1.24)	(0.04)
EGARCH	1.25	0.70	0.058	5.76	7.23	6.25
	(0.66)	(0.14)		(1.76)	(2.11)	(−1.09)
Brazil						
GARCH	1.27	0.73	0.113	3.08	7.45	3.02
	(0.57)	(0.11)				
GJR	0.54	0.83	0.136	2.98	7.28	2.81
	(0.67)	(0.12)		(4.72)	(7.50)	(2.34)
EGARCH	−0.34	0.98	0.134	2.96	7.17	2.69
	(0.74)	(0.14)		(3.27)	(5.79)	(2.97)
Chile						
GARCH	0.20	0.78	0.101	11.52	1.48	2.95
	(0.21)	(0.17)				
GJR	0.18	0.79	0.104	11.49	1.48	2.93
	(0.21)	(0.17)		(2.75)	(1.39)	(2.90)
EGARCH	−0.18	1.05	0.105	11.44	1.47	3.08
	(0.29)	(0.23)		(0.35)	(1.49)	(−1.58)
Mexico						
GARCH	1.43	0.71	0.075	3.22	5.12	5.86
	(0.75)	(0.21)				
GJR	−0.11	1.09	0.147	2.96	4.92	4.94
	(1.07)	(0.30)		(3.51)	(4.30)	(1.79)
EGARCH	−1.91	1.61	0.156	3.01	4.79	4.82
	(1.45)	(0.42)		(3.22)	(5.12)	(2.06)
Korea						
GARCH	0.67	0.94	0.153	7.18	8.05	4.23
	(1.29)	(0.25)				
GJR	−0.15	1.07	0.180	6.99	7.95	3.97
	(1.59)	(0.29)		(2.34)	(2.79)	(1.62)
EGARCH	0.09	1.01	0.171	7.09	8.05	3.70
	(1.53)	(0.27)		(0.36)	(−0.04)	(1.73)
Malaysia						
GARCH	1.56	0.70	0.060	10.53	6.27	10.79
	(0.67)	(0.20)				
GJR	2.11	0.61	0.043	10.83	6.17	10.18
	(0.51)	(0.16)		(−1.17)	(1.49)	(0.69)
EGARCH	0.78	0.95	0.054	10.42	5.90	9.84
	(0.57)	(0.22)		(0.53)	(3.87)	(1.24)
Philippines						
GARCH	1.20	0.62	0.033	2.26	3.84	22.13
	(0.43)	(0.14)				
GJR	0.63	0.81	0.037	2.23	3.75	21.54
	(0.45)	(0.16)		(1.09)	(4.21)	(1.18)
EGARCH	−0.21	1.11	0.047	2.25	3.67	19.53
	(0.62)	(0.23)		(0.12)	(4.07)	(1.03)
Taiwan						
GARCH	0.36	0.85	0.076	0.48	3.90	3.72
	(0.33)	(0.11)				
GJR	0.28	0.88	0.086	0.48	3.88	3.91
	(0.35)	(0.11)		(1.07)	(2.09)	(−0.63)
EGARCH	0.40	0.84	0.086	0.48	3.91	3.74
	(0.42)	(0.13)		(0.55)	(−0.43)	(−0.11)
Thailand						
GARCH	1.33	0.73	0.094	3.31	6.37	8.97
	(0.69)	(0.15)				
GJR	0.83	0.84	0.097	3.27	6.24	10.19
	(0.69)	(0.16)		(0.72)	(4.52)	(−1.00)
EGARCH	0.01	1.03	0.096	3.28	6.15	10.75
	(0.83)	(0.19)		(0.29)	(4.04)	(−1.08)

Notes: The table presents results from the evaluation of one-step ahead out-of-sample volatility forecasts. The second to fourth columns contain estimates of the regression (7), where figures in brackets below b_j, $j = 0, 1$ are heteroskedasticity-consistent standard errors. Figures in brackets below MSPE, MAE and HMSPE in columns 5-7 are Diebold-Mariano statistics of equal forecast accuracy, comparing the relevant model with the GARCH model, where negative values indicate that the GARCH model is more accurate. The forecast evaluation period covers January 2, 1993-December 31, 2002 ($R \approx 2600$).

according to the corresponding Diebold-Mariano statistic. This is confirmed by the results from the regression (7), which show that the estimates of b_0 and b_1 generally are closer to their target values for the nonlinear models. Note though that in many cases the forecasts appear to be biased as $\hat{b}_0$ and $\hat{b}_1$ are significantly different from 0 and 1, respectively. The R^2 from this regression is in the typical range between 0.05 and 0.15, confirming the results of (Andersen and Bollerslev, 1998). Finally, it is difficult to distinguish between the GJR and EGARCH models based on these results, in the sense that neither model dominates the other, although the EGARCH model appears to perform slightly better on average, also based on the results from the regression (7).

The out-of-sample forecast evaluation results discussed so far depend on the use of the squared innovation $\hat{\varepsilon}^2_{t+1}$ as a proxy for the true unobserved volatility h_{t+1}. As discussed in the previous section, this proxy is known to be inaccurate and, hence, the results from Table 2 need to be interpreted with caution. In the following we evaluate the volatility forecasts indirectly by means of probability forecasts. As one of the main uses of volatility forecasts is constructing estimates of Value-at-Risk [VaR], defined as the minimum return which is not exceeded with a certain given probability, we consider probability forecasts which resemble VaR measures, see also (Lopez, 1999). In particular, we consider the event that the return at day $t+1$ is smaller than the VaR at 10%, 5% and 1%. In terms of the lower and upper bounds $L_{t+1|t}$ and $U_{t+1|t}$ in (8), we set $L_{t+1|t} = -\infty$ and $U_{t+1|t}$ equal to the 0.10, 0.05 or 0.01 quantile of the conditional distribution of r_{t+1}. Table 3 presents QPS-values for the corresponding probability forecasts and the associated Diebold-Mariano statistics. Again, the GJR and EGARCH models outperform the linear GARCH model, where the EGARCH model seems to have a slight advantage relative to the GJR model. Note that the number of significant differences in QPS values declines as the quantile that is considered becomes smaller, that is the difference between the different GARCH models becomes less pronounced as we move further into the left tail of the conditional distribution of r_{t+1}.

Table 3. Out-of-Sample Probability Forecast Evaluation, January 1993-December 2002

	10%	5%	1%
Argentina			
GARCH	0.153	0.091	0.033
GJR	0.151	0.080	0.030
	(1.00)	(3.42)	(2.24)
EGARCH	0.142	0.085	0.030
	(2.88)	(1.51)	(1.29)
Brazil			
GARCH	0.163	0.083	0.027
GJR	0.159	0.078	0.024
	(1.34)	(1.81)	(1.63)
EGARCH	0.152	0.080	0.025
	(2.66)	(0.69)	(1.00)
Chile			
GARCH	0.147	0.076	0.026
GJR	0.147	0.077	0.025
	(0.00)	(−0.58)	(1.00)
EGARCH	0.143	0.077	0.025
	(1.40)	(−0.30)	(0.58)
Mexico			
GARCH	0.176	0.096	0.036
GJR	0.168	0.092	0.030
	(1.72)	(1.13)	(1.94)
EGARCH	0.164	0.090	0.030
	(2.54)	(1.48)	(2.00)
Korea			
GARCH	0.162	0.095	0.027
GJR	0.154	0.087	0.023
	(2.42)	(2.53)	(1.67)
EGARCH	0.153	0.084	0.023
	(2.22)	(2.61)	(1.21)
Malaysia			
GARCH	0.141	0.086	0.031
GJR	0.133	0.079	0.028
	(1.95)	(2.30)	(1.27)
EGARCH	0.122	0.071	0.022
	(4.54)	(4.02)	(3.00)
Philippines			
GARCH	0.157	0.090	0.030
GJR	0.152	0.084	0.028
	(1.88)	(2.18)	(0.82)
EGARCH	0.153	0.081	0.028
	(1.15)	(2.50)	(0.71)
Taiwan			
GARCH	0.153	0.094	0.030
GJR	0.154	0.087	0.025
	(−0.69)	(2.36)	(2.12)
EGARCH	0.153	0.080	0.023
	(−0.18)	(3.09)	(2.33)
Thailand			
GARCH	0.153	0.080	0.027
GJR	0.146	0.077	0.026
	(2.40)	(0.94)	(1.41)
EGARCH	0.135	0.075	0.027
	(4.54)	(1.46)	(0.45)

Notes: The table presents results from the evaluation of one-step ahead out-of-sample probability forecasts, for the event that the return at day $t+1$ is smaller than the VaR at 10%, 5% and 1%, defined as the minimum return which is not exceeded with the specified probability. The table entries are the values of the QPS (9), where figures in brackets are Diebold-Mariano statistics of equal forecast accuracy as given in (10), comparing the relevant model with the GARCH model, where negative values indicate that the GARCH model is more accurate. The forecast evaluation period covers January 2, 1993-December 31, 2002 ($R \approx 2600$).

6. CONCLUSIONS

In this paper we have examined the usefulness of nonlinear GARCH models for forecasting daily volatility in a number of Asian and Latin American emerging equity markets. Two of the most popular nonlinear GARCH specifications, the GJR model and the EGARCH model, were shown to outperform a linear GARCH model in terms of one-day ahead out-of-sample volatility forecasts. This conclusion holds both when volatility forecasts are evaluated directly by means of traditional criteria (such as MSPE) that rely upon a proxy for unobserved volatility or by means of indirect probability forecasts.

Topics for future research include examining the forecasting performance of alternative nonlinear GARCH specifications and stochastic volatility models as well as the performance of nonlinear GARCH models for other emerging equity markets. Evaluating and comparing volatility forecasts by means of techniques recently developed by (Diebold *et al.*, 1998), among others, is also of interest but beyond the scope of this paper.

REFERENCES

Aggarwal, R., C. Inclan and R. Leal (1999). Volatility in emerging stock markets. *Journal of Financial and Quantitative Analysis* **34**, 33–55.

Andersen, T.G. and T. Bollerslev (1998). Answering the critics: Yes, standard volatility models do provide accurate volatility forecasts. *International Economic Review* **39**, 885–905.

Bekaert, G. and C.R. Harvey (1997). Emerging equity market volatility. *Journal of Financial Economics* **43**, 29–78.

Bekaert, G. and C.R. Harvey (2003). Emerging markets finance. *Journal of Empirical Finance* **10**, 3–55.

Bekaert, G., C.B. Erb, C.R. Harvey and T.E. Viskanta (1998). Distributional characteristics of emerging market returns and asset allocation. *Journal of Portfolio Management* **24**, 102–116.

Bollerslev, T. (1986). Generalized autoregressive conditional heteroscedasticity. *Journal of Econometrics* **31**, 307–327.

Bollerslev, T., R.Y. Chou and K.F. Kroner (1992). ARCH modeling in finance: a review of the theory and empirical evidence. *Journal of Econometrics* **52**, 5–59.

Diebold, F.X., T.A. Gunther and A.S. Tay (1998). Evaluating density forecasts with applications to financial risk management. *International Economic Review* **39**, 863–883.

Engle, R.F. (1982). Autoregressive conditional heteroscedasticity with estimates of the variance of United Kingdom inflation. *Econometrica* **50**, 987–1007.

Franses, P.H. and D. van Dijk (2000). *Nonlinear Time Series Models in Empirical Finance.* Cambridge University Press. Cambridge.

Glosten, L.R., R. Jagannathan and D.E. Runkle (1993). On the relation between the expected value and the volatility of the nominal excess return on stocks. *Journal of Finance* **48**, 1779–1801.

Lopez, J.A. (1999). Mthods for evaluating Value-at-Risk estimates. Federal Reserve Bank of San Francisco *Economic Review* pp. 3–17.

Lopez, J.A. (2001). Evaluating the predictive accuracy of volatility models. *Journal of Forecasting* **20**, 87–109.

Nelson, D.B. (1991). Conditional heteroskedasticity in asset returns: a new approach. *Econometrica* **59**, 347–370.

Richards, A.J. (1996). Volatility and predictability in national markets: how do emerging and mature markets differ?. *IMF Staff Papers* **43**, 461–501.

Santis, G. De and S. Imrohoroğlu (1997). Stock returns and volatility in emerging stock markets. *Journal of International Money and Finance* **16**, 561–579.

IFAC Publications
www.elsevier.com/locate/ifac

STOCHASTIC PROPERTIES OF MULTIVARIATE TIME SERIES EQUATIONS WITH EMPHASIS ON ARCH

Anders Rahbek

* *Department of Applied Mathematics and Statistics, University of Copenhagen, Universitetsparken 5, DK-2100 Copenhagen Ø, Denmark*

Abstract: Markov chain theory is applied to the nonlinear modelling of conditional variance with focus on the in financial econometrics widely applied class of multivariate autoregressive conditional heteroscedastic (ARCH) processes. The multivariate socalled BEKK-ARCH of Engle and Kroner (1995) as well as other multivariate ARCH processes in the literature are discussed. The results show that an essential regularity condition for the existence of moments is that the largest modulus of the eigenvalues or equivalently, that the spectral radius of a certain matrix Φ parametrizing the conditional heteroscedasticity in the ARCH process is smaller than one. Due to the fact that multivariate systems are considered it is demonstrated that an important step in the derivations is based on changing the measure of size of the matrix Φ from norm to spectral radius. *Copyright © 2003 IFAC*

Keywords: Multivariate ARCH, Nonlinear processes, Geometric Ergodicity, Spectral Radius, Markov Chain, Drift Criteria, Asymptotics.

1. INTRODUCTION

Nonlinearities in multivariate discrete time series models in financial econometrics fundamentally appear in the conditional mean and the conditional variance. Stochastic properties of these such as stationarity, ergodicity and existence of moments of the processes are vital for understanding the dynamics of the process as well as for obtaining appropriate asymptotic theory for obtained extremum estimators of the parameters in the models. Solutions, and hence stochastic properties, of linear multivariate time series such as the vector autoregressive (VAR) processes can be derived explicitly based on well-known recursions of the system. In the VAR case the stochastic properties are fully characterised by the eigenvalues of the matrix parametrising the conditional mean. This includes stationary solutions, but also the case of non-stationary behaviour such as error or equilibrium correction encountered in cointegration analysis. However, in general nonlinear multivariate processes do not have an explicit solution and instead properties of the solutions to the equations as proposed here can be derived from general Markov chain theory.

Specifically, this paper introduces an operational drift criterion from Rahbek *et al.* (2002) based on Markov chain theory in Meyn and Tweedie (1993). Under regularity conditions the drift criterion implies not only stationarity and ergodicity, but also importantly existence of appropriate moments for the process considered. In addition it implies that limit theory such as laws of large numbers and classic and functional central limit theory holds, see Rahbek *et al.* (2002). The drift criterion provides sufficient conditions for the mentioned properties but also necessary conditions for existence of moments are discussed.

The theory is applied to the modelling of conditional variance with focus on the in financial econometrics widely applied class of multivariate autoregressive conditional heteroscedastic (ARCH) processes. The multivariate ARCH class originally introduced was the socalled BEKK-ARCH of Engle and Kroner (1995) which is studied here but also other multivariate ARCH processes in the literature are discussed: The BEKK-GARCH (as opposed to BEKK-ARCH) of Engle and Kroner (1995) for which asymptotic inference in the stationary case is treated in Comte and Lieberman (2001); the constant correlation ARCH process suggested in Bollerslev (1990) and for which inference is discussed in Jeantheau (1998); and the diagonal ARCH process. The results show that an essential regularity condition for the existence of second order moments is that the largest modulus of the eigenvalues or equivalently, that the spectral radius of a certain matrix Φ parametrizing the conditional heteroscedasticity in the ARCH process is smaller than one. Due to the fact that multivariate systems are considered it is demonstrated that an essential step in the derivations is based on changing the measure of size of the matrix Φ from norm to spectral radius. The fact that Φ – or any matrix – can have an arbitrarily large norm but small spectral radius is fundamental for the difference between multivariate nonlinear (ARCH) processes and, say, univariate (ARCH) processes.

The general idea of using drift criteria to show existence of moments as well as geometric ergodicity for non linear processes is by now well-established in the econometric literature; see for instance Carrasco and Chen (2002) for a recent application to univariate GARCH processes. With respect to the multivariate BEKK-ARCH processes it appeared initially in Hansen and Rahbek (1998), while the applied idea of drift criteria similar to Hansen and Rahbek (1998) appears in Tjøstheim (1990) derived under different assumptions.

Some notation is used throughout: First regarding norms, with x a vector in $\mathbb{R}^n$, $\|x\|$ denotes the euclidean norm, $\|x\|^2 = \Sigma_{i=1}^n x_i^2$. Denote the space of linear maps from $\mathbb{R}^n \to \mathbb{R}^n$ by $L(\mathbb{R}^n)$. Then with A an $n\times n$ matrix, A represents an element in $L(\mathbb{R}^n)$ and has norm, $\|A\|$ which is chosen as $\|A\|^2 = tr(A'A)$. The spectral radius $\rho(A)$ is the maximal modulus of the eigenvalues of A, noting $\|A\| \geq \rho(A)$. At one or two occasions the operator norm on matrices, as defined by $\|A\|_{\text{op}} := \sup_{\|x\|\neq 0} \frac{\|Ax\|}{\|x\|} = \sqrt{\rho(A'A)}$, is used. Also note that $\|A\|_{\text{op}} \geq \rho(A)$, with equality provided A is symmetric. For linear mappings $\phi : L(\mathbb{R}^n) \to L(\mathbb{R}^m)$ use the operator norm defined by $\|\phi\| := \sup_{\|X\|\neq 0} \frac{\|\phi(X)\|}{\|X\|}$. As an example, $\phi = (A \otimes A)$ is a linear map from $L(\mathbb{R}^n) \to L(\mathbb{R}^n)$ defined by $(A \otimes A)(X) = AXA'$ for X in $L(\mathbb{R}^n)$. Note that the latter definition closely resembles the well-known practice of using the $vec(\cdot)$ operator.

2. DRIFT CRITERION

As mentioned Markov chain theory and drift criteria applied to ARCH processes from Rahbek *et al.* (2002) is used here. For applications of drift criteria on multivariate chains in general see inter alia Feigin and Tweedie (1985), Pham (1986), Rahbek and Shephard (2002) and Tjøstheim (1990).

Let $\mathbf{X}_0, \mathbf{X}_1, \mathbf{X}_2, \ldots$ be a time-homogenous Markov process on the state space $\mathbb{R}^q$ induced with the Borel σ-algebra $\mathbb{B}^q$.

Assumption 1. Consider the time-homogenous Markov chain $\mathbf{X}_t$ on $(\mathbb{R}^q, \mathbb{B}^q)$. Assume that for all sets $A \in \mathbb{B}^q$ and for some integer $m \geq 1$, that the m step transition density with respect to the Lebesgue measure, $f(\cdot|\cdot)$ as defined by

$$P(\mathbf{X}_t \in A|\, \mathbf{X}_{t-m} = x) = \int_A f(y|x)dy$$

is strictly positive and bounded on compact sets.

This assumption is made in order to circumvent some of the difficulties of Markov chain theory on general state spaces. From an abstract Markov chain viewpoint, Assumption 1 is strong. Nonetheless, it is frequently satisfied in the context of multivariate time-series as illustrated below. An important implication of Assumption 1 is as shown in Rahbek *et al.* (2002) that the Markov chain is *irreducible* with respect to the Lebesgue measure μ, it is *aperiodic* and compact sets $C \subset \mathbb{R}^q$ are *small*. Given these regularity conditions are satisfied, the (k-step) drift criterion from Hansen and Rahbek (1998) can be applied. A drift function v is any function defined on the state space $\mathbb{R}^q$ which takes values in $\mathbb{R}$ greater than or equal to one (but not identically ∞). An example of a typical choice of drift function is

$$v(x) = 1 + \|x\|^2. \tag{1}$$

Theorem 1. Let $(\mathbf{X}_t)_{t=0,1,..}$ be a time homogeneous Markov chain which satifies Assumption 1. Let $v : \mathbb{R}^p \mapsto [1,\infty]$ be some drift function. Assume there exists an integer $k \geq 1$, a compact set $C \subset \mathbb{R}^p$ and constants $0 < \gamma < 1$, $g > 0$ such that

$$E\left(v(\mathbf{X}_{t+k})|\, \mathbf{X}_t = x\right) \leq \gamma v(x) \tag{2}$$

for x in C^c, while $E\left(v(\mathbf{X}_{t+k})|\, \mathbf{X}_t = x\right)$ is bounded by g on C.

Then $\mathbf{X}_t$ is geometrically ergodic and $\mathbf{X}_0$ can be given an initial distribution such that $\mathbf{X}_t$ is stationary. Furthermore, all moments bounded by $v(\cdot)$ exist.

3. BEKK-ARCH

Consider the BEKK-ARCH process of Engle and Kroner (1995) as given by

$$X_t = \Omega_t^{1/2}\varepsilon_t, \tag{3}$$

$$\Omega_t = \Omega + \Sigma_{i=1}^m \Sigma_{j=1}^{\tilde{m}} A_{ij} X_{t-i} X_{t-i}' A_{ij}', \tag{4}$$

where ε_t is assumed to be $p-$dimensional i.i.d. with mean zero, variance I_p and positive continuous density $f(\cdot)$. As a simple and natural example ε_t can be i.i.d. Gaussian, $N(0, I_p)$. The variance parameter Ω is symmetric and positive definite and the A_{ij} matrices are real $p \times p$ matrices.

With X_t defined in (3) consider the induced Markov chain

$$\mathbf{X}_t = (\mathbf{X}_t', \mathbf{X}_{t-1}', ..., \mathbf{X}_{t-m+1}')' \quad (5)$$

on $\mathbb{R}^q$, $q = pm$ endowed with the Borel σ-algebra. By the factorization of the density of X_{t+m} conditional on $\mathbf{X}_t$,

$$g(\mathbf{X}_{t+m}|\, \mathbf{X}_t) = \Pi_{i=1}^m f(\mathbf{X}_{t+i}|\, \mathbf{X}_{t-1+i,...,}\mathbf{X}_{t-m+i}), \quad (6)$$

the m-step transition density is continuous and positive as ε_t is assumed to be i.i.d. $(0, I_p)$. Hence Assumption 1 holds and Theorem 1 can be applied.

Using the drift function in (1), sufficient conditions for stationarity and existence of second order moments are given in Theorem 2 below, while necessity is discussed seperately in Theorem 3 afterwards. Finally, fourth and higher order moments are discussed in connection with Theorem 4.

Theorem 2. Consider the BEKK-ARCH process X_t defined in (3). Then a sufficient condition for geometric ergodicity, stationarity and existence of second order moments of X_t is that $\rho(\Phi) < 1$. Here the $mp^2 \times mp^2$ matrix Φ is defined by

$$\begin{aligned} &\Phi = (\Phi_{ij})_{i,j=1,2} \quad (7) \\ &\Phi_{11} = \\ &\quad \Sigma_{j=1}^{\tilde{m}}(A_{1j} \otimes A_{1j}) \cdots \Sigma_{j=1}^{\tilde{m}}(A_{m-1j} \otimes A_{m-1j}), \\ &\Phi_{12} = \Sigma_{j=1}^{\tilde{m}}(A_{mj} \otimes A_{mj}), \\ &\Phi_{21} = (I_{p(m-1)} \otimes I_{p(m-1)}) \\ &\Phi_{22} = (0_{p(m-1)\times p} \otimes 0_{p(m-1)\times p}). \end{aligned}$$

Before turning to the proof note that in the case where $m = \tilde{m} = 1$, then $\Phi = (A_{11} \otimes A_{11})$, while if $m = 2$ and $\tilde{m} = 1$,

$$\Phi = \begin{pmatrix} A_{11} \otimes A_{11} & A_{21} \otimes A_{21} \\ I_p \otimes I_p & 0 \otimes 0 \end{pmatrix}$$

Proof of Theorem 2 (Rahbek *et al.*, 2002): As noted above the focus is on the role of the measure of matrix size by spectral radius and norm respectively.

For an element X in $L(\mathbb{R}^{pm})$ identify it with the $pm \times pm$ matrix divided into m^2 blocks of size $p \times p$, $X = (X_{ik})_{i,k=1,..,m}$ and define the important mapping $\phi : L(\mathbb{R}^{pm}) \to L(\mathbb{R}^{pm})$ by

$$\begin{aligned} &\phi(X) = (\phi_{ij})_{i,j=1,2} \quad (8) \\ &\quad \phi_{11} = \Sigma_{i=1}^m \Sigma_{j=1}^{\tilde{m}} A_{ij} X_{ii} A_{ij}', \ \phi_{12} = \phi_{21} = 0 \text{ and} \\ &\quad \phi_{22} = (I_{p(m-1)}, 0) X (I_{p(m-1)}, 0)' \end{aligned}$$

With this definition turn to sufficiency of the condition $\rho(\phi) < 1$: Applying the drift function $v(\cdot)$ in (1) along with the definition of x_t in (5) immediately leads to

$$\begin{aligned} &E(v(\mathbf{X}_{t+1})|\, \mathbf{X}_t) \\ &= 1 + E\left(tr\left\{\mathbf{X}_{t+1}\mathbf{X}_{t+1}'\right\}\middle|\, \mathbf{X}_t\right) \\ &= 1 + tr\left\{\phi(\mathbf{X}_t\mathbf{X}_t')\right\} + tr\{\tilde{\Omega}\} \end{aligned}$$

where $\tilde{\Omega}$ =blockdiag$(\Omega, 0_{p(m-1)\times p(m-1)})$. Clearly this can be used inductively by successive conditioning to obtain

$$\begin{aligned} &E(v(\mathbf{X}_{t+k})|\, \mathbf{X}_t = x) \\ &= 1 + E\left(tr\left\{\mathbf{X}_{t+k}\mathbf{X}_{t+k}'\right\}\middle|\, \mathbf{X}_t = x\right) \quad (9) \\ &= 1 + tr\left\{\phi^k(xx')\right\} + \Sigma_{i=0}^{k-1} tr\{\phi^i(\tilde{\Omega})\}. \quad (10) \end{aligned}$$

The central argument for (2) to hold concerns finding an upper bound for the term $tr\left\{\phi^k(xx')\right\}$. In terms of the operator norm

$$\begin{aligned} &tr\left\{\phi^k(xx')\right\} \\ &= \left|tr\left\{\phi^k(xx')\right\}\right| \leq \kappa \left\|\phi^k(xx')\right\| \\ &\leq \kappa \left\|\phi^k\right\| \|xx'\| = \kappa \left\|\phi^k\right\| \|x\|^2 \end{aligned}$$

for some positive constant κ. It has been used that $\phi(xx')$, and hence $\phi^k(xx')$, are symmetric and positive semidefinite.

Obviously, in the first step, where $k = 1$, $tr\{\phi(xx')\}$ is bounded from above by $\gamma \|x\|^2$ by choosing the norm of ϕ small. However, this will lead to a too strong restriction on the parameters. Instead the relevant measure for ϕ is the spectral radius

$$\lim_{k\to\infty} \left\|\phi^k\right\|^{1/k} = \rho(\phi),$$

see Pedersen (1988, Theorem 4.1.13). Hence if $\rho(\phi) < 1$, then for some k large enough the norm of ϕ^k is arbitrarily small irrespectively of the size of $\|\phi\|$.

The proof of sufficiency is now concluded by choosing k such that $\kappa \left\|\phi^k\right\| = \tilde{\gamma} < 1$. Next, define $c(\tilde{\gamma}) = 1 - \tilde{\gamma} + \Sigma_{i=0}^{k-1} tr\{\phi^i(\tilde{\Omega})\}$ and choose $\tilde{\gamma} < \gamma < 1$. Then

$$E(v(\mathbf{X}_{t+k})|\, \mathbf{X}_t = x) \leq \gamma v(x) \ \text{ for } x \in K^c$$

where the compact set C is defined by

$$C = \{x|\, v(x) \leq c(\tilde{\gamma})/(\gamma - \tilde{\gamma})\}.$$

For computational purposes note that $\rho(\phi) = \rho(C_0 \otimes C_0 + \Sigma_{i=1}^m \Sigma_{j=1}^{\tilde{m}}(C_{ij} \otimes C_{ij}))$ where the $pm \times pm$ matrices C_{ij} are given by

$$C_{ij} = \begin{pmatrix} 0 \cdots 0 \ A_{ij} \ 0 \cdots 0 \\ 0 \cdots \quad\quad \cdots 0 \end{pmatrix}$$

for $i = 1, 2, .., m$ and $j = 1, 2, ..., \tilde{m}$, such that C_{ij} is the zero matrix except for the entries in the first m rows and columns i to $i + m$ which contain the A_{ij} matrix. C_0 is the zero matrix except for the lower left corner which contains the $p(m-1)$-dimensional identity matrix. Finally note that

$$\rho(C_0 \otimes C_0 + \Sigma_{i=1}^m \Sigma_{j=1}^{\tilde{m}}(C_{ij} \otimes C_{ij})) < 1$$

is equivalent to $\rho(\Phi) < 1$ with Φ defined in (7). This ends the proof of Theorem 2. □

Next turn to necessity of the conditions for existence of second order moments:

Theorem 3. Consider the the BEKK-ARCH process $\mathbf{X}_t$ defined in (3). For $\mathbf{X}_t$ to have a finite and well-defined finite second order moment, $E(\mathbf{X}_t\mathbf{X}_t') = \Sigma$, and thereby be weakly stationary, it is necessary that the spectral radius $\rho(\Phi|_{sym}) < 1$. The spectral radius $\rho(\Phi|_{sym})$ is the maximal absolute value of the eigenvalues of Φ corresponding to (the vec) of symmetric matrices, while Φ is defined in Theorem 2.

The proof is given in the Appendix.

Recall that in the case of a BEKK-ARCH(1) process with $m = \tilde{m} = 1$ Φ will be identical to $(A \otimes A)$ for some A and note that in this case $\rho(\Phi|_{sym}) = \rho(\Phi)$.

Next, turn to the existence of fourth order moments:

Theorem 4. Consider the BEKK-ARCH process X_t defined in equation (3) with $m = \tilde{m} = 1$. Then a sufficient condition for the existence of fourth order moments of X_t is that $\rho(\Phi) < 1/\sqrt{3}$, where Φ is defined in Theorem 2.

Proof of Theorem 4: Define the drift function

$$v(x) = 1+(x\otimes x)'(x\otimes x) = 1+(x'x)^2 = 1+\|x\|^4,$$

and denote A_{11} by A. It follows that

$$E\left(v(X_t)|\,X_{t-1}=x\right) = 1+2tr\left\{\Omega_x\Omega_x\right\}+tr^2\left\{\Omega_x\right\} \tag{11}$$

where $\Omega_x = \Omega + Axx'A'$ and, with Q any positive definite matrix, the identity,

$$Var\left(tr\left\{\varepsilon_t\varepsilon_t'Q\right\}\right) = 2tr\left\{QQ\right\},$$

has been used. Ignoring terms of lower order than $\|x\|^4$, the right hand side of (11) is given by

$$\begin{aligned}&2tr\left\{Axx'A'Axx'A'\right\}+tr^2\left\{Axx'A'\right\}\\&=3\left(x'A'Ax\right)^2.\end{aligned}$$

Simple recursion shows that $E\left(v(X_{t+k})|\,X_t=x\right)$ apart from lower order terms is given by

$$3^k\left(x'A^{k\prime}A^kx\right)^2 \leq \left\|\left(3^{1/4}\right)^kA^k\right\|^4\|x\|^4, \tag{12}$$

and hence the condition that $\rho(A\otimes A) < \frac{1}{\sqrt{3}}$ emerges as in the proof of Theorem 2. □

Regarding generalizations of Theorem 4, note first as a conjecture that the result in Theorem 4 holds for general m and $\tilde{m}$ by applying similar arguments as in the proof above. As to existence of moments of higher order than fourth order, note that the ARCH process may be represented as a multivariate random coefficient autoregressive process. In terms of this representation conditions for existence of these higher order moments can be stated as in Feigin and Tweedie (1985, Theorem 5) where a different drift function is applied.

3.1 *Other ARCH processes*

Asymptotic theory for the BEKK-GARCH process suggested in Engle and Kroner (1995) has recently been studied in Comte and Lieberman (2001). The BEKK-GARCH is given by a natural extension of the conditional variance in (4) as

$$\begin{aligned}&\xi_t = \Omega_t^{1/2}\varepsilon_t,\ \Omega_t = \\&\quad\Omega+\Sigma_{i=1}^m\Sigma_{j=1}^{\tilde{m}}(A_{ij}\xi_{t-i}\xi_{t-i}'A_{ij}'+B_{ij}\Omega_{t-i}B_{ij}')\end{aligned} \tag{13}$$

where as before ε_t is i.i.d. $(0, I_p)$, the variance parameter Ω is symmetric and positive definite and A_{ij} and B_{ij} are real $p\times p$ matrices. Consider Φ as given by (7) but with each entry $(A_{ij}\otimes A_{ij})$ replaced by

$$(A_{ij}\otimes A_{ij})+(B_{ij}\otimes B_{ij}).$$

One finds, provided the drift criterion can be applied, that $\rho(\Phi)$ should be smaller than one for existence of second order moments. In this case ξ_t will also be geometrically ergodic. Comte and Lieberman (2001) quote results in Boussama (1998) where in fact a similar condition appears for geometric ergodicity.

As for the diagonal ARCH process in Li, Ling and Wong (2001) and diagonal GARCH process used for simulations in Boswijk, Lucas and Taylor (2000) these are straightforward examples of the BEKK-ARCH and BEKK-GARCH processes. Specifically, Ω_t in (4) and (13) is assumed to be diagonal with each element on the diagonal, σ_t^{ii} say, a univariate ARCH or GARCH process respectively.

Finally, Jeantheau (1995) and Ling and McAleer (2003) study the Constant Correlation GARCH process suggested in Bollerslev (1990). In this case the p-dimensional conditional variance in (13) is given by the entries

$$(\Omega_t)_{ij} = \begin{cases}\sigma_{it}\sigma_{jt}\sigma_{ij}^2 & \text{for } i\neq j\\ \sigma_{it}^2 & \text{for } i=j\end{cases}$$

where σ_{ij}^2 is strictly positive and σ_{it}^2 is a univariate GARCH process

$$\sigma_{it}^2 = \sigma_i^2+\Sigma_{j=1}^m(a_{ij}\xi_{t-j}^2+b_{ij}\sigma_{it-j}^2),$$

where a_{ij} and b_{ij} are greater than or equal to zero. Note that by definition the conditional correlation between ξ_{it} and ξ_{jt} is given by σ_{ij}^2. Define the $p\times p$ diagonal matrices A_j =diag$(a_{ij})_{i=1,\dots,p}$ and B_j =diag$(b_{ij})_{i=1,\dots,p}$ for $j = 1,..,m$. Then the Φ for which $\rho(\Phi) < 1$ is the relevant condition can be written in the form

$$\begin{aligned}\Phi &= \begin{pmatrix}\Phi_0 & (A_m+B_m)\\ I_{p(m-1)\times p(m-1)} & 0_{p(m-1)\times p}\end{pmatrix}\\ \Phi_0 &= (A_1+B_1)\cdots(A_{m-1}+B_{m-1}).\end{aligned}$$

For the case of no lagged variances, i.e. all $B_j = 0$, the Constant Correlations ARCH process, or rather the implied Markov chain, satisfies the regularity conditions for applying our drift criterion in Theorem 1. Applying simple algebra will lead to the condition that $\rho(\Phi) < 1$ for the just given Φ matrix. Due to the nature of the state space inherited from the univariate specification of the $\sigma_{it}'s$ the drift criterion can also be applied for the case of Constant Correlation GARCH, see also Carrasco and Chen (2002) and Ling and McAleer (2003).

4. REFERENCES

Bollerslev, T. (1990). Modelling the Coherence in Short-Run Nominal Exchange Rates: A Multivariate Generalized ARCH Model. *Review of Economics and Statistics,* **72**, 498-505.

Boswijk, H.P. and A.Lucas (2002). Semi-Nonparametric Cointegration Testing. *Forthcoming in Journal of Econometrics.*

Boussama, F. (1998). Ergodicity, Mixing and Estimation in Multivariate GARCH Processes. *Ph.D. dissertation, University Paris 7.*

Carrasco, M. and X.Chen (2002). Mixing and Moment Properties of Various GARCH and Stochastic Volatility Models. *Econometric Theory,* **18**, 17-39.

Comte, F. and O.Lieberman (2001). Asymptotic Theory for Multivariate GARCH Processes. *Working paper, CREST-ENSAE.*

Engle, R.F. and K.F. Kroner (1995). Multivariate Simultaneous Generalized ARCH. *Econometric Theory,* **11**, 122-150.

Feigin, P.D. and R.L.Tweedie (1985). Random Coefficient Autoregressive Processes: A Markov Chain Analysis of Stationarity and Finiteness of Moments. *Journal of Time Series Analysis,* **6**, 1-14.

Hansen, E. and A. Rahbek (1998). Stationarity and Asymptotics of Multivariate ARCH Time Series with an Application to Robustness of Cointegration Analysis. *Preprint, Department of Theoretical Statistics, University of Copenhagen. CAF working paper series no.22.*

Jeantheau, T. (1998). Strong Consistency of Estimators for Multivariate GARCH Models. *Econometric Theory,* **14**, 70-86.

Li, W.K., S.Ling and H.Wong (2001). Estimation for Partially Nonstationary Multivariate Autoregressive Models with Conditional Heteroscedasticity. *Biometrika,* **88**, 1135-1152.

Ling, S. and M. McAleer (2003). Asymptotic Theory for a Vector ARMA-GARCH Model. *Econometric Theory*, **19**, 280-310.

Meyn, S.P. and R.L. Tweedie (1993). *Markov Chains and Stochastic Stability. Communications and Control Engineering Series, Springer-Verlag.*

Pedersen, G.K. (1988). *Analysis Now, Springer-Verlag.*

Pham, D.T. (1986). The Mixing Property of Bilinear and Generalised Random Coefficient Autoregressive Models. *Stochastic Processes and their Applications,* **23**, 291-300.

Rahbek, A. and N. Shephard (2002). Inference and Ergodicity in the Autoregressive Conditional Root Model. *Working paper, Oxford University.*

Rahbek, A., J.G. Dennis and E. Hansen (2002). ARCH Innovations and their Impact on Cointegration Rank. *Working paper, Department of Statistics and Operations Research, University of Copenhagen.*

Rudin, W. (1991). *Functional Analysis, McGraw-Hill Inc., New York*

Tjøstheim, D. (1990). Non-linear Time Series and Markov Chains. *Advances in Applied Probability,* **22**, 587-611.

Appendix A. PROOF OF THEOREM 3

Proof (Rahbek *et al.*, 2002): Consider first the case of $m = \tilde{m} = 1$. From (3) it follows that $\Sigma = V(\xi_t) = E(\xi_t\xi_t')$ satisfies

$$\Sigma = \phi(\Sigma) + \Omega \quad \text{(A.1)}$$

where the mapping ϕ in this case is defined by $\phi(\Sigma) = A\Sigma A'$, see equation (8) and Ω is positive definite, $\Omega > 0$.

With B some $p \times p$ positive semi-definite matrix, $B \geq 0$, then by Lemma 1 below there exists a small positive real number γ for which

$$\Omega \geq \gamma B.$$

This gives the inequalities

$$\Omega \geq \Omega - \gamma B \geq 0$$

which as ϕ is linear and maps positive semi-definite matrices into positive semi-definite matrices gives

$$\phi^n(\Omega) \geq \phi^n(\Omega) - \gamma\phi^n(B) \geq 0. \quad \text{(A.2)}$$

By iterating in (A.1) it follows that for all m,

$$\Sigma_{n=0}^{m}\phi^n(\Omega) \leq \Sigma.$$

This implies that

$$x'\phi^n(\Omega)x \to 0 \text{ as } n \to \infty \text{ and for all } x \in \mathbb{R}^p$$

and hence, using (A.2),

$$x'\phi^n(B)x \to 0 \text{ as } n \to \infty \text{ and for all } x \in \mathbb{R}^p$$

which means that $\phi^n(B)$ converges to zero in the weak operator topology, see Pedersen (1988, section 4.6.1). By Rudin, (1991, Theorem 1.21) there is only one vector space topology on any finite dimensional vector space and therefore in particular for the operator norm we have,

$$\|\phi^n(B)\| \to 0 \text{ as } n \to \infty. \quad \text{(A.3)}$$

Next, it remains to show that the same holds for any symmetric matrix C. To do so, decompose C as the sum of two positive semi-definite matrices

$$C = B^+ - B^-, \quad B^+ \geq 0, B^- \geq 0.$$

It follows that,

$$\|\phi^n(C)\| = \left\|\phi^n(B^+) - \phi^n(B^-)\right\| \to 0 \quad \text{(A.4)}$$

as $n \to \infty$. The convergence in (A.4) for any symmetric C means that $\phi^n(\cdot)$ when restricted to the space of symmetric matrices converges in the strong operator topology, see Pedersen (1988, Section 4.6.1). Similar to before introducing the operator norm,

$$\left\| \phi|_{sym}^{n} \right\| := \sup_{C symmetric} \frac{\|\phi^n(C)\|}{\|C\|} \tag{A.5}$$

it holds by Rudin (1991, Theorem 1.21) that,

$$\left\| \phi|_{sym}^{n} \right\| < 1 \text{ for some } n \text{ large enough.}$$

Next, using submultiplicativity of the operator norm in (A.5) then by Pedersen (1988, 4.1.13),

$$\begin{aligned} \rho\left(\phi|_{sym}\right) &= \lim_{m\to\infty} \left\| \phi|_{sym}^{m} \right\|^{1/m} \\ &= \lim\ \sup_{m\to\infty} \left\| \phi|_{sym}^{mn} \right\|^{1/mn} \\ &\leq \lim\ \sup_{m\to\infty} \left\| \phi|_{sym}^{n} \right\|^{1/n} < 1 \end{aligned}$$

Identifying $\rho(\Phi)$ with $\rho(\phi)$ as in the proof of Theorem 2 finishes the proof for the case of $m = \tilde{m} = 1$.

For general $m, \tilde{m}$ use that $\tilde{\Sigma} = E(x_t x_t')$ with x_t defined in (5) satisfies

$$\begin{aligned} \tilde{\Sigma} &= \phi(\tilde{\Sigma}) + \tilde{\Omega} = \phi^m(\tilde{\Sigma}) + \Sigma_{i=0}^{m-1}\phi^i(\tilde{\Omega}) \\ &:= \phi^*(\tilde{\Sigma}) + \Omega^* \end{aligned}$$

where ϕ is defined in (8) and Ω^* is positive definite, see (6). Using the arguments from before we see that

$$\|\phi^{mn}\| < 1 \text{ for some } n \text{ large enough.}$$

and therefore that $\rho(\phi|_{sym}) < 1$. □

Lemma 1. Assume that Ω is a symmetric and positive definite $n \times n$ matrix. Then for any $n \times n$ positive semi-definite matrix B there exists a constant $\gamma > 0$ such that

$$\Omega - \gamma B > 0$$

Proof: For any $x \neq 0$ in $\mathbb{R}^n$,

$$\begin{aligned} x'(\Omega - \gamma B)x &= \left(\tfrac{x'\Omega x}{x'x} - \gamma \tfrac{x'Bx}{x'x}\right)x'x \\ &\geq \left[\lambda_{\min}(\Omega) - \gamma\lambda_{\max}(B)\right] \|x\|^2 > 0 \end{aligned}$$

for γ small enough. Here $\lambda_{\min}$ and $\lambda_{\max}$ denote the minimal, respectively maximal, modulus of the eigenvalues of the matrix involved. □

IFAC Publications
www.elsevier.com/locate/ifac

A RATIONAL PROBABILITY DENSITY APPROACH TO STOCHASTIC VOLATILITY ESTIMATION

B. Hanzon *

* *Mathematical Institute, Leiden University, P.O. Box 9512, 2300 RA Leiden, The Netherlands, E-mail:*
`bhanzon@math.leidenuniv.nl`

Abstract: In the area of financial time series the Black-Scholes model is often used. However, it is well-known that although the volatility is assumed to be constant in the Black-Scholes model, in practice it is varying in time. This has led to the investigation of more general models in which the volatility is allowed to vary. In one class of such models the dynamic behavior of volatility is described by some stochastic process. A problem with such models is that it is generally very difficult to solve the volatility estimation problem for such models: the calculation of the conditional density of the volatility at some point in time, given the observations up till that point in time, is usually a difficult task for which there are no closed form expressions. In the present paper a class of models of the same type is presented, which however has the advantage that for these models the volatility estimation problem can be solved exactly. In our models all disturbances have a rational probability density function on the real line.

Keywords: Non-linear filtering, stochastic volatility, rational densities, estimation, realization theory.

1. INTRODUCTION

In the area of financial time series the Black-Scholes model is often used for modelling the behaviour of the price of stocks, exchange rates and other financial time series. This is also the basis for much of the literature on pricing of derivative financial instruments such as options. However it is considered to be a well-known fact that although the volatility is assumed to be constant in the Black-Scholes model, in practice it is varying. This has led to the investigation of more general models in which the volatility is allowed to vary. One can broadly distinguish between two types of generalizations. One is the type of model in which the volatility is varying over time and its dynamic behaviour is described by some stochastic process. A problem with such models is that it is generally difficult to solve the *volatility estimation problem* for such models: the calculation of the conditional density of the volatility at some point in time, given the observations up till that same point in time, is usually a difficult task for which there are no closed form expressions. In the literature there are several proposals to approximate the conditional density, cf. e.g. (A. Harvey, 1994),(Mahieu and Schotman, 1998),(Brigo and Hanzon, 1998).The other, second type of model that is used is the ARCH model and its generalizations ((Engle, 1982), see also e.g. (Gouriéroux, 1997)), as applied to financial time series. These models have the advantage that the volatility is again time varying, and the conditional volatility (also called conditional heteroskedasticity in this context) is in fact prescribed by the model as a deterministic function of the past observations. By construction the problem of estimating the stochastic volatility has been solved in these models. However one could argue that this is at the expense of a less transparent model for the underlying data generating process. In the present paper a model of the first type will be presented, however with the advantage that for this model the volatility estimation problem can be solved, as we will show. Apart from the volatility to be time-varying another feature of financial time series that is often reported is that it has *fat tails*. In the literature there are many studies that try to deal with this phenomenon by specifying non-Gaussian disturbances. This goes back to the work of (Mandelbrot, 1963) who suggested to consider the class of stable distributions as possible distributions for the disturbances. An important example of stable distributions is given by the Cauchy distributions. More recent studies seem to favor other distributions, including Student t-distributions (cf e.g. (J.Y. Campbell, 1997), (Lucas, 1996)). In the approach followed in the present paper all disturbances are allowed which have a *rational probability density function* on the real line. This includes the Cauchy distributions and Student t-distributions with odd number of degrees of freedom. In fact it is well-known that the Gaussian distribution can be approximated by a Student t-distribution of sufficiently high number of degrees of freedom. Therefore in a sense the corresponding Gaussian model is a limiting case of the class of models presented here. It should perhaps be stressed from the start that there is a price to be paid in the form of high complexity if one wants to use rational densities of higher (McMillan) degree. From the point of view of complexity in fact the estimation problem is easiest when the disturbances have Cauchy density. In a previous paper a matrix calculus was developed for performing various calculations with rational probability density functions ((Hanzon and Ober, 2001)) and applied to a filtering problem for a class of linear dynamical models. Here we extend this calculus and show how it can be fruitfully applied to the non-linear filtering problem of volatility estimation, in a specific class of stochastic volatility models. The class of stochastic volatility models considered is reasonably flexible, but the flexibility may have to be weighed against the complexity of the solutions of the estimation problem that results. In the matrix calculus that we use for calculations with rational probability densities the complexity can be measured in terms of the sizes of the matrices that are required to represent the various rational probability density functions. At the same time with the present developments in computer technology the bottleneck in computing is often much more the effort to develop software than the size of the matrices or other data structures. If this is the criterion then the solution presented here is easy because each of the steps involved consists of a relatively simple operation on some matrices and vectors.

2. THE MODEL CLASS

Stochastic volatility models that we will consider are of the following form:

$$\begin{aligned} X_{t+1} &= aX_t + W_t \\ Y_t &= V(X_t)U_t \end{aligned} \tag{1}$$

where for each $t \in T := \{0, 1, 2, \dots\}$, the random variables X_t, W_t, Y_t, U_t take their values in the real numbers, and where $V(x)$ is a real-valued, nonnegative polynomial function of $x \in \mathbf{R}$; $\{W_t,\ t \in T\}$ and $\{U_t,\ t \in T\}$ are sequences of jointly stochastically independent real valued random disturbances with time-invariant probability density functions: for each $t \in T$, W_t has rational probability density function $p_W(w)$, U_t has rational probability density function $p_U(u)$. The initial state X_0 has rational probability density function $p_{X_0}(x)$. The parameter a is a real number that will be assumed to be unequal to zero for ease of exposition. In financial applications, the Y_t usually stand for the returns $Y_t = \log(S_{t+1}/S_t)$ of some price process $\{S_t,\ t \in T\}$.
We consider the following nonlinear filtering problem: Estimate at each time $t \in T$ the volatility $V(X_t)$ from the sequence of observations $Y_0^t := \{Y_s,\ s \in T, s \leq t\}$. The solution of such a problem consists of finding for each $t \in T$ the conditional probability density function of X_t given Y_0^t, and deriving the desired estimate of $V(X_t)$ from this.

A number of remarks can be made about this model class.

(i) The family of rational probability density functions is a very rich class. It contains the stable class of Cauchy densities, it contains the Student distributions with odd number of degrees of freedom. Under relatively mild conditions, probability density functions can be approximated by rational probability density functions, as follows from results of the theory of rational approximation. It is well-known that the Gaussian probability density functions can be approximated for example by the Student t-distribution of sufficiently high number of degrees of freedom, therefore the Gaussian case appears in a certain sense as a limiting case of our model class.

(ii) The function V is a non-negative polynomial, i.e. for all $x \in \mathbf{R}$, $V(x) \geq 0$. Here this is required for technical reasons. In the literature one finds other positive functions as specifications for V as well, for example an exponential function $V(x) = \exp(\frac{x+\gamma}{2})$ (cf. e.g. (Taylor, 1986)). If desired one can approximate the exponential function on any given finite interval by a positive polynomial.

(iii) The parameters in the model as well as in the rational probability density functions of W_t and V_t, $t = 0, 1, 2, \dots$, are assumed to be constants here. However they could be taken time-varying if desired. The resulting filter equations for that case form a straightforward extension of the filter equations presented in this paper.

3. STATE-SPACE CALCULUS FOR RATIONAL PROBABILITY DENSITY FUNCTIONS

In the paper (Hanzon and Ober, 2001) a nonnegative integrable rational density is represented by a triple of complex-valued matrices (A, b, c), where A is a matrix with all eigenvalues in the open left half-plane. Calculations that one wants to perform on rational densities are translated into operations in terms of such triples. A concise summary of the various operations will be given in the presentation. For the application to the problem of stochastic volatility estimation a few more operations are needed. These are the operation which maps a rational function $g(s)$ to the rational function $h(s) = g(s^2)$ and more generally, if $p(s)$ is a polynomial, the operation which maps the rational function $g(s)$ to the rational function $h_p(s) = g(p(s))$. Furthermore it will be necessary to perform the operation of additive decomposition into a stable and an anti-stable part of a rational function, again in terms of associated triples of matrices. This step requires a certain block-diagonalization of a matrix. To make it a bit easier to discuss the notions of rational probability density function and unnormalized rational density function, and to distinguish densities on the real line and on the imaginary axis, we introduce the following notation: Let $\mathcal{U}(\mathbf{R})$ denote the class of unnormalized rational density functions on the real line, not including the zero function, which are nonnegative and integrable on the real line. Let $\mathcal{U}_0(\mathbf{R})$ denote the class of nonnegative rational density functions, i.e. $\mathcal{U}_0(\mathbf{R}) = \mathcal{U}_0(\mathbf{R}) \cup \{0\}$ where 0 here denotes the zero function on the real line. Let $\mathcal{R}(\mathbf{R})$ denote the class of rational probability density functions on the real line, i.e. consisting of all elements of $\mathcal{U}(\mathbf{R})$ which integrate to one. Let $\mathcal{U}(i\mathbf{R})$ denote the class of unnormalized rational density functions on the imaginary axis, not including the zero function, which are nonnegative and integrable on the imaginary axis. Hence $\mathcal{U}(i\mathbf{R}) = \{\mathbf{\Phi} | \exists \rho \in \mathcal{U}(\mathbf{R}) : \forall \mathbf{x} \in \mathbf{R} : \mathbf{\Phi}(\mathbf{ix}) = \rho(\mathbf{x})\}$. And similarly let $\mathcal{U}_0(i\mathbf{R})$ denote the set of all nonnegative, integrable rational density functions on the imaginary axis. Hence this set consists of all the functions in $\mathcal{U}(i\mathbf{R})$ together with the zero function. Let $\mathcal{R}(i\mathbf{R})$ denote the class of rational probability density functions on the imaginary axis, i.e. consisting of all elements of $\mathcal{U}(i\mathbf{R})$ which integrate to one.

Now consider a rational density function $\rho \in \mathcal{U}(\mathbf{R})$. With it we associate a complex rational function Φ which is specified on the imaginary axis by $\Phi(ix) = \rho(x)$ for every $x \in \mathbf{R}$. This function can be considered as the sum $\Phi(s) = Z(s) + Z^*(-s)$ of the complex rational functions $Z(s)$ and $Z^*(-s)$, where $Z(s)$ has all its poles in the open left half plane $\{s | \Re(s) < 0\}$ and Z^* is the complex rational function which is given on the imaginary axis by $Z^*(-ix) = \overline{Z(ix)}$, where $\overline{z}$ denotes the complex conjugate of z for any complex number $z \in \mathbf{C}$. The complex rational function $Z(s)$ can be written as $Z(s) = c(sI - A)^{-1}b$, where the triple (A, b, c) consists, for some positive integer n, of an $n \times n$ matrix A, an $n \times 1$ column vector b and a $1 \times n$ row vector c. The triple (A, b, c) is referred to as a *state-space realization* of Z. (This terminology stems from system theory). The matrix A can be taken such that it has all its eigenvalues in the open left half plane and if A satisfies this property the triple (A, b, c) will be called a *stable* realization. In (Hanzon and Ober, 2001) the following result concerning the normalization constant and the moments of a rational density from $\mathcal{U}$ was presented. Let X be a real random variable with unnormalized rational density function $\rho \in \mathcal{U}$, with corresponding density summand Z, hence $\Phi(ix) := \rho(x) = Z(ix) + Z^*(-ix)$, and let (A, b, c) be a stable state-space realization of Z. Then cb is real and positive and the density function $\frac{\rho}{2\pi cb}$ is the rational *probability* density function $\in \mathcal{R}(\mathbf{R})$ corresponding to ρ. If X is a random variable with rational probability density function $\frac{\rho}{2\pi cb}$, the moments $E(X^l)$ of X exist for $l = 0, \ldots, k-2$, where k is the smallest integer larger than or equal to two such that $cA^{k-1}b \neq (-1)^{k-1}(cA^{k-1}b)^*$, and the moments are given by $E(X^l) = (-i)^l \frac{cA^l b}{cb}$, $l = 0, 1, 2, \ldots, k-2$. In (Hanzon and Ober, 2001) it was shown that the operations of translation, scaling, multiplication and convolution of rational density functions can be translated into linear algebra operations on corresponding state-space realizations.

The key step in the present paper is the construction of a realization of the spectral summand of the rational density function given by $\rho(\frac{y}{V(x)})/V(x)$, where y is a fixed non-zero real number and $V(x)$ is a nonnegative polynomial on the real line, V not equal to the zero polynomial, if a realization of the spectral summand of ρ is known. Without loss of generality it will be assumed that $V = x^k + v_1 x^{k-1} + v_2 x^{k-2} \ldots + v_{k-1}x + v_k$ is a monic polynomial.

How such a realization can be found is the content of the following Proposition.

Let $\rho \in \mathcal{U}(\mathbf{R})$, let $\Phi \in \mathcal{U}(i\mathbf{R})$ be given by $\Phi(ix) = \rho(x)$. Let $\Phi(ix) = Z(ix) + Z^*(ix)$ and let (A, b, c) be a stable state-space realization of Z, i.e. all the eigenvalues of A lie in the open left half-plane. Let y be a non-zero real number and let V be a real polynomial which is non-negative on the real line and which is not the zero polynomial, i.e. $V(x) \geq 0$ for all real values of x and $\exists x \in \mathbf{R} : V(x) > 0$. Let q be the rational function given by $q(x) = \rho(\frac{y}{V(x)})/V(x)$. Then $q \in \mathcal{U}(\mathbf{R})$. A realization of the spectral summand of q can be found by performing the following five steps:

(i) Let (A_Φ, b_Φ, c_Φ) be the realization of Φ, given by $A_\Phi = \begin{pmatrix} A & 0 \\ 0 & -A^* \end{pmatrix}$, $b_\Phi = \begin{pmatrix} b \\ c^* \end{pmatrix}$, and $c_\Phi = \begin{pmatrix} c & -b^* \end{pmatrix}$.

(ii) Let (A_H, b_H, c_H) be given by $A_H = iy.A_\Phi^{-1}$, $b_H = b_\Phi$ and $c_H = -c_\Phi A_\Phi^{-1}$.

(iii) Let (A_N, b_N, c_N) be given as follows. A_N is a block companion matrix with $k \times k$ blocks $A_N(i,j)$, $i, j = 1, \ldots, k$, of size $2n \times 2n$ each, specified by:
$A_N(i, i+1) = I_{2n}$ for $i = 1, \ldots, k-1$,

$A_N(k,1) = (A_H - v_k I)$,
$A_N(k,j) = -v_{k+1-j} I_{2n}$, for $j = 2,\ldots,k$,
and all other blocks are zero.
b_N is a block column with k blocks $b_N(i)$, $i = 1,\ldots,k$ each of size $2n \times 1$, specified by:
$b_N(k) = b_H$, and all other blocks are zero.
Finally: c_N is a block row with k blocks $c_N(j)$, $j = 1,2,\ldots,k$, each of size $1 \times 2n$, specified by
$c_N(1) = c_H$ and all other blocks are zero.

(iv) Let (A_F, b_F, c_F) be given by: $A_F = iA_N$, $b_F = ib_N$ and $c_F = c_N$.

(v) There exists a state transformation T such that TA_FT^{-1} is a block-diagonal matrix with two blocks on the diagonal, which we denote by A_l and A_r respectively, the first one with all its eigenvalues in the open left half-plane and the other one with all its eigenvalues in the open right half-plane. Let $(A_D, b_D, c_D) = (TA_FT^{-1}, Tb_F, c_FT^{-1})$ and let $b_D = (b_l^T, b_r^T)^T$ be partitioned into the two vectors b_l and b_r each having equal number of entries. Similarly let $c_D = (c_l, c_r)$ be partitioned into the two vectors c_l and c_r each having equal number of entries.

Then a realization of the spectral summand of q is given by (A_l, b_l, c_l).

4. THE FILTER

The filter consists of a set of recursive equations by which one can calculate the conditional probability density function of the state X_t given the observations $Y_0^t = \{Y_0, Y_1, \ldots, Y_t\}$. The filter consists of a prediction step and an update step. In the prediction step one calculates the conditional density of X_{t+1} given the observations Y_0^t starting from the conditional density of X_t given Y_0^t. Because the state equation of our model is linear and the probability densities involved are all rational, the prediction step is the same as in (Hanzon and Ober, 2001). Therefore here we will discuss the update step. In the update step one calculates the conditional density of X_t given the observations Y_0^t from the observation Y_t and the conditional density of X_t given Y_0^{t-1}, using Bayes' rule. Suppose the conditional probability density function $\rho_{X_t|Y_0^{t-1}}$ of X_t given Y_0^{t-1} is known and the observation Y_t becomes available. The joint density of (X_t, Y_t) can be obtained from the joint density of (X_t, U_t) by a change of variables:

$$\begin{pmatrix} X_t \\ Y_t \end{pmatrix} = \begin{pmatrix} X_t \\ V(X_t).U_t \end{pmatrix}.$$

The inverse Jacobian determinant of this change of variables is $\frac{1}{V(X_t)}$, which is a positive because V is a positive polynomial. It follows that the joint density of (X_t, Y_t) is given by $\rho_{X_t,Y_t|Y_0^{t-1}}(x,y) = \rho_{X_t,U_t|Y_0^{t-1}}(x, \frac{y}{V(x)})\frac{1}{V(x)} = \rho_{X_t|Y_0^{t-1}}(x)\rho_{U_t}(\frac{y}{V(x)})\frac{1}{V(x)}$.

Substituting $y = Y_t$, we obtain the following expression for the (unnormalized) density of $X_t|Y_0^t$:

$$\rho_{X_t|Y_0^t} = \rho_{X_t|Y_0^{t-1}}(x)\rho_{U_t}(\frac{Y_t}{V(x)})\frac{1}{V(x)}.$$

Based on the results on realization of the factors $\rho_{X_t|Y_0^{t-1}}(x)$ and $\rho_{U_t}(\frac{Y_t}{V(x)})\frac{1}{V(x)}$ a realization of the summand of $\rho_{X_t|Y_0^t}$ can be obtained. The results imply that $\rho_{X_t|Y_0^t}, \rho_{X_{t+1}|Y_0^t}$ have rational probability density functions for $t = 0, 1, 2, \ldots$

5. CONCLUSION

The exact filter for a class of stochastic volatility models is derived. The standard stochastic volatility model can be viewed as a limiting case. For models with disturbances with rational probability density with higher McMillan degree and with volatility polynomial $V(x)$ of high degree, the complexity of the filter increases in the sense that the matrices that are used to represent the rational probability density functions tend to grow quickly in that case.

6. REFERENCES

A. Harvey, E. Ruiz, N. Shephard (1994). Multivariate stochastic variance models. *Review of Economic Studies* **61**, 247–264.

Brigo, D. and B. Hanzon (1998). On some filtering problems arising in mathematical finance. *Insurance: Mathematics and Economics* **22**, 53–64.

Engle, R. (1982). Autoregressive conditional heteroskedasticity with estimates of the variance of uk inflation. *Econometrica* **50**, 987–1008.

Gouriéroux, C. (1997). *ARCH Models and Financial Applications*. Springer. New York.

Hanzon, B. and R.J. Ober (2001). A state-space calculus for rational probability density functions and applications to non-gaussian filtering. *SIAM J. Control and Optimization* **40**, 724–740.

J.Y. Campbell, A.W. Lo, A. C. MacKinlay (1997). *The Econometrics of Financial Markets*. Princeton University Press. Princeton, New Jersey.

Lucas, A. (1996). *Outlier robust unit root analysis (Ph D Thesis)*. Thesis Publishers. Amsterdam.

Mahieu, R. and P. Schotman (1998). An empirical application of stochastic volatility models. *Journal of Applied Econometrics* **13**, 333=360.

Mandelbrot, B. (1963). The variation of certain speculative prices. *J. Business* **36**, 394–419.

Taylor, S. (1986). *Modelling Financial Time Series*. John Wiley and Sons. London.

IFAC Publications
www.elsevier.com/locate/ifac

SNIPPETS OF SYSTEM IDENTIFICATION IN COMPUTER VISION

Stefano Soatto [*,1] **Alessandro Chiuso** [**,2]

[*] *University of California, Los Angeles - CA 90095, soatto@ucla.edu*
[**] *Università di Padova - Italy 35131, chiuso@dei.unipd.it*

Abstract: In this paper we illustrate the use of identification-theoretic techniques in computer vision, and hint at some open problems.

Keywords: System identification, computer vision, dynamic scene analysis, visual recognition, dynamic textures, human gaits, dynamic vision.

1. INTRODUCTION

The sense of vision plays a crucial role in the life of primates, allowing them to infer spatial properties of the environment and perform crucial tasks for survival. Primates use vision to explore unfamiliar surroundings, negotiate physical space with one another, detect and recognize a prey at a distance, fetch it, all with seemingly little effort. To this day, engineered systems are far from exhibiting similar capabilities, and endowing a computer with a "sense of vision" is proving to be a formidable task: [3] How can we take a collection of digital images (i.e. arrays of positive numbers) as shown in Table 3, and tell whether we are looking at an apple or a person, and whether she is wearing a hat?

This is no easy task, since the measured images depend upon the geometry of the scene (which is unknown), its reflectance properties (also unknown), its motion (unknown) and the illumination of the scene (unknown). While some of these unknowns are not necessarily of interest, they all affect the measurements, and therefore the inference process has to be invariant with respect to these "nuisance" variables. It is easy to convince oneself that from images alone, no matter how many, it is impossible to recover a physically correct model of the geometry (shape), photometry (reflectance) and dynamics (motion) of the scene. In this sense, visual perception *per se* is an ill-

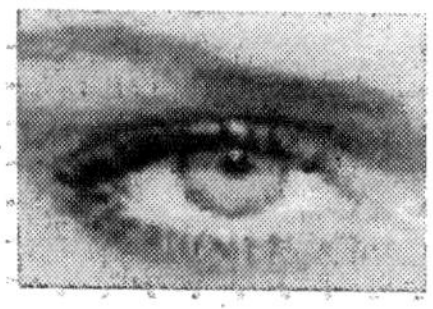

188	186	188	187	168	130	101	99	110	113	112	107	117	140	153	153
189	189	188	181	163	135	109	104	113	113	110	109	117	134	147	152
190	190	188	176	159	139	115	106	114	123	114	111	119	130	141	154
190	188	188	175	158	139	114	103	113	126	112	113	127	133	137	151
191	185	189	177	158	138	110	99	112	119	107	115	137	140	135	144
193	183	178	164	148	134	118	112	119	117	118	106	122	139	140	152
185	181	178	165	149	135	121	116	124	120	122	109	123	139	141	154
175	176	176	163	145	131	120	118	125	123	125	112	124	139	142	155
170	170	172	159	137	123	116	114	119	122	126	113	123	137	141	156
171	171	173	157	131	119	116	113	114	118	125	113	122	135	140	155
174	175	176	156	128	120	121	118	113	112	123	114	122	135	141	155
176	174	174	151	123	119	126	121	112	108	122	115	123	137	143	156
175	169	168	144	117	117	127	122	109	106	122	116	125	139	145	158
179	179	180	155	127	121	118	109	107	113	125	133	130	129	139	153
176	183	181	153	122	115	113	106	105	109	123	132	131	131	140	151
180	181	177	147	115	110	111	107	107	105	120	132	133	133	141	150
181	174	170	141	113	111	115	112	113	105	119	130	132	134	144	153
180	172	168	140	114	114	118	113	112	107	119	128	130	134	146	157
186	176	171	142	114	114	116	110	108	104	116	125	128	134	148	161
185	178	171	138	109	110	114	110	109	97	110	121	127	136	150	160

Table 1. *An "image" (top) is an array of positive numbers (bottom, subsampled) whose values are influenced by many "nuisance factors" including illumination and material properties of the scene.*

---

[1] Supported by AFOSR F49620-03-1-0095 and ONR N00014-02-1-0720.
[2] Supported by ASI and MURST.
[3] Although primates appear to be able to "see" effortlessly, it is interesting to notice that about half of their cerebral cortex is devoted to processing visual information (Felleman and van Essen, 1991).

posed problem. However, the inference problem may become well-posed within the context of a specific *task*. For instance, while one cannot infer "the" (physically correct) model of the scene in Figure 2.3, one can infer a *representation* of the scene that can be sufficient to support, for instance, control tasks, or recognition tasks. After all, even if we cannot infer the correct model of steam, or smoke or fire – no matter what computational device we have available, it be a computer or a brain – we sure know how to recognize smoke or fire when we see it.

In this paper we seek to infer models of dynamic visual processes for the purpose of classification and recognition tasks. For instance, from a number of sequences of images of fire, smoke or steam, we want to identify a model that can be used to then recognize, say, fire in a new sequence. The same goes for human motion: how can we infer a model of various gaits, such as walk, run, jump, limp, so that we can for instance detect a limping person from afar? We will first address the simplest possible classes of models, working under the assumption that the underlying data exhibit some form of stationarity. Here the current body of knowledge in system identification has already played a key role in a number of applications, as we describe in Section 2. Even for such simple models, however, recognition and classification tasks remain largely an open problem, which we discuss in Section 3. Imposing simpler models and inferring the model parameters as well as the domain where they are satisfied within a prescribed accuracy leads to a segmentation problem, which is described in Section 4. Extensions to more complex models are countless, and so are their applications. We therefore highlight some open problems in system identification that are motivated by extensions of the research described in this paper in Section 5.

2. MODELING DYNAMIC VISUAL PROCESSES

Since a physically correct model of the *geometry*, *photometry* and *dynamics* of a scene cannot be recovered from visual information alone, it is customary to make *assumptions* on some of the unknowns, in order to infer the others. For instance, in stereo reconstruction one often exploits the assumption that the scene exhibits Lambertian reflection in order to recover shape (see (Ma *et al.*, 2003) and references therein). Naturally, if the assumption is violated, the resulting inference is meaningless, which results in visual illusions that both biological and artificial systems are subject to (Chiuso *et al.*, 2000).

Therefore, here we take a different approach: rather than making prior assumptions and attempt to use visual information to recover a physical model of the scene, we seek to recover a *statistical* model of visual data directly at the outset. Hopefully, this model will support recognition and classification tasks, which we discuss in Section 3. To this end, we represent visual data as the output of a dynamical system driven by realizations of a process drawn from an unknown distribution. Inferring a model of the scene then consists of inferring the model parameters as well as the input distribution. In the next two subsections we illustrate the simplest possible model applied first to pixel intensity, resulting in so-called "dynamic textures", and then to the position of a number of landmark "features", as a model of human gaits.

2.1 Dynamic textures

Let $\{I(t) \in \mathbb{R}^{k \times l}\}_{t=1\dots\tau}$ be a sequence of images. Suppose that at each instant of time t we can measure a noisy version of the image, $y(t) = I(t)+w(t)$ where $w(t)$ is an independent and identically distributed (IID) sequence drawn from a distribution $p_w(\cdot)$ resulting in a positive measured sequence $y(t) \in \mathbb{R}^m$, $t = 1\dots\tau$[4], where $m = k \times l$. We say that *the sequence* $\{I(t)\}$ *is a (linear) dynamic texture* if there exists a set of n spatial filters ϕ_α, $\alpha = 1\dots n$ and a stationary distribution $q(\cdot)$ such that, calling $z(t) \doteq \phi(I(t))$, $z(t)$ can be modeled as an ARMA process excited by the white noise $v(t)$, distributed according to $q(\cdot)$. Therefore, a dynamic texture is associated to (a second-order stationary process and, therefore) a state space model with unknown input distribution

$$\begin{cases} x(t+1) = Ax(t) + Bv(t) \\ z(t) = Cx(t) + Dv(t) \\ y(t) = \psi(z(t)) + w(t) \end{cases} \quad (1)$$

with $x(0) = x_0$, $v(t) \overset{IID}{\sim} q(\cdot)$ unknown, $w(t) \overset{IID}{\sim} p_w(\cdot)$ given, and $I(t) = \psi(z(t))$ where $\psi(\phi(I)) = I$. One can obviously extend the definition to an arbitrary non-linear model of the form $x(t+1) = f(x(t), v(t))$, leading to the concept of *non-linear dynamic texture.*

The definition of dynamic texture above, which was proposed in (Doretto *et al.*, 2003), entails a choice of filters ϕ_α, $\alpha = 1\dots n$. These filters are also inferred as part of the identification process for a given dynamic texture. There are several criteria for choosing a suitable class of filters, ranging from biological motivations to computational efficiency. In the trivial case, we can take ϕ to be the identity, and therefore look at the

[4] This distribution can be inferred from the physics of the imaging device.

dynamics of individual pixels $x(t) = I(t)$ in (1). However, in texture analysis the dimension of the signal is huge (tens of thousands components) and there is a lot of redundancy. Hence we view the choice of filters as a dimensionality reduction step, and seek for a decomposition of the image in the simple (linear) form

$$I(t) = \sum_{i=1}^{n} x_i(t)\theta_i \doteq Cx(t) \tag{2}$$

where $C = [\theta_1, \ldots, \theta_n]$ and $\{\theta\}$ can be an orthonormal or overcomplete basis of $\mathcal{L}^2$, a set of principal components, or a wavelet filter bank. In the simples yet effective solution C can be estimated using linear techniques such as principal component analysis. The advantage of this approach, besides simplicity, is that it allows to effectively reduce complexity via a data-tailored construction of basis function. Experimental results show that 20 to 30 principal components yield synthesized textures which are practically indistinguishable from the original ones. Standard approaches based on filter banks require more coefficients to obtain comparable results; for instance, dynamic textures based on Fourier and Gabor filters have been shown in (Zhu and Wang, 2002).

2.2 Human gaits

Instead of modeling the pixel intensities, one could model the position in space of a number of landmark points, for instance the joints of an articulated body. We start from the assumption that a sequence of joint angle trajectories $y(t)$, $t = 1 \ldots \tau$ is a realization from a second-order stationary stochastic process, i.e.

$$\begin{cases} x(t+1) = Ax(t) + Bv(t) \\ y(t) = Cx(t) + Dv(t) \end{cases} \tag{3}$$

where $v(t)$ is a normalized white noise sequence. Although this may seem like a severely restrictive assumption, we show that it is sufficient to characterize models that are general enough for the purpose of recognition.

2.3 Inference

The problem of going from data to models is the usual system identification problem. Several approaches have been proposed in the literature ranging from standard prediction error methods (Ljung, 1987), to iterative solutions based on expectation-maximization which seem to have obtained a fair amount of attention in the learning community, to the more recent subspace methods (Overschee and Moor, 1993).

For the case of dynamic textures, due to the dimension of the signal (76,800 for video at half-resolution), one cannot apply standard identification algorithms. A dimensionality reduction step is a must as we have discussed at the end of section 2.1. Even after this reduction step the signal is high dimensional and therefore subspace methods seems to be best way to go. If we restrict ourselves to first-order AR processes, the following algorithm yields a simple and yet effective solution: Let $Y_1^\tau \doteq [y(1), \ldots, y(\tau)] \in \mathbb{R}^{m\times\tau}$ with $\tau > n$, and similarly for X_1^τ and W_1^τ, and notice that

$$Y_1^\tau = CX_1^\tau + W_1^\tau; \quad C \in \mathbb{R}^{m\times n}; \; C^T C = I \tag{4}$$

by our assumptions. Now let $Y_1^\tau = U\Sigma V^T$; $U \in \mathbb{R}^{m\times n}$; $U^T U = I$; $V \in \mathbb{R}^{\tau\times n}$, $V^T V = I$ be the singular value decomposition (SVD) with $\Sigma = \mathrm{diag}\{\sigma_1, \ldots, \sigma_n\}$, and consider the problem of finding the best estimate of C in the sense of Frobenius: $\hat{C}(\tau), \hat{X}(\tau) = \arg\min_{C, X_1^\tau} \|W_1^\tau\|_F$ subject to (4). It follows immediately from the fixed-rank approximation property of the SVD (Golub and Loan, 1989) that the unique solution is given by

$$\hat{C}(\tau) = U \quad \hat{X}(\tau) = \Sigma V^T \tag{5}$$

and $\hat{A}$ can be determined uniquely, again in the sense of Frobenius, by solving the following linear problem: $\hat{A}(\tau) = \arg\min_A \|X_1^\tau - AX_0^{\tau-1}\|_F$ which is trivially done in closed form using an estimate of X from (5):

$$\hat{A}(\tau) = \Sigma V^T D_1 V (V^T D_2 V)^{-1} \Sigma^{-1} \tag{6}$$

where $D_1 = \begin{bmatrix} 0 & 0 \\ I_{\tau-1} & 0 \end{bmatrix}$ and $D_2 = \begin{bmatrix} I_{\tau-1} & 0 \\ 0 & 0 \end{bmatrix}$. Notice that $\hat{C}(\tau)$ is uniquely determined up to a change of sign of the components of C and x. Also note that

$$E[\hat{x}(t)\hat{x}^T(t)] \equiv \lim_{\tau\to\infty} \frac{1}{\tau} \sum_{k=1}^{\tau} \hat{x}(t+k)\hat{x}^T(t+k) \simeq \Sigma^2 \tag{7}$$

which is diagonal. Finally, the sample input noise covariance Q can be estimated from

$$\hat{Q}(\tau) = \frac{1}{\tau} \sum_{i=1}^{\tau} \hat{v}(i)\hat{v}^T(i) \tag{8}$$

where $\hat{v}(t) \doteq \hat{x}(t+1) - \hat{A}(\tau)\hat{x}(t)$. In the algorithm above we have assumed that the number of principal components n was given. In practice, this needs to be inferred from the data. In practice this is done from the singular values $\sigma_1, \sigma_2, \ldots$, by choosing n as the cutoff where the value of σ drops below a threshold. A threshold can also be imposed on the difference between adjacent singular values.

Identifying a model from a sequence of 100 frames takes about 5 minutes in MATLAB on a 1GHz pentium® III PC. Synthesis can be performed in

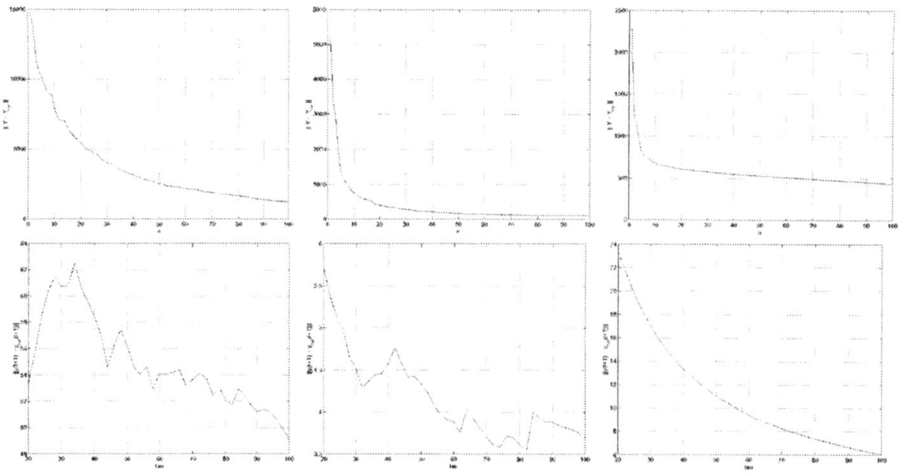

(a) (b) (c)

Fig. 1. *Compression error as a function of the dimension of the state space n (top row), and extrapolation error as a function of the length of the training set τ (bottom row). Column (a)* `river` *sequence, column (b)* `smoke` *sequence, column (c)* `toilet` *sequence.*

(a)

(b)

(c)

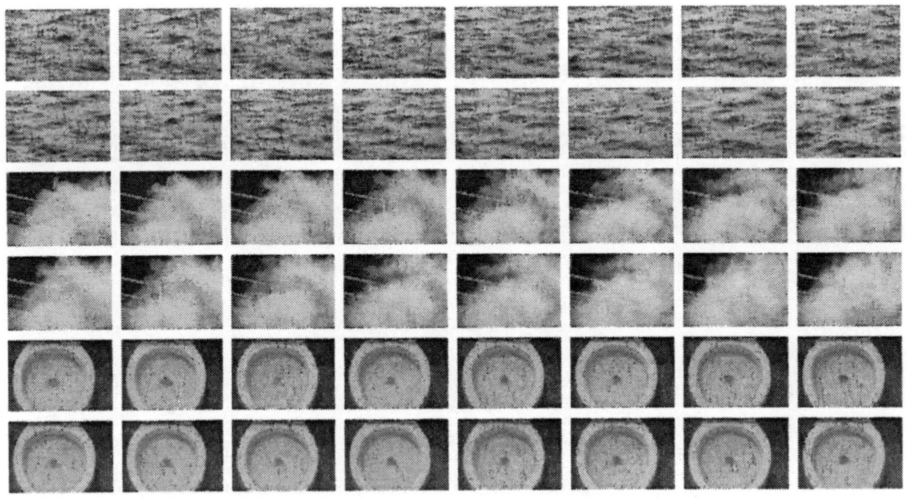

Fig. 2. *(a) river sequence, (b) smoke sequence, (c) toilet sequence. For each of them the top row are samples of the original sequence, the bottom row shows samples of the extrapolated sequence. All the data are available on-line at* `http://www.vision.ucla.edu/projects/dynamic-textures.html`

real time. In our implementation we have used τ between 50 and 150, n between 20 and 50 and k between 10 and 30. Figures 2.3 to 2.3 show the behavior of the algorithm on a representative set of experiments. In each case of Figure 2.3, on the first row we show a few images from the original dataset, on the second row we show a few extrapolated samples. Figure 2.3 shows the overall compression error as a function of the dimension of the state space (top row) as well as the prediction error as a function of the length of the learning set (bottom row). The prediction error is computed by using the first τ images to identify a model, then using the model to predict the image at $\tau + 1$, finally comparing the result with the actual image measured at $\tau + 1$. The plot shows the error between the predicted and the measured images at $\tau + 1$ as a function of τ. A simple histogramming of the residual shows that it deviates considerably from a Gaussian density, with considerable weight at the tails.

3. CLASSIFICATION AND RECOGNITION OF DYNAMIC VISUAL PROCESSES

One approach to classification and recognition is to look directly to the data sequences. These can be regarded as samples from some distribution and therefore classical concepts of discrepancy measures such as Kullback-Leibler divergence, Battacharyya or Hellinger, could be employed. The space of probability distributions, suitably restricted, can be given the structure of a differentiable manifold. One could endow these manifolds with a Riemannian structure, and define distances and probability distributions, which are the basis of standard techniques such as Bayes classification or likelihood ratios. This, however, has several drawbacks. First, one has to take into account that different datasets may have different length. In fact, the probability distribution of an n-vector lives on a different space than that of an m-vector if $m \neq n$. Therefore, it seems more appropriate to look at invariants of the process itself. As we are working with second-order stochastic processes, these are the spectrum, the covariance function or a spectral factor or, in other words, an ARMA model.

ARMA models, learned from data as described in the previous section, do not live on a linear space. The matrix A is constrained to be stable, the matrix C has non-trivial geometric structure after a canonical realization is chosen (for instance, its columns may form an orthogonal set). More in general, the model lives in the quotient space of coordinate transformations of the state space. That leads us to endowing Grassmann and Stiefel manifolds with a metric and probabilistic structure that, although possible, is not straightforward (although see the work of (Hannan and Deistler, n.d.) for references on how to endow transfer functions with the structure of a differentiable manifold.)

While a simple probabilistic approach to classification based on likelihood ratios with respect to a simple probability distribution on Stiefel manifolds has been proposed in (Saisan *et al.*, 2001), a rigorous treatment of this matter has yet to come. In this expository paper, we restrict ourselves to describing an approach to classification and recognition based on the metric structure of the space of models, for instance nearest-neighbor classification or k-means clustering (Duda and Hart, 1973).

Suppose a set of samples $C_1, C_2, \ldots$ is given, where each sample is labelled as belonging to one of c classes λ_j. Given a new sample C, the label λ_m is chosen by taking a vote among the k nearest samples. That is, λ_m is selected if the majority of the k nearest neighbors have label λ_m, which happens with probability

$$\sum_{i=(k+1)/2}^{k} \binom{k}{i} P(\lambda_m|C)^i (1 - P(\lambda_m|C))^{k-i}. \quad (9)$$

It can be shown (Cover and Hart, 1967) that if k is odd the large-sample 2-class error rate is bounded above by the smallest concave function of P^* – the optimal error rate – greater than

$$\sum_{i=0}^{(k-1)/2} \binom{k}{i} \left(P^{*i+1}(1 - P^*)^{ki} + P^{*k-i}(1 - P^*)^{i+1} \right). \quad (10)$$

Note that the analysis holds for k fixed as $n \to \infty$, and that the rule approaches the minimum error rate for $k \to \infty$. For small samples, there are no known results except negative counter-examples that show that an arbitrarily bad error rate can be achieved.

Distances between identified models (which we have inferred using the implementation of the N4SID algorithm (Overschee and Moor, 1993) in the Matlab System Identification Toolbox), can be computed in a number of ways. First, we have computed the "naive" distances (the 2-norm of the difference between corresponding system matrices, without taking the geometry of the manifold into account) and the geodesic distance between models. Not surprisingly these led to quite disappointing results. Then we have computed two metrics between observability subspaces - also taking into account the geometry of the manifold : the Finsler (Weinstein, 1999) and a generalization of the Martin distance, defined in (Martin, 2000) for scalar models. We computed the principal angles between observability subspaces using the algorithm proposed in (Coch and Moor, 2000), where it is formulated for the scalar case but it can be naturally extended to MIMO systems which are innovation models of full-rank vector processes. Then we have calculated the Finsler distance as the maximum subspace angle, and extended the Martin distance d_M by using its relation to the subspace angles θ_i in the scalar case: $d_M = -\ln \prod_i \cos^2 \theta_i$. These two metrics gave similar results, with a slight advantage for the latter. To the distance between learned zero-mean models we added the norm of the difference between the the means of the joint configurations, weighted by a scale factor whose value was set empirically.

For the purpose of illustration, we show the pictorial result of an experiment with three classes of motions that result in similar gaits: walking, running and going up and down a staircase. Notice that these three gaits are quite similar to each other (as opposed, say, to dancing or jumping), and yet the algorithm proposed is capable of correct classification in most cases. These are admittedly preliminary results, and significant work in this area is needed. In Figure 3 we show sample frames from the training datasets. Figure 3 shows the pairwise distance between each model in the dataset. As it can be seen, similar gaits result in smaller distances, with a few outliers. Although this is a very restricted database, it suffices to test our hypothesis.

We have then chosen a few sample sequences for each category as a test sequence. For each of the sequences we have estimated a model by first preprocessing the sequence (after manual initialization) using the ideas described in (Bregler, 1997) to extract joint coordinates, and finally compared the models using a nearest neighbor criterion. A sample frame from the test sequence is shown in

Fig. 3. *Sample frames from the dataset of the gaits: waking, running and walking a staircase.*

Figure 3, while the first two corresponding nearest neighbors are shown to the right. Although this dataset is quite small, the discriminating power of the model as a representation of the dynamic sequence is visible.

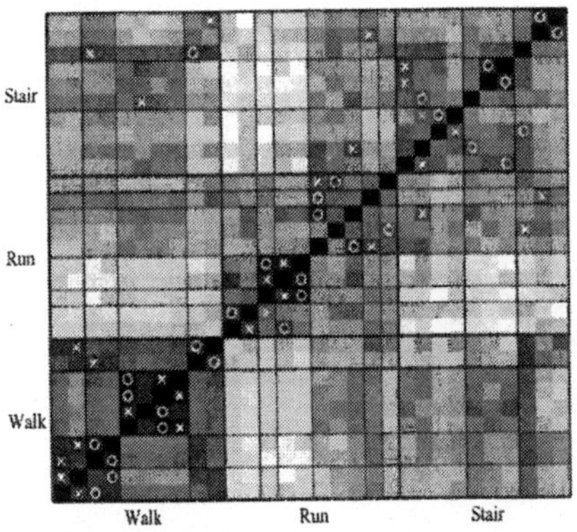

Fig. 4. *The pairwise distance between each sequence in the dataset is displayed in this plot. Each row/column of a matrix represents a sequence, and sequences correspondence to similar gaits are grouped in block rows/columns. Dark indicates a small distance, light a large distance. The minimum distance is of course along the diagonal, and for each row the next closest sequence is indicated by a circle, while the second nearest is indicated by a cross.*

We shall now briefly recall the basics of the distance concept introduced by Martin and elaborated upon by (Coch and Moor, 2000). Let $M_1 \doteq (A_1, C_1, K_1, \Lambda_1)$ and $M_2 \doteq (A_2, C_2, K_2, \Lambda_2)$ be two ARMA models; we define the extended observability matrix

$$\mathcal{O}_\infty(M_i) \doteq \left[C_i^T \ A_i^T C_i^T \ \ldots \ (A_i^T)^n C_i^T \ \ldots\right]^T.$$

It turns out that, as shown in (Coch and Moor, 2000), the concept of distance defined by Martin can be expressed directly in terms of angles between the (extended) observability spaces. In fact, being M_i^{-1} the inverse of the model M_i (recall that M_i is minimum-phase), the distance between M_1 and M_2 can be expressed in terms of the subspace angles between the column spaces of $\left[\mathcal{O}_\infty(M_1) \ \mathcal{O}_\infty(M_2^{-1})\right]$ and $\left[\mathcal{O}_\infty(M_2) \ \mathcal{O}_\infty(M_1^{-1})\right]$. If we denote by θ_i the ith canonical angle between

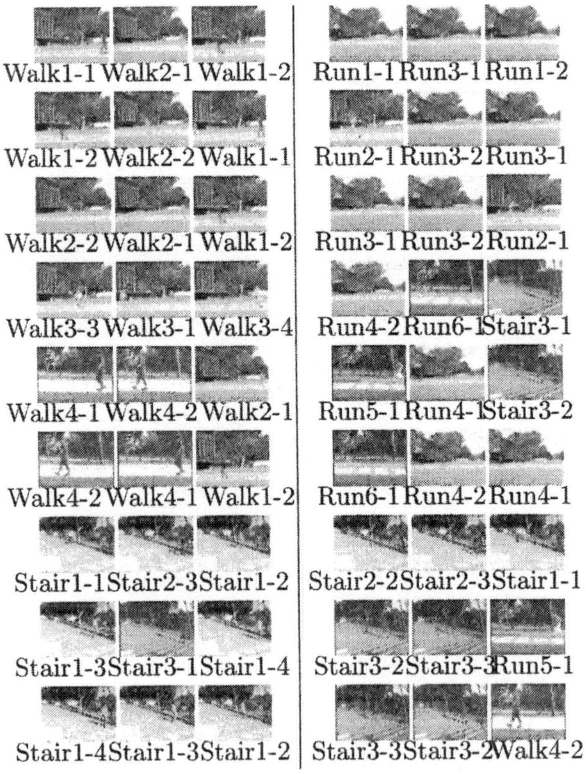

Fig. 5. *For each gait we have chosen a few sample sequences (left) and computed the distance to every other sequence in the dataset. The closest sequence is shown in the central column, while the second nearest is shown in the right column. With a few exceptions, the nearest neighbor belongs to the same gait as the test sequence. Notice that all gaits are quite similar; similar experiments performed on much more diverse gaits such as jumping or dancing return correct classifications.*

these spaces the distance defined by Martin is given by

$$d_M(M_1, M_2)^2 = \ln \prod_{i=1}^{2n} \frac{1}{\cos^2\theta_i}, \qquad (11)$$

where n is the model order. Although this is true for scalar ARMA processes, one can measure the distance between multivariable ARMA models using the same concept. In this case the link with the cepstrum is clearly lost. In order to compute subspace angles between models, we proceed as follows

- Compute the solution $Q = \begin{pmatrix} Q_{11} & Q_{12} \\ Q_{21} & Q_{22} \end{pmatrix}$ of the Ljapunov equation

  $$\mathcal{A}^T Q \mathcal{A} - Q = -\mathcal{C}^T \mathcal{C}$$

  where

  $$\mathcal{A} = \begin{pmatrix} A_1 & 0 & 0 & 0 \\ 0 & A_2 - K_2 C_2 & 0 & 0 \\ 0 & 0 & A_2 & 0 \\ 0 & 0 & 0 & A_1 - K_1 C_1 \end{pmatrix}$$

  and

$$\mathcal{C} \doteq \begin{pmatrix} C_1 & -C_2 & C_2 & -C_1 \end{pmatrix}.$$

- Compute the $2n$ largest eigenvalues $\lambda_1, \ldots, \lambda_{2n}$ of

$$\begin{pmatrix} 0 & Q_{11}^{-1}Q_{12} \\ Q_{22}^{-1}Q_{21} & 0 \end{pmatrix}.$$

- $\lambda_i = \cos\theta_i$

We would like to stress that the distance just defined is independent of the basis chosen in the state space. The only requirement is that the model be in innovation canonical form, which is guaranteed by construction in our representation.

It is also worthwhile emphasizing that, as we have anticipated, the innovation variance Λ_i does not enter into the distance computation. This implies that data which differ just by the innovation variance will be treated as deriving from the same model, at least as far as classification/recognition is concerned.

4. SEGMENTATION OF DYNAMIC VISUAL PROCESSES

In modeling the spatio-temporal statistics of a process one is always faced with the conundrum of whether to use a complex model that captures the global statistics, or choose a simple class of models and then partition the scene into regions where the model fits the data within a specified accuracy. In the next section, we discuss a simple model for partitioning the scene into regions where the spatio-temporal statistics, represented by a model identified as we have described above, is constant.

4.1 Spatial segmentation

In this section, which follows (Cremers *et al.*, 2003), we discuss the problem of partitioning a sequence of images into regions that can be described by an ARMA model. Let $\Omega_i \subset \mathbb{R}^2$, $i = 1, \ldots, N$ be a partition of the image into N (unknown) regions[5]. We assume that each pixel contained in the region Ω_i is a Gauss-Markov process. In particular, we assume that there exist (unknown) parameters $A_i \in \mathbb{R}^{n\times n}, C_i \in \mathbb{R}^{m_i\times n}$, covariance matrices $Q_i \in \mathbb{R}^{n\times n}, R_i \in \mathbb{R}^{m_i\times m_i}$, white, zero-mean Gaussian processes $\{v(t)\} \in \mathbb{R}^n$, $\{w(t)\} \in \mathbb{R}^{m_i}$ and a process $\{x(t)\} \in \mathbb{R}^n$ such that the pixels in each region at each instant of time, $y(t)$, obey a model of the type (3). Note that we allow the number of pixels m_i to be different in each region, as long as $\sum_{i=1}^N m_i = m$, the size of the entire image, and that we require that neither the regions nor the parameters change over time, $\Omega_i, A_i, C_i, Q_i, R_i, x_{i,0} = \text{const}$.

Given this generative model, one way to formalize the problem of segmenting a sequence of images is the following: *Given a sequence of images $\{y(t) \in \mathbb{R}^m,\ t = 1, \ldots, T\}$ with two or more distinct regions Ω_i, $i = 1, \ldots, N \geq 2$ that satisfy the model (3), estimate both the regions Ω_i, the unknown state $\{x(t)\}$ in each region and the "signature" of each region, namely the parameters A_i, C_i, the initial state $x_{i,0}$ and the covariance of the driving process Q_i (the covariance R_i is uninformative and therefore excluded from the signature).* Assuming that the parameters A_i, C_i, Q_i, $x_{i,0}$ have been inferred for each region, in order to set the stage for a segmentation procedure, one has to define a discrepancy measure among regions. This has been discussed in the previous section. Now, if the boundaries of each region were known, one could easily estimate a simple model of the spatio-temporal statistics within each region. Unfortunately, in general one does not know the boundaries of each region, and this is one of the goals of the inference process. On the other hand, if the dynamic signature associated with each pixel was known, then one could easily determine the regions by thresholding or by other grouping or segmentation technique. Unfortunately, the model we wish to infer is not a point process, and therefore one cannot pre-compute the signature of each pixel and convert the problem into a static segmentation at the outset. Therefore, we have a classic "chicken-and-egg" problem: If we knew the regions, we could easily identify the dynamical models, and if we knew the dynamical models we could easily segment the regions. Unfortunately, we know neither.

In order to address this problem, one can set up an alternating minimization procedure, starting with an initial guess of the regions, $\hat{\Omega}_i(0)$, estimating the models within each region, $\hat{A}_i(0), \hat{C}_i(0), \hat{Q}_i(0)), \hat{x}_{i,0}$, and then at any given time t seeking for the modification of the regions $\hat{\Omega}_i(t)$, and the update of the models $\hat{A}_i(t), \hat{C}_i(t), \hat{Q}_i(t)), \hat{x}_{i,0}$ so as to minimize a chosen cost functional. For instance, one can minimize the norm of the innovation, integrated in space and time. For the sake of example, in the case of two regions Ω_1, Ω_2, one could seek for[6] $\hat{\Omega}_i(t+1)$ that solves the following optimization problem:

$$\arg\min_{\Omega_i} \sum_{i=1,2} \int_{\Omega_i} \sum_{k=1}^{T} \|y(\xi,k) - \hat{y}_i(\xi,k|k-1)\| d\xi \tag{12}$$

where $y_i(\xi, k|k-1)$ is the predictor of $y(\xi, k)$ based on model i. Once the partition has been estimated one can update $\hat{A}_i(t+1), \hat{C}_i(t+1), \hat{Q}_i(t+1), \hat{x}_0(t+1)$ using for instance subspace identification techniques.

---

[5] That is, $\Omega = \cup_{i=1}^N \Omega_i$ and $\Omega_i \cap \Omega_j = \emptyset, i \neq j$.

[6] We use the notation $y(\xi, t)$ to indicate the value of a pixel at location $\xi \in \Omega$ at time t.

Under the assumption that each pixel in a region obeys the same dynamical model, the minimum of the corresponding functional is attained when the distance between the two models (one for region Ω_1, the other one for its complement Ω_2) is maximized. Therefore, it is tempting to formulate the problem by simultaneously finding the regions Ω_i and the models M_i by maximizing the distance (11) subject to the constraint that models M_i is identified from data in region Ω_i. A suboptimal approach for this task has been proposed in (Cremers *et al.*, 2003).

For each pixel ξ we generate a local spatio-temporal signature given by the cosines of the subspace angles $\{\theta_j(\xi)\}$ between M_ξ and a reference model, M_{ξ_0}:

$$s(x) = \big(\cos\theta_1(\xi), \ldots, \cos\theta_n(\xi)\big). \tag{13}$$

With the above representation, the problem of dynamic texture segmentation can be formulated as one of grouping regions of similar spatio-temporal signature. We propose to perform this grouping by reverting to the Mumford-Shah functional (Mumford and Shah, 1989). A segmentation of the image plane Ω into a set of pairwise disjoint regions Ω_i of constant signature $s_i \in \mathbb{R}^n$ is obtained by minimizing the cost functional

$$E(\Gamma, \{s_i\}) = \sum_i \int_{\Omega_i} \big(s(\xi) - s_i\big)^2 d\xi + \nu\,|\Gamma|, \tag{14}$$

simultaneously with respect to the region descriptors $\{s_i\}$ modeling the average signature of each region, and with respect to the boundary Γ separating these regions (an appropriate representation of which will be introduced in the next section). The first term in the functional (14) aims at maximizing the homogeneity with respect to the signatures in each region Ω_i, whereas the second term aims at minimizing the length $|\Gamma|$ of the separating boundary.

Let the boundary Γ in (14) be given by the zero level set of a function $\phi : \Omega \to \mathbb{R}$:

$$\Gamma = \{\xi \in \Omega \,|\, \phi(\xi) = 0\}. \tag{15}$$

With the Heaviside function

$$H(\phi) = \begin{cases} 1 \text{ if } \phi \geq 0 \\ 0 \text{ if } \phi < 0 \end{cases}, \tag{16}$$

the functional (14) can be replaced by a functional on the level set function ϕ:

$$E(\phi, \{s_i\}) = \int_\Omega \big(s(\xi) - s_1\big)^2 H(\phi)\, d\xi + \int_\Omega \big(s(\xi) - s_2\big)^2 \big(1 - H(\phi)\big)\, d\xi + \nu\,|\Gamma|. \tag{17}$$

We minimize the functional (17) by alternating the two fractional steps of:

- Estimating the mean signatures.

  For fixed ϕ, minimization with respect to the region signatures s_1 and s_2 amounts to averaging the signatures over each phase:

$$s_1 = \frac{\int s\, H(\phi)\, d\xi}{\int H(\phi)\, d\xi}, \quad s_2 = \frac{\int s\,(1-H(\phi))\, d\xi}{\int (1-H(\phi))\, d\xi}. \tag{18}$$

- Boundary evolution.

  For fixed region signatures $\{s_i\}$, minimization with respect to the embedding function ϕ can be implemented by a gradient descent given by:

$$\frac{\partial \phi}{\partial t} = \delta(\phi)\left[\nu \nabla\left(\frac{\nabla\phi}{|\nabla\phi|}\right) + (s - s_2)^2 - (s - s_1)^2\right],$$

In Figure 6 we show a few snapshots of the contour evolution, starting from a circle. Notice that the final contour is the contour of an "average" region obtained by combining the different regions in time. Therefore, our approach shows robustness also to changes in the original hypotheses that dynamic textures were spatially stationary.

4.2 Temporal segmentation: filtering and identification of hybrid systems

In addition to partitioning the spatial domain into regions of constant statistics, one can partition the temporal domain, thereby segmenting a continuous process into discrete "events". For instance, the trajectory of joint positions and angles can be partitioned into segments each corresponding to a particular gait, so that we can detect when a walking person begins running or limping.

This is very much related to filtering and identification on hybrid systems and in particular Jump Linear Systems. We have addressed some of the issues related to identifying and filtering linear systems of this kind in (Vidal *et al.*, 2003*a*; Vidal *et al.*, 2002; Vidal *et al.*, 2003*b*) where also one can find a formalization of the problem. We use the techniques developed there to detect spatio-temporal events from live video, for instance the inception of a fire, or an explosion, or the transition from walking to running of a subject in the field of view.

5. EXTENSIONS

All that we have discussed above pertains to systems that are either linear or piecewise linear. However, as we have anticipated, if we identify a model within a linear segment and then plot a histogram of the residual, it is far from Gaussian. An optimality criterion for inference of model and

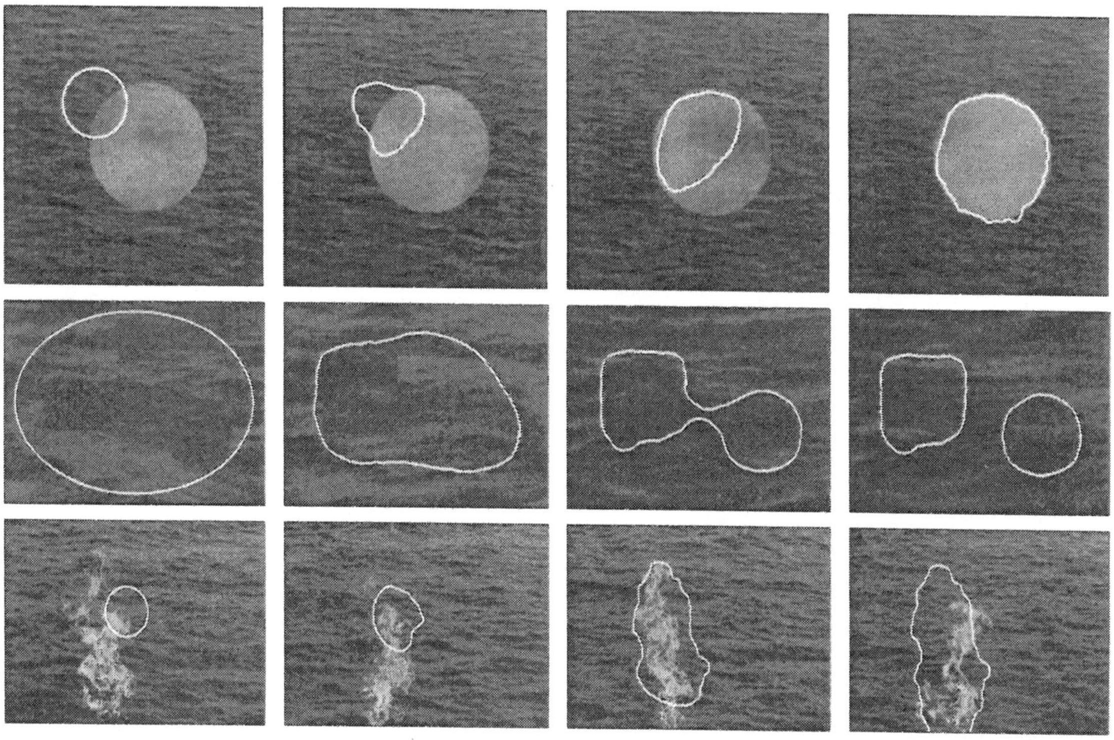

Fig. 6. In this experiment, courtesy of (Cremers et al., 2003), we show the segmentation of scenes based on the dynamic content on different regions. The last example is particularly challenging since the regions change over time. Animation of these results can be downloaded from `http://www.cs.ucla.edu/~doretto/projects/dynamic-segmentation.html`.

input descriptions can be constructed by requiring the estimated input sequence $\hat{v}(t)$ to be a realization from a stochastic process that has maximally independent components. Independence can be expressed in terms of the mutual information among input components, which in turn can be written in terms of the Kullback-Leibler divergence. This approach results in a semi-parametric statistical inference problem, where one has to simultaneously infer the (finite-dimensional) model parameters as well as the (infinite-dimensional) input distribution q. This is essentially an independent component analysis (ICA) problem.

In its conventional static from, ICA attempts to decompose a random vector into a linear combination of statistically independent components. If we call $\mathbf{y} \in \mathbb{R}^m$ the random vector, then ICA looks for a matrix $C \in \mathbb{R}^{m \times n}$ with $n \leq m$ and a random vector $\mathbf{x} \in \mathbb{R}^n$ with independent components, $p_{\mathbf{x}}(x_1, \ldots, x_n) = p_1(x_1) \ldots p_n(x_n)$ such that

$$\mathbf{y} = C\mathbf{x}. \tag{19}$$

The unknowns C and p_i can be estimated by minimizing the mutual information $I(\mathbf{y}||C\mathbf{x}) \doteq \int p_{\mathbf{y}} \log \frac{p_{\mathbf{y}}}{p_{C\mathbf{x}}} d\mathbf{y}$, computed or approximated using a number of independent and identically distributed (IID) samples from $p_{\mathbf{y}}$: $\mathbf{y}(1), \ldots, \mathbf{y}(t) \overset{IID}{\sim} p_{\mathbf{y}}$. Typically the process $\mathbf{y}$ is assumed to be ergodic, and therefore a time series is used in lieu of a fair sample. What we have here, however, is a dynamic ICA problem of separating independent components mixed by linear dynamical (state-space) systems.

Let us rewrite the output of the model at time t:

$$\mathbf{y}(t) = \left[CA^t,\ CA^{t-1}B, \ldots, CB\right] \begin{bmatrix} \mathbf{x}(0) \\ \mathbf{v}(0) \\ \vdots \\ \mathbf{v}(t-1) \end{bmatrix} \doteq \tilde{\mathcal{C}}^t \tilde{\mathbf{V}} \tag{20}$$

and stack the observations $\mathbf{y}(1), \ldots, \mathbf{y}(t)$ into a vector $\mathbf{Y}^t$ to obtain

$$\mathbf{Y}^t = \tilde{\mathcal{C}}^t \tilde{\mathbf{V}}. \tag{21}$$

One may be tempted to invoke the independence of the components of $\mathbf{V}$ – based on the assumptions that $\mathbf{v}(t)$ is white (time samples are independent) and has independent components – and use standard ICA to estimate the mixing matrix $\tilde{\mathcal{C}}^t$. This, however, does not work because it is not possible to use time realizations as independent samples of $\mathbf{Y}$ due to the initial condition $\mathbf{x}(0)$. One can may conjecture that if t is large enough and A is stable the effect of initial condition will wane; therefore, the assumption may not be as restrictive. Under this assumption, the problem of dynamic ICA can be posed as follows. Consider $\mathbf{Y}^t(k) = [\mathbf{y}((k-1)t)^T, \ldots, \mathbf{y}(kt-1)^T]^T$, and similarly for $\mathbf{V}^t(k)$. Furthermore, let $\mathcal{C}^t$ be the matrix obtaining by completing, in the sense of Toeplitz, the following matrix

$$\begin{bmatrix} CB & & & \\ CAB & CB & & \\ \vdots & \vdots & \ddots & \\ CA^{t-1}B & CA^{t-2}B & \dots & CB \end{bmatrix}. \quad (22)$$

Then $\hat{A}, \hat{B}, \hat{C}$ can be found sub-optimally by first estimating the mixing matrix $\mathcal{C}^t$ having the particular structure above from a set of independent samples $\mathbf{Y}^t(1), \dots, \mathbf{Y}^t(k)$ (notice that $\mathbf{Y}^t(i)$ and $\mathbf{Y}^t(j)$ do not share components $\mathbf{y}(k)$):

$$\hat{\mathcal{C}}^t(A, B, C) = \arg\min_{\mathcal{C}^t} I(\mathbf{Y}^t(i) || \mathcal{C}^t \mathbf{V}^t(i)) \quad (23)$$

A suboptimal algorithm for identification based on maximization of input independence has been proposed in (Bissacco and Saisan, 2002). It is based on Amari's natural gradient flow for semiparametric statistical problems (Amari and Cardoso, 1997). Sampling techniques can also be used to perform inference, in a particle filtering framework.

6. DISCUSSION

We have presented a handful of examples where current algorithms for system identification can be successfully employed to address modeling, synthesis and recognition problems in computer vision. These include modeling dynamic textures and human gaits for the purpose of synthesis and classification or recognition. In addition, we have indicated several directions where further work in system identification is needed in order to address difficult tasks of modeling non-Gaussian, non-linear, non-stationary processes for detection, classification, recognition and segmentation.

Acknowledgments

We wish to thank Alessandro Bissacco, Daniel Cremers, Gianfranco Doretto, Paolo Favaro, Payam Saisan, and Ying-Nian Wu for discussions and experimental material used in this paper.

REFERENCES

Amari, S. and F. Cardoso (1997). Blind source separation– semiparametric statistical approach. *IEEE Trans. Signal Processing* **45(11)**, 2692–2700.

Bissacco, A. and P. Saisan (2002). Modaling human gaits with subtleties. In: *Workshop on Dynamic Scene Analysis.*

Bregler, C. (1997). Learning and recognizing human dynamics in video sequences. In: *Proc. of the Conference on Computer Vision and Pattern Recognition.* pp. 568–574.

Chiuso, A., R. Brockett and S. Soatto (2000). Optimal structure from motion: local ambiguities and global estimates. *Intl. J. of Computer Vision* **39**(3), 195–228.

Coch, K. De and B. De Moor (2000). Subspace angles and distances between arma models. *Proc. of the Intl. Symp. of Math. Theory of Networks and Systems.*

Cover, T. and P. Hart (1967). Nearest neighbor pattern classification. *IEEE Trans. on Information Theory* **13**, 21–27.

Cremers, D., G. Doretto, P. Favaro and S. Soatto (2003). Dynamic texture segmentation. In: *UCLA CSD-TR030014.*

Doretto, G., A. Chiuso, Y. Wu and S. Soatto (2003). Dynamic textures. *Intl. J. of Comp. Vis.* **51(2)**, 91–109.

Duda, R. O. and P. E. Hart (1973). *Pattern classification and scene analysis.* Wiley and Sons.

Felleman, D. J. and D. C. van Essen (1991). Distributed hierarchical processing in the primate cerebral cortex. *Cerebral Cortex* **1**, 1–47.

Golub, G. and C. Van Loan (1989). *Matrix computations.* 2 ed.. Johns Hopkins University Press.

Hannan, E. J. and M. Deistler (n.d.). *The statistical theory of linear systems.* Wiley and Sons.

Ljung, L. (1987). *System Identification: theory for the user.* Prentice Hall.

Ma, Y., S. Soatto, J. Kosecka and S. Sastry (2003). *An invitation to 3D vision, from images to models.* Springer Verlag.

Martin, R. (2000). A metric for arma processes. *IEEE Trans. on Signal Processing* **48(4)**, 1164–1170.

Mumford, D. and J. Shah (1989). Optimal approximations by piecewise smooth functions and associated variational problems. *Comm. on Pure and Applied Mathematics* **42**, 577–685.

Overschee, P. Van and B. De Moor (1993). Subspace algorithms for the stochastic identification problem. *Automatica* **29**, 649–660.

Saisan, P., G. Doretto, Y. Wu and S. Soatto (2001). Dynamic texture recognition. In: *Proc. IEEE Conf. on Comp. Vision and Pattern Recogn..* pp. II 58–63.

Vidal, R., A. Chiuso and S. Soatto (2002). Observability and identifiability of jump-linear systems. In: *Proc. of the Intl. Conf. on Decision and Control.*

Vidal, R., A. Chiuso, S. Soatto and S. Sastry (2003*a*). Observability of linear hybrid systems. In: *Proc. of the Hybrid Systems Computation and Control.*

Vidal, R., S. Soatto and S. Sastry (2003*b*). An algebraic geometric approach to the identification of linear hybrid systems. In: *IEEE Conf. on Decision and Control.* p. (submitted).

Weinstein, A. (1999). Almost invariant submanifolds for compact group actions.

Zhu, S. C. and Y. Z. Wang (2002). A generative method for textured motion: analysis and synthesis. In: *Proc. of the Eur. Conf. on Comp. Vision.*

IFAC Publications
www.elsevier.com/locate/ifac

INTERVAL ANALYSIS FOR GUARANTEED NONLINEAR PARAMETER ESTIMATION

Eric Walter and Michel Kieffer

Laboratoire des Signaux et Systèmes
CNRS – Supélec – Université Paris-Sud
Plateau de Moulon, F-91192 Gif-sur-Yvette, France
{walter, kieffer}@lss.supelec.fr

Abstract: Interval analysis was initially developed to analyze and control numerical errors in computers. It can now be used to solve problems that are at the core of nonlinear parameter estimation, such as the minimization of possibly nonconvex cost functions or the characterization of sets defined by nonlinear inequalities. The solutions obtained are approximate but guaranteed, in the sense that no solution can be lost, a definite advantage over the usual local iterative techniques. After recalling basic concepts of interval analysis, this introductory paper describes algorithmic tools that can be used for nonlinear parameter estimation and applies them to simple illustrative examples drawn from compartmental modeling, a formalism widely used in biology. Guaranteed numerical integrators and properties of cooperative systems make it possible to deal with differential models, even when they do not admit closed-form solutions. Pointers to freely downloadable software are provided. *Copyright © 2003 IFAC*

Keywords: compartmental models, cooperativity, global optimization, identifiability, interval analysis, nonlinear estimation, parameter bounding.

1. INTRODUCTION

When the vector $\mathbf{p}$ of the parameters of a model has to be estimated from experimental data by minimizing a cost function that is quadratic in $\mathbf{p}$, for instance because this cost function is quadratic in an error that is itself affine in $\mathbf{p}$, it is well known that an explicit expression for the optimal estimate $\widehat{\mathbf{p}}$ as a function of the data can be derived, and that, under suitable hypotheses on measurement noise, it is possible to compute the covariance matrix of this estimate and thus to assess its uncertainty. Although this is an important special case, it is far from covering all situations of interest. Most knowledge-based models, for instance, have outputs that are nonlinear in their parameters, which usually makes it impossible to obtain such explicit estimates. Iterative local optimization procedures are then employed, which can provide no guarantee as to their results.

The purpose of this paper is to show how interval analysis, initially developed to analyze and control numerical errors in computers (Moore, 1959), allows such guaranteed results to be obtained in nonlinear parameter estimation. Interval analysis can indeed be used to solve sets of nonlinear equations or inequalities or to minimize nonconvex cost functions (Moore, 1979), (Ratschek and Rokne, 1988), (Neumaier, 1990), (Hansen, 1992), (Hammer *et al.*, 1995), (Kearfott, 1996). Numerical solutions are provided under the form of sets guaranteed to contain *all* actual solutions of the initial mathematical problem, which is a considerable advantage over the usual numerical methods that deliver point estimates by iterative local refinement of some initial guess. The basic concepts of interval analysis will be presented first. Some tools based on interval analysis and especially relevant for nonlinear parameter estimation will then be described and applied, before providing pointers to freely downloadable software. To simplify presentation and save space, many details will be skipped, for instance on how interval computation can be implemented in a guaranteed way even when a floating-point representation of real numbers is used. Much more information may be found in (Jaulin *et al.*, 2001*b*).

2. BASIC CONCEPTS

An *interval* is a connected subset of $\mathbb{R}$. Here, it will be assumed to be closed and bounded, but open-ended and unbounded intervals may also be considered. Let $\mathbb{IR}$ be the set of all closed intervals. Denote the *lower bound* of $[x] \in \mathbb{IR}$ by $\underline{x}$ or $\text{lb}([x])$, and its *upper bound* by $\overline{x}$ or $\text{ub}([x])$, so

$$[x] = \{x \in \mathbb{R} \mid \underline{x} \leqslant x \leqslant \overline{x}\}.$$

The *width* of $[x]$ is

$$w([x]) = \overline{x} - \underline{x},$$

and its *midpoint* (or *center*) is

$$\text{mid}([x]) = \frac{\underline{x}+\overline{x}}{2}.$$

When $\underline{x}$ and $\overline{x}$ are equal, $[x]$ is said to be *degenerate*. Any real number can thus be represented as a degenerate interval. A closed interval is entirely defined by its lower and upper bounds, so these intervals have a dual nature: they may be viewed as *sets*, to which set-theoretic operations such as intersection, union and Cartesian product apply, and as *pairs of real numbers*, on which an arithmetic can be built.

2.1 *Operations on intervals*

The *intersection* of two intervals

$$[x] \cap [y] = \{z \in \mathbb{R} \mid z \in [x] \text{ and } z \in [y]\}$$

is always an interval (provided that the empty set is considered as one). The *union* of two intervals

$$[x] \cup [y] = \{z \in \mathbb{R} \mid z \in [x] \text{ or } z \in [y]\}$$

is not necessarily an interval, hence the definition of the *interval union* of $[x]$ and $[y]$, denoted by $[x] \sqcup [y]$, which is the smallest interval that contains $[x] \cup [y]$. The Cartesian product of intervals is an *interval vector* (or *box*), see the next column.

Addition, subtraction, multiplication and division can be extended to intervals according to

$$[x] \diamond [y] = [\{x \diamond y \in \mathbb{R} \mid x \in [x] \text{ and } y \in [y]\}], \quad (1)$$

where $\diamond \in \{+, -, \cdot, /\}$ and where $[\mathbb{S}]$ is the *interval hull* of the set $\mathbb{S}$, *i.e.*, the smallest interval that contains it. In (1), $[x]$ and $[y]$ are treated as corresponding to independent variables x and y, which is a major source of pessimism. For instance, $[x] - [x]$ is not equal to $[0, 0]$, unless $[x]$ is degenerate. Addition and multiplication remain associative and commutative, but multiplication is no longer distributive with respect to addition. Instead

$$[x] \cdot ([y] + [z]) \subset [x] \cdot [y] + [x] \cdot [z],$$

a property known as *subdistributivity*. As a result, it is recommended to factorize expanded forms as much as possible, to make the width of the interval results as small as possible. For nonempty intervals,

$$[x] + [y] = [\underline{x} + \underline{y}, \overline{x} + \overline{y}],$$

$$[x] - [y] = [\underline{x} - \overline{y}, \overline{x} - \underline{y}],$$

$$[x] \cdot [y] = [\min\{\underline{x}\underline{y}, \underline{x}\overline{y}, \overline{x}\underline{y}, \overline{x}\overline{y}\}, \max\{\underline{x}\underline{y}, \underline{x}\overline{y}, \overline{x}\underline{y}, \overline{x}\overline{y}\}],$$

and if $0 \notin [y]$ then

$$[x]/[y] = [x] \cdot [1/\overline{y}, 1/\underline{y}].$$

(Specific formulas involving unbounded intervals are available for division by an interval containing zero.) When applied to degenerate intervals, these rules simplify into those of real arithmetic, of which interval arithmetic can thus claim to be an extension.

The interval counterpart $[f]^*$ of a function f from $\mathbb{R}$ to $\mathbb{R}$ satisfies

$$[f]^*([x]) = [\{f(x) \mid x \in [x]\}].$$

For any continuous function, $[f]^*([x])$ is thus equal to the image set $f([x])$. Elementary interval functions can be expressed in terms of bounds. This is especially simple for monotonic functions. For instance, for any nonempty $[x]$,

$$[\exp]^*([x]) = [\exp(\underline{x}), \exp(\overline{x})].$$

For nonmonotonic functions, the situation is more complicated. For example, $[\sin]^*([0, \pi]) = [0, 1]$, which differs from $[\sin(0), \sin(\pi)] = [0, 0]$, and specific algorithms had to be built for trigonometric and hyperbolic functions.

An *interval vector* (or *box*) $[\mathbf{x}]$ is a Cartesian product of intervals. It will be written indifferently as

$$[\mathbf{x}] = [x_1] \times [x_2] \times \cdots \times [x_n],$$

or

$$[\mathbf{x}] = ([x_1], [x_2], \ldots, [x_n])^{\mathrm{T}}.$$

The set of all n-dimensional boxes will be denoted by $\mathbb{IR}^n$. Nonempty boxes of $\mathbb{IR}^n$ are n-dimensional axis-aligned parallelepipeds. Most notions introduced for intervals extend without difficulty to boxes. The *lower bound* of $[\mathbf{x}] \in \mathbb{IR}^n$ is the vector of the lower bounds of its interval components:

$$\underline{\mathbf{x}} = (\underline{x}_1, \underline{x}_2, \cdots, \underline{x}_n)^{\mathrm{T}}.$$

Similarly, the *upper bound* of $[\mathbf{x}]$ is

$$\overline{\mathbf{x}} = (\overline{x}_1, \overline{x}_2, \cdots, \overline{x}_n)^{\mathrm{T}}.$$

The *width* of $[\mathbf{x}]$ is

$$w([\mathbf{x}]) = \max_{1 \leqslant i \leqslant n} w([x_i]).$$

Its *midpoint* (or *center*) is

$$\text{mid}([\mathbf{x}]) = (\text{mid}([x_1]), \ldots, \text{mid}([x_n]))^{\mathrm{T}}.$$

The *interval hull* $[\mathbb{S}]$ of a subset $\mathbb{S}$ of $\mathbb{R}^n$ is the smallest box of $\mathbb{IR}^n$ that contains it.

The intersection of the boxes $[\mathbf{x}]$ and $[\mathbf{y}]$ of $\mathbb{IR}^n$ satisfies

$$[\mathbf{x}] \cap [\mathbf{y}] = ([x_1] \cap [y_1]) \times \ldots \times ([x_n] \cap [y_n]),$$

provided that it is nonempty. Most often the union of two boxes $[\mathbf{x}]$ and $[\mathbf{y}]$ is not a box. Its interval hull can be computed as

$$[\mathbf{x}] \sqcup [\mathbf{y}] = ([x_1] \sqcup [y_1]) \times \ldots \times ([x_n] \sqcup [y_n]).$$

The test for the inclusion of $[\mathbf{x}]$ in $[\mathbf{y}]$ can be performed componentwise, since

$$[\mathbf{x}] \subset [\mathbf{y}] \Leftrightarrow \begin{cases} [x_1] \subset [y_1], \\ \vdots \\ [x_n] \subset [y_n]. \end{cases}$$

Classical operations on vectors trivially extend to interval vectors. For instance, if $\alpha \in \mathbb{R}$,

$$\begin{aligned} \alpha[\mathbf{x}] &= (\alpha[x_1]) \times \cdots \times (\alpha[x_n]), \\ [\mathbf{x}]^{\mathrm{T}} \cdot [\mathbf{y}] &= [x_1] \cdot [y_1] + \cdots + [x_n] \cdot [y_n], \\ [\mathbf{x}] + [\mathbf{y}] &= ([x_1] + [y_1]) \times \cdots \times ([x_n] + [y_n]). \end{aligned}$$

Interval matrices can be similarly defined, and the notions of lower and upper bounds, width, midpoint and interval hull extend as for boxes. If $[\mathbf{A}]$ and $[\mathbf{B}]$ are intervals, interval vectors or interval matrices of appropriate dimensions and if $\diamond$ is a binary operator, then

$$[\mathbf{A}] \diamond [\mathbf{B}] = [\{\mathbf{A} \diamond \mathbf{B} \mid \mathbf{A} \in [\mathbf{A}] \text{ and } \mathbf{B} \in [\mathbf{B}]\}].$$

For example, if $[\mathbf{A}]$ and $[\mathbf{B}]$ are in $\mathbb{IR}^{n \times n}$ then

$$[\mathbf{A}] \cdot [\mathbf{B}] = \left(\sum_{k=1}^{n} [a_{ik}] \cdot [b_{kj}] \right)_{1 \leqslant i \leqslant n, 1 \leqslant j \leqslant n}.$$

More sophisticated operations such as matrix inversion and the computation of eigenvalues and eigenvectors raise difficulties that go beyond this introductory paper. See (Neumaier, 1990) for details.

2.2 *Inclusion functions*

Let $\mathbf{f}$ be a function from $\mathbb{R}^n$ to $\mathbb{R}^m$, which may be defined by an algorithm or even by a differential equation. The interval function $[\mathbf{f}]$ from $\mathbb{IR}^n$ to $\mathbb{IR}^m$ is an *inclusion function* for $\mathbf{f}$ if

$$\forall [\mathbf{x}] \in \mathbb{IR}^n, \ \mathbf{f}([\mathbf{x}]) \subset [\mathbf{f}]([\mathbf{x}]).$$

The image set $\mathbf{f}([\mathbf{x}])$ may have any shape. It may be nonconvex, or even disconnected if $\mathbf{f}$ is discontinuous. Whatever this shape, an inclusion function $[\mathbf{f}]$ of $\mathbf{f}$ makes it possible to compute a box $[\mathbf{f}]([\mathbf{x}])$ guaranteed to contain $\mathbf{f}([\mathbf{x}])$. Actually this box may offer a very pessimistic vision of $\mathbf{f}([\mathbf{x}])$, but since it is far easier to manipulate boxes than generic sets, this is a very interesting tool for studying functions. One of the main challenges of interval analysis is to provide inclusion functions that can be evaluated quickly while keeping $[\mathbf{f}]([\mathbf{x}])$ as small as possible.

An inclusion function $[\mathbf{f}]$ for $\mathbf{f}$ is *convergent* if, for any sequence of boxes $[\mathbf{x}]_k$,

$$\lim_{k \to \infty} w([\mathbf{x}]_k) = 0 \Rightarrow \lim_{k \to \infty} w([\mathbf{f}]([\mathbf{x}]_k)) = 0.$$

It is *minimal* if for any $[\mathbf{x}]$, $[\mathbf{f}]([\mathbf{x}])$ is the smallest box that contains $\mathbf{f}([\mathbf{x}])$. The minimal inclusion function for $\mathbf{f}$ is unique and will be denoted by $[\mathbf{f}]^*$, as it corresponds to the interval counterpart of $\mathbf{f}$. $[\mathbf{f}]$ is *inclusion monotonic* if

$$[\mathbf{x}] \subset [\mathbf{y}] \Rightarrow [\mathbf{f}]([\mathbf{x}]) \subset [\mathbf{f}]([\mathbf{y}]).$$

A minimal inclusion function is inclusion monotonic but not necessarily convergent (because $\mathbf{f}$ may be discontinuous). A convergent inclusion function may not be inclusion monotonic.

The construction of inclusion functions for $\mathbf{f}$ can be cast into that of inclusion functions for each of its coordinate functions. This is why we shall focus attention on getting inclusion functions for real-valued functions $f : \mathbb{R}^n \to \mathbb{R}$. The first idea that comes to mind is to perform two optimizations to compute the *infimum* and *supremum* of f over the box $[\mathbf{x}]$ of interest. At least in principle, one should thus get the smallest interval containing $f([\mathbf{x}])$, denoted by $[f]^*([\mathbf{x}])$. However, these optimization problems themselves turn out to be far from trivial in general. An alternative and much more tractable approach is as follows. Consider a function f expressed as a finite composition of the operators $+, -, \cdot, /$ and elementary functions sin, cos, exp, sqrt... An inclusion-monotonic inclusion function $[f] : \mathbb{IR}^n \to \mathbb{IR}$ for f is obtained by replacing each real variable x_i by an interval variable $[x_i]$ and each operator or elementary function by its interval counterpart. The result is called the *natural inclusion function* of f. If f involves only continuous operators and continuous elementary functions, then $[f]$ is convergent. If, moreover, each of the variables $x_1, \ldots, x_n$ occurs at most once in the formal expression of f then $[f]$ is minimal. Unfortunately, natural inclusion functions are not minimal in general, and their performance strongly depends on the formal expression of f, as illustrated by the next example.

Consider the following three formal expressions of the same function f:

$$\begin{aligned} f_1(x) &= x(x+2), \\ f_2(x) &= x^2 + 2x, \\ f_3(x) &= (x+1)^2 - 1. \end{aligned}$$

At $[x] = [-1, 1]$, their natural inclusion functions take the values

$$\begin{aligned} [f_1]([x]) &= [x]([x]+2) = [-3, 3], \\ [f_2]([x]) &= [x]^2 + 2[x] = [-2, 3], \\ [f_3]([x]) &= ([x]+1)^2 - 1 = [-1, 3]. \end{aligned}$$

$[f_2]$ is less pessimistic than $[f_1]$ because it takes advantage of the fact that the lower bound of a square cannot be negative. Since x occurs only once in f_3 and f_3 is continuous, $[f_3]$ is minimal. Thus $[f_3]([x]) = f([x]) = [-1,3]$.

Natural inclusion functions are not always to be recommended. Their efficiency depends strongly on the number of occurrences of each variable, which is often difficult to reduce. An important field of investigation of interval analysis (Ratschek and Rokne, 1984) has thus been to propose other types of inclusion function that might provide less pessimistic results, such as those now briefly described.

Assume that f is differentiable over $[\mathbf{x}]$, and denote $\text{mid}([\mathbf{x}])$ by $\mathbf{m}$. The mean-value theorem then implies that

$$\forall \mathbf{x} \in [\mathbf{x}], \exists \mathbf{z} \in [\mathbf{x}] \text{ such that}$$
$$f(\mathbf{x}) = f(\mathbf{m}) + \mathbf{g}^{\mathrm{T}}(\mathbf{z}) \cdot (\mathbf{x} - \mathbf{m}),$$

where $\mathbf{g}$ is the gradient of f. Thus,

$$\forall \mathbf{x} \in [\mathbf{x}], f(\mathbf{x}) \in f(\mathbf{m}) + [\mathbf{g}^{\mathrm{T}}]([\mathbf{x}]) \cdot (\mathbf{x} - \mathbf{m}),$$

where $[\mathbf{g}^{\mathrm{T}}]$ is an inclusion function for $\mathbf{g}^{\mathrm{T}}$, so

$$f([\mathbf{x}]) \subset f(\mathbf{m}) + [\mathbf{g}^{\mathrm{T}}]([\mathbf{x}]) \cdot ([\mathbf{x}] - \mathbf{m}).$$

Therefore, the interval function

$$[f_c]([\mathbf{x}]) = f(\mathbf{m}) + [\mathbf{g}^{\mathrm{T}}]([\mathbf{x}]) \cdot ([\mathbf{x}] - \mathbf{m})$$

is an inclusion function for f, called its *centered inclusion function*. This function becomes especially interesting when the width of $[\mathbf{x}]$ is small, because the pessimism resulting from the interval evaluation of $[\mathbf{g}]([\mathbf{x}])$ is reduced by the scalar product with $[\mathbf{x}] - \mathbf{m}$, which is a small interval centered on zero. The centered inclusion function can be significantly improved at the cost of a slightly more complicated formulation (Hansen, 1968). Iterating the reasoning that led to the centered inclusion function, one may think of using a Taylor series expansion to approximate f at a higher order. This leads to the second-order *Taylor inclusion function*

$$[f]_{\mathrm{T}}([\mathbf{x}]) = f(\mathbf{m}) + \mathbf{g}^{\mathrm{T}}(\mathbf{m}) \cdot ([\mathbf{x}] - \mathbf{m}) + \tfrac{1}{2}([\mathbf{x}] - \mathbf{m})^{\mathrm{T}} \cdot [\mathbf{H}]([\mathbf{x}]) \cdot ([\mathbf{x}] - \mathbf{m}),$$

where $[\mathbf{H}]$ is an inclusion function for the *Hessian matrix* of f.

The convergence rate of a natural inclusion function is at least linear, whereas those of the centered and Taylor inclusion functions are at least quadratic provided that the Taylor inclusion function is based on an expansion to order $k \geqslant 2$. Quadratic convergence looks more interesting than linear convergence, but it should be remembered that it only means that more accurate results will be obtained in the case of infinitesimal boxes. Nothing similar can be said on the behavior of these inclusion functions for large boxes. To the contrary, when the box involved is large enough, the natural inclusion function is often more satisfactory than the centered and Taylor inclusion functions. No approach to building an inclusion function can thus claim to be uniformly the best, and a compromise between complexity and efficiency must often be struck. One may also use several inclusion functions and take the intersection of their image sets to get a better approximation of the image set of the original function.

2.3 *Subpavings*

Intervals and boxes are not by themselves general enough to describe all sets of interest, which include, for instance, unions of disconnected subsets. This motivates the introduction of the notion of subpaving. A *subpaving* of a box $[\mathbf{x}]$ is a union of nonoverlapping subboxes of $[\mathbf{x}]$ with nonzero width. The set $\mathbb{S}$ of interest will be bracketed between inner and outer approximations by computing subpavings $\underline{\mathbb{S}}$ and $\overline{\mathbb{S}}$ such that

$$\underline{\mathbb{S}} \subset \mathbb{S} \subset \overline{\mathbb{S}}.$$

The distance between $\underline{\mathbb{S}}$ and $\overline{\mathbb{S}}$ is indicative of the quality of the approximation of $\mathbb{S}$. Computation on subpavings allows approximate computation on compact sets, and forms the basic ingredient of the parameter estimation algorithms to be presented below. The special class of *regular subpavings* simplifies representation and manipulation by computers as they can be represented by binary trees. A subpaving of $[\mathbf{x}]$ is regular if each of its boxes can be obtained from $[\mathbf{x}]$ by a finite succession of bisections and selections.

2.4 *Contractors*

Consider a vector $\mathbf{x}$ of n real variables x_i, linked by relations (or constraints) that can be written in vector form as

$$\mathbf{f}(\mathbf{x}) = \mathbf{0}. \tag{2}$$

Assume that the prior domain for $\mathbf{x}$ is the box

$$[\mathbf{x}] = [x_1] \times \cdots \times [x_n].$$

Solving (2) for $\mathbf{x}$ in $[\mathbf{x}]$ is a *constraint satisfaction problem* (CSP) (Davis, 1987), (Hyvönen, 1992), (Sam-Haroud and Faltings, 1996), (van Hentenryck *et al.*, 1998), the solution set of which is

$$\mathbb{S} = \{\mathbf{x} \in [\mathbf{x}] \mid \mathbf{f}(\mathbf{x}) = \mathbf{0}\}.$$

Inequality constraints can be dealt with within this framework by introducing *slack variables*. Instead of looking for $\mathbb{S}$, which is an NP-complete problem, *contractors* aim more modestly at reducing the size of the prior domain without loosing any solution or performing any bisection. Many contractors are available (Jaulin *et al.*, 2001*b*) and we shall only present one of them, namely the *interval Newton contractor* classically used for global optimization. Assume that $\mathbf{f}$ is once differentiable, and let $\mathbf{J_f}$ be its Jacobian matrix and $\mathbf{m}$ be the midpoint of $[\mathbf{x}]$. The mean-value theorem implies that for any $\mathbf{x} \in [\mathbf{x}]$ there exists $\mathbf{z} \in [\mathbf{x}]$ such that

$$\mathbf{f}(\mathbf{x}) = \mathbf{f}(\mathbf{m}) + \mathbf{J_f}(\mathbf{z}) \cdot (\mathbf{x} - \mathbf{m}). \tag{3}$$

Now assume that $\widehat{\mathbf{x}} \in [\mathbf{x}]$ is a solution of (2), so $\mathbf{f}(\widehat{\mathbf{x}}) = \mathbf{0}$. Assume further that the number of constraints is equal to the number of variables, so $\mathbf{J_f}$ is square. Provided that it is also invertible, (3) implies that

$$\widehat{\mathbf{x}} = \mathbf{m} - \mathbf{J_f}^{-1}(\mathbf{z}) \cdot \mathbf{f}(\mathbf{m}).$$

Now $\mathbf{z}$ is known to belong to $[\mathbf{x}]$, so

$$\widehat{\mathbf{x}} \in \mathbf{m} - \mathbf{J_f}^{-1}([\mathbf{x}]) \cdot \mathbf{f}(\mathbf{m}).$$

Since $\widehat{\mathbf{x}}$ is also assumed to belong to $[\mathbf{x}]$, it must belong to

$$[\mathbf{x}_r] = [\mathbf{x}] \cap \left(\mathbf{m} - \mathbf{J_f}^{-1}([\mathbf{x}]) \cdot \mathbf{f}(\mathbf{m})\right). \quad (4)$$

The interval Newton contractor thus replaces the prior box $[\mathbf{x}]$ by a possibly much smaller box $[\mathbf{x}_r]$, which may even be empty if $[\mathbf{x}]$ does not contain any solution. This is obtained without bisection and thus escapes the curse of dimensionality. The often unrealistic assumption that $\mathbf{J_f}([\mathbf{x}])$ is invertible can be relaxed by replacing (4) by the computation of an outer approximation of the set of all solutions for $\widehat{\mathbf{x}}$ in $[\mathbf{x}]$ of the *linear* system of equations.

$$\mathbf{f}(\mathbf{m}) + \mathbf{J_f}([\mathbf{x}]) \cdot (\widehat{\mathbf{x}} - \mathbf{m}) = \mathbf{0}.$$

Specific methods involving *preconditioning* are used for this purpose (Hammer *et al.*, 1995).

3. BASIC TOOLS FOR NONLINEAR ESTIMATION

3.1 *Global optimization*

The problem to be considered now is the minimization of a cost function c over a compact set $\mathbb{P} \subset \mathbb{R}^{n_p}$:

$$\min_{\mathbf{p} \in \mathbb{P}} c(\mathbf{p}).$$

It is, of course, always possible to transform a maximization problem into a minimization problem, for instance by multiplying $c(\mathbf{p})$ by -1. The *global minimum* will be denoted by $\widehat{c}$, and the set of all corresponding *global minimizers* by $\mathbb{S}$. Although interval analysis may also be used for constrained or minimax problems, we shall limit ourselves to *unconstrained minimization*. This means that $\mathbb{P}$ is only a (possibly very large) domain of interest, and that the minimizers are not expected to lie on its boundary. The main ideas of Hansen's algorithm will be presented. As this algorithm forms the main subject of his book, this will be an oversimplification and the reader is urged to consult (Hansen, 1992) for more details. For simplicity, the search domain $\mathbb{P}$ will be assumed to be a box $[\mathbf{p}]_0$ in parameter space. Let $\mathscr{L}$ be a list of boxes included in $\mathbb{P}$ that may contain global minimizers. Initially, $\mathscr{L}$ contains only $[\mathbf{p}]_0$. The basic structure of the algorithm is as follows, where $\varepsilon > 0$ is a tolerance parameter to be chosen by the user.

While $\mathscr{L}$ contains at least one box with width greater than ε, do {

(1) Pop the first such box $[\mathbf{p}]$ out of $\mathscr{L}$,
(2) Update the upper bound $\text{ub}(\widehat{c})$ of the global minimum $\widehat{c}$ over $[\mathbf{p}]_0$.
(3) Try to eliminate $[\mathbf{p}]$ via the midpoint, monotonicity and convexity tests.
(4) Should this fails, try to eliminate or at least reduce $[\mathbf{p}]$ with the interval Newton contractor.
(5) If $[\mathbf{p}]$ resists elimination and $w([\mathbf{p}]) > \varepsilon$ then bisect it and push the two resulting boxes into $\mathscr{L}$. Otherwise $[\mathbf{p}]$ is deemed too small to be bisected and will be pushed back as is into $\mathscr{L}$. }

At last eliminate all boxes of $\mathscr{L}$ that no longer pass the midpoint test.

Except during the first iteration of the algorithm, when $\mathscr{L}$ contains only one box, the rule for selecting the box to be considered at Step 1 must be chosen carefully as it has much impact on performance. A good strategy is to take one of the boxes with the largest potential, *i.e.*, a box $[\mathbf{p}]$ such that for any other box $[\mathbf{q}]$ in $\mathscr{L}$ with width greater than ε

$$\text{lb}([c]([\mathbf{p}]) \leqslant \text{lb}([c]([\mathbf{q}]).$$

To update the upper bound of $\widehat{c}$ as required by Step 2, it suffices to compute $c(\text{mid}([\mathbf{p}]))$, which is of course an upper bound of the optimal value of the cost, and to compare it with the best upper bound available so far.

Three methods for eliminating the box $[\mathbf{p}]$ under consideration are tried at Step 3. The *midpoint test* requires the evaluation of the image of $[\mathbf{p}]$ by an inclusion function $[c]$ of the cost function. If

$$\text{lb}([c]([\mathbf{p}])) > \text{ub}(\widehat{c}),$$

then no vector $\mathbf{p}$ in $[\mathbf{p}]$ can be a global minimizer of c over $\mathbb{P}$, so $[\mathbf{p}]$ can be eliminated. The *monotonicity test* takes advantage of the fact that if c is once differentiable then any unconstrained minimizer corresponds to a stationary point at which the gradient $\mathbf{g}$ takes the value zero. Thus if an inclusion function $[\mathbf{g}]$ is available for $\mathbf{g}$ and if $[\mathbf{p}]$ is such that

$$\mathbf{0} \notin [\mathbf{g}]([\mathbf{p}]),$$

then no vector $\mathbf{p}$ in $[\mathbf{p}]$ can be an unconstrained minimizer so $[\mathbf{p}]$ can be eliminated. Finally the *convexity test* is based on the fact that if c is twice differentiable, then a necessary condition for $\mathbf{p}$ to be an unconstrained minimizer is that the Hessian $\mathbf{H}$ of c be nonnegative definite at $\mathbf{p}$. A necessary condition for this to be true is that all diagonal entries of $\mathbf{H}(\mathbf{p})$ satisfy

$$h_{ii}(\mathbf{p}) \geqslant 0, i = 1, \ldots, n_p.$$

So if inclusion functions $[h_{ii}]$ are available for the diagonal entries of the Hessian and if $[\mathbf{p}]$ is such that

$$\exists i \in \{1, \ldots, n_p\} \mid [h_{ii}]([\mathbf{p}]) < 0,$$

then no $\mathbf{p}$ in $[\mathbf{p}]$ can be an unconstrained minimizer and $[\mathbf{p}]$ can be eliminated.

Automatic differentiation can be used to compute inclusion functions for $\mathbf{g}$ and $\mathbf{H}$ from the code evaluating $c(\mathbf{p})$ (Rall, 1981), (Griewank and Corliss, 1991),

(Hammer *et al.*, 1995), (Rall and Corliss, 1999), (Jaulin *et al.*, 2001*b*). Note that if the cost function c is not differentiable, or if it turns out that inclusion functions for its gradient or Hessian cannot be obtained, then parts of Step 3 may have to be dropped. In the limit, Step 3 may be based only on the midpoint test.

Step 4 uses an interval Newton contractor to reduce the size of $[\mathbf{p}]$ without having to bisect it. The problem is to find a box guaranteed to contain all solutions in $[\mathbf{p}]$ of

$$\mathbf{g}(\mathbf{p}) = \mathbf{0},$$

so $\mathbf{g}$ plays the role of $\mathbf{f}$ in Section 2.4, and the role of the Jacobian matrix of $\mathbf{f}$ is played by the Hessian of c. Step 4 can of course be implemented only if c is twice differentiable.

The simplest strategy for bisection at Step 5 is to cut $[\mathbf{p}]$ perpendicularly to one of its edges of maximum width, but more sophisticated policies taking into account information provided by the gradient may also be considered.

Since the tolerance parameter ε is strictly positive, the algorithm stops after a finite number of steps and the set $\mathbb{S}$ of all global minimizers of c is included in the union of the boxes in $\mathscr{L}$, each of which has now a width smaller than ε. It is also possible to compute an interval $[\widehat{c}]$ guaranteed to contain $\widehat{c}$, with its upper bound the best upper bound obtained at Step 2 and its lower bound given by

$$\underline{\widehat{c}} = \min_{[\mathbf{p}]\in\mathscr{L}} \operatorname{lb}([c]([\mathbf{p}]).$$

The algorithm can also be made to stop when

$$w([\widehat{c}]) < \eta,$$

where η is another tolerance parameter to be set by the user.

3.2 *Set inversion*

Let $\mathbf{f}$ be a possibly nonlinear function from $\mathbb{R}^{n_\mathrm{p}}$ to $\mathbb{R}^{n_\mathrm{y}}$ and $\mathbb{Y}$ be a subpaving of $\mathbb{R}^{n_\mathrm{y}}$. *Set inversion* is the characterization of the reciprocal image of $\mathbb{Y}$

$$\mathbb{S} = \{\mathbf{p} \in \mathbb{R}^{n_\mathrm{p}} \mid \mathbf{f}(\mathbf{p}) \in \mathbb{Y}\} = \mathbf{f}^{-1}(\mathbb{Y}). \qquad (5)$$

Using an inclusion function $[\mathbf{f}]$ for $\mathbf{f}$, two regular subpavings $\underline{\mathbb{S}}$ and $\overline{\mathbb{S}}$ such that

$$\underline{\mathbb{S}} \subset \mathbb{S} \subset \overline{\mathbb{S}}$$

can be obtained with the algorithm SIVIA, for Set Inverter Via Interval Analysis (Jaulin and Walter, 1993).

A (possibly very large) search box $[\mathbf{p}]_0$ to which $\overline{\mathbb{S}}$ is guaranteed to belong must be provided by the user. It will be used to grow a subpaving by successive bisections and selections. Consider a box $[\mathbf{p}]$ of this subpaving. Four cases may be encountered.

(1) If $[\mathbf{f}]([\mathbf{p}])$ has a nonempty intersection with $\mathbb{Y}$, but is not entirely in $\mathbb{Y}$, then $[\mathbf{p}]$ may contain a part of $\mathbb{S}$ without being included in $\mathbb{S}$ and is said to be *undetermined*. If it has a width greater than a prespecified precision parameter ε, then this undetermined box should be bisected into the boxes $L[\mathbf{p}]$ and $R[\mathbf{p}]$, each of which should be investigated by the algorithm.
(2) If $[\mathbf{f}]([\mathbf{p}])$ has an empty intersection with $\mathbb{Y}$, then $[\mathbf{p}]$ has an empty intersection with $\mathbb{S}$ and should be discarded.
(3) If $[\mathbf{f}]([\mathbf{p}])$ is included in $\mathbb{Y}$, then $[\mathbf{p}]$ is included in $\mathbb{S}$, and should be stored in $\underline{\mathbb{S}}$ and in $\overline{\mathbb{S}}$.
(4) If the box considered is undetermined, but its width is lower than ε, then it is deemed too small to be bisected further and is stored as is in the outer approximation $\overline{\mathbb{S}}$ of $\mathbb{S}$.

The resulting recursive algorithm is given in Table 1, where the subpavings $\underline{\mathbb{S}}$ and $\overline{\mathbb{S}}$ should be initialized as empty before the first call.

Algorithm SIVIA(in: $\mathbf{f}, \mathbb{Y}, [\mathbf{p}], \varepsilon$; inout: $\underline{\mathbb{S}}, \overline{\mathbb{S}}$)

if $[\mathbf{f}]([\mathbf{p}]) \cap \mathbb{Y} = \emptyset$ return;
if $[\mathbf{f}]([\mathbf{p}]) \subset \mathbb{Y}$ then
$\{\underline{\mathbb{S}} := \underline{\mathbb{S}} \cup [\mathbf{p}]$; $\overline{\mathbb{S}} := \overline{\mathbb{S}} \cup [\mathbf{p}]$; return;};
if $w([\mathbf{p}]) < \varepsilon$ then $\{\overline{\mathbb{S}} := \overline{\mathbb{S}} \cup [\mathbf{p}]$; return;};
SIVIA($\mathbf{f}, \mathbb{Y}, L[\mathbf{p}], \varepsilon, \underline{\mathbb{S}}, \overline{\mathbb{S}}$);
SIVIA($\mathbf{f}, \mathbb{Y}, R[\mathbf{p}], \varepsilon, \underline{\mathbb{S}}, \overline{\mathbb{S}}$).

Table 1. SIVIA

The *uncertainty layer* $\Delta\mathbb{S}$ consisting of all boxes of $\overline{\mathbb{S}}$ that are not in $\underline{\mathbb{S}}$ is a regular subpaving, all boxes of which have a width smaller than ε.

3.3 *Guaranteed numerical integration*

As seen above, Hansen's algorithm and SIVIA both require the availability of inclusion functions. When no closed-form expression or finite algorithm is available for the evaluation of $\mathbf{f}$, the construction of $[\mathbf{f}]$ becomes much more difficult. Such a situation often occurs, for example, when the evaluation of $\mathbf{f}$ requires the solution of some ordinary differential equation

$$\mathbf{x}' = \mathbf{g}(\mathbf{x}, \mathbf{p}, t), \text{ with } \mathbf{x}(0) = \mathbf{x}_0(\mathbf{p}), \qquad (6)$$

where

$$\mathbf{x}' = \frac{\mathrm{d}\mathbf{x}}{\mathrm{d}t}$$

and $\mathbf{p}$ is some parameter vector only known to belong to a box $[\mathbf{p}]$. At a given time instant t, let $\mathbf{f}(\mathbf{p},t)$ be the solution of (6) for a given $\mathbf{p} \in [\mathbf{p}]$ and let $\mathbf{f}([\mathbf{p}],t)$ be the set of all values that can be reached by the solution of (6) for all $\mathbf{p}$ in $[\mathbf{p}]$.

Several numerical integrators based on interval analysis are readily available to compute boxes guaranteed to contain $\mathbf{f}([\mathbf{p}],t)$, *e.g.*, AWA (Lohner, 1992), COSY (Hoefkens *et al.*, 2001) or VNODE (Nedialkov and Jackson, 2001). The techniques employed usually consist of two steps. First, the existence of a solution for (6) over a given time interval is ensured using

Brouwer's fixed-point theorem. As a result, a box containing $\mathbf{f}([\mathbf{p}],t)$ is obtained. This box is then refined using a Taylor expansion of the solution or other similar schemes to get a smaller box $[\mathbf{f}]([\mathbf{p}],t)$ satisfying

$$\mathbf{f}([\mathbf{p}],t) \subset [\mathbf{f}]([\mathbf{p}],t).$$

Guaranteed numerical integration thus provides an inclusion function for $\mathbf{f}(.,t)$. This technique is well suited for systems such as (6), where $[\mathbf{p}]$ is a degenerate box with zero width. For large boxes as needed in the context of parameter estimation, the enclosure for $\mathbf{f}([\mathbf{p}],t)$ may become *very* pessimistic. However, when (6) can be bounded between *cooperative systems*, it remains possible to obtain an accurate inclusion function for $\mathbf{f}([\mathbf{p}],t)$, see (Gouzé *et al.*, 2000).

Definition 1. The dynamical system

$$\mathbf{x}' = \mathbf{g}(\mathbf{x},t),$$

where $\mathbf{x} \in \mathscr{D} \subset \mathbb{R}^n$, is *cooperative* over $\mathscr{D}$ if

$$\frac{\partial g_i(\mathbf{x},t)}{\partial x_j} \geqslant 0 \text{ for all } i \neq j,\ t \geqslant 0 \text{ and } \mathbf{x} \in \mathscr{D}.$$

The following theorem, which is a reformulation of a result described in (Smith, 1995), indicates how an enclosure of the solution of (6) can be obtained with cooperative systems when $\mathbf{p}$ is only known to belong to $[\mathbf{p}]$.

Theorem 1. If there exists a pair of cooperative systems

$$\underline{\mathbf{x}}' = \underline{\mathbf{g}}(\underline{\mathbf{x}},\underline{\mathbf{p}},\overline{\mathbf{p}},t) \text{ and } \overline{\mathbf{x}}' = \overline{\mathbf{g}}(\overline{\mathbf{x}},\underline{\mathbf{p}},\overline{\mathbf{p}},t)$$

satisfying

$$\underline{\mathbf{g}}(\mathbf{x},\underline{\mathbf{p}},\overline{\mathbf{p}},t) \leqslant \mathbf{g}(\mathbf{x},\mathbf{p},t) \leqslant \overline{\mathbf{g}}(\mathbf{x},\underline{\mathbf{p}},\overline{\mathbf{p}},t)$$

for all $\mathbf{p} \in [\underline{\mathbf{p}},\overline{\mathbf{p}}]$, $t \geqslant 0$ and $\mathbf{x} \in \mathscr{D}$, and if these cooperative systems can be given initial conditions $\underline{\mathbf{x}}_0(\underline{\mathbf{p}},\overline{\mathbf{p}})$ and $\overline{\mathbf{x}}_0(\underline{\mathbf{p}},\overline{\mathbf{p}})$ such that

$$\underline{\mathbf{x}}_0(\underline{\mathbf{p}},\overline{\mathbf{p}}) \leqslant \mathbf{x}_0(\mathbf{p}) \leqslant \overline{\mathbf{x}}_0(\underline{\mathbf{p}},\overline{\mathbf{p}})$$

for all $\mathbf{p} \in [\underline{\mathbf{p}},\overline{\mathbf{p}}]$, then the solution of (6) satisfies

$$\underline{\mathbf{x}}(t) \leqslant \mathbf{x}(t) \leqslant \overline{\mathbf{x}}(t), \text{ for all } t \geqslant 0,$$

where $\underline{\mathbf{x}}(t) = \underline{\varphi}(\underline{\mathbf{p}},\overline{\mathbf{p}},t)$ is the flow associated with

$$\{\underline{\mathbf{x}}' = \underline{\mathbf{g}}(\underline{\mathbf{x}},\underline{\mathbf{p}},\overline{\mathbf{p}},t), \underline{\mathbf{x}}(0) = \underline{\mathbf{x}}_0(\underline{\mathbf{p}},\overline{\mathbf{p}})\}$$

and $\overline{\mathbf{x}}(t) = \overline{\varphi}(\underline{\mathbf{p}},\overline{\mathbf{p}},t)$ is the flow associated with

$$\{\overline{\mathbf{x}}' = \overline{\mathbf{g}}(\overline{\mathbf{x}},\underline{\mathbf{p}},\overline{\mathbf{p}},t), \overline{\mathbf{x}}(0) = \overline{\mathbf{x}}_0(\underline{\mathbf{p}},\overline{\mathbf{p}})\}.$$

■

For any given $t \geqslant 0$, the box-valued function

$$[\varphi](\underline{\mathbf{p}},\overline{\mathbf{p}},t) = \left[\underline{\varphi}(\underline{\mathbf{p}},\overline{\mathbf{p}},t), \overline{\varphi}(\underline{\mathbf{p}},\overline{\mathbf{p}},t)\right]$$

is thus an inclusion function for the solution of (6). Usually, however, no closed-form expressions are available for $\underline{\varphi}(\underline{\mathbf{p}},\overline{\mathbf{p}},t)$ and $\overline{\varphi}(\underline{\mathbf{p}},\overline{\mathbf{p}},t)$, but guaranteed integration makes it possible to compute enclosures for these flows at any $t \geqslant 0$ as

$$\left[\underline{\varphi}(\underline{\mathbf{p}},\overline{\mathbf{p}},t)\right] = \left[\underline{\underline{\varphi}(\underline{\mathbf{p}},\overline{\mathbf{p}},t)}, \overline{\underline{\varphi}(\underline{\mathbf{p}},\overline{\mathbf{p}},t)}\right]$$

and

$$\left[\overline{\varphi}(\underline{\mathbf{p}},\overline{\mathbf{p}},t)\right] = \left[\underline{\overline{\varphi}(\underline{\mathbf{p}},\overline{\mathbf{p}},t)}, \overline{\overline{\varphi}(\underline{\mathbf{p}},\overline{\mathbf{p}},t)}\right].$$

The function

$$[\Phi](\underline{\mathbf{p}},\overline{\mathbf{p}},t) = \left[\underline{\underline{\varphi}(\underline{\mathbf{p}},\overline{\mathbf{p}},t)}, \overline{\overline{\varphi}(\underline{\mathbf{p}},\overline{\mathbf{p}},t)}\right]$$

is then such that

$$[\mathbf{f}]([\mathbf{p}],t) \subset [\Phi](\underline{\mathbf{p}},\overline{\mathbf{p}},t) \tag{7}$$

and is therefore an inclusion function for the solution of (6), which can be numerically evaluated for any $t \geqslant 0$.

Note that the right-hand side of (7) no longer involves boxes, contrary to its left-hand side. This allows guaranteed numerical integration to be used on degenerate boxes, with much more accurate results.

4. PARAMETER ESTIMATION

The interval tools previously presented will now be applied to two types of nonlinear parameter estimation problems, namely parameter optimization and parameter bounding.

4.1 *Parameter optimization*

Nonlinear parameter estimation often boils down to the optimization of a cost function, which may for instance result from information or hypotheses about the noise corrupting the data. This cost function is not convex in general, and local optimization techniques cannot be guaranteed to locate its global minimum and all corresponding global minimizers. Since computing time is finite, no global search method based on random exploration can offer any guarantee either. This makes deterministic global optimization algorithms such as that presented in Section 3.1 particularly attractive.

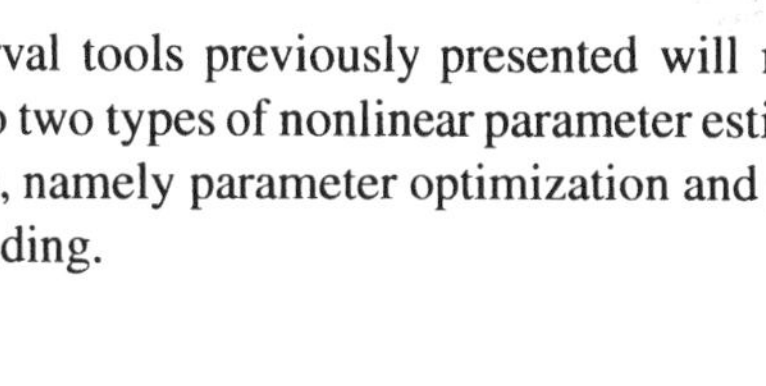

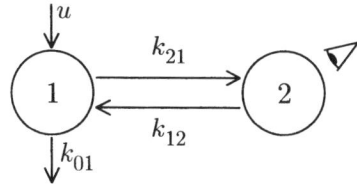

Fig. 1. Two-compartment model

To illustrate their potential and limitation, consider the compartmental model described by Figure 1. Such models are widely used in biology. They consist of finite sets of homogeneous subsystems, called compartments and represented by circles, which may exchange material as indicated by arrows. The equations describing the behavior of the compartmental model are readily obtained by writing down conservation equations, under the form of a state equation. Let $\mathbf{x} = (x_1, x_2)^{\mathrm{T}}$ be the vector of the (positive) quantities of material in the two compartments of the model of Figure 1 and assume, for the time being, that all flows of material are proportional to the quantity of material in the origin compartment. The evolution of $\mathbf{x}$ is then governed by the linear state equation

$$\mathbf{x}' = \mathbf{g}(\mathbf{x}, \mathbf{p}, u), \tag{8}$$

where

$$\mathbf{p} = (k_{01}, k_{12}, k_{21})^{\mathrm{T}}$$

and

$$\mathbf{g}(\mathbf{x}, \mathbf{p}, u) = \begin{pmatrix} -(k_{21}+k_{01})x_1 + k_{12}x_2 + u \\ k_{21}x_1 - k_{12}x_2 \end{pmatrix}.$$

Assume that the quantity x_2 of material in Compartment 2 is observed, so the model output is

$$y_{\mathrm{m}}(\mathbf{p}, t_i) = x_2(\mathbf{p}, t_i),\ i = 1, ..., n_{\mathrm{y}}. \tag{9}$$

Assume further that there is no input ($u \equiv 0$) and that the initial condition is $\mathbf{x}_0 = (1, 0)^{\mathrm{T}}$. Then, for $t_i \geqslant 0$,

$$y_{\mathrm{m}}(\mathbf{p}, t_i) = \alpha(\mathbf{p}) \left(\mathrm{e}^{\lambda_1(\mathbf{p})t_i} - \mathrm{e}^{\lambda_2(\mathbf{p})t_i} \right), \tag{10}$$

where

$$\alpha(\mathbf{p}) = \frac{k_{21}}{\sqrt{(k_{01} - k_{12} + k_{21})^2 + 4k_{12}k_{21}}}, \tag{11}$$

$$\lambda_{1,2}(\mathbf{p}) = -\frac{1}{2}[(k_{01} + k_{12} + k_{21}) \pm ((k_{01} - k_{12} + k_{21})^2 + 4k_{12}k_{21})^{1/2}]. \tag{12}$$

Although the model is linear, its output thus depends nonlinearly on its parameters.

The least-square estimate of $\mathbf{p}$ is obtained by minimizing the cost function

$$c(\mathbf{p}) = \sum_{i=1}^{n_{\mathrm{y}}} (y(t_i) - y_{\mathrm{m}}(\mathbf{p}, t_i))^2, \tag{13}$$

where $y(t_i)$ is the quantity of material measured in Compartment 2 at time t_i. Results obtained with Hansen's algorithm on this problem where reported in (Kieffer and Walter, 1998) and will only be summarized here. With sixteen data points, the search domain $\mathbb{P} = [0.01, 2.0] \times [0.05, 3.0] \times [0.05, 3.0]$ and a precision parameter $\varepsilon = 10^{-9}$, it took about a day on a Pentium at 200MHz to enclose the set of all global minimizers of (13) in the union of two very small boxes in parameter space. These two solutions are due to the fact that the parameters of this model are only locally identifiable, see, *e.g.*, (Walter, 1982), because the values of k_{12} and k_{01} can be exchanged without modifying input-output behavior. Note that no identifiability study was needed to reach this conclusion here. The disappointing length of these computations is due to multioccurrences of the parameters in the natural inclusion function for $y_{\mathrm{m}}(\mathbf{p}, t_i)$ and *a fortiori* in the natural inclusion function for c. Note that $\mathbf{p}$ appears n_{y} times in the formal expression of the cost function. Optimization time can be drastically decreased to about 90s by using an intermediate parametrization of $y_{\mathrm{m}}(\mathbf{p}, t_i)$ in terms of α, λ_1 and λ_2 and then solving (11) and (12) for $\mathbf{p}$ using an interval Newton solver.

4.2 *Parameter bounding*

Parameter bounding, see, *e.g.*, (Milanese *et al.*, 1996), represents an attractive alternative to parameter optimization. Instead of looking for an optimal value of $\mathbf{p}$, one looks for the *set* of all parameter vectors that are consistent (in a sense to be specified) with the experimental data, model structure and error bounds. It is assumed that each experimental datum $y(t_i)$ corresponds to a known interval $[\underline{e}_i, \overline{e}_i]$ of acceptable errors. A parameter vector $\mathbf{p} \in [\mathbf{p}]_0$ is deemed acceptable if $\underline{e}_i \leqslant y(t_i) - y_{\mathrm{m}}(\mathbf{p}, t_i) \leqslant \overline{e}_i$ for all $i = 1, \ldots, n_{\mathrm{y}}$. Parameter estimation then amounts to the characterization of the set

$$\begin{aligned} \mathbb{S} &= \{\mathbf{p} \in [\mathbf{p}]_0 \mid y(t_i) - y_{\mathrm{m}}(\mathbf{p}, t_i) \in [\underline{e}_i, \overline{e}_i], \\ &\qquad i = 1, \ldots, n_{\mathrm{y}}\} \\ &= \{\mathbf{p} \in [\mathbf{p}]_0 \mid \mathbf{y}_{\mathrm{m}}(\mathbf{p}) \in [\mathbf{y}]\}, \end{aligned} \tag{14}$$

with

$$[\mathbf{y}] = [y(t_1) - \overline{e}_1, y(t_1) - \underline{e}_1] \times \cdots \times [y(t_{n_{\mathrm{y}}}) - \overline{e}_{n_{\mathrm{y}}}, y(t_{n_{\mathrm{y}}}) - \underline{e}_{n_{\mathrm{y}}}]$$

and

$$\mathbf{y}_{\mathrm{m}}(\mathbf{p}) = (y_{\mathrm{m}}(\mathbf{p}, t_1), \ldots, y_{\mathrm{m}}(\mathbf{p}, t_{n_{\mathrm{y}}}))^{\mathrm{T}}.$$

One of the advantages of this approach from the point of view of interval analysis is that the multioccurrences of $\mathbf{p}$ due to cost functions such as (13) are avoided, thus decreasing the pessimism of inclusion functions and the number of bisections required to reach a satisfactory conclusion.

A guaranteed enclosure of $\mathbb{S}$ can be obtained using SIVIA, described in Section 3.2, as (14) is a set-inversion problem similar to (5), with $\mathbb{Y} = [\mathbf{y}]$. Solving (14) with SIVIA requires an inclusion function for $\mathbf{y}_{\mathrm{m}}(\mathbf{p})$ to be provided. Two approaches may be considered depending on whether a closed-form expression is available for $y_{\mathrm{m}}(\mathbf{p}, t_i)$. They will be illustrated by considering again the model of Figure 1.

4.2.1. *Set inversion using a closed-form expression*

Assume that

$$y(t_i) = x_2(\mathbf{p}^*, t_i)(1 + b(t_i)),\ i = 1, ..., 16, \tag{15}$$

where $\mathbf{p}^*$ is the true value of the parameter vector and $b(t_i) \in [\underline{b}, \overline{b}]$. Noisy data have been obtained as follows. First the noise-free response $x_2(\mathbf{p}^*, t_i)$ was computed by simulating (8) with the true value of the parameters to be estimated given by $(k_{01}^*, k_{12}^*, k_{21}^*) = (1, 0.25, 0.5)$. This noise-free response was then corrupted according to (15) with realizations of a pseudorandom noise $b(t_i)$ belonging to $[-0.05, 0.05]$. The resulting data points are indicated by dots on Figure 2.

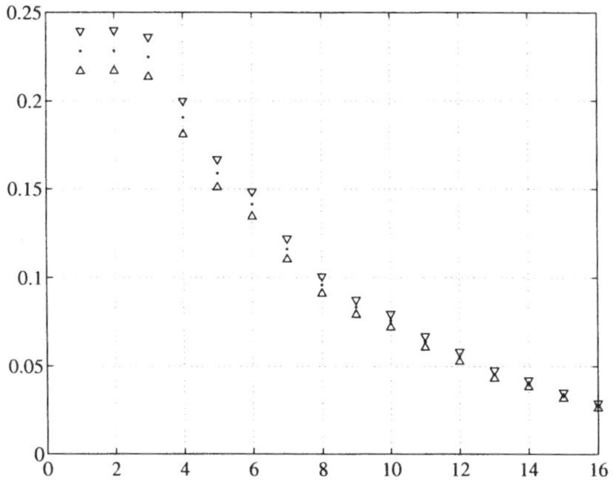

Fig. 2. Measured output (·), upper bounds (▽) and lower bounds (△) of the tolerance intervals (the true system is linear)

The intervals $[\underline{e}_i, \overline{e}_i]$ must be chosen such that

$$x_2(\mathbf{p}^*, t_i) \in [y(t_i) - \overline{e}_i, y(t_i) - \underline{e}_i], \; i = 1, ..., 16. \quad (16)$$

Equation (15) implies that

$$x_2(\mathbf{p}^*, t_i) \in \frac{1}{1 + [-0.05, 0.05]} y(t_i) \subset [0.95238, 1.05264] y(t_i),$$

so taking

$$\underline{e}_i = -0.05264 y(t_i) \quad (17)$$

and

$$\overline{e}_i = 0.04762 y(t_i) \quad (18)$$

ensures that (16) is satisfied.

Using SIVIA with the initial search box $[\mathbf{p}]_0 = [0, 5]^{\times 3}$ and various values of the precision parameter ε leads to the results summarized in Table 2 for computations on an Athlon 1800+.

ε	0.005	0.0025	0.00125
Comput. time (s)	9	14	24
Volume of $\overline{\mathbb{S}}$	$1.7 \cdot 10^{-3}$	$4 \cdot 10^{-4}$	$1.2 \cdot 10^{-4}$

Table 2. Estimation results using a closed-form expression

Figure 3 presents the projections of $\overline{\mathbb{S}}$ onto the (k_{01}, k_{12}) and (k_{12}, k_{21}) planes when $\varepsilon = 0.0025$. $\mathbb{S}$ consists of two disconnected subsets. Again, k_{01} and k_{12} can be exchanged without modifying the input-output behavior of the model.

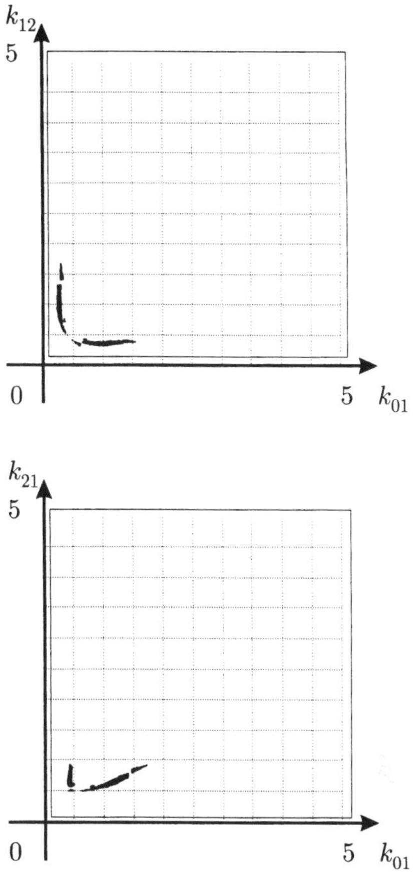

Fig. 3. Projections onto the (k_{01}, k_{12}) and (k_{01}, k_{21}) planes of $\overline{\mathbb{S}}$ found using a closed-form expression with $\varepsilon = 0.0025$

4.2.2. *Set inversion using guaranteed integration* With this approach, a closed-form expression for $y_m(\mathbf{p}, t_i)$ is no longer needed.

For a given box $[\mathbf{p}] = [\underline{\mathbf{p}}, \overline{\mathbf{p}}]$ in parameter space such that $\underline{\mathbf{p}} \geqslant \mathbf{0}$, with

$$\underline{\mathbf{p}} = (\underline{k}_{01}, \underline{k}_{12}, \underline{k}_{21})^{\mathrm{T}} \text{ and } \overline{\mathbf{p}} = (\overline{k}_{01}, \overline{k}_{12}, \overline{k}_{21})^{\mathrm{T}},$$

it is possible to enclose $\mathbf{g}(\mathbf{x}, \mathbf{p}, u)$ in (8) between

$$\underline{\mathbf{g}}(\mathbf{x}, \underline{\mathbf{p}}, \overline{\mathbf{p}}, u) = \begin{pmatrix} -(\overline{k}_{21} + \overline{k}_{01}) x_1 + \underline{k}_{12} x_2 + u \\ \underline{k}_{21} x_1 - \overline{k}_{12} x_2 \end{pmatrix}$$

and

$$\overline{\mathbf{g}}(\mathbf{x}, \underline{\mathbf{p}}, \overline{\mathbf{p}}, u) = \begin{pmatrix} -(\underline{k}_{21} + \underline{k}_{01}) x_1 + \overline{k}_{12} x_2 + u \\ \overline{k}_{21} x_1 - \underline{k}_{12} x_2 \end{pmatrix}$$

As $\underline{\mathbf{p}} \geqslant \mathbf{0}$, it is easy to show that

$$\mathbf{x}' = \underline{\mathbf{g}}(\mathbf{x}, \underline{\mathbf{p}}, \overline{\mathbf{p}}, u) \text{ and } \mathbf{x}' = \overline{\mathbf{g}}(\mathbf{x}, \underline{\mathbf{p}}, \overline{\mathbf{p}}, u)$$

are cooperative. The technique presented in Section 3.3 thus makes it possible to obtain an inclusion function for $y_m(\mathbf{p}, t_i)$. Since the initial condition of the model does not depend on $\mathbf{p}$, the bounding cooperative systems are given the same initial condition

$$\underline{\mathbf{x}}_0 = \mathbf{x}_0 = \overline{\mathbf{x}}_0.$$

For $[\underline{e}_i, \overline{e}_i]$, $i = 1, \ldots, 16$, the same intervals are taken as in Section 4.2.1. Using SIVIA and the guaranteed integration toolbox VNODE with the same initial search box in parameter space now leads to the results summarized in Table 3. The shape and volume of the set obtained for $\varepsilon = 0.005$ are similar to those obtained in Section 4.2.1 for $\varepsilon = 0.0025$.

ε	0.01	0.005
Comput. time (s)	1300	1600
Volume of $\overline{\mathbb{S}}$	$2.5 \cdot 10^{-3}$	$6 \cdot 10^{-4}$

Table 3. Estimation results using guaranteed integration

Two remarks are in order when comparing Tables 2 and 3. First, with the same value of the precision parameter ε, the volume of $\overline{\mathbb{S}}$ is smaller when using guaranteed integration than with the closed-form expression used in Section 4.2.1. This is due to a less pessimistic inclusion function. However, if results of the same accuracy are compared, the computing time using guaranteed integration is more than 100 times larger than with the closed-form expression.

4.2.3. *Nonlinear system* Assume now that k_{01} in (8) depends on x_1 according to

$$k_{01}(x_1) = \frac{a}{1+bx_1}.$$

This corresponds to a Michaelis-Menten nonlinearity (Godfrey, 1983).

The evolution of the quantities of material in the two compartments of the model of Figure 1 is now described by the nonlinear state equation

$$\mathbf{x}' = \mathbf{h}(\mathbf{x}, \mathbf{p}, u), \qquad (19)$$

where

$$\mathbf{p} = (a, b, k_{12}, k_{21})^{\mathrm{T}}$$

and

$$\mathbf{h}(\mathbf{x}, \mathbf{p}, u) = \begin{pmatrix} -k_{21}x_1 - \dfrac{ax_1}{1+bx_1} + k_{12}x_2 + u \\ k_{21}x_1 - k_{12}x_2 \end{pmatrix}.$$

Again, Compartment 2 is observed, with input and initial condition as in Sections 4.2.1 and 4.2.2. The observation equation is as in (15), with the same hypothesis about the $b(t_i)$s. An inclusion function based on guaranteed numerical integration will be employed.

For any $\mathbf{p} \in [\underline{\mathbf{p}}, \overline{\mathbf{p}}]$ such that $\underline{\mathbf{p}} \geqslant \mathbf{0}$, it is possible to bound $\mathbf{h}(\mathbf{x}, \mathbf{p}, u)$ in (19) between

$$\begin{pmatrix} -\left(\overline{k}_{21} + \dfrac{\overline{a}}{1+\underline{b}x_1}\right)x_1 + \underline{k}_{12}x_2 + u \\ \underline{k}_{21}x_1 - \overline{k}_{12}x_2 \end{pmatrix}$$

and

$$\begin{pmatrix} -\left(\underline{k}_{21} + \dfrac{\underline{a}}{1+\overline{b}x_1}\right)x_1 + \overline{k}_{12}x_2 + u \\ \overline{k}_{21}x_1 - \underline{k}_{12}x_2 \end{pmatrix}.$$

The resulting systems are cooperative, as $\underline{\mathbf{p}} \geqslant \mathbf{0}$. It is thus again simple to built an inclusion function for $y_{\mathrm{m}}(\mathbf{p}, t_i)$.

Two sets of data points have been considered. In the first one, the data points $y(t_i)$ and intervals $[\underline{e}_i, \overline{e}_i]$, $i = 1, \ldots, 16$, are as in Sections 4.2.1 and 4.2.2. SIVIA has been used here with the initial search box

$$[\mathbf{p}]_0 = [0,5] \times [0,5] \times [0.25, 0.25] \times [0.5, 0.5],$$

which means that k_{12} and k_{21} are now treated as known *a priori*. When $\varepsilon = 0.001$, in less than 55s on an Athlon 1800+, the subpaving $\overline{\mathbb{S}}$ presented on Figure 4 is obtained.

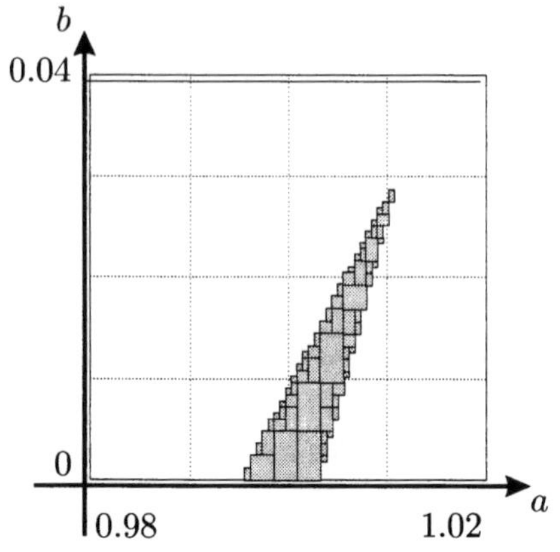

Fig. 4. Outer approximation of the solution set for (a, b) when the true system is linear

By projecting it on the axes of parameter space, one gets the following parameter uncertainty intervals

$$[a] = [0.9955, 1.0114]$$

and

$$[b] = [0, 0.02930].$$

Since the data have been generated with a linear model, it comes as no surprise that the parameter uncertainty interval for b includes 0.

The second set of data points has been obtained by corrupting the noise-free response $x_2(\mathbf{p}^*, t_i)$ of (19) where $(a^*, b^*, k_{12}^*, k_{21}^*) = (1, 4/3, 0.25, 0.5)$ with realizations of a pseudorandom noise $b(t_i)$ belonging to the interval $[-0.05, 0.05]$ according to (15), see Figure 5.

The intervals for $[\underline{e}_i, \overline{e}_i]$, $i = 1, \ldots, 16$, are computed as before, using (17) and (18), and the initial search box is as for the first set of data points.

When $\varepsilon = 0.01$, in less than 4mn, the subpaving $\overline{\mathbb{S}}$ presented in Figure 6 is obtained. As b cannot be zero, the data represented on Figure 5 could not have been generated by a linear model, given our hypotheses.

5. SOFTWARE

We shall limit ourselves to tools that can be downloaded freely, at least for academic use.

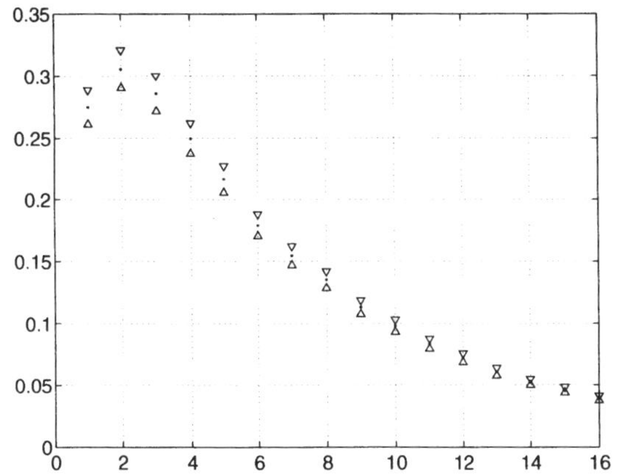

Fig. 5. Measured output (·), upper bounds (▽) and lower bounds (△) of the tolerance intervals (the true system is nonlinear)

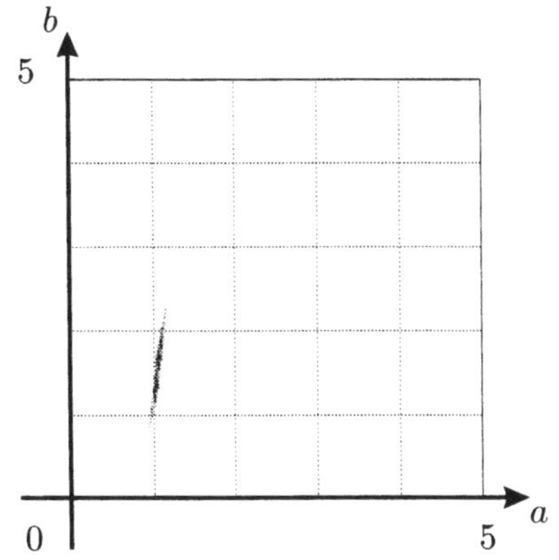

Fig. 6. Outer approximation of the solution set for (a,b) when the true system is nonlinear

Those familiar with MATLAB will find INTLAB, a library developed by S. M. Rump, particularly useful for experimenting with interval techniques. INTLAB runs under WINDOWS, UNIX and LINUX and may be downloaded from

`http://www.ti3.tu-harburg.de /english/index.html`

A number of interval algorithms cannot be implemented efficiently enough with an interpreted language such as MATLAB; this is why interval libraries built in compiled languages should also be useful.

C-XSC 2.0, a C++ class library developed by many people over more than a decade of an effort started at the University of Karlsruhe, is available at

`http://www.math.uni-wuppertal.de /~xsc/xsc/download.html`

together with a C++ toolbox for verified computing that works with C-XSC 2.0.

PROFIL/BIAS, also written in C++ and developed by O. Knüppel and S. M. Rump, is available at

`ftp://ti3sun.ti3.tu-harburg.de /pub/profil/`

Chapter 11 of (Jaulin *et al.*, 2001*b*) shows how a library such as PROFIL/BIAS could be built and how subpavings and algorithms such as SIVIA can then be implemented. The corresponding C++ source code is available at

`http://www.lss.supelec.fr /books/intervals`

A FORTRAN 90 software developed by R. B. Kearfott for global optimization is available at

`http://interval.louisiana.edu /GlobSol/download_GlobSol.html`

All of these products have benefited considerably from object-oriented programming and operator overloading, which allow intervals and boxes to be manipulated about as simply as standard data types.

Tools for guaranteed numerical integration have been developed in various languages. AWA uses PASCAL-XSC and may be found at

`ftp://ftp.iam.uni-karlsruhe.de /pub/awa/`

COSY was written in FORTRAN 90 and is available at

`http://cosy.pa.msu.edu/`

VNODE has been developed in C++ using PROFIL/BIAS and may be downloaded from

`http://www.cas.mcmaster.ca/~nedialk /Software/VNODE/VNODE.shtml`

A regularly updated list of pointers to software implementing interval analysis can be found at

`http://www.cs.utep.edu /interval-comp/main.html`

6. CONCLUSIONS

When nonlinear parameter estimation involves minimizing possibly nonconvex cost functions or solving sets of nonlinear inequalities, global deterministic methods based on interval analysis have definite advantages over more conventional local iterative methods that are unable to provide any guarantee as to their results. Structural identifiability studies can be bypassed since all solutions are provided, including (but not limited to) those due to a lack of global identifiability. Examples drawn from compartmental modeling have shown that it is possible to estimate the parameters of models defined by (possibly nonlinear) differential equations. The main challenge is to increase the complexity of the problems that can be considered as much as possible. Two allies in this endeavor have been briefly presented; the first one is the notion of contractor, which allows boxes to be reduced and sometimes eliminated without bisection, and the second is the property of cooperativity, which allows efficient inclusion functions to be derived for important classes of differential models.

The ideas presented here in the context of parameter identification readily extend to state estimation or parameter tracking, see (Kieffer *et al.*, 1998), (Jaulin *et al.*, 2001*a*) and (Jaulin, 2002).

7. REFERENCES

Davis, E. (1987). Constraint propagation with interval labels. *Artificial Intelligence* **32**(3), 281–331.

Godfrey, K. (1983). *Compartimental Models and Their Application*. Academic Press. London.

Gouzé, J. L., A. Rapaport and Z. M. Hadj-Sadok (2000). Interval observers for uncertain biological systems. *Journal of Ecological Modelling* (133), 45–56.

Griewank, A. and Corliss, G. F., Eds.) (1991). *Automatic Differentiation of Algorithms: Theory, Implementation and Applications*. SIAM. Philadelphia, PA.

Hammer, R., M. Hocks, U. Kulisch and D. Ratz (1995). *C++ Toolbox for Verified Computing*. Springer-Verlag. Berlin.

Hansen, E. R. (1968). On solving systems of equations using interval arithmetic. *Mathematical Computing* **22**, 374–384.

Hansen, E. R. (1992). *Global Optimization Using Interval Analysis*. Marcel Dekker. New York, NY.

Hoefkens, J., M. Berz and K. Makino (2001). Efficient high-order methods for ODEs and DAEs. In: *Automatic Differentiation : From Simulation to Optimization* (G. Corliss, C. Faure and A. Griewank, Eds.). Springer-Verlag. New-York, NY. pp. 341–351.

Hyvönen, E. (1992). Constraint reasoning based on interval arithmetic; the tolerance propagation approach. *Artificial Intelligence* **58**(1-3), 71–112.

Jaulin, L. (2002). Nonlinear bounded-error state estimation of continuous-time systems. *Automatica* **38**, 1079–1082.

Jaulin, L. and E. Walter (1993). Set inversion via interval analysis for nonlinear bounded-error estimation. *Automatica* **29**(4), 1053–1064.

Jaulin, L., M. Kieffer, I. Braems and E. Walter (2001*a*). Guaranteed nonlinear estimation using constraint propagation on sets. *International Journal of Control* **74**(18), 1772–1782.

Jaulin, L., M. Kieffer, O. Didrit and E. Walter (2001*b*). *Applied Interval Analysis*. Springer-Verlag. London.

Kearfott, R. B. (1996). *Rigorous Global Search: Continuous Problems*. Kluwer. Dordrecht, the Netherlands.

Kieffer, M. and E. Walter (1998). Interval analysis for guaranteed nonlinear parameter estimation. In: *MODA 5-Advances in Model-Oriented Data Analysis and Experiment Design* (A. C. Atkinson, L. Pronzato and H. P. Wynn, Eds.). pp. 115–125. Physica-Verlag. Heidelberg.

Kieffer, M., L. Jaulin and E. Walter (1998). Guaranteed recursive nonlinear state estimation using interval analysis. In: *Proceedings of the 37th IEEE Conference on Decision and Control*. Tampa, FL. pp. 3966–3971.

Lohner, R. (1992). Computation of guaranteed enclosures for the solutions of ordinary initial and boundary value-problem. In: *Computational Ordinary Differential Equations* (J. R. Cash and I. Gladwell, Eds.). Clarendon Press. Oxford. pp. 425–435.

Milanese, M., Norton, J., Piet-Lahanier, H. and Walter, E. (Eds.) (1996). *Bounding Approaches to System Identification*. Plenum Press. New York, NY.

Moore, R. E. (1959). Automatic error analysis in digital computation. Technical Report LMSD-48421 Lockheed Missiles and Space Co, Palo Alto, CA.

Moore, R. E. (1979). *Methods and Applications of Interval Analysis*. SIAM. Philadelphia, PA.

Nedialkov, N. S. and K. R. Jackson (2001). Methods for initial value problems for ordinary differential equations. In: *Perspectives on Enclosure Methods* (U. Kulisch, R. Lohner and A. Facius, Eds.). Springer-Verlag. Vienna. pp. 219–264.

Neumaier, A. (1990). *Interval Methods for Systems of Equations*. Cambridge University Press. Cambridge, UK.

Rall, L. B. (1981). *Automatic Differentiation: Techniques and Applications*. Vol. 120 of *Lecture Notes in Computer Science*. Springer-Verlag. Berlin.

Rall, L. B. and G. F. Corliss (1999). Automatic differentiation: Point and interval AD. In: *Encyclopedia of Optimization* (P. M. Pardalos and C. A. Floudas, Eds.). Kluwer. Dordrecht, the Netherlands.

Ratschek, H. and J. Rokne (1984). *Computer Methods for the Range of Functions*. Ellis Horwood. Chichester, UK.

Ratschek, H. and J. Rokne (1988). *New Computer Methods for Global Optimization*. Ellis Horwood. Chichester, UK.

Sam-Haroud, D. J. and B. Faltings (1996). Consistency techniques for continuous constraints. *Constraints* **1**(1-2), 85–118.

Smith, H. L. (1995). *Monotone Dynamical Systems: An Introduction to the Theory of Competitive and Cooperative Systems*. Vol. 41 of *Mathematical Surveys and Monographs*. American Mathematical Society. Providence, RI.

van Hentenryck, P., L. Michel and F. Benhamou (1998). `Newton`: Constraint programming over nonlinear constraints. *Science of Computer Programming* **30**(1–2), 83–118.

Walter, E. (1982). *Identifiability of State Space Models*. Springer-Verlag. Berlin.

IFAC
Publications
www.elsevier.com/locate/ifac

Online Detection of Tyre Pressure Deflation in Passenger Cars

Jitendra Shah*, Marcus Börner+, Rolf Isermann+, Srinivasa Y G*

* Precision Engineering and Instrumentation Laboratory
Indian Institute of Technology Madras
India

+ Institute of Automatic Control
Darmstadt University of Technology
Germany

Abstract: Monitoring of tyre a pressure in passenger vehicle is a major aspect of improved active vehicle safety. In this contribution, a new velocity compensated method for monitoring tyre pressure using vertical body acceleration signals is presented. It is found that the wheel hop frequency is strongly correlated with damping coefficient and nature of road excitation. Limitations of using the shift of wheel hop frequency are presented. To avoid the influence of nature of road excitation, a virtual transfer function has been derived between front and rear body acceleration signals based on certain predefined event of the vehicle. The delay in transfer function is strongly related to velocity of the vehicle. The characteristic features are generated from the frequency response of the virtual transfer function. The sensitivity analysis presents the effects of the variation of mass of the vehicle and the spring stiffness has no effect on the frequency response but the damping coefficient has an influence. A simple threshold is found from experiment data for classification of relative tyre deflation. The tyre pressure diagnostic algorithm is implemented online and tested for different relative tyre pressure deflations. *Copyright © 2003 IFAC*

Keywords: Tyres; Fault detection; Spectral analysis; Identification; Automotive.

1. INTRODUCTION

Many accidents have been recorded due to air leakage in pneumatic tyres. It is imperative to monitor the air pressure in tyres to properly maneuver the vehicle. Vehicles operating with underinflated tyres pose significant safety problems. According to the Rubber Manufactures Association, there were 647 fatalities in 1999 that involved "tyre related factors" [RMA (2001)]. International Independent Research has shown that on an average, only 17% of tyres are correctly inflated. The minor underinflations of only 140 kPa results in an increase in tyre wear of at least 20% and increase in fuel consumption by 5%. This is quoted in a report published National Highway Traffic Safety Administration, USA, [Grygier *et al.* (2001)].

On November 1, 2000, the Transportation, Recall, Enhancement, Accountability and Documentation (TREAD) [Grygier *et al.* (2001)] Act became law in United States of America. One section of this Act deals with above-mentioned problem through the following requirement:

SEC. 13. "*Tyre Pressure Warning: No later than one year after enactment the National Highway Traffic Safety Administration must complete a rulemaking to require a warning system in new motor vehicles to indicate to the operator when a tyre is significantly underinflated. The rule must be effective within two years after completing the rulemaking*". [Grygier *et al.* (2001)]

A considerable amount of literatures exists on indirect sensing of tyre pressure from the vehicle behaviours. Different ideas were proposed using wheel speeds and body acceleration sensors. Bußhardt and Isermann (1992) carried out simulations and experiments on a quarter-car stand and estimated the stiffness of the tyre at different tyre pressure in offline basis. However, the results were limited only to laboratory stage and couldn't be implemented in test vehicles. He gave an idea of spectral approach to fault diagnosis of tyre pressure. Weispfenning (1996,1997) stated that the resonant frequency of wheel vibration could be used for tyre pressure monitoring. Mayer (1995) proposed an idea of classification of tyre pressure by estimation of virtual transfer function between the vertical accelerations of front axle and rear axle. The approach is not suitable for on-line implementation as the time delay is estimated by cross correlation of measured data and delay was not normalised for different velocities. So, a compensation of vehicle states was not done suitably. This new approach compensates delay in excitations and illustrates the powerful approach of frequency domain estimation. The selection of frequency domain classification in this paper has a specific advantage over time domain classification is improved envelope of threshold.

2. EFFECTS OF TYRE PRESSURE LOSS

Tyre pressure is one of the major factors that determines the comfort, stability, safety and economic driving of the vehicle. The major parameters that govern tyre pressure are type of the road, temperature of the climate and degree of comfort. The total damage of tyre after running at very low pressure due to a puncture is shown in Fig.1

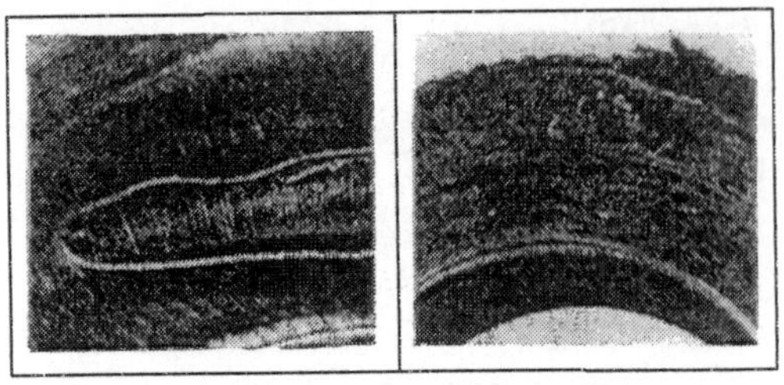

Fig. 1 Total tyre failure after running at very low pressure [French (1989)]

Natural diffusion is one of the causes of deflation of the tyre when the vehicle is not used for long time. It is found that diffusion rate is 0.02 bar/month at 0°C and at higher temperature such as 40°C is 0.35 bar/month.

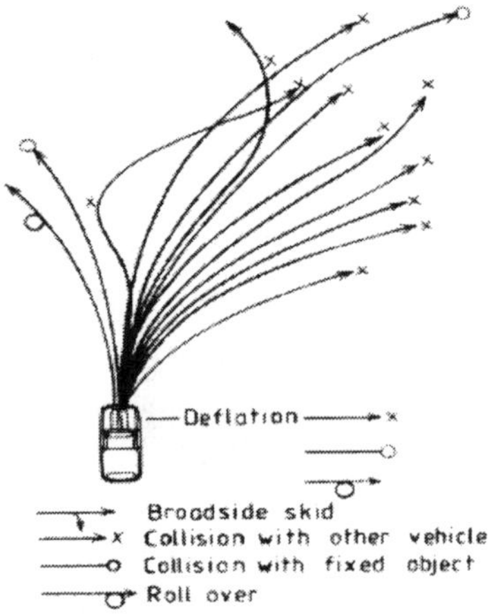

Fig. 2 Effects of front right tyre burst and deviation from path [French (1989)]

Grogan [French(1989)] showed that sudden change in pressure of tyre caused severe stability problems and caused inadequate control over lateral dynamics of vehicle. Fig.2 shows deviation from desired trajectory during normal lane change when front right tyre is deflated. Climatic temperature variation is one of the causes for variation of tyre pressure. The behaviors of the tyre change during winter and summer.

3. MODELLING AND SIMULATION

The spectral analysis of the vertical body acceleration has shown many limitations of the earlier works done at laboratory level [Bußhardt and Isermann (1992)]. In the University of Technology Darmstadt at laboratory level researches, static test has been carried out on a tyre test stand. It gives good and satisfactory results, as it is free from the tyre unbalance, engine vibration and road excitation. The damped natural frequency of the wheel-tyre vibration is a linear function of tyre pressure. There is a shift in the damped natural frequency with a change in the tyre pressure. The shift of damped natural frequency is very sensitive to the type of road excitation and other suspension parameters of the vehicle. However, tyre pressure is not the only factor that causes a shift in the damped natural frequency.

3.1 Calculation of Wheel-Hop-Frequency

A linear quarter car model of two degrees of freedom is assumed for the analysis. Usually, the quarter car model is thought as a single independent suspension with a part of the sprung masses on it. To calculate the wheel hop frequency the sprung mass m_A is hooked to sky or assumed of infinite mass is shown in Fig.3.

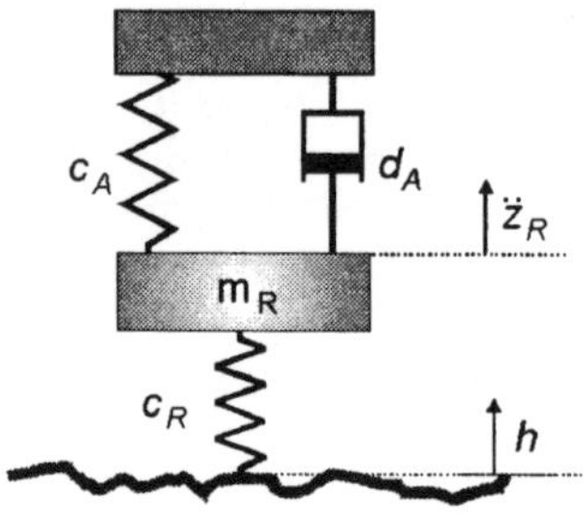

Fig. 3 Estimation of damped natural frequency of wheel

$$\omega = \sqrt{\frac{c_A + c_R}{m_R}} \sqrt{1 - 2\frac{d_A^2}{4m_R(c_A + c_R)}} \qquad \textbf{(1)}$$

ω denotes damped natural frequency.

It is observed from the Eq.1 that the wheel hop frequency ω is dependent on tyre stiffness c_R and suspension damper coefficient d_A.

Another interesting effect of change in road excitation h is simulated using a linear quarter car model. It is observed that the damped natural frequency ω is increasing for an equal increase in tyre stiffness c_A in Fig.4. The effect of different types of road excitation h for same tyre stiffness c_R is also shown in Fig.4. In simulation, the statistical properties of the excitation signal are kept constant, only the road irregularities are changed to visualise the effects of the road irregularities on the damped natural frequency ω. With changes in road irregularities the change in wheel hop frequency is quite random. It is only possible in simulation to keep the road excitation repeatable. Hence, it cannot be assumed that the road surface will be similar for all driving situations.

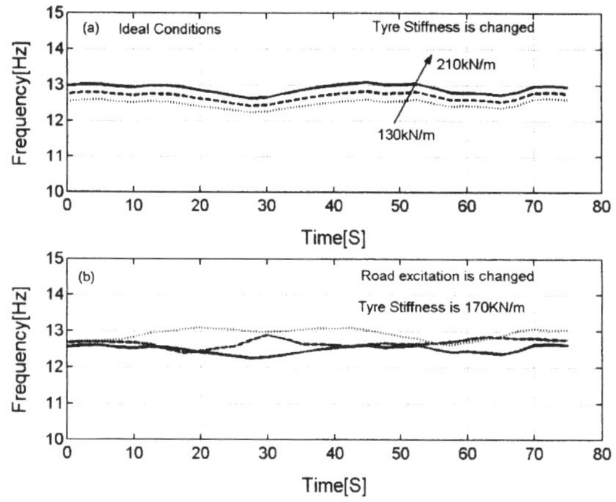

Fig. 4 (a) Effects of change in tyre stiffness c_R on damped natural frequency ω
(b) Effects of change in road excitation h on damped natural frequency ω

3.2 Concept of Virtual Transfer Function

The concept of the virtual transfer function is derived from the fact that both wheels of one side of vehicle track are rolling over same road surface. There is always a delay in road excitation between the wheels on one side of the vehicle. Hence, it is assumed that rear axle body vibration is delayed and transformed form of front axle body vibration. The vertical body vibrations of the rear axle and the front axle are used to estimate the parameters of the virtual transfer function $G_v(s)$.

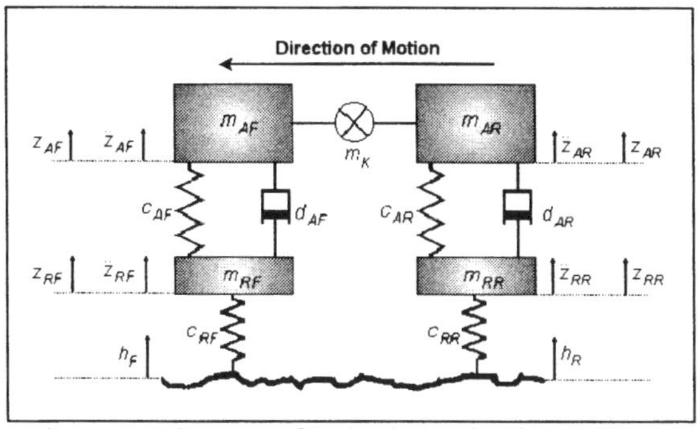

Fig. 5 Diagram of a single track of a vehicle

The model as shown in Fig.5 is comprised of two quarter-car model and is connected with a coupling mass m_k between them and the effect of coupling mass to transfer of body vibration is not of great importance as compared to axle vibrations. There is a time delay τ_d between front road excitation $h_F(t)$ and rear wheel road excitation $h_R(t)$. [Würtenberger (1997)]

$$h_R(s) = h_F(s)e^{-s\tau_d} \qquad \textbf{(2)}$$

The following equation describe linear transfer function $G_B(s)$ between road excitation and body vibration for rear(R) axle and front (F) axle.

$$\begin{aligned} s^2 Z_{AF}(s) &= G_{BF}(s)h_F(s) \\ s^2 Z_{AR}(s) &= G_{BR}(s)h_R(s) \end{aligned} \qquad \textbf{(3)}$$

The virtual transfer function is written as

$$G_V(s) = \frac{s^2 Z_{AR}(s)}{s^2 Z_{AF}(s)} = \frac{G_{BR}(s)}{G_{BF}(s)}e^{-s\tau_d} \qquad \textbf{(4)}$$

The $G_v(s)$ is found by dividing the body acceleration signal of front and rear quarter car model [4].

$$G_V(s) = \frac{1 + b_1 s + b_2 s^2 + b_3 s^3 + b_4 s^4 + b_5 s^5}{1 + a_1 s + a_2 s^2 + a_3 s^3 + a_4 s^4 + a_5 s^5}$$

where

$$b_1 = \left(\frac{d_{AR}}{C_{AR}} + \frac{d_{AF}}{C_{AF}}\right);$$

$$b_2 = \left(\frac{d_{AF}d_{AR}}{C_{AF}C_{AR}} + \frac{m_R C_{AF} + m_A(C_{AF} + C_{RF})}{C_{AF}C_{RF}}\right)$$

$$b_3 = \left(\frac{m_R C_{AF} d_{AR} + m_A d_{AR}(C_{AF} + C_{RF})}{C_{AF}C_{RF}C_{AR}}\right)$$

$$b_4 = \left(\frac{m_R d_{AF} d_{AH} + m_A d_{AF} d_{AR}}{C_{AF}C_{RF}C_{AR}} + \frac{m_A m_R}{C_{AF}C_{RF}}\right)$$

$$b_5 = \left(\frac{m_A m_R m_{AR}}{C_{AF}C_{RF}C_{AR}}\right)$$

$$a_1 = \left(\frac{d_{AF}}{C_{AF}} + \frac{d_{AR}}{C_{AR}}\right)$$

$$a_2 = \left(\frac{d_{AR}d_{AF}}{C_{AR}C_{AF}} + \frac{m_R C_{AR} + m_A(C_{AR} + C_{RR})}{C_{AR}C_{RR}}\right) \qquad \textbf{(5)}$$

$$a_3 = \left(\frac{m_R C_{AR} d_{AF} + m_A d_{AF}(C_{AF} + C_{RF})}{C_{AR}C_{RR}C_{AF}} + \frac{m_R d_{AR} + m_A d_{AR}}{C_{AR}C_{RR}}\right)$$

$$a_4 = \left(\frac{m_R d_{AR} d_{AF} + m_A d_{AR} d_{AF}}{C_{AR}C_{RR}C_{AF}} + \frac{m_A m_R}{C_{AR}C_{RR}}\right)$$

$$a_5 = \left(\frac{m_A m_R d_{AF}}{C_{AR}C_{RR}C_{AF}}\right)$$

The frequency response of the virtual transfer function $G_v(s)$ derived from the above formulas is shown in Fig.6. In the simulation the rear tyre stiffness is changed while the front tyre stiffness is kept constant to simulate the effect of relative change in tyre pressure between front tyre and rear tyre.

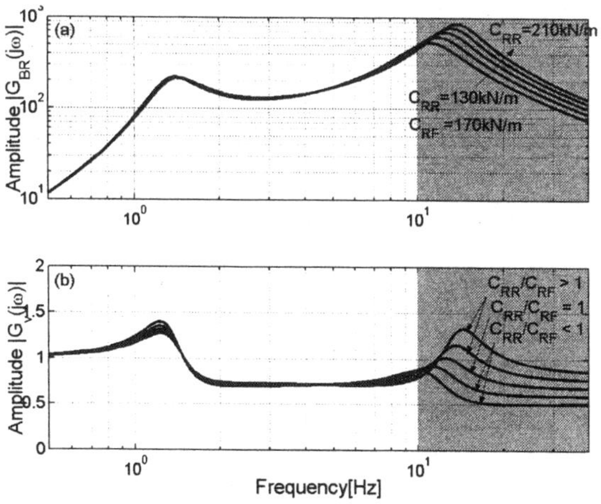

Fig. 6 (a) Effects of change of rear tyre stiffness on frequency response of $G_{BR}(s)$
(b) Frequency response of $G_v(s)$ for relative change in tyre pressure.

By analysing the frequency spectra, only relative pressure difference could be detected. Different pressure ratio line is normalised with respect to unity pressure line. The possibility to detect relative pressure difference is visible in Fig. 7 and symptom generation criteria are illustrated in Table.1.

Table. 1 Symptom generation criteria

$\frac{P_R}{P_F} > 1$	Front tyre pressure is reduced compared to rear tyre pressure.
$\frac{P_R}{P_F} = 1$	All tyres have relatively equal pressure.
$\frac{P_R}{P_F} < 1$	Rear tyre pressure is reduced compared to front tyre pressure.

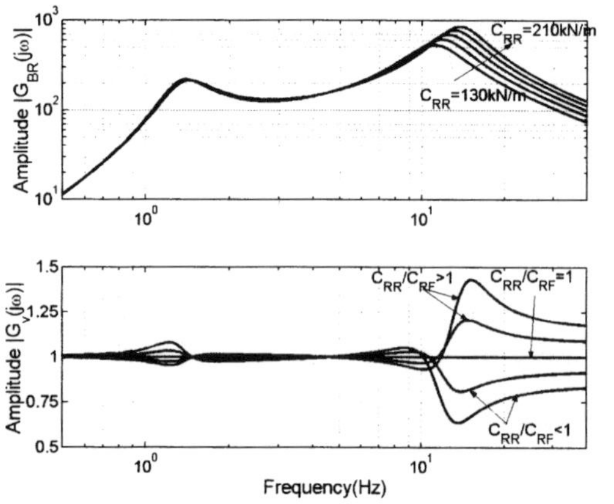

Fig. 7 Normalised around unity pressure ratio

4. SENSITIVITY ANALYSIS

It is certain that not only the change in tyre pressure, which is strongly correlated with C_R, will cause a variation in the coefficient of $G_V(s)$, but also the changes in suspension parameters and more over different mass of chassis resulting from different load affects the frequency response.

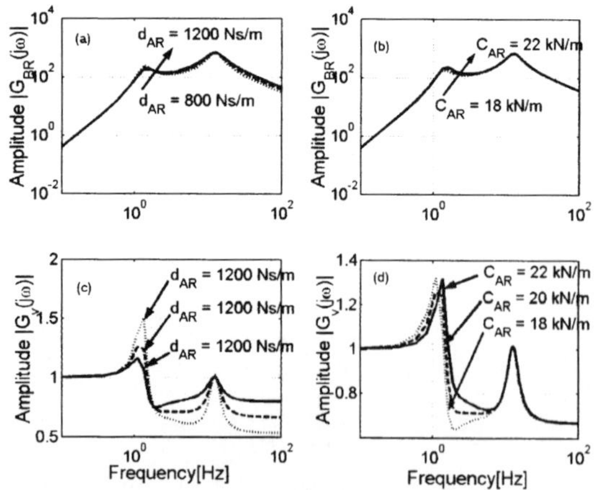

Fig.8 Effects of parameter change of rear suspension on $G_V(s)$
(a & c) rear suspension damping coefficient
(b & d) rear suspension spring stiffness

The effects of parameter variation are shown in Fig.8 and Fig.9. The effects of change of rear suspension spring stiffness on the frequency response of the virtual transfer function are nil. But, the effects of parameter variation of suspension damper have considerable effect above 10 Hz. The dampers, which are used in most of the passenger vehicle, are passive and changes in damping coefficient with time are very less. The change of frequency caused by sensitive parameters is mitigated by careful selection and filtering of identification zone.

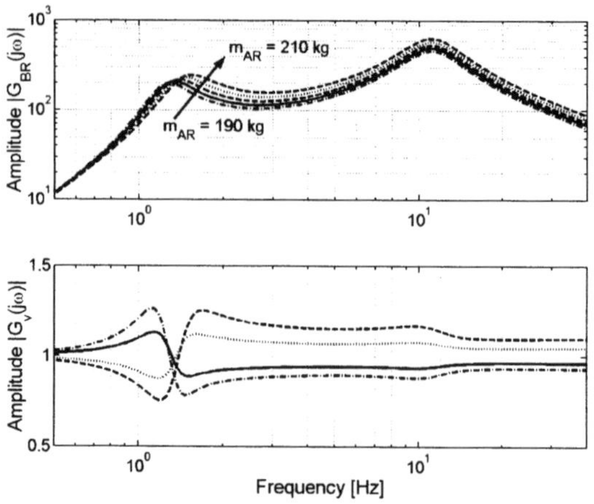

Fig.9 Effect of variation of sprung mass of vehicle

5. MODEL REDUCTION

The order of the virtual transfer function is five which is much higher for parameter estimation. A model of virtual transfer function which is responsible for the frequency response after 10 Hz is needed. The pole and zero analysis can give some clue for the model reduction.

$$G_V(s) = k + G_{V1}(s) + G_{V2}(s) + G_{V3}(s) \qquad \textbf{(6)}$$

The G_{V1} is composed of one pole pair with high imaginary parts whereas the G_{V2} is characterised by one pole pair with relatively lower imaginary parts and G_{V3} is denoted by one pole on real axis. The term k is the gain of the virtual transfer function. The reduced transfer functions are represented below.

$$G^r_{V1} = k + G_{V1}(s)$$

$$G^r_{V1} = \frac{b_{0rv1} + b_{1rv1}s + b_{2rv1}s^2}{1 + a_{1rv1}s + a_{2rv1}s^2}$$

$$G^r_{V2} = k + G_{V2} + G_{V3} \qquad \textbf{(7)}$$

$$G^r_{V2} = \frac{b_{0rv2} + b_{1rv2}s + b_{2rv2}s^2 + b_{3rv2}s^3}{1 + a_{1rv2}s + a_{2rv2}s^2 + a_{3rv2}s^3}$$

Generally, the wheel hop frequency range lies within 10 Hz to 15 Hz. So, the frequency response of the reduced virtual transfer function after 10 Hz is of prime importance for detection of tyre pressure drop.

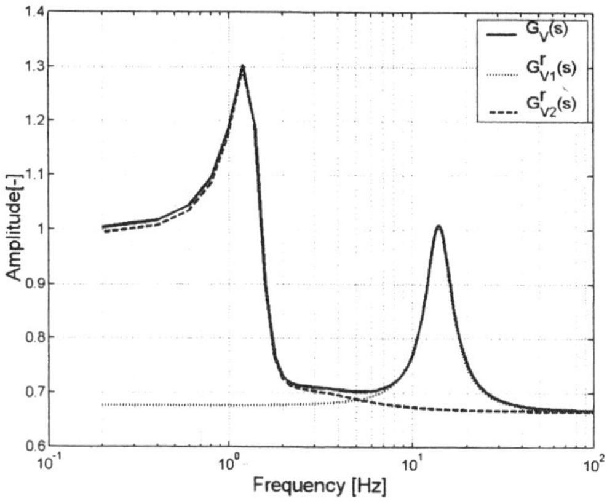

Fig. 10 Frequency response of original and two reduced virtual transfer functions

The frequency responses of original and reduced virtual transfer functions are shown in Fig.10. The frequency response of $G^r{}_{V2}(s)$ is characterising the behaviour of transfer function below 5 Hz. This model represents the existence of the natural frequency of chassis that lies between 1 Hz to 5 Hz. The characteristic for tyre pressure loss detection can only be represented by reduced transfer function of order two, i.e.$G^r{}_{V1}(s)$ because it shows the characteristic peak after 10 Hz. The body acceleration estimated by reduced transfer function $G^r{}_{V1}(s)$ is compared with the original rear axis body acceleration in Fig.11 to validate the new model.

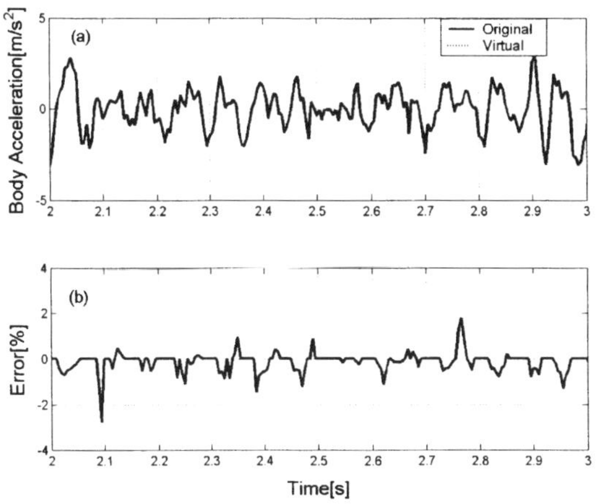

Fig. 11 (a) Error between original rear body acceleration and reduced transfer function $G^r{}_{v1}(s)$ output
(b) Percentage of error in estimation

6. EXPERIMENTAL VERIFICATION

The experimental data of body acceleration is collected for different driving manoeuvres. The nature of data is a normal uncorrelated Gaussian distribution with zero mean and 0.1[m²/s⁴] standard deviation. The shape of the histogram is symmetrical and data is mesokurtic. The cross correlation technique is used to estimate delay in road excitation of front wheel and rear wheel of one side of vehicle.

$$\ddot{Z}_{AR}(t) = k \cdot \ddot{Z}_{AF}(t - \tau_d) \quad \textbf{(8)}$$

k is dynamic modification factor.

The cross correlation is calculated between rear right body acceleration and front right body acceleration. The maximum value of cross correlation gives delay. In real-time an enabled block is used to estimate the parameters of reduced order virtual transfer function is shown in Fig.12. The well versed recursive least square algorithm [Isermann(1992),Juang(1994)] with forgetting factor is used. The minimum value of the forgetting factor is taken as 0.99, as the road excitation is less for estimation of five parameters from two signals. The estimation starts when the velocity of vehicle is nearly constant to reduce extra efforts to find time delay of excitation. The reduction of complicacy of time delay estimation makes this algorithm more suitable for real time online implementation.

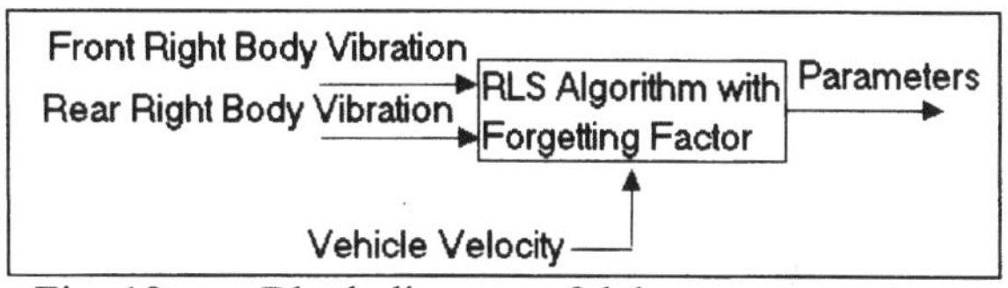

Fig. 12 Block diagram of delay compensator

The frequency response of estimated parameters gives simple threshold values to classify pressure differences. The threshold values are calculated after recording many datasets and finding the frequency response for particular sets of tyre deflation. The observed thresholds from this test vehicle could be different for other models of car. The envelope of confidence is obtained after a careful observation on the frequency response of virtual transfer function. The algorithm is sensitive to a pressure difference of 0.2 bar. The algorithm could detect a change in relative pressure ratio of 10%. Table.2 shows a simple classification criteria based on experimental datasets.

Table. 2 Pressure classification and threshold value

$\frac{P_R}{P_F} > 0.2$	Front tyre is relatively deflated compared to the rear tyre
$\frac{P_R}{P_F} < 0.2 \;\&\; > 0.1$	All tyres are at equal pressure.
$\frac{P_R}{P_F} < 0.1$	Rear tyre is relatively deflated compared to the front tyre

The reliability of the fault detection system is tested with few runs for same settings and is shown in Fig.13. Two trial runs are made for each setting and algorithm is able to classify relative tyre pressure difference. One of the two test runs is made blind to detect accuracy and reliability of the algorithm. It is found that in most of cases the differences in tyre pressure detected are true.

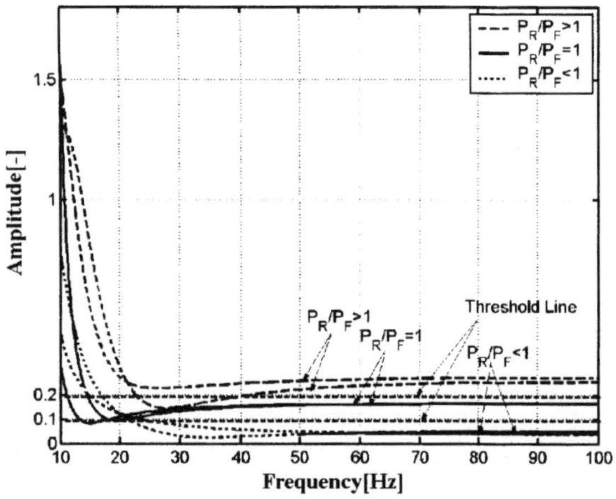

Fig.13 Frequency response of estimated virtual transfer functions

7. CONCLUSIONS

A robust model based tyre pressure deflation detection system for passenger vehicle is developed. In this paper the major drawbacks of spectral analysis of body acceleration to distinguish tyre pressure are mentioned. As wheel hop frequency is sensitive to other suspension parameters and road surface irregularities and type of excitation. The signal is filtered between 10Hz to 20Hz by second order band pass filter. The effects of road surface and other similar interfering disturbances to both front and rear suspension of one side of vehicle are eliminated by introducing a virtual transfer function between front body acceleration and rear body acceleration. The delay in road excitation is dependent upon the speed of the vehicle. The wrong estimation of delays will cause incorrect parameter estimation of virtual transfer function that in turn will cause wrong diagnosis of tyre pressure loss. The complexity of time delay estimation is totally removed in this approach. The simple thresholds for classification of tyre pressure deflation are set from experimental results. The envelope of threshold is found from observation. So, It could be different in other vehicle parameter settings. The compensation for vehicle velocity and online detection of faults are unique contribution of this paper.

REFERENCES

Bußhardt, J. and Isermann, R. (1992). "Adaptive and semi-active shock absorbers based on real-time parameter estimation", *24 FISTA Congress, London*, 1-7.

French, T. (1989). "*Tyre Technology*", Adam Hilger, Newyork.

Grygier, P., Garrott, R. W. and Mazzae N. E. (2001). "An evaluation of existing tire monitoring systems" *Publication of National Highway Traffic Safety Administration, U.S. Department of Transportation.* July 2001.

Isermann, R (1992). "*Identifikation Dynamischer System*", **2**, Springer-Verlag.

Juang Jer-Nan (1994). "*Applied System Identification*", Prentice Hall PTR, New Jersey

Mayer, H. (1995), "Model based detection of tyre pressure deflation by estimation of a virtual transfer function", *Proceedings of 4th IEEE Conference on Control Applications, Albany*, 285-290.

Rubber Manufacturer Association, RMA (2001). "Tire and autosafety facts", [Online] available at http://www.rma.org/tiresafety/auto_safety_facts.html.

Weispfenning, T. and Isermann, R. (1997). "Fault detection of vehicle suspensions", *IFAC Symposium SAFEPROCESS'97*, Hull, England.

Weispfenning, T., (1996). "Fault detection and diagnosis of components of the vehicle vertical dynamics", *1st International Conference on Control and Diagnostics in Automotive Applications,* Geneva, Italy.

Würtenberger, M. (1997), Modellgestützte Verfahren zur Überwachung des Fahrzustands eines PKW. Fortschr.-Ber. VDI Reihe 12, No. 314 Düsseldorf, Deutschland, 1997.

IFAC
Publications
www.elsevier.com/locate/ifac

A SUBSPACE-BASED IDENTIFICATION APPROACH FOR THE ANALYSIS OF ROAD VEHICLES YAW DYNAMICS AROUND STEERING-PAD CONDITIONS

Sergio M. Savaresi*, Enrico Silani*, Sergio Bittanti*, Francesco Farachi†

Dipartimento di Elettronica e Informazione, Politecnico di Milano, Piazza L. da Vinci, 32, 20133 Milano, ITALY.
†Ferrari S.p.A., Via Abetone inf., 4, 41053 Maranello (Modena), ITALY.

Abstract. The topic of this paper is the identification of the yaw dynamics of a road vehicle, around a steering-pad condition. Unfortunately, a linearized model cannot be easily computed from the simulator since the working condition is not an equilibrium for the vehicle. Hence a black-box identification approach has been used. The generator of Input/Output data is a detailed simulator of the vehicle (high-order, non-linear). The I/O data-set is constituted by impulse responses from the steering angle (Input) to the slip-side angle of the vehicle (Output). Subspace-Based System Identification methods have been used. This work is a sub-task of a larger project regarding the analysis of the improvements on the maximum lateral acceleration achievable on a modern road vehicle by means of active yaw control and active suspensions.

Keywords. Road vehicles; lateral acceleration; yaw control; active suspensions; subspace identification; steering-pad.

1. INTRODUCTION

The topic of this paper is the identification of the yaw dynamics of a road vehicle, around a steering-pad condition. Unfortunately, a linearized model cannot be easily computed from the simulator since the working condition is not an equilibrium for the vehicle. Hence a black-box identification approach has been used. The generator of Input/Output data is a detailed simulator of the vehicle (high-order, non-linear). The I/O data-set is constituted by impulse responses from the steering angle (Input) to the slip-side angle of the vehicle (Output). Subspace-Based System Identification methods have been used. This work is a part of a work with the wider scope of analyzing and developing a new concept of "performance-oriented" controller of yaw and roll dynamics.

The notion of "performance-oriented" yaw and roll active control is quite new and unexplored (see Silani *et al.*, 2002, Valtolina *et al.*, 2001). It is very appealing for sport GT-cars manufacturers since it can help to overcome traditional trade-offs of "passive" vehicles, so providing, at the same time, race-like performance with a smooth and easy-to-drive behavior. Note that this concept can be interesting also in a completely different realm, namely in the development of ultra-light low-consumption vehicles: the improvement of lateral forces via active control may allow the use of narrower and lighter wheels an tires, so reducing rolling forces, car weight, and aerodynamic drag forces.

The work presented in this paper is a sub-task of a more general project having the following goals:

- develop a complete model of the car specifically developed and suited for the analysis and design of performance-oriented yaw and roll dynamics control systems;
- develop a quantitative analysis on the improvements on the maximum lateral acceleration achievable by means of active yaw control and active suspensions.

The identification task presented in this paper solves the problem of the dynamic analysis of the handling behavior of the car. This analysis is complicated by the fact that it must be done along a steering-pad (constant-radius turn) working condition. A Black-Box Subspace-Based System Identification approach has quickly and effectively solved the problem of estimation the yaw dynamics around this very peculiar working condition.

The outline of the paper is the following: in Section 2 the structure of the car model is briefly outline and the steering-pad analysis is introduced. Section 3 is devoted to the analysis of the yaw dynamics around the steering-pad conditions: this requires the estimation of a linear model, which is solved using Subspace-Based System Identification methods. Section 4 ends the paper with some conclusive remarks.

2. CAR MODEL AND STEERING-PAD QUASI-STATIC ANALYSIS

For the development of the concept of "performance-oriented" yaw and roll active control, building an accurate mathematical model of the car is mandatory. A software simulator developed in a Matlab-Simulink® environment has been done, tested and validated. This model represents a good compromise between modeling accuracy and complexity, and it has been developed bearing in mind the ultimate goal of this work, namely the quantitative analysis of the benefits and critical issues of performance-oriented yaw and roll active control. The structure of the model is displayed in Fig.1. It is a continuous-time non-linear model, characterized by 28 state variables. A detailed description of this model is in Silani *et al.*, 2002.

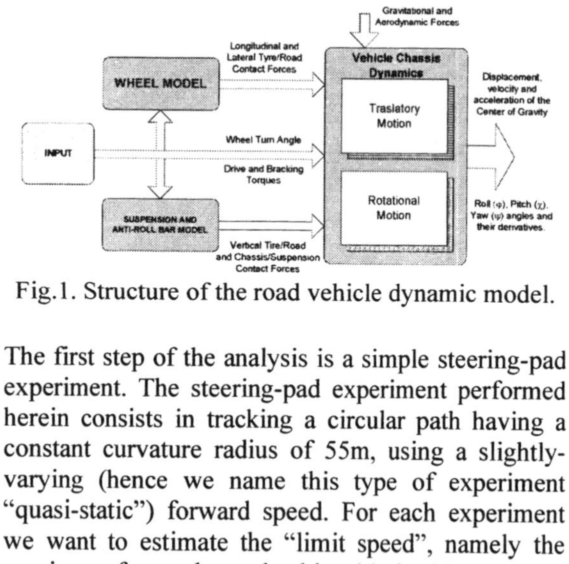

Fig.1. Structure of the road vehicle dynamic model.

The first step of the analysis is a simple steering-pad experiment. The steering-pad experiment performed herein consists in tracking a circular path having a constant curvature radius of 55m, using a slightly-varying (hence we name this type of experiment "quasi-static") forward speed. For each experiment we want to estimate the "limit speed", namely the maximum forward speed achievable in this constant-radius circular circuit. The analysis consists in determining the influence of the roll-bars (Front and Rear – see Fig.2 and Bosch, 2000, Gillespie, 1992, and Kienke and Nielse, 2000) on such limit speed.

All the experiments have been done using the Simulink model. The parameters used in the simulator do not refer to any specific car. Standard (average) characteristics and parameters of a mid-size modern sedan have been used.

In order to compute the limit speed in each experiment, the throttle has been fed (in open-loop) with a slowly-varying increasing ramp. This results in a slightly increasing forward speed. According to the speed rise, the steering angle must be varied in order to track exactly the 55m-radius circular path. To this end, the current curvature radius is measured and compared with its set-point. Using a simple PID controller (tuned so as to provide a narrow bandwidth) this curvature-radius error has been used to suitably modify the steering angle.

Fig.2. An example of roll-bar in an old F1 car (right).

A set of 8 steering-pad experiments have been done; they differ on the stiffness coefficients of the Front (K_{b_F}) and the Rear (K_{b_R}) roll-bars (see Table 1).

Experiment #	K_{b_F}	K_{b_R} / K_{b_F}
1	16.000 N/m	0.5
2	16.000 N/m	0.75
3	16.000 N/m	1
4	16.000 N/m	1.5
5	18000 N/m	0.5
6	18000 N/m	0.75
7	18000 N/m	1
8	18000 N/m	1.5

Table 1. Roll-bars tuning in the steering-pad experiments.

For each roll-bar tuning, there is a "limit" forward speed: when the car reaches such speed it is no longer possible to track the 55-m radius circle, and the curvature radius of the car increases. The limit speed has been conventionally defined as the forward speed measured when the car curvature radius overshoots the limit of 56m. This can be appreciated in Fig.3, where some results abridged from Experiment #3 are displayed. In Fig.3 an eye-bird view of the *Center of Gravity* (*CoG*) trajectory in a steering-pad experiment is plotted.

The results of this quasi-static analysis are condensed in Fig.4, where the limit speed is plotted as a function of the roll-bars tuning. This plot is very explicative and many interesting remarks can be drawn:

- Using the same K_{b_R} / K_{b_F} ratio, the limit speed

decreases as the roll-bar stiffness increases; this shows a trade-off between the goal of reducing rolling movements of the car (obtained using stiffer roll-bars) and increasing the maximum lateral acceleration.

- Increasing the K_{b_R} / K_{b_F} ratio improves the maximum lateral acceleration. This means that using stiffer roll-bars on the Front than on the Rear inherently reduces the maximum lateral acceleration of a car.

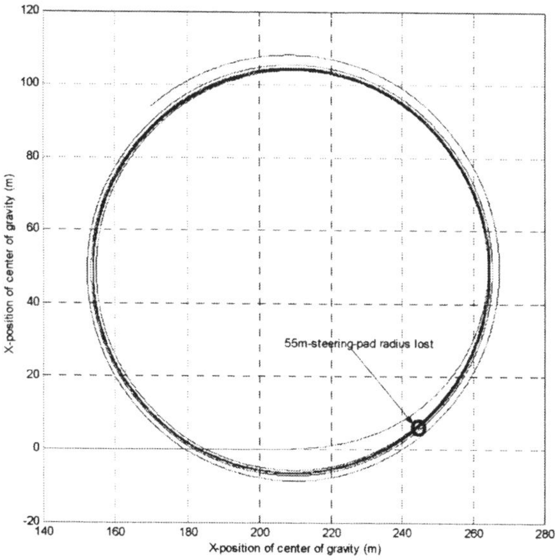

Fig.3. Eye-bird view of the *CoG* trajectory in a steering-pad experiment.

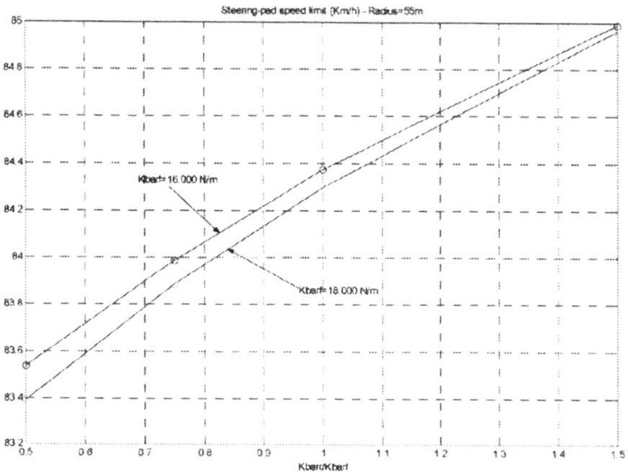

Fig.4. The limit speed as a function of roll-bars tuning.

From this quasi static analysis one might conclude that using a large K_{b_R} / K_{b_F} ratio is the best tuning choice. This is false. In the next section a dynamic analysis will show that K_{b_R} / K_{b_F} may lead to unstable lateral dynamics. Henceforth there is a strong trade-off between dynamic and static behavior.

3. MODEL IDENTIFICATION AND DYNAMIC ANALYSIS

The simplest way of analyzing the lateral dynamics of a car (handling characteristics) is to compute the transfer function associated to the dynamics (assumed linear and time-invariant) from the steering angle δ_w to the body side-slip angle β. The analysis of the poles-zeros location of such transfer function can provide a clear picture of the lateral dynamic behavior of the car.

Unfortunately, the computation of such a transfer function is not trivial. As a matter of fact the overall system is non-linear and high-order (it is characterized by 28 state variables); hence, a linearization around an equilibrium point is strictly required. However an equilibrium point is not well-defined, since the transfer function we are searching for is in the neighborhood of a steering-pad working condition; this inherently makes its computation difficult since some state variables always increase even when the car moves at constant forward speed around a circle.

Due to this very peculiar features, the computation of the linear transfer function from the steering angle δ_w to the body side-slip angle β cannot be easily done (although possible) using standard analysis tools typically available in the simulation package.

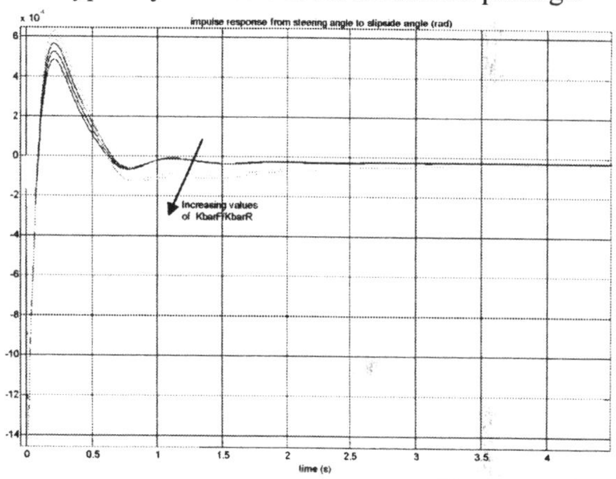

Fig.5. Impulse responses (from $\delta_w - \bar{\delta}_w$ to $\beta - \bar{\beta}$).

An unusual but very effective approach to perform such a linear analysis is to identify the transfer function from Input/Output signals. In other words, even if we have a complete knowledge of the mathematical model of the system, it turns out to be more practical to ignore it and to use the simulator just as an I/O signals generator. Among the many identification approaches one can use (see e.g. Bittanti and Picci, 1996), Subspace-based System State-Space Identification (4SID) methods are very appealing since they allow a one-shot non-parametric identification. The simplest way of estimating the I/O dynamics using 4SID is to use its output when the input is an impulse. Using this rationale, the following experiment has been repeated for different roll-bars tuning:

- The car is trimmed to a stationary steering-pad condition with a curvature radius of 55m and a forward speed of 80 Km/m (this speed is always lower than the limit speed).

- The constant steering angle corresponding to the steering-pad condition has been added with a unitary discrete-time impulse of 10ms. The corresponding value of β has been measured.
- The transfer function has been estimated using the signals $\delta_w - \bar{\delta}_w$ and $\beta - \bar{\beta}$ ($\bar{\delta}_w$ and $\bar{\beta}$ are the constant values of steering angle and slip-side angle in the nominal steering-pad working condition).

The impulse-response excitation has been repeated in 5 experiments: Experiments#1-4 (see Table 1) plus an additional Experiment with $K_{b_F} = 16.000N/m$ and $K_{b_R}/K_{b_F} = 1.1$. The results are plotted in Fig.5.

A preliminary analysis of such impulse responses reveals that, for increasing values of K_{b_R}/K_{b_F}, the lateral dynamics become less damped; a quantitative analysis however calls for a precise estimation of the model.

The origin of the 4SID methods goes back to the so-called "realisation theory" (Ho and Kalman, 1996, Levin, 1960); the impressive growth of the interest in such methods, which has been registered in the last decade, is essentially due to the introduction of the Singular Value Decomposition (SVD) as a highly numerically stable tool for the noise reduction in the measured data set. The method (see Bayard, 1992, Bittanti *et al.*, 2000, Liu and Skelton, 1992, Lovera *et al.*, 2000, Maciejowski and Ober, 1988, Van Overschee and de Moor, 1994, and Viberg, 1994 for a more detailed overview) is based on a state-space representation of a SISO time-invariant discrete-time system.

The most distinctive features of the subspace-based methods:

Numerical stability and efficiency. The most critical and computationally expensive calculus of the subspace-based methods is the SVD decomposition of the extended Hankel matrix. The computation of the SVD of an high-dimensional matrix is a numerically stable operation, which can be efficiently performed. This feature is extremely appealing, especially if one compares with the standard parametric identification methods. These parametric methods usually require the minimisation of a multi-variable nonlinear loss function, which calls for iterative optimisation methods, so that one encounters non-trivial convergence and numerical stability problems.

No initialisation requirement. A distinctive feature of the subspace-based methods is that no initialisation, nor a-priori knowledge of the structure of the system to be identified is required.

This method has been applied on the five impulse responses depicted in Fig.5. Note that, thanks to the fact that they have been generated using a mathematical simulator, the impulse responses are virtually noise-free. Note, however, that in this case the noise can be interpreted as model-error (constituted by the non-linear part of the system).

After computing the Hankel Matrix and the corresponding SVD, an interesting intermediate step is the analysis of the singular values. They are plotted, in descending order, in Fig.6.

The analysis of Fig.6 reveals that a good liner model of the $\delta_w - \beta$ dynamics can have an order much lower than 28. The choice of the model order can be made empirically by direct inspection of Fig.6. A good compromise between accuracy and complexity is the choice n=8. Note, however, that the main dynamics could be captured even with a model of order 5 or even 3. In Fig.7 the "true" impulse response of the simulator is compared with the impulse response of the estimated linear low-order models. Note that the 8th-order model perfectly fits the impulse-response; the 3rd-order model roughly captures the main dynamics only.

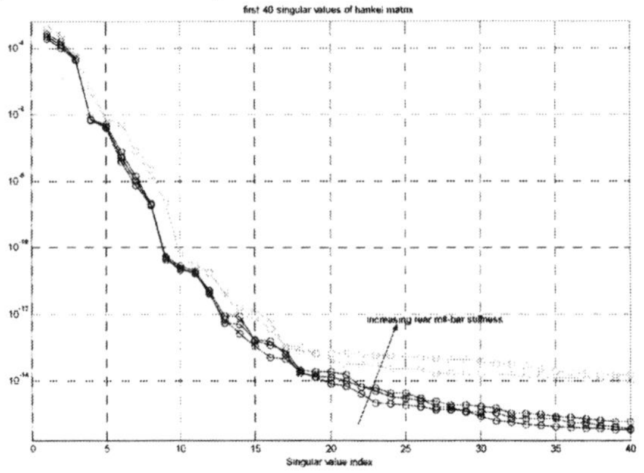

Fig.6a. The first 40 singular values of the Hankel matrices.

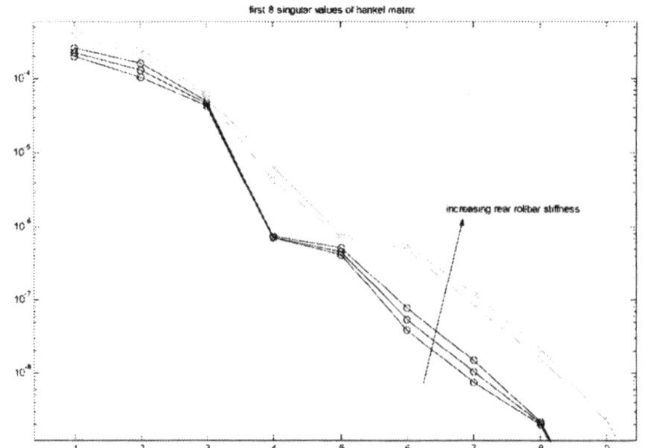

Fig.6b. The first 8 singular values of the Hankel matrices (zoom of Fig.6a).

The poles location of the 8th-order models estimated in correspondence to 5 different roll-bars tuning are displayed in Fig.8. The poles-location structure is as follows:

- in all 5 cases there are 3 pairs of complex-conjugate poles; one of this pair of poles has a dominant effect, since they are not attenuated by nearby zeros.
- In all 5 cases there is a real pole, asymptotically stable, very close to the unit circles; also this pole is attenuated by a nearby zero;

- In the case $K_{b_R} / K_{b_F} \leq 1$ the 8th pole is a real pole very close to the origin (hence its effect on the dynamics is negligible); in the case of $K_{b_R} / K_{b_F} > 1$ it is interesting to see that the 8th pole is still a real pole but it is very close to the unit circle; this effect can be clearly seen in the impulse responses of Fig.5.

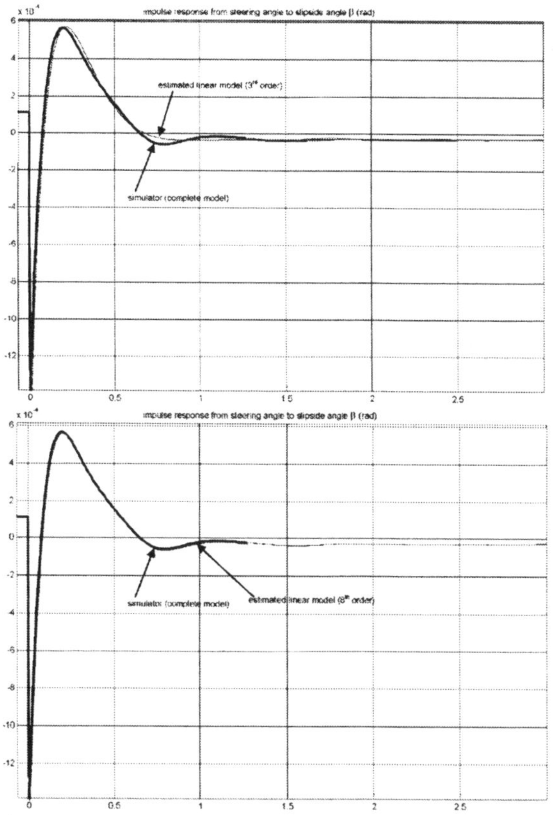

Fig.7. Step-responses of the simulator and the linear models. Left: 3rd-order model; right: 8th-order model.

The dynamic analysis of the linear transfer function from $\delta_w - \overline{\delta}_w$ to $\beta - \overline{\beta}$ around a steering-pad condition can be summarized as follows:

- It is characterized by two dominant poles; their damping significantly decreases as K_{b_R} / K_{b_F} increases. This can be clearly appreciated in Fig.8, where the distance from the origin of the dominant poles is displayed as a function of K_{b_R} / K_{b_F}.
- When $K_{b_R} / K_{b_F} > 1$ a real pole moves from the neighborhood of the origin to the neighborhood of the unit circle; this give rise to comparatively small but very slow transient behavior.

This analysis have clearly shown the trade off between the maximum lateral acceleration (increasing with K_{b_R} / K_{b_F}) and quality of handling (decreasing with K_{b_R} / K_{b_F}). In practice $K_{b_R} / K_{b_F} \leq 1$ is always used. In a performance-oriented tuning (racing cars) K_{b_R} / K_{b_F} is close to 1; in a standard family sedan K_{b_R} / K_{b_F} is significantly lower than 1.

Finally, this analysis shows that a "performance-oriented" active control of yaw and lateral dynamics could be explored: using K_{b_R} / K_{b_F} close to 1 (or even >1) the car enlarges its lateral acceleration limit; however, in order to be easily handled by non-professional drivers, yaw and lateral dynamics must be always assisted by an automatic controller.

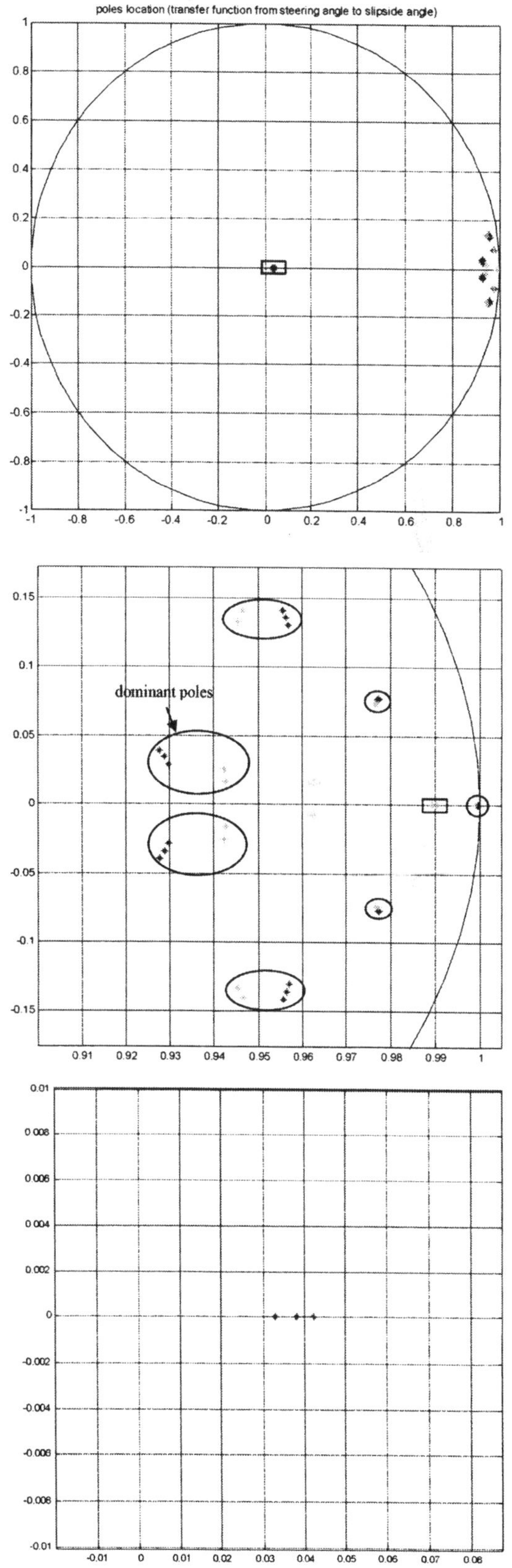

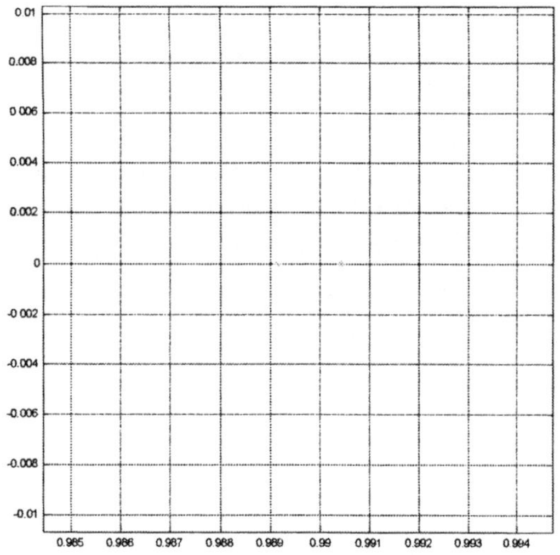

Fig.8. Poles location (8-poles model) of the 5 estimated transfer functions. From top: overall poles location; dominant poles; real pole in the experiments with $K_{b_R} / K_{b_F} \leq 1$; real pole in the experiments with $K_{b_R} / K_{b_F} > 1$.

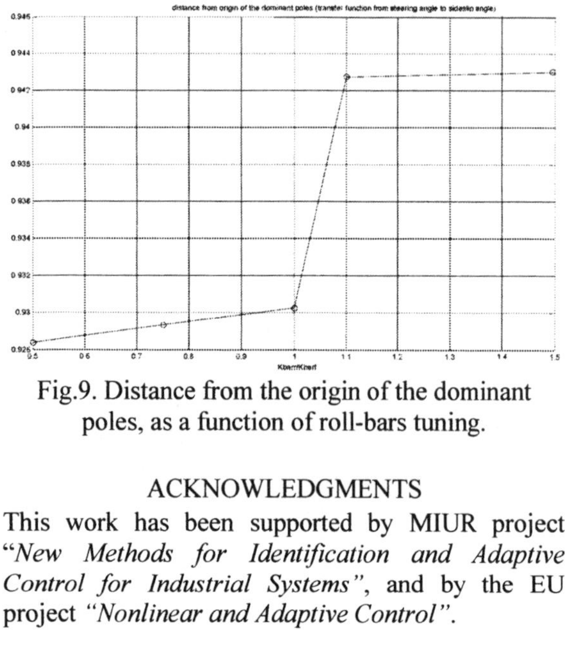

Fig.9. Distance from the origin of the dominant poles, as a function of roll-bars tuning.

ACKNOWLEDGMENTS

This work has been supported by MIUR project "*New Methods for Identification and Adaptive Control for Industrial Systems*", and by the EU project "*Nonlinear and Adaptive Control*".

REFERENCES

Bayard D.S. (1992). "An Algorithm for State-Space Frequency Domain Identification Without Windowing Distorsions". Proceedings of the 31st Conference on Decision and Control. Tucson, Arizona. pp. 1707-1712.

Bittanti S., Gatti E., Ripamonti G., Savaresi S.M. (2000). Poles identification of an analog filter for nuclear spectroscopy via subspace-based techniques. IEEE Transactions on Control System Technology, Vol.8, n.1, pp.127-137.

Bittanti, S., G. Picci (Eds.) (1996). Identification, Adaptation, Learning - The Science of Learning Models from Data. Computer and Systems Sciences Series, Springer-Verlag, Berlin.

Bosch (2000). Automotive Handbook, 5th Edition. BOSCH Gmbh.

Gillespie, T. D. (1992). Fundamentals of Vehicle Dynamics. Society of Automotive Engineers Inc.

Ho, B., and R.E. Kalman (1966). "Efficient construction of Linear State Variable Models from Input/Output functions", Regelungstechnik, 14, pp.545-548.

Kiencke, U. and L. Nielsen (2000). Automotive Control Systems. Springer-Verlag. Berlin.

Levin M.J. (1960). "Optimum Estimation of Impulse Response in the Presence of Noise". IRE Transactions on Circuit Theory. Vol. CT7, pp.50-56.

Liu K., R. Skelton (1992). "Identification of Linear systems from their Pulse Response". Proceedings of American Control Conference, pp.1243-1247.

Lovera M., T. Gustafsson, M. Verhaegen (2000). "Recursive subspace identification of linear and nonlinear Wiener type models", Automatica, Vol. 36, N. 11, pp.1639-1650.

Maciejowski, J.M., and Ober, R.J. (1988). "Balanced Parametrizations and Canonical Forms for System Identification". Proceedings 8th IFAC Symp. on Identification and Parameter Estimation, Beijing, China.

Silani E., S.M. Savaresi, S. Bittanti, A. Visconti, F. Farachi (2002). The concept of performance-oriented yaw-control systems: vehicle model and analysis. *2002 SAE Automotive Dynamics & Stability Conference*, Detroit, MI, May 7-9, pp.235-245.

Valtolina E., S.M. Savaresi, S. Bittanti, A. Visconti, A. Longhi (2001). A Co-ordinate Approach for the Control of Road Vehicles. *6th European Control Conference*, Porto, Portugal, Sptember 4-7. pp.629-634.

Van Overschee P., B. de Moor (1994). "N4SID: Subspace Algorithms for the Identification of Combined Deterministic-Stochastic Systems". Automatica, Vol. 30-1, pp.75-93.

Viberg M. (1994). "Subspace Methods in System Identification". Proceedings of the Symposium on System Identification, Copenhagen, pp. 1-12.

IFAC Publications
www.elsevier.com/locate/ifac

IDENTIFICATION AND FAULT DETECTION OF AN ACTIVE VEHICLE SUSPENSION

Daniel Fischer[*], Markus Zimmer[+], Rolf Isermann[*]

[*]*Institute of Automatic Control, Darmstadt University of Technology*
Landgraf-Georg-Str. 4, 64283 Darmstadt, Germany; Phone: +49-6151-16-7412
{dfischer, risermann}@iat.tu-darmstadt.de
[+]*DaimlerChrysler Research and Technology, Goldsteinstr. 235*
60528 Frankfurt, Germany
markus.g.zimmer@daimlerchrysler.com

Abstract: After a short introduction into the topic of active vehicle suspension systems, a mathematical model of the considered active vehicle suspension, which is presented in a test rig, is derived. It is shown how the unknown parameters can be obtained experimentally by parameter estimation. The results of parameter estimation are used for model based fault detection and identification, in order to obtain reliable knowledge of the system's state. All results are shown for measurements from an active suspension on a test rig.

Keywords: Parameter Estimation, Fault Detection, Active Vehicle Suspension, Recursive Least Square, Damper, Springs, Hydraulic Actuators

1 INTRODUCTION

Automobiles show an increasing number of mechatronic integration of mechanical functions and intelligent control (Isermann, 2001). Introducing such technologies in the handling systems of a vehicle requires good fault detection and diagnosis systems. The electronic systems are complex due to interaction between many systems like the steering system, brake system, and suspension system (see Fig. 1).
One of the newest advances is the Global Chassis Control System (GCC), Active Body Control System (ABC), and Electronic Suspension System (ESS), combining the vertical and lateral systems of a vehicle. The aim of this system is to equalize the driving performance, comfort, and safety to a common standard (Börner *et al.*, 2001).

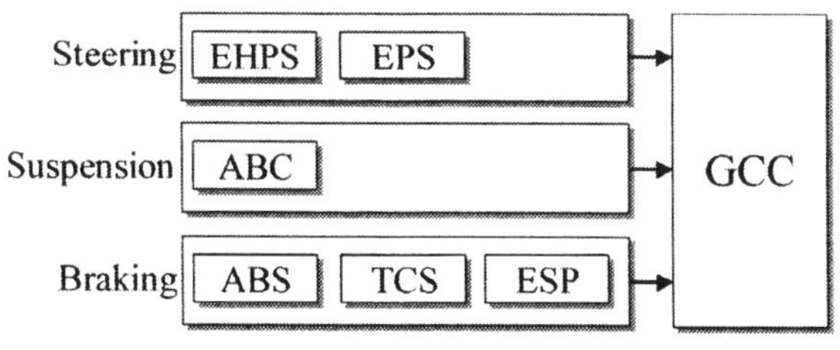

Fig. 1 Development of Global Chassis Control (GCC) with Electro-Hydraulic Power Steering (EHPS), Electronic Power Steering (EPS), Antilock Braking System (ABS), and Traction Control System (TCS).

In the literature, information about the electronic stability program (ESP) or vehicle dynamic control (VDC) can be found (i.e. van Zanten, et al. (1996), Ding (1999)), which controls the lateral dynamic and information about the vertical dynamics (Weispfenning, 1996, Isermann and Bußhard, 1993). Concerning active vehicle suspension systems today's developments are accompanied by the integration of mechatronic systems. This comprises the application of hydraulic, pneumatic and/or electric actuators with the aim of active roll and pitch stabilization and an increase of driving safety and comfort in general. Supplementary, the integration of these actuators and the involved sensors and electronic control units enables and demands process supervision, fault detection and identification for the whole system.

Until now, the identification and fault detection of active vehicle suspension systems is only incipiently investigated for example in Rjamani (1997) or Tan (1997), whereas such fault detection and diagnosis methods have been applied in detail to passive and semi-active suspensions (e.g. Weispfenning, 1996; Majjad, 1997; Börner *et al.*, 2000; Isermann *et al.* 2002). Even a variety of publications exists dealing with control strategies for active suspension systems (e.g. Alleyne A. and J.K. Hedrick, 1995; Oya *et al.*, 1998; Chen, H. and K. Guo, 2001). As a result, this paper presents a model based fault detection and diagnosis approach for an active vehicle suspension system on a test rig.

In general, faults are unexpected and unpermitted deviations from acceptable/usual/standard conditions, which are subjected to components, sensors and actuators in physical systems. Faults can cause the loss of the overall performance of a physical system, which may present hazards to personnel or lead to unacceptable economic loss. Therefore many research efforts in the field of process supervision, fault detection and diagnosis (FDI) have been made (Isermann and Ballè, 1996; Patton, 1994; Isermann, 1994 and 1997; Frank, 1990; Spreitzer and Ballé, 2000). The objective of FDI is to not only determine if a fault is present in a system (fault detection), but also the determination of kind and location (fault isolation) or determination of size and time varying behavior (fault identification) of the fault. Most schemes consist of two levels, a symptom generation part and a diagnostic part. In the first one, symptoms are generated which indicate the state of the process and in the second one, the relation between symptoms and faults is established. It is the task to find significant symptoms, which are robust against noise, disturbances and changes of the setpoint. One approach is the parity space approach. It is based on the deviations of estimated process states or outputs from measured ones, the so-called residuals. The method is based on process models. Another approach is the parameter estimation approach. Here, deviations of online estimated physical parameters of the process from fault free parameters are used to detect faults. In the following, this method is applied to an active vehicle suspension, in order to detect various faults.
For this purpose, algorithms are developed to identify a quarter car test rig, which is equipped with a hydraulic fully loaded active suspension. This system is equivalent to the Active-Body-Control-System known from Mercedes cars (e.g. Mercedes CL-Class; Streiter, 1996).

2 ACTIVE SUSPENSION SYSTEMS

The vehicle suspension system is responsible for driving comfort and safety. Active suspension systems enable the suspension system to adapt to various driving conditions consequently leading to an improvement in comfort and safety compared to a passive suspension. Active suspension systems provide an active force intervening directly in the dynamics of the vehicle with the aim of compensating for road disturbances and mainly pitch and roll motions of the body. So, active suspension systems enable the suspension system to adapt to various driving conditions consequently leading to an improvement in comfort and safety compared to a passive suspension. The control frequencies usually lie between the natural frequencies of the body and the tire. The energy demand is relatively high and increases quadratically with the frequency. Depending on the system's frequency partially and fully active systems can be distinguished. Furthermore, there are non-, partially- and fully loaded suspension systems, depending on whether the controlled system contributes to carrying the body. Fig. 2 gives a short survey of active suspension systems.

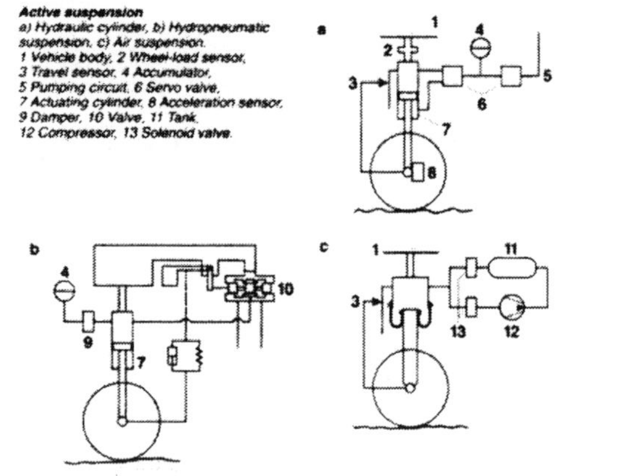

Fig. 2 Active suspension systems (Bosch 1996)

3 MODEL AND TEST RIG OF THE ACTIVE SUSPENSION

The investigated active suspension is a fully loaded hydraulic system. It consists of a hydraulic plunger, which is connected in series to a steel spring, whereas the damper directly connects the body and the wheel mass. The controlled actuation system provides a specific pressure and hydraulic liquid in/out-flow to the hydraulic cylinder. Whereas Fig. 3 represents a schematic sketch of the suspension, the real test rig reproduces the exact geometry of a real car by considering suspension arms. However, it can be shown that the force and distance transmission of the suspension arms do not have to be taken into consideration for the modeling of the system. Neglecting the suspension arms leads to an effective spring stiffness and damper ratio including the transmission of the suspension arms (Würtenberger, 1997).

Using the balance of forces Eq. (1) and (2) can be derived,

$$m_B \ddot{z}_B = F_{hyd} + F_D - F_B \tag{1}$$

$$m_W \ddot{z}_W = F_{zdyn} - F_S - F_D \tag{2}$$

where F_{hyd} describes the plunger's action of force to the body mass. The steel spring and the wheel can be approximated by a linear spring in their operating point, leading to the following forces:

$$F_S = c_B \cdot (z_W - z_B + z_P) \tag{3}$$

$$F_{zdyn} = c_W \cdot (r - z_W) \tag{4}$$

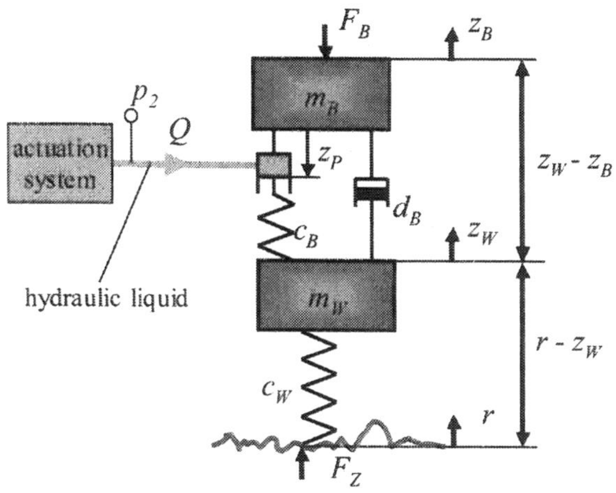

Fig. 3 Model of the active suspension

The strongly degressive course of the damper's characteristic curve can be described non-linearly (5) or segmentic linearly (6) with $i = 1 .. n$ segments.

$$F_D = d_{Bl} \cdot (\dot{z}_W - \dot{z}_B) + F_C \cdot sign(\dot{z}_W - \dot{z}_B) + .. \quad d_{Bnl} \cdot \left|\dot{z}_W + \dot{z}_B\right|^{\frac{2}{1+2m}} \cdot sign(\dot{z}_W - \dot{z}_B) \tag{5}$$

$$F_D = d_{Bi} \cdot (\dot{z}_W - \dot{z}_B) + F_{Ci} \tag{6}$$

The constant m is usually set to m = *1.5* (Majjad 1997).

The volumetric flow out of the actuation system is outputted by the electronic control unit of the system. Additionally, the pressure of the hydraulic liquid of the output of the actuation system is measured. In order to obtain the force F_{hyd} the pressure in the plunger has to be calculated. Neglecting the plunger and oil mass in the plunger, the forces F_{hyd} and F_S are the same and can be determined with the effective cross-sectional plunger area A:

$$F_{hyd} = F_S = p_P \cdot A \tag{7}$$

On account on the pressure loss in the pipe from the pressure sensor to the plunger, the pressure p_P has to be calculated using p_2. Taking into consideration a laminar and a turbulent resistance R_1 and R_2 of the pipe and mass effects of the moving liquid Eq. (8) can be derived:

$$p_P(t) = p_2(t) - R_1 Q(t) - R_2 Q^2(t) - L\dot{Q}(t) \tag{8}$$

In view of estimation purpose and FDI, two equations for the body acceleration can be obtained. Combining Eqs. (1), (6), (7) and (8) results in a flow-based and the combination of Eqs. (1), (3) and (6) in a plunger-position-based description for the body acceleration. The turbulent resistance of the pipe is abandoned according to experimental results, compare chapter 4.

$$\ddot{z}_B(t) = \frac{A}{m_B} \cdot p_2(t) - \frac{AR}{m_B} \cdot Q(t) - \frac{AL}{m_B} \cdot \dot{Q}(t) + \frac{d_{Bi}}{m_B} \cdot (\dot{z}_R - \dot{z}_A)(t) + \frac{F_{Ci}}{m_B} \tag{9}$$

$$\ddot{z}_B(t) = \frac{c_B}{m_B} \cdot (z_W - z_B + z_P)(t) + \frac{d_{Bi}}{m_B} \cdot (\dot{z}_R - \dot{z}_A)(t) + \frac{F_{Ci}}{m_B} \tag{10}$$

Neglecting the elasticity of the hydraulic liquid, the volume balance leads to the following equation:

$$\dot{z}_P(t) = \frac{1}{A} \cdot Q(t) \tag{11}$$

Applying the balance of forces to the plunger mass m_P an additional equation can be derived. Again, considering the pressure loss in the pipe Eq. (12) can be found.

$$\ddot{z}_P(t) = \frac{A}{m_P} \cdot p_2(t) - \frac{AR}{m_P} \cdot Q(t) - \frac{AL}{m_P} \cdot \dot{Q}(t) - \frac{c_B}{m_P} \cdot (z_W - z_B + z_P)(t) \tag{12}$$

4 IDENTIFICATION OF THE ACTIVE SUSPENSION

Based on Eqs. (9), (10), (11) and (12) (neglecting m_P) the following estimation equations are derived.

$$\dot{z}_P(t) = \frac{1}{A} \cdot Q(t) + \frac{Q_0}{A} \tag{13}$$

$$p_2(t) = \frac{c_B}{A} \cdot (z_W - z_B + z_P)(t) + R \cdot Q(t) + L \cdot \dot{Q}(t) + p_{2,0} \tag{14}$$

$$\ddot{z}_B(t) = \frac{A}{m_B} \cdot p_2(t) - \frac{AR}{m_B} \cdot Q(t) + \frac{d_{Bi}}{m_B} \cdot (\dot{z}_W - \dot{z}_B)(t) + \frac{F_C}{m_B} \cdot sign((\dot{z}_W - \dot{z}_B)(t)) \tag{15}$$

$$\ddot{z}_B(t) = \frac{c_B}{m_B} \cdot (z_W - z_B + z_P)(t) + \frac{d_{Bi}}{m_B} \cdot (\dot{z}_W - \dot{z}_B)(t) + \frac{F_C}{m_B} \cdot sign((\dot{z}_W - \dot{z}_B)(t)) \tag{16}$$

The damper is modeled with a two-segmented linear characteristic with a coulomb friction. The coulomb friction is assumed equal for compression and decompression. The segmentation realizes a specific damping ratio for both, compression and decompression motion, which is indicated with $i = +, -$. Here, $d_{B,+}$ for example denotes damping in positive direction of $\dot{z}_W - \dot{z}_B$ referring to compression mode.
As it is shown in the following, the influence of the inductive resistance of the pipe is marginal, which is why this resistance is abandoned in Eq. (15).

For the excitation of the test rig, the base plate of the wheel at the test rig is excited with a measured street profile. The power density spectrum of the wheel movement and the wheel and body mass acceleration is shown in Fig. 4. This demonstrates that all typical frequencies for a vehicle are excited. Therefore, the presented fault detection is demonstrated for realistic conditions. For all presented experiments, these measurements are applied.

Measured data from the test rig is applied to Eqs. (13) - (16). These models are identified using the DSFI-algorithm (discrete square root filter in information form). This recursive algorithm is based on the least squares method and provides a fast converging estimation with low calculation effort. The forgetting factor was chosen to $\lambda = 0.999$. The required derivatives of the suspension deflection and the hydraulic flow are determined using a state variable filter due to noise reduction (Moseler, 2001). The identification result is exemplarily demonstrated in Fig. 5 and Fig. 6. As it can be seen, the estimation converges fast and stable. This indicates correct models, which contain all relevant physical effects. Using the estimation results of Eq. (15) the plunger cross sectional area A can be calculated. With this known quantity all other physical parameters can be calculated successively without any previous knowledge. Additionally, the mentioned transmission of the suspensions level arms is considered for the calculation of the physical parameters, so that the values refer to the real components' properties.
The comparison of the estimated values with reference values indicates an accurate estimation and substantiates the correct modeling of the process. The physical quantities determined by two different estimation equations (such as c_B, $d_{B,+}$, $d_{B,-}$, F_C, R) are estimated with deviations between each other less than 8%. In general, the relative errors compared to reference values are between 6% and 13%.
The comparison of the hydraulic resistant R and the hydraulic inductivity L shows a 10^2 times smaller value for L. This justifies the neglecting of this effect for Eq. (15).

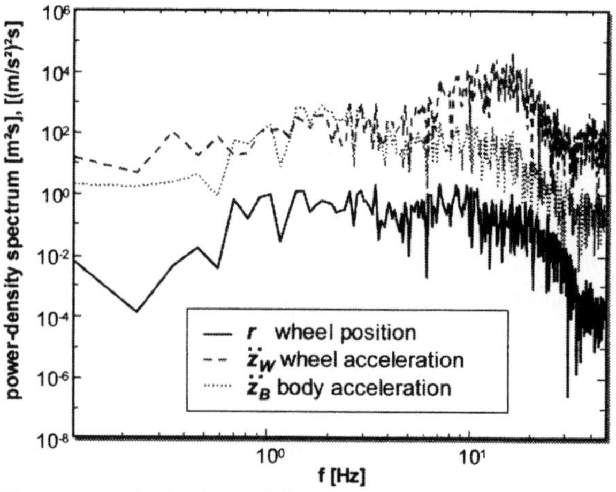

Fig. 4 Excitation of the test rig; power-density spectrum of measured wheel spectrum, wheel and body acceleration

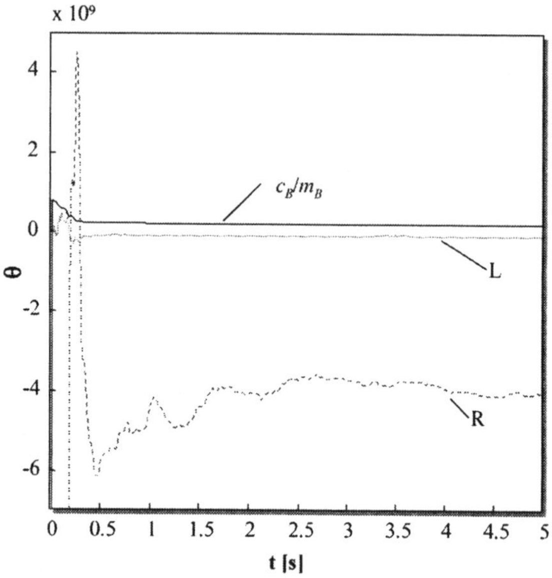

Fig. 5 Parameter Estimation results of Eq. (14)
—— c_B/A, - - - - R, ……… L

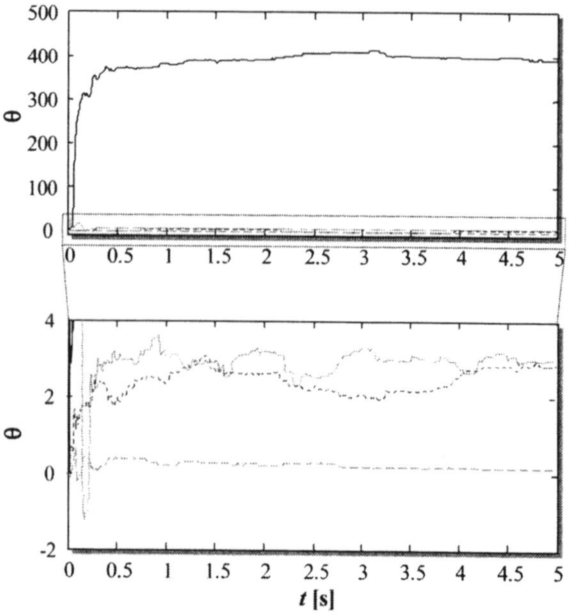

Fig. 6 Parameter Estimation results of Eq (16)
—— c_B/m_B, - - - - $d_{B,comp}/m_B$
……… $d_{B,decomp}/m_B$, F_C/m_B

5 FAULT DETECTION OF THE ACTIVE SUSPENSION SYSTEM USING ESTIMATED PARAMETERS

Sensor faults and process faults are reflected by changing parameter estimations. This is used to detect faults. Therefore, symptoms are generated comparing the online estimated parameters with fault free values. In order to classify different faults, Table 1 shows how each fault affects the estimated parameters. In case of deflections of the symptoms above certain thresholds, the matching fault pattern can be classified. Isermann (1998) proposes a Fuzzy logic approach for the evaluation of the fault symptom scheme.

In the first left column process or sensor faults are given, e. g. "Friction" stands for increased friction. The influence on the estimated variables is shown by the other columns. The symbols -, 0 or + refers to a negative, zero or positive deflection of the symptom, respectively. Generally, all symptom deflections are considered for a positive/greater-one occurrence of the faults. The indices refers to their respective estimation equations. The separate identification of offset and gain faults of the sensors without using variances of residuals is enabled by the presented estimation equation design. Offset faults only affect the DC component of the estimation equation (such as F_C, $p_{1,0}$, Q_0), whereas gain faults directly influences the estimation parameter affected by the faulty sensor signal. Due to the estimation order of the physical parameters, some faults also influence other physical parameters, which are normally not affected by this fault. The symptoms $d_{B,+}$ and $d_{B,-}$, are combined to a single symptom for Eq. (15) and (16) each, as these quantities have the same significance relating the considered faults.

In general, the estimation of the parameters requires an adequate excitation of the system. This is ensured by an excitation-indicator enabling the estimation only for an adequate excitation (Isermann *et al.*, 2002).

Table 1 shows that there is a different pattern for nearly every fault. Only friction and an offset fault of the body acceleration sensor cannot be distinguished. This fault detection scheme can be supplemented by a model-based approach (Fischer *et al.*, 2003). This enables a more unambiguous and detailed sensor fault detection.

6 CONCLUSION

The presented fault detection and isolation approach provides reliable fault detection for an active suspension system distinguishing 15 sensor and process faults such as sensor offset and gain faults as well as friction, leakage and clogging. For this, a model-based approach based on parameter estimation is presented. Using the DSFI algorithm, 13 physical parameters of the system are estimated with measured data with minor deviations to reference values between 6% and 13%. The resulting fault symptom classification shows an almost isolated nature. As a result, at least all process faults – except friction and body acceleration sensor offset – can unambiguously be distinguished.

Table 1 Fault Symptom Classification

	Online estimated parameters										
Fault	$d_{B+-,15}$	$d_{B+-,16}$	$c_{B,14}$	$c_{B,16}$	$F_{C,15}$	$F_{C,16}$	A	m_B	R	$p_{2,0}$	Q_0
Clogging	0	0	0	0	0	0	0	0	+	0	0
Leakage	0	0	0	0	0	0	-	0	0	0	+
Friction	0	0	0	0	+	+	0	0	0	0	0
Damping	+	+	0	0	0	0	0	0	0	0	0
Spring stiffness	0	0	+	+	0	0	0	0	0	0	0
Offset z_B''	0	0	0	0	+	+	0	0	0	0	0
Gain z_B''	+	+	0	+	+	+	+	-	-	0	0
Offset z_W-z_B	0	0	0	0	-	-	0	0	0	+	0
Gain z_W-z_B	-	-	-	-	0	0	-	+	+	0	0
Offset p_2	0	0	0	0	+	+	0	0	0	+	0
Gain p_2	0	0	+/-	+	0	0	0	+	+	0	0
Offset z_P	0	0	0	0	0	+	0	0	0	+	+
Gain z_P	+	+	0	+	+	+	-	+	0	0	0
Offset Q	0	0	0	0	+	0	0	0	0	+	+
Gain Q	-	-	+	-	-	-	+	-	-	0	0

REFERENCES

Alleyne, A. and J.K. Hedrick (1995). Non-linear Adaptive Control of Active Suspensions, *IEEE Transactions on Control Systems Technology*, **3**, No 1.

Börner, M., M. Zele and R. Isermann (2001). Comparison of different fault detection algorithms for active body control components: automotive suspension system. *American Control Conference, Arlington*, USA.

Börner, M., H. Straky, T. Weispfenning and R. Isermann (2000). Model based fault detection of vehicle suspension and hydraulic brake systems, 1st *IFAC Conference on Mechatronic systems*, Darmstadt, Germany.

Bosch (1996). *Automotive Handbook*. Robert Bosch GmbH, Stuttgart.

Chen, H. and K. Guo (2001). An LMI approach to multiobjective RMS gain control for active suspensions, *American Control Conference*, Arlington VA.

Ding, E.L. (1999). Modellgestützte Sensorüberwachung eines ESP-Systems. ATP 41, No. 7, pp. 35-42.

Fischer, D, E. Kaus and R. Isermann (2003). Fault detection for an active suspension. *IFAC Symposium on Fault Detection, Supervision and Safety of Technical Processes,* Washington, USA.

Frank, P. M. (1990). Fault diagnosis in dynamic systems using analytical and knowledge-based redundancy. *Automatica* **26**, 459–474.

Isermann R. and J. Bußhardt (1993). Parameter adaptive semi-active shock absorbers based on nonlinear models. *2nd European Control Conference*, Groningen, Netherlands.

Isermann, R (1994). *Überwachung und Fehlerdiagnose*. VDI Verlag, Düsseldorf.

Isermann, R. and Peter Ballé (1996). Trends in the application of model based fault detection and diagnosis of technical processes. In: *13th IFAC World Congress*.

Isermann, R. (1997). Supervision, fault-detection and fault-diagnosis methods – an introduction. *Control Engineering Practice*, Vol. **5**, No. 5, p. 639-652.

Isermann, R. (1998). On Fuzzy Logic Applications for Automatic Control, Supervision, and Fault Diagnosis. *IEEE Transactions on Systems, Man, and Cybernetics*. **8**, No. 2.

Isermann, R (2001). Diagnosis Methods for Electronic Controlled Vehicles. *Vehicle System Dynamics,* Vol. 36, **2-3**. Swets&Zeitlinger.

Isermann, R, M. Börner and D. Fischer (2002). Mechatronic Semi-active vehicle suspensions. 1st *International Symposium on Mechatronic Systems (ISOM)*, Chemnitz, Germany.

Majjad, R. (1997) "*Estimation of Suspension Parameters*", IEEE International Conference on Control Applications Hartford, CT.

Moseler, O (2001). *Mikrocontrollerbasierte Fehlererkennung für mechatronische Komponenten am Beispiel eines elektronischen Stellantriebs*. VDI Verlag, Düsseldorf.

Oya, M., Y. Araki and H. Harada (1998). Robust Control of Active Automotive Suspensions with Model Uncertainties, *AVEC* pp.467-472.

Patton, R. J. (1994). Robust model- based fault diagnosis, the state of the art. In: *IFAC-Symposium SAFEPROCESS'94*

Rajamani, R and J.K. Hedrick (1995). Adaptive Observers for Active Automotive Suspensions: Theory and Experiment. IEEE Transactions on Control Systems Technology, 3, No. 1.

Streiter, R. (1996). *Entwicklung und Realisierung eines analytischen Regelkonzepts für eine aktive Federung*, Dissertation, TU Berlin.

Tan, H-S (1997). Model identification of an automotive hydraulic active suspension system. *American Control Conference*. Albuquerque, USA.

Van Zanten, A.T., Erhard R., Pfaff, G., Kost F., Hartmann, U. and Ehret, T. (1996). Control Aspects of the Bosch-VCD. *International Symposium on Advanced Vehicle Control (AVEC 1996),* Aachen, Germany.

Weispfenning, T (1996) „*Fault Detection of Components of the Vehicle Vertical Dynamics*", 1st Int. Conference on Control and Diagnostics in Automotive Applications, Genova.

Würtenberger, M (1997). *Modellgestützte Verfahren zur Überwachung des Fahrzustandes eines PKW*. VDI-Verlag, Düsseldorf.

SYMBOLS

Symbol	**Description**
A	plunger cross sectional area
c_B	spring coefficient body
c_W	spring coefficient of the tire
d_B	damper coefficient body
F_B	force to the body
F_C	Coulomb force
F_D	damper force
F_{hyd}	hydraulic force of the plunger
F_S	spring force
F_Z	tire load
L	hydraulic inductivity
λ	forgetting factor DSFI
m_B, m_W	body mass, wheel mass
m_P	plunger mass
p_2	hydraulic pressure at actuation system
p_P	plunger pressure
Q	hydraulic flow
r	road height
R_1, R_2	laminar, turbulent hydraulic resistance
z_P	plunger position
z_B	body height
z_W	wheel mass height

IFAC Publications
www.elsevier.com/locate/ifac

NON-ADAPTIVE NEURAL AUTOMOTIVE SIDESLIP VIRTUAL SENSOR

M. Battipede[†], D. Danesin[‡], P. Krief[‡], G. Sassi*, M. Velardocchia[†]

[†]*Department of Mechanics - Politecnico di Torino - Torino – Italy*
* *Department of Chemical Engineering - Politecnico di Torino - Torino - Italy*
[‡] *FIAT Auto - Alfa Romeo Performance Competence Center Handling Ride & Brake - Balocco (VC) -Italy*

Abstract: A virtual sideslip sensor has been developed for automotive applications, to improve active chassis control systems devoted to increase vehicle safety and performance. High costs and scarce reliability of conventional sideslip sensors do not allow their employment on passenger vehicles, while approximated estimation methods are effective only when the vehicle lateral dynamics is confined in the linear field. The proposed virtual sensor consists in a cluster-structure neural system, trained and tested over a very wide range of maneuvers, performed on different road surface conditions.

Keywords: Neural Networks, Identification, Automotive

1. INTRODUCTION

The increasing general demand for car passenger safety and performance has determined the application of a number of vehicle chassis control systems. Nowadays, the most diffused chassis control systems regard brakes (e.g. Antilock Braking Systems and Electronic Stability Program), suspensions (Continuous Damping Control and Active Roll Control) and traction (All Wheel Drive). From the current letterature it comes out that most of these chassis control systems would greatly take advantage from the sideslip angle feedback, to predict critical situations, caused by unfavourable road conditions or vehicle wear.

So far, the means to measure the sideslip angle aboard are mainly two. One way consists in measuring the longitudinal and lateral ground speed components with a non-contact velocimeter; the sideslip angle is thus obtained through a vector composition. However, as the non-contact velocimeter uses the road irregularities for speed detection, some difficulties may arise on the mirror-like surface such as a waterhole, where the method usually fails or becomes scarcely accurate. Moreover, the sensor itself is too expensive to be employed on passenger vehicles. The second method consists in calculating the sideslip angle rate by using the vehicle lateral acceleration, the longitudinal speed and the yaw rate, and then integrating the resulting signal to obtain the sideslip angle. Though this way is widespread, it has several drawbacks which make the method unreliable for control purpose. Firstly, sensor signals can be affected by off-set inaccuracies, that may be uninfluential on the measured variables, but may become critical when these signals are combined to calculate a function to integrate. Secondly, this method does not account for road surface conditions, which may affect results significantly. Empirical methods have proven unsuited to fill this gap as the friction parameter, that should reflect the road friction conditions, can not be measured online.

In the last decade, the increasing reliability of the nonlinear identification techiques, as those based on fuzzy systems or neural networks (NN), have paved the way of the virtual sensor employment. Sasaki and Nishimaki (2000) developed a simple virtual sideslip angle sensor using a two-hidden-layer MISO NN for an all-wheel-steering vehicle. Training and testing of this sensor are performed through simple ISO maneuvers, featuring a limited range of steering patterns. Although results are promising, their virtual sensor show scarce reliability beyond the training speed range.

In this paper the authors adopt a similar approach, making use of a very exhaustive set of experimental maneuvers, performed on different kind of tracks and road surfaces. To fill the gap of missing information, two different expedients have been adopted: first the velocity variable have been introduced in the input vector through the longitudinal acceleration signal; second the NN has been extended to a MIMO structure, where the extra outputs are used to supply the system with a supporting trend for the sideslip angle. Both the expedients are expected to strengthen the neural sensor reliability outside the training range, more than to improve performance whitin its boundaries. The longitudinal speed, in fact, has a very strong influence on the vehicle handling behaviour and a wide domain which may range from 0 to beyond 60 m/s; in this case the longitudinal acceleration signal can be effectively used to compensate for gaps in the longitudinal speed range of the training set.

2. PHISICAL FRAMEWORK

1.1 The lateral vehicle dynamics

The sideslip angle β is defined as the angle between the vehicle longitudinal axis and the velocity vector measured in the center of gravity (Gillespie, 1992)
In steady state maneuvers, e.g. a ramp steer, the sideslip angle β increases linearly with the lateral acceleration a_y only in the linear vehicle operating range. In the same operating conditions a_y can be considered a linear function of the steering wheel angle δ_{SW}. This trend is not affected by the road surface condition, neither by the vehicle speed. For high lateral acceleration values, the relationships between β, δ_{SW} and a_y become strongly non linear, while the influence of some state variables increases. Moreover, the road surface condition becames critical, affecting dramatically the lateral vehicle dynamics, through the tire-terrain friction coefficient μ.. This coefficient orientatively ranges between *0.1* and *1*, where the higher values are used for dry road conditions and the lower for icy road surfaces. As shown in Figure 1, during cornering on a slippery surface a higher steering wheel angle must be used to develop the same lateral acceleration, meaning that the understeering vehicle behaviour is stressed.

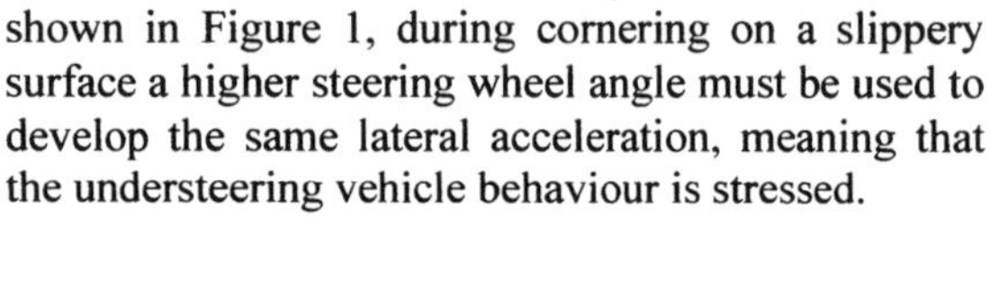

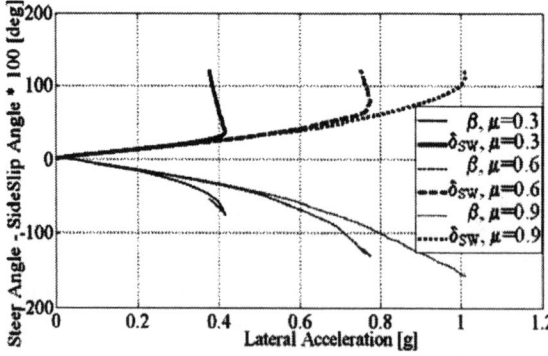

Fig. 1. Vehicle lateral behaviour during cornering

At the same time, the sideslip angle increases due to the combined effect of μ and of the greater steering wheel angle required, according to the following relation (single track vehicle model):

$$\beta = \delta_{SW} \left[\frac{b}{L} - \frac{V^2}{\left(L + K_{US} V^2\right) g C_c} \right] \quad (1)$$

where the understeering gradient K_{US} is a function of μ.. The other parameters are typical vehicle parameters. This phenomenon is sketched in Figure 2. Morover, according to (1), the sideslip angle range is expected to decrease for higher values of the forward speed V.

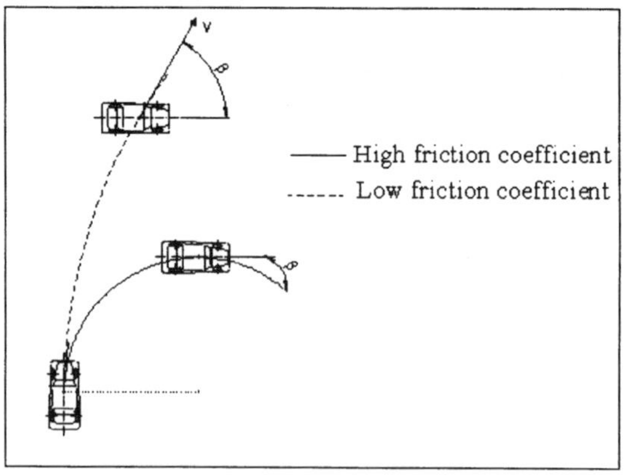

Fig. 2. Friction coefficient μ. influence on the vehicle lateral dynamics.

1.2 Experimental data set

The above described phenomena are cleary more evident for a race car, but can be easily detected by any driver, dealing with a mixed driving route. As variables affecting the sideslip angle calculation are numerous, a thorough analysis of a β function requires a great amount of experimental data, acquired on a very wide range of roads and surface conditions.
Tests have been performed with a high-powered sedan car. The road test tracks cover an ehaustive collection of real situations, with nominal and alterated load conditions as well as asymmetric friction conditions, brought about also through different tire pressure. Precisely, the experimental data set consist of *27* time histories, each collecting of *14* sensors signals, sampled during time periods ranging from *20* to *60* seconds. As shown in Table 1, the time histories include generic road tests (named Handling1, Countryside, Handling2, English Track, White Track, High Speed Lap) as well as standard handling maneuvres (ramp steer, step steer with or without throttle release).
Each manouvre has a sampling frequency of *100 Hz* and is postprocessed to be depurated from the sensor offset. Three different values of the friction coefficient are considered, corresponding to dry (μ=*0.9*), wet (μ=*0.7*) and icy asphalt, simulated through a particular kind of gravel (μ=*0.4*). Almost all the maneuvers are performed with the single driver, while alterated load conditions are simulated with four people aboard. Finally, the data set also contains maneuvers performed with different tire pressure conditions, as pointed in Table 4.

Track Name	Main Features
Handling	Mixed route, high/low speed
Countryside	Countryside route (60-110km/h)
City	Low speed (50km/h), 90° curves
White Track	Gravel road surface
High speed Lap	High speed (160 km/h)
ramp steer	Steady maneuver
step steer	Transient maneuver
step steer with throttle release	Transient maneuver

Table 1 maneuver description

3. STRUCTURE OF THE NN ESTIMATOR

The friction factor μ is the external parameter which affects the dynamic lateral dynamics, reducing the lateral acceleration range while increasing the corresponding sideslip angle, according to a nonlinear relation. Moreover μ cannot be measured on board and, consequently, it cannot be used as an input to the neural system. For this reason a cluster approach has been used, based on a manual switch which allows the driver to choose between *dry&wet* or *snow&ice* asphalt conditions. Thus, according to the driver selection, the sensor signals are processed by one of the two NNs: the first trained with maneuvers performed on dry and wet asphalt (medium-high μ), the second trained on gravel or icy road surface (low μ). Thus, the two single-hidden-layer NNs identify two nonlinear multivariable dynamic systems, whose features are representative of a given range of the friction factor. The hidden and the output neurons of each NN have bipolar hyperbolic tangent activation functions.

The NNs are trained off-line to match the output $\hat{\beta}$ with the input regression vector, according to the time series approssimation approach of the linear system theory. For each NN two different architectures have been tested and compared on the basis of the tracking error *rms* and of the robustness criterion, as will be explained in the results section. The two architectures differ for the output vector dimension; precisely, the first is MISO while the second is MIMO.

The MISO neural system features a regressor vector of order n, arranged as follows:

$$\phi = \left(u_1(k)\ u_1(k-1) \cdots u_1(k-n) \cdots u_m(k)\ u_m(k-1) \cdots u_m(k)\right)^T$$

where $\mathbf{u}$ is the input vector and m is its dimension. This structure can be generalized and assimilated to a NARX or NOE model (Ljung, 1987), where the output feedback signal is a zero order. Actually, this means that the NN output is not fed-back, to avoid instability problems.

The MIMO system is derivated from the respective MISO through the addition of one or two output signals, selected among the variables which show greater correlation with the sideslip angle. The choice of these extra outputs is constrained by the on board sensor availability, meaning that the extra outputs are not fed-back but simply estimated, according to the NARX model. Thus, once the NN is trained, the correlated extra outputs are used only to supply a trend for the sideslip angle estimation.

The choice of the input/output data sets is suggested by the degree of correlation among the variables, as well as by their functional connection. Strictly speaking, the input of the real vehicle system is the steering wheel angle, at least as far as the lateral dynamics is concerned. However, the dynamic features of the mechanical transmission introduce further nonlinearities, as well as time delays. For this reason the steering wheel angle, or the analogous steering wheel torque, is not considered as a preferential input signal and is included in the input vector only if strenghtened by the presence of the beta-correlated signals. According to one of the most simple vehicle model (single-track), in fact, the rate of the sideslip angle could be expressed by the equation:

$$\dot{\beta} = \frac{a_y}{V} - \dot{\psi} \qquad (2)$$

which is approximately valid for linear and steady vehicle conditions, i.e. for low values of the lateral and longitudinal accelerations. To take into account unsteady contributions, the longitudinal acceleration should be included in the input data set. Thus, according to (2) and to the previous considerations the input signals for the MISO structure are:

$$\mathbf{u} = (\dot{\psi}\ a_y\ a_x)^T$$

For the MIMO structure, one or more of these variables are used as output signals, as will be shown in the following section.

4. RESULTS

Results concernes different strucures which are exactly duplicated for the two clusters *dry&wet* e *snow&ice*. This means that MISO1 of Table 4 and MISO1 of Table 5 have the same structure, but are trained with different time histories. The data set contains only three maneuvers performed on ice, and for this reason the NNs of the second cluster are trained just on a single maneuver, and tested on the other two. On the countrary, the great amount of data concerning the first cluster has enabled a massive training and testing activity. A brief description of each maneuver is provided in Table 4 and Table 5, while neural systems are labeled by names which identify their stucture (MISO or MIMO) as well as the particular training.

	MISO1	MISO2	MISO3	MIMO1
inputs	$\dot{\psi}$ ay	$\dot{\psi}$ ay	$\dot{\psi}$ ay ax	ay ax Δω
outpus	β	β	β	$\dot{\psi}$ β
$train_{W\&D}$	M5	M5+M2	M5+M2	M5+M2
$train_{S\&I}$	G2	-	G2	G2

Table 2: NNs structure

As shown in Table 2, MISO1 has 2 inputs and 1 output, and recalls the neural sensor developed by Sasaki and Nishimaki (2000). It has been training with a single maneuver (M5), featuring the same narrow range of longitudinal velocity of the manuevers used by Sasaki and Nishimaki.

	MISO1	MISO2	MISO3	MIMO1
# of input neurons	4	4	6	8
# of hidden neurons	8	8	10	10
# of ouput neurons	1	1	1	2
# param	49	49	81	112

Table 3: NNs dimensions

MISO2 has the same structure of MISO1, but training is performed on two maneuvers (M5 and M2), which allows the sensor to enlarge the velocity field. M5 and M2 have been selected since they contain very detailed information concerning the velocity range *0÷35 m/s*, where β is supposed to vary to a greater extent. Actually, has shown in Fig. 3, the maximum value of the error function $\beta_{act} - \hat{\beta}$ is by far lower for MISO2, apart for M9 where neither are able to identify the dynamics of a rash maneuver on wet road.

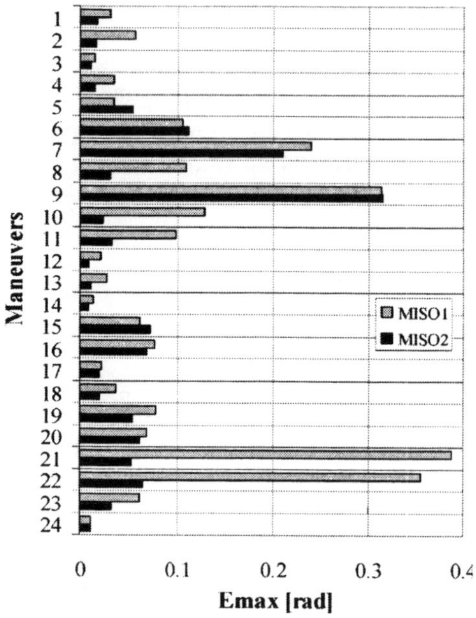

Fig. 3. Comparison between MISO1 and MISO2

MISO3 has 3 inputs and 1 output: the longitudinal acceleration a_x has been included in the input vector to account for unsteady contributions. Alhough this extra input brings about only slight improvements, in terms of maximum and rms error values, the β trend is estimated with greater precision during rush braking or acceleration, as shown in the first tract of Fig. 4.

Analogous conclusions can be drawn by analyzing the performance of MIMO1, which has 3 inputs and 2 outputs. Precisely, the yaw rate has been used as an extra output, to increase the number of connections and perform the gradient descent on a multidimensional error function.

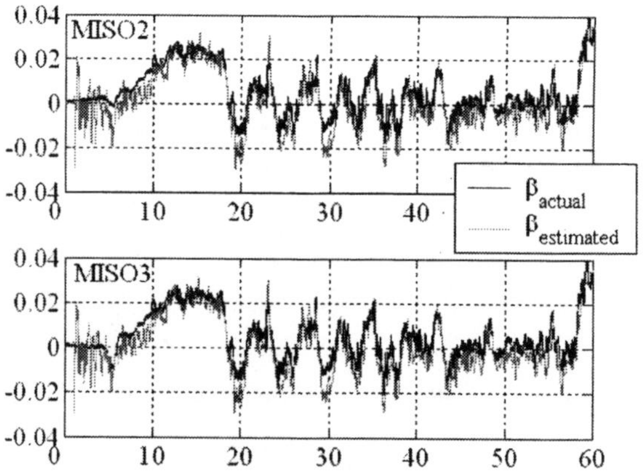

Fig. 4. β[rad] estimation on M17

The third input is a proper combination of the wheels linear velocities, that should account for the slippery of the rear axle towards the front axle:

$$\Delta\omega = \frac{1}{2}(V_{FR} + V_{FL}) - \frac{1}{2}(V_{RR} + V_{RL}) \qquad (3)$$

Still, information provided by this variable slightly improve the sensor precision for almost all the maneuvers (apart from M13, where a sudden V_{FR} sensor failure can be detected from the maneuver analysis).

All the proposed neural systems are reliable both on dry and wet road surface. Conversely, the asymmetric tire pressure condition is always considered as a pathological situation and β is consequently badly estimated. Further training on M6, for example, improves the β prediction on M7, but when the tire symmetry is restored a new training session should be performed.

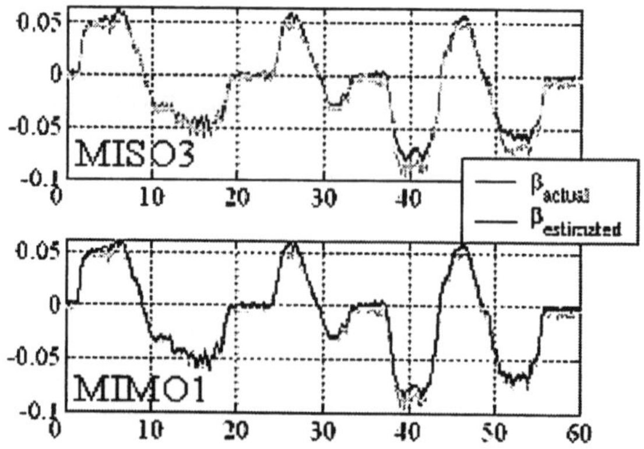

Fig. 5. β[rad] estimation on M12

As far as the second cluster is concerned, the three structures MISO1, MISO3 and MIMO1 has comparable performance even if testing is restricted to only two maneuvers. However, the real interesting point is the feasibility of the cluster-based sensor, which should enable reliable estimation in the whole range of the the external parameter μ.

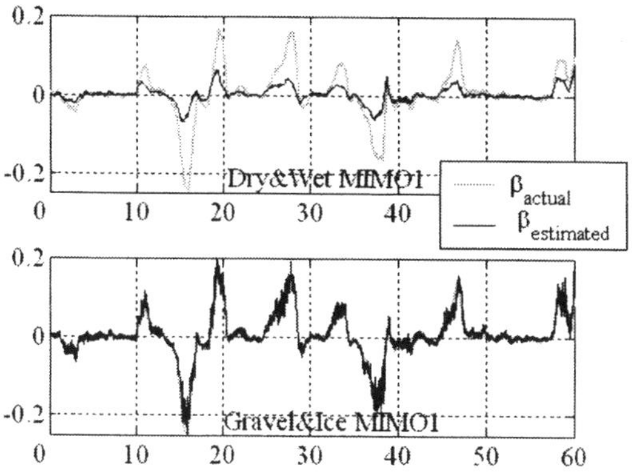

Fig. 6. β [rad] estimation on icy surface performed with a 1st cluster (upper) and a 2nd cluster (bottom) NN.

5. CONCLUSIONS

A neural virtual sideslip sensor has been developed for large scale employment on passenger vehicles.
A cluster approach has been used, based on a manual switch which allows the driver to choose between *dry&wet* or *snow&ice* asphalt conditions. Results clearly shows the sensor feasibility as well as its reliability over a very wide range of driving conducts (normal/sport), load and road surface conditions. Different architectures have been tested, with an increasing number of input and output parameters. Precision is improved mostly by a massive training activity on medium/slow and high speed maneuvers. Although this method is effective even when the lateral vehicle dynamics has a strongly non linear behaviour, it fails for non nominal pressure conditions, which imply radical dynamics alterations.

ACKNOWLEDGEMENTS

Support for the authors has been provided through a grant from the *Alfa Romeo Performance Competence on Handling Ride & Brake Center*.

REFERENCES

Gillespie T.D. (1992), Fundamental of Vehicle Dynamics, Society of Automotive Engineers, Warrendale

Ljung L. (1987), *System Identification: Theory for the User*, *Prentice-Hall Inc.*, Englewood Cliffs, New Jersey, pp. 71-72.

Sasaki H. and T. Nishimaki (2000), A Side-Slip Angle Estimation Using Neural Network for a Wheeled Vehicle. *In Proc. of the SAE 2000 World Congress,* SAE Int., Warrendale, PA

Dry & Wet		**MISO1**		**MISO2**		**MISO3**		**MIMO1**	
	Maneuver Features	rms_E [rad]	E_{max} [rad]	rms_E [rad]	E_{max} [rad]	rms_E [rad]	E_{max} [rad]	rms_E [rad]	E_{max} [rad]
M1	Handling 1, 1st part, full load	0.0043	0.0297	0.0049	0.0175	0.0046	0.0176	0.0044	0.0175
M2	Handling 1, 1st part	0.0121	0.0552	0.0038	0.0162	0.0036	0.0168	0.0036	0.0172
M3	Handling 1, Lesmo curves	0.0041	0.0144	0.0043	0.0107	0.0043	0.0104	0.0045	0.0105
M4	Handling 1, 2nd part	0.0041	0.0337	0.0059	0.0144	0.0056	0.0148	0.0051	0.0150
M5	Handling 1, sport	0.0056	0.0344	0.0092	0.0534	0.0089	0.052	0.0091	0.0454
M6	Handling 1, sport, defl. tyres	0.0179	0.1053	0.0251	0.1119	0.0244	0.1108	0.0241	0.1093
M7	Handling 1, 1st part , sport, defl. tyres	0.0287	0.2402	0.0343	0.2097	0.0337	0.2094	0.0341	0.2087
M8	Handling 1, 2nd part,Wet	0.0171	0.1086	0.0106	0.0300	0.0106	0.0302	0.0103	0.0313
M9	Handling 1, 2nd part, sport,Wet	0.0482	0.3145	0.0462	0.3146	0.0455	0.3134	0.0453	0.3125
M10	City	0.0328	0.1279	0.0119	0.0225	0.0126	0.0214	0.012	0.0202
M11	City	0.0323	0.0985	0.0248	0.0319	0.0251	0.0288	0.0238	0.0287
M12	Countryside, 1st part	0.0050	0.0209	0.0078	0.0082	0.0071	0.0079	0.0054	0.0080
M13	Countryside,3rd part	0.0053	0.0264	0.0059	0.0106	0.0056	0.0101	0.0112	0.0341
M14	Countryside, 2nd part	0.0040	0.0138	0.0048	0.0084	0.0046	0.0086	0.0044	0.0086
M15	Handling 2, 1st part, sport	0.0115	0.0600	0.0135	0.0716	0.0131	0.0701	0.0135	0.0667
M16	Handling 2, 2nd part, sport	0.0124	0.0769	0.0141	0.0673	0.0135	0.0677	0.0133	0.0708
M17	Eng. track, 1st part	0.0057	0.0224	0.0066	0.0199	0.0065	0.0200	0.0065	0.0201
M18	Eng. track, 2nd part, bumping	0.0079	0.0367	0.0082	0.0194	0.0079	0.0382	0.0078	0.033
M19	Ramp Steer	0.0224	0.0778	0.0437	0.0532	0.0153	0.0494	0.0139	0.0484
M20	Step Steer, Wet	0.0221	0.0679	0.0111	0.0611	0.0103	0.0612	0.0105	0.0610
M21	Step Steer	0.0880	0.3879	0.1023	0.0516	0.1004	0.0507	0.0997	0.0496
M22	Step Steer, throttle release	0.0755	0.3548	0.0925	0.0640	0.0925	0.0615	0.0918	0.0600
M23	High speed ring	0.0079	0.0604	0.0046	0.0320	0.0047	0.0333	0.0046	0.0333
M24	High speed ring	0.0035	0.0107	0.0032	0.0111	0.0032	0.0080	0.0032	0.0102

Table 4 Rms and maximum value of the error signal for the dry&wet cluster

	Gravel & Ice Maneuver Features	MISO1		MISO3		MIMO1	
		rms_E [rad]	E_{max} [rad]	rms_E [rad]	E_{max} [rad]	rms_E [rad]	E_{max} [rad]
G1	White tr, 2nd part, sport	0.0175	0.1097	0.018	0.0805	0.0144	0.0732
G2	White tr, 1st part, sport	0.0311	0.1585	0.0296	0.1472	0.0223	0.2022
G3	White tr, 2nd part, sport	0.021	0.2042	0.0216	0.2122	0.0433	0.6069

Table 5 Rms and maximum value of the error signal for the gravel&ice cluster

IFAC Publications
www.elsevier.com/locate/ifac

PARAMETRIC IDENTIFICATION OF THE CAR DYNAMICS

Gentiane Venture* Maxime Gautier**
Wisama Khalil Philippe Bodson***

**P.S.A. Peugeot Citroën – Direction Plates-formes, Techniques et Achats*
Route de Gisy – 78943 Vélizy-Villacoublay – France
venture@mpsa.com bodson@mpsa.com

***Institut de recherche en Communications et Cybernétique de Nantes (IRCCyN)*
1, rue de la Noë – B.P. 92101 – 44321 Nantes Cedex 3 – France
Wisama.Khalil@irccyn.ec-nantes.fr Maxime.Gautier@irccyn.ec-nantes.fr

Abstract: The aim of this paper is to give a general and unifying presentation of the identification issues of a passenger car using the Denavit and Hartenberg robotic formalism for the modelling. The desired parameters are the dynamic parameters of the car and the parameters of the suspension and the tires. The kinematics and dynamic models are automatically calculated using the software package SYMORO+. The identification is based on the use of the inverse dynamic model. Parameters are estimated using a weighted least squares method. The weighting procedure is developed. Practical results are given for a 406 Peugeot car.

Key words: parameter identification, automobile, dynamic modelling, least squares.

1. INTRODUCTION

The aim of this paper is to present a general and unifying methodology of modelling a car and to identify its dynamic parameters.

Several techniques of derivation of kinematics and dynamic models for mobile robots are available in the literature (Tilbury *et al.*, 1994, Zodiac, 1996), but the usual approach considers systems made up of a rigid cart and rigid wheels, moving in a horizontal plane, with the constraint of rolling without slipping. Meanwhile real working conditions of motion do not satisfy such hypothesis. Consequently, it is necessary that the dynamic model takes into account the 3D motion and the forces between the wheels and the soil and the aerodynamics.

With such a complexity, a systematic method of modelling, based on the modified Denavit & Hartenberg geometric description of a multi-body system (Khalil and Kleinfinger, 1986), facilitates the derivation of the dynamic and identification models. The car is considered as a tree structure made with 42 real or virtual bodies, where the four wheels are considered as the terminal links. This description allows to automatically calculate the symbolic expression of the geometric, kinematic and dynamic models by using robotics techniques or even by a symbolic software package like SYMORO+ (Symbolic Modelling of Robots) (Khalil and Creusot, 1997). The car suspensions, the anti-roll bars and the vertical behaviour of the tires are modelled with elasticities and with dampers if needed (Khalil and Gautier, 2000). Such a model allows to calculate the inverse dynamic model which is linear in the dynamic parameters, and to identify those parameters with a weighted least squares method, (Khosla, 1986, Khalil and Dombre, 2002, Guillo and Gautier, 2001).

2. DESCRIPTION OF A PASSENGER CAR USING MODIFIED DENAVIT & HARTENBERG NOTATIONS

The modified Denavit and Hartenberg (MDH) (Khalil and Kleinfinger 1986) notations can be applied to obtain the geometric parameters of a car. (Venture, and al. 2002).

Let R_0 be a fixed reference frame attached to the ground. The reference link C_r is the chassis, it corresponds to the body whose location ζ (i.e. position & orientation) gives the system posture in the frame R_0: $\zeta = [q_1\ q_2\ q_3\ q_4\ q_5\ q_6]^T$ for a movement in the 3D space. In that case the model of a car can be composed of a tree structure with 42 links C_j such that (Fig.1):

- C_0 is the base attached to the ground,
- $C_1, C_2, \ldots, C_5$ are virtual links used to define the car posture. Their variables $q_1\ q_2\ q_3\ q_4\ q_5$ correspond to the posture position variables $x\ y\ z$ and the angles $\theta\ \phi$, of the posture orientation variables respectively.
- C_6 is the chassis, its variable is the roll $q_6 = \psi$.
- C_9, C_{18}, C_{27} and C_{36} are the dampers, represented by linear springs and dampers.
- C_{14} and C_{23} are the rear wheels
- C_{32} and C_{41} are the front wheels

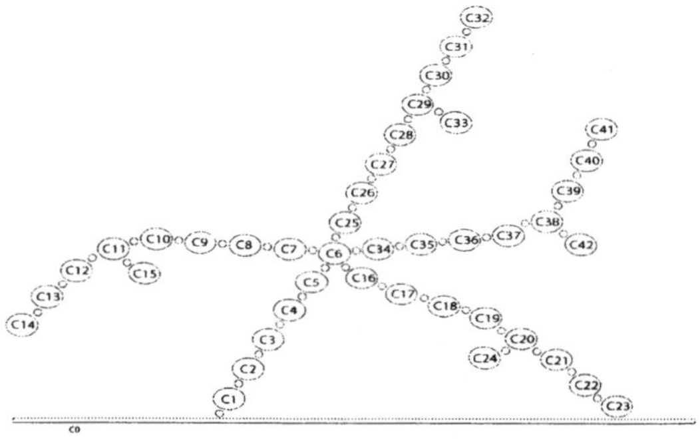

Figure 1: multi-bodies description of a car

- $q_7, q_{16}, q_{25}, q_{34}$ are the ½ track width,
- $q_8, q_{17}, q_{26}, q_{35}$ are the ½ wheelbase,
- $q_9, q_{18}, q_{27}, q_{36}$ are the dampers clearances,
- $q_{10}, q_{19}, q_{28}, q_{37}$ are the steering angles,
- $q_{11}, q_{20}, q_{29}, q_{38}$ are the camber angles,
- $q_{13}, q_{22}, q_{31}, q_{40}$ are the kingpin angles
- $q_{14}, q_{23}, q_{32}, q_{41}$ are the rotation of the wheels
- $q_{15}, q_{24}, q_{33}, q_{42}$ are the tire deflection.

This description allows calculating the geometric, kinematic and dynamic models automatically with the help of SYMORO+, software of symbolic calculations developed by the robotics team of the IRCCyN. (Khalil and Creusot, 1997)

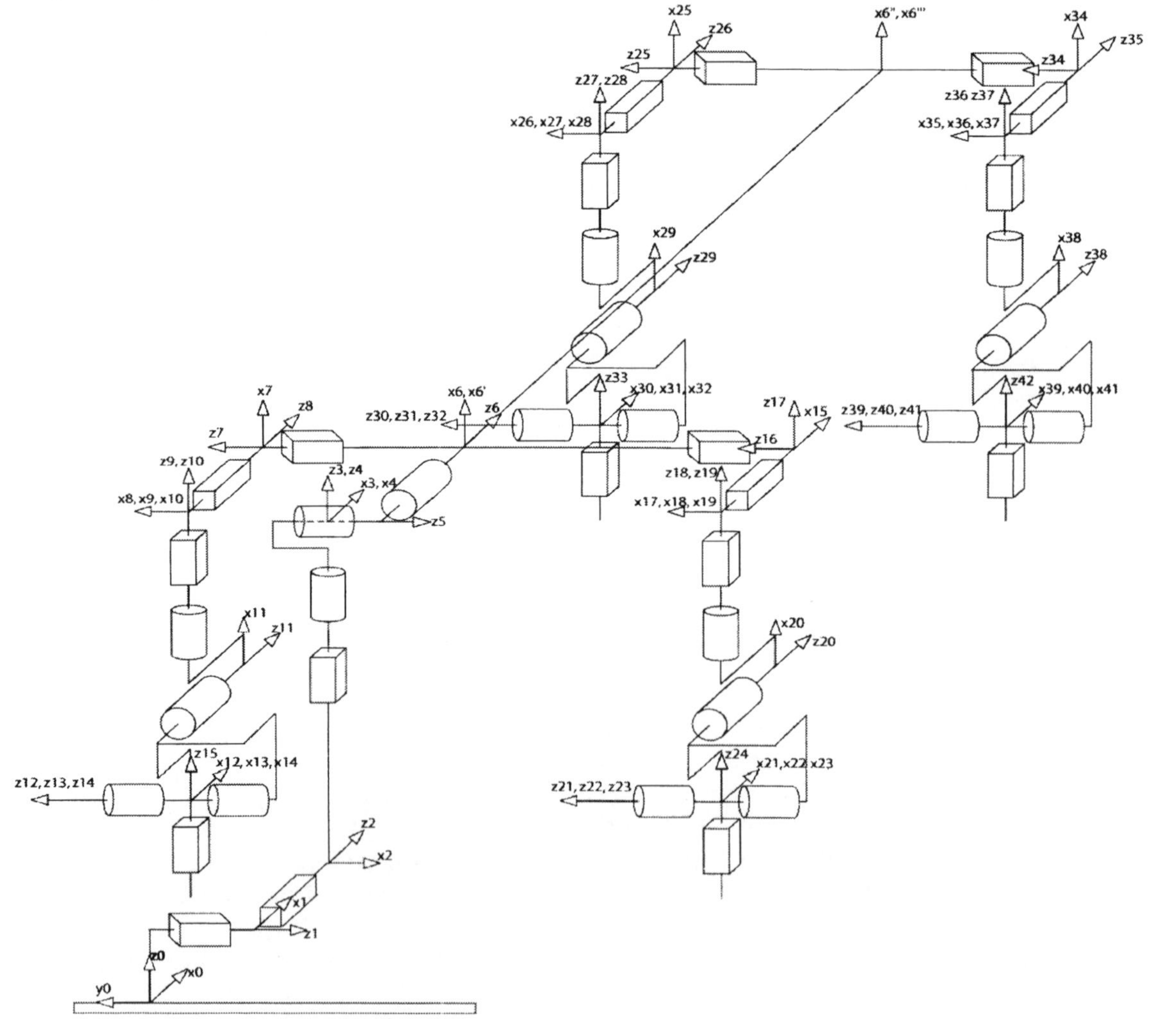

Figure 2: geometric description of a car

3. COMPUTATION OF THE DYNAMIC MODEL

The Inverse Dynamic Model (IDM) gives the joint torques as a function of the joint positions, velocities and accelerations. Let the IDM be written as:

$$\mathbf{M}(\mathbf{q})\ddot{\mathbf{q}} + \mathbf{H}(\mathbf{q},\dot{\mathbf{q}}) = \mathbf{L} + \mathbf{Q}^c + \mathbf{Q}^a \quad (1)$$

- $\mathbf{M}(\mathbf{q})$ is the mass matrix of the system Σ
- $\mathbf{H}(\mathbf{q},\dot{\mathbf{q}})$ is the vector of centrifugal, Coriolis and gravity terms.
- **L** is the vector of the internal forces between the vehicle bodies: motor torque, friction, elasticity
- $\mathbf{Q}^c$ is the vector of the contact forces between the ground and the wheels projected on the joint axes.
- $\mathbf{Q}^a$ is the vector of aerodynamic forces projected on the joint axes.

3.1. Internal forces: L

The internal forces vector is composed of:

- the elastic forces

If j is an elastic joint, the j-component of L^e_j, the elastic forces vector can be written:

$$L^e_j = -k_j q_j - h_j qp_j$$

where q_j is the joint displacement w.r.t. the steady state position, k_j is the stiffness of the j-joint and h_j the damping coefficient.

- the friction vector component, is simplified as:

$$L^f_j = -F_{vj}\dot{q}_j - F_{sj} sign(\dot{q}_j)$$

F_{vj} is the viscous friction coefficient.
F_{sj} is the Coulomb friction force.

3.2. Aerodynamic forces: $\mathbf{Q}^a$

The aerodynamic forces have to be taken into account for high-speed (> 100 km/h) tests. They are given by tables and applied on body C_6.

$$\mathbf{Q}^a = \mathbf{J_a}^T\mathbf{F}^a$$

where $\mathbf{F}^a$ is the aerodynamic forces and Ja is the Jacobian matrix corresponding to its projection on the joint axes.

3.3. Forces between the wheels and the ground: $\mathbf{Q}^c$

In the experiments, four dynamometric wheels measure the contact forces. They give the 6 directional forces and torques expressed in the frames R_8, R_{11}, R_{15}, R_{19}, attached to the wheel axes.

$$\mathbf{Q}^c = \Sigma \mathbf{J_{cj}}^T\mathbf{F}^c_j$$

where $\mathbf{F_j}^c$ is the contact forces on wheel j and $\mathbf{J_{cj}}$ is the Jacobian matrix corresponding to its projection on the joint axes.

3.4. Modelling of the anti-roll bars

The constraint equation (2) is simpler to add it by hand than to model it with the MDH notation.

$$F_{ar} = k_{ar}\,(q - q_{op}) \quad (2)$$

Where :

- F_{ar} is the anti-roll bar force,
- k_{ar} the anti-roll bar stiffness,
- q the damper clearance of the considered wheel
- q_{op} the damper clearance of the opposite wheel of the same axle.

It gives the complete equation for links 9, 18, 27, 36:

$$F_i = ks_i.q_i + h_i.qp_i + fs_i.\text{sign}(qp_i) + off_i + k_{ar}\,(q_i - q_{opi}) \quad (3)$$

4. DYNAMIC PARAMETERS

This is the parameters which are used to write the dynamic equations of the system.

4.1. Standard inertial parameters

For each link there are 10 standard inertial parameters (Gautier and Khalil, 1990), composed of :

- $[XX_j\ XY_j\ XZ_j\ YY_j\ YZ_j\ ZZ_j]$: the 6 elements of the inertia matrix of link j with respect to frame j,
- $[MX_j\ MY_j\ MZ_j]$: the 3 first moments of link j around the origin of frame j,
- M_j, the mass of link j.

Beside the inertial parameters, the dynamic parameters include for some joints the spring stiffness k_i, and dampers coefficients h_i, offset off_i, friction coefficients fs_i.

4.2. Base inertial parameters

The base inertial parameters are defined as the minimum inertial parameters that can be used to obtain the dynamic model. They represent the set of inertial parameters that can be identified using the dynamic model.

They are obtained from the standard inertial parameters by eliminating those that have no effect on the dynamic model and by grouping some others. There are two techniques to obtain those parameters: a symbolic one (Gautier and Khalil, 1990), or a numerical one (Gautier, 1991).

In the model concerned the base parameters are the standard parameters, there is neither regrouping relation nor elimination of parameters.

4.3. Base dynamic parameters

The base dynamic parameters, defining the vector of the dynamic parameters, X, to be identified, consists of:

- For the chassis (link C_6):
$\mathbf{X_6} = [XX_6\ XY_6\ XZ_6\ YY_6\ YZ_6\ ZZ_6\ MX_6\ MY_6\ MZ_6\ M_6]^T$
- For the wheels (links C_i, i = 14, 23, 32, 41):
$\mathbf{X_i} = [XX_i\ YY_i\ ZZ_i\ M_i]^T$
- For the suspensions (joints i = 9, 18, 27, 36):
$\mathbf{X_i} = [ks_i\ h_i\ fs_i\ off_i]^T$
- For the anti roll bars: $\mathbf{X_{ar}} = [k_{arf}\ k_{arr}]^T$
- For the tire deflection (joints i = 15, 24, 33, 42):
$\mathbf{X_i} = [kt_i]^T$

And finally X, the vector of identifiable parameters has the following expression:

$$\mathbf{X} = [\mathbf{X_6}^T\ \mathbf{X_9}^T\ \mathbf{X_{14}}^T\ \mathbf{X_{15}}^T\ \mathbf{X_{18}}^T\ \mathbf{X_{23}}^T\ \mathbf{X_{24}}^T\ \mathbf{X_{27}}^T\ \mathbf{X_{32}}^T\ \mathbf{X_{33}}^T\ \mathbf{X_{36}}^T\ \mathbf{X_{41}}^T\ \mathbf{X_{42}}^T\ \mathbf{X_{ar}}^T]^T$$

Let n_p be the number of the parameters to identify:

$$n_p = \text{length}(\mathbf{X}) \quad (4)$$

5. IDENTIFICATION

5.1. Identification model

The dynamic model (1) can be expressed as a linear relation w.r.t. the vector of identifiable parameters. It can be written as:

$$\mathbf{y} = \mathbf{D}(\mathbf{q}, \dot{\mathbf{q}}, \ddot{\mathbf{q}}).\mathbf{X} \tag{5}$$

5.2. Identification method : weighted least squares

X can be estimated, from the sampling of the dynamic model, as the least squares (L.S.) solution $\hat{\mathbf{X}}$, of the following linear system:

$$\mathbf{Y} = \mathbf{W}\mathbf{X} + \rho \tag{6}$$

- **W** is the observation matrix (n_{exp} x n_p), it represents the concatenation of the **D** matrix on the different samples of the trajectory.
- **Y** the (n_{exp} x 1) vector of joint and reaction forces
- ρ the (n_{exp} x 1) vector of model errors
- n_{exp} is the (number of samples*42)

The method allows computing a standard deviation for each parameter $\sigma_{\hat{X}i}$. Classically it is estimated considering **W** to be a deterministic matrix and ρ to be zero mean and independent noise, with standard deviation σ_ρ such that: $\mathbf{C}_{\rho\rho} = \sigma_\rho^2.\mathbf{I}_{nexp}$

The variance-covariance matrix can be calculated as:

$\mathbf{C}_{\hat{X}} = \sigma_\rho^{\,2}\left[\mathbf{W}^T.\mathbf{W}\right]^{-1}$, and $\sigma_{\hat{X}i}^{\,2} = C_{\hat{X}ii}$ is the ith diagonal coefficient of $\mathbf{C}_{\hat{X}}$. So the relative standard deviation $\sigma_{\hat{X}ri\%}$ is given by :

$$\sigma_{\hat{X}ri\%} = \frac{\sigma_{\hat{X}i}}{\hat{X}_i} \tag{7}$$

But, in fact, $\underline{\mathbf{Y}}$ and **W** are obtained by concatenation of 42 equations that correspond of (j =1,42) linear systems with different standard deviation $\hat{\sigma}_\rho^j$. They can be calculated using the minimal 2-norm of errors calculated on each linear system:

$$\hat{\sigma}_\rho^j = \frac{\left\|Y^j - W^j.\hat{X}^j\right\|}{\sqrt{n_p^j - p^j}} \tag{8}$$

X^j is the (p^j x 1) vector of the minimal parameters of equation j

n_p^j is the number of equations of the linear system j

The inverse of these standard deviations give the weighting matrix **P**:

$$\mathbf{P} = \begin{bmatrix} S_1 & & \\ & \ddots & \\ & & S_{42} \end{bmatrix} \text{with } S_j = (1/\hat{\sigma}_\rho^j).I_{n_p^j} \tag{9}$$

Equation (6) weighted by P gives the following system:

$$\mathbf{Y_p} = \mathbf{W_p}\mathbf{X} + \rho_\mathbf{p} \tag{10}$$

where: $\mathbf{Y_p} = \mathbf{P}\,\mathbf{Y}$, $\mathbf{W_p} = \mathbf{P}\,\mathbf{W}$ *et* $\rho_\mathbf{p} = \mathbf{P}\,\rho$.

The solution $\hat{\mathbf{X}}_\mathbf{W}$ and the standard deviations $\hat{\sigma}_{\hat{X}Wi}$ are those given by the L.S. solution of (10). This weighting procedure allows improving both the estimation of the parameters and the estimation of the standard deviations.

6. EXPERIMENTAL SET UP

For the data acquisition, a real car is equipped with many sensors, which allows estimating the joint variables needed for the identification model and the contact forces with the ground.

6.1. Sensors and measurements

The sensors used in our case are composed of: four dynamometric wheels, an inertial unit, a laser sensor for vertical position, video camera "Zimmer", and a speed sensor "Correvit".

Table 1 : sensors characteristics

sensor	measurement	accuracy	Bandwidth	resolution
Inertial unit Sagem	Yaw speed *qp4*	0.01°/s	50°/s 10Hz	0.005°/s
	Roll speed *qp6*	0.01°/s	50°/s 10Hz	0.005°/s
	Pitch speed *qp5*	0.01°/s	50°/s 10Hz	0.005°/s
	Lateral acceleration rear axle *qdp1*	0.1m/s²	15m/s² 10Hz	0.01m/s²
	Longitudinal acceleration rear axle *qdp2*	0.02m/s²	15m/s² 10Hz	0.01m/s²
	Vertical acceleration rear axle *qdp3*	0.02m/s²	15m/s² 10Hz	0.01m/s²
	Absolute roll angle *q6*	0.1°	50° 5Hz	0.05°
	Absolute pitch angle *q5*	0.1°	50° 5Hz	0.05°
Speed sensor Correvit	Longitudinal speed *qp2*	1km/h	150km/h 5Hz	
	Lateral speed *qp1*		15m/s 5Hz	0.075m/s
Video camera Zimmer	Ground clearances *q9, q18, q27, q36*	1mm on smooth ground	125mm 200Hz	0.062mm
	Steering angles *q10, q19, q28, q37*	0.033°	5° 30 Hz	0.011°
	Camber angles *q11, q20, q29, q38*	0.042°	2.5° 30Hz	0.014°
Dynanometric wheel igel	Angular speed axles *q14, q23, q32, q41*		30tr/s 150Hz	
	Wheel total forces and torques	25N	20kN 150Hz	

Other joint coordinates are tabulated using the damper clearances and the steering angle. Tables of data are given by the car manufacturer (PSA) as a result of measurements carried out on a special test bench.

6.2. Filtering and joint coordinates estimation

All the measurements (position, velocity or acceleration) are filtered in order to estimate the position, the velocity and the acceleration for each joint. The derivatives are estimated with a zero-phase band pass filter whose transfer function is the product of a forward and reverse low pass butterworth filter (filtfilt matlab function) with a central difference algorithm. The order (between 5 and 8) and the cut off frequency (around 3Hz) are chosen according to the order of the derivatives and the dynamics to be identified (around 1Hz). The integrations are estimated with a trapezoidal method without phase shift. Once they are estimated it is possible to compute Y and W for the considered test.

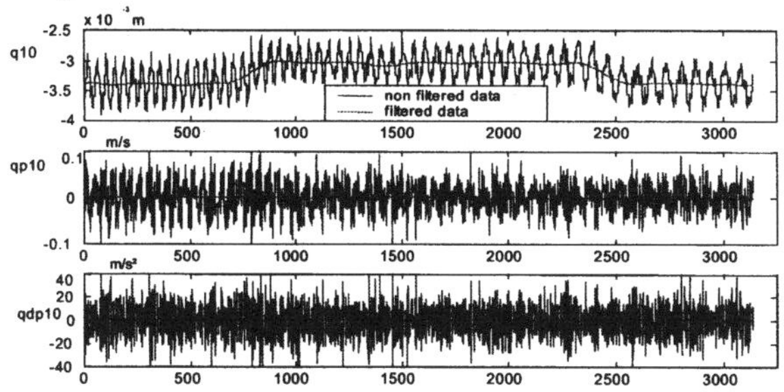

Figure 3 : filtering and derivation

7. RESULTS

7.1. Results

Results are obtained using standard tests such as sinus steering, straight line braking and spiral steering tests. That allow to limit the cost. The sample time is 0.0016 second, the number of samples: n_{exp}, depends on the tests. The estimated values and the relative standard deviations are shown in tables 2,3 and 4. Parameters are given in IS units and in the modelling frames.

Table 2: Inertial parameters of the chassis

parameter	*A priori*	estimated	%standard deviation
XX6	*6810.44*	6877.46	0.30
XZ6	*-445.9*	-590.92	0.56
YY6	*6567*	16697.76	1.72
ZZ6	*668.56*	714.26	1.01
MX6	*537.24*	357.43	0.29
MY6	*5.81*	-2.06	3.26
MZ6	*2368.21*	2446.61	0.07
M6	*1685*	1644.11	0.10

M6 is the mass of the whole car (spring mass and non-sprung mass). The position of the centre of mass in frame R6 is given as:

$X_{G/R6} = MX6/M6$, $Y_{G/R6} = MY6/M6$,
$Z_{G/R6} = MZ6/M6$,

Table 3: dynamic parameters of the suspensions

parameter	*A priori*	estimated	%standard deviation
fs9		12.60	7.95
off9		589.08	0.38
ks9	*22000*	24425	1.42
h9	*4200*	4131.7	1.17
fs18		30.41	3.31
off18		-14.58	13.38
ks18	*22000*	20798	1.73
h18	*4200*	4445.2	1.13
karr	*19185*	21662	0.82
fs27		28.56	4.49
off27		420.94	1.67
ks27	*20000*	27478	1.32
h27	*3200*	3143.2	1.54
fs36		35.87	3.34
off36		1018.87	0.80
ks36	*20000*	26073	1.53
h36	*3200*	3869.1	1.36
karf	*19780*	18342	1.05

Table 4: vertical stiffness of the tires

parameter	*A priori*	estimated	%standard deviation
kt15	*200000*	216502.50	0.32
kt24	*200000*	217816.21	0.31
kt33	*200000*	232581.41	0.37
kt42	*200000*	230549.08	0.37

7.2. Interpretation and validation

A rule of thumb is to consider that the parameters whose standard relative deviations are greater than 20 times the smallest one are poorly identified. Most of all parameters are well identified with the given data. Another step of validation consists in comparing Y to the reconstructed vector $W.X$. (Figures 4,5,6,7,8).

Figures 4,5 and 6 give the results of a direct validation (carried out on the trajectories used in the identification). Figures 7 and 8 give the results of a cross validation, where the validation trajectory has not been used for the estimation: the test used for the cross validation is a braking in straight line while the identification test was a 90km/h sinus steering.

Abscises are $n_{exp} * n_{joint\ represented}$

Ordinates represent the joint force or the joint torque.

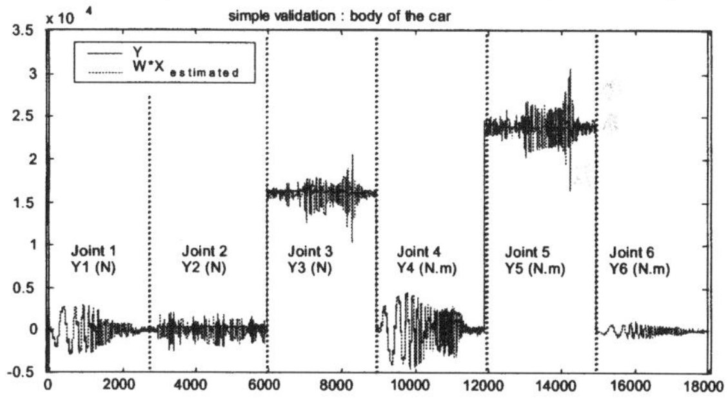

Figure 4: direct validation for the chassis

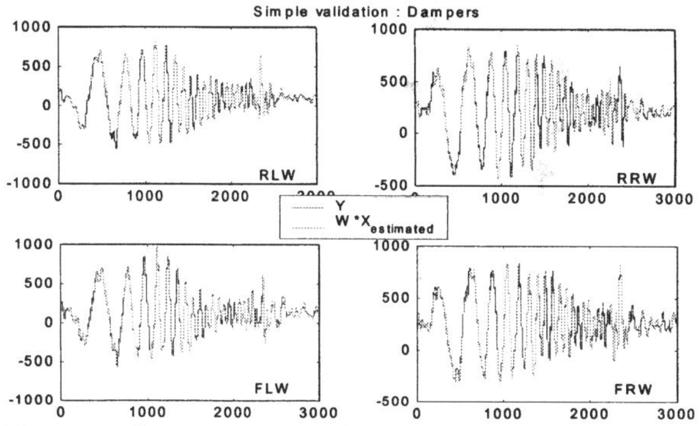

Figure 5: direct validation for the suspension

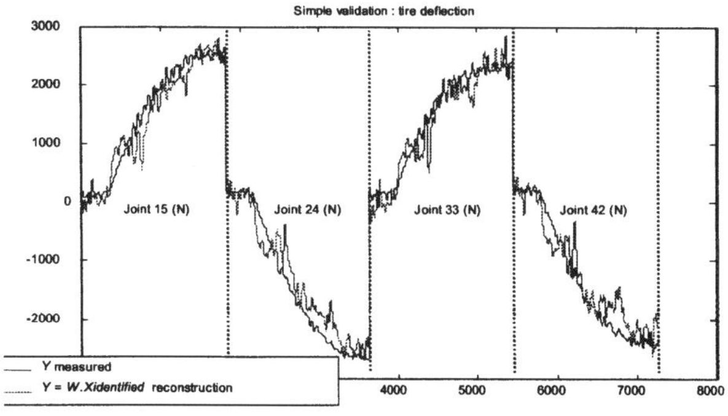

Figure 6: direct validation for the vertical stiffness of the tires

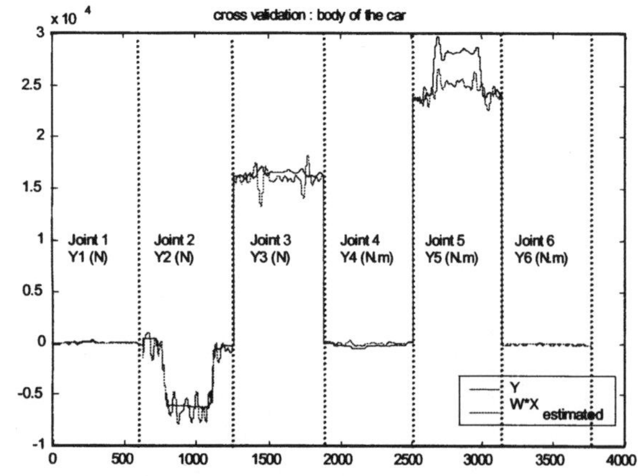

Figure 7: cross validation for the chassis

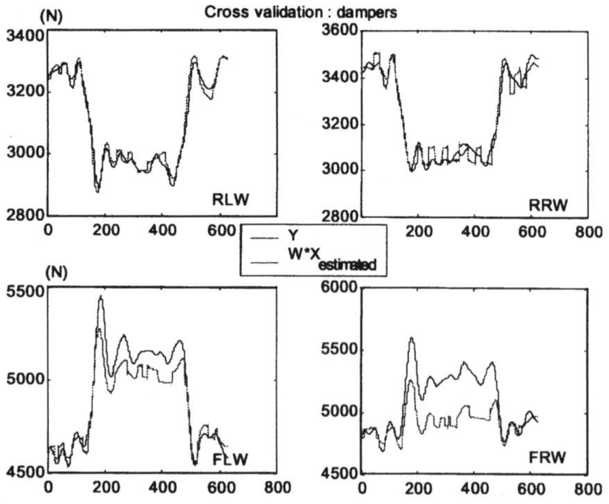

Figure 8: cross validation for the suspension

8. CONCLUSION

Results presented in this paper show the possibility of extending robotics formalism to the mobile robots and to the ground vehicles such as passenger cars. This formalism and the use of appropriated tools allow to compute automatically the dynamic models. The inverse dynamic model can be written as a linear relation w.r.t. the dynamic parameters, and the use of a weighted least squares technique is not time consuming and straight forward. Good results are obtained: they have small standard deviations, they are close to the a priori values, and they are confirmed by direct and cross validation.

The model has been validated and compared with respect to other dynamic models used for simulations.

Nevertheless, it is important to notice that the suspension stiffness and damping coefficients are supposed to be constant. This is a rough approximation particularly for the damping coefficients. It could be better to tabulate them from data obtained by special test measurements.

Further work is to extend the set of tires parameters in order to estimate tires characteristics.

REFERENCES

Gautier M., Khalil W (1990), Direct calculation of minimum set of inertial parameters of serial robots, *IEEE Trans. on Robotics and Automation*, Vol. RA-6(3), 1990, p. 368-373.

Gautier M. (1991), Numerical calculation of the base inertial parameters, *J. of Robotic Systems*, Vol. 8 (4), August 1991, p. 485-506.

Gautier M (1997), Dynamic Identification of Robots with Power Model, *Proc. of IEEE International Conference on Robotics and Automation* . pp.1922-1927, Albuquerque, NM, USA.

Jacquinot E. (2000), Simulink/car : modèle dynamique simplifié de véhicule, document interne PSA

Khalil W., Creusot D. (1997), SYMORO+: a system for the symbolic modelling of robots, *Robotica*, Vol. 15, 1997, p. 153-161.

Khalil W. and Dombre E. (2002), Modelling, identification and control of robots, *Hermès Penton, London & Paris.*

Khalil W., Gautier M (2000), Modelling of mechanical systems with lumped elasticity, *Proc. IEEE Int. Conf. on Robotics and Automation*, pp. 3965-3970, San Francisco, CA, USA.

Khalil W., Kleinfinger J.F.(1986), A new geometric notation for open and closed loop robots, *Proc. IEEE on robotics and automation*, pp. 1174-1180, San Francisco, CA, USA.

Khosla P.K (1986), Real-time control and identification of direct drive manipulators, Ph. D. Thesis, Carnegie Mellon University, Pittsburgh, USA.

Milliken W.F. and D.L (1998)., Race car vehicle dynamics, *SAE international, ISBN 1-56091-526-9.*

Pacejka H.B., Besselink I.J. (1997), Magic formula tyre model with transient properties, *Vehicle system dynamics supplement*, n°27, pp 234-249

C. Canudas de Wit, B. Siciliano, G. Bastin (1996), Theory of robot control, Eds., Springer-Verlag, Berlin, Germany, 1996.

Venture G., Khalil W., Gautier M, Bodson P.(2002) Dynamic modelling and identification of a car, *Proc. IFAC 2002*, Barcelona, Spain.

IFAC
Publications
www.elsevier.com/locate/ifac

SIMULATING ENERGY CONSUMPTION OF AUXILIARY UNITS IN HEAVY VEHICLES[1]

Niklas Pettersson, Karl Henrik Johansson

Department of Signals, Sensors & System, Royal Institute of Technology, Sweden
niklas.pettersson@s3.kth.se, kallej@s3.kth.se

Abstract: Models that can be used to analyse the fuel saving potential of electrically driven auxiliaries in heavy vehicles are presented. With the purpose of evaluating the influence on fuel consumption from various concepts and control principals, a model library is developed in the modelling language Modelica. The library contains a mixture of models developed from physical principles and models fitted to collected data. Modelling of the cooling system is described in some detail. Simulation results are compared with measurement data from tests in a wind tunnel. *Copyright © 2003 IFAC*

Keywords: Automotive Control, Energy Management Systems, Computer Simulation, Modelica.

1. INTRODUCTION

Improving fuel efficiency is central when developing heavy vehicles. The cost for fuel constitutes a major part of the total costs for transport of goods. Thus, a fuel-efficient truck will be economically attractive to own and can be sold at a higher price. In the same time, fuel efficiency is directly related to the environmental impact of road transports. More and more focus is put on environmental issues, and in the end, only the ones that can adapt to the demands on reduced environmental impact from their business will survive. Over the years, huge efforts have been put on improving efficiency of the combustion engine. However, the marginal cost for achieving a certain improvement of the engine is increasing. Therefore, it is getting increasingly more rewarding spending development effort on optimising the efficiency of the auxiliary units and subsystems in the vehicle. Resent advances within the area of electrical drive systems, and alternative power production devices such as fuel cells, may offer a possibility to utilise electricity powered auxiliaries where it former was non-practicable or non-economical. In the process of introducing new technologies, working with models and computer simulations is essential. Simulations studies are needed both to evaluate novel concepts and to increase the knowledge of how the behaviour of today's design.

This paper presents the work of developing vehicle models that can be used to evaluate alternative architectures for the electrical system in heavy vehicles. With aid of the simulation models, the potential energy savings of new designs can be assessed. Here the ideas behind development and maintenance of a comprehensive model library are presented. The Modelica language is used to build models with a modular structure. Figure 1 shows composition of the model at the highest level.

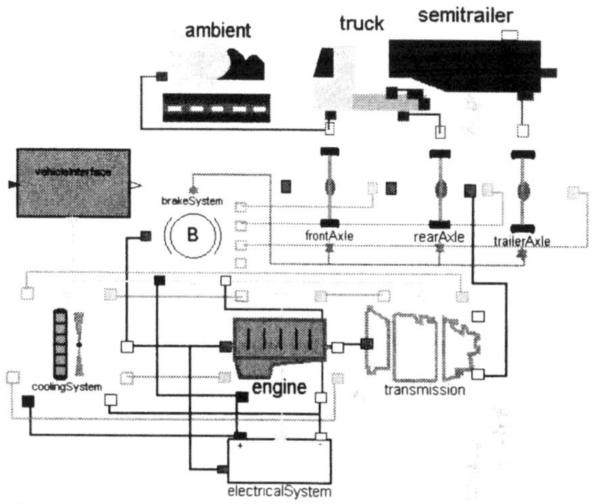

Fig 1. Modules of the simulation model.

In the simulations, the vehicle is set to drive a road with varying topology and speed limit that have been obtained from recordings of real roads. The vehicle is assumed to run on cruise control and with computer-controlled gear shifting (automated manual transmission). Algorithms from the production version of the control are incorporated in the simulation model. The vehicle model has been validated with respect to the energy consumption of the combustion engine and losses such as rolling resistance and air drag, (Sandberg, 2001). Influences from the, sub-systems, the cooling system, and the electrical network, were only included as a lumped effect on the net fuel consumption. This work refines the description of the auxiliary units. The paper describes the modelling of the cooling system in some detail. Sub-models are built from physical principles, resulting in grey-box models with

[1] This work is supported by Scania CV AB and Vinnova.

parameters identified from various tests in a laboratory environment. The sub-models are assembled into a model of the complete vehicle. Measurements collected from tests in a wind tunnel are used to tune the performance of the total model. The model is validated against data recorded from a dynamic driving cycle in the wind tunnel.

2. SELECTION OF MODEL STRUCTURE

The prime goal with the vehicle model is to serve as a tool for study effects on the fuel economy from different designs of sub-systems. The main quantity that is studied is flow of energy between the parts of the vehicle. To give an accurate estimate of the energy balance, the model must cover the whole vehicle and describe processes involved in the energy conversion with a significant level of detail. Besides description of physical phenomena, it contains control software and various look-up tables. We refer to this type of model as *simulation models* since it (in contrast to simpler models of sub-systems or black-box models of complete processes) might not be possible to summarise the model in a set of mathematical equations. In the literature this type of models sometimes are referred to as software models (cf., Ljung, 1999). Although the model might be comprehensive, by necessity it must in many parts be a rough approximation in order to keep the complexity on a reasonable level. Fast dynamics and other effects not relevant for the study may be neglected.

Models used in vehicle development can be classified in three groups according to which level of detail they use to describe the physical world, shown in figure 2. The most detailed models, referred to as *special purpose models*, are used designing hardware components. In this category one find models of the combustion process, 3-D models of the suspension etc. On the other extreme we find models used for control design. Since their limited extent allows for a mathematical description we denote these models *mathematical models*. Often mathematical models represent an abstract image of the physical world. Parts may be lumped into a single quantity, e.g., total inertia felt in one motion, blurring out the interfaces between components in the system described in the model. Similarly to special purpose models, mathematical models typically cover a limited part of the vehicle.

2.1 Simulation models

It has become more and more important to perform simulations covering the complete behaviour of the vehicle as control functions develop to be more integrated. Beside this, overall performance parameters as fuel consumption, electric energy balance and cooling capacity have become more critical in vehicle design. Models integrating relevant descriptions of major parts of the vehicle tend to grow in complexity.

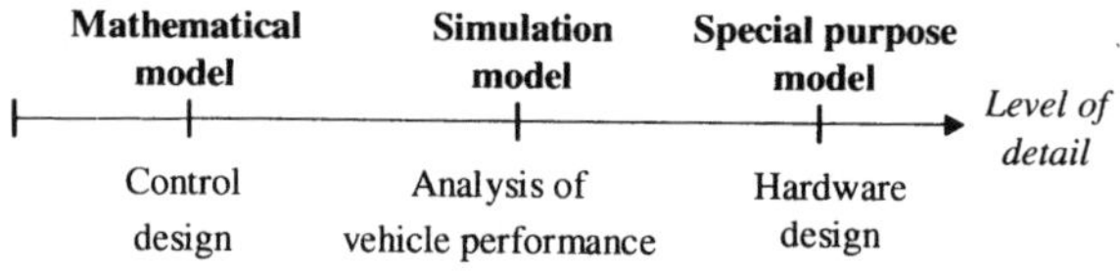

Fig 2. Classes of models used in the automotive industry.

To assemble the model in a closed form system of equation might not be possible. Stability of complex systems may not be possible to analyse mathematically. To study stability with simulation models heuristically may be the only possible alternative. In general, not only the input-output behaviour of model is of interest. Internal signals may be just as important to study. Therefore it is preferable to split up the model in components corresponding to physical objects and derive sub-models from first principles. Structured in this way, simulation models are intrinsically multi-domain models.

Although all the classes of models play a specific part, it is advantageous to organise the modelling effort in an integrated way. Clearly, there exist a possibility to utilise results gained working with one type of model also for other type of models. From the perspective of simulation models, an outline of benefits with integrated modelling efforts might include: (1) time will be saved assembling a simulation models if modelling paradigms and measurement data can be adopted from the teams working with special purpose models. (2) Mathematical and simulation models can share databases for parameter values. In case either of them work with a more lumped definition of parameters, it may be possible to implement translation rules for transferring on type of parameter set to the other. (3) Control algorithms developed based on a mathematical model can be tested out together with a simulation model containing more details before the control is applied in a real vehicle. (4) In the work with complex simulation models one often have to resort to a simplified mathematical model in order to gain insight in dependencies between parts of the model or to do sensibility analysis of parameter influences. (5) The work with simulation models can give useful input to the hardware design. Results from a complete vehicle simulation can be used to define operating profiles for the components such as average speed, time spent on a certain load, etc.

3. MODEL LIBRARY

A library of simulation models is developed according to the principles described in the previous section. The library is developed in Modelica™, (Modelica Association, 2000). Modelica is an object oriented modelling language well suited to describe behaviour of complex systems containing parts from different engineering disciplines, e.g., mechanics and electronics.

3.1 Library structure.

In contrast to the Modelica Standard Library, (Modelica Association, 2002), the library is not organised in different engineering disciplines. Instead it is organised after the parts of the truck. The library, named Scania Modelica Library, SML, consist of four main branches:

1. Interfaces
2. Components
3. Modules
4. Examples

The principal structure of the library can be viewed in figure 3.

The Interface branch contains classes describing connections between model components. Although the library relies heavily on connector classes defined in the Modelica Standard Library, some unique connectors are defined. One example is the CAN connector, used to mimic the information flow between control units in the truck. Further, under the Interfaces sub-library Media, base classes for thermodynamic and hydraulic models are found. These base classes are mainly used in models of components in the cooling system. In the thermodynamic and hydraulic base classes many of the modelling ideas used are adopted from the Modelica library ThermoFluid developed by Thummescheit, et al. (2000). However, here a somewhat simpler structure and less extensive description of media properties are used. In the Components branch models of all physical parts needed to build up the complete model of a truck are gathered. Modules, in the next branch, are a higher level of abstraction, and contain more compound models. The idea is to define a set of generic modules with well-defined interfaces that can be used to for simulations with various purposes. In the last branch a number of working examples is built that can be used directly for simulations.

Figure 4 illustrate how the models are parameterised to obtain modules that correspond to physical modules. Each component contains a placeholder for a set of parameters of a defined structure. Parameter sets with values describing various versions of the components are gathered in special sub-libraries. When modules are put together, illustrated with the cooling module, the generic placeholders are replaced with the parameter set of the current versions of components.

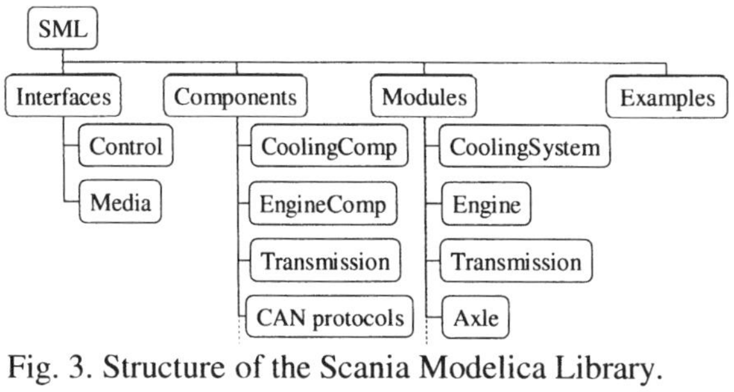

Fig. 3. Structure of the Scania Modelica Library.

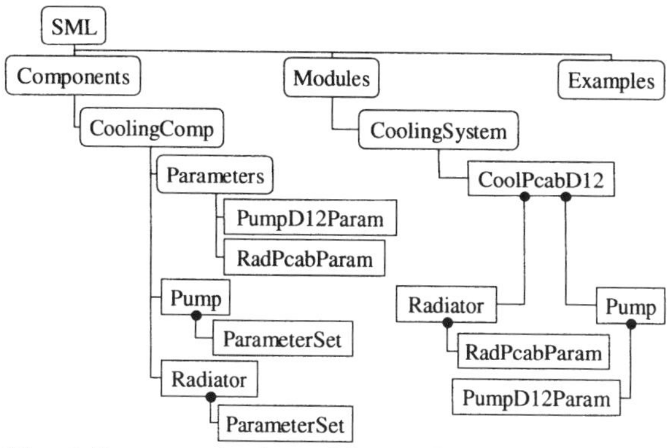

Fig. 4. Parameterisation of the model exemplified with the cooling module.

4. COOLING SYSTEM MODULE

The cooling system is one of the modules of the vehicle model. Energy consumers in the cooling system are primarily the cooling fan and the water pump. In heavy vehicles, these units normally are mechanically driven. The model corresponds the current design of a Scania truck where the water pump is directly driven from the crankshaft while the cooling fan is connected to the shaft via a viscous clutch enabling a passive speed control. However, the basic structure allows for changing the model to describe other ways of driving and controlling these auxiliaries.

4.1 Cooling system components

The main parts of the cooling system are modelled, using the thermodynamic and hydraulic base classes. In figure 5 the structure of the cooling system is depicted. The model mainly consists of two adjoining flows of mass and energy: the flow of coolant fluid and the airflow.

The pump drives the flow of coolant fluid through the engine and the retarder. The retarder is a hydraulic brake mounted on the secondary side of the gearbox. When used, it produces heat that is emitted to the cooling system. The temperature of the coolant is controlled with the thermostat by splitting up the coolant flow into one part passing the radiator and one part flowing in a by-pass pipe.

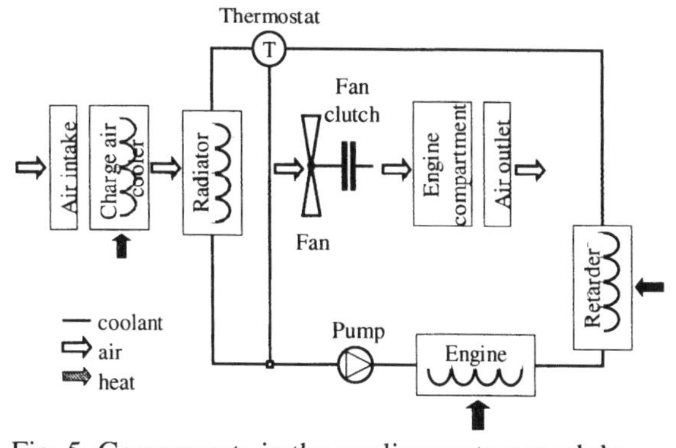

Fig. 5. Components in the cooling system module.

The air enters in the air intake at the front of the truck cab and exits at the air outlet at the rear. The airflow is partly driven by the fan and partly by the pressure build up caused by the wind speed at the intake and outlet. The air is used to cool down both the turbo charged intake air to the engine, and the coolant fluid. The charge air cooler and the radiator are connected in series so that the cooling air first passes the charge air cooler and then the radiator. Both the charge air cooler and the radiator are cross directional heat exchangers, i.e., the hot and cool media streams are perpendicular to each other.

The models of the coolant and the air streams are built up with alternating control volumes and flow models. In the control volumes, mass and energy balances are defined, while in the flow models, relations between the pressure drop and the flow are determined. The control volumes describe the dynamic behaviour and are parameterised purely with geometrical quantities and properties of the contained media. The flow models describe pressure drops, heat transfer and consumed power based on empirical relations. No explicit identification of the parameters of the control volumes is needed, since they could be found in the technical specification of the components. The parameters of the flow models, however, typically have to be estimated from data.

4.2 Dynamics of the cooling system

For the control volumes it is possible to select which state representation that should be used. The transformation of state variables from the primary mass and energy balances to the selected states is dependent on the properties of the media inside the volume. The modelling of the control volumes is rather standard. Here it essentially follows the principles in ThermoFluid (Tummescheit et al. 2000)

For the airflow, pressure, p, and temperature, T, are chosen as state variables. The transformed balance equations then become

$$\begin{aligned} m\frac{\partial u}{\partial p}\dot{p}+m\frac{\partial u}{\partial T}\dot{T} &= \dot{U} \\ V\frac{\partial \rho}{\partial p}\dot{p}+V\frac{\partial \rho}{\partial T}\dot{T}+\rho\dot{V} &= \dot{m} \end{aligned} \quad (1)$$

Here $\dot{U}$ and $\dot{m}$ denote the net flow of energy and mass into the control volume while m and V are the mass trapped in the volume and the size of the volume, respectively. Additionally, the air is regarded as an ideal gas yielding the following expressions for the density, ρ, and the partial derivatives in equation (1)

$$\begin{aligned} &\rho = \frac{pM}{TR} \\ &\frac{\partial u}{\partial p} = 0, \qquad \frac{\partial \rho}{\partial p} = \frac{M}{TR} \\ &\frac{\partial u}{\partial T} = c_v, \qquad \frac{\partial \rho}{\partial T} = -\frac{pM}{T^2R} \end{aligned} \quad (2)$$

where M denotes the molar mass and c_v the specific heat capacity at constant volume, respectively, while R is the molar gas constant.

Similar expressions are used for the state derivatives of the coolant media, although only the temperature is chosen as state variable. The pressure of the coolant is determined purely from static hydraulic relationships.

4.3 Parameters of the flow models

For the airflow, pressure drops in the components along the flow path are modelled as an exponential friction loss

$$\Delta p = c\,|\,q\,|\,\dot{m}^e \quad (3)$$

The frictional pressure losses in the coolant path is modelled with second order polynomials

$$\Delta p = c_2\,|\,q\,|\,q + c_1 q \quad (4)$$

The pressure rise in the pump and the fan depend on the flow through the components and the angular velocity of the shaft. In the model the following equations are used to describe the operation of the pump and the fan, respectively

$$\Delta p = R_1\,|\,\omega\,|\,\omega + 2R_2\omega q - R_3\,|\,q\,|\,q \quad (5)$$

$$\Delta p = R_1\rho\,|\,\omega\,|\,\omega + 2R_2\omega\dot{m} - R_3\,|\,q\,|\,\dot{m} \quad (6)$$

In equation (3)–(6), q and $\dot{m}$ denotes volume flow rate and mass flow rate, respectively, while ω denotes the angular velocity of the pump or the fan.

The wind speed gives rise to a differential pressure at the air intake and outlet relative the ambient pressure. In the model, the pressure difference depends on the wind speed, v, the air density, ρ, and the non-dimensional coefficient CD according to

$$\Delta p = CD\frac{\rho}{2}v^2 \quad (7)$$

In order to find the parameter values of the sub-models, experimental data is collected from tests on individual components in a laboratory environment. For each component, a small identification problem is solved trying to fit the predicted output from the models to the measurements. The model predictions can in the most cases be expressed as linear regressions from which the parameters estimates that give the best fit in a least square sense can be found. Essentially parameters of equation (3)–(7) and other characteristics are identified for each component depicted in the overview of the cooling module in figure 5. Table 1 summarises which parameters that are identified and what data that are used.

Input from other parts of the total model is primarily heat losses that need to be cooled away. The engine emits heat to the cooling system both directly into the engine block, which is heated up by the combustion, and through the cooling of the charge air. The amount of heat depends on the current torque and speed of the engine. In the model this calculated from a look-up table. The table is obtained from measurements done in test cells. The heat emitted to the cooling system from the retarder is directly proportional to the braking power. In some sub-models, the parameters solely represent basic quantities such as mass or volume that are found from the data sheet of the corresponding component.

The tests are performed in the laboratory under well-controlled conditions. As a result the obtained prediction errors are very small as can be seen by the example in figure 6, showing the pressure drops in the airflow path.

5. ASSEMBLING THE TOTAL MODEL

The modelling errors in the sub-models are very small. However, when they are assembled to a full model, effects that are not handled in the sub-models may play an important role. It may be effects from the installation the truck cab such as the piping between the components. Non-linearities may amplify small errors in the sub-models when these are connected and new feedback paths are closed. It can be shown, using a simplified model of the cooling system, that the change of temperature of the coolant in steady state due to a small perturbation of the airflow is proportional to the squared inverse of the airflow. Thus, the simulated temperature will be very sensitive to modelling errors influencing the airflow.

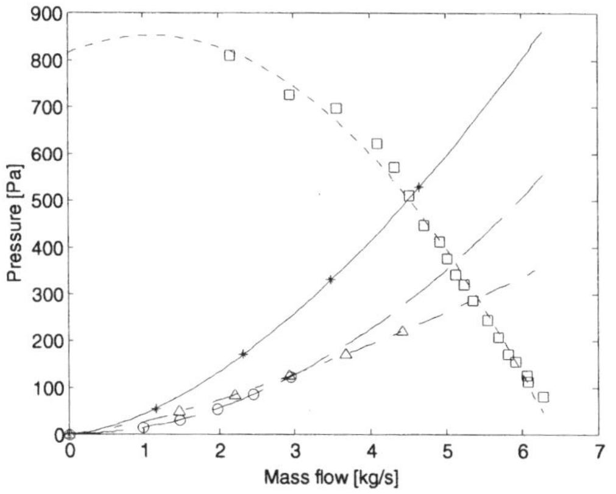

Fig. 6. Pressure drop as a function of airflow for the charge air cooler model (solid) compared with measurements (stars). Corresponding drops for radiator (dashed and triangles), and engine compartment (dash-dotted and circles). Pressure rise of the fan model (dotted) at 1400 rpm compared with measurements (squares).

Further, for the pressure build-up due to the wind speed there exist no practicable experiment on a component level. Therefore, the result of the total model is verified through comparison with experimental data collected in a wind tunnel. In the wind tunnel, the vehicle is driven on a dynamometer with a defined load and speed of the engine. Fans are used to simulate the wind speed. Results from nine steady-state tests and two step-response tests are used to tune the model parameters. A number of the parameters in the sub-models are assigned as slack parameters that are adjusted to fit the behaviour of the total model to the measurements. In table 1 the choice of slack parameters is indicated in the last column. In figures 7 and 8 the cooling temperature obtained with the tuned model are compared with measurements.

Table 1 Summary of model components in the cooling module.

Component	Characteristic	Data source	Slack
Pump	- Pressure rise	Rig test	s
	- Power consumption	Rig test	
Engine	- Flow resistance	Rig test	
	- Heat capacitance	Data sheet	s
	- Heat emission to coolant	Rig test	
	- Heat emission from charge air	Rig test	
Retarder	- Flow resistance	Rig test	
	- Heat capacitance	Data sheet	s
	- Heat emission	None	
Thermostat	- Opening characteristic	Rig test	
	- Flow resistance	Rig test	
	- Dynamic response	Rig test	
Radiator	- Flow resistance coolant	Rig test	
	- Flow resistance air	Rig test	
	- Operating characteristics	Rig test	
	- Heat capacitance	Data sheet	
Air intake	- Pressure build-up	None	s
Charge air cooler	- Flow resistance	Rig test	s
Fan	- Pressure rise	Rig test	
	- Power consumption	Rig test	
Fan clutch	- Slip characteristics	Rig test	
Engine compartment	- Flow resistance	Rig test	
Air outlet	- Pressure build-up	None	s

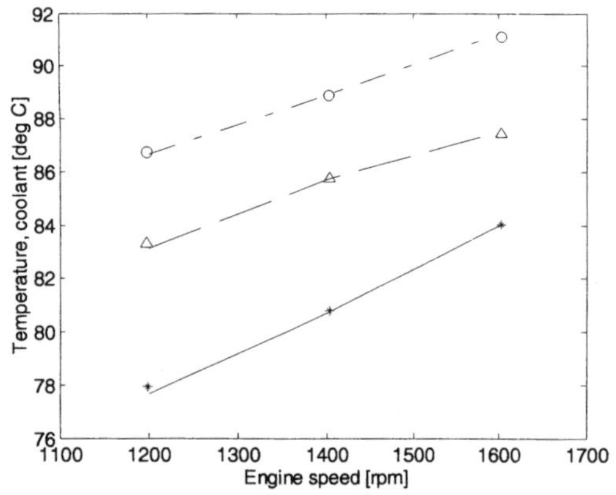

Fig. 7. Simulated temperature of the coolant in steady state at 80 km/h with full load and different speeds on the engine (solid) compared with measurements (stars). Corresponding at 60 km/h (dashed and triangles) and at 40 km/h (dash-dotted and circles).

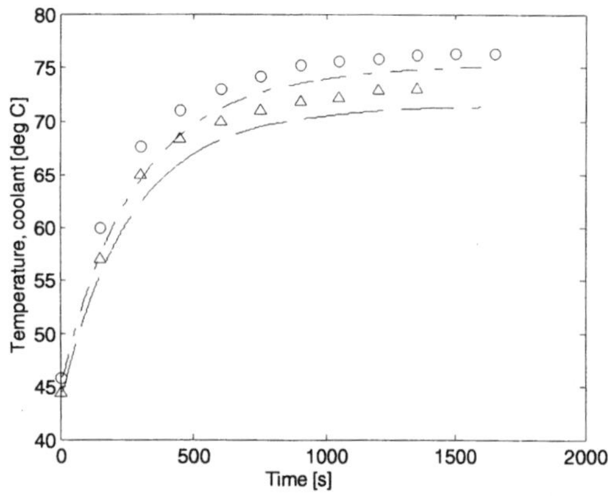

Fig. 8. Simulated response of the coolant temperature on a step in the engine load at 60 km/h with engine speed 1400 rpm (dashed), compared with measurements (triangles). Corresponding at 40 km/h (dash-dotted and circles).

As a last step of the parameter identification, a validation of the total model is performed. Validation data is recorded during a dynamic drive cycle in the wind tunnel, where the load and speed of the dynamometer is programmed to follow a cycle corresponding to a specified road. In figure 9 the simulation result is compared with measurements where the dynamometer follows the profile of a 57 km section of the road between the cities Koblenz and Trier in Germany. The validation shows that the model is capable to capture the main dynamics of the cooling system while it does not describe the small oscillations observed in the measurements. The oscillations around 80° C most likely have its origin in the complex dynamics of the thermostat. The model of the thermostat is a rather rough approximation and do not give raise to corresponding oscillations around the opening temperature. Despite the observed differences, the model should be sufficient to evaluate the energy consumption of the auxiliary units in the cooling system.

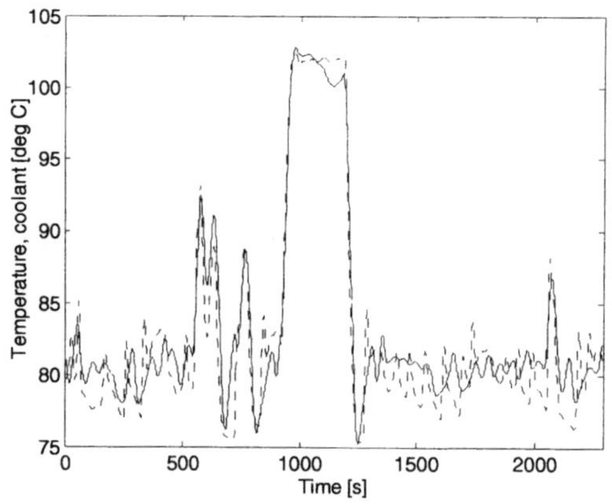

Fig. 9. Simulated coolant temperature (solid) during a dynamic driving cycle compared with measurements (dotted).

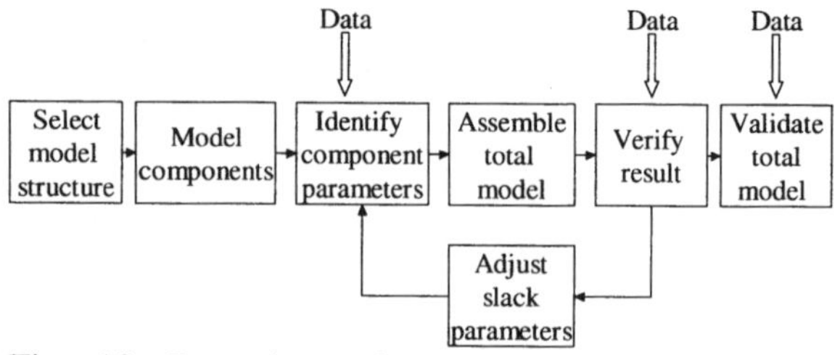

Fig. 10. Procedure of sequential modelling and identification.

6. SUMMARY

The procedure of building a model consisting of sub-models with physical interpretation and performing sequential identification of parameters may be summarised with the illustration in figure 10. The cascaded identification allows for keeping the physical structure of the model while the total behaviour of the model can be tuned to give a good fit to measurement data.

REFERENCES

Ljung, L., (1999). *System Identification, Theory for the User.* Prentice Hall, Upper Saddle River, New Jersey 07458.

Modelica Association, (2002). Modelica™ - A Unified Object-Oriented Language for Physical Systems Modeling. *Language Specification Ver2.0.* http://www.modelica.org/, 2002.

Sandberg, T., (2001). *Heavy Truck Modeling for Fuel Consumption Simulations and Measurements.* Licentiate thesis, department of Electrical Engineering, Linköping University Sweden.

Tummesheit, H., Eborn J. and Wagner FJ. Development of a Modelica Base Library for Modeling of ThermoHydraulic Systems. *Proceedings of Modelica Conference 2000.*

IFAC
Publications
www.elsevier.com/locate/ifac

PRIOR CHARACTERIZATION OF THE PERFORMANCE OF SOFTWARE SENSORS

Isabelle Braems * **Michel Kieffer** * **Eric Walter** *

* *L2S - CNRS - Supélec - Université Paris-Sud*
Plateau de Moulon, F-91192 Gif-sur-Yvette, France
{braems, kieffer, walter}@lss.supelec.fr

Abstract: Sensor performance is usually evaluated *a posteriori* after numerous essays, in a probabilistic framework were unknwon quantities are modeled by random variables. This paper addresses the problem of evaluating *a priori* the limits of the performance that can be achieved with a software sensor in a given range of operation. This is done in a context of bounded-error estimation, worst-case design and MinMax optimization. An algorithm based on interval analysis is used to obtain guaranteed results, and the procedure advocated is illustrated on a simple example of saturating vapour pressure thermometers. *Copyright © 2003 IFAC*

Keywords: Bounded-error estimation, MinMax optimization, Performance analysis, Software sensors, Worst-case design

1. INTRODUCTION

Bounded-error parameter estimation (see, *e.g.*, (Milanese *et al.*, 1996) and the references therein) computes a set guaranteed to contain all parameter vectors consistent with measurements and hypotheses on the noise. As a by-product, this set estimate provides a posterior measure of the accuracy with which the parameters of interest are estimated. When designing a measuring device, it might be useful to assess *a priori* (*i.e.*, before taking any measurement) the accuracy that can be achieved so as to compare it with the design specifications. This paper focuses on the prior evaluation of this accuracy in the case of a software sensor, *i.e.*, an experimental device consisting of a transducer in series with a processing unit. This topic pertains to the class of sensor performance evaluation problems, which are classical in metrology (ISO, 1993*a*). It is usually considered *a posteriori* in a probabilistic framework where unknown quantities are modelled by random variables. Thus, many essays must be conducted for the sensor to be evaluated. However, such a framework is not useful for prior evaluation, when no measurements are available yet. In this paper, only prior information is assumed available on the model of the transducer and on the bounds of the measurement errors.

In Section 2, some metrological vocabulary is briefly recalled and extended to the context considered here. Some solutions to the problem of prior characterization of the performance of a software sensor are proposed in Section 3. The resulting methodology is finally illustrated on a simple example in Section 4.

2. SOFTWARE SENSORS

The estimation of a quantity x (the *measurand*) by a *software sensor* is based on the measurement of an intermediate quantity y (usually of an electrical nature) produced by a *transducer*. The estimate $\widehat{x}$ of x is then obtained by processing y. The *response characteristic* of the transducer is the relation between x and y, whereas the relation between y and $\widehat{x}$ is called the *calibration curve*. Usually, the calibration curve is assumed to be the inverse of the response characteristic, and only one of them is provided by sensor manufacturers.

As many types of transducers may be considered for a given physical quantity x, many types of software sensor may be considered. For example, a temperature can be measured by evaluating the dilatation of a liquid, the variation of the resistance of some metal, a pressure of saturating vapour, *etc.* To compare the performances of sensors, norms such as (ISO, 1993*b*) have defined some characteristics which will now be briefly recalled and extended to prior evaluation in a bounded-error context.

The actual value of the measurand will be denoted by x^*, and its estimate by $\hat{x}$. In a bounded-error context, a measure is an interval $[\widehat{x}] = [\underline{\widehat{x}}, \overline{\widehat{x}}]$, with *center* $c([\widehat{x}]) = (\overline{\widehat{x}} + \underline{\widehat{x}})/2$ and *radius* $r([\widehat{x}]) = (\overline{\widehat{x}} - \underline{\widehat{x}})/2$.

The *uncertainty of measurement* is the upper bound $\delta x\,(x^*)$ of the absolute value of the measurement error, and

$$\widehat{x} - \delta x\,(x^*) \leq x^* \leq \widehat{x} + \delta x\,(x^*)\,.$$

In a bounded-error context, $\delta x\,(x^*)$ becomes

$$\delta x(x^*) = \max(|\underline{\widehat{x}} - x^*|, |\overline{\widehat{x}} - x^*|). \qquad (1)$$

The *bias error* $b\,(x^*)$ is the systematic component of the error

$$b\,(x^*) = E\,(\widehat{x}) - x^*,$$

which translates in a bounded-error context into

$$b(x^*) = c\,([\widehat{x}]) - x^*. \qquad (2)$$

The *repeatability error* Δx is the random component of the error, evaluated as

$$\begin{aligned}\Delta x\,(x^*) &= \sqrt{E\left((\widehat{x} - E\,(\widehat{x}))^2\right)} \\ &= \sqrt{E\left((\widehat{x} - x^* - b\,(x^*))^2\right)}.\end{aligned}$$

This quantity can also be extended to a bounded-error context as

$$\begin{aligned}\Delta x(x^*) &= \max\left(|\underline{\widehat{x}} - x^* - b(x^*)|\,, \left|\overline{\widehat{x}} - x^* - b(x^*)\right|\right) \\ &= |\underline{\widehat{x}} - c\,([\widehat{x}])| = \left|\overline{\widehat{x}} - c\,([\widehat{x}])\right|.\end{aligned}$$

Thus,

$$\Delta x(x^*) = r\,([\widehat{x}])\,. \qquad (3)$$

A sensor is *accurate* for the measurand x^* when it provides measurements that are free from bias and with low repeatability error. All the quantities previously introduced are summarized in Figure 1, where $\mathbb{X}$ represents the measuring range.

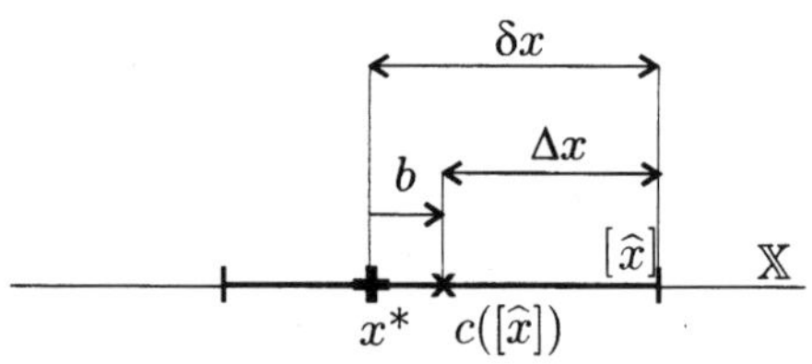

Fig. 1. Bias error b, repeatability error Δx and measurement uncertainty δx of a sensor when the actual value of the measurand is x^*

3. PRIOR CHARACTERIZATION OF A SOFTWARE SENSOR

In the sequel the relation between the mesurand x and the output $y\,(x)$ of the transducer embedded into the software sensor is assumed to be

$$y\,(x) = f(x, \mathbf{e}) + b\,(x)\,, \text{ for any } x \in \mathbb{X}, \quad (4)$$

where $f\,(\cdot,\cdot)$ is a known continuous function. The model and bias errors are both represented by the error vector $\mathbf{e} \in [\underline{\mathbf{e}}, \overline{\mathbf{e}}]$, where the bounds $\underline{\mathbf{e}}$ and $\overline{\mathbf{e}}$ are assumed to be known. The repeatability error of the transducer is represented by $b\,(x) \in [-\Delta y\,(x)\,, \Delta y\,(x)]$. A continuous calibration curve

$$\widehat{x} = g(y), \qquad (5)$$

surjective for all $x \in \mathbb{X}$ is also assumed to be available. Note that f and g are usually nonlinear.

Two types of problems will be considered. The *direct* problem is the prior characterization of a software sensor when the characteristic $\Delta y\,(x)$ of the embedded transducer is known. The *inverse* problem is the determination of the minimum performance that a transducer has to reach for the software sensor based on this transducer to be capable of reaching some prespecified performance on $\delta x\,(x)$.

3.1 *Direct problem*

From (4) , the set of all possible outputs of the transducer can be computed for any given value x^* of the measurand as

$$[y]\,(x^*) = f\,(x^*, [\underline{\mathbf{e}}, \overline{\mathbf{e}}]) + \Delta y\,(x^*)\,[-1, 1]\,,$$

where $\Delta y\,(x^*)$ is supposed to be known for the direct problem. Define

$$\begin{aligned}h^{\mathrm{d}} : \mathbb{X} \times [-1,1] \times [\underline{\mathbf{e}}, \overline{\mathbf{e}}] &\longmapsto \mathbb{R} \\ (x^*, \varepsilon, \mathbf{e}) &\longrightarrow g\,(f\,(x^*, \mathbf{e}) + \varepsilon \Delta y\,(x^*))\,.\end{aligned}$$

The set of all values that the output of the software sensor can take is

$$[\widehat{x}]\,(x^*) = h^{\mathrm{d}}(x^*, [-1,1], [\underline{\mathbf{e}}, \overline{\mathbf{e}}]).$$

From now on, two situations will be considered.

3.1.1. *Assume that* $[\underline{\mathbf{e}}, \overline{\mathbf{e}}] = \mathbf{0}$ Under this assumption, the transducer is free from bias. For any $x^* \in \mathbb{X}$, one may then evaluate :

the bias error of the software sensor

$$\begin{aligned} b(x^*) &= c([\widehat{x}](x^*)) - x^* \\ &= \frac{1}{2}\Bigg(\min_{\varepsilon \in [-1,1]} h^{\mathrm{d}}(x^*, \varepsilon, \mathbf{0}) - x^* \\ &\quad + \max_{\varepsilon \in [-1,1]} h^{\mathrm{d}}(x^*, \varepsilon, \mathbf{0}) - x^* \Bigg), \quad (6) \end{aligned}$$

its repeatability error

$$\begin{aligned} \Delta x(x^*) &= r([\widehat{x}](x^*)) \\ &= \frac{1}{2}\Bigg(\max_{\varepsilon \in [-1,1]} h^{\mathrm{d}}(x^*, \varepsilon, \mathbf{0}) \\ &\quad - \min_{\varepsilon \in [-1,1]} h^{\mathrm{d}}(x^*, \varepsilon, \mathbf{0}) \Bigg), \quad (7) \end{aligned}$$

and its uncertainty of measurement

$$\begin{aligned} \delta x(x^*) &= \max(|\underline{\widehat{x}}(x^*) - x^*|, |\overline{\widehat{x}}(x^*) - x^*|) \\ &= \max\Bigg(- \min_{\varepsilon \in [-1,1]} \left(h^{\mathrm{d}}(x^*, \varepsilon, \mathbf{0}) - x^*\right), \\ &\quad \max_{\varepsilon \in [-1,1]} \left(h^{\mathrm{d}}(x^*, \varepsilon, \mathbf{0}) - x^*\right) \Bigg). \quad (8) \end{aligned}$$

However, except for sensor calibration, the actual value x^* of the mesurand is unknown. Worst-case characteristics have thus to be evaluated over all possible x^* in the measuring range $\mathbb{X}$.

Thus, the worst bias error is

$$\overline{b} = \max\left(- \min_{x^* \in \mathbb{X}} b(x^*), \max_{x^* \in \mathbb{X}} b(x^*) \right), \quad (9)$$

the worst repeatability error is

$$\overline{\Delta x} = \max_{x^* \in \mathbb{X}} \Delta\widehat{x}(x^*), \quad (10)$$

and the worst uncertainty of measurement is

$$\overline{\delta x} = \max_{x^* \in \mathbb{X}} \delta x(x^*). \quad (11)$$

For the sake of brevity, the expressions of $\overline{b}$, $\overline{\Delta x}$ and $\overline{\delta x}$ are not detailed here (see (Braems, 2002)), but one may easily check that their evaluation requires solving MinMax optimization problems.

3.1.2. *Assume that* $[\underline{\mathbf{e}}, \overline{\mathbf{e}}] \neq \mathbf{0}$ Now, for any $x^* \in \mathbb{X}$, only an enclosure of the bias error may be obtained

$$b(x^*) \in \left[\min_{\mathbf{e} \in [\underline{\mathbf{e}}, \overline{\mathbf{e}}]} b(x^*, \mathbf{e}), \max_{\mathbf{e} \in [\underline{\mathbf{e}}, \overline{\mathbf{e}}]} b(x^*, \mathbf{e}) \right], \quad (12)$$

where

$$\begin{aligned} b(x^*, \mathbf{e}) &= \frac{1}{2}\Bigg(\min_{\varepsilon \in [-1,1]} h^{\mathrm{d}}(x^*, \varepsilon, \mathbf{e}) \\ &\quad + \max_{\varepsilon \in [-1,1]} h^{\mathrm{d}}(x^*, \varepsilon, \mathbf{e}) \Bigg) - x^*. \end{aligned}$$

The uncertainty of measurement $\delta x(x^*)$ may also only be enclosed

$$\delta x(x^*) \in \left[\min_{\mathbf{e} \in [\underline{\mathbf{e}}, \overline{\mathbf{e}}]} \delta x(x^*, \mathbf{e}), \max_{\mathbf{e} \in [\underline{\mathbf{e}}, \overline{\mathbf{e}}]} \delta x(x^*, \mathbf{e}) \right], \quad (13)$$

with

$$\begin{aligned} \delta x(x^*, \mathbf{e}) &= \max\Bigg(\left| \min_{\varepsilon \in [-1,1]} h^{\mathrm{d}}(x^*, \varepsilon, \mathbf{e}) - x^* \right|, \\ &\quad \left| \max_{\varepsilon \in [-1,1]} h^{\mathrm{d}}(x^*, \varepsilon, \mathbf{e}) - x^* \right| \Bigg) \\ &= \max_{\varepsilon \in [-1,1]} \left| h^{\mathrm{d}}(x^*, \varepsilon, \mathbf{e}) - x^* \right|. \end{aligned}$$

On the other hand, for any $x^* \in \mathbb{X}$, the repeatability error of the software sensor can still be evaluated

$$\begin{aligned} \Delta x(x^*) &= r([\widehat{x}](x^*)) \\ &= \frac{1}{2}\Bigg(\max_{\mathbf{e} \in [\underline{\mathbf{e}}, \overline{\mathbf{e}}], \varepsilon \in [-1,1]} h^{\mathrm{d}}(x^*, \varepsilon, \mathbf{e}) \\ &\quad - \min_{\mathbf{e} \in [\underline{\mathbf{e}}, \overline{\mathbf{e}}], \varepsilon \in [-1,1]} h^{\mathrm{d}}(x^*, \varepsilon, \mathbf{e}) \Bigg), \quad (14) \end{aligned}$$

Again, maximum values of $b(x^*)$, $\delta x(x^*)$ and $\Delta x(x^*)$ over the measuring range can be evaluated by solving MinMax problems.

3.2 *Inverse problem*

This section will focus on the characterization of the largest value of the repeatability error $\overline{\Delta y}$ of the transducer such that the uncertainty of measurement of x^* remains below a prespecified value δx. Moreover, this constraint has to be satisfied for all $x^* \in \mathbb{X}$. This problem is significantly more complicated than the direct one and again only results will be provided. For detailed calculations, see (Braems, 2002).

$\overline{\Delta y}$ is the upper bound of the set of all Δy such that

$$\forall x^* \in \mathbb{X},\ g([y](x^*)) \subset x^* + [-\delta x, \delta x],$$

with

$$[y](x^*) = f(x^*, [\underline{\mathbf{e}}, \overline{\mathbf{e}}]) + \Delta y [-1, 1].$$

The solution, when it exists, is given by

$$\overline{\Delta y} = \min\left(-\max_{x^* \in \mathbb{X}} \min_{\varepsilon \in [-\delta x, \delta x]} \max_{\mathbf{e} \in [\underline{\mathbf{e}}, \overline{\mathbf{e}}]} h^{\mathrm{i}}(x^*, \varepsilon, \mathbf{e}), \right.$$
$$\left. \min_{x^* \in \mathbb{X}} \max_{\varepsilon \in [-\delta x, \delta x]} \min_{\mathbf{e} \in [\underline{\mathbf{e}}, \overline{\mathbf{e}}]} h^{\mathrm{i}}(x^*, \varepsilon, \mathbf{e}) \right), \quad (15)$$

where

$$h^{\mathrm{i}} : \mathbb{X} \times [-\delta x, \delta x] \times [\underline{\mathbf{e}}, \overline{\mathbf{e}}] \to \mathbb{R}$$
$$(x^*, \varepsilon, \mathbf{e}) \mapsto g^{-1}(x^* + \varepsilon) - f(x^*, \mathbf{e}).$$

As $\Delta y(x^*)$ has to be positive, if

$$\max_{x^* \in \mathbb{X}} \min_{\varepsilon \in [-\delta x, \delta x]} \max_{\mathbf{e} \in [\underline{\mathbf{e}}, \overline{\mathbf{e}}]} h^{\mathrm{i}}(x^*, \varepsilon, \mathbf{e}) > 0$$

or if

$$\min_{x^* \in \mathbb{X}} \max_{\varepsilon \in [-\delta x, \delta x]} \min_{\mathbf{e} \in [\underline{\mathbf{e}}, \overline{\mathbf{e}}]} h^{\mathrm{i}}(x^*, \varepsilon, \mathbf{e}) < 0$$

or equivalently if there exists some $(x^*, \mathbf{e}) \in \mathbb{X} \times [\underline{\mathbf{e}}, \overline{\mathbf{e}}]$ such that

$$f(x^*, \mathbf{e}) \notin g^{-1}(x^* + [-\delta x, \delta x]),$$

then $\overline{\Delta y} < 0$. This means that no value of the repeatability error may be such that the desired uncertainty of measurement δx is satisfied. In such situations, the models f and g are not compatible with the requested δx. A more accurate model of the calibration curve has to be constructed, the transducer has to be changed or the requirement on δx has to be relaxed.

The direct and inverse problems both require continuous worst-case optimization. The partitioning algorithm MINIMAX, described in (Jaulin *et al.*, 2001), page 126, can be used to compute *guaranteed enclosures* of the solution set of such problems, based on interval analysis.

4. STUDY OF THERMOMETERS

When software sensors are designed, in practice, no exact model of the response characteristic of the transducer is available. The calibration curve and response characteristic are then built experimentally from some measurements or from commonly considered behavioural models. Two problems may be noticed. First, model errors and measurement errors are difficult to take into account. Second, the effect of a calibration curve that is not the exact inverse of the actual response characteristic is difficult to evaluate. The aim of this section is to show on a simple example how both problems can be addressed using the prior estimation framework presented in the previous sections.

Consider the measurement of temperatures using saturating vapour pressure thermometers. For such thermometers, the saturating vapour pressure, denoted by P_s, is measured and the temperature T is then deduced. The flash curve, consisting of a (T, P_s)-plot for the considered material represents the response characteristic which can only be obtained experimentally. No knowledge-based model is able to describe it over the whole temperature range to be considered. However, some behavioural models have been proposed. When $T \in [90, 300]\,^\circ C$, the Duperray model is usually employed

$$P_s(T) = \left(\frac{T}{100}\right)^4, \quad (16)$$

where P_s is in atmosphere. The relative error between this model and the experimental flash curve varies from 0.12% to 7.7%. The Rankine model

$$\ln P_s(T) = 13.7 - \frac{5120}{T + 273.16}, \quad (17)$$

is used when $T \in [5, 140]\,^\circ C$. Here, the relative error is from 0.39% to 4.1%. To emphasize the differences between these two models, two saturation vapour pressure thermometers $\mathcal{T}_1$ and $\mathcal{T}_2$ will be considered over the measurement range

$$\mathbb{T} = [90, 170]\,^\circ C.$$

For $\mathcal{T}_1$, a response characteristic based on the Rankine model will be considered

$$P_s = f_1(T, e_1)$$
$$= \exp\left(13.7 - \frac{5120}{T + 273.16}\right)(1 + e_1), \quad (18)$$

with $e_1 \in [-0.041, 0.041]$, whereas for $\mathcal{T}_2$ the Duperray model will serve as a basis

$$P_s = f_2(T, e_2) = \left(\frac{T}{100}\right)^4 (1 + e_2), \quad (19)$$

with $e_2 \in [-0.077, 0.077]$. The parameters e_1 and e_2 represent the modelling error of the response characteristic. For both thermometers, a calibration curve g is built using the inverse of $f_1(\cdot)$

$$\widehat{T} = g(P_s) = \frac{5120}{13.7 - \ln P_s} - 273.16. \quad (20)$$

The thermometer $\mathcal{T}_1$, for which $f_1(.,0) = g^{-1}(.)$, serves to illustrate the influence of a modelling error e_1 on the response characteristic, whereas $\mathcal{T}_2$ will be helpful to illustrate the additional influence of a response characteristic differing from the calibration curve. All computions have been performed on an Athlon 1800XP-based computer.

4.1 *Direct problem*

For the direct problem, only the case $e_1 = e_2 = 0$ will be considered for the sake of brevity. The value of $\overline{\delta T}$, the maximum of the uncertainty of measurement of T over $\mathbb{T}$

$$\overline{\delta T} = \max_{T^* \in \mathbb{T}} \delta T\,(T^*)\,, \tag{21}$$

with $\delta T\,(T^*)$ evaluated using (11), has been computed for 21 equally-spaced values of $\Delta P_s \in [0, 0.1]$ atm with the MINIMAX algorithm presented in (Jaulin *et al.*, 2001). Figure 2 presents an enclosure of $\overline{\delta T}$ as a function of ΔP_s between plain and dashed curves for both thermometers. These results have been obtained in 10 mn for $\mathcal{T}_1$ and 15 mn for $\mathcal{T}_2$. For any given value of ΔP_s, the uncertainty of measurement is lower when $\mathcal{T}_1$ is used, because in this case, $f_1 = g^{-1}$. For $\mathcal{T}_2$, it is not possible to obtain a certain measurement, even if $\Delta P_s = 0$, because $f_2 \neq g^{-1}$.

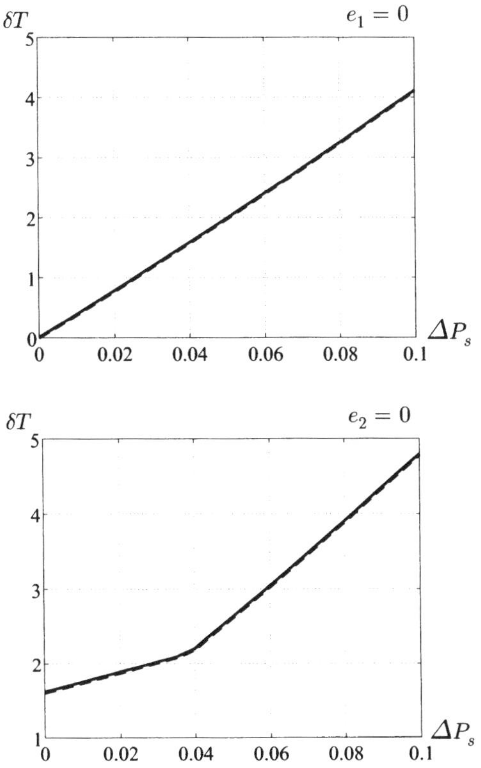

Fig. 2. δT as a function of ΔP_s when no model error is taken into account, for thermometers $\mathcal{T}_1$ (top) et $\mathcal{T}_2$ (bottom)

4.2 *Inverse problem*

The problem is now to find the maximum value $\overline{\Delta P_s}$ of the repeatability error ΔP_s such that a measurement of $T \in \mathbb{T}$ can be obtained with an uncertainty of measurement lower than δT.

4.2.1. *Thermometer $\mathcal{T}_1$* Consider the first thermometer $\mathcal{T}_1$, and assume for the time being that $e_1 = 0$. The repeatability error ΔP_s and uncertainty of measurement δP_s are then equal. Figure 3 presents the enclosure of $\overline{\Delta P_s}$ obtained in less than 4s by the MINIMAX algorithm applied on (15) for 40 equally-spaced values of δT between $0°C$ and $1°C$. For example, to achieve an accuracy of measurement δT that is better than $0.5°C$ for any $T \in \mathbb{T}$, ΔP_s has to be lower than 0.0123 atm. As could be expected, $\overline{\Delta P_s}$ increases with δT; the larger the acceptable uncertainty of measurement, the larger the repeatability error that can be tolerated.

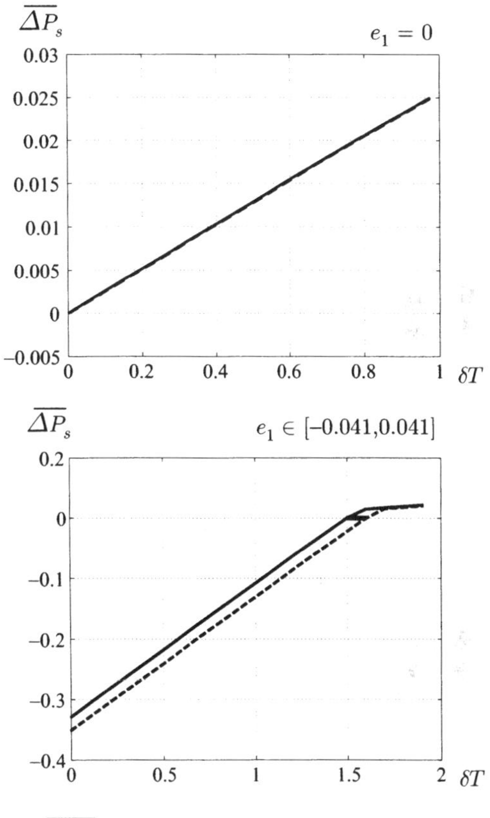

Fig. 3. $\overline{\Delta P_s}$ as a function of δT for $\mathcal{T}_1$ when $e_1 = 0$ (top) and $e_1 \in [-0.041, 0.041]$ (bottom)

If there exists a model error $e_1 \in [-0.041, 0.041]$ in (18), then δP_s and ΔP_s are no more equal. An enclosure of ΔP_s for 20 equally-spaced values of δT over $[0, 2]\,°C$ is again computed using the MINIMAX algorithm applied to (15). The enclosures obtained in 10mn are represented in Figure 3. Using the MINIMAX algorithm, it is possible to prove that for $\delta T = 1.5°C$, $\overline{\Delta P_s} < 0$, whereas for $\delta T = 1.6°C$, $\overline{\Delta P_s} > 0$. Thus, using $\mathcal{T}_1$, the presented technique has proved that it was not possible to achieve an accuracy of measurement better than $1.5°C$, even if the repeatability error of the pressure measurement is zero (see Section 3.2). On the other hand, accuracies of measurement δT no better than $1.6°C$ may be obtained provided that the pressure measurement is accurate enough.

4.2.2. *Thermometer $\mathcal{T}_2$* The same study can be conducted with the second thermometer $\mathcal{T}_2$. When $e_2 = 0$, the MINIMAX algorithm can again be used to bound the maximum value of ΔP_s as a function of the required δT, see Figure 4. Here, 80 equally-spaced values of δT between $0°C$ and $4°C$ have been considered. The total computing time is 38s. Now, even if $e_2 = 0$, it is not possible to achieve an accuracy of measurement better than $1.55°C$. This is due to the difference between the considered response characteristic and calibration curve. However, accuracies of measurement no better than $1.70°C$ could be obtained.

To derive more precise conclusions, the MINIMAX algorithm could be used with increased accuracy of the enclosure of $\overline{\Delta P_s}$ for each δT. This would easily be done at the price of an increased computing time.

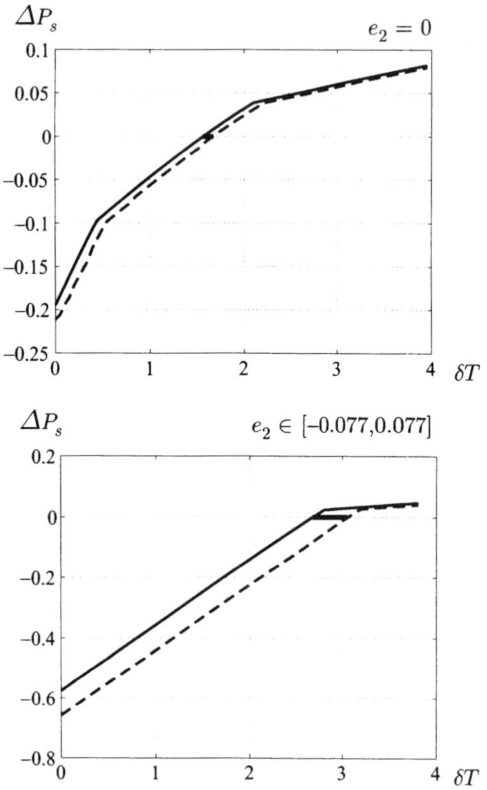

Fig. 4. $\overline{\Delta P_s}$ as a function of δT for $\mathcal{T}_2$ when $e_2 = 0$ (top) and $e_2 \in [-0.077, 0.077]$ (bottom)

Assume now that $e_2 \in [-0.077, 0.077]$. Enclosures of ΔP_s for 20 equally-spaced values of $\delta T \in [0, 4]\,°C$ are described by Figure 4. The computing time is now 13mn. Figure 4 implies that an accuracy of measurement δT better than $2.6°C$ cannot be obtained wheras it is possible to obtain $\delta T \geqslant 3.2°C$. Again, more precise conclusions would require more computing time.

Remark 1. The conclusion obtained for the inverse problem a consistent with that obtained for the direct problem. For example, for $\mathcal{T}_2$ with $e_2 = 0$, the solution of the direct problem indicates that when $\Delta P_s = 0$, $\delta T \approx 1.60°C$, whereas the solution of the inverse problem indicates that when $\delta T < 1.55°C$, $\overline{\Delta P_s} < 0$ and when $\delta T > 1.7°C$, $\overline{\Delta P_s} > 0$.

5. CONCLUSIONS

In this paper, the prior analysis of software sensors consisting of transducers followed by processing units has been studied. Two types of problems have been considered. The direct problem is the evaluation of the performance of the software sensor based on available knowledge on the performance of the transducer. The inverse problem is the assessment of the largest value of the repeatability error of the transducer that can be tolerated if the measurement uncertainty must be lower than some pre-specified value.

Both studies can be conducted before any actual measurement. As conclusions have to be reached for any possible value of the mesurand in a given range, both problems are treated as worst-case optimization problems. Using the MINIMAX algorithm based on interval arithmetic, is is possible to obtain an approximate, but guaranteed, enclosure of the solutions. The resulting methodology for solving direct and inverse problems has been illustrated on two saturating vapour pressure thermometers.

Preliminary results have also been obtained for multiple sensors working in parallel or one sensor working under varying conditions, see (Braems, 2002). The next step is to use this framework of prior sensor performance evaluation for the design of experiments.

6. REFERENCES

Braems, I. (2002). Méthodes ensemblistes garanties pour l'estimation de grandeurs physiques. PhD thesis. Université Paris-Sud. Orsay, France.

ISO (1993*a*). *Guide to the Expression of Uncertainty in Measurement.* ISO. Geneva, Switzerland.

ISO (1993*b*). *International Vocabulary of Basic and General Terms in Metrology : Vocabulaire international des termes fondamentaux et généraux de métrologie.* ISO. Geneva, Switzerland.

Jaulin, L., M. Kieffer, O. Didrit and E. Walter (2001). *Applied Interval Analysis.* Springer-Verlag. London.

Milanese, M., Norton, J., Piet-Lahanier, H. and Walter, E., Eds.) (1996). *Bounding Approaches to System Identification.* Plenum Press. New York, NY.

IFAC Publications
www.elsevier.com/locate/ifac

MODEL BASED SOURCE LOCALISATION BY DISTRIBUTED SENSORS FOR POINT SOURCES AND DIFFUSION

Jörg Matthes * Lutz Gröll *

* *Institute for Applied Computer Science, Forschungszentrum Karlsruhe, P.O. Box 3640, 76021 Karlsruhe, Germany, matthes/groell@iai.fzk.de*

Abstract: The inverse problem of locating a point source of an emission based on measurements from spatial distributed sensors is studied for isotrope diffusion. Equivalent localisation problems concern heat sources or pollution sources. A new two-step approach with the steps: estimation of a scalable sensor-source-distance for each sensor and estimation of the source position using these distances is presented. In contrast to conventional one-step approaches for solving the nonlinear least squares output error problem by iterative algorithms, the new approach is not sensitive to local minima. *Copyright © 2003 IFAC*

Keywords: spatial distributed sensors, source localisation, electronic nose, inverse problem

1. INTRODUCTION

1.1 *Problem*

Spatial distributed electronic noses allow online concentration measurements of different air admixtures. From these measurements the emission source is to be localised. This task leads to an ill-posed, inverse problem of causation, which seeks for the cause to a given model and given measurements (Groetsch, 1993). Applications for the source localisation problem are the emission detection in environmental monitoring and the leakage detection in industrial monitoring.

Here, the case of isotrope diffusion in a semi-infinite medium with a point source placed on the impermeable surface is studied. The source position is denoted by $\mathbf{x}_0 = (x_0, y_0, 0)^T$ and the sensor positions are $\mathbf{x}_i = (x_i, y_i, z_i)^T$; $i = 1, .., m$. Further it is assumed, that the source rate $q(t) = q_0\, 1(t - t_0)$ is a step-function. The behavior of the concentration $C(\mathbf{x}, t)$ at any position $\mathbf{x} = (x, y, z)^T$ and at any time t is described by the inhomogeneous diffusion equation (Crank, 1975; Carslaw and Jaeger, 1959)

$$\frac{\partial C}{\partial t} - K\left(\frac{\partial^2 C}{\partial x^2} + \frac{\partial^2 C}{\partial y^2} + \frac{\partial^2 C}{\partial z^2}\right) = q_0 \cdot 1(t - t_0) \cdot \delta(x - x_0) \cdot \delta(y - y_0) \cdot \delta(z - 0) \quad (1)$$

where K denotes the isotrope diffusion coefficient. Taking into account the initial condition $C(\mathbf{x}, t) \equiv 0$ for $t < t_0$ and arbitrary $\mathbf{x}$ the solution for $t \geq 0$ is given by

$$C(\mathbf{x}, t) = \frac{q_0 \cdot \operatorname{erfc}\left(\frac{\|\mathbf{x} - \mathbf{x}_0\|_2}{2\sqrt{K(t - t_0)}}\right) \cdot 1(t - t_0)}{2\pi K \|\mathbf{x} - \mathbf{x}_0\|_2}. \quad (2)$$

In this paper a new approach for estimating the source position $\mathbf{x}_0$, the constant source rate q_0, the source start time t_0 and the diffusion coefficient K based on spatial distributed measurements of $C(\mathbf{x}_i, t)$ is presented.

For the sake of a better graphical presentation of the results, the sensors are placed on the surface ($z_i = 0$) in the given examples.

In this paper all variables are dimensionless. For practical applications a consistent system of units (e. g. SI-System) must be applied. Estimates are denoted by a hat ($\hat{}$). Henceforth, $C_i(t)$ is the shortcut for the concentration of sensor i at time t, calculated according to (2). $C_{i,k}$ is the measured concentration at sensor i and discrete point in time $t_{i,k}$ ($k = 1, .., N_i$).

1.2 *Problem classification*

Approaches for source localisation can be classified into

- Approaches with mobile sensors (Duckett *et al.*, 2001; Lilienthal *et al.*, 2002) and
- Approaches with stationary sensors (Alpay and Shor, 2000; Khapalov, 1994; Nievergelt, 1998; Wacholder *et al.*, 1995)

Mobile sensors are mostly transported by autonomous robots. From the measured concentrations the robot decides by a search strategy, which direction to go, in order to find the source.

Approaches with stationary sensors can be divided into three groups:

- *Classification approaches* use measured time responses of the concentration $C_{i,k}$ and reference signals. These reference signals are obtained by experiment or simulation for different source positions. In the monitoring phase the source position is classified by comparing the measurements with the references (Alpay and Shor, 2000; Wacholder *et al.*, 1995). The large number of necessary simulations or experiments is the main disadvantage of these approaches.
- *Approaches with discrete models* use a state-space-model, which results from space discretisation. It is assumed, that sensors and sources are located at lattice points only. By an observer formulation the unknown input (source position and rate) is estimated from the measured sensor data (Alpay and Shor, 2000). The large order of the state-space-system in contrast to the small number of measured states (number of sensors) is a disadvantage of these approaches.
- *Approaches with continuous models* (Khapalov, 1994; Nievergelt, 1998; Jeremić and Nehorai, 2000) use analytical solutions of the advection-diffusion equation and formulate a nonlinear least squares output error problem for the source position, source rate and source start time. This optimisation problem has to be solved by iterative algorithms which can fail due to local minima of the objective function (cp. Section 2). The approach presented in this paper is ranged in this group. It overcomes the problem of local minima by a two-step approach.

2. PROBLEMS OF ONE-STEP APPROACHES USING OUTPUT ERROR MODEL

The analytical solution of the diffusion equation (2) is used to estimate the source position $\mathbf{x}_0$, source rate q_0, diffusion coefficient K and start time t_0 from

$$J = \sum_{i=1}^{m} \sum_{k=1}^{N_i} (C_{i,k} - C_i(t_{i,k}))^2 \stackrel{!}{=} \underset{\mathbf{x}_0, q_0, K, t_0}{\text{Min}} \quad . \qquad (3)$$

Usually, for solving this nonlinear optimisation problem gradient methods or Newton's methods are applied. Dependent on the choice of the unknown initial values these methods can abort in local minima and thus lead to wrong solutions.

In order to show this problem, the objective function (3) is plotted over $\mathbf{x}_0$ for the noisefree case, where q_0, K, t_0 are fixed at the true values. In the underlying example (Fig. 1) four sensors at the positions $\mathbf{x}_1 = (0,0,0)^T, \mathbf{x}_2 = (4,0,0)^T, \mathbf{x}_3 = (2,2,0)^T, \mathbf{x}_4 = (2,4,0)^T$ and a source at $\mathbf{x}_0 = (1,2,0)^T$ yields to a local minimum at $\mathbf{x} \approx (3,2,0)^T$ and the global minimum at the true source position $\mathbf{x}_0 = (1,2,0)^T$. Besides the

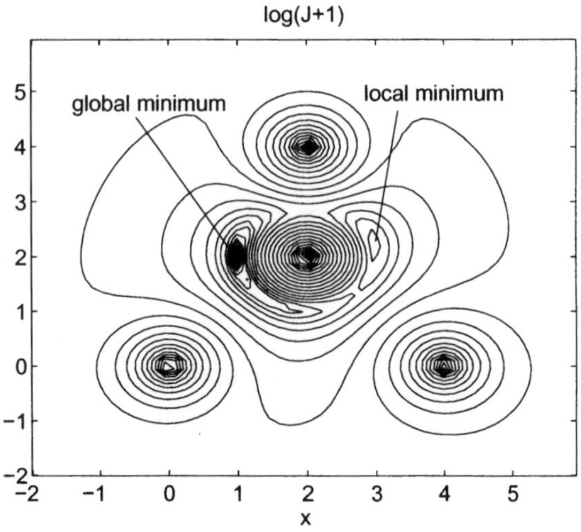

Fig. 1. Logarithmic contour plot of the squared output error dependent on the source position $\mathbf{x}_0$

global and local minimum Fig. 1 shows local maxima at each sensor position. Furthermore, the narrow valleys in the topology of the objective function shown in Fig. 1, understated by the logarithmic representation, cause slow convergence of the iterative algorithms.

The plot becomes even worse with regards to the iteration if source rate q_0, diffusion coefficient K and start time t_0 have to be estimated, too. This is caused by the nonlinear connection of the parameters and the fact, that t_0 and K can not be estimated consistently. The inconsistence leads asymptotically to valleys in the topology of the objective function. The problem of inconsistence is unavoidable, since the persistence excitation condition is violated for a constant source rate q_0. However, the problem can be relieved by a decomposition into:

(1) estimation of scalable sensor-source-distances ρ_i for each sensor i and estimation of the start time t_0,

(2) determination of the source position $\mathbf{x}_0$, of the source rate q_0 and of the diffusion coefficient K using the estimated scalable sensor-source-distances ρ_i.

3. ESTIMATION OF THE SCALABLE SENSOR-SOURCE-DISTANCES AND OF THE START TIME

3.1 *Introduction*

The scalable sensor-source-distance ρ_i for each sensor i is defined by

$$\rho_i = \gamma \cdot \|\mathbf{x}_i - \mathbf{x}_0\|_2 = \gamma \cdot r_i. \tag{4}$$

Here, the scaling factor γ is independent of i, but it depends on, whether ρ_i is determined by output or equation error approach.

3.2 *Output error approach*

The start time t_0 is estimated by minimising

$$J_i = \sum_{k=1}^{N_i} (C_{i,k} - C_i(t_{i,k}))^2 \stackrel{!}{=} \underset{\alpha_i,\beta_i,t_0}{\text{Min}}\ , \tag{5}$$

where i denotes the sensor with the highest concentration. Note, that due to the summation the measurement information is exploited not only for the current point in time, but for the whole observation time.

Using the shortcuts

$$\alpha_i = \frac{q_0}{2\pi K\|\mathbf{x}_i - \mathbf{x}_0\|_2}, \quad \beta_i = \frac{\|\mathbf{x}_i - \mathbf{x}_0\|_2}{2\sqrt{K}} \tag{6}$$

(2) can be written as

$$C(\mathbf{x}_i, t) = \alpha_i \cdot \text{erfc}\left(\frac{\beta_i}{\sqrt{t - t_0}}\right) \cdot 1(t - t_0). \tag{7}$$

Note, that the time dependent term in (7) tends to 1:

$$\lim_{t\to\infty} \text{erfc}\left(\beta_i \cdot \frac{1}{\sqrt{t - t_0}}\right) \cdot 1(t - t_0) = 1. \tag{8}$$

As a consequence, the information content for t_0 decreases in the measurements of $C(\mathbf{x}_i, t)$, and thus the estimation of t_0 is inconsistent.

Next, $\hat{\alpha}_i$ and $\hat{\beta}_i$ are estimated for each sensor i according to (5), but now t_0 is fixed by the estimate $\hat{t}_0$ corresponding to the sensor with the highest concentration.

Obviously, due to the inconsistence of $\hat{t}_0$ the estimation of $\hat{\beta}_i$ is inconsistent, too. Also if $\hat{t}_0$ would be the true start time, then $\hat{\beta}_i$ is inconsistent, since (8) causes a decrease in the information content. In contrast, the estimation of α_i improves with increasing measurement time. It becomes a consistent mean estimation, due to (8). Therefore, the calculation of the scalable sensor-source-distance $\hat{\rho}_i$ should be based on $\hat{\alpha}_i$:

$$\hat{\rho}_{i,\alpha} = \frac{1}{\hat{\alpha}_i}. \tag{9}$$

With (4) and (6) the scaling factor follows

$$\gamma = \gamma_\alpha = \frac{2\pi K}{q_0}. \tag{10}$$

A special case of this estimation arises from a very large measurement time. Then, the concentration reaches the stationary values at each sensor position (cp. (2) and (8))

$$C(\mathbf{x}_i, \infty) = \frac{q_0}{2\pi K\|\mathbf{x}_i - \mathbf{x}_0\|_2}. \tag{11}$$

An estimate $\hat{C}(\mathbf{x}_i, \infty)$ is given by a measurement $C_{i,k}$ with a k corresponding to a large t. Using this, another scalable sensor-source-distance can be defined

$$\hat{\rho}_{i,\alpha,\infty} = \frac{1}{\hat{C}(\mathbf{x}_i, \infty)}. \tag{12}$$

The advantage of the output error approach is its robustness with respect to disturbances in the concentration measurement. Especially the estimation of the scalable sensor-source-distances with α_i leads to a consistent estimation of $\mathbf{x}_0$. However, an iterative algorithm for solving (5) is necessary. In contrast to the one-step approach, the estimation of ρ_i is unproblematic with respect to iterative algorithms (Fig. 2). Regarding the same example as described in Section 2, Fig. 2 shows the sum of squared output errors J_i over the sensor-source-distance r_i (not scaled, because q_0 and K are given) for each sensor i. Here, no trouble with local minima occurs.

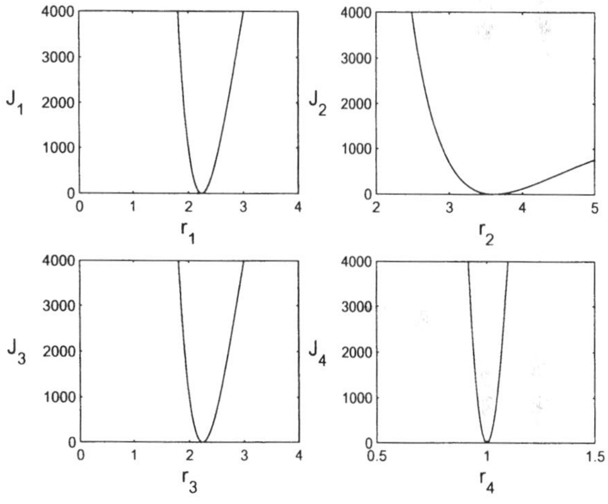

Fig. 2. Squared output error with respect to sensor-source-distance for each sensor

3.3 *Equation error approach*

It can be shown, that the concentration $C_i(t)$ at each sensor i satisfies the second-order ordinary differential equation

$$\ddot{C}_i(t) + \dot{C}_i(t)\left[\frac{3}{2(t - t_0)} - \frac{\|\mathbf{x}_i - \mathbf{x}_0\|_2^2}{4K(t - t_0)^2}\right] = 0. \tag{13}$$

In order to get a parameter linear estimation problem for a scalable sensor-source-distance ρ_i and start time t_0 from (13), ρ_i is defined by

$$\rho_{i,Eq} = \frac{\|\mathbf{x}_i - \mathbf{x}_0\|_2}{2\sqrt{K}} \tag{14}$$

and with

$$\gamma = \gamma_{Eq} = \frac{1}{2\sqrt{K}}. \tag{15}$$

Thus, with the parameter vector

$$\theta = \left(\rho_{1,Eq}^2, \quad \ldots, \quad \rho_{m,Eq}^2, \quad t_0, \quad t_0^2\right)^T \tag{16}$$

the optimisation problem

$$||\mathbf{A}\theta - \mathbf{b}||_2^2 \stackrel{!}{=} \underset{\theta}{\text{Min}} \quad \text{s.t.} \quad \theta_{m+1}^2 = \theta_{m+2} \tag{17}$$

with

$$\begin{aligned}
\mathbf{A} &= \begin{pmatrix} \mathbf{a}_1 & \mathbf{0} & \mathbf{0} & \mathbf{0} & \mathbf{A}_1 \\ \mathbf{0} & \mathbf{a}_2 & \mathbf{0} & \mathbf{0} & \mathbf{A}_2 \\ \cdots & \cdots & \cdots & \cdots & \cdots \\ \mathbf{0} & \mathbf{0} & \mathbf{0} & \mathbf{a}_m & \mathbf{A}_m \end{pmatrix}, \\
\mathbf{a}_i &= \left(-2\dot{C}_{i,1}, \quad \ldots, \quad -2\dot{C}_{i,N_i}\right)^T, \\
\mathbf{A}_i &= \begin{pmatrix} -4\ddot{C}_{i,1}t_{i,1} - 3\dot{C}_{i,1} & 2\ddot{C}_{i,1} \\ \vdots & \vdots \\ -4\ddot{C}_{i,N_i}t_{i,N_i} - 3\dot{C}_{i,N_i} & 2\ddot{C}_{i,N_i} \end{pmatrix}, \\
\mathbf{b}_i &= \begin{pmatrix} -2\ddot{C}_{i,1}t_{i,1}^2 - 3\dot{C}_{i,1}t_{i,1} \\ \vdots \\ -2\ddot{C}_{i,N_i}t_{i,N_i}^2 - 3\dot{C}_{i,N_i}t_{i,N_i} \end{pmatrix}
\end{aligned} \tag{18}$$

results. (Gröll, 2002) describes a method for a quasi analytical solutions of quadratically constrained least squares problems. Up to a root finding, all steps in the algorithm are straight forward, which requires little computation time.

However, problems arise with noisy signals, since the first and second derivatives of the concentration have to be computed. For this reason, the equation error approach is only applicable, if the noise is low. Furthermore, the derivatives of the concentration with respect to time $\dot{C}_i(t)$ and $\ddot{C}_i(t)$ tend to 0 for $t \to \infty$, and thus, only zero-rows are added to the matrix $\mathbf{A}$.

4. ESTIMATION OF SOURCE POSITION, DIFFUSION COEFFICIENT AND SOURCE RATE

4.1 *Estimation of* $\mathbf{x}_0$

Rewriting (4) yields

$$\mathbf{x}_i^T\mathbf{x}_i - 2\cdot\mathbf{x}_i^T\mathbf{x}_0 + \mathbf{x}_0^T\mathbf{x}_0 - \frac{\rho_i^2}{\gamma^2} = 0; \quad i = 1,..,m. \tag{19}$$

Since only the estimated scalable sensor-source-distances are available, a least squares problem for the unknowns $\mathbf{x}_0$ and γ^2 has to be formulated.

Recall, $z_0 = 0$, (19) gives with the parameter vector

$$\theta = \left(x_0, \quad y_0, \quad x_0^2 + y_0^2, \quad \frac{1}{\gamma^2}\right)^T \tag{20}$$

a quadratically constrained least squares problem

$$||\mathbf{A}\theta - \mathbf{b}||_2^2 \stackrel{!}{=} \underset{\theta}{\text{Min}} \quad \text{s.t.} \quad \theta_1^2 + \theta_2^2 = \theta_3 \tag{21}$$

with

$$\begin{aligned}
\mathbf{A} &= \begin{pmatrix} -2x_1 & -2y_1 & 1 & -\hat{\rho}_1^2 \\ \vdots & \vdots & \vdots & \vdots \\ -2x_m & -2y_m & 1 & -\hat{\rho}_m^2 \end{pmatrix}, \\
\mathbf{b} &= \begin{pmatrix} -\mathbf{x}_1^T\mathbf{x}_1 \\ \vdots \\ -\mathbf{x}_m^T\mathbf{x}_m \end{pmatrix}.
\end{aligned} \tag{22}$$

Note, that the problem generally possesses a local and a global minimum. In contrast to (17), here, the local minimiser is also in the focus of interest, especially in that cases, where no a-priori knowledge is available to exclude the local minimiser as irrelevant. Both minima can be found with the approach described in (Gröll, 2002).

A strong view at the parameter estimation problem signalises, that no constraint concerning the positivity of $\hat{\theta}_4$ or equivalent $1/\gamma^2 > 0$ were met. Nevertheless, no real application and no simulated problem gave a negative $\hat{\theta}_4$.

Obviously, from $\hat{\theta}$ follows

$$\hat{\mathbf{x}}_0 = (\hat{\theta}_1, \quad \hat{\theta}_2, \quad 0)^T \tag{23}$$

and the estimate of the scaling factor is

$$\hat{\gamma} = \sqrt{1/\hat{\theta}_4}. \tag{24}$$

Depending on whether $\hat{\rho}_{i,\alpha}$ or $\hat{\rho}_{i,Eq}$ were used in (22), the estimates of $\mathbf{x}_0$ and γ are consistent or inconsistent, since the consistence property of $\hat{\rho}_i$ is inherited over (21) and (24).

4.2 *Estimation of K and* q_0

There are two ways to estimate K and q_0, namely by using $\hat{\alpha}_i$ and $\hat{\beta}_i$ or by using $\hat{\gamma}_{Eq}$.

In the first way both equations in (6) are combined to the following in generally overdetermined set of equations

$$\underbrace{\begin{pmatrix} -1 & 2\pi\hat{\alpha}_1 r_1 \\ \vdots & \vdots \\ -1 & 2\pi\hat{\alpha}_m r_m \\ 0 & 4\hat{\beta}_1^2 \\ \vdots & \vdots \\ 0 & 4\hat{\beta}_m^2 \end{pmatrix}}_{\mathbf{A}} \underbrace{\begin{pmatrix} q_0 \\ K \end{pmatrix}}_{\theta} \cong \underbrace{\begin{pmatrix} \mathbf{0}_m \\ \hat{r}_1^2 \\ \vdots \\ \hat{r}_m^2 \end{pmatrix}}_{\mathbf{b}}. \tag{25}$$

An appropriate criterion for the fit is

$$||\mathbf{A}\theta - \mathbf{b}||_2^2 \stackrel{!}{=} \underset{\theta}{\text{Min}} \tag{26}$$

with its explicit solution $\hat{\theta} = (\mathbf{A}^T\mathbf{A})^{-1}\mathbf{A}^T\mathbf{b}$.

It is worth to note, that K and q_0 can not be calculated separately if $\rho_{i,\alpha}$ is estimated from the stationary concentration $C(\mathbf{x}_i, \infty)$. Then, only the ratio q_0/K is determinable.

In the second way $\hat{K}$ is obtained by rewriting (15)

$$\hat{K} = \frac{1}{4\hat{\gamma}_{Eq}^2}. \tag{27}$$

Further, by (2) each measurement $C_{i,k}$ is related to

$$(q_0)_{i,k} = \frac{2\pi\hat{K}||\mathbf{x}_i - \hat{\mathbf{x}}_0||_2}{\mathrm{erfc}\left(\frac{||\mathbf{x}_i-\hat{\mathbf{x}}_0||_2}{2\sqrt{\hat{K}(t_{i,k}-\hat{t}_0)}}\right) C_{i,k}}. \tag{28}$$

Thus, the weighted average of $(q_0)_{i,k};\ i = 1,..,m;\ k = 1,..,N_i$ is an appropriate estimate for q_0.

5. ANALYSIS OF THE SOLUTION

In the foregoing sections the structural identifiability of the unknown parameters was taken for granted. Since all variants of the two-step approach have to solve problem (21) the identifiability of this problem is analyzed at first. In this part the information of each sensor has been reduced to the scalable sensor-source-distance $\hat{\rho}_i$. So it is clear, that at least three sensors are necessary in order to determine the three parameters x_0, y_0 and γ by (21).

If only three sensors are used, then the source localisation problem possesses two solutions: the true source position and a false one. Both solutions have the same value of the objective function (21) independently whether the measured signals are disturbed or not. Evidently, they are global minimisers.

Besides the algebraic interpretation of the two solutions, here, a geometric one is given for $z_i = 0$ as an intersection problem of three circles. Fig. 3 illustrates this situation with the two solutions $\mathbf{x}_{0,I}$ and $\mathbf{x}_{0,II}$, where each circle around the sensor position $\mathbf{x}_i$ with radius $\hat{\rho}_i$ represents all potential source positions from the view of sensor i for a fixed γ. There exist two γ, namely γ_I and γ_{II}, for which the three circles intersect in a common point.

An interesting feature of the two solutions $\mathbf{x}_{0,I}$ and $\mathbf{x}_{0,II}$ is their location related to the circumcircle U of the sensor positions $(x_i,y_i,0)^T$. They are polar reciprocal points with respect to U, that acts as the inversion circle. This means algebraically

$$||\mathbf{x}_{0,I} - \mathbf{x}_U||_2 \cdot ||\mathbf{x}_{0,II} - \mathbf{x}_U||_2 = R^2 \tag{29}$$

and for the associated scaling factors

$$\frac{\gamma_I^2}{\gamma_{II}^2} = \frac{||\mathbf{x}_{0,II} - \mathbf{x}_U||_2}{||\mathbf{x}_{0,I} - \mathbf{x}_U||_2}, \tag{30}$$

where $\mathbf{x}_U$ denotes the centre and R the radius of U.

To proof this statement, observe that the minima of (21) is zero, since $\mathbf{A}\theta = \mathbf{b}$ is underdetermined. The solution set for (x_0,y_0) of this equation system is a line in the (x,y)-plane. Incorporating the constraint, the two solutions are specified at this line. By showing, that $\mathbf{x}_U$ lies at this line, too, the proof is completed straightforward by some algebra.

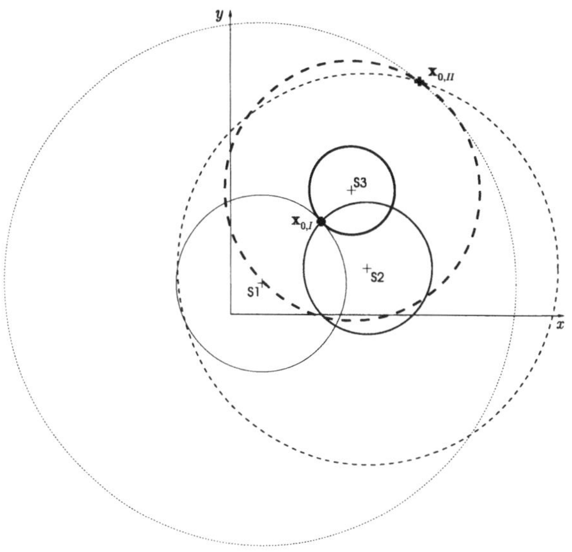

Fig. 3. Two solutions for the source position using three sensors S1, S2, S3

As a consequence of the above considerations, a unique solution of the source localisation problem can only be obtained, if at least four sensors are used. However, it is to assure, that not all sensor positions $(x_i,y_i,0)^T$ have a common circumcircle. Would they be at the circumcircle and would all $\hat{\rho}_i$ be undisturbed, then an argumentation about rank defect in (22) reveals, that it is impossible to distinguish true and false source position. Although in this critical case the least squares criterion yields a global and a local minimum for the disturbed case, and thus a distinction between true and false source positions seems possible, it is meaningless in the statistical sense.

Note, that in the case of four sensors, it is often sufficient to check, whether the sensors do not form a chord quadrangle in the (x,y)-plane.

Independently on the more algebraic arguments, there is an interesting illustration of the circumcircle problem. Therefore the special case of stationary concentrations is considered, see Fig. 4. If the concentration profiles according to (11) are plotted above the (x,y)-plane for $\mathbf{x}_0 = \mathbf{x}_{0,I}$ and $\mathbf{x}_0 = \mathbf{x}_{0,II}$ it can be seen, that the concentrations for the true and the false source are identically at U. In other words, the projection of the intersection curve in the concentration profiles onto the (x,y)-plane is the circumcircle U. Consequently, no further information for the distinction of the true and the false source can be obtained from measurements at U.

For the case $z_i \geq 0$ the circum-hemisphere with the centre in the (x,y)-plane and all three sensors at the surface has to be regarded. In this case, the inversion circle for the two solutions (polar reciprocal points) is the intersection curve of this circum-hemisphere with the (x,y)-plane. To distinguish the true and the false solution the fourth sensor must not lie at the circum-hemisphere.

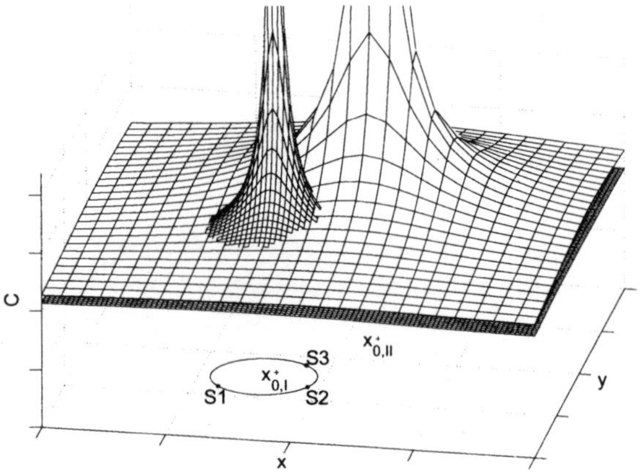

Fig. 4. Identical concentrations at the circumcircle U for the true and the false source

Finally, it should be remarked, that the problem of two solutions (true and false solution), which arises with the two-step approach when only three sensors are applied, is also relevant for the one-step approach. For a large observation time the time dependent part of (2) tends to 1 and thus the output error (3) is dominated by the time independent part of (2). This leads to an additional local minimum at the false source position, in which iterative algorithms can abort.

6. CONCLUSIONS

The localisation problem for a point source has been studied. It was assumed, that concentration measurements from spatially distributed sensors are available. The presented two-step approach is based on the analytical solution of the diffusion equation for a semi-infinite, isotropic medium.

The new two-step approach bypasses the problem of local minima, which occurs in common one-step approaches and makes trouble in iterative algorithms. In the first step a scalable sensor-source-distance for each sensor is estimated. In the second step, these estimates are used to determine the source position by a quadratically constrained least squares (QCLS) fit. In difference to the one-step approach, local and global minima are obtained directly by a QCLS-solver. Further, the approach gives more insight in the structure of the localisation problem, and thus, statements concerning the consistence, the identifiability and the sensor placement can be derived.

The geometric interpretation of the solutions has been illustrated for that cases, where all sensors are located at the impermeable surface. Additionally, the geometric interpretation for an arbitrary three-dimensional sensor placement was briefly discussed.

Since most of the results are found both algebraically and geometrically, there is a straightforward way to extend them to the infinite medium problem. For it, the analytical solution of the diffusion equation is well-known and its structure is equal to that of the considered case.

A current field of activities is the improvement of the statistical properties of the estimation problems by reformulating the criterions. Thereby, the online applicability is to keep in mind.

Future research is directed to include an anisotropic medium and the existence of advection. Another interesting question is: How to place sensors optimally with respect to the source localisation problem?

7. REFERENCES

Alpay, M. E. and M. H. Shor (2000). Model-based solution techniques for the source localization problem. *IEEE Transactions on Control Systems Technology* **8**(6), 893–902.

Carslaw, H. S. and J. C. Jaeger (1959). *Conduction of Heat in Solids*. Clarendon Press. Oxford.

Crank, J. (1975). *The Mathematics of Diffusion*. Clarendon Press. Oxford.

Duckett, T., M. Axelsson and A. Saffiotti (2001). Learning to locate an odour source with a mobile robot. In: *IEEE International Conference on Robotics and Automation*. Seoul, Korea. pp. 4017–4021.

Groetsch, C. W. (1993). *Inverse Problems in the Mathematical Sciences*. Vieweg. Braunschweig, Wiesbaden.

Gröll, L. (2002). LS-Identifikation mit einer quadratischen Restriktion. In: *Workshop "Theoretische Verfahren der Regelungstechnik" des GMA-FA 1.4, Interlaken, 22.-25.09.2002*. Johannes Kepler Universität Linz, Abteilung für Regelungstechnik und Prozessautomatisierung. pp. 172–180.

Jeremić, A. and A. Nehorai (2000). Landmine detection and localization using chemical sensor array processing. *IEEE Transactions on Signal Processing* **48**, 1295–1305.

Khapalov, A. Y. (1994). Localization of unknown sources for parabolic systems on the basis of available observations. *Int. Journal Systems Sci.* **25**(8), 1305–1322.

Lilienthal, A., A. Zell, M. Wandel and U. Weimar (2002). Detektion und Lokalisation einer Geruchsquelle mit einem autonomen mobilen Roboter. In: *VDI-Berichte 1679*. VDI/VDE-Gesellschaft Mess- und Automatisierungstechnik. pp. 689–694.

Nievergelt, Y. (1998). Solution to an inverse problem in diffusion. *SIAM Rev.* **40**(1), 74–80.

Wacholder, E., E. Elias and Y. Merlis (1995). Artificial neural networks optimization method for radioactive source localization. *Nuclear Technology* **110**, 228–237.

IFAC Publications
www.elsevier.com/locate/ifac

CONTINUOUS-TIME MODEL IDENTIFICATION BY USING ADAPTIVE OBSERVER

Kenji Ikeda * Yoshio Mogami * Takao Shimomura *

* *Department of Information Science and Intelligent Systems*
University of Tokushima,
Tokushima 770-8506 JAPAN
e-mail: {ikeda,moga,simomura}@is.tokushima-u.ac.jp

Abstract: This paper proposes a continuous-time model identification from sampled I/O data by using an adaptive observer. The boundedness of the parameter estimate and the exponential convergence of the parameter estimate error to 0 under the PE assumption are guaranteed. In order to identify the plant from a finite number of the I/O data, an adaptive observer of a backward system is also proposed. *Copyright © 2003 IFAC*

Keywords: Continuous-Time Model, Identification, Sampled-Data System, Adaptive Observer

1. INTRODUCTION

In recent years, the continuous-time model identification has been raised by several authors. The importance of the continuous-time model identification is owing to the fact that discrete-time models do not have a consistency with the physical parameters and that many of the control system synthesis methods are based on continuous-time models. Indirect method, which translates the identified discrete-time model to a continuous-time one, introduces a numerically ill-conditioned problem or difficulty in transforming the zeros of the discrete-time model. From these reasons, direct method, which directly identifies a continuous-time model from the sampled input-output data, has been received much attentions (Young, 1981; Unbehauen and Rao, 1990; Sinha and Rao, 1991).

In the study of direct method, the points are how to construct derivatives of the signals from the sampled data and how to introduce a discrete-time model equivalent to a continuous-time model. In Refs.(Sagara and Zhao, 1990; Kowalczuk and Kozlowski, 2000), the plant is modeled by an integral equation and the integral of the signals are approximated by a numerical method such as trapezoidal method or Simpson's method. In Refs.(Söderström and Mossberg, 2000; Huang and Katayama, 2001), the differential operator is approximated by a difference of the signals e.g. by using δ-operator. In these approximation approaches, a continuous-time model will be obtained by limiting the sampling period h to 0. The use of pre-filters is another approach to a continuous-time model identification. By inserting Poisson moment filter (Garnier *et al.*, 2000), Laguerre filter(Haverkamp *et al.*, 1995; Yang, 1998), more generalized filter utilizing test functions in the random distribution (Ohsumi *et al.*, 2002), etc before sampling the signals, an equivalent discrete-time model will be obtained, in which the parameters of the continuous-time model appear explicitly. Although this approach does not require the infinitesimally small sampling period, it requires pre-filters, which must be implemented in continuous-time (analog) circuits. From the practical viewpoint, the use of pre-filters will be restricted to the case when a fast sampling is possible. Another approach is to restrict the signals to be periodic (Shiotsuki and Kimura, 2001). By expanding the signals by sinusoidal functions, strict derivatives of the signals can be obtained. However, periodicity of the signals may not be assumed in a practical use. Adaptive approach (Kreisselmeier, 1977; Unbehauen and Rao, 1990) is also considered to be one of the main approaches. However, the adaptive systems for the continuous-time model are based on the continuous-time input-output signals. Thus, they are not appropriate for the present framework of digital control systems.

In this paper, a continuous-time model identification, which does not require pre-filters, infinitesimally small sampling periods, nor periodicity of the signals, is considered. The difficulty of the approximation approach is due to that the derivatives of the signals must be estimated without the inter-sampling data. On the other hand, the plant state of inter-sampling period can be estimated if the plant parameters are known. Kreisselmeier's adaptive observer(Kreisselmeier, 1977) is to estimate the plant state as well as the plant parameters by using state variable filters (SVF). If we can discretize this adaptive observer, the inter-sampling output can be estimated by using the plant parameter estimates and to improve the parameter estimates by using the inter-sampling output estimate. Though the state variable filter must be implemented as a continuous-time system, a continuous-time simulation such as Runge-Kutta method achieves a sufficient accuracy for our purpose so that it is easily implemented in a digital computer.

This paper proposes an adaptive observer which identifies a continuous-time model from a finite number of sampled input-output data. The problem is formulated in section 2. A Luenberger observer for a sampled-data system and its equivalent but structurally different observer using a state variable filter are presented in sections 3 and 4, respectively. An adaptive observer is presented in section 5. In order to estimate the model from a finite number of data, an adaptive observer of a backward system is proposed in section 6. Finally, numerical examples are shown to illustrate the proposed method.

Notation: Let h be a sampling period throughout the paper. Let S_h denote a sampler with 0 order hold, *i.e.* the following equation holds for any signal $x(t)$:

$$(S_h x)(t) = x(\lfloor t/h \rfloor h) \tag{1}$$

where $\lfloor x \rfloor$ is the maximum integer less than or equal to x. A function of continuous time t will be denoted by $x(t)$ while its sampled value $x(kh)$ will be denoted by $x[k]$. Piecewise continuous signals are assumed to be right continuous, *i.e.* $\lim_{h \downarrow 0} x(t+h) = x(t)$ for any signal $x(t)$. The left limit of a signal is denoted by $x(t-0)$ *i.e.* $x(t-0) = \lim_{h \uparrow 0} x(t+h)$.

2. PROBLEM STATEMENT

Consider a single-input single-output (SISO) continuous-time system:

$$\dot{x}(t) = Ax(t) + bu(t) \tag{2}$$

$$y(t) = c^{\mathrm{T}} x(t) \tag{3}$$

where $u(t) \in R$, $y(t) \in R$, and $x(t) \in R^n$ are the input, the output, and the state of the system. A, b, and c are the system matrices of appropriate dimensions. Assume the followings for this system.

(A1) (A, c) is observable,
(A2) the dimension n is known.

Without loss of generality, (A, b, c) is assumed to be the observer canonical form:

$$A = \begin{pmatrix} -a_1 & 1 & & \\ \vdots & & \ddots & \\ \vdots & & & 1 \\ -a_n & & & \end{pmatrix} \quad b = \begin{pmatrix} b_1 \\ \vdots \\ b_n \end{pmatrix}$$

$$c^{\mathrm{T}} = (1 \quad 0 \quad \dots \quad 0)$$

Furthermore, the sampled data of the input and the output is available for the identification:

(A3) the output $y(t)$ can be measured at the discrete-time instances $t = kh$ $(k = 0, 1, 2, \dots, N)$.
(A4) the input $u(t)$ is an output of the 0 order hold, *i.e.*,

$$u(t) = u(\lfloor t/h \rfloor h), \quad \forall t \in [0, Nh].$$

Problem formulation: System identification requires the determination of $\{a_i, b_i\}$, $i = 1, \dots, n$ from the sampled input-output data $\{u[k], y[k]\}$ for $k = 0, \dots, N$ and to estimate the state variable $x(t)$ for $t \in [0, Nh]$.

3. LUENBERGER OBSERVER

A Luenberger observer for a sampled-data system will be defined by

$$\dot{\hat{x}}(t) = F\hat{x}(t) + hu(t) + g\left[S_h y(t) + (I - S_h)\hat{y}(t)\right] \tag{4}$$

$$\hat{y}(t) = c^{\mathrm{T}} \hat{x}(t) \tag{5}$$

where F, g, and h are the observer matrices of the form:

$$F = \begin{pmatrix} -f_1 & 1 & & \\ \vdots & & \ddots & \\ \vdots & & & 1 \\ -f_n & & & \end{pmatrix}, \tag{6}$$

$$g = (g_1, \dots, g_n)^{\mathrm{T}}, \tag{7}$$

$$h = (h_1, \dots, h_n)^{\mathrm{T}}. \tag{8}$$

Let the observer parameters be $g_i = f_i - a_i$ and $h_i = b_i$, then the state estimate error $e_x(t) = \hat{x}(t) - x(t)$ must satisfy the following equation.

$$\dot{e}_x(t) = (A - gc^{\mathrm{T}} S_h) e_x(t). \tag{9}$$

At the sampling instances kh, $k = 0, 1, 2, \dots, N-1$, state estimate error $e_x[k]$ can be written explicitly as:

$$e_x[k+1] = \left[e^{Ah} - \int_0^h e^{At} dt g c^{\mathrm{T}} \right] e_x[k]. \quad (10)$$

Since the difference between the sampled-data system and the continuous time system becomes

$$\left[e^{Ah} - \int_0^h e^{At} dt g c^{\mathrm{T}} \right] - e^{(A - gc^{\mathrm{T}})h} = O(h^2), \quad (11)$$

the state estimate error $e_x[k]$ goes to 0 as $k \to \infty$ if f_i's are the coefficients of a Hurwitz polynomial and the sampling period h is sufficiently small. Note that the sampling period h is not required to be infinitesimally small for the convergence of the state estimate error $e_x[k]$.

The addition of the inter-sampling estimate $(I - S_h)\hat{y}(t)$ to the input of the Luenberger observer (4) plays an important role. If there are no inter-sampling estimation, the state estimate $\hat{x}(t)$ does not converge to the plant state $x(t)$ in general.

In order for the regressor vector to behave like a continuous-time signals, the step size of the ordinary differential equation solver (e.g. Euler method) must be taken less than or equal to h/n. Otherwise, the influence of the input at time k does not spread over the state at time $k+1$.

4. REALIZATION OF THE OBSERVER BY USING STATE VARIABLE FILTER

In order for the output estimate error to be a linear function of unknown parameters g and h, the Luenberger observer explained in the previous section will be realized by using State Variable Filter (SVF):

$$\dot{\zeta}_1(t) = F^{\mathrm{T}}\zeta_1(t) + c[S_h y(t) + (I - S_h)\hat{y}(t)], \quad (12)$$

$$\dot{\zeta}_2(t) = F^{\mathrm{T}}\zeta_2(t) + c u(t). \quad (13)$$

Let $z(t) = (\zeta_1^{\mathrm{T}}(t), \zeta_2^{\mathrm{T}}(t))^{\mathrm{T}}$ and $\hat{p} = (\hat{g}^{\mathrm{T}}, \hat{h}^{\mathrm{T}})^{\mathrm{T}}$, then the output estimate becomes

$$\hat{y}(t) = z^{\mathrm{T}}(t)\hat{p}. \quad (14)$$

Furthermore, the estimate of the plant state becomes

$$\hat{x}(t) = (\hat{g}_1 T_1 + \cdots + \hat{g}_n T_n)\zeta_1(t) + (\hat{h}_1 T_1 + \cdots + \hat{h}_n T_n)\zeta_2(t) \quad (15)$$

where $T_i \in \mathbf{R}^{n \times n}$ is a constant matrix such that

$$(sI - F)^{-1} e_i = T_i (sI - F^{\mathrm{T}})^{-1} c, \quad (16)$$

in which e_i is a i-th unit vector. Define an ideal state variable filter as

$$\dot{\zeta}_1^*(t) = F^{\mathrm{T}}\zeta_1^*(t) + c y(t), \quad (17)$$

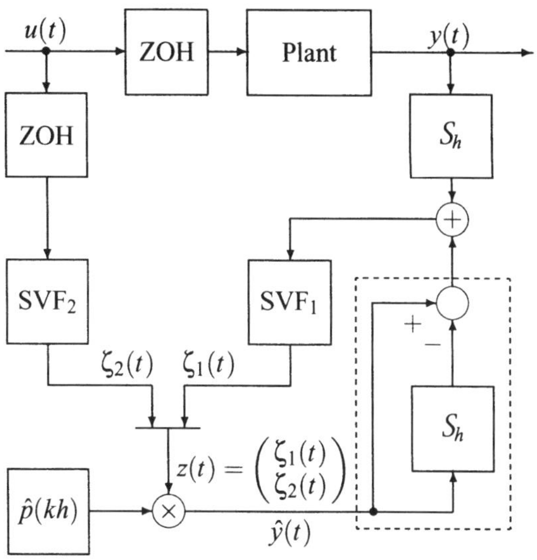

Fig. 1. Observer using SVF

then the output $y(t)$ has an equivalent form:

$$y(t) = p^{*\mathrm{T}} z^*(t) + \varepsilon(t) \quad (18)$$

where $z^*(t) = (\zeta_1^{*\mathrm{T}}(t), \zeta_2^{\mathrm{T}}(t))^{\mathrm{T}}$, $p^* = (g^{\mathrm{T}}, h^{\mathrm{T}})^{\mathrm{T}}$ is a true value of the parameter, and $\varepsilon(t)$ is an exponentially vanishing term due to the initial state. From eqs.(14) and (18), the output estimate error $e_y(t) = \hat{y}(t) - y(t)$ becomes

$$e_y(t) = z^{\mathrm{T}}(t)\tilde{p}(t) + g^{\mathrm{T}}\tilde{\zeta}_1(t) - \varepsilon(t). \quad (19)$$

where $\tilde{p}(t) = \hat{p}(t) - p^*$ is a parameter estimate error, $\tilde{\zeta}_1(t) = \zeta_1(t) - \zeta_1^*(t)$ is the error caused by the use of the inter-sampling estimation instead of the true output $y(t)$.

The error $\tilde{\zeta}_1(t)$ must obey the following equation:

$$\dot{\tilde{\zeta}}_1(t) = F^{\mathrm{T}}\tilde{\zeta}_1(t) + c[I - S_h]e_y(t). \quad (20)$$

From eq.(20), $\tilde{\zeta}_1(t)$ is proportional to the sampling period h and is considered to be small compared to $z(t)$. Thus, the following assumption is made.

(A5) The term $\tilde{\zeta}_1(t)$ is negligible, *i.e.*, the output estimate error $e_y(t)$ can be approximated by

$$e_y(t) = z^{\mathrm{T}}(t)\tilde{p}(t) - \varepsilon(t). \quad (21)$$

5. ADAPTIVE OBSERVER

Since the output estimate error becomes a linear function of the unknown parameters, we can introduce a parameter adjustment law. Discrete-time adaptive laws will be appropriate for the adjustment of this sampled-data system, since the output $y(t)$ can be measured only at the sampling instances kh. Thus, the parameter estimate $\hat{p}(t)$ is piecewise constant and changes its value only on the sampling instances kh.

i.e. $\hat{p}(t) = \hat{p}([t/h]h)$. An ordinary discrete-time parameter adjustment law is adopted as:

$$\hat{p}[k] = \hat{p}[k-1] - \frac{Gz[k]}{1+z[k]^{\mathrm{T}}Gz[k]} e_y(kh-0) \quad (22)$$

where $G = G^{\mathrm{T}} > 0$ is an adaptation gain and $e_y(kh-0)$ is an output estimate error just before the parameter changes:

$$e_y(kh-0) = \hat{y}(kh-0) - y(kh-0) \quad (23)$$

$$= z[k]^{\mathrm{T}}\hat{p}[k-1] - y[k]. \quad (24)$$

For the adaptive observer defined above, it can be stated that the parameter estimates are bounded and that the parameter estimate error converges exponentially to 0 under the assumption of Persistently Excitation (PE) property of the signals.

Theorem 1. For the adaptive observer system defined by the eqs. (2), (3), (12), (13), (14), and (22), with the assumptions (A1) to (A5), define a sum of the square of the parameter estimate error as

$$W[k] = \tilde{p}[k]^{\mathrm{T}}G^{-1}\tilde{p}[k]. \quad (25)$$

Then, $W[k]$ is monotonically non-increasing. Furthermore, if there exists $k_{\min} > 0$ such that

$$0 < k_{\min}I \le \sum_{1}^{N} \frac{z[k]z[k]^{\mathrm{T}}}{1+z[k]^{\mathrm{T}}Gz[k]}, \quad (26)$$

then the following inequality holds

$$\frac{W[0]-W[N]}{W[0]} \ge \frac{k_{\min}\lambda_{\min}(G)}{1+N^2\frac{\lambda^2_{\max}(G)}{\lambda^2_{\min}(G)}}. \quad (27)$$

By virtue of the assumption (A5), the proof of this theorem is reduced to the case of an ordinary discrete-time adaptive observers, so omitted here.

6. ADAPTIVE OBSERVER OF BACKWARD SYSTEM

In general, it takes a long time for the convergence of the parameter estimate in the adaptive observer. From the practical point of view, it is not preferable because there are cases when it is not possible to obtain the long term I/O data of the system. Thus, it is necessary to develop a method to identify the system from a finite number of I/O data. For this purpose, it might be necessary to utilize the given data repeatedly for the adaptation.

Consider the backward system:

$$\dot{x}'(t) = -Ax'(t) - bu'(t) \quad (28)$$

$$y'(t) = c^{\mathrm{T}}x'(t) \quad (29)$$

where $u'(t) = u(Nh-t)$, $y'(t) = y(Nh-t)$, $x'(t) = x(Nh-t)$. In the rest of the paper, variables and parameters of the backward system will be denoted by putting ($'$).

Define the matrices of the state variable filter for this backward system as:

$$F' = \begin{pmatrix} (-1)^n f_1' & -1 & & \\ \vdots & & \ddots & \\ f_{n-1}' & & & -1 \\ -f_n' & & & \end{pmatrix} \quad (30)$$

$$\hat{g}'(t) = (\hat{g}_1' \quad \ldots \quad \hat{g}_n'(t))^{\mathrm{T}} \quad (31)$$

$$\hat{h}'(t) = (\hat{h}_1' \quad \ldots \quad \hat{h}_n'(t))^{\mathrm{T}}, \quad (32)$$

then the state variable filter (12) and (13), the output estimate (14), the state estimate (15), and the parameter adjustment law (22) are used as the adaptive observer of the backward system. In eq. (15), T_i are replaced by T_i' which satisfies

$$(sI-F')^{-1}e_i = T_i'(sI-(F')^{\mathrm{T}})^{-1}c. \quad (33)$$

The true value of the parameters $g_i'^*$ and $h_i'^*$ are defined

$$g_i'^* = a_i + (-1)^{n-i}f_i', \quad h_i'^* = -b_i.$$

Both the monotonically non-increasing property and the exponential convergence of the parameter estimate error can be proved in the same manner in the previous section.

In order to succeed the parameter estimate of the forward system, the initial value of the parameter estimate of the backward system $\hat{p}'(0) = (\hat{g}_1'(0), \ldots, \hat{g}_n'(0), \hat{h}_1'(0), \ldots, \hat{h}_n'(0))^{\mathrm{T}}$ is defined as follows:

$$\hat{g}_i'(0) = -\hat{g}_i(Nh) + f_i + (-1)^{n-i}f_i', \quad (34)$$

$$\hat{h}_i'(0) = -\hat{h}_i(Nh). \quad (35)$$

The initial state of the SVF of the backward system should be defined such that $\hat{x}'(0) = \hat{x}(Nh)$. Let G and H be as follows:

$$G = \hat{g}_1'(0)T_1 + \cdots + \hat{g}_n'(0)T_n$$

$$H = \hat{h}_1'(0)T_1 + \cdots + \hat{h}_n'(0)T_n$$

and let a singular value decomposition of the matrix (G,H) as

$$(G,H) = U\Sigma V^{\mathrm{T}} \quad (36)$$

where U and V are unitary matrices with appropriate dimentions and Σ is a diagonal matrix. Then, the initial value of the SVF of the backward system will be defined as

$$z'(0) = V(:,1:n)\Sigma(:,1:n)^{-1}U^{\mathrm{T}}\hat{x}[N] \quad (37)$$

where $\hat{x}[N]$ is defined by

$$\hat{x}[N] = (\hat{g}_1[N]T_1 + \cdots + \hat{g}_n[N]T_n)\zeta_1[N], \\ +(\hat{h}_1[N]T_1 + \cdots + \hat{h}_n[N]T_n)\zeta_2[N].$$

In the adaptation of the backward system, similar results as Theorem 1 are hold. Define $W'[k]$ as

$$W'[k] = \tilde{p}'[k]^{\mathrm{T}} G^{-1} \tilde{p}'[k], \tag{38}$$

then $W[k]$ is monotonically non-increasing. Furthermore, a similar inequality as eq. (27) holds if the signals are persistently exciting.

When the adaptation in the backward system stops at $t = Nh$, switch to the forward system again, where the initial value of the parameter estimate and the initial state of the SVF are determined in the same fashion as in eqs. (34), (35), and (37). Switching the forward/backward system repeatedly, define W_i be the value of $W[0]$ or $W'[0]$ at the i-th switching, then under the PE condition, there exists $\rho > 0$ s.t.

$$1 \geq \frac{W_i - W_{i+1}}{W_i} \geq \rho.$$

By solving this inequality, we obtain

$$W_i \leq (1-\rho)^i W_0.$$

This means the parameter estimate error decreases exponentially.

7. NUMERICAL EXAMPLE

7.1 *Illustration of the Proposed Method*

In order to illustrate how the proposed adaptive observer works, a numerical example of the identification of 3rd order system is presented. The system parameters are $(a_1, a_2, a_3) = (1, 4, 2)$ and $(b_1, b_2, b_3) = (0, 1, 1)$, the parameters of the SVF are defined as $(f_1, f_2, f_3) = (f_1', f_2', f_3') = (4, 8, 8)$, and the adaptation gain is defined as $G = I$. The input and output data of the system is presented in Fig.2, in which the axis of abscissa is a time in second and the ordinate is the input and the output. Fig.3 shows the sum of the square of the parameter estimate errors $W(t) = \tilde{p}(t)^{\mathrm{T}}\tilde{p}(t)$. It is observed that $W(t)$ is monotonically non-decreasing not only at $t \in (0, Nh)$ but at the instance of the forward/backward switch. Fig.4 is the plot of $\log_{10} W_i$ at i-th forward/backward switching instance. From the fact that the graph is almost linear, we can confirm the exponential convergence of the parameter estimate errors.

For the comparison study, the simulation result of a method without an inter-sampling estimation is also presented in Fig.4. As is expected, it does not achieve the convergence to the true value.

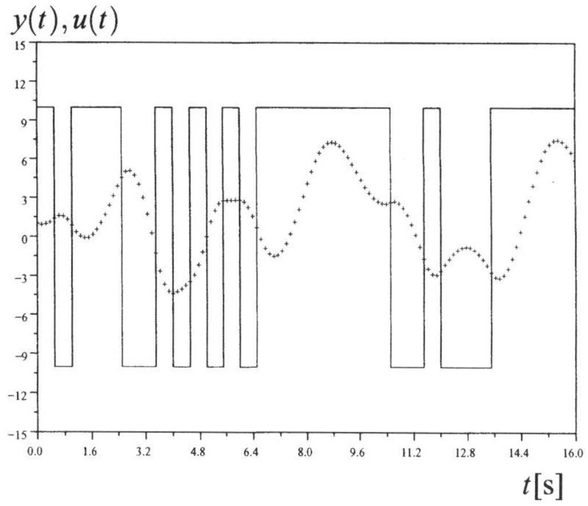

Fig. 2. Input and Output Data

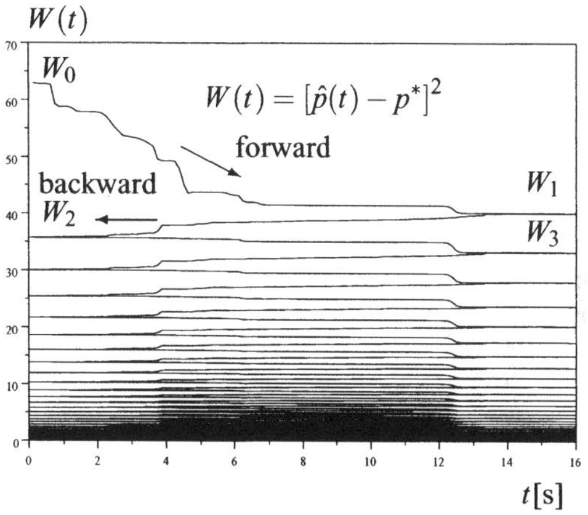

Fig. 3. Sum of the Square of the Parameter Estimate Error

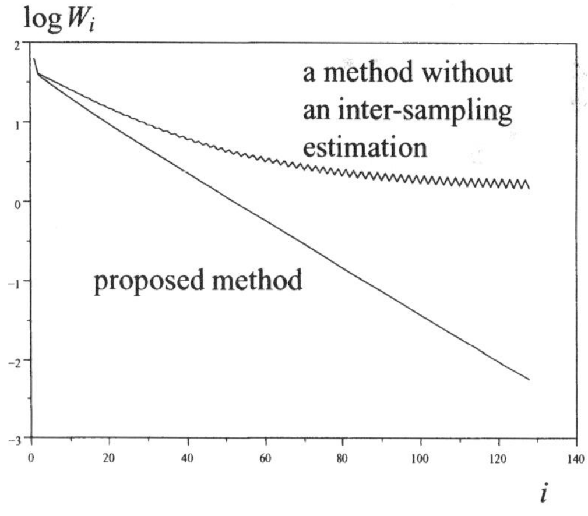

Fig. 4. Sum of the Square of the Parameter Estimate Error at the Forward/Backward Switch (in the log scale)

7.2 *Effect of the Measurement Noise*

The effect of the measurement noise is investigated by numerical simulations. The parameters of the plant to be estimated are $(a_1, a_2, a_3) = (5, 10, 8)$ and $(b_1, b_2, b_3) = (6, 7, 8)$. The input $u(t)$ to the system is generated to be a filtered random noise, where the filter is a second order Butterworth filter with cut-off frequency equal to 10 [rad/sec]. The number of data is

150 and the simulation time is 15 [sec]. Random noise is added to each output data with 0% to 10% noise to signal ratio (NSR) defined as

$$\text{NSR} = \frac{\|\text{noise}\|_2}{\|\text{signal}\|_2} \times 100(\%).$$

Twenty sets of input/output data are generated for each NSR and the accuracy of the estimated parameters for each NSR is defined as an average of 20 results measured by relative root square error (RRSE) as

$$\text{RRSE} = \frac{\|p^* - \hat{p}\|_2}{\|p^*\|_2}.$$

Fig.5 shows the results of the identifications. In order to compare the existing subspace methods, numerical simulations based on the methods in (Yang, 1998) are also presented. SVF is the result of the direct subspace identification by using the state variable filter with its design parameter $\alpha = 2$ while ω-operator is the result of the direct subspace identification by using ω-operator with its design parameter $\alpha = 3$. When NSR is greater than 3%, RRSE of the proposed method is greater than SVF and ω-operator. However, when NSR is less than 2%, RRSE of the proposed method is less than those of the other two methods. RRSE of the proposed method is almost zero if there are no noise while RRSE's of SVF and ω-operator remain positive even if there are no noise.

In order for the proposed method to be applicable for a practical situation, a more robust algorithm for the noise and disturbances must be investigated. However, the unbiased estimation of the proposed method in noise free case will be an advantage in continuous-time model identification.

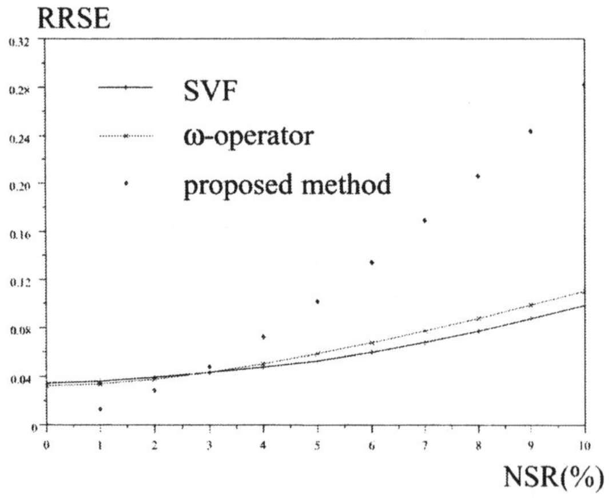

Fig. 5. Results of the SVF, the ω-operator, and the proposed method

8. CONCLUSION

This paper proposed a continuous time model identification method by using an adaptive observer. The estimated parameters are bounded and the parameter estimate error converges to 0 exponentially. In order to utilize the sampled data repeatedly, the adaptive observer of a backward system is proposed. The selection of the initial value of the parameter estimate and the initial state of the SVF are given.

9. REFERENCES

Garnier, H., M. Gilson and W. X. Zheng (2000). A bias-eliminated least-squares method for continuous-time model identification of closed-loop systems. *Int. J. Control* **70**(1), 38–48.

Haverkamp, B. R. J., C. T. Chou, M. Verhaegen and R. Johansson (1995). Identification of continuous-time mimo state space models from sampled data,in the presence of process and measurement noise. In: *Proc. of the 35th CDC*. pp. 1539–1544.

Huang, D. and T. Katayama (2001). A subspace-based method for continuous-time model identification by using δ-operator model. *Trans. of ISCIE* **14**(1), 1–9.

Kowalczuk, Z. and Janusz Kozlowski (2000). Continuous-time approaches to identification of continuous-time systems. *Automatica* **36**, 1229–1236.

Kreisselmeier, G. (1977). Adaptive observers with exponential rate of convergence. *IEEE Tr. on Automat. Contr.* **AC-22**(1), 2–8.

Ohsumi, A., K. Kameyama and K. Yamaguchi (2002). Subspace identification for continuous-time stochastic systems via distribution-based approach. *Automatica* **38**, 63–79.

Sagara, S. and Z. Zhao (1990). Numerical integration approach to on-line identification of continuous-time systems. *Automatica* **26**, 63–74.

Shiotsuki, T. and A. Kimura (2001). Direct identification method of continuous-time system model from time-series (in japanese). In: *SICE 1st Annual Conference on Control Systems*. pp. 413–416.

Sinha, N. K. and G. P. Rao (1991). *Identification of Continuous-Time Systems*. Kluwer.

Söderström, T. and M. Mossberg (2000). Performance evaluation of methods for identifying continuous-time autoregressive processes. *Automatica* **36**, 53–59.

Unbehauen, H. and G. P. Rao (1990). Continuous-time approaches to system identification–a survey. *Automatica* **26**(1), 23–35.

Yang, Z. (1998). Subspace model identification of continuous-time systems with the aid of the *w*-operator (in japanese). *Trans. of SICE* **34**(6), 546–554.

Young, P. (1981). Parameter estimation for continuous-time models–a survey. *Automatica* **17**(1), 23–39.

IFAC Publications
www.elsevier.com/locate/ifac

OPTIMAL FILTERING OF NONLINEAR SYSTEMS BASED ON PSEUDO GAUSSIAN DENSITIES

Uwe D. Hanebeck

Institute of Computer Design and Fault Tolerance
Universität Karlsruhe
76128 Karlsruhe, Germany
Uwe.Hanebeck@ieee.org

Abstract: We consider the problem of estimating the state of a discrete–time dynamic system comprising a linear system equation and a nonlinear measurement equation based on measurements corrupted by non–Gaussian noise. The problem is solved by recursively calculating the complete posterior density of the state given the measurements. For representing the resulting non–Gaussian posterior, a new exponential type density, the so called pseudo Gaussian density, is introduced. By converting the original nonlinear system to an equivalent linear representation in a higher–dimensional space, the parameters of the pseudo Gaussian posterior are obtained by means of a linear estimator operating in the higher–dimensional space. The resulting filtering algorithms are easy to implement and always guarantee valid posterior densities.

Keywords: Estimators, Filtering Theory, Mathematical systems theory, Non–Gaussian processes, Nonlinear systems, Optimal filtering, Recursive estimation, Stochastic systems

1. INTRODUCTION

Filtering consists of estimating parameters of one process, the system state sequence, given uncertain information from another related process, the measurement sequence. When the measurements are related nonlinearly to the system state, this estimation problem is in general difficult to solve. Usually, a linearization is performed to permit application of filtering methods derived for linear systems (Anderson and Moore, 1979). Of course, this only works for certain type of nonlinearities. In addition, the presence of non–Gaussian measurement noise further limits the applicability of linear methods.

More advanced methods for providing state estimates in the nonlinear case have been developed by keeping nonlinear terms in a Taylor series expansion of the nonlinearity, see (Bohn and Unbehauen, 2000) for an elegant derivation. However, in this paper the focus is on calculating the *complete* posterior density of the unknown system state given all the measurements. A parametric closed–form density description is desired, which is defined by a finite number of parameters. In addition, the density representation should allow for recursive application and should not suffer from a permanently growing number of description parameters with an increasing number of available measurements.

A grid representation of densities for numerical nonlinear filtering based on quantization of the state space has been introduced in (Bucy and Senne, 1971), but has proven to be useful only for a limited state vector dimension (Bergman *et al.*, 1999). Monte Carlo techniques (Doucet *et al.*, 2000; Liu and Chen, 1998) use stochastic samples to represent density functions in order to numerically solve the filtering problem.

Closed–form representations of densities include the Edgeworth expansion, i.e., a Gaussian density times a sum of Hermite polynomials, which has been proposed in (Sorenson and Stubberud, 1968). A method for updating this type of density numerically is described in (Challa *et al.*, 2000). The approach has the disadvantage that truncated Edgeworth expansions are not themselves valid density functions and may give negative values (Jazwinski, 1970). A Gaussian mixture representation has been proposed in (Alspach and Sorenson, 1972), which always provides valid density functions. However, each term is individually updated based on linearization, which results in a bank of parallel extended Kalman filters.

The simplest form of the measurement update seems to be obtained when using exponential type densities (Kulhavý, 1992). In addition, these densities are always positive. However, depending on the exponent function, e.g. polynomials, numerical inaccuracies during the update recursion may lead to densities that are not integrable, i.e., the integral over the density does not give a finite value.

In this paper, a new type of exponential density, the so called pseudo Gaussian density, is proposed. It is defined by a standard Gaussian function in a hyperspace S^* related to the original state space S via a nonlinear transformation. Because of its special structure, pseudo Gaussians are always valid density functions even in the presence of numerical inaccuracies. In addition, it will be shown that under certain assumptions, this type of density can be updated by means of a linear filter operating in the hyperspace S^*.

The nonlinear filtering problem is formulated in Section 2. The concept of pseudo Gaussian densities is explained in detail in Section 3. Section 4 provides a generic conversion of a nonlinear system to an equivalent system, which is linear in a higher–dimensional space. The corresponding filtering algorithm is given in Section 5. An example in Section 6 illustrates the proposed approach.

2. PROBLEM FORMULATION

We consider estimating the state of a linear dynamic system, which may evolves in discrete time according to

$$\underline{x}_{k+1} = \mathbf{A}_k \, \underline{x}_k + \hat{\underline{u}}_k \ , \tag{1}$$

where $\hat{\underline{u}}_k$ is a known input sequence. The system state $\underline{x}_k$ is not directly observable, but will instead be deduced from measurements of the system output. Measurements are assumed to be taken sequentially at discrete time steps $k = 1, 2, \ldots$ and are corrupted by white non–Gaussian noise.

An M–dimensional measurement $\hat{\underline{y}}_k$ [1] at time step k is related to the N–dimensional system state $\underline{x}_k$ via the *nonlinear time–variant* measurement equation

$$\hat{\underline{y}}_k = \underline{h}_k(\underline{x}_k) + \underline{v}_k \tag{2}$$

and is corrupted by *additive* white noise $\underline{v}_k$ from a possibly non–Gaussian noise density $p_k^v(\underline{v}_k)$.

Instead of providing point estimates of the unknown state $\underline{x}_k$, an estimator is used to construct the complete conditional density of the state

$$p_k^e(\underline{x}_k) = p(\underline{x}_k | \hat{\underline{y}}_k, \hat{\underline{y}}_{k-1}, \ldots, \hat{\underline{y}}_1)$$

given all observations up to time step k. A recursive estimation procedure is preferred, which calculates a state estimate based on the estimate at the previous time step and hence, does not require to store all measurements. A suitable time update procedure produces a predicted density

$$p_k^p(\underline{x}_k) = p(\underline{x}_k | \hat{\underline{y}}_{k-1}, \ldots, \hat{\underline{y}}_1)$$

by propagating the previous estimate $p_{k-1}^e(\underline{x}_{k-1})$ through the system model. Although not strictly required, an initial density $p_0^e(\underline{x}_0)$ is assumed to be given.

Arbitrary characteristic values of the estimate such as mean, covariance matrix, mode, or median can be derived once the estimated density is available.

3. PSEUDO GAUSSIANS

The key idea is to represent complicated probability density functions in the N–dimensional original state space S_x by simpler densities in a higher–dimensional space S_x^*. Points $\underline{x}_k$ in S_x are related to points $\underline{x}_k^*$ in S_x^* via a nonlinear transformation $\underline{T}_x(.)$ according to

$$\underline{x}_k^* = \underline{T}_x(\underline{x}_k) = [T_1(\underline{x}_k), \ldots, T_{L_x}(\underline{x}_k)]^T \ ,$$

where L_x denotes the dimension of space S_x^*. Hence, the original space S_x is transformed by $\underline{T}_x(.)$ to an N–dimensional manifold U_x^* in the L_x–dimensional space S_x^*.

---

[1] We use a hat to indicate that the given measurement at time step k is a non–random quantity and is used as an estimate of the true measurement.

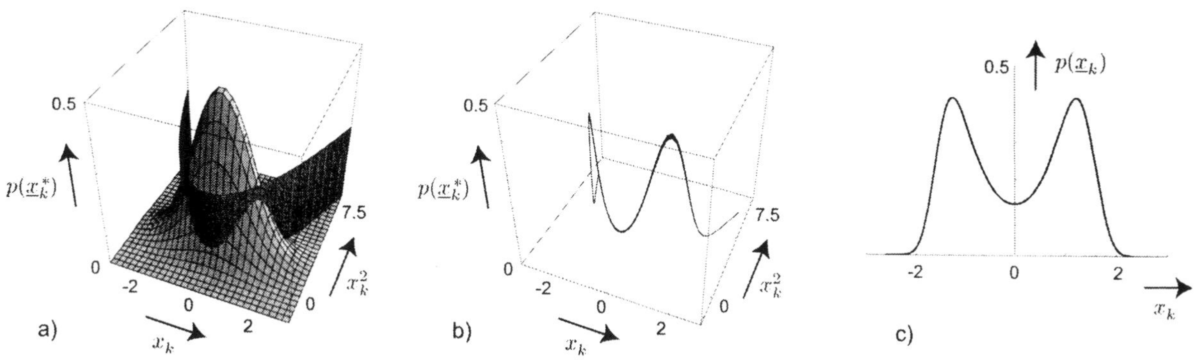

Fig. 1. Example for demonstrating the concept of pseudo Gaussians with scalar state x_k and twodimensional hyperspace S_x^*. a) Pseudo Gaussian in hyperspace S^* with mean and covariance matrix according to example 3.1. b) Parts of the pseudo Gaussian density lying on the manifold U_x^*. c) Corresponding density in the original space S_x.

In S_x^*, L_x–dimensional Gaussian probability density functions are defined according to

$$p(\underline{x}_k^*) = c_k^x \exp\left\{ -\frac{1}{2}(\underline{x}_k^* - \hat{\underline{x}}_k^*)^T (\mathbf{C}_x^*)^{-1} (\underline{x}_k^* - \hat{\underline{x}}_k^*) \right\}$$

with mean $\hat{\underline{x}}_k^*$, symmetric positive definite covariance matrix $\mathbf{C}_k^{x,*}$, and normalizing constant c_k^x. Densities of this type will be called pseudo Gaussian in the following, because the components of $\underline{x}_k^*$ are not independent.

The intersection of a pseudo Gaussian $p(\underline{x}_k^*)$ with the manifold U_x^* defines a non–Gaussian, e.g. multimodal, probability density function in the original space S_x.

REMARK 3.1. A non–Gaussian density in the original space S_x is defined by *both* the transformation $\underline{T}_x(.)$ *and* the mean $\hat{\underline{x}}_k^*$ and covariance matrix $\mathbf{C}_k^{x,*}$ of the pseudo Gaussian $p(\underline{x}_k^*)$.

EXAMPLE 3.1. A scalar state x_k is considered, which is related to a two–dimensional state $\underline{x}_k^*$ via

$$\underline{x}_k^* = \underline{T}_x(\underline{x}_k) = [x_k, x_k^2]^T \ .$$

An example of a pseudo Gaussian density defined in the space S_x^* with mean

$$\hat{\underline{x}}_k^* = \begin{bmatrix} 0 & 2 \end{bmatrix}^T$$

and covariance matrix

$$\mathbf{C}_k^{x,*} = \begin{bmatrix} 1 & 0 \\ 0 & 1 \end{bmatrix}$$

is shown in Fig. 1 a) together with the manifold U_x^*. Fig. 1 b) then shows that part of the pseudo Gaussian density lying on the manifold U_x^*, which defines the density in the original space shown in Fig. 1 c).

The selection of the functions $T_i(\underline{x}_k)$, $i = 1, \ldots, L_x$ depends on the type of nonlinearity considered. However, multidimensional Bernstein polynomials appear to be a good choice in many cases, e.g. polynomial nonlinearities. They are defined on the basis of one–dimensional Bernstein polynomials, which on the interval $[l, r]$ are given by

$$H_i^n(x) = \binom{n}{i} \left(\frac{l-x}{l-r}\right)^i \left(\frac{r-x}{r-l}\right)^{n-i}$$

for $i = 0, \ldots, n$. With

$$\underline{x}_k = \begin{bmatrix} x_k^1 & x_k^2 & \ldots & x_k^N \end{bmatrix}^T \ ,$$

the above transformation is defined by

$$T_i(\underline{x}_k) = \prod_{j=1}^{N} H_{i_j}^{L_j - 1}(x_k^j) \ ,$$

for $i_j = 0, \ldots, L_j - 1$, $j = 1, \ldots, N$, $L_x = \prod_{j=1}^N L_j$, and $i = \sum_{j=1}^N i_j$. For example, in two dimensions this gives

$$T_i(\underline{x}_k) = H_{i_1}^{L_1 - 1}(x_k^1)\, H_{i_2}^{L_2 - 1}(x_k^2) \ ,$$

for $i_1 = 0, \ldots, L_1 - 1$, $i_2 = 0, \ldots, L_2 - 1$, $L_x = L_1 L_2$, and $i = i_1 + i_2$.

4. SYSTEM TRANSFORMATION

The original nonlinear system given by the linear system equation (1) and the nonlinear measurement equation (2) will now be converted to an equivalent linear representation in a higher–dimensional space.

For that purpose, the nonlinear measurement equation is transformed according to

$$\underline{T}_v(\hat{\underline{y}}_k - \underline{v}_k) = \underline{T}_v(\underline{h}_k(\underline{x}_k)) \ . \qquad (3)$$

The left hand side is then converted into an affine function of $\underline{v}_k^*$

$$\underline{T}_v(\hat{\underline{y}}_k - \underline{v}_k) = -\mathbf{G}_k^* \underline{v}_k^* + \hat{\underline{y}}_k^* \ ,$$

where the term $\hat{\underline{y}}_k^*$ does not depend on elements of $\underline{v}_k^*$. Of course, $\mathbf{G}_k^*$ and $\hat{\underline{y}}_k^*$ are polynomial functions

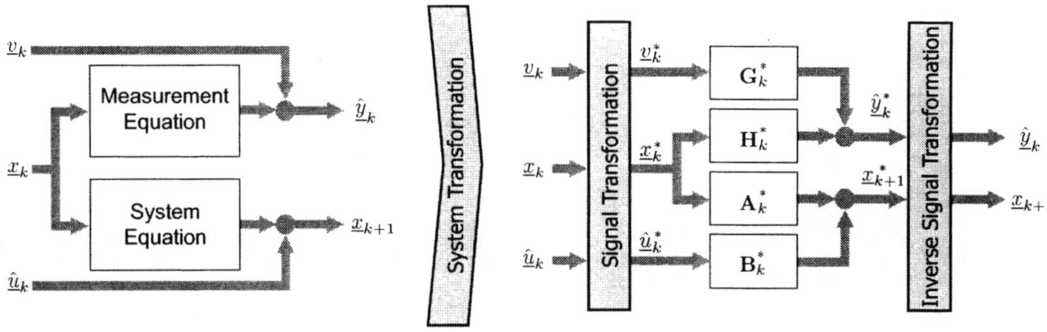

Fig. 2. Block diagram of the system transformation: The original system comprising a linear system equation and a nonlinear measurement equation is converted to a higher–dimensional representation, which is linear in a higher-dimensional space.

of the measurements $\hat{\underline{y}}_k$. The right hand side of (3) is expanded into a linear function of $\underline{x}_k^*$

$$\underline{T}_v(\underline{h}_k(\underline{x})) = \mathbf{H}_k^* \underline{x}_k^*$$

with

$$\underline{x}_k^* = \underline{T}_x(\underline{x}_k)$$

and $L_x \geq \max(N, L_v)$. This expansion is exact for a polynomial measurement nonlinearity $\underline{h}_k(.)$. Finally, we obtain a *linear* measurement equation

$$\hat{\underline{y}}_k^* = \mathbf{H}_k^* \underline{x}_k^* + \mathbf{G}_k^* \underline{v}_k^*$$

in S_x^* with $\hat{\underline{y}}_k^* \in \mathbb{R}^{L_v}$, $\underline{x}_k^* \in \mathbb{R}^{L_x}$, $\underline{v}_k^* \in \mathbb{R}^{L_v}$.

In addition, the system equation given by (1) is transformed according to

$$\underline{T}_x(\underline{x}_{k+1}) = \underline{T}_x(\mathbf{A}_k \, \underline{x}_k + \hat{\underline{u}}_k) \ ,$$

which is rewritten as

$$\underline{x}_{k+1}^* = \mathbf{A}_k^* \, \underline{x}_k^* + \mathbf{B}_k^* \, \hat{\underline{u}}_k^* \ .$$

5. FILTERING

Given the linear representation from Section 4, the desired densities are obtained by a linear filter operating in a higher–dimensional space S_x^* with state dimension L_x, provided the noise density $p_k^v(\underline{v}_k)$ is given as a pseudo Gaussian

$$p_k^v(\underline{v}_k^*) = c_k^v \exp\left\{-\frac{1}{2}(\underline{v}_k^* - \hat{\underline{v}}_k^*)^T (\mathbf{C}_k^{v,*})^{-1} (\underline{v}_k^* - \hat{\underline{v}}_k^*)\right\}$$

in a space S_v^* with dimension L_v and the initial state is characterized by a pseudo Gaussian density $p_0^e(\underline{x}_0)$ defined by $\hat{\underline{x}}_0^{e,*}$ and $\mathbf{C}_0^{e,*}$. The prediction step is given by

$$\begin{aligned} \hat{\underline{x}}_{k+1}^{p,*} &= \mathbf{A}_k^* \, \hat{\underline{x}}_k^{e,*} + \mathbf{B}_k^* \, \hat{\underline{u}}_k^* \ , \\ \mathbf{C}_{k+1}^{p,*} &= \mathbf{A}_k^* \, \mathbf{C}_k^{p,*} \, (\mathbf{A}_k^*)^T \ , \end{aligned} \qquad (4)$$

the filter step is given by

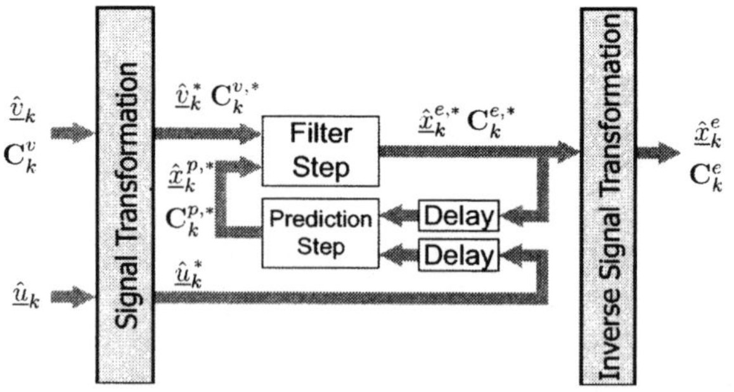

Fig. 3. Block diagram of the proposed new estimator: Estimation is performed by a linear estimator, e.g. a Kalman filter, in the higher–dimensional space. Please note that the recursion is completely performed in the higher–dimensional space. The inverse transformation is done outside of the recursion loop.

$$\begin{aligned} \hat{\underline{x}}_k^{e,*} &= \hat{\underline{x}}_k^{p,*} + \mathbf{C}_k^{p,*}(\mathbf{H}_k^*)^T \left\{\mathbf{G}_k^* \mathbf{C}_k^{v,*} (\mathbf{G}_k^*)^T \right. \\ &\quad \left. + \mathbf{H}_k^* \mathbf{C}_k^{p,*} (\mathbf{H}_k^*)^T \right\}^{-1} (\hat{\underline{y}}_k^* - \mathbf{G}_k^* \hat{\underline{v}}_k^* - \mathbf{H}_k^* \hat{\underline{x}}_k^{p,*}) \ , \\ \mathbf{C}_k^{e,*} &= \mathbf{C}_k^{p,*} - \mathbf{C}_k^{p,*}(\mathbf{H}_k^*)^T \left\{\mathbf{G}_k^* \mathbf{C}_k^{v,*} (\mathbf{G}_k^*)^T \right. \\ &\quad \left. + \mathbf{H}_k^* \mathbf{C}_k^{p,*} (\mathbf{H}_k^*)^T \right\}^{-1} \mathbf{H}_k^* \mathbf{C}_k^{p,*} \ , \end{aligned} \qquad (5)$$

where nonzero mean measurement noise has been considered. However, to ensure symmetry and positive definiteness of the covariance matrix $\mathbf{C}_k^{e,*}$, square–root forms of the Kalman filter (Park and Kailath, 1995; Sayed and Kailath, 1994) are a better choice.

The recursion is completely performed in the higher–dimensional space, only the calculation of characteristic values of the estimate is done outside of the recursion loop. Typically, the mean $\hat{\underline{x}}_k^e$ and the covariance matrix $\mathbf{C}_k^e$ are provided, which in general requires numerical computation. However, an efficient algorithm for calculating moments of general exponential densities including pseudo–Gaussian densities by means of differential algebraic equations is given in (Rauh and Hanebeck, 2003).

6. EXAMPLE

To illustrate the proposed filtering algorithm, the following dynamic system with scalar state x_k is considered, which evolves in discrete time according to the linear system equation

$$x_{k+1} = a\, x_k + \hat{u}_k \ , \tag{6}$$

with a known scalar input sequence $\hat{u}_k$. Measurements $\hat{y}_k$ of the system output are related to the system state x_k via the nonlinear measurement equation

$$\hat{y}_k = x_k^2 + v_k \ . \tag{7}$$

The noise distribution is given by a two-dimensional pseudo Gaussian $p_k^v(\underline{v}_k^*)$, i.e., $L_v = 2$, with mean and covariance matrix

$$\hat{\underline{v}}_k^* = \begin{bmatrix} 0.5 \\ 2 \end{bmatrix} \ , \ \mathbf{C}_k^{v,*} = \begin{bmatrix} 1 & 0.5 \\ 0.5 & 2 \end{bmatrix} \ ,$$

where $\underline{v}_k^*$ is selected as

$$\underline{v}_k^* = \begin{bmatrix} v_k \\ v_k^2 \end{bmatrix}$$

for illustration purposes. $p_k^v(v_k)$ is visualized in Fig. 4.

Transformation of the original nonlinear measurement equation (7) and the system equation (6) according to Section 4 yields a linear measurement equation

$$\underbrace{\begin{bmatrix} \hat{y}_k \\ \hat{y}_k^2 \end{bmatrix}}_{\hat{\underline{y}}_k^*} \underbrace{\begin{bmatrix} 0 & 1 & 0 & 0 \\ 0 & 0 & 0 & 1 \end{bmatrix}}_{\mathbf{H}_k^*} \underbrace{\begin{bmatrix} \hat{x}_k \\ \hat{x}_k^2 \\ \hat{x}_k^3 \\ \hat{x}_k^4 \end{bmatrix}}_{\underline{x}_k^*} + \underbrace{\begin{bmatrix} 1 & 0 \\ 2\,\hat{y}_k & -1 \end{bmatrix}}_{\mathbf{G}_k^*} \underbrace{\begin{bmatrix} v_k \\ v_k^2 \end{bmatrix}}_{\underline{v}_k^*}$$

and a linear system equation

$$\underbrace{\begin{bmatrix} x_{k+1} \\ x_{k+1}^2 \\ x_{k+1}^3 \\ x_{k+1}^4 \end{bmatrix}}_{\underline{x}_{k+1}^*} = \underbrace{\begin{bmatrix} a & 0 & 0 & 0 \\ 2a\hat{u}_k & a^2 & 0 & 0 \\ 3a\hat{u}_k^2 & 3a^2\hat{u}_k & a^3 & 0 \\ 4a\hat{u}_k^3 & 6a^2\hat{u}_k^2 & 4a^3\hat{u}_k & a^4 \end{bmatrix}}_{\mathbf{A}_k^*} \underbrace{\begin{bmatrix} x_k \\ x_k^2 \\ x_k^3 \\ x_k^4 \end{bmatrix}}_{\underline{x}_k^*} + \underbrace{\begin{bmatrix} 1 & 0 & 0 & 0 \\ 0 & 1 & 0 & 0 \\ 0 & 0 & 1 & 0 \\ 0 & 0 & 0 & 1 \end{bmatrix}}_{\mathbf{B}_k^*} \underbrace{\begin{bmatrix} \hat{u}_k \\ \hat{u}_k^2 \\ \hat{u}_k^3 \\ \hat{u}_k^4 \end{bmatrix}}_{\hat{\underline{u}}_k^*}$$

in a higher-dimensional space with $L_x = 4$ and $L_u = 4$.

The state of the system given by (7) and (6) can now be estimated by means of a Kalman filter in the L_x-dimensional space according to Fig. 3 with the prediction step given by (4) and the filter step given by (5).

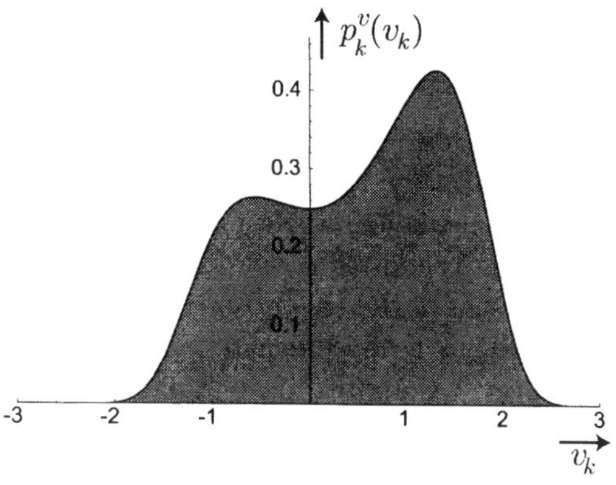

Fig. 4. Pseudo Gaussian noise density $p_k^v(v_k)$ discussed in the example.

Numerical simulations of a scalar system comprising a cubic measurement equation and a linear system equation are given in (Hanebeck, 2001*a*).

7. CONCLUSIONS

Estimating the state of a nonlinear dynamic system from measurements corrupted by non-Gaussian measurement noise is reformulated as a linear filtering problem in a higher-dimensional space. This is similar to the concept of support vector machines (Schölkopf, 1998), which perform nonlinear classification by means of linear hyperplane classifiers in a higher-dimensional space nonlinearly related to the input or problem space.

For that purpose, a specific type of exponential probability density function, the so-called pseudo Gaussian density, is introduced for representing the resulting non-Gaussian posterior densities. Furthermore, the measurement nonlinearity is converted into a linear measurement equation in a higher-dimensional space. Hence, proven linear filtering techniques can be employed for solving the nonlinear estimation problem.

A similar approach has been proposed for the case of nonlinear set-theoretic estimation in the case of unknown-but-bounded noise descriptions (Hanebeck, 2001*b*) and evaluated extensively in applications (Horn *et al.*, 2002; Briechle and Hanebeck, 2003).

8. EXTENSIONS

So far, an exact expansion of the measurement nonlinearity was assumed to exist. In that case a sufficient statistic is provided by the mean vectors and the covariance matrices of the pseudo

Gaussians used to represent the posterior densities. However, in many practical applications it is not possible to use an exact expansion of the nonlinearity. In addition, an approximation may be desirable to keep the dimensions of the hyperspace low even when an exact expansion is known. The resulting pseudo Gaussian densities are then approximations of the true posterior densities and are described by a nonsufficient or reduced statistic. However, an approximate expansion can be selected in such a way that a certain distance, e.g. the Kullback–Leibler distance, between the approximate and the exact posterior is minimized.

The proposed technique can also be applied to more complex additive noise descriptions, for example colored noise or noise with partially known statistics. These problems can be solved analogously by applying the appropriate linear filter in the higher–dimensional space, e.g. (Hanebeck *et al.*, 1999; Hanebeck and Horn, 2000; Julier and Uhlmann, 1997).

9. REFERENCES

Alspach, D. L. and H. W. Sorenson (1972). Nonlinear Bayesian Estimation Using Gaussian Sum Approximation. *IEEE Transactions on Automatic Control* **AC–17**(4), 439–448.

Anderson, B. D. O. and J. B. Moore (1979). *Optimal Filtering.* Prentice–Hall.

Bergman, N., L. Ljung and F. Gustafsson (1999). Terrain Navigation Using Bayesian Statistics. *IEEE Control Systems Magazine* **19**(3), 33–40.

Bohn, C. and H. Unbehauen (2000). The Application of Matrix Differential Calculus for the Derivation of Simplified Expressions in Approximate Non–Linear Filtering Algorithms. *Automatica* **36**(10), 1553–1560.

Briechle, K. and U. D. Hanebeck (2003). Localization of Mobile Robots Using Relative Bearing Measurements. *to appear in: IEEE Transactions on Robotics and Automation.*

Bucy, R. S. and K. D. Senne (1971). Digital Synthesis of Non–linear Filters. *Automatica* **7**, 287–298.

Challa, S., Y. Bar-Shalom and V. Krishnamurthy (2000). Nonlinear Filtering via Generalized Edgeworth Series and Gauss–Hermite Quadrature. *IEEE Transactions on Signal Processing* **48**(6), 1816–1820.

Doucet, A., S. Godsill and C. Andrieu (2000). On Sequential Monte Carlo Sampling Methods for Bayesian Filtering. *Statistics and Computing* **10**(3), 197–208.

Hanebeck, U. D. (2001*a*). Optimal Filtering for Polynomial Measurement Nonlinearities with Additive Non-Gaussian Noise. In: *Proceedings of the American Control Conference (ACC'2001), Arlington, Virginia.*

Hanebeck, U. D. (2001*b*). Recursive Nonlinear Set–Theoretic Estimation Based on Pseudo–Ellipsoids. In: *Proceedings of the IEEE Conference on Multisensor Fusion and Integration for Intelligent Systems (MFI'2001), Baden–Baden.* pp. 159–164.

Hanebeck, U. D. and J. Horn (2000). Fusing Information Simultaneously Corrupted by Uncertainties with Known Bounds and Random Noise with Known Distribution. *Information Fusion, Elsevier Science* **1**(1), 55–63.

Hanebeck, U. D., J. Horn and G. Schmidt (1999). On Combining Statistical and Set Theoretic Estimation. *Automatica* **35**(6), 1101–1109.

Horn, J., U. D. Hanebeck, K. Riegel, K. Heesche and W. Hauptmann (2002). Nonlinear Set–Theoretic Position Estimation of Cellular Phones. *Ortung und Navigation, Deutsche Gesellschaft für Ortung und Navigation e.V. (DGON)* (1), 93–99.

Jazwinski, A. H. (1970). *Stochastic Processes and Filtering Theory.* Academic Press.

Julier, S. and J. K. Uhlmann (1997). A Nondivergent Estimation Algorithm in the Presence of Unknown Correlations. In: *Proceedings of the American Control Conference (ACC'1997).*

Kulhavý, R. (1992). Recursive Nonlinear Estimation: Geometry of a Space of Posterior Densities. *Automatica* **28**(2), 313–323.

Liu, J. S. and R. Chen (1998). Sequential Monte Carlo Methods for Dynamic Systems. *Journal of American Statistical Association* **93**, 1032–1043.

Park, P. and T. Kailath (1995). New Square–Root Algorithms for Kalman Filtering. *IEEE Transactions on Automatic Control* **40**(5), 895–899.

Rauh, A. and U. D. Hanebeck (2003). Calculating Moments of Exponential Densities Using Differential Algebraic Equations. *to appear in: IEEE Signal Processing Letters.*

Sayed, A. H. and T. Kailath (1994). A State–Space Approach to Adaptive RLS Filtering. *IEEE Signal Processing Magazine* **11**(3), 18–70.

Schölkopf, B. (1998). Support Vector Machines – A Practical Consequence of Learning Theory. In: M. A. hearst, B. Schölkopf, S. Dumais, E. Osuna, J. Platt. Trends and Controversies – Support Vector Machines. *IEEE Intelligent Systems* **13**(4), 18–28.

Sorenson, H. W. and A. R. Stubberud (1968). Non–Linear Filtering by Approximation of the A Posteriori Density. *International Journal of Control* **8**(1), 33–51.

IFAC Publications
www.elsevier.com/locate/ifac

A TOTAL LEAST SQUARES APPROACH TO SENSOR CHARACTERISATION

Peter C.F. Hung*, Dr. Seán McLoone#, Prof. George Irwin*, Dr. Robert Kee*

*Virtual Engineering Centre**
The Queen's University of Belfast
Belfast, Northern Ireland
United Kingdom, BT9 5HN

#Department of Electronic Engineering
National University of Ireland, Maynooth
Maynooth, Co. Kildare
Ireland

E-mail: p.hung@qub.ac.uk

Abstract: The use of robust, low-bandwidth sensors makes exhaust gas temperature variations difficult to measure in internal combustion engines. One common solution involves measuring gas temperature using two thermocouples with different time-constants and estimating the time-constants from the resulting signals. This assumes that the ratio of the thermocouple time-constants is invariant and known *a priori*. In addition they are generally subject to singularities and sensitive to noise. This paper presents a novel total least squares (TLS) *difference equation* based characterisation method. It makes no such assumption and is potentially superior to existing methods in terms of time-constant estimation accuracy and noise tolerance. *Copyright © 2003 IFAC*

Keywords – Least Squares, Sensor, System Identification

1. INTRODUCTION

Gas temperature in combustion engines is an important signal for gaining insights into combustion phenomena, and evaluating engine performance and as a potential input to advanced engine management systems (Kee, *et al.*, 1999). Thermocouples are widely used as temperature measurement devices due to their low cost, robustness, ease of installation and reliability, but their design, a compromise between robustness and speed of response, poses major problems when measuring high frequency temperature fluctuations. The bandwidth of a thermocouple is dependent on its diameter according to the equation

$$w_B = \kappa d^{m-2} v^m \qquad (1)$$

where κ and m are constants (approximately) arising from thermodynamic considerations, d is the diameter of the thermocouple wire and v is the velocity of the gas. Large diameter thermocouples are usually required to withstand the harsh environment in combustion engine systems, resulting in low bandwidth sensors, typically less than 1 Hz. Temperature variations, such as those in the exhaust of a reciprocating internal combustion engine, are usually 2 to 3 orders of magnitude faster than this. This leads to raw signal measurements that are severely attenuated and lagged. For such signals, some form of dynamic compensation must therefore be employed to reconstruct the actual gas temperature. In theory, assuming a sensor with first order dynamics and time constant τ, the true gas temperature ($T_g(t)$) can be reconstructed from the measured temperature ($T_m(t)$) and its derivative with respect to time ($dT_m(t)/dt$) using the equation

$$T_g(t) = T_m(t) + \tau \frac{dT_m(t)}{dt}. \qquad (2)$$

In practice this approach is infeasible as the time constant τ, being a function of the physical properties of the gas and its velocity as well as the geometrical and physical properties of the thermocouple, is generally time varying and unknown *a priori*. Furthermore, measurements are generally corrupted by noise, making $dT_m(t)/dt$ difficult to compute accurately.

To carry out thermocouple sensor characterization, its time constant should be accurately measured. Traditionally, this involves studying the thermocouple cooling curve, which can be obtained by applying a step temperature change to the junction. The time constant can then be estimated from the exponentially decreasing thermocouple temperature profile. Unfortunately this method becomes impractical for high bandwidth thermocouples, as their cooling curves fall off too quickly for accurate characterizations.

Several time and frequency domain approaches have been developed to tackle the problem of time constant estimation (sensor characterisation) and subsequent temperature reconstruction on the basis of measurements taken from two or more thermocouples with different time-constants (Kee, *et al.*, 1999; Tagawa and Ohta, 1997; Forney and Fralick, 1994). These two-thermocouple probe methods rely on the restrictive assumption that the ratio of the thermocouple time-constants is invariant and known *a priori*. They are also subject to singularities and sensitive to noise.

A novel *difference equation* based method is proposed, which makes no such assumption. Yet it is an improvement to existing methods in terms of time constant estimation accuracy and noise tolerance. It has been tested in simulation on a composite signal similar to exhaust gas fluctuations and shown to give very good time constant estimates at low noise levels (Hung, 2002). Unfortunately the basic method, which involves identifying the parameters of a specially constructed ARX model, produces increasingly biased estimates as the noise level increases. Biased estimation in the presence of noise, even if it is zero mean and Gaussian, is a well-known problem with least squares identification of ARX models. The problem is particularly severe here as both the input and output of the ARX model to be identified are subject to noise corruption. Instrumented variable (IV) and output error (OE) methods, which address the problem of output signal noise, were investigated in Hung (2002) with some success. However, the capabilities of all these approaches are limited in the presence of input noise.

This paper investigates the use of total least squares (TLS) identification to produce improved sensor characterizations. TLS specifically addresses the problem of parameter estimation in the presence of input and output noise and produces unbiased estimates when the noise is zero mean and Gaussian.

The remainder of the paper is organised as follows. Section 2 gives a brief overview of existing two-thermocouple probe (TTP) reconstruction techniques. In particular, *difference equation* based sensor characterisation is introduced. The new TLS approach is discussed in Section 3. Simulation results are then presented in Section 4 comparing the new approach to three existing time-constant estimation methods. Finally Section 5 gives a summary and conclusions.

2. EXISTING TWO-THERMOCOUPLE PROBE (TTP) RECONSTRUCTION TECHNIQUES

The basis for TTP temperature reconstruction techniques is that both thermocouples are subject to the same environmental conditions, i.e. the same gas temperature $T_g(t)$ and gas velocity v. It therefore follows from Equation (1) that the ratio of their time constants, given by

$$\alpha = \frac{\tau_1}{\tau_2} = \frac{\omega_{B2}}{\omega_{B1}} = \frac{\kappa d_2^{m-2} v^m}{\kappa d_1^{m-2} v^m} = \left(\frac{d_1}{d_2}\right)^{2-m}, \quad \alpha < 1, \tag{3}$$

is a function of thermocouple geometry only and therefore approximately invariant. Here subscripts 1 and 2 are used to distinguish between the two thermocouples and the corresponding thermocouple models are given by

$$T_g(t) = T_{m1}(t) + \tau_1 \dot{T}_{m1}(t) \tag{4}$$

and

$$T_g(t) = T_{m2}(t) + \tau_2 \dot{T}_{m2}(t) \tag{5}$$

respectively. Noting that

$$\frac{T_g(t) - T_{m1}(t)}{T_g(t) - T_{m2}(t)} = \alpha \frac{\dot{T}_{m1}(t)}{\dot{T}_{m2}(t)} \tag{6}$$

and assuming knowledge of α, $T_g(t)$ can be estimated directly using

$$T_g(t) = \frac{T_{m1}(t)\dot{T}_{m2}(t) - \alpha \dot{T}_{m1}(t) T_{m2}(t)}{\dot{T}_{m2}(t) - \alpha \dot{T}_{m1}(t)}. \tag{7}$$

This formula, although simple, requires instantaneous derivative estimates and has a singularity due to the form of the denominator leading to numerical difficulties when reconstructing $T_g(t)$ from noisy measurements.

2.1 Time Domain Reconstruction (TDR)

Tagawa and Ohta (1997) describe a method for estimating τ_1 and τ_2 based on minimising the time-averaged, mean squared difference between the reconstructions generated by (4) and (5), that is

$$\tau_1, \tau_2 = \arg\{ \min_{\tau_1,\tau_2} [J(\tau_1, \tau_2)] \} \tag{8}$$

where

$$J(\tau_1, \tau_2) = \frac{1}{N} \sum_{i=1}^{N} \begin{bmatrix} T_{m1}(t_i) + \tau_1 \dot{T}_{m1}(t_i) \\ -T_{m2}(t_i) - \tau_2 \dot{T}_{m2}(t_i) \end{bmatrix}^2 \tag{9}$$

for N data sample points. The resulting time constant formulae

$$\tau_1 = \frac{E[\dot{T}_{m2}^2]E[\dot{T}_{m1}\Delta T_m] - E[\dot{T}_{m1}\dot{T}_{m2}]E[\dot{T}_{m2}\Delta T_m]}{E[\dot{T}_{m1}^2]E[\dot{T}_{m2}^2] - E[\dot{T}_{m1}\dot{T}_{m2}]^2},$$
$$\tau_2 = \frac{E[\dot{T}_{m1}\dot{T}_{m2}]E[\dot{T}_{m1}\Delta T_m] - E[\dot{T}_{m1}^2]E[\dot{T}_{m2}\Delta T_m]}{E[\dot{T}_{m1}^2]E[\dot{T}_{m2}^2] - E[\dot{T}_{m1}\dot{T}_{m2}]^2}, \tag{10}$$

where $\Delta T_m = T_{m2} - T_{m1}$, avoid the singularity problems of (7), do not require knowledge of α and work well on noise free data. (Here $E[.]$ indicates time averaged values computed over N data points). In practice, however, the formulae have been found to be unreliable when applied to noisy data, underestimating time constants and in some cases generating infeasible results such as negative values (Kee, *et al.*, 1999; Tagawa, *et al.*, 1998).

Kee, *et al.* (1999) propose a modification to the Tagawa-Ohta (1997) approach which results in a much more robust formula. By introducing the time constant ratio α, expression (8) reduces to a function of τ_2 only and is minimised when

$$\tau_2 = \frac{E[(T_{m2} - T_{m1})(\alpha\dot{T}_{m1} - \dot{T}_{m2})]}{E[(\alpha\dot{T}_{m1} - \dot{T}_{m2})^2]}, \tag{11}$$
$$\tau_1 = \alpha\tau_2$$

(Hung, *et al.*, 2002).

While singularities are still possible in this method, experience has shown it to be more robust than other TDR techniques. It should be noted that measurement signals $T_{m1}(t)$ and $T_{m2}(t)$ are pre-filtered before application of any TDR approaches. Kee, *et al.* (1999) recommend the use of polynomial smoothing on a sliding data window, as this avoids introducing phase shift and facilitates the estimation of signal derivatives.

2.2 Frequency Domain Reconstruction

Forney and Fralick (1994) propose the use of Fast Fourier Transforms (FFT) as a means of avoiding the use of derivatives in the reconstruction of $T_g(t)$. In the frequency domain, the real gas temperature can be expressed as:

$$\overline{T}_g = \frac{\overline{T}_{m1} \cdot \overline{T}_{m2}(\alpha - 1)}{\alpha\overline{T}_{m1} - \overline{T}_{m2}}, \tag{12}$$

where $\overline{T}$, denotes FFT($T(t)$). Thus, $T_g(t)$ can be found by simply taking the inverse FFT:

$$T_g(t) = FFT^{-1}(\overline{T}_g). \tag{13}$$

This method is not suitable for temperature reconstruction when inputs are not purely sinusoidal. Moreover, singularities can still occur due to noise. As a result, large amplitude non-existent oscillations can occur in the reconstructed EGT (Kee, *et al.*, 1999).

2.3 Difference Equation Reconstruction (DER)

Hung, *et al.* (2002) proposed an approach based on discrete-time system identification techniques. This method has the advantage that it makes no assumption of time constant ratio being time-invariant and known *a priori*.

The first-order difference equation equivalent to the continuous-time, single thermocouple model (2) is given by

$$T_m(k) = aT_m(k-1) + bT_g(k-1). \tag{14}$$

Assuming ZOHs and a sample rate τ_s, the parameters of the discrete and continuous models are related by

$$a = \exp(-\frac{\tau_s}{\tau}),\ b = 1 - a. \tag{15}$$

Equation (14) is the conventional ARX model structure (Ljung, 1999) and its parameters can be estimated using standard linear, least-squares system identification techniques if an appropriate set of input-output samples is provided. Of course in this application $T_g(k-1)$ is unknown, hence a and b cannot be determined directly. However, a TTP based identification strategy can be developed as follows. The ARX models for the two thermocouples defined in Equations (4) and (5) are given by

$$T_{m1}(k) = a_1 T_{m1}(k-1) + (1-a_1)T_g(k-1) \quad (16)$$

and

$$T_{m2}(k) = a_2 T_{m2}(k-1) + (1-a_2)T_g(k-1) \quad (17)$$

respectively. Using (16), $T_g(k-1)$ can be eliminated from (17) yielding the following difference equation relationship between the thermocouple outputs:

$$\begin{aligned} T_{m2}(k) &= a_2 T_{m2}(k-1) + \frac{(1-a_2)}{(1-a_1)} T_{m1}(k) \\ &- \frac{a_1(1-a_2)}{(1-a_1)} T_{m1}(k-1) \end{aligned} \quad . \quad (18)$$

By minimising the mean-squared prediction error

$$J(a_1, a_2) = \frac{1}{N} \sum_{k=1}^{N} (T_{m2}(k) - \hat{T}_{m2}(k))^2 \quad (19)$$

where $\hat{T}_{m2}(k)$ is the prediction generated by (18), parameters a_1 and a_2 can now be identified. However, since this would require the use of non-linear optimisation techniques, a better strategy is to convert (18) into a three-parameter linear ARX model by defining

$$\gamma_1 = a_2,\ \gamma_2 = \frac{1-a_2}{1-a_1},\ \gamma_3 = -a_1 \frac{1-a_2}{1-a_1}. \quad (20)$$

Equation (20) then becomes

$$\begin{aligned} T_{m2}(k) &= \gamma_1 T_{m2}(k-1) + \gamma_2 T_{m1}(k) \\ &+ \gamma_3 T_{m1}(k-1) \end{aligned} \quad . \quad (21)$$

By choosing $T_{m2}(k)$ as the output variable and $T_{m1}(k)$ as the input variable, the ARX structure illustrated in Figure 1 is obtained. Conventional linear system identification tools can now be used to determine the parameters γ_1, γ_2 and γ_3.

Thus, the difference equation reconstruction (DER) technique can be summarised as follows: Estimate parameters γ_1, γ_2 and γ_3. Then compute a_1 and a_2 using (20). Having determined a_1 and a_2, temperature reconstruction can be performed using (16) or (17) or some combination of both. Alternatively the corresponding time constants τ_1 and τ_2 can be calculated using (15) and $T_g(t)$ reconstructed using one of the TDR formulae described in Section 2.

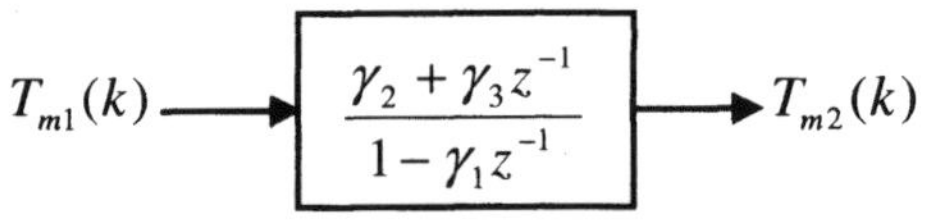

Fig. 1. Equivalent ARX model for TTP parameter identification

2.4 Problems with existing DER

As discussed in the introduction, DER is characterised by an ARX identification problem in which both the input and output signals are subject to noise (assumed to be zero mean Gaussian white noise) leading to biased estimates with conventional least squares. Other models, which take account of coloured noise on the output, such as ARMAX and OE were evaluated in Hung, *et al.* (2002) and indeed lead to significant reductions in estimation bias. However, as already noted, these methods assume output noise only and are therefore limited in capability. To further improve DER sensor characterizations, both input and output noise must be dealt with simultaneously. A total least squares (TLS) solution to this problem is presented in the next section.

3. TOTAL LEAST SQUARES (TLS) PARAMETER ESTIMATION

TLS is proposed because theoretically it gives unbiased estimates in the presence of zero-mean Gaussian white noise on both input and output signals (Rao and Principe, 2002).

Rewriting (21) in vector-matrix form gives:

$$\begin{bmatrix} T_{m2}(k-1) \\ T_{m1}(k) \\ T_{m1}(k-1) \end{bmatrix}^T \cdot \begin{bmatrix} \gamma_1 \\ \gamma_2 \\ \gamma_3 \end{bmatrix} = [T_{m2}(k)]. \quad (22)$$

where $k = 1, 2, \cdots N$. One method to obtain the numerical TLS solution involves computing a SVD (singular value decomposition) of the augmented $N \times 4$ data matrix $S = \begin{bmatrix} T_{m2}(k-1) \\ T_{m1}(k) \\ T_{m1}(k-1); \\ T_{m2}(k) \end{bmatrix}^T$. This can be very computationally intensive when there are a large number of parameters. However, since only three parameters are to be estimated in DER (20), the computational load is not significant. In addition, SVD is desirable as it can deal effectively with ill-conditioned matrices.

The SVD of matrix S is

$$S = U \Sigma V^T, \tag{23}$$

where

$$\begin{aligned} U &= [\mathbf{u}_1, \mathbf{u}_2, \mathbf{u}_3, \cdots \mathbf{u}_N], \quad U^T U = I_N \\ V &= [\mathbf{v}_1, \mathbf{v}_2, \mathbf{v}_3, \mathbf{v}_4], \quad V^T V = I_4 \quad , \\ \Sigma &= \begin{bmatrix} diag(\sigma_1, \sigma_2, \sigma_3, \sigma_4) \\ O_{(N-4)\times 4} \end{bmatrix} \end{aligned} \tag{24}$$

with

$$\sigma_1 \geq \sigma_2 \geq \sigma_3 \geq \sigma_4, \quad \sigma_4 > 0. \tag{25}$$

In order to obtain a solution, the rank of S is reduced from 4 to 3 by making $\sigma_4 = 0$. The TLS parameter estimates $\hat{\gamma}_i$ are then computed as:

$$\begin{bmatrix} \hat{\gamma}_1 \\ \hat{\gamma}_2 \\ \hat{\gamma}_3 \end{bmatrix} = -\frac{1}{v_{4,4}} \cdot \mathbf{v}_4 \tag{26}$$

where $v_{4,4}$ is the last element of the minor eigenvector $\mathbf{v}_4$. The parameters a_1 and a_2 can then be determined using (20).

Provided the variance of the noise on both the input and output signals are the same, this implementation of TLS produces unbiased estimates (Rao and Principe, 2002).

4. RESULTS

To test the feasibility of the proposed method, simulation studies were carried out in MATLAB©. The two-wire thermocouple system was represented as first-order transfer functions fed by the same input (Figure 2) and with time constants τ_1 and τ_2 equal to 0.6 seconds and 0.9 seconds respectively. A composite signal consisting of a sum of sinusoids and a transient pulse was used to represent the fluctuating input exhaust gas temperature (EGT) (Figure 3). Zero-mean, Gaussian white noise was added to each thermocouple model output to represent measurement noise, and the resulting signals sampled every 0.1 seconds over a 50 second interval.

Using this platform, the Difference Equation (DER) and Time Domain Reconstruction (TDR-Kee) techniques described in Section 2 were compared to the TLS presented in Section 3 for a range of noise levels. The noise level, K, which is the same for both thermocouples, is defined as:

$$K = \frac{N_{RMS}}{S_{RMS}} \times 100\% \tag{27}$$

where S_{RMS} and N_{RMS} are the RMS values of the signal and noise respectively.

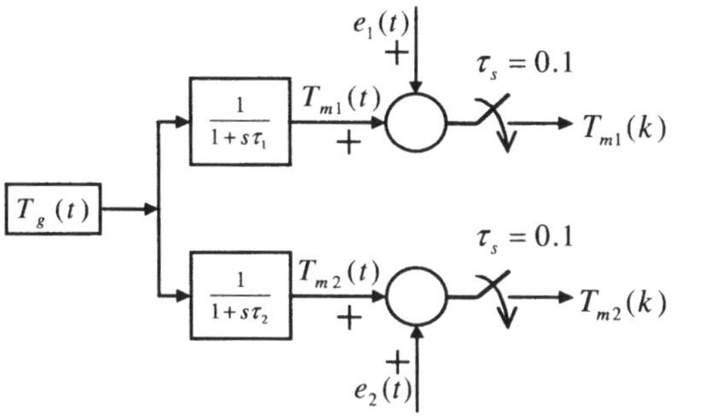

Fig. 2. Block diagram representation of the TTP measurement system

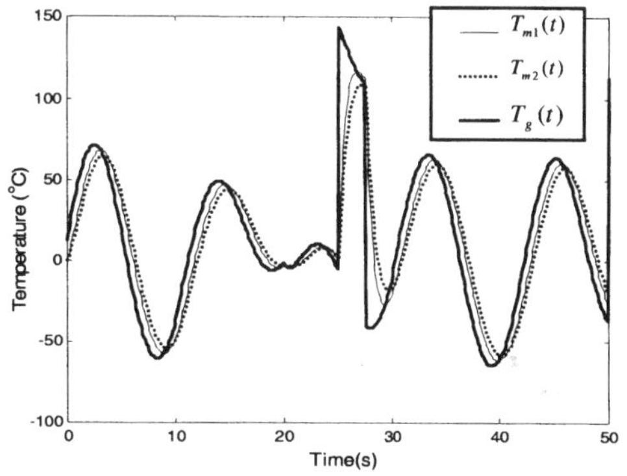

Fig. 3. TTP response for a composite signal

Both the TDR-Kee and TLS algorithms were implemented using standard MATLAB functions. In contrast, the ARX model was estimated using the linear least-squares based *arx* command from MATLAB's System Identification Toolbox. Similarly, the OE model was estimated via the PEM (Prediction Error Method) based Toolbox command *oe*.

For a given noise level, the performance of each method was assessed in terms of the percentage absolute time-constant estimation error ($e_{\tau i}$)

$$e_{\tau i} = \frac{\tau_i - \hat{\tau}_i}{\tau_i} \times 100\% \tag{28}$$

and averaged over a hundred runs. Here τ_i is the true value and $\hat{\tau}_i$ is the estimate.

The means and standard deviations of the percentage time constant errors for τ_i obtained with each approach are given in Figure 4. Table 1 gives the corresponding 1% error bounds.

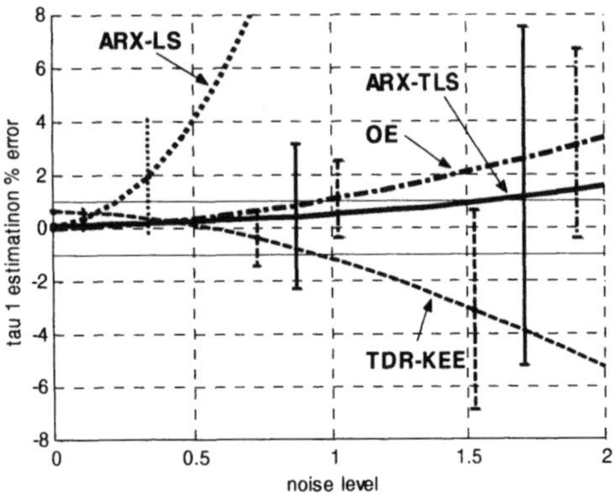

Fig. 4. Variation in mean and standard deviation of $e_{\tau 1}$ with noise level for various methods

Table 1. 1% error bounds on means and standard deviations of $e_{\tau 1}$ for various methods

Method	Mean bound	S. D. bound
TDR-Kee	0.95	0.65
ARX-LS	0.23	0.13
OE	0.95	0.52
ARX-TLS	1.55	0.27

At low noise levels (K<0.5), most algorithms work well, except ARX-LS. All other DER algorithms perform better than TDR-Kee. The graph shows the biased DER estimates, a well-known failing of least squares identification (Section 2.4). It should be emphasized that the value of the time constant ratio (α), a critical TDR input parameter, is *not required* for DER characterisations. This implies that TDR is sensitive to poor α estimations, while it will not have any impact on DER performance.

Employing more sophisticated identification formulations such as Output-Error (OE) does improve the estimations. However, using TLS reduces the bias even more. This shows that the use of TLS to tackle both noisy input and output in the ARX models correctly addresses the biased issue. At high noise levels (K>2), the tendency for TLS to give less consistent estimates may be due to heavy corruption of inputs and TLS is unable to distinguish signals from both thermocouples.

5. CONCLUSIONS

A novel TLS *difference equation* based method for two-thermocouple sensor characterisation has been presented. Unlike most existing techniques, DER makes no assumptions about the time constants or their ratio. The use of TLS in the simple ARX model significantly reduces the bias in estimations. Simulation results indicate that DER is also superior to existing methods in terms of time constant estimation accuracy and noise tolerance when appropriate identification procedures are used.

Future research will include the development of recursive real-time implementations of DER and validating all thermocouple characterisation techniques using actual gas measurements from a test rig and an engine test bed.

6. ACKNOWLEDGEMENTS

The first author wishes to acknowledge the financial support of the Virtual Engineering Centre, Queen's University of Belfast, *http://www.vec.qub.ac.uk.*

REFERENCES

Forney, L.J., G.C. Fralick (1994). Two wire thermocouple: Frequency response in constant flow. *Rev. Sci. Instrum.*, **65**, pp 3252-3257.

Hung, P., S. McLoone, R. Kee and G. Irwin (2002). A Novel Approach to Two-wire Thermocouple Temperature Reconstruction. *Proc. Irish Signals and Systems Conference*, **2002**, pp 193-198.

Hung, P.C.F. (2002). Two-wire thermocouple sensor characterization. *Proc. UKACC Postgraduate Symposium*, **2002**, pp 13-18.

Kee R.J, P.G. O'Reilly, R. Fleck and P.T. McEntee (1999). Measurement of Exhaust Gas Temperature in a High Performance Two-Stroke Engine. *SAE Trans. J. Engines*, **107**, SAE Paper No. 983072.

Ljung, L. (1999). *System identification: Theory for the user, second edition.* Prentice-Hall, New Jersey.

Rao, Y.N., J.C. Principe (2002). Efficient Total Least Squares Method for System Modeling Using Minor Component Analysis. *Proc. IEEE Workshop on Neural Networks for Signal Processing XII*, **2002**, pp 259-267.

Tagawa, M. and Y. Ohta (1997). Two-Thermocouple Probe for Fluctuating Temperature Measurement in Combustion – Rational Estimation of Mean and Fluctuating Time Constants. *Combustion and Flame*, **109**, pp 549-560.

IFAC
Publications
www.elsevier.com/locate/ifac

Estimation and Validation of Semi-Parametric Dynamic Nonlinear Models.

Yves ROLAIN, Wendy VAN MOER and Johan SCHOUKENS
Vrije Universiteit Brussel (VUB, Dept. ELEC/TW); Pleinlaan 2; B-1050 Brussels (Belgium)
Phone: +32.2.629.29.44; Fax: +32.2.629.28.50; e-mail: Yves.Rolain@vub.ac.be

Abstract

An approach for measurement based modelling of nonlinear devices is proposed that extends the mixture of parametric models and nonparametric verification, that is common for LTI systems, to a class of nonlinear systems. The applicability of the method is illustrated on the baseband modelling of an radio-frequency amplifier over a wide power and frequency range. Copyright © 2003 IFAC

keywords: identification, nonlinear systems, Volterra models.

I. Introduction

During the last years, there is an increasing interest in modelling of the nonlinear behavior of subsystems that are 'close' to be linear. Mainly, these systems were designed to be linear. However, a description of the deviation from this ideal behavior is vital for the evaluation or simulation of the performance of the global system to which the considered system belongs. In telecommunication applications, for example, the data error rate is linked to the in-band distortion of the power amplifier used in the transmitter. As a consequence, a nonlinear model for the operation of the device in the baseband (the neighborhood of the fundamental frequency) is a needed by the RF system designers to analyze, optimize and tune the system.

For such a system, there is a lot of prior model information available. The model class that is considered here is restricted further to the class of the systems that, when excited by a periodic excitation, produce a periodic output with the same period. This class of systems is called the NICE systems from now on. It is clear that this definition excludes 'nasty' behavior such as chaos or bifurcation.

In applications such as modelling for telecommunication, the class of signals that is used during device operation is known in advance. The class of narrowband modulated signals (signals with a modulation bandwidth of a few percent proportional to the carrier frequency) covers most practical applications and will be used here.

To obtain a model that is useful in a design context, one needs more than 'just a model'. It should be validated in the frequency band and the power range where the model will be used later on. Therefore, a robust visualization and validation tool to compare the measured system response to the estimated model response is mandatory. In this paper, a measurement based, non-parametric validation tool based on the Volterra theory is proposed.

Parametric models for NICE systems based on measured input/output characteristics were already obtained earlier, following two main tracks: the polynomial models (including Volterra models) and the behavioral models such as neural nets or wavelet models ([2],[3]). The approach used here is a combination of both approaches: it extends Volterra type models with neural net like kernel functions to obtain a parsimonious model for devices operating under hard nonlinear conditions.

The proposed three step approach consisting of experiment design, measurement visualization and finally model extraction and validation is illustrated experimentally on the identification of a power amplifier in a frequency band from 500 to 2500 MHz and for an input power range from -15 to 5 dBm.

II. Measurement and Experiment design.

The signal has to persistently excite the device over the user specified frequency and power range to be eligible as an excitation signal for model identification. Both spectrally rich signals and sine waves can be used to meet this.

Spectrally rich excitations - such as random multisines - are used extensively for baseband devices up to a few MHz. These signals enable a fast device characterization, over the analysis frequency band in 1 single take. Inclusion of a power sweep realizes the requested experiment. Such signals can also be used for the narrow-band modulation considered here. For the considered application, noise excitation is not an option as the measurement setup requires periodic excitation signals to allow measurement.

Using sine waves, both the frequency and the power have to be swept. This results in much higher number of experiments to be performed, as 1 full experiment is required for each (power, frequency) pair.

The information that is obtained by both excitation signals is more or less the same in both experiments, due to the bandpass nature of the excitation signal. Consider an narrowband modulated signal as in Figure 1 where $U(\omega)$ is the applied input and $Y(\omega)$ the system response.

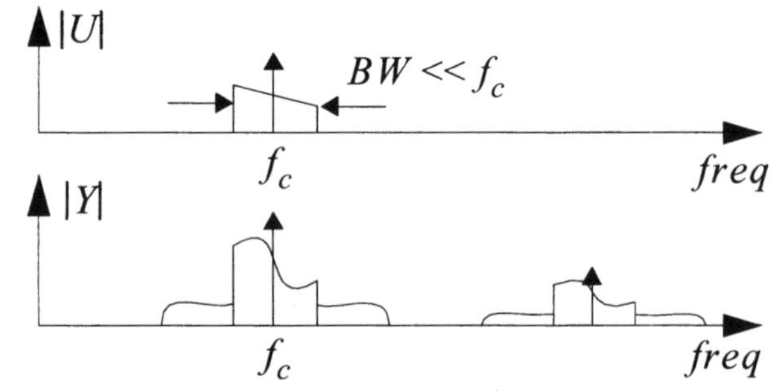

Figure 1 : The excitation spectrum

The bandpass input signal ($BW << f_c$) is formalized as a periodic signal, viz.

$$\begin{aligned} U(f_k) &= U_k & & \\ U(-f_k) &= \overline{U}_k & f_k = f_c + k\Delta f & \quad -N < k \le N \\ U(f) &= 0 & elsewhere & \end{aligned} \tag{1}$$

where $N = 0$ for a sine wave excitation.

A multicarrier signal is a first choice whenever it is technologically realizable. However, in the RF and microwave example considered here, it is almost impossible to generate an arbitrary, wide band signal. Sine wave generators are the most common signal sources in the GHz frequency range. A first brand of arbitrary waveform generators covering the frequency band from DC to 2GHz became available only very recently.

III. MEASUREMENT VISUALIZATION

Once the measurements are taken at n_F frequencies and n_P power levels, one possesses $n_F n_P$ input and output spectra describing the device under test (DUT). The major question now is how to extract the information out of this data. To this end, a nonparametric approach comparable to the frequency response function (FRF) of the LTI systems is used.

Assume that the Volterra theory gives a qualitatively valid description of the frequency mixing. This is a very weak hypothesis as the system is assumed to be a NICE system.

A sine wave excitation will be used as an example to illustrate the proposed method. Consider the response at frequency f_0:

$$\begin{aligned} Y(f_0) &= \sum_{k=0}^{L} K_k(f_0) U^{k+1}(f_0) \overline{U}^k(f_0) \\ &= U(f_0) \left\{ \sum_{k=0}^{L} K_k(f_0) \| U(f_0) \|_2^{2k} \right\} \end{aligned} \tag{2}$$

The last factor of the right hand side is a polynomial with complex coefficients in the (real) power of the input as an independent variable. Dividing this equation by $U(f_0)$:

$$ETF_1(f_0, |U(f_0)|) = \frac{Y(f_0)}{U(f_0)} = \left\{ \sum_{k=0}^{L} K_k(f_0) |U(f_0)|^{2k} \right\} \tag{3}$$

This quantity is defined here as an energy transfer function, and would be equal to the frf if the system were linear. Due to the nonlinearity of the system, $ETF_1(f, |U(f_0)|)$ is no longer a constant but rather a smooth complex surface that is phase coherent with the input signal. The shape of this surface (both magnitude and phase) gives a lot of insight in the behavior of the device, as can be shown in Figure 2 and 3.

This approach nicely extends to multicarrier excitation signals or contributions that shift the frequency of the output. As an example, consider that a second harmonic ($3f_0$) is also present in the input signal. The output at the fundamental frequency becomes:

$$Y(f_0) = U(f_0)\Phi_1(U) + U(3f_0)\overline{U}(f_0)\overline{U}(f_0)\Phi_2(U) \tag{4}$$

Again, the functions Φ_1 and Φ_2 are functions of a real variable. The difference with the first example is that besides power terms like $U(f_0)\overline{U}(f_0)$, these functions also contain pairs of terms like $U(3f_0)\overline{U}^3(f_0) + \overline{U}(3f_0)U^3(f_0)$, whose sum is again a real variable. A smooth surface that is phase coherent with the input spectrum results again. Evaluating $Y(f_0)/U(f_0)$ and $Y(f_0)/(U(3f_0)\overline{U}(f_0)\overline{U}(f_0))$

only yields a coherent surface in the ranges of (input power, frequency) where the corresponding energy transfer function (Φ_1 or Φ_2) dominates the other contributions. This yields a lot of insight in the operation of the device.

These ideas can be extended to a general response with more than 2 excited lines, at the cost of a complex mathematical formalism that is out of the scope of the paper.

IV. IDENTIFICATION AND MODEL VALIDATION

In the context of this paper, an output error noise model is used. This ensures consistency of the estimates. Using repeated experiments, the variance of the noise source can be used to improve the estimators'efficiency.

For the considered class of NICE systems, a Volterra-like description is flexible enough to yield appropriate models (in mean square sense), but fails to describe device saturation (also called compression). This is very similar to the observation that a polynomial can not give a high quality description for a static system that is driven into deep saturation. Since the systems used in this context are driven up to deep saturation - or high compression levels - the plain Volterra series are bound to the impossible.

To get around this problem, one can start from the equations derived in the previous paragraph. Reading in between the lines, a simple and accurate model should be obtained by replacing the polynomial function Φ by some other real function which matches the effects of amplitude saturation adequately. This new set of functions should behave like a polynomial for moderate power levels, but saturate outside the measured power range.

Here, the basis $U(f)$ is replaced by $U(f)/(g+|U(f)|)$, with g a real gain factor. For the first example used above, the model used in the identification of the semi-parametric identification at test frequency f_l becomes:

$$Y(f_l) = \sum_{k=0}^{L} K_{lk}\left\{\frac{U(f_l)}{g_l+|U(f_l)|}\right\}^{k+1}\left\{\frac{\bar{U}(f_l)}{g_l+|U(f_l)|}\right\}^{k} \quad (5)$$

To extend this model to the general case discussed earlier, each non-zero input spectral line is extended with one separate gain factor.

Three different modelling approaches are tried out here.

- In a first step, a parametric model is estimated for the n_F test frequencies separately. This model is from now on called non-parametric versus frequency.
- The second step proposes a parametric model for g versus frequency. The $(L+1)n_F$ parameters K_{lk} and the parametric representation of g are now estimated in one step over the $n_F n_P$ experiments. This model is called semi-parametric versus frequency.
- Finally, a parametric LTI model is also used to model the K_l versus frequency. The parametric models for all parameters are extracted together for the $n_F n_P$ measurements.

A. Non-Parametric versus frequency.

The model proposed is parametric in the input power but non-parametric in the frequency f_l. One model is therefore evaluated at each measured input frequency f_l. The estimation for each frequency boils down to the optimization of the following cost function:

$$L_N(Y(f_l), U(f_l), \theta) = e_w^H e_w \qquad e_{wk} = \frac{Y_k(f_l) - M(\theta, U_k(f_l))}{W_k(\theta_p)} \quad (6)$$

with N the number of power levels measured at each frequency, $\theta = [K_{l1}, \ldots, K_{lL}, g_l]$ the parameter vector and $M(\theta, U(f_l))$ the model equation as defined in (5). The weight $W_k(\theta)$ is set equal to the experimental standard deviation of the equation error $Y_k(f_l) - M(\theta_p, U_k(f_l))$ obtained from repeated experiments. Note that this model is linear in K_{lj}, g_l is the only parameter that enters the model in a nonlinear way. The model is hence a semi-linear model as considered in [4]. This simplifies the optimization significantly, a 1 dimensional nonlinear optimization remains to be solved after the elimination of the linear parameters. The estimates for this first step, which is non-parametric over the frequency, are:

$$\theta_l = \underset{\theta}{\text{argmin}}(L_N(Y(f_l), U(f_l), \theta)) \quad (7)$$

B. Semi-Parametric versus frequency

In this second step, the parameters g_l are replaced by a parametric polynomial model.

$$g(f) = \sum_{m=0}^{G} \gamma_m f^m \quad (8)$$

The optimization is now performed over the n_F frequencies together, and is still linear in the parameters K_{lk}. Using the elimination scheme of the non-parametric case above for the parameters that appear linearly in the model equation, only the parameters γ_m need to be optimized.

C. Parametric versus frequency

Finally, the non-parametric kernel values K_{lk} are also replaced by a parametric LTI model $K_l(f)$. The parameters that appear linearly in the model are again eliminated as earlier, leaving a reasonable amount of parameters to be optimized. Of course, the resulting set of parameters involves the estimation of Linear Time Invariant systems. This problem is nonlinear in the parameters, but is readily solved using the techniques in [4].

For the validation of the model, the same technique is used as for the visualization of the measurement, but this time modeled and measured validation output are displayed on the same surface plots together with the magnitude of the complex error between model and measurement.

V. EXAMPLE SYSTEM.

A GSM band power amplifier of type MAR6 (Mini Circuits) is modelled. The supply voltage is set to 4V, while the device output is terminated in a 50Ω load. Absolutely calibrated incident and reflected wave spectra at both ports of the DUT are measured by the nonlinear vectorial network analyzer (hp85120a-k60) [1].

The amplifier input is excited by a sine wave. The frequency is stepped from 600 to 1500 MHz in 20 MHz steps. The input power is stepped from -20 to -8 dBm in steps of 0.2 dB. At -8 dBm, the 2 dB compression point is reached in the pass band of the device. Sample variances for the measured spectra are obtained using 5 repeated measurements. The relation

$$ETF_1(f_0, |U(f_0)|) = \frac{Y(f_0)}{U(f_0)} \tag{9}$$

is analyzed. The model equation hence is given by (5).

8736 spectral measurements are now condensed in 1 measured complex ETF_1 surface: magnitude and phase of the surface are shown in Figure 2 and 3. Note that the magnitude of the response has important gain at both 900 and 1800 MHz. The gain compression is much higher close to 900 MHz than to 1800 MHz. The phase dependency shows important phase distortion close to 1800 MHz.

A non-parametric model containing 4 terms is estimated for the 96 test frequencies.

$$Y(f_l) = \sum_{k=0}^{3} K_{lk}\left\{\frac{U(f_l)}{g_l + \|U(f_l)\|^2}\right\}^{k+1}\left\{\frac{\overline{U}(f_l)}{g_l + \|U(f_l)\|^2}\right\}^{k} \tag{10}$$

The model residual is shown in Figure 4. Over the whole band, the relative complex modelling error is below 35 dB. The K_{lk} parameters estimated for all frequencies are shown in Figure 6. Even if the model is very good, the behavior of the gain factor and the linear kernel K_{l0} contain several jumps and can hardly be interpreted in a physical context. This potentially indicates that gain and linear kernel are not totally independent.

Finally, a semi-parametric model with a parametric gain $g = \gamma_0$ is extracted. The resulting modelling error is shown in Figure 5. Note only a slight increase in the residual when compared to the non-parametric case of Figure 4. The K_{lk} parameters estimated for all frequencies are shown in Figure 7. In this plot, the linear kernel K_{l0} clearly gets close to the linear behavior of the DUT, both in magnitude and phase.

VI. CONCLUSION

A 3 step approach is proposed to model nonlinear systems in the frequency domain. In a first step, a set of experiments is designed and a large amount of data is acquired. In the second step, energy transfer functions are introduced to condense the data in a few characteristics that allow to gain insight in the device operation using measured data only. Building up this measurement based knowledge is a vital - and often overlooked - part in nonlinear modelling. Finally, a modified Volterra-type model is proposed that allows to model systems in deep compression approporiately. Validation of the proposed models is performed using the visual tools. The modelling approach is illustrated on the identification of an RF power amplifier operating between 600 and 2500 MHz and between -15 and 5 dBm input power.

VII. REFERENCES

[1] T. Van den Broeck, J. Verspecht, "Calibrated Vectorial Nonlinear Network Analyser", *IEEE-MTT-S*, San-Diego, USA, 1994, pp. 1069-1072.

[2] Y. Rolain, W. Van Moer, Ph. Vael, "Measuring the Characteristics of Modulated Non-Linear Devices", *53rd ARFTG Conference - Nonlinearity Characterization*, Anaheim, California, June 17-18, 1999.

[3] W. Van Moer, Y. Rolain and A. Geens, "Measurement Based Nonlinear Modeling of Spectral Regrowth" IEEE Transactions on Instrumentation and Measurement Technology, Vol. 50, No 6, in press,2001

[4] R.Pintelon and J. Schoukens, "System Identification. A Frequency Domain Approach", IEEE Press, New York, 2001

Figure 2 : Phase of the measured $ETF_1(f_0, |U(f_0)|)$ as a function of input power and frequency.

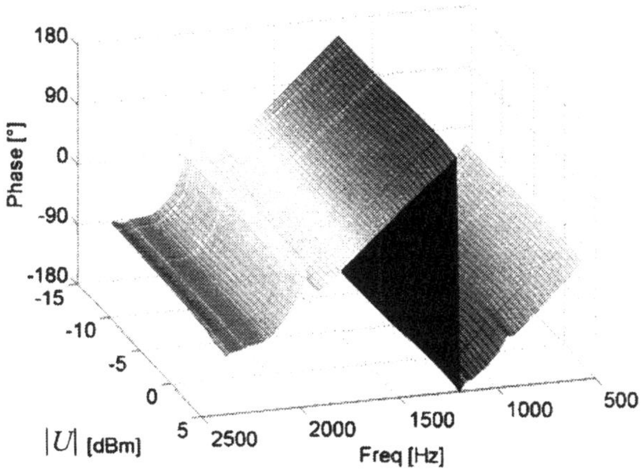

Figure 3 : Magnitude of the measured $ETF_1(f_0, |U(f_0)|)$ as a function of input power and frequency

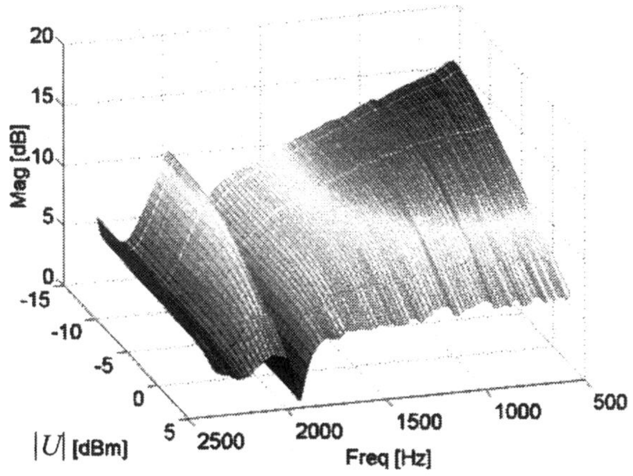

Figure 4 : Non-parametric model: comparison of model and measurement

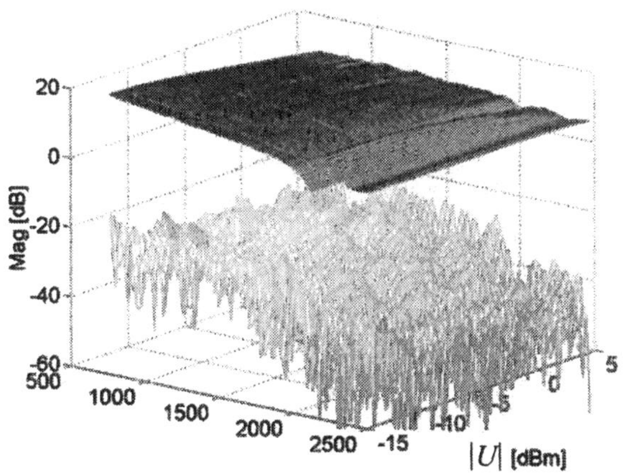

Figure 5 : Semi-parametric model: comparison of model and measurement

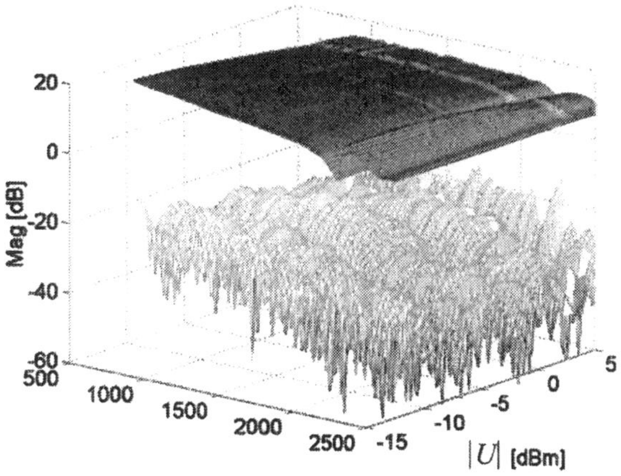

Figure 6 : Estimated K_{lk} and g_l for the non-parametric model.

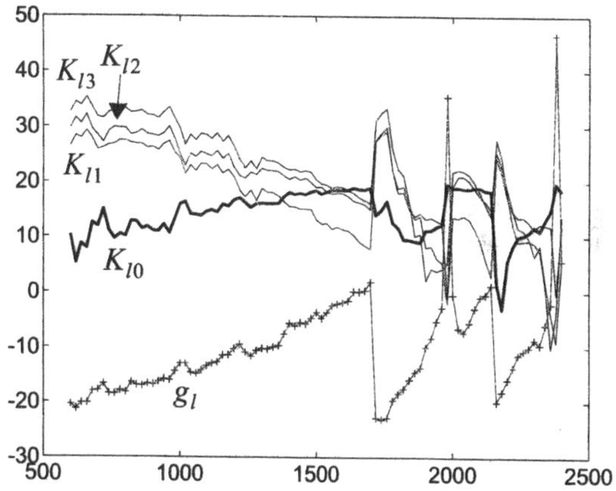

Figure 7 : Estimated K_{lk} and g_l for the semi-parametric model.

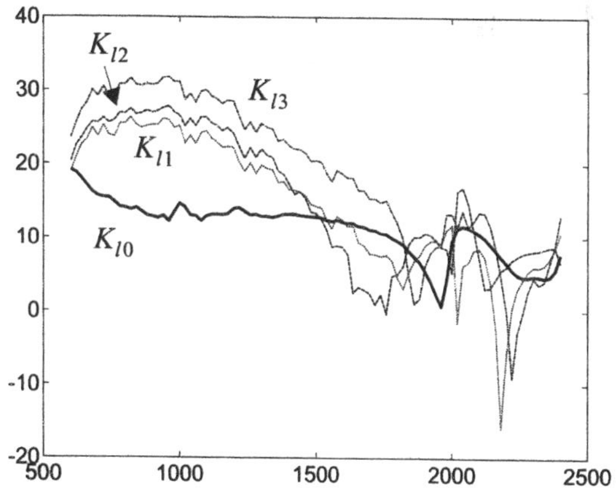

IFAC Publications
www.elsevier.com/locate/ifac

NONLINEAR SYSTEM MODELLING USING THE RBF NEURAL NETWORK-BASED REGRESSIVE MODEL

Hui Peng*, Tohru Ozaki, Yukihiro Toyoda †, Kazushi Nakano††**

** College of Information Science & Engineering, Central South University, Changsha 410083, China. Currently a visiting researcher at the Institute of Statistical Mathematics, 4-6-7 Minami Azabu, Minato-ku, Tokyo 106-8569, Japan. E-mail: peng@ism.ac.jp*
*** The Institute of Statistical Mathematics, 4-6-7 Minami Azabu, Minato-ku, Tokyo 106-8569, Japan.*
† Niihama National College of Technology, 7-1 Yagumo-cho Niihama, Ehime 792-0805, Japan
†† The University of Electro-Communications, 1-5-1 Chofu-ga-oka, Chofu, Tokyo 182-8585, Japan

Abstract: This paper presents an off-line identification-based modelling method for a class of smooth SISO nonlinear systems for the purposes of process control, output prediction or dynamics reconstruction. A pseudo-linear ARX (RBF-ARX) model with system operating-point dependent Gaussian RBF network-type coefficients is built to characterize the system. The RBF-ARX model is off-line identified, and the structured nonlinear parameter optimization method (Peng, *et al.* 2002) is applied to estimate its parameters. Output stability of the estimated RBF-ARX model is investigated in dynamics reconstruction problem. Comparisons of the RBF-ARX model and the on-line estimated linear time-varying ARX model are illustrated in case study. *Copyright © 2003 IFAC*

Keywords: Nonlinear systems, modelling, radial basis function networks, ARX model, prediction problems, stability.

1. INTRODUCTION

In nonlinear system modelling problems for control, the off-line identified piecewise linearization models based approach, and the on-line parameter estimation based modeling technique have been widely applied. For piecewise linearization approach (*e. g.* Prasad *et al.* 1998, Bloemen *et al.* 2001), although the off-line identified piecewise linearized models are easy to use for control design, the identification-data-making for estimation of many linear models being valid only in each small region is not easy in real application due to high cost etc. For on-line parameter estimation approaches, if the nonlinear system dynamics vary with time rapidly, the modelling precision may not be satisfactory due to some common problems such as too slow convergence rate for parameter-tracking, and even estimation failure.

Furthermore, if the estimated model is a purely nonlinear model which is used to design a controller, there usually may result in on-line solving a higher order nonlinear optimization problem, which is still an unsolved problem. Therefore, it is a very meaningful work to look for a model, which can effectively describe the nonlinear behavior of a system, and can be also easily used to design a controller. In fact, a large number of nonlinear processes may be regarded as this kind of systems whose working point varies with time, and which can be locally linearized at any fixed working point. For this kind of nonlinear systems mentioned above, in this paper, a hybrid pseudo-linear (RBF-ARX) model built on the basis of the Gaussian radial basis function (RBF) networks and ARX structure is proposed for the purposes of controller design, output prediction (for process monitoring), or dynamics reconstruction.

The RBF-ARX model is built as a global model, and to avoid on-line estimating time-varying parameters, the model parameters are estimated off-line using a quickly-convergent structured nonlinear parameter optimization method (SNPOM) proposed in Peng *et al.* (2002). The RBF-ARX model may be treated as a linear time-invariant ARX model at any working point; therefore the estimated model may be easily used to design controller. Although the system working-point dependent RBF-ARX model may be treated as a linear ARX model at each sampling interval, the approximated locally-linear ARX model parameters still vary with time because the system working point varies with time due to nonlinearity. Therefore, the RBF-ARX model behaves like a time-varying linear ARX model. In this paper, under the assumption that the process input and measurable disturbance are bounded, the output bounded stability of the estimated RBF-ARX model is investigated for the output dynamics reconstruction problem. Comparisons of the on-line estimation accuracy between the RBF-ARX model proposed and the on-line estimated linear time-varying ARX model is also shown in Section 4.

2. THE RBF-ARX MODEL

2.1 Derivation of the RBF-ARX model

Consider the smooth nonlinear SISO system with working point dependent dynamics, which can be described by the following NARX model

$$y(t) = f(\mathbf{W}(t-1)) + \zeta(t) \tag{1}$$

$$\begin{aligned}\mathbf{W}(t-1) = [&y(t-1),\cdots,y(t-n_y),u(t-1),\\ &\cdots,u(t-n_u),v(t-1),\cdots,v(t-n_v)]^{\mathrm{T}}\end{aligned} \tag{2}$$

where $y(t)$ is the output, $u(t)$ is the input, $v(t)$ is the measurable disturbance, and $\zeta(t)$ is the modelling error generally assumed as a white noise. If regarding $\mathbf{W}(t-1)$ in (2) as the state variables representing system working-point state, then system (1) may be described by a state $\mathbf{W}(t-1)$ dependent ARX model as is shown in Theorem 2.1.

Theorem 2.1: *If assuming that the function $f(\bullet)$ in (1) is analytic, then nonlinear system (1) may be approximated by the state $\mathbf{W}(t-1)$ dependent ARX model given by*

$$\begin{aligned}y(t) = {}&\pi_0(\mathbf{W}(t-1)) + \sum_{i=1}^{n_y}\pi_i^y(\mathbf{W}(t-1))y(t-i)\\ &+\sum_{i=1}^{n_u}\pi_i^u(\mathbf{W}(t-1))u(t-i) + \sum_{i=1}^{n_v}\pi_i^v(\mathbf{W}(t-1))v(t-i) + \xi(t)\end{aligned} \tag{3}$$

where π_0, π_i^y, π_i^u and π_i^v are the state-dependent function-style coefficients, and $\xi(t)$ is the modelling error.

Proof: Since $f(\bullet)$ is analytic, one may use a Taylor expansion to approximate it around any fixed time point, say t_0. If truncating this expansion to first order, for model (1), yields that

$$\begin{aligned}y(t) = {}&f(\mathbf{W}(t_0)) + \sum_{i=1}^{n_y}(y(t-i)-y(t_0-i))f_{y,i}(\mathbf{W}(t-1))\\ &+\sum_{i=1}^{n_u}(u(t-i)-u(t_0-i))f_{u,i}(\mathbf{W}(t-1))\\ &+\sum_{i=1}^{n_v}(v(t-i)-v(t_0-i))f_{v,i}(\mathbf{W}(t-1)) + \xi(t)\end{aligned} \tag{4}$$

where $$\begin{cases} f_{y,i}(\mathbf{W}(t-1)) = \left.\dfrac{\partial f(\mathbf{W}(t-1))}{\partial y(t-i)}\right|_{y(t-i)=y(t_0-i)} \\ f_{u,i}(\mathbf{W}(t-1)) = \left.\dfrac{\partial f(\mathbf{W}(t-1))}{\partial u(t-i)}\right|_{u(t-i)=u(t_0-i)} \\ f_{v,i}(\mathbf{W}(t-1)) = \left.\dfrac{\partial f(\mathbf{W}(t-1))}{\partial v(t-i)}\right|_{v(t-i)=v(t_0-i)} \end{cases}$$

and $\xi(t)$ is the modelling error including noise. For any choice of t_0 we can rewrite (4) in the form, *i.e.* (3) with

$$\begin{aligned}\pi_0(\mathbf{W}(t-1)) = {}&-\sum_{i=1}^{n_y}y(t_0-i)f_{y,i}(\mathbf{W}(t-1))\\ &-\sum_{i=1}^{n_u}u(t_0-i)f_{u,i}(\mathbf{W}(t-1))\\ &-\sum_{i=1}^{n_v}v(t_0-i)f_{v,i}(\mathbf{W}(t-1)) + f(\mathbf{W}(t_0))\end{aligned} \tag{5a}$$

$$\begin{cases}\pi_i^y(\mathbf{W}(t-1)) = f_{y,i}(\mathbf{W}(t-1))\\ \pi_i^u(\mathbf{W}(t-1)) = f_{u,i}(\mathbf{W}(t-1))\\ \pi_i^v(\mathbf{W}(t-1)) = f_{v,i}(\mathbf{W}(t-1))\end{cases} \tag{5b}$$

□

Model (3) has an autoregressive structure similar to a linear ARX model, and its state dependent coefficients make the model's dynamics change with system state, so it may be called the state dependent ARX model or the functional-coefficient ARX model as was done in time series field (Priestly 1980, Chen and Tsay 1993). Model (3) may be applied to implement the local linearization of general nonlinear system (1) by fixing $\mathbf{W}(t-1)$ at time t; however, the problem is how to determine the functional coefficients. The one way is to approximate each partial derivative as well as the state-dependent local mean in (3) and (5) by an expansion of a limited number of RBF neural networks, because of the universal approximation capability of RBF nets; this is always possible to any degree of accuracy, as far as the partial derivatives and the state-dependent local mean in (5) are continuous (Poggio and Girosi 1990, Chen *et al.* 1990). According to Theorem 2.1, and using Gaussian RBF networks to construct the coefficients of the state-dependent ARX model (3), the RBF-ARX model can then be built as follows

$$\begin{cases} y(t)=\phi_0(\mathbf{W}(t-1))+\sum_{i=1}^{n_y}\phi_{y,i}(\mathbf{W}(t-1))y(t-i) \\ +\sum_{i=1}^{n_u}\phi_{u,i}(\mathbf{W}(t-1))u(t-i)+\sum_{i=1}^{n_v}\phi_{v,i}(\mathbf{W}(t-1))v(t-i)+\xi(t) \\ \phi_0(\mathbf{W}(t-1))=c_0^0+\sum_{k=1}^{m}c_k^0\exp\{-\lambda_k\left\|\mathbf{W}(t-1)-\mathbf{Z}_k^y\right\|_2^2\} \\ \phi_{j,i}(\mathbf{W}(t-1))=c_{i,0}^j+\sum_{k=1}^{m}c_{i,k}^j\exp\{-\lambda_k\left\|\mathbf{W}(t-1)-\mathbf{Z}_k^j\right\|_2^2\} \\ \mathbf{Z}_k^j=\left(z_{k,1}^j,z_{k,2}^j,\cdots,z_{k,n_w}^j\right)^{\mathrm{T}},\ j=y,u,v \end{cases} \tag{6}$$

where n_y, n_u, n_v, m, and n_w are the orders; $\mathbf{Z}_k^j(k=1,2,\cdots,m)$ are the centers of RBF networks; $\lambda_k\,(k=1,2,\cdots,m)$ are the scaling parameters; $c_{i,k}^j\,(i=1,2,\cdots,n_j\ ;\ j=y,u,v;\ k=0,1,2,\cdots,m)$ and $c_k^0(k=0,1,2,\cdots,m)$ are the scalar weighting coefficients; and $\|\bullet\|_2$ denotes the vector 2-norm. In fact, to use (6) as a more general model, the state $\mathbf{W}(t-1)$ in RBF-ARX model (6), which makes the model coefficients vary with working-point, may be extended to be an output signal, an input signal, an external signal, a computed signal based on measured signals, or a composition of the above two or more signals. On the other hand, many nonlinear industrial processes may be characterized by linear models at certain fixed working point, and at different working point the dynamics of systems may be described by the different linear models. Therefore, the signal(s) which govern system's working point are more suitable to be used as $\mathbf{W}(t-1)$ in (6). Based on this opinion, model (6) is called the system's working-point dependent RBF-ARX model.

The RBF-ARX model (6) is a rather general form of working-point dependent ARX style-model by adding a local mean (offset term) $\phi_0(\mathbf{W}(t-1))$, which is necessary to describe a non-stationary process by a global model. One can see that the local linearization of the model is a linear ARX model at any working point by fixing $\mathbf{W}(t-1)$ in (4) at time t. It has a natural and appealing interpretation as a locally linear ARX model in which the evolution of the process at time $(t-1)$ is governed by a set of autoregressive coefficients $\{\phi_{y,i}\ ,\ \phi_{u,i}\ ,\ \phi_{v,i}\}$, and a local mean ϕ_0, all of which depend on the 'working point' of the process at time $(t-1)$. Using the working point dependent functional-coefficients, especially due to the satisfactory properties of RBF networks in function approximation and in learning local variations, the RBF-ARX model may effectively represent the system behavior at each working point. The RBF-ARX model has the advantages of the state-dependent ARX model in nonlinear dynamics description and the RBF networks in function approximation. In general, it does not require too many RBF centers compared with a single RBF network model, because the complexity of the model is dispersed into the lags of the autoregressive parts of the model. In order to avoid some potential problems caused by on-line parameter estimation, such as estimation failure, all parameters of the RBF-ARX model are here estimated by off-line approach. For the nonlinear systems with working point dependent dynamics, the RBF-ARX model identified off-line may exhibit satisfactory fitting precision because of its capability of globally representing the nonlinear dynamics as is shown in Section 4.

2.2 Identification of the RBF-ARX model

Identification of RBF-ARX model (6) includes order selection and estimation of all parameters. The orders (n_y, n_u, n_v, m, and n_x) may be selected by comparing the estimated AIC (Akaike Information Criterion) (Akaike 1974) values under different orders and the model dynamics. First we must have a good parameter estimation method, and then we repeat the method for the models under different orders to select final model. Here main concern is focused on the parameter estimation, which is an off-line nonlinear parameter optimization problem. In general cases, the number of linear weights is larger than that of nonlinear centers and scaling parameters in a RBF-ARX model, so applying some classic methods to estimate all parameters simultaneously regardless of the feature of them, such as Gauss-Newton method (GNM) and Lenvenburg-Marquardt method (LMM), may spend many computation time and may not obtain a satisfactory result. In this paper, the structured nonlinear parameter optimization method (SNPOM) proposed by Peng *et al.* (2002) is applied to estimate the RBF-ARX model. This is a hybrid method combining the LMM for nonlinear parameter estimation and the Least Squares method (LSM) for linear parameter estimation, but it is not a variable rotation method (VRM) (*i.e.* rotationally fix partial variables to optimize other variables). Therefore the SNPOM could largely accelerate the computational convergence of parameter optimization search process, especially for the RBF-ARX model with more linear weights and less nonlinear parameters.

3. STABILITY ANALYSIS

In this paper, we give the stability analysis for the off-line estimated RBF-ARX model based output dynamics reconstruction problem. To this end, assume that the process under consideration is now described by the RBF-ARX model below

$$\begin{aligned} A_t(q^{-1})y(t)&=a_0(t-1)+B_t(q^{-1})u(t-1) \\ &\quad+D_t(q^{-1})v(t-1)+e(t) \end{aligned} \tag{7}$$

$$
\text{where} \quad \begin{cases} A_t(q^{-1}) = 1 + \sum_{i=1}^{n_y} a_i(t-1) q^{-i} \\ B_t(q^{-1}) = \sum_{i=1}^{n_u} b_i(t-1) q^{-i+1} \\ D_t(q^{-1}) = \sum_{i=1}^{n_v} d_i(t-1) q^{-i+1} \end{cases} \tag{8}
$$

$$
\begin{cases} a_0(t-1) = c_0^0 + \sum_{k=1}^{m} c_k^0 \exp[-\lambda_k^y \|\mathbf{W}(t-1) - \mathbf{Z}_k^y\|_2^2] \\ a_i(t-1) = c_{i,0}^y + \sum_{k=1}^{m} c_{i,k}^y \exp[-\lambda_k^y \|\mathbf{W}(t-1) - \mathbf{Z}_k^y\|_2^2] \\ b_i(t-1) = c_{i,0}^u + \sum_{k=1}^{m} c_{i,k}^u \exp[-\lambda_k^u \|\mathbf{W}(t-1) - \mathbf{Z}_k^u\|_2^2] \\ d_i(t-1) = c_{i,0}^v + \sum_{k=1}^{m} c_{i,k}^v \exp[-\lambda_k^v \|\mathbf{W}(t-1) - \mathbf{Z}_k^v\|_2^2] \\ \mathbf{W}(t-1) = [w(t-1), w(t-2), \cdots, w(t-n_w)]^{\mathrm{T}} \\ \mathbf{Z}_k^j = \left(z_{k,1}^j, z_{k,2}^j, \cdots, z_{k,n_w}^j\right)^{\mathrm{T}}, \ j = y, u, v \end{cases} \tag{9}
$$

and $e(t)$ is the modelling error including the unmodeled dynamics. Assume that the process input $u(t)$ and the measurable disturbance $v(t)$ are all bounded. The model used for process dynamics reconstruction is then obtained from (7) as follows

$$
y(t) = a_0(t-1) + (1 - A_t(q^{-1}))y(t) + B_t(q^{-1})u(t-1) + D_t(q^{-1})v(t-1) \tag{10}
$$

Define the state vector $\boldsymbol{\alpha}(t)$ below, and then model (10) can be written as

$$
\boldsymbol{\alpha}(t) = \mathbf{A}(t-1)\boldsymbol{\alpha}(t-1) + \mathbf{B}(t-1)\mathbf{U}(t-1) + \mathbf{D}(t-1)\mathbf{V}(t-1) + \mathbf{L}a_0(t-1) \tag{11}
$$

where $\boldsymbol{\alpha}(t) = [y(t), y(t-1), \cdots, y(t-n_y+1)]^{\mathrm{T}}$ (12)

$$
\mathbf{U}(t-1) = [u(t-1), u(t-2), \cdots, u(t-n_u)]^{\mathrm{T}}
$$

$$
\mathbf{V}(t-1) = [v(t-1), v(t-2), \cdots, v(t-n_v)]^{\mathrm{T}}
$$

$$
\mathbf{A}(t) = \begin{bmatrix} -a_1(t) & -a_2(t) & \cdots & -a_{n_y-1}(t) & -a_{n_y}(t) \\ 1 & 0 & \cdots & 0 & 0 \\ 0 & 1 & \cdots & 0 & 0 \\ \vdots & \vdots & \ddots & \vdots & \vdots \\ 0 & 0 & \cdots & 1 & 0 \end{bmatrix}
$$

$$
\mathbf{B}(t) = \begin{bmatrix} b_1(t) & b_2(t) & \cdots & b_{n_u}(t) \\ 0 & 0 & \cdots & 0 \\ \vdots & \vdots & & \vdots \\ 0 & 0 & \cdots & 0 \end{bmatrix}
$$

$$
\mathbf{D}(t) = \begin{bmatrix} d_1(t) & d_2(t) & \cdots & d_{n_v}(t) \\ 0 & 0 & \cdots & 0 \\ \vdots & \vdots & & \vdots \\ 0 & 0 & \cdots & 0 \end{bmatrix}, \quad \mathbf{L} = \begin{bmatrix} 1 \\ 0 \\ \vdots \\ 0 \end{bmatrix}
$$

To derive the results given in Theorem 3.1, we have now to introduce the following.

Lemma 3.1: *For RBF-ARX model (10) or (11) with the RBF network-style parameters, the parameters are bounded, and the parameter variations are slow, i.e. there exist positive constants* c_1 *and smaller* c_2*, such that*

$$
\|\boldsymbol{\theta}(t)\| \le c_1 \ , \forall t \tag{13}
$$

$$
\|\boldsymbol{\theta}(t) - \boldsymbol{\theta}(t-1)\| \le c_2 \ , \forall t \tag{14}
$$

$$
\boldsymbol{\theta}(t) = [a_0(t), \cdots, a_{n_y}(t), b_1(t), \cdots, b_{n_u}(t), d_1(t), \cdots, d_{n_v}(t)]^{\mathrm{T}}
$$

Proof: For the off-line estimated RBF-ARX model (10), assume the scaling factors in (9) are positive, *i.e.*

$$
\lambda_k^j > 0, \quad (k = 1, \cdots, m \ ; \ j = y, u, v)
$$

This is able to be guaranteed, for which one can apply the heuristic approach to compute λ_k^j, such that

$$
\begin{cases} \lambda_k^j = -\log \varepsilon_k / \max_{\forall t} \{\|\mathbf{W}(t-1) - \mathbf{Z}_k^j\|_2^2\} \\ \varepsilon_k \in [0.1 \sim 0.0001] \end{cases} \tag{15}
$$

after updating the RBF center $\mathbf{Z}_k^j$ at each search iteration in the model parameter optimization process (see Peng *et al.* 2002). Thus, from (15), yields

$$
0 < \exp[-\lambda_k^j \|\mathbf{W}(t-1) - \mathbf{Z}_k^j\|_2^2] \le 1 \ , \forall \mathbf{W} \ , \forall \mathbf{Z}_k^j
$$

and then from (9) one can see that all the estimated linear weights in (9) are finite constants. It implies the boundedness of $\boldsymbol{\theta}(t)$. Further, it is easy to confirm that for an exponential function below

$$
y = \exp(-\lambda_k^j x^2) \ , \ \lambda_k^j > 0 \ , x \in (-\infty \ , +\infty)
$$

its maximal differential is given by $\max_x |\dot{y}| = \sqrt{2\lambda_k^j} e^{-\frac{1}{2}}$, which is a constant. Furthermore, from the computation formula of λ_k^j in (15), yields that λ_k^j is typically smaller. This implies that the parameter variations of model (10) are slow. □

Theorem 3.1: *For an open-loop stable system described by model (7), if properly choosing the sampling period and model order to estimate the model, make all the eigenvalues of* $\mathbf{A}(t-1)$ $(\forall t)$ *in (11) lie inside the unit circle, i.e.*

$$
\max_{j,t} \{|\lambda_j[\mathbf{A}(t-1)]|\} \le 1 - 2\rho < 1, \ \rho > 0
$$

there then exist constants $c_3 > 0$, $c_4 > 0$ *and* $0 < \eta < 1$ *such that, the zero-input-response of system (11) is exponentially stable. Furthermore, the signal* $y(t)$ *in (10) is then uniformly bounded for any initial state, any bounded input signal* $u(t)$, *and any bounded disturbance signal* $v(t)$.

Proof: From Lemma 3.1, one can see that there exists an $c_3 > 0$ and an smaller $c_4 > 0$, such that

$$\|\mathbf{A}(t-1)\| \le c_3 \ , \forall t$$
$$\|\mathbf{A}(t) - \mathbf{A}(t-1)\| \le c_4 \ , \forall t$$

If properly choosing sampling period and model order to estimate model (7), make all eigenvalues of $\mathbf{A}(t-1)$ $(\forall t)$ in (11) lie strictly inside the unit circle. Thus, one can conclude that the zero-input-response of system (11) is exponentially stable (Desoer 1970), where the explicit parameter relations can be given as follows (see Desoer 1970)

$$\begin{cases} c_4 = \dfrac{(1-(1-\rho)^2)^2}{2\vartheta^4 c_3}(1-\eta) \\ \vartheta = (1-\rho)\dfrac{(1-\rho)+c_3^{n_\alpha -1}}{\rho^{n_\alpha}} \\ 0 < \eta < 1 \ , \ n_\alpha = \dim\{\boldsymbol{\alpha}(t)\} \end{cases}$$

Furthermore, in view of the boundedness of the model parameters in (11), the input signal $u(t)$, the measurable disturbance signal $v(t)$ and the local mean value $a_0(t-1)$, from the output-dynamics-reconstructed system (11), we obtain

$$\|\boldsymbol{\alpha}(t)\| \le \beta_0 \|\boldsymbol{\alpha}(0)\| + \beta_1 \qquad (16)$$

for any t and suitable finite nonnegative constants β_0 and β_1. From the definition of $\boldsymbol{\alpha}(t)$ in (12), one can see that $y(t)$ is bounded. □

4. CASE STUDY

Fig.1 shows a set of sample data from a Nitrogen Oxide (NOx) decomposition (de-NOx) process in thermal power plants. This process is nonlinear non-stationary, whose dynamics depend on the load demand $x(t)$ of power plants (Matsumura *et al.* 1997). In fact, working point of the process is dependent on the load $x(t)$, and the process dynamics may be represented by different linear model at different load $x(t)$, thus this process may be described by a load $x(t)$ dependent RBF-ARX model (7), in which v_2, v_1, x are in place of u, v, w respectively. In de-NOx process control, the most commonly used technology is the selective catalytic reduction method (Matsumura *et al.* 1997), which is to use ammonia (NH3) to decompose NOx. For comparison, the linear time-varying ARX (TVLARX) model (17) given below is also applied to modelling the sample data.

$$\tilde{A}_t(q^{-1})y(t) = \tilde{a}_{t,0} + \tilde{B}_t(q^{-1})v_2(t-1) + \tilde{D}_t(q^{-1})v_1(t-1) + \xi(t)$$
$$\tilde{A}_t(q^{-1}) = 1 + \sum_{i=1}^{n_y} \tilde{a}_{t,i} q^{-i} \ , \ \tilde{B}_t(q^{-1}) = \sum_{i=1}^{n_u} \tilde{b}_{t,i} q^{-i+1}$$
$$\tilde{D}_t(q^{-1}) = \sum_{i=1}^{n_v} \tilde{d}_{t,i} q^{-i+1} \qquad (17)$$

The estimated result of RBF-ARX model (7) using the SNPOM (Peng *et al.* 2002) for the data is given in Fig.2, where the model orders are chosen as, $n_y = 4$, $n_{v_2} = 6$, $n_{v_1} = 6$, $m = 1$, $n_x = 1$, and the sampling period is 10 sec. The on-line estimated result of model (17) with the same order as model (7) is shown in Fig.3, where the function '*RARX*' using the recursive forgetting factor least squares algorithm (Ljung 1999) in Matlab System Identification Toolbox is applied to implement on-line estimation of model (17), and the used forgetting factor is 0.985, the initial scaled covariance matrix of the parameters is 100 times the identity matrix. Fig.2 shows that the RBF-ARX model (7) has better modelling precision than model (17), and the larger residual sometimes occurred in the result by model (17) as is seen in Fig.3. Figs.4-5 shows the estimated model parameters varying with time, which are the characteristic polynomial coefficients $(a_{t,1}, a_{t,2}, \cdots, a_{t,n_y})$ of model (7) or (17). It is clear that the parameter variation of RBF-ARX mode (7) is much slower than that of TVLARX model (17). The parameters of model (17) are always fluctuating due to system nonlinearity, working-point varying, slower parameter tracking rate, and noise effect etc. The larger fluctuation of parameters of model (17) leads that the model dynamics change fastly and un-predictablely as seen in Fig.7, which shows the poles of model (17) during modelling process. Fig. 6 shows the pole variation of the estimated RBF-ARX model (7), which is smooth and are all inside the unit circle.

5. CONCLUSIONS

For a class of smooth nonlinear system whose dynamics changes with working point and may be linearized at each working point, the off-line estimated RBF-ARX model could be applied as a global model to characterize the system over a large operation range. The RBF-ARX model may be used for controller design, output prediction, or dynamics reconstruction. Compared to the modelling approach using on-line estimated time-varying linear model, the off-line estimated RBF-ARX model can avoid the risk such as on-line estimation failure, and may provide a more precise model with smooth and stabile dynamic characteristics.

REFERENCES

Akaike, H., 1974, A new look at the statistical model identification. *IEEE Transactions on Automatic Control*, **19**, 716-723.

Bloemen, H. H. J., Van Den Boom, T. J. J., and

Verbruggen, H. B., 2001, Model-based predictive control for Hammerstein-Wiener systems. *International Journal of Control*. **74**, 482-485.
Chen, S., Billings, S. A., Cowan, C. F. N., and Grant, P. M., 1990, Non-linear systems identification using radial basis functions. *International Journal of Systems Science*, **21**, 2513-2539.
Chen, R., and Tsay, R. S., 1993, Functional- coefficient autoregressive models. *Journal of the American Statistical Association*, **88**, 298-308.
Desoer, C. A., 1970, Slowly varying discrete system $x_{i+1} = A_i x_i$. *Electronics Letters*, **6**(11), 339-340.
Ljung, L., 1999, *System Identification: Theory for the Users (2nd edition)*. Addison-Wesley, Reading, MA.
Matsumura, S., Iwahara, T., Ogata, K., Fujii, S., and Suzuki, M., 1997, Improvement of de-NOx device control performance using software sensor. *Proceedings of the 11th IFAC Symposium on System Identification*, Fukuoka, Japan, pp. 1507-1512.
Peng, H., Ozaki, T., Haggan-Ozaki, V., and Toyoda, Y., 2002, A parameter optimization method for the radial basis function type models. *IEEE Trans. on Neural Networks*, **14**, 432-438.
Poggio, T., and Girosi, F., 1990, Networks for approximation and learning. *Proceedings of IEEE*, **78**, 1481-1497.
Prasad, G., Swidenbank, E., and Hogg, B.W., 1998, A local model networks based multivariable long-range predictive control strategy for thermal power plants. *Automatica*, **34**, 1185-1204.
Priestley, M. B., 1980, State dependent models: a general approach to nonlinear time series analysis. *Journal of Time Series Analysis*, **1**, 57-71.

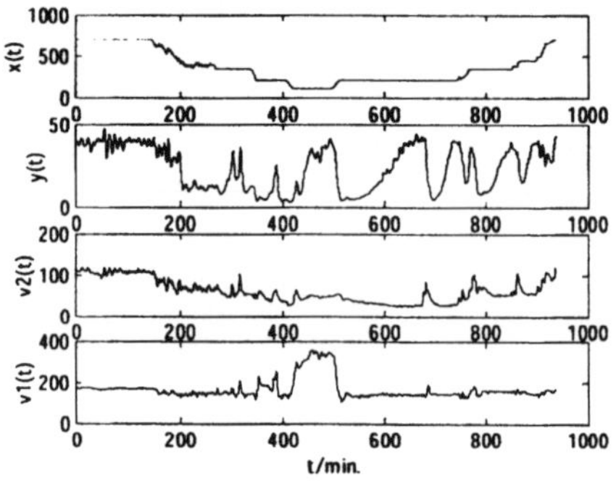

Fig. 1. Sample data, where $x(t)$ is the load demand of power plant; $y(t)$ is the NOx concentration in exhaust gas; $v_2(t)$ is the injected NH3 flow; and $v_1(t)$ is the NOx concentration in the de-NOx device inlet, which is measurable disturbance.

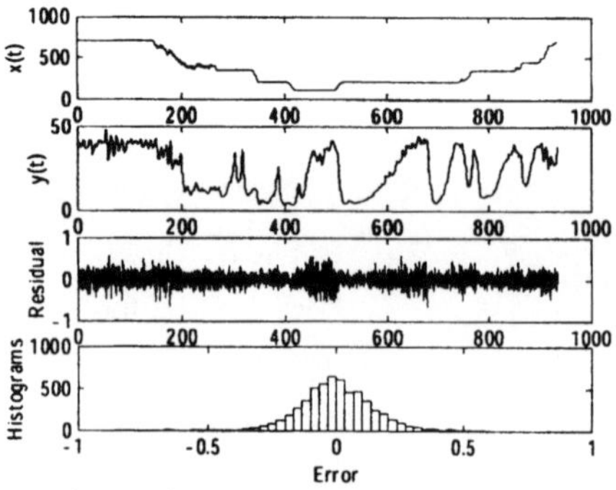

Fig. 2. Estimated (one-step) residual and histogram of RBF-ARX model (7).

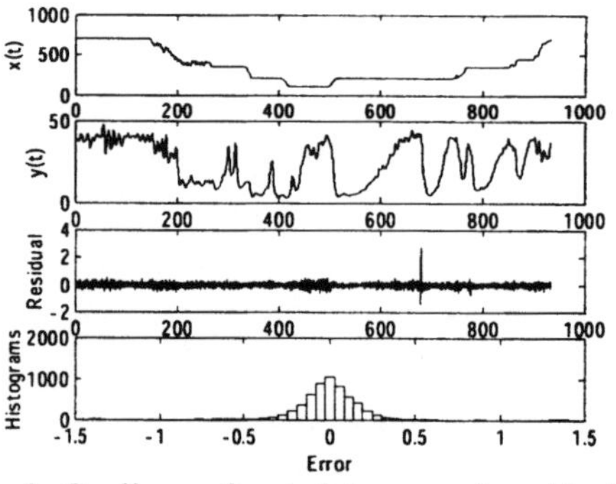

Fig. 3. On-line estimated (one-step) residual and histogram of linear time-varying ARX model (17).

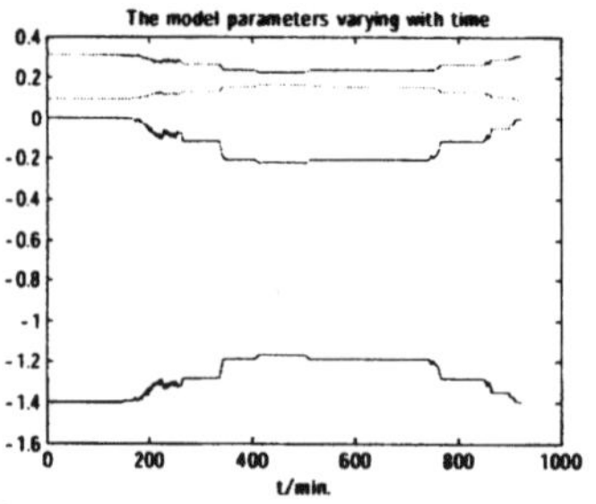

Fig. 4. Off-line estimated characteristic polynomial coefficients of model (7) varying with time.

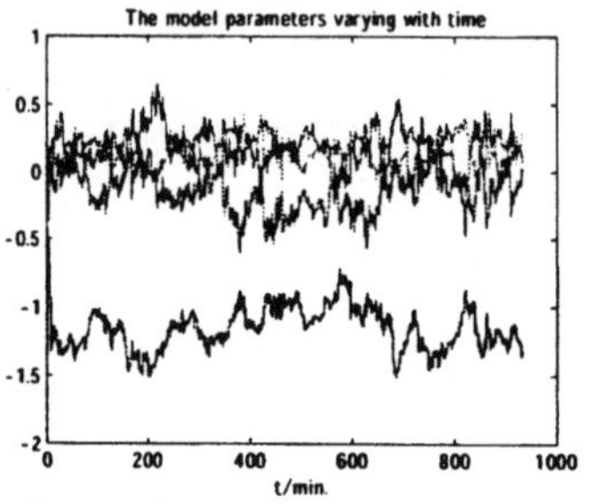

Fig. 5. On-line estimated characteristic polynomial coefficients of model (17) varying with time.

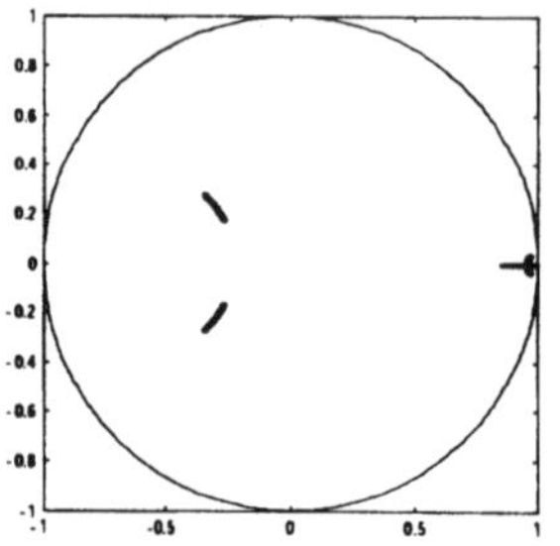

Fig. 6. Eigenvalues of the estimated RBF-ARX model (7) varying with time.

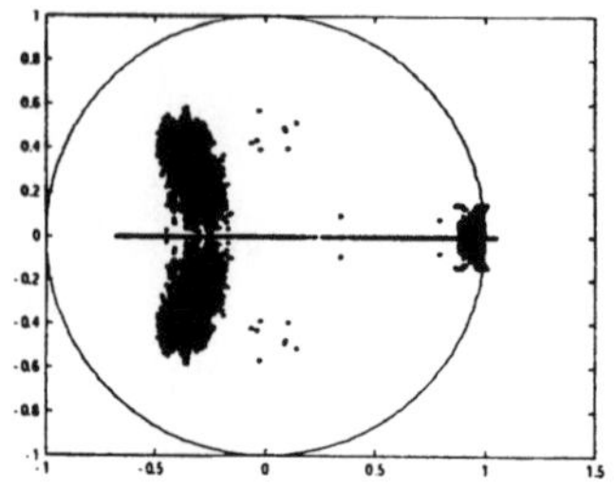

Fig. 7. Eigenvalues of the on-line estimated linear time-varying ARX model (17) varying with time.

IFAC Publications
www.elsevier.com/locate/ifac

MODELING AND LINEARIZATION OF NONLINEAR DYNAMIC SYSTEMS

József G. Németh (*) and Johan Schoukens (**)

(*): Budapest Univ. of Technology & Econ., Dep. MIS,H-1521 Budapest, Hungary
(**): Vrije Univ. Brussel, Dep. ELEC, Pleinlaan 2, B-1050 Brussels, Belgium
email: nemeth@mit.bme.hu

Abstract: In this paper a powerful method is presented for the modeling and the linearization of even highly dynamic nonlinear systems. The method approximates Volterra-series expansion, and factors out part of the system dynamics into linear structures that can be identified with minimal effort. The method is illustrated on experimental measured data.

Keywords: nonlinear, identification, instrumentation

1. INTRODUCTION

Linearization and equalization of data acquisition channels are common problems in many fields of engineering. The goal is to achieve the linear behavior of the overall system and that the resulting transfer function has flat magnitude and linear phase characteristic.

Methods for the removal of linear distortions are usually based on identifying the transfer function first, which is then used to design a digital filter having an inverse characteristic within a known delay (Pintelon *et al.* 1990, Kollar *et al.* 1991). In a similar fashion, dynamic nonlinear compensation is feasible through the identification of nonlinear transfer functions based on Volterra kernels and the computation of the p-th order inverse (Schetzen 1980, Mathews and Sicuranza 2000).

A different approach may consist in approximating directly the inverse characteristic in the frequency range of interest. I.e., the output of the system is used as the input to the model, while the (delayed) input of the system is used as the desired output. An important detail needs to be stressed, though: outside the frequency band of interest, the gain of the compensating model should be controlled because too high gains would significantly amplify the out-of-band noise. In this paper, this latter approach is applied in combination with an approximate Volterra kernel estimation technique (Nemeth *et al.* 2002, Nemeth and Schoukens 2002).

The paper is organized as follows. First, the setup is described. Second, the model structure is introduced. Next, the identification procedure is presented. Finally, the method is tested on a nonlinear resonator possessing a 20 dB gain at its resonance frequency. The identification of the transfer function (TF) and the identification of the equalizer are both presented.

2. PROBLEM SETUP

The basic setup is given in Fig. 1. A nonlinear device $y(t) = g_{\mathrm{NL}}(y(t))$ should be compensated starting from the sampled output $y(k)$. For this purpose, a nonlinear digital filter $z(k) = g_{\mathrm{C}}(y(k))$ is designed such that within the frequency band of interest, the output $z(k)$ is a delayed copy of the input:

$$z(k) = u(k - \tau) \tag{1}$$

In practice, the choice of the delay τ will be part of the design problem.

The first author is member of the HAS-BUTE Embedded Information Technology Research Group. This work was also supported by the Hungarian National Fund for Scientific Research (OTKA: T 033053), the Flemish government (BIL99/18, and GOA-ILiNos) and the Belgian government (IUAP V/22).

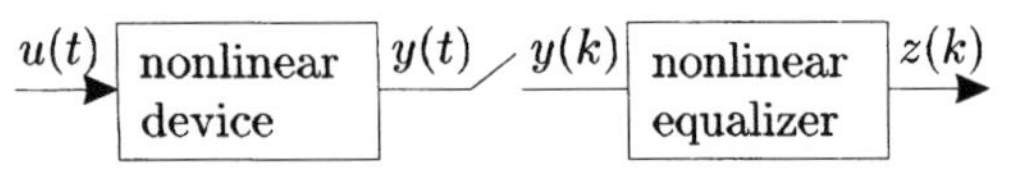

FIGURE 1. Basic setup of equalization.

3. THE MODEL

A Volterra series (Schetzen 1980) provides a description for dynamic systems in a similar way as the Taylor series does for static input/output relationships, i.e., the system output is split into linear, quadratic, cubic, etc. contributions:

$$y(t) = \sum_{\alpha=0}^{\infty} G^{(\alpha)}[u(t)] = \sum_{\alpha=0}^{\infty} y^{(\alpha)}(t), \qquad (2)$$

each Volterra operator $G^{(\alpha)}$ having the property that for any scalar c:

$$G^{(\alpha)}[c \cdot u(t)] \equiv c^{\alpha} \cdot G^{(\alpha)}[u(t)]. \qquad (3)$$

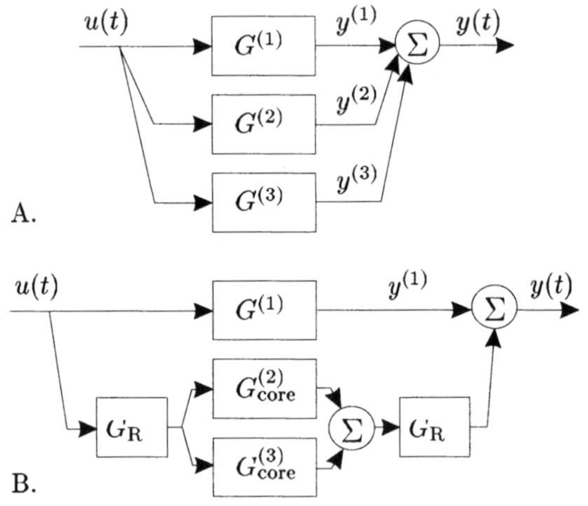

FIGURE 2. A. Third order Volterra model. B. Modified model containing pre- and post-filters G_R

The time-domain and frequency-domain Volterra kernels that describe each operator are α-dimensional impulse response and frequency response functions, respectively. Also, if an α-order Volterra filter with N memory is to be identified or an α-order frequency-domain Volterra kernel is sampled by N harmonics, the number of unknowns amounts to $O(N^{\alpha})$. In most problems such a high degree of freedom is unnecessary, especially if the higher-order contributions are dominated by the linear one, hence their accuracy has less impact on the overall quality of the model.

Therefore, an approximate method for the identification of third-order Volterra models was introduced (Nemeth *et al.* 2002): the second and third order frequency-domain kernels are estimated in the bases of B-spline functions, which results in smooth kernels. This method aims at tailoring the kernel complexity to the identification task by leveling the different error components. As a result, the required measurement duration can be reduced by an important factor compared to nonparametric kernel estimation methods (Kim and Powers 1988, Nam and Powers 1994).

Nevertheless, as in the case of most other black box modeling techniques, systems with high dynamics would require a nonlinear model with long memory, hence a large number of coefficients. In consequence, the identification would require long measurement records and would be computationally expensive. The most powerful solution would be to use a dedicated model. However, this might be very time-consuming since the whole modeling process should be restarted for each new design.

In order to factor out the major part of the system dynamics from the nonlinear kernels, linear pre- and post-filters are introduced into the model. See Fig. 2/B. G_R is the identified best linear approximation to the nonlinear device. Often the kernels $G^{(2)}_{\text{core}}$ and $G^{(3)}_{\text{core}}$ that remain to be estimated are close to being static nonlinearities. The motivation behind the proposed filtering is explained in the following.

3.1 *Motivation for the pre- and post-filters*

Pre- and post-filtering can be effective for dynamic systems containing the nonlinearity in the feedback path. In order to keep the discussion simple, the model in Fig. 3/A is considered: the feedforward branch is linear, whereas the feedback branch is a single (cubic) Volterra operator. (The more general case where both branches are general Volterra operators is treated in (Schoukens *et al.* 2003a) and in Section 8.5 of (Schetzen 1980).)

By inverting the input and the output of this feedback structure, a feedforward structure is obtained as shown in Fig. 3/B. The 3rd order p-inverse of this structure is indicated in Fig. 3/C. The inversion can be verified by cascading the two systems (B and C) and by computing the first, second and third-order contributions, which yield u, 0 and 0, respectively. The operator $\mathcal{I}$ in the figure denotes the higher order contributions produced, which are not accounted for by the compensation.

The important observation to be made concerning Fig. 3/C is that $G^{(1)}$ figures (twice) in $G^{(3)}$. Hence, if $G^{(1)}$ contains high resonances then $G^{(3)}$ is bound to contain strong dynamics, too, which makes the identification of the Volterra model (Fig. 2/A) difficult. On the other hand, the best linear approximation G_R can be easily estimated and used for filtering as shown in Fig. 2/B. The kernel $G^{(3)}_{\text{core}}$ that remains to be estimated proves to contain much less dynamics in most cases.

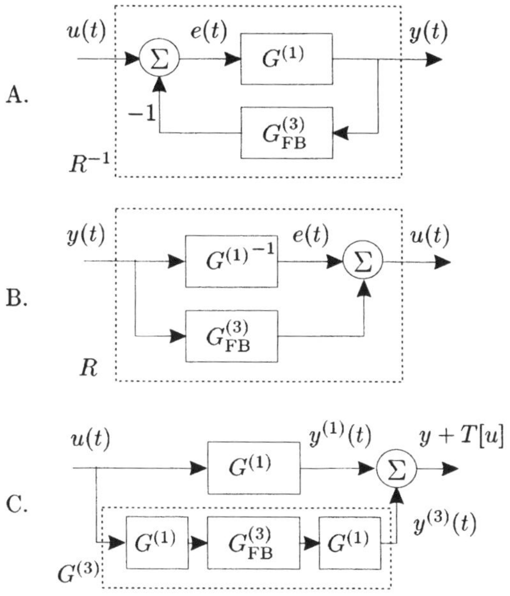

FIGURE 3. A. Feedback structure: R^{-1} B. Feedforward structure: R C. The p-inverse of R

3.2 *Alternatives*

The factorization of the system dynamics into linear structures, as proposed by the pre- and post filtering technique, can be taken further to have *static* nonlinear blocks in Fig. 2/B. Since in this case, each nonlinear block introduces a single unknown (being the coefficient of a polynomial), the model can be extended towards higher nonlinear orders economically.

The general nonlinear structure proposed in (Schoukens *et al.* 2003a) combines such a model by other types of nonlinear branches (Wiener and Hammerstein branches and static nonlinear corrections). This general nonlinear structure can encompass a smaller class of system dynamics, then the method proposed in this paper, but accounts for the most common types of nonlinear behavior, thus can direct the model design towards more specific and complex models. Since its identification does not require excessive measurements or computation, it can be applied in a trial and error manner with no extra price. Its use for linearization is presented in (Schoukens *et al.* 2003b).

4. IDENTIFICATION PROCEDURE

The identification of the parameterized Volterra kernels requires measuring a number of independent realizations of the system response using random phase multisine excitation or periodic noise. The use of non-periodic excitation is possible, but introduces leakage effects. The product between the number of the exciting harmonics N and the number of gathered realizations will determine the affordable model complexity and the discrete-time length of the measurements. The parameter estimation is a linear least-squares problem. For the details, see (Nemeth *et al.* 2002).

If pre- and post-filtering is to be applied then the best linear approximation must be identified in a preliminary step based on the same measurement data.

When identifying the inverse characteristic all signals are oversampled and the selected frequency-band of interest is the band of the excitation. The out-of-band output components are omitted, which is equivalent to placing a low-pass filter at the front-end of the nonlinear equalizer (Fig. 1). The desired output for the equalizer can be forced to zero outside the band of interest or left without constraint during the identification. If the second choice is taken then a low-pass filter is necessary at the output of the equalizer when running the compensation. The price to be paid for the low-pass filters is that they will introduce a (known) delay into the system.

A stable equalizer can be obtained by fitting a finite impulse response filter onto the linear component $G^{(1)}(f)$ of the inverse characteristic. The estimated frequency-domain kernels automatically yield finite length time-domain kernels, because B-splines have near to optimal time-frequency bandwidth product TFBP (Unser 2000).

Since the FIR filters obtained for each kernel extrapolate the frequency response in an uncontrolled way outside the band of interest, the low-pass filtering of the equalizer *input* is always necessary. The design of these low-pass filters is dependent on the extrapolated frequency-domain kernels.

5. EXPERIMENTAL ILLUSTRATION

The method was tested on a nonlinear electrical circuit that is ideally described by the following nonlinear 2nd order differential equation:

$$m\frac{\mathrm{d}}{\mathrm{d}^2t} + d\,\frac{\mathrm{d}}{\mathrm{d}t}y(t) + a\,y(t) = u(t) - b\,y^3(t) \qquad (4)$$

This corresponds exactly to the model in Fig. 3/A, with $G^{(3)}_{\mathrm{FB}}$ being a static nonlinearity: $b\,y^3(t)$. Of course, the actual realized circuit is not in perfect agreement with (4). For instance, the presence of a small quadratic term $y^2(t)$ could also be noticed in the measurements. A detailed study on this nonlinear system can be found in (Pintelon and Schoukens 2001).

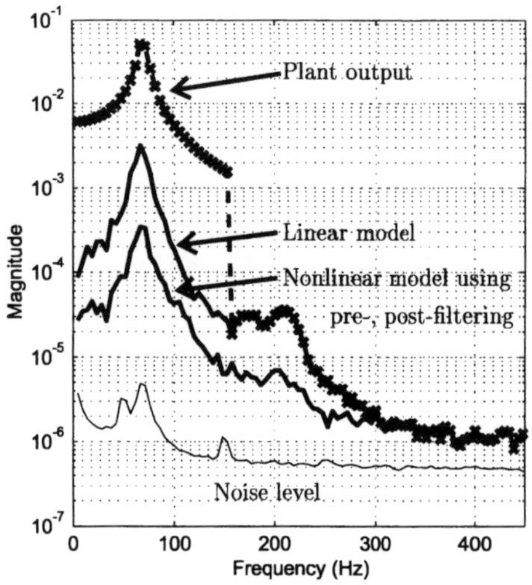

FIGURE 4. Validation results $|Y - Y_{\text{model}}|$ when modeling the transfer function. (Pre- and post-filtering is used.)

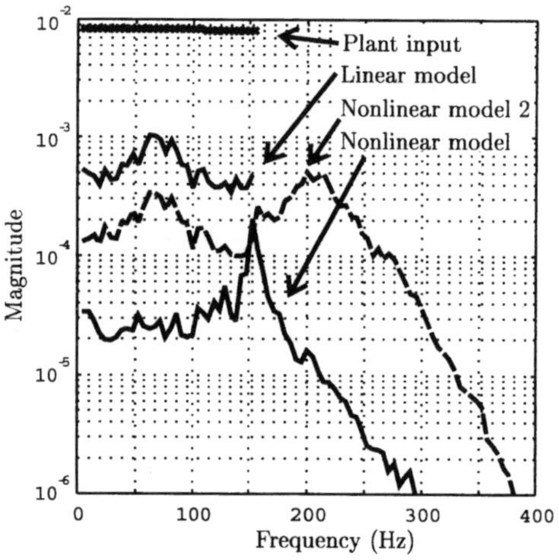

FIGURE 5. Validation results $|U - Z|$ of the compensation. (Pre- and post-filtering is *not* used.)

The system is driven with a random multisine having a flat amplitude spectrum and $N = 32$ harmonics. The output spectrum, depicted in Fig. 4, exhibits a 20 dB resonance peak. The measurements were repeated with 40 realizations of the random multisine excitation. 30 experiments were used for identification and 10 were reserved for the validation. When static kernels were used only 2 realizations were necessary for the estimation.

Although the modeling of the system is not necessary for the design of the equalizer, it illustrates well the use of pre- and post-filtering. Three models were identified. The best linear approximation had a normalized mean-square error (NMSE) of about -19 dB, the nonlinear model (with pre- and post-filtering) achieved a 14 dB improvement (i.e., NMSE $= -33$ dB). Using static kernels did not degrade the result significantly, which means that the pre- and post-filters could well encompass the system dynamics. Meanwhile, without this filtering the model failed completely. These results are in perfect agreement with the theory.

When identifying the equalizer, the band-limited output of the system was taken as the input of the model and the input (having flat magnitude spectrum) was the desired output for the equalizer. The validation results are depicted in Fig. 5. The linear compensation yielded NMSE $=$ -21 dB, whereas the nonlinear model achieved NMSE $= -43$ dB. Inferior results were obtained when using static kernels (see Nonlinear Model 2 in Fig. 5). The latter result is in agreement with the results published in (Schoukens *et al.* 2003b), and illustrates that there is a trade-off between the simplicity and the accuracy of the modeling approach.

The pre- and post- filtering technique was not used in this case, since the inverted system corresponds to a feed-forward structure Fig. 3/B.

Again, this experimental result is in agreement with the theoretical expectations. It should also be noted that the pre- and post-filtering could be required for identifying the equalizer if the system behavior corresponded to the scheme of Fig 3/B This situation would be complementary to the case presented here.

The equalizer was also validated on non-periodic random data. The amplitude of the excitation was increased gradually. In Fig. 6 the input signal to be approximated by the equalizer is depicted together with the error (once in the same amplitude scale, once with zoom). The error obtained with the linear compensation is depicted in grey for comparison. The signal power around $t = 25 \ldots 27$sec corresponds to the power level used during the identification. Around this level the spikes in the grey plot become very frequent, because the linear equalizer cannot extrapolate the compensation of the system towards higher amplitude levels. Meanwhile the nonlinear model seems to be robust in this respect.

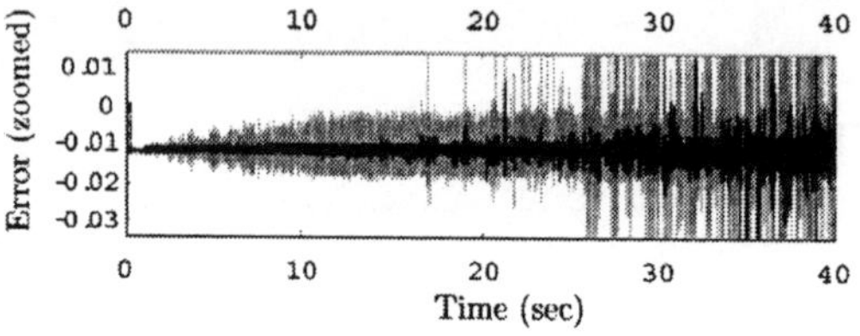

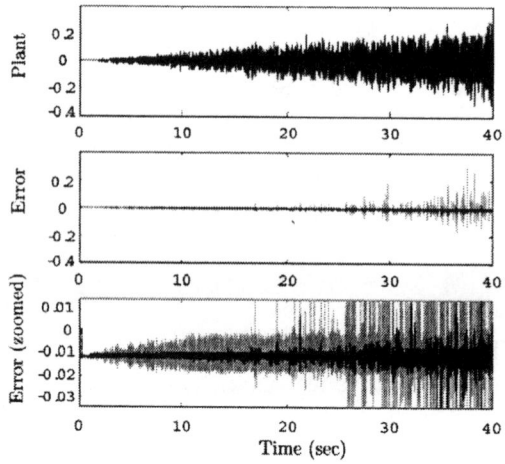

FIGURE 6. Validation results: y and $u - z$ for non-periodic random excitation as the input power is gradually increased.

6. CONCLUSION

An approximate Volterra-kernel identification method was used for the modeling and the linearization of a highly dynamic, nonlinear physical device.

It was shown that the nonlinear dynamics coming from a feedback can be "factored out" from the kernel estimation problem by means of pre- and post-filtering with the best linear approximation G_R.

The system was compensated within a -43 dB NMSE, which is an improvement of env. 20 dB over the linear compensation.

7. REFERENCES

K. I. Kim and E. J. Powers, Nov. 1988. A digital method of modeling quadratically nonlinear systems with a general random input. *IEEE Trans. Acoust. Speech Signal Process.*, 36 (1988), 1758–1769.

I. Kollar, R. Pintelon, Y. Rolain, and J. Schoukens, June 1991. Another step towards an ideal data acquisition channel. *IEEE Trans. Instrum. Meas.*, 40 (1991), 659–660.

J. V. Mathews and L. G. Sicuranza, 2000. *Polynomial signal processing.* Wiley-Interscience.

S. W. Nam and E. J. Powers, July 1994. Application of higher order spectral analysis to cubically nonlinear system identification. *IEEE Trans. Signal Proc.*, 42-7 (1994), 1746–1765.

J. G. Nemeth, I. Kollar, and J. Schoukens, Aug. 2002. Identification of Volterra Kernels Using Interpolation. *IEEE Transactions on Instrumentation and Measurement*, 51-4 (2002), 770–775. *Best paper award for 2002 in the IM Transactions.*

J. G. Nemeth and J. Schoukens, May 2002. Efficient Identification of Third Order Volterra Models Using Interpolation Techniques. In *19th IEEE Instrumentation and Measurement Technology Conference*, Anchorage, AK, USA.

R. Pintelon, Y. Rolain, M. Vanden Bossche, and J. Schoukens, Feb. 1990. Towards an ideal data acquisition channel. *IEEE Trans. Instrum. Meas.*, 39 (1990), 116–120.

R. Pintelon and J. Schoukens, 2001. *System Identification: a Frequency Domain Approach.* IEEE Press.

M. Schetzen, 1980. *The Volterra and Wiener Theories of Nonlinear Systems.* Wiley-Interscience.

J. Schoukens, J. G. Nemeth, P. Crama, Y. Rolain, and R. Pintelon, August 2003a. Fast Approximate Identification of Nonlinear Systems. In *13th IFAC Symposium on System Identification (SYSID 2003)*, The Netherlands.

J. Schoukens, J. G. Nemeth, G. Vandersteen, P. Crama, and R. Pintelon, May 2003b. Linearization of nonlinear dynamic systems. In *IEEE Instrumentation and Measurement Technology Conference*, USA.

M. Unser, April 2000. Sampling – 50 years after Shannon. *Proceedings of the IEEE*, 88-4 (2000).

IFAC Publications
www.elsevier.com/locate/ifac

LINEAR PARAMETER ESTIMATION AND PREDICTIVE CONSTRAINED CONTROL OF WIENER/HAMMERSTEIN SYSTEMS

**Krzysztof J. Latawiec,† Czesław Marciak,† Ryszard Rojek†
and Gustavo H.C. Oliveira‡**

† *Department of Electrical Engineering and Automatic Control
Technical University of Opole, Opole, Poland; lata@po.opole.pl*
‡ *LAS/CCET/PUCPR, Curitiba, Brazil; oliv@rla01.pucpr.br*

Abstract. A new, analytical, orthonormal basis functions (OBF)-based design methodology for adaptive predictive constrained control of open-loop stable, possibly nonminimum phase, time-varying Wiener and Hammerstein systems is presented. A linear adaptive least-squares parameter estimation algorithm is applied both to a nonlinear static part and a linear dynamic, OBF-modeled factor of the Wiener/ Hammerstein system. A notion of inverse systems is crucial for linear estimation of both Wiener and Hammerstein systems, with inverses of the nonlinear or linear parts respectively involved. The adaptive estimator is coupled with a simple but robust, predictive control strategy called Extended Horizon Model Algorithmic Control, with input/output constraints handled in a trivial way. Simulation examples demonstrate computational and numerical effectiveness of the new adaptive nonlinear constrained control approach.

Keywords: Adaptive control, adaptive systems, nonlinear models, least-squares estimation, time-varying systems, predictive control, nonlinear control.

1. INTRODUCTION

Simple structures of nonlinear Wiener and Hammerstein models make them attractive in control applications. In particular, when a nonlinear static characteristic is invertible, the effect of nonlinearity can be essentially reduced, or theoretically eliminated, by an inverse serial compensator. Therefore, Wiener and Hammerstein models have attracted a considerable interest both from academic and industrial environments. A number of various solutions have been offered to estimate Wiener and Hammerstein models (Bai, 1998; Garulli *et al.*, 2002; Greblicki, 1997, 2002; Hasiewicz, 1999; Norquay *et al.*, 1999; Pearson and Pottman, 2000; Wigren, 1993; Zhao *et al.*, 2001; Zhu, 2000). Those numerical estimation methodologies are rather computationally involving and can hardly be used in the adaptive control environment, in particular for 'fast' systems requiring frequent sampling. This also concerns the attractive techniques involving the inverse static nonlinearity (Kalafatis *et al.*, 1995; Pearson and Pottman, 2000). Recently, an interesting recursive least-squares scheme for linear parameter estimation of ARX-based Wiener model has been presented (Janczak, 2001). The method is based on the assumption that a nonlinear static characteristic is invertible, in which case a polynomial approximation of the inverse characteristic can be exploited. With well-known limitations in application of the classical polynomial approximation (Zhu, 2000), one disadvantage may present prohibitive burdens in adaptive estimation/control: the parameter estimation problem may often be numerically ill conditioned. On the other hand, ARX-based Wiener modeling (e.g. by Janczak, 2001) involves excessively large number of parameters to be estimated (due to the lack of separation of linear and nonlinear parts of the model). The impact of the two drawbacks can be essentially reduced when using a new, adaptive least-squares, low-order, *o*rthonormal *b*asis *f*unctions (OBF)-based Wiener modeling method (Marciak *et al.*, 2001), which employs the inverse static nonlinearity concept without the necessity to invert a linear part of the system. The method takes advantage of the output-error structure of OBF models, which is particularly magnified for nonlinear Wiener modeling. On the other hand, employing the inverse model concept to a linear dynamic part of the Hammerstein model can, again, lead to the linear parameter estimation problem, with the separation of linear and nonlinear parts of the model and without getting into the bilinearity issue. Therefore, making use of our OBF-based Hammerstein modeling offers, again, an essential reduction of computational burden as compared with the ARX-based one.

This new, inverse model-based, adaptive, linear estimation methodology for Wiener/Hammerstein systems is here combined with a simple predictive control strategy called Extended Horizon Model Algorithmic Control (EHMAC) (Latawiec, 1997; Latawiec and Rojek, 2000; Latawiec *et al.*, 2001). IMC-structured EHMAC is redesigned to control open-loop stable nonminimum phase Wiener/ Hammerstein systems, with input/output constraints handled in an a trivial way. In another interesting, robustness-oriented, reduced-order, OBF-based, nonlinear identification and predictive control method (Oliveira and Latawiec, 2001), input/output constraints have been dealt with using the classical, sequential quadratic programming tools.

The remainder of this paper is organized as follows. Section 2 compares new, ARX- and OBF-based modeling solutions for Wiener systems. The adaptive parameter estimation issue, both for linear dynamic and nonlinear static parts, is presented in the linear regression framework in Section 3. A linear parameter estimation solution is presented in a similar way for the Hammerstein system in Section 4. In Section 5, predictive constrained EHMAC is extended to nonlinear Wiener/Hammerstein models. Simulation examples are presented in Section 6 and new results of the paper are summarized in conclusions of Section 7.

2. ARX- VERSUS OBF-BASED WIENER MODELS

In a single-input single-output Wiener system, a linear dynamic part is cascaded with a nonlinear static element. The discrete-time output $y(t)$ can be expressed as $y(t)=f[G(q)u(t)+e(t)]$, where f is a nonlinear function, $G(z)$ is a Z-transfer function of a linear part, q is the forward shift operator, $u(t)$ is the system input and $e(t)$ is the disturbance term. It is assumed that 1) a linear dynamic system is asymptotically stable and 2) a nonlinear function f is continuous and invertible. It is well known that the first assumption can be weakened, so that linear systems with their pole(s) on the unit circle can be easily tractable. Also, open-loop unstable minimum phase plants can be effectively dealt with (Latawiec *et al.*, 2003). On the other hand, the invertibilty condition of a nonlinear part can be relaxed, e.g. making use of constrained control strategies.

The output $\hat{y}(t)$ of the Wiener model, or the system output predictor, can be calculated as

$$\hat{y}(t) = \hat{f}[\hat{G}(q)u(t)] \tag{1}$$

where the estimate $\hat{G}(z)$ of a transfer function of a linear part is given either as

$$\hat{G}(z) = \hat{B}(z^{-1}) / \hat{A}(z^{-1}) \tag{2}$$

or (Bokor *et al.*, 1999; Latawiec *et al.*, 2000)

$$\hat{G}(z) = \sum_{i=1}^{M} \hat{c}_i L_i(z^{-1}) \tag{3}$$

for ARX and OBF models, respectively, with the familiar ARX-model polynomials $\hat{B}(z^{-1})$ and $\hat{A}(z^{-1})$ being of orders n, whereas the weighting parameters $\hat{c}_i$, i=1,...,M, and orthonormal transfer functions $L_i(z^{-1})$ characterize an OBF model.

In case of discrete Laguerre models to be exploited hereinafter, the orthonormal transfer functions

$$L_i(z) = \frac{\sqrt{1-p^2}}{z-p}\left[\frac{1-pz}{z-p}\right]^{i-1} \qquad i=1,...,M \tag{4}$$

consist of a first-order low-pass factor and (i-1)th-order all-pass filters. Assuming that the dominant pole p is known (or tuned, Latawiec *et al.*, 2000), the estimation problem reduces to determination of parameter estimates $\hat{c}_i$, i=1,...,M, a number M of which is quite low as compared to a number of components of another familiar orthonormal model, that is finite impulse response (FIR).

3. PARAMETER ESTIMATION PROBLEM FOR THE WIENER SYSTEM

Assuming that a nonlinear static characteristic is invertible we can rewrite equation (1) in form

$$\hat{f}^{-1}[\hat{y}(t)] = \hat{G}(q)u(t) \tag{5}$$

The function $\hat{f}^{-1}[\hat{y}(t)]$ can be approximated with e.g. a polynomial expansion

$$\hat{f}^{-1}[\hat{y}(t)] = \hat{a}_1\hat{y}(t) + \hat{a}_2\hat{y}^2(t) + ... + \hat{a}_{\mathrm{m}}\hat{y}^{\mathrm{m}}(t) \tag{6}$$

Without loss of generality, the leading coefficient can be put equal to unity. In fact, the gain of a linear contributor to the polynomial approximation can be estimated from a linear dynamic part of the Wiener model.

We will continue now with the OBF-based Wiener model. Detailed derivations for the nonlinear ARX-based model have been given by Janczak (2001). Combining equations (3), (5) and (6) while replacing, in a standard manner (Janczak, 2001), parameter estimates with parameters themselves and the model output with the system output itself (wherever appropriate) we arrive at the linear regression function

$$\hat{y}(t) = \sum_{i=1}^{M} c_i L_i(q^{-1})u(t) - a_2 y^2(t) - \ldots - a_m y^m(t) \quad (7)$$

which can be presented in the familiar form

$$\hat{y}(t) = \boldsymbol{\varphi}^{\mathrm{T}}(t)\boldsymbol{\theta} \quad (8)$$

with $\boldsymbol{\theta}^{\mathrm{T}} = [c_1\ c_2 \ldots c_M\ a_2\ a_3 \ldots a_m]$ and $\boldsymbol{\varphi}^T(t) = [v_1(t)\ v_2(t) \ldots v_M(t)\ -y^2(t)\ -y^3(t) \ldots -y^m(t)]$, where the OBF-related regressor elements $v_i(t)=L_i(q^{-1})u(t)$, $i=1,\ldots,M$, can be conveniently calculated in a recursive way.

Recall that in case of similar, ARX-based Wiener modeling (Janczak, 2001) the regressor is of form $\boldsymbol{\varphi}^T(t)=[-y(t-1) \ldots -y(t-n)\ u(t-1) \ldots u(t-n)\ -y^2(t) \ldots -y^m(t)\ -y^2(t-1) \ldots -y^m(t-1) \ldots -y^2(t-n) \ldots -y^m(t-n)]$. Thus, in that equation-error modeling approach as many as $2n+(n+1)(m-1)$ parameters have to be estimated, of which only $2n+m-1$ estimates are finally utilized in the ARX-based model (1) of the Wiener system. In fact, the remaining $n(m-1)$ parameters are 'needlessly' estimated, thus increasing dimensions of the covariance matrix and so the variance error of a parameter estimator. The reason is obvious: the lack of separation of linear and nonlinear parts of the model. In contrast, only $M+m-1$ parameters have to be estimated in the output error-related model of the OBF-structured Wiener system, and all these estimates are included in the model (1).

In addition to lower variance errors, the bias error of an estimator can also be expected to be lower in case of parameter estimation of the Wiener system based on OBF models. In fact, parameter estimates for the output error model structure are unbiased, even for correlated additive disturbances. Moreover, numerical conditioning of the OBF-based estimation problem is essentially better than that for the ARX-based one (Bokor *et al.*, 1999; Latawiec *et al.*, 2000; Latawiec *et al.*, 2001). This results from the orthonormality property, which is *not* associated with ARX modeling, however. It can be concluded that the superiority of OBF models over ARX ones is still magnified when using the former models in the nonlinear, Wiener system environment, especially in terms of reduced variance errors of an estimator. This will be confirmed in simulations.

3.1. Adaptive estimation of OBF-Wiener models.

An unknown parameter vector $\boldsymbol{\theta}$ can be estimated by one of available estimation algorithms. In order to examine the behaviour of OBF-Wiener models under time-varying plant parameters we employ adaptive algorithms. Specifically, we use a standard recursive least-squares estimation scheme with exponential forgetting, to be referred to as an Adaptive Least-Squares (ALS) algorithm (Latawiec *et al.*, 2000). It is interesting that, owing to the improved numerical conditioning properties of the OBF-based estimation scheme, there is no need to use a more sophisticated (and complex) ALS algorithm developed for the ARX model (Latawiec, 1998).

It is also worth mentioning that, owing to the orthonormality property improving the numerics of the estimator, it is not necessary to introduce e.g. square root update of the covariance matrix into the algorithm. The convergence of the algorithm is controlled by means of the trace of the ALS system transition matrix, that is $\mathrm{tr}\{[\boldsymbol{I} - \boldsymbol{K}(t)\boldsymbol{\varphi}^T(t)]\}$, where $\boldsymbol{K}(t)$ is the Kalman gain of the estimator.

4. PARAMETER ESTIMATION PROBLEM FOR THE HAMMERSTEIN SYSTEM

In the Hammerstein system, a nonlinear static part is cascaded with a linear dynamic element. Assuming e.g. a polynomial approximation of the nonlinear part, the output $\hat{y}(t)$ of the Hammerstein model can be expressed as

$$\hat{y}(t) = \hat{G}(q) \sum_{i=1}^{m} a_i u^i(t) \quad (9)$$

which can be rewritten as

$$\hat{G}^{-1}(q)\hat{y}(t) = \sum_{i=1}^{m} a_i u^i(t) \quad (10)$$

thus yielding, again, the separation of the linear and nonlinear components of the model.

Without loss of generality, the coefficient a_1 can be put equal to unity. Assuming again that the linear part is asymptotically stable one can approximate its dynamics by FIR, with the components g_i, $i=0,1,\ldots,N-1$. Alternatively, one can operate with the components r_j, $j=0,1,\ldots,L-1$, of FIR of the inverse $R(q)=\hat{G}^{-1}(q)$, with g_is and r_js easily recalculated from each other. (Note that $\hat{G}(q)$ need not be stable now.) The only problem is that for systems with time delay d one can obtain $\hat{G}^{-1}(q) = = r_0 q^d + r_1 q^{d-1} + \ldots + r_d + r_{d+1} q^{-1} + \ldots + r_{L-1} q^{-L+d+1}$.

However, one can easily rewrite equation (10) as

$$\hat{y}(t)(1+\underline{r}_1 q^{-1}+...+\underline{r}_{L-1} q^{-L+1})=\sum_{i=1}^{m}\alpha_i u^i(t-d) \quad (11)$$

with $\underline{r}_j = r_j/r_0$, $j=1,\ldots,L-1$, and $\alpha_1=1/r_0$, $\alpha_i=a_i/r_0$, $i=2,\ldots,m$. After simple manipulations, this is just the equation (8), with $\boldsymbol{\theta}^{\mathrm{T}}=[\underline{r}_1\ \underline{r}_2\ \ldots\ \underline{r}_{L-1}\ \alpha_1\ \alpha_2\ \ldots\ \alpha_m]$ and $\boldsymbol{\varphi}^{\mathrm{T}}(t)=[-y(t-1)\ -y(t-2)\ \ldots\ -y(t-L+1)\ u(t-d)\ u^2(t-d)\ \ldots u^m(t-d)]$.

Now, the unknown parameters g_j, $j=0,1,\ldots,N-1$, and a_i, $i=1,\ldots,m$, can be easily recovered from the ALS estimated $\underline{r}_j$s and α_is. Also, an inverse characteristic $\hat{f}^{-1}[\hat{y}(t)]$ can be constructed.

Instead of using the FIR model one can immediately employ e.g. the Laguerre model as in equations (3) and (4), receiving $\boldsymbol{\theta}^{\mathrm{T}}=[c_1\ c_2\ \ldots\ c_M\ \alpha_1\ \alpha_2\ \ldots\ \alpha_m]$ and $\boldsymbol{\varphi}^{\mathrm{T}}(t)=[-v_1\ -v_2\ \ldots\ -v_M\ u(t-d)\ u^2(t-d)\ \ldots\ u^m(t-d)]$, with $M \ll L$ (and v_is driven by $y(t)$). Notice how the bilinearity involved in e.g. ARX-based modeling is avoided here, thus contributing (like for OBF-based Wiener modeling) to reduction of computational burden. It is the authors' belief that the above is a new effective solution to the parameter estimation problem for the Hammerstein system, which can be an attractive alternative to the one presented by Ninness and Gibson (2002). Additional advantages of our inverse OBF-based solution are (Latawiec *et al.*, 2003): 1) open-loop unstable minimum phase systems can be OBF-modeled, 2) Laguerre filters can be effectively used to model oscillatory systems (instead of Kautz functions).

5. NONLINEAR EHMAC

The EHMAC strategy, being a simple combination of Extended Horizon Predictive Control and Model Algorithmic Control, has proved to be an effective tool to control open-loop stable, possibly nonminimum phase systems (Latawiec, 1997; Latawiec and Rojek, 2000; Latawiec *et al.*, 2001). Having estimated an inverse static characteristic of the Wiener/Hammerstein model, we can immediately use it to form the nonlinear EHMAC law

$$\Gamma(q^{-1})\,u(t)=\hat{f}^{-1}[\hat{y}(t)]\,F(q^{-1})\,[y_{ref}-y(t)+\hat{y}(t)] \quad (12)$$

where y_{ref} is the output reference, $F(z^{-1}) = [(1-\beta z^{-1})^{-1}(1-\beta)]^{\kappa}$ represents the simple IMC filter (with $0\le\beta<1$ and κ being an integer to be further equal to unity) and the control polynomial Γ is of form

$$\Gamma(q^{-1})=\gamma_0+\sum_{i=k-d+1}^{N-1} g_i q^{-i+k-d} \quad (13)$$

where, e.g. for the EHPC1 version of EHPC (Latawiec and Rojek, 2000)

$$\gamma_0=\sum_{i=0}^{k-d} g_i \quad (14)$$

with k being the prediction horizon ($k\ge d$), and g_i, $i=0,1,\ldots,N-1$, are the FIR components of the model of a linear part of the Wiener/Hammerstein system. The two control tuning parameters, i.e. k and β, provide a range of either accuracy/speed- or robustness-oriented performances of the controller (Latawiec and Rojek, 2000).

It is interesting to note that the Γ polynomial represents the 'best' (in the sense of the EHPC criterion) minimum phase approximation to the nonminimum phase model of a linear part of the system, so that Γ^{-1} is the 'best' approximate inverse of that part.

In Wiener/Hammerstein-related EHMAC, components of the FIR model, used in the EHMAC law (12), can be recalculated from estimated parameters c_i, $i=1,\ldots,M$, of the OBF model

$$\bar{g}_i=\sum_{i=1}^{M} c_i L_i(q^{-1})\delta(t) \qquad t=0,\ldots,N+d \quad (15)$$

$$g_i=\bar{g}_{i+d+1} \qquad i=0,\ldots,N-1 \quad (16)$$

where $\delta(t)$ is the unit impulse. Thus, a number of FIR components need not be estimated, nor is recalculation from possibly estimated ARX model parameters necessary, the latter being ineffective due to the equation-error model structure involved (Latawiec *et al.*, 2000; Marciak *et al.*, 2001).

5.1. Constrained EHMAC

Predictive control algorithms are often required to operate under input and/or output signal constraints. Usually, available constrained predictive control methods are computationally involved, as they typically employ quadratic programming machinery. This may create computational problems, especially in multivariable and/or adaptive control of 'fast' plants.

A solution to the constrained control problem is trivial for EHMAC. Since only current input/output signals are subject to constraints (as opposed to eg. GPC), the constrained (and yet optimal) signals simply enter directly the control law (12). Thus, the simple structure of EHMAC is particularly useful in the constrained control environment.

6. SIMULATION RESULTS

Of a number of simulation runs performed, we have selected two representative examples of ALS estimation and control of Wiener systems, for which the above methodology is particularly useful.

Example 1.
Consider the Wiener system consisting of a linear plant

$$\hat{G}(z) = \frac{0.01z^{-1} - 0.1z^{-2} + 0.24z^{-3} - 0.155z^{-4} + 0.03z^{-5}}{1 - 2.98z^{-1} + 3.4z^{-2} - 1.85z^{-3} + 0.48z^{-4} - 0.048z^{-5}}$$

coupled with a nonlinear static element described by the characteristic

$$f[G(q)u(t)] = \text{atan}\,[G(q)u(t)]$$

First we compare estimation performance for ARX- and OBF-based Wiener models under PRBS excitation, which is known not to be the optimal one (Zhu, 2000). For the case of constant parameters we set the forgetting factor λ equal to unity. An order m of the approximating polynomial is chosen to be equal to 13. Thus, the number of estimated parameters of the ARX-based Wiener model is 82, of which only 22 (!) estimates are finally utilized in the model (1). Fig. 1 presents the plots of estimated and original characteristics of a nonlinear part of the Wiener system. The characteristics can be considered indistinguishable and so are FIRs of the linear part of the system and its model.

For the Laguerre-based Wiener model with m=13, choosing M=14 is sufficient for estimates both of the static characteristic and the linear part to become indistinguishable from the original ones. Thus, estimating only 26 model parameters is sufficient to achieve high estimation accuracy.

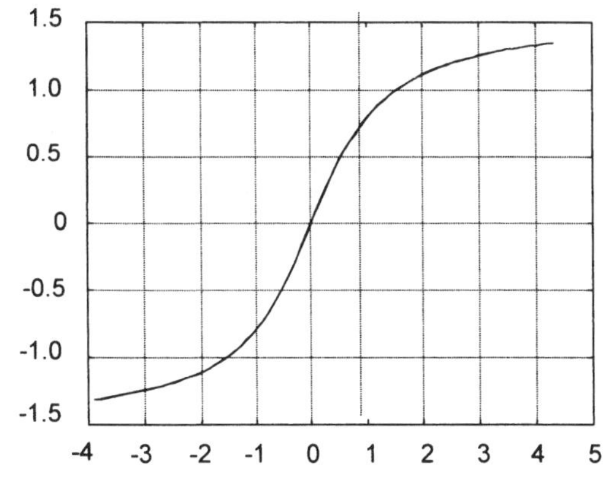

Fig. 1. Nonlinear static characteristic and its estimate, Example 1.

Definitely, an excessively high number of estimated parameters of the ARX-based Wiener model leads to increased variance errors of the estimator, in addition to a bias error in case of correlated disturbances. We have examined the ALS estimation accuracy in the stochastic case of additive correlated noise $e(t)$. Estimator variances for ARX-based modeling were at least twice as large as for OBF-based identification of the Wiener system. Note that the inefficiency of ARX-based modelling of Wiener systems, apparently magnified as compared to the linear identification case, results from the specific equation error structure involved.

Example 2.
A nonlinear part of the Wiener system is the same as before, whereas a linear part is of form

$$G(z) = \frac{-0.1(z-0.2)(z-0.5)(z-1.5)}{(z-0.25)(z-0.45)(z-0.55)(z-0.8)(z-0.9)}$$

At time t=350 one zero of the plant is changed from 0.2 to 0.5, at time t=600 the largest pole is raised from 0.9 to 0.93 and at time t=800 the original time delay equal to 2 is increased up to 4. The order M is selected as 8, which is fairly low, considering the nature of the plant under study. In fact, it does not provide estimated FIR of the linear part to be indistinguishable from the original one. Until t=300 the plant is excited with PRBS, after which feedback control with nonlinear adaptive OBF-based EHMAC is started, with y_{ref}=10 for $t\in(300,500]$, y_{ref}=20 for $t\in(500,700]$ and y_{ref}=10 for $t\geq 700$. The forgetting coefficient λ is set to 0.98. An order m is selected as 13. The prediction horizon k is equal to 8. Until $t<700$, the EHMAC algorithm is run without any constraint and then the constraint is imposed on the control signal $u(t)>1.3$. Fig. 2 presents plots of the reference, plant output (top) and input (bottom) variables. Observe the correct performance of the adaptive control algorithm under poor excitation resulting from only occasional set-point changes.

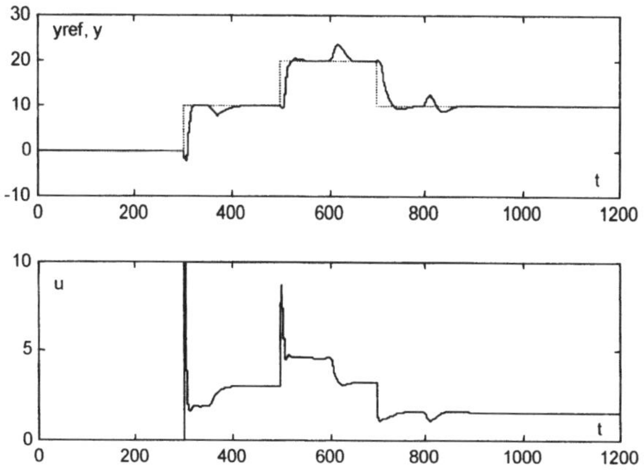

Fig. 2. Performance of nonlinear adaptive EHMAC; Example 2.

7. CONCLUSION

This paper presents a new analytical solution to the problem of nonlinear adaptive predictive constrained control. A new OBF-based adaptive estimation methodology provides effective means for reconstruction both of an inverse nonlinear static characteristic and a transfer function of open-loop stable nonminimum phase linear part of the Wiener system. The OBF-based Wiener modeling approach has proved in simulations to be much more effective than its ARX-based counterpart, both in terms of essentially lower computational burden and better numerical conditioning of the ALS parameter estimator. Similarly, a new analytical solution to the parameter estimation problem has also been proposed for the Hammerstein system, with inverse OBF modeling effectively employed. The crucial concept of the inverse modeling approach for either nonlinear or linear part of the Wiener/Hammerstein system contributes to the separation of the two parts of the model, while making use of the OBF modeling tools enables to get rid of the bilinearity problem and to reduce computational burden.

The new adaptive estimators for Wiener/ Hammerstein systems have been coupled with a simple nonlinear version of a predictive constrained control algorithm called EHMAC. Simulation results have demonstrated the effectiveness of the new nonlinear estimation/control approach.

It is worth mentioning that the EHMAC algorithm has been effectively implemented in the MATLAB/ SIMULINK/DSPACE environment on a laboratory-scale electric motor.

REFERENCES

Bai, E.-W. (1998). An optimal two-stage identification algorithm for Hammerstein-Wiener nonlinear systems. *Automatica*, **34**, 333-338.

Bokor, J., P. Heuberger, B. Niness, T.O. e Silva, P. Van den Hof and B. Wahlberg (1999). Modelling and identification with orthogonal basis functions. *Proc. Preconference Workshop, 14th IFAC World Congress*, Beijing, P.R. China.

Garulli, A., L. Giarre and G. Zappa (2002). Identification of approximated Hammerstein models in a worst-case setting. *IEEE Trans. Automat. Contr.*, **47**, 2046-2050.

Greblicki, W. (1997). Nonparametric approach to Wiener system identification. *IEEE Trans. Circuits and Systems*, **44**, 538-545.

Greblicki, W. (2002). Stochastic approximation in nonparametric identification of Hammerstein systems. *IEEE Trans. Automat. Contr.*, **47**, 1800-1810.

Hasiewicz, Z. (1999). Hammerstein system identification by the Haar multiresolution approximation. *Int. J. Adapt. Contr. and Signal Processing*, **13**, 191-217.

Janczak, A. (2001). On identification of Wiener systems based on a modified serial-parallel model. *Proc. European Control Conference ECC 2001,* Porto, Portugal, 1852-1857.

Kalafatis, A., N. Arifin, L. Wang and W.R. Cluett (1995). A new approach to the identification of pH processes based on the Wiener model. *Chemical Engineering Science*, **50**, 3693-3701.

Latawiec, K.J. (1997). Extended horizon adaptive model algorithmic control. *Proc. 11th IFAC Symp. on System Identification (SYSID'97)*, Kitakyushu, Japan, **1**, 297-302.

Latawiec, K.J. (1998). Towards robust adaptive least-squares parameter estimation with internal feedback, *Proc. IFAC Workshop on Adaptive Systems in Control and Signal Processing*, Glasgow, Scotland, 442-447.

Latawiec, K.J. and Rojek R. (2000). EHMAC - a new simple tool for robust linear multivariable control. *Int. J. of Applied Mathematics and Computer Science*, **10**, 101-116.

Latawiec, K.J., C. Marciak, R. Rojek and G.H.C. Oliveira (2003). Modeling and parameter estimation of open-loop unstable systems by means of orthonormal basis functions. To be presented at the 9th *IEEE MMAR Conference*, Miedzyzdroje, Poland.

Latawiec, K.J., R. Rojek, G.H.C. Oliveira and C. Marciak (2000). Adaptive parameter estimation of OBF models. *Proc. 6th MMAR Conference*, Miedzyzdroje, Poland, 927-932.

Latawiec, K.J., R. Rojek, C. Marciak and G.H.C. Oliveira (2001). Robust adaptive OBF-based constrained EHMAC: simple and effective. *Proc. 5th World Multiconference on Systemics, Cybernetics and Informatics*, Orlando, Florida, **IX**, 265-270.

Marciak, C., K.J. Latawiec, R. Rojek and G.H.C. Oliveira (2001). Adaptive least-squares parameter estimation of OBF-based Wiener models. *Proc. 7th IEEE MMAR Conference*, Miedzyzdroje, Poland, **2**, 965-999.

Ninness, B. and S. Gibson (2002). Quantifying the accuracy of Hammerstein model estimation. *Automatica*, **38**, 2037-2051.

Norquay, S.N., A. Palazoglu and J.A. Romagnoli (1999). Application of Wiener model predictive control (WMPC) to a pH neutralization experiment. *IEEE Trans. on Automatic Control*, **7**, 437-445.

Oliveira, G.H.C. and K.J. Latawiec (2001). Methodology for identification and control of nonlinear systems using Laguerre-Volterra models. *Proc. 7th IEEE MMAR Conference*, Miedzyzdroje, Poland, **2**, 921-926.

Pearson, R.K. and M. Pottman (2000). Gray-box identification of block-oriented nonlinear models. *Journal of Process Control*, **10**, 301-315.

Wigren, T. (1993). Recursive prediction error identification algorithm using the nonlinear Wiener model. *Automatica*, **29**, 1011-1025.

Zhao, H. and J. Guiver, R. Neelakantan and L.T. Biegler (2001). A nonlinear industrial model predictive controller using integrated PLS and neural net state-space model. *Control Engineering Practice*, **9**, 125-133.

Zhu, Y. (2000). Identification of Hammerstein models for control using ASYM. *Int. Journal of Control*, **73**, 1692-1702.

IFAC Publications
www.elsevier.com/locate/ifac

IDENTIFICATION OF WIENER SYSTEMS USING REDUCED COMPLEXITY VOLTERRA MODELS

Rıfat Hacıoğlu * **Geoffrey A. Williamson** **

* *Dept. of Elec.&Electronics Engr., Zonguldak Karaelmas University, Zonguldak 67100 TURKEY*
e-mail: hacirif@iit.edu
** *Dept. of Elec.&Computer Engr., Illinois Institute of Technology, Chicago, IL 60616 USA*
e-mail: williamson@iit.edu

Abstract: In this paper we study the identification problem of single-input, single-output Wiener systems with polynomial nonlinearities. Wiener systems can be represented by a cascade of a dynamic linear subsystem followed by a static nonlinearity. Our approach is to use a reduced complexity Volterra model structure called fixed pole expansion technique (FPET) to estimate the products of the coefficients of the nonlinearity and the linear subsystems coefficients. We then present a method using the singular value decomposition to extract the coefficients of the nonlinearity and of the dynamic linear subsystem. *Copyright © 2003 IFAC*

Keywords: Nonlinear system identification, Wiener systems, Volterra models, orthonormal functions, Laguerre functions, Kautz functions, singular value decomposition.

1. INTRODUCTION

The identification of nonlinear systems has attracted considerable interest in system theory, control design and signal processing. The reason for such an interest is that most natural or man-made systems are inherently nonlinear. There is a wide variety of nonlinear system structures so that the first task is to determine the most appropriate structure for the underlying problem. Several approaches have been developed for this problem, including black box and grey box techniques (Bai, 1998; Greblicki, 1992; Hunter and Korenberg, 1986; Hacıoğlu and Williamson, 2001; Lacy and Bernstein, 2002; Ljung, 1987; Marmarelis, 1993). The Volterra model representation (Schetzen, 1980) has been a popular model in the black box approach. The gray box approaches include the identification of the block structured models such as Hammerstein model (a nonlinear memoryless system is followed by a linear dynamic one) and Wiener model (the same subsystems connected in reverse order) (Greblicki, 1992; Hunter and Korenberg, 1986; Lacy and Bernstein, 2002). We in this paper study the identification of Wiener systems using a reduced complexity Volterra model structure (Hacıoğlu and Williamson, 2001).

The Wiener system identification problem has been considered in (Greblicki, 1992; Hunter and Korenberg, 1986) under the assumption that the nonlinearity is invertible. The invertibility assumption simplifies the problem. However, we in this paper consider Wiener system identification in which the nonlinear block is both unknown and not necessarily invertible. In this case the goal is to identify both the linear subsystem and static nonlinearity coefficients.

The Volterra model structure can be used to identify a wide class of nonlinear systems (Boyd and Chua, 1985; Schetzen, 1980). Wiener systems, which have a linear dynamic system followed by a static nonlinearity, each have a unique Volterra model representation. We use this property to extract the Wiener blocks from the Volterra model coefficients. Due to the high memory requirement in some nonlinear systems, the Volterra model structure requires a huge number of parameters. Reducing the parametric complexity

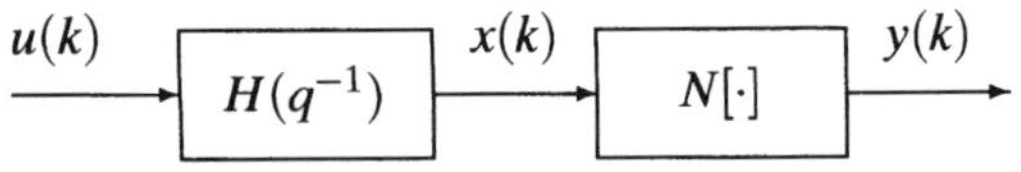

Fig. 1. Wiener System

has been studied in (Hacıoğlu and Williamson, 2001) using a fixed pole expansion technique (FPET). The Volterra-FPET approach is investigated in this paper in order to identify the Wiener systems with a reduced parametric complexity.

This paper is organized as follows. In section 2, the Volterra model representation of Wiener systems is described. The reduction of parametric complexity of the Volterra model structure is considered in section 3 and the FPET approach is presented. In section 4, we introduce a fixed pole selection approach for FPET. We present a solution to identify the coefficients of linear and nonlinear blocks in the Wiener model using the FPET model via a multi-dimensional singular value decomposition. In section 6, we give an illustrative example to discuss the performance.

2. WIENER BLOCK STRUCTURE

We study here the identification of single-input, single-output, linear time invariant dynamic system whose output is measured through a static nonlinearity. The Wiener system is represented in Fig.1. Here we assume that the nonlinearity is written as a finite power series. If that is not satisfied, then an approximation is considered. The polynomial nonlinearity is unknown and not necessarily invertible.

We consider the Wiener model

$$x(k) = \sum_{l=0}^{\infty} h(l)u(k-l) \qquad (1)$$

$$y(k) = \sum_{n=0}^{N} c_n x(k)^n \qquad (2)$$

where $u(k)$ is the input to the system, $x(k)$ is the unmeasured output of the linear subsystem, and $y(k)$ is the measured output of the nonlinearity. We may write the input-output relationship for an Nth order Wiener system as

$$\begin{aligned} y(k) &= c_0 + \sum_{l_1=0}^{\infty} c_1 h(l_1)u(k-l_1) \\ &+ \sum_{l_1=0}^{\infty}\sum_{l_2=0}^{\infty} c_2 h(l_1)h(l_2)u(k-l_1)u(k-l_2) + \ldots \\ &+ \sum_{l_1=0}^{\infty}\cdots\sum_{l_N=0}^{\infty} c_N h(l_1)\cdots h(l_N)u(k-l_1)\cdots u(k-l_N) \end{aligned} \qquad (3)$$

where $h(l)$ is the impulse response of the linear dynamic system

$$H(q^{-1}) = \frac{B(q^{-1})}{A(q^{-1})} = \sum_{l=0}^{\infty} h(l)q^{-l} \qquad (4)$$

with q^{-1} the delay operator ($q^{-l}u(k) = u(k-l)$).

There is a straightforward relationship between (3) and Volterra model structure

$$\begin{aligned} y(k) &= h_0 + \sum_{l_1=0}^{\infty} h_1(l_1)u(k-l_1) \\ &+ \sum_{l_1=0}^{\infty}\sum_{l_2=0}^{\infty} h_2(l_1,l_2)u(k-l_1)u(k-l_2) + \cdots \\ &+ \sum_{l_1=0}^{\infty}\cdots\sum_{l_N=0}^{\infty} h_N(l_1,\ldots,l_N)u(k-l_1)\cdots u(k-l_N) \end{aligned} \qquad (5)$$

where $h_n(l_1,\ldots,l_n)$ is the nth order Volterra kernel of the system. The output $y(k)$ at a particular discrete-time k can be seen as depending, in a nonlinear way, on all values of the input at times prior to k, i.e. $y(k)$ depends on $u(k-l_i)$ for all $l_i \geq 0$. All the infinite sums are of course truncated in practice to obtain an approximate representation.

3. FIXED POLE EXPANSION TECHNIQUE

For higher memory length and/or model order, estimation of Volterra kernel parameters becomes much more difficult. We suggest an approach to solve this problem: the Fixed Pole Expansion Technique (FPET). The Volterra kernels are expanded on a fixed pole basis that transforms (5) into the multinomial power series expression

$$\begin{aligned} y(k) &= b_0 + \sum_{l_1=0}^{M} b_1(l_1)x_{l_1}(k) \\ &+ \sum_{l_1=0}^{M}\sum_{l_2=0}^{M} b_2(l_1,l_2)x_{l_1}(k)x_{l_2}(k) + \cdots \\ &+ \sum_{l_1=0}^{M}\cdots\sum_{l_N=0}^{M} b_N(l_1,\ldots,l_N)x_{l_1}(k)\cdots x_{l_N}(k). \end{aligned} \qquad (6)$$

In the matrix form, we may write

$$y(k) = \varphi^T(k)\theta \qquad (7)$$

where the parameter vector θ contains the weight parameters $b_n(l_1,\ldots,l_n)$, and the regressor vector $\varphi(k)$ contains not only $x_i(k)$ but also their cross terms at time k such that

$$\theta = [b_0, b_1(0), \ldots, b_1(M), \ldots, b_N(M,\ldots,M)]^T, \qquad (8)$$

$$\varphi(k) = [1, x_0(k), \ldots, x_M(k), \ldots, x_0^N(k), \ldots, x_M^N(k)]^T. \tag{9}$$

We will estimate the unknown weight parameters θ from input-output data sets by using linear regression of the output data $y(k)$ on the terms of the multinomial expansion of (6), as long as the data length is finite. The signals $x_i(k)$ derive from the input $u(k)$ via

$$x_i(k) = \sum_{l=0}^{\infty} g_i(l)u(k-l),\ i = 0, 1, \ldots, M \tag{10}$$

using impulse response functions $g_i(l)$ of filters $G_i(z)$. Our goal is to choose $G_0(z), \ldots, G_M(z)$ so that a good model may be obtained with small M. To this end we consider the set of orthonormal basis functions. We specify the orthonormal basis functions for either real poles or complex-conjugate pole pairs.

Definition 3.1. Fixed real pole basis functions can be defined as

$$G_k(z) = \frac{\sqrt{1-\alpha_k^2}}{(z-\alpha_k)} \prod_{i=1}^{k-1} \frac{(1-\alpha_i z)}{(z-\alpha_i)}, \tag{11}$$

$k = 0, 1, \cdots, M$ if the poles are real numbers in the unit circle. Also, $G_0(z) = 1$.

Definition 3.2. The sequence of fixed complex pole basis functions

$$G_{2k-1}(z) = K_{2k-1} \frac{(z-1)}{(z-\beta_k)(z-\beta_k^*)} \prod_{i=1}^{k-1} H_i(z), \tag{12}$$

$$G_{2k}(z) = K_{2k} \frac{(z+1)}{(z-\beta_k)(z-\beta_k^*)} \prod_{i=1}^{k-1} H_i(z) \tag{13}$$

forms an orthonormal set by the choice of

$$K_{2k-1} = \sqrt{\frac{(1+\beta_k)(1+\beta_k^*)(1-\beta_k\beta_k^*)}{2}}, \tag{14}$$

$$K_{2k} = \sqrt{\frac{(1-\beta_k)(1-\beta_k^*)(1-\beta_k\beta_k^*)}{2}} \tag{15}$$

$k = 1, \cdots, M/2$, and

$$H_i(z) = \frac{(1-\beta_i z)(1-\beta_i^* z)}{(z-\beta_i)(z-\beta_i^*)} \tag{16}$$

where the complex conjugate pole pairs, (β_k, β_k^*) are in the unit circle. Also, $G_0(z) = 1$.

Note here that if $\{\alpha_k\}$ are a simple real pole $\{\alpha\}$ in (11) then $\{G_j(z)\}$ are called Laguerre functions (Marmarelis, 1993; Wahlberg, 1994), and also if $\{\beta_k\}$ are a simple complex-conjugate pole pair in (12)-(16), then $\{G_j(z)\}$ are called Kautz functions (Wahlberg, 1994). Note also that if $\{\alpha_k\}$ are all zero then (6) becomes the FIR model (Ljung, 1987).

The FPET approach naturally arises from the Volterra model structure. It has been shown in (Williamson and Zimmerman, 1996) that an M pole fixed pole system can realize any transfer function in the form (4) by choice of weight parameters such as

$$\begin{aligned} H(q^{-1}) &= \sum_{l=0}^{M} b(l)G_l(q^{-1}) \\ &= \sum_{l=0}^{M} b(l) \sum_{k=0}^{\infty} g_l(k)q^{-k}. \end{aligned} \tag{17}$$

Based on the Wiener block structure representation, (6) can be written as

$$\begin{aligned} y(k) = c_0 &+ \sum_{l_1=0}^{M} c_1 b(l_1) x_{l_1}(k) \\ &+ \sum_{l_1=0}^{M} \sum_{l_2=0}^{M} c_2 b(l_1) b(l_2) x_{l_1}(k) x_{l_2}(k) + \cdots \\ &+ \sum_{l_1=0}^{M} \cdots \sum_{l_N=0}^{M} c_N b(l_1) \cdots b(l_N) x_{l_1}(k) \cdots x_{l_N}(k). \end{aligned} \tag{18}$$

This implies from (6) and (18) that

$$\begin{aligned} b_0 &= c_0 \\ b_1(l_1) &= c_1 b(l_1) \\ b_2(l_1, l_2) &= c_2 b(l_1) b(l_2) \\ &\vdots \\ b_N(l_1, \ldots, l_N) &= c_N b(l_1) \cdots b(l_N). \end{aligned} \tag{19}$$

4. FIXED POLE SELECTION BASED ON MODIFIED PRONY APPROACH

Selection of fixed pole location in FPET approach is an important issue to have an appropriate estimate of the nonlinear system. An adaptive gradient algorithm is introduced in (Hacıoğlu and Williamson, 2001) for online estimation of pole locations. We address here a fixed pole selection approach based on availability of *a priori* knowledge about the unknown system.

Prony's method is a technique to fit exponentials to a time series. It can be used to fit a pole/zero transfer function model to a given impulse response. An Mth order rational function can be uniquely recovered from its first $2M+1$ impulse response coefficients $h(0), \ldots, h(2M)$.

We proceed as follows. Assume that we have available the Nth order Volterra kernels, $h_1(l), \ldots, h_N(l_1, \ldots, l_N)$, $l_i = 0, \ldots, L$ for a suitably large L. We then fit a sum of exponentials to these kernel functions via modified Prony's approach. We first define certain prescribed functions from the Volterra kernels. These predescribed functions play the same role as the impulse response did in the Prony's method. These predescribed functions can be written as

$$h^{(1)}(\ell) = h_1(\ell)$$
$$h^{(2)}(\ell) = h_2(\ell,0)$$
$$\vdots$$
$$h^{(L+2)}(\ell) = h_2(\ell,L)$$
$$\vdots$$
$$h^{(S)}(\ell) = h_N(\ell,L,\ldots,L)$$

where $\{S = \frac{(L)^N-1}{L-1}\}$.

The iterative algorithm of (Williamson and Zimmerman, 1996), translated to the notation we use here, has the following steps.

1) Given the jth function $h^{(j)}(l)$, generate the $(L-M-1)\times 1$ vector

$$h^{(j)} = \left[h^{(j)}(M+1),\ldots,h^{(j)}(L)\right]^T,$$

and the $(L-M-1)\times M$ matrix

$$H^{(j)} = \begin{bmatrix} h^{(j)}(M) & \ldots & h^{(j)}(1) \\ h^{(j)}(M+1) & \ldots & h^{(j)}(2) \\ \vdots & & \vdots \; \vdots \\ h^{(j)}(L-1) & \ldots & h^{(j)}(L-M-1) \end{bmatrix},$$

where M is the number of pole location, and L is large enough to capture most of the energy in each predescribed function.

2) Initialize the parameter vector

$$a(0) = -\left[\sum_{j=1}^{S}(H^{(j)})^T H^{(j)}\right]^{-1}\left[\sum_{j=1}^{S}(H^{(j)})^T h^{(j)}\right]$$

where

$$a = \left[a_1,\ldots,a_M\right].$$

3) Construct the $(L-1)\times(L-M-1)$ matrix

$$A(a) = \begin{bmatrix} a_M & \cdots & a_1 & 1 & & & 0 \\ & a_M & \cdots & a_1 & 1 & & \\ & & & & \ddots & & \\ 0 & & & a_M & \cdots & a_1 & 1 \end{bmatrix}^T.$$

4) For step i, find

$$F^{(i-1)} = \left[A^T(a(i-1))A(a(i-1))\right]^{-1},$$

and

$$a(i) = -\left[\sum_{j=1}^{S}(H^{(j)})^T F^{(i-1)} H^{(j)}\right]^{-1}\left[\sum_{j=1}^{S}(H^{(j)})^T F^{(i-1)} h^{(j)}\right]$$

with the obtainable performance measure

$$J(a(i)) = \sum_{j=1}^{S}(H^{(j)}a(i)+h^{(j)})^T F^{(i)}(H^{(j)}a(i)+h^{(j)}).$$

5) Continue iteration until the performance measure $J(a(i))$ no longer decreases, and then find the fixed pole locations as the roots of the polynomial

$$1 - a_1 z^{-1} - \ldots - a_M z^{-M}.$$

The procedure described above computes the M pole locations directly from the Volterra kernel parameters. We call the above algorithm the modified Prony's approach.

5. SVD APPROACH

In this approach, we will arrange the components of the estimated weight parameter vector $\hat{\theta}$ in a matrix form that is also an outer product of b and d vectors. Then we compute the singular value decomposition of this matrix to find $\hat{b}$ and $\hat{d}$. Finally, we extract $\hat{c}$ from $\hat{d}$. We proceed as follows.

First, $\hat{c}_0$ can be estimated directly,

$$\hat{c}_0 = \hat{b}_0 = \hat{\theta}(1) \tag{20}$$

Then, we define the elements of $\hat{\theta}$ into several tensors

$$b_1 = c_1 b$$
$$b_2 = c_2 b^T \otimes b$$
$$\vdots$$
$$b_N = c_N b^T \otimes \cdots \otimes b. \tag{21}$$

where $\otimes$ denotes the kronoker tensor product operator and

$$b = [b(0),\ldots,b(M)]^T \tag{22}$$

Next we arrange these results in

$$B = [b_1,\ldots,b_N] = bd^T = U_B S_B V_B^T \tag{23}$$

and use a singular value decomposition to estimate $\hat{b}$ and $\hat{d}$

$$\hat{b} = \sigma S_B(1,1)U_B(:,1), \tag{24}$$
$$\hat{d} = \frac{1}{\sigma}V_B(:,1) \tag{25}$$

Table 1. NMSE values for Volterra-FPET models

Model Class Selection	fixed pole(s)	NMSE (in dB)
Volterra (M=4)	0	-19.08
FPET (M=4)	$0.72 \pm 0.13j, 0.42 \pm 0.16j$	-39.58
Volterra (M=20)	0	-40.06

where the choice of the scaler σ selects the normalization constraint. The nonlinearity coefficients $\hat{c}$ can be found from $\hat{d}$ and $\hat{b}$ using least squares estimation.

Notice that normalization constraint gives a unique Wiener model from Volterra representation. We also remark that non-invertibility of the nonlinearity does not violate identifiability of the Wiener system using Volterra approach because we have access to the input sequence as well as the output.

6. EXAMPLE

In this section we consider an illustrative example to apply the Volterra-FPET approach. The nonlinear system to be identified is depicted by the Wiener block structure (linear filter following by memoryless nonlinearity) shown in Fig.1. The linear dynamic subsystem has the repeated poles at $\alpha_{1,2} = 0.5$, and $\alpha_{3,4} = 0.65$ and two repeated zeros at $z = 0.82 \pm 0.15j$ and $z = 0.06 \pm 0.21j$. The memoryless nonlinearity has the coefficients $c = [0.5, 1.0, -0.25, 0.4]$. Note that this system may be exactly described by an FPET of the form of (6) with $N = 3$, and $M = 4$, using only two fixed poles at $z = 0.5$ and $z = 0.65$. Fig. 2 shows the impulse response of the linear system, and the input-output relation of the memoryless nonlinear block. An independent, identically distributed (i.i.d.), zero mean, white Gaussian 15,000 samples input signal with unit variance is used to generate the output data set in this example. With this setup, the memoryless nonlinear system is operated in the nonlinear region most of the time. The measured output contains an additive white Gaussian noise signal such that the signal to noise ratio is 40 *dB*.

The performance measure, a normalized mean square error (NMSE), is defined as

$$10\log_{10}\frac{\frac{1}{K}\sum_{k=0}^{K-1}(y(k)-\hat{y}(k))^2}{\frac{1}{K}\sum_{k=0}^{K-1}(y(k))^2}. \qquad (26)$$

We estimate the weight parameters to find the performance of the FPET approach so that the NMSE results are shown in Table 6. The table shows for three model structures with the pole parameters, and the achievable NMSE values. If two optimal fixed poles are chosen for the FPET, then we will have a reduction of approximately $20dB$ compared to truncated Volterra with the same parameter complexity.

The Volterra kernel parameters become negligible for this example when M reaches about 20, and the

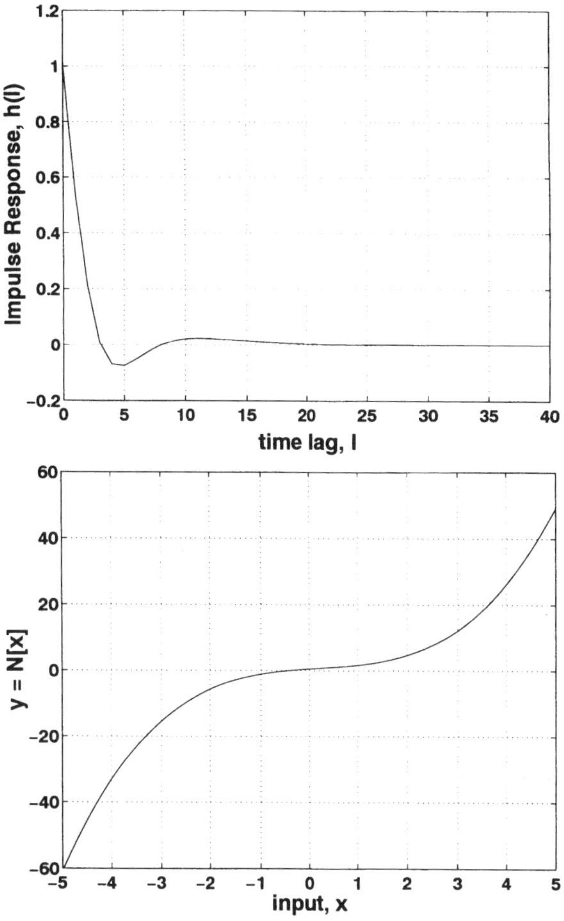

Fig. 2. Dynamic linear and memoryless nonlinear block of Wiener system

achievable reduction in NMSE reaches $40dB$, comparable to the NMSE for the two pole, $M = 4$ FPET case. It is illustrative to compare the parametric complexity of the $M = 20$ Volterra model with the $M = 4$ two pole FPET. In the case of the $M = 20$ Volterra model, we have 2,024 parameters to identify, while there are only 56 parameters in the FPET case. Hence the reduction in the parametric complexity is significant when we use FPET approach in this example.

In order to extract coefficients of the linear and nonlinear blocks, we arrange the components of the estimated weight vector $\hat{\theta}$ into several matrices

$$b_0 = c_0 = \theta(1)$$

$$b_1 = c_1 b = \begin{bmatrix} \theta(2) \\ \vdots \\ \theta(M_1) \end{bmatrix}$$

$$b_2 = c_2 b^T \otimes b = \begin{bmatrix} \theta(M_2) & \dots & \theta(M_4) \\ & \vdots & \\ \theta(M_3) & \dots & \theta(M_5) \end{bmatrix}$$

$$b_3 = c_N b^T \otimes b \otimes b = \begin{bmatrix} \theta(M_6) & \dots & \theta(M_8) \\ & \vdots & \\ \theta(M_7) & \dots & \theta(M_9) \end{bmatrix}$$

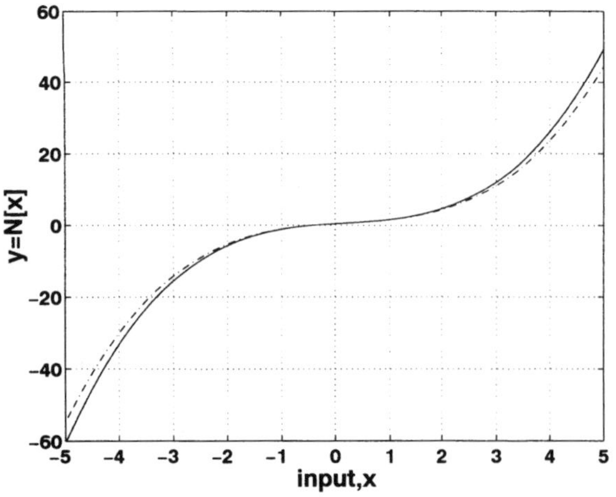

Fig. 3. Estimated memoryless nonlinearity of Wiener system: Solid line actual, dash-dotted line Volterra ($M = 4$), dotted-dotted line Volterra ($M = 20$), dash-dashed line Volterra-FPET ($M = 4$)

Table 2. NMSE values for estimated Wiener system

Model Class Selection	fixed pole(s)	NMSE (in dB)
Volterra (M=4)	0	-19.00
FPET (M=4)	$0.72 \pm 0.13j$,$0.42 \pm 0.16j$	-39.48
Volterra (M=20)	0	-38.65

where the dummy variables are $M_1 = M+2$, $M_2 = M+3$, $M_3 = 2(M+2)$, $M_4 = (M+1)(M+2)+1$, $M_5 = (M+2)^2$, $M_6 = (M+2)^2+1$, $M_7 = (M+3)(M+2)$, $M_8 = (M+2)((M+1)^2+1)$, $M_9 = (M+2)^2+(M+1)^3$. Next, we arrange these matrices as an outer product of b and d. Then, we compute the singular value decomposition of this matrix

$$B = [b_1, b_2, b_3] = bd^T = U_B S_B V_B^T. \quad (27)$$

We select the normalization constraint by setting $\sigma = V_B(1,1)$ resulting in $\hat{c}_1 = 1$. The nonlinearity coefficients $\hat{c}_2$ and $\hat{c}_3$ were found from $\hat{d}$ and $\hat{b}$ using least squares estimation. We plot the memoryless nonlinearity in Fig. 3. Estimated nonlinear curve using FPET approach is on the top of the actual system while Volterra model with $M = 4$ shows observable distinction. Finally we in Table 6 show the NMSE values for the estimated Wiener system using three model structures when we have only $M+5$ coefficients. FPET structure requires only 9 parameters to represent this Wiener system.

7. CONCLUSION

Identification of Wiener systems with polynomial nonlinearities has been presented based on the reduced complexity Volterra model, i.e. FPET approach. First, the weight parameters in the Volterra-FPET model structure were estimated using least squares approach. Then singular value decomposition method were presented to separate the linear system coefficients and the polynomial nonlinearity coefficients from each other. Modified Prony's approach were introduced to select the fixed pole location for FPET. Since we need to identify a Volterra model to start for the pole selection procedure, FPET is just as complex as a Volterra model approach as far as identification is concerned. However, we then reduce the parameter complexity of the model via the FPET and Wiener extraction. To illustrate the implementation, we applied Volterra-FPET approach and SVD method to an example and discussed the parametric complexity reduction. A significant reduction in the parametric complexity has been observed. This may be important, for instance, in the overall complexity of a model predictive control scheme.

8. REFERENCES

Bai, E.-W. (1998). An optimal two-stage identification algorithm for hammerstein-wiener nonlinear systems. *Automatica* **34**, 333–338.

Boyd, S. and L.O. Chua (1985). Fading memory problem of approximating nonlinear operators with volterra series. *IEEE Trans. on Circuit and Systems* **CAS–32**, 1150–1161.

Greblicki, W. (1992). Nonparametric idenrification of wiener systems. *IEEE Trans. on Information Theory* **38**, 1487–1493.

Hacıoğlu, R. and G.A. Williamson (2001). Reduced complexity volterra models for nonlinear system identification. *EURASIP Journal on Applied Signal Processing* **2001:4**, 257–265.

Hunter, I.W. and M.J. Korenberg (1986). The identification of nonlinear biological systems: Wiener and hammerstein cascade models. *Biological Cybernetics* **55**, 135–144.

Lacy, S.L. and D.S. Bernstein (2002). Identification of fir wiener systems with unknown, noninvertibele, polynomial nonlinearities. *Proceedings of the American Control Conference* pp. 893–898.

Ljung, L. (1987). *System Identification, Theory for the User*. Prentice Hall. New Jersey.

Marmarelis, V.Z. (1993). Identification of nonlinear biological systems using laguerre expentions of kernels. *Annals of Biomedical Engineering* **21**, 573–589.

Schetzen, M. (1980). *The Volterra and Wiener Thories of Nonlinear Systems*. John Wiley. New York.

Wahlberg, B. (1994). Laguerra and kautz models. *10th IFAC Symposium on System Identification* **3**, 3.1–3.12.

Williamson, G.A. and S. Zimmerman (1996). Global convergent adaptive iir filters based on fixed pole locations. *IEEE Trans. on Signal Processing* **44**, 1418–1427.

IFAC Publications
www.elsevier.com/locate/ifac

STRUCTURE SELECTION FOR POLYNOMIAL NARX MODELS BASED ON SIMULATION ERROR MINIMIZATION

L. Piroddi and W. Spinelli

Dipartimento di Elettronica e Informazione, Politecnico di Milano, Piazza Leonardo da Vinci 32, 20133, Milano (Italy), tel. ++39-2-23993556, fax. ++39-2-23993412, e-mail: piroddi@elet.polimi.it

Abstract: This paper investigates the problem of model structure selection for polynomial NARX models. In particular it discusses how classical identification approaches based on prediction error minimization may lead to an incorrect evaluation of the importance of the regressors within the model, with the consequent inclusion of spurious terms in the model. The paper suggests an alternative approach, in which the model structure is selected based on the minimization of the simulation error. The approach is shown to be particularly effective when the identification data are not adequately exciting or oversampled or when the model family is under-parameterized. *Copyright © 2003 IFAC*

Keywords: identification algorithms, nonlinear models, NARX models, model structure selection, simulation, prediction error

1. INTRODUCTION

In the field of black-box nonlinear modeling, the nonlinear autoregressive moving average with exogenous variables (NARMAX) representation has gained considerable interest in applications (Loh and Duh, 1996; Palumbo and Piroddi, 2000; Chiras *et al.*, 2001; Palumbo *et al.*, 2001; Leva and Piroddi, 2002). In particular, polynomial models are often employed, because they are linear-in-the-parameters and polynomial terms are more easily interpretable than, say, neural networks. Also, the less general NARX model family is often preferred for simplicity reasons.

What makes the polynomial NARX approach difficult is not so much parameter estimation, which can be easily performed using linear regression methods. Rather, the model structure selection problem is critical, since the number of possible terms in polynomial expansions grows rapidly. In the lines of prediction error minimization (PEM) methods for model identification, an effective criterion for structure selection is the Error Reduction Ratio (ERR), which is used in the popular Forward-Regression Orthogonal Estimator (FROE) (Billings *et al.*, 1989). The ERR criterion rates the significance of model terms, by evaluating their ability to explain the output variance (in prediction).

However, the PEM paradigm can yield misleading results in terms of model structure selection, especially when the available data are not adequately exciting for the system to be identified (e.g. insufficient amplitude range, low frequency data, oversampled data) or when the model family is under-parameterized. In these cases, often encountered in usual practice, the model structure selection is hard to solve with the ERR, since many different model structures can have comparable prediction performance. Also, autoregressive terms are typically inserted in the model regardless of the original model structure, due to their effectiveness for prediction. Redundant models are often obtained where many spurious compensation terms are included for every "exact" regressor missing. This is not without consequences in applications where the ultimate use of the model is for interpretation, simulation or long-range prediction purposes, since structural model errors severely affect the simulation performance of identified models.

This paper introduces a model structure selection criterion for polynomial NARX models, based on the minimization of the simulation error. It is shown how this criterion can be more effective in selecting an appropriate model structure in non ideal identification conditions. Therefore, it can be useful for NARX identification, especially when long-range prediction or simulation models are required.

The paper is organized as follows. The polynomial NARX representation and the FROE are briefly recalled in Section 2. The limitations of the PEM approach for model structure selection are discussed in Section 3. The novel structure selection criterion is developed in Section 4, and its performance compared to that of the ERR. Section 5 summarizes the main points of the paper.

2. NARX MODELS AND THE FORWARD-REGRESSION ORTHOGONAL ESTIMATOR

A quite general representation of nonlinear models is given by the NARX model (Leontaritis and Billings 1985a, b):

$$y(t) = f(y(t-1),...,y(t-n_y),u(t-1),...,u(t-n_u)) + \xi(t), \quad (1)$$

where $u(\cdot), y(\cdot)$ are the model input and output, n_y, n_u are the respective maximum lags, $\xi(\cdot)$ is a white noise term, and $f(\cdot)$ is a suitable nonlinear function. In the following, $f(\cdot)$ is assumed to be a polynomial function with maximum nonlinearity degree $l \in Z^+$. Therefore, (1) is a classical linear regression of the type:

$$y(t) = \psi^T(t-1)\,\vartheta + \xi(t), \quad (2)$$

where $\psi^T(t-1)$ is the regressor vector that contains linear and nonlinear combinations of the input and output signals, and ϑ is the parameter vector containing the coefficients of the polynomial terms. Terms with the same degree of nonlinearity in the $u(\cdot)$ and $y(\cdot)$ elements form a *cluster* (Aguirre and Billings, 1995). For example, both $y(t-1)^2$ and $y(t-2)y(t-3)$ belong to the same cluster, denoted Ω_{y^2}.

One of the most efficient algorithms for the identification of NARX models is the Forward-Regression Orthogonal Estimator (FROE) (Billings *et al.* 1989). In this approach, the model structure is iteratively incremented until a specified prediction accuracy is obtained, starting from an empty model. Orthogonal Least Squares (OLS) are employed to decouple the estimation of additional parameters from that of the parameters already included in the model. In this way, at each step the significance of each candidate regressor can be separately evaluated, by computing the Error Reduction Ratio (ERR) criterion:

$$[\mathrm{ERR}]_i = [\hat{g}_i^2 \sum_{t=1}^{N} w_i^2(t)] \,/\, [\sum_{t=1}^{N} y^2(t)]$$

where w_i is the i-th auxiliary orthogonal regressor and $\hat{g}_i$ is the corresponding estimated parameter. The ERR actually represents the reduction of the mean square prediction error (MSPE), due to the inclusion of an additional regressor, expressed as a fraction of the total MSPE (equal to the output variance):

$$[\mathrm{ERR}]_i = [\mathrm{MSPE}(M_j) - \mathrm{MSPE}(M_{j+1})] \,/\, [\sum_{t=1}^{N} y^2(t)], \quad (3)$$

where M_j is the model obtained at the j-th iteration and M_{j+1} is the candidate model at the subsequent iteration, with the inclusion of the i-th regressor. At each iteration, the regressor with the highest ERR value is added to the model. At the end of the procedure the estimated parameters for the original, non orthogonal regressors can be recovered from the auxiliary regressors and parameters. Notice that both parameter estimation and structure selection in the FROE are based on prediction error minimization.

It is well known that the order in which parameters are progressively included in the model influences the model selection process. Also, it is not the overall MSPE that is minimized, but rather its reduction at each iteration: therefore, there is no guarantee that the model obtained be optimal in any sense (Korenberg *et al.* 1987, Billings *et al.* 1989).

3. STRUCTURE SELECTION BASED ON THE ERR CRITERION

Structure selection becomes particularly critical when models are to be used for long-range prediction or simulation purposes. In fact, while in short-term prediction the effect of an incorrect or redundant structure is reduced to the minimum, since at each stage the use of measured past output values in the model can help keep the prediction close to the real output, in long-term prediction an error in the structure tends to accumulate in time, affecting all the more severely the prediction accuracy the more the simulation is protracted.

To see that correct structure selection is much more important for the identification of models intended for simulation, consider the following system:

$$S: \quad w(t) = u(t-1)u(t-2)^2 + 0.4\,u(t-1)w(t-1)$$
$$y(t) = w(t) + \xi(t)\,, \quad \xi(\cdot) \sim \mathrm{WN}(0, 0.02)$$

A set of I/O data is generated using a low frequency AR(2) input signal with a mean value of 1.5 and a 0.5 variance. The NARX model family with $n_y = n_u = 2$ and $l = 3$ (35 possible regressors) is considered for identification. All the possible two regressor models (including the correct one) are identified (with LS), and for each identified model the MSPE and the mean square simulation error (MSSE) are evaluated. The MSPE is shown in the bottom right section of figure 1 (the performance index is symmetrical with respect to the main diagonal), while the MSSE of the same models is shown in the top left section. Dark zones correspond to low values of the index, whereas white areas denote high values.

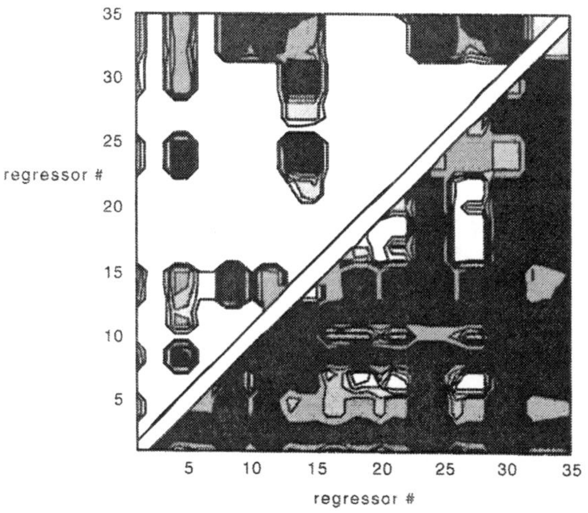

Figure 1. MSPE (bottom right) and MSSE (top left) for all two regressor models

The MSPE index is almost flat, showing that a large number of models share a comparable prediction performance: an incorrect model structure does not necessarily prevent from obtaining acceptable 1-step ahead predictions. On the other hand, the MSSE plot displays narrow local minima, which specify a much smaller set of models with comparable simulation performance (which mainly differ by way of regressors of the same clusters). Notice also that many of the models which are classified as nearly optimal for prediction are in fact no good at all for simulation purposes (many turn out to be even unstable). On the contrary, good simulation models are generally also acceptable in prediction.

In general, there is no relationship between the model prediction and simulation performance. It is therefore important to understand the limitations in simulation performance of models identified with PEM methods. The optimal 1-step ahead predictor of a NARX model can be easily obtained as:

$$\hat{y}(t|t-1) = f(y(t-1), \ldots, y(t-n_y), u(t-1), \ldots, u(t-n_u)). \quad (4)$$

Notice that (4) is always stable. The optimal simulation $\hat{y}_S(t)$ can be computed on the basis of the optimal predictor, by setting $y(t) = \hat{y}_S(t)$ and $\hat{y}(t|t-1) = \hat{y}_S(t)$ (Sjöberg *et al.*, 1995):

$$\hat{y}_S(t) = f(\hat{y}_S(t-1), \ldots, \hat{y}_S(t-n_y), u(t-1), \ldots, u(t-n_u)). \quad (5)$$

The stability of (5) depends on the original process model. Notice that the simulation error $\varepsilon_S(t) = y(t) - \hat{y}_S(t)$ is not white, in general. If the simulator model is computed on the basis of the *optimal* predictor, it is optimal too, i.e. it minimizes the MSSE. However, only sub-optimal predictors are obtained in practice, due to the mis match between the real system and the abstract model family (1), the excitation characteristics of identification data and the non-optimality of the FROE. Unfortunately, the simulation model computed on the basis of a non optimal predictor may perform in a significantly different way from the optimal simulator, even if the predictor model is nearly optimal.

4. STRUCTURE SELECTION BASED ON SIMULATION ERROR MINIMIZATION

A workaround for these problems can be envisaged in an algorithm which tries to assess the model structure on the basis of the simulation error. The simplest way to obtain such an algorithm is to modify the selection criterion in the FROE, thus minimizing the mean square simulation error. This is obtained by substituting the MSPE with the MSSE in equation (3). The resulting criterion is denoted *Simulation error Reduction Ratio* (*SRR*):

$$[\mathrm{SRR}]_i = [\mathrm{MSSE}(M_j) - \mathrm{MSSE}(M_{j+1})] \,/\, [\sum_{t=1}^{N} y^2(t)]\,, \quad (6)$$

This criterion can be used like the ERR in the identification algorithm, *i.e.* at each iteration the regressor with the highest SRR is included in the model. Notice that since the MSSE performance index is employed for structure selection, it would be natural to use it also for parameter estimation. However, this would involve an unacceptable computational load since an iterative estimation algorithm would have to be executed for each model structure tested by the identification algorithm. A satisfactory compromise solution consists in performing parameter estimation still by LS, thereby minimizing the MSPE in the identification of a *given* model structure, whereas estimated models of different structures are compared in terms of MSSE for structure selection.

4.1. Regressor weighting

Consider the following system:

$$S: w(t) = 0.75w(t-2) + 0.25u(t-1) - 0.2w(t-2)u(t-1)$$
$$y(t) = w(t) + \xi(t), \quad \xi(\cdot) \sim \mathrm{WN}(0, 0.02)$$

where the input signal is a low frequency AR(2) process (poles in 0.9 and 0.95), with 0 mean and variance 0.25. Assume that the NARX model family with $n_y = n_u = 2$ and $l = 3$ is employed in identification. The following tables 1-2 show the values of the two criteria for each regressor in the first 3 iterations. The values in bold are related to the selected regressors.

Table 1 ERR values in the first 3 iterations

regressor	1st iteration	2nd iteration	3rd iteration
1	3.56E-02	3.40E-04	2.69E-03
y(t-1)	**8.50E-01**	-	-
y(t-2)	8.20E-01	2.10E-02	**2.28E-02**
u(t-1)	5.94E-01	**2.96E-02**	-
u(t-2)	6.53E-01	2.38E-02	7.09E-05
y(t-1)·y(t-1)	1.67E-01	6.08E-04	5.48E-03
y(t-1)·y(t-2)	1.84E-01	9.93E-04	6.78E-03
y(t-2)·y(t-2)	1.62E-01	1.33E-03	7.20E-03
y(t-1)·u(t-1)	2.97E-02	8.13E-04	7.01E-03
y(t-1)·u(t-2)	2.72E-02	6.92E-04	6.70E-03
y(t-2)·u(t-1)	3.13E-02	6.56E-04	6.58E-03
y(t-2)·u(t-2)	2.83E-02	4.22E-04	5.94E-03
u(t-1)·u(t-1)	5.41E-03	1.18E-04	3.71E-03
u(t-1)·u(t-2)	6.62E-03	7.17E-05	4.13E-03
u(t-2)·u(t-2)	6.39E-03	2.36E-05	4.21E-03
y(t-1)·y(t-1)·y(t-1)	5.70E-01	1.40E-03	1.26E-04
y(t-1)·y(t-1)·y(t-2)	5.92E-01	1.20E-06	9.30E-04
y(t-1)·y(t-2)·y(t-2)	5.84E-01	1.23E-03	3.80E-03
y(t-2)·y(t-2)·y(t-2)	5.44E-01	4.58E-03	7.54E-03
y(t-1)·y(t-1)·u(t-1)	5.75E-01	1.72E-03	1.22E-03
y(t-1)·y(t-1)·u(t-2)	5.66E-01	3.32E-04	1.56E-03
y(t-1)·y(t-2)·u(t-1)	6.18E-01	6.45E-03	9.42E-05
y(t-1)·y(t-2)·u(t-2)	6.16E-01	3.46E-03	1.74E-05
y(t-2)·y(t-2)·u(t-1)	5.74E-01	1.22E-02	1.93E-03
y(t-2)·y(t-2)·u(t-2)	5.73E-01	8.46E-03	1.55E-03
y(t-1)·u(t-1)·u(t-1)	4.43E-01	4.75E-03	2.60E-03
y(t-1)·u(t-1)·u(t-2)	4.39E-01	3.29E-03	3.22E-03
y(t-1)·u(t-2)·u(t-2)	4.35E-01	1.94E-03	3.23E-03
y(t-2)·u(t-1)·u(t-1)	4.54E-01	1.13E-02	1.54E-05
y(t-2)·u(t-1)·u(t-2)	4.49E-01	8.68E-03	4.91E-05
y(t-2)·u(t-2)·u(t-2)	4.49E-01	6.44E-03	1.13E-04
u(t-1)·u(t-1)·u(t-1)	2.63E-01	1.16E-02	1.44E-03
u(t-1)·u(t-1)·u(t-2)	2.74E-01	9.74E-03	1.84E-03
u(t-1)·u(t-2)·u(t-2)	2.83E-01	7.99E-03	2.14E-03
u(t-2)·u(t-2)·u(t-2)	2.91E-01	6.28E-03	2.23E-03

The following facts can be observed from tables 1-2:

- The ERR values are always positive (the prediction error always decreases with the inclusion of a new regressor), whereas SRR values can have positive or negative sign: some regressors are explicitly marked as unsafe. This is reasonable, since a new regressor can introduce significant variations in the model dynamics, even worsening the simulation performance.
- With the exception of the first iteration the SRR values of high degree terms (not present in the system) are very low or negative, whereas the ERR values are comparably high. This shows that the SRR criterion can be effective in rejecting redundant terms.
- Typically, regressors with persistently negative values of SRR during iterations belong to spurious clusters. On the other hand, the negative SRR values of the terms of cluster Ω_y at the first iteration point out that they are of no use *by themselves alone* in characterizing the system dynamics.

Table 2 SRR values in the first 3 iterations

regressor	1st iteration	2nd iteration	3rd iteration
1	5.68E-02	5.52E-02	5.45E-02
y(t-1)	-1.87E-02	2.26E-01	-1.51E-03
y(t-2)	-3.01E-02	**2.48E-01**	-
u(t-1)	**5.58E-01**	-	-
u(t-2)	4.73E-01	8.65E-05	7.29E-05
y(t-1)·y(t-1)	7.08E-04	unstable	6.53E-02
y(t-1)·y(t-2)	6.97E-04	1.67E-01	6.33E-02
y(t-2)·y(t-2)	6.58E-04	1.23E-01	5.54E-02
y(t-1)·u(t-1)	7.01E-04	6.68E-02	1.16E-01
y(t-1)·u(t-2)	6.98E-04	7.25E-02	1.12E-01
y(t-2)·u(t-1)	6.88E-04	1.00E-01	**1.19E-01**
y(t-2)·u(t-2)	6.86E-04	1.09E-01	1.13E-01
u(t-1)·u(t-1)	3.73E-02	5.71E-02	9.23E-02
u(t-1)·u(t-2)	3.90E-02	6.05E-02	9.24E-02
u(t-2)·u(t-2)	3.71E-02	5.89E-02	8.92E-02
y(t-1)·y(t-1)·y(t-1)	6.76E-04	-2.51E-02	-5.09E-02
y(t-1)·y(t-1)·y(t-2)	6.82E-04	-2.14E-02	-5.03E-02
y(t-1)·y(t-2)·y(t-2)	6.84E-04	-1.61E-02	-4.48E-02
y(t-2)·y(t-2)·y(t-2)	6.86E-04	-9.28E-03	-3.49E-02
y(t-1)·y(t-1)·u(t-1)	6.78E-04	unstable	-9.12E-02
y(t-1)·y(t-1)·u(t-2)	6.80E-04	unstable	-7.95E-02
y(t-1)·y(t-2)·u(t-1)	6.83E-04	unstable	-7.07E-02
y(t-1)·y(t-2)·u(t-2)	6.84E-04	unstable	-5.47E-02
y(t-2)·y(t-2)·u(t-1)	6.85E-04	unstable	-3.53E-02
y(t-2)·y(t-2)·u(t-2)	6.86E-04	-1.13E-01	-2.92E-02
y(t-1)·u(t-1)·u(t-1)	6.82E-04	-1.11E+00	-4.70E-02
y(t-1)·u(t-1)·u(t-2)	6.83E-04	-1.79E+00	-5.64E-02
y(t-1)·u(t-2)·u(t-2)	6.83E-04	-1.31E+00	-4.66E-02
y(t-2)·u(t-1)·u(t-1)	6.86E-04	-3.01E-01	-1.16E-02
y(t-2)·u(t-1)·u(t-2)	6.85E-04	-4.04E-01	-1.44E-02
y(t-2)·u(t-2)·u(t-2)	6.85E-04	-3.34E-01	-1.29E-02
u(t-1)·u(t-1)·u(t-1)	2.89E-01	1.31E-03	3.16E-05
u(t-1)·u(t-1)·u(t-2)	3.17E-01	1.52E-03	-1.10E-04
u(t-1)·u(t-2)·u(t-2)	3.34E-01	1.49E-03	-5.71E-05
u(t-2)·u(t-2)·u(t-2)	3.46E-01	1.27E-03	1.37E-04

- Both criteria yield comparable values for terms of the same cluster: these are “confused” because of the input signal mostly excites the low frequency dynamics. However, only the SRR criterion selects terms of the three correct clusters.
- The SRR criterion gives indirect indication of the fact that the optimal model is selected: at the fourth iteration most of the remaining regressors have negative or extremely low SRR, so that no significant increase can be obtained in the model simulation accuracy.

4.2. Regressor selection ordering

With reference to the example introduced in subsection 4.1, table 3 reports the regressors actually selected according to the SRR criterion in the first three iterations, together with the corresponding ERR values. Notice that while regressors $u(t-1)$ and $y(t-2)$ have comparable importance for both criteria, the third regressor is not considered significant by the ERR criterion: its elimination worsens the prediction performance by 1%, whereas a 10% decrease in simulation accuracy would be experienced.

Table 3 Regressors selected in the first 3 iterations according to the SRR criterion

Iteration	Regressor	SRR	ERR
1	**u(t-1)**	5.58 E–01	5.57 E–01
2	y(t-2)	2.48 E–01	3.42 E–01
3	**y(t-2)·u(t-1)**	1.19 E–01	1.43 E–02

If the regressors are chosen according to the ERR criterion (table 4), the 1st term has a negative SRR (the autoregressive term *alone* is useless in simulation) compensated by the 2nd regressor: summing the corresponding SRR values one obtains a simulation accuracy (0.784) nearly equal to that achieved by the first two regressors selected according to the SRR (0.806). In fact, at the 2nd iteration the models identified with the two criteria are equivalent at least in terms of clusters. However, the ERR criterion wrongly picks out $y(t-2)$ as third regressor, instead of the much more significant nonlinear term $y(t-2)u(t-1)$.

Table 4 Regressors selected in the first 3 iterations according to the ERR criterion

Iteration	Regressor	SRR	ERR
1	y(t-1)	-1.87 E–02	8.50 E–01
2	**u(t-1)**	8.03 E–01	2.96 E–01
3	**y(t-2)**	2.05 E–02	2.28 E–02
	[y(t-2)·u(t-1)]	[1.13 E–01]	[6.58 E–03]

4.3. Model family

It is important to investigate the performance of the two selection criteria with respect to variations in the model family in which the search is performed.

Consider the following system:

$$S : y(t) = 0.5\, y(t-1) - 1.2\, u(t-1) - 0.1\, y(t-2)^2 + \\ - 0.5\, u(t-2)^2 + 1.5\, u(t-1)^3,$$

where the input signal is a low frequency AR(2) process (poles in 0.7 and 0.8), with 0 mean and variance 0.25. 500 data are generated for identification and other 500 for validation. If a NARX model family with $n_y = n_u = 2$ and $l = 4$ is employed in the identification procedure, both criteria ultimately succeed in obtaining the correct structure (which belongs to the model family). The difference lies in the evolution of the MSSE with the algorithm iterations (figure 2): only the SRR criterion yields a uniform convergence towards the final model, whereas unstable models (denoted with the symbol × in figure 2) are occasionally selected with the ERR criterion.

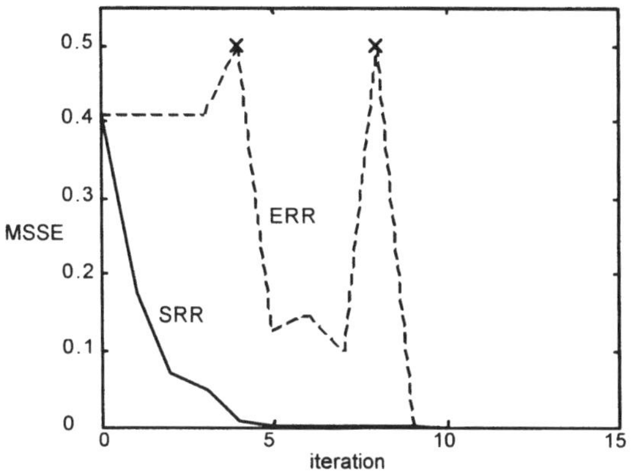

Figure 2. Evolution of the MSSE during algorithm iterations, with the ERR and SRR criteria.

If the nonlinearity degree is reduced ($l = 3$), the model family used in the identification procedure is under-parameterized. Therefore, the algorithm has to compensate for the absence of the correct terms. The ERR criterion requires as many as 11 iterations to achieve a satisfactory performance in simulation (MSSE = 7.128E-03 on identification data, MSSE = 7.387E-03 on validation data). On the other hand, a comparable simulation performance is obtained with a much smaller 6-term model using the SRR (MSSE = 7.589E-03 on identification data, MSSE = 7.412E-03 on validation data). The obtained models are reported in tables 5 and 6, respectively.

Table 5 Model identified with the ERR with an under-parameterized model family

Iter.	Regressors	Estimated parameters	3·STD %	ERR
1	y(t-1)	1.121 E+00	8.0%	9.762 E–01
2	y(t-2)	-3.555 E–01	17.6%	1.597 E–02
3	$y(t-1)^2 \cdot u(t-1)$	-5.047 E–02	74.9%	5.301 E–04
4	u(t-1)	-8.586 E–01	12.4%	2.252 E–04
5	$y(t-1) \cdot u(t-2)^2$	-3.516 E–01	18.4%	8.197 E–04
6	$u(t-1)^3$	1.889 E+00	19.6%	1.546 E–04
7	$u(t-1)^2 \cdot u(t-2)$	-1.455 E+00	28.0%	4.634 E–04
8	u(t-2)	3.143 E–01	43.3%	1.275 E–03
9	$u(t-1)^2$	1.656 E+00	12.7%	1.274 E–04
10	u(t-1)·u(t-2)	-1.372 E+00	13.8%	1.094 E–03
11	y(t-1)·u(t-1)	3.621 E–01	17.5%	1.170 E–03

4.4. Structure selection and sampling time

In system identification, the effects of oversampling are usually associated with numerical problems during parameter estimation, because of the conditioning of the regression matrix. In practice, however, data oversampling will also pose problems for model structure selection (Billings and Aguirre, 1995).

Table 6 Model identified with the SRR with an under-parameterized model family

Iter.	Regressors	Estimated parameters	3·STD %	ERR
1	u(t-2)	-1.980 E+00	4.2%	5.669 E–01
2	$u(t\text{-}2)^3$	1.852 E+00	4.6%	2.588 E–01
3	$u(t\text{-}1)^2$	7.701 E–01	6.8%	7.676 E–03
4	$y(t\text{-}2)^2$	-3.469 E–01	16.3%	9.013 E–02
5	y(t-1)·u(t-2)	6.436 E–01	15.2%	3.864 E–02
6	$y(t\text{-}2)^2 \cdot u(t\text{-}2)$	3.395 E–01	37.4%	1.938 E–02

A common experience, for example, is that the autoregressive term $y(t-1)$ will usually be selected as first regressor in the model by the ERR criterion, almost independently of the real system structure. In fact, it is easy to show that the value of ERR associated to $y(t-1)$ tends to 1 as $T_S \rightarrow 0$, since $y(t) \approx y(t-1)$ (Billings and Aguirre, 1995).

In practice, T_S will never be zero, but if the sample time is too short, then $[ERR]_1 \approx 1$. Therefore, since the sum of the ERR values of the selected regressors cannot be greater than 1, all the other terms will have very small ERR values. Consequently, it becomes difficult to select the structure in such a situation. On the contrary, the SRR does not suffer from such a limitation. In fact, if $y(t-1)$ were selected as first regressor, the corresponding parameter, estimated with LS, would be computed as

$$\hat{\theta} = \left(\sum_{t=1}^{N} y^2(t-1) \right)^{-1} \sum_{t=1}^{N} y(t)y(t-1),$$

and $\hat{\theta} \approx 1$ for $T_S \rightarrow 0$. The corresponding simulation model would be $\hat{y}_S(t) \approx \hat{y}_S(t-1)$, which would result in a SRR almost equal to zero.

5. CONCLUSIONS

A critical analysis of classical PEM approaches for NARX model structure selection has pointed out some unacceptable features for the use of identified models in simulation (or long range prediction).

To overcome these limitations, a simulation error based criterion for regressor selection has been proposed. The SRR criterion has been experimentally compared to the ERR, and has been found to be effective for model structure selection, especially when the available data for identification are not adequately exciting or oversampled.

ACKNOWLEDGEMENTS

Paper supported by MURST project "Identification and Control of Industrial systems".

REFERENCES

Aguirre, L.A. and S.A. Billings (1995). Improved structure selection for nonlinear models based on term clustering. *Int. J. of Control*, **62**, 569-587.

Billings, S.A. and L.A. Aguirre (1995). Effects of the sampling time on the dynamics and identification of nonlinear models. *Int. J. of Bifurcation and Chaos*, **5** (6), 1541-1556.

Billings, S.A., S. Chen and M.J. Korenberg (1989). Identification of MIMO non-linear systems using a forward-regression orthogonal estimator. *Int. J. of Control*, **49**, 2157-2189.

Chiras, N., C. Evans and D. Rees (2001). Nonlinear Gas Turbine Modeling Using NARMAX structures. *IEEE Trans. on Instrumentation and Measurement*, **50** (4), 893-898.

Korenberg, M., S.A. Billings, Y.P. Liu and P.J. McIlroy (1987). Orthogonal parameter estimation algorithm for non-linear stochastic systems. *Int. J. of Control*, **48**, 193-210.

Leontaritis, I.J., and S.A. Billings (1985a). Input-output parametric models for non-linear systems - Part I: deterministic non-linear systems. *Int. J. of Control*, **41**, 303-328.

Leontaritis, I.J., and S.A. Billings (1985b). Input-output parametric models for non-linear systems - Part II: stochastic non-linear systems. *Int. J. of Control*, **41**, 329-344.

Leva, A. and L. Piroddi (2002). NARX-based technique for the modeling of magneto-rheological damping devices. *Smart Materials and Structures*, **11**, 79-88.

Loh, C.H. and J.Y. Duh (1996). Analysis of nonlinear system using NARMA models. *JSCE - Structural Eng./Earthquake Eng.*, **13**, 11-21.

Palumbo, P. and L. Piroddi (2000). Seismic behaviour of buttress dams: nonlinear modelling of a damaged buttress based on ARX/NARX models. *J. of Sound and Vibration*, **239**, 405-422.

Palumbo, P., L. Piroddi, S. Lancini and F. Lozza (2001). NARX Modelling of Radial Crest Displacements of the Schlegeis Arch Dam. *6th Benchmark Workshop on Numerical Analysis of Dams*, Salzburg (Austria), 17-19 October.

Sjöberg, J., Q. Zhang, L. Ljung, A. Benveniste, B. Delyon, P. Glorennec, H. Hjalmarsson and A. Juditsky (1995). Nonlinear black-box modeling in system identification: a unified overview. *Automatica*, **12**, 1691-1724.

IFAC
Publications
www.elsevier.com/locate/ifac

NONLINEAR IDENTIFICATION OF A TWO LINK ROBOTIC SYSTEM USING DYNAMIC NEURAL NETWORKS

S. Torres * **V.M. Becerra** **

* *Departamento de Fisica, Electronica y Sistemas. Universidad de La Laguna. 38271. Tenerife. Spain. E–mail: storres@ull.es*
** *The University of Reading. Department of Cybernetics Whiteknights, Reading RG6 6AY. United Kingdom E–mail: v.m.becerra@reading.ac.uk*

Abstract:
The aim of this work is the identification of a two link robotic system using dynamic neural networks. The identified system was a mechanical leg formed by two revolute links. The theoretical model of the system corresponds to a two-link kinematic chain. However, a number of nonlinearities which are present in the real system are difficult to include in the theoretical model. An empirical model of the system was obtained instead using dynamic neural network. New training and validation techniques that assure a good performance of the empirical model have been applied. They consist of including the initial state of the hidden neurons in the decision vector associated with the optimisation problem that is solved for training the network. Once a network has been suitably trained, the initial states of the hidden neurons are also optimised based on the validation data.

Keywords: system identification, dynamic neural networks, robotics

1. INTRODUCTION

The dynamics of a robot composed by an articulate chain of rigid elements joined by links can be normally obtained from the matrix Lagrange-Euler formulation. Other formulations, like the recursive Lagrange or the generalised d'Alembert-Lee, can be used in the same way. Using any of them, a set of nonlinear coupled equations, called the inverse dynamics of the robot, is obtained (Angeles, 1997).

These equations relate the torques acting on each link with the generalized coordinates of the robot, angles in the case of revolution links or displacements in the case of prismatic links. The terms depending of these coordinates are the inertia terms (reaction forces induced by the acceleration that a link produces on other link), centrifugal and Coriolis terms and, finally, the gravity terms due to the mass of the elements of the articulate chain.

Those equations may contain terms related with frictions. These terms contain parameters of difficult evaluation, such as viscous or coulombian friction coefficients. Thus, frictions add uncertainty in the model. Moreover, there exist other nonlinearities that become apparent appear when the mechanic prototype is constructed. These include, for instance, link looseness and dead zones. These characteristics are difficult to model theoretically. Hence, an empirical model may be a good alternative to a theoretical model.

Neural networks have become the predominant family of structures for nonlinear system identification. Their ability to model complex nonlinear behaviour together with their suitability to be trained using experimental data are some of the reasons for their success.

The attention of researches in nonlinear system identification was first focused on static networks such as multilayer perceptrons (Narendra and Parthasarathy, 1990). The inputs to these static networks are usually delayed values of the inputs and outputs of the plant. This approach, however, has some disadvantages. First, the input structure is not easy to choose. Moreover, the discrete time model requires re–training when the sampling time is changed. Furthermore, if the models are to be employed as part of a nonlinear control scheme, methods for discrete time non–linear control are not as well developed as continuous time nonlinear control methods.

The continuous time Hopfield dynamic neural networks (DNN) (Hopfield, 1982) and their variations (Hopfield, 1984; Hopfield and Tank, 1985; Hopfield and Tank, 1986; Koiran, 1994), do not present these disadvantages and they are capable of approximating multivariable dynamic systems. Input affine DNNs can approximate autonomous nonlinear dynamic systems (Funahashi and Nakamura, 1993; Kimura and Nakano, 1998), multivariable control affine systems (Garces *et al.*, 1999; Kambhampati *et al.*, 2000; Garces *et al.*, 2003) and furthermore general nonlinear systems (Garces, 2000; Garces *et al.*, 2003). Stability conditions for DNNs have also been analysed (Matsuoka, 1992; Sanchez and Perez, 1999).

Recently, Becerra *et al* (2002) and Garces *et al* (2003) have introduced techniques to initialise the state of the hidden neurons both for training and validation purposes. Particularly, it has been proposed to include the hidden states in the decision vector of the optimisation problem associated with training. Also, it has been proposed to initialise these hidden states by solving an optimisation problem when the trained network is been used or validated.

Using these techniques, one of the problems associated with dynamic neural networks is solved. Previous works on the use of dynamic neural networks for system identification carried out the initialization of the network's hidden states with zeros or random values, which is not always suitable.

In this paper, the idea of adjusting the hidden initial states proposed by (Becerra *et al.*, 2002; Garces *et al.*, 2003) is explored and applied for the identification of a robot system consisting of a two DOF mechanic leg. In order to improve the identification process, a regularisation term was added to the cost function associated with the training problem.

The paper is organized as follows. Section 2 introduces dynamic neural networks (DNNs). Section 3 describes the problem of DNN training. Section 4 describes the method for initialising the DNNs. Section 5 describes a cross-validation method for selecting the best model structure. Section 6 describes the two

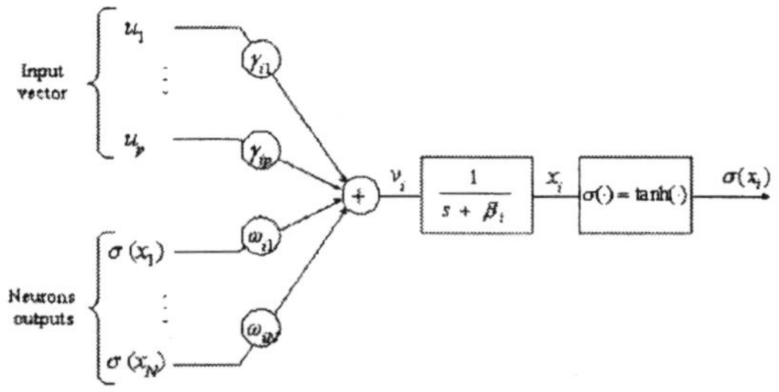

Fig. 1. Diagram of a dynamic neuron

link robotic system employed in this work. Section 7 presents training and validation results. Finally, Section 8 gives final remarks about this research work.

2. DYNAMIC NEURAL NETWORKS

Recurrent networks were introduced in the context of associative or content addressable memory (CAM) problems (Kohonen, 1989; Hopfield, 1984). The introduction of feedback or internal dynamics into a feedforward neural network architecture produces a state space dynamic model. The model is defined by a one-dimensional array of neurons; each unit can be described as follows,

$$\frac{\mathrm{d}x_i}{\mathrm{d}t} = -\beta_i x_i + \sum_{j=1}^{n} \omega_{ij}\sigma(x_j) + \sum_{j=1}^{p} \gamma_{ij}u_j \qquad (1)$$

where β_i, ω_{ij} and γ_{ij} are adjustable weights, with $1/\beta_i$ as positive time constant and $p \leq n$, x_i the activation state of unit i, and $u_1,\ldots,u_m$ the input signals. The function $\sigma(\cdot)$ is typically a nonlinear sigmoid-type function like the hyperbolic tangent function. The structure of a dynamic neuron is illustrated in Figure 1.

The DNN is formed by a single layer of n units. The output of the network is often taken as the first p units of the state vector x, leaving $n-p$ units as hidden neurons. The network is defined in equation (2) by the vectorised expression of equation (1),

$$\begin{aligned}\frac{\mathrm{d}x}{\mathrm{d}t} &= -\beta x + \omega\sigma(x) + \gamma u \\ \hat{y} &= C_n x\end{aligned} \qquad (2)$$

where $x \in \Re^n$ is a state vector, $u \in \Re^m$ is the input vector, $\hat{y} \in \Re^p$ is the output vector, $\omega \in \Re^{n\times n}$, $\sigma(x) = [\sigma(x_1),\ldots,\sigma(x_n)]^T$, $\gamma \in \Re^{n\times m}$, $C_n = [I_{p\times p} 0_{p\times(n-p)}]$, and $\beta \in \Re^{n\times n}$ is a diagonal matrix with elements $\{\beta_1,\ldots,\beta_n\}$ in the diagonal.

It is assumed that $n > p$, so that the network has $n-p$ hidden units. Hidden units are used to increase the dynamic mapping potential of the network (Funahashi and Nakamura, 1993). The state vector of the DNN can thus be partitioned into the output states x_o and the hidden states x_h:

$$x = \begin{bmatrix} x_o \\ x_h \end{bmatrix} \qquad (3)$$

where $x_o \in \Re^p$ and $x_h \in \Re^{n-p}$.

3. TRAINING DYNAMIC NEURAL NETWORKS

Suppose that data have been collected from a real system that is to be modelled by means of a dynamic neural network. Consider a training data set with N input–output pairs and sampling time T_s:

$$Z_N = [y(t_k), u(t_k)]_{k=1,N} \quad (4)$$

where $y \in \Re^p$ is the measured output, $u \in \Re^m$ is the input variable, and k is a sampling index. Then the problem of training the DNN to learn the dynamics from data set Z_N may be written as an optimisation problem.

Let be $\theta \in \Re^{n_\theta}$ the decision vector where n_θ is the number of parameters to be optimised. If the loss function given by the mean squarre error is chosen,

$$V_N(\theta, Z_N) = \frac{1}{2N} \sum_{k=1}^{N} ||y(t_k) - \hat{y}(t_k|\theta)||^2 \quad (5)$$

where $y(t_k|\theta)$ is the output vector of the network (2) at time t_k given the decision vector θ, the nonlinear identification problem can be described by the following unconstrained optimisation problem:

$$\min_{\theta} \quad V_N(\theta, Z_N)$$

In (Becerra *et al.*, 2002) and (Garces *et al.*, 2003) it is proposed that the decision vector θ be augmented to include the initial values of hidden states of the DNN, $x_h(t_0)$:

$$\theta = \begin{bmatrix} \beta_d \\ \text{vec}(\omega) \\ \text{vec}(\gamma) \\ x_h(t_0) \end{bmatrix} \quad (6)$$

where β_d is a vector with the diagonal elements of β and vec$(\cdot)$ is a vector created with the stacked columns of an argument matrix $(\cdot)$.

In this form, the potential bias that would arise on the parameters of the DNN with an incorrect initialisation of the hidden states is avoided. This is particularly true for data sets of limited length in time compared with the dynamics of the system. As (Becerra *et al.*, 2002) demonstrates, fixing the initial values of the hidden states to zero (as usual) leads to an unnecessary adjustment of the DNN parameters in order to reduce the differences between the output states of the DNN with the measured outputs in the early stages of the training record.

A variation of the identification problem that help to the DNN training is to choose the following modified loss function:

$$V'_N(\theta, Z_N) = \frac{1}{2N} \sum_{k=1}^{N} ||y(t_k) - \hat{y}(t_k|\theta)||^2 + \rho ||x_h(t_0)||^2 \quad (7)$$

where $\rho > 0$ weights the squared norm of the initial values of the hidden states. In this form, by the addition of a regularisation term, unnecessarily high values of these states are avoided and only those hidden states whose values are relevant would be non-zero at the solution.

A way of allowing for the presence of local minima in the error surface defined by the objective function is to run the DNN training procedure from different randomly selected initial decision variables. Another possibility is using genetic algorithms (Garces *et al.*, 2003).

4. INITIALIZING DYNAMIC NEURAL NETWORKS

If a further data set not used for training purposes is available, the trained neural network can be validated using this unseen data set. The initial values of the hidden states obtained with the training procedure are in general not valid for the validation data set, so a simple procedure has been developed to initialize the DNN when it is to be used or validated (Garces *et al.*, 2003). It consists of initialising the output states of the DNN with the values of the system's output:

$$x_o(t_0) = y(t_0) \quad (8)$$

and the hidden states by solving the following optimisation problem:

$$x_h(t_0) = \arg \min_{x_h(t_0)} \left\| \frac{\mathrm{d}y(t_0)}{\mathrm{d}t} - \frac{\mathrm{d}x_o(t_0)}{\mathrm{d}t} \right\|^2 \quad (9)$$

where $dy(t_0)/dt$ can be calculated from known data and $dx_o(t_0)/dt$ can be calculated from equation (1), once $x_0(t_0)$, $x_h(t_0)$, β, ω and γ are known. By using this procedure, it is ensured that the initial value of the hidden state vector $x_h(t_0)$ is such that the initial output derivative of the DNN, $\mathrm{d}x_o(t_0)/\mathrm{d}t$, is as close as possible to the required value $\mathrm{d}y(t_0)/\mathrm{d}t$.

5. CROSS-VALIDATION AND MODEL STRUCTURE SELECTION

Cross-validation is an approach to aid in the model structure selection using an extra data set. Given a training data set Z_N and a validation data set $\hat{Z}_N$, a sequence of models of increasing complexity $\{\mathcal{M}_1, \mathcal{M}_2, \ldots, \mathcal{M}_k\}$ is identified based on the training data set Z_N. This gives rise to a decreasing trend in the training objective function $V_N(\theta^{(k)}, Z_N)$ for increasing complexity. However, when the objective function is evaluated using the validation data set, $V_N(\theta^{(k)}, \hat{Z}_N)$, its value will initially decrease with increasing model complexity until some critical k^*, and then it will start to increase. The model $\mathcal{M}_{k^*}$ that provides the minimum value of $V_N(\theta^{(k)}, \hat{Z}_N)$ is usually chosen as the best model for the system that generated the data.

In the case of the dynamic neural networks used in this work, the model complexity index is the number of neurons in the network; this is, the model order n.

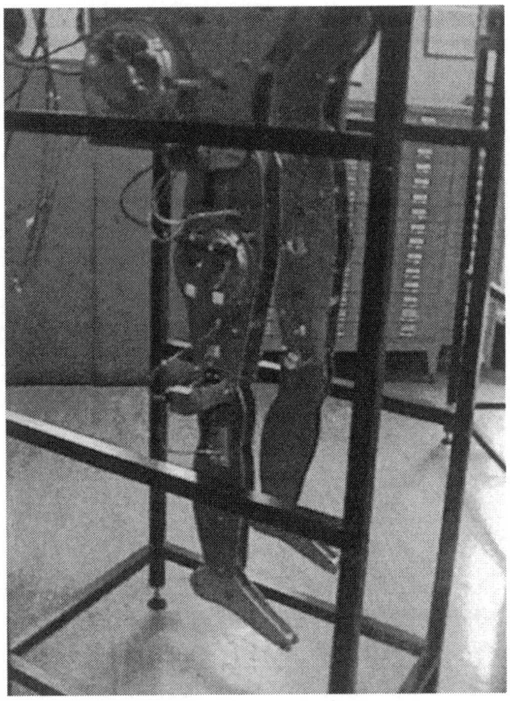

Fig. 2. View of the legs

6. CASE STUDY

A dynamic neural network model was trained using the techniques described above to model a real two-link robotic leg which is shown in Figure 2. The complete system is composed of a pair of decoupled legs, but only one of them used for identification experiments described in this paper.

The system consists of two rigid elements actuated by two DC motors situated on both revolute links: hip and knee. The final element of the leg is the foot, which is articulated but it is not actuated, so the leg is considered as a 2–DOF system.

The angles of each link are measured by two potentiometers located in the internal part of the leg. The corresponding velocities were computed by digital filtering of angular signals. The manipulated inputs were the two voltages applied to the power amplifiers that drive the DC motors. It is not difficult to notice that nonlinearities such as different types of friction, dead-zones and some looseness are present in this system.

7. RESULTS

A DNN with the form given by equation 2 was used, applying the training and validation aspects described in the previous sections.

Both the training and validation data sets were obtained experimentally and had 1595 input–output samples each. A sum of sinusoids with different amplitudes and frequencies was chosen as inputs, both inputs and outputs covering a large range of values which the real system can vary in. These values have been scaled to have zero mean and variance 1.

The optimisation was carried out using a Quasi-Newton unconstrained optimisation algorithm implemented in function *fminunc*, which is part of the Optimisation Toolbox in MATLAB version 6.1.

The DNN is intended to provide the angles of the two links, so the first two neurons were chosen as the outputs, with the remaining neurons being hidden neurons. The best number of neurons $n = 4$ was selected using the cross-validation method described in Section 5.

The values of the parameters obtained were the following:

$$\beta_d = \begin{bmatrix} 5.0856 \\ 2.9197 \\ -3.3450 \\ 2.0491 \end{bmatrix} \tag{10}$$

$$\omega = \begin{bmatrix} 3.6484 & -4.1867 & 6.1201 & -4.0089 \\ -1.2547 & 1.7313 & -3.2064 & -4.3606 \\ -0.4706 & -0.9042 & -5.7358 & -2.5743 \\ 0.7214 & 0.3866 & 0.5750 & 1.7928 \end{bmatrix} \tag{11}$$

$$\gamma = \begin{bmatrix} -0.3534 & 0.0060 \\ -0.3835 & -0.7599 \\ 0.0249 & -1.0831 \\ -0.3994 & -0.8308 \end{bmatrix} \tag{12}$$

The results obtained for both links corresponding to the training data are shown in Figures 3 and 4. As it can be seen, the behavior of the DNN is very close to the real system behaviour.

In results obtained for the validation data are shown in Figures 5 and 6. The DNN presents a good performance compared with the real data measured.

8. CONCLUSIONS

The aim of this work was the identification of a robotic system using a dynamic neural network. The system consists of a robotic 2-DOF leg, which has nonlinearities that are difficult to model theoretically. A method for the initialisation of the hidden states both during training and validation was applied in order to improve the quality of the model. A cross-validation method was employed to select the best model structure. The results obtained show that a four neuron DNN approximates in a satisfactory manner the dynamic behaviour of the real system.

REFERENCES

Angeles, J. (1997). *Fundamentals of Robotic Mechanical Systems. Theory, Methods and Algorithms.* Springer-Verlag. New-York, Inc.

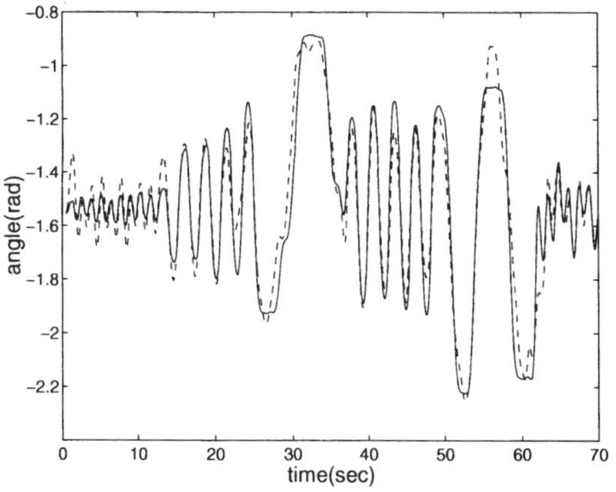

Fig. 3. Angles of link 1 (solid line) and DNN output. Training data

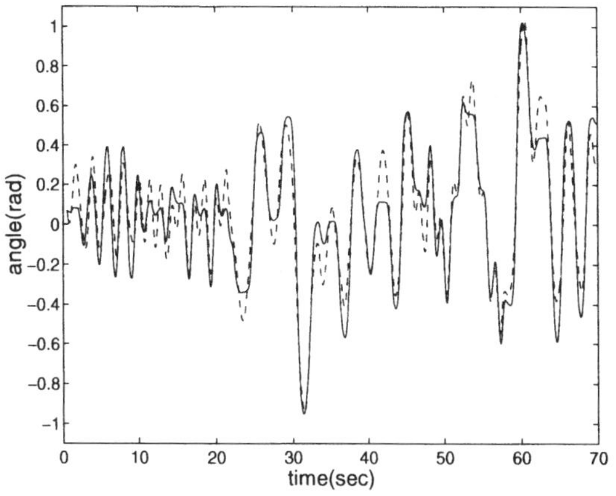

Fig. 4. Angles of link 2 (solid line) and DNN output. Training data

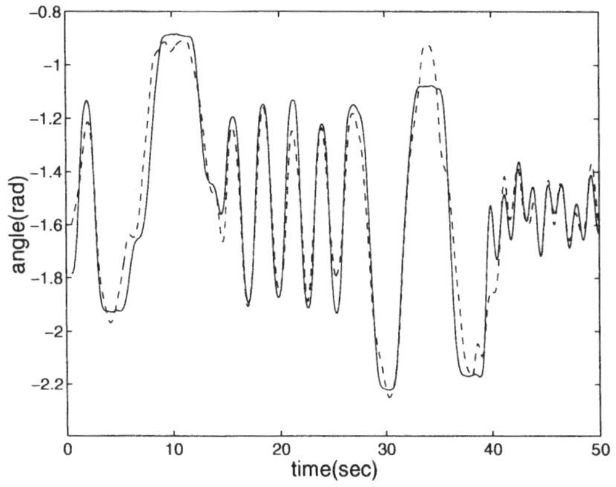

Fig. 5. Angles of link 1 (solid line) and DNN output. Validation data

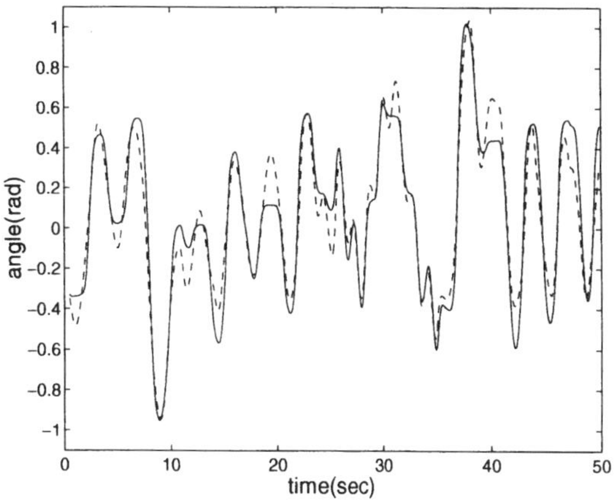

Fig. 6. Angles of link 2 (solid line) and DNN output. Validation data

Becerra, V.M., J.M.F. Calado, P.M. Silva and F. Garces (2002). System identification using dynamic neural networks: training and initialization aspects. In: *15th World Congress of the IFAC*. Barcelona. Spain.

Funahashi, K. and Y. Nakamura (1993). Approximation of dynamical-systems by continuous-time recurrent neural networks. *Neural Networks* **6**(6), 801–806.

Garces, F. (2000). Dynamic neural networks for approximate input-output linearisation-decoupling of dynamic systems. PhD Thesis. University of Reading.

Garces, F., C. Kambhampati and K. Warwick (1999). Dynamic recurrent neural networks for identification of a multivariable nonlinear evaporator system. In: *International Conference on Dynamic Control Systems, DYCONS99*. World Scientific. Ottawa.

Garces, F., V.M. Becerra, C. Kambhampati and K. Warwick (2003). *Strategies for feedback linearisation: a dynamic neural network approach*. Springer. London.

Hopfield, J. J. (1982). Neural networks and physical systems with emergent collective computational abilities. *Proceedings of the National Academy of Sciences of the United States of America-Biological Sciences* **79**(8), 2554–2558.

Hopfield, J. J. (1984). Neurons with graded response have collective computational properties like those of 2-state neurons. *Proceedings of the National Academy of Sciences of the United States of America-Biological Sciences* **81**(10), 3088–3092.

Hopfield, J. J. and D. W. Tank (1985). Neural computation of decisions in optimization problems. *Biological Cybernetics* **52**(3), 141–152.

Hopfield, J. J. and D. W. Tank (1986). Computing with neural circuits - a model. *Science* **233**(4764), 625–633.

Kambhampati, C., F. Garces and K. Warwick (2000). Approximation of non-autonomous dynamic systems by continuous time recurrent neural networks. In: *Neural networks* (S. I. Amari, Ed.). IEEE International Conference on Neural Networks. Como, Italy. pp. I–64–I–69.

Kimura, M. and R. Nakano (1998). Learning dynamical systems by recurrent neural networks from orbits. *Neural Networks* **11**(9), 1589–1599.

Kohonen, T. (1989). *Self-organization and associative memory*. 3rd ed.. Springer-Verlag. Berlin ; New York.

Koiran, P. (1994). Dynamics of discrete-time, continuous state Hopfield networks. *Neural Computation* **6**(3), 459–468.

Matsuoka, K. (1992). Stability conditions for nonlinear continuous neural networks with asymmetric connection weights. *Neural Networks* **5**(3), 495–500.

Narendra, K. S. and K. Parthasarathy (1990). Identification and control of dynamical systems using neural networks. *IEEE Transactions on Neural Networks* **NN1**, 4–27.

Sanchez, E. N. and J. P. Perez (1999). Input-to-state stability (ISS) analysis for dynamic neural networks. *IEEE Transactions on Circuits and Systems I-Fundamental Theory and Applications* **46**(11), 1395–1398.

IFAC Publications
www.elsevier.com/locate/ifac

NEURAL NETWORK SYSTEM IDENTIFICATION FOR A LOW PRESSURE NON-LINEAR DYNAMICAL SUBSYSTEM ONBOARD THE ALICIA II CLIMBING ROBOT

Domenico Longo, Giovanni Muscato, Giuseppe Nunnari

Dipartimento di Ingegneria Elettrica Elettronica e dei Sistemi, Università degli Studi di Catania, viale A. Doria 6, 95125 Catania Italy
e-mail: dlongo@diees.unict.it, gmuscato@diees.unict.it, gnunnari@diees.unict.it

Abstract: In this work, a 'black box' non-linear dynamic model for the low pressure subsystem onboard the base module Alicia II robot has been computed by using Artificial Neural Network methodology. The obtained model can be useful to implement and tune a control algorithm for the pressure inside the cup of the robot, also by using Neural Network, to prevent it to fall down. *Copyright © 2003 IFAC*

Keywords: Neural Network, System identification, Robotics.

1. INTRODUCTION

Climbing robots are useful devices that can be adopted in a variety of applications for solving problems of maintenance, inspection and safety in the process and construction industries. These systems are mainly adopted in all those places where direct access by a human operator is very expensive (because of the need of scaffolding) or very dangerous (due to the presence of a hostile environment) [1]-[4]. Several different examples of climbing robots and their applications can be found in the proceedings of the CLAWAR conferences [5]-[7], in the Walking Machine Catalogue [8] and in previous work [9]-[11].

The proposed identification methodology is applied to the robot module Alicia II. This robot is designed for the inspection of non-porous vertical surfaces like those of petrochemical storage tanks. For this application, due to the extremely corrosive substances the tanks may contain, it is very important to perform periodic inspection, as standardised by the American Petroleum Institute [12].

The structure of the Alicia II module prototype, shown in Fig. 1 and Fig. 2, is actually composed by a PVC cup with an internal diameter of 20 cm and 8 cm height. The adhesion to the wall is guaranteed by a negative pressure inside the cup. To generate a suitable vacuum level onboard the robot, a centrifugal air aspirator has been used. The aspirator uses a universal motor (230Vac - 1200W) and it is capable of a maximum Δp of 200 mbar when no appreciable gap between the cup and the wall is present. The whole structure has been designed to contain two wheels with two independent DC motors/gearboxes. Two 5 cm diameter wheels have been used to allow the structure moving along all directions, while the third contact point is guaranteed by a passive wheel.

Fig. 1. The Alicia II robot module.

Fig. 2. The Alicia II robot module in action.

Such a kind of system requires a different behaviour for the cup with respect to a standard fixed suction cup. In this case, the cup can not adhere to the surface with high friction, so a particular kind of seal between the wall and the robot needs to be realised. The seal must guarantee the internal negative pressure and should allow the robot to pass over small obstacles (lower than 1 cm height) like screws or welding traces. Actually a 2 cm height sandwich of rubber/bristles sealing is under testing with good results. A maximum Δp of 100 mbar corresponding to a maximum normal force of about 300 N was measured.

2. NEEDS FOR A PRESSURE CONTROL

By using this kind of movement and sealing method, it is possible, due to unexpected small obstacles on the surface, to have some air leakage in the cup. This leakage can cause the internal negative pressure to rise up, and in this situation the robot could fall down.

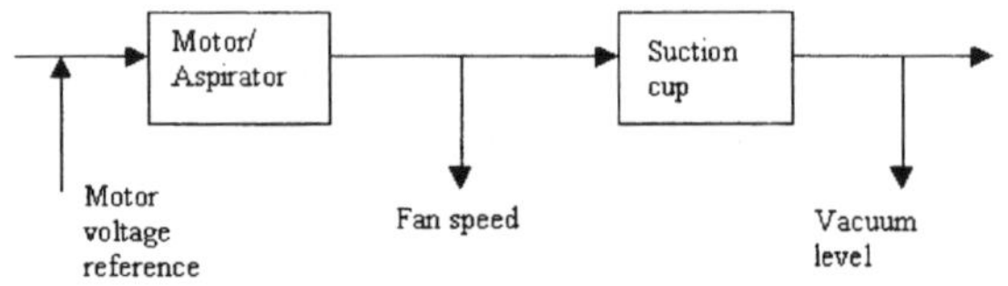

Fig. 3. Used open loop scheme.

On the other side if the internal pressure is too low (high Δp), a very big normal force is applied to the system. As a consequence, the friction can increase in such a way to not allow robot movements. This problem can be solved by introducing a control loop to regulate the pressure inside the chamber to a suitable value to sustain the system. This control loop can also optimise the power consumption of the aspirator. The open loop system and the most easily accessible system variables has been schematised in Fig. 3; in this scheme the first block includes the electrical and the mechanical subsystem and the second block includes the pneumatic subsystem. The used variables are the *Motor voltage reference* (the input signal that fixes the motor power), the *Fan speed* (the velocity of the aspirator fan) and the *Vacuum level* (the negative pressure inside the chamber).

Since it is very difficult to have a reliable analytical model of that system, because of the big number of parameters involved, it has been decided to identify a black box dynamic model of the system by using input/output measurements. This model was designed with two purposes. It will be adopted to compute a suitable control strategy and to implement a simulator for tuning the control parameters.

3. BLACK BOX SYSTEM IDENTIFICATION

An experimental setup was realised, as represented in Fig. 4, by using the DS1102 DSP board from Dspace, in order to acquire the input/output variables. Each of the system variables has been measured with a suitable sensor. Since the aspirator is actuated by an AC universal motor, a power interface has been realised; this block, based on a ST52E430 MCU, uses phase partialisation method to translate in power the voltage reference signal for the AC motor coming from a DAC channel of the DS1102 board. The aspirator velocity has been acquired by using a reflective optical sensor and a frequency to voltage converter, while the vacuum level inside the chamber has been measured by using a piezoresistive pressure sensor. For these sensors a suitable conditioning block has been realised and the signals have been acquired with two analog inputs of the DS1102. The software running on the DSpace DSP board, simply generates an exciting motor voltage reference signal (pseudo random, ramp or step signals) and acquires the two analog inputs with a sampling time of 0.1 s, storing the data in its internal SRAM. This software has been written in the Matlab/Simulink environment.

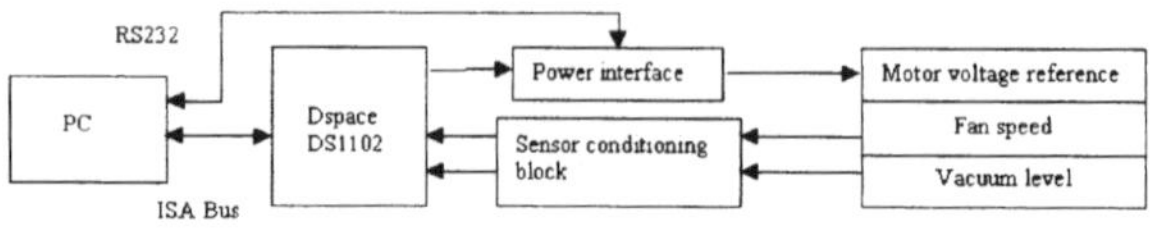

Fig. 4. I/O variable acquisition scheme.

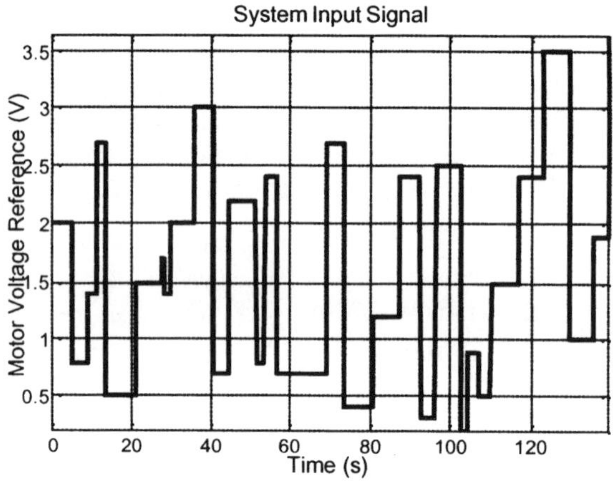

Fig. 5. Typical input signal.

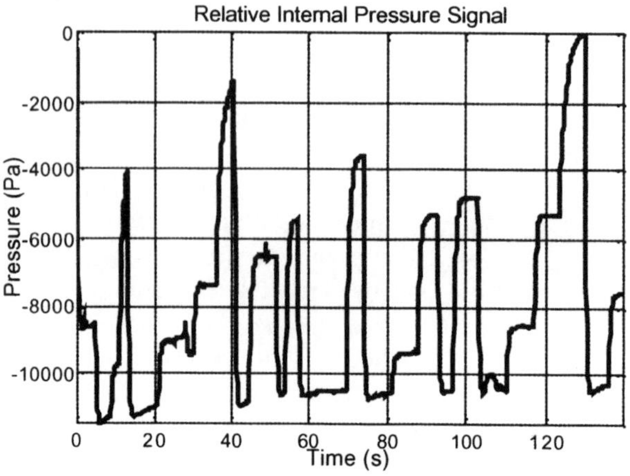

Fig. 6. Typical output signal.

4. MODEL ESTIMATION ALGORITHMS

4.1 Linear approximation.

Input/Output measurement represented in Fig. 5 and Fig. 6, suggest that the system can be modelled as a linear first order system. Also, as first approximation, it is possible to consider an electric equivalent of the pneumatic subsystem of the Alicia II. In Fig. 7 there is an RC equivalent network where R_{pump} is the pressure drop inside the aspirator, *Cup* is the suction cup of the robot, and *Leakage* is an equivalent resistor for the lateral leakage of the sealing.

Based on these considerations, a first trial with a linear regression has been done. The used model is the ARX one, as in the equation (1).

$$y(k) = a_0 u(k) + a_1 u(k-1) + \ldots + b_1 y(k-1) + b_2 y(k-2) + \ldots \quad (1)$$

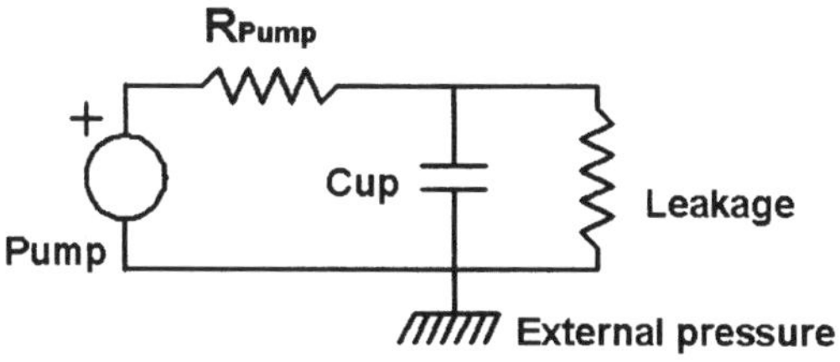

Fig. 7. Linear RC equivalent network.

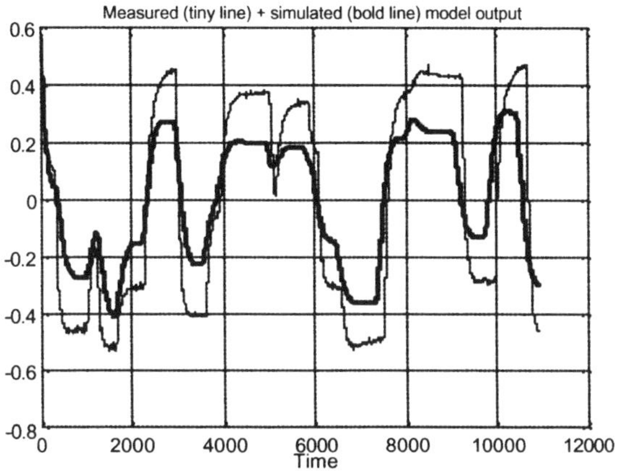

Fig. 8. Some results for the ARX algorithm.

In Fig. 8 is represented the simulated output and the real output of the system, where the simulation is generated by a ten order ARX equation with nine zeroes. It is possible to see as the results are very useless. System non-linearity is not negligible so a non-linear model is required.

4.2 Non-linear model.

In order to obtain better results, the system needs to be considered as non-linear. A NARMAX model has been used is in the form of equation (2), where *f* is a non linear function [13], [14].

$$y(k) = f(u(k), u(k-1), \ldots; y(k-1), y(k-2), \ldots) \quad (2)$$

To implement this kind of non-linearity, it was decided to use Artificial Neural Network (ANN). As the ANN realise a static mapping of the input pattern over the output one, the system dynamic has been introduced in the input vector with some I/O regression terms. Using the I/O data from the real system, a suitable network structure and training algorithm, the net training phase can be schematised as in Fig. 9.

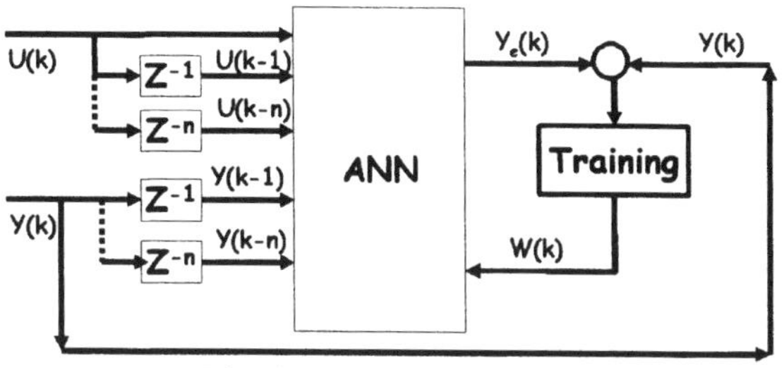

Fig. 9. Net training phase.

Once the net has been trained to a suitable mean square error, it has been simulated giving it as input the real input measurement only (parallel identification model) [13]. With this methodology, equation (2) can be modified in order to obtain equation (3).

$$\tilde{y}(k) = f(u(k), u(k-1), \ldots; \tilde{y}(k-1), \tilde{y}(k-2), \ldots) \quad (3)$$

where $\tilde{y}$ is the estimated system output. In Fig. 10 the testing phase of the net has been represented; Y_e is the estimated output.

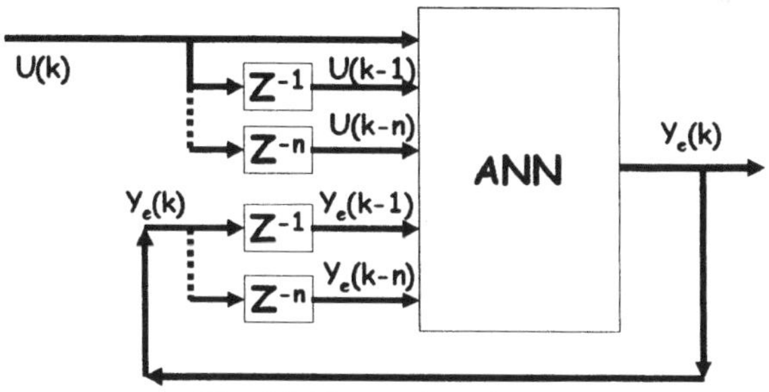

Fig. 10. Net testing phase.

At the moment the best simulation results have been obtained with a single layer perceptron network with a number of hidden neurons between 4 and 7, 3 output regressions and 2 input regressions. The used training algorithm is the standard Levenberg-Marquardt.

In Fig. 11, Fig. 12, Fig. 13 and Fig. 14 some simulation results are reported. In Fig. 11 and Fig. 13 there are input time series for the system. In Fig. 12 and Fig. 14 output measures (tiny lines) are compared with the simulated output of the system (bold line). This simulation was obtained with an ANN trained over the data in Fig. 5 and Fig. 6 with six neurons in the hidden layer.

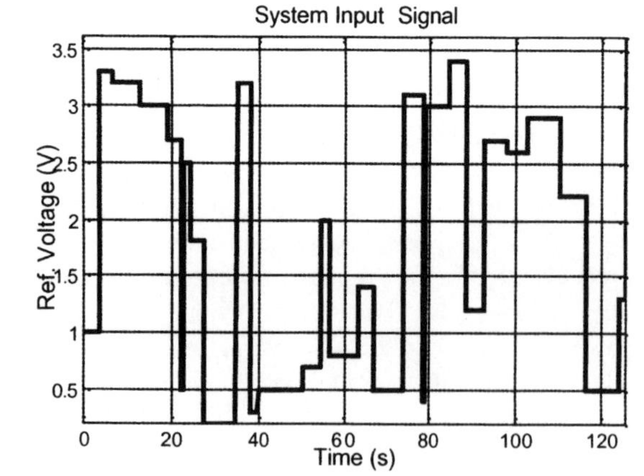

Fig. 11. Input data: case 1.

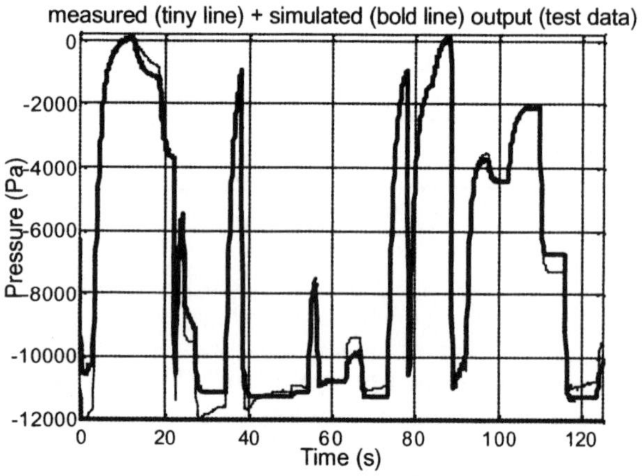

Fig. 12. Output data + simulated output: case 1.

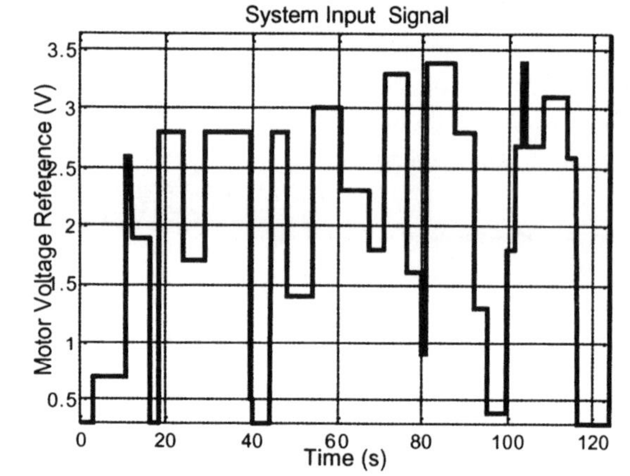

Fig. 13. Input data: case 2.

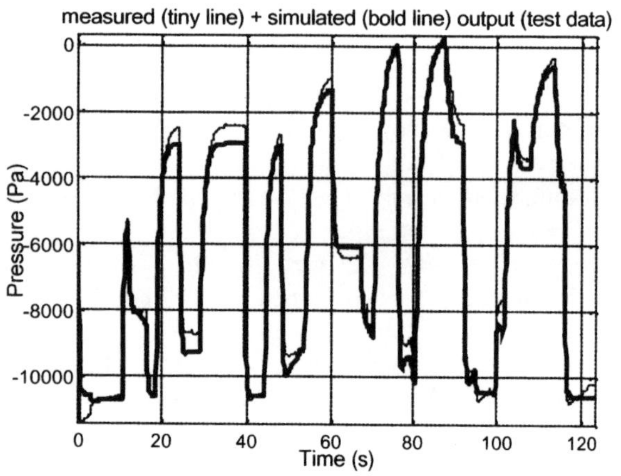

Fig. 14. Output data + simulated output: case 2.

5. SOME RESULTS

Once the neural model has been tested against the real system, it is possible to derive a control algorithm starting from the system emulator [15] [16].

As shown in Figure 15, a closed loop control system has been formed by using the plant model block and a controller block. The first block contains the previously trained network.

The main idea to obtain the controller equation is to write the plant model in the NARMA-L2 approximate form, as shown by equation (3).

$$y(k+d) = f[y(k), y(k-1), \dots, y(k-n+1), u(k), u(k-1), \dots, u(k-n+1)] + g[y(k), \dots, y(k-n+1), u(k), \dots u(k-n+1)]u(k+1)$$

(3)

$d \geq 2$

When the plant is modeled by means of the equation (3), the controller equation is simply derived from the plant model and is in the form of equation (4).

$$u(k+1) = \frac{y_r(k+d) - f[y(k), \dots, y(k-n+1), u(k), \dots u(k-n+1)]}{g[y(k), \dots, y(k-n+1), u(k), \dots, u(k-n+1)]}$$

(4)

$d \geq 2$

In this equation, the y_r signal is the reference signal after a filtering phase, in order to impose an output dynamics. In our simulation, this transfer function (reference model) has been taken as a constant, because we are interested in solving a regulation problem and not a tracking one.

The NARMA-L2 form of the plant has been obtained once again by approximating the two non-linear functions f and g included in equation (3) and equation (4) with two separate neural networks. At this point it is possible to simulate the closed loop behavior of the system.

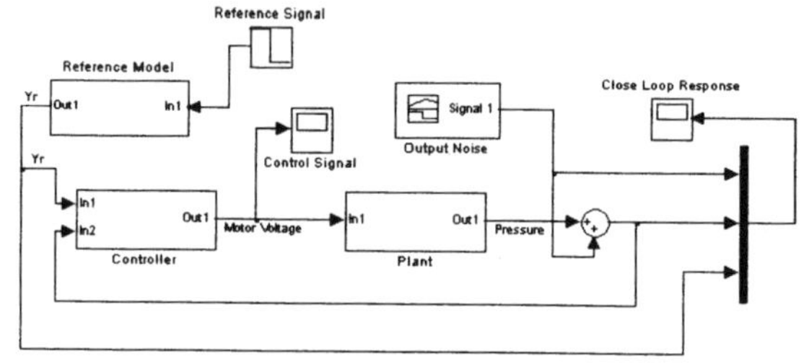

Fig. 15. Simulink diagram for the NARMA-L2 controller simulation

In Figure 16 a simulation results has been reported. The reference signal is a pressure step that starts from the atmospheric one and goes to a lower level at the step time. The pressure values in the y axes are normalized in the range [-1, 1]. Moreover some noise is added to the system output (dashed line). As can be seen from Figure 17, the controller is able to filter the noise on the output pressure with only a brief transient. In Figure 18 the control signal generated by the controller is reported. Also a steady-state leakage (the final part of the plot in Figure 16) has been compensated by modulating the motor voltage applied to the aspirator.

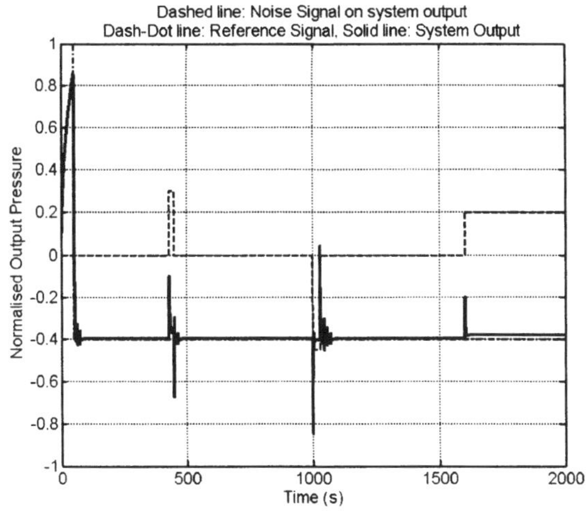

Fig. 16. A Simulation result for the NARMA-L2 Neuro Controller.

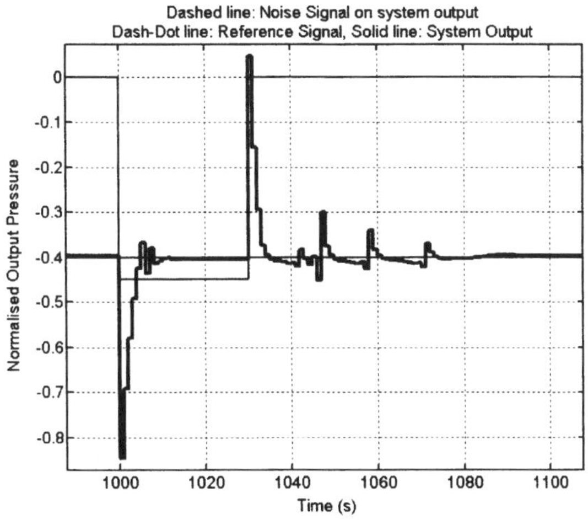

Fig. 17. A detail of the simulation.

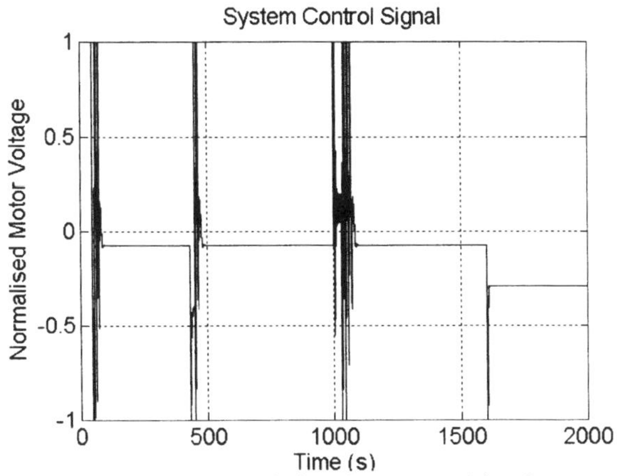

Fig. 18. The control signal generated by the controller.

Finally, in Fig. 19, a picture of the final version of the Alicia II module is reported. It is an improved version of the prototype shown in Fig. 1 and Fig. 2 in what is concerning robustness, modularity for easy maintenance and weight.

Fig. 19. The final version of the Alicia II module during a test.

6. CONCLUSION

In this work an Artificial Neural Network identification methodology has been applied to the low pressure pneumatic subsystem of the Alicia II module robot. This subsystem is responsible for the robot to stay attached to non-porous walls. If for any reason the sealing between the robot and the wall has some leakages, the internal pressure can rise up to the external value and the robot can fall down.

To solve this problem a suitable control algorithm for the internal pressure of the cup is required. The system composed by the cup and the aspirator can be thought as a first order linear RC network, so in first approximation a linear model can be utilised. Some trials in that direction have shown that system non-linearity can't be neglected, so a NARMAX model has been implemented by using ANN.

The trained network is capable to do an n-step prediction of the system output over some testing pattern not used during the training phase, which means that the trained network has good generalising properties. It must also be considered that the real system has some repeatability problem due to a random behaviour of the sealing materials (Teflon/bristle/rubber).

The model that has been found out has been used to realise a system emulator (dynamical model for the pneumatic subsystem of the Alicia II module) and to derive a control system to regulate the internal pressure of the cup. Some results have been reported.

REFERENCES

[1] Y. Wang, X. Zhao, D. Xu – "Design of a wall cleaning robot with a single suction cup" – Robot Research Institute of Harbin Institute of Technology , China. Proceedings of 2nd International Conference on Climbing and Walking Robot – Pag. 405 – 1999

[2] T. S. White, D. S. Cooke – "Robosense – Robotic delivery of sensors for seismic risk assessment"– Portsmouth Technology Consultant Limited, UK Proceedings of 3rd International Conference on Climbing and Walking Robot – Pag. 847 – 2000

[3] Y. Wang, X. Zhao, D. Xu – "The study and application of wall-climbing robot for cleaning" – Robot Research Institute of Harbin Institute of Technology , China. Proceedings of 3rd International Conference on Climbing and Walking Robot – Pag. 789 – 2000

[4] G. La Rosa, M. Messina, G. Muscato, R. Sinatra – "A low cost lightweight climbing robot for the inspection of vertical surfaces" – Mechatronics 12 (2002) 71-96 - Pergamon

[5] Proceedings of the First International Symposium Climbing and Walking Robots CLAWAR '98, Brussels, 26-28 November 1998.

[6] Proceedings of the Second International Conference on Climbing and Walking Robots CLAWAR 99, Portsmouth, U.K., 14-15 September1999, Professional Engineering Publishing.

[7] Proceedings of the Third International Conference on Climbing and Walking Robots CLAWAR 2000, Madrid (Spain), 2-4 October 2000, Professional Engineering Publishing.

[8] K. Berns, "The Walking Machine Catalogue", http://www.fzi.de/ipt/WMC/walking_machines_katalog/walking_machines_katalog.html

[9] L. Fortuna, A. Gallo, G. Giudice, G. Muscato, "Sensor Fusion to Improve the Control of a Mobile Walking Robot for Automatic Inspection: The ROBINSPEC System", Proceedings of the 6th IMEKO International Symposium on Measurement and Control in Robotics ISMCR96, Brussels (Belgium), 9-11 May 1996, pp.376-380.

[10] L. Fortuna, A. Gallo, G. Giudice, G. Muscato, "ROBINSPEC: A Mobile Walking Robot for the Semi-Autonomous Inspection of Industrial Plants", in Robotics and Manufacturing: recent trends in research and applications, Vol. 6, ASME PRESS New York (USA), pp. 223-228, May 1996.

[11] G. Muscato, G. Trovato, "Motion control of a pneumatic climbing robot by means of a fuzzy processor", First International Symposium CLAWAR '98 Climbing and Walking Robots, Brussels, 26-28 Novembre 1998.

[12] American Petroleum Institute – "Tank Inspection, Repair, Alteration and Reconstruction", Standard 653, January 1992, API, 1220, L. St. NW, Washington, DC, 20005.

[13] K. S. Narendra, K. Parthasarathy, "Identification and control of dynamical system using neural networks", IEEE Transaction on Neural Network, Vol. 1 no. 1, pp 4-27, March 1990.

[14] S. Chen, S. A. Billings, C. F. N. Cowant, P. M. Grant, "Practical identification of NARMAX models using radial basis functions", Int. j. Control, Vol. 52, no. 6, pp. 1327-1350, 1990.

[15] D. H. Nguyen, B. Widrow – "Neural Networks for self learning Control Systems" – IEEE Control System Magazine – pp 18-23 – 1990.

[16] K. S. Narendra, S. Mukhopadhyay – "Adaptive Control Using Neural Networks and Approximate Models" – IEEE Transaction on Neural Networks – Vol. 8, No 3, pp 475-485 – 1997.

IFAC Publications
www.elsevier.com/locate/ifac

MEASUREMENT OF YOUNG'S MODULUS VIA MODAL ANALYSIS EXPERIMENTS: A SYSTEM IDENTIFICATION APPROACH

R. Pintelon[(1)], P. Guillaume[(2)], K. De Belder[(1)], and Y. Rolain[(1)]

[(1)]Vrije Universiteit Brussel, Department ELEC, Pleinlaan 2, 1050 Brussel, Belgium
[(2)]Vrije Universiteit Brussel, Department WERK, Pleinlaan 2, 1050 Brussel, Belgium

Abstract: The stress-strain relationship of linear visco-elastic materials is characterized by a complex-valued, frequency dependent elastic modulus $E(j\omega)$ (Young's modulus). Using system identification techniques it is shown in this paper how $E(j\omega)$ can be measured accurately in a broad frequency band from forced flexural (transverse) and longitudinal vibration experiments on a beam under free-free boundary conditions. The advantages of the proposed method are (i) it takes into account the disturbing noise and the nonlinear distortions, (ii) $E(j\omega)$ is delivered with an uncertainty bound, (iii) the low sensitivity to non-idealities of the experimental set up, and (iv) the ability to measure lowly damped materials. The approach is illustrated on plexiglass and copper. *Copyright © 2003 IFAC*

Keywords: Young's modulus, modal analysis, frequency domain.

1. INTRODUCTION

The complex modulus $E(j\omega)$ is a widely used and powerful tool for describing the linear dynamic elastic and damping properties of visco-elastic materials. For example, knowledge of $E(j\omega)$ in the audio frequency range is needed by acousticians when developing and characterizing new materials for noise and vibration control. The idea of measuring the complex moduli (Young's, shear, ...) from vibration tests (flexural, longitudinal, and torsional) is not new and dates from the early sixties (see the references in [1]). From those early days till now a continuing attention has been paid to this measurement problem in the literature: see [1-6] and the references therein for the longitudinal vibration tests, and [1], [7-10] and the references therein for the flexural vibration tests. The longitudinal (wave) experiments assume that the material is homogeneous and isotropic [8-10], while the flexural (transverse or bending) experiments can also handle layered structures [8], [10]. In the wave experiments the beam is sometimes loaded by an end mass (e.g. for measuring rubber, plastic foams and mineral wools) [3, 4], while in the bending experiments the beam is often clamped at one [8-10] or two [7] sides. These boundary conditions are always non-ideal and should be avoided especially when measuring stiff and lowly damped materials.

According to the way Young's modulus is obtained from the measured frequency response function (FRF), the different approaches can be split into two classes: the first class uses the poles (resonant frequencies) of the FRF and the partial differential equation (PDE) model only [1, 4, 8, 10], while the second class uses the whole FRF and PDE model [5, 6, 9]. It is clear that the second class will be much more sensitive to non-idealities of the experimental set up such as misalignment creating other vibration modes than those described by the PDE model. These errors are easily eliminated in the first class by selecting the appropriate resonances (see also Section 3). The advantage of the second class is the much higher frequency resolution of the $E(j\omega)$ measurement. On the other hand, except for highly damped materials, the resonant frequencies (poles) of the FRF contain most information about the material damping.

This work is supported by the Belgium National Fund for Scientific Research; the Belgian Government (IUAPV/22); and the Flemish Community (GOA-IMMI).

Except [6], none of the existing methods take into account the disturbing noise when estimating Young's modulus, and no uncertainty bound is given. In [6], Young's modulus is determined using strain measurements from longitudinal impulse response experiments (hammer and air gun) on a beam under free-free boundary conditions. Uncertainty bounds on $E(j\omega)$ are calculated assuming that the disturbing noise is spatially and temporarily white.

This paper uses frequency domain system identification techniques to measure the elastic modulus $E(j\omega)$ of homogenous isotropic materials in a broad frequency band from forced flexural and longitudinal vibration experiments on a beam under free-free boundary conditions. Special designed broadband periodic excitation signals are used for the measurement procedure. It allows to calculate simultaneously the frequency response function (FRF), the noise level, and the level of the nonlinear distortions. This information is used to estimate the poles (and their uncertainty) of the measured FRF. The final result is Young's modulus together with its uncertainty due to the disturbing noise and/or the nonlinear distortions. The presented method takes into account the disturbing noise on the output and the input, the noise colouring, the correlation between the input and output disturbances, and the non-linear distortions.

The outline of the paper is as follows. First the theoretical resonances of the longitudinal (Love model) and the flexural (Euler-Bernoulli and Timoshenko models) vibration experiments are calculated (Section 2). Next the identification procedure for estimating Young's modulus is explained step by step and illustrated on a wave experiment (Section 3). Finally some practical aspects of the experimental setup are discussed and a comparison between the wave and the transverse experiments is made (Section 4).

2. THEORETICAL RESONANCE FREQUENCIES

2.1. Longitudinal (wave) vibrations

Consider a longitudinal vibration experiment on a uniform (along its length) beam composed of a linear, homogeneous and isotropic visco-elastic material (see Fig. 1). According to the Love model [2] the poles s_k of the force $u(t)$ to longitudinal displacement $y(t, x)$ transfer function are related to Young's modulus E by

$$E(s_k) = -\rho\left(\frac{Ls_k}{k\pi}\right)^2\left[1+\left(\frac{k\pi\nu(s_k)J}{L}\right)^2\right] \tag{1}$$

with $k = \pm1, \pm2, \ldots$. L is the length of the beam, ρ the density of the material, $\nu(s_k)$ Poisson's

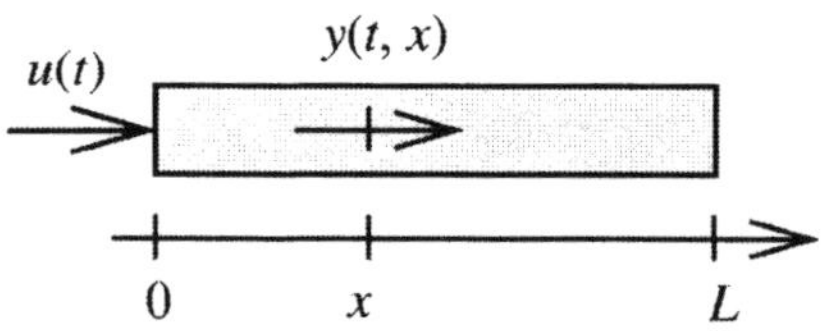

Fig. 1: Longitudinal (wave) vibration experiment of a beam with length L (side view): $u(t)$ is the applied force per unit area and $y(t, x)$ the longitudinal displacement at position x.

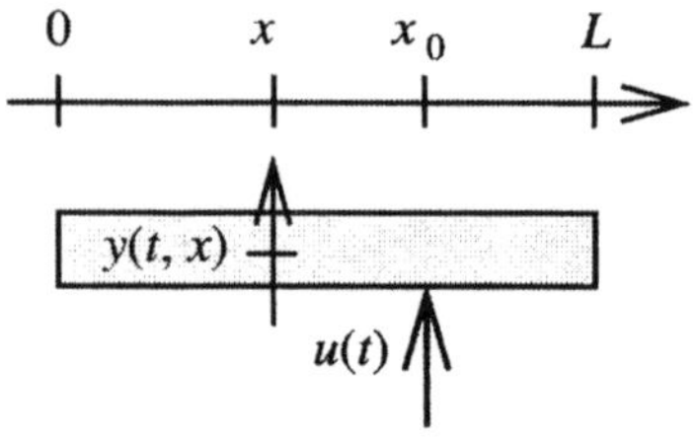

Fig. 2: Flexural (transverse or bending) vibration experiment on a beam with length L (top view): $u(t)$ is the applied force and $y(t, x)$ the transverse displacement as a function of the position x.

coefficient, and J the radius of gyration of the cross section of the beam about the x-axis ($J = r/\sqrt{2}$ for a circle with radius r; and $J = \sqrt{(h_y^2 + h_z^2)/12}$ for a rectangle with sides h_y and h_z [11]).

2.2. Flexural (transverse or bending) vibrations

Consider a flexural vibration experiment on a uniform (along its length) beam composed of a linear, homogeneous and isotropic visco-elastic material without axial loads (see Fig. 2). According to the Euler-Bernoulli model [12] the poles of the force $u(t)$ to the transverse displacement $y(t, x)$ transfer function are related to Young's modulus E by

$$E(s_k) = -\frac{\rho A L^4 s_k^2}{I\zeta_k^4} \tag{2}$$

with $k = \pm1, \pm2, \ldots$. I is the moment of inertia of the cross-section of the beam about the z-axis (axis perpendicular to the x- and y- axes; $I = \pi r^4/4$ for a circle with radius r, and $I = h_z h_y^3/12$ for a rectangle with sides h_z and h_y in the z- and y- directions respectively [11]), and ζ_k the wave number of the k th resonance frequency

$$\cosh(\zeta_k)\cos(\zeta_k) = 1 \tag{3}$$

Numerical values of ζ_k are tabulated in the literature (see, for example, Table 6.4 of [12]): $\zeta_1 = 4.730041$, $\zeta_2 = 7.853205$, $\zeta_3 = 10.995608$, and $\zeta_k = (2k+1)\pi/2$ for $k \geq 4$ within a relative error smaller than 1.2×10^{-7}.

The Euler-Bernoulli theory leading to (2) ignores the

effects of shear deformation and rotary inertia, and is accurate for thin beams only [12]. In practice (2) can be used to model the first few resonance frequencies corresponding to the simple mode shapes. Timoshenko's theory [12] considers the effects of shear deformation and rotary inertia, giving the following implicit relationship between the poles s_k and Young's modulus E

$$\cosh(b_1(s_k)L)\cos(b_2(s_k)L) - 1 + \frac{b_2^2(s_k) - b_1^2(s_k)}{2b_1(s_k)b_2(s_k)}\sinh(b_1(s_k)L)\sin(b_2(s_k)L) = 0 \quad (4)$$

where

$$b_i^2(s) = (-1)^{(i+1)}c(s) + \sqrt{c^2(s) + a(s)}$$

$$\begin{cases} c(s) = \dfrac{\rho}{2E(s)}(1 + \gamma(s))s^2 \\ a(s) = -\left(\dfrac{\rho A}{E(s)I}s^2 + \dfrac{\rho^2\gamma(s)}{E^2(s)}s^4\right) \end{cases} \quad (5)$$

A is the cross section area of the beam, and $\gamma(s)$ depends on the shape of the cross section and Poisson's coefficient $\nu(s)$ ($\gamma(s) = (12 + 11\nu(s))/5$ for a rectangle, and $\gamma(s) = (7 + 6\nu(s))/3$ for a circle [13]).

2.3. Discussion

Equations (1), (2) and (4) establish a relationship between the poles s_k of the longitudinal and flexural vibration experiments and Young's modulus $E(s_k)$ when the beam (L, A, I, J) and the material (ρ, $\nu(s)$) properties are known. Fortunately, the cross section dimensions of the beam can always be chosen such that the terms in (1) and (4) depending on Poisson's coefficient $\nu(s)$ are correction terms. Hence, only a rough guess of $\nu(s)$ is required. This is the basic idea used for identifying Young's modulus $E(s)$ in Section 3.

3. THE IDENTIFICATION PROCEDURE

The procedure consists of the follow main steps. (i) Choice of the broadband periodic excitation signals and measurement of the frequency response function (result: FRF with its uncertainty). (ii) Approximation of the measured FRF by a rational form in the Laplace variable s (result: poles and their uncertainty). (iii) Selection of the poles of the rational form corresponding to the longitudinal or flexural vibration modes and calculation of Young's modulus $E(s_k)$ via eq. (1), (2) or (4) (result: Young's modulus $E(s_k)$ and its uncertainty at the values of the poles s_k). (iv) Approximation of $E(s_k)$ by a rational form in the Laplace variable s (result: Young's modulus $E(j\omega)$ and its uncertainty in the frequency band of interest). These four main steps are explained in detail in the sequel of this section, and each step is illustrated on the longitudinal vibrations of a plexiglass beam.

3.1. Measurement of the frequency response function

The frequency response function (FRF) of the system is measured using random phase multisines [14, 15]. These are periodic signals consisting of the sum of harmonically related sine waves with user defined amplitudes and random phases. The measurement procedure of [16] is followed to estimate the FRF and its uncertainty. It consists of the following basic steps:

1 Choose the amplitude spectrum and the frequency resolution of the random phase multisine.
2 Make a random choice of the phases of the random phase multisine, and calculate the corresponding time signal.
3 Apply the excitation to the system and measure $P \geq 2$ periods of the steady state response $u(t)$, $y(t)$.
4 Repeat steps 2 and 3 $M \geq 6$ times.
5 Calculate the DFT spectra of the input $u(t)$, and output $y(t)$ signals for each period of each experiment at the excited DFT frequencies.

From these $M \times P$ sets of noisy input/output spectra, one can calculate for each experiment the average FRF $\hat{G}^{[m]}$ over the P periods and its sample variance $\hat{\sigma}^2_{\hat{G}^{[m]}}$ ($m = 1, \ldots, M$). An additional averaging over m gives the final FRF $\hat{G}$ of the whole measurement procedure

$$\hat{G} = \sum_{m=1}^{M}\frac{\hat{G}^{[m]}}{M}, \ \hat{\sigma}^2_{\hat{G}} = \sum_{m=1}^{M}\frac{|\hat{G}^{[m]} - \hat{G}|^2}{M(M-1)} \quad (6)$$

together with its sample variance $\hat{\sigma}^2_{\hat{G}}$. If the system is linear, then $\hat{\sigma}^2_{\hat{G}}$ should be equal to the mean value of $\hat{\sigma}^2_{\hat{G}^{[m]}}$ divided by M

$$\hat{\sigma}^2_{\hat{G}n} = \frac{1}{M^2}\sum_{m=1}^{M}\hat{\sigma}^2_{\hat{G}^{[m]}} \quad (7)$$

(the extra factor M copes with the averaging over the M experiments). Indeed, in the linear case the variability of the FRF measurement over the M experiments with different excitation signals can be explained by the disturbing input/output noise only. In that case $\hat{\sigma}^2_{\hat{G}}$ (6) and $\hat{\sigma}^2_{\hat{G}n}$ (7) are both a measure of the influence of the disturbing input/output noise on the FRF. If $\hat{\sigma}^2_{\hat{G}}$ (6) is larger than $\hat{\sigma}^2_{\hat{G}n}$ (7), then this is an indication that the systems behaves nonlinearly. Indeed, the difference

$$\hat{\sigma}^2_{\hat{G}S} = \hat{\sigma}^2_{\hat{G}} - \hat{\sigma}^2_{\hat{G}n} \quad (8)$$

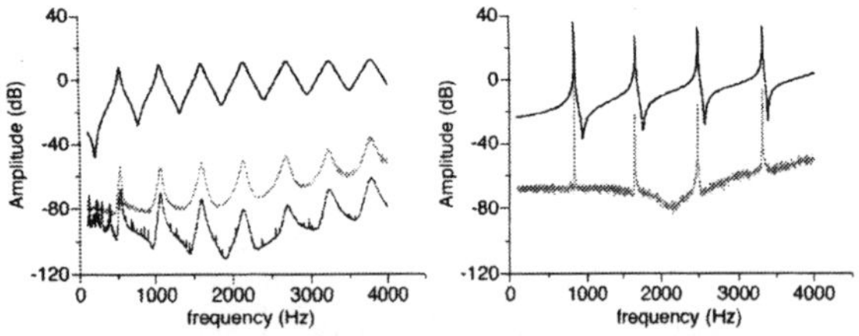

Fig. 3: Force-to-acceleration FRF (bold black line) and its standard deviation (thin black line: (7) noise errors only, and gray line: (6) noise errors + nonlinear distortions) of the longitudinal vibration experiments. Left: plexiglass ($L = 1.983$ m), right: copper ($L = 2.209$ m).

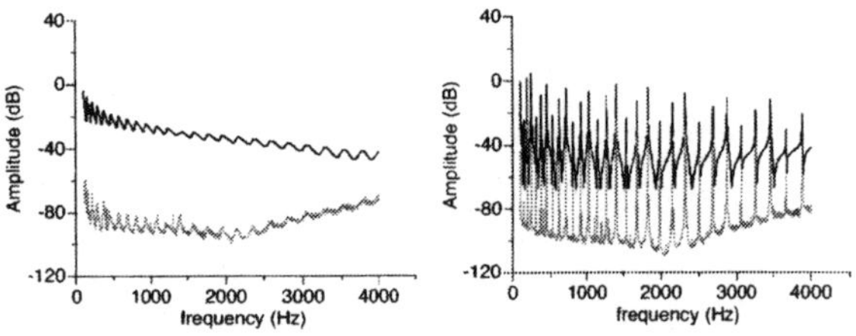

Fig. 4: Force-to-velocity FRF (bold black line) and its standard deviation (6) (gray line) of the flexural vibration experiments. Left: plexiglass ($L = 1.372$ m), right: copper ($L = 2.209$ m).

quantifies the contribution of the stochastic non-linear distortions to the FRF measurement; while the non-linear bias contributions G_B are bounded by

$$\frac{1}{\gamma}\sqrt{M}\hat{\sigma}_{\hat{G}S} \leq |G_B| \leq \gamma\sqrt{M}\hat{\sigma}_{\hat{G}S} \tag{9}$$

where γ typically lies between two and ten (see [15, 17 and 18] for the details).

The measurement procedure is illustrated in Figures 3 and 4. For each measurement $P = 10$ and $M = 25$. It can be seen that the stochastic nonlinear distortions are mostly dominant, and that the signal-to-noise ratio $|\hat{G}|/\hat{\sigma}_{\hat{G}}$ of the FRF measurement (6) varies between 40 to 60 dB.

3.2. Approximation of the FRF by a rational form

According to the Mittag-Leffler theorem [19] the infinite dimensional transfer functions shown in Figures 3 and 4 can be approximated arbitrary well by a rational form of finite order in a particular frequency band

$$G(s, \theta) = \frac{B(s, \theta)}{A(s, \theta)} = \frac{\sum_{n=0}^{n_b} b_n s^n}{\sum_{n=0}^{n_a} a_n s^n} \tag{10}$$

The parameters θ of the rational form are found by minimizing the sample maximum likelihood cost function (see [15] for the details). The result is an

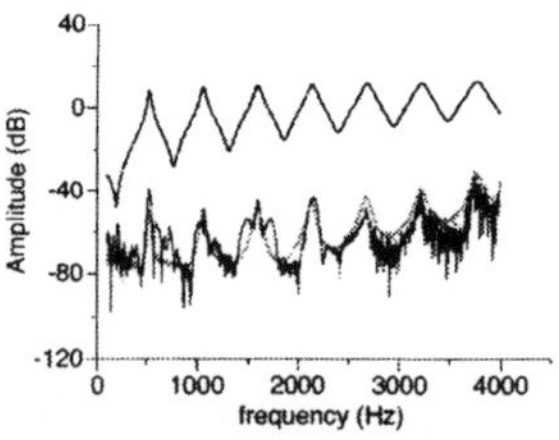

Fig. 5: Comparison between the measured and modelled FRF of the longitudinal vibration experiment on the plexiglass beam. Bold black line: $\hat{G}(j\omega)$ and $G(j\omega, \hat{\theta})$, gray line: the 95% confidence bound of the FRF measurement ($\sqrt{3}\hat{\sigma}_{\hat{G}}$), and black line: the complex error $\hat{G}(j\omega) - G(j\omega, \hat{\theta})$.

estimate $\hat{\theta}$ together with its covariance matrix $\mathrm{Cov}(\hat{\theta})$. Finally, from $\hat{\theta}$ and $\mathrm{Cov}(\hat{\theta})$ the poles and their uncertainty are calculated.

Fig. 5 illustrates the procedure on the longitudinal vibration experiment of a plexiglass beam. Using a rational form (10) of order $n_a = n_b = 34$, the approximation error is about at the level of the uncertainty of the FRF measurement (between -50 dB and -60 dB).

3.3. Calculation of Young's modulus at the poles

The first step consists in selecting the poles s_k of $G(s, \hat{\theta})$ (10) corresponding to the longitudinal or flexural vibration modes. This can easily be done by comparing $|s_k|/(2\pi)$ to the resonance frequencies of the FRF. Next the index k of each observed resonance peak (pole) is determined. This is done by comparing the first few peaks of the measured FRF to the resonance frequencies predicted using eq. (1), (2) or (4), the beam parameters, and a rough guess of Young's modulus and Poisson's coefficient. Finally Young's modulus is calculated at the values of the poles s_k via eq. (1), (2) or (4), where (1) and (4) need a rough guess of $\nu(s_k)$ ($\nu = 0.33$ for plexiglass and $\nu = 0.3$ for copper). By an appropriate choice of the beam cross section dimensions, the correction term in (1) depending on $\nu(s_k)$ can often be neglected. Equation (4) is a nonlinear algebraic equation in $E(s_k)$, and is solved via the Newton-Raphson root finding algorithm [20]. As starting value we use the Euler solution (2) or the previous solution $E(s_{k-1})$ of (4). Uncertainty bounds on $E(s_k)$ are obtained from the uncertainty of the poles through a first order sensitivity analysis of eq. (1), (2) and (4).

Fig. 6 gives the result for the longitudinal vibration experiments. The following standard deviations are found

$$\begin{aligned} &\text{plexiglas: } \mathrm{std}(E(s_k)) = 1\times10^5 \text{ N/m}^2 \\ &\text{copper: } \mathrm{std}(E(s_k)) = 1\times10^6 \text{ N/m}^2 \end{aligned} \tag{11}$$

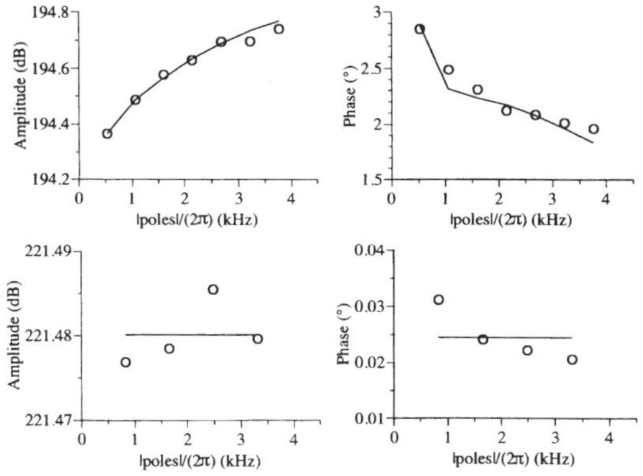

Fig. 6: Comparison between the measured and the modelled elastic modulus (N/m^2) at the poles s_k of the longitudinal vibration experiments: plexiglass (row 1), and copper (row 2). Circles: measurement $E(s_k)$ (1), and solid line: model $E(s_k, \hat{\theta})$ (row 1 (10), row 2 (12))

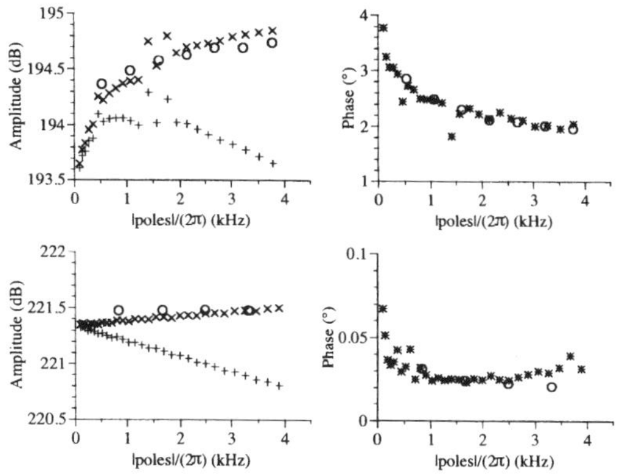

Fig. 7: Comparison between the elastic moduli (N/m^2) obtained from the longitudinal (o: Love model (1)) and the flexural (+: Euler model (2), x: Timoshenko model (4)) vibration experiments.

3.4. Modelling of Young's modulus

In Section 3.3 Young's modulus E and its variance var(E) are obtained at the value of the poles of the longitudinal or flexural vibration modes. In practise Young's modulus should be known along the $j\omega$-axis, which requires an additional modelling step. Based on the spring-dashpot representations of linear visco-elastic materials [21] Young's modulus is modelled as a rational from $E(s, \theta)$ (10). For metals Young's modulus is modelled as a complex, frequency independent constant

$$E(s, \theta) = R + Ij \tag{12}$$

The parameters θ in (10) and (12) are found by minimizing the following cost function w.r.t. θ

$$\sum_{k=1}^{K} \frac{|E(s_k) - E(s_k, \theta)|^2}{\text{var}(E(s_k))} \tag{13}$$

The procedure is illustrated in Fig. 6 on the longitudinal vibration experiments shown in Fig. 3 For plexiglass a rational form (10) with $n_a = n_b = 2$ is sufficient to model the elastic modulus. The poles and zeroes of the estimated second order model have multiplicity one, lie on the negative axis, and are alternating. This proves that $E(s, \hat{\theta})$ can be represented by an RC-network [22] or an equivalent spring-dashpot scheme [21]. Although models (10) and (12) are quite accurate (model errors less than 0.05 dB = 5‰ for plexiglass, and less than 0.008 dB = 0.8‰ for copper), the model errors are significantly larger than the uncertainty of $E(s_k)$ (compare eq. (11) and Fig. 6). It can be due to the non-idealities of the experimental set up and the test specimen (non-homogeneity of the material, non-uniform cross section beam, ...), and/or the fact that the true material behaviour cannot be described perfectly by (10) or (12).

4. EXPERIMENTAL RESULTS

The following test beams are used for the experiments: plexiglass ($\rho = 1200$ kg/m^3, rectangular cross section with sides $h_y = 10.3$ mm and $h_z = 20.0$ mm, $L = 1.372$ m and $L = 1.983$ m), and copper ($\rho = 8900$ kg/m^3, rectangular cross section with sides $h_y = 10.0$ mm and $h_z = 30.0$ mm, $L = 2.209$ m). The beams are hung by two or three nylon threads perpendicular to the excitation direction. They are excited in the longitudinal or transverse direction by a mini-shaker. A stinger rod connects the mini-shaker to the impedance head, which is attached as close as possible to the test specimen (distance less than 7 mm).

The results of the longitudinal and flexural vibration experiments are compared in Fig. 7. It can be seen that the amplitude of Young's modulus obtained from the Timoshenko (4) and Love (1) models coincide (errors of less than 0.1 dB = 1%), while that of the Euler model (2) is accurate for the first few (five) values only. The phase of Young's modulus matches very well for the three models. This can be explained by the fact that Timshenko's solution (4) does almost not change the phase of the poles. Note, however, that the differences between the Timoshenko and Love models are larger than the uncertainty of the $E(s_k)$ measurement (compare eq. (11) with Fig. 7). For plexiglass this can be due to the variability of the material properties over the two test beams, and/or a temperature difference between the two experiments; while for the copper beam it can be due to non-isotropy caused by the rolling process (the side walls are stiffer).

5. CONCLUSION

A system identification approach for measuring the elastic modulus of homogeneous visco-elastic materials from longitudinal and flexural vibration experiments has been presented. The final result is a parametric model for Young's modulus over a broad frequency band together with an uncertainty bound. The uncertainty bound takes into account the disturbing input/output (measurement) noise and the stochastic nonlinear distortions. As such it is a measure for the repeatability of the experiment over the class of random excitation signals with the same power spectrum. It can be used to study the influence (non-idealities) of the measurement set up, the temperature, the nonlinear material behaviour, the variability of the material over different test specimen, the value of Poisson's coefficient ...

Note that the same ideas could be used to measure the shear modulus $G(j\omega)$ from torsional vibration experiments.

REFERENCES

[1] ASTM Standard E 1876-99 (1999). Standard Test Method for Dynamic Young's Modulus, Shear Modulus, and Poisson's Ratio by Impulse Excitation of Vibration, *Annual book of ASTM Standards*, American Society for Testing and Materials: West Conshohocken (PA).

[2] T. Pritz (1981). Apparent complex Young's modulus of a longitudinally vibrating viscoelastic rod, *Journal of Sound and Vibration*, vol. 77, no. 1, pp. 93-100.

[3] T. Pritz (1982). Transfer function method for investigating the complex modulus of acoustic materials: rod-like specimen, *Journal of Sound and Vibration*, vol. 81, no. 3, pp. 359-376.

[4] T. Pritz (1994). Dynamic Young's modulus and loss factor of plastic foams for impact sound isolation, *Journal of Sound and Vibration*, vol. 178, no. 3, pp. 315-322.

[5] L. Hillström, M. Mossberg, and B. Lundberg (2000). Identification of complex modulus from measured strains on an axially impacted bar using least squares, *Journal of Sound and Vibration*, vol. 230, no. 3, pp. 689-707.

[6] M. Mossberg, L. Hillström, and T. Söderström (2001). Non-parametric identification of viscoelastic materials from wave propagation experiments, *Automatica*, vol. 37, no. 4, pp. 511-521.

[7] T. Miyazaki, J. Muroi, Y. Yamamoto, and T. Uyemura (1994). Measurement of complex elastic modulus by holography - effect of moment caused under incorrect driving conditions, *Int. J. Japan Soc. Prec. Eng.*, vol 28. no. 3, pp. 243-248.

[8] Y. C. Xu, and A. Nashif (1996). Measurement, analysis and modeling of the dynamic properties of materials, *Sound and Vibration*, vol. 30, no. 7, pp. 20-23.

[9] R. Caracciolo, A. Gasparetto, and M. Giovagnoni (2000). Measurement of the isotropic dynamic Young's modulus in a seismically excited cantilever beam using a laser sensor, *Journal of Sound and Vibration*, vol. 231, no. 5, pp. 1339-1353.

[10] C. L. Sisemore and C. M. Darvennes (2002). Transverse vibration of elastic-viscoelastic-elastic sandwich beams: compression-experimental and analytical study, *Journal of Sound and Vibration*, vol. 252, no. 1, pp. 155-167.

[11] R. D. Blevins (2001). Formulas for Natural Frequency and Mode Shape, Krieger Publishing Compagny, Malabar (Fl., USA).

[12] D. J. Inman (1996). *Engineering Vibration*, Prentice Hall, Englewood Cliffs (N.J., USA).

[13] G. R. Cowper (1966). The shear coefficient in Timoshenko's beam theory, *Transactions of the ASME: Journal of Applied Mechanics*, vol. 33, pp. 335-340.

[14] J. Schoukens, T. Dobrowiecki and R. Pintelon (2002). Linear modeling in the presence of nonlinear distortions, *IEEE Trans. Instrum. and Meas.*, vol. 51, no. 4, pp. ?-?.

[15] R. Pintelon, and J. Schoukens (2001). *System Identification: A Frequency Domain Approach*, IEEE Press, Piscataway (N.J., USA).

[16] R. Pintelon, J. Schoukens, W. Van Moer and Y. Rolain (2001). Identification of linear systems in the presence of nonlinear distortions, *IEEE Trans. Instrum. and Meas.*, vol. 50, no. 4, pp. 855-863.

[17] J. Schoukens, T. Dobrowiecki, and R. Pintelon (1998). Parametric Identification of Linear Systems in the Presence of Nonlinear Distortions. A Frequency Domain Approach, *IEEE Trans. Autom. Contr.*, vol. AC-43, no. 2, pp. 176-190.

[18] R. Pintelon, and J. Schoukens (2002). Measurement and modeling of linear systems in the presence of non-linear distortions. *MSSP*, vol. 16, no. 5, pp. 785-801.

[19] P. Henrici (1974). *Applied and Computational Complex Analysis*, vol. 1, John Wiley & Sons, New York (USA).

[20] Fletcher, R. (1991). *Practical Methods of Optimization (2nd ed.)*, Wiley, New York (USA).

[21] W. Flügge (1967). *Viscoelasticity*, Blaisdell Publishing Compagny, London (UK).

[22] N. Balabanian (1964). *Network Synthesis*, Prentice-Hall, Englewood Cliffs (USA).

IFAC Publications
www.elsevier.com/locate/ifac

A NOVEL ALGORITHM FOR FULLY AUTONOMOUS STAR IDENTIFICATION

S. Bittanti *, **E. De Marchi** **, **M. Giranzani** *, **M. Lovera** *,
B. Lübke-Ossenbeck *** and **E. Silani** *

* *Dipartimento di Elettronica e Informazione, Politecnico di Milano,
Piazza Leonardo da Vinci 32, 20133 Milano, Italy,
Tel. +39-02-23993650, Fax +39-02-23993412,
E-mail: [bittanti,lovera,silani]@elet.polimi.it*
** *Tecnomare S.p.A., S. Marco 3584, 30124 Venezia, Italy,
Tel. +39-041-796711, Fax +39-041-796800,
E-mail: demarchi.e@tecnomare.it*
*** *OHB System, Universitätsallee 27-29 28359, Bremen, Germany,
E-mail: luebke@ohb-system.de*

Abstract: In this paper a novel approach to the problem of star identification for spacecraft attitude determination based on data provided by a star camera is proposed. The algorithm is based on pattern matching ideas and provides improved certainty with respect to similar methods available in the literature. The achievable performance is demonstrated via simulation results. *Copyright © 2003 IFAC*

Keywords: Space vehicles, Attitude algorithms, Pattern identification, Sensor systems, Spacecraft autonomy.

1. INTRODUCTION

Accurate attitude determination is becoming more and more important for the success of space missions, as the requirements on spacecraft attitude and angular rate control become more stringent. Attitude information can be derived from a wide range of measurements (see, e.g., (Wertz, 1978)), however it is generally recognized that star sensors represent the best available solution to the problem (Liebe, 2002; Dungate and Airey, 2002; Davies and Holt, 2002).

Star sensors operate according to the following, general principle: a camera, usually with a very small field of view (FOV), provides images of a portion of the celestial sphere. Such images are processed and compared with a star catalog stored on board, in order to *match* measured stars with the stars from the catalog (star identification problem). Knowledge of the position of stars both in the camera reference frame (as provided by the sensor) and in an inertial reference frame (as provided by the catalog) constitutes the input for the actual attitude determination problem, for which a number of solutions are available in the literature (see, e.g., (Markley, 1993; Mortari, 1998)).

Existing fully autonomous star identification algorithms can roughly be grouped into two classes (Padgett *et al.*, 1997). Both classes make use of a database consisting of a catalog of known star characteristics (e.g., location and apparent brightness) and data structures that aid in pairing sensor stars to the appropriate catalog stars. The major difference in the two classes arises from their respective approach in identifying the sensor sky-field.
The first class of algorithms, see e.g., (Liebe, 1992), tends to approach star identification as an instance of subgraph isomorphism. In this case, the known stars are treated as vertices in an undirected graph G, the edges of which correspond to the angular separation between neighboring stars that could possibly share the same star-sensor FOV. The set of points extracted from the sensor forms an undirected graph G_S. An identification arises when the resultant graph obtained from the sensor image (or a portion of it) is uniquely identified with a portion of the database. The problem of star identification is thus

transformed into the problem of finding a subgraph of G that is isomorphic to G_S. Typically, the data structures employed by the algorithms that make use of this formulation include lists of star pair distances or polygons (mainly triangles) from the catalog. As sensor accuracy is limited, there are often numerous pairs (or triples) that match a given sensor pair (or triple). In this case, employing also the brightness information can be useful. Anyway, as more parts of the sensor graph match, many of the database subgraphs can be eliminated until (in principle) only a single subgraph remains.
The algorithms which belong to the second class, see e.g., (Padgett and Kreutz-Delgado, 1997), tend to approach the star identification more in terms of pattern recognition or best match. In this case, each star on the onboard catalog is associated with a well-defined pattern or signature that can (and should) be determined only by the surrounding star field (at least to some degree). This allows similar patterns to be generated from the sensor image. As each star now has an individual signature, finding the nearest neighbor pattern is sufficient for star identification, provided that the patterns are close "enough". The data structures used to implement this kind of algorithms often include lookup and hash tables to facilitate finding the best matching pattern.
Herein, we propose a novel star identification algorithm that cannot be exactly classified into the first or the second class of algorithms. Precisely, we resort to a "mixed" strategy: given a sensor image, the best possible candidates for a sensor stars are found by means of a pattern matching formulation and then an unambiguous identification is obtained solving a subgraph isomorphism problem.

2. ALGORITHM DESCRIPTION

To facilitate analysis and understanding of the implementation details given in the following sections, we introduce some terms for stars in different reference frames or that have special uses:

- *Visible stars* (**V**): those stars or objects that can actually be imaged by the sensor.
- *Reference stars* (**R**): a subset of **V** contained in the catalog used on-board the spacecraft. The star locations in this set are in a standard reference frame and are used in attitude estimation.
- *Sensor stars* (**S**): the set of detected objects on the sensor image plan. This includes a subset of **V** and spurious non-star items such as sensor noise or effects of direct radiation. Locations are defined in terms of the sensor reference frame.

The set of **V** is meant to indicate the real sky and all the objects in it that the sensor might recognize as stars. We can approximate **V** with the stars found in a standard sky catalog (e.g., the *Hipparcos* star catalog developed by the European Space Agency) whose apparent brightness (when corrected for the sensor) is greater then the minimum sensitivity of the sensor (the apparent magnitude below which a star will not be recognized).

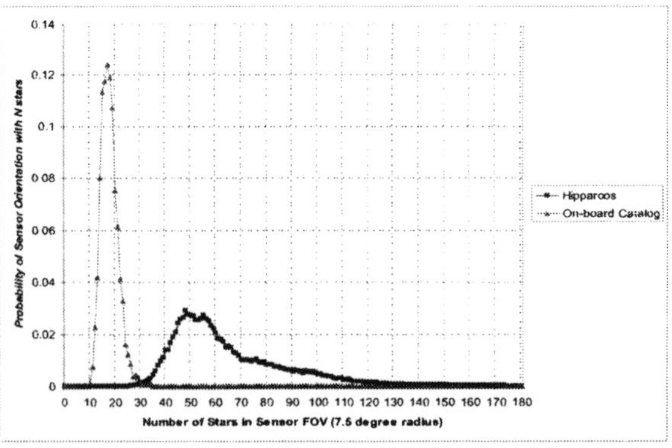

Figure 1. Star distribution over the celestial sphere. Plot points represent number of 7 magnitude stars or brighter in a circular 7.5 deg FOV sampled at 1 deg increments.

2.1 *On-board catalog generation*

The stars in our approximation of the set **V** of the visible stars (Hipparcos star catalog) constitute the base set for selecting the set of known reference stars **R**. Generally, some of the stars in **V** are not suitable for navigation purposes. Many of them will be binary stars or have variable brightness which can cause difficulties during the identification process. More importantly, the distribution of stars in **V** varies significantly over the celestial sphere (see Figure 1), and therefore incorporating all the suitable stars in **R** could bias the identification routine and seriously degrade its performance. In order to implement an unbiased identification algorithm for a fully autonomous attitude determination system, **R** should be chosen so that it is relatively uniform over the celestial sphere (see Figure 1). Finally, for memory and performance reasons, **R** should contain as few stars as possible to achieve the desired recognition rate.

The approach we take in constructing **R** is to determine a minimum number χ of reference stars that we require to be imaged in any arbitrary sensor orientation. Even if only two identified stars are required for attitude estimation, in order to achieve an unambiguous star identification, χ has to be at least three. The actual minimum sensitivity of the sensor is in almost every case sufficient to have more than 3 imaged stars in every possible FOV. However, even assuming that the sensor will support at least χ stars per sensor orientation, a number of other factors will influence the size of the parameter χ, namely:

(1) the algorithm used for the identification,
(2) the expected level of noise in the image (higher levels of noise, that degrade both the position and the brightness measurements of the imaged stars, could seriously cause problems for identification),
(3) the confidence level desired from the star recognition routine and the desired level of accuracy from the attitude determination algorithm.

The result of these factors is that the actual value for χ is likely to be significantly higher than the strict minimum value ($\chi = 3$) required for an unambiguous star identifi-

cation, since a larger number of stars in **R** for a section of sky will increase the likelihood that at least three can be identified correctly. Of course the value of χ cannot be too large due to memory and performance constraints.
Any number of methods could be used for determining which stars to include in **R**. For our purpose however, a relatively simple procedure is used. After discarding binary stars whose neighbor may degrade the image and variable stars, an incremental uniform scan (with a 1 deg step) is made across the celestial sphere and the brightest χ stars within the pattern radius (PR) are added at each orientation if they are not already elements of **R**. The only constraint placed on the selection of stars was to require that they be separated by greater than a fixed value, (ε), from any other viewable star (this was done for the same reason that unsuitable binary stars are not included).
Finally, the generated on-board catalog **R** is memorized as a table in which every record corresponds to a reference star. For each star in **R** the following information are stored: an unique identification code, the star position given in right ascension (α) and declination (δ), and the visual magnitude (M_v). The on-board catalog is ordered by increasing value of declination. The reason for this will be made clear in the following.

2.2 *Pattern generation*

The star identification algorithm which we propose involves generating a set of patterns for the selected group of reference stars in **R**, whose locations are known in a standard reference frame. This pattern set constitutes a database which is used to compare patterns derived in a similar way from the sensor image. As each star has its own signature, finding a suitably close match to a pattern is equivalent to pair the two stars for the purpose of identification. For convenience we assume that the index of a particular pattern is the star from which the pattern was generated. Given the reference star R_i, its pattern is generated in the following manner:

(a) *Determination of the "neighbors" set.*
First we build the set of all the stars in **R** "close" to R_i, i.e., the reference stars whose angular distances from R_i are less than the pattern radius PR and greater than a fixed value D_{MIN}. Thus, given the generic element R_j in **R**, with position $[\alpha_j, \delta_j]$ in right ascension and declination, the angular distance ϑ_{ij} between R_j and R_i is:

$$\vartheta_{ij} = \arccos[\sin(\delta_i)\sin(\delta_j) + \cos(\delta_i)\cos(\delta_j)\cos(\alpha_i - \alpha_j)]. \quad (1)$$

and R_j is a neighbor of R_i if and only if the following condition holds,

$$D_{MIN} \le \vartheta_{ij} \le PR. \quad (2)$$

Remark 1. In order to determine the set of the neighbors of R_i, (1) and (2) should be verified (in principle) for all the stars in **R**. Since the on-board catalog is ordered by increasing values of declination, the two above conditions must be verified only for the reference stars whose declinations are in the range $[\delta_i - PR, \delta_i + PR]$.

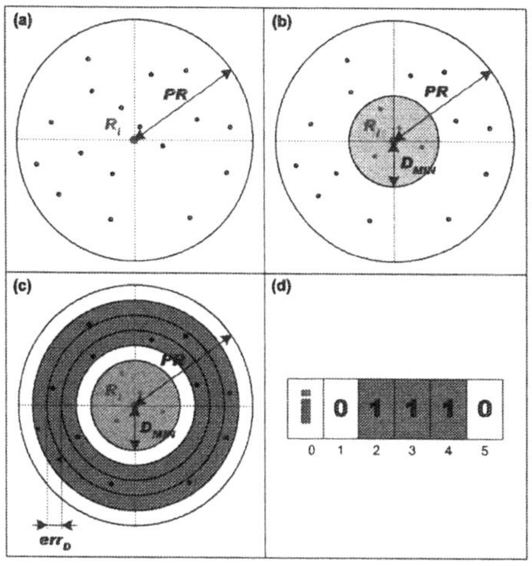

Figure 2. Panels (a)-(d) show how pattern vector is derived for the reference star R_i

(b) *Angular distances discretization and bar-code generation.*
If the reference star R_j belongs to the set of the "neighbors" of R_i, we knows by definition that the angular distance $\vartheta_{ij} \in [D_{MIN}, PR]$. By dividing the range $[D_{MIN}, PR]$ in smaller intervals of length err_D, we can discretize the angular distance ϑ_{ij} into,

$$\hat{\vartheta}_{ij} = \text{fix}\left(\frac{\vartheta_{ij} - D_{MIN}}{err_D}\right) + 1$$

where the function fix(x) rounds the elements of x to the nearest integers towards zero.
Note that $\hat{\vartheta}_{ij} \in [1, N_{max}]$, where

$$N_{max} = \text{fix}\left(\frac{PR - D_{MIN}}{err_D}\right) + 1$$

Thus, for the star R_i we can derive a N_{max} length binary vector (bar code), so that for each neighbor star R_j, whose discretized angular distance from R_i is $\hat{\vartheta}_{ij} = Y$, there is 1 on the bit Y.
Finally, the pattern generated from the reference star R_i is a $N_{max}+1$ length vector whose first element is the index i and the last N_{max} elements are the bits of the binary vector derived above. Figures 2.(a) - 2.(d) show how the pattern vector is derived for the reference star R_i.

2.3 *Database generation*

Once the on-board catalog **R** has been constructed, a pattern is generated for each of its elements. The set of all pattern vectors and the reference stars of **R** constitute the database **D**.
The structure of pattern vector as described in Section (2.2) is unsuitable for an efficient matching operation and requires more memory than necessary (assuming that N_{max} is much greater than the average number of stars per pattern). Instead of a vector, the patterns are incorporated into a lookup table **LT**. The original binary vector locations serve as the table index (see Figure 3).

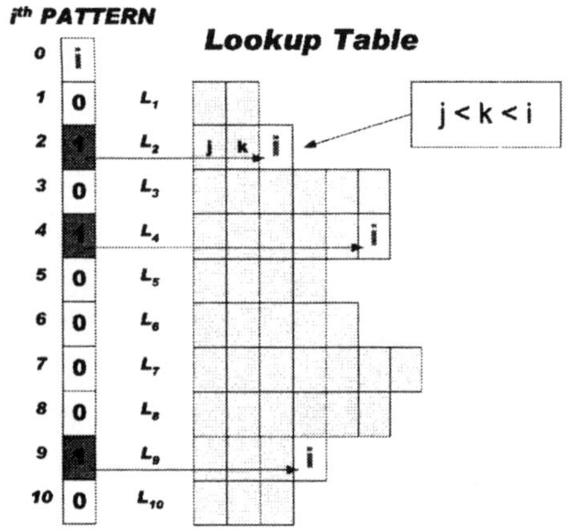

Figure 3. Lookup table structure

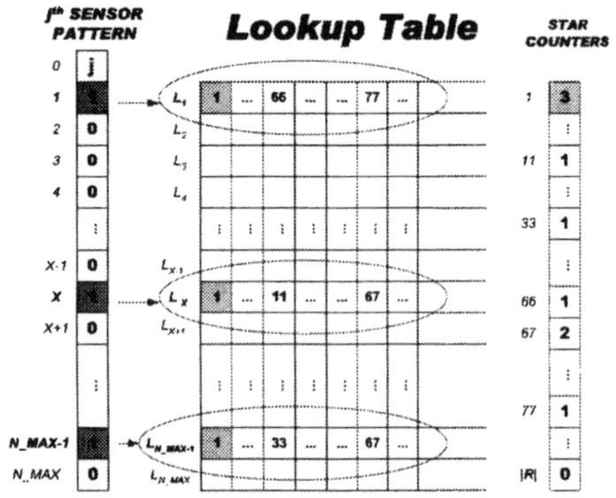

Figure 4. Location of 1 bits in the j-th sensor pattern are used to index entries of lookup table. Entries consists of database stars that also contain a 1 in the same binary vector locations. For instance, reference stars 1, 11, 67 all contain a 1 in location X. Each indexed entry in **LT** is a match and associated star counters are each incremented by one. After completing the process for all binary vector locations, the star counters with the highest values are the best matches (the best candidates for the associated sensor star).

For each neighbor star R_j in the pattern associated to the reference star R_i (a 1 in the pattern vector in the position corresponding to the discretized angular distance $\hat{\vartheta}_{ij}$ between R_i and R_j), the index i is entered into the lookup table at the row number corresponding to the binary vector location $\hat{\vartheta}_{ij}$. At the end of this process, the m-th (with $1 \leq m \leq N_{max}$) entry in the table contains a sorted (by increasing star index values) list of the indexes of all the stars in **R** that have the m-th element in their associated pattern vector set at value 1.
In order to find and calculate the best matches for a sensor pattern (the best candidates for the associated sensor star), we simply examine the table at each bit location in the sensor pattern where a 1 occurs and increment a counter for each reference star listed there. At the end of this procedure, the reference star counters with the highest values are the patterns in **D** closest to the sensor pattern (see Figure 4).

Remark 2. It is easy to see that the representation of the patterns in **LT** and in the original vector form are equivalent (see Figure 3)

2.4 *Polestar algorithm*

Once the database **D** has been constructed the actual identification process is quite simple. The input to the Polestar algorithm is a sensor image **S**. A sensor star j in **S** has information regarding its position on the sensor and its apparent brightness. For convenience we assume that **S** is ordered with star j being of a greater or equal brightness to star $j+1$. Ideally, the first χ sensor stars in **S** would be those whose signature is included in **D**. In this case, testing all the sensor stars would increase the probability of generating 'spurious' matches (matches of sensor stars which are either not in **D** or identified with the wrong star). Due to noise however, some of the stars shared by **S** and **D** may no longer be among the χ brightest stars so that some confidence factor $c \geq 1$ is employed (typical values of c are in the range 3 to 5). Testing $c\chi$ sensor stars improves the likelihood that we will find most of the stars shared by **S** and **D**.
The Polestar algorithm can be divided into the following steps:

(1) *Sensor patterns generation*
The identification process starts generating a pattern for each of the $c\chi$ stars in **S** (following the procedure described in Section 2.2). The number of 1 contained in each of the sensor patterns is directly related to its amount of useful information (for the aim of identification). This number can largely varies from a sensor pattern to another one due to the associated star position on the sensor image (see Figure 5). Thus, the $c\chi$ sensor patterns are ordered by increasing number of elements set at 1 and only the first N_{pat} (with $3 \leq N_{pat} \leq c\chi$) patterns are considered.

(2) *Best candidates determination*
For each of the N_{pat} sensor patterns, the best candidates are determined first generating the associated star counters (following the procedure described in Section 2.3) and then selecting from the on-board catalog **R** the stars whose counters are greater than a minimum threshold (β) and that have distance from the maximum less than one. Thus, if we indicate with $cont_i$ the value of the i-th star counter, the i-th reference star in **R** is one of the best candidates for the considered sensor pattern if and only if $\beta \leq cont_i \leq \max(cont_j) - 1$, where $j = 1, 2, \ldots, |\mathbf{R}|$.

Remark 3. Of course, if $\max(cont_j) < \beta$, $j = 1, 2, \ldots, |\mathbf{R}|$, the set of the best candidates for the considered sensor star is empty.

(3) *Candidates selection and identification*
The aim of this step is to obtain an unambiguous identification of at least three of the stars in **S** associated to the N_{pat} sensor patterns. This is done using a strategy similar to the first class of star identification algorithms (see Section 1).
The N_{pat} sensor stars and their catalog candidates are used to construct first pair, then triangles and finally polygons (with a number $n \geq 3$ of edges). An unambiguous identification arises when we find

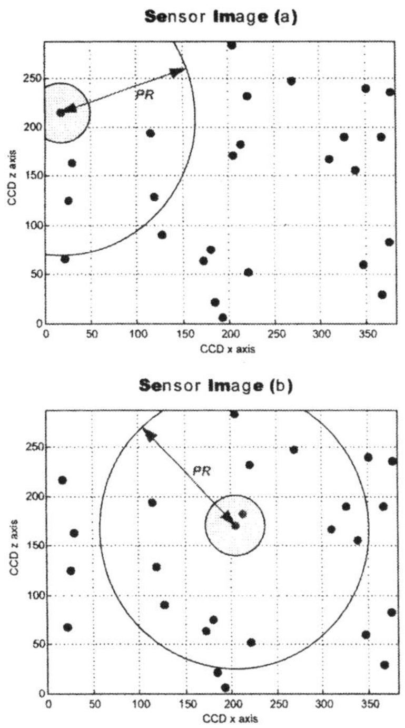

Figure 5. Usually, the "neighbors" set of a sensor star close to the edges of the sensor image (case (a)) has cardinality less than the one of a sensor star close to the center of the sensor image (case (b)). Testing the stars in **S** close to the edges could increase the risk of generating spurious matches.

a unique match between a triangle obtained from the sensor stars and a triangle generated from the candidates and no matches between polygons with a number $n \geq 3$ of edges, or when we find a unique match between a 4-edges polygon obtained from the sensor stars and 4-edges polygon generated from the candidates and no matches between polygons with a number of edges greater than 4, and so on. Any other possible situation is marked as ambiguous and no identification is provided.

Table 2.4 summarizes all the parameters used in the Polestar algorithm. The value of the pattern radius *PR* doesn't need to be the same size as the FOV. Typically, we use the value of the smallest FOV dimension as the pattern radius. This increases the number of stars required for the on-board catalog **R** but reduces the likelihood that the reference stars are all located close to the edges of the sensor, where accuracy is lower and only a portion of the star pattern can be generated (see Figure 5.a). The minimum match value (β) is used to limit the number of spurious star identifications and reduce the likelihood of misidentifications. Only a database pattern with at least this number of matches can be identified with a sensor pattern.

3. SIMULATIONS

To evaluate the performance of the Polestar algorithm, a number of simulations were conducted to measure the identification rate under a variety of different noise

Polestar Algorithm Parameters		
Parameter	Name	Brief Description
PR	pattern radius	Viewable stars within this distance from the central star are included in that star pattern.
D_{MIN}	minimum distance	The minimum angular distance considered in the pattern generation process between the central star and any other reference star.
err_D	discretization error	Length of the discretization interval considered in the pattern generation process.
χ	known star density	The minimum number of reference stars in an arbitrary orientation with radius *PR*.
ε	separation distance	The minimum distance (in pixel) between two reference stars
c	confidence factor	The total number of the brightest sensor objects that we attempt to identify.
N_{pat}	number of patterns	The total number of the sensor patterns that we attempt to associate to a known reference star.
β	minimum match	The minimum number of match between a sensor pattern and a database pattern required for identification.

Table 1. Polestar algorithm parameters and brief description

conditions. The sensor configuration used for the simulations reported here made use of a 7.5×10 degree FOV with an image plane consisting of 288×384 pixels. The minimum sensitivity of the sensor was set at 7 units apparent stellar magnitude. Any observed star whose apparent brightness (including noise) falls below this threshold was not imaged. The addiction of random Gaussian noise to the stars (with nominal standard deviation $\hat{\sigma}_M = 0.323$ units apparent stellar magnitude, which correspond to a nominal error of ± 1 in the observed magnitude determination process) could allow dimmer stars to actually appear during the simulation and for stars brighter than the minimum sensitivity to be lost. The number of stars actually imaged was limited to 30.
Along with the changes made to observed brightness for each star, positional noise was included in the imaged section of the celestial sphere. This source of noise is due to the optical properties of the lens of the sensor and the star extraction algorithm used to derive the location of the centroid of an object. Also in this case, random Gaussian noise was added to the projected locations of the imaged stars on the sensor plane (with nominal standard deviation $\hat{\sigma}_P = 0.062$ pixels, which correspond to a nominal error of ± 0.01 deg in the centroid locations determination process).
The on-board catalog **R** was generated by a uniform 1 deg increment scan of the star catalog **V**, selecting the $\chi = 10$ brightest stars within a pattern radius $PR = 7.5$ deg of each orientations provided that they are not closer together than $\varepsilon = 5$ pixels (when projected onto the sensor image plane). This resulted in 4127 stars in **R** or about one third of the stars from **V** that were brighter than 7 units magnitude apparent brightness.
For our simulation and testing of the algorithm, we set $D_{MIN} = 0.2$ deg, $err_D = 0.02$ deg, $c = 3$, $\beta = 6$ and $N_{pat} = 10$. In order to evaluate the identification rate

A	B	C
0.3 %	99.7 %	0 %

Table 2. Nominal identification rate ($\sigma_M = \hat{\sigma}_M$ and $\sigma_P = \hat{\sigma}_P$).

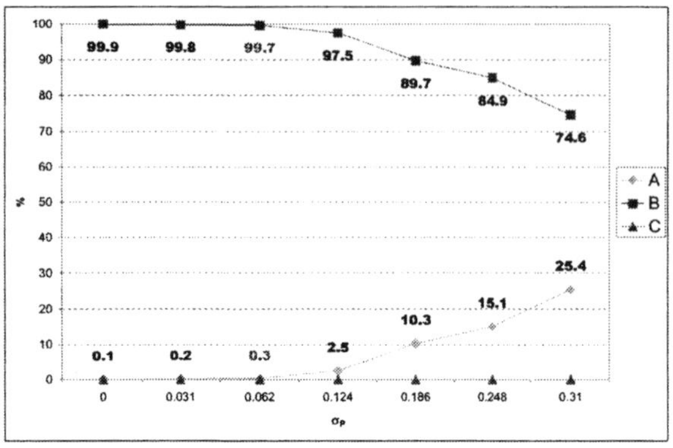

Figure 6. Identification rate as a function of the position accuracy of the sensor ($\sigma_M = \hat{\sigma}_M$).

of the Polestar algorithm the following distinction were made:

(A) No star identification is provided, since the result of the algorithm is "ambiguous" (not a unique match between database polygons and sensor polygons was obtained).

(B) Correct star identification: the result of the algorithm is a set of N identified imaged stars and all of them are correct.

(C) Wrong star identification: the result of the algorithm is a set of N identified imaged stars but some of them are not correct.

Of course, the last one is the worse case that should be avoided (it is better to provide no information than a misidentification).
Finally, the performance of the algorithm are evaluated in terms of occurrences of the above cases for three different simulation conditions:

(1) both apparent brightness and position noises at their nominal values;

(2) apparent brightness noise at its nominal value and position noise varying from standard deviation equal to zero to $\sigma_P = 0.31$ pixels;

(3) position noise at its nominal value and apparent brightness noise varying from standard deviation equal to zero to $\sigma_M = 0.969$ units apparent stellar magnitude.

The results of the above simulations are summarized in Table 3 and in Figures 6 and 7, respectively and the value for each test is taken over 10,000 uniform random sensor orientations.

4. CONCLUDING REMARKS AND ACKNOWLEDGMENTS

In this paper a novel approach to the problem of star identification for spacecraft attitude determination based on data provided by a star camera is proposed. The

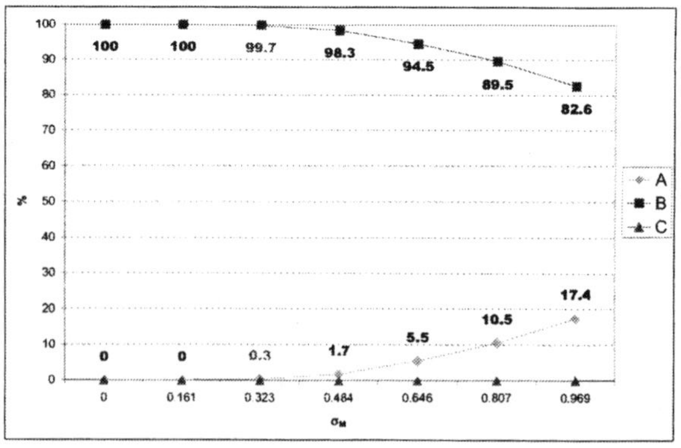

Figure 7. Identification rate as a function of the apparent brightness accuracy of the sensor ($\sigma_P = \hat{\sigma}_P$).

algorithm is based on pattern matching ideas and provides robust star identification over a wider range of sensor noise without parameter readjustment. The algorithm compares also favorably with current published star identification techniques in terms of accuracy and performance.
Paper supported by MURST project "Identification and Control of Industrial systems".

5. REFERENCES

Davies, A.R. and A.P. Holt (2002). Use of autonomous star trackers in modern attitude and orbit control systems. In: 5th *ESA International Conference on Spacecraft Guidance, Navigation and Control Systems.*

Dungate, D. and S.P. Airey (2002). Star sensor specification standard. In: 5th *ESA International Conference on Spacecraft Guidance, Navigation and Control Systems.*

Liebe, C.C. (1992). Pattern recognition of star constellation for spacecraft applications. *IEEE Aerospace and Electronics Systems Magazine* **7**(6), 34–41.

Liebe, C.C. (2002). Accuracy performance of star trackers - a tutorial. *IEEE Transactions on Aerospace and Electronic Systems* **38**(2), 587–599.

Markley, F.L. (1993). Attitude determination using vector observations: a fast optimal matrix algorithm. *The Journal of the Astronautical Sciences* **41**(2), 261–280.

Mortari, D. (1998). Euler-q algorithm for attitude determination from vector observations. *Journal of Guidance, Control and Dynamics* **21**(2), 328–334.

Padgett, C. and K. Kreutz-Delgado (1997). A grid algorithm for autonomous star identification. *IEEE Transactions on Aerospace and Electronic Systems* **33**(1), 202–213.

Padgett, C., K. Kreutz-Delgado and S. Udomkesmalee (1997). Evaluation of star identification techniques. *Journal of Guidance, Control, and Dynamics* **20**(2), 259–267.

Wertz, J. (1978). *Spacecraft attitude determination and control.* D. Reidel Publishing Company.

IFAC
Publications
www.elsevier.com/locate/ifac

FAST MODEL UPDATES AND SIMULATION FOR EFFICIENT FLIGHT CONTROL SOFTWARE DESIGN

Dr.-Ing. Holger Friehmelt, Dipl.-Ing. Detlef Rohlf

German Aerospace Center DLR
Institute of Flight Systems
Lilienthalplatz 7, 38108 Braunschweig, Germany
Fax: +49 (531) 295-2845

Abstract: The development, verification, and validation of flight control software for aircraft is characterized by high demands on accuracy, reliability and efficiency. Such processes cover all stages from first simulations to final flight test. The respective environment is described here on the basis of the experimental aircraft X-31A. Besides the applied hard- and software, special emphasis is given to simulation models and their updates, being marked by two new approaches: the use of nearly identical simulation software during the whole development and verification process and update of integrated simulation models due to special flight tests by means of global model system identification. *Copyright © 2003 IFAC*

Keywords: nonlinear, simulation, identification, aerospace systems

1. INTRODUCTION

During the X-31 EFM Program, an experimental aircraft of the famous 'X-Series' has been realized with international partnership (Francis, 1995). Two X-31A aircraft, built by Boeing International (formerly Rockwell International) and EADS (formerly MBB) (Robinson and Herbst, 1990) have demonstrated the concept of enhanced fighter maneuverability impressively in more than 550 flights since their maiden flights in Southern California in 1990 (Eubanks *et al.*, 1995). Final highlight of that program was the participation at the 1995 Paris Le Bourget Airshow.

The X-31A is a naturally unstable delta wing-canard configuration (Fig. 1) with extremely high maneuverability and outstanding flying qualities. Using a thrust vectoring system (Georg, 1994) and an innovative flight control design (Beh and Hofinger, 1994), the X-31A aircraft can be flown within the so-called 'poststall regime' (PST), i.e. well beyond the conventional stall barrier up to 70° angle of attack. Although the aerodynamic control devices lack efficiency in this low-speed range, the aircraft remains fully controllable and maneuverable because of the thrust vectoring system which deflects the engine exhaust flow using temperature-resistant composite vanes. Installed power plant is a General Electric F404-GE-400 engine with afterburner and typical take-off weight of this aircraft is approximately 7000 kg.

Fig. 1. The X-31A in EFM Configuration

Fig. 2. The X-31A in VECTOR Configuration

After completion of the EFM project, the X-31A aircraft was mothballed but reactivated again for the VECTOR program to demonstrate extremely short take-off and landing and to test a novel flush mounted air data system (FADS) (Friehmelt and Huber, 2001; Friehmelt and Hahn, 2002). Several modifications were necessary to accomplish these tasks. This included a flight control software redesign with autothrottle integration and the installation of a new inertial navigation system augmented by dGPS. The outward changes are the relocation of the noseboom, the reshaping of the nose cone to house the FADS and a new painting (Fig. 2). The X-31 VECTOR version was first flown on May 17, 2002, and has performed more than 50 sorties since then, including automatic approaches to a virtual runway at 5000 ft altitude.

The DLR Institute of Flight Systems has participated in both X-31 programs under contracts from BMVg/BWB and EADS (Friehmelt *et al.*, 1995; Friehmelt *et al.*, 2002). Besides contributions to general flight test data analysis (Huber and Galleithner, 1992), the DLR Institute of Flight Systems was involved in the system identification efforts with special attention on the determination of aerodynamic stability and control derivatives. This task generally includes the development of adequate model structures and estimation of the associated parameters and is based on special designed flight tests. If the results of such flight tests do not validate the predictions from CFD and wind-tunnel within certain tolerances, aerodynamic updates turn out to be necessary. Subsequently, these updates can require further control law adjustments and renewed stability analysis.

2. SIMULATION HARDWARE

Starting with a development simulation on PC basis for conceptual design, parametric variations and initial quantitative assessment (e.g. stability analysis), the next steps require real-time capabilities of the simulation with RTADS (Real-Time All Digital Simulation). This simulation type allows operation with 'Cockpit-in-the-Loop' and thus first quantitative statements about flying qualities. The most complex simulation FHILS (Flight-Hardware-in-the-Loop-Simulation, Fig. 3) is mainly used for verification & validation and final flying qualities assessment. These are prerequisites to gain the flight certification. For this simulation task, the simulation computer is connected via Scramnet with various peripheral systems. The 'System Test Console' (STC) houses four flight control computers, which execute the flight control software. System status and error message are displayed on the so-called 'Status & Test Panel' (STP). The cockpit computer serves as link between the different input and output devices in the cockpit, e.g. stick, pedals, throttle lever, displays, and various dials and switches.

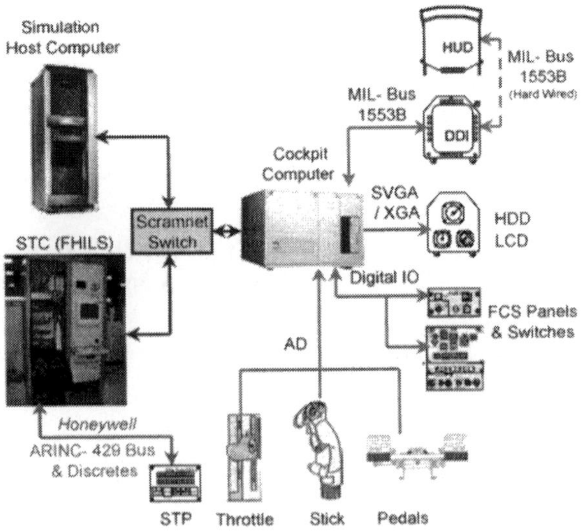

Fig. 3. Flight-Hardware-in-the-Loop-Simulation

3. SIMULATION SOFTWARE

During the VECTOR program, the simulation environment 'CASTLE' provided by the US Navy is used (Fig. 4). The abbreviation stands for 'Control Analysis and Simulation Test Loop Environment' and already implies the key application. CASTLE is a nonlinear simulation of the aircraft motion with various integrated hard- and software modules which represent aircraft subsystems, aircraft characteristics and aircraft environment.

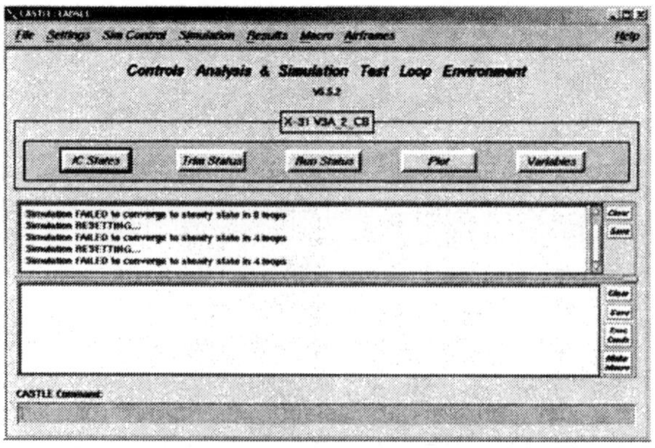

Fig. 4. Graphical User Interface of CASTLE

CASTLE thus represents all relevant characteristics of and around the aircraft. The core process is a cycle of calls of the individual modules with its order given by the flow of events (Fig. 5). To simplify the picture, only the main modules are depicted here, like electric system components, sensors, hydraulics, flight control laws, engine, aerodynamics, and the 'surfaces' module which embraces atmospheric data (with gusts and turbulence) and runway data as well.

Various steps are required for verification & validation in support of the flight test certification (Fig. 6). Offline simulations yield data for linear and nonlinear assessment, e.g. stability, robustness, and accuracy of the control laws. Manned simulations are used to derive handling qualities data. Additionally, time critical events and moding are checked, where moding is the transition of one flight control mode into another. During such transitions, the transients are to retain within small limits and all parameters must be transmitted correctly. Since the different modules are executed at individual cycle rates, simulation runs have to verify the absence of frame overruns.

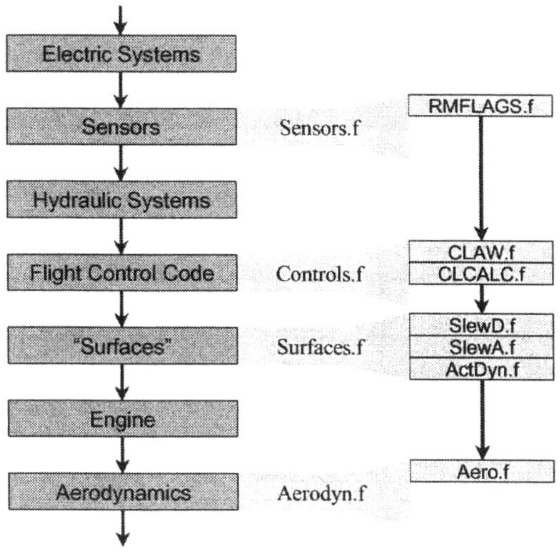

Fig. 5: Schematic Process Order of CASTLE

CASTLE provides a common simulation environment using most of the integrated simulation modules unaltered for the three different types of simulations. Exchange or update of such modules changes all three development simulations more or less simultaneously preventing possible coding errors and thus leading to an efficient model update for the PC-based, the real-time, and the flight-hardware-in-the-loop simulation.

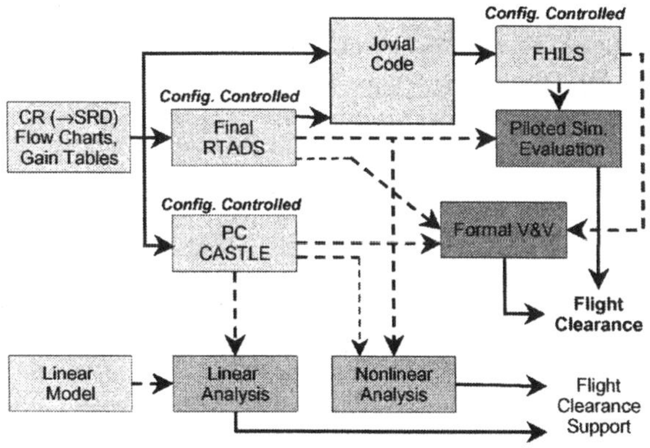

Fig. 6. Schematic View of the Simulation for Flight Clearance

4. IDENTIFICATION PROCEDURE AND MODEL

Each of the described simulations is able to fulfill its particular task only if sufficient exact models are applied. These models are to be verified and – if necessary – to be updated due to flight test results. The system identification has proven to be an effective tool for this task, combined here with an innovative global identification model. In addition to the complex aircraft aerodynamics, the global model represents e.g. engine and landing gear dynamics (Fig. 7). Thus, the procedure remains not only restricted to flight tests with relatively small deviations around distinct reference conditions. Furthermore, the entire flight regime – including take-off and landing – of the experimental aircraft X-31A is covered, i.e. from the low subsonic flight regime with extreme high angles of attack (max. 70°) up to the low super sonic flight regime (max. Mach number = ~1.3).

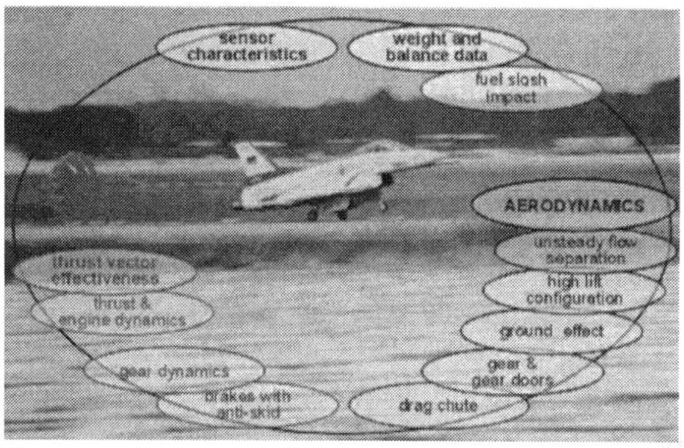

Fig. 7. Global Identification Model

The global identification model (Rohlf, 1999) contains the original database used in the simulations (e.g. aircraft aerodynamics derived from wind tunnel tests and a table-based thrust model) which is adapted and supplemented by increments with means of a regression method (Fig. 8). Therefore, flight-measured aircraft states, control surface deflections, etc. are used to re-simulate the flight-measured aircraft accelerations with sufficient small deviations. It is indispensable to apply actual values for aircraft weight, center of gravity and moments of inertia which are changing in course of flight because of the decreasing fuel quantity. Depending on center of gravity are the lever arms of the various sensors such as acceleration sensors and airflow sensors located in nose and noseboom of the aircraft. The center of gravity and the moments of inertia arc additionally affected by the landing gear extension.

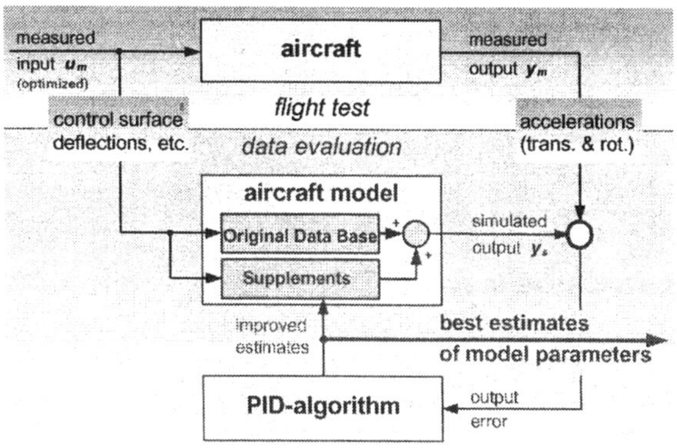

Fig. 8. Procedure for the Identification of Aerodynamic Increments

The regression method was chosen mainly for two reasons, (1) because it is applicable to unstable systems and (2) because it consumes relatively short computer run time even for large amount of data. Presumption for a successful application of the regression method is a good data quality and compatibility which has to be proven - or if necessary to be established by re-calibration of opposing signals - with means of a flight path reconstruction (FPR) procedure.

5. INPUT SIGNALS, GENERATION AND CHARACTERISTICS

For the identification of the aerodynamics of an highly augmented aircraft, it is indispensable to excite the system in its characteristic frequencies (determination of stability derivatives) and, in addition, to stimulate each individual control surface separately (determination of control derivatives). Therefore, special designed and optimized input signals are necessary, which – in the case of X-31 VECTOR – are generated in a supplementary part of the flight control software (Fig. 9) and which can be selected by the pilot during the flight test. The input in the pilot command path runs through the flight control laws just as an regular pilot control input and is distributed to the individual control surfaces, i.e. blended depending on flight conditions and corresponding control surface effectiveness. The separated surface excitation (SSE), however, bypasses the flight control laws and results in un-correlated deflections of the selected control surface.

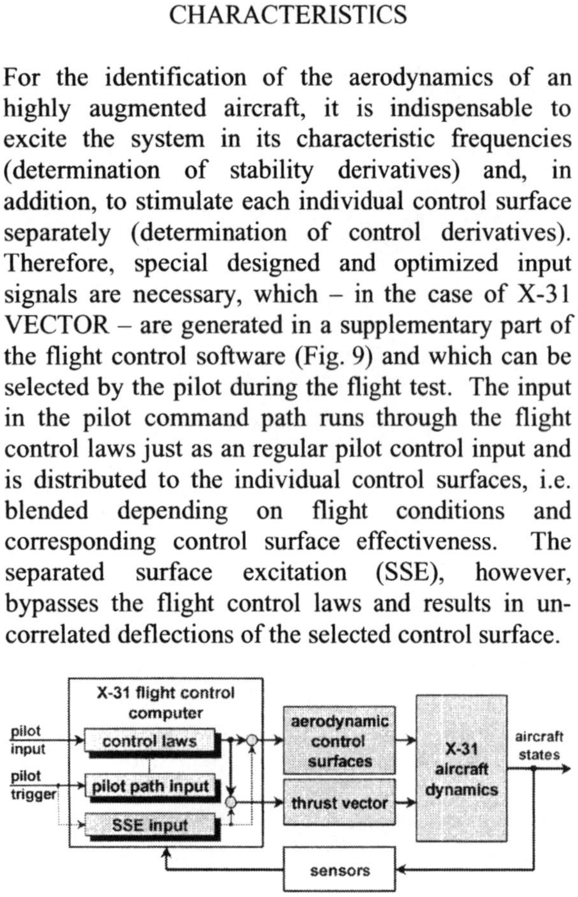

Fig. 9. Integration of SysId-Input Sequences into the Flight Control System

To generate an effective excitation, although not to large (e.g. due to avoid structural overloads), the amplitudes of the input signals are calculated depending on the dynamic pressure occurring actually during the flight test. Fig. 10 shows the measured time histories of such a typical input signal sequence. In the pilot path, a modified 1123-signal is used to get a relatively broad-banded excitation of the different characteristic motion frequencies. 1123 is a sequence of numbers which represents the bar sequence: two short, one doubled and one tripled bar with alternate signs, designed with balanced positive and negative areas. For the determination of the control surface effectiveness, a shorter 121-signal is used because the main purpose is to get sharp-edged aircraft reaction which can be attributed to each individual control surface deflection. To avoid that the electro-hydraulic actuators runs at their rate limits without special need, the 121-signal slope is limited.

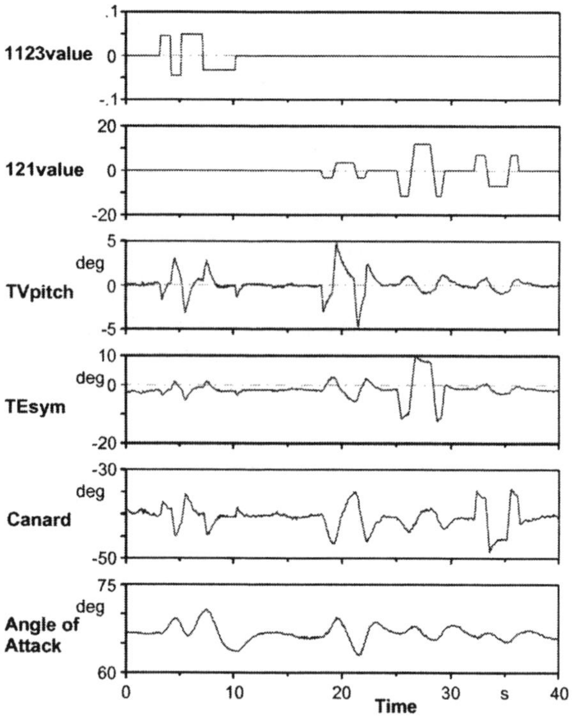

Fig. 10. SysId-Input Sequence for the Excitation of the Longitudinal Aircraft Motion

6. IDENTIFICATION RESULTS

As mentioned above, a good fit of flight-measured and re-simulated time histories is a presumption for reliable SysId results. Such a comparison is shown in Fig. 11. The reactions to input sequences as shown in Fig. 10 are depicted at a reference angles of attack of about 15° and 65° (time slices 1 and 2) and additionally two so-called 'Large-Amplitude-Maneuvers' (time slices 3 and 4) which starts at relatively small angles of attack and are flown far into the poststall regime.

The measured accelerations (solid line) show noticeable more noise at high angles of attack which can be related to both, the increased activities of the flight control computer to stabilize the commanded flight conditions and to the partial separated airflow. As the resulting airflow turbulence is not included in the model, it is treated like measurement noise. Besides the noise, the flight-measured accelerations are well matched by the simulation results. Dealing with the simulation of the poststall maneuvers, it is necessary to include hysteresis effects in the build-up of the aerodynamic forces. At fast increasing angles of attack, the resulting lift is higher than afterwards at the angle of attack decrease. This is caused by the delayed separation and re-attachment of the airflow. During the envelope expansion flights, the pilot commands standard maneuvers (e.g. doublets in the pitch axes). Such maneuvers and those with single surface excitation can be evaluated using linear

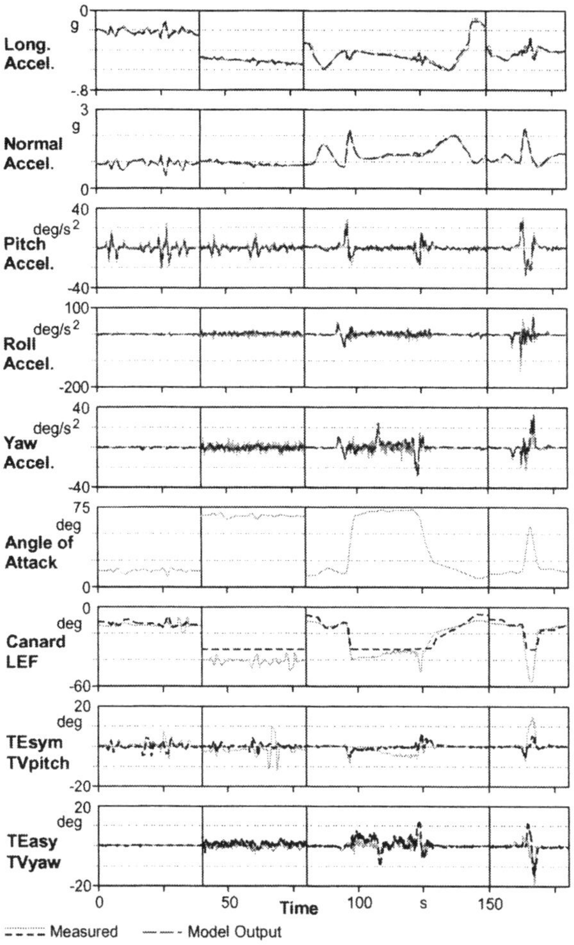

Fig. 11. SysId-Sequences and Poststall Maneuvers

models at the respective reference point (single-point evaluation). This leads, among others, to the lift coefficient CLtrim as compared in the left part of Fig. 12 to the predictions of the database. On the right side of the figure, the result of the global model identification is presented which follows from the evaluation of all standard and large amplitude maneuvers as a continuos curve (multi-point evaluation). Not shown is the above mentioned hysteresis effect which has to be superposed to the quasi-static lift coefficient. Differences between the both methods are additionally due to the influence of engine thrust which can be identified and updated in parallel to the aerodynamics only in the global model evaluation using tests with throttle transients. Thus, possible deviations in the thrust model do not affect the identified aerodynamic parameters.

Fig. 13 shows the derivative of the dihedral coefficient Clβ (rolling moment due to angle of sideslip), again the result of both methods compared to the predictions. At angles of attack between 30° and 45°, the original database has a deep trough which is not validated by the flight test results. The database was therefor updated, at that time on the basis of single maneuver evaluation (left side). The later performed global model identification leads to good corresponding results, only the peak at about 45° angle of attack seems to be too high. This may be due to the model structure which uses - to keep the total amount of parameters to be estimated (presently 1200) within handy limits – a relative small number of breakpoints (presently 12 for angle of attack and 6 for Mach number) with linear interpolation in between. The aerodynamic grid is possibly to rough to satisfy the dependency of each of the various derivatives. Looking at the increment ΔClβ? it is obvious that an additional breakpoint between 38° and 50° is missing to compensate for the big changes in the original database.

7. SIMULATION UPDATE

The results of the global model identification, i.e. in particular the aerodynamic increments, are written into multi dimensional tables which can be converted into 'Real-Functions' at the end of the evaluation run. These functions can be easily implemented into the simulations. It is highly important to take great pains over the modeling of the transients from updated to non-updated flight regimes being not covered during the flight test. Only under well defined circumstances, the identified increments may be extrapolated. In most of the cases, it is suitable that the updates are faded out within a transition regime. This is valid especially when the identification is an integrated part of a successive flight regime expansion where each new flight has to be prepared and flown in advance without any failures in the – possibly updated – simulation.

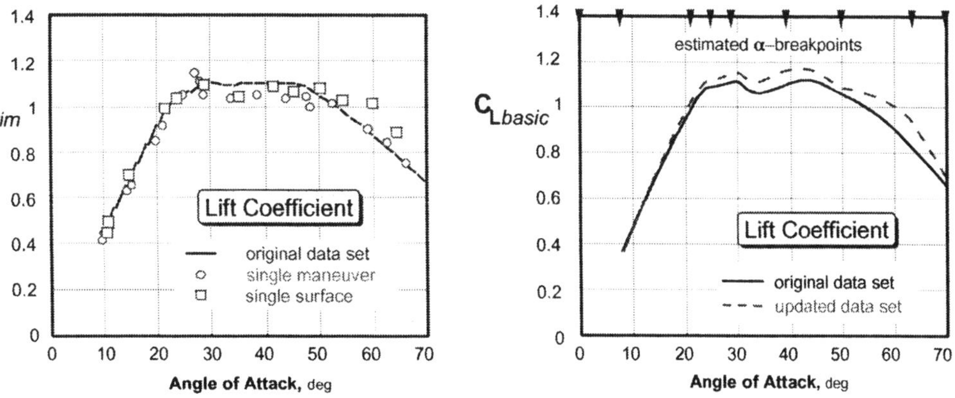

Fig. 12. Lift Coefficient, Single Point Evaluation versus Global Model Evaluation

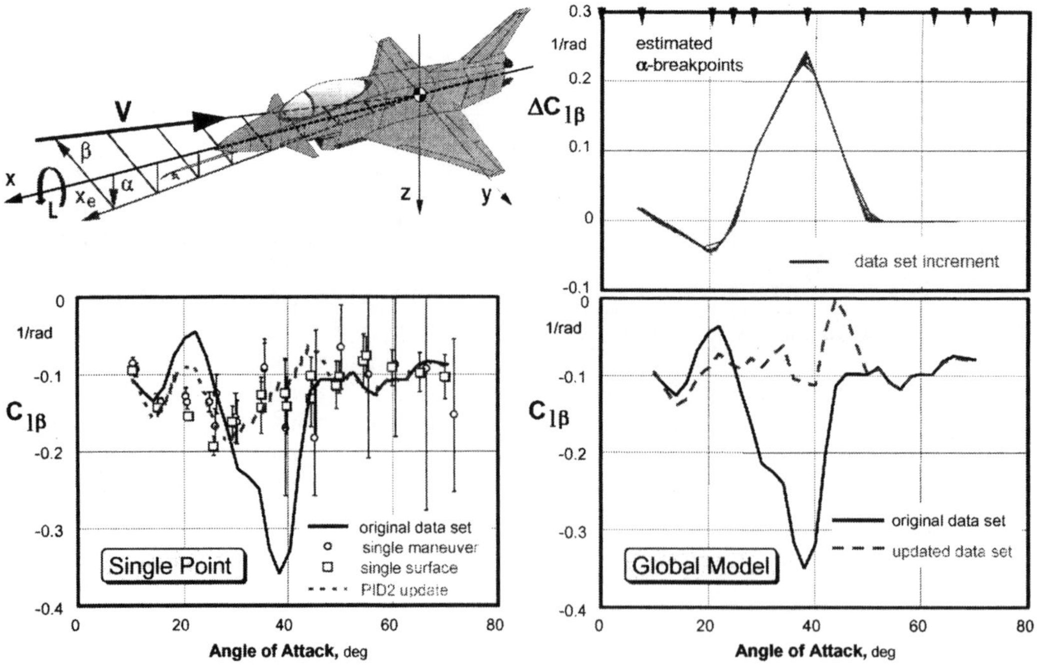

Fig. 13. Dihedral Effect, Single Point Evaluation versus the Global Model Evaluation

8. CONCLUSIONS

Simulation and update of the integrated models by means of system identification during the evaluation of flight control software of modern aircraft has been described on the basis of the experimental aircraft X-31A. The design and validation process uses several steps starting with increasingly complex simulations - with on the whole identical software modules - and finally ending up in the flight test. In the case of the occurrence of significant deviations between simulation and flight test, the integrated simulation models must be updated. With respect to the aerodynamics, special designed flight tests are necessary which are evaluated by means of system identification applying an innovative global model. The global model leads to aerodynamic increment tables for supplementation of the original database. The tables are converted into 'Real-Functions' which are easily to be implemented into the simulation software.

REFERENCES

Beh, H. and G. Hofinger (1994). Control Law Design of the Experimental Aircraft X-31A. *19th Congress of the International Council of the Aeronautical Sciences,* Paper ICAS 94-7.2.1. AIAA, Annaheim, CA, USA.

Eubanks, D., R. Gütter and B. Lee (1995). X-31 CIC Flight Test Results. *4-PWR SM TG Workshop on Full Envelope Agility,* pp. 31 – 47. Eglin AFB, USA.

Francis, M.S. (1995). X-31: An International Success Story. *Aerospace America*, Feb., pp. 22-27 & 32.

Friehmelt, H., S. Weiß, D. Rohlf and E. Plaetschke (1995). Using Single Surface Excitation for X-31 System Identification at High AoA. *26th Annual SFTE Symposium.* Society of Flight Test Engeneers, Berlin, Germany.

Friehmelt, H. and P. Huber (2001). VECTOR – Die X-31A fliegt zu neuen bahnbrechenden Technologiedemonstrationen. *DGLR Annual Conference,* Paper No. 2001-090. Hamburg, Germany.

Friehmelt, H., T. Grohs and D. Rohlf (2002). New Challenges and Opportunities for System Identification during VECTOR. *RTA SCI Panel Symposium 'Challenges in Dynamics, System Identification, Control and Handling Qualities for Land, Air, Sea, and Space Vehicles',* Paper 1. Berlin, Germany.

Friehmelt, H. and M. Hahn (2002). VECTOR – Ein kooperatives Experimentalprogramm als Schmelztiegel der Disziplinen und beteiligten Partner. *DGLR Annual Conference,* Paper No. 2002-048. Stuttgart, Germany.

Georg, H.-U. (1994). Aerodynamic Development and Effectiveness Evaluation of the X-31 Thrust Vectoring System. *Fourth High Alpha Conference, NASA CP-10143, Vol. 2.* NASA Dryden, USA.

Huber H. and H. Galleithner (1992). Control Laws / Flying Qualities and Flight Test Results. *High-Angle-of-Attack Projects and Technology Conference, NASA-CP-3137, Vol. 1,* pp. 171-188. NASA Dryden, USA.

Robinson, M.R. and W.B. Herbst (1990). The X-31A and Advanced Highly Maneuverable Aircraft. *17th ICAS* Congress, *Proceedings Vol. I,* pp. LV – LXIV. Stockholm, Sweden.

Rohlf, D. (1999). Direct Update of a Global Simulation Model with Increments via System Identification. *RTO-MP-11,* Paper No. 28.

IFAC Publications
www.elsevier.com/locate/ifac

CONTINUOUS-TIME IDENTIFICATION OF FIRST-ORDER PLUS DEAD-TIME MODELS FROM STEP RESPONSE IN CLOSED LOOP

Flávio S. Coelho and Péricles R. Barros

Dep. de Eng. Elétrica, Univ. Fed. de Campina Grande,
Cx.P. 10105, Campina Grande, PB – BRAZIL.
E-mail: flcoelho@dee.ufcg.edu.br, prbarros@dee.ufcg.edu.br

Abstract: In this work the identification of first-order plus dead-time models from closed loop step response experiments is considered. Existing approaches for closed loop identification are examined and simple algorithm are proposed for dealling with the dead-time. Simulation examples are used to illustrate the techniques. *Copyright © 2003 IFAC*

Keywords: Continuous time identification; Process identification; Closed-loop identification; Time delay process.

1. INTRODUCTION

Methods for identification in closed-loop are very attractive to industrial applications. The closed-loop identification doesn't cause stops in system operation, unlike open-loop identification. Other reasons which can be listed are demands on safety in process operation, unstable processes and restrictions in production (Forssell and Ljung, 1999) and (Ljung, 1999). An additional consideration to perform experiments in closed-loop is that if the process model is of restricted complexity (as is usual with PI and PID controllers) the dynamics exhibited by the plant with the old controller is relevant to the new controller design (den Hof and Schrama, 1998), (Gevers, 1995) and (Landau and Karimi, 1997). Despite the developments in closed-loop identification, most of the theory has considered the discrete-time case and very little has been reported for the continuous-time models. One reason for this is the expectation that most results from discrete-time will directly extend to the continuous-time case.

The estimation of continuous-time models from sampled data has received some attention in the last years, motivated by the need of such models to recover physical parameters or to use design techniques developed for continuous-time controllers. An extensive list of references on the subject can be found in (Mensler, 1999), where a detailed survey discusses the advantages of a direct approach in relation to the indirect estimation of a discrete-time model plus a later transformation into a continuous-time model. Invited sessions have been organized in recent conferences (for instance, IFAC World Triannual Conference 2002) to report new developments and applications. Several studies are based on the use of the toolbox package CONTSID (see (Garnier, 2002)). It should be noticed that most results are of identification in open loop, with a few exceptions (Garnier et al., 2000).

The continuous-time results reported in the literature mainly address finite-dimensional systems. But dead-time is present in several industrial processes so that first and second order dead-time continuous time models are widely used to tune industrial controllers. Usually, a model is estimated from the process open-loop step response (Åström and Hägglund, 1995) and used to design PI and PID controllers. In these designs the process model that receives most attention is first-order plus dead-time model (FOPDT) (Sudaresan and Krishnaswamy, 1977). There are a few methods of estimate parameters for this model. Among them one can mention the graphics and the area methods (Åström and Hägglund, 1995). A method less sensitive to noise is proposed in (Wang et al., 1999) which uses least-squares method to estimate the parameters

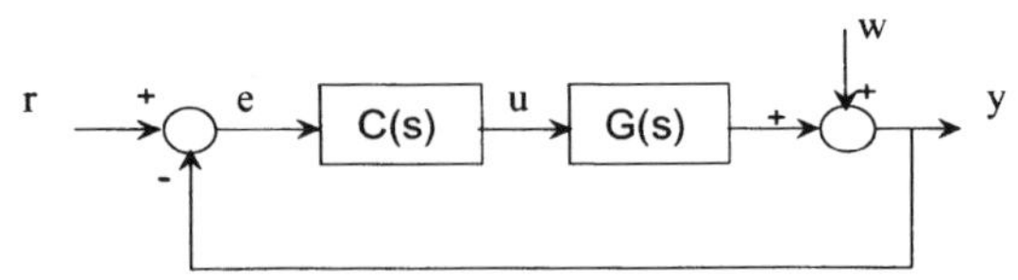

Fig. 1. Closed-loop system.

of FOPDT model. Variants of this methods are used in (Wang and Zhang, 2001) and (Wang et al., 2000). A close look reveals that these methods are extensions of continuous-time estimation algorithms (*Integral Methods* - see (Mensler, 1999)) to dead-time models when a step excitation is applied at the process input. For such simple models the results are remarkably good and motivated the present work. Identification in closed-loop from step responses can be found in (Ananth and Chidambaram, 1999), which uses approximations on the infinite dimensional closed loop to estimate parameters for unstable processes, and (Wang et al., 2001), which uses FFT/IFFT to estimate the process step response from closed loop signals.

In this paper continuous-time identification from discrete-time measurements and closed loop identification are combined to estimate FOPDT models. The excitation is restricted to come from a closed loop step experiment. A simple algorithm is proposed to capture dead-times which are not multiple of the sampling period. Three closed loop estimation structures are compared using the proposed algorithm. As in the open-loop case (Wang et al., 2000), similar results can be obtained for second-order models and are omitted here.

This paper is organized as follows. In Section 2, continuous-time identification of FOPTD models is discussed and the algorithm proposed. The algorithm is applied in the framework of closed-loop system identification in Section 3, resulting in three different structures. In Section 4 the proposed structures are illustrated using simulations of examples and, finally, conclusions are presented in Section 5.

2. IDENTIFICATION OF DEAD-TIME SYSTEMS

In this Section continuous-time system identification is presented for FOPTD models. The Integral Method (Mensler, 1999) is chosen as it is simple one which requires little *a-priori* information.

2.1 *The Closed Loop*

Consider the closed-loop system shown in Fig. 1, with r is the setpoint signal, u is the control signal or process input, y is the process output and w is the disturbance signal.

In this paper a first-order dead-time model

$$G(s) = \frac{b}{s+a} e^{-Ls} \tag{1}$$

is chosen to represent the process behavior. The process is in closed loop with the known controller $C(s)$, assumed to be PI or PID. The data at the process input and output are continuous-time, although the data available to the estimation is sampled.

2.2 *Identification of FOPTD Model from a Step Input*

The process is assumed to be at steady-state at $t = 0$ so that, without loss of generality, $u(t) = 0$ for $t < 0$ and zero initial conditions are assumed. For $t \geq L$ the process is assumed to satisfy the differential equation

$$\dot{y}(t) + ay(t) = bu(t-L), \tag{2}$$

where the disturbance term has been discarded.

Integrating Eq.(2) from $\tau = 0$ to $\tau = t$ yields

$$y(t) = -a\int_0^t y(\tau)\,d\tau + b\int_0^t u(\tau - L)\,d\tau. \tag{3}$$

In continuous-time identification, techniques which use such models (with $L = 0$) are named *Integral Methods* (Mensler, 1999). An integral method has been used in (Wang et al., 2000), with the process in open loop and under a step input with amplitude h applied at $t = 0$. For this case the model also satisfy

$$y(t) = -a\int_0^t y(\tau)\,d\tau + bht - bhL, \tag{4}$$

from which a regression model can be obtained on model parameters $\{a, b, L\}$. The estimate can be computed in one step using either least-squares or instrumental variable methods.

2.3 *Identification of FOPTD Model Under a Step Setpoint*

In closed-loop, the procedure used in (Wang et al., 2000) cannot be applied, so that some additional processing must be made. Under the same conditions (steady-state at $t = 0$, $u(t) = 0$ for $t < 0$ and zero initial conditions) the process model can be written as

$$y(t) = -a\int_0^t y(\tau)\,d\tau + b\int_0^{t-L} u(\tau)\,d\tau. \tag{5}$$

It can also be rewritten as

$$y(t) = -a\int_0^t y(\tau)\,d\tau + b\int_0^t u(\tau)\,d\tau - b\int_{t-L}^t u(\tau)\,d\tau. \tag{6}$$

Define

$$\phi(t) = \left[-\int_0^t y(\tau)\,d\tau \;\; \int_0^t u(\tau)\,d\tau \;\; -\int_{t-L}^t u(\tau)\,d\tau \right]^T,$$

$$\theta = \left[a \;\, b_1 \;\, b_2 \right]^T. \tag{7}$$

and Eq.(6) can be written in regression form

$$y(t) = \phi(t)\theta.$$

Assume that the setpoint $r(t)$ is a step applied at $t = 0$ to the closed loop formed by the process with controller $C(s)$. The data is collected until the closed loop system reaches a new steady state. Assuming the closed loop stable and that an integrator is present in the controller, the new steady state output is equal to the step amplitude. For this closed loop experiment the plant input signal $u(t)$ is not constant. Furthermore, the value of L is not known so that, unlike the open-loop case, the third integral of Eq.(6) can not be computed. In this case, a staightforward procedure is to search for the best fit among several values of L. But, this procedure will yield an estimate for L which is a multiple of the sampling period. The following algorithm can be used to overcome this limitation. Its motivation comes from the fact that $b_1 = b_2$ for the true value of L.

2.3.1. *Iterative Algorithm for Dead-time Identification*

(1) Perform a step test on the closed loop and store the data until the closed loop reaches a new steady-state ($t = 0$ to $t = (N-1)T$, with T the sampling period).
(2) Choose a range for the dead-time, say $[L_{\min}, L_{\max}]$, with $L_{\min} = k_{\min}T$ and $L_{\max} = k_{\max}T$, from observation of the signals or, automatically, by computing the time the output exceeds a certain percentage of the step amplitude.
(3) Use the regression model and estimate parameters for each value of k in $[k_{\min}, k_{\max}]$. For each value $i = k - k_{\min} + 1$ compute the estimate $\hat{\theta}^i$

$$\begin{bmatrix} a^i \\ b_1^i \\ b_2^i \end{bmatrix} = \begin{bmatrix} \hat{\theta}^i(1) \\ \hat{\theta}^i(2) \\ \hat{\theta}^i(3) \end{bmatrix}. \tag{8}$$

(4) Compute

$$L_1 = \hat{k}T \quad \text{with} \quad \hat{k} = \min_i \left| b_1^i - b_2^i \right|.$$

(5) Apply the estimator to the regression vector

$$\phi(t) = \begin{bmatrix} -\int_0^t y(\tau)\,d\tau \\ \int_0^t u(\tau)\,d\tau \\ -\frac{1}{L_1}\int_{t-L_1}^t u(\tau)\,d\tau \end{bmatrix}, \tag{9}$$

$$\theta = \left[a \; b \; \beta\right]^T, \tag{10}$$

to recover the final estimate $\left\{\hat{a}, \quad \hat{b}, \quad \hat{L} = \hat{\beta}/\hat{b}\right\}$.

It should be remarked that the best the *a-priori* information on the bounds $\{L_{\min}, L_{\max}\}$ the less iterations are needed.

3. CLOSED LOOP ESTIMATORS

Now estimation closed loop is considered. Closed-loop estimation structures are presented in several references, such as (den Hof and Schrama, 1998), (Gevers, 1995), (Landau and Karimi, 1997) and (Albertos and Sala, 2002). Due to the infinite dimensional nature of the process and the little excitation from a step setpoint, several of the methods found in the references cannot be applied to the identification problem considered here. For instance, the two-stage method (den Hof and Schrama, 1993) can not be used as a step will not provide enough excitation to get a very good estimate of the sensitivity function (which is also infinite-dimensional). The closed loop output error method of (Landau and Karimi, 1997) can not used because recursion is not applicable to the dead-time estimation.

The following estimators can be applied to the experimental setup of this paper and were chosen: The direct least squares estimator, the direct Steiglitz-McBride method and a closed loop instrumental variable method. In this Section the methods willl be described and in the following Section simulation examples are used to illustrate the use of the structures together with the proposed iterative dead-time algorithm.

3.1 *Least Squares Method*

For the least squares method, form matrices

$$Y = \left[\; y(0) \quad y(T) \quad \cdots \quad y((N-1)T) \;\right]^T$$

and

$$\Phi = \left[\; \phi(0) \quad \phi(T) \quad \cdots \quad \phi(((N-1)T)) \;\right]^T.$$

Compute the least-squares estimate

$$\hat{\theta}_{LS} = \left(\Phi^T\Phi\right)^{-1}\Phi^T Y \tag{11}$$

and the corresponding model $\hat{G}_{LS}(s)$.

The estimate will be biased as the data is collected in closed loop and the disturbance is neglected.

3.2 *Steiglitz-McBride Method*

Here the algorithm proposed in (Steiglitz and McBride, 1965) is applied to the data. Compute the least squares estimate $\hat{\theta}_{LS}$. Take the original data and filter through $F(s) = \hat{b}/(s+\hat{a})$ to obtain y_f and u_f. Apply now least squares to the filtered data to obtain a new estimate $\hat{\theta}_{SM}$ using the algorithm from previous Section. Repeat the procedure with the new estimate until $\hat{\theta}_{SM}$ converges. Construct the related process model G_{SM}. It should be noticed that this algorithm can not be used to unstable processes.

3.3 *Closed Loop Instrumental Variable Method*

Depart again from the least squares estimate $\hat{\theta}_{LS}$. Simulate the auxiliar closed-loop system composed by the estimated model $\hat{G}_{LS}(s)$ and the controller $C(s)$, using the same step setpoint. The data from this simulation is now named $\bar{u}(t)$ and $z(t)$ for the process input and output, respectively. Form vector

$$\psi(t) = \left[-\int_0^t z(\tau)d\tau \quad \int_0^t \bar{u}(\tau)d\tau \quad -\int_{t-L}^t \bar{u}(\tau)d\tau \right] \tag{12}$$

as before, for the chosen L, and matrix

$$Z = \left[\begin{array}{cccc} \psi(0) & \psi(T) & \cdots & \psi((N-1)T) \end{array} \right]^T.$$

Finally apply the algorithm from last Section to obtain the instrumental variable estimate $\hat{\theta}_{IV}$ from

$$\hat{\theta}_{IV} = \left(Z^T \Psi\right)^{-1} Z^T Y, \tag{13}$$

and recover G_{IV}. The instrumental variable procedure can be repeated with the new G_{IV} until the estimate converges.

4. SIMULATION EXAMPLES

In this section the closed loop identification algorithms are applied to several processes in closed loop. The cost function used to compare the estimates is

$$\varepsilon = \frac{1}{N}\sum_{k=0}^{N-1}\left[x(kT) - \hat{x}(kT)\right]^2$$

where $x(kT_s)$ is the actual process output (with noise), while $\hat{x}(kT_s)$ is the estimated process output from a closed loop simulation with the same controller and under the same step setpoint.

In all experiments $T = 0.1s$, $k_{\max} = 50$ samples $(= 5s)$, $k_{\min} = 5$ samples $(= 0.5s)$, and the regression vector was formed for $t \geq 30$ samples $(= 3s)$. The simulation runs for $40s$.

For the first simulation the controller used is $C_1 = 1 + \frac{0.1}{s}$. White noise is added only to the output of the process G_1, with variance $\sigma = 0.01$. The Process and the results are shown below.

$G_1 + C_1$	Estimate	ε	$\hat{k}$
Process	$\frac{0.14}{s+0,12}e^{-0.95s}$	—	—
G_{LS}	$\frac{0.1413}{s+0.1205}e^{-0.9682s}$	0.0104	10
G_{SM}	$\frac{0.1368}{s+0.1158}e^{-0.8083s}$	0.0105	8
G_{IV}	$\frac{0.1366}{s+0.1162}e^{-0.7661s}$	0.0105	8

In this example the LS estimate seems the best. The controller is changed to $C_1 = 6 + \frac{0.5}{s}$ in order to yield a more oscillatory response. The results are:

$G_1 + C_2$	Estimate	ε	$\hat{k}$
Process	$\frac{0.14}{s+0,12}e^{-0.95s}$	—	—
G_{LS}	$\frac{0.1431}{s+0.1221}e^{-1.135s}$	0.0133	11
G_{SM}	$\frac{0.1374}{s+0.1169}e^{-0.8596s}$	0.0116	9
G_{IV}	$\frac{0.1407}{s+0.1200}e^{-0.9953s}$	0.0110	10

Now, the best estimate comes from the IV structure.

Now, an unmodelled pole is added to the process and the noise variance is reduced to $\sigma = 0.0001$. The controller is again $C_1 = 1 + \frac{0.1}{s}$. The resulting estimates are shown below with the SM estimator giving the best result.

$G_2 + C_1$	Estimate	ε	$\hat{k}$
Process	$\frac{0.14}{(s+0,12)(s+0.6)}e^{-0.95s}$	—	—
G_{LS}	$\frac{0.2268}{s+0.1158}e^{-2.5296s}$	0.0011	25
G_{SM}	$\frac{0.2174}{s+0.1105}e^{-2.2295s}$	0.0006	22
G_{IV}	$\frac{0.2263}{s+0.1156}e^{-2.5142s}$	0.0010	25

In the two following example, the process order is increased, with repeated poles. The process and the noise variance is $\sigma = 0.0001$ and the controller is again $C_1 = 1 + \frac{0.1}{s}$. The resulting estimates are shown in the following tables.

$G_3 + C_1$	Estimate	ε	$\hat{k}$
Process	$\frac{1}{(s+1)^4}$	—	—
G_{LS}	$\frac{0.3522}{s+0.3518}e^{-1.1859s}$	0.0017	12
G_{SM}	$\frac{0.4606}{s+0.4594}e^{-1.8745s}$	0.0011	19
G_{IV}	$\frac{0.4216}{s+0.4211}e^{-1.6525s}$	0.0011	25

and

$G_4 + C_1$	Estimate	ε	$\hat{k}$
Process	$\frac{1}{(s+1)^8}$	—	—
G_{LS}	$\frac{0.2913}{s+0.2877}e^{-4.8240s}$	0.0029	48
G_{SM}	$\frac{0.2840}{s+0.2804}e^{-4.7117s}$	0.0037	47
G_{IV}	$\frac{0.2899}{s+0.2865}e^{-4.7968s}$	0.0032	48

Again, it is hard to point out the best result.

Finally an unstable plant is considered, taken from (Ananth and Chidambaram, 1999) with $C_3 = 0.350\left(1 + \frac{1}{29.412s}\right)$ and no noise is added. Their estimate is $\hat{G} = \frac{0.9857e^{-2.0s}}{(s-0.2458)}$. In this case, the Steiglitz-McBride method cannot be applied. The results are very precise, with

$G_5 + C_3$	Estimate	ε	$\hat{k}$
Process	$\frac{1}{s-0.25}e^{-2s}$	—	—
G_{LS}	$\frac{0.9999}{s-0.2500}e^{-2.0003s}$	1.1242×10^{-6}	20
G_{IV}	$\frac{0.9999}{s-0.2500}e^{-2.0003s}$	1.1239×10^{-6}	20

In the Figures the outputs for some process and estimated models (closed loop) are compared as well as the Bode and Nyquist plots (no controller).

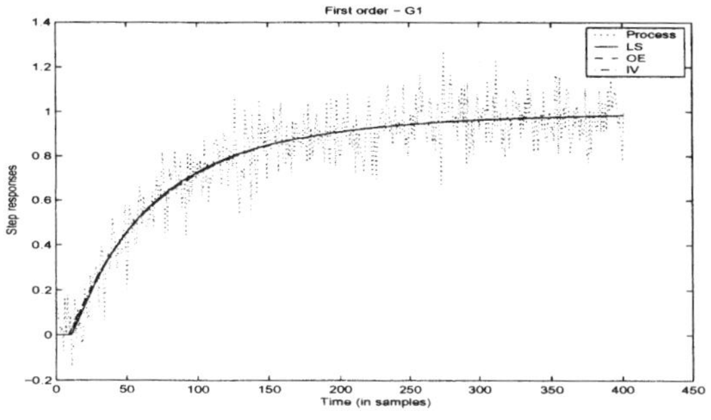

Fig. 2. Step responses - Process G_1 Controller C_1 .

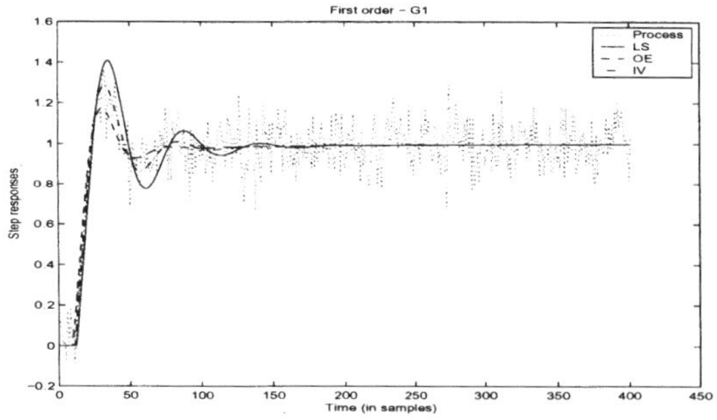

Fig. 3. Step responses - Process G_1 Controller C_2 .

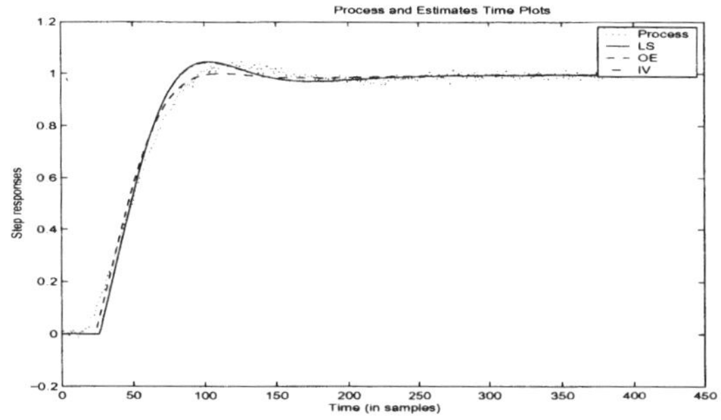

Fig. 4. Step responses - Process G_2 Controller C_1 .

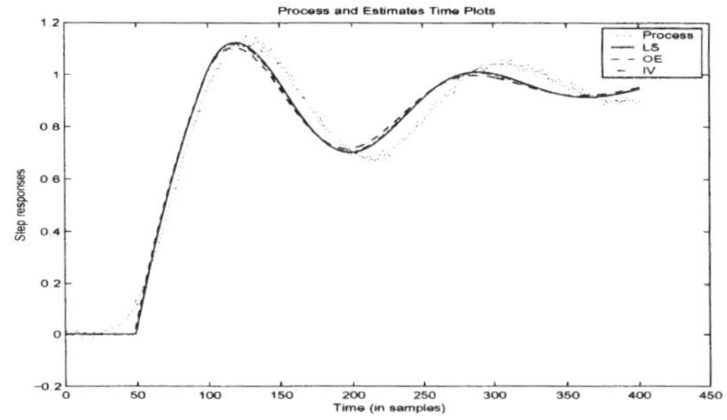

Fig. 5. Step responses - Process G_4 Controller C_1 .

As can be seen from the results shown above, the estimates are remarkably good. On the other side, the simulations does not indicate which one is the best.

5. CONCLUSIONS

In this paper continuous-time FOPDT identification from closed loop step response was presented. An algorithm for dealing with dead-time was presented. Structured for identification in closed loop were also discussed. Simulation examples illustrated the capa-

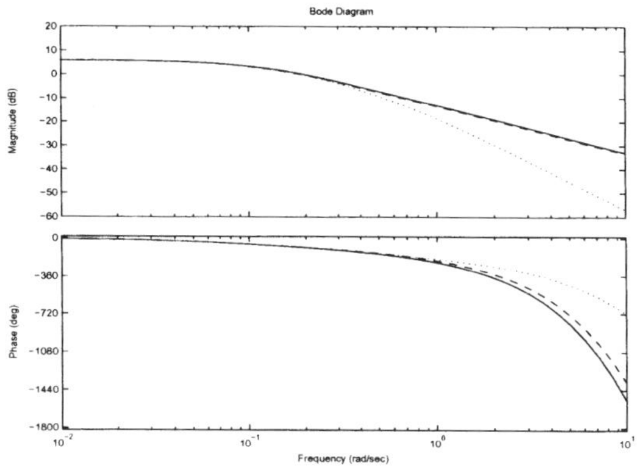

Fig. 6. Bode Plot - Process G_2 -$(\cdots)$ Controller C_1 : LS(–), OE(- -), IV(- .).

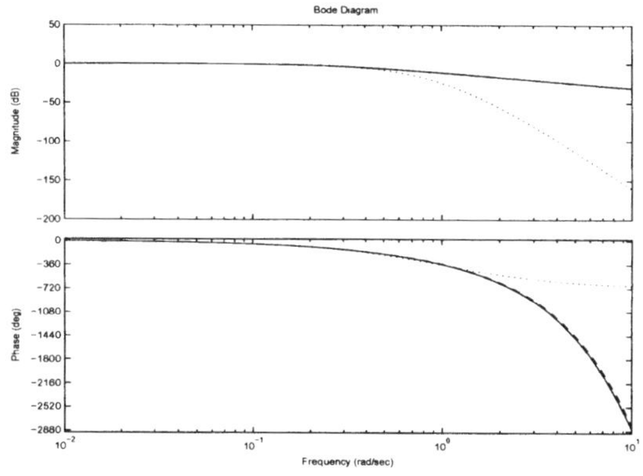

Fig. 7. Bode Plot Process G_4 - $(\cdots)$ Controller C_1 : LS(-), OE(- -), IV(- .).

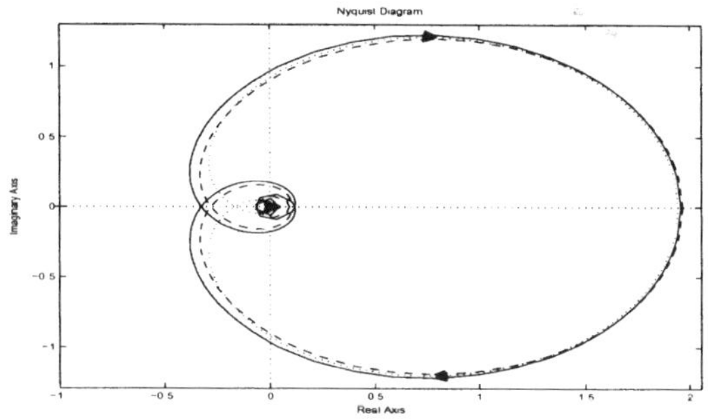

Fig. 8. Nyquist Plot Process G_2 - $(\cdots)$ Controller C_1 : LS(-), OE(- -),IV(- .).

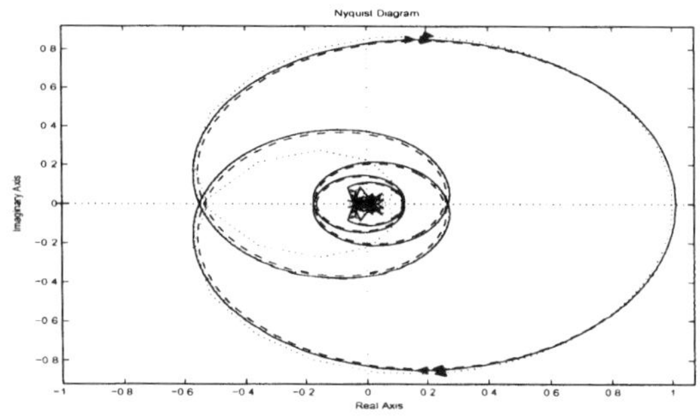

Fig. 9. Nyquist Plot Process G_4 - $(\cdots)$ Controller C_1 : LS(-), OE(- -), IV(- .).

bilities of the methods. In fact, for the step setpoint excitation and for a FOPTD model, the results were very good, but so close that one can not point out a better structure to be used.

6. REFERENCES

Albertos, P. and Sala, A. (2002). *Iterative Identification and Control*, Advances in Theory and Application, Springer-Verlag, London, Grat Britain.

Ananth, I. and Chidambaram, M. (1999). Closed-loop identification of transfer function model for unstable systems, *Journal of the Franklin Institute* **336**: 1055–1061.

Åström, K. J. and Hägglund, T. (1995). *PID Controllers: Theory, Design and Tuning*, 2nd edn, Instrument Society of America, Research Triangle Park, North Carolina.

den Hof, P. M. J. V. and Schrama, R. J. P. (1993). An indirect method for transfer function estimation from closed loop data, *Automatica* **29**: 1523–1527.

den Hof, P. M. J. V. and Schrama, R. J. P. (1998). Closed loop issues in system identification, *Annual Reviews in Control* **22**: 173–186.

Forssell, U. and Ljung, L. (1999). Closed-loop identification revistited, *Automatica* **35**(7): 1215–1241.

Garnier, H. (2002). Identification of real-life processes with the contsid toolbox, 15th IFAC Triennial World Congress, Barcelona (Spain).

Garnier, H., Gilson, M. and Zheng, W. (2000). A bias-eliminated least-squares method for continuous-time model identification of closed-loop systems, *International Journal of Control* **73**(1): 38–48.

Gevers, M. (1995). Identification for control, *Plenary Talk, 5th IFAC Symposium on Adaptive Systems in Control and Signal Processing*, Budapest, Hungary.

Landau, I. D. and Karimi, A. (1997). Recursive algorithms for identification in closed loop: A unified approach and evaluation, *Automatica* **33**(8): 1499–1523.

Ljung, L. (1999). *System Identification: Theory for the User*, 2nd edn, Prentice Hall, Upper Saddle River, NJ.

Mensler, M. M. (1999). *Analyse et étude comparative de méthodes d'identification des systèmes à représéntation continue. Développement d'une boîte à outilis logicielle*, Thèse du doctorad, Université Henry Poincaré Nancy 1, Nancy,France.

Steiglitz, K. and McBride, L. E. (1965). A technique for the identification of linear systems, *IEEE Transactions on Automatic Control* **AC-10**: 461–464.

Sudaresan, K. and Krishnaswamy, P. (1977). Estimation of time delay time constant parameters in time, frequency, and laplace domains, India.

Wang, Q., Bi, Q., Cai, W., Lee, E., Hang, C. and Zhang, Y. (1999). Robust identification of first-order plus dead-time model from step response., *Control Engineering Practice* (7): 71–77.

Wang, Q.-G., Hwang, B. and Guo, X. (2000). Auto-tuning of tito decoupling controllers from step tests, *ISA Transactions* **39**: 407–418.

Wang, Q. and Zhang, Y. (2001). Robust identification of continuous systems with dead-time from step responses, *Automatica* **37**(3): 377–390.

Wang, Q., Zhang, Y. and Guo, X. (2001). Robust closed-loop identification with application to auto-tuning, *Journal Process Control* **11**: 519–530.

IFAC
Publications
www.elsevier.com/locate/ifac

IDENTIFICATION OF SIMPLE CONTINUOUS-TIME MODELS FROM RELAY FEEDBACK

Gustavo H. M. de Arruda and Péricles R. Barros

*Dep. de Eng. Elétrica, Univ. Fed. de Campina Grande,
Cx.P. 10105, Campina Grande, PB – BRAZIL.
E-mail: arruda@dee.ufcg.edu.br, prbarros@dee.ufcg.edu.br*

Abstract: In this paper it is presented a procedure for the estimation of a first order plus dead-time (FOPDT) model using a relay test and a relay test with an integrator. Exact expressions for the limit cycle in each case are obtained. The time constant and dead-time are computed by solving two equations from Poincaré map analysis of time-delay systems. The process gain is then estimated using three different approaches. Simulation examples illustrates the procedure in the presence of unmodelled dynamics and noise. *Copyright © 2003 IFAC*

Keywords: Relay tests; first order plus dead-time; Poincaré maps;

1. INTRODUCTION

Relay feedback systems are a special class of non-linear systems that have proven to be very useful for process identification and on-line controller tuning (Åström and Hägglund, 1995). Analysis of relay feedback systems has early been performed in the frequency domain by describing function analysis, which yields approximate results. These results have been used in process modelling, controller tuning and performance evaluation, among others (see Yu (1999) and Tan *et al.* (1999)). Exact conditions have been obtained by time-domain analysis and operator theory, which also provided a deeper understanding of the problem (Åström, 1995; Varigonda and Georgiou, 2001).

There exists many references regarding estimation of first order plus dead-time (FOPDT) models from relay feedback data (see early work from Luyben (1987)). In the majority of the cases, describing function approximation is used to obtain the model, which results in errors even though the process structure matches the model structure. Recently, some effort has been made in order to obtain more accurate results from relay feedback experiments. In Luyben (2001), the notion of shape factor is introduced, and the model is obtained by taking into account the behavior of the output. However, it is assumed that the frequency of the oscillation is the critical frequency, which is only an approximation by describing function analysis. In Kaya and Atherton (2001), stable or unstable FOPDT and SOPDT parameters are computed exactly, assuming model matching and no measurement errors, and using data from an asymmetrical limit cycle from a relay feedback. The procedure is based on the A-locus method, which gives exact solutions for the limit cycle frequency and amplitude. Time domain approach for a stable FOPDT model is presented in Wang *et al.* (1997).

In this paper it is presented a FOPDT modelling procedure based on two relay feedback experiments: a standard relay test and a standard relay test with an integrator. In sections 2 and 3, the two expressions relating the model dead-time and the time constant are obtained from Poincaré map analysis. The advantage of the procedure is that the model fits better in the frequency range where the process phase angle is between -90° and -180°, in the case of unmodelled dynamics. Also, no asymmetry in the relay amplitude is required for estimation of the process gain, which prevents errors caused by disturbances since the output must remain symmetric during the tests. Three approaches for the estimation of the process gain are presented in section 4. Simulations examples are presented in section 5 to evaluate the procedure in the presence of unmodelling errors and noise.

2. RELAY FEEDBACK DYNAMICS

Consider a linear system with dead-time D represented in state space form,

$$\dot{x}(t) = Ax(t) + Bu(t-D) , \quad y(t) = Cx(t) , \tag{1}$$

connected in feedback with a relay

$$u = rel(y) = \begin{cases} -1 , & y > 0 \text{ and} \\ 1 , & y < 0 . \end{cases} \tag{2}$$

where $x \in \mathscr{R}^n$. For now, the relay amplitude can be considered unitary in the subsequent analysis, as it will only influence the estimation of the process gain, which will be discussed later. The dead-time is considered at the input of the linear system, without loss of generality, so that the feedback structure can be represented as shown in Fig. 1.

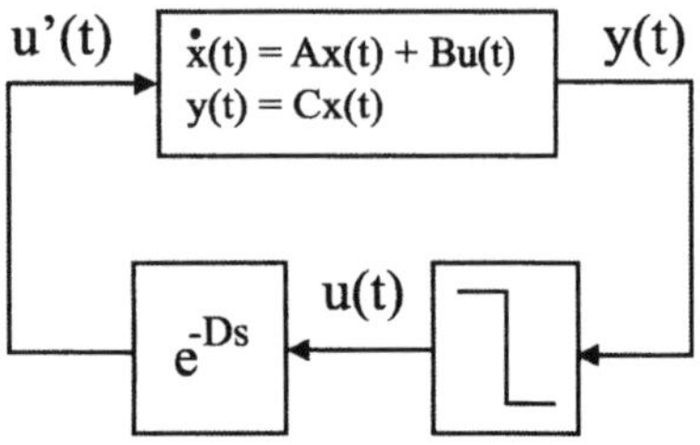

Fig. 1. Relay feedback of linear system with time delay.

A limit cycle γ is the limit set of the nontrivial periodic orbit defined by the solution of the nonlinear system,

$$\dot{x}(t) = Ax(t) + B \cdot rel(Cx(t-D)) . \tag{3}$$

The limit cycle is *symmetric* if the periodic solution of (3) satisfies $x(t+h^*) = -x(t)$, where $2h^*$ is the period of the oscillation. It is called *unimodal* if the relay switches twice, and only twice, per oscillation period. From Eq. (2), the switching surface is defined as

$$\mathscr{S}_n = \{x \in \mathscr{R}^n : Cx = 0\} . \tag{4}$$

When the trajectory of the linear system hits $\mathscr{S}_n$, the relay output changes its state.

The following theorems, which state necessary and sufficient conditions for the existence of limit cycles in the relay feedback system without dead-time ($D = 0$), are based on obtaining the Poincaré map from a state belonging to $\mathscr{S}_n$, to the next state that hits $\mathscr{S}_n$ again (see Åström (1995) and Varigonda and Georgiou (2001)).

Theorem 1. Consider a linear system in state space form

$$\dot{x} = Ax + Bu , \quad y = Cx , \tag{5}$$

connected in feedback with the relay given by (2). There exists a symmetric unimodal limit cycle with period $2h^*$ in the relay feedback system if and only if the following applies:

$$(i)\ f(h^*) = C(e^{Ah^*} + I)^{-1} \int_0^{h^*} e^{A(h^*-s)} ds B = 0, \tag{6}$$

$$(ii)\ y(t) = C(e^{At}x^* - \int_0^t e^{A(h^*-s)} ds B) > 0 , \tag{7}$$

$\forall t \in (0, h^*)$, where

$$x^* = \left(e^{Ah^*} + I\right)^{-1} \int_0^{h^*} e^{A(h^*-s)} ds B , \tag{8}$$

is the initial condition $x(0) = x^* \in \mathscr{S}_n$ that yields the periodic solution.

The use of a linear system with dead-time to represent a process as shown in Fig. 1 allows a more accurate approximation of the process dynamics, in general. Furthermore, many chemical processes can be successfully modelled by such class of models.

The next theorem provides necessary and sufficient conditions for the existence of limit cycles in linear systems with dead-time. The proof will be omitted here, as it is not essential for the purpose of this work. It follows the same ideas of the proof of the Theorem 1, taking into account the delay in the input of the linear system, $u'(t)$.

Theorem 2. Consider a linear system with dead-time (Eq. (1)) connected in feedback with a relay (Eq. (2)), as shown in Fig. 1. There exists a symmetric unimodal limit cycle with period $2h^*$, $h^* > D$, in the relay feedback system if and only if the following applies:

$$(i)\ f_D(h^*) = C\left(e^{Ah^*} + I\right)^{-1} \left[-\int_0^D e^{A(h^*-s)} ds + \int_D^{h^*} e^{A(h^*-s)} ds \right] B = 0 , \tag{9}$$

$$(ii)\ y(t) = Cx(t) > 0 , \forall t \in (0, h^*) , \tag{10}$$

with

$$x^* = \left(e^{Ah^*} + I\right)^{-1} \left[-\int_0^D e^{A(h^*-s)} ds + \int_D^{h^*} e^{A(h^*-s)} ds \right] B , \tag{11}$$

being the state where the trajectory of the linear system intercepts the switching surface $\mathscr{S}_n = \{x \in \mathscr{R}^n : Cx = 0\}$. The periodic solution is obtained by specifying an initial trajectory and relay output, $x(\tau)$, $u(\tau) = -1$, $-D \leq \tau < 0$ with $x(0) = x^*$.

Proof. See Åström (1995). ■

3. PROPERTIES OF FOPDT SYSTEMS

The following lemma is obtained from Theorem 2.

Lemma 3. Consider a FOPDT model,

$$G(s) = \frac{K}{Ts+1}e^{-Ds}, \quad (12)$$

connected in feedback with a relay as shown in Fig. 1. If the oscillation period of the limit cycle is $2h_1$, then

$$D = T\ln\left(\frac{e^{h_1/T}+1}{2}\right). \quad (13)$$

Proof. A state space representation of the model is

$$\dot{x}(t) = (-1/T)x(t) + (K/T)u(t-D),$$
$$y(t) = x(t).$$

Then from Theorem 2 a limit cycle exists at h_1, such that

$$\begin{aligned} f_D(h_1) =& C\left(e^{Ah_1}+I\right)^{-1} \\ & \left[-\int_0^D e^{A(h_1-s)}ds + \int_D^{h_1} e^{A(h_1-s)}ds\right]B \\ =& \left(e^{-h_1/T}+1\right)^{-1}(-e^{-(h_1-D)/T}+e^{-h_1/T} \\ & +1-e^{-(h_1-D)/T})(-1/T)^{-1}\frac{K}{T} = 0, \end{aligned}$$

which gives

$$K\frac{\left(1+e^{-h_1/T}-2e^{(D-h_1)/T}\right)}{\left(e^{-h_1/T}+1\right)} = 0.$$

Since $K/\left(e^{-h_1/T}+1\right) \neq 0$, then

$$1+e^{-h_1/T}-2e^{(D-h_1)/T} = 0$$

yielding Eq. 13. ■

This result indicates that the family of FOPDT models that have a limit cycle with period $2h_1$ is given by Eq. (13). This is an important result, since most of the FOPDT estimation procedures does not yield models capable of reproducing such oscillation when applied to a standard relay test.

Remark 4. Note that

$$\lim_{T\to 0} D(T) = h_1 \text{ and } \lim_{T\to\infty} D(T) = h_1/2,$$

which implies that in a FOPDT model in relay feedback, if the limit cycle period is $2h_1$, then the dead-time is restricted to the interval $[h_1/2; h_1]$.

Remark 5. On the other hand, this also means that relay feedback of a FOPDT process with large D/T ratio oscillates with a period close to twice the dead-time ($2h_1 \cong 2D$), where the equality is obtained in the limit $T \to 0$, which is the case of pure delay systems. For the case of a FOPDT process with small D/T ratio, the oscillation period is close to the dead-time ($2h_1 \cong D$), and the equality is also obtained in the limit $T \to \infty$, which is the case of the integrator plus dead-time model. Therefore, by just measuring the oscillation period, one is able to guess the total delay of the linear system, also taking in account the shape of the output signal (see Luyben (2001))

Consider now the case of a relay test with an integrator. If an integrator is added to the feedback path of Fig. 1, then the frequency of the oscillation will be close to the frequency where $\angle G(j\omega)$ is $-\pi/2$. The following lemma is also obtained from Theorem 2.

Lemma 6. Consider the FOPDT model in Eq. 12 connected in feedback with a relay and an integrator, such that the linear part is

$$\frac{G(s)}{s} = \frac{K}{s(Ts+1)}e^{-Ds}. \quad (14)$$

If the oscillation period of the limit cycle is $2h_2$, then

$$\begin{aligned} & 2T\left(1-2e^{(D-h_2)/T}+e^{-h_2/T}\right) \\ & +\left(1+e^{-h_2/T}\right)(2D-h_2) = 0 \end{aligned} \quad (15)$$

Proof. A state space representation of the linear system is now

$$\begin{bmatrix}\dot{x}_1(t)\\ \dot{x}_2(t)\end{bmatrix} = \begin{bmatrix}-1/T & 0\\ 1 & 0\end{bmatrix}\begin{bmatrix}x_1(t)\\ x_2(t)\end{bmatrix} + \begin{bmatrix}K/T\\ 0\end{bmatrix}u(t-D),$$
$$y(t) = \begin{bmatrix}0 & 1\end{bmatrix}\begin{bmatrix}x_1(t)\\ x_2(t)\end{bmatrix}.$$

If this system is connected in a relay feedback, as shown in Fig. 1, then from Theorem 2 a limit cycle exists at h_2, such that

$$\begin{aligned} f_D(h_2) =& C\left(\exp\left(\begin{bmatrix}-1/T & 0\\ 1 & 0\end{bmatrix}h_2\right)+I_{2\times 2}\right)^{-1} \\ & \left[-\int_0^D \exp\left(\begin{bmatrix}-1/T & 0\\ 1 & 0\end{bmatrix}(h_2-s)\right)ds \right. \\ & \left. +\int_D^{h_2} \exp\left(\begin{bmatrix}-1/T & 0\\ 1 & 0\end{bmatrix}(h_2-s)\right)ds\right]B. \end{aligned}$$

Since

$$e^{At} = \begin{bmatrix} e^{-t/T} & 0\\ T\left(1-e^{-t/T}\right) & 1\end{bmatrix},$$

and $f_D(h_2) = 0$, then the result follows after tedious manipulation. ■

4. FOPDT MODELLING

In this section the procedure for obtaining a FOPDT model from a standard relay test and a standard relay test with an integrator is presented. Clearly the Lemmas 3 and 6 can be combined together to yield a solution for D and T using just the measurements of the period of the limit cycle in both tests. Inserting (13) into (15) yields

$$\begin{aligned} 0 =& 2T(1-e^{(h_1-h_2)/T}) \\ & +(1+e^{-h_2/T})(2T\ln\left(\frac{e^{h_1/T}+1}{2}\right)-h_2). \end{aligned} \quad (16)$$

The roots of Eq. (16) are computed using numeric routines, and when a root $\hat{T}$ is found, the corresponding dead-time, $\hat{D}$, is obtained from Eq. (13). Note that the roots of Eq. (16) can be validated by observing that the corresponding value of $\hat{D}$ should be at the interval $[h_1/2; h_1]$, according to remark 4. The parameters computed this way give a family of models of the form

$$\hat{G}(s) = \frac{\hat{K}}{\hat{T}s+1}e^{-\hat{D}s}. \tag{17}$$

Remark 7. It may however not exist a solution if there is strong unmodelled dynamics. However, this situation can be predicted only by observing the values of h_1 and h_2, which will be discussed in future work.

It remains now to determine the process direct gain. It will be considered three alternatives for the estimate of the process gain, K:

(1) using a step test in open loop;
(2) measuring the peak of the output signal of the process;
(3) computing a least squares estimate using only output data.

These alternatives are discussed next, where it will be required now the knowledge of the relay amplitude, d. Note the all three alternatives will give the same result if the process is FOPDT and noise effects are neglected.

4.1 *Using a step test*

If a step is applied in open loop, than an estimate of the process gain is obtained as

$$\hat{K}_1 = \frac{y(\infty) - y(0)}{u(\infty) - u(0)}. \tag{18}$$

This gives a good magnitude fitting at very low frequencies, at the expense of a modelling error of the magnitude at the critical point. If the model is intended to obtain controller parameters, such as PID tuning procedures, low frequency modelling errors are tolerated. Also in all cases the model is adjusted using two frequency points, which increases its accuracy.

4.2 *Measuring the output peak*

The information of the maximum value of the process output signal can also be used to estimate process gain. The output of the process with the standard relay test, and relay amplitude d, during one half-period is given by

$$y(t) = \begin{cases} Kd(1-e^{-t/T}), & 0 < t \leq D \\ Kd(2e^{(D-t)/T} - e^{-t/T} - 1), & D < t \leq h_1 \end{cases}. \tag{19}$$

It is possible to show that the maximum achieved by $y(t)$ is $\max_t y(t) = Kd\left(1-e^{-D/T}\right)$. If this value is compared to the measured output maximum amplitude, than an estimate for the process gain can be obtained as

$$\hat{K}_2 = \frac{y_{peak}}{d\left(1-e^{-\hat{D}/\hat{T}}\right)}. \tag{20}$$

This gives a model that resembles the same peak of the output signal when applied to a standard relay feedback test. The great advantage of this estimate is that no input or output data storing is required for the estimation procedure, except the measurements of the oscillation periods in both experiments, and the peak of the output signal during the standard relay test. It may however be necessary to consider the effect of the noise in data, when measuring the output peak.

4.3 *Using least squares*

Finally, a third estimate can be obtained from least squares estimation, using the process output data during standard relay test, with relay amplitude d, and sampling period T_s. The output of the model in Eq. (12) during standard relay test can be written in regressor form as

$$\hat{y}_1(kT_s) = \theta\phi_1(kT_s), \tag{21}$$

where $\theta = K$ and $\phi_1(kT_s)$ is the output of the FOPDT model with unitary gain and parameters given by $\hat{D}$ and $\hat{T}$, to an input signal of the form

$$u(k) = \begin{cases} +d, & 0 < kT_s \leq \hat{D} \\ -d, & \hat{D} < kT_s \leq h_1 \end{cases}, \tag{22}$$

with initial condition $\hat{y}_1(0) = 0$, so that the limit cycle is obtained. For the case of the standard relay feedback test with an integrator, the output is also written as

$$\hat{y}_2(kT_s) = \theta\phi_2(kT_s), \tag{23}$$

where $\theta = K$ and $\phi_2(kT_s)$ is the output of the FOPDT model plus an integrator with unitary gain and parameters given by $\hat{D}$ and $\hat{T}$, to an input signal of the same form as in Eq. (22), with initial condition given as in Eq. (11). The least squares estimate is then given by (see Ljung (1999))

$$\hat{K}_3 = \theta_N^{LS} = \left(\sum_{k=1}^{N}\phi^2(k)\right)^{-1}\sum_{k=1}^{N}\phi(k)y(k), \tag{24}$$

where the data set $[\phi, y]$ can be $[\phi_1, y_1]$, $[\phi_2, y_2]$ or a combination of both. These estimates gives better approximations at certain frequency ranges than the previous ones. If $[\phi_1, y_1]$ is chosen, a better approximation of the process magnitude and phase around its critical frequency is obtained, and if $[\phi_2, y_2]$ is chosen, the model is better at the frequency where the phase of the process is $-\pi/2$. If the vectors are combined, then the obtained model is close to the process frequency response in the frequency range where the process phase is between $-\pi/2$ and $-\pi$, which is a relevant range for control.

5. SIMULATION EXAMPLES

In this section, simulation examples are presented. It will be discussed the effectiveness of the estimation procedure in the presence of unmodelled dynamics. Since this is not the case with true FOPDT processes, their analysis will be postponed to the end of the section, including the effects of noise.

5.1 *Second Order plus Dead-time Processes*

Consider a SOPDT process given by

$$G(s) = \frac{1}{(s+1)^2} e^{-Ds} \, , \, D = 0.1; 1; 10 \, .$$

The standard relay test and the standard relay test with an integrator are applied to this set of process, and the obtained oscillation periods are presented in Table 1.

	Dead-time		
	0.1	1	10
h_1 (sec)	0.7429	2.3878	11.6783
h_2 (sec)	3.5284	5.7721	23.9999

Table 1. Limit cycle measurements for example 5.1

With the measurements of h_1 and h_2, the estimates $\hat{D}$ and $\hat{T}$ are obtained. The process gain estimate is computed using the three different approaches presented in the previous section. The results are presented in the first row of the Table 2. The estimate using the least squares technique is performed with the combined regressor vector, with data from the two relay feedback tests, using $N = 1000$ points per oscillation period. The Bode plot of the process and the models for the case $D = 0.1$ is presented in Fig. 2. The three models have the same phase angle curve, since it depends only on $\hat{D}$ and $\hat{T}$. Note the good phase matching of the model with the process in both frequencies close to the ones where the phase of the processes is $-\pi/2$ and $-\pi$. This is due to the modelling procedure, where these frequency points are used to compute exactly which dead-time and time constant are required to match them in a FOPDT model. The relay output during standard relay test is presented in Fig. 3. Note that while the model using the step response method gives a larger oscillation amplitude at the critical frequency, the model using the peak estimate gives the same peak amplitude of the process, and the model using LS estimate gives an intermediate value.

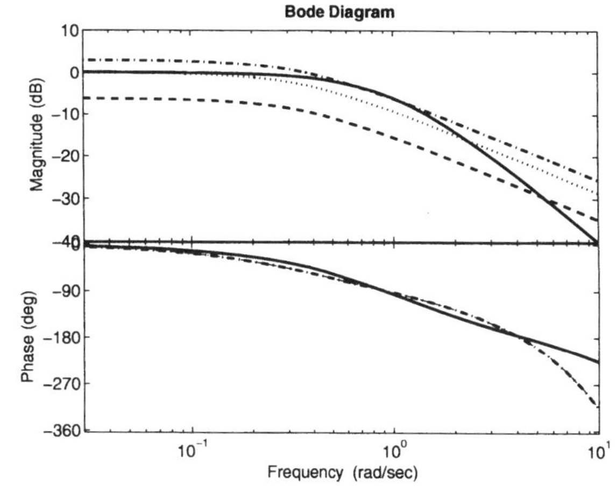

Fig. 2. Bode plot for example 5.1 (case $D/T = 0.1$): (−) process, (·) using step test, (−−) using output peak and (·−) using least squares.

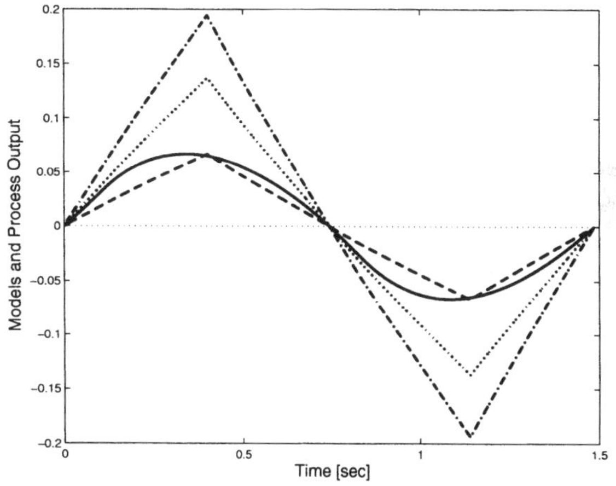

Fig. 3. Limit cycle for example 5.1 (case $D/T = 0.1$): (−) process, (·) using step test, (−−) using output peak and (·−) using least squares.

5.2 *Nonminimum Phase Processes*

Consider now the following inverse response model,

$$G(s) = \frac{1-s}{(s+1)^2} e^{-Ds} \, , \, D = 0.1; 1; 10 \, .$$

The results from the relay tests are presented in Table 3. Again, the obtained models are presented in the second row of the Table 2, and the least squares estimate is computed with the combined regression vector with data from the two relay feedback tests, using $N = 1000$ points per oscillation period. Note that in each estimate the dead-time is much greater than the real one. This can be seen as caused by the inverse response behavior, which resembles a delay in the output signal.

	Dead-time		
	0.1	1	10
h_1 (sec)	2.3068	3.3648	12.4773
h_2 (sec)	6.0446	7.9343	26.0000

Table 3. Limit cycle measurements for example 5.2

5.3 *Noise Effects*

Consider the following processes

$$G(s) = \frac{1}{s+1} e^{-Ds} \, , \, D = 0.1; 1; 10 \, .$$

Process		Estimates				
Structure	Dead-time	$\hat{D}$	$\hat{T}$	$\hat{K}_1$ (step)	$\hat{K}_2$ (peak)	$\hat{K}_3$ (LS)
	0.1	0.3969	2.6964	1.0000	0.4875	1.4158
$\frac{1}{(s+1)^2}e^{-Ds}$	1	1.6658	1.3458	1.0000	0.7024	0.9324
	10	10.9516	1.0484	1.0000	0.9998	0.9907
	0.1	1.4997	1.8024	1.0000	1.4005	1.2291
$\frac{(1-s)}{(s+1)^2}e^{-Ds}$	1	2.4140	1.6767	1.0000	1.4368	1.1284
	10	11.2988	1.7018	1.0000	1.4279	1.0152
	0.1	0.101	0.9927	0.9972	1.2098	0.9997
$\frac{1}{s+1}e^{-Ds}$	1	1.001	0.9991	0.9975	1.1943	0.9998
	10	9.994	1.0063	0.9974	1.2900	1.0005

Table 2. Results for examples 5.1, 5.2 and 5.3

The processes are now simulated including the effects of the noise at the process output, in order to verify the accuracy of the proposed procedure. The noise is added to the output of the process. In order to have the same signal-to-noise ratio, the relay amplitude is adjusted according to the desired output level. The output is recorded for three consecutive cycles, and the period is computed as an average of the obtained curve. The measurements are presented in Table 4. One period of the limit cycle for each case is shown in Fig. 4. The estimation results are presented in the third row of the Table 2.

	Dead-time		
	0.1	1	10
h_1 (sec)	0.1918	1.4908	10.6918
h_2 (sec)	1.0750	3.7512	22.0012

Table 4. Limit cycle measurements for example 5.3

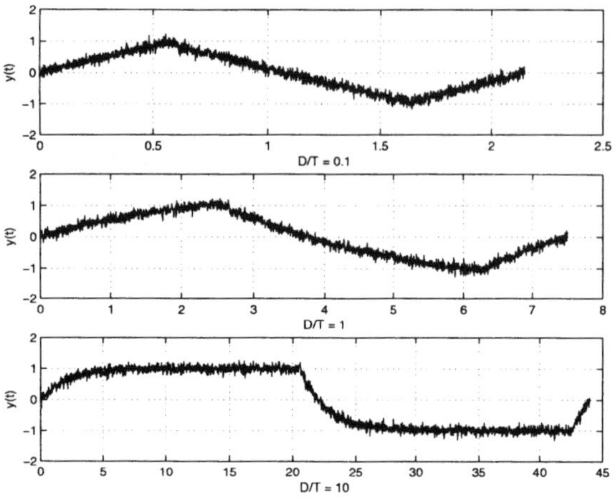

Fig. 4. Relay output for example 5.3.

6. CONCLUSIONS

Exact FOPDT modelling can be accomplished with relay feedback data if time domain analysis is applied. The results point out that in a standard relay feedback test, the dead-time and time constant are related to each other at the frequency of the limit cycle, if the model supposed to reproduce the process critical frequency in a relay feedback simulation. By performing a second relay feedback test with an integrator, one is able to obtain a pair of estimates of the dead-time and time constant which yields a family of models that matches the process at the frequencies where the phase is approximately $-\pi/2$ and $-\pi$. The model is finally obtained with an estimate of the process direct gain. Three alternatives for obtaining this estimate were presented and discussed, according to accuracy and implementation complexity. Simulation examples are used to illustrate the properties of the procedure in the presence of unmodelled dynamics and noise. Results point out that a good modelling is achieved in a frequency range that is relevant for process control.

7. REFERENCES

Åström, K. J. (1995). Oscillations in systems with relay feedback. In: *IMA Vol. Math. Appl.: Ad. Cont., Filt., Sig. Proc.*. pp. 1–25.

Åström, K. J. and T. Hägglund (1995). *PID Controllers: Theory, Design and Tuning*. 2nd ed.. Instrument Society of America. Research Triangle Park, North Carolina.

Kaya, I. and D. P. Atherton (2001). Parameter estimation from relay autotuning with asymmetric limit cycle data. *J. of Proc. Cont.* **11**, 429–439.

Ljung, L. (1999). *System Identification: Theory for the User*. 2nd ed.. Prentice Hall. Upper Saddle River, NJ.

Luyben, W. L. (1987). Derivation of transfer functions for highly nonlinear distillation columns. *Ind. and Eng. Ch. Res.* **26**, 2490–2495.

Luyben, W. L. (2001). Getting more information from relay-feedback tests. *Ind. and Eng. Ch. Res.* **40**, 4391–4402.

Tan, K. K., Q.-G. Wang, C. C. Hang and T. J. Hägglund (1999). *Advances in PID Control*. Springer-Verlag. London.

Varigonda, S. and T. T. Georgiou (2001). Dynamics of relay relaxation oscillators. *IEEE Trans. on Aut. Cont.* **46**(1), 65–77.

Wang, Q.-G., C.-C. Hang and B. Zou (1997). Low order modelling from relay feedback. *Ind. and Eng. Ch. Res.* **36**, 375–381.

Yu, C.-C. (1999). *Autotuning of PID Controllers: Relay Feedback Approach*. Springer-Verlag. London.

IFAC Publications
www.elsevier.com/locate/ifac

CONTINUOUS-TIME MODEL IDENTIFICATION OF SYSTEMS OPERATING IN CLOSED-LOOP

Marion Gilson and Hugues Garnier

Centre de Recherche en Automatique de Nancy (CRAN)
CNRS UMR 7039
Université Henri Poincaré, Nancy 1, BP 239
F-54506 Vandoeuvre-lès-Nancy Cedex, France
email: [marion.gilson, hugues.garnier]@cran.uhp-nancy.fr

Abstract: Schemes for system identification based on closed-loop experiments have attracted considerable interest in the last two decades. Most of the existing methods have been developed for discrete-time models. In this paper, various instrumental variable-based methods for identifying continuous-time models of systems operating in closed-loop are proposed and their performances are compared on the basis of numerical simulations. *Copyright © 2003 IFAC*

Keywords: continuous-time model identification, closed-loop identification, linear estimators, instrumental variable, bias-eliminated least-squares.

1. INTRODUCTION

System identification is an established field in the area of system analysis and control. It aims at determining particular models for dynamical systems based on observed inputs and outputs. Although dynamical systems in the physical world are native to continuous-time (CT) domain, most system identification schemes have been based in the past on discrete-time (DT) models without concern for the merits of the native continuous-time models. The development of continuous-time model based system identification techniques originated in the middle of the last century but was overshadowed by the overwhelming developments of discrete-time methods. This was mainly due to the 'go completely digital' trend that was spurred by parallel developments in digital computers. Interest in continuous-time approaches to system identification has however been growing in the last decade (see Sinha and Rao (1991) and Pintelon *et al.* (2000), Söderström and Mossberg (2000), Garnier *et al.* (2004) for more recent references).

The relevance of CT model identification methods has been recently illustrated with extensive simulation examples (Rao and Garnier 2002). However, there are many issues which have not received adequate attention so far in the case of CT model identification. One such issue is the identification of closed-loop systems.

Many results were established during the last decade in this area (Van den Hof 1998, Forssell and Ljung 1999). These methods can be broadly classified into three main approaches: direct, indirect and joint input/ouput identification. In this paper, a particular attention will be payed to the indirect approach which consists in firstly estimating the closed-loop model and secondly computing the process model by using the knowledge of the controller. It is well-known that the conventional least-squares (*ls*) method fails to produce consistent estimates in this approach when disturbances act in the loop. Further methods have therefore been developed to cope with this bias problem, as the instrumental variable type of estimators, or equivalently estimators based on the bias correction of *ls* estimates. They provide asymptotically unbiased plant parameter estimates in the closed-loop identification framework of systems subject to colored disturbances.

Moreover, they are computationally economical since they are based on linear regressions.

In this paper, instrumental variable (*iv*) and bias-eliminated least-squares (*bels*) methods are used to handle the identification problem of continuous-time models of linear dynamic systems operating in closed-loop. Moreover, a new but simple approach is proposed to eliminate the artificially introduced pre-filter in a recently developed *bels* method (Garnier *et al.* 2000*b*). The performances of *iv* based methods and proposed bias-compensating techniques are then compared on the basis of simulation examples.

2. PRELIMINARIES

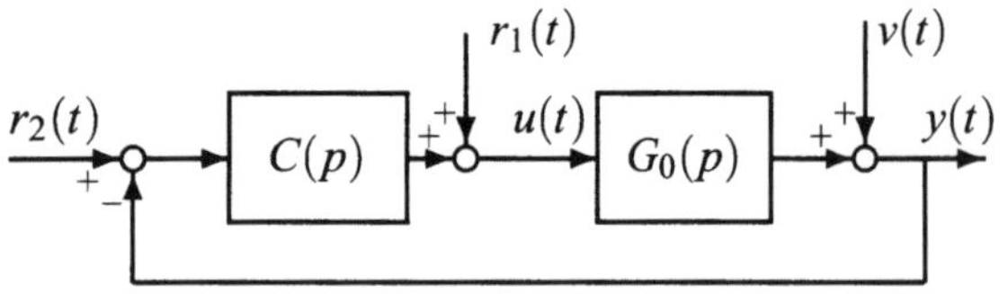

Fig. 1. Closed-loop configuration.

Consider a stable linear SISO closed-loop system shown in figure 1. The process is denoted with $G_0(p)$ and the controller with $C(p)$ where p is the differentiation operator ($p = d/dt$); $u(t)$ describes the process input signal, $y(t)$ the process output signal and $v(t)$ describes the colored disturbance acting on the loop. The external signals $r_1(t)$, $r_2(t)$ are assumed to be uncorrelated with the noise disturbance $v(t)$. For ease of notation we also introduce the reference signal $r(t) = r_1(t) + C(p)r_2(t)$. With this notation, the data generating system is assumed to be given by the relations

$$\mathscr{S} : \begin{cases} y(t) = G_0(p)u(t) + v(t) \\ u(t) = r(t) - C(p)y(t) \end{cases} \tag{1}$$

with $G_0(p) = B_0(p)/A_0(p)$. A parametrized process model is considered

$$\mathscr{G} : G(p,\theta) = \frac{B(p,\theta)}{A(p,\theta)} = \frac{b_0 p^{n_b} + b_1 p^{n_b-1} \cdots + b_{n_b}}{p^{n_a} + a_1 p^{n_a-1} + \cdots + a_{n_a}} \tag{2}$$

where n_b, n_a denote the orders of the numerator and denominator of the process respectively and with the pair (A,B) assumed to be coprime. The process model parameters are stacked columnwise in the parameter vector

$$\theta = \left[a_1 \ \cdots \ a_{n_a} \ b_0 \ \cdots \ b_{n_b}\right]^T \in \mathbb{R}^{n_a+n_b+1}. \tag{3}$$

The numerator and denominator orders n_b and n_a are supposed to be known. The controller $C(p)$ is also assumed to be known and given by

$$C(p) = \frac{Q(p)}{P(p)} = \frac{q_0 p^{n_q} + q_1 p^{n_q-1} \cdots + q_{n_q}}{p_0^{n_p} + p_1 p^{n_p-1} + \cdots + p_{n_p}} \tag{4}$$

with the pair (P,Q) assumed to be coprime. The closed-loop transfer function stemmed from (1) can be written as

$$y(t) = \frac{G_0(p)}{1 + C(p)G_0(p)} r(t) + v(t) \tag{5}$$

or in polynomial fraction form

$$y(t) = \frac{B^0_{cl}(p)}{A^0_{cl}(p)} r(t) + \frac{1}{A^0_{cl}(p)} \xi(t) \tag{6}$$

with $\xi(t) = A_0(p)P(p)v(t)$. The polynomials $B^0_{cl}(p)$ and $A^0_{cl}(p)$ have orders $n_\beta = n_b + n_p$ and $n_\alpha = n_a + n_p$ respectively. For parametrizing the closed-loop transfer function, the following model structure is used

$$y(t) = \frac{B_{cl}(p,\Theta)}{A_{cl}(p,\Theta)} r(t) + \frac{1}{A_{cl}(p,\Theta)} \xi(t) \tag{7}$$

$$B_{cl}(p,\Theta) = \beta_0 p^{\bar{n}_\beta} + \beta_1 p^{\bar{n}_\beta - 1} + \cdots + \beta_{\bar{n}_\beta} \tag{8}$$

$$A_{cl}(p,\Theta) = p^{n_\alpha} + \alpha_1 p^{n_\alpha - 1} + \cdots + \alpha_{n_\alpha} \tag{9}$$

and the closed-loop parameters are collected in the parameter vector

$$\Theta = \left[\alpha_1 \cdots \alpha_{n_\alpha} \ \beta_0 \cdots \beta_{\bar{n}_\beta}\right]^T \in \mathbb{R}^{n_\alpha + \bar{n}_\beta + 1}. \tag{10}$$

For $\bar{n}_\beta \geq n_\beta$ the closed-loop model structure will be flexible enough to exactly represent the reference to output transfer function in the closed-loop system (6). In the following, the closed-loop system is assumed to be asymptotically stable and $r(t)$ is a persistently exciting of sufficient high order signal.

3. CT MODEL IDENTIFICATION

There are two ways to obtain a CT model. The first is to estimate from the sampled data an initial DT model and then convert it into a CT model. The second approach consists in identifying directly a CT model from the DT data. In comparison with the DT counterpart, CT model identification raises several technical issues. The first point is related to implementation. Unlike the difference equation model, the differential equation model is not a linear combination of samples of only the measurable process input and output signals. It also contains input and output time-derivatives which are not available as measurement data in most practical cases. Various types of continuous-time filters have been devised to circumvent the need to reconstruct these time-derivatives (Sinha and Rao 1991), (Garnier *et al.* 2004). The CONtinuous-Time System IDentification (CONTSID) toolbox has been developed on the basis of these methods (Garnier *et al.* 2003).

Suppose that a causal stable analog filter with Laplace transfer function $F(s)$ of minimal order n_α is selected for the identification procedure. By passing both reference and output measurements $r(t)$ and $y(t)$ through this filter, the time-derivatives of the filtered signals may be obtained. This operation when applied to model (6) for time-instant $t = t_k$ yields

$$\sum_{i=0}^{n_\alpha} \alpha_i y_f^{(i)}(t_k) = \sum_{i=0}^{\bar{n}_\beta} \beta_i r_f^{(i)}(t_k) + \varepsilon(t_k) \tag{11}$$

where $\varepsilon(t_k)$ denotes the equation error and

$$y_f^{(i)}(t) = \mathscr{L}^{-1}[s^i F(s)Y(s)] \quad (12)$$

$$r_f^{(i)}(t) = \mathscr{L}^{-1}[s^i F(s)R(s)] \quad (13)$$

with $\mathscr{L}^{-1}$ denoting the inverse Laplace transform. For simplicity, it has been assumed that the differential equation model (6) is initially at rest. Note however that in the general case the initial condition terms do not vanish in equation (11). Whether they require estimation or they can be neglected depends upon the selected pre-processing method. There is a multitude of choice for the pre-filter. Four typical filters are as follows (Garnier *et al.* 2004):

$$F_1(s) = \left(\frac{\beta}{s+\lambda}\right)^{n_\alpha} \qquad F_2(s) = \left(\frac{\beta}{s+\lambda}\right)^{n_\alpha+1} \quad (14)$$

$$F_3(s) = \left(\frac{1}{s}\right)^{n_\alpha} \qquad F_4(s) = \left(\frac{1-e^{-lT_s s}}{s}\right)^{n_\alpha} \quad (15)$$

where $F_1(s)$ and $F_2(s)$ represent the filters used in the case of the minimal order multiple filter method and generalised Poisson moment Functional (*gpmf*) approach respectively; $F_3(s)$ denotes the usual multiple integral operation while $F_4(s)$ is referred to a linear integral filter (*lif*).

4. INDIRECT APPROACHES FOR CLOSED-LOOP IDENTIFICATION

The problem of indirect closed-loop identification can be stated as follows; under the assumptions described in section 2 and using available sampled data $Z^N = \{r(t_k); y(t_k)\}_{k=1}^N$, estimate consistently the CT parameters of the closed-loop system and then use the known regulator to compute the CT open-loop parameters of primary interest. In this second step, use is made of the linear relation connecting (open-loop) process parameters and closed-loop parameters

$$\Theta = M\theta + \rho \quad (16)$$

where ρ is a known vector and M is a known full-column rank matrix, given by

$$M = \begin{pmatrix} P_c & Q_c \\ 0 & \bar{P}_c \end{pmatrix} \in \mathbb{R}^{(n_\alpha+\bar{n}_\beta+1)\times(n_a+n_b+1)} \quad (17)$$

$$\rho = (p_1 \cdots p_{n_p}\ 0 \cdots 0)^T \in \mathbb{R}^{(n_\alpha+\bar{n}_\beta+1)} \quad (18)$$

$P_c \in \mathbb{R}^{n_\alpha \times n_a}$, $\bar{P}_c \in \mathbb{R}^{(\bar{n}_\beta+1)\times(n_b+1)}$, $Q_c \in \mathbb{R}^{n_\alpha \times (n_b+1)}$ are Sylvester matrices expanded by $\begin{bmatrix}1 & p_1 & \cdots & p_{n_p}\end{bmatrix}^T$ and $\begin{bmatrix}q_0 & q_1 & \cdots & q_{n_q}\end{bmatrix}^T$ respectively, e.g.

$$P_c = \begin{bmatrix} 1 & 0 & \cdots & 0 \\ p_1 & \ddots & \ddots & \vdots \\ \vdots & \ddots & \ddots & 0 \\ p_{n_p} & \vdots & \ddots & 1 \\ 0 & \ddots & \vdots & p_1 \\ \vdots & \ddots & \ddots & \vdots \\ 0 & \cdots & 0 & p_{n_p} \end{bmatrix} \quad \bar{P}_c = \begin{bmatrix} P_c \\ 0_{(\bar{n}_\beta+1-n_\alpha)\times(n_b+1)} \end{bmatrix}.$$

Three indirect approaches are presented in this section to estimate the CT parameters of a system operating in closed-loop. The first two consist in applying to the closed-loop case commonly used open-loop identification estimators in the first step of the indirect approach. The third proposed estimator is only devoted to the closed-loop case, since the knowledge of the controller is also required and taken into account for estimating the closed-loop parameters.

4.1 *Least-squares estimator*

To estimate the closed-loop parameters Θ, equation (11) can be reformulated using the filtered variables into standard linear regression form as

$$y_f^{(n_\alpha)}(t_k) = \phi_f^T(t_k)\Theta + \varepsilon(t_k) \quad (19)$$

with

$$\phi_f^T(t_k) = \left[-y_f^{(n_\alpha-1)}(t_k) \cdots -y_f(t_k)\, r_f^{(\bar{n}_\beta)}(t_k) \cdots r_f(t_k)\right] \quad (20)$$

From N available samples of the reference and output signals, the least-squares (*ls*) estimate that minimizes the sum on the squared errors is given by

$$\hat{\Theta}_{ls} = \left[\sum_{i=1}^{N} \phi_f(t_k)\phi_f^T(t_k)\right]^{-1} \sum_{i=1}^{N} \phi_f(t_k) y_f^{(n_\alpha)}(t_k) \quad (21)$$

provided that the inverse exists. The open-loop parameters can then be computed by solving equation (16) in a least-squares sense.

4.2 *Instrumental variable estimator*

It is however well known that the conventional least-squares method delivers biased estimates in presence of general cases of measurement noise. One of the simplest solutions to the asymptotic bias problem associated with the basic *ls* algorithm is to use instrumental variable (*iv*) methods since they do not require *a priori* knowledge of the noise statistics. A bootstrap estimation of *iv* type where the instrumental variable is built up from an auxiliary model can be considered as a first solution (Young 1970). The instrument is given by

$$\hat{\phi}_f^T(t_k) = \left[-\hat{y}_{r,f}^{(n_\alpha-1)}(t_k) \cdots -\hat{y}_{r,f}(t_k)\, r_f^{(\bar{n}_\beta)}(t_k) \cdots r_f(t_k)\right]$$

where

$$\hat{y}_{r,f}(t_k) = F(p)\hat{y}_r(t_k), \quad r_f(t_k) = F(p)r(t_k) \quad (22)$$

and $\hat{y}_r(t_k)$ is the noise-free output calculated from

$$\hat{y}_r(t_k) = \frac{B_{cl}(p, \hat{\Theta}_{ls})}{A_{cl}(p, \hat{\Theta}_{ls})} r(t_k). \quad (23)$$

The *iv*-based estimated parameters are then given by

$$\hat{\Theta}_{iv} = \left[\sum_{i=1}^{N} \hat{\phi}_f(t_k)\phi_f^T(t_k)\right]^{-1} \sum_{i=1}^{N} \hat{\phi}_f(t_k) y_f^{(n_\alpha)}(t_k) \quad (24)$$

provided that the inverse exists. As previously, the open-loop parameters can then be computed by solving (16) in a least-squares sense.

4.3 *Bias-eliminated least-squares estimator*

The bias-eliminated least-squares method (*bels*) for closed-loop identification as discussed in (Garnier *et al.* 2000*b*) is designed to provide an unbiased estimate for the CT process model $G(p,\theta)$, by only using linear regression techniques without noise modelling. The method comprises the following main steps:

- Estimate a CT model for the closed-loop system (6) on the basis of sampled data Z^N; this estimate is denoted by $\hat{\Theta}_{ls}$ and is given by equation (21).
- This estimate is biased due to the fact that $\varepsilon(t_k)$ in (19) will not be white noise; however the bias on $B_{cl}(p,\hat{\Theta}_{ls})/A_{cl}(p,\hat{\Theta}_{ls})$ can be estimated and subtracted from the closed-loop estimate.
- The corrected closed-loop parameter is converted to an open-loop process parameter by solving (16) in a least-squares (LS) sense.

In short

$$Z^N \underset{ls}{\longrightarrow} \hat{\Theta}_{ls} \underset{Computation}{\longrightarrow} \hat{\Theta}_{corr} \underset{LS}{\longrightarrow} \hat{\theta}_{bels}$$

The bias correction principle is based on the following reasoning. If the differential model structure is rich enough to capture all dynamics of the closed-loop system, then

$$\hat{\Theta}_{ls}(N) = \Theta_0 + \hat{R}^{-1}_{\phi_f\phi_f}(N)\hat{R}_{\phi_f\varepsilon}(N) \qquad (25)$$

where Θ_0 is the parameter vector of the true closed-loop plant, and $\hat{R}_{\phi_f\varepsilon}(N) = \frac{1}{N}\sum_{k=1}^{N}\phi_f(t_k)\varepsilon(t_k)$. Then, under minor regularity conditions on the data, the *ls* estimate $\hat{\Theta}_{ls}(N)$ is known to converge for $N \to \infty$ with probability 1 to

$$\Theta^*_{ls} = \Theta_0 + R^{-1}_{\phi_f\phi_f}R_{\phi_f\varepsilon}$$

with $R_{\phi_f\phi_f} = \bar{E}\phi_f(t_k)\phi_f^T(t_k)$ and $R_{\phi_f\varepsilon} = \bar{E}\phi_f(t_k)\varepsilon(t_k)$, where the notation $\bar{E}[.] = \lim_{N\to\infty}\frac{1}{N}\sum_{k=1}^{N}E[.]$ is adopted (Ljung 1999). As the noise disturbance $\varepsilon(t_k)$ is assumed to be uncorrelated with the reference signal $r(t_k)$, the bias in the asymptotic estimate is given by

$$\Delta^* := R^{-1}_{\phi_f\phi_f}R_{\phi_f\varepsilon} = R^{-1}_{\phi_f\phi_f}\begin{bmatrix} I_{n_\alpha} \\ 0 \end{bmatrix} R_{y_f\varepsilon} \qquad (26)$$

with $R_{y_f\varepsilon} = \bar{E}\{[-y_f^{(n_\alpha-1)}(t_k) \quad \cdots \quad -y_f(t_k)]^T \cdot \varepsilon(t_k)\}$. Based on this expression, an estimate for Δ^* is obtained by

$$\hat{\Delta}(N) = \hat{R}^{-1}_{\phi_f\phi_f}(N)\begin{bmatrix} I_{n_\alpha} \\ 0 \end{bmatrix}\hat{R}_{y_f\varepsilon}(N). \qquad (27)$$

The unknown $\hat{R}_{y_f\varepsilon}(N)$ in this relation can be obtained as follows. As matrix M in (16) has full column rank, there exists a full column rank matrix $H \in \mathbb{R}^{(n_\alpha+\bar{n}_\beta+1)\times(\bar{n}_\beta+n_p-n_b)}$ that satisfies $H^TM = 0$. Multiplying equation (25) by H^T and using equation (16) for Θ_0, it follows that

$$H^T\hat{R}^{-1}_{\phi_f\phi_f}(N)\begin{bmatrix} I_{n_\alpha} \\ 0 \end{bmatrix}\hat{R}_{y_f\varepsilon}(N) = H^T(\hat{\Theta}_{ls}(N) - \rho). \qquad (28)$$

This constitutes a set of $(\bar{n}_\beta + n_p - n_b)$ equations with n_α unknowns in $\hat{R}_{y_f\varepsilon}(N)$, requiring $\bar{n}_\beta \geq n_a + n_b$ to have at least as many equations as unknowns. There are two situations to be distinguished

- $n_p \geq n_a$. $\bar{n}_\beta$ is chosen according to $\bar{n}_\beta = n_\beta = n_b + n_p$, and equation (28) is an overdetermined set of equations ($2n_p$ equations, $n_\alpha = n_a + n_p$ unknowns) that is solved in a least-squares sense, leading to

$$\hat{\Delta}(N) = \hat{R}^{-1}_{\phi_f\phi_f}(N)\begin{bmatrix} I_{n_\alpha} \\ 0 \end{bmatrix}\left[H^T\hat{R}^{-1}_{\phi_f\phi_f}(N)\begin{bmatrix} I_{n_\alpha} \\ 0 \end{bmatrix}\right]^+ H^T\left[\hat{\Theta}_{ls}(N) - \rho\right] \qquad (29)$$

  with $[\cdot]^+$ denoting the matrix pseudo-inverse.
- $n_p < n_a$. By choosing $\bar{n}_\beta = n_\beta$ as previously, the number of equations in (28) is not sufficient to uniquely determine $\hat{\Delta}$. In (Garnier *et al.* 2000*b*) this is solved by applying an artificial prefilter to the reference signal such that a system with higher numerator degree is obtained. Then, the artificially augmented model estimation is achieved through a combined use of the known information of the controller and the designed prefilter. However, this solution denoted as *belsf* in the following (*f* stands for filtering), suffers from the drawback concerning the choice of this artificially introduced filter. To overcome this drawback, a solution consists in deliberately enlarging the number of parameters in the numerator of the closed-loop model, as recently suggested in (Gilson and Van den Hof 2001, Zheng 2001). Then the resulting augmented closed-loop system includes some known zero parameters. This consists in choosing $\bar{n}_\beta = n_a + n_b$, thus obtaining the situation that (28) is uniquely solvable for $\hat{R}_{y_f\varepsilon}(N)$. An estimate $\hat{\Delta}(N)$ can then be constructed according to

$$\hat{\Delta}(N) = \hat{R}^{-1}_{\phi_f\phi_f}(N)\begin{bmatrix} I_{n_\alpha} \\ 0 \end{bmatrix}\left[H^T\hat{R}^{-1}_{\phi_f\phi_f}(N)\begin{bmatrix} I_{n_\alpha} \\ 0 \end{bmatrix}\right]^{-1} H^T\left[\hat{\Theta}_{ls}(N) - \rho\right]. \qquad (30)$$

This solution will be denoted as *bels* in the following. The bias elimination can now be performed by constructing the corrected closed-loop parameter vector

$$\hat{\Theta}_{corr}(N) = \hat{\Theta}_{ls}(N) - \hat{\Delta}(N). \qquad (31)$$

Finally the plant parameter estimate $\hat{\theta}_{bels}$ is obtained by solving (16) in a least-squares sense

$$\hat{\theta}_{bels}(N) = (M^TM)^{-1}M^T(\hat{\Theta}_{corr}(N) - \rho). \qquad (32)$$

This resulting parameter estimate has been shown to be asymptotically unbiased.

Remarks.

- It has been proved in (Gilson and Van den Hof 2001) that in the DT case, this method is equivalent to a tailor-made instrumental variable. This result also holds in the CT case. The *bels* algorithm can thus be seen as a way to find the instrument to well-estimate the process parameters.

- Unlike several standard identification methods, the *bels* technique is able to provide satisfactory models for unstable plants while only using linear regressions and without modelling the noise contribution. Moreover, the version which does not use any prefilter for the bias estimation greatly simplifies the presentation, the understanding and the implementation of the algorithm.

5. IV METHOD FOR CLOSED-LOOP IDENTIFICATION

The closed-loop instrumental variable (*cliv*) method considered in this section is an adaptation to the CT model identification case of the method presented in (Söderström *et al.* 1987). It attempts to estimate the CT (open-loop) process parameters in one step, by using the available sampled data $Z^N = \{r(t_k); u(t_k); y(t_k)\}_{k=1}^N$.

The instrumental variable $\zeta_f(t_k)$ is defined as

$$\zeta_f^T(t_k) = \left[r_f^{(n_a+n_b)}(t_k) \cdots r_f(t_k) \right] \in \mathbb{R}^{n_a+n_b+1} \quad (33)$$

where $r_f(t_k) = F(p)r(t_k)$, with $F(p)$ of minimal order $n_a + n_b$. The estimation of the open-loop parameter is then determined by

$$\hat{\theta}_{cliv} = \left[\sum_{i=1}^{N} \zeta_f(t_k)\psi_f^T(t_k) \right]^{-1} \sum_{i=1}^{N} \zeta_f(t_k) y_f^{(n_a)}(t_k) \quad (34)$$

provided that the inverse exists and where the regressor is given by

$$\psi_f^T(t_k) = \left[-y_f^{(n_a-1)}(t_k) \cdots -y_f(t_k) u_f^{(n_b)}(t_k) \cdots u_f(t_k) \right]$$

The estimated parameters $\hat{\theta}_{cliv}$ are consistent since the instrument $\zeta_f(t_k)$ is uncorrelated with the disturbance but correlated with $y_f(t_k)$ and $u_f(t_k)$.

6. SIMULATION RESULTS

The numerical example with the same simulation conditions described in (Garnier *et al.* 2000*b*) has been used to compare the performances of the proposed approaches. From the comparative studies recently presented (Garnier *et al.* 2004), the generalized Poisson moment functionals (*gpmf*) approach can be considered as one of the more efficient method to handle the time-derivative problem. This latter has been therefore associated with the five estimators presented above. To illustrate the effectiveness of the algorithms and to investigate their performances, some Monte Carlo simulations of 200 runs with about 4000 data points have been performed. Note that in the case of the *belsf* method, since the controller is one order lower than the open-loop plant, a first-order prefilter $L(p) = 10/(p+10)$ has been chosen. The *gpmf* transform of minimal order 3 has been applied and the Poisson filter coefficients have been set to $\lambda = \beta = 1$ (see $F_2(s)$ in equation (14)).

Tables 1 and 2 report the mean and the standard deviations of the estimated closed and open-loop (OL) model parameters respectively. It can be seen from both tables that the parameter estimates obtained with the *lsgpmf* techniques, but also and more surprisingly with the *ivgpmf* methods are not very accurate with quite high standard deviation values. It is also seen that the two *bels*-based *gpmf* methods are slightly more accurate than the *clivgpmf* algorithm. The *belsgpmf* estimates are desirably accurate and are similar to those of the *belsfgpmf* method. These results demonstrate that without filtering, unbiased parameter estimates can be obtained by the proposed *bels* method. Bode diagrams of the 200 estimated CT models for the *clivgpmf* and the *belsgpmf* methods are plotted in figure 2 to 3. These plots in the frequency domain make it possible to further analyse the performances of the evaluated methods and confirm the slight superiority of the *bels*-based method.

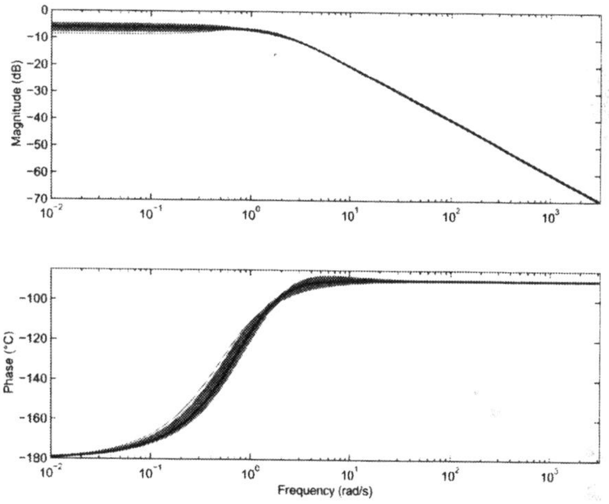

Fig. 2. Bode diagrams of the OL *clivgpmf* models

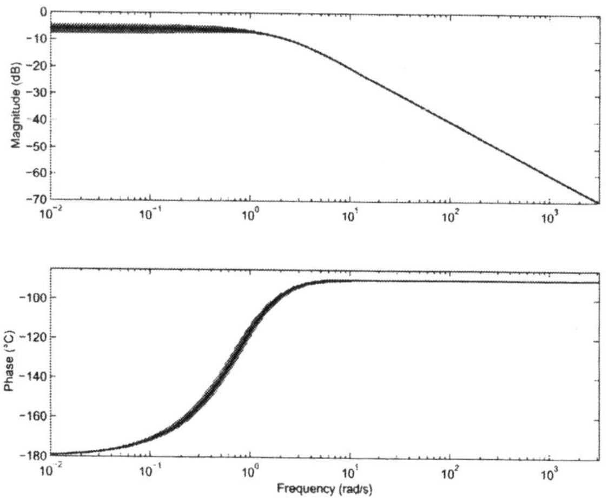

Fig. 3. Bode diagrams of the OL *belsgpmf* models

7. CONCLUSION

In this paper, the identification problem of continuous-time models of linear dynamic systems operating in

method	$\hat{\beta}_0 \pm \sigma_{\hat{\beta}_0}$	$\hat{\beta}_1 \pm \sigma_{\hat{\beta}_1}$	$\hat{\beta}_2 \pm \sigma_{\hat{\beta}_2}$	$\hat{\alpha}_1 \pm \sigma_{\hat{\alpha}_1}$	$\hat{\alpha}_2 \pm \sigma_{\hat{\alpha}_2}$	$\hat{\alpha}_3 \pm \sigma_{\hat{\alpha}_3}$
lsgpmf	0.998 ± 0.008	0.76 ± 1.03	0.009 ± 0.029	10.74 ± 1.11	20.61 ± 10.12	11.85 ± 14.09
ivgpmf	1.000 ± 0.013	1.09 ± 3.02	0.001 ± 0.078	11.11 ± 3.24	23.78 ± 29.53	16.30 ± 41.13
clivgpmf	0.995 ± 0.022	1.00 ± 0.24	0.000 ± 0.000	10.95 ± 0.58	22.92 ± 2.34	14.98 ± 3.53
belsfgpmf	1.000 ± 0.008	1.01 ± 0.10	0.000 ± 0.000	11.02 ± 0.17	23.08 ± 0.94	15.13 ± 1.49
belsgpmf	1.000 ± 0.008	1.01 ± 0.10	0.000 ± 0.000	11.02 ± 0.17	23.08 ± 0.94	15.13 ± 1.49
true value	1	1	0	11	23	15

Table 1. Mean and standard deviations of the closed-loop parameter estimates for 200 runs

method	$\hat{b}_0 \pm \sigma_{\hat{b}_0}$	$\hat{b}_1 \pm \sigma_{\hat{b}_1}$	$\hat{a}_1 \pm \sigma_{\hat{a}_1}$	$\hat{a}_2 \pm \sigma_{\hat{a}_2}$
lsgpmf	0.9976 ± 0.0080	0.7897 ± 0.9400	0.7666 ± 1.0642	-2.2462 ± 0.7721
ivgpmf	1.0003 ± 0.0130	1.0865 ± 2.7432	1.1046 ± 3.1307	-2.0926 ± 1.9971
clivgpmf	0.9947 ± 0.0221	0.9987 ± 0.2353	1.0076 ± 0.3617	-1.9869 ± 0.3334
belsfgpmf	1.0002 ± 0.0077	1.0089 ± 0.0995	1.0136 ± 0.1265	-2.0111 ± 0.1085
belsgpmf	1.0002 ± 0.0081	1.0087 ± 0.0991	1.0133 ± 0.1250	-2.0108 ± 0.1068
true value	1	1	1	-2

Table 2. Mean and standard deviations of the open-loop parameter estimates for 200 runs

closed-loop has been addressed by using closed-loop dedicated methods based on the instrumental variable and the bias-eliminated least-squares techniques. A new but simple approach has been proposed to eliminate the artificially introduced pre-filter in the *belsf* method. This simplifies the use of the *bels* based algorithm as the user is relieved from the task of designing a prefilter and pre-filtering the sampled data. The performances of the developed methods have been compared on the basis of a simulation example. The comparative analysis has shown that the three dedicated closed-loop methods give a very good accuracy of the parameter estimates. An interesting future research topic concerns the development of direct *bels* closed-loop identification approaches which could deliver accurate estimates without the knowledge of the controller.

8. REFERENCES

Forssell, U. and L. Ljung (1999). Closed-loop identification revisited. *Automatica* **35**(7), 1215–1241.

Garnier, H., M. Gilson and E. Huselstein (2003). Developments for the Matlab CONTSID toolbox. In: *SYSID 2003*. Rotterdam - Netherlands.

Garnier, H., M. Gilson and W.X. Zheng (2000*b*). A bias-eliminated least-squares method for continuous-time model identification of closed-loop systems. *International Journal of Control* **73**(1), 38–48.

Garnier, H., M. Mensler and A. Richard (2004). Continuous-time model identification from sampled data. Implementation issues and performance evaluation. *International Journal of Control*. To appear.

Gilson, M. and P. Van den Hof (2001). On the relation between a bias-eliminated least-squares (BELS) and an IV estimator in closed-loop identification. *Automatica* **37**(10), 1593–1600.

Ljung, L. (1999). *System identification : theory for the user - Second Edition*. Prentice-Hall.

Pintelon, R., J. Schoukens and Y. Rolain (2000). Box-Jenkins continuous-time modeling. *Automatica* **36**(7), 983–991.

Rao, G.P. and H. Garnier (2002). Numerical illustrations of the relevance of direct continuous-time model identification. In: *15th Triennial IFAC World Congress on Automatic Control*. Barcelona (Spain).

Sinha, N.K. and G.P. Rao (ed.) (1991). *Identification of continuous-time systems. Methodology and computer implementation*. Kluwer Academic Press. Dordrecht.

Söderström, T. and M. Mossberg (2000). Performance evaluation of methods for identifying continuous-time autoregressive processes. *Automatica* **36**, 53–59.

Söderström, T., P. Stoica and E. Trulsson (1987). Instrumental variable methods for closed-loop systems. In: *10th IFAC World Congress*. Munich - Germany. pp. 363–368.

Van den Hof, P.M.J. (1998). Closed-loop issues in system identification. *Annual Reviews in Control* **22**, 173–186.

Young, P.C. (1970). An instrumental variable method for real-time identification of a noisy process. *Automatica* **6**, 271–287.

Zheng, W.X. (2001). Parametric identification of linear systems operating under feedback control. *IEEE Transactions on Circuits and Systems - I: Fundamental theory and applications* **48**(4), 451–458.

IFAC Publications
www.elsevier.com/locate/ifac

MULTIVARIABLE CLOSED-LOOP SYSTEM IDENTIFICATION OF PLANTS UNDER MODEL PREDICTIVE CONTROL

E. de Klerk and I.K. Craig

*Department of Electrical, Electronic and Computer Engineering,
University of Pretoria, Pretoria, 0002, South Africa*

Abstract: This paper discusses a simulation where a multivariable plant under Model-based Predictive Control was identified from closed-loop data. A motivation for closed-loop system identification in this context is given and an identification methodology is proposed. To evaluate the consistency of the methodology, the plant was identified for different controller settings and different added disturbances. Different methods to ensure identifiability were also investigated. *Copyright © 2003 IFAC*

Keywords: simulation; closed-loop identification; multivariable; model-based predictive control; evaluation; validation.

1. INTRODUCTION

Model-based predictive control (MPC) constitutes a class of control algorithms that make direct use of a process model (Maciejowski, 2002). In recent times, MPC has become one of the dominant methods of advanced industrial process control.

Central to the success of the MPC technique is the derivation of accurate process models. It has been found that model accuracy degrades with time after the MPC controller has been commissioned. Consequently, controller performance is adversely affected. Therefore, periodic re-identification of the process models is necessary to ensure optimal long-term controller performance (Shouche, *et al.*, 1998).

Industrial project experience has shown that the most difficult and time-consuming work in MPC projects is modelling and identification (Zhu and Butoyi, 2002). In practice, during controller design, the process models are obtained by conducting open-loop step testing on the relevant process units (Zhu, 1998). Widespread applications of MPC technology call for a more effective and efficient method of multivariable system identification (SID).

The open-loop step testing approach works for stable processes, but the cost is very high. Stepping the manipulated variables may disturb product quality and long time constants consume much manpower and make production planning difficult. Furthermore, the tests are done manually, which dictate extremely high commitment of engineers and operators (Zhu, 1998). When the process is non-linear, ill-conditioned or sensitive then it is also very difficult to carry out open-loop tests. The process may even be unstable in open-loop, which makes open-loop tests undesirable (Forssell and Ljung, 1999).

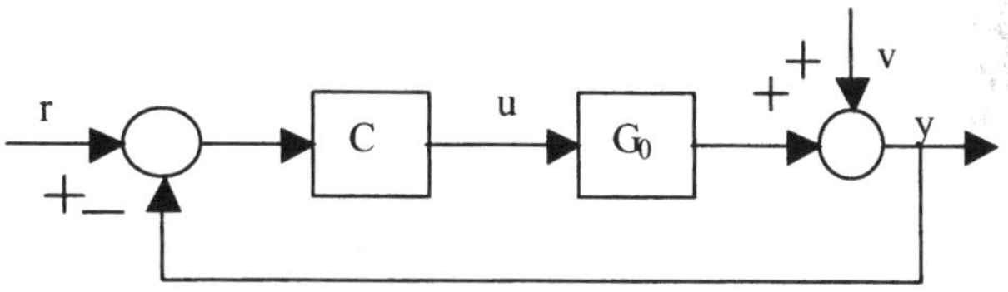

Fig. 1. A closed-loop system.

Closed-loop identification of the process models may well address some of these issues (De Klerk and Craig). This SID technique is less intrusive and may reduce re-identification time considerably (Forssell, 1999). According to Landau (2001), plant models identified in closed-loop may also provide better models for controller design.

2. CLOSED-LOOP SYSTEM IDENTIFICATION

Closed-loop SID refers to the process in which plant models are identified using data collected from closed-loop experiments, where the underlying process is fully or partly under feedback control, as in Fig. 1 (Ljung, 1999). In this configuration the signal r is the set-point. This signal is uncorrelated with v, which is a filtered white noise disturbance, i.e. $v(t)=He(t)$. G_0 and C are the plant and controller respectively. The closed-loop system equations are:

$$y(t) = G_0 C S_0 r(t) + S_0 v(t), \qquad (1)$$

$$u(t) = S_0 C r(t) - C S_0 v(t), \qquad (2)$$

with the sensitivity function $S_0=(1+CG_0)^{-1}$.

2.1 System Identification Steps

SID deals with the construction of models from data. According to Forssell (1999), the SID problem can be divided into a number of subproblems:

- experiment design,
- data collection,
- model structure selection,
- model estimation, and
- model validation.

These steps are also applicable to closed-loop SID. However, most of the standard open-loop estimation methods fail when applied directly to closed-loop data. The main reason for this is the correlation between the input u and the additive output noise v, which models all disturbances in the system. The prediction error method (PEM) is one of the few methods that work in the presence of this correlation (Forssell, 1999).

Ljung (1999) states that a PEM method will consistently estimate a system if the data are informative and the model set contains the true system, irrespective if the data have been collected under feedback or not.

Although called conventional by many researchers, the PEM method is a more powerful methodology than many newly proposed schemes for use in closed-loop SID for MPC (Zhu, 1998). Thus, this estimation method is used in the proposed SID methodology.

2.2 Closed-loop Identification Approach

The possible closed-loop SID approaches can be divided into three categories (Ljung, 1999):

Direct Approach: In the direct approach the basic PEM method is applied in a straightforward manner. The output y of the process and the input u are used in the same way as for open-loop SID, ignoring any possible feedback, and not using the reference signal r to estimate the model.

Indirect Approach: The closed-loop system is identified from reference input r to output y. From this identified closed-loop system the open-loop system (plant) is retrieved making use of the known controller.

Joint Input-Output Approach: The output y of the process and the input u of the plant are considered as outputs of a system driven by the reference input r and noise v. Knowledge of the system and the controller is recovered from this joint model.

3. IDENTIFICATION OF MPC CONTROLLED PLANTS

In this section the different options regarding the identification methodology are considered.

3.1 Identification Approach

Common assumptions in closed-loop identification are that the existing controller is linear and the process is single variable. Practical MPC controllers are usually non-linear, because, invariably, plant inputs and outputs are constrained. Also, plants are often multivariable, making many closed-loop SID results unsuitable for MPC applications (Zhu and Butoyi, 2002).

Forssell (1999) states that the indirect and the joint input-output approaches are typically only used with linear feedback. Non-linear controllers will cause considerably more work. MPC is also complex, since it computes the control action at each sampling interval, using constrained optimisation (Zhu and Butoyi, 2002). This complexity makes these methods even less attractive for MPC.

One variant of the joint input-output approach that does work with non-linear controllers is the projection method (Forssell and Ljung, 2000). However, this method depends on the derivation of non-casual Finite Impulse Response (FIR) models, which are nonparametric. According to Zhu *et al.* (2000), parametric models are better for use in industrial process identification, since they are more user friendly, more accurate, and require shorter test times than nonparametric models. Thus, for industrial processes, which are the type of processes usually controlled by MPC, the projection method is not ideal.

The direct closed-loop SID approach is the obvious choice when MPC controllers are involved. According to Ljung (1999), this approach should be seen as the natural approach to closed-loop data analysis. The main reasons are:

- this method works regardless of the complexity of the controller,
- no special algorithms and software are required,
- consistency and optimal accuracy are obtained if the model structure contains the true system (including the noise properties), and
- unstable systems can be handled without problems, as long the closed-loop system is stable and the predictor is stable.

A stable predictor means that any unstable poles of the plant model G must be shared by the noise model H. Parametric models like AutoRegressive with eXternal Input (ARX) and ARMAX satisfy this constraint (Ljung, 1999).

The only drawback with the direct approach is that a good noise model is needed in order to prevent a bias in the estimated plant model (Forssell and Ljung, 1999).

3.2 Guarantee of Identifiability

The purpose of feedback is to make the sensitivity function So small; especially at frequencies where the disturbance signals v have energy. Feedback will, thus, worsen the measured data's information about the system at these frequencies (Ljung, 1999).

In closed-loop an experiment will, however, still be informative in each of the following situations (Van den Hof, 1998):

- the reference signal is persistently exciting (PE),
- the controller is of sufficiently high order, or
- the controller switches between several settings during the experiment.

Although a multivariable MPC changes its structure as it deals with changes in the active constraints, this is unpredictable under normal operation and cannot guarantee a given number of changes in the controller settings. Making deliberate changes to the MPC may have unforeseen consequences for process operation, and it does not provide the same control over the signal-to-noise ratio (SNR) as adding external test signals (Doma, *et al.*, 1996). This SNR influences the size of the bias. The best way to satisfy the identifiability condition is, thus, to ensure that the reference signal is PE.

An approach to simultaneous constrained MPC and identification (MPCI) was developed by Shouche *et al.* (1998). In this approach, a persistent identification criterion is used as an additional constraint in the standard on-line optimisation of MPC.

Case studies were done by applying: MPCI; MPC with external dithering (Pseudo-Random Binary Signals (PRBS) were added to control signals); and MPC with no external input (constant references). The first two cases gave good results due to the process excitation by the controller in MPCI, and the external dithering signals in MPC. However, the case of the MPC without excitation gave very poor parameter estimates due to the lack of any information about the dynamics of the process (Shouche, *et al.*, 1998). This also demonstrates the need for PE reference signals.

Just recently, a novel approach to direct closed-loop identification, called output inter-sampling has been introduced. It is claimed that by using the inter-sampled plant input-output data, traditional restrictive identifiability conditions are removed (Sun *et al.*, 2001). A simulation study was conducted, which showed that, in general, this method delivers a large variance in the model.

3.3 Model Structure

For closed-loop identification, the choice of model structure depends on three, often conflicting, issues (Zhu, 1998):

- the compactness of the model, i.e. number of parameters needed,
- the numerical complexity of the parameter estimation, and
- the consistency of the model in closed-loop identification.

Parametric models, such as the ARX and the ARMAX, are much more compact than nonparametric models, such as FIR models. Although the ARMAX model is also more compact than the ARX model, the numerical complexity when estimating ARMAX models is much higher than with ARX models, since non-linear optimisation routines are needed. These optimisation routines often suffer from local minima and convergence problems when identifying multivariable processes (Zhu, 1998).

As already stated, a parametric model is a better choice in the modelling of industrial processes. Since many of these processes may also be unstable, it is natural to choose from these models the ARX or ARMAX model, because these models give stable predictors.

Since the ARMAX model might suffer from local minima and convergence problems, the ARX model is the best choice, provided that the noise model is accurate for the process to be modelled.

3.4 Model Validation

Model validation deals with the question of whether the best model is also *good enough* for its intended use (Forssell, 1999).

Standard validation tools are *residual analysis* and *cross-validation*, where the model is simulated using validation data and where the output is compared to measured output data (Ljung, 1999). These tools can also be used in closed-loop. The validation data should be a set of measured closed-loop data, not used in the estimation.

Since the aim of this closed-loop SID methodology is to substitute the open-loop approach, the identified models should also be compared to models identified in open-loop: e.g. visual comparison in the time and frequency domain; and comparison of *residual analysis* and *cross-validation* results.

The closed-loop identified model can be more accurate than the open-loop identified model, since the models' frequency weighting are very different (Ljung, 1999). Thus, when the true model is known, the open-loop and closed-loop identified models should also be compared, by evaluating how close these models are to the true model. A numerical value that can be used for comparison is the relative norm of the frequency magnitude responses:

$$frekfit = \left\| \frac{\left|\hat{G}_N(\omega)\right| - \left|G_0(\omega)\right|}{\left|G_0(\omega)\right|} \right\|_2 , \qquad (3)$$

$$\|x\|_2 = \sqrt{\sum_{i=1}^{n} |x_i|^2} . \qquad (4)$$

Here $\left|\hat{G}_N(\omega)\right|$ is the magnitude of the estimated plant model in the frequency domain and $G_0(\omega)$ is the true plant model. Similarly, in the time domain, the relative norm of the step response coefficients can be computed:

$$stepfit = \sqrt{\sum_{i=1}^{n} \left| \frac{\hat{s}_i - s_{0i}}{s_{0i}} \right|^2} . \qquad (5)$$

Here $\hat{s}_i$ is the ith estimated step response coefficient and s_{0i} is the true value. The model that gives the smallest value for Eqns. 3 and 5 has the best fit.

The goal of model validation is to test whether the model is good enough for its purpose (Zhu and Butoyi, 2000). The purpose of the model in question is to design an acceptable MPC controller, which in turn will ensure good closed-loop control. Therefore, the test is to demonstrate the acceptable performance of the controller (Ljung, 1999).

4. SIMULATION SET-UP

A multivariable MPC controlled plant was identified from simulated closed-loop data. To evaluate the consistency of the methodology, the plant was identified for different controller settings and different added disturbances. Different methods to ensure identifiability were also investigated.

4.1 The Plant

The plant, given in Eqn. 6, is a benchmark example used in the MATLAB MPC toolbox (Morari and Ricker, 1995). This linear two-input two-output plant aids in demonstrating the proposed methodology.

$$\begin{bmatrix} y_1(s) \\ y_2(s) \end{bmatrix} = \begin{bmatrix} \dfrac{12.8e^{-s}}{16.7s+1} & \dfrac{-18.9e^{-3s}}{21.0s+1} \\ \dfrac{6.6e^{-7s}}{10.9s+1} & \dfrac{-19.4e^{-3s}}{14.4s+1} \end{bmatrix} \begin{bmatrix} u_1(s) \\ u_2(s) \end{bmatrix} \quad (6)$$

4.2 Controller

Table 1 gives the final parameters chosen for the MPC controller. The MPC toolbox function, *mpccon*, was used for the design of the unconstrained controllers and *cmpc* was used for the design of the constrained controllers. These functions make use of quadratic optimisation. The cmpc function solves the quadratic problem iteratively, which results in a non-linear controller (Morari and Ricker, 1995).

4.3 Case Scenarios

Data sets from the following different scenarios were used for identification:

Unconstrained Control Law:

- case 1: no disturbances,
- case 2: pulse disturbance added to $u_1(1)$,
- case 3: output pulse disturbance added to $y_1(1)$,
- case 4: step disturbance added to $u_1(1)$,
- case 5: output step disturbance added to $y_1(1)$,
- case 6: saturation limits on the manipulated variables,
- case 7: no disturbances to the system with $\mathbf{r_i(t)=0}$,
- case 8: output step disturbance added to $y_1(1)$ with $\mathbf{r_i(t)=0}$,
- case 9: saturation limits on the manipulated variables with output step disturbance added to $y_1(1)$, $\mathbf{r_i(t)=0}$ and the **output inter-sampled,**

Table 1: Controller parameters.

Execution Time	1s
Output Weights	[1 1]
Input Weights	[1 1]
Prediction Horizon	6
Number of control moves	2
Saturation limits (cases 6, 9)	$\Delta u_1<0.1$, $\Delta u_2<0.05$,
Input constraints (case 10)	$u_i \in [-0.5,0.5], i=1,2$
Output constraints (cases 11, 12, 13)	$y_i \in [-1.5,1.5], i=1,2$

Constrained Control Law:

- case 10: enforced hard bounds on the manipulated variables
- case 11: enforced hard bounds on the output variables,
- case 12: enforced hard bounds on the output variables with $\mathbf{r_i(t)=0}$, and
- case 13: enforced hard bounds on the output variables with output step disturbance added to $y_1(1)$ and $\mathbf{r_i(t)=0}$.

Cases 1-6 and 10-11 were used for evaluation of the consistency of the proposed methodology: the plant was identified for different settings in the controller (cases 1, 6, 10 and 11), as well as for different added disturbances (cases 1-5). Structured tests were performed, by adding external test signals to the reference inputs. This ensured good SNRs and PE reference signal. SNR refers to the ratio between the noise and the plant input signal *u*.

In cases 7-9 and 12-13 no structured tests were performed. The reference signals were zero and not PE. Other methods to ensure identifiability were considered. In cases 7 and 8, no identifiability condition was satisfied. In case 9 the outputs were inter-sampled and the plant was, thus, identifiable. In cases 12 and 13 the plant was also identifiable, since the controller was constrained and, thus, non-linear. In case 13 a disturbance was added, to evaluate the influence of the SNR with non-linear feedback.

5. IDENTIFICATION STEPS

In this section the proposed methodology is summarised with a discussion of the five SID steps.

Since the direct closed-loop SID approach do not involve any reference signals, or controller information, the MATLAB SID Toolbox (Ljung, 1995), without any custom-written algorithms, was suitable for the implementation of the methodology.

5.1 Experiment Design

Adding a PRBS to the control signal and to the reference signal can have different effects due to the integrator inherent in MPC, e.g. PRBS added to the control signal will result in zero SNR at steady state making estimation of the steady state gain difficult.

To guarantee informative data, PE PRBS test signals were added to the reference signals. The period of the PRBS signals was taken as a tenth of the slowest time constant, i.e. 2s. In the cases where other guarantees for identifiabality were evaluated, the reference signals were kept at zero.

5.2 *Data Collection*

The data were sampled every second. In case 9 where the inter-sampling method was evaluated, the sampling time was 0.5s. Since the direct approach was implemented, the collected data sets consisted of manipulated variables u_i and output variables y_i.

5.3 *Model Structure Selection*

The ARX model structure was used. Since this methodology will be used in re-identification, one can assume that an old model exists, which gives an indication of the model order. Thus, the known order (first order) of the model was used. In cases where the order is unknown, it is better to identify higher order models and then reduce the order later on (Zhu, 1998).

5.4 Model Estimation

The *idarx* command in MATLAB, which estimates multivariable ARX models, was used. This function uses the least-squares estimation method, which is a PEM method (Ljung, 1995).

5.5 Model Validation

For model validation the standard validation methods were used: *residual analysis* and *cross-validation.* For validation of the methodology the followings tests were used:

Comparison with Open-Loop Identified Model:

- Bode magnitude, Bode phase, step and impulse responses were *visually* compared,
- *residual analysis* results were compared, and
- *cross-validation* results were compared.

Comparison of Both Models with True Model: The *frekfit* and *stepfit* values were computed using Eqns. 3 and 5.

Examination of Closed-Loop System: An MPC controller was designed from each of the identified models and the closed-loop responses were evaluated. The stability of the closed-loop systems were also evaluated by determining if all the discrete poles were within the unit circle.

6. RESULTS

In all the cases where structured tests with PE reference signals and good SNRs were used (cases 1-6 and 10-11) satisfactory models were identified: In the *residual analysis,* the functions of the auto-correlation and cross-correlation of the errors with the outputs stayed within the confidence bounds, similarly to the open-loop identified model. This means that the errors are white and the models are unbiased (Ljung, 1999). The *cross-validation* analysis showed that the simulated outputs followed the true outputs closely. The percentages of fit for output y_1 are shown in Fig. 2. The percentages of fit were similar for the open-loop identified model. The *visual comparison* also showed that these models corresponded very well in both time and frequency domain with the open-loop identified model. Figs. 3 and 4, in which the *frekfit* and *stepfit* values are plotted, show that for cases 1-6 and 10-11 these values are low and the models are, thus, close to the true model. In Fig. 5 the *system response* for a controller designed from the model identified in case 11 is shown. Cases 1-6, and 10 delivered similar results. This shows that these models ensured good *controller performance*.

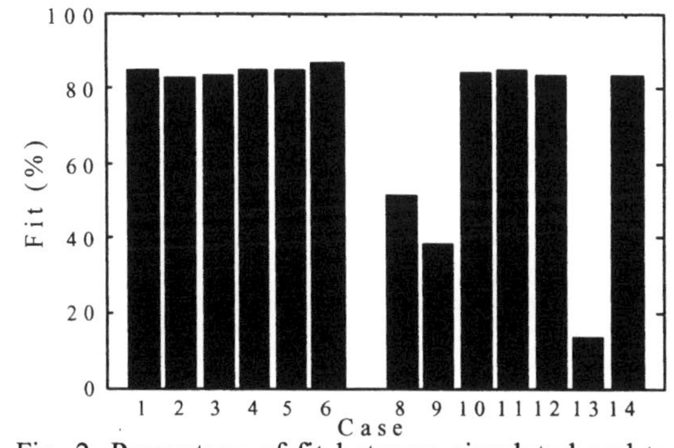

Fig. 2. Percentage of fit between simulated and true output y_1. Case 14 is the open-loop identified model.

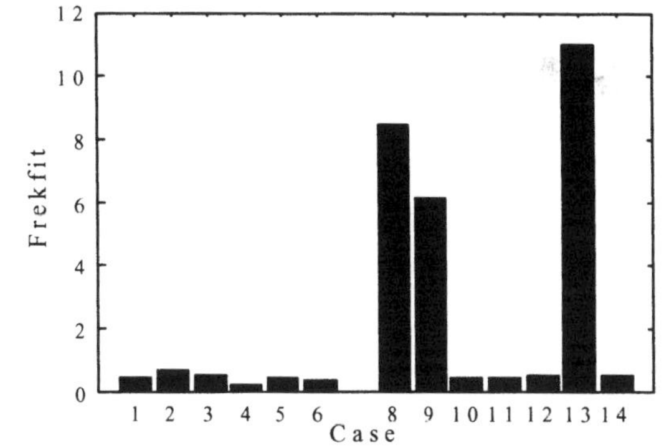

Fig. 3. The added *frekfit* values of the SISO magnitude responses. Case 14 is the open-loop identified model.

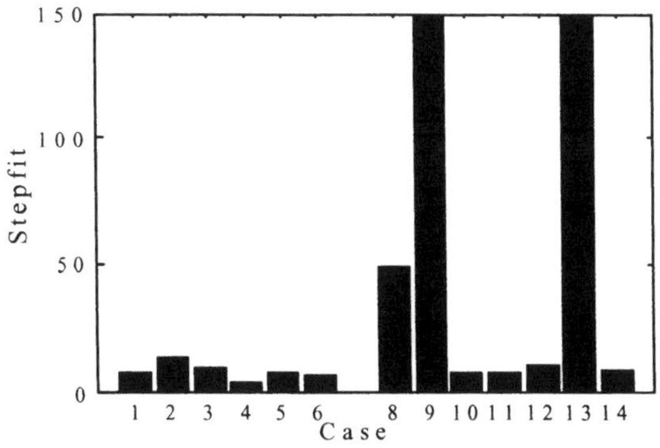

Fig. 4. The added *stepfit* values of the SISO step responses. The maximum value was taken as 150. Case 14 is the open-loop identified model.

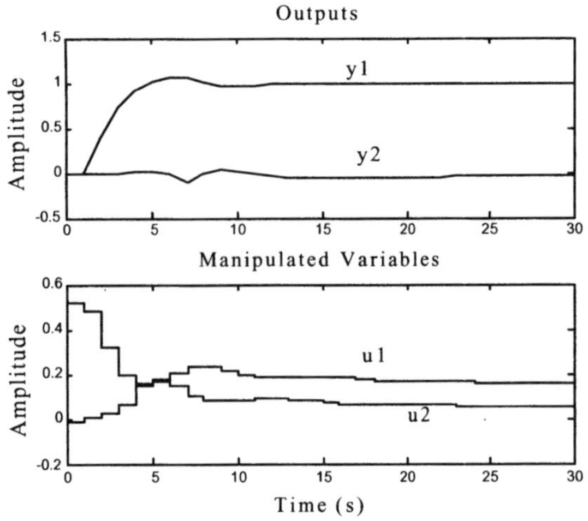

Fig. 5. The closed-loop response for a controller designed from the closed-loop identified model in case 11 with $r_1(t)=1$ and $r_2(t)=0$.

All the cases that delivered unsatisfactory results had $r_i(t)=0$ (cases 7, 8, 9 and 13). In case 7, with the linear controller and no disturbances, zero input-output signals resulted; therefore, no model could be identified. In case 8 the data were not informative enough. In case 9, where the inter-sampling method ensured identifiability, an imprecise model was identified, as expected from the variance simulation study. The only exception was case 12 with the constrained controller and no disturbance. Here the non-linearity of the controller ensured identifiability and the good SNR ensured a precise model. Since the changes in the input-output signals were very small, the SNR became unacceptable in case 13 with the added disturbance and an imprecise model resulted. Fig. 2 shows that cases 8, 9 and 13 have low percentages of fit, while case 12 has a high percentage. Figs. 3 and 4 also show that cases 8, 9 and 13 have large *frekfit* and *stepfit* values and are, thus, very different from the open-loop identified model, while case 12 still has low values.

7. CONCLUSION

From these simulation results, it can be concluded that the proposed closed-loop SID methodology gives reliable results for multivariable MPC controlled plants, irrespective of system disturbances and constraints, as long as the reference signals are PE and the SNRs are good.

Other methods that ensure identifiability, e.g. inter-sampling and non-linear controllers, do not guarantee precise models, if the SNR is not good, which is possible when no structured tests are performed. Structured tests should, thus, be conducted to ensure good SNRs.

It would be helpful in future to investigate how the choices of MPC tuning parameters and the presence of constraints impact the goodness of fit in the identification procedure.

REFERENCES

De Klerk, E. and I.K. Craig (2003). A laboratory experiment to teach closed-loop system identification. Accepted for publication in *IEEE Transactions on Education*.

Doma, M.J., P.A. Taylor and P.T. Vermeer (1996). Closed loop identification of MPC models for MIMO processes using genetic algorithms and dithering one of the variables at a time: application to an industrial distillation tower. *Computers & Chemical Engineering*, **Vol. 20**, pp. 1034-1040.

Forssell, U. and L. Ljung (2002). Projection method for closed-loop identification. *IEEE Transactions on Automatic Control*, **Vol. 45**, pp. 2101-2106.

Forssell, U. and L. Ljung (1999). Closed-loop identification revisited. *Automatica*, **Vol. 35**, pp.1215-1241.

Forssell, U. (1999). *Closed-loop identification methods, theory, and applications, dissertation No. 566*. Department of Electrical Engineering, Linköping University, Linköping, Sweden.

Landau, I.D. (2001). Identification in closed loop: a powerful design tool (better design models, simpler controllers). *Control Engineering practice*, **Vol. 9**, pp. 51-65.

Ljung, L. (1999). *System identification theory for the user, second edition*, chapter 14. Prentice Hall, New Jersey, USA.

Ljung, L. (1995). *System identification toolbox: for use with MATLAB*. The MathWorks Inc., Mass, USA.

Maciejowski, J.M. (2002). *Predictive control with constraints First Edition*, chapter 1. Prentice Hall, New York, USA.

Morari, M. and N.L. Ricker (1995). *Model predictive control toolbox: for use with MATLAB*. The MathWorks Inc., Mass, USA.

Shouche, M., H. Genceli, P. Vuthandam and M. Nikolaou (1998). Simultaneous constrained model predictive control and identification of DARX processes. *Automatica*, **Vol. 34**, pp. 1521-1530.

Sun, L., H. Ohmori and A. Sano (2001). Output intersampling approach to direct closed-loop identification. *IEEE Transactions on Automatic Control*, **Vol. 46**, pp. 1936 - 1941.

Van den Hof, P. (1998). Closed-loop issues in system identification. *Annual Reviews in Control*, **Vol. 22**, pp. 173-186.

Zhu, Y.C. and F. Butoyi (2002). Case studies on closed-loop identification for MPC, *Control Engineering Practice*, **Vol. 10**, pp. 403-417.

Zhu, Y.C., E. Arrieta, F. Butoyi and F. Cortes (2000). Parametric versus nonparametric models in MPC process identification. *Hydrocarbon Processing*, **Vol. 79**.

Zhu, Y.C. (1998). Multivariable process identification for MPC: the asymptotic method and its applications. *Journal of Process Control*, **Vol. 8**, pp. 101-115.

IFAC Publications
www.elsevier.com/locate/ifac

DEAD TIME MEASUREMENT OF CLOSED LOOP SYSTEM BY WAVELET

Tetsuya Tabaru * and Seiichi Shin *

* *Graduate School of Information Science and Technology,
the University of Tokyo, 7-3-1 Hongo, Bunkyo-ku, Tokyo, 113-8656, JAPAN
Phone: +81-3-5841-7670, E-mail: baru@axis.t.u-tokyo.ac.jp*

Abstract: This paper shows that the wavelet based dead time measurement method, which has been already studied for open loop systems, is also applicable to closed loop systems. The method uses a wavelet transform of a cross correlation function between an input and an output. To achieve our objective, the cross correlation function is derived for the closed loop case and its wavelet transform is analyzed under a certain condition. *Copyright © 2003 IFAC*

Keywords: wavelet analysis, dead time, closed loop system, cross correlation

1. INTRODUCTION

We have proposed a dead time (pure time delay, transport delay) measurement method by wavelet (Tabaru et al., 1997; Tabaru et al., 2000). The method measures a dead time of a linear system from a wavelet transform of a cross correlation function between an input and an output of the system. It has some advantages over conventional dead time measurement methods.

(1) Measurable even if a degree and a relative degree of a system are unknown.
(2) Special inputs aren't required, such as step input and white noise, which are used in some conventional methods.
(3) Robust to disturbances if they have no correlation with inputs.

Therefore it is efficient to use the method as the first step of system identification(Nakano et al., 2002), especially when there is few information known priorly about a measured target.

The proposed method is benetifical even when other method (*e.g.* prediction error based methods) is applied for a dead time measurement. Our method's estimate will be good prior information. Moreover, it will be useful to compare the estimate with other method's one in the validation process.

The preceding studies (Tabaru et al., 1997; Tabaru et al., 2000) have provided the reason why the method can measure the dead time of the system when it is a open loop system. However, no analysis has been given for the case of a closed loop system regardless of its importance. Since there are many systems that can be operated under only closed loop, the analysis should be developed for such a case.

This paper analyzes a cross correlation function between an input and an output and its wavelet transform for a closed loop system. The result proves that our dead time measurement method is also efficient for the closed loop case. It means that the same method is applicable to both open and closed loop systems. This is an additional merit since it is not necessary to know whether a target system is a part of a closed loop system or not.

Section 2 gives a brief explanation about wavelet transform and a relation between a wavelet transform of a cross correlation function and a cross spectrum density. Section 3 introduces our method and review the analysis of the method for the open loop case. In section 4, an analysis for the closed loop case will be provided to show that the method can be applied to closed loop systems also. In section 5, numerical simulations show validity of our discussion. Section 6 concludes this paper.

Notation: The set of real numbers will be represented by $\boldsymbol{R}$ and the set of complex numbers by $\boldsymbol{C}$. The asterisk $*$ denotes complex conjugation and $\angle$ denotes a phase (argument) value. The inner product in time domain is defined by $\langle x(t), y(t)\rangle = \int_{-\infty}^{\infty} x^*(t)y(t)dt$. The Fourier transform of a signal $x(t)$ is defined by $X(\omega) = \int_{-\infty}^{\infty} x(t)e^{-j\omega t}dt$. The cross correlation function between $u(t) \in \boldsymbol{R}$ and $y(t+b) \in \boldsymbol{R}$ is denoted by $\phi_{u(t),y(t+b)}(\tau)$. If $b=0$, it is simply denoted by $\phi_{uy}(\tau)$. Thus $\Phi_{u(t),y(t+b)}(\omega)$, which is the Fourier transform of $\phi_{u(t),y(t+b)}(\tau)$, is the cross spectrum density between

$u(t)$ and $y(t+b)$. It is also denoted by $\Phi_{uy}(\omega)$ if $b=0$. Auto correlation function and power spectrum density are defined similarly by replacing $y(t)$ with $u(t)$.

2. WAVELET TRANSFORM OF CORRELATION FUNCTION AND CROSS SPECTRUM

The objective of this section is to introduce a property of a wavelet transform of a cross correlation function, in particular, a relation with a cross spectrum density. This relation plays an important role to analyze our proposed method. At first, a brief explanation is given about the wavelet transform (Kaiser, 1994; Meyer, 1993), then the property will be described.

2.1 *Basics of Wavelet Transform*

The *wavelet transform* is defined by a family of functions that are dilation and translation of a unique function $\psi(t)$. This mother function $\psi(t)$ is called an *analyzing wavelet* and the family of functions is given by

$$\psi_{a,b}(t)=\frac{1}{\sqrt{a}}\psi\left(\frac{t-b}{a}\right),\quad a,b\in\boldsymbol{R},a>0. \quad (1)$$

To normalize the norm of $\psi_{a,b}(t)$, $1/\sqrt{a}$ is inserted.

The wavelet transform is defined by this family of functions as the following.

Definition 1. The wavelet transform of $x(t)$ is defined by

$$\tilde{x}(a,b)=\langle\psi_{a,b}(t),x(t)\rangle = \frac{1}{\sqrt{a}}\int_{-\infty}^{\infty}x(t)\,\psi^{*}\left(\frac{t-b}{a}\right)dt. \quad (2)$$

■

The parameter a is called the *dilation parameter* and it corresponds to frequency. The parameter b is called the *shift parameter* and it corresponds to time.

An analyzing wavelet can be chosen to be arbitrary as long as $\int_{-\infty}^{\infty}\psi(t)dt=0$.

If an analyzing wavelet is a complex function, a wavelet transform of a signal is also a complex function. Our method needs a complex analyzing wavelet since it uses an phase (argument) value of the wavelet transform.

2.2 *Wavelet Transform of Cross-correlation Function and Its Relation with Cross Spectrum*

In this paper, we will focus the following analyzing wavelet.

$$\psi(t)=w(t)e^{j\omega_p t},\ w(t)=w(-t)\in\boldsymbol{R} \quad (3)$$

Here $w(t)$ is a real and even function and ω_p is a center frequency of the analyzing wavelet. This limitation on the analyzing wavelet makes it easy to analyze our dead time measurement method. One of the example of this analyzing wavelet is the Gabor function, which is widely used in wavelet applications. The Gabor function and its Fourier transform are

$$\psi(t)=\frac{1}{\pi^{1/4}}\frac{\omega_p}{\gamma}\exp\left(-\frac{\omega_p^2t^2}{2\gamma^2}\right)\exp(-j\omega_p t) \quad (4\text{-}1)$$

$$\Psi(\omega)=\pi^{1/4}(\gamma/\omega_p)\exp(-\gamma^2(\omega-\omega_p)^2/(2\omega_p^2)). \quad (4\text{-}2)$$

Now, consider a wavelet transform of a cross correlation function. It is defined by

$$\tilde{\phi}_{uy}(a,b)=\langle\psi_{a,b}(\tau),\phi_{uy}(\tau)\rangle. \quad (5)$$

The next theorem states a basic relation between the wavelet transform of the cross correlation function and the corresponding cross spectrum.

Theorem 2. Assume that $\Phi_{uy}(\omega)$ is n-times differentiable and $\Phi^{(n)}_{u(t),y(t+b)}(\omega)$ is bounded for all ω. Then

$$\tilde{\phi}_{uy}(a,b)=\frac{\phi^{*}(0)}{\sqrt{a}}\Phi_{u(t),y(t+b)}\left(\frac{\omega_p}{a}\right)+\sum_{k=1}^{n-1}Q_{2k}+R_n \quad (6)$$

where R_n is bounded and

$$Q_l=\frac{\sqrt{a}}{2\pi l!}\Phi^{(l)}_{u(t),y(t+b)}\left(\frac{\omega_p}{a}\right)\int_{-\infty}^{\infty}\lambda^l\Psi^{*}(\omega_p+a\lambda)d\lambda. \quad (7)$$

Proof : The equation (5) can be transformed into $\tilde{\phi}_{uy}(a,b)=\langle\psi_{a,0}(\tau'),\phi_{uy}(\tau'+b)\rangle$ by applying $\tau'=\tau-b$. According to the property of inner products, $\int_{-\infty}^{\infty}f^{*}(\tau)g(\tau)d\tau=\frac{1}{2\pi}\int_{-\infty}^{\infty}F^{*}(\omega)G(\omega)d\omega$. Therefore

$$\tilde{\phi}(a,b)=\frac{\sqrt{a}}{2\pi}\int_{-\infty}^{\infty}\Psi^{*}(a\omega)\Phi_{u(t),y(t+b)}(\omega)d\omega$$

since $\Psi_{a,0}(\omega)=\sqrt{a}\Psi(a\omega)$. This equation is transformed as follows by applying $\omega=\omega_p/a+\lambda$.

$$\tilde{\phi}(a,b)=\frac{\sqrt{a}}{2\pi}\int_{-\infty}^{\infty}\Psi^{*}(\omega_p+a\lambda)\Phi_{u(t),y(t+b)}\left(\frac{\omega_p}{a}+\lambda\right)d\lambda \quad (8)$$

The Taylor series of $\Phi_{u(t),y(t+b)}(\omega_p/a+\lambda)$ at ω_p/a is $\Phi_{u(t),y(t+b)}(\omega_p/a)+\lambda\Phi'_{u(t),y(t+b)}(\omega_p/a)+\cdots+\lambda^n\Phi^{(n)}_{u(t),y(t+b)}(\omega_p/a+\xi(\lambda))$, where $0<|\xi|<|\lambda|$. Consequently, the series and (8) lead to (6). The assumption on $w(t)$ eliminates the terms Q_{2k+1} since $\int_{-\infty}^{\infty}\lambda^{2k+1}\Psi^{*}(\omega_p+a\lambda)d\lambda=0$. It can be derived by showing that $W(\omega)$ is also even function if $w(t)$ is even and $\Psi(\omega_p+\lambda)=W(\lambda)$. The first term's coefficient is obtained by using $\frac{1}{2\pi}\int_{-\infty}^{\infty}\Psi^{*}(\omega_0+a\lambda)d\lambda=\psi^{*}(0)/a$. The assumption that $\Phi^{(n)}_{u(t),y(t+b)}$ is bounded guarantees R_n to be bounded. ■

The following theorem ensures that Q_{2k} is minimized by choosing ω_p as center frequency.

Theorem 3. Suppose that $\psi(t)$ is defined by (3) and $\Psi(\omega)\geq 0$ for all ω. Then $\int_{-\infty}^{\infty}\lambda^{2k}\Psi^{*}(\omega_0+a\lambda)d\lambda$ has a minimum value at $\omega_0=\omega_p$.

Proof : Trivial because of a moment property (Note that $\Psi(\omega_p+a\lambda)=\Psi(\omega_p-a\lambda)$ from (3)). ■

Above discussions allow us to approximate

$$\tilde{\phi}_{uy}(a,b)\approx\frac{\psi^{*}(0)}{\sqrt{a}}\Phi_{u(t),y(t+b)}\left(\frac{\omega_p}{a}\right). \quad (9)$$

This implies that $\tilde{\phi}_{uy}(a,b)$ corresponds to the cross spectrum between $u(t)$ and $y(t+b)$ at ω_p/a. Since $\sqrt{a}$ and $\psi^{*}(0)$ is real (since $\psi(0)=w(0)$),

$$\angle\tilde{\phi}_{uy}(a,b)\approx\angle\Phi_{u(t),y(t+b)}(\omega_p/a). \quad (10)$$

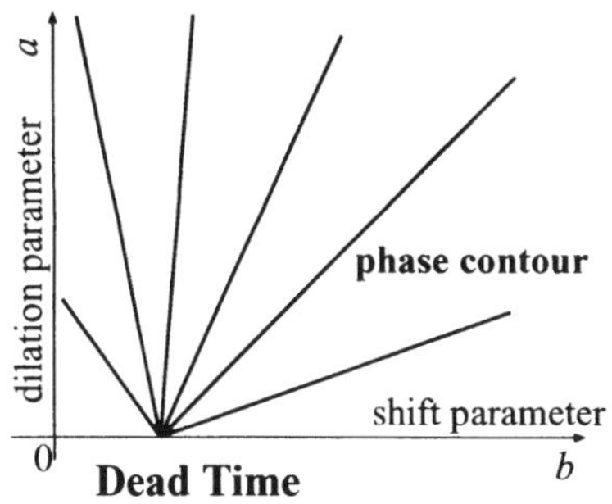

Fig. 1. The dead time measuring method from phase contour plots of wavelet transform of a cross correlation function.

3. MEASUREMENT METHOD

This section states the procedure of our dead time measurement method and reviews the analysis for the open loop case (Tabaru et al., 1997; Tabaru et al., 2000).

The open loop system treated in our studies is a SISO (single-input single-output) LTI (linear time-invariant) system whose transfer function is

$$G(s) = G_R(s)e^{-Ls} \tag{11}$$

where $G_R(s)$ is a strictly proper rational transfer function with real coefficients and L is a dead time of the system. An input and an output of the system are denoted by $u(t)$ and $y(t)$ respectively. Assume that the system is stable to keep $y(t)$ bounded.

The procedure of the method consists of three steps.

(1) Calculate $\tilde{\phi}_{uy}(a,b)$ with a complex analyzing wavelet (*e.g.* Gabor function).
(2) Plot phase (argument) contour lines of $\tilde{\phi}_{uy}(a,b)$. (see Fig. 1).
(3) Then, the contour lines concentrate at the dead time of the system as $a \to 0$.

The reason why the method can measure the dead time can be explained by the property described in the section 2.2 (Tabaru et al., 2000). From (10) and $\Phi_{u(t),y(t+b)} = G(j\omega)e^{j\omega b}\Phi_{uu}(\omega)$, it follows that $\angle\tilde{\phi}_{uy}(a,b) = \angle G_R(j\omega_p/a) + (b-L)\omega_p/a$. This equation can be rewritten as

$$b = \frac{\angle G_R(j\omega_p/a) - \angle\tilde{\phi}_{uy}(a,b)}{\omega_p}a + L. \tag{12}$$

Suppose that $\angle G_R(j\omega_p/a)$ is constant. On a phase contour line of $\tilde{\phi}_{uy}(a,b)$, $\angle\tilde{\phi}_{uy}(a,b)$ is also constant. Consequently, the phase contour's equation is

$$b = sa + L, \quad s \in \boldsymbol{R}. \tag{13}$$

This equation implies that any contour line concentrates at $b = L$ as $a \to 0$, since $\angle G_R(j\omega_p/a)$ becomes almost constant as $a \to 0$.

We also note that this phase contour's property can be derived from a self-similarity property of an impulse response around a dead time (Tabaru et al., 1997).

4. CLOSED LOOP CASE

This section gives a main result in this paper. After problem description, calculations follow to obtain a cross correlation function of the closed loop system. Then coarse analysis will be provided based on the calculations and more precise analysis will be developed with some conditions to simplify the problem.

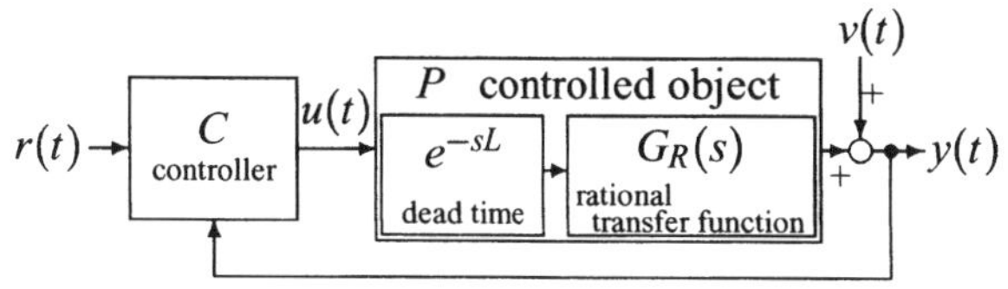

Fig. 2. Closed loop system considered in this paper.

4.1 *Settings and Problem Description*

Figure 2 shows a system considered in this paper. It consists of a controller C and a controlled object P, which is the same as $G(s)$ in the previous section. Both of them are linear and time-invariant. An input and an output of the overall system are $r(t) \in \boldsymbol{R}$ and $y(t) \in \boldsymbol{R}$ respectively. An output of the controller is $u(t) \in \boldsymbol{R}$, which is connected to an input of the controlled object. A disturbance $v(t) \in \boldsymbol{R}$ is added to the output of P. Assume that the overall system is stable and $v(t)$ has no correlation with $r(t)$ (*i.e.* $\phi_{rv}(\tau) = 0$).

The state variable equations of the controlled object are as follows

$$\dot{\boldsymbol{x}}_p(t) = A_p\boldsymbol{x}_p(t) + \boldsymbol{b}_p u(t-L) \tag{14-1}$$

$$y(t) = \boldsymbol{c}_p^T\boldsymbol{x}_p(t) + v(t) \tag{14-2}$$

where $\boldsymbol{x}_p(t)$ is its state vector. This system is assumed to be controllable and observable. The state variable equations of the controller are

$$\dot{\boldsymbol{x}}_c(t) = A_c\boldsymbol{x}_c(t) + [\boldsymbol{b}_{c1}\ \boldsymbol{b}_{c2}]\begin{bmatrix} r(t) \\ y(t) \end{bmatrix} \tag{15-1}$$

$$u(t) = \boldsymbol{c}_c^T\boldsymbol{x}_c(t) + [d_{c1}\ d_{c2}]\begin{bmatrix} r(t) \\ y(t) \end{bmatrix} \tag{15-2}$$

where $\boldsymbol{x}_c(t)$ is controller's state vector. Now, define $\boldsymbol{x}(t) = [\boldsymbol{x}_p(t)\ \boldsymbol{x}_c(t)]^T$ and $\boldsymbol{f}_{rv}(t) = [r(t)\ v(t)]^T$ respectively. Therefore, the overall state variable equations are the following.

$$\dot{\boldsymbol{x}}(t) = A_1\boldsymbol{x}(t) + A_2\boldsymbol{x}(t-L) + B_1\boldsymbol{f}_{rv}(t) + B_2\boldsymbol{f}_{rv}(t-L) \tag{16-1}$$

$$u(t) = C_u\boldsymbol{x}(t) + D_u\boldsymbol{f}_{rv}(t) \tag{16-2}$$

$$y(t) = C_y\boldsymbol{x}(t) + v(t) \tag{16-3}$$

where

$$A_1 = \begin{bmatrix} A_p & 0 \\ \boldsymbol{b}_{c2}\boldsymbol{c}_p^T & A_c \end{bmatrix}, A_2 = \begin{bmatrix} \boldsymbol{b}_p d_{c2}\boldsymbol{c}_p^T & \boldsymbol{b}_p\boldsymbol{c}_c^T \\ 0 & 0 \end{bmatrix} \tag{17-1}$$

$$B_1 = \begin{bmatrix} 0 & 0 \\ \boldsymbol{b}_{c1} & \boldsymbol{b}_{c2} \end{bmatrix}, B_2 = \begin{bmatrix} \boldsymbol{b}_p d_{c1} & \boldsymbol{b}_p d_{c2} \\ 0 & 0 \end{bmatrix} \tag{17-2}$$

$$C_u = [d_{c2}\boldsymbol{c}_p^T\ \boldsymbol{c}_c^T], C_y = [\boldsymbol{c}_p^T\ 0] \tag{17-3}$$

$$D_u = [d_{c1}\ d_{c2}] \tag{17-4}$$

This is a retarded system (Hale, 1977). The initial conditions are assumed that $\boldsymbol{x}(t) = \boldsymbol{f}_{rv}(t) = 0$ for $t < 0$. These assumptions make it simple to calculate responses without loss of generality since their initial responses don't affect correlation functions if stable.

The system involves various configurations of a controller for a SISO system. For example, a rational transfer function controller, a two degrees of freedom controller, and a PID controller whose derivative action is implemented with a low-pass filter for noise attenuation. Thus, the analysis in this section is meaningful for many real systems.

The measurement procedure is the same as the open loop case, that is calculate $\tilde{\phi}_{uy}(a,b)$ and plot its phase contour. This is an additional advantage of our method and it is not necessary to know whether a target is an open loop system or a part of a closed loop system before a measurement.

4.2 *Calculation of Cross Correlation Function*

Now, let us obtain $\phi_{uy}(\tau)$ for the closed loop system. The calculation consists of the following steps. The first is to investigate responses of state variables of a retarded system (Lemma 4). The second is to obtain responses of $u(t)$ and $y(t)$ respectively (Lemma 5). The last is to represent the cross correlation function in terms of $\phi_{rr}(\tau)$ and $\phi_{vv}(\tau)$ (Lemma 6).

The lemma described below is about responses of state variables of a retarded system.

Lemma 4. Consider the following retarded system.

$$\dot{\boldsymbol{x}}(t) = A_1\boldsymbol{x}(t) + A_2\boldsymbol{x}(t-L) + \boldsymbol{f}(t) \qquad (18)$$

The response of $\boldsymbol{x}(t)$ of the system is

$$\boldsymbol{x}(t) = \sum_{k=0}^{\infty}\int_0^t g_k(t-\xi)\boldsymbol{f}(\xi - kL)d\xi, \qquad (19)$$

where $g_k(t)$ is defined recursively as follows.

$$g_k(t) = \begin{cases} e^{A_1 t}, \ k=0, t \geq 0 \\ \int_0^t e^{A_1(t-\xi)}A_2 g_k(\xi)d\xi, \ k \geq 1, t \geq 0 \\ 0, \ t < 0 \end{cases} \qquad (20)$$

Proof: For $t \geq 0$, the response of (18) is (Hale, 1977)

$$\boldsymbol{x}(t) = \int_0^t e^{A_1(t-\xi)}(\boldsymbol{f}(\xi) + A_2\boldsymbol{x}(\xi - L))d\xi.$$

The lemma can be proved by applying this equation recursively. ■

The previous lemma leads us to the next lemma, which shows responses of $u(t)$ and $y(t)$.

Lemma 5. The responses of $u(t)$ and $y(t)$ of (16) are

$$u(t) = \sum_{k=0}^{\infty} g_{u,k}(t)\boldsymbol{f}_{rv}(\xi - kL)d\xi + D_u\boldsymbol{f}_{rv}(t) \quad (21\text{-}1)$$

$$y(t) = \sum_{k=1}^{\infty} g_{y,k}(t)\boldsymbol{f}_{rv}(\xi - kL)d\xi + v(t) \qquad (21\text{-}2)$$

where $g_{u,k}(t)$ and $g_{y,k}(t)$ are defined by

$$g_{u,k}(t) = [g_{ur,k}(t)\ g_{uv,k}(t)] = \begin{cases} C_u g_k(t)B_1 & k=0 \\ C_u(g_k(t)B_1 + g_{k-1}(t)B_2), & k \geq 1 \end{cases} \quad (22\text{-}1)$$

$$g_{y,k}(t) = [g_{yr,k}(t)\ g_{yv,k}(t)] = C_y(g_k(t)B_1 + g_{k-1}(t)B_2). \qquad (22\text{-}2)$$

Proof: The response of (16-1) is the sum of responses of the following two retarded systems.

$$\dot{\boldsymbol{x}}(t) = A_1\boldsymbol{x}(t) + A_2\boldsymbol{x}(t-L) + B_1\boldsymbol{f}_{rv}(t) \qquad (23\text{-}1)$$

$$\dot{\boldsymbol{x}}(t) = A_1\boldsymbol{x}(t) + A_2\boldsymbol{x}(t-L) + B_2\boldsymbol{f}_{rv}(t-L) \quad (23\text{-}2)$$

From lemma 4, the response of these systems are

$$\boldsymbol{x}(t) = \sum_{k=0}^{\infty}\int_0^t g_k(t-\xi)B_1\boldsymbol{f}_{rv}(\xi - kL)d\xi \qquad (24\text{-}1)$$

$$\boldsymbol{x}(t) = \sum_{k=1}^{\infty}\int_0^t g_{k-1}(t-\xi)B_2\boldsymbol{f}_{rv}(\xi - kL)d\xi \quad (24\text{-}2)$$

respectively. Hence the response of (16-1) is the sum of (24-1) and (24-2). The results are obtained by substituting the response into (16-2) and (16-3). Note that the index of (21-2) begins at $k=1$ since there is a dead time element to go from $r(t)$ to $y(t)$. This is also implied by $C_y g_0(t)B_1 = C_y e^{A_1 t}B_1 = 0$. ■

Note that $g_{u,k}(t)$ and $g_{y,k}(t)$ are impulse responses corresponding to rational transfer functions, *i.e.* they include no dead time element.

The cross correlation function $\tilde{\phi}_{uy}(\tau)$ is given by the next lemma for the considered closed loop system.

Lemma 6. Let $\phi_{rr,vv}(\tau) = [\phi_{rr}(\tau)\ \phi_{vv}(\tau)]^T$. Then

$$\phi_{uy}(\tau) = \sum_{k=-\infty}^{+\infty}\int_{-\infty}^{\infty} \boldsymbol{h}_k^T(\tau - t)\phi_{rr,vv}(t-kL)dt + \sum_{k=-\infty}^{0}\int_{-\infty}^{\infty} g_{uv,-k}(t-\tau)\phi_{vv}(t-kL)dt + d_{c2}\phi_{vv}(\tau) \qquad (25)$$

where $\boldsymbol{h}_k(t) = [h_{r,k}(t)\ h_{v,k}(t)]^T$ and $h_{r,k}(t)$ and $h_{v,k}(t)$ are defined as follows.

$$h_{r,k}(t) = \begin{cases} \sum_{l=1-k}^{+\infty}\int_0^{\infty} g_{ur,l}(\xi - t)g_{yr,k+l}(\xi)d\xi, & k \leq 0 \\ d_{c1}g_{yr,k}(t) + \sum_{l=0}^{+\infty}\int_0^{\infty} g_{ur,l}(\xi - t)g_{yr,k+l}(\xi)d\xi, & k \geq 1 \end{cases} \qquad (26)$$

$$h_{v,k}(t) = \begin{cases} \sum_{l=1-k}^{+\infty}\int_0^{\infty} g_{uv,l}(\xi - t)g_{yv,k+l}(\xi)d\xi, & k \leq 0 \\ d_{c2}g_{yv,k}(t) + \sum_{l=0}^{+\infty}\int_0^{\infty} g_{uv,l}(\xi - t)g_{yv,k+l}(\xi)d\xi, & k \geq 1 \end{cases} \qquad (27)$$

Proof: Let $z_1(t) = \int_{-\infty}^{\infty} g_1(t-\xi)f_1(\xi - k_1L)d\xi$ and $z_2(t) = \int_{-\infty}^{\infty} g_2(t-\xi)f_2(\xi - k_2L)d\xi$. Then

$$\phi_{z_1z_2}(\tau) = \int_{-\infty}^{\infty} h(\tau - t)\phi_{f_1f_2}(t - (k_1 - k_2)L)dt$$

where $h(t) = \int_0^t g_1(\xi - t)g_2(\xi)d\xi$. The lemma can be derived by using this propery and $\phi_{rv}(\tau) = 0$. ■

Note that every $h_k(t)$ includes no dead time element and $h_k(t) \neq 0$ for $t < 0$. The latter means that $h_k(t)$ are not causal if they are considered as impulse responses.

4.3 *Coarse Analysis of Phase Contour Line*

The obtained correlation function has very complicated form for the closed loop case. On the other hand, the one for the open loop can be represented simply as

$$\phi_{uy}(\tau) = \int_0^t g_R(\tau - t)\phi_{uu}(t - L)dt \qquad (28)$$

where $g_R(t)$ is the impulse response corresponding to $G_R(s)$. The main difference between them is that $\phi_{uy}(\tau)$ for the closed loop case is written in terms of $\phi_{rr}(\tau - kL)$ and $\phi_{vv}(\tau - kL)$ with $k = \cdots, -1, 0, 1, 2, \cdots$, while in terms of only $\phi_{uu}(\tau - L)$ for the open loop case. It causes that contour lines concentrate at multiple points, that is every $b = kL$ (k is integer) for the closed loop case (Fig. 3).

Because of difficulties of complete and precise analysis, we only pointed out that the concentration at $b = kL$ corresponds the terms concerned with $\phi_{rr}(\tau - kL)$ or $\phi_{rr}(\tau - kL)$. Instead of general analyses, the following description gives more analysis in detail under certain conditions.

4.4 *More Analysis under Loop Gain Condition*

For simplicity, we ignore the second and third terms of (25). Then the wavelet transform of $\phi_{uy}(\tau)$ can be represented by

$$\tilde{\phi}_{uy}(a,b) = \sum_{k=-\infty}^{+\infty} \tilde{\phi}_{uy,k}(a,b) \qquad (29)$$

$$\tilde{\phi}_{uy,k}(a,b) = \int_{-\infty}^{\infty} \psi^*_{a,b}(\tau) \int_{-\infty}^{\infty} \boldsymbol{h}_k^T(\tau - t)\phi_{rr,vv}(t - kL)dtd\tau. \qquad (30)$$

We also assume that an analyzing wavelet has the form (3). Then the result of the section 2.2 allows us to approximate $\tilde{\phi}_{uy,k}(a,b)$ as

$$\tilde{\phi}_{uy,k}(a,b) \approx \frac{\psi^*(0)}{\sqrt{a}} e^{j\omega_p(b-kL)/a} \times \boldsymbol{H}_k^T(j\omega_p/a)\Phi_{rr,vv}(\omega_p/a) \qquad (31)$$

where $\Phi_{rr,vv}(\omega) = [\Phi_{rr}(\omega)\ \Phi_{vv}(\omega)]^T$. This equation shows that $\boldsymbol{H}_k(j\omega)$ governs $\tilde{\phi}_{uy,k}(a,b)$. Hence we will represent $\boldsymbol{H}_k(j\omega)$ in terms of transfer functions in the considered closed loop system.

Let $G_k(s)$ denote the Laplace transform of $g_0(t)$, so that $G_0(s) = (sI - A_1)^{-1}$ and $G_{k+1}(s) = G_0(s)A_2G_k(s)$ from (20). Now define $\boldsymbol{b}_{p0} \in \boldsymbol{R}^4$ by $[\boldsymbol{b}_p^T\ \boldsymbol{0}^T]^T$. The transfer function of a rational part of P can be represented by $G_R(s) = C_yG_0(s)\boldsymbol{b}_{p0}$ and it is the same as $G_R(s)$ of the open loop case. Similary, the transfer function from $r(t)$ and $y(t)$ to $u(t)$ is $G_C(s) = [G_{C1}(s)\ G_{C2}(s)] = C_uG_0(s)B_1 + D_u$ and the one from the input of $G_R(s)$ to $u(t)$ is $G_{LP}(s) = C_uG_0(s)\boldsymbol{b}_{p0}$. The latter is also the loop transfer function except for the dead time element.

Theorem 7. For ω such that $|G_{LP}(j\omega)| < 1$,

$$\boldsymbol{H}_k(j\omega) = \begin{cases} \dfrac{G_R(j\omega)}{1 - |G_{LP}(j\omega)|^2}[|G_{C1}(j\omega)|^2\ |G_{C2}(j\omega)|^2], & k = 1 \\ (G^*_{LP}(j\omega))^{1-k}H_1(j\omega), & k < 1 \\ (G_{LP}(j\omega))^{k-1}H_1(j\omega), & k > 1. \end{cases} \qquad (32)$$

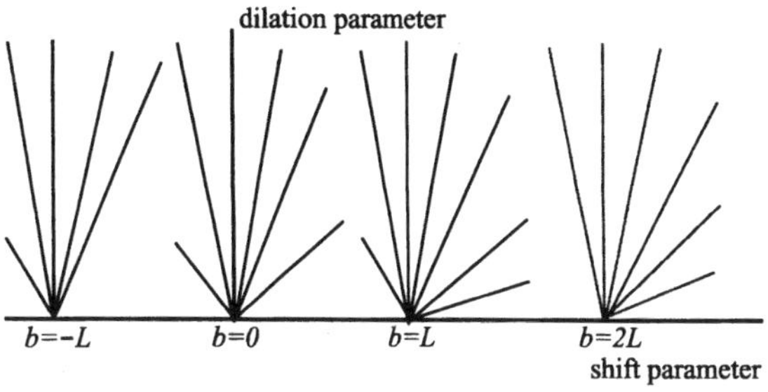

Fig. 3. Phase contour line plots of the closed loop case. The contour lines may concentrate at every $b = kL$ ($k = \cdots, -1, 0, 1, 2, \cdots$) as $a \to 0$.

Proof: From (26) and (27), for $k \leq 0$

$$H_{r,k}(\omega) = \sum_{l=1-k}^{+\infty} G^*_{ur,l}(j\omega)G_{yr,k+l}(j\omega)$$

$$H_{v,k}(\omega) = \sum_{l=1-k}^{+\infty} G^*_{uv,l}(j\omega)G_{yv,k+l}(j\omega)$$

and for $k \geq 1$

$$H_{r,k}(\omega) = d_{c1}G_{y,k}(j\omega) + \sum_{l=0}^{+\infty} G^*_{ur,l}(j\omega)G_{yr,k+l}(j\omega)$$

$$H_{v,k}(\omega) = d_{c2}G_{y,k}(j\omega) + \sum_{l=0}^{+\infty} G^*_{uv,l}(j\omega)G_{yv,k+l}(j\omega)$$

respectively. From (22) and the definition of $G_C(s)$,

$$G_{u,k}(s) = [G_{ur,k}(s)\ G_{uv,k}(s)] = \begin{cases} G_C(s) - D_u & k = 0 \\ C_u(G_k(s)B_1 + G_{k-1}(s)B_2), & k \geq 1 \end{cases}$$

$$G_{y,k}(s) = [G_{yr,k}(s)\ G_{yv,k}(s)] = C_y(G_k(s)B_1 + G_{k-1}(s)B_2).$$

Since $A_2 = \boldsymbol{b}_{p0}C_u$, $C_uG_{k+1}(s) = G_{LP}(s)C_uG_k(s)$ and $C_yG_{k+1}(s) = G_R(s)C_uG_k(s)$. These relations derive

$$G_{u,k}(s) = \begin{cases} G_C(s) - D_u & k = 0 \\ (G_{LP}(s))^kG_C(s), & k \geq 1 \end{cases} \qquad (33\text{-}1)$$

$$G_{y,k}(s) = G_R(s)(G_{LP}(s))^{k-1}G_C(s). \qquad (33\text{-}2)$$

Substitution of (33) and some caculations simplify $H_{r,k}(j\omega)$ for $k \geq 1$ as follows.

$$G_R(j\omega)|G_{C1}(j\omega)|^2(G_{LP}(j\omega))^{k-1}\sum_{l=0}^{\infty}|G_{LP}(j\omega)|^{2l}$$

If $|G_{LP}(j\omega)| < 1$, the summation converge and the one of the result is obtained. Other results can be derived similarly. ■

From this theorem, for a such that $|G_{LP}(j\omega_p/a)| < 1$

$$|\tilde{\phi}_{uy,k}(a,b)| = |G_{LP}(j\omega_p/a)|^{|k-1|}|\tilde{\phi}_{uy,1}(a,b)|.$$

It implies that $|\tilde{\phi}_{uy,k}(a,b)| \ll |\tilde{\phi}_{uy,1}(a,b)|$ for such a. Hence $\tilde{\phi}_{uy}(a,b)$ can be approximated as

$$\tilde{\phi}_{uy}(a,b) \approx \tilde{\phi}_{uy,1}(a,b)$$

for a such that $|G_{LP}(j\omega)|$ is small enough. Then

$$\tilde{\phi}_{uy}(a,b) \approx \frac{\psi^*(0)}{\sqrt{a}} \frac{G_R(j\omega)}{1 - |G_{LP}(j\omega)|^2} e^{j\omega_p(b-L)/a} \times [|G_{C1}(j\omega)|^2\ |G_{C2}(j\omega)|^2]\Phi_{rr,vv}(\omega_p/a)$$

and $\angle\tilde{\phi}_{uy}(a,b) \approx \angle G_R(j\omega_p/a) + (b - L)\omega_p/a$ since $\psi^*(0)$ is real by (3). Consequently, the equation of contour lines is the same as (12) and they concentrate at only $b = L$ as $a \to 0$.

5. NUMERICAL EXAMPLES

This section gives two numerical simulations to illustrate validity of discussions in the previous section.

Figure 4 shows a configuration of a controller C and a controlled object P in the both simulations. The controller C was a SISO LTI system without a dead time element and P was the same except that it had a dead time on its input. We added exogenous inputs $r(t)$ and got $u(t)$ and $y(t)$, which were inputs and outputs of the controlled object respectively. The purpose of the simulation was to measure the dead time of P from the wavelet transform of the cross correlation function between $u(t)$ and $y(t)$.

Example 1 : The transfer functions of P and C were

$$P(s) = \frac{1}{36s-2}e^{-5s},\ C(s) = \frac{1.97}{10s+1}. \quad (34)$$

The dead time of P was 5 seconds. Note that the controlled object needs a feedback controller to measure the dead time due to its instability. The number of samples was set to 8192 and the sampling rate was set to 0.1 second. We generated the input signal $r(t)$ by applying the filter whose transfer function was $1/(5s+1)$ to normally distributed random numbers with variance 1.0. An analyzing wavelet was chosen to be the Gabor function. In this simulation, ω_p and γ were set to 1 (rad/sec) and 2π respectively. This simulation was carried out under disturbance free.

Figure 5 is a phase contour plot. The horizontal axis represents shift parameter (time) and vertical axis represents a dilation parameter, which corresponds to a frequency. Those lines are the phase contour lines of $\tilde{\phi}_{uy}(a,b)$, which is the wavelet transform of the cross correlation function between the input $u(t)$ and the output $y(t)$. The contour lines concentrate around $b=5$, where the dead time is located. Thus the method can measure the dead time in this case.

Example 2 : The transfer functions of P and C were

$$P(s) = \frac{5}{s+5}e^{-2s},\ C(s) = 0.98 + \frac{0.1}{s}. \quad (35)$$

The dead time was 2 seconds. The number of samples, the sampling rate, the input signal $r(t)$, and the analyzing wavelet were the same as Example 1. This simulation was also disturbance free.

Figure 6 is a phase contour plot of $\tilde{\phi}_{uy}(a,b)$. Contour lines become complicated in this example. They concentrate at $b = -4, -2, 0, 2, \cdots$, every two seconds. The dead time is estimated to be two seconds although the estimation may be more difficult than Example 1. It is obvious that this example doesn't satisfy the loop gain condition in the section 4.4.

6. CONCLUSION

This paper presents that the dead time measurement method based on wavelet is available for not only open loop systems but also closed loop systems. We derive a correlation function between an input and an output for a closed loop system and point out that contour lines may concentrate at every $b = kL$ (k is integer).

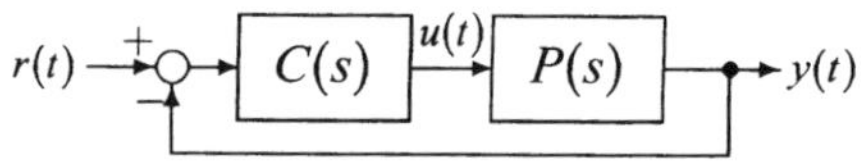

Fig. 4. Configuration of numerical simulations.

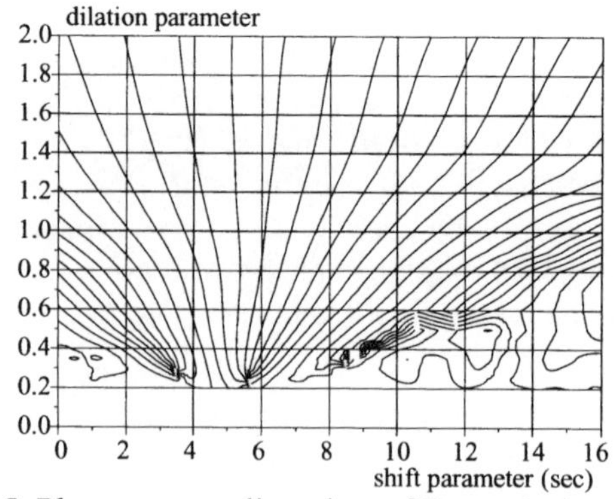

Fig. 5. Phase contour line plots of Example 1.

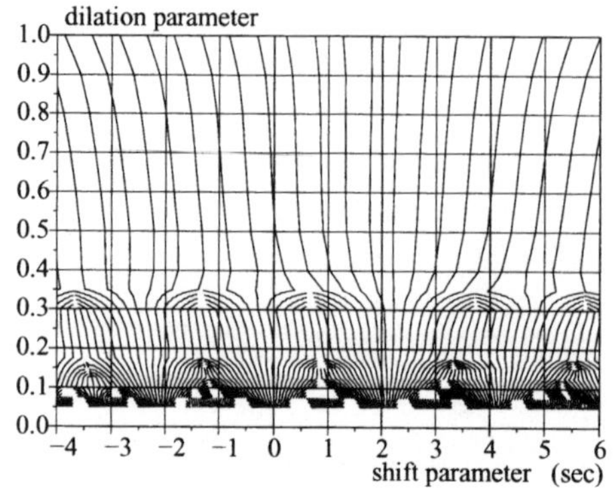

Fig. 6. Phase contour line plots of Example 2.

On the other hand, the contour lines will be the same as the open loop case under the loop gain condition. Anyway, the dead time is measurable by the method for both cases. Numerical simulations illustrate that our analysis is true.

More precise and general analysis are expected since it will make required conditions clearer for the dead time measurement of the closed loop system.

REFERENCES

G. Kaiser (1994). *A Friendly Guide to Wavelets*, Birkhäuser, 1994.

J. Hale (1977). *Theory of Functional Differential Equations*, Springer-Verlag.

Y. Meyer (1993). *Wavelets, Algorithms & Applications*, SIAM.

K. Nakano, T. Tabaru, S. Shin and Y. Toyoda (2002). "Wavelet-based Identification for Control of Water-Tube Drum Boiler," *The 15th IFAC World Congress on Automatic Control*, Barcelona, Spain, 21-26 July 2002.

T. Tabaru and S. Shin (1997). "Dead Time Detection Based on Wavelet Analysis of Cross Correlation Data," *11th IFAC Symp. SYSID'97*, Kitakyushu, Japan, 8-11 July 1997, **1**, pp. 33-38.

T. Tabaru and S. Shin (2001). "Reconsideration of Dead Time Measurement by Wavelet from Phase Property of Frequency Response," *IFAC Symp. SYSID 2000*, Santa-Barbara, U.S.A., 21-23 June 2000, **2**, pp. 775-779.

IFAC Publications
www.elsevier.com/locate/ifac

CLOSED LOOP IDENTIFICATION METHOD USING A SUBSPACE APPROACH

Pouliquen Mathieu * **M'Saad Mohammed** *

* *Control Group, GREYC CNRS UMR 6072*
ENSICAEN, 06 Bd du Maréchal Juin - 14050 Caen Cedex, France

Abstract: This paper presents a free model reduction closed loop subspace identification method for multivariable systems operating in a well posed closed loop environment. This allows to determine the impulse response of the system from the identification of the control system sensitivity functions. The rational behind the proposed method is twofold. Firstly, the general purpose design features of the subspace system identification approach and its inherent numerical robustness. Secondly, the model reduction performed in the available subspace identification is removed by a proper order selection procedure. A bias analysis is carried out to emphasize the robustness features. *Copyright © 2003 IFAC*

Keywords: closed loop system identification, multivariable systems, impulse response, state space realization, subspace methods.

1. INTRODUCTION

There has been increasing interest in the closed loop identification as well as the subspace identification over the last decade. The former alleviated the interplay between the identification and control while the latter provided the relevant engineering features of an experimental identification methodology for multivariable systems. Comprehensive informative overviews can be found in the open literature ((Gevers, 1993), (Van Den Hof and Schrama, 1995), (Ljung, 1999), (Van Overschee and De Moor, 1996)).

Three approaches have been pursued to address the closed loop identification problem. The first one is referred to as the direct approach. The identification is performed as in a usual open loop context up to a suitable data processing. The indirect approach is mainly based on an open loop identification of the control system sensitivity function using the system output and an external excitation input. It requires knowledge of the regulator transfer function. The joint input/ouput approach uses the system input/output behavior together with an external excitation input. Two methods are worth to be mentioned in this context. The three steps method performs two open loop identification experiments of the control system sensitivity functions. These sensitivity functions are used to recover the input/output behavior of the system in the free disturbances framework. This input/output behavior recovery is used to identify the system as it is done in the open loop context ((Ebert *et al.*, 1997)). In (Verhaegen, 1993) and (Katayama *et al.*, 2002) a high order model is directly determined from the control system sensitivity functions.

This paper presents a new joint input/output identification method using the subspace identification approach. This consists in partially identifying the impulse response of the control system sensitivity functions to determine the impulse response of the system. A realization problem is then solved to determine the model of the system with an appropriate order selection. It is worth mentioning that it is possible to separate the system dynamics in order to take into account some prior knowledge.

2. PROBLEM STATEMENT AND NOTATIONS

Let consider the problem of identifying a linear time invariant multivariable system operating in a closed

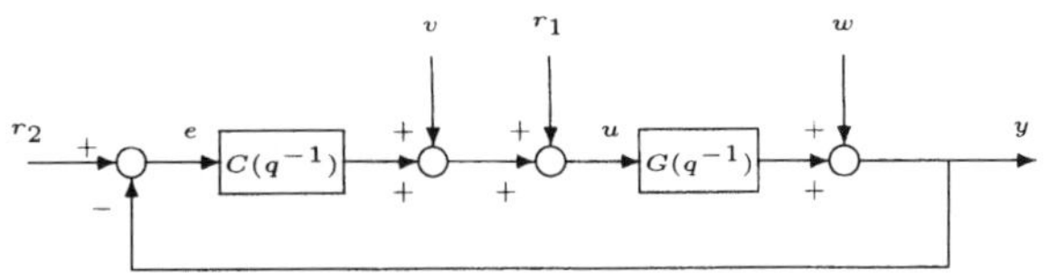

Fig. 1. Closed loop

loop environment. The control system configuration is shown in Fig. 1. $u(t) \in \mathbb{R}^m$ and $y(t) \in \mathbb{R}^p$ are respectively the input and the output of the system that can be described by its transfer matrix $G(q^{-1})$. $v(t) \in \mathbb{R}^m$ and $w(t) \in \mathbb{R}^p$ are respectively input and output noise measurements. $r_1(t) \in \mathbb{R}^m$ and $r_2(t) \in \mathbb{R}^p$ are external excitation inputs which are introduced so as to satisfy the persistent excitation condition. $\{r_1(t)\}$, $\{r_2(t)\}$, $\{u(t)\}$, $\{y(t)\}$, $\{v(t)\}$ and $\{w(t)\}$ are assumed to be jointly quasi stationary zero mean bounded sequences. Moreover the external excitation input sequences are assumed to be uncorrelated with the noise vector sequences.

The relationship between the different sequences and transfer functions can be easily derived from the control system depicted in Fig. 1, namely

$$\begin{cases} u(t) = C(q^{-1})\,(r_2(t) - y(t)) + r_1(t) + v(t) \\ y(t) = G(q^{-1})u(t) + w(t) \end{cases} \tag{1}$$

For notational convenience considerations the shift operator q^{-1} will be omitted when clear from the context.

Of fundamental importance, the control system of Fig. 1 is assumed to be well posed in the sense that the outputs are uniquely determined by the states and the external inputs. Obviously the control system is also assumed to be asymptotically stable. From (1), one gets

$$s(t) = G_{cl}r(t) + \nu(t)$$

with

$$G_{cl} = \begin{pmatrix} \mathbb{S}_{r_1u} & \mathbb{S}_{r_2u} \\ \mathbb{S}_{r_1y} & \mathbb{S}_{r_2y} \end{pmatrix} \tag{2}$$

$$s(t) = \begin{pmatrix} u(t) \\ y(t) \end{pmatrix},\ r(t) = \begin{pmatrix} r_1(t) \\ r_2(t) \end{pmatrix} \text{ and } \nu(t) = \begin{pmatrix} v'(t) \\ w'(t) \end{pmatrix}$$

where $\mathbb{S}_{r_1u}$, $\mathbb{S}_{r_2u}$, $\mathbb{S}_{r_1y}$ and $\mathbb{S}_{r_2y}$ denote the usual sensitivity functions of the control system given by

$$\begin{cases} \mathbb{S}_{r_1u} = (I_m + CG)^{-1} \\ \mathbb{S}_{r_2u} = (I_m + CG)^{-1}C \\ \mathbb{S}_{r_1y} = G(I_m + CG)^{-1} \\ \mathbb{S}_{r_2y} = G(I_m + CG)^{-1}C \end{cases}$$

The components of the disturbing signal $\nu(t)$ are given by

$$\begin{cases} v'(t) = \mathbb{S}_{r_1u}v(t) - \mathbb{S}_{r_2u}w(t) \\ w'(t) = \mathbb{S}_{r_1y}v(t) + (I_p - \mathbb{S}_{r_2y})\,w(t) \end{cases}$$

The system G can be described by its state space representation as follows

$$\begin{cases} x(t+1) = Ax(t) + Bu(t) \\ y(t) \ \ \ \ = Cx(t) + Du(t) + w(t) \end{cases}$$

where $A \in \mathbb{R}^{n_G \times n_G}$, $B \in \mathbb{R}^{n_G \times m}$, $C \in \mathbb{R}^{p \times n_G}$ and $D \in \mathbb{R}^{p \times m}$. (A, C) is assumed to be observable and (A, B) is assumed to be controllable.

Before presenting the solution to this identification problem, let us define the following matrices

$$\Gamma_{G/i} = \begin{pmatrix} C \\ \cdots \\ CA^{i-1} \end{pmatrix} \text{ and } \Delta_{G/i} = \begin{pmatrix} A^{i-1}B & \cdots & B \end{pmatrix}$$

which are respectively the extended observability matrix ($i > n_G$) and the reversed extended controllability matrix, and

$$H_{G/i} = \begin{pmatrix} D & 0 & \cdots & 0 \\ CB & D & 0 & \cdots \\ \cdots & \cdots & \cdots & \cdots \\ CA^{i-2}B & \cdots & CB & D \end{pmatrix}$$

which is a lower triangular Toeplitz matrix containing the first i Markov parameters of the system. Notice that

$$H_{G/2i} = \begin{pmatrix} H_{G/i} & 0 \\ \Gamma_{G/i}\Delta_{G/i} & H_{G/i} \end{pmatrix} \tag{3}$$

Furthermore $\Gamma_{G/i}\Delta_{G/i}$ will be noted $\Gamma\Delta_{G/i}$ for convenience consideration. One will use the same notations for the controller and the closed loop system, i.e. $\Gamma_{C/i}$, $\Delta_{C/i}$, $H_{C/i}$ and $\Gamma_{cl/i}$, $\Delta_{cl/i}$, $H_{cl/i}$.

3. IDENTIFICATION METHOD

The identification method splits into three steps. The first step consists in identifying set of the first Markov parameters of the control system sensitivity functions. Such an identification can be performed in different ways. In the second step the impulse response of the system is identified from the previous result. The third step consists in determining the order together with a state space realization of the system.

3.1 *Impulse response of* G_{cl}

The aim of this part is the identification of some Markov parameters of G_{cl}. Unlike the solutions presented in (Verhaegen, 1993) and (Katayama *et al.*, 2002) which identify a state space realization of the tracking dynamic of the control system, one only determines the $2i-1$ first Markov parameters of the G_{cl}, i.e. $H_{cl/2i}$. This is achieved using an appropriate subspace identification method with a relevant external excitation sequence ((Verhaegen, 1994), (Mc Kelvey and Akcay, 1995)). Notice that the involved identification can be performed using the prediction error method appropriately initialized using the N4SID method.

3.2 *Impulse response of G*

In this part, one will focus on the estimation of a finite number of Markov parameters of the system G, i.e. the product $\Gamma\Delta_{G/i}$. The following result provides the relationship between the system G, the controller C and the control system G_{cl}, namely useful expressions of $\Gamma_{cl/i}$, $\Delta_{cl/i}$ and $H_{cl/i}$.

Theorem 1. - Let consider the control system of Fig. 1 and its transfer matrix G_{cl} defined by (2). Then $\Gamma_{cl/i}$, $\Delta_{cl/i}$ and $H_{cl/i}$ are of the form

$$\Gamma_{cl/i}=L_i^T\begin{pmatrix} -W_i^{-1}H_{C/i}\Gamma_{G/i} & W_i^{-1}\Gamma_{C/i} \\ (I_{ip}-H_{G/i}W_i^{-1}H_{C/i})\Gamma_{G/i} & H_{G/i}W_i^{-1}\Gamma_{C/i} \end{pmatrix}T^{-1} \quad (4)$$

$$\Delta_{cl/i}=T\begin{pmatrix} \Delta_{G/i}W_i^{-1} & \Delta_{G/i}W_i^{-1}H_{C/i} \\ -\Delta_{C/i}H_{G/i}W_i^{-1} & \Delta_{C/i}(I_{ip}-H_{G/i}W_i^{-1}H_{C/i}) \end{pmatrix}L_i \quad (5)$$

$$H_{cl/i}=L_i^T\begin{pmatrix} W_i^{-1} & W_i^{-1}H_{C/i} \\ H_{G/i}W_i^{-1} & H_{G/i}W_i^{-1}H_{C/i} \end{pmatrix}L_i \quad (6)$$

where $W_i = (I_{im}+H_{C/i}H_{G/i})$, T is a non singular $n_{cl}\times n_{cl}$ transformation matrix and L_i is a non singular $i(m+p)\times i(m+p)$ matrix of permutation.

Proofs of all theorems are given in the full version of the paper ((Pouliquen and M'Saad, 2002)). The following remarks are worth to be mentioned.

Remark 2. W_i is a lower triangular Toeplitz matrix with $(I_m + D_C D)$ on its diagonal. The well posed assumption of the control system ensures that W_i is non singular.

Remark 3. From (4) and (5), it follows

$$\begin{pmatrix} \Gamma\Delta_{G/i} & 0 \\ 0 & \Gamma\Delta_{C/i} \end{pmatrix} = \begin{pmatrix} -H_{G/i} & I_{ip} \\ I_{im} & H_{C/i} \end{pmatrix} L_i\Gamma\Delta_{cl/i}L_i^T\begin{pmatrix} I_{ip} & -H_{C/i} \\ H_{G/i} & I_{im} \end{pmatrix} \quad (7)$$

Let us introduce the following form of $L_{2i}H_{cl/2i}L_{2i}^T$

$$L_{2i}H_{cl/2i}L_{2i}^T=\begin{pmatrix} H_{11} & 0 & H_{13} & 0 \\ H_{21} & H_{11} & H_{23} & H_{13} \\ H_{31} & 0 & H_{33} & 0 \\ H_{41} & H_{31} & H_{43} & H_{33} \end{pmatrix}=\begin{pmatrix} H_{1:2/1:2} & H_{1:2/3:4} \\ H_{3:4/1:2} & H_{3:4/3:4} \end{pmatrix} \quad (8)$$

From theorem 1 and remark 3, one can obtain several different ways to determine the matrix $\Gamma\Delta_{G/i}$ depending on the composition of the external signal $r(t)$

- The case where only $r_1(t)$ is present. One can derive the following expression of $H_{G/2i}$ from theorem 1.

$$H_{G/2i}=H_{3:4/1:2}H_{1:2/1:2}^{-1}$$

  where $H_{1:2/1:2}$ and $H_{3:4/1:2}$ are respectively the impulse response of the sensitivity functions $\mathbb{S}_{r_1u}$ and $\mathbb{S}_{r_1y}$. $\Gamma\Delta_{G/i}$ is simply extracted from $H_{G/2i}$ using (3).
- The case where only $r_2(t)$ is present. As previously $H_{G/2i}$ is determined as follows

$$H_{G/2i}=H_{3:4/3:4}H_{1:2/3:4}^{-1} \quad (9)$$

  where $H_{1:2/3:4}$ and $H_{3:4/3:4}$ are respectively the impulse response of the sensitivity function $\mathbb{S}_{r_2u}$ and $\mathbb{S}_{r_2y}$. (9) holds provided that $H_{1:2/3:4}$ is non singular or equivalently there is no delay in the controller.
- The case where $r_1(t)$ as well as $r_2(t)$ are present. The determination of $\Gamma\Delta_{G/i}$ can be performed in various manners. As before $H_{G/2i}$ can be obtained from the solution of the equation

$$H_{G/2i}H_{1:2/1:4}=H_{3:4/1:4}$$

  and $\Gamma\Delta_{G/i}$ is extracted from $H_{G/2i}$ thanks to (3) or estimated from (7) as follows

$$\Gamma\Delta_{G/i}=\begin{pmatrix} -H_{G/i} & I_{ip} \end{pmatrix}\begin{pmatrix} H_{21} & H_{23} \\ H_{41} & H_{43} \end{pmatrix}\begin{pmatrix} I_{ip} \\ H_{G/i} \end{pmatrix}$$

A usual problem in system identification is the use of a priori known dynamics to improve the parameter estimation consistency. This feature can be easily incorporated as shown in (Pouliquen and M'Saad, 2002).

3.3 *Order and state space realization*

The results presented above allow to deal with the estimation problem of the impulse response of the system. The order of the minimal realization of the system can be determined from the following well known result.

Lemma 4. - Assume that $i \geq n_G$ and consider the Singular Value Decomposition (SVD) of $\Gamma\Delta_{G/i}$

$$\Gamma\Delta_{G/i}=\begin{pmatrix} U_1 & U_2 \end{pmatrix}\begin{pmatrix} S_1 & 0 \\ 0 & 0 \end{pmatrix}\begin{pmatrix} V_1^T \\ V_2^T \end{pmatrix}$$

then one has the following properties

P1. $\Gamma\Delta_{G/i}$ is of order n_G which is equal to the number of singular value different from zero.

P2. $\Gamma_{G/i}$ and $\Delta_{G/i}$ are respectively of the form

$$\Gamma_{G/i}=U_1S_1^{1/2}T^{-1} \text{ and } \Delta_{G/i}=TS_1^{1/2}V_1^T$$

where T is a non singular $n_G\times n_G$ transformation matrix.

$\Gamma_{G/i}^T\Gamma_{G/i}$ and $\Delta_{G/i}\Delta_{G/i}^T$ are respectively truncated observability and controllability grammian on a horizon i. The ellipsoids $\left\{x\in\mathbb{R}^n, x(\Gamma_{G/i}^T\Gamma_{G/i})x^T\leq 1\right\}$ and $\left\{x\in\mathbb{R}^n, x(\Delta_{G/i}\Delta_{G/i}^T)^{-1}x^T\leq 1\right\}$ show how a state is observable and controllable on a horizon i.

The choice $T = I_{n_G}$ leads to an internally balanced realization. A low order model can be obtained by simply neglecting the modes corresponding to the singular values that could be considered as sufficiently small from both observability and controllability point of views. This idea is the corner stone of the balanced truncation reduction model method ((Moore, 1981)).

The determination of the state space realization can be carried out using the available algorithms ((Zeiger and Mc Ewen, 1974), (Kung, 1978)).

4. ANALYSIS

This section deals with a bias analysis of the parameter estimation process presented above. The bias problem is a natural consequence of those ubiquitous noise measurements and/or modelling errors in the real world life. The underlying disturbing effects have some impacts on the sensitivity functions estimation accuracy and henceforth the quality of the estimated state realization of the model $(\hat{A}, \hat{B}, \hat{C}, \hat{D})$.

Let $\widehat{\begin{pmatrix} \mathbb{S}_{r_1u} \\ \mathbb{S}_{r_1y} \end{pmatrix}}$ and $\widetilde{\begin{pmatrix} \mathbb{S}_{r_1u} \\ \mathbb{S}_{r_1y} \end{pmatrix}}$ be the estimate sensitivity transfer functions and their corresponding estimation error

$$\widetilde{\begin{pmatrix} \mathbb{S}_{r_1u} \\ \mathbb{S}_{r_1y} \end{pmatrix}} = \begin{pmatrix} \mathbb{S}_{r_1u} \\ \mathbb{S}_{r_1y} \end{pmatrix} - \widehat{\begin{pmatrix} \mathbb{S}_{r_1u} \\ \mathbb{S}_{r_1y} \end{pmatrix}}$$

From these definitions and the description (8) of the control system, one introduces the matrix error

$$\begin{pmatrix} \widetilde{H}_{1:2/1:2} \\ \widetilde{H}_{3:4/1:2} \end{pmatrix} = \begin{pmatrix} H_{1:2/1:2} \\ H_{3:4/1:2} \end{pmatrix} - \begin{pmatrix} \widehat{H}_{1:2/1:2} \\ \widehat{H}_{3:4/1:2} \end{pmatrix}$$

where $\left(\widetilde{H}_{1:2/1:2}\right)$ and $\left(\widetilde{H}_{3:4/1:2}\right)$ contains respectively the estimation error on the $2i-1$ first Markov parameters of the sensitivity functions $\mathbb{S}_{r_1u}$ and $\mathbb{S}_{r_1y}$.

The following result provides a bound on the error between $\Gamma\Delta_{G/i}$ and its estimate $\widehat{\Gamma\Delta}_{G/i}$ in a Frobenius norm sense. This bound may be interpreted as a measure of the disturbing effects on the estimation of sensitivity functions $\mathbb{S}_{r_1u}$ and $\mathbb{S}_{r_1y}$.

Lemma 5. - Let $\widetilde{\Gamma\Delta}_{G/i} = \Gamma\Delta_{G/i} - \widehat{\Gamma\Delta}_{G/i}$ be the error on $\Gamma\Delta_{G/i}$, one has

$$\|\widetilde{\Gamma\Delta}_{G/i}\|_F \leq \mathbf{E}_{cl}$$

with

$$\mathbf{E}_{cl}=\left(\|\widetilde{H}_{3:4/1:2}\|_F+\|H_{G/2i}\|_F\|\widetilde{H}_{1:2/1:2}\|_F\right)\|(\widehat{H}_{1:2/1:2})^{-1}\|_F$$

The aim of the SVD used in lemma 4 is to estimate the rank of the matrix $\widehat{\Gamma\Delta}_{G/i}$. If $\widetilde{\Gamma\Delta}_{G/i} = 0$ then only n_G singular value are different from zero. For $\widetilde{\Gamma\Delta}_{G/i} \neq 0$ it's generally no more the case. The SVD of $\widehat{\Gamma\Delta}_{G/i}$ is commonly written under the following form

$$\widehat{\Gamma\Delta}_{G/i} = \begin{pmatrix} \widehat{U}_1 & \widehat{U}_2 \end{pmatrix} \begin{pmatrix} \widehat{S}_1 & 0 \\ 0 & \widehat{S}_2 \end{pmatrix} \begin{pmatrix} \widehat{V}_1^T \\ \widehat{V}_2^T \end{pmatrix}$$

with
$\widehat{S}_1 = diag_{1\leq k\leq n_G}(\widehat{\sigma}_k)$ and $\widehat{S}_2 = diag_{n_G+1\leq k\leq i}(\widehat{\sigma}_k)$

.

Some basic perturbation bounds for singular values of a matrix are given in (Golub and Van Loan, 1983) and (Stewart, 1973). This makes it possible to derive the following result

Lemma 6. - Consider the SVD of $\Gamma\Delta_{G/i}$ and $\widehat{\Gamma\Delta}_{G/i}$. Together with the estimation error $\mathbf{E}_{cl}$ given in theorem (5), one has

$$\sqrt{\sum_{k=1}^{i}(\sigma_k - \widehat{\sigma}_k)^2} \leq \mathbf{E}_{cl}$$

This result clearly shows that $\mathbf{E}_{cl}$ is an upper bound on the estimation error for each singular value

$$|\sigma_k - \widehat{\sigma}_k| \leq \mathbf{E}_{cl} \; ; \; \forall\, k \in [1; i]$$

The estimation of the system order depends hence on the smallest non zero singular value σ_{n_G} and the noise level through $\mathbf{E}_{cl}$. If σ_{n_G} is important enough then $\widehat{\sigma}_{n_G}$ and $\widehat{\sigma}_{n_G+1}$ will be different enough. Thus, it will be easy to separate the system singular values from those due to the disturbing effects $\widetilde{\Gamma\Delta}_{G/i}$. Otherwise, if σ_{n_G} is relatively small with respect to $\mathbf{E}_{cl}$, then it will be difficult to determine n_G.

The choice of the order leads to a low rank approximation of $\widehat{\Gamma\Delta}_{G/i}$ as follows

$$\widehat{\Gamma}_{G/i}\widehat{\Delta}_{G/i} = \widehat{U}_1\widehat{S}_1\widehat{V}_1^T$$

with

$$\widehat{\Gamma}_{G/i} = \widehat{U}_1\widehat{S}_1^{1/2}T^{-1} \text{ and } \widehat{\Delta}_{G/i} = T\widehat{S}_1^{1/2}\widehat{V}_1^T$$

The following lemma provides a bound on this low rank approximation.

Lemma 7. -

$$\|\Gamma\Delta_{G/i} - \widehat{\Gamma}_{G/i}\widehat{\Delta}_{G/i}\|_F \leq 2\mathbf{E}_{cl}$$

From realization theory, the estimation of the state space realization (A, B, C, D) depends on the left and right singular vector. An error on the impulse response of the control system transfer function $\begin{pmatrix} \mathbb{S}_{r_1u} \\ \mathbb{S}_{r_1y} \end{pmatrix}$ induces a variation on the singular subspace spanned by the singular vectors. Thus, it can be interesting to determine a bound between singular subspace spanned by $\Gamma_{G/i}$ and $\widehat{\Gamma}_{G/i}$, $\Delta_{G/i}$ and $\widehat{\Delta}_{G/i}$. The notion of canonical angle between subspace is interesting in order to compare singular subspace ((Wedin, 1983)). Let X and Y be two full column rank matrices, the angle matrix $\theta(X, Y)$ between X and Y is given by

$$\theta(X,Y)=arccos\left((X^TX)^{-1/2}X^TY(Y^TY)^{-1}Y^TX(X^TX)^{-1/2}\right)^{1/2}$$

The canonical angles between the subspace spanned by the columns of X and the columns of Y are defined to be the singular value of matrix $\theta(X,Y)$. Two subspaces are said to be closed each other if the largest canonical angle is small. From this definition, one has

$$\theta(\widehat{U}_1, U_1) = arccos\left(\widehat{U}_1^T U_1 U_1^T \widehat{U}_1\right)^{1/2}$$

and the following formulas is well known ((Li, 1999))

$$\|sin(\theta(\widehat{U}_1, U_1))\|_F = \|U_2^T \widehat{U}_1\|_F$$

Similar relation can be established for the angle $\theta(\widehat{V}_1, V_1)$.

The following result describes how the identification error on the closed loop system transfer matrix $\begin{pmatrix} \mathbb{S}_{r_1 u} \\ \mathbb{S}_{r_1 y} \end{pmatrix}$ influences the angle between singular subspaces while providing the perturbations on the matrices $\widehat{U}_1$ and $\widehat{V}_1$.

Lemma 8. - Consider the angle matrices $\theta(\widehat{U}_1, U_1)$ and $\theta(\widehat{V}_1, V_1)$. If $\sigma_{n_G} > 0$, then one has

$$\sqrt{\|sin(\theta(\widehat{U}_1, U_1))\|_F^2 + \|sin(\theta(\widehat{V}_1, V_1))\|_F^2} \leq \frac{2\sqrt{2n_G}}{\sigma_{n_G}}$$

Moreover $\widehat{U}_1$ and $\widehat{V}_1$ can be written as

$$\widehat{U}_1 = U_1 T_u + U_2 Q_u$$

with

$$\begin{cases} \|Q_u\|_F = \|sin(\theta(\widehat{U}_1, U_1))\|_F \\ \|T_u\|_F = \sqrt{n_G - \|sin(\theta(\widehat{U}_1, U_1))\|_F^2} \end{cases}$$

and

$$\widehat{V}_1^T = T_v V_1^T + Q_v V_2^T$$

with

$$\begin{cases} \|Q_v\|_F = \|sin(\theta(\widehat{V}_1, V_1))\|_F \\ \|T_v\|_F = \sqrt{n_G - \|sin(\theta(\widehat{V}_1, V_1))\|_F^2} \end{cases}$$

This result shows the continuity of the singular subspace estimate with respect to the estimation of the closed loop system transfer function $\begin{pmatrix} \mathbb{S}_{r_1 u} \\ \mathbb{S}_{r_1 y} \end{pmatrix}$. Let suppose that T_u is non singular (U_1 and $\widehat{U}_1$ are close enough) then one obtains

$$\widehat{U}_1 = (U_1 + U_2 Q_u T_u^{-1}) T_u$$

The matrix $P_u = Q_u T_u^{-1}$ describes how the subspace spanned by $\widehat{U}_1$ is close to the subspace spanned by U_1 and the matrix T_u can be interpreted as a transformation matrix. A similar remark can be made about $\widehat{V}_1$.

5. EXAMPLE

In this section simulation results are included to show the applicability of the proposed closed loop identification algorithm. The system under consideration is a discrete time model of a laboratory plant setup of two circular paltes rotated by an electrical servo motor with flexible shafts ((Hakvoort, 1990)). This example is also used in others papers about closed loop subspace identification such that (Verhaegen, 1993), (Van Overschee and De Moor, 1997) and (Katayama *et al.*, 2002).

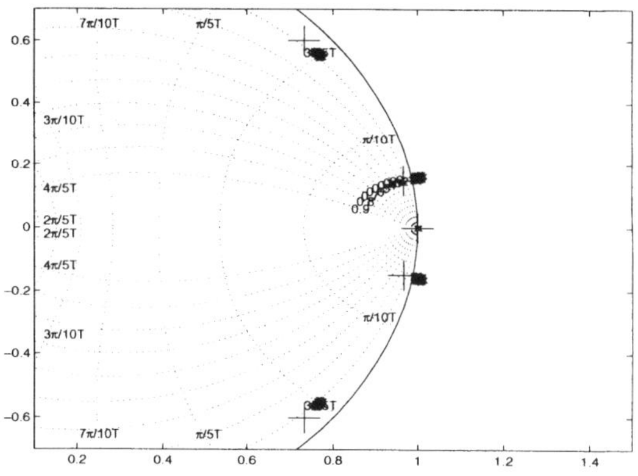

Fig. 2. Estimates of the poles obtained with the direct identification

$\mathbf{E}_{cl}$Only the external excitation signal $r_1(t)$ is used. This consists in a pseudo random binary sequence with unit variance. The model of the system and the controller are respectively given by

$$G(z) = \frac{10^{-3}(0.98z^4 + 12.99z^3 + 18.59z^2 + 3.30z - 0.02)}{z^5 - 4.4z^4 + 8.09z^3 - 7.83z^2 + 4z - 0.86}$$

$$C(z) = \frac{0.61z^4 - 2.03z^3 + 2.76z^2 - 1.83z + 0.49}{z^4 - 2.65z^3 + 3.11z^2 - 1.75z + 0.39}$$

Notice that the plant has an integrator. $v(t) = 0$ and $w(t) = H(q^{-1})e(t)$ is a Gaussian white noise sequence with variance $1/9$ filtered by a linear filter described by

$$H(z) = \frac{10^{-2}(2.89z^2 + 11.13z^1 + 2.74)}{z^3 - 2.7z^2 + 2.61z - 0.9}$$

The identification of a 5th order state space model of the plant has been performed over 1200 samples. The experiment was repeated 100 times with a fixed sequence $r_1(t)$ and different noise sequence $v(t)$. Two experiments have been carried out:

- Direct identification using PEM algorithm initialized with N4SID estimate.
- Closed loop identification using the proposed method in a PEM identification framework with and adequate initialization.

The fact that the plant contains an integrator has been taken into account in the closed loop identification. Some results are shown in Fig. 2 and 3. The latter show the eigenvalues of the state matrix, denoted by $+$, and its estimate, denoted by $\times$. Notice that the direct identification method leads to biased model on high and low frequency poles; e.g. the estimations of the low frequency poles are outside the unit circle. On the contrary the closed loop identification method performs very well from the modes identification point of view.

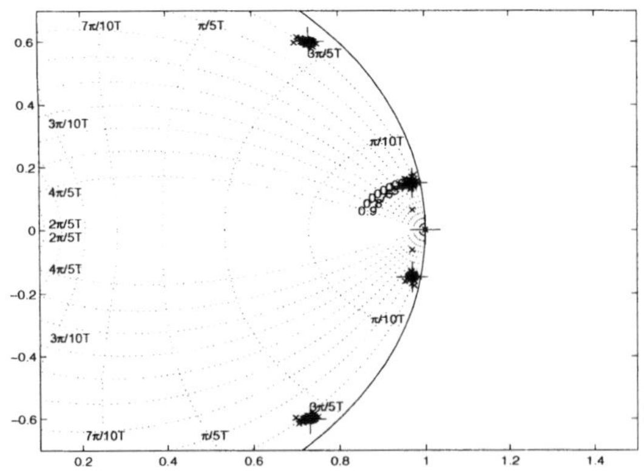

Fig. 3. Estimates of the poles obtained with the closed loop identification

6. CONCLUSION

The motivation of this paper was to develop a closed loop identification method for multivariable systems from the subspace identification culture. A free model reduction algorithm has been proposed for this end from the joint input/output closed loop identification approach. The estimation bias has been investigated in the case of noise measurement and finite data records. An illustrative problem has been addressed to demonstrate the performances of the proposed method with respect to the direct approach.

7. REFERENCES

Ebert, W., M. M'Saad and J. Chebassier (1997). Three stage procedure for closed loop identification. *Proceedings European Control conference - Brussels.*

Gevers, M. (1993). *Essays on Control : Perspectives in the Theory and its Applications*. Chap. Towards a joint design of identification and control, pp. 111–151. Birkhauser, Boston. Editors : L. Trentelman and J. C. Willems.

Golub, G.H. and C. Van Loan (1983). *Matrix computations*. Baltimore MD, John Hopkis University Press.

Hakvoort, R. (1990). Approximate identification in the controller design problem. Master thesis. Delft University of Technology.

Katayama, T., H. Kawauchi, T. Inoue and G. Picci (2002). Subspace identification of closed loop systems by orthogonal decomposition method. technical report 2002-003. Kyoto university-Kyoto.

Kung, S.Y. (1978). A new identification and model reduction algorithm via singular value decomposition. 12th *Asimolar conference on Circuits, Systems and Computers, Pacific Grove.*

Li, R. (1999). Relative perturbation theory: II. Eigenspace and singular subspace variations. *SIAM Journal of matrix analysis and applications* **20**(2), 471–492.

Ljung, L. (1999). *System identification: theory for the user*. Prentice Hall.

Mc Kelvey, T. and H. Akcay (1995). Subspace based system identification with periodic excitation signal. *Systems and control letters* **26**, 349–361.

Moore, B.C. (1981). Principal component analysis in linear systems: controllability, observability and model reduction. *IEEE Transactions on automatic control* **26**(1), 17–31.

Pouliquen, M. and M. M'Saad (2002). A closed loop identification method using subspace approach. technical report. Laboratoire d'Automatique de Procédés - ISMRA.

Stewart, G.W. (1973). *Introduction to matrix computations*. Academic Press, New York.

Van Den Hof, P.M.J. and R.J.P. Schrama (1995). Identification and control - closed-loop issues. *Automatica* **31**(12), 1751–1770.

Van Overschee, P. and B. De Moor (1996). *Subspace identification for linear systems. Theory, implementation applications*. Kluver academic Publishers.

Van Overschee, P. and B. De Moor (1997). Closed loop subspace system identification. *Proceedings* 36th *IEEE Control and Decision Conference - San Diego* pp. 1848–1853.

Verhaegen, M. (1993). Application of a subspace model identification technique to identify lti systems operating in closed loop. *Automatica* **29**(4), 1027–1040.

Verhaegen, M. (1994). Identification of the deterministic part of mimo state space models given in innovations form from input output data. *Automatica* **30**(1), 61–74.

Wedin, P.A. (1983). *Matrix pencils - B. Kargstrm and A. Ruhe*. Chap. On angles between subspaces, pp. 263–285. Springer-Verlag, New York.

Zeiger, H.P. and A.J. Mc Ewen (1974). Approximate linear realizations of given dimension via ho's algorithm. *IEEE transaction on automatic control* **19**, 153.

IFAC Publications
www.elsevier.com/locate/ifac

MODEL IDENTIFICATION OF A MULTIVARIABLE INDUSTRIAL FURNACE

Marta Barreras, Mario García-Sanz

Automatic Control and Computer Science Department
Public University of Navarre
31006, Pamplona . SPAIN
E-mail: mgsanz@unavarra.es

Abstract: This paper presents the modeling and identification of a real industrial furnace (900 kW, 38 m) used to cure large laminated composite pieces. The system exhibits a multivariable configuration with seven inputs (heat resistors and fans) and seven outputs (temperature sensors). A very complex MIMO dynamic model is carried out by solving the heat transfer equations obtained from analogous electrical circuits that represent the multi-zone heat exchange. A set of experiments have to be designed in the mentioned industrial furnace in order to generate a useful real data collection which reliably allows to identify the parameters of the multivariable model.

Keywords: Multivariable, identification, process control systems, manufacturing.

1. INTRODUCTION

The thermal dynamic behaviour of a multi-zone furnace under several load conditions is a complex process to model (Astigarraga, 1994), identify (Ljung, 1987, 1997), (Pintelon and Shoukens 2001) and control (Houpis and Lamont 1985), (D'azzo and Houpis, 1995) due to the multivariable characteristics of the system, the important loops coupling involved, and the low input excitation and low sensor accuracy. Dealing with all these difficulties, this article develops a model of a real furnace used to cure large laminated composite pieces (Areal, 1987).

The goal of this work is to identify the MIMO system and develop an accurate model that represents the real furnace behavior under every working condition. In the future, the model obtained will help to improve the present furnace controller, which will be capable of shortening the curing time in order to manufacture high quality pieces at the lowest possible cost.

The following sections are organized as follows. In section 2 the real system under study is introduced. The furnace and all its characteristics, features and elements are described. In section 3 the analytical model is presented and the heat transfer processes that occur inside are investigated. A mathematical model is carried out by solving the heat transfer equations obtained from analogous electrical circuits that represent the heat exchange. In section 4, the set of experiments designed in order to generate a useful real data collection and the estimation method are described. In addition, temperature simulations are performed and the final model obtained is shown.

2. FURNACE DESCRIPTION

The system under study is a real resistance furnace manufactured by Siflexa and located at the M.Torres company. The furnace (see Figure 1) is used to cure large laminated composite pieces. These pieces are made of a fiber-reinforced epoxy-matrix composite and are destined for supporting aerodynamic structures.

Figure 1 – Siflexa's industrial furnace for composite manufacturing located at M.Torres

2.1.-Heating chamber description

The furnace is a long stainless steel room which has seven *virtual* areas of approximately 5 metres long, except the first and the last one whose lengths are 4.8 and 7.8 metres respectively. These zones are called virtual because neither walls nor panels separate them. Due to the piece length, the heating chamber must be big enough in order to process a whole load at a time. Furnace cross section and dimensions are shown in Figure 2.

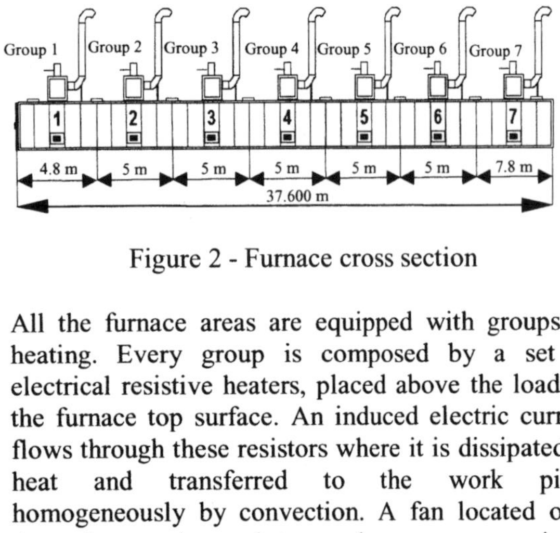

Figure 2 - Furnace cross section

All the furnace areas are equipped with groups of heating. Every group is composed by a set of electrical resistive heaters, placed above the load, at the furnace top surface. An induced electric current flows through these resistors where it is dissipated as heat and transferred to the work piece homogeneously by convection. A fan located over these heaters is used to produce a pressure-jump which creates the forced convection by rising the air velocity and therefore the heat transfer. In addition, there is a small gate located on the side that allows the entry of external air during the cooling process.

Several standard thermocouples attached to the walls, one for each zone, measure the temperature inside the furnace and twenty more sensors evenly placed over the load, give its surface temperature.

A summary of the furnace most important features, dimensions and data can be seen in Table 1.

Table 1 – Furnace characteristics

Heating chamber size (W x H x L)		4,030 m x 1,925 m x 37,600 m
Total Power		900 kW
Voltage		380 V
Zones		7
Gates		7
Heating Elements	Number	7
	Name	Resistor
	Power	85 kW
Fans	Number	7
	rpm	1500
Thermocouples		7 (on the walls)
		20 (on the load)

2.2.-Control

The manufacturing technique used in the furnace (Areal, 1987) exposes the composite to the required pressures and temperatures, about 120 Celsius degrees, for a predetermined time which depends on the cure cycle (Areal, 1987) and the load size and weight.

Nowadays, if the load distribution is homogeneous, the temperature is acceptably controlled in each zone by seven different digital PIDs which have all the same parameters. As can be seen in Figure 3, the power input is represented by OP_i signal, besides there is a feedback of the output temperature (Ti) which is compared with the reference temperature, fixed according to the curing program specifications.

The main problem with these PIDs is the physical coupling of the zones. Underloaded zones transfer heat to those which have more load. That leads to couplings between several inputs obtaining, as a result, a complex model which is difficult for the PIDs to handle.

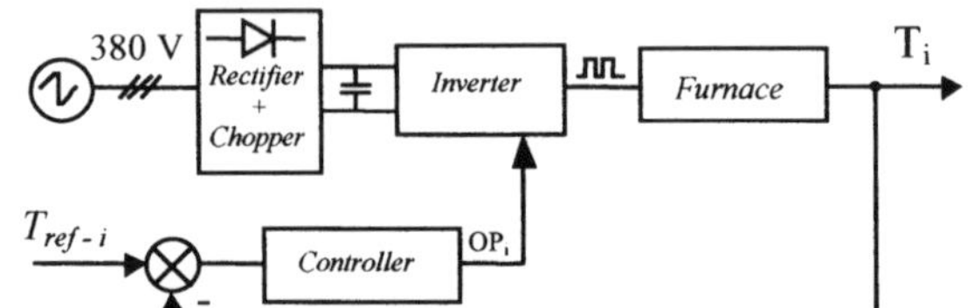

Figure 3 – Block diagram of control loop and furnace

Due to the above reasons, the present control loop needs to be improved by taking into account the multivariable characteristic of the process. In addition the new controller will have to be able to optimise the curing process considering the load weight and its distribution inside the furnace.

However, this future improvement cannot be achieved without a good model of the system which gathers the important features of the process behaviour. In that context, the following sections describe the analytical and experimental model of the furnace and estimate the main process parameters.

3. FURNACE MODEL

3.1 Multivariable electrical model

This section develops the furnace model describing the multivariable system as an electrical pi-model (Wellstead, 1979), represented by the scheme shown in Figure 4.

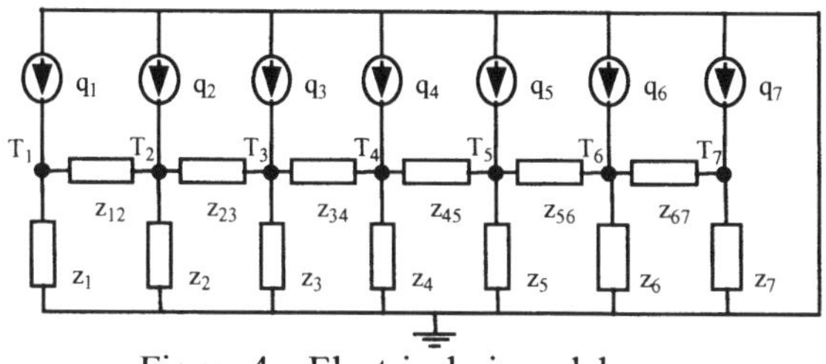

Figure 4 – Electrical pi-model

Kirchoff's current and voltage laws will be used to determine heat fluxes and temperatures respectively. By applying a node electrotechnical analysis, the following system is obtained:

$$\mathbf{G}(s)\,\mathbf{T}(s) = \mathbf{Q}(s) \tag{1}$$

Where **T**(s) is a 7x1 vector that represents the air temperatures read by the thermocouples in each zone and **Q**(s) is a 7x1 vector which symbolises the heat flux supplied by the resistors,

$$\mathbf{T}(s) = \begin{bmatrix} T_1(s) \\ T_2(s) \\ T_3(s) \\ T_4(s) \\ T_5(s) \\ T_6(s) \\ T_7(s) \end{bmatrix} ; \quad \mathbf{Q}(s) = \begin{bmatrix} q_1(s) \\ q_2(s) \\ q_3(s) \\ q_4(s) \\ q_5(s) \\ q_6(s) \\ q_7(s) \end{bmatrix} \tag{2}$$

G(s) is a 7x7 tri-diagonal matrix as can be seen in equation (3). Its elements are a sum of conductances, so that,

$$g_i(s) = \frac{1}{z_i(s)} \quad ; \quad g_{ij}(s) = \frac{1}{z_{ij}(s)} \tag{4}$$

where $z_i(s)$ and $z_{ij}(s)$ represent the thermal impedance in one zone and in the adjoining zone respectively.

Consequently, the equation which connects heat fluxes and temperatures is:

$$\mathbf{T}(s) = \mathbf{G}(s)^{-1}\,\mathbf{Q}(s) \tag{5}$$

The elements of $\mathbf{G}^{-1}(s)$ are extremely complex. Hence, further research is needed in order to estimate these parameters by applying another complementary method.

3.2 Analytical zone model

The analytical model of each zone is based on a heat transfer study (Chapman, 1990) that allows to obtain simple expressions which will be compared with the experimental results.

By a thorough study of the process (Chapman, 1990), it can be noticed that forced convection is the dominant heat exchange mechanism inside the furnace. Although radiation contributes to the energy exchange, below 700°C it can be disregarded (Astigarraga, 1994), (Chapman, 1990). Conduction heat transfer is considered through the walls and load.

Quite a few basic heat transfer equations will be used to get the model that describes the furnace behaviour. First of all, several hypotheses will be set in order to reduce complexity and shorten the operations.

- It has been assumed that the material properties are uniform in the whole part.
- As the records of the temperature measurements are long, (9 hours) the slight effects of the external environment should be taken into consideration. In this model, this effect is assumed a constant external temperature because its variation is very small.

Apart from the previous assumptions, the uniformity of surface load temperature facilitates the calculations.

Experimental temperatures measured in different points of the load surface are plotted in Figure 5. As can be seen, they are all very similar, not only in the initial time but also during all the cure cycle. This uniformity is really helpful to the analysis because makes it quicker and more straightforward.

$$\mathbf{G}(s) = \begin{bmatrix} g_1(s)+g_{12}(s) & -g_{12}(s) & 0 & 0 & 0 & 0 & 0 \\ -g_{12}(s) & g_{12}(s)+g_2(s)+g_{23}(s) & -g_{23}(s) & 0 & 0 & 0 & 0 \\ 0 & -g_{23}(s) & g_{23}(s)+g_3(s)+g_{34}(s) & -g_{34}(s) & 0 & 0 & 0 \\ 0 & 0 & -g_{34}(s) & g_{34}(s)+g_4(s)+g_{45}(s) & -g_{45}(s) & 0 & 0 \\ 0 & 0 & 0 & -g_{45}(s) & g_{45}(s)+g_5(s)+g_{56}(s) & -g_{56}(s) & 0 \\ 0 & 0 & 0 & 0 & -g_{56}(s) & g_{56}(s)+g_6(s)+g_{67}(s) & -g_{67}(s) \\ 0 & 0 & 0 & 0 & 0 & -g_{67}(s) & g_{67}(s)+g_7(s) \end{bmatrix} \tag{3}$$

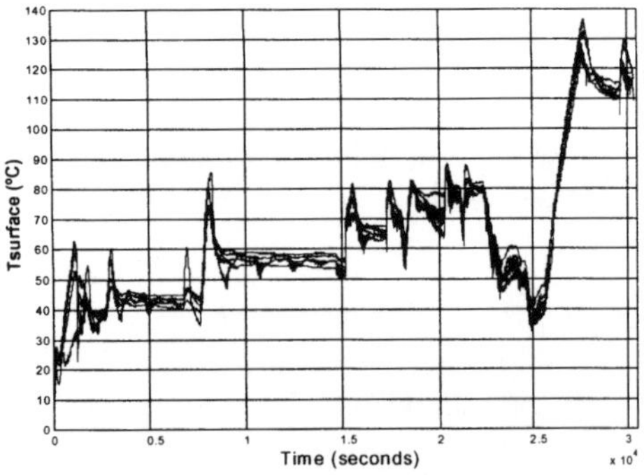

Figure 5 – Surface temperatures

Once all the energy balances that occur inside the furnace have been established (Wellstead, 1979), (Astigarraga, 1994), (Chapman, 1990), (García-Sanz, 1997), an electrical circuit is used to model the heat transmission differential equations that describe every zone of the furnace (see Figure 6).

The thermal characteristics of the furnace vary continuously in time and space. Therefore an accurate model would be described with distributed parameters. As this furnace exhibits high thermal capacity, a lumped approach can be used to build a thermal capacitance model (Hudson, 1999).

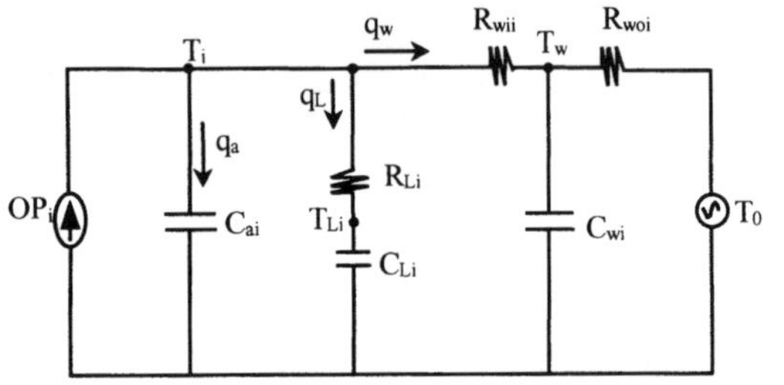

Figure 6 – Thermal - electric analogy in one zone

The analogous equivalencies shown in Figure 6 are as follows: voltages represent temperatures, currents are heat fluxes, electrical resistances correspond to heat transmission resistances and finally electrical capacities symbolise thermal capacities (Wellstead, 1979), (Chapman, 1990),(García-Sanz, 1997). The physical meaning of every element is shown in the Nomenclature section.

Taking advantage of the extra information about the system, obtained from physical insight and experimental observations, and using the analogous electrical circuit, the thermodynamic equations are formulated. Obviously the previous hypotheses will be applied.

First of all the non-steady state heat balance equations that govern the process are formulated from the electric analogy point of view (Astigarraga, 1994), (García-Sanz, 1997). Global heat balance or heat energy conservation is given by equation (6). The input heat flux (OP_i), provided in each zone, is equal to that absorbed by the work piece represented in equation (8), and by the air - equation (7) - and the portion lost through the walls of the chamber, (9).

$$OP_i = q_a + q_L + q_w \tag{6}$$

$$q_a = C_{ai}\frac{dT_a}{dt} \tag{7}$$

$$q_L = C_{Li}\frac{dT_L}{dt} = \frac{T_i - T_L}{R_{Li}} \tag{8}$$

$$q_w = C_{wi}\frac{dT_w}{dt} = \frac{T_i - T_0}{R_{w\,ii}} \tag{9}$$

Applying the lumping capacitance approach (Houpis, 1985), the differential equation that describes the furnace can be written, so that,

$$OP_i = (C_{wi} + C_{ai})R_{Li}C_{Li}\frac{d^2T_L}{dt^2} + (C_{wi} + C_{ai} + C_{Li})\frac{dT_L}{dt} \tag{10}$$

The above expression is translated into the s-domain using Laplace transform (Houpis, 1985). As a result the following equation is obtained. It relates the input heat flux in one zone to the temperature measured in the same zone.

$$\frac{T_i(s)}{OP_i(s)} = z_i(s) = \frac{R_{Li}C_{Li}\ s+1}{s\big((C_{ai}+C_{wi})R_{Li}C_{Li}\ s+(C_{ai}+C_{wi}+C_{Li})\big)} \tag{11}$$

Right hand term in the equation (11) represents the thermal impedance within one zone.

A new mathematical expression, similar to the previous one, can be developed with the aim of describing the adjoining zone thermal impedance.

$$\frac{T_{ij}(s)}{OP_{ij}(s)} = z_{ij}(s) = \frac{R_{Lij}C_{Lij}\ s+1}{s\big((C_{aij}+C_{wij})R_{Lij}C_{Lij}\ s+(C_{aij}+C_{wij}+C_{Lij})\big)} \tag{12}$$

Parameters in equation (12) are calculated as followed,

$$R_{ij} = \frac{R_i + R_j}{2} \qquad C_{ij} = \frac{C_i + C_j}{2} \tag{13}$$

Equations (12) and (13) describe the effect that the heat flux provided by the heaters located in one area has on the temperature measured in the contiguous area. The impedance between two zones is the same whether the heat is transferred from one zone or from the other, consequently $R_{ij} = R_{ji}$. All these results are substituted in equations (4) and (5). The final expression obtained is extremely complex to operate with because further calculations reveal that the 49 elements of $\mathbf{G}^{-1}(s)$ have a very high order (70/70).

Next step will estimate the parameters of the obtained model by using the real data measured in different experiments.

4 ESTIMATION METHOD

4.1 Main steps of the estimation method

The previous results yield important knowledge about the structure of the transfer functions that represent the system behaviour. The knowledge of the model order is a really precious information to use in identification toolboxes (Ljung, 1997). The estimation method described in this section can be divided into two steps:

Firstly $z_i(s)$ parameters are estimated (Ljung, 1997) by using the information provided by the data measured during a real experiment. The complexity of the equation (5) leads to a really difficult identification problem. For this reason a new set of experiments is proposed where the air temperature of every zone must be the same (this is achieved by the PIDs controllers). Now if each zone has the same temperature, as shows Figure 7a, no heat transfers between the zones will occur. Thus, the multivariable problem is reduced to a monovariable one where only relations between the heat flux supplied in one zone and the measurements in the same zone $z_i(s)$ are considered.

Secondly $z_{ij}(s)$ elements are determined from a different set of experiments. This time, the present controller accomplishes to get the same temperature in all the heating chamber zones except one (the first one in this application), as can be seen in Figure 7b. Now the $z_{12}(s)$ transfer function is excited and estimated.

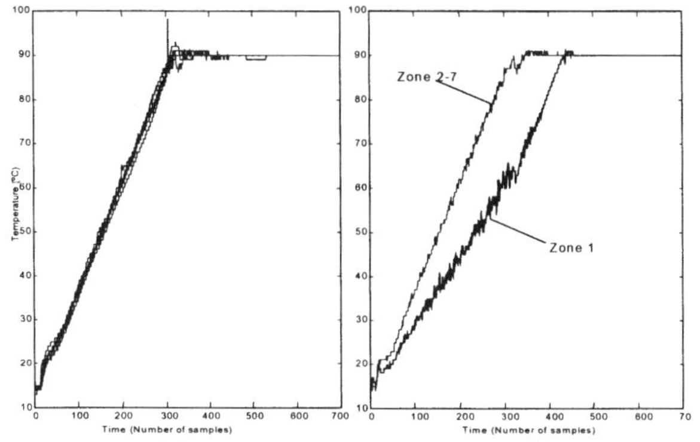

(a) (b)

Figure 7 – Output temperature measurements

As the analytical matrix element structure is known, the rest of the parameters can be calculated by extrapolating the previous results following the size, geometry and weight of the work piece.

The estimated parameters calculated for the multivariable electrical pi-model and the electrical thermal analogy are shown in Table 2 and Table 3.

Viewing the fit between the real system output along with the identified one (for instance, zones 5 & 7 in Figure 8), it can be concluded that this method finds a good and acceptable model which shows a good behaviour.

Table 2 – Parameters of the zone model

	ZONE i						
	1	**2**	**3**	**4**	**5**	**6**	**7**
R_{Li}	41.75	12.95	8.62	6.48	5.05	4.12	2.5
C_{Li}	99.11	616.3	1380.7	2456.1	4227	5970.5	12003.7
C_{ai}	45.01	46.34	45.53	44.38	42.5	40.65	34.24
C_{wi}	4432.4	5046	5046	5046	5046	5046	5512.1

(Units according to Nomenclature section)

Table 3 – Parameters of the coupling model

	ZONE ij					
	12	**23**	**34**	**45**	**56**	**67**
R_{Lij}	27.35	10.78	7.55	5.76	4.58	63.31
C_{Lij}	357.7	998.5	1918.4	3341.8	5100	8987
C_{aij}	45.67	45.93	45	43.44	41.57	37.44
C_{wij}	4740	5046	5046	5046	5046	5280

(Units according to Nomenclature section)

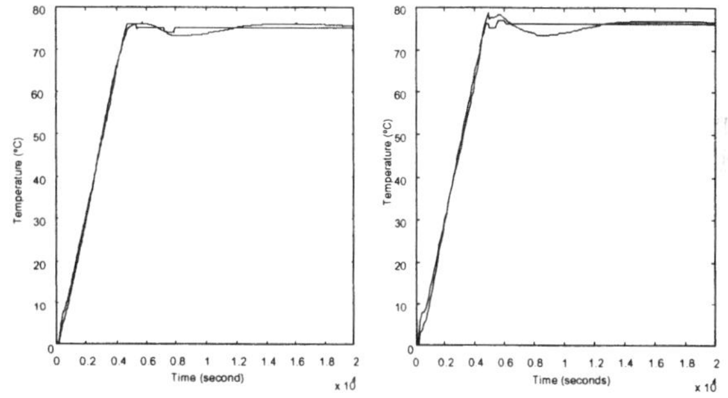

(a) (b)

Figure 8 - Measured data vs estimated data
(a) Zone 5, (b) Zone 7

4.2 Identification problems

Finally it is interesting to mention some of the main problems found during the identification process.

First of all, the feedback in data. It makes the furnace difficult to identify. The existing closed loop correlates the output with the input. A possible remedy could be to eliminate the close loop. However in this application it is no possible at all, because an uncontrolled temperature may deteriorate the composite. Consequently the identification must be handled with the closed loop.

Secondly, the effect of oversampling. It is often overlooked but its consequences are severe. In this application the data were sampled every 15 seconds. That is quite a fast sampling for this thermal process and for the sensor precision. In this case the precision is 1°C. Figure 9 shows one example where the measured temperature does not change until the power increase of nearly 33% takes place. As a conclusion the data set was resampled and the new sampling interval was considered equal to 120 seconds.

Finally it is important to mention the lack of excited input signals in this process. This has a strong impact on the identification problem and adds an extra difficulty. However it is not possible to modify them because they only depend on the load and the treatment chosen to cure it.

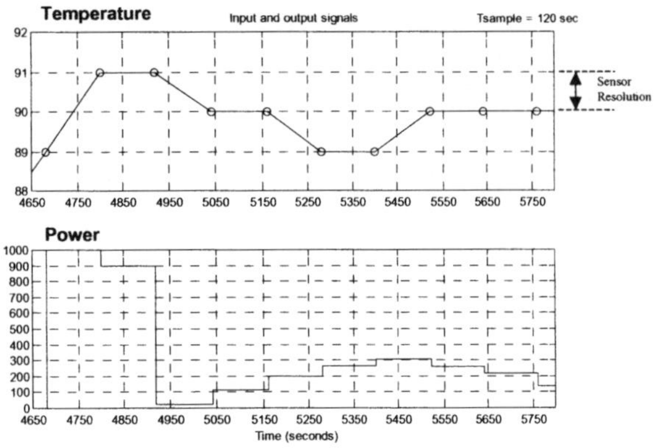

Figure 9 – Sensor precision

5. CONCLUSIONS

Thermal dynamic behaviour of a multi-zone furnace under several load conditions is a complex process challenging to model, identify and control. In that context, the present paper has shown the modeling and identification of a real industrial furnace (900 kW, 38 m long) used to cure large laminated composite pieces which exhibits a multivariable performance with seven inputs (heat resistors and fans) and seven outputs (temperature sensors).

A very complex MIMO dynamic model has been obtained by solving heat transfer equations from analogous electrical circuits that represent the multi-zone heat exchange. To identify the parameters of the multivariable model, special experiments have been performed in the existing furnace in order to generate a collection of useful real data. The final model has been validated with a new set of experiments that excite the seven zones of the furnace simultaneously.

ACKNOWLEDGEMENTS

The authors gratefully appreciate the support given by Siflexa and M.Torres companies and the Spanish 'Comisión Interministerial De Ciencia y Tecnología' (CICYT) under grant DIP2000-0785.

REFERENCES

Areal, A.(1987). *An introduction to composite materials*. Cambridge University Press.

Astigarraga, J. (1994). *Hornos industriales de resistencias: teoría, cálculo y aplicaciones*. McGraw-Hill.

Chapman, A.J., (1990). *Transmisión de calor*. 3rd Edition. Bellisco.

D'azzo, J.J, Houpis, C.H., (1995). Linear Control. System Analysis and design. 4th ed. McGraw-Hill.

García-Sanz, M. (1997). *A reduced model of central heating systems as a realistic scenario for analyzing control strategies.* Appl. Math. Modelling. Vol **21**, September, pp.535-545.

Houpis, C.H., Lamont, G.B. (1985). *Digital Control Systems: Theory, Hardware, Software*. 2nd Edition, McGraw-Hill.

Hudson, G.,Underwood,C.P. (1999). *A simple building modeling procedure for Matlab/Simulink. Proceedings of the 6th ICBPS.* pp: 777-783.

Ljung, L. (1987). *System Identification:theory for the user*. Prentice Hall Information and System Science Series.

Ljung, L.(1997). *System Identification Toolbox User's Guide.* The Mathworks, Inc.

M.Torres Diseños Industriales, S.A. Crta. Pamplona-Huesca, Km. 9. 31119 Torres de Elorz (Spain).

Pintelon, R., Shoukens, J. (2001). *System Identification: a frequency domain approach.* IEEE Press

Siflexa Tecnología Tratamiento. Polígono .Industrial. Masia D'Espi 46930 (Spain)-

Wellstead, P.E. *Introduction to Physical System Modelling*. Academic Press, New York, 1979.

NOMENCLATURE

Temperature - Units = [K]

T_L = Load internal temperature

Tw = Wall internal temperature

T_i = Inside air furnace temperature

T_0 = Outside air temperature

Heat flux per zone - Units = [Kw]

OP = Total heat flux from a resistance group

q_a = Heat flux in the air inside the furnace

q_c = Load heat flux

q_p = Conduction losses through the walls

Resistance - Units = [m² °C/w]

$R_L = (1/hA_L)+(Ln(Re/Ri)/2\pi kL)$=Load thermal resistance

R_{wi} = Input wall heat conduction resistance

R_{wo} = Output wall heat conduction resistance

Capacity - Units =[KJ/ °C]

$C_w = V_w\ \rho_w\ c_{pw}$ = Wall heat capacity

$C_a = V_a\ \rho_a\ c_{pa}$ = Internal air thermal capacity

$C_L = V_L\ \rho_{aL} c_{pL}$ = Load thermal capacity

Film coefficient - Units = [w/m² °C]

h =Furnace air convection coefficient

Density - Units=[Kg/m³]

ρ_a = 1.2 = air density

$\rho_L = 1.6\ 10^3$ = Load density

ρ_w = wall material density

Specific heat - Units=[KJ/ Kg°C]

c_{pair} = 1.01 = air specific heat

$C_{pc} = 712\ 10^{-3}$ = Load specific heat

c_{pw} =wall specific heat

Volume, area & length

V_a = Air volume [m³]

V_L = Load volume [m³]

L=Load length [m]

A_L = Load surface area [m²]

Re=External load radio [m]

Ri=Internal load radio [m]

***Condutivity* - Units=[w/ m°C]**

k = Load thermal conductivity [w/ m°C]

IFAC Publications
www.elsevier.com/locate/ifac

EXTENDED FUZZY GK CLUSTERING WITH APPLICATION TO IDENTIFICATION OF AN AUTOMATIC VOLTAGE REGULATION LOOP DYNAMICS

Lixin Ren, G. W. Irwin

Intelligent Systems and Control Group
School of Electrical and Electronic Engineering
Queen's University, Belfast

Abstract: A new strategy for nonlinear identification which combines an extended fuzzy Gustafson-Kessel (GK) clustering technique using competitive agglomeration with locally weighted least-squares is proposed. This is applied to nonlinear identification of the highly nonlinear loop dynamics of an automatic voltage regulation (AVR) loop across a wide operating range. A Takagi-Sugeno fuzzy model is automatically constructed from plant input-output response data set. The validity of the resultant TS fuzzy model representation is confirmed on a validated simulation of a 3kVA laboratory micro-machine. *Copyright © 2003 IFAC*

Keywords: fuzzy clustering, Takagi-Sugeno fuzzy system, nonlinear identification, turbogenerator AVR

1. INTRODUCTION

In addition to analytical approaches, models for complex and nonlinear systems can also be constructed from data. Such approaches are the so-called data-driven modelling techniques. Recently, data-driven fuzzy modelling of nonlinear dynamic systems using Takagi-Sugeno (TS) fuzzy models (Zhao *et al.*, 1994; Babuska, 1998) offers a general and practical alternative, which captures the plant nonlinearity in a smooth and elegant way through fuzzy set representation. In this paper, an extended fuzzy GK clustering technique using competitive agglomeration (Frigui and Krishnapuram, 1997) combined with a locally weighted least-squares technique for nonlinear identification is proposed and applied to identification of the AVR loop dynamics for a turbogenerator using a TS fuzzy system.

Turbogenerators are the important components of most electric power systems and their performance is directly related to the security and stability of the power system operation. A turbogenerator is both complex and inherently highly nonlinear with dynamic characteristics that vary as operating conditions change (Anderson and Fouad, 1977; Hogg, 1981). The increasing complexity of the deregulated electrical power systems, and their increased interconnection make the electric power plant unit operate in wide operating conditions, which drives the turbogenerator operated outside its linear region. This increases the need to investigate effective nonlinear modelling in order to achieve wider control range for such machine.

Based on the previous study (Ren, *et al.,* 2002), an extended fuzzy GK clustering technique is proposed in this paper and used to achieve automatic construction from plant input-output data. A two-input, single-output nonlinear AutoRegressive with eXgenous output (NARX) model of the dynamics of a turbogenerator AVR loop is automatically constructed from four Tagaki-Sugeno linear ARX sub-models. It takes account of the influences of both excitation input and turbine governor input to the terminal voltage response for wide range operation. The paper includes results from tests on a validated simulation of a 3 kVA laboratory turbogenerator. These confirm the efficacy of the approach for nonlinear identification of the nonlinear AVR loop. Section 2 describes the extended GK clustering technique using competitive agglomeration in details and its use for TS fuzzy model construction. Section 3 then illustrates the turbogenerator system and the associated nonlinear AVR loop. TS fuzzy model identification of the AVR loop dynamics using the extended GK clustering technique and a locally-weighted least-squares technique is addressed in Section 4 followed by discussion and conclusions.

2. EXTENDED FUZZY GK CLUSTERING FOR TS FUZZY MODEL CONSTRUCTION

2.1 Fuzzy product space clustering for nonlinear regression

Identification of nonlinear dynamical systems can be transformed into a static nonlinear regression problem ((Leonaritis and Billings, 1985). For input-output modelling, a relationship between the past input-output data and the future output need to be determined, which leads to the NARX model structure. Consider a multiple input single output (MISO) system with n_i inputs and one output. A finite number of past inputs $\boldsymbol{u}(k)$ and outputs $y(k)$ are collected into the regression vector

$$\phi(k) = [y(k), y(k-1),..., y(k-n+1), u_1^m(k), u_2^m(k),..., u_{n_i}^m(k)]^T \tag{1}$$

where $\boldsymbol{u}_l^m(k)=[\boldsymbol{u}_l(k),\boldsymbol{u}_l(k-1),...,\boldsymbol{u}_l(k-m+1)]^T$ $l=1..n_i$, n and m are the orders of the output and input respectively.

Suppose the smooth nonlinear system that generated the data is described by

$$y(k) = f(\phi(k))+\varepsilon \tag{2}$$

where the deterministic nonlinear function f captures the dependence of y on $\phi(k)$ and ε is the additive stochastic component which reflects the dependence of y on quantities other than $\phi(k)$.

From a geometrical viewpoint, the unknown nonlinear function $y(k) = f(\phi(k))$ represents a nonlinear hypersurface. This is a subspace of dimension $m \times n_i + n + 1$, called the regression surface, in the product space of the regression vector $\phi(k)$ and the output variable y.

Fuzzy product space clustering partitions the product space of regression matrix and output data set of the dynamic systems into overlapping groups such that the clusters produced describe the underlying structure within the data. It allows for imprecision in the data. The technique can be used for nonlinear regression, i.e. modelling the dependence of the output variable $y(k) \in \mathrm{Y} \subset \mathrm{R}$ on the regression vector $\phi(k)$ over some domain $\phi(k) \in \mathrm{D} \subset \mathrm{R}^p$ containing the available data.

2.2 Extended Fuzzy GK Clustering Algorithm

GK Clustering Algorithm The aim of using fuzzy clustering is to determine the rule antecedent parameters of a Takagi-Sugeno fuzzy model, which approximates the nonlinear hypersurface in a piecewise manner by hyperplanes. The Gustafson-Kessel (GK) clustering algorithm (Gustafson and Kessel, 1979), which is capable of detecting clusters in a set of feature attribute vectors formed from plant input-output data sets, can be used to achieve this goal. It has therefore been intensively studied for dynamical system modelling (Babuska, 1998).

Assume that necessary data for identification has been collected and a set of N data pairs $(\phi(k), y(k)), k = 1,..., N$ is available and denoting $\mathbf{x} = [\phi(k), y(k)]^T$, a q-dimensional feature vector where $q = p+1, p = m \times n_i + n$, the data set can then be written in a matrix form. A set of N feature vectors can then be represented as a $q \times N$ data matrix

$$\mathrm{X} = \begin{bmatrix} \mathbf{x}_1 \\ \mathbf{x}_2 \\ \vdots \\ \mathbf{x}_N \end{bmatrix} = \begin{bmatrix} x_{11} & x_{12} & \cdots & x_{1q} \\ \cdots & \cdots & \cdots & \cdots \\ x_{N1} & x_{N2} & \cdots & x_{Nq} \end{bmatrix} \tag{3}$$

Fuzzy clustering partitions the data set X into K fuzzy clusters, forming a fuzzy partition in X denoted by a partition matrix U, whose elements $\mu_{ik} \in [0,1]$ represent the membership degree of $\mathbf{x}_k$ to the cluster i relative to all other clusters. The GK clustering algorithm finds the partition matrix U by minimising the distance of the datum $\mathbf{x}_k$ to the prototypes, which are represented by an objective function of the following form:

$$\mathrm{J(X,V,U)} = \sum_{i=1}^{M}\sum_{k=1}^{N}\mu_{ik}^l \mathrm{d}^2(\mathbf{x}_k, \mathbf{v}_i) \tag{4}$$

subject to:

$$\sum_{i=1}^{M}\mu_{ik} = 1 \quad k = 1,..., N \tag{5}$$

$$0 < \sum_{k=1}^{N}\mu_{ik} < 1, \quad i = 1,...,M \tag{6}$$

Here l>1 is a parameter that controls fuzziness of the clusters, with higher values of l the clusters overlap more. Typically l=2 is used. The function $\mathrm{d}(\mathbf{x}_k, \mathbf{v}_i)$ is the distance of the data vector $\mathbf{x}_k$ from the cluster prototype $\mathbf{v}_i$. The constraint (5) avoids the trivial solution U=0 and the constraint (6) guarantees that clusters are neither empty nor contain all the points to degree 1. The clustering is implemented by minimise fuzzy objective function (4) subject to (5) and (6). The shape of clusters is determined by the particular distance measure $\mathrm{d}(\mathbf{x}_k, \mathbf{v}_i)$ involved. The GK algorithm employs an adaptive distance measure:

$$d^2(\mathbf{x}_k, \mathbf{v}_i) = (\mathbf{x}_k - \mathbf{v}_i)A_i(\mathbf{x}_k - \mathbf{v}_i)^T \qquad (7)$$

where A_i is a positive definite symmetric matrix called the *distance inducing matrix* adapted according to the actual shapes of the individual clusters approximately described by the cluster covariance matrices P_i:

$$P_i = \frac{\sum_{k=1}^{N} \mu_{i,j}^{l}(\mathbf{x}_j - \mathbf{v}_i)(\mathbf{x}_j - \mathbf{v}_i)^T}{\sum_{k=1}^{N} \mu_{ik}^{l}}, 1 \leq i \leq M \qquad (8)$$

The distance-inducing matrix A_i is an optimisation variable in (4) and calculated as a normalised inverse of the cluster covariance matrix:

$$A_i = \det(P_i)^{\frac{1}{d+1}} P_i^{-1} \qquad (9)$$

Normalisation by the determinant of P_i is involved in order to avoid a trivial solution of (4) for $A_i = 0$.

Extended Fuzzy GK Algorithm Besides the advantages of using conventional GK clustering algorithm for identification, a major problem for conventional GK algorithm is that the number of clusters has to be pre-specified in advance, which may be difficult when there is little knowledge available for the plant. This problem could be solved using GK clustering through competitive agglomeration (Frigui and Krishnapuram, 1997). Clustering through competitive agglomeration was first proposed for use in image segmentation. In competitive agglomeration clustering, the objective function J in (4) is modified to the following form by adding a regularisation term α (l=2),

$$J_{CA}(X, U, V) = \sum_{i=1}^{M}\sum_{k=1}^{N}(\mu_{ik})^2 d^2(\mathbf{x}_k, \mathbf{v}_i) - \alpha \sum_{i=1}^{M}\left[\sum_{k=1}^{N}\mu_{ik}\right]^2 \qquad (10)$$

The modified objective function in (10) consists of two components. The first is the sum of squared distances from data points to the prototypes weighted by its constrained memberships. This component allows control of the shapes and sizes of the clusters by selection of different distance measures. A global minimum of zero for this component is achieved when the number of clusters M is equal to the number of samples N, i.e., each cluster contains a single data point. The second term in (10) is the sum-of-squares of the cardinalities of the clusters, which controls the number of clusters. The global minimum of this term is achieved when all data points are lumped in a single cluster and all other clusters are empty. When both components are combined and α is chosen properly, the final partition will minimise the sum of intra-cluster distances, while partitioning the data set into the smallest possible number of clusters.

To minimise J_{CA} with respect to U, Lagrange multipliers λ_k are added to remove the constraints, which gives

$$J_{CA}(X, U, V) = \sum_{i=1}^{M}\sum_{k=1}^{N}(\mu_{ik})^2 d^2(\mathbf{x}_k, \mathbf{v}_i) - \alpha \sum_{i=1}^{M}\left[\sum_{k=1}^{N}\mu_{ik}\right]^2 - \sum_{k=1}^{N}\ddot{e}_k\left(\sum_{k=1}^{M}\mu_{ik} - 1\right) \qquad (11)$$

The updating equation for the memberships μ_{ik} is subsequently obtained by solving the set of equations with fixed V.

$$\frac{\partial J_{CA}}{\partial \grave{\imath}_{ik}} = 2\mu_{ik} d^2(\mathbf{x}_k, \mathbf{v}_i) - 2\alpha \sum_{k=1}^{N}\mu_{ik} - \ddot{e}_k = 0, \; for \; i \in \{1, ..., M\}, k \in \{1, ..., N\} \qquad (12)$$

Equation (12) consists a set of $N \times M$ linear equations in the $N \times M + N$ unknown variables μ_{ik} and $\ddot{e}_k$. These $N \times M + N$ variables can be found using (12) in conjunction with the N equations resulting from the constraints in (6). The solution can be simplified considerably with the reasonable assumption that the membership values do not change significantly between two successive iterations. The term $\sum_{k=1}^{N}\mu_{ik}$ in (12) is then calculated using the membership values from the previous iteration to give

$$\mu_{ik} = \frac{2\alpha \times N_i + \ddot{e}_k}{2d^2(\mathbf{x}_k, \mathbf{v}_i)} \qquad (13)$$

where $$N_i = \sum_{k=1}^{N}\mu_{ik} \qquad (14)$$

Here N_i is called the cardinality of the cluster i. Substituting (14) into the constraints (6) then gives

$$\alpha \sum_{i=1}^{M}\frac{N_i}{d^2(\mathbf{x}_k, \mathbf{v}_i)} + \ddot{e}_k \sum_{i=1}^{M}\frac{1}{2d^2(\mathbf{x}_k, \mathbf{v}_i)} = 1 \qquad (15)$$

Solving (15) for $\ddot{e}_k$ produces the result

$$\ddot{e}_k = \frac{1 - \alpha \sum_{i=1}^{M}\frac{N_i}{d^2(\mathbf{x}_k, \mathbf{v}_i)}}{\sum_{i=1}^{M}\frac{1}{2d^2(\mathbf{x}_k, \mathbf{v}_i)}} \qquad (16)$$

Substituting for $\ddot{e}_k$ in (13) from (16), results in the following update equation for the membership of the data point $\mathbf{x}_k$ in cluster $\mathbf{v}_i$.

$$\mu_{ik} = \frac{\frac{1}{2d^2(\mathbf{x}_k,\mathbf{v}_i)}}{\sum_{i=1}^{M}\frac{1}{2d^2(\mathbf{x}_k,\mathbf{v}_i)}} + \frac{\alpha}{d^2(\mathbf{x}_k,\mathbf{v}_i)}\left(N_i - \frac{\sum_{i=1}^{M}\frac{N_i}{2d^2(\mathbf{x}_k,\mathbf{v}_i)}}{\sum_{i=1}^{M}\frac{1}{2d^2(\mathbf{x}_k,\mathbf{v}_i)}}\right) \tag{17}$$

Agreeably, the first term in (17) is the membership term for the fuzzy C-Means algorithm, which considers only the relative distances of the data point to all clusters. The second component in (17) is a signed bias term, which depends on the difference between the cardinality of the cluster of interest and the weighted average of the cardinalities from the viewpoint of the data points $\mathbf{x}_k$.

This bias term is positive for clusters with cardinality higher than the average, thus appreciating the membership value. On the other hand, the negative bias term for lower cardinality clusters will depreciate their membership values. Moreover, since the bias term in (17) is also inversely proportional to the distance of the data point $\mathbf{x}_k$ to the cluster of interest $\mathbf{v}_i$, it serves as an amplification factor, which will heavily reduce the membership of the data point for low-cardinality clusters when their distance to such clusters is low.

When the cardinality of a cluster drops below a pre-specified threshold, the cluster is discarded and the number of clusters is updated. The algorithm starts with an over- specified number of clusters, such that each data point is approximated by many small clusters at the beginning. As the algorithm proceeds, the second term in (17) causes each cluster to expand and include as many data points as possible. At the same time, the constraint in (6) causes adjacent clusters to compete. As a result, only a few clusters will survive, while the others will shrink and eventually vanish.

The choice of α reflects the balance of the second term relative to the first in the modified objective function J_{CA} given in (10). If α is too small, the second term will be neglected, and the number of clusters will not be reduced. If α is too large, the first term will be neglected and all the data points will be lumped into just one cluster. The value of α should therefore be chosen such that both terms are of the same order of magnitude. It can usefully be set to be proportional to the ratio of the two terms (Frigui and Krishnapuram, 1997), such that

$$\alpha \propto \frac{\sum_{i=1}^{M}\sum_{k=1}^{N}(\mu_{ik})^2 d^2(\mathbf{x}_k,\mathbf{v}_i)}{\sum_{i=1}^{M}\left[\sum_{k=1}^{N}\mu_{ik}\right]^2} \tag{18}$$

2.3 TS Fuzzy Model Identification

For a Takagi-Sugeno fuzzy model, each local sub-model or fuzzy consequent regression structure is valid for a certain range of operating conditions. An interpolative scheduling mechanism in the shape of fuzzy antecedent structure then combines the outputs of the local models into a global output. Construction of a Takagi-Sugeno fuzzy model requires identification of the antecedent and consequent structure, identification of the membership functions for different operating regions and estimation of the consequent regression parameters. While the latter task can be solved using a linear estimation technique, the construction of the membership functions constitutes a nonlinear optimisation problem.

Fuzzy clustering can be used to determine the regions in the product space of the input and output variables within which the system can be approximated locally by simple model, such as linear sub-models. By means of extended fuzzy GK clustering, the available data set is hence partitioned into fuzzy subsets, or fuzzy clusters that can be well approximated locally by linear regression models. The parameters of these models can then be estimated by a least-squares technique.

After generating the parameters of rule antecedents by the extended GK clustering, the consequent parameters are obtained using a weighted least-squares method with the premise being fixed. The identification data and the membership degrees of the fuzzy partition are arranged in the following matrices:

$$U_i = \begin{bmatrix} \mu_{i1} & 0 & \cdots & 0 \\ 0 & \mu_{i2} & \cdots & 0 \\ \vdots & \vdots & \ddots & \vdots \\ 0 & 0 & \cdots & \mu_{iM} \end{bmatrix}, i = 1,..,M \tag{19}$$

Denote the consequent parameters as a single parameter vector

$$\grave{e}_i = \left[a_i^T \quad b_i\right]^T \; i = 1,..,M \tag{20}$$

Since each cluster represents a local linear description of the system, the consequent parameter vectors θ_ι, i=1,...,M can be estimated by a weighted least-squares method. The membership degree μ_{ik} of the fuzzy partition serves as the weighting factor for the data point $\mathbf{x} = [\phi_k(k), y_k(k)]^T$ to that local model. Consequently, the parameters θ_ι can be obtained as

$$\grave{e}_i = \left[X^T U_i X\right]^{-1} X^T U_i y \tag{21}$$

which is the least-square solution of $y = X_e\theta_\iota + \varepsilon$ where the data point $\mathbf{x}_k$ is weighted by μ_{ik} and $X_e = [X \vdots 1]$. The parameters a_i^T and b_i are given by:

$$a_i^T = [\theta_1,...,\theta_p],\; b_i = \theta_q \tag{22}$$

3. NONLINEAR IDENTIFICATION OF AVR LOOP DYNAMICS

This section addresses the Takagi-Sugeno fuzzy model identification of a turbogenerator using fuzzy product clustering technique from the plant input-output response data set. The dotted line block in Fig. 1 shows a schematic of a turbogenerator AVR loop. The control of the generator bus voltage is achieved mainly by adjusting the excitation current via an automatic voltage regulator (AVR). In this work, the open loop identification of a turbogenerator AVR loop dynamics from process response data is studied using a first principles model (Hogg, 1981) as a benchmark. The purpose is to develop an effective, yet efficient, approach for building a fuzzy type representative model of such plant, which can be applied in situations where the first-principle model or its model parameters are not available. Since the nonlinear dynamics of the terminal voltage response V_T is dependent on both the generator excitation input and the turbine governor input, the plant is modelled as a nonlinear ARX model of the following form:

$$y(k) = f\,(y(k-1), y(k-2), u_1(k-1), u_2(k-1)) \qquad (23)$$

where f is to be approximated by an affine TS model locally. The model is a behavioural model, which has the same dynamic response as the plant. Its parameters do not have any physical meaning relating to the physical plant parameters.

The TS fuzzy model orders and the required sampling rate are chosen using 'a priori' experience and knowledge of the application. Simulated data that covers the complete machine operational range is used for automatic construction of the global fuzzy TS model representation of the AVR loop dynamics.

Fig. 2 shows the training and validation data for constructing the TS fuzzy model representation of the turbogenerator AVR loop dynamics where the system is driven over a wide operating region (P = 0.2-0.8 p.u., Q = 0.0-0.5 p.u.) by varying of the excitation input and the governor input. The two input signals didn't show here due to the limited space. The data set from 250 seconds to 500 seconds was used as a training data set for model construction, and the data from 0-250 seconds was used as a validation data set.

In this case, four fuzzy clusters are finally obtained using the extended GK clustering with 20 initial clusters, which provide good performance with a small number of rules. This is equivalent to using four local sub-models. Each local sub-model structure is second-order of the following form.

$$y(k)=a_1 y(k-1)+a_2 y(k-2)+a_3 u_1(k-1)+a_4 u_2(k-1)+b \qquad (24)$$

The local model order is chosen to be two based on previous linear identification experiments at some single operating points (Ren, *etal.*, 2001). This is kept as low as possible to reduce the model complexity while maintaining its representational ability.

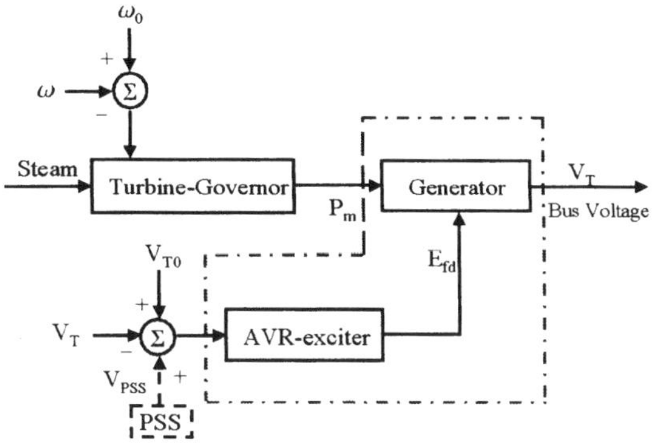

Fig. 1 Turbogenerator AVR loop

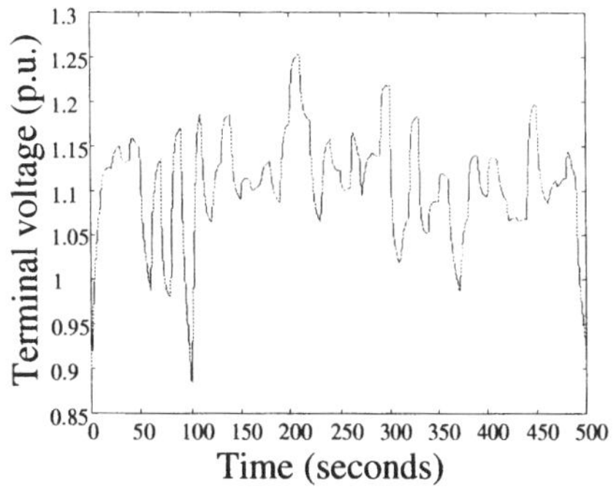

Fig. 2 Training and validation data

4. RESULTS AND DISCUSSION

Applying the extended fuzzy GK clustering to the identification data set produces the following TS fuzzy model rules.

Rule 1:
IF $y(k-1)$ is A_{11} AND $y(k-2)$ is A_{12} AND $u_1(k-1)$ is A_{13} AND $u_2(k-1)$ is A_{14} THEN $y(k)=1.75y(k-1)-0.753(k-2)+3.76u_1(k-1)-0.00319u_2(k-1)+0.00491$

Rule 2:
IF $y(k-1)$ is A_{21} AND $y(k-2)$ is A_{22} AND $u_1(k-1)$ is A_{23} AND $u_2(k-1)$ is A_{24} THEN $y(k)=1.60y(k-1)-0.64y(k-2)+9.62u_1(k-1)-0.00427u_2(k-1)-0.0295$

Rule 3:
IF $y(k-1)$ is A_{31} AND $y(k-2)$ is A_{32} AND $u_1(k-1)$ is A_{33} AND $u_2(k-1)$ is A_{34} THEN $y(k)=1.72y(k-1)-0.744y(k-2)+7.23u_1(k-1)-0.00391u_2(k-1)+0.0189$

Rule 4:
IF $y(k-1)$ is A_{41} AND $y(k-2)$ is A_{42} AND $u_1(k-1)$ is A_{43} AND $u_2(k-1)$ is A_{44} THEN $y(k)=1.36y(k-1)-0.431y(k-2)+1.66u_1(k-1)-0.00475u_2(k-1)+0.053$

where A_{ij} (i=1,2,3,4 and j=1,2,3,4) in each rule represent the fuzzy domains for different regression variables. The cluster centers produced in terms of the regression variables for each rule are given in Table 1.

The variation of the coefficients in the rule consequent highlight the nonlinearity of the plant, it can also be seen that the dynamics of the local models is mainly influenced by input u_1, i.e. excitation input, which has large values in this case.

Table1 Fuzzy cluster centers

Rule	$y(k\text{-}1)$	$y(k\text{-}2)$	$u_1(k\text{-}1)$	$u_2(k\text{-}1)$
1	1.04	1.04	0.00121	0.586
2	1.09	1.09	0.00136	0.446
3	1.11	1.11	0.00146	0.659
4	1.15	1.15	0.00155	0.518

To assess the quality of the resultant model, the VAF (percentile *variance accounted for*) (Babuska, 1998) is used as a performance index for comparing the model output with the plant output. The VAF between N sample values of two signals y_1 and y_2 is defined as,

$$\mathrm{VAF} = 100 \cdot \left[1 - \frac{\mathrm{cov}(y_1 - y_2)}{\mathrm{cov}(y_1)} \right] \% \qquad (25)$$

where cov is the covariance operator. The VAF between the fuzzy TS model response in parallel (or long range prediction) mode and the plant response for unseen validation data was 92.0173 % while that VAF between the fuzzy TS model response in parallel mode and the plant response for training data was 96.19%.

Fig. 3 shows the TS fuzzy model response represented by the dashed line in parallel mode versus the turbogenerator plant response shown as solid line for the unseen validation data. Fig. 4 shows the corresponding error curves. It can be seen that the TS fuzzy model representation with 4 sub-models provides excellent prediction performance for the global plant response. The better model response for the training data (VAF1>VAF2) is because the fuzzy product space clustering is based on the training data set rather than the validation data set. The validation data set is used as a model generalisation test. The combination of the membership function on each regression variables determines the effective region of each local model.

5. CONCLUSION

A new strategy for nonlinear dynamical system identification using an extended fuzzy GK clustering algorithm through competitive agglomeration with a locally-weighted least-squares technique has been proposed and applied for TS fuzzy model identification of a Automatic voltage regulation loop dynamics. The proper number of clusters can be automatically detected from the training data set. The efficacy of this new strategy has been confirmed by nonlinear identification of the highly nonlinear turbogenerator AVR loop dynamics based on a validated simulation of a turbogenerator system.

REFERENCES

Babuska R., (1998) *Fuzzy Modelling for Control*, Kluwer Academic Publishers.

Frigui H. and Krishnapuram R., (1997) "Clustering by competitive agglomeration", *Pattern Recognition*, **Vol. 30**, No. **7**, pp. 1223-1232.

Gustafson, D. and W. C. Kessel (1979) "Fuzzy clustering with a fuzzy covariance matrix", *Proceedings of IEEE CDC*, San Diego, CA. pp. 761-766.

Hogg B. W., (1981) "Representation and control of turbogenerators in electric power system", Chapter 5, in H. Nicholson *"Modelling of Dynamic Systems-***Vol. 2**", P. Peregrinus.

Johansen, T. A. and Foss, B. A. (1993) "Constructing NARMAX models using ARMAX models", *Int. J. Control*, **Vol. 58**, No. 5, pp. 1125-1153.

Leonaritis and A. Billings (1985) "Input-output parametric models for nonlinear systems", *International Journal of Control*, **Vol. 41**, pp. 303-344.

Ren Lixin, G. W. Irwin, and D. Flynn (2001) "Multiple Model identification of AVR Loop Dynamics", *Proceedings of the Irish Signals and Systems Conference*, June 25-27, National University of Ireland, Maynooth, pp 64-69.

Ren Lixin, G. W. Irwin, and D. Flynn (2002) "Nonlinear Identification of Turbogenerator AVR Loop Dynamics using Fuzzy Clustering", *Proceedings of the IEEE International Conference on Power System Technology*, Kunming, China, October 13-17, pp. 1503-1508.

Takagi, T. and M. Sugeno (1985) "Fuzzy identification of systems and its application to modelling and control", *IEEE Transactions on Systems, Man and Cybernetics*, **Vol. 15**, No.1, pp. 116-132.

Zhao J., V. Wertz and R. Gorez, (1994) "A fuzzy clustering method for the identification of fuzzy models for dynamical systems", *Proceedings of IEEE Symposium on Intelligent Control*, Columbus, pp.172-177.

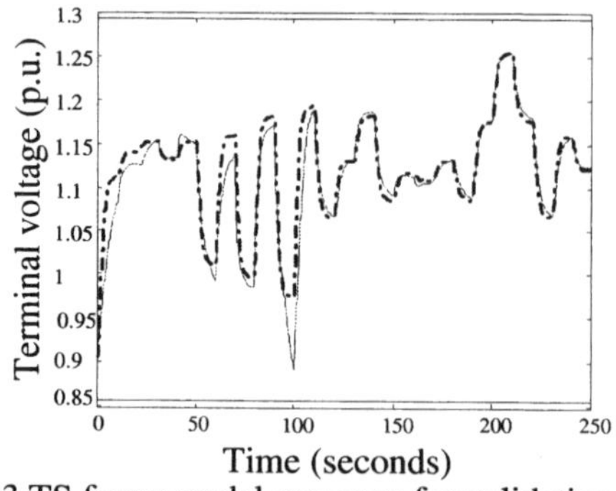

Fig. 3 TS fuzzy model response for validation data

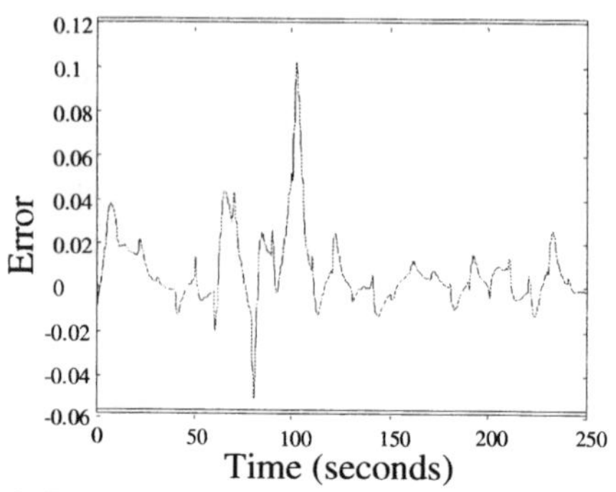

Fig. 4 Error between TS fuzzy model response and validation data

IFAC Publications
www.elsevier.com/locate/ifac

ON SIMPLIFIED MODELLING APPROACHES TO SMB PROCESSES

V. Grosfils[a], C. Levrie[b], M. Kinnaert[a], A. Vande Wouwer[b]

[a] *Service d'Automatique et d'Analyse des Systèmes, Université Libre de Bruxelles, CP 165/55, 50, Av. F. D. Roosevelt, B-1050 Brussels, Belgium*
Fax: 32-2-650.26.77;E-mail: valerie.grosfils@ulb.ac.be
[b]*Service d'Automatique, Faculté Polytechnique de Mons, Boulevard Dolez, 31, 7000 Mons, Belgium (E-mail: {Caroline.Levrie, Alain.VandeWouwer}@fpms.ac.be)*

Abstract: The Simulated Moving Bed (SMB) technology is important in various fields, from sugar to enantiomer separation, and operating conditions must be carefully selected and regulated, based on an appropriate process model. Basically, two modelling approaches exist, i.e. True Moving Bed (TMB) and SMB models. Both approaches show advantages and drawbacks. In this work, an attempt is made to develop time-varying-velocity TMB models, which retain the simplicity of the original TMB model and its modest computational load, while capturing the essential features of the cyclic steady state reproduced by the SMB model. Alternative modelling approaches are compared and their use is discussed. *Copyright © 2003 IFAC*

Keywords: mathematical models, chemical processes, process control, chromatography, simulated moving bed

1. INTRODUCTION

Conventional batch chromatography is relatively inefficient in terms of adsorbent and solvent consumption and significant benefits can be achieved by performing separation of high-added value products, such as enantiomers produced in the pharmaceutical industry, with a simulated moving bed (SMB) process. The SMB process allows a counter-current movement of the liquid and the solid to be achieved in order to increase the exchange capabilities between both phases. (For further details about the process see (Ruthven and Ching, 1989)).

However, the transfer of the SMB technology, used industrially for hydrocarbons and sugars separation, to the separation of fine chemicals is not immediate. Indeed, the conditions and requirements (product quantities and purities, characteristics of the phases, interactions, ...) are very different. The main issues are the selection of optimal operating conditions and the process control, problems which require the development of a model of the process.

Two modelling approaches are commonly applied to SMB processes. The first one is called TMB (true moving bed) and assumes an equivalent counter-current movement of the solid phase (see Fig. 1), whereas the solid movement is achieved in practice by periodically switching inlet and outlet valves in the direction of the liquid flow. The second, more rigorous approach, called SMB, considers the system as an arrangement of static chromatographic columns and takes the discrete nature of the solid movement into account. In the literature, it is generally agreed that the TMB model represents the average behaviour of the SMB and that the correspondence between TMB and SMB becomes better and better when the number of columns in the SMB increases. The TMB model brings significant reduction of the computational complexity (Haag *et al*, 2001; Ruthven and Ching, 1989) and can be used for a first analysis in design, optimization and control. However, the TMB model, as opposed to the SMB model, does not show the cyclic steady-state of the process and

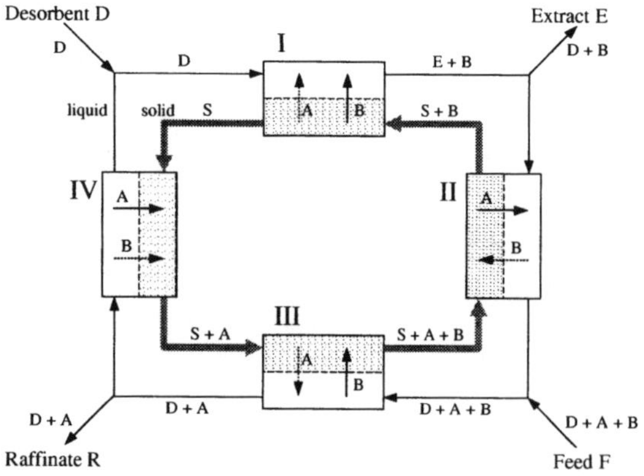

Fig. 1. Equivalent counter-current representation of a simulated moving bed process for separation of a mixture with two species A and B – Material flows and adsorption-desorption phenomena in each section

introduces modelling inaccuracies. Indeed, in the vicinity of the feed point, the concentrations are higher than those predicted by the SMB model. This is due to the fact that the flow rates in the TMB model are always smaller than in the SMB model, leading to smaller dilution. Moreover, when there is only one column in zone 2 and 3, differences between TMB and SMB models may be large (Pais *et al.*, 1998).

The objective of this study is to build a model which combines the advantages of the TMB and SMB approaches without having their drawbacks. This means a model with computational efficiency for use in identification and control, giving concentration profiles close to those predicted by the SMB model and reproducing the cyclic steady-state. To this end, original approaches are discussed in this paper. The first approach considers a modification of the conventional TMB model, in which a periodic solid velocity (sinusoidal or pulse velocity) is introduced. The second approach is based on the assumption that the TMB model represents the SMB at 50% of the switching period, and makes use of a transformation (translation) of the TMB internal concentration profiles.

2. TMB AND SMB MODELS

In this section, the SMB and TMB models are described. The model formulation chosen is the LDF model which assumes that the mass transfer kinetics are described by linear driving forces. Moreover, it is assumed that velocities and concentrations are radially homogeneous. Only a brief introduction to the models is given. For more details about the initial and boundary conditions and about the numerical solution procedures, see (Haag *et al.*, 2001).

2.1 SMB model

In this approach, it is considered that the solid phase does not move between two switching times. Following the assumptions described before, the mass balance in the liquid phase for component i is given by:

$$D_{Li}\frac{\partial^2 c_{i,k}}{\partial z^2} - u_k\frac{\partial c_{i,k}}{\partial z} = \frac{\partial c_{i,k}}{\partial t} + \frac{1-\varepsilon}{\varepsilon}\frac{\partial q_{i,k}}{\partial t} \qquad (1)$$

where k numbers the columns, c_i is the concentration of component i in the liquid phase, q_i, the corresponding concentration in the solid phase. D_{Li} represents the axial dispersion coefficient. ε is the bed porosity, and u is the fluid velocity. t denotes the time and z, the axial coordinate.
For the solid phase, the mass balance equation is given by

$$\frac{\partial q_{i,k}}{\partial t} = k_{FSi}(q_{i,k}^{*} - q_{i,k}) \qquad (2)$$

with k_{FSi}, the overall effective mass transfer coefficient, q^*_i, the adsorbed equilibrium concentration. The latter is given by the multicomponent adsorption equilibrium isotherm:

$$q^*_{i,k} = f_i(c_{A,k}, c_{B,k}) \qquad (3)$$

where i = A, B refers to the species in the mixture. Note that Langmuir adsorption isotherms are considered in this study.

Valve switching is taken into account by considering that the concentration profiles in column k at the beginning of a switching interval are equal to the profiles obtained in the column k+1 at the end of the previous period:

$$c_{i,k}(t_p = 0, z_k) = c_{i,k+1}(t_{p-1} = \Delta t, z_{k+1})$$

where the index p denotes the current switching period and Δt is the switching period.

2.2. TMB model

As explained in the introduction, the TMB model assumes an equivalent counter-current movement of the solid phase. Following the assumptions presented before, the mass balance for each component i in the liquid phase can be expressed as:

$$D_{L,i}\frac{\partial^2 c_{i,j}}{\partial z^2} - u_{cc,j}\frac{\partial c_{i,j}}{\partial z} + \frac{1-\varepsilon}{\varepsilon}u_S\frac{\partial q_{i,j}}{\partial z} = \frac{\partial c_{i,j}}{\partial t} + \frac{1-\varepsilon}{\varepsilon}\frac{\partial q_{i,j}}{\partial t} \qquad (4)$$

j = 1,.., 4
where $u_{cc,j}$ represents the velocity of the liquid in zone j. u_s is the equivalent solid phase velocity and is given by $u_s = L/\Delta t$ with L, the length of one column and Δt, the switching period. In order to get the equivalence between the SMB and the TMB models,

$$u_{cc,j} = u_j - u_s. \qquad (5)$$

Moreover, the mass balance for component i in the solid phase is written as

$$\frac{\partial q_{i,j}}{\partial t} = u_S \frac{\partial q_{i,j}}{\partial z} + k_{FSi}(q^*_{i,j} - q_{i,j}) \qquad (6)$$

3. FROM TMB TO SMB

In order to model the cyclic steady state of the SMB process with a modified TMB model, two approaches are presented. The first one attempts to reproduce the cyclic behaviour of the process by incorporating a periodic solid velocity in a classical TMB model. At this stage, different formulations of the solid velocity are considered, i.e. a simple sinusoid, a combination of sinusoids and a rectangular pulse. In the second approach, the classical TMB model is used and the cyclic behaviour is simulated by translating the concentration profiles along the z-axis during a switching period. The two options are described below and simulation results are compared with those obtained with the SMB model for a 8 columns SMB process.

3.1 TMB with a periodic solid velocity

Simple sinusoid

The TMB model is used with a sinusoidal solid velocity,

$$u_s^P(t) = u_s\,(1 + A\sin(\omega t)) \qquad (7)$$

with u_s, the usual velocity of the solid in a TMB, A, the amplitude of the sinusoidal variation, and the pulsation, $\omega = 2\pi / \Delta t$. Equation (5) describing the fluid velocity is adapted: $u_{cc,j}(t) = u_j - u^P_s(t)$. Note that for values of A larger than 1, the solid velocity becomes negative, which is non physical. Some simulation results are shown in Figure 2. In order to get the maximum amplitudes possible for the extract and raffinate signals, and to approach those produced by the SMB model, A has to be chosen equal to 1. However, it is not sufficient, and the amplitudes of the extract and the raffinate signals remain too small in comparison with those produced by the SMB model. The obtained raffinate and extract signals are also symmetric during a switching period which is not the case in the SMB simulation.

Sum of sinusoids

Here, the TMB model is used with a solid velocity equal to

$$u_s^P(t) = u_s(1 + C\sum_{i=1}^{n} A_i \sin \omega_i t), \qquad (8)$$

expression in which u_s is the solid velocity in a classical TMB model, C is a constant that has to be estimated, A_i and ω_i are the amplitudes and pulsations of the sinusoids determined from a FFT analysis of the extract signal simulated with the SMB model. The fluid velocity is calculated from $u_{cc,j}(t) = u_j - u^P_s(t)$. C is chosen in order to have the maximum amplitude of the solid velocity while keeping the velocity positive. Some results are shown in Figure 3. The same conclusions as for the simple sinusoidal velocity can be drawn, except that the extract and raffinate are no longer symmetric during a switching period. However, the abrupt change of the signals at the switching time is not observed because of the filtering action of the process.

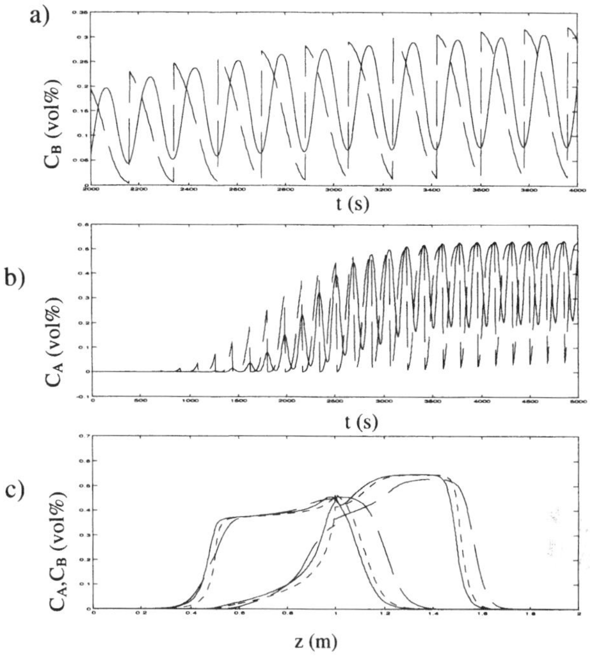

Fig. 2 : TMB model with sinusoidal solid velocity; a) extract; b) raffinate; c) internal concentration profile in steady-state at 50% of the switching period (__ modified TMB; __ __ SMB; - - classical TMB)

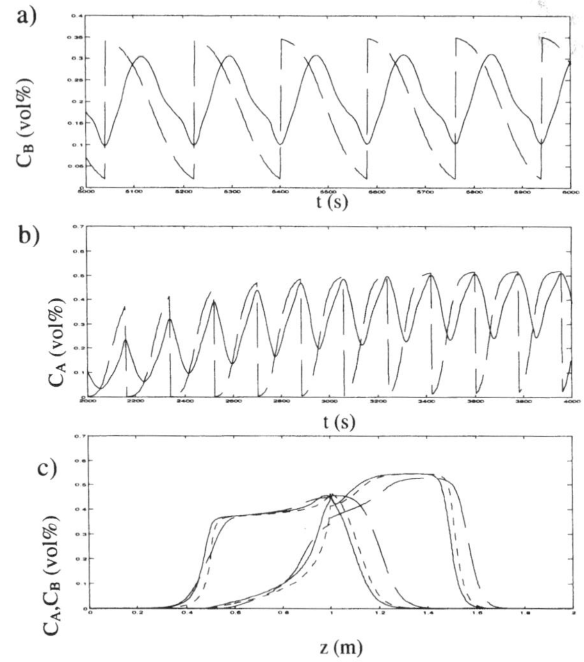

Fig 3 : sum of sinusoids solid velocity; a) extract; b) raffinate; c) internal concentration profile in steady-state at 50% of the switching period (__ modified TMB; __ __ SMB; - - classical TMB)

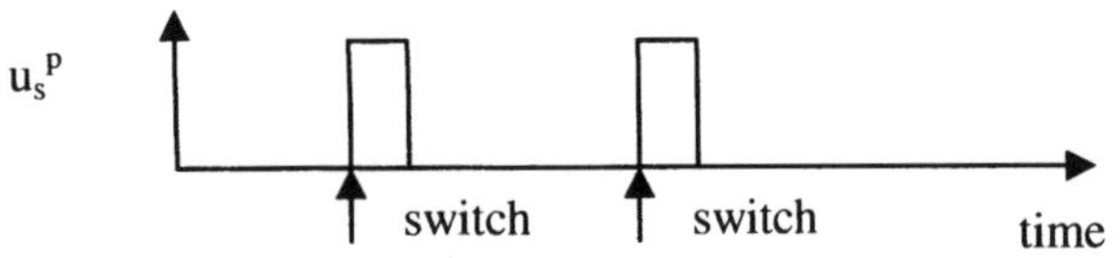

Fig. 4 : rectangular pulse velocity

Rectangular pulse velocity

As shown in Figure 4, the solid velocity of the TMB model is equal to a narrow rectangular pulse occurring just after valve switching. The mean value of the velocity $u_s^P(t)$ calculated during a switching period is equal to the value of the solid velocity in a classical TMB model, u_s, and $u_{cc,j}(t) = u_j - u^P{}_s(t)$. Hence the sole parameter is the ratio of the amplitude to the duration of the pulse. The length of the pulse determines the duration of the switching phenomenon. Moreover, delaying the pulse introduces a lag in the extract and raffinate concentration signals. Thus, it is possible, as can be seen in Figure 5, to reproduce the time delay and the smoothing of the curves which are physically due to hold-up and mixing elements as described in (Strube and Schmidt-Traub, 1998) and (Beste *et al.*, 2000). This is an important advantage that nor SMB nor classical TMB models have.

Results in Figure 6 show that the raffinate and the extract signals of the modified TMB and SMB models are in very good agreement. Particularly, in Figure 6.c the displacement of the concentration profile along the z-axis during a switch, as reproduced by the modified TMB model, is close to the prediction of the SMB model. In Fig. 6.d, significant improvements of the shape of the profile are obtained by using the modified TMB approach rather than a classical TMB model. As shown in Table 1, the simulation time is shorter than for the SMB model and becomes even shorter with a longer pulse because the maximum step size allowed for time integration of the model equations may be increased. However, the modified TMB model with a pulse-like velocity (other types of pulses, with smoother temporal evolutions, have also been considered) is more difficult to solve numerically, and further investigation of this issue is needed.

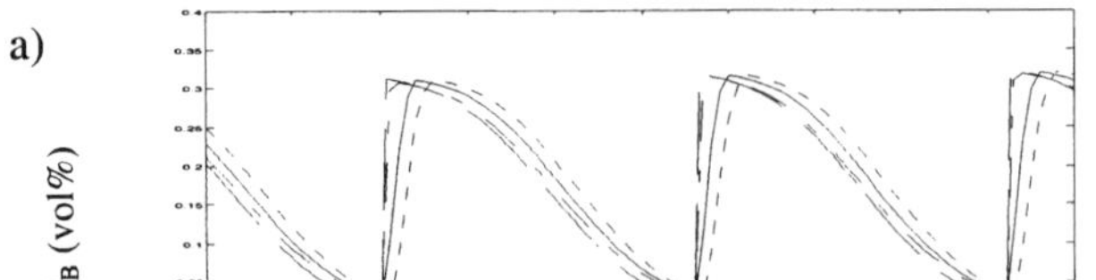

Fig. 5: extract signal (__ __ SMB model, _ __ rectangular pulse velocity with a pulse length of 6 s; __ rectangular pulse velocity with a pulse length of 20 s; - - delayed rectangular pulse velocity with a pulse length of 20 s)

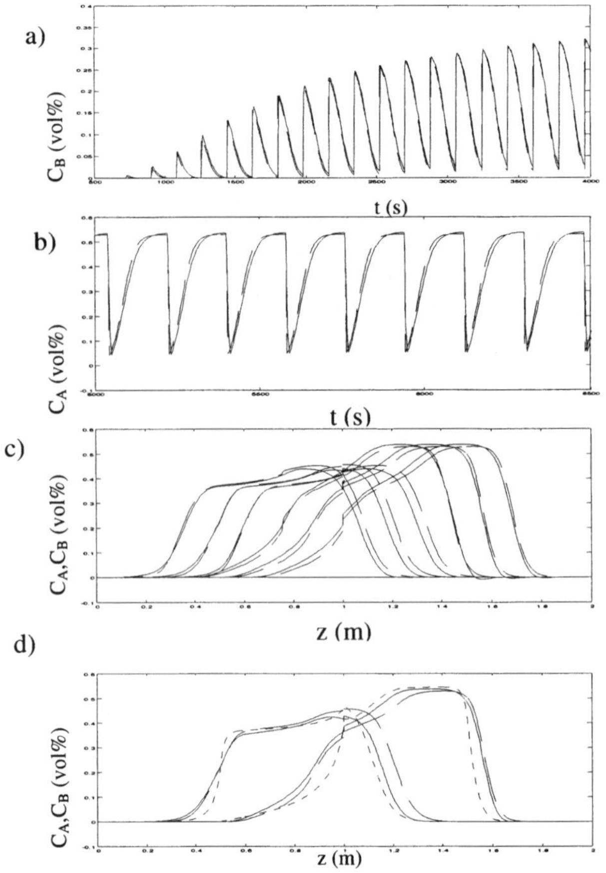

Fig. 6 : rectangular pulse solid velocity; a) extract; b) raffinate; c) concentration profiles in steady-state at 0%, 50% and 100% of Δt ; d) internal concentration profile in steady-state at 50% of the switching period (__ modified TMB; __ __ SMB; - - classical TMB)

Table 1: comparison of r, the ratio of the simulation time with the modified TMB model on the simulation time with the SMB model, for a 8 columns SMB process

Modified TMB model	r
u_s^P=sin	0.12
u_s^P = sum of sin	0.24
u_s^P = rectangular pulse (delay 6s)	0.67
u_s^P = rectangular pulse (delay 20 s)	0.55
translated TMB	0.23

3.2 Translation of the TMB profile

This approach is based on the assumption that the TMB model represents the SMB at 50% of the switching period. Moreover, it is based on the fact that the temporal variations of the extract and raffinate concentrations during one switching period are due to the movement of the spatial concentration profiles through the columns in the direction of the fluid flow as shown in Fig. 6.c.

Indeed, the components move through the columns as

nonlinear waves (Hellferich and Carr, 1993). As shown in fig.7, a SMB profile is composed of adsorption waves (labelled "A") and desorption waves (labelled "D"). In the case of Langmuir adsorption isotherms, the former are spread whereas the latter keep their constant pattern (Hellferich and Carr, 1993). These properties influence the wave velocities, i.e., for desorption waves, the velocity of component i in zone j is given by:

$$v_{i,j}^{D} = \frac{u_j}{1+\frac{1-\varepsilon}{\varepsilon}\frac{dq_{i,j}}{dc_{i,j}}} \qquad (9)$$

while for adsorption waves, the velocity of component i in zone j is defined as:

$$v_{i,j}^{A} = \frac{u_j}{1+\frac{1-\varepsilon}{\varepsilon}\frac{\Delta q_{i,j}}{\Delta c_{i,j}}} \qquad (10)$$

where Δ stands for the difference between the downstream and upstream sides of the wave. These equations are derived from the ideal chromatography theory (no diffusion, no mass transfer resistance).

The basic idea is therefore to translate the TMB profile along the z-axis as a function of the time elapsed since the last switch by taking into account these wave velocities. In practice, the concentration at the position p is calculated according to the following equation:

$$C_i^{translated\ \ TMB}(p,t) = C_i^{TMB}(p + tr_i(t)) \qquad (11)$$

where C_i, is the concentration of component i, p is the position along the z-axis, and tr_i, the translation of which the general shape is shown in Figure 8. The slope of the translation factor is equal to the corresponding wave velocity (equation (9) and (10)). In our case, $dq_{i,j}/dc_{i,j}$ is approximated by the slope of the isotherm at infinite dilution and $\Delta q_{i,j}/\Delta c_{i,j}$ is calculated from the difference between the maximum and minimum values of the concentrations in the wave.

Moreover, it is also possible to improve the results by identifying the diffusion coefficients of the TMB model by minimizing the deviations between the

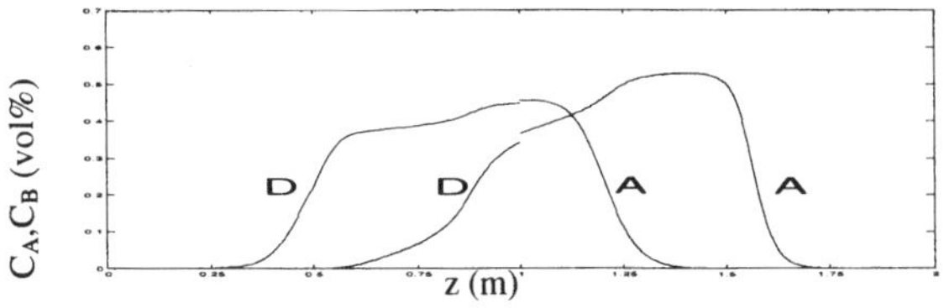

Fig. 7: SMB profile - D : desorption wave; A : adsorption wave

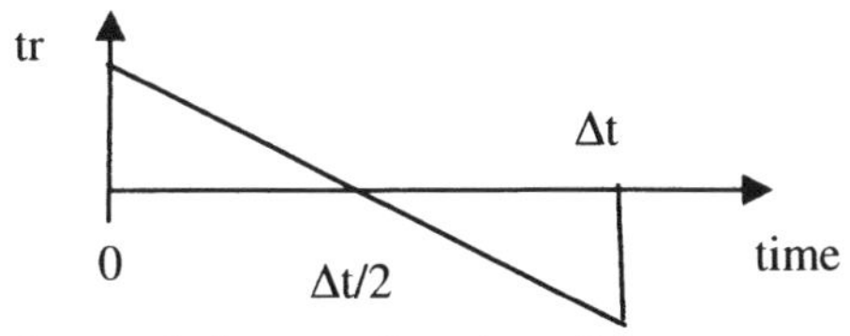

Fig. 8: translation as a function of time

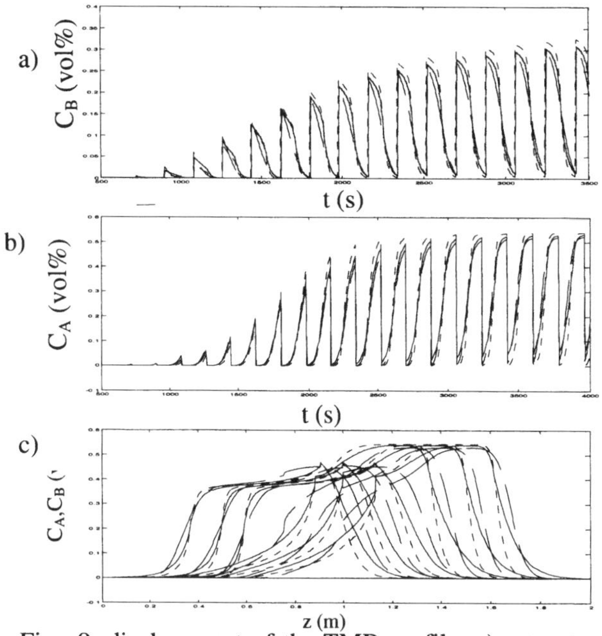

Fig. 9: displacement of the TMB profile; a) extract; b) raffinate c) evolution of the profile along the z-axis during a switching period (__ modified TMB with tuning of diffusion coefficients;
- - translated TMB without tuning of diffusion coefficients;__ __ SMB);

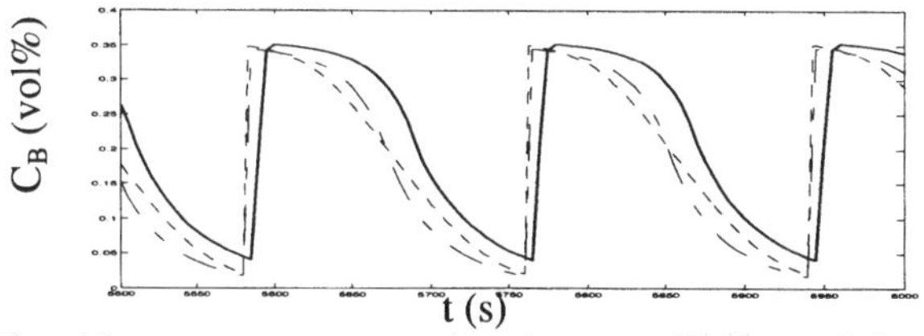

Fig. 10: extract concentration (__ __ SMB model, _ __ translated TMB with tuning of diffusion coefficients; __ translated TMB with time delay (20 s) and with tuning of diffusion coefficients)

internal concentration profiles generated by the modified TMB and SMB models. Figure 9 shows that the raffinate and extract concentrations produced by the "translated" TMB are close to those produced by the SMB model, especially those simulated after tuning of the diffusion parameters.

It is also possible to reproduce the time delay and the smoothing of the curves as shown in Figure 10.

4. DISCUSSIONS

In this section, the different modelling approaches are compared and their use is discussed.

Extract and raffinate signals obtained in Section 3 show that the TMB with rectangular pulse and the translated TMB give results which are in good agreement with those obtained with the SMB model.

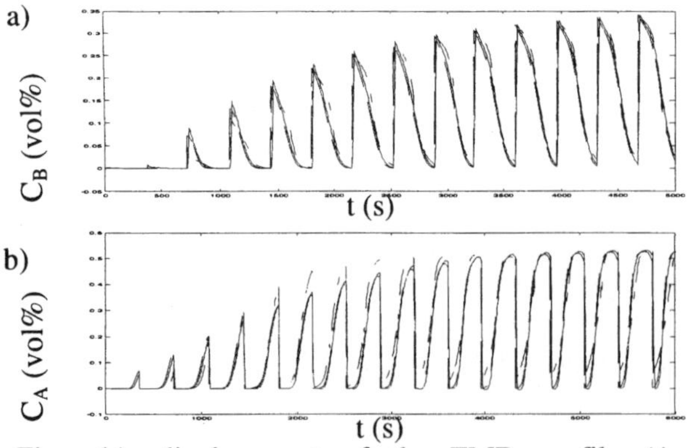

Fig. 11: displacement of the TMB profile (4 columns); a) extract; b) raffinate; (_ __ translated TMB; _ rectangular pulse velocity; _ _ SMB)

On the contrary, the simulations performed with sinusoidal velocities are quite different from those obtained with the SMB model.

If the TMB with rectangular pulse in Fig. 6.c and the translated TMB in Fig.9.c are compared, it is seen that results produced with the former are in closer agreement with those produced by the SMB model but the latter is less computationally demanding. In further work, calibration of the diffusion coefficients or the mass transfer coefficients could be achieved in order to improve the results of the translated TMB.

An important advantage of our models is the possibility to reproduce delays and dispersion introduced by connecting lines and valves.

As it is well known, the differences between SMB and TMB models become more significant when the number of columns decreases. In our study, the quality of the simulations obtained with the modified TMB is not significantly affected when the number of columns decreases as shown in Fig.11. However, in the extreme case where there is only 1 column per section, the simulation time of the SMB model could be extremely short, even shorter than the time required by the TMB with rectangular pulse velocity. Only the translated TMB model may be useful in this extreme case.

5. CONCLUSION

Simulated moving bed processes are usually represented by simplified models (so called TMB models), in which the solid movement is assumed to occur in a continuous way, or by more rigorous models (so called SMB models), in which the discrete nature of the solid movement is taken into account. Of course, the hybrid dynamics of the SMB models makes them more computationally demanding than their TMB counterparts. The main advantages of SMB models are that they are able to reproduce the cyclic steady state observed in real-life operation and bring less model inaccuracies.

The contribution of this study is to propose several modified TMB models, which incorporate a time-varying velocity of the solid phase. At this stage several model variations are presented and thoroughly analysed. These models retain the original TMB model simplicity, and therefore their relatively modest computational load, while capturing the essential features of the SMB cyclic steady state. Rectangular-pulse-velocity TMB and "translated" TMB models appear of particular value with this respect.
Classical TMB models have already demonstrated their usefulness in optimising the operation of SMB processes (Kloppenburg and Gilles, 1999), based on the assumption that TMB models represent well the average behaviour of the real process. It is expected that the proposed modified TMB models convey more information on the process dynamics, and therefore have potentials (to be explored in future studies) in process optimization and control.

ACKNOWLEDGEMENTS

This work is performed in the framework of Movida project (contracts n°0114843 and 0114961) funded by the Walloon Region (Belgium).

REFERENCES

Beste Y. A., M. Lisso, G. Wozny, W. Arlt, (2000), Optimization of simulted moving bed plants with low efficient stationary phases: separation of fructose and glucose, *Journal of Chromatography* A, **86**, 169-188

Haag J., A. Vande Wouwer, S. Lehoucq, P. Saucez, (2001), Modelling and simulation of a SMB chromatographic process designed for enantioseparation, *Control Engineering Practice*, **9**, 921-928

Kloppenburg E., E. D. Gilles, (1999), Automatic control of the simulated moving bed process for C_8 aromatics separation using asymptotically exact input/output-linearization, *Journal of Process Control,* 9, 41-50

L.S. Pais, J M Loureiro, A. E. Rodrigues, (1998), Modeling Strategies for enantiomers separation by SMB chromatography, *AIChE Journal*, **vol. 44**, no 3

D. M. Ruthven and C.B. Ching, , (1989),Counter-current and simulated counter-current adsorption separation processes, *Chemical Engineering Science*, **vol. 44**, n°,5, pp.1011-1038

Strube J. and H. Schmidt-Traub, , (1998), Dynamic simulation of simulated moving-bed chromatographic processes, *Computers Chemical Engineering*, **vol. 22**, n° 9, pp. 1309 – 1317

IFAC Publications
www.elsevier.com/locate/ifac

OPTIMAL FILTERING FOR BILINEAR SYSTEMS AND ITS APPLICATION TO TERPOLYMERIZATION PROCESS STATE IDENTIFICATION

Michael Basin * **Maria Aracelia Alcorta-Garcia** *

* *Autonomous University of Nuevo Leon, Mexico*

Abstract: The paper presents the optimal nonlinear filter for bilinear state and linear observation equations confused with white Gaussian disturbances. The general scheme for obtaining the optimal filter in case of polynomial state and linear observation equations is announced. The obtained bilinear filter is applied to solution of the state identification problem for the bilinear terpolymerization process and compared to the optimal linear filter available for the linearized model and to the mixed filter designed as a combination of those filters. *Copyright © 2003 IFAC*

Keywords: Bilinear system, Filtering, Process identification

1. INTRODUCTION

It is virtually the common opinion that the optimal nonlinear finite-dimensional filter exists and can be obtained in a closed form only in the case of linear state and observation equations. This famous construction is called the linear Kalman-Bucy filter (Kalman and Bucy, 1961), referring to the scientists who derived it in 1960s. However, it is much less known that the optimal nonlinear finite-dimensional filter can be obtained in many other cases, if, for example, the state vector can take only a finite number of admissible states (Wonham, 1965) or if the observation equation is linear and the drift term in the state equation satisfies the Riccati equation $\frac{df}{dx}+f^2=x^2$ (see (Benes, 1981)). Moreover, the complete classification of the "general situation" cases (this means that there are no special assumptions on the structure of state and observation equations) when the optimal nonlinear finite-dimensional filter exists is given in (Yau, 1994).

This paper would like to attract attention to relatively simple (but important in practical applications, see (Ogunnaike, 1994)) cases when the optimal nonlinear finite-dimensional filter can be obtained in a closed form. Indeed, if the observation equation is linear and the observation matrix is invertible, then, as shown below in the paper, it is possible to obtain the optimal finite-dimensional filter for a polynomial state equation, provided that the system coefficients depend on time only. In the case of a bilinear state equation, the corresponding filtering equations are derived in the paper directly. The possibility to derive similar results for an arbitrary polynomial state equation is underlined.

The paper is organized as follows. Section 2 briefly reminds the linear Kalman-Bucy filter for reference purposes, considers the case of nonlinear state and linear observation equations, establishes the procedure to obtain a closed system of the filtering equations for polynomial state and linear observation equations, and gives the optimal filter for bilinear system states and linear observations in the explicit form. In Section 3, the obtained bilinear filter is applied to solution of the state identification problem for the bilinear terpolymerization process and compared to the optimal linear filter available for the linearized model and to the mixed filter designed as a combination of those filters. The simulation results show an advantage of the optimal bilinear filter in comparison to the other filters. It should be noted that there are a number of papers addressing a near optimal state and parameter identification for polynomial processes (see, for example, (Zhou and Blanke, 1989)).

2. OPTIMAL FILTERING FOR POLYNOMIAL STATE EQUATION

2.1 *Linear Kalman-Bucy filter*

It is well known that the linear optimal filter (Kalman and Bucy, 1961) can be designed in a closed form, if the state and observation equations of a dynamic system are linear. Let an unobservable random process $x(t)$ satisfy a linear equation

$$dx(t) = (a_0(t) + a(t)x(t))dt + b(t)dW_1(t), \quad x(t_0) = x_0, \tag{1}$$

and linear observations are given by

$$dy(t) = (A_0(t) + A(t)x(t))dt + B(t)dW_2(t). \tag{2}$$

Here, $W_1(t)$ and $dW_2(t)$ are Wiener processes, whose weak derivatives are Gaussian noises and which are assumed independent of each other and of the initial value x_0. The last equation can also be written in the algebraic form:

$$\dot{y}(t) = A_0(t) + A(t)x(t) + B(t)\psi(t). \tag{3}$$

where $\psi(t)$ is a white Gaussian noise (a weak derivative of $W_2(t)$).

The estimation problem is to find the best estimate for the real process $x(t)$ at time t based on the observations $Y(t) = \{y(s), t_0 \leq s \leq t\}$, that is the conditional expectation $m(t) = E(x(t) \mid Y(t))$ of the real process $x(t)$ with respect to the observations $Y(t)$. Let $P(t) = E((x(t) - m(t)(x(t) - m(t))^T \mid Y(t))$ be the estimate variance (correlation function).

The solution to this problem is given by the following system of filtering equations, which is closed with respect to the introduced variables, $m(t)$ and $P(t)$:

$$dm(t) = (a_0(t)+a(t)m(t))dt+P(t)A^T(t)(B(t)B^T(t))^{-1}\times$$
$$[dy(t) - (A_0(t) + A(t)m(t))dt], \tag{4}$$
$$m(t_0) = E(x(t_0) \mid Y(t_0))$$
$$dP(t) = (a(t)P(t) + P(t)a^T(t) + b(t)b^T(t))dt- \tag{5}$$
$$P(t)A^T(t)(B(t)B^T(t))^{-1}A(t)P(t)dt,$$
$$P(t_0) = E((x(t_0) - m(t_0)(x(t_0) - m(t_0))^T \mid Y(t_0).$$

The advantages of the Kalman-Bucy filter are very well known: the equations are simple, the variance equation is independent of the observations $y(t)$ and can be solved off-line, the estimate equation is linear and the variance one is quadratic of the Riccati type.

2.2 *Nonlinear filtering equation*

In the case of nonlinear state and observation equations, the problem is more complicated. Let an unobservable random process $x(t)$ satisfy a nonlinear equation

$$dx(t) = f(x(t))dt + b(t)dW_1(t), \quad x(t_0) = x_0, \tag{6}$$

and nonlinear observations are given by

$$dy(t) = h(x(t))dt + B(t)dW_2(t). \tag{7}$$

There exist two principal results related to this case (Mitter, 1996). First, as in the previous linear case, the innovations process $\vartheta(t) = y(t) - \int_{t_0}^{t} E(h(x(s)) \mid Y(s))ds$ is a Wiener process and, second, contains the same new information as the observation process $y(t)$ itself. The first result means that for every fixed t, the random variable $\vartheta(t)$ is Gaussian and the second one implies that for every function $\varphi(x)$ depending on the real unobservable vector $x(t)$, the expectations with respect to the observation and innovations processes are the same: $E(\varphi(x(t)) \mid Y(t)) = E(\varphi(x(t)) \mid \{\vartheta(s), t_0 \leq s \leq t\})$, in particular, if $\varphi(x) = x$, then $m(t) = E(x(t) \mid Y(t)) = E(x(t) \mid \{\vartheta(s), t_0 \leq s \leq t\})$.

Using these basic properties, it is possible to obtain the equation for the optimal estimate $m(t) = E(x(t) \mid Y(t))$, the so-called nonlinear filtering equation, first derived by Kushner (Kushner, 1964), in the form

$$dm(t) = E(f(x(t)) \mid Y(t))dt+$$
$$[E(h(x(t))x^T(t) \mid Y(t)) - E(h(x(t)) \mid Y(t))m^T(t)]^T\times$$
$$(B(t)B^T(t))^{-1}[dy(t) - E(h(x(s)) \mid Y(t))dt],$$
$$m(t_0) = E(x(t_0) \mid Y(t_0)). \tag{8}$$

However, the computation of $m(t)$ requires computing the functions in the right-hand side of this equation, which, in turn, requires computing the quantities: $E(f(x(t)) \mid Y(t))$, $E(h(x(t))x(t) \mid Y(t))$, and $E(h(x(t)) \mid Y(t))$. Each of them is a nonlinear function of x and, as a consequence, a non-Gaussian random variable. Thus, one has to solve a nonlinear stochastic differential equation for each of these variables, which involves higher moments of these variables in its right-hand side. Hence, an infinite-dimensional system of nonlinear stochastic equations should be obtained as the optimal filter. In other words, the optimal filter cannot be obtained in a closed form, i.e., with respect to a finite number of filtering variables (there are two, $m(t)$ and $P(t)$, in the linear Kalman-Bucy filter), or one can say that the optimal finite-dimensional filter does not exist. Actually, there are only a few number of examples where the optimal finite-dimensional filter exists for a nonlinear model of state and observation processes (Wonham, 1965; Benes, 1981; Yau, 1994) in the "general situation."

2.3 *Polynomial state and linear observation equations*

Nonetheless, it should be possible to obtain the optimal finite-dimensional filter in a closed form in the following case. Let a unobserved random process $x(t)$ satisfy a nonlinear equation

$$dx(t) = f(x(t))dt + b(t)dW_1(t), \quad x(t_0) = x_0, \tag{9}$$

and linear observations are given by

$$dy(t) = (A_0(t) + A(t)x(t))dt + B(t)dW_2(t), \tag{10}$$

where the function $f(x(t)) = a_0(t) + a_1(t)x + a_2(t)x^2 + \ldots$ is a polynomial and the observation matrix $A(t)$ is invertible, i.e., the inverse matrix $A^{-1}(t)$ exists.

Since the observation equation is linear, the first result of nonlinear filtering implies that the innovations process $\vartheta(t) = y(t) - \int_{t_0}^{t}(A_0(s) + A(s)m(s))ds = \int_{t_0}^{t}(A_0(s) + A(s)x(s))ds + \int_{t_0}^{t} B(s)dW_2(s) - \int_{t_0}^{t}(A_0(s) + A(s)m(s))ds = \int_{t_0}^{t} A(s)(x(s)-m(s))ds + \int_{t_0}^{t} B(s)dW_2(s)$ is a Wiener process, and, since $\int_{t_0}^{t} B(s)dW_2(s)$ is also a Wiener process, the random variable $A(t)(x(t)-m(t))$ is Gaussian for every fixed t. If the inverse matrix $A^{-1}(t)$ exists, then the random vector $(x(t) - m(t))$ is also Gaussian (Pugachev, 1984).

Moreover, in this case, the second term in the nonlinear filtering equation is equal to

$$[E(h(x(t))x^T(t) \mid Y(t)) - E(h(x(t)) \mid Y(t))m^T(t)]^T \times$$
$$(B(t)B^T(t))^{-1}[dy(t) - A(t)m(t)dt] =$$
$$[E(x(t)x^T(t)A^T \mid Y(t)) - m(t)E(x^T(t)A^T(t) \mid Y(t))] \times$$
$$(B(t)B^T(t))^{-1}[dy(t) - A(t)m(t)dt] =$$
$$[E(x(t)x^T(t) \mid Y(t))A^T(t) - m(t)E(x^T(t) \mid Y(t))A^T(t)] \times$$
$$(B(t)B^T(t))^{-1}[dy(t) - A(t)m(t)dt] =$$
$$[E(x(t)x^T(t) \mid Y(t)) - m(t)m^T(t)]A^T(t) \times$$
$$(B(t)B^T(t))^{-1}[dy(t) - A(t)m(t)dt] =$$
$$P(t)A^T(t)(B(t)B^T(t))^{-1}[dy(t) - A(t)m(t)dt].$$

Hence, the nonlinear filtering equation for the optimal estimate $m(t)$ takes the form:

$$dm(t) = E(f(x(t)) \mid Y(t))dt +$$
$$P(t)A^T(t)(B(t)B^T(t))^{-1}[dy(t) - A(t)m(t)dt],$$

$$m(t_0) = E(x(t_0) \mid Y(t_0)). \quad (11)$$

Let us note now that if the function $f(x(t)) = a_0(t) + a_1(t)x + a_2(t)x^2 + ...$ is a polynomial, it should be possible to compute a finite-dimensional filter in a closed form for variables $m(t)$ and $P(t)$, using the fact that the random variable $(x(t) - m(t))$ is Gaussian. Since all the system coefficients in (9), (10) do not depend on state $x(t)$ and observations $y(t)$, the conditional moments of $(x(t) - m(t))$ with respect to observations $y(t)$ coincide with the unconditional ones. This implies that all odd central conditional moments of this Gaussian variable $\mu_1 = E((x(t) - m(t)) \mid Y(t))$, $\mu_3 = E((x(t) - m(t))^3 \mid Y(t))$, $\mu_5 = E((x(t) - m(t))^5 \mid Y(t)), ...$ are equal to 0, and all even central conditional moments $\mu_2 = E((x(t) - m(t))^2 \mid Y(t))$, $\mu_4 = E((x(t) - m(t))^4 \mid Y(t))$, $\mu_6 = E((x(t) - m(t))^6 \mid Y(t)), ...$ can be represented as functions of the variance $P(t)$. For example, $\mu_2 = P$, $\mu_4 = 3P^2$, $\mu_6 = 15P^3, ...$ Thus, all higher moments of $(x(t)-m(t))$ can be expressed using $P(t)$, and this yields additional relations for representing every higher initial moment of $x(t)$ and, finally, the possibility to obtain the optimal filter in a closed form, i.e., the optimal finite-dimensional filter should exist in the polynomial-linear case.

For example, if the function

$$f(x) = a_0(t) + a_1(t)x + a_2(t)xx^T \quad (12)$$

is a bilinear polynomial, where x is now an n-dimensional vector, a_1 is an $n \times n$ - matrix, and a_2 is a 3D tensor of dimension $n \times n \times n$, the system of filtering equations is as follows

$$dm(t) = (a_0(t) + a_1(t)m(t) + a_2(t)m(t)m^T(t) + a_2(t) \times$$
$$P(t))dt + P(t)A^T(t)(B(t)B^T(t))^{-1}[dy(t) - A(t)m(t)dt],$$
$$m(t_0) = E(x(t_0) \mid Y(t_0)), \quad (13)$$
$$dP(t) = (a_1(t)P(t) + P(t)a_1^T(t) +$$
$$2a_2(t)m(t)P(t) + 2P(t)m^T(t)a_2^T(t) + \quad (14)$$
$$b(t)b^T(t))dt - P(t)A^T(t)(B(t)B^T(t))^{-1}A(t)P(t)dt,$$
$$P(t_0) = E((x(t_0) - m(t_0))(x(t_0) - m(t_0))^T \mid Y(t_0)),$$

since the third central moment μ_3 is equal to 0, and the third initial moment of $x(t)$ can be expressed using its second and first moments, i.e., $P(t)$ and $m(t)$. In this bilinear-linear case, the variance equation is also independent of the observations $y(t)$, but has the bilinear terms $m(t)P(t)$ in its right-hand side and depends on $m(t)$, thus making both the equations interconnected. The estimate equation is bilinear with respect to m, as expected. It should be noted that the questions of asymptotic stability of the obtained polynomial filter estimate, as well as its stability with respect to parameter variations, are not addressed in this paper.

3. APPLICATION

The obtained optimal filter for bilinear system states and linear observations is applied to solution of the terpolymerization process state identification problem in the presence of direct linear observations. The mathematical model of terpolymerization process given by Ogunnaike (Ogunnaike, 1994) is reduced to ten equations for the concentrations of input reagents, the zeroth live moments of the product molecular weight distribution (MWD), and its first bulk moments. These equations are intrinsically nonlinear (bilinear), so their linearization leads to large deviations from the real system dynamics, as it could be seen from the simulation results. Of course, the assumption that the MWD moments can be measured in the real time is artificial, since this can be done only with large time delays, however, at this step, the objective is to verify the performance the obtained nonlinear filtering algorithm for a nonlinear system and compare it with other filtering algorithms based on the linearized model. Taking into account delays in some of the observation components would be the subject of subsequent papers.

Let us rewrite the bilinear state equations (9),(12) and the linear observation equations (10) in the component form using index summations

$$dx_k(t)/dt = a_{0k}(t) + \sum_i a_{1ki}(t)x_i(t) + \tag{15}$$

$$\sum_{ij} a_{2kij}(t)x_i(t)x_j(t) + \sum_i b_{ki}(t)\psi_{1i}(t), \quad k = 1, n,$$

$$y_k(t) = \sum_i A_{ki}(t)x_i(t) + \sum_i B_{ki}(t)\psi_{2k}(t),$$

where $\psi_1(t)$ and $\psi_2(t)$ are white Gaussian noises. Then, the filtering equations (13),(14) can be rewritten in the component form as follows:

$$dm_k(t)/dt = (a_{0k}(t) + \sum_i a_{1ki}(t)m_i(t) + \tag{16}$$

$$\sum_{ij} a_{2kij}(t)m_i(t)m_j(t) + \sum_{ij} a_{2kij}(t)P_{ij}(t))dt +$$

$$\sum_{ijlps} P_{kj}(t)A_{jl}^T(t)(B_{lp}(t)B_{ps}(t)))^{-1} \times$$

$$[dy_s - \sum_r A_{sr}(t)m_r(t)dt],$$

with

$$m_k(t_0) = E[x_k(t_0) \mid Y(t_0)],$$

$$dP_{ij}(t) = \sum_k a_{1ik}(t)P_{kj}(t) + \sum_j P_{kj}(t)a_{1jk}(t) + \tag{17}$$

$$2\sum_{kl} a_{2ikl}(t)m_l(t)P_{kj} + 2\sum_{kl} a_{2jkl}(t)m_l(t)P_{ki}(t) +$$

$$\sum_k b_{ik}(t)b_{kj}(t) -$$

$$\sum_{klpsr} P_{ik}(t)A_{kl}^T(t)(B_{lp}(t)B_{ps}(t)))^{-1}A_{sr}(t)P_{rj}(t),$$

with

$$P_{ij}(t_0) = E[(x_i(t_0) - m_i(t_0))(x_j(t_0) - m_j(t_0))^T \mid Y(t_0)].$$

The terpolymerization process model reduced to 10 bilinear equations selected from (Ogunnaike, 1994) is given by

$$dC_{m1}/dt = [(1/V)d\Delta_{m1}/dt - ((1/\theta) + K_{L1}C^* + \tag{18}$$
$$K_{11}\mu_P^o + K_{21}\mu_Q^o + K_{31}\mu_R^o)C_{m1};$$
$$dC_{m2}/dt = (1/V)d\Delta_{m2}/dt - ((1/\theta) + K_{L2}C^* +$$
$$K_{12}\mu_P^o + K_{22}\mu_Q^o)C_{m2};$$
$$dC_{m3}/dt = (1/V)d\Delta_{m3}/dt - ((1/\theta) + K_{13}\mu_P^o)C_{m3};$$
$$dC_{m4}/dt = (1/V)d\Delta_{m^*}/dt - ((1/\theta) +$$
$$K_d + K_{L1}C_{m1} + K_{L2}C_{m2})C^*;$$
$$d\mu_P^o/dt = (-1/\theta - K_{t1})\mu_P^o + K_{L1}C_{m1}C^* -$$
$$(K_{12}C_{m2} + K_{13}C_{m3})\mu_P^o + K_{21}C_{m1}\mu_Q^o + K_{31}C_{m1}\mu_R^o;$$
$$d\mu_Q^o/dt = (-1/\theta)\mu_Q^o + K_{L2}C_{m2}C^* -$$
$$(K_{21}C_{m1} + K_{t2})\mu_Q^o + K_{12}C_{m2}\mu_P^o;$$
$$d\mu_R^o/dt = (-1/\theta)\mu_R^o - (K_{31}C_{m1} + K_{t3})\mu_R^o + K_{13}C_{m3}\mu_P^o;$$
$$d\lambda_1^{100}/dt = (-1/\theta)\lambda_1^{100} + K_{L1}C_{m1}C^* + K_{L2}C_{m2}C^* +$$
$$K_{11}C_{m1}\mu_P^o + K_{21}C_{m1}\mu_Q^o + K_{31}C_{m1}\mu_R^o;$$
$$d\lambda_1^{010}/dt = (-1/\theta)\lambda_1^{010} + K_{L1}C_{m1}C^* + K_{L2}C_{m2}C^* +$$
$$K_{12}C_{m2}\mu_P^o + K_{22}C_{m2}\mu_Q^o;$$
$$d\lambda_1^{001}/dt = (-1/\theta)\lambda_1^{001} + (K_{L1}C_{m1} + K_{L2}C_{m2})C^* +$$
$$K_{13}C_{m3}\mu_P^o;$$

Here, the state variables are: C_{m1}, C_{m2}, and C_{m3} are the reagent (monomer) concentrations, C^* is the active catalyst concentration; μ_P^o, μ_Q^o, and μ_R^o are the zeroth live moments of the product MWD, and λ_1^{100}, λ_1^{010}, and λ_1^{001} are its first bulk moments. The reactor volume V and residence time θ, as well as all coefficients K's, are known parameters, and $\Delta_{m1}, \Delta_{m2}, \Delta_{m3}, \Delta_{m^*}$ stand for net molar flows of the reagents and active catalyst into the reactor.

The identification (filtering) problem is to find the optimal estimate for the unobservable states (18) assuming that the direct observations Y_i mixed with Gaussian noises ψ_2's are provided for each of the ten state components x_i

$$y_i = x_i + \psi_{2i}.$$

Here, x_1 denotes C_{m1}, x_2 denotes C_{m2}, and so on up x_{10}. In this situation, the bilinear filtering equations (16) for the vector of the optimal estimates $m(t)$ take the form

$$dm_1(t)/dt = (1/V)d\Delta_{m1}/dt - \tag{19}$$
$$((1/\theta) + K_{L1}m_4(t) + K_{11}m_5(t) +$$
$$K21m_6(t) + K_{31}m_7(t))m_1(t) - K_{L1}P_{14}(t) - K_{11}P_{15}(t) -$$
$$K_{21}P_{16}(t) - K_{31}P_{17}(t) + \sum_j P_{1j}[dy_j/dt - m_j]$$

$$dm_2(t)/dt = (1/V)d\Delta_{m2}/dt -$$
$$((1/\theta) + K_{L2}m_4(t) + K_{12}m_5(t) +$$
$$K_{22}m_6(t))m_2(t) - K_{L2}P_{24}(t) - K_{12}P_{25}(t) - K_{22}P_{26}(t) +$$
$$\sum_j P_{2j}[dy_j/dt - m_j]$$

$$dm_3(t)/dt = (1/V)d\Delta_{m3}/dt - ((1/\theta) + K_{13}m_5(t))m_3(t) -$$
$$K_{13}P_{35}(t) + \sum_j P_{3j}[dy_j/dt - m_j]$$

$$dm_4(t)/dt = (1/V)d\Delta_{m^*}/dt - ((1/\theta) + K_d +$$
$$K_{L1}m_1(t) + K_{12}m_2(t))m_4(t) - K_{L1}P_{14}(t) -$$
$$K_{12}P_{24}(t) + \sum_j P_{4j}[dy_j/dt - m_j]$$

$$dm_5(t)/dt = (-1/\theta - K_{t1})m_5(t) + K_{L1}m_4(t)m_1(t) -$$
$$K_{12}m_2(t)m_5(t) + K_{21}m_6(t)m_1(t) +$$
$$K_{31}m_7(t)m_1(t) - K_{13}m_5(t)m_3(t) +$$
$$K_{L1}P_{14}(t) + K_{21}P_{16}(t) + K_{31}P_{17}(t) - K_{12}P_{25}(t) -$$
$$K_{13}P_{35}(t) + \sum_j P_{5j}[dy_j/dt - m_j]$$

$$dm_6(t)/dt = (-1/\theta - K_{t2} - K_{21}m_1(t))m_6(t) +$$
$$K_{L2}m_4(t)m_2(t) + K_{12}m_5(t)m_2(t)$$
$$-K_{21}P_{16}(t) + K_{L2}P_{24}(t) + K_{12}P_{25}(t) +$$
$$\sum_j P_{6j}[dy_j/dt - m_j]$$

$$dm_7(t)/dt = (-1/\theta - K_{t3} - K_{31}m_1(t))m_7(t) + K_{13}m_5(t)m_3(t) - K_{31}P_{17}(t) + K_{13}P_{35}(t) + \sum_j P_{7j}[dy_j/dt - m_j]$$

$$dm_8(t)/dt = (-1/\theta)m_8(t) + (K_{L1}m_4(t) + K_{11}m_5(t) + K_{21}m_6(t) + K_{31}m_7(t))m_1(t) + K_{L2}m_4(t)m_2(t) + K_{L1}P_{14}(t) + K_{11}P_{15}(t) + K_{21}P_{16}(t) + K_{31}P_{17}(t) + K_{L2}P_{24}(t) + \sum_j P_{8j}[dy_j/dt - m_j]$$

$$dm_9(t)/dt = (-1/\theta)m_9(t) + K_{L1}m_4(t)m_1(t) + K_{L2}m_4(t)m_2(t) + K_{12}m_5(t)m_2(t) + K_{22}m_6(t)m_2(t) + K_{L1}P_{14}(t) + K_{L2}P_{24}(t) + K_{12}P_{25}(t) + K_{22}P_{26}(t) + \sum_j P_{9j}[dy_j/dt - m_j];$$

$$dm_{10}(t)/dt = (-1/\theta)m_{10}(t) + K_{L1}m_4(t)m_1(t) + K_{L2}m_4(t)m_2(t) + K_{13}m_5(t)m_3(t) + K_{L1}P_{14}(t) + K_{L2}P_{24}(t) + K_{13}P_{35}(t) + \sum_j P_{10j}[dy_j/dt - m_j].$$

Here, $m_1(t)$ is the optimal estimate for C_{m1}, $m_2(t)$ for C_{m2}, and so on up to $m_{10}(t)$. The fifty-five variance component equations are similarly generated by the equations (17), however are not given here due to place shortage.

In the simulation process, the initial conditions at $t = 0$ are equal to zero for the state variables $C_{m1}, ..., \lambda_1^{001}$, to 0.5 for the estimates $m_1(t), ..., m_{10}(t)$, to 1 for the diagonal entries of the variance matrix, and to zero for its other entries. The system parameter values are all set to 1: $V = 1; d\Delta_{m1}/dt = 1; K_{L1} = 1; K_{11} = 1; K_{21} = 1; K_{31} = 1; K_{32} = 1; d\Delta_{m2}/dt = 1; d\Delta_{m3}/dt = 1; d\Delta_{m^*}/dt = 1; K_{L2} = 1; K_{L3} = 1; K_{12} = 1; K_{13} = 1; K_{22} = 1; K_d = 1; K_{t1} = 1; K_{t2} = 1; K_{t3} = 1; \theta = 1$. The white Gaussian noises in the equations (19) are realized as sinusoidal signals: $\psi_i = \sin t$ for $i = 1, 10$.

In Figure 1, the obtained values of the state variables $C_{m1}, ..., \lambda_1^{001}$ are given in the blue, and the values of the bilinear optimal filter estimates $m_1(t), ..., m_{10}(t)$ are depicted in the red.

The performance of the optimal bilinear filter (16),(17) is compared to the performance of the optimal linear Kalman-Bucy filter available for the linearized system. This linear filter consists of only the linear terms and innovations processes in the equations (16) (or (19)) for the optimal estimates and the Riccati equations for the variance matrix components corresponding to the equations (17):

$$dm_k(t)/dt = (a_{0k}(t) + \sum_i a_{1ki}(t)m_i(t) + \sum_{jlps} P_{kj}(t)A_{jl}^T(t)(B_{lp}B_{ps}))^{-1}(t)[dy_s - \sum_r A_{sr}(t)m_r(t)dt] \quad (20)$$

with

$$m_k(t_0) = E[x_k(t_0) \mid Y(t_0)];$$

$$dP_{ij}(t)/dt = \sum_k a_{1ik}(t)P_{kj}(t) + \sum_k P_{ki}(t)a_{1jk}(t) + \sum_k b_{ik}(t)b_{kj}(t) - \sum_{klpsr} P_{ik}(t)A_{kl}^T(t)(B_{lp}B_{ps}))^{-1}A_{sr}P_{rj}(t). \quad (21)$$

with

$$P_{ij}(t_0) = E[(x_i(t_0) - m_i(t_0))(x_j(t_0) - m_j(t_0))^T \mid Y(t_0)].$$

The graphs of the estimates obtained using this linear Kalman-Bucy filter are shown in Figure 1 in the green.

Finally, the performance of the optimal bilinear filter (16),(17) is compared to the performance of the mixed filter designed as follows. The estimate equations in this filter coincide with the equations (16) (or (19)) from the optimal bilinear filter, and the variance equations coincide with the equations (21) from the linear Kalman-Bucy filter. The graphs of the estimates obtained using this mixed filter are shown in Figure 1 in the black. The initial conditions and white Gaussian noise realizations remain the same for all the filters involved in the simulation.

Upon comparing all simulation results given in Figure 1, it can be concluded that the optimal bilinear filter gives the best estimate in comparison to two other filters. Although this conclusion follows from the developed theory, the numerical simulation serves as a convincing illustration.

4. REFERENCES

Benes, V.E. (1981). Exact finite-dimensional filters for certain diffusions with nonlinear drift. *Stochastics* **5**, 65–92.

Kalman, R. E. and R. S. Bucy (1961). New results in linear filtering and prediction theory. *ASME Journal of Basic Engineering, Ser. D* **83**, 95–108.

Kushner, H. J. (1964). On differential equations satisfied by conditional probability densities of Markov processes. *SIAM J. Control* **2**, 106–119.

Mitter, S. K. (1996). Filtering and stochastic control: A historic perspective. *Control Systems* **16**, 67–76.

Ogunnaike, B. A. (1994). On-line modeling and predictive control of an industrial terpolymerization reactor. *Int. J. Control* **59**, 711–729.

Pugachev, V. S. (1984). *Probability Theory and Mathematical Statistics for Engineers.* Pergamon Press. London.

Wonham, W.M. (1965). Some applications of stochastic differential equations to nonlinear filtering. *SIAM J. Control* **2**, 347–369.

Yau, S.S.-T. (1994). Finite-dimensional filters with nonlinear drift i: a class of filters including both kalman-bucy and benes filters. *J. Math. Systems, Estimation, and Control* **4**, 181–203.

Zhou, W.W. and M. Blanke (1989). Identification of a class of non-linear state space models using rpe techniques. *IEEE Trans. Automat. Contr.* **34**, 312–316.

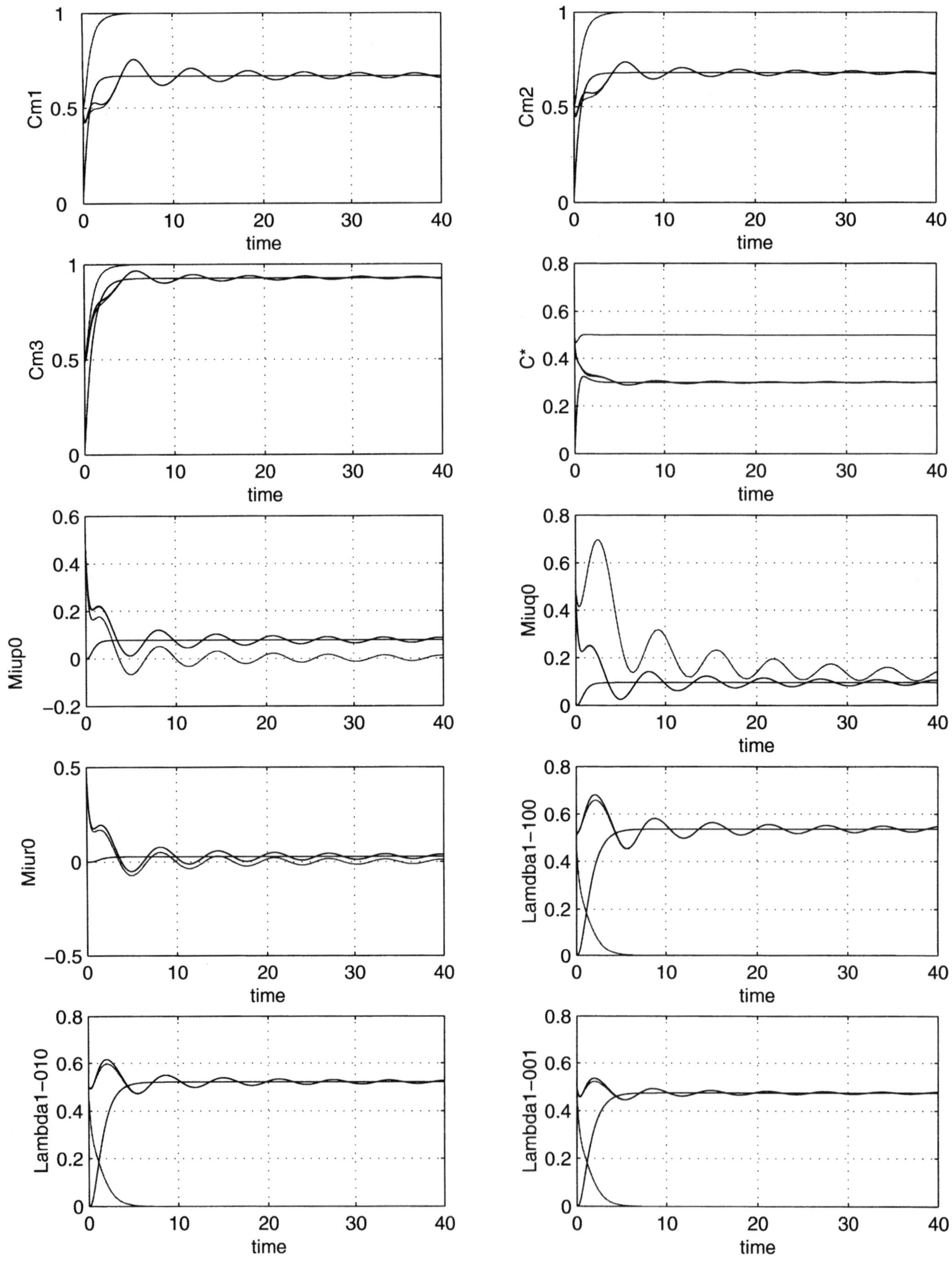

Fig. 1. Graphs of the ten state variables (18) (blue), the estimates given by the optimal bilinear filter (16),(17) (red), the estimate given by the linear Kalman-Bucy filter (20),(21) (green), the estimates given by the mixed filter (16),(21) (black)

IFAC Publications
www.elsevier.com/locate/ifac

NEURAL PREDICTION OF CYLINDER AIR MASS FOR AFR CONTROL IN SI ENGINE

G. Bloch*, Y. Chamaillard, G. Millerioux* and P. Higelin****

* : *Centre de Recherche en Automatique de Nancy (CRAN, UMR CNRS 7039)*
ESSTIN, 2, rue Jean Lamour, 54519 Vandoeuvre les Nancy, France.
gerard.bloch@esstin.uhp-nancy.fr

** : *Laboratoire de Mécanique et d'Energétique (LME)*
ESEM, 8, rue Léonard de Vinci, 45072 Orléans Cédex 2, France.
yann.chamaillard@univ-orleans.fr

Abstract: As parsimonious and flexible universal approximator, the one hidden layer perceptron can be used for non linear prediction. An application is described in the framework of Air-Fuel Ratio (AFR) control in spark-ignition engines, a critical point to satisfy pollutant emission legislation. AFR control depends essentially on the prediction of the air mass to be admitted in cylinder. The building of an air mass predictive neural network is described and its performances are evaluated. Compared to classical solutions based on static mappings, the neural predictor allows for reduction of AFR excursions on rapid torque transients. *Copyright © 2003 IFAC*

Keywords: identification, neural networks, prediction problems, control oriented model, applied neural control, engine modeling, engine control, automotive emissions, Air-Fuel Ratio.

1. INTRODUCTION

Artificial neural networks have been the focus of a great deal of attention during the last two decades, due to their capabilities to solve non-linear problems by learning from data. Although a broad range of neural network architectures can be found, multilayer perceptrons (MLP) and radial basis function networks (RBFN) are the most popular neural models, particularly for system modeling and identification (Chen, and Billings, 1992; Sjöberg, *et al.*, 1995).

In the second part, a particular form of MLP, the one hidden layer perceptron with linear output unit, is briefly recalled. With an adequate choice of the regression vector, such a black box model can be used as Non linear Output Error model. An application is then described in the framework of Air-Fuel Ratio (AFR) control in spark-ignition engines, a critical point to satisfy pollutant emission legislation. Section 3 is devoted to AFR control. The physical problem and the associated control problems are described. Then, the AFR control method is detailed and the air mass prediction issue developed. In part 4, a solution using a model-based neural air mass predictor in addition with transient fuel film compensation is proposed. Compared to classical solutions based on static mappings, the neural predictor allows for system dynamics modeling and so for reduction of AFR excursions on rapid torque transients.

2. MLP FOR NON LINEAR MODELLING

Because of their ability to represent complex non-linear mappings with good flexibility and accuracy, neural networks have become popular to model various subsystems as discrete black boxes (Narendra, and Parthasarathy, 1990; Chen, and Billings, 1992; Norgaard, *et al.*, 2000).

As parsimonious (Barron, 1993) and flexible universal approximator, the one hidden layer perceptron with linear output unit is used here. Its form is given, for a single output f, by:

$$f=\sum_{i=1}^{n} w_i^2 \, g\left(\sum_{j=1}^{p} w_{ij}^1 \varphi_j + b_i^1\right) + b^2 \qquad (1)$$

where φ_j, $j=1,\cdots,p$, are the inputs of the network, w_{ij}^1 and b_i^1, $i=1,\cdots,n$, $j=1,\cdots,p$, are the weights and biases of the hidden layer, the activation function g is a sigmoid function, chosen here as often as the hyperbolic tangent, w_i^2, $i=1,\cdots,n$, and b^2 are the weights and bias of the output neuron or node.

The restriction to only one hidden layer and to a linear activation function at the output brings the general perceptron closer to other non linear models, neural or not (Sjöberg, *et al.*, 1995). Indeed, the one hidden layer perceptron corresponds to a unique particular choice, the sigmoid function, for the basis function g_i, and to a "ridge" construction for the inputs in a function expansion:

$$f(\varphi,\theta)=\sum_{i=1}^{n} \alpha_i \, g_i(\varphi,\beta_i) \qquad (2)$$

where $\varphi=\left[\varphi_1 \cdots \varphi_p\right]^T$ is the regression vector and the parameter vector θ is the concatenation of all the weights w and biases b, with $\alpha_i = w_i^2$, $i=1,\cdots,n$ and $\beta_i=\left[w_{i1}^1 \cdots w_{ip}^1 \, b_i^1\right]^T$.

The process of approximating a non-linear relationship from data can be decomposed in several steps:

- determining the structure of the regression vector φ or selecting the inputs of the network,
- choosing the non-linear mapping f or, in the neural network terminology, selecting an internal network architecture,
- estimating the parameter vector θ, i.e. (weight) "learning" or "training".

A review of the methods for selecting the network architecture and estimating the parameter vector can be found in (Bloch, *et al.*, 2003). For dynamical systems in discrete-time *k*, different models can be derived, depending on the choice of the regressors in φ, which then is $\varphi(k)$. Particularly, the regression vector of a Non linear (or Neural) Output Error (NOE) Multiple Input Single Output (MISO) model will contain delayed inputs $u_i(k-l)$ and outputs simulated from past inputs u only $\hat{y}_u(k-l|\theta)$. Such a model can be used to replace a complex non linear simulation model for real time control purposes, as in that follows.

3. AFR CONTROL

3.1 Physical problem and control

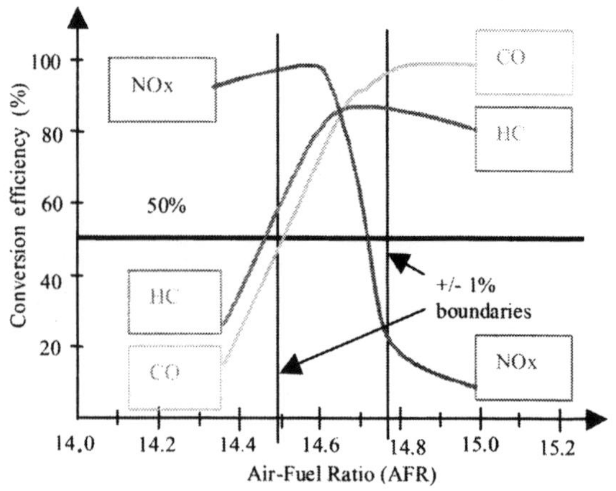

Fig. 1. Catalytic converter efficiency.

In today's spark ignition engines, three-way catalysts are used to reduce the exhaust emission of the three main pollutants that are: unburned hydrocarbons (HC), carbon monoxide (CO) and nitrogen oxides (NO_x). The Air-Fuel Ratio (AFR) is the ratio between the air mass and the fuel mass admitted into the cylinder. The optimization of the three-way catalyst efficiency requires the cylinder AFR to be kept in a narrow band which corresponds to the stoichiometric conditions (Heck and Farrauto, 2001). Figure 1 describes the catalytic conversion efficiency for the three main pollutants versus the in-cylinder mixture AFR. Even a small deviation from stoichiometric conditions can result in a dramatic degradation of the conversion efficiency.

In recent spark ignition engines, the actuators and sensors related to the AFR consist of electronic throttle, electronic injector, electronic ignition, intake manifold pressure sensor, engine speed sensor and UEGO sensor for AFR measurement (see figure 2).

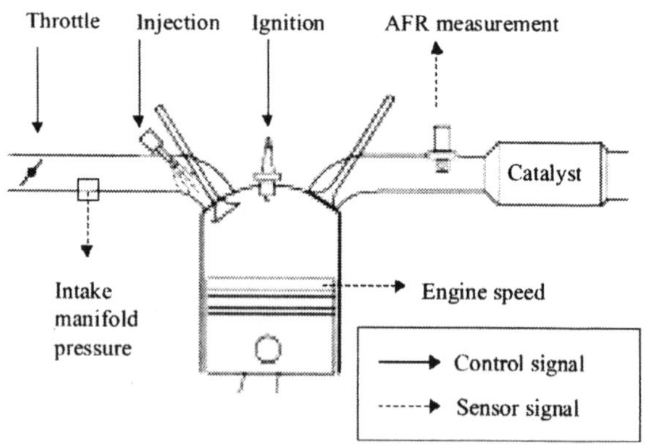

Fig. 2. Basic SI engine.

A modern engine control unit, as the common ones installed on new vehicles, handles the AFR regulation task very well under steady state conditions (Chamaillard and Perrier, 2001). It provides the injection controller with a prediction of the air mass to be admitted in the cylinder and uses a Universal Exhaust Gas Oxygen (UEGO) sensor in the exhaust flow for the AFR measurement that permits to correct an eventual bias by feedback.

The control problem becomes more difficult in transient phases owing to the more difficult prediction of the air mass, the fuel flow dynamics and the inherent delay in the feedback system. This results in AFR excursions during fast transients, and so increased pollutant emissions.

Lots of researches deal with the air mass prediction (Weeks, and Moskwa, 1994; Jankovic, and Magner, 2001; Magner, and Jankovic, 2002) and with fuel film dynamics (Hendricks, *et al.*, 1993; Behnia, and Milton, 2001; Arsie, *et al.*, 2001) to improve AFR control.

The two variables involved in the AFR, i.e. the air and fuel masses admitted into the cylinder, are not accessible for measurement but are essentially dependant (through dynamic systems) of the throttle angle reference and the injection duration.

Accurate simulation models are available for describing the fuel and air dynamics. These dynamics are fairly complex and non linear. On the other hand, a close look at the engine processes shows that the operations divide the physical processes into four distinct regimes corresponding to the four events: intake, compression, power and exhaust, and suggests an event-based approach according to the crank angle.

As a result, the characteristic behavior of an engine consists of a combination of two types of dynamics: time-based and event-based. Event-based dynamics are described in the crank angle domain. From the engine control point of view, only one value of AFR exists at each cycle for each cylinder, and the outputs of an engine control system are synchronous with crank angle. So, the fundamental sampling period T_e (constant in the crank-angle domain but varying in the time domain) corresponds to the rate of occurrence of intake event and is defined in seconds by:

$$T_e = \frac{120}{Ne\, n_{cyl}} \qquad (3)$$

where n_{cyl} is the number of cylinders and Ne the engine speed (rpm).

In the crank angle domain, the AFR at the end of the exhaust phase is defined by:

$$AFR(k) = \frac{M_{air}(k-3)}{M_{fuel}(k-3)} \qquad (4)$$

where $M_{air}(k-3)$ and $M_{fuel}(k-3)$ are respectively the air and fuel masses admitted in the cylinder at the end of the intake phase. The delay in the AFR corresponds to the delay inherent in a four-stroke engine between the end of the intake and the exhaust phase.

3.2 AFR control strategy

There are two main kinds of AFR control systems, fuel control and air control (Chang, *et al.*, 1993). The fuel injection control system is the most common one and regulates the fuel flow according to the airflow. The AFR control is here an element of a global torque control strategy that first controls the air mass in order to satisfy the torque reference and then adjusts the injection duration to be in desired conditions.

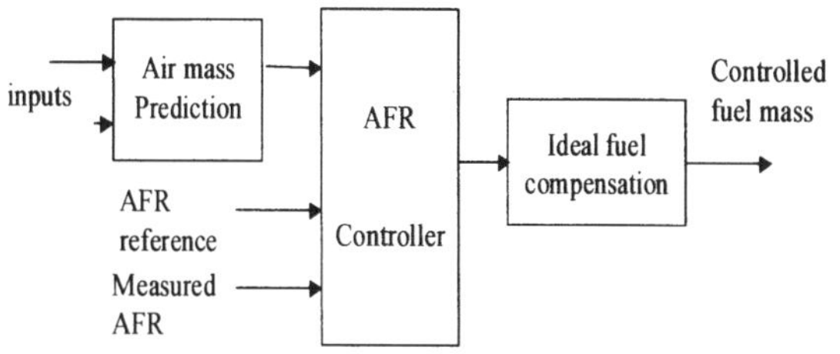

Fig. 3. AFR control.

The general scheme is described in figure 3. A Luenberger observer is built to estimate the manifold pressure. This observer, obtained from physical equations after appropriate linearization and discretization, is given by:

$$\hat{p}(k+1) = a\,\hat{p}(k) + b + k_o(k)\big(p(k) - c\,\hat{p}(k)\big) \qquad (5)$$

where $\hat{p}(k)$ is the prediction at discrete time k of the pressure $p(k)$, a, b, c are resulting scalar constants, and $k_o(k)$ the time varying observer gain.

The admitted air mass is then predicted by a volumetric efficiency map from the estimated manifold pressure and engine speed, and, in turn, the fuel mass to be injected is calculated from this prediction and the engine speed. A fuel film dynamics compensation is also achieved.
When the admission occurs and the intake valve closes, the AFR is already determined. However, its measurement is only possible after the exhaust phase to correct the injection duration. Classically, the feedback uses a Proportional Integral (PI) controller that is able to correct a bias on the air estimation thanks to its integral action, but cannot manage the transient errors.

As stated in Chang, *et al.* (1993), several sources of AFR excursions have been identified during engine transients including manifold air filling, air mass prediction error, fuel flow dynamics, feedback delay and back-flow. The feedback delay, the sensor dynamics, slow with respect to the variation to be detected, and the fuel flow dynamics lead to inefficiency of the feedback scheme during transients. Nevertheless, the main problem to be tackled lies in the estimation of the air mass that will be admitted in the cylinder. As in port fuel injection engines the injection system requires time to dispense fuel, the injection should be completed before the intake valve opens and the injection controller must get a prediction of the air mass to be admitted before a direct estimation is available. The problem is due to the fact that during transient, operating conditions change between the instant the estimation is done and the admission phase, resulting in prediction errors. This prediction is essential in AFR regulation. A solution for air prediction is proposed in the following section.

4. NEURAL AIR MASS PREDICTION

4.1 The neural model

As the measurement of the AFR presents a delay and the sensor dynamics is slow with respect to the variation to be detected during transient phases, a feedforward control seems to be the solution during transients. For such a scheme, the quality of the AFR regulation depends essentially on the prediction of the air mass to be admitted. The goal is to obtain a discrete event-based model of the air admission in order to predict the air mass flow to be admitted in the cylinder. The delay between the angle reference and the effective throttle position can be used in this way to develop an air charge anticipation algorithm. Magner and Jankovic (2002) develop such a solution using a neural network predictor. Other works (Majors, *et al.*, 1994; Li, *et al.*, 1999) already use neural networks to optimize AFR control.

As the variable to estimate, here the air mass, is not measured, a simulation model, involving outputs predicted by the model in the regression vector, is needed. Hence, a Neural Output Error (NOE) model is used. To predict $\hat{y}(k)$, the air mass to be admitted at discrete time *k*, the following regressors have been chosen:

Air mass prediction at $k-1, \hat{y}(k-1)$,
Manifold pressure $p(k-1)$ and $p(k-2)$,
Engine speed $Ne(k-1)$,
Throttle angle reference $Thr_{ref}(i)$, $i = k-1, \cdots, k-6$.

This choice is based on physical equations, which involve as dynamical inputs the manifold pressure, effective throttle section and engine speed. Including $p(k-1)$ and $p(k-2)$ reflects the presence of the manifold pressure time derivative in these equations. The engine speed, beyond its role in air admission model, permits to handle the variable sampling period *Te* issue. The last regressors allow the prediction thanks to the delay present in the throttle actuator which is around 30 ms. At 6000 rpm, the sampling period is 5 ms and 6 samples are then necessary. Magner and Jankovic (2002) made the same choice. The use of a rapid throttle can reduce the delay and so the number of regressors.

The experiments are made on a non-linear fuel-injected, mean-value and event-based model. Computation is performed at each Top Dead Center. The engine model includes the engine (fuel flow dynamics, airflow dynamics, combustion and delays inherent in four–stroke engine), actuators, sensors and a dynamic model of the load. The model used is representative of a PSA engine.

Training was performed by minimizing the mean squared error function, with the Levenberg-Marquardt method implemented in a specific Matlab toolbox (Norgaard, 1995). The different signals involved in training the network should have been scaled in order to avoid saturation. A hidden layer of $n = 14$ neurons (see eq. 1) was selected to reach good prediction accuracy. The training data set was obtained by simulating the engine on a large range of operation. The torque reference signal consisted of steps of random length and size, to which was added up another random step signal with length and amplitude divided by 10, as shown in figure 4.

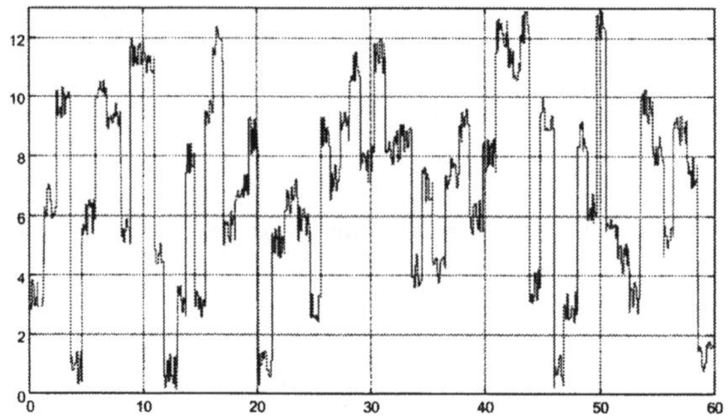

Fig. 4. Engine torque reference (daN.m) vs. time (sec).

The whole engine operation range for different engine speeds was covered. The speed reference signal was varying from 1000 to 6000 rpm by step of 1000 rpm. As the sampling period depends on the engine speed, the step duration varied with the speed reference to keep the same learning points number for each level.

4.2 The results

Two simulation scenarios can be considered for the validation: the engine speed scenario and the torque scenario. Chamaillard and Perrier (2001) showed that transients in torque are the most disturbing. So the torque scenario is used here for comparison. Different simulations have been performed to test the neural air mass predictor. The torque reference represented on figure 5 is chosen to generate fast throttle angle variations and thus rapid transient phases. That signal is used with different engine speed references from 1000 to 6500 rpm by 500 rpm step to compare the results with data similar but different from the learning set.

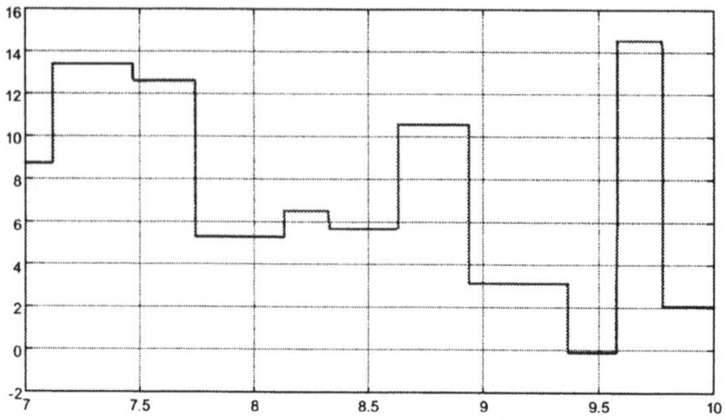

Fig. 5. Torque reference (daN.m) vs. time (sec).

The test results in table 1. The engine speed reference value used with the torque reference is reported in the first row. The Root Mean Square Error (RMSE) values for the air mass prediction with a traditional method (Air_t) (prediction by a volumetric efficiency map from estimated manifold pressure, given by (5), and engine speed) and for the neural prediction (Air_nn) are reported in the second and third rows. The last two rows give the RMSE on AFR control results with traditional (AFR_t) and neural (AFR_nn) predictions. For the control results of these two last rows, the error is simply the difference between the actual AFR and the reference, which is equal to 14.7. The simulations have been done with ideal fuel film compensation and a PI controller on AFR measurement (in the AFR controller - figure 3) to avoid bias.

Table 1 Results

Rpm	Air_t	Air_nn	AFR_t	AFR_nn
1000	360.75	8.59	76.9	0.42
1500	84.94	12.79	29.4	0.93
2000	40.70	3.07	2.81	0.32
2500	10.40	3.71	1.03	0.35
3000	7.52	2.37	0.91	0.18
3500	4.72	2.18	0.70	0.25
4000	4.24	1.29	0.67	0.24
4500	3.47	2.0	0.53	0.27
5000	3.33	1.77	0.56	0.30
5500	2.75	2.16	0.40	0.29
6000	2.51	3.52	0.41	0.44
6500	1.37	4.96	0.29	0.58

The results show that the neural prediction leads to a very significant improvement in AFR control thanks to its better prediction of the air mass to be admitted. The neural network interpolates the learning data very well, but, for extrapolation, the performances fall down compared to traditional method (at 6500 rpm for example).

Results at 3500 rpm are shown in figures 6 and 7, during only 3 seconds to better illustrate the differences. Figure 6 shows the neural air mass prediction compared to the real (simulated) air mass. It can be noticed that the prediction error is very weak and the real and predicted air masses are difficult to distinguish

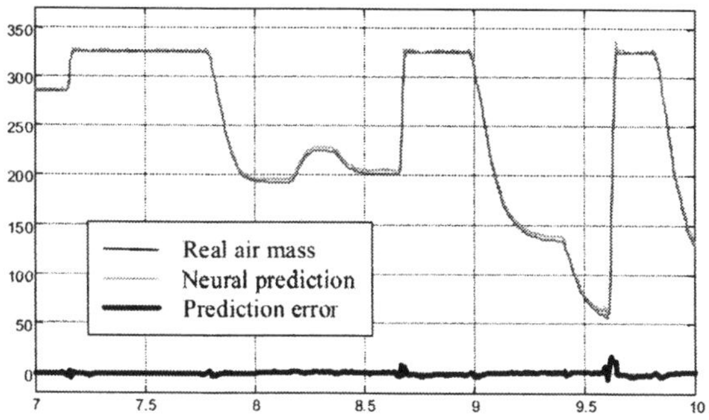

Fig. 6. Predicted and real air mass (mg) vs. time (sec).

As previously mentioned, the good accuracy of the air mass neural predictor allows to significantly enhance the AFR control. Figure 7 shows a comparison between the AFR excursions with the traditional air mass flow predictor (AFR_t) and the neural one (AFR_nn). In all cases the AFR excursions are reduced (by 50%) especially for high excursions, which are the most problematic for consumption, pollution and agreement. However the static error compensation is rather slower because the feedback controller has not been redefined.

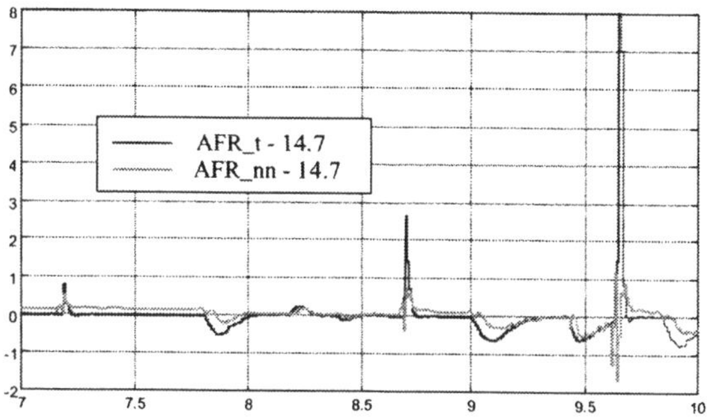

Fig. 7. Comparison of the AFR error vs. time (sec).

5. CONCLUSION

In the framework of AFR control, a non-linear NOE model has been built from a detailed and accurate physical model to predict the air mass to be admitted in cylinder. Using this predictor in addition with transient fuel film compensation allows to significantly reduce the AFR excursions on rapid torque transients if the inputs/outputs of the air admission can be correctly collected. The neural model takes advantage of the delay in the throttle actuator and uses a variable time sampling period. Compared to classical solutions based on static mappings, the neural predictor allows for system dynamics modeling which can be embedded in the control scheme.

From AFR control results it appears that the feedback controller must be redefined to optimize static error compensation. Neural prediction of cylinder air mass permits a clear improvement of the AFR control, but some issues must be investigated for further application, like the data collection from real engine and the system non-stationarity with time.

REFERENCES

Arsie, I., C. Pianese, G. Rizzo and V. Cioffi (2001). An adaptive estimator of fuel film dynamics in the intake port of a spark ignition engine. 3rd IFAC Workshop Advances in Automotive Control, Karlsruhe, 293-298.

Barron, A.R. (1993) Universal approximation bounds for superpositions of a sigmoidal function. *IEEE Trans. Inform. Theory*, **39**(3), 930-945.

Behnia, M. and B.E. Milton (2001). Fundamentals of fuel film formation and motion in spark ignition engine induction systems. *Energy Conversion and Management*, **42**, 1751-1768.

Bloch, G. and T. Denoeux (2003). Neural networks for process control and optimization: two industrial applications. *ISA Transactions*, **41**(2), 39-51.

Chamaillard, Y. and C. Perrier (2001). Air-fuel ratio control by fuzzy logic, preliminary investigation. 3rd IFAC Workshop Advances in Automotive Control, Karlsruhe, 221-226.

Chang, C.F., N.P. Fekete and J.D. Powell (1993). Engine air-fuel ratio control using an event-based observer. SAE International Congress and Exposition, SAE Paper 930766, Detroit, MI.

Chen, S. and S. Billings (1992). Neural networks for non-linear dynamic systems modelling and identification. *International Journal of Control*, **56**, 319-346.

Heck, R.M. and R.J. Farrauto (2001). Automobile exhaust catalysts. *Applied Catalysis, A: General*, **221**, 443-457.

Hendricks, E., T. Vesterholm, P. Kaidantzis, P. Rasmussen and M. Jensen (1993). Nonlinear Transient Fuel Film Compensation (NTFC). SAE International Congress and Exposition, SAE Paper 930767, Detroit, MI.

Jankovic, M. and S. Magner (2001). Cylinder air-charge estimation for advanced intake valve operation in variable cam timing engines. *JSAE*, **22**, 445-452.

Li, N., K. Li and S. Thompson (1999). Employing a new type of neural network to optimise power plant air-fuel ratio. 14th IFAC Triennial World Congress, Beijing, 333-338.

Magner, S. and M. Jankovic (2002). Delta air charge anticipation for mass air flow and electronic throttle control based systems. Proc. of American Control Conference, Anchorage, 1407-1412.

Majors, M., J. Stori and D. Cho (1994). Neural network control of automotive fuel injection systems. IEEE International Symposium on Intelligent Control, 31-36.

Narendra, K.S. and K. Parthasarathy (1990). Identification and control of dynamical systems using neural networks., *IEEE Trans. on Neural Networks*, **1**(1), 4-27.

Norgaard, M., O. Ravn, N.K. Poulsen and L.K. Hansen (2000). *Neural networks for modeling and control of dynamic systems*, Springer-Verlag.

Norgaard, M. (1995). Neural Network Based System Identification Toolbox. *Technical Report 95-E-77*, Institute of Automation, Technical University of Denmark.

Sjöberg, J., Q. Zhang, L. Ljung, A. Benveniste, B. Delyon, P.Y. Glorennec, H. Hjalmarsson and A. Juditsky (1995). Nonlinear black-box modeling in system identification: A unified overview. *Automatica*, **31**(12), 1691-1724.

Weeks, R.W. and J.J. Moskwa (1994). Transient airflow rate estimation in a natural gas engine using a non-linear observer. SAE Paper 940759.

IFAC Publications
www.elsevier.com/locate/ifac

CONTRIBUTION TO IDENTIFICATION OF THERMO-MECHANIC INTERACTION AT VIBRATING RUBBER-LIKE MATERIALS

Pešek Luděk, Půst Ladislav, Vaněk František

Institute of Thermomechanics, Academy of Science of the Czech Republic, Dolejškova 5, CZ 182 00, Prague 8, Czech Republic.

Abstract: Effect of interaction between the vibration of a damped mechanical system and thermal processes is investigated. This interaction, which has in general non-linear character, is important for systems containing springs made from elastomers and other rubber-like materials with higher inner damping and marked dependence of mechanical properties, such as stiffness and damping, on temperature. Forms of response curves, often used at identification of dynamic systems, depend on a lot of mechanical and thermal parameters. In the paper, two relationships are studied: influence of decrease of stiffness with temperature and change of heat transfer on the form of response curves. Elaborated method of solution of thermo-mechanic interaction will be used for identification of dynamic systems with dissipation layers. *Copyright © 2003 IFAC*

Keywords: non-linear, modelling, numerical methods, thermo-mechanic interaction, dissipation, passive control

1. INTRODUCTION

In the frame of our grant task GA CR no. 101/02/0241 "Vibroacoustic problems of mechanical systems with a single dissipation layer" we deal with the mathematical modelling of elastomers. This material is often used as passive damping elements for diminishing of vibration in mechanical structures. Our investigation is aimed at a mathematical description of dynamic behaviour by use of rheological models with consideration of non-linear terms in their both stiffness and damping characteristics. We develop also new methods for the identification of the parameters of the non-linear rheological models (Pešek *et al.*, 2002). In the latest we are engaged also by the interaction of mechanical and thermal processes in the vibrating damped structures (Pešek *et al.*, 2003). The effect of this interaction is studied herein by means of the numerical simulations in frequency domain.

Vibrating mechanical systems, in which a disippation of energy is realised mainly in the elements of plastic materials, change their dynamic behaviour with a temperature change e.g. (Mark *et al.*, 1988, Mead, 1999). This effect is substantial particularly at rubber-like components, which have a high inner damping and hence a high absorption of vibrating energy. This energy is inside a body transformed into heat. An increase of temperature causes backwardly changes of elastic and damping behaviours and that consequently changes dynamic properties of a whole system.

Higher temperature of the component has also an effect of higher heat transfer into environment. At stationary regime the temperature proportions stabilize after certain time so that the temperature rise, which is proportional to energy dissipated by the inner damping, is in equilibrium with the temperature drained to environment.

In this contribution, the simple computational model is developed for analysis of both the dissipated energy by inner damping and the arise of heat and temperature including their influence on the mechanical behaviour of a whole system. For simplicity these phenomena are investigated on 1 DOF system exited by a harmonic force.

2. PHYSICAL AND MATHEMATICAL MODEL

The scheme of a system with a thermo-mechanical interaction is depicted in Fig. 1. At a vibration the energy E_b is dissipated in the component $g(\dot{x},\theta)$ and is all transformed to heat $E_T + E_{T0}$. The energy E_{T0} is drained from the system into environment either by heat transfer, radiation or convection of surrounding media. The energy of acoustic waves can be also included into this part of energy. The rest of energy E_T remains in the compliant component and increases its temperature θ. The increase of temperature $\Delta\theta$ is proportional to heat input E_T. From logic consideration that heat output E_{T0} will increase with an increase of temperature, it results, that for stationary vibration these properties stabilize after some time. It is valid both for temperature $\theta \to \theta_{stac}$ and the heat output, when $E_b = E_{T0}$ and $E_T = 0$.

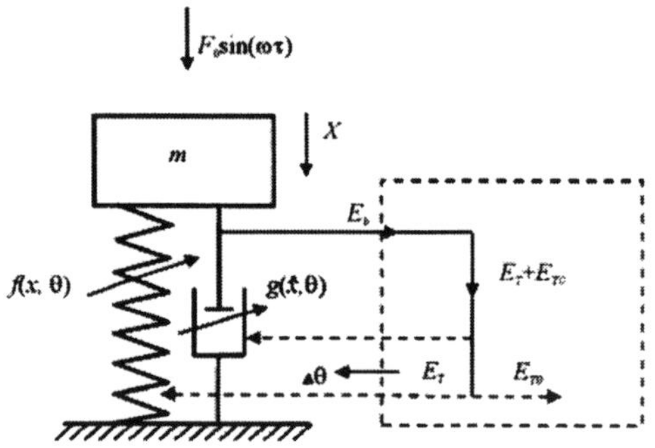

Fig. 1 Scheme of interaction of mechanical system and heat

The equation of motion of the mechanical system is

$$m\ddot{x} + f(x,\theta) + g(\dot{x},\theta) = F_0 \sin\omega t\,, \qquad (1)$$

where generally non-linear functions $f(x,\theta)$ represents reversible force, dependent on displacement and x and temperature θ and $g(\dot{x},\theta)$ represents damping force, that depends on velocity $\dot{x}$ and temperature θ.

The temperature arise in one period $T = 2\pi/\omega$ of the excitation force is given by lost energy in the damping component in the same time period

$$E_b = \oint g(\dot{x},\theta)dx = \int_0^T g(\dot{x},\theta)\dot{x}dt\,. \qquad (2)$$

For simplicity we assume herein a linear dependence of these quantities f and g on displacement x and velocity $\dot{x}$, respectively. So that

$$m\ddot{x} + k(\theta)x + b(\theta)\dot{x} = F_0 \sin\omega t\,. \qquad (3)$$

Since at ordinary materials the lost energy is very small for one period, it can be assumed, that the coefficient *b(θ)* will be during 1 or several few period constant. Therefore, it is sufficient to determine a lost energy for time of *n* periods, i.e. for time $T_n = n\dfrac{2\pi}{\omega}$, $n \le 5$.

$$E_b = \int_0^{n2\pi/\omega} b(\theta)\dot{x}^2 dt\,. \qquad (4)$$

The loss energy can be evaluated during calculations in different ways. We can come either from the relations (2), (4) or from an equivalent linearization assuming that the movement is close to harmonic. In latter case for $n = 1$ we get

$$E_b = b_e(\theta)\int_0^{2\pi}(\omega x_{\max}\cos\omega t)^2 dt = \pi\, b_e(\theta)\omega\, x_{\max}^2\,.$$

For numerical simulations of resonant region passages the modified relations (2) or (4) are the most suitable.

The equation (2) will be first arranged into the non-dimensional form

$$X'' + B(\theta)X' + K(\theta) = \sin\varphi\,, \qquad (5)$$

where $X = \dfrac{xk_0}{F_0}$, $B(\theta) = \dfrac{b(\theta)}{\sqrt{k_0 m}}$, $\tau = t\sqrt{k_0/m}$,
$K(\theta) = k(\theta)/k_0$.

The quantities $k_0[\mathrm{kgs}^{-2}]$, $b_0[\mathrm{kgs}^{-1}]$ are values $k(\theta)$, $b(\theta)$ for $\theta = 0$ in time $t = \tau = 0$. The angle φ can be expressed as

$$\varphi = \omega_1 t + k_1 t^2/2 \text{ or } \varphi = \eta_0\tau + k_2\tau^2/2\,. \qquad (6)$$

The temporal angular velocity is defined as $\omega = \omega_0 + k_1 t$, where ω_0 is an initial angular velocity in time $t = 0$. The quantity k_1 is an angular acceleration [rad/s^2].

In a non-dimensional form there is an angular velocity $\eta = \sqrt{m/k_0}$ expressed

$$\eta = \eta_0 + k_2\tau\,, \qquad (7)$$

where $k_2 = k_1 m/k_0$ and $\varphi = \int_0^\tau \eta d\tau$.

For numerical evaluation of the resonant curves in the programming language Matlab 6.0 the equation of motion is arranged into a form

$$X_1' = X_2$$
$$X_2' = -X_1 - BX_2 + \sin(\varphi)\,. \qquad (8)$$

These equations are solved discretely in a time domain by the Runge-Kutta method of the 4th order.

If a number of steps $l = \frac{2\pi}{\Delta\tau}$ belongs to one period at a selected integration step $\Delta\tau$, there is a contribution of loss energy due to a dissipation after n periods according to (4) in a non-dimensional form

$$\Delta E_b = \sum_{i=0}^{nl} B(\theta) * X_2 * (X_{1i} - X_{1(i-1)}). \qquad (9)$$

After nl steps the computation is interrupted, the evaluated ΔE_b is added to the energy E_b, which is stored in the compliant component of the system in a form of heat E_{TC}.

This heat is inside the system longer time, however, it escapes from the system by conduction according to an exponential law. Therefore at each addition of new energetic contribution ΔE_b the energy E_{TC-1} stored in the body must be decreased by a certain part that is proportional to an actual amount of heat E_{TC-1} and also proportional to temperature θ

$$E_{TC} = (1-\alpha)E_{TC-1} + \beta\Delta E_b. \qquad (10)$$

The coefficient α of heat output giving a decrease of system temperature by heat conduction to surroundings within n periods of exciting force is proportional to a number of measured periods n and is relatively very small. It depends on material behaviours, on a shape of a deformed component and also on temperature and conduction conditions of surroundings. It can be chose for example 1.10^{-3}, i.e. $(1-\alpha) = 0.999$.

The heat E_{TC} remains within n periods inside the deformed component and it determines a component's temperature θ with respect to a temperature capacity of the component. The temperature θ is numerically measured from the initial temperature, e.g. a room temperature.

Let's assume, that the deformed component has a temperature capacity coefficient $c_T = 1\text{J}/^0C = 1\text{kgm}^2\text{s}^{-2}(^0\text{C}^{-1})$. Then a numerical value of the heat change ΔE_T [$\text{kgm}^2\text{s}^{-2}$] is equivalent with a value of the temperature change $\Delta\theta$ [^{0}C]. It holds

$$\Delta\theta = 1.\Delta E_{TC}, \text{ or } \theta = E_{TC}. \qquad (11)$$

This temperature θ influences the quantity of stiffness $K(\theta)$ and damping $B(\theta)$. These dependencies are generally non-linear. Let's define these dependencies as

$$K(\theta) = K_0/(1+K_1\theta) \text{ and } B(\theta) = B_0 + B_1\theta. \qquad (12)$$

The next $(k+1)^{\text{th}}$ interval of the computation begins with the new values of $K(\theta)$, $B(\theta)$ and $\theta = \theta_{k+1}$ and it continues next n periods of the excitation. It means, that the equations (7), (8) are gradually solved and the values X_{1i} and X_2 are substituted in the equation (9). After n steps, i.e. time $(n\Delta\tau)$, the equation (9) is enumerated and substituted in the (10), (11). For the obtained value θ the new coefficients $K(\theta)$ and $B(\theta)$ are calculated from the equation (12) and the whole cycle is repeated.

3. EXAMPLES

The interaction between mechanical and thermal processes is influenced by a lot of physical parameters and properties. As examples of the above-mentioned computational method, let us present selected cases of resonant curves of maximal amplitudes and corresponding temperatures obtained by passing through the resonance zones. They were computed for such values of system parameters at which the basic phenomena evoked by mechanical and heat coupling were emphasised. Dependencies of stiffness and damping on temperature correspond to the relations in (12). The following figures show an effect of the stiffness dependence

$$K(\theta) = K_0/(1+K_1\theta), \qquad (13)$$

where the parameters are K_0=1, K_1=0.0 (0.01, 0.02, 0.03), B_0=0.04, B_1=0.01, α=0.005, $\beta=1/c_T$=0.001.

Figure 2 shows the courses of temperature and of amplitude for slowly increasing and decreasing frequency $\eta = \omega\sqrt{m/k}$. The arrows show directions of passages.

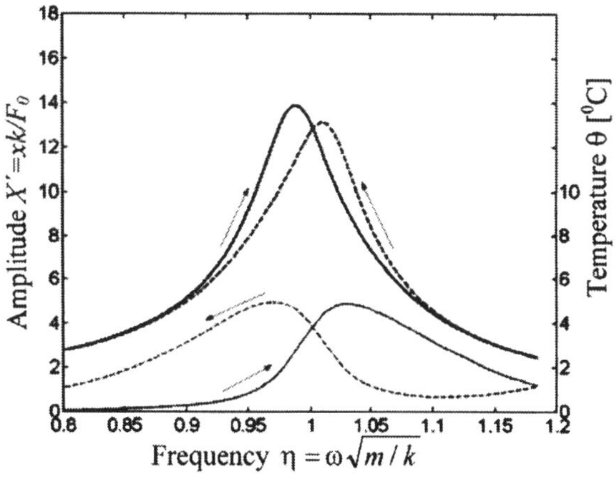

Fig. 2 Amplitude (thick) and temperature (thin lines) characteristics for K_1=0, B_1=0.01, α=0.005, β=0.001.

The velocity of resonance passages was 1280 periods of excitation on the range η=0.8-1.2. No backward influence of temperature on the stiffness is supposed $K(\theta)=K_0$=constant, K_1=0. The temperature influences only damping, which increases according to law

$$B(\theta)=0.04+0.01\theta\,. \qquad (14)$$

It is well known that the acceleration (deceleration) of frequency η shifts the resonance peaks to higher (lower) frequency. But the damping dependence on temperature according (14) overturn the order of peaks as it is seen in Fig. 2.

If the stiffness decreases with temperature $K(\theta)=1/(1+0.01\theta)$ see Fig. 3, the shift of peaks does not change, but the response peak at decreasing frequency changes its form and becomes broader. The larger extension of resonance peak occur at $K_1=0.02$ $K(\theta)=1/(1+0.02\theta)$ as shown in Fig. 4.

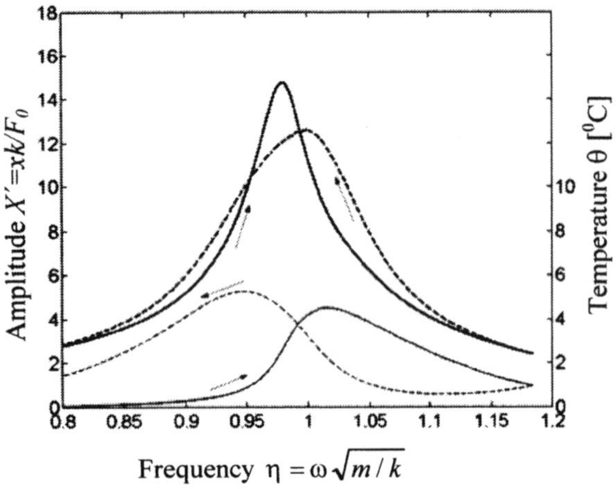

Fig. 3 Amplitude (thick) and temperature (thin lines) characteristics for K_1=0.01, B_1=0.01, α=0.005, β=0.001.

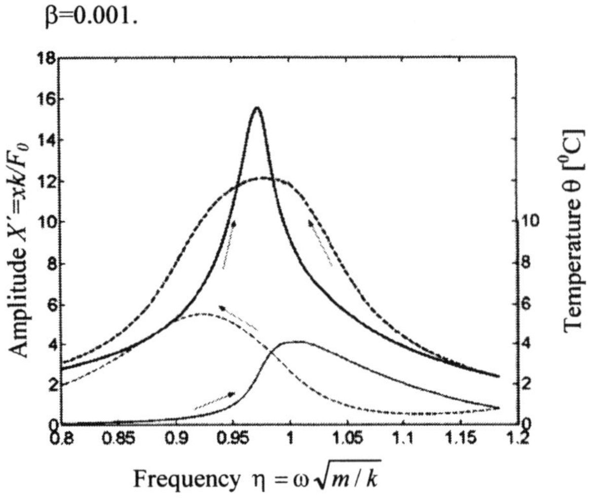

Fig. 4 Amplitude (thick) and temperature (thin lines) characteristics for K_1=0.02, B_1=0.01, α=0.005, β=0.001.

Response of a system containing more sensitive material having the relation $K(\theta)=1/(1+0.03\theta)$ is shown in Fig. 5. The response curve at this stronger stiffness-temperature relation (K_1=0.03) is characterized by further decrease of amplitudes at running down with frequency η and by a very wide resonant peak.

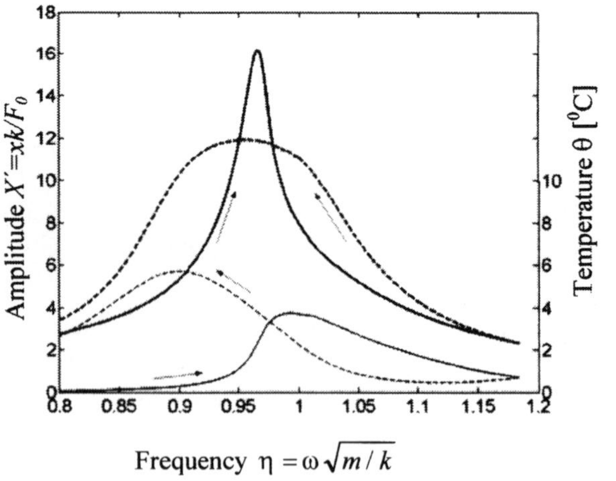

Fig. 5 Amplitude (thick) and temperature (thin lines) characteristics for K_1=0.03, B_1=0.01, α=0.005, β=0.001.

The influence of heat removal (parameter α) is drawn in Fig. 4, 6 and 7. The coefficient α of heat drain was changed in these cases to: α = 0.005, 0.002, 0.001. All other parameters have the same constant magnitudes as in previous case: K_0 = 1, K_1 = 0.002, B_0 = 0.04, B_1 = 0.01, $\beta = 1/c_T = 0.001$.

Fig.4 corresponds to the system having very good cooling and good removal of heat developed by internal material damping: α = 0.005.Decrease of α on value α=0.002 caused by poorer heat removal results in increase of temperature at running frequency down, in further fall of response peak and in its widening (Fig. 6).

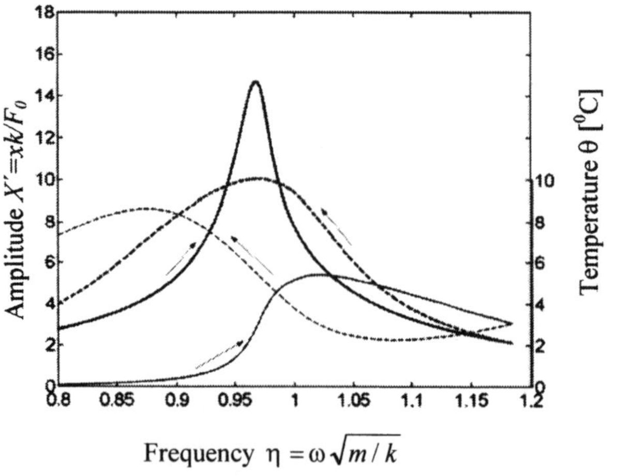

Fig. 6 Amplitude (thick) and temperature (thin lines) characteristics for K_1=0.02, B_1=0.01, α=0.002, β=0.001.

All these phenomena are intensified at stronger isolation and worse heat removal. The influence of twice smaller heat drift, described by α=0.001, on the course of response curve and temperature-frequency plot can be seen from Fig. 7.

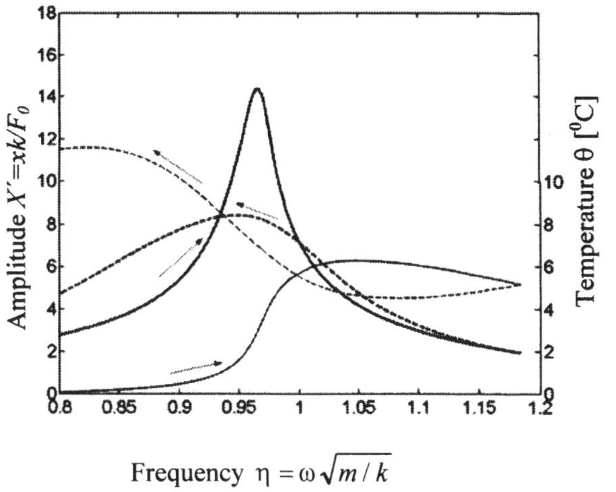

Fig. 7 Amplitude (thick) and temperature (thin lines) characteristics for K_I=0.02, B_I=0.01, α=0.001, β=0.001.

The analyses of effects of other parameters as well as application of this method for identification of rubberlike materials is a subject of current research.

4. CANTILEVAR RUBBER BEAM

For experimental research of the thermo-mechanical interaction of rubber-like materials the pattern of soft rubber beam was dynamically loaded and measured (Fig. 8). The beam parameters were: width a=40 mm, height b=16 mm, length l=200 mm, density ρ =1.477.10^{3}kg/m^3, Young modulus E'_{DYN} = 60.6 MPa, damping ratio $\varsigma \cong 0.1$, E'_{STAT} = 32 MPa, the first bending resonance f_1=11 Hz (Vaněk *et al.*, 2003).

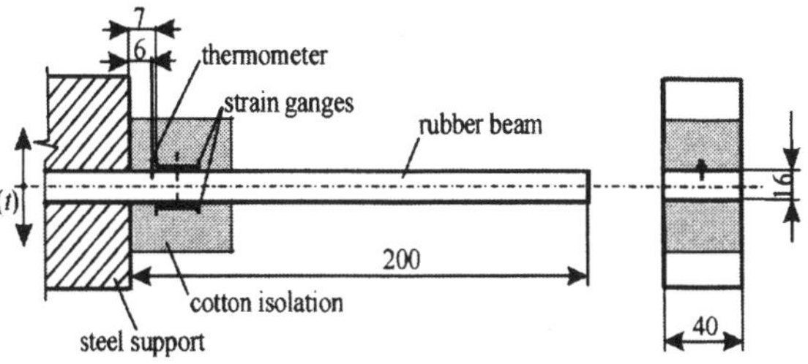

Fig.8 Drawing of the measured rubber beam with transducer placement

Semiconductor transducers sensitive to deformation and temperature measured the physical quantities such as strains and temperatures, respectively. In case of strains two resistive silicon strain gauges of type AP150-6-100 (coefficient of deformation sensitivity $K_T \doteq 150$), produced by VTS Zlín, CR, were stuck directly without a pad to the beam opposite surfaces and connected into semi-bridge. As to temperature measurements we tested two solutions of thermometers: the special differential thermometer with sensitivity 1.9mV/^{0}C and single ended bead-type thermistor with sensitivity 8V/^{0}C. The latter one proved better and hence it is used. The thermistor was stabbed in a vicinity of the measuring strain-gauges. PSD detectors measure the amplitudes of the beam end and the support movement optically. Power electro-dynamic vibrator B&K 4817 controlled by Simulink RTW, The MathWorks Inc., performed the sweep harmonic dynamic loading. Analog measured signals were acquired by A/D converter NI PCI-6035E and processed by Data Acquisition Toolbox and developed programs in Matlab6. The parameters of the bilateral sweep regime were: minimal frequency 6Hz, maximal frequency 25Hz.

Two examples of preliminary measurements are presented in Fig. 9a,b. Both are realized on the same system (Fig.8) but they differ by the rate of frequency sweep: 0.2Hz/s (Fig. 9a) and 0.1Hz/s (Fig.9b). From the plot temperature θ versus frequency f, seen in bottom parts of figures 9a,b, it is

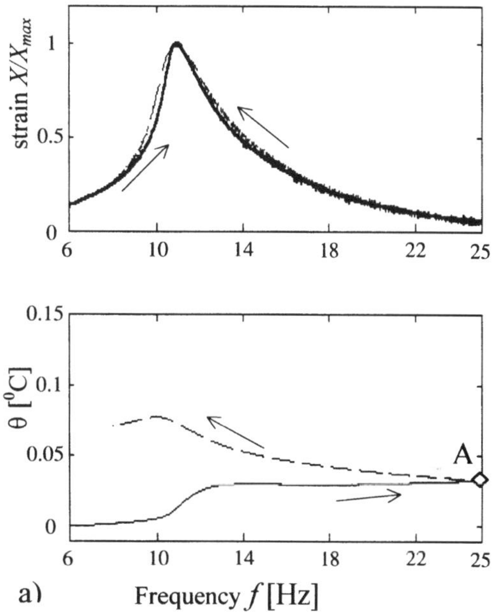

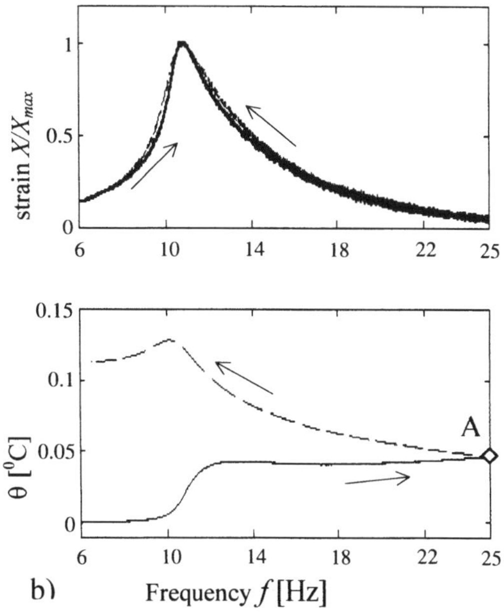

Fig. 9a,b Experimental amplitude (above) and temperature (lower) characteristics

evident, that the differences of the temperatures at running up and down of excitation frequency are inversely proportional to the rate of frequency sweep. Maximum slopes $d\theta / df$ of temperature curves occur near resonance frequencies but temperatures rise further and their maximum values lie behind these frequencies. Curves $\theta(f)$ at return point A (25Hz) form a sharp edge without any loop, which is seen in Figures 2÷7. It can be deduced from the property, that the energy output from the loaded body into environment is negligible; heat output coefficient is very small, $\alpha \to 0$. In spite of the fact that the measurement of temperature on the surface and inside of rubber-like elements due to its small relative changes is very difficult and is strongly influenced by systematic and random deviations coming from environmental and thermometer influences, the response curves in combination with measurement of temperature give good bases for identification of main thermo-mechanical parameters.

Till now only qualitative agreement between data of measurements and those gained by numerical simulations were achieved. Next, we are elaborating appropriate identification methods for achieving also quantitative results from experiments.

5. REMARK TO STATIONARY STATE

Above presented method is focused on transient phenomena, but the stationary state of temperature and material properties can be simply calculated as well. If the system oscillates for a long time at a constant frequency the temperature and heat output increases. System comes nearer to the state, when the heat output E_{out} equals the energy input of damping loss. Supposing linear properties of this phenomena we get:

$$\frac{1}{2}b(\theta)a^2\omega^2 = \alpha\theta \qquad (15)$$

The stationary temperature can be ascertained from this equation. If damping coefficient depends linearly on temperature, i.e. $b = b_0 + b_1\theta$, a simple formula is obtained from (15)

$$\theta = \frac{b_0 a^2\omega^2}{2\alpha - b_1 a^2\omega^2}. \qquad (16)$$

Measurement of temperature θ, at various amplitude a and frequency ω enable to ascertain unknown parameters α, b_1, b_0. Similar procedure can be used also for more complicated (nonlinear) dependences of damping on temperature, frequency and amplitude, as well as for nonlinear properties of heat output (drain) to environment e.g.

$$E_{out} = \alpha\theta + \alpha^2\theta^2 + \ldots \qquad (17)$$

Identification procedure based on measurement of stationary state doesn't enables to identify temperature capacity of deformed component. For these purposes, the nonstationary course of frequency and temperature has to be measured and treated.

6. CONCLUSION

In this contribution the method for numerical computations of a mutual interaction between heat and mechanical events in a vibrating system was elaborated. This method was applied for investigation of resonant curves of the 1 DOF's mechanical system containing a complaint component of material sensitive to temperature.

Influence of increased temperature shows itself mainly in resonances, when lost energy due to internal material damping is the highest. The drain of heat into surroundings is also respected. The higher temperature the lower is stiffness of the complaint component and hence the eigenfrequency of the system. Its damping coefficient increases at the same time.

During passages of resonant regions, the maximum of temperature is always delayed after the maximum of displacement. Influence of temperature on resonant curve shapes is the higher the slower is a resonant region cross over and the lower is a heat drain. Deformations of resonant peaks are relatively small at increasing of excitation frequency. However at decreasing these frequencies the deformations are much higher and they show themselves by both a higher shift of the resonant peak and its widening and lowering.

This method of numerical solution will be used for processing of measurement and identification of stiffness and damping characteristics of mechanical systems with dissipation layer, as outlined herein on the example of cantilever rubber beam.

REFERENCES

Mark J. E. , B. Erman (1988)
Rubberlike elasticity a molecular primer, *John Wiley & Sons*.

Mead J. Denys (1999)
Passive Vibration Control, *John Wiley & Sons*.

Pešek, L, J. Veselý (2002)
Parametric identification of non-linear models of an internal dissipation layer in mechanical systems. In: *Proceedings of ISMA2002*. (Ed.: Sas, P. - Van Hal B.), Leuven, KU, 2002, 471-479.

Pešek, L., L. Půst (2003)
Numerical solution of mechanical and thermal processes at vibrations, In: *Proceedings of National Colloquium Dynamics of Machines'2003*, Prague, pp.147-154.

Vaněk, F., Pešek, L., Cibulka, J. (2003)
Strain gauge measurements of deformations on rubber patterns, In: *Proceedings of National conference Engineering Mechanics 2002*, Svratka, 2003, in press, (in Czech).

This research has been solved in a frame of the grant GA CR no.101/02/0241.

IFAC Publications
www.elsevier.com/locate/ifac

IDENTIFICATION OF A HIGH EFFICIENCY BOILER BY SUPPORT VECTOR MACHINES WITHOUT BIAS TERM

Michael Vogt * Karsten Spreitzer * Vojislav Kecman **

** Darmstadt University of Technology, Institute of Automatic Control,*
Landgraf-Georg-Strasse 4, 64283 Darmstadt, Germany,
E-mail: mvogt@iat.tu-darmstadt.de
*** University of Auckland, School of Engineering, Private Bag 92019*
Auckland, New Zealand

Abstract: This paper considers the application of support vector machines for the identification of the nonlinear dynamic behavior of a high efficiency boiler. A new algorithm for the computation of support vector machines without bias term is proposed. Whereas the advantages of this concept are known in classification, it has been hardly made use of for regression. The main intention is to provide a simulation tool for the development engineer.

Keywords: identification, modelling, high efficiency boiler, support vector machines

1. INTRODUCTION

The development of modern control strategies for *building energy management systems* requires an integrated approach. To find an optimal control solution the interactions of the building, the heating system and the heat generating unit have to be taken into account. The major requirements for the control strategy are satisfaction of the inhabitant's thermal comfort demands and driving the heat generating unit in an optimal operating point. High efficiency boilers especially depend on the operating point. Their operating point is determined by the temperature of the heating system and the mass flow of the heating water. Hence, the heating system (e.g. radiators, floor heating) and the energy demands of the building have to be included in the development of new control strategies.

The investigation of new control strategies requires the consideration of an entire heating period (September through April in Central Europe). Performing these investigations on real systems is both time and cost intensive. Simulations on the other hand offer the major advantage of running much faster than real time experiments and of being reproducible. Hence, the effects of changes in the control strategy can be examined in a fraction of the time it would take on a real system. This shortens the development time and reduces the development costs.

Accurate modelling of all the different components of the system is essential to obtain reasonable results from a simulation. Basically, there are two ways to obtain those models: theoretical modelling and modelling by identification.

Theoretical modelling of the dynamic behavior of a thermal plant (Schwamberger, 1991; Pfannstiel, 1991; Pfannstiel, 1992) like a boiler remains a difficult task due to the inherent uncertainties, e.g., inaccurate heat and mass transfer coefficients, vague information about the hydro- and aerodynamic conditions. Many simplifications have to be made which result in mediocre model performance.

In the recent years, system engineers have been involved in the subject of dynamic nonlinear system identification using neural networks and dynamic local linear neuro-fuzzy models (Narendra and Parthasarathy, 1990; Chen *et al.*, 1990; Ayoubi *et al.*, 1995; Nelles, 2001). One of the latest developments is the concept of *support vector machines* (SVMs) that has its origin in the 1960's but has become practical only several years ago.

The main benefits of SVMs (e.g., compared to neural networks) are their better generalization capabilities and their solid theoretical foundation in the statistical learning theory (Schölkopf and Smola, 2002). Whereas SVMs are widely used for *classification* tasks, much less *regression* applications can be found so far. Additionally, there are very few regression algorithms to compute SVMs without bias term.

In section 2 the boiler test stand and the signals used for identification are described. Section 3 explains the necessary theory of SVMs and provides a new regression method for SVMs without bias term based on Platt's SMO algorithm. In section 4 this method is used to generate the boiler model. The paper finishes with the conclusions in section 5.

2. HIGH EFFICIENCY BOILER TEST STAND

2.1 *Description of the high efficiency boiler*

This section describes the high efficiency boiler under investigation and the test stand for acquiring the necessary measurement data. The measurement data was gathered at a test stand for boilers which is shown in its basic structure in Fig. 1. This test stand consists of the high efficiency boiler, a water-water heat exchanger (WWHE) and a PC for controlling and monitoring the system.

The outlet temperature of the boiler (T_{31}) is controlled by a built-in PI controller. This control task is performed by adjusting the speed of the fan which feeds the combustion air to the boiler depending on the difference between the set point $T_{31\text{d}}$ and the actual value T_{31} of the outlet temperature. Due to mechanical restrictions, the continuous operation mode is limited to the range of 35% to 100% of the nominal thermal output. If the load is below 35% the controller operates in an on-and-off mode.

The thermal output of the premix burner cannot be measured directly, but the control system (CS) provides access to the control signal of the fan which is a measure of the thermal output. The signal provided

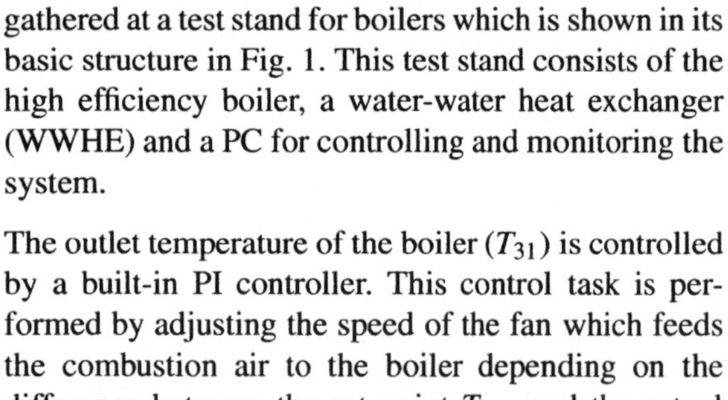

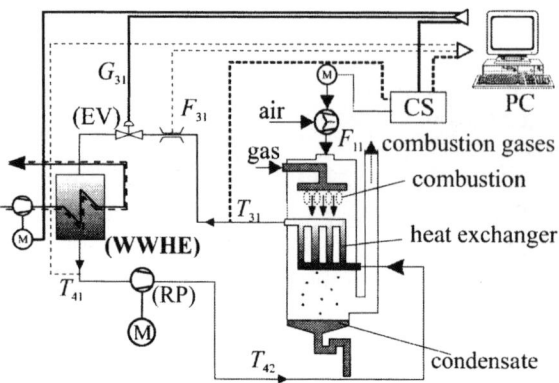

Fig. 1. Basic structure of the boiler and the test stand.

by the CS is in the range of 0% to 100% which corresponds to 0% to 100% of the nominal thermal output.

Identification of the boiler requires direct excitation of the thermal output of the boiler. Unfortunately, this is not possible because influencing the fan's control signal is not possible. To solve this problem, a second controller was implemented to control the thermal output by means of calculating the set point for the outlet temperature $T_{31\text{d}}$, see Fig. 1.

Depending on the combustion air flow, natural gas is added to the air to obtain an air-gas-mixture which is burned in the furnace. This combustion process produces hot combustion gases which are cooled down by cold water in the boiler's heat exchanger. The special feature of the high efficiency boiler is that the combustion gases are cooled down to the dew point of the steam which is a component of the combustion gases. This causes the steam to condensate so that the evaporation heat is released and can be used by the heating system. Using this effect, the efficiency of the boiler can be improved up to 11%. The condensate flows down the pipes of the heat exchanger and flows out at the bottom of the heat exchanger.

The water of the heating system enters the boiler having the temperature T_{42}. The heated water is lifted by a radial rotary pump (RP) and leaves the boiler with the temperature T_{31}. Within the WWHE the water is cooled down from the temperature T_{31} to the temperature T_{41} by a cooling water flow which is controlled by an electrically driven valve (EV). The valve adjustment is determined by a manipulating voltage in the range of 0 V to 10 V.

Finally a PC is used to calculate the control signals, set point values and for monitoring the system.

Fig. 2 shows a block diagram of the system with three inputs and one output. The inputs are the heating water

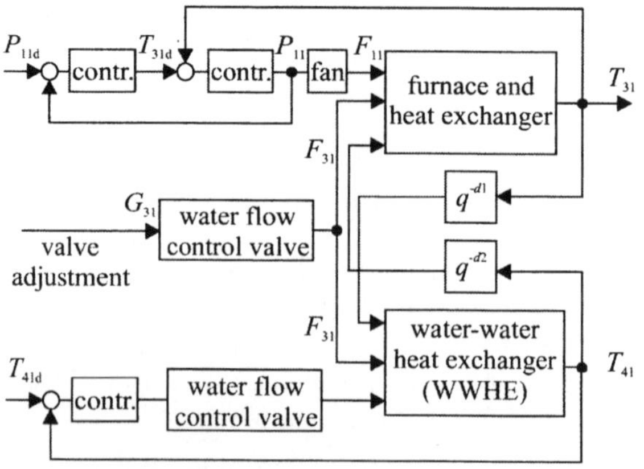

Fig. 2. Simplified block diagram of the thermal plant as a multi-input/single-output nonlinear process.

mass flow F_{31} [m^3/h], the desired thermal output of the boiler P_{11} [%] and the desired water temperature $T_{41\text{d}}$ after the WWHE. The outlet temperature of the WWHE T_{41} is assumed approximately equal to T_{42}. The output of the system is the outlet temperature T_{31}.

The dynamic behavior of the plant depends significantly on several variables. Firstly, the inlet and outlet water temperatures T_{42} and T_{31} are considerably correlated and influenced by the load of the boiler and the water flow. Secondly, the dead times depend on the operating point. For $F_{31} = 1\,\mathrm{m}^3/\mathrm{h}$, a dead time of approximately 100 s can be measured. Finally, the step response time slightly depends on the step direction (dynamic nonlinearity) and significantly depends on the steady state with respect to P_{11} and F_{31} (static nonlinearity).

2.2 *Generation of an Excitation-Sequence*

Identification of a process requires an appropriate excitation signal (Isermann, 1992*a*; Isermann, 1992*b*). Without proper excitation, estimation of the process-behavior is not possible. This is especially the case with nonlinear processes (Nelles, 2001). They require excitation of both their dynamic and static properties in all relevant operating points.

The high efficiency boiler has three inputs and one output and is a so-called MISO-process: $T_{31} = f(P_{11}, F_{31}, T_{42})$. Hence, identifying a model for the boiler requires dynamic and static excitation of all the input signals to assure excitation of all relevant process dynamics in all operating points. This is achieved by exciting one input dynamically and leaving the other two inputs at constant values. This leads to a division of the excitation sequence into several static and dynamic parts. The static parts consist of piecewise-constant functions while an **a**mplitude modulated **p**seudo **r**andom **b**inary **s**ignal (APRBS) represents the dynamical parts. The excitation sequence for identifying the boiler model is shown in Fig. 3. To avoid numerical problems (ill-conditioned regressor-matrices) all input signals have been normalized.

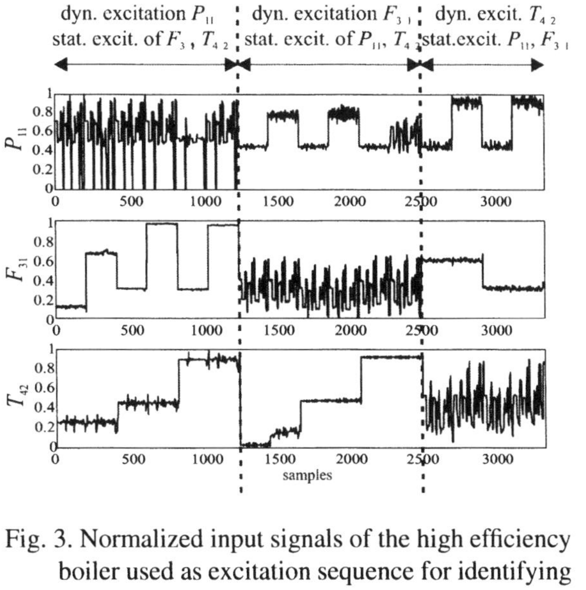

Fig. 3. Normalized input signals of the high efficiency boiler used as excitation sequence for identifying the boiler model.

3. SUPPORT VECTOR MACHINES

For building models from data, usually methods like *least squares estimation* are employed which try to minimize the error between the model output and the measured process output on a given training data set. However, these techniques usually do not consider the generalization behavior of the model, so that there is always the danger of *overfitting*. Support vector machines try to overcome this problem by optimizing the *flatness* of the model while accepting a certain model error. This concept was originally used for classification and then transferred to regression problems.

3.1 *Support vector machine regression*

For linear regression problems, a linear model $f(\mathbf{x}) = \langle \mathbf{w}, \mathbf{x} \rangle + b$ (with the normal vector $\mathbf{w}$) and a certain precision ε are assumed, see Fig. 4. Following the

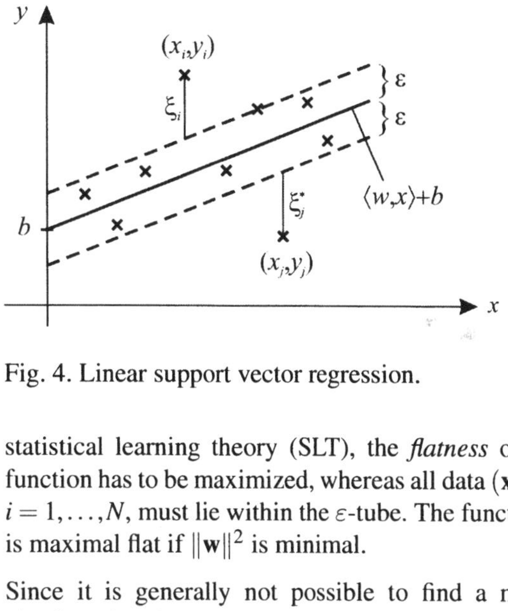

Fig. 4. Linear support vector regression.

statistical learning theory (SLT), the *flatness* of the function has to be maximized, whereas all data $(\mathbf{x}_i, y_i)$, $i = 1, \ldots, N$, must lie within the ε-tube. The function f is maximal flat if $\|\mathbf{w}\|^2$ is minimal.

Since it is generally not possible to find a model obeying the above restrictions for a given ε, *slack variables* $\xi_i \geq 0$ and $\xi_i^* \geq 0$ are introduced for each sample i to make the problem feasible. Each violation of the original constraints (i.e., $\xi_i + \xi_i^* > 0$) is punished in the objective function by the factor C. This leads to the following (primal) optimization problem:

$$\begin{aligned} \text{minimize} \quad & L_\mathrm{p} = \frac{1}{2}\|\mathbf{w}\|^2 + C \cdot \sum_{i=1}^{N} (\xi_i + \xi_i^*) \\ \text{subject to} \quad & \langle \mathbf{w}, \mathbf{x}_i \rangle + b - y_i \leq \varepsilon + \xi_i \\ & y_i - \langle \mathbf{w}, \mathbf{x}_i \rangle - b \leq \varepsilon + \xi_i^* \\ & \xi_i, \xi_i^* \geq 0, \quad i = 1, \ldots, N \end{aligned}$$

Introducing the Lagrange multipliers α_i and α_i^*, and exploiting the saddlepoint condition, the *dual* optimization problem is given by

$$\text{minimize} \quad L_\text{d} = \frac{1}{2}\sum_{i=1}^{N}\sum_{j=1}^{N}(\alpha_i-\alpha_i^*)(\alpha_j-\alpha_j^*)\langle\mathbf{x}_i,\mathbf{x}_j\rangle - \sum_{i=1}^{N}(\alpha_i-\alpha_i^*)y_i+\varepsilon\sum_{i=1}^{N}(\alpha_i+\alpha_i^*)$$

$$\text{subject to} \quad \sum_{i=1}^{N}(\alpha_i-\alpha_i^*)=0$$

$$0\leq\alpha_i,\alpha_i^*\leq C, \quad i=1,\ldots,N\,.$$

This is a *QP problem*, which can be solved by standard QP solvers if N is small. However, the main advantage of the dual formulation is its easy extendability to *non-linear* SVMs. For that, the input vector $\mathbf{x}$ is mapped into a *feature space* by a nonlinear function $\mathbf{\Phi}(\mathbf{x})$. If the linear SVM is applied to the feature space, the objective function to be minimized is solely dependent from $\langle\mathbf{\Phi}(\mathbf{x}_i),\mathbf{\Phi}(\mathbf{x}_j)\rangle$, not from $\mathbf{\Phi}(\mathbf{x})$ itself. The map $\mathbf{\Phi}(\mathbf{x})$ may be extremely high-dimensional since it never needs to be computed. Hence, the scalar product $\langle\mathbf{x}_i,\mathbf{x}_j\rangle$ has only to be substituted by the *kernel function*

$$k(\mathbf{x}_i,\mathbf{x}_j)=\langle\mathbf{\Phi}(\mathbf{x}_i),\mathbf{\Phi}(\mathbf{x}_j)\rangle\,.$$

There are many possible kernel functions, e.g., polynomials or Gaussian bells. The output of a nonlinear SVM is calculated as

$$f(\mathbf{x})=\sum_{i=1}^{N}(\alpha_i-\alpha_i^*)\cdot k(\mathbf{x}_i,\mathbf{x})+b\,.$$

Only those vectors $\mathbf{x}_i$ with $\alpha_i>0$ or $\alpha_i^*>0$ take effect on $f(\mathbf{x})$: The *support vectors*. For nonlinear SVMs, the above *support vector expansion* is used instead of the coefficients $\mathbf{w}$, since these would require the computation of $\mathbf{\Phi}(\mathbf{x})$.

The Lagrange multipliers α_i and α_i^* are computed by QP optimization, whereas b is calculated separately and is also responsible for the equality constraint of the QP problem. If b could be omitted, the problem structure would be much simpler. This concept is referred to as *SVMs without bias term*.

The bias term b can be omitted if the kernel function k provides an *implicit* bias. E.g., the *inhomogeneous* polynomial kernel

$$\mathbf{\Phi}(\mathbf{x})=(x_1^2,\sqrt{2}x_1x_2,x_2^2,\sqrt{2}x_1,\sqrt{2}x_2,1)^\text{T}$$
$$\Rightarrow\quad k(\mathbf{x},\mathbf{y})=(\langle\mathbf{x},\mathbf{y}\rangle+1)^2$$

provides an implicit bias, since the according non-linear map $\mathbf{\Phi}(\mathbf{x})$ includes a constant term (last component), whereas the *homogeneous* polynomial kernel

$$\mathbf{\Phi}(\mathbf{x})=(x_1^2,\sqrt{2}x_1x_2,x_2^2)^\text{T}$$
$$\Rightarrow\quad k(\mathbf{x},\mathbf{y})=\langle\mathbf{x},\mathbf{y}\rangle^2$$

requires an *explicit* bias b. In general, each positive definite kernel function can be used (even if $\mathbf{\Phi}(\mathbf{x})$ is not explicitly known) if C is chosen large enough.

3.2 *A SMO regression algorithm for SVMs without bias term*

Unfortunately, the size of SVM optimization problems is dependent from the number of samples. For regression, $2N$ parameters have to be determined if the data set consists of N samples, and the QP problem needs to store $4N^2$ matrix elements. Consequently, standard QP solvers can only be used for small and medium N (on current PC hardware up to $N=1000\ldots2000$).

For large data sets, algorithms with a memory consumption smaller than $\mathcal{O}(N^2)$ are desired. The most familiar algorithm of this category is Platt's *Sequential Minimal Optimization* (SMO), see (Platt, 1999). SMO has also been applied to regression problems by Schölkopf and Smola (2002).

Caused by the explicit bias b, the solution must obey the equality constraint $\sum(\alpha_i-\alpha_i^*)=0$. This is the reason why SMO optimizes *two* parameters in each iteration step. For a SVM without bias term, the equality constraint vanishes, so that only *one* parameter needs to be optimized in each step. This option is mentioned in (Platt, 1999) for classification. In the following, an SMO algorithm for SVMs without bias term is shown based on the SMO regression algorithm given in (Schölkopf and Smola, 2002). A detailed description of this new algorithm can be found in (Vogt, 2002).

As pointed out above, L_d has to be minimized with respect to only *one* variable. However, for regression problems L_d has to be minimized with respect to both α_l and α_l^*, but at least one of them is zero. The parameters to be optimized are chosen by Platt's "first choice heuristic" in each iteration step. To find an analytical solution of the resulting 2-dimensional optimization problem, all terms containing α_l and α_l^* have to be separated in L_d:

$$L_\text{d} = \Big(E_l-(\alpha_l^\text{old}-\alpha_l^{*\,\text{old}})k_{ll}+\varepsilon\Big)\alpha_l - \Big(E_l-(\alpha_l^\text{old}-\alpha_l^{*\,\text{old}})k_{ll}-\varepsilon\Big)\alpha_l^* + \frac{1}{2}k_{ll}\alpha_l^2+\frac{1}{2}k_{ll}\alpha_l^{*\,2}+\text{const.}$$

with the kernel function value $k_{ll}=k(\mathbf{x}_l,\mathbf{x}_l)$, the old parameters α_l^old and $\alpha_l^{*\,\text{old}}$ (which are known from the last step) and the cached error value

$$E_l=\sum_{j=1}^{N}(\alpha_j^\text{old}-\alpha_j^{*\,\text{old}})k_{lj}-y_l\,.$$

The *unconstrained* minimum is found by solving $\partial L/\partial\alpha_l=0$, and $\partial L/\partial\alpha_l^*=0$, respectively:

$$\alpha_l=\alpha_l^\text{old}-\alpha_l^{*\,\text{old}}-\frac{E_l+\varepsilon}{k_{ll}}$$
$$\alpha_l^*=-(\alpha_l^\text{old}-\alpha_l^{*\,\text{old}})+\frac{E_l-\varepsilon}{k_{ll}} \quad = \quad -\alpha_l-\frac{2\varepsilon}{k_{ll}}$$

Since there is no equality constraint, the solution of the *constrained* optimization problem is found by simply clipping α_l and α_l^* to $[0, C]$:

$$\alpha_l^{\text{new}} = \min\{\max\{\alpha_l, 0\}, C\}$$
$$\alpha_l^{*\text{new}} = \min\{\max\{\alpha_l^*, 0\}, C\}.$$

Before optimizing α_l and α_l^*, the algorithm checks the *Karush-Kuhn-Tucker (KKT) conditions*, and computes an update of α_l and α_l^* only if

$$\begin{aligned} &\alpha_l < C \ \wedge\ \varepsilon + E_l < -\tau \\ \text{or}\quad &\alpha_l > 0 \ \wedge\ \varepsilon + E_l > \tau \\ \text{or}\quad &\alpha_l^* < C \ \wedge\ \varepsilon - E_l < -\tau \\ \text{or}\quad &\alpha_l^* > 0 \ \wedge\ \varepsilon - E_l > \tau, \end{aligned}$$

where τ is a certain precision (e.g., $\tau = y_{\max} \cdot 10^{-3}$). Like in standard SMO, also an error cache is employed which is updated in each step using

$$E_i = E_i^{\text{old}} + t_l k_{il} \qquad \text{for all unbound } \alpha_i$$

with the auxiliary variable $t_l = \alpha_l - \alpha_l^* - \alpha_l^{\text{old}} + \alpha_l^{*\text{old}}$, which is independent from i.

The new algorithm has been compared to the standard SMO regression algorithm as described in (Schölkopf and Smola, 2002). Both algorithms were implemented as MATLAB MEX-files. A data set was generated based on the test function $f(x) = \frac{10}{x} \cdot \sin(\frac{x}{10})$ evaluated on the interval $x \in [-1, 1]$, see Fig. 5. Gaussian bells

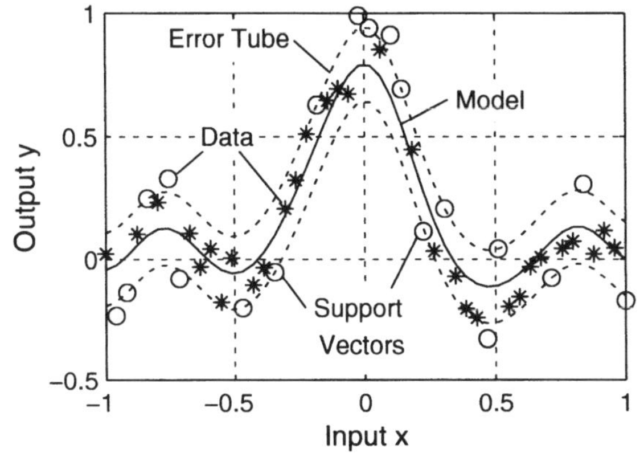

Fig. 5. Simulation example ($N = 50$, $\varepsilon = 0.1$). The support vectors are indicated by circles.

with standard deviation s are used as kernel functions; ε is set to 0.1 for all simulations. Tab. 1 shows the computation time (on a 800 MHz Pentium-III PC) and the support vector ratio for $C = 1000$ and $N = 100$. It is not reasonable to compare the number of iterations

Table 1. SMO simulations, variation of s: Computation time and support vector ratio.

Std. Dev. s	0.01	0.1	0.5
Standard SMO	37.5 s / 82%	2.91 s / 37%	0.67 s / 12%
SMO w. b. t.	0.14 s / 63%	0.38 s / 31%	0.21 s / 11%

since a single step is much easier for SMO without bias term and has effect on only *one* parameter instead of *two*. Tab. 1 shows that SMO without bias term is much faster (particularly for small s) and produces less support vectors.

If C becomes smaller, the number of support vectors with $\alpha_i = C$ or $\alpha_i^* = C$ increases, i.e., more samples lie outside the ε-tube. Tab. 2 shows for $N = 500$ that the SMO algorithm without bias term copes better with this situation than standard SMO.

Table 2. SMO simulations, variation of C: Computation time and support vectors with $\alpha_l = C$ or $\alpha_l^* = C$.

Constant C	0.001	0.1	10
Standard SMO	18.69 s / 61%	1.53 s / 5.8%	1.20 s / 0%
SMO w. b. t.	0.42 s / 58%	0.83 s / 6.2%	0.75 s / 0%

In Tab. 3, s is fixed to 0.5, whereas N is variable. In particular for small N the same result as in Tab. 1 and Tab. 2 is observed: SMO without bias term works faster and produces less support vectors.

Table 3. SMO simulations, variation of N: Computation time and support vector ratio.

Samples N	50	200	500
Standard SMO	0.06 s / 44%	0.45 s / 15%	1.18 s / 6.4%
SMO w. b. t.	0.01 s / 40%	0.09 s / 14%	0.76 s / 6.4%

A further discussion on algorithms for SVMs without bias term can be found in (Kecman *et al.*, 2003).

4. MODELLING OF THE BOILER

The SMO algorithm without bias term has been applied to build a model of the boiler described in section 2.1 using the signals from section 2.2 as process excitation. The training data set consists of 3344 samples and the generalization data set of 2926 samples. The sample time is 30 s. The boiler can be described as a second order dynamic process, so the discrete model

$$\begin{aligned} T_{31}(k) = f(&P_{11}(k), P_{11}(k-1), P_{11}(k-2), \\ &F_{31}(k), F_{31}(k-1), F_{31}(k-2), \\ &T_{42}(k), T_{42}(k-1), T_{42}(k-2), \\ &T_{31}(k-1), T_{31}(k-2)) \end{aligned}$$

employing 11 regressors is used for regression. The SVM parameters were chosen as $\varepsilon = 10^{-2}$, $C = 1000$, $s = 0.5$ and $\tau = 10^{-3}$. The choice of ε results from the measurement errors of the used sensors, C and s have been found by experiment, and τ is a standard setting. The solution was found after approximately 260 s (again on a 800 MHz Pentium-III PC) and consists of 231 support vectors, which is 6.9% of the data set. There were no support vectors with $\alpha_l = C$ or $\alpha_l^* = C$, i.e., all support vectors lie on the edge of the ε-tube.

Fig. 6 shows a comparison between the measured and predicted output for both the training and the generalization data set. To compare the performance

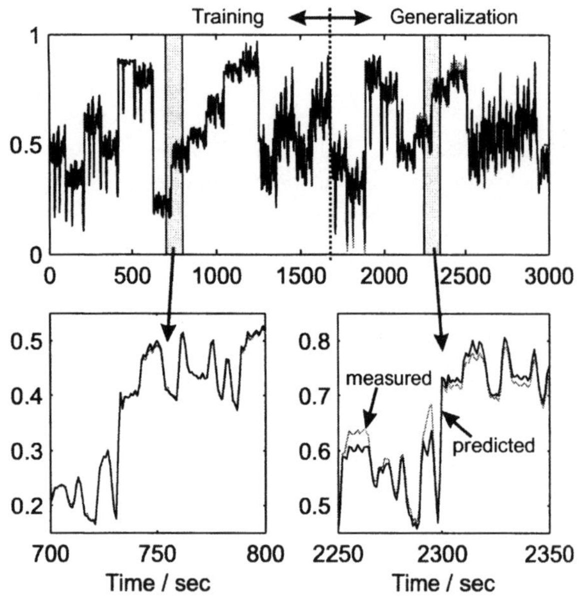

Fig. 6. Measured and predicted output of the boiler in both training and generalization phase.

on both data sets, the *root-mean-square error*

$$\mathrm{RMSE} = \sqrt{\frac{1}{N}\sum_{k=1}^{N}(y(k) - \hat{y}(k))^2}$$

has been calculated:

Training data set:	RMSE = 0.05
Generalization data set:	RMSE = 0.18

As expected, the RMSE on the generalization data set is higher but the model still shows a reliable performance.

5. CONCLUSIONS

Simulations are useful for designing new control strategies for building energy management systems. Since most components of the building and heating system show complex nonlinear behavior, data driven modelling seems to be a good approach.

In this paper a new SMO regression algorithm for support vector machines without bias term has been used to build a model of the high efficiency boiler. The algorithm was derived and tested in MATLAB simulations. It has been shown that it works faster than the standard SMO regression algorithm and produces less support vectors. The user only has to make sure to choose a kernel that does not require an explicit bias.

Further research will compare the new algorithm to more advanced methods like SVM^{light} (Joachims, 1999) or implementations using Keerthi's improvements (Keerthi *et al.*, 2001).

6. REFERENCES

Ayoubi, Mihiar, M. Schaefer and S. Sinsel (1995). Dynamic neural units for nonlinear dynamic systems identification. In: *International Workshop on Artificial Neural Networks IWANN'95.*

Chen, Sheng, S. Billings and P. Grant (1990). Nonlinear sytem identification using neural networks. *International Journal of Control* **51**(6), 1191–1214.

Isermann, Rolf (1992*a*). *Identifikation Dynamischer Systeme*. Vol. 1. 2 ed.. Springer Verlag.

Isermann, Rolf (1992*b*). *Identifikation Dynamischer Systeme*. Vol. 2. 2 ed.. Springer Verlag.

Joachims, Thorsten (1999). Making large-scale SVM learning practical. In: *Advances in Kernel Methods – Support Vector Learning* (Bernhard Schölkopf, Christopher J. C. Burges and Alexander Smola, Eds.). Chap. 11. MIT Press. Cambridge, MA.

Kecman, Vojislav, Michael Vogt and Te Ming Huang (2003). On the equality of kernel AdaTron and Sequential Minimal Optimization in classification and regression tasks and alike algorithms for kernel machines. In: *Proceedings of the 11th European Symposium on Artificial Neural Networks (ESANN 2003).* Bruges, Belgium.

Keerthi, S. Sathiya et al. (2001). Improvements to Platt's SMO algorithm for SVM classifier design. *Neural Computation* **13**, 637–649.

Narendra, Kumpati S. and Kannan Parthasarathy (1990). Identification and control of dynamic systems using neural networks. *IEEE Transactions on Neural Networks* **1**(1), 4–27.

Nelles, Oliver (2001). *Nonlinear System Identification*. Springer Verlag.

Pfannstiel, Dieter (1991). In circuit simulation of industrial processes with personal computers. In: *Mediterranean Electrotechnical conference Melecon '91.*

Pfannstiel, Dieter (1992). *Modellbildung, Simulation und digitale Regelung eines ölbefeuerten Heizkessels mit kleiner Leistung*. Wärmetechnik / Kältetechnik. VDI-Verlag. Düsseldorf.

Platt, John C. (1999). Fast training of support vector machines using sequential minimal optimization. In: *Advances in Kernel Methods – Support Vector Learning* (Bernhard Schölkopf, Christopher J. C. Burges and Alexander Smola, Eds.). Chap. 12. MIT Press. Cambridge, MA.

Schölkopf, Berhhard and Alexander J. Smola (2002). *Lerning with Kernels*. The MIT Press. Cambridge, MA.

Schwamberger, Klaus (1991). *Modellbildung und Regelung von Gebäudeheizungsanlagen mit Wärmepumpen*. Energieerzeugung. VDI-Verlag. Düsseldorf.

Vogt, Michael (2002). SMO algorithms for support vector machines without bias. Technical report. Darmstadt University of Technology, Institute of Automatic Control. Darmstadt.

IFAC
Publications
www.elsevier.com/locate/ifac

IMPLEMENTING GA-BASED PREDICTIVE CONTROLLER FOR ON-LINE CONTROL OF A PROCESS MINI-PLANT

Yul Y. Nazaruddin[1] and Fajrih Maulana

Industrial Instrumentation Laboratory
Department of Engineering Physics - Institut Teknologi Bandung
Jl. Ganesa 10 Bandung 40132, INDONESIA
Phone/Fax: +62-22-2508138
e-mail : [1]yul@tf.itb.ac.id

Abstract: This paper is concerned with a development of an alternative intelligent control strategy, which is an integration between Predictive Control Technique and Neuro-Fuzzy as well as Genetic Algorithm (GA) approaches, and its real-time implementation for controlling a process mini-plant. Generalized Predictive Control (GPC), which is considered as universal method for model-based predictive control, was integrated with a neuro-fuzzy approach for the plant identification/modelling. In this strategy, GA was employed, firstly, as an optimization method for parameter learning in neuro-fuzzy based plant modelling, and secondly, in determining the optimal control signal value by minimizing the cost function of the GPC. The implementation demonstrates the applicability and the performance of the proposed control strategy to handle nonlinear as well as changing plants characteristics in real-time environment. *Copyright © 2003 IFAC*

Keywords: fuzzy models, experimentation, process control systems

1. INTRODUCTION

Generalized Predictive Control (GPC), which is considered as universal method for model-based predictive control, is proven to be successful in handling various kind of processes and has also been successfully applied in various industries. GPC can be used either to control a simple plant with little prior knowledge or a complex plant such as non-minimum phase, open-loop unstable and a process having variable dead-time [Garcia *et al.*, 1989]. A very critical step toward the success of the implementation of GPC is the availability of a reliable process model as an accurate plant model is necessary to derive a set of future plant output close to its corresponding reference signal sequence. As most processes in industry have nonlinear behavior, then the modelling process is even more difficult

In recent years many efforts have been made to combine neural-network and fuzzy system methodologies in control system design. The primary concern is the integration of the strength of both methodologies in order to achieve learning and adaptation capability and knowledge representation via fuzzy if-then rules, producing the so-called neuro-fuzzy systems. The main advantage of the neuro-fuzzy methodologies is its learning capability from numerical data obtained from the measurement and hence no mathematical model of the plant to be control required, which is very beneficial in dealing with nonlinear plants which its mathematical models are usually very difficult to derive.

The successfulness of neuro-fuzzy based modelling strongly depends on the optimization method used in parameter learning inside the neuro-fuzzy structure. In the previous investigation [Nazaruddin and

Tjandrakusuma, 2001], gradient-based Backpropagation was employed, but the result could not reach an optimal one since a local maximum or minimum solution of the objective function could be obtained.

An alternative optimization technique as a potential solution for the neuro-fuzzy learning algorithm is the Genetic Algorithm (GA). GA, also known as population-based optimization, is one of the derivative-free optimization methods. GA is inspired by evolutionary and natural selection process, where the best individual is the only candidate that can survive. In the proposed intelligent control strategy, GA will also be employed in the GPC control algorithm. In order to derive the future plant outputs $y(t+j)$ close to the desired set-point $w(t+j)$ in some sense, GA is implemented not only to optimize the required control signal sequence but also to minimize the increments in control signal.

The overall control scheme was tested in real-time environment to control the level of a process mini-plant which is assumed to have a strongly inherent mechanical nonlinearities. The experiments will show the real application of the integration of the strength of GPC, neuro-fuzzy and GA techniques in real-time environment.

2. PREDICTIVE CONTROL LAW

Suppose that a future set-point or reference sequence $[w(t+j); j=1,2,...]$ is available. In most cases $w(t+j)$ will be constant w equal to current set-point $w(t)$, though sometimes (as batch process control or robotic) future variations in $w(t+j)$ would be known. The objective of the predictive control law is then to derive the future plant outputs $y(t+j)$ close to $w(t+j)$ in some sense, bearing in mind that the control activity required to do so. This is done using a receding-horizon approach for which at each sample-instant t :

(1). the future set-point sequence $w(t+j)$ is calculated;

(2). the prediction model is used to generate a set of predicted outputs $\hat{y}(t+j\mid t)$ with corresponding predicted errors $e(t+j)=w(t+j)-\hat{y}(t+j\mid t)$, by noting that $\hat{y}(t+j\mid t)$ for $j>k$ depends in part on the future control signals $u(t+j\mid t)$ which are to be determined using genetic algorithm optimization method;

(3). an appropriate quadratic function of the future error and control is minimized, provide a suggested sequence of future controls $u(t+j\mid t)$;

The quadratic function, also known as the cost function, is defined as

$$J=\sum_{j=N_1}^{N_2}[\hat{y}(t+j)-w(t+j)]^2+\sum_{J=1}^{N_U}\lambda(j)[\Delta u(t+j-1)]^2 \quad (1)$$

where N_2, N_1, N_u and $\lambda(j)$ is the horizon maximum $(1\le N_2)$, the horizon minimum $(1\le N_1\le N_2)$, the control horizon $(1\le N_u\le N_2)$ and a control-weighting sequence, respectively. Further, the way of the optimization of control signal using genetic algorithms will be explained in part 3.

3. NEURO-FUZZY BASED MODELLING AND OPTIMIZATION OF CONTROL SIGNAL USING GENETIC ALGORITHM

3.1 Plant/Process Modelling

Ideally, a plant/process model should be derived from physical and chemical consideration. However, in many cases, this approach of modelling is not favorable as the lack of plant/process knowledge contributes mostly to the difficulties. Therefore, an empirical plant/process modelling approach is used in many cases in which the dynamics can be inferred from the measured plant data directly. A parametric model of plant/process identifications is favorable to be used in industrial practice. Since most of the process models in industrial control show a strongly nonlinear behavior, a popular Nonlinear Auto-Regressive with eXogenous Variable (NARX) parametric model form is widely used to represent nonlinear systems. In this model, the output is a nonlinear function of previous outputs and inputs of the system, or

$$y(t)=F(y(t-1),..,y(t-n),u(t-d-1),..,u(t-d-n))+e(t) \quad (2)$$

Here y(t) and u(t) are the sampled plant/process output and input at time instant t respectively, *e(t)* is the equation error, *n* denotes the order of the process, *d* represents the process dead time as an integer number of samples and *F(.)* is an unknown nonlinear function to be identified.

A technique which is successfully used for nonlinear plant modelling is based on neuro-fuzzy approach [Nazaruddin and Tjandrakusuma, 2001]. An architecture called Adaptive Neuro-Fuzzy Inference System (ANFIS) [Jang *et al.*, 1997], which is an integration between neural network and fuzzy inference system has been implemented. The ANFIS architecture consists of five layers with different function in each layer. The adaptive network is manifested only in the first and fourth layer. In the first layer, the adaptive parameters are the parameters of the membership function of the input fuzzy set, which are nonlinear function of the system output, also known as premise parameters. The parameter in the fourth layer are the linear function of the system output assuming that the parameters of the membership function are fixed. These parameters are then determined using hybrid learning, which involves backward pass for nonlinear parameters and

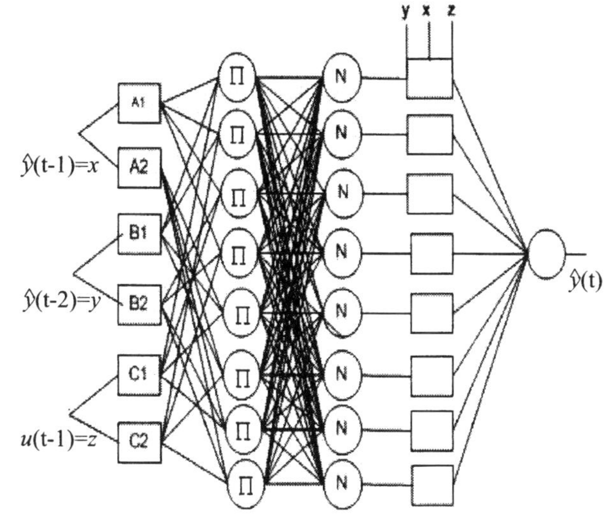

Fig. 1. Structure of ANFIS with 3 inputs

forward pass learning for linear parameters. Due to the linear relationship with respect to the output parameters, then a least-square estimator (RLSE) can be applied for the learning process. Whereas the modelling and learning process for the nonlinear parameters employs Genetic Algorithm that will be explained below.

Assuming that the plant under consideration can be represented as a second order process, then the NARX representation becomes

$$y(t) = F(y(t-1), y(t-2), u(t-d-1)) + e(t) \qquad (3)$$

Using the neuro-fuzzy approach, the function F(.) is obtained, so that

$$\hat{y}(t) = F(y(t-1), y(t-2), u(t-d-1)) \qquad (4)$$

where $\hat{y}(t)$ is the model predicted output.

The configuration of the neuro-fuzzy system being used in this investigation was as follows :

- Fuzzy system : 1st order Takagi-Sugeno-Kang model
- Input : $y(t-1), y(t-2), u(t-d-1)$
- Output : *y(t)*
- Number of fuzzy set per input : 2
- Number of rule fuzzy : 8
- Number of premise parameters : 18
- Number of consequent parameters : 32

The overall structure of ANFIS using the above configuration can be seen in Fig. 1.

3.2 Backward Learning and Optimization of Control Signal Using Genetic Algorithm

The basic procedure of genetic algorithm is shown in Fig. 2. The initial population P(t) is built which contains chromosomes, representing the solution of optimization problem [Gen and Cheng, 1997]. The chromosomes evolve through successive iterations, called generations. During each generation, the chromosome is evaluated, using some measures of fitness where an evaluation function plays the role of the environment, and it rates individual in terms of their fitness.

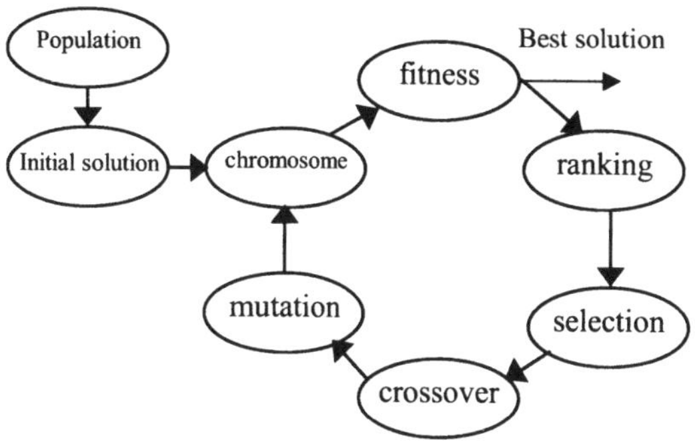

Fig. 2. Basic procedure of GA

In the learning case, the population contains of N individual with 18 chromosomes as many as number of premise parameters which is encoded in real number. For this case, the fitness is a function of RMSE (Root Means Square Error) of each individual for N data sequence. The fitness function is then defined as

$$f(i) = \frac{1}{1+e(i)} \qquad (5)$$

where *e(i)* is RMSE of each individual. To create the next generation, new chromosome, called offspring, are formed by merging two chromosomes from current generation using crossover operator or modifying a chromosome using a mutation operator. A new generation is formed by selecting, according to the fitness values, some of the parents and offspring and rejecting others so as to keep the population size constant. After several generations, the algorithms converge to the best chromosome, which hopefully represents the optimum or sub optimal solutions to the problem.

The same optimization procedure is implemented in determining of the optimal control signal, in which the chromosome is the future control signal sequence $[u(t+j\,|\,t), j = 0,1,...N_u - 1]$ and the fitness functions is a function of the cost function *J*. Genetic operation is conducted at N time genetic iteration for a parent population, where after N genetic iteration, a best individual in term of its fitness is obtained. The first element $u(t)$ of the best chromosome sequence is then applied and the algorithm is repeated at next sample instant.

4. EXPERIMENTAL RESULTS AND EVALUATION

4.1 Process Mini-plant Description

The process mini-plant basically consists of two tanks containing fluid which its level will be controlled, and real industrial scaled components, such as differential pressure transmitter, control valve, I/P converter, so that it resembles almost real-

Fig. 3. View of the process mini-plant

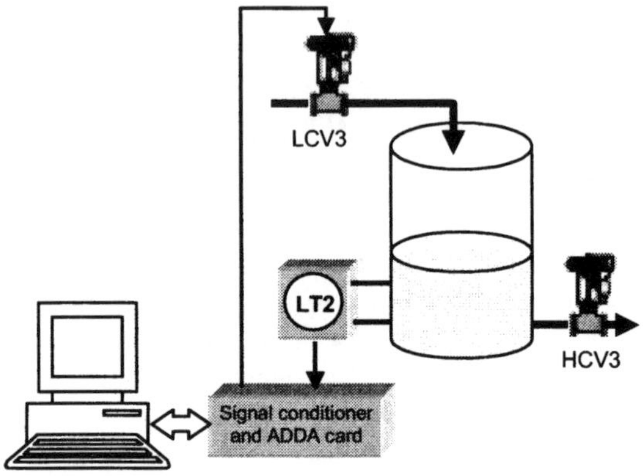

Fig. 4. Experimental set-up

plant characteristics. View of the process mini-plant is shown in Fig. 3. In the experiment only one tank was used to be controlled with the configuration as shown in Fig. 4. The controller output signal manipulates the flow rate of fluid entering the top tank. The measured plant variable is fluid level. Fluid drains through a pipe and is adjusted with a throttling valve and can be considered as disturbance flow. Due to its mechanical components, the plant is assumed to have strongly inherent mechanical nonlinearities. The proposed intelligent control scheme was implemented as a real-time control software developed using graphical-based programming language LabVIEW [LabVIEW, 1998]. The software was connected on-line to the process mini-plant through an AD/DA card and a signal conditioner.

4.2 Results of Control

Real-time control was conducted on the process mini-plant to see the capability and performance of the proposed intelligent control strategy in real-time environment, either using off-line or on-line learning mode, which configuration is shown in Fig. 5 and 6. Observation was also done to see the ability of the algorithm in handling the plant dynamics changes.

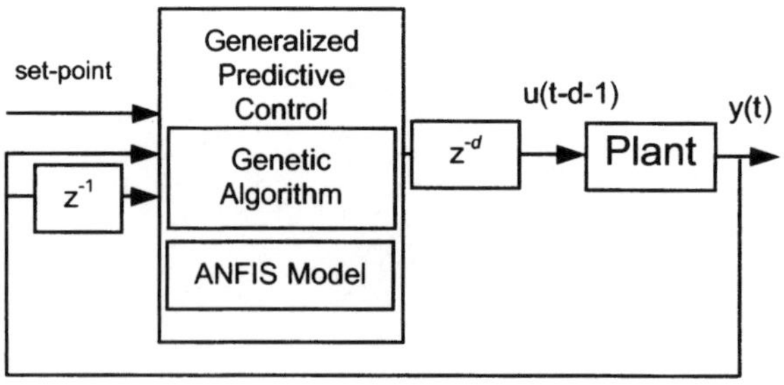

Fig. 5. Control structure using off-line learning mode

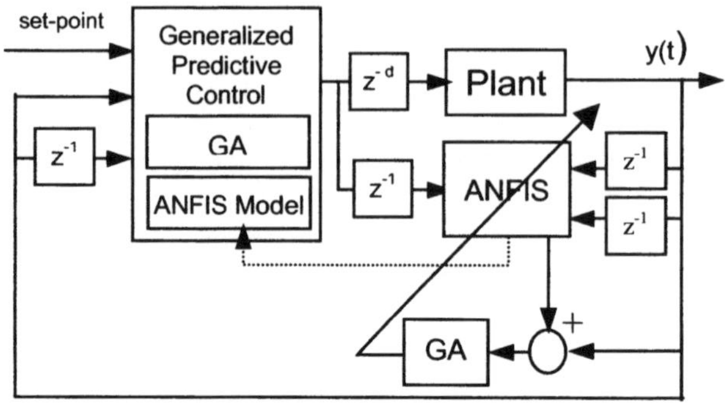

Fig. 6. Control structure using on-line learning mode

The configurations of GA parameters used in the experiment were as follows :

Number of genetic iterations : 100
Number of populations : 20
Ranking method : exponential ranking, q=0.2
Crossover method : Non-uniform, p_m= 0.2
Mutation method : arithmetics, $\lambda_1 = 0.4$ and $\lambda_2 = 0.6$
Elitism Strategy : number of elite is 1 and the best individual from one epoch iteration.

Fig. 7 shows the results of control using off-line mode. It can be seen that unsatisfactory performance of the controller is shown, which is reflected by the remaining offset higher than the criteria 5%. This result demonstrates that the model obtained from off-line learning does not represent the process dynamics quite well. The control system was then switched into on-line learning mode at sampling time k = 76. Since on-line learning mode was introduced, a better plant response was observed, even after the desired set-point was changed to 60 cm (level of the water in the tank)

Investigations were also made to see the control performance for the various setting of the GPC controller parameters applied to the proposed control scheme, which results are shown in Fig. 8. Observations revealed that increasing N_2 results in shorter overshoot of the response and smoother response of the control signal. On the other hand, if N_1 was increased then the response fluctuated as it gets closer to the desired set-point and the control signal was also more reactive.

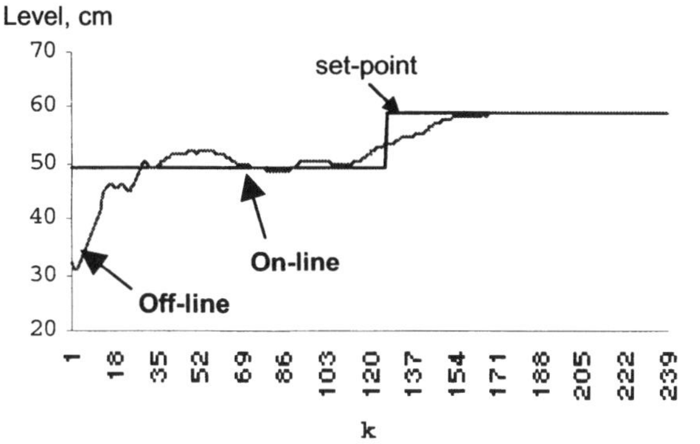

Fig 7. Result of the on-line control using off-line and on-line learning mode, with N_2=20, N_1=1,N_u=1.

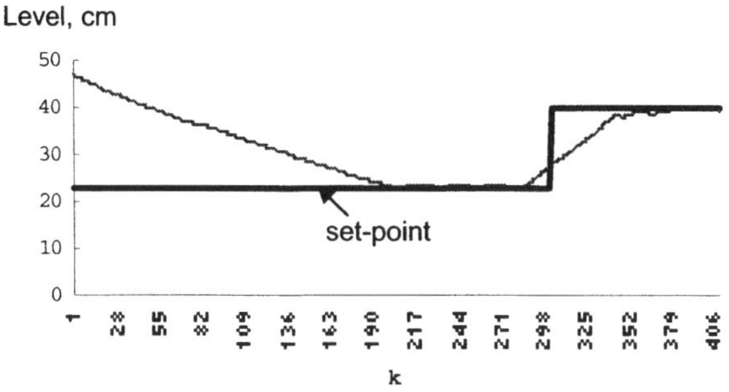

(a). N_2 =5, N_1 =1,N_u =1

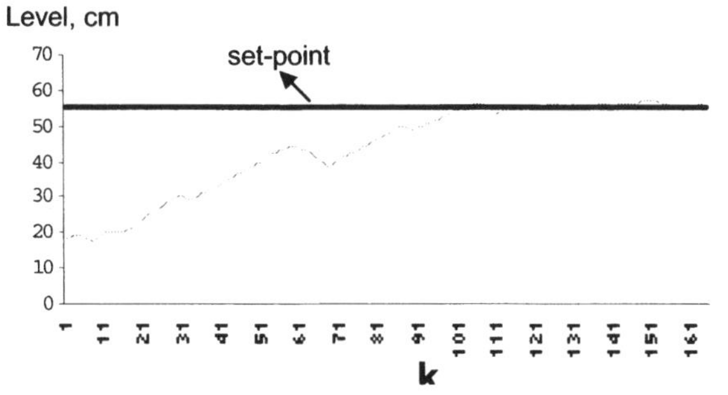

(b). N_2=10, N_1=1,N_u=1

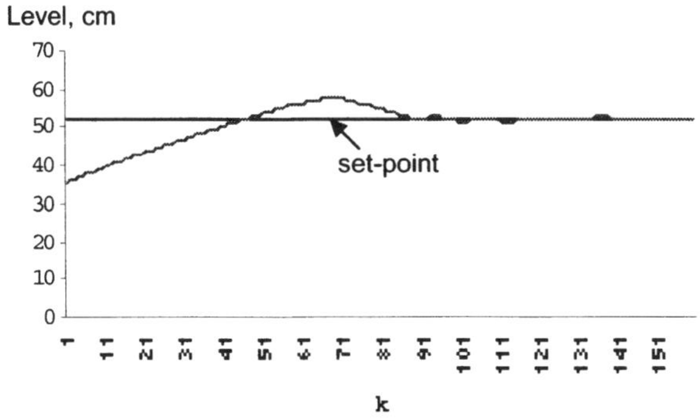

(c). N_2=15, N_1=5,N_u=1

Fig 8. Result of the control using on-line learning mode, with different value of controller parameters.

To investigate whether the learning system is adaptive to the plant disturbances and dynamics changes, the valve HCV3 was closed more firm at sampling instant k=70 causing the instability to the process, as can be seen from Fig. 9.a, where the level of water in the tank increased suddenly. However this sudden change could be anticipated by the controller so that the response returns to the set-point after a while. With its on-line learning mode, the control scheme acts in response to the new data and the GPC controller responded accordingly. After that the set-point was also changed to the level of 40 cm, and at sampling instant of k = 1470 , the valve HCV3 was open more loosely so that the level in the tank decreased rapidly. Again the controller reacted and the level in the tank returned to the desired set-point. Satisfactorily response is also shown to the set-point change subsequently.

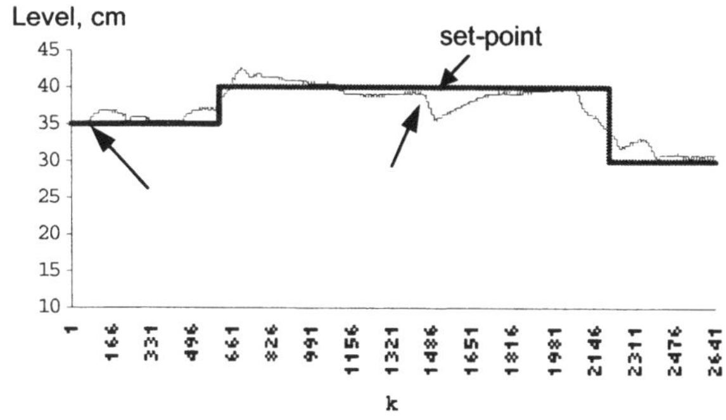

(a)

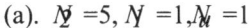

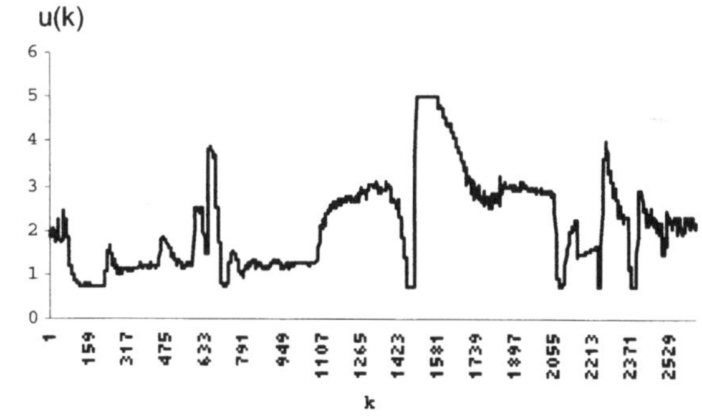

(b)

Fig. 9. Response of plant due to plant disturbances and dynamics changes, using control with on-line learning mode and N_2=20, N_1=1,N_u=1.

For the purpose of performance comparison, the experiment was also conducted using the conventional Proportional + Integral (PI) type controller. Similar behaviour can also be observed on the plant response using PI controller. However, to obtain a satisfactory response, an optimal tuning (through an extensive trial and error or using Ziegler-Nichols approach, which was often time consuming) was needed for PI controller. In addition, by using the standard PI controller, on-line plant dynamics change can not be handled satisfactorily.

5. CONCLUSIONS

An alternative intelligent control scheme which is an integration between the GPC scheme, neuro-fuzzy approach and genetic algorithm based optimization has been proposed and tested on a real-time environment for on-line control of a process mini-plant process. The results show the effectiveness and the performance of the method in controlling nonlinear systems either using *off-line* and *on-line* learning approaches. The experimental result demonstrated that the on-line learning mode gives a better result, especially when there is a disturbance or dynamics change of the system.

REFERENCES

Garcia, C. E., D. M. Prett and M. Morari (1989), Model Predictive Control - Theory and Practice - A Survey, *Automatica*, **Vol. 25**, pp. 335-348

Gen, M, and R. Cheng (1997), *Genetic Algorithms & Engineering Design*, John Wiley & Sons, New York, 1997

Jang, J.S.R., C.T. Sun, C.T. and E. Mizutani (1997), *Neuro-fuzzy and Soft Computing*, Prentice-Hall Inc., Englewood Cliffs, New Jersey

LabVIEW (1998), *User Manual*, National Instruments

Nazaruddin, Y.Y. and Paula F. Tjandrakusuma (2001), On-line Adaptive Predictive Control of a Process Mini-Plant Using Neuro-Fuzzy Based Modelling, *Proceed. of the IEEE International Conference on Mechatronics and Machine Vision in Practice* (M^2VIP'2001), Hong Kong, August 27-29.

IFAC Publications
www.elsevier.com/locate/ifac

LONG-RANGE OPTIMAL MODEL AND MULTI-STEP-AHEAD PREDICTION IDENTIFICATION FOR PREDICTIVE CONTROL

R. Haber*, U. Schmitz*, R. Bars**

**Department of Plant and Process Engineering, Laboratory of Process Control, University of Applied Science Cologne, D-50679 Köln, Betzdorfer Str. 2, Germany fax: +49-221-8275-2836 and e-mail: robert.haber@fh-koeln.de, us@fh-av.av.fh-koeln.de*
***Department of Automation and Applied Informatics, Control Research Group of the Hungarian Academy of Sciences, Budapest University of Technology and Economics, Hungary; e-mail : bars@aut.bme.hu*

Abstract: Long-range optimal prediction algorithms use the predicted output for several steps ahead. The prediction based on traditionally estimated model parameters does not result in an optimal prediction if the measurements are noisy or/and model structure differs from real process structure. In this paper two different identification schemes are presented and compared: long-range predictive single-model identification and simultaneous multi-step-ahead prediction identification. It is shown that the first method is easier to realize but the second one leads to more accurate results. Both methods are derived for a first-order model in details. Simulation runs and a level control example illustrate the algorithms presented. *Copyright © 2003 IFAC*

Keywords: Process identification, prediction, parameter estimation, predictive control

1. INTRODUCTION

Predictive control algorithms require the prediction of the controlled signal several steps ahead. There are different ways to calculate the predicted values:

- Transformation of the process model equation to a predictive equation solving a Diophantine equation or
- Iterative simulation several steps ahead by using the process model (Clarke *et. al.* 1989).

With long-range optimal control the controlled output has to be predicted several steps ahead simultaneously, thus a multi-step prediction is required. If the process model is exactly known, then the prediction is biasfree. There are, however, cases when the process model has to be estimated:

- either because of noisy measurements,
- or because of model order reduction.

In both cases the process model, or more precisely the prediction equations have to be estimated. The traditional process identification results in one-step-ahead optimal prediction, which may not be optimal for more prediction steps simultaneously. There are two ways to estimate multi-step optimal prediction equations:

- *LRPI (Long-Range Predictive Identification)*: parameter estimation of single model, which is optimal for the prediction in a long-range horizon,
- *MSPI (Multi-Step-ahead Prediction Identification)*: estimation of several prediction equations in the long-range horizon simultaneously.

The paper shows both methods for a first-order process model, although the method works for higher-order processes, as well.

In a former paper (Haber et. al., 1999) on adaptive non-linear predictive control it was already shown, that the estimation of a prediction equation is prior to the prediction based on an estimated process model.

In the sequel the two estimation methods (long-range and multi-step) are presented.

The process model is a linear one with the pulse-transfer function

$$y(k)=\frac{B(q^{-1})}{A(q^{-1})}u(k-(d+1))$$

where k is the discrete time, d the discrete-time (mathematical) delay, $u(k)$ the input and $y(k)$ the output signal, further

$$B(q^{-1})=b_1q^{-1}+...+b_nq^{-n}$$
$$A(q^{-1})=1+a_1q^{-1}+...+a_nq^{-n}$$

2. LONG-RANGE OPTIMAL MODEL EQUATION IDENTIFICATION

The cost function of the long-range predictive horizon optimal parameter estimation is:

$$J_{ident}=\sum_{k=1}^{N}\sum_{n_p=n_{p1}}^{n_{p2}}\left(y(k)-\hat{y}(k|k-d-n_p)\right)^2$$

where $y(k)$ is the measured and $\hat{y}(k|k-d-n_p)$ is the $d+n_p$ steps earlier predicted output, n_{p1} is the least prediction step, n_{p2} is the greatest prediction step and N is the number of input/output measurements.

Figure 1 illustrates the idea of a long-range optimal model equation identification.

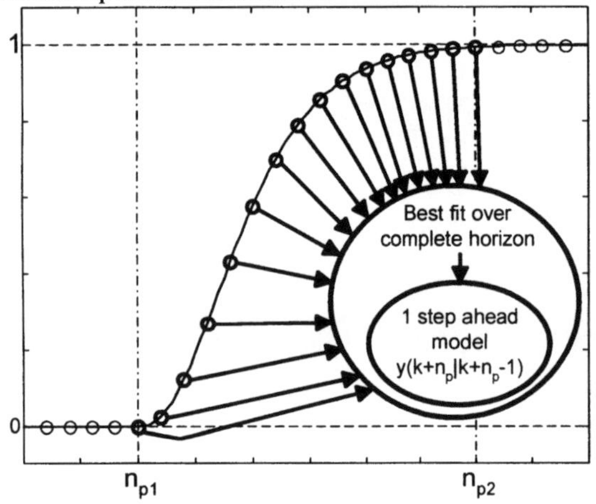

Fig. 1: The idea of long-range optimal model equation identification

Shook et. al. (1991) have shown that the minimization of the long-range cost function can be performed by the following iterative way:

1. Let $L(q^{-1})=1$
2. Filter the measured input and output data by $L(q^{-1})$
3. Estimate the parameters of the process model by least squares method.
4. Solve

$$L(q^{-1})L(q)=\sum_{j=d+n_{p1}}^{d+n_{p2}}E_j(q^{-1})E_j(q)$$

where $E_j(z^{-1})$ is the solution of the Diophantine equation

$$1=E_j(q^{-1})A(q^{-1})+q^{-j}F_j(q^{-1})$$

5. Go to Step 2 until the estimated parameters converge.

3. MULTI-STEP-AHEAD PREDICTION EQUATION IDENTIFICATION

An alternative way to the long-range prediction optimal single model identification is the estimation of the coefficients of all prediction equations in the prediction horizon separately (see e.g., Rossiter, 2000). Figure 2 illustrates this idea.

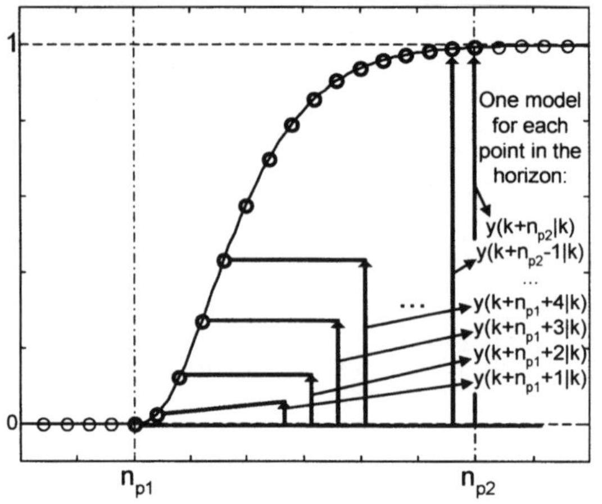

Fig. 2: The idea of multi-step prediction equation identification

For the case of a first-order model

$$\hat{y}(k)=a_1y(k-1)+b_1u(k-1)$$

the prediction equations - till the maximal prediction of 5 steps ahead - are as follows

$$\hat{y}(k+1|k)=a_1y(k)+b_1u(k)$$
$$\hat{y}(k+2|k)=a_1^2y(k)+a_1b_1u(k)+b_1u(k+1)$$
$$\hat{y}(k+3|k)=a_1^3y(k)+a_1^2b_1u(k)+a_1b_1u(k+1)$$
$$+b_1u(k+2)$$
$$\hat{y}(k+4|k)=a_1^4y(k)+a_1^3b_1u(k)+a_1^2b_1u(k+1)$$
$$+a_1b_1u(k+2)+b_1u(k+3)$$
$$\hat{y}(k+5|k)=a_1^5y(k)+a_1^4b_1u(k)+a_1^3b_1u(k+1)$$
$$+a_1^2b_1u(k+2)+a_1b_1u(k+3)$$
$$+b_1u(k+4)$$

All prediction equations are linear in the parameters, that means the output signal can be written as a scalar product of a memory vector with measured values or their known (e.g. delayed) functions and a parameter vector of the unknown coefficients, if not the original model parameters (a_1 and b_1) but the coefficients of the prediction equation (α_i and β_i) are estimated.

$$\hat{y}(k+1|k)=\alpha_0^{(1)}y(k)+\beta_0^{(1)}u(k)$$
$$\hat{y}(k+2|k)=\alpha_0^{(2)}y(k)+\beta_0^{(2)}u(k)+\beta_1^{(2)}u(k+1)$$

$$\hat{y}(k+3\mid k)=\alpha_0^{(3)}y(k)+\beta_0^{(3)}u(k)+\beta_1^{(3)}u(k+1) + \beta_2^{(3)}u(k+2)$$
$$\hat{y}(k+4\mid k)=\alpha_0^{(4)}y(k)+\beta_0^{(4)}u(k)+\beta_1^{(4)}u(k+1) + \beta_2^{(3)}u(k+2)+\beta_3^{(3)}u(k+3)$$
$$\hat{y}(k+5\mid k)=\alpha_0^{(5)}y(k)+\beta_0^{(5)}u(k)+\beta_1^{(5)}u(k+1) + \beta_2^{(5)}u(k+2)+\beta_3^{(5)}u(k+3) + \beta_4^{(5)}u(k+4)$$

Therefore, the LS (Least Squares) parameter estimation can be performed in one step. Of course, alternatively, any other usual parameter estimation algorithms can be used.

As only the actual and older signals can be measured or stored, the actual output signal (not the future output signals) has to be predicted from older measured values for the purpose of parameter estimation:

$$\hat{y}(k\mid k-1)=\alpha_0^{(1)}y(k-1)+\beta_0^{(1)}u(k-1)$$
$$\hat{y}(k\mid k-2)=\alpha_0^{(2)}y(k-2)+\beta_0^{(2)}u(k-2)+\beta_1^{(2)}u(k-1)$$
$$\hat{y}(k\mid k-3)=\alpha_0^{(3)}y(k-3)+\beta_0^{(3)}u(k-3)+\beta_1^{(3)}u(k-2) + \beta_2^{(3)}u(k-1)$$
$$\hat{y}(k\mid k-4)=\alpha_0^{(4)}y(k-4)+\beta_0^{(4)}u(k-4)+\beta_1^{(4)}u(k-3) + \beta_2^{(3)}u(k-2)+\beta_3^{(3)}u(k-1)$$
$$\hat{y}(k\mid k-5)=\alpha_0^{(5)}y(k-5)+\beta_0^{(5)}u(k-5)+\beta_1^{(5)}u(k-4) + \beta_2^{(5)}u(k-3)+\beta_3^{(5)}u(k-2) + \beta_4^{(5)}u(k-1)$$

The coefficients of the prediction equations have to be estimated from this fit. The simultaneous parameter estimation can be simplified because of common components in several equations.

The algorithm of the LS parameter estimation is well known and is not given here. It is assumed, that additive noise leads to a white noise equation error and to a biasfree estimation.

4. LONG-RANGE OPTIMAL MODEL V.S. MULTI-STEP-AHEAD PREDICTION IDENTIFICATION

The main features of both estimation methods are:

- *long-range optimal single model identification*:
  - only one model has to be estimated,
  - model has relatively few parameters,
  - estimation with repeated LS algorithm,
- *multi-step-ahead prediction identification:*
  - several models have to be estimated,
  - models have relatively many parameters,
  - estimation with a one-shot LS algorithm.

Example 1 *Long-prediction range optimal estimation of a third-order process*

The transfer function of the process is

$$G(s)=\frac{1}{[1+(1/3)s]^3}.$$

The excitation was a PRTS (Pseudo Random Ternary Signal) with levels –2, 0 and 2, period of 26 steps. The sampling time was ΔT=0.5 sec .

Figs. 3 to 5 show the results of three methods for an approximating PT1 model

- one-step-ahead optimal model identification,
- long-range optimal single model identification,
- multi-step-ahead prediction identification

for different prediction horizons and minimal switching times.

Fig 3. shows the estimation for the prediction horizon 1 to 5 steps ahead. The minimal switching time of the PRTS was equal to the sampling time.

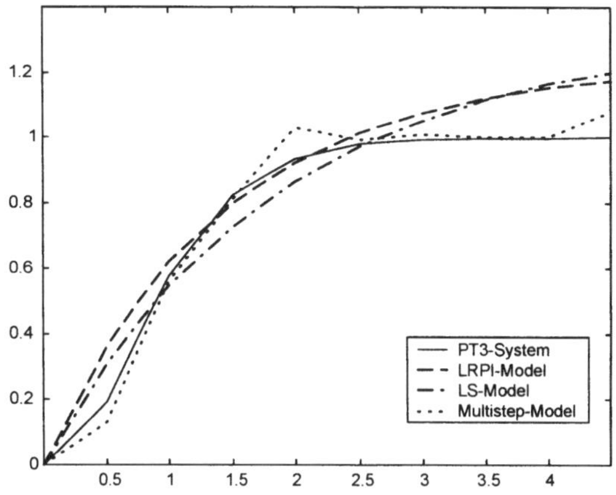

Fig 3: Step response of one-step-ahead optimal (dash-dotted line), long-range single model optimal (dashed line) and multi-step-ahead prediction identification (dotted line) with an identification horizon of 5. 0.5 sec = 2.5 sec

Fig. 3 shows that the prediction based on a first-order model approximated well a third-order process model and that the multi-step prediction identification worked best.

Fig 4. shows the estimation for the prediction horizon 1 to 25 steps ahead. The minimal switching time of the PRTS was equal to the sampling time.

The large deviation with great prediction length (i.e. in the steady-state) are caused by the too short minimal switching time of the test signal (PRTS) used for identification. Better fit can be achieved by doubling of the minimal switching time, as seen in Fig. 5.

In the next the difference between one-step-ahead and multi-step-ahead parameter estimation is shown by an example.

Example 2 *Difference between one-step-ahead and multi-step-ahead parameter estimation*

Fig. 6 shows the input and output records of the identification for the same process as in Example 1.

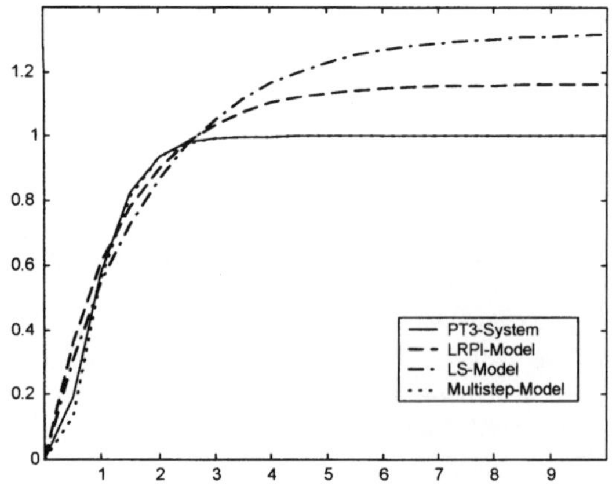

Fig 4: Step response of one-step-ahead optimal (dash-dotted line), long-range single model optimal (dashed line) and multi-step-ahead prediction identification (dotted line) with an identification horizon of 25. 0.5 sec = 12.5 sec and minimal switching time equal to the sampling time

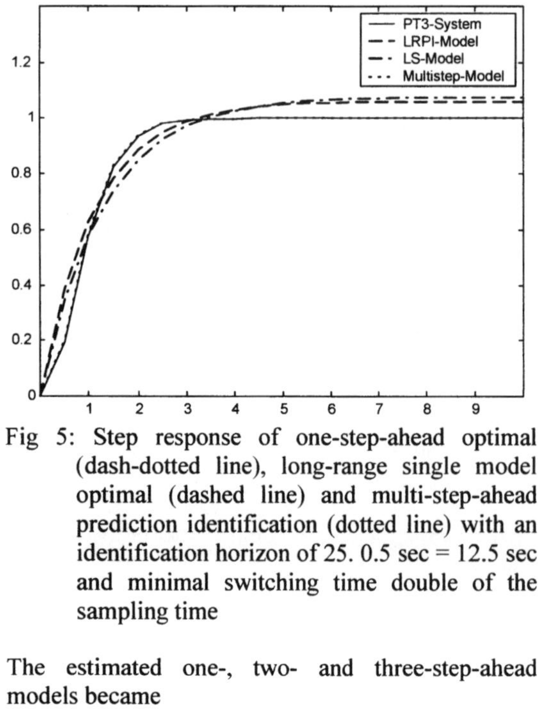

Fig 5: Step response of one-step-ahead optimal (dash-dotted line), long-range single model optimal (dashed line) and multi-step-ahead prediction identification (dotted line) with an identification horizon of 25. 0.5 sec = 12.5 sec and minimal switching time double of the sampling time

The estimated one-, two- and three-step-ahead models became

$$\hat{y}(k|k-1) = 0.679y(k-1) + 0.3462u(k-1)$$
$$\hat{y}(k|k-2) = 0.3303y(k-2) + 0.488u(k-2) + 0.2003u(k-1)$$
$$\hat{y}(k|k-3) = 0.1311y(k-3) + 0.2967u(k-3) + 0.3858u(k-2) + 0.1955u(k-1)$$

As one can see, the estimated two- and three-step-ahead models are not equal to the prediction equations calculated from the one-step-ahead estimated model.

5. GPC (GENERALIZED PREDICTIVE CONTROL)

The control aim of a predictive controller is to

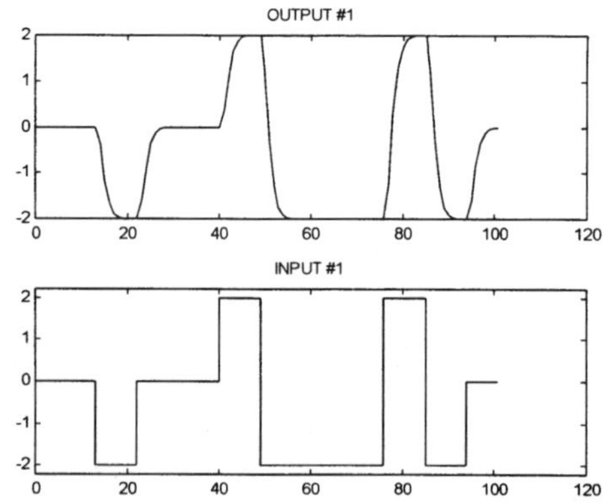

Fig. 6 Input and output signals for the identification

minimize the cost function:

$$J_{control} = \sum_{n_e=n_{e1}}^{n_{e2}} [y_r(k+d+n_e) - \hat{y}(k+d+n_e|k)]^2 +$$
$$+\lambda_u \sum_{j=1}^{n_u} \Delta u^2(k+j-1) \Rightarrow \underset{\Delta u}{MIN}$$

where k is the actual discrete time point, d is the discrete-time delay (mathematical), $\Delta u(k)$ the manipulated variable, $\hat{y}(k+d+n_e|k)$ the controlled signal and $y_r(k+d+n_e|k)$ the reference signal predicted $n_e + d$ steps ahead the actual time point k

The controller setting parameters are as follows:

- n_{e1} extension of the start of the control error prediction horizon over the mathematical dead time,
- n_{e2} extension of the end of the control error prediction horizon over the mathematical dead time,
- n_u the number of the subsequent predicted changes of the manipulated variable,
- λ_u weighting factors of the manipulated variable.

Example 3 *GPC of a third-order process based on a first-order model using long-range optimal single model identification*

The process and the sampling time are as in Example 1. The controller parameters are n_{e1}=0, n_{e2}=N_2-1=4, n_u=1, λ_u=0.8. The control was based on a first-order estimated model. Fig. 7 shows the controlled output (y) and the manipulated variable (u) of GPC for a reference signal step $y_r = 1 \cdot 1(t-1)$ and a stepwise disturbance $dist = -1 \cdot 1(t-6)$ at the process input for both identification methods:

- one-step-ahead optimal model identification,
- long-range optimal single model identification .

As is seen, the control is quicker and the oscillation as the amplitude of the oscillation is less with the long-range optimal identification than with the one-step-ahead estimation.

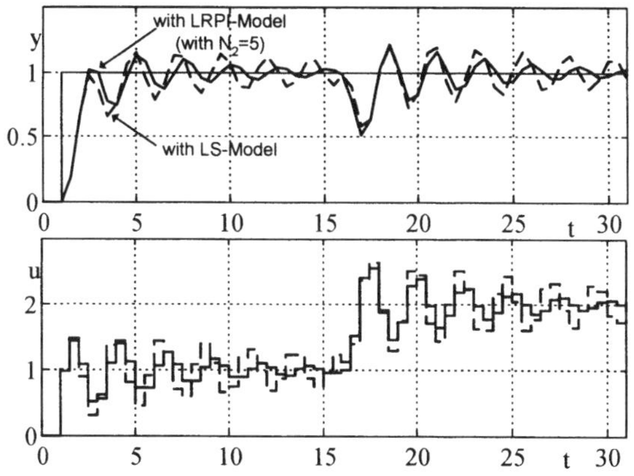

Fig 7. GPC of a third-order process based on first-order model one-step-ahead LS (dashed line) and LRPI (solid line) identification.

Example 4 GPC *(with matrix inversion) of a third order process based on a first-order model using long-range identification*

The process model, sampling time and the controller parameters are as in Example 3. The control was based on a first-order estimated model and the identification methods were:

- long-range optimal single model identification,
- multi-step-ahead prediction identification.

Fig. 8 compares the two identification methods. There is no big difference between the two methods. The control based on the multi-step predictive identification is a little better (quicker and less oscillation and amplitude of the oscillation) than the long-range optimal single model identification.

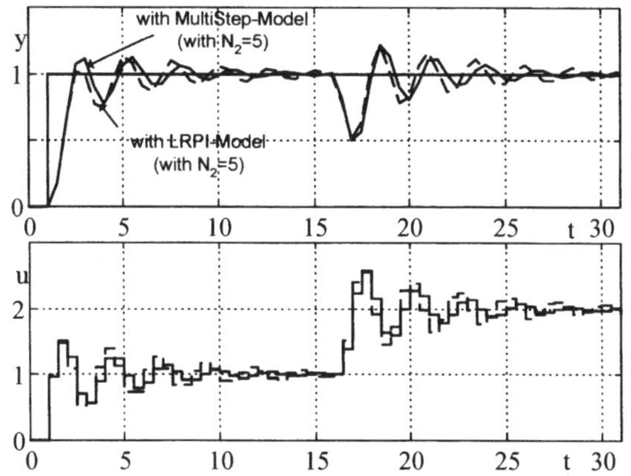

Fig 8. GPC of a third-order process based on first-order model with LRPI (dashed line) and MSPR (solid line) identification.

6. LEVEL CONTROL OF A TWO-TANK SYSTEM

At the Department of Automation and Applied Informatics of the Budapest University of Technology and Economics a level control rig has been built. The pilot plant consists of two tanks with free flow out (Fig. 9.). The upper tank is filled through a pump and the water stream is controlled by a valve. The water leaving the upper tank fills the lower tank.

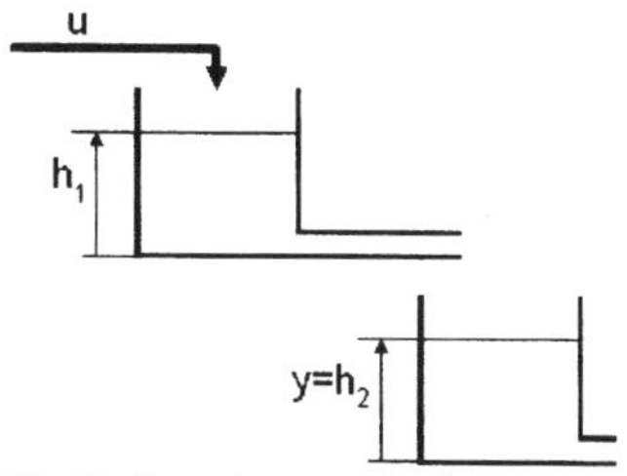

Fig.9. Level control pilot plant with two tanks.

The system is non-linear. For small changes in the control signal linear proportional models with two lags were identified. Their parameters are summarized in Table 2 for different levels in the lower tank.

Table 2. Linearized models of the level pilot plant

Level [%]	Gain	T_1 [sec]	T_2 [sec]
30	0.886	415	298
40	1.024	480	344
50	1.144	536	385
60	1.254	587	421

The process was identified by a PRTS (Pseudo Random Ternary Signal) in the working point 27 %. with the signal amplitude +/-3 %. The parameters of the second-order model were estimated by

- one-step ahead identification and
- long-range identification till $n_{e2} = 9$ prediction steps.

The GPC algorithm was applied with sampling time ΔT=1 sec. and with the following controller parameters: n_{e1}=0, n_{e2}=9, n_u=3, λ_u=0.5 in different working points.

Fig 10 and 11 show the control in working-point of 27% with a model, identified in the working-point of 27% As is seen, the controller outputs are very similar, but the manipulated variable is smoother in the case of long-range identification. The advantage of the long-range optimal identification over the one-step-ahead one can be observed better, if the working points of parameter estimation and control differ from each other, as illustrated in Figs. 12 and 13.

The process was identified again by a PRTS in the working point 5 % with the signal amplitude +/-0.5 %. The parameters of the second-order model were estimated by

- one-step ahead identification and
- long-range identification till $n_{e2} = 9$ prediction steps.

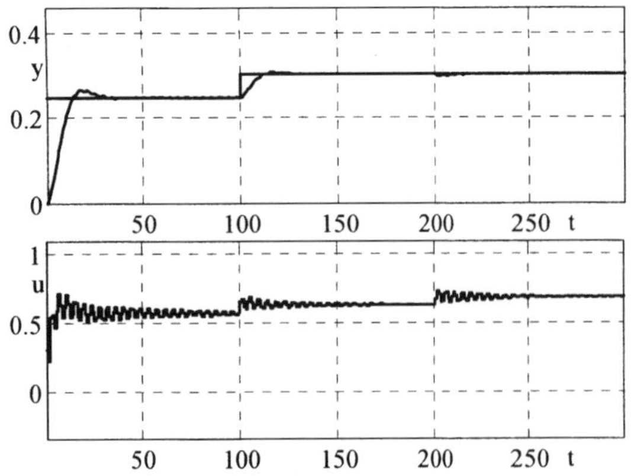

Fig 10. GPC of the two-tank-system based on one-step-optimal estimated model (working-point of control: 27 %, working-point of identification 27 %)

The controller parameters are the same as in working point 27 %.

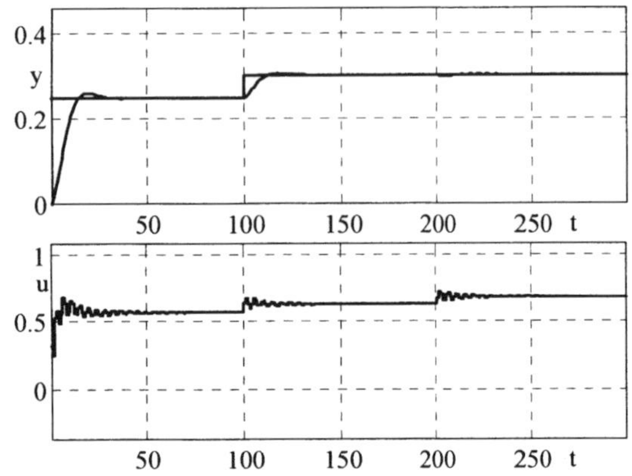

Fig 11. GPC of the two-tank-system based on long-range optimal estimated model (working-point of control: 27 %, working-point of identification 27 %)

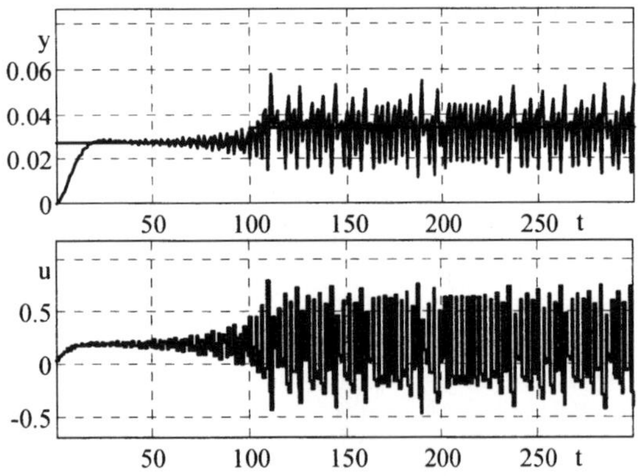

Fig 12. GPC of the two-tank-system based on one-step-optimal estimated model (working-point of control: 27 %, working-point of identification 5 %)

7. CONCLUSION

If the model, used for predictive control, does not fit perfectly the real process model then the prediction equations cannot be calculated biasfree from a one-step-ahead optimal estimated process model. Long-

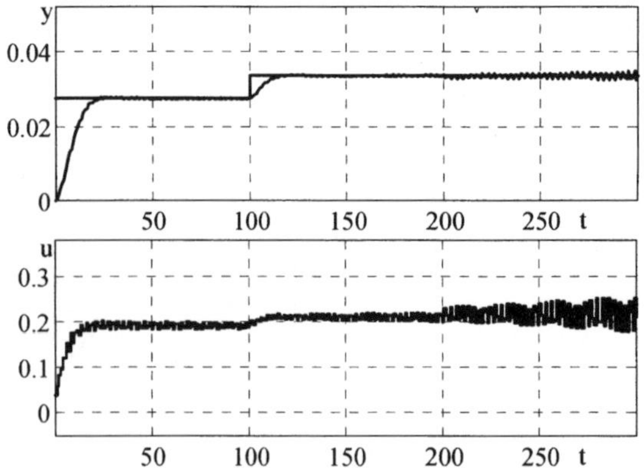

Fig 13. GPC of the two-tank-system based on long-range optimal estimated model (working-point of control: 27 %, working-point of identification 5 %)

range optimal identification fits an equivalent process model, which leads to optimally fitting prediction equations in the whole horizon length. An alternative way, estimating all prediction equations individually but simultaneously was recommended and illustrated by several simulation runs and a level control example. As it was expected and was shown by several simulations, the long-range simultaneous multi-step-ahead prediction (MSPR) resulted in better control behavior than the long-range predictive optimal single-model identification (LRPI) .

ACKNOWLEDGMENTS

The work has been supported earlier by the Ministry of Science and Research of NRW (FRG) in the program „Support of the European Contacts of the Universities / Förderung der Europafähigkeit der Hochschulen“ and now is supported by the University of Applied Science Cologne in the program “Advanced Process Identification for Predictive Control” and by the program of EU-Socrates. The third author’s work was also supported by the fund of the Hungarian Academy of Sciences for control research and partly by the OTKA fund T042741. All supports are kindly acknowledged.

REFERENCES

Clarke, D.W. and C. Mohtadi (1989). Properties of generalised predictive control. *Automatica*, **25 (6)**, 859-875.

Shook, D.S., C. Mohtadi, S.L. Shah (1991). Identification for long-range predictive control, *IEE Proceedings-D*, Vol. **138 (1)**, 75-84.

Rossiter, J.A. (2000). Identification of models for predictive control. *IEE Seminar on Model Validation for Plant Control and Condition Monitoring*. London, UK, 7/1-7/7.

Haber, R., R. Bars and O. Lengyel (1999). Three extended horizon adaptive nonlinear predictive control schemes based on the parametric Volterra model. *ECC'99*, Paper No.F625

IFAC Publications
www.elsevier.com/locate/ifac

PREDICTIVE CONTROL OF FLOW QUANTITY AND SLOSHING-SUPPRESSION DURING BACK-TILTING OF A LADLE FOR BATCH-TYPE CASTING POURING PROCESSES

Kazuhiko Terashima*, Ken'ichi Yano*, Motoki Kaneko*

* *Department of Production Systems Engineering*
Toyohashi University of Technology
Hibarigaoka 1-1, Tempaku-cho, Toyohashi, 441-8580 Japan
FAX : +81-532-44-6690
E-mail : {terasima, yano, kaneko}@procon.tutpse.tut.ac.jp

Abstract: The purpose of this paper presents a method to predict the molten metal quantity overspilt from a ladle at the back-tilting period of a ladle acted to finish the pouring, and also to give a method to suppress the sloshing(liquid vibration)during back-tilting. By making use of predictive control of flow quantity, a supervisory control system is given to suitably switch without overflow of the fluid in a sprue cup from a level controller to a back-tilting controller with sloshing-suppresion. A Hybrid-Shape Approach with a notch-filter under consideraion of the time and the frequency characteristics is applied to suppress the sloshing. The effectiveness of the proposed method is demonstrated through experiments. *Copyright © 2003 IFAC*

Keywords: predictive control, supervisory control, process control, sloshing, casting, automatic pouring.

1. INTRODUCTION

In the recent years, the major reasons for installing the batch-type automatic molten metal pouring systems are increasing productivity, cutting scrap loss, reducing operating costs, impoving working conditions in the foundry.Because the process of pouring the molten metal into molds is potentially dangerous and chaotic, the establishment of reliable automation system is of great important from the standpoints of both safety and hygiene of the foundry environment.

Figure 1 shows illustration of the tilting-type automatic pouring machines used for the rotation of a ladle(tank). The ladel is tilted forward for pouring, and moves slightly backward after pouring each mold. The molds stop at fixed location during pouring. The main objects of pouring are the following(Burditt and Bralower, 1989).

(1) To pour the molten metal precisely into the sprue cup or the pouring basin of a mold.

(2) To keep the sprue cup filled to the specified level with metal from the start of the pour to the end, in order to avoid trapping dust and a vortex in the mold, which induces casting defects.

(3) To pour the amount of metal per mold needed to fill the mold and avoid over-pouring exactly by predicting the flow quantity overspilt from a ladle at the back-tilting.

(4) To suppress sloshing(liquid vibration)when the ladle is filled backward when finishing the pouring in the specified time, and especially to eliminate the residual viberation of molten metal at the end point.

High-speed transfer control with suppressing the sloshing has been largely studied by authors

(K.Terashima and K.Yano, 2001; K.Yano and K.Terashima, 2001). Liquid level control in the mold has been reported for the continuous casting by H_∞ control(H.Kitada and K.Sasame, 1998). Studies on liquid level control in a sprue cup have few been reported for the batch casting pocesses. Authors presented reasonable results for it, by means of two-degree-of freedom control(Y.Sugimoto, et al, 2001).

With respect to (3) of the above items, there is no scientific report, and it has been carried out by the human operator's skill or feedback of liquid level. Then, overflow of liquid has been often occured. Then, it is indispensable to start the back-tilting of a ladle, when the present poured weight plus the predicted weight to be still further flowed during back-tilting are equal to the necessary weight to completely fill the mold with metal.

With respect to (4), the authors have already reported the study using the feedback of sloshing. However, it is difficult to feedback the sloshing of molten metal in real time due to the sensor's problem in the actual industries.

Therefore, in order to know the starting time of back-tilting of a ladle, this paper presents a method to predict the flow quantity overspilt from a ladle during back-tilting. Further, this paper gives a sensorless control method without using feedback of sloshing in real time while tilting a ladle backward to finish the pouring.

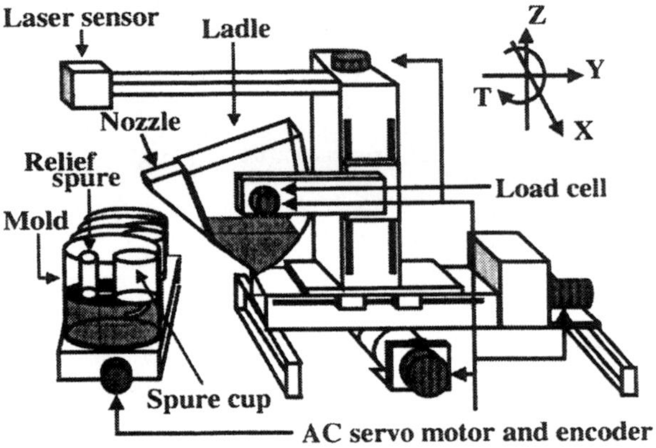

Fig. 1. Illustration of Automatic Pouring Robot (APR)

2. AUTOMATIC POURING ROBOT

The laboratory experimental apparatus used in this paper is shown in Fig. 1 and Fig.2. The rotary direction of the T-axis is driven by an AC servomotor. The driving force of the AC servomotor can be amplified by reducing the gear ratio. The center of the ladle's rotation shaft is placed near the ladle's center of gravity. When the ladle is rotated around the center of gravity, the tip of the ladle nozzle (or mouth) moves in a

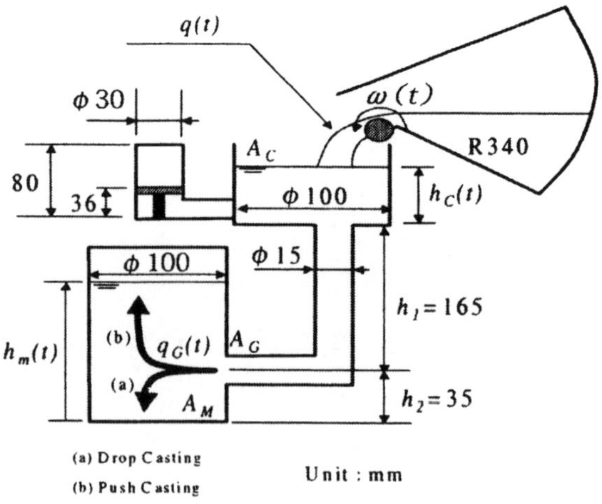

Fig. 2. Illustration of a pouring model.

circular trajectory. It is then difficult to pour the molten metal into a mold if the pouring mouth is moved by tilting. Then, the position of the tip of the ladle nozzle is controlled invariable during pouring, by means of a synchronous control of Y- and Z-axes for rotational motion around T-axis of a ladle.

The rotation angle is measured by an encoder installed in AC servo motor. X-, Y-, Z- and T-axes are also driven by AC servo motors. But, the driving force of each of these motors is amplified through the ball and screw mechanism. Each axis can be independently moved.

The liquid level in a sprue cup is measured by a laser sensor above the relief sprue. This laser sensor measures the distance by triangulation. The liquid level is measured by using a relief sprue that has a float inside it.This float reflects the laser beam. The laser sensor used in this paper can measure liquid levels higher than 0.036[m] from the bottom of a cup (0[m]). The weight of fluid in a ladle is measured by a load cell located under the pouring machine.

The measured value from the laser sensor, encoder of each axis, and the load cell are inputted to a computer by A/D converter. A control input is sent to a motor driver via a D/A converter, driving the AC servomotor. The amount of sloshing in the ladle is measured by using an electric resistance-type sensor comprised of two stainless bars, where this sensor is not used for the purpose of feedback. A control instruction, a calculation of the control law and data processing are carried out in the DSP through the AD/DA converter and an up/down counter.

3. SUPERVISORY CONTROL

Systematic pouring operations that fill up a mold with molten metal and hold the liquid level constantly in a sprue cup are in high demanded. The present process is a batch casting pouring process

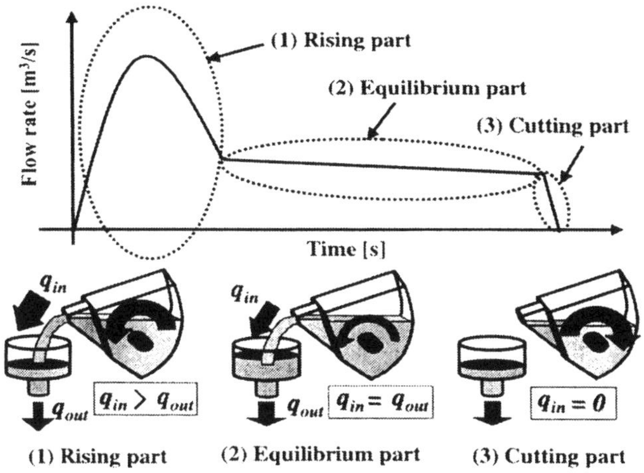

Fig. 3. Pouring flow rate curve.

that handles the liquid from a transient state to a steady state.

An ideal pouring pattern has three parts as drawn in Fig. 3. First, in the Rising Part, the liquid level in a sprue cup must be raised to the set point as quickly as possible. Second, the liquid level must be held constant in the sprue cup while metal is filling into mold. Finally, in the Cutting part of the pouring pattern, backward tilting after pouring must be carried out quickly without residual vibration. A series of process shown in Fig. 3 is needed to maintain the liquid level in a sprue cup to a set point. This research has been already done by authors(Y.Sugimoto, K.Yano and K.Terashima, 2002), except for the cutting part of (3). Cutting part is required to finish the pouring. When the molten metal has been completely filled into mold, the pouring must be finished to avoid the overflow of the liquid from a sprue cup.

The required quantity to completely fill the liquid into a mold can be calculated. Then, if the weight of liquid in a ladle is measured by the real time sensor of a load cell, the poured weight of the liquid can be measured. In principle, when the poured weight agrees with the required weight to fill the liquid into a mold, the pouring is demanded to be finished, and therefore the ladel is filled backward. However, when a ladle is tilted backward, some amount of liquid is still poured for a while from a ladle to a mold, due to the inertia by back-tilting of a ladle. Therefore, without predicting this amount of the poured weight during back-tilting, the overflow of the liquid from a sprue cup is occured. On the other hand, if cutting part is started in the inadequate timing, the molten metal may not be completely filled into a mold, and then the casting defects will be generated.

So,the pouring weight from a ladel to a mold during back-tilting must be predicted. Then, when the total weight of the poured metal weight from a

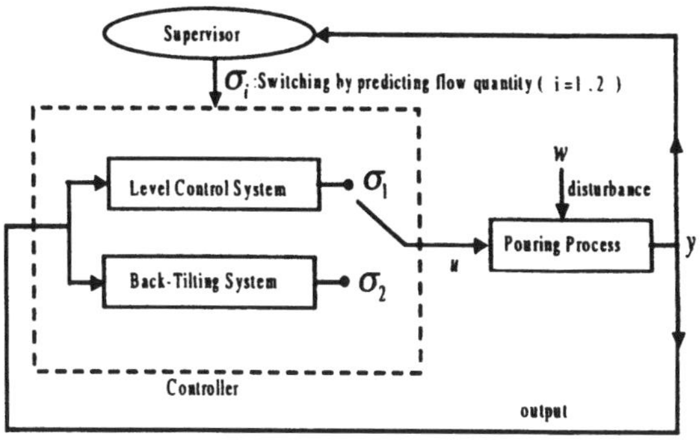

Fig. 4. Structure of supervisory control system

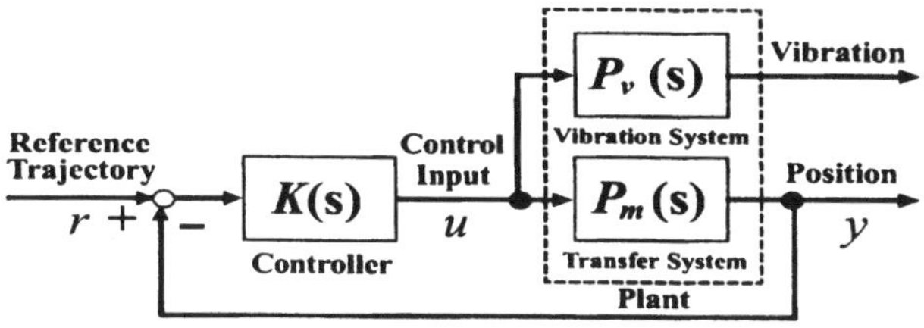

Fig. 5. Target closed loop system

ladle measured in the load cell and the predicted metal weight to be overspilt during back-tilting agrees with the metal weight to compleately fill the mold, the back-tilting of a ladle must be satrted. In previous researches of authors, level control in (1) Rising part and (2) Equilibrium part has been achieved, and we will omit it in this paper due to the paper limitation.

In this paper, the adequate switching control from Level Control System to Back-Tilting Control System is conducted by supervisory control system as shown in Fig. 4. The present supervisory control is simple. Supervisor is always calculating the total amount of metal weight comprised of the poured weight up to the present and the predicted weight to be poured in future by back-tilting from the present time, and the supervisor gives a singal to be switched from a level control mode to a back-tilting control mode, when the total amount of metal weight agrees with the metal weight to completely fill the mold.

4. BACK-TILTING CONTROL WITH SLOSHING SUPPRESSION

In order to suppress sloshing, this system is designed using Hybird Shape Approach(K.Yano, N.Oguro and K.Terashima, 2001). This approach is a design method for starting control and vibration damping, and satisfies hybrid specifications in both frequency domain(gain margin, phase margin, vibration characteristics, and so on)and time domain(transient response, settling time, overshoot, restriction of control input and so on)by using real-time feedback of only the container's position data.

Table 1. Sloshing natural frequencies

	1st mode	3rd mode
1st pouring	1.27[Hz] (7.98[rad/s])	-
2nd pouring	1.17[Hz] (7.35[rad/s])	-
3rd pouring	1.07[Hz] (6.72[rad/s])	-
4th pouring	0.97[Hz] (6.09[rad/s])	3.02[Hz]

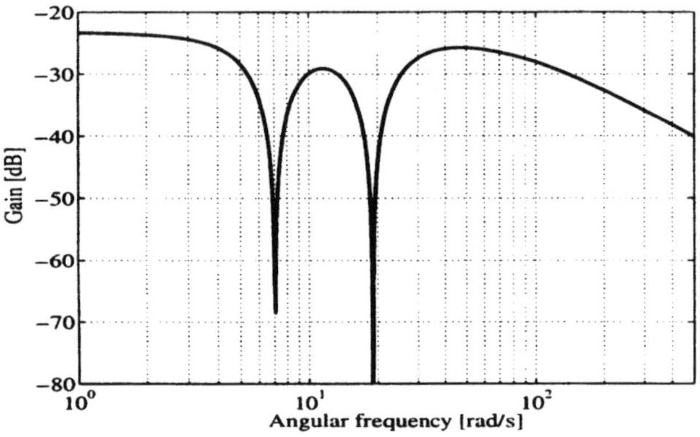

Fig. 6. Frequency response of the controller

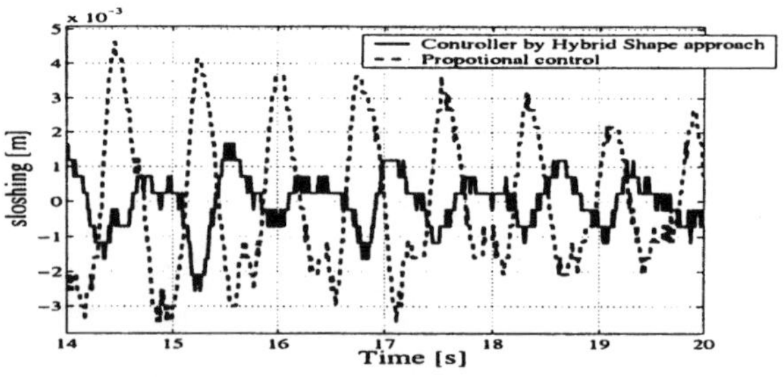

Fig. 7. Sloshing after backward tilting

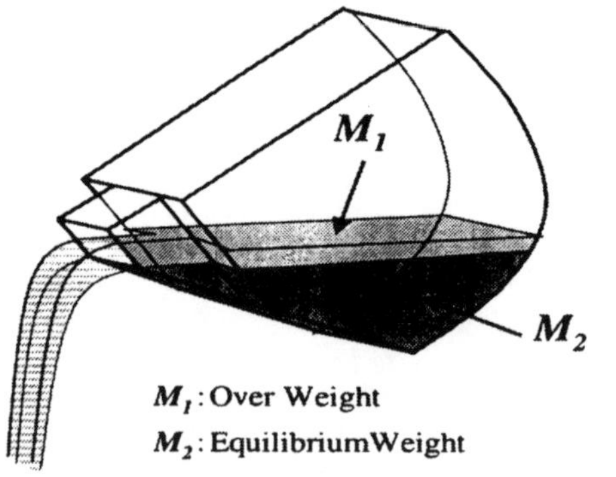

Fig. 8. Over weight and equilibrium weight.

In fact, in a liquid container transfer system, sloshing is caused by self-transfer movement under circumstances in which there is no disturbance. To suppress sloshing, it is important to shape the frequency response of the controller to be notch-typed at the resonance frequency. Therefore, it is theoretically possible to suppress vibration by controlling acceleration. Therefore, dynamical model of sloshing is not necessarily required in the control design. Firure. 5 shows target closed loop system. In this approach, sloshing can be damped by furnishing the controller with a notch filter at the resonance frequency of sloshing. The details of the method is omitted(K.Yano, N.Oguro and K.Terashima, 2001). In order to obtain the controller not to exicte sloshing, it is necessary to specify the natual frequency of the sloshing.

Table 1 shows natual frequencies of the sloshing after each pouring. By considering this results, the controller obtained by Hybird Shape Approach is as follows:

$$K(s) = \frac{K_p}{T_l s + 1} \prod_{j=1}^{2} \frac{s^2 + 2\zeta\omega_{nj}s + \omega_{nj}^2}{s^2 + \omega_{nj}s + \omega_{nj}^2} \quad (1)$$

where $\omega_{n1} = 7.04[rad/s], \omega_{n2} = 18.97[rad/s], \zeta = 0.0001$, and $K_p = 0.0807, T_l = 0.052$. K_p and T_l of the controller were determined by a simplex method such as minimizes the settling time of back-tilting of a ladle.

Figure 6 shows the frequency response of the controller. Controller gain becomes lower in notch-shape around the natural frequency of sloshing. Figure 7 shows sloshing after backward tilting. Sloshing by the Hybrid Shape Approach is suppressed better than it is by proportional control.

5. PREDICTIVE CONTROL OF FLOW QUANTITY

Residual Pouring Quantity(RPQ) is defined as the liquid weight flowing out of a ladle from the start of cutting part to the end during back-tilting of a ladle. In order to exactly conduct the control of flow quantity from a ladle to a mold, it is necessary to predict the residual pouring quantity. In order to predict RPQ, let us consider the state of the liquid inside a ladle as shown in Fig.8. Then, the liquid inside a ladle is considered as two parts, comprised of the upper part of the liquid above the nozzle and the lower part under the nozzle. Upper part is called "Over Weight Part, M_1", and lower part "Equilibrium Weight Part, M_2". Total weight inside a ladle becomes M_1 plus M_2.

When the pouring is stopped in the state as shown in Fig.9, the metal weight in a ladle becomes only equilibrium weight finally.

Let us consider the decrease of over weight M_1 with the increase of time. The relationship between the tilting angular velocity of the ladle and the input voltage to the motor is described in the following equations:

$$\frac{d\omega(t)}{dt} = -\frac{1}{T_{tm}}\omega(t) + \frac{K_{tm}}{T_{tm}}u(t) \quad (2)$$

where $\omega[rad/s]$ is the tilting angular velocity, u[V] is the input voltage for T-axis, $K_{tm}[rad/(sV)]$ is the gain of the motor, $T_{tm}(s)$ is the time constant of the motor, where $K_{tm} = 24.3[deg/sV]$ and $T_{tm} = 0.005[s]$.

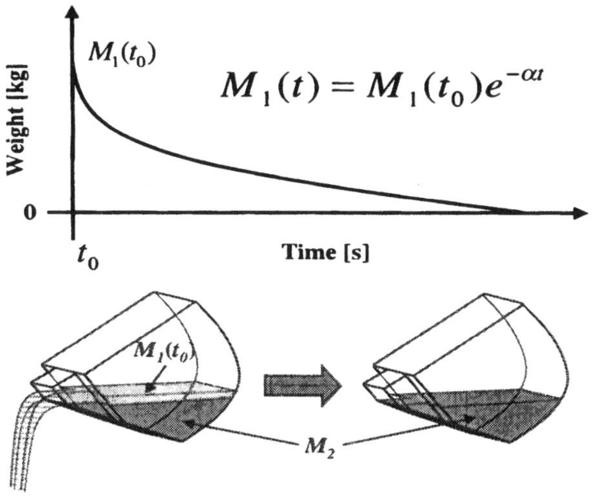

Fig. 9. Static outflow of the over weight in a ladle.

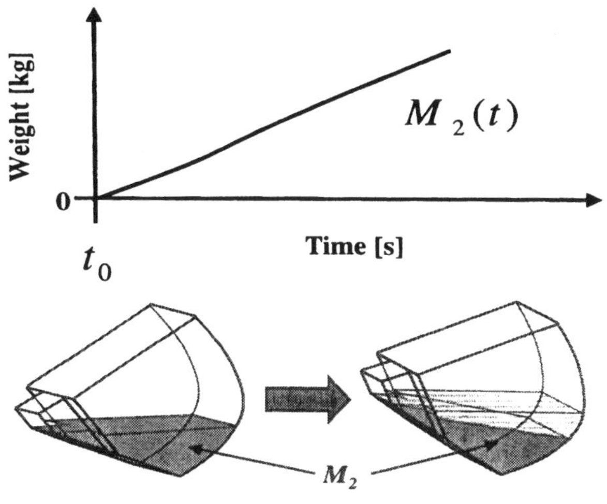

Fig. 10. Change of the over weight by the back-tilting motion.

In this paper, a fan-type ladle is used for the pouring. Controlling the tip of the ladle nozzle invariably, the surface area of liquid in the ladle becomes constant while the ladle is tilted. Therefore, it can be considered that the pouring flow rate is constant when a ladle is rotated at a constant angular velocity. Hence the flow rate model can be described by a first-order transfer function as follows:

$$\frac{dq(t)}{dt} = -\frac{1}{T_f}q(t) + \frac{K_f}{T_f}\omega(t) \tag{3}$$

where $K_f[m^3/rad]$ and $T_f[s]$ are the gain and the time constant of the flow rate model, where $K_f = 2.39 \times 10^{-}4[m^3/deg]$, and $T_f = 1.75[s]$.

Then, when $\omega(t) = 0$, over wejght is considered to become

$$M_1(t) = M_1(t_0)e^{-\alpha t}, \tag{4}$$

Equation (4) is transformed into the following discrete-time equation:

$$M_1[k+j] = C_{d1}A_{d1}^j M_1[k] \tag{5}$$

where $A_{d1} = e^{-\alpha T}$, $\alpha = T_f$, $C_{d1} = 1$, T is a sampling period(T=0.01[s]).

Nextly, let us consider the change of M_2 with the increase of time. During back-tilting, the ladle is tilted backward at the center of gravity of a ladle for reducing the sloshing because of decreasing inertia. As shown in Fig.10, the equilibrium weight increases with back-tilting. $M_2(t)$ can be calculated by the tilting angle $\theta(t)$ of a ladle, as shown in Fig. 10. The ladle is tilted backward by Hybrid-Shape controller described in the previous section.

In Fig.5, transfer system $P_m(s)$ of plant is expressed due to the motor model as follows:

$$P_m(s) = \frac{\theta(s)}{u(s)} = \frac{K_{tm}}{(1+T_{tm}s)s}, \tag{6}$$

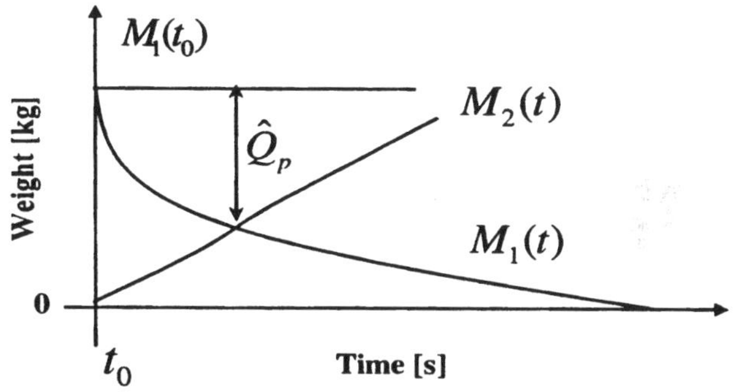

Fig. 11. Predictive value of the outflow.

where $u(s)$ is control input($L(u(t)) = u(s)$; L: laplace operator), and $\theta(s)$ is output of tilting angle($y(s) = \theta(s)$). And,controller $K(s)$ is expressed by Eq.(1). Then, a realization of the closed-loop transfer function from $r(s) = \theta_r(s)$ to $y(s) = \theta(s)$ becomes

$$\dot{x} = Ax(t) + B\theta_r(t) \tag{7}$$

$$y(t) = Cx(t) \tag{8}$$

where,

$$A = \begin{bmatrix} 0 & 1 & 0 & 0 & 0 & 0 & 0 \\ 0 & -\frac{1}{T_{tm}} & g_5 & g_4 & g_3 & g_2 & g_1 \\ 0 & 0 & 0 & 1 & 0 & 0 & 0 \\ 0 & 0 & 0 & 0 & 1 & 0 & 0 \\ 0 & 0 & 0 & 0 & 0 & 1 & 0 \\ 0 & 0 & 0 & 0 & 0 & 0 & 1 \\ -1 & 0 & -k_5 & -k_4 & -k_3 & -k_2 & -k_1 \end{bmatrix}$$

$B = [0 \ 0 \ 0 \ 0 \ 0 \ 0 \ 1]^T$

$C = [1 \ 0 \ 0 \ 0 \ 0 \ 0 \ 1]$, and $g_i(i = 1, \cdots, 5)$, $k_i(i = 1, \cdots, 5)$ are constructed by the parameters K_{tm}, T_{tm} in Eq.(6) and $K_p, T_l, \omega_{n1}, \omega_{n2}$ in Eq.(1).

Then, the discrete-time state equation from Eq.(7) becomes

$$x_{d2}[k+1] = A_{d2}x_{d2}[k] + B_{d2}u_{d2}[k] \tag{9}$$

$$Q[k] = C_{d2}x_{d2}[k] \tag{10}$$

Then,it follows that

$$Q(k+1) = C_{d2}x_{d2}[k+1]$$

$$= C_{d2}\left(A_{d2}x_{d2}[k] + B_{d2}u_{d2[k]}\right) \tag{11}$$

Further, it follows that

$$Q[k+j] = C_{d2}A_{d2}^{j}x_{d2}[k]$$

$$+\sum_{-\infty}^{\infty} C_{d2}A_{d2}^{j-1-i}B_{d2}\theta_r[k+1] \tag{12}$$

Because the increment $M_2[k+j][Kg]$ of Equilibrium weight can be calculated by multilpying the trajectory $\theta[k+j]$ during back-tilting with flow gain $1000K_f[Kg/deg]$, it follows that

$$M_2[k+j] = K_f(C_{d2}A_{d2}^{j}x_{d2}[k]$$

$$+\sum_{i=1}^{j-1} C_{d2}A_{d2}^{j-1-i}B_{d2}\theta_r[k+i]) \tag{13}$$

Liquid weight over the nozzle in a ladle becomes $M_1(t)$ minus $M_2(t)$ while back-tilting as shown in Fig.11. $M_1(t)$ equals with $M_2(t)$ at time t_n. Therefore, after t_n, the fluid is not flowed out of a ladle, even if a ladle is further tilted backward. Hence, the prediction value $\hat{Q}_p(t_0)$ of RPQ at time t_0 becomes

$$\hat{Q}_p(t_0) = M_1(t_0) - M_1(t_n) \tag{14}$$

If the following condition is satisfied, a supervisor sends a signal σ to a controller, and controller is switched from a level controller to a back-tilting controller.

Switching condition:

$$Q_{ref} = \hat{Q}_p(t_0) + M(t_0), \tag{15}$$

where $M(t_0)$ is a total metal weight flowed out of a ladle up to time t_0 which can be measured by a load cell, and Q_{ref} is the reference pouring weight.

Figure 12 shows experimental results by the supervisory control with prediction of residual pouring quantity(RPQ). Supervisory control with prediction of flow quantity shows better results than the results without prediction. For reference weight of 1.274[Kg], the proposed method 1.282[Kg], and the conventional method without prediction 1.304[Kg]. A lot of experiments have been done by various tilting velocity, and angles. In any cases, the proposed method showed good results in the similar precison with Fig.12.

6. CONCLUSION

In this paper, a supervisory control was proposed to suitably switch from a level control to a back-tilting control with sloshing-supression. In order to avoid the overflow of liquid from a sprue cup during back-tilting, a method to predict the flow quantity flowing out of a laldle during back-tilting is proposed. By supervising the predictions information of the flow quantity in real time, it has been shown that the switching from a level control system to a back-tilting control system with damping of sloshing is well executed. The effectiveness of the proposed approach was demonstrated through experiments.

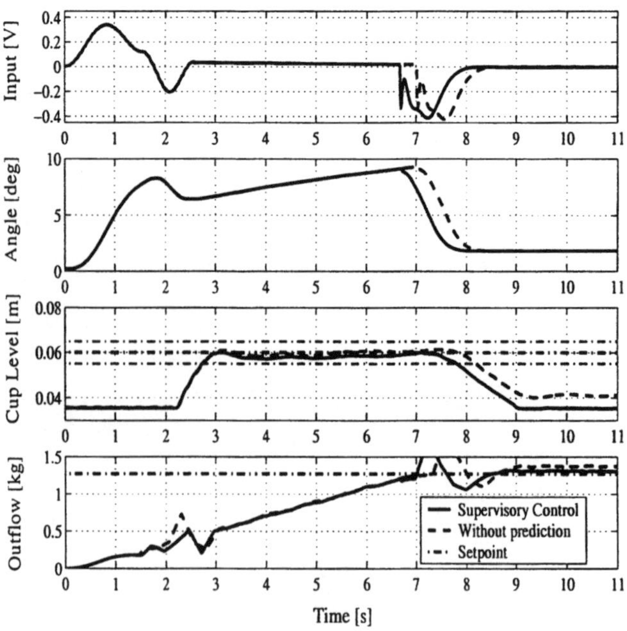

Fig. 12. Experimental result by the flow rate control.

7. REFERENCES

H.Kitada, O.Konko, H.Kasachi and K.Sasame (1998). H_∞ control of molten steel level in continuous caster. *IEEE Trans. Systems Technology*, pp.200-207

K.Terashima and K.Yano (2001). Sloshing analysis and suppression control of tilting-type automatic pouring machine, *Control Engineering Practice*, 9, pp.607-620.

K.Yano and K.Terashima (2001). Robust liquid container transfer control for complete sloshing suppression, *IEEE Trans. or Control System Technology*, Vol.9, No.3, pp.483-493.

Y.Sugimoto, K.Yano and K.Terashima (2002), Liquid level control of automatic pouring robust by two-degree-of-freedom control, *IFAC 15th Triennial World Congress, Barcelona.*

K.Yano, N.Oguro and K.Terashima (2001), Starting control with vibration damping by Hybrid Shape approach considering time and frequency specitictions, *Trans of SICE*, Vol.37, No.5, pp.403-410

IFAC
Publications
www.elsevier.com/locate/ifac

OPTIMAL PREFILTERING IN ITERATIVE FEEDBACK TUNING [1]

R. Hildebrand,* A. Lecchini, G. Solari,** M. Gevers****

* *Center for Operations Research and Econometrics (CORE)*
Université Catholique de Louvain
Voie du Roman Pays, 34
B-1348 Louvain-la-Neuve, Belgium
{hildebrand}@core.ucl.ac.be
** *Center for Systems Engineering and Applied Mechanics*
(CESAME)
Université Catholique de Louvain
Bâtiment Euler, 4 Av. Georges Lemaître,
B-1348 Louvain-la-Neuve, Belgium
{lecchini, solari, gevers}@auto.ucl.ac.be

Abstract: Iterative Feedback Tuning (IFT) is a widely used procedure for controller tuning. It is a sequence of iteratively performed special experiments on the plant interlaced with periods of data collection under normal operating conditions. In this paper we derive the asymptotic convergence rate of IFT for disturbance rejection, which is one of the main fields of application. Further we present a method to improve the convergence of IFT by prefiltering the input data for the special experiment. At each iteration step the optimal prefilter is computed from data collected under normal operating conditions of the plant.

Keywords: Iterative Feedback Tuning, filter design

1. INTRODUCTION

Iterative Feedback Tuning (IFT) is a data based method for the tuning of restricted complexity controllers. It has proved to be very effective in practice and is now widely used in process control, often for disturbance rejection. Following the original formulation of the method in (Hjalmarsson *et al.*, 1998) many improvements and modifications of IFT have been suggested. The reader is referred to (Hjalmarsson, 2002) for a recent overview.

The objective of IFT is to minimize a quadratic performance criterion. IFT is a stochastic gradient descent scheme in a finitely parameterized controller space. The gradient of the cost function at each step is estimated from data. These data are collected with the actual controller in the loop. Under suitable assumptions the algorithm converges to a local minimum of the performance criterion. One of the advantages of IFT is that most data are collected while the process runs under normal operating conditions. These data are then used to design a special experiment, which yields a noisy, but unbiased, estimate of the cost function gradient. This gradient estimate is used to perform the next descent step in controller space. For more details of the procedure see (Hjalmarsson *et al.*, 1998).

In this and in the companion paper (Hildebrand *et al.*, 2002) we focus on IFT for disturbance rejection. We provide an analytic expression for the asymptotic convergence rate of the algorithm, as the number of data collected in each experiment tends to infinity.

[1] Paper supported by the the Belgian Programme on Interuniversity Poles of Attraction initiated by the Belgian State, Prime Minister's Office for Science, Technology and Culture and the European Research Network on System Identification (ERNSI) funded by the European Union. The scientific responsibility rests with its authors.

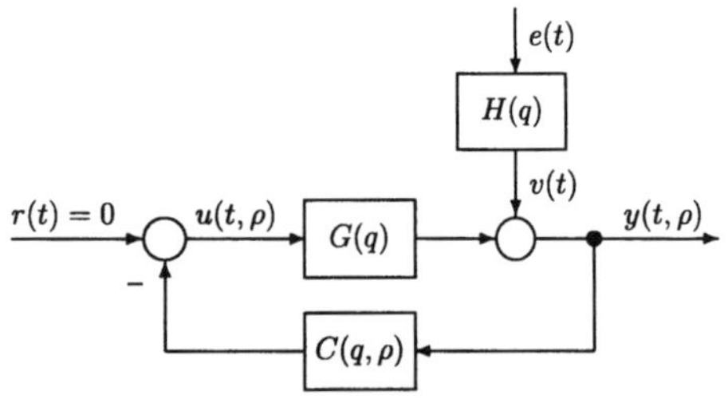

Fig. 1. The control system under normal operating conditions.

The convergence rate depends on the covariance of the gradient estimates. Therefore, the calculation of this covariance is a part of our analysis.

In (Hjalmarsson *et al.*, 1998) it was proposed to reduce the error in the gradient estimate by prefiltering the reference input data for the special experiment. Our second contribution is to optimize the corresponding prefilter with respect to both the convergence speed of the procedure and the accuracy of the finally obtained value. An analytical expression for the optimal prefilter is given. It depends on certain characteristics of the unknown process. However, in the spirit of IFT, these characteristics can be estimated from data collected under normal operating conditions. Thus the computation of the optimal prefilter does not necessitate any special experiment on the process and hence does not impose any additional cost.

The remainder of the paper is structured as follows. In the next section we summarize the details of the IFT algorithm for disturbance rejection. In Section 3 we derive an expression for the asymptotic convergence rate dependent on the covariance of the gradient estimates. In Section 4 the asymptotic expression of this covariance is calculated. This enables us in Section 5 to establish a design criterion for the optimization of the algorithm. We compute the optimal prefilter which minimizes the design criterion. In Section 6 we demonstrate the gains in convergence speed and accuracy in a simulation example. Finally, we draw some conclusions in the last section.

2. IFT FOR DISTURBANCE REJECTION

In this section we review the IFT method for the disturbance rejection problem with a classical LQ criterion. For a more general and detailed presentation of IFT the reader is referred to (Hjalmarsson *et al.*, 1998).

Consider a SISO discrete time system described by

$$y(t) = G(q)u(t) + v(t)\,, \qquad (1)$$

where $y(t)$ is the output, $u(t)$ is the input, $G(q)$ is a linear time-invariant transfer function, with q being the shift operator, and $v(t) = H(q)e(t)$ is the process disturbance. Here $H(q)$ is a monic, stable and inversely stable transfer function and $e(t)$ is zero mean white noise with variance σ^2. The transfer functions $G(q)$ and $H(q)$ are unknown.

Consider the feedback loop around $G(q)$ depicted in Figure 1, where $C(q,\rho)$ is a one-degree-of-freedom controller belonging to a parameterized set of controllers with parameter $\rho \in \mathbf{R}^n$. The transfer function from $v(t)$ to $y(t,\rho)$ is named sensitivity function and is denoted by $S(q,\rho)$. We assume that in the control system of Figure 1 the reference signal $r(t)$ is set at zero under normal operating conditions. Our goal is to tune the controller $C(q,\rho)$ so that the variance of the noise-driven closed loop output is as small as possible subject to a penalty on the control effort. Thus we want to find a minimizer for the cost function

$$J(\rho) = \frac{1}{2}\mathbf{E}\left[y(t,\rho)^2 + \lambda u(t,\rho)^2\right]\,, \qquad (2)$$

where $\lambda \geq 0$ is chosen by the user.

The IFT method yields an approximate solution to the above problem. IFT is based on the possibility of obtaining an unbiased estimate of the gradient $\frac{\partial J}{\partial \rho}(\rho)$ of the cost function at $\rho = \rho_n$ from data collected from the closed-loop system with the controller $C(\rho_n)$ operating on the loop. The cost function $J(\rho)$ can then be minimized with an iterative stochastic gradient descent scheme of Robbins-Monro type (Blum, 1954). In the scheme a sequence of controllers $C(q,\rho_n)$ is computed and applied to the plant. In the n-th iteration step, data obtained from the system with the controller $C(\rho_n)$ operating on the loop are used to construct the next parameter vector ρ_{n+1}.

The data based iterative procedure is as follows.

IFT procedure

1. Collect a sequence $\{u^1(t,\rho_n),\, y^1(t,\rho_n)\}$ with $t = 1,\ldots,N$ of input-output data under normal operating conditions, i.e. without reference signal.

2. Collect a sequence $\{u^2(t,\rho_n),\, y^2(t,\rho_n)$ with $t = 1,\ldots,N$ of input-output data by performing a special experiment with reference signal

$$r_n^2(t) = -K_n(q)y^1(t,\rho_n)$$

where $K_n(q)$ is any stable minimum-phase prefilter.

3. Construct the estimates of the gradients of $u^1(t,\rho_n)$ and $y^1(t,\rho_n)$ as

$$est\left[\frac{\partial u^1}{\partial \rho}(t,\rho_n)\right] = \frac{1}{K_n(q)}\frac{\partial C}{\partial \rho}(q,\rho_n)\,u^2(t,\rho_n)\,,$$

$$est\left[\frac{\partial y^1}{\partial \rho}(t,\rho_n)\right] = \frac{1}{K_n(q)}\frac{\partial C}{\partial \rho}(q,\rho_n)\,y^2(t,\rho_n)\,.$$

4. Form the estimate of the gradient of $J(\rho)$ at ρ_n as

$$est_N\left[\frac{\partial J}{\partial \rho}(\rho_n)\right] = \frac{1}{N}\sum_{t=1}^{N}\left[y^1(t,\rho_n)\right.$$

$$\left.\times est\left[\frac{\partial y^1}{\partial \rho}(t,\rho_n)\right] + \lambda u^1(t,\rho_n) est\left[\frac{\partial u^1}{\partial \rho}(t,\rho_n)\right]\right]$$

5. Calculate the new parameter vector ρ_{n+1} according to

$$\rho_{n+1} = \rho_n - \gamma_n R_n^{-1}\, est_N\left[\frac{\partial J}{\partial \rho}(\rho_n)\right]$$

where γ_n is a positive step size and R_n is a positive definite matrix.

In the above procedure we assume negligible measurement noise and independence between the disturbance realizations $v_n^1(t)$, in the first experiment, and $v_n^2(t)$, in the second experiment. Under these assumptions the estimate of the gradient turns out to be unbiased (Hjalmarsson *et al.*, 1998). The sequences γ_n and R_n are basically left to the choice of the user. The matrix R_n should be an approximation of the Hessian of the cost function in ρ_n. A biased estimate of the Hessian, obtained from data, has been proposed in (Hjalmarsson *et al.*, 1998).

3. ANALYSIS OF CONVERGENCE RATE IN IFT

In this section we quantify the effect of the variability of the gradient estimate on the asymptotic convergence rate of the algorithm.
The proposition below describes the asymptotic behavior of the sequence ρ_n. It follows from a general proposition on the convergence rate of for Robbins-Monro processes as can be found in (Nevelson and Khasminskii, 1976).

Proposition 1. Assume that the sequence ρ_n converges to a local isolated minimum $\bar{\rho}$ of $J(\rho)$ (the reader is referred to (Hildebrand *et al.*, 2002) for the conditions of convergence). Let H be the Hessian of $J(\rho)$ at $\rho = \bar{\rho}$. Suppose further that the following conditions hold.

1. The sequence γ_n of step sizes is given by $\gamma_n = \frac{a}{n}$, where a is a positive constant. There exists an index $\bar{n}$ and a matrix R such that $R_n = R$ for all $n > \bar{n}$.
2. The matrix $A = \frac{1}{2}I - aR^{-1}H$ is stable, i.e. the real parts of its eigenvalues are negative.
3. The covariance matrix $\mathbf{Cov}\left[est_N\left[\frac{\partial J}{\partial \rho}(\rho)\right]\right]$ at $\rho = \bar{\rho}$ is positive definite.

Then the sequence of random variables $s_n = \sqrt{n}(\rho_n - \bar{\rho})$ converges in distribution to a normally distributed zero mean random variable with covariance matrix

$$\Sigma = \int_0^\infty e^{At} R^{-1}\mathbf{Cov}\left[est_N\left[\frac{\partial J}{\partial \rho}(\bar{\rho})\right]\right] \times R^{-1} e^{A^T t} dt,$$

i.e. $\sqrt{n}(\rho_n - \bar{\rho}) \xrightarrow{D} \mathcal{N}(0, \Sigma)$. □

Proposition 1 shows that the asymptotic accuracy of the estimate crucially depends on the distribution of the error on the gradient. This distribution in turn can be influenced by the prefilters $K_n(q)$. Before turning to the question of designing the filters $K_n(q)$ for optimal accuracy, we analyze in detail how the covariance of the gradient estimate depends on $K_n(q)$. This will be done in thc next section.

4. THE COVARIANCE OF THE GRADIENT ESTIMATE

This section is devoted to finding an explicit expression for the covariance of $est_N\left[\frac{\partial J}{\partial \rho}(\rho_n)\right]$. We will show that this covariance can be written as the sum of two terms. These two contributions originate in the variability of the noise realizations in the first and second experiment of iteration n, respectively. Consequently, the first term is independent of the prefilter $K_n(q)$, because the filter is applied only to the reference signal for the second experiment. However, the second term can be influenced by the choice of this prefilter.
It can be shown that the estimates of the gradients of $u^1(t, \rho_n)$ and $y^1(t, \rho_n)$ obtained in Step 3 of the IFT procedure are corrupted by the realization $v_n^2(t)$ of the noise in the second experiment as follows

$$est\left[\frac{\partial u^1}{\partial \rho}(t, \rho_n)\right] = \frac{\partial u^1}{\partial \rho}(t, \rho_n) - \frac{S(q, \rho_n)}{K_n(q)} C(q, \rho_n) \frac{\partial C}{\partial \rho}(q, \rho_n)\, v_n^2(t),$$

$$est\left[\frac{\partial y^1}{\partial \rho}(t, \rho_n)\right] = \frac{\partial y^1}{\partial \rho}(t, \rho_n) + \frac{S(q, \rho_n)}{K_n(q)} \frac{\partial C}{\partial \rho}(q, \rho_n)\, v_n^2(t).$$

Therefore we can separate $est_N\left[\frac{\partial J}{\partial \rho}(\rho_n)\right]$ as

$$est_N\left[\frac{\partial J}{\partial \rho}(\rho_n)\right] = S_N(\rho_n) + E_N(\rho_n),$$

$$S_N(\rho_n) = \frac{1}{N}\sum_{t=1}^{N}\left[y^1(t, \rho_n)\frac{\partial y^1}{\partial \rho}(t, \rho_n) + \lambda u^1(t, \rho_n)\frac{\partial u^1}{\partial \rho}(t, \rho_n)\right]$$

$$E_N(\rho_n) = \frac{1}{N}\sum_{t=1}^{N}\left[y^1(t, \rho_n)\left[\frac{S(q, \rho_n)}{K_n(q)} \frac{\partial C}{\partial \rho}(q, \rho_n)\, v_n^2(t)\right] + \lambda u^1(t, \rho_n) \times \left[-\frac{C(q, \rho_n) S(q, \rho_n)}{K_n(q)} \frac{\partial C}{\partial \rho}(q, \rho_n)\, v_n^2(t)\right]\right].$$

The term $S_N(\rho_n)$ corresponds to the sampled estimate of the gradient of $J(\rho)$. This term is entirely dependent on the realization $v_n^1(t)$ of the noise in the first experiment. The second term $E_N(\rho_n)$ is an error due to the corruption of the estimates of the gradients of $u^1(t,\rho_n)$ and $y^1(t,\rho_n)$ by $v_n^2(t)$. The estimate $est_N\left[\frac{\partial J}{\partial \rho}(\rho_n)\right]$ turns out to be unbiased under the assumption that the two experiments in the algorithm are sufficiently separated in time. In fact, under this assumption, the realization $v_n^2(t)$ can be considered as being independent of the signals coming form the first experiment and therefore the mean of $E_N(\rho_n)$ is zero. The dispersion of $est_N\left[\frac{\partial J}{\partial \rho}(\rho_n)\right]$ is described in the following proposition.

Proposition 2.
1. The following relation holds

$$\mathbf{Cov}\left[est_N\left[\frac{\partial J}{\partial \rho}(\rho_n)\right]\right] = \mathbf{Cov}\left[S_N(\rho_n)\right] + \mathbf{Cov}\left[E_N(\rho_n)\right].$$

2. The following asymptotic frequency-domain expression of $\mathbf{Cov}\left[E_N(\rho_n)\right]$ holds

$$\lim_{N\to\infty} N\mathbf{Cov}\left[E_N(\rho_n)\right] = \frac{\sigma^4}{2\pi}\int_{-\pi}^{\pi} \frac{1}{|K_n(e^{j\omega})|^2}|S(e^{j\omega},\rho_n)H(e^{j\omega})|^4 \left[1 + \lambda|C(e^{j\omega},\rho_n)|^2\right]^2 \times \frac{\partial C}{\partial \rho}(e^{j\omega},\rho_n)\frac{\partial C^*}{\partial \rho}(e^{j\omega},\rho_n)\, d\omega.$$

3. Under the additional assumption that the 4th order cumulants of the noise v are zero (e.g the noise is normally distributed), the following asymptotic frequency-domain expression of $\mathbf{Cov}\left[S_N(\rho_n)\right]$ holds

$$\lim_{N\to\infty} N\mathbf{Cov}\left[S_N(\rho_n)\right] = 2\cdot\frac{\sigma^4}{2\pi}\int_{-\pi}^{\pi} |S(e^{j\omega},\rho_n)H(e^{j\omega})|^4 \times \mathcal{R}e\left\{\left[G(e^{j\omega}) - \lambda\bar{C}(e^{j\omega},\rho_n)\right]S(e^{j\omega},\rho_n)\frac{\partial C}{\partial \rho}(e^{j\omega},\rho_n)\right\} \times \mathcal{R}e\left\{\left[G(e^{j\omega}) - \lambda\bar{C}(e^{j\omega},\rho_n)\right]S(e^{j\omega},\rho_n) \times \frac{\partial C}{\partial \rho}(e^{j\omega},\rho_n)\right\}^T d\omega.$$

Proof See (Hildebrand *et al.*, 2002). □

In Proposition 2 it has been shown that the covariance of the gradient estimate can be represented as the sum of the covariances of the separate contributions $S_N(\rho_n)$ and $E_N(\rho_n)$ (i.e. $S_N(\rho_n)$ and $E_N(\rho_n)$ are uncorrelated). Both $\mathbf{Cov}\left[S_N(\rho_n)\right]$ and $\mathbf{Cov}\left[E_N(\rho_n)\right]$ decay asymptotically proportionally to $1/N$ as the number of data tends to infinity. Their asymptotic frequency domain expressions as $N\to\infty$ have been given.

By Proposition 1, $\mathbf{Cov}\left[est_N\left[\frac{\partial J}{\partial \rho}(\rho_n)\right]\right]$ enters linearly into the asymptotic covariance Σ of the estimated controller parameter value. Hence this covariance can as well be expressed as a sum of two terms, dependent on the covariances of $S_N(\rho_n)$ and $E_N(\rho_n)$, respectively. Since $S_N(\rho_n)$ does not depend on $K_n(q)$, the corresponding term in Σ can be regarded as constant for the purpose of designing the prefilter $K_n(q)$. This observation leads us to the design of an optimal prefilter.

5. DESIGN OF THE OPTIMAL PREFILTER

We are now ready to specify the criterion for the design of the prefilter $K_n(q)$. In this section we state this criterion and deliver the expressions of the corresponding optimal prefilter. We assume that the current controller is near the optimal one. Then the convergence rate of the procedure is measured by the accuracy of the estimate. Therefore, in order to construct a design criterion for the prefilter, one can employ the asymptotic results on the accuracy given in Section 3.

5.1 *The design criterion*

Let the sequence γ_n of step lengths in the IFT procedure be proportional to $1/n$, i.e. $\gamma_n = \frac{a}{n}$. Define $\Delta\bar{\rho}_n = \rho_n - \bar{\rho}$, where $\bar{\rho}$ is the optimal parameter. Let us take $\mathbf{E}[J(\rho_n)] - J(\bar{\rho})$ as a measure of quality of the controller $C(\rho_n)$. Expanding $J(\rho)$ into a Taylor series around $\bar{\rho}$ and retaining only terms up to the second order, we obtain $\mathbf{E}\left[\Delta\bar{\rho}_n^T\mathbf{H}(\bar{\rho})\Delta\bar{\rho}_n\right]$ as an approximation of $\mathbf{E}[J(\rho_n)] - J(\bar{\rho})$. Here $\mathbf{H}(\rho)$ denotes the Hessian of $J(\rho)$.
Following Proposition 1, $\sqrt{n}\Delta\bar{\rho}_n$ is asymptotically normally distributed with zero mean and covariance

$$\Sigma = a^2\int_0^\infty e^{At}R^{-1}\mathbf{Cov}\left[est_N\left[\frac{\partial J}{\partial \rho}(\bar{\rho})\right]\right] \times \left[R^{-1}\right]^T e^{A^T t}dt,$$

where $R = \lim_{n\to\infty} R_n$ and $A = \frac{1}{2}I - aR^{-1}\mathbf{H}(\bar{\rho})$. Let us now assume $R = \mathbf{H}(\bar{\rho})$, i.e. we consider a Gauss-Newton scheme. Then we obtain $A = \left(\frac{1}{2} - a\right)I$ and

$$\Sigma = \frac{a^2}{2a-1}R^{-1}\mathbf{Cov}\left[est_N\left[\frac{\partial J}{\partial \rho}(\bar{\rho})\right]\right]\left[R^{-1}\right]^T \quad (3)$$

This yields

$$\lim_{n\to\infty} n\mathbf{E}\left[\Delta\bar{\rho}_n^T\mathbf{H}(\bar{\rho})\Delta\bar{\rho}_n\right] = \frac{a^2}{2a-1} \times \mathrm{Trace}\left[\mathbf{Cov}\left[est_N\left[\frac{\partial J}{\partial \rho}(\bar{\rho})\right]\right]\left[R^{-1}\right]^T\right].$$

We shall take this expression as the criterion to be minimized for the design of the optimal prefilter.
There are different methods to satisfy the condition $R = \mathbf{H}(\bar{\rho})$. A classical method to obtain a sequence of estimated matrices R_n which converges to the Hessian is to fit a regression model using the gradient estimates obtained in the previous iterations. The reader is referred to (Wei, 1985; Yin, 1988).
In practice, in order to use (4) as a criterion for the design of the optimal filter, at the iteration n one has to replace the optimal parameter $\bar{\rho}$ on the right-hand side by the current parameter ρ_n, since (as it will be shown in Subsection 5.2) the prefilter is estimated from data obtained under the current operating conditions. In the same way, R has to be replaced by the current estimate of the Hessian R_n. This estimate could be the (biased) data-based estimate of the Hessian proposed in (Hjalmarsson *et al.*, 1998) which is constructed, at each step, with the data of the first and second experiment. However, in order to not violate a certain condition for the convergence of ρ_n (see (Hildebrand *et al.*, 2002)), the estimated R_n has to be uncorrelated with the noise realizations $v_n^1(t)$ and $v_n^2(t)$. Therefore it has to be calculated by using the data of iteration $n-1$. These approximations are reasonable, because we assume that the current controller is near the optimal controller.

5.2 The optimal prefilter

The quantity that has to be optimized in the design of the prefilter is thus (4) with $\bar{\rho}$ replaced by ρ_n. The optimal prefilter minimizes the weighted trace of the covariance of the gradient estimate. In order to obtain a bounded solution we have to restrict the gain of the prefilter. A straightforward constraint is a bound on the energy of the reference signal $r_n^2(t)$, i.e. on the input of the second experiment. This bound represents the level of acceptable perturbation to the normal operating conditions during the second experiment at each step. We thus arrive at the following optimization problem:

$$K_n^{opt} = \arg\min_K \text{Trace}\left[R_n^{-1}\mathbf{Cov}\left[est_N\left[\frac{\partial J}{\partial \rho}(\rho_n)\right]\right]\right]$$
$$\text{subject to} \quad \mathbf{Var}\left[r_n^2(t)\right] \leq \alpha,$$

where α is selected by the user. By Proposition 2, and recalling that $\mathbf{Cov}\left[S_N(\rho_n)\right]$ does not depend on the prefilter, we can rewrite the problem as follows:

$$K_n^{opt} = \arg\min_K \text{Trace}\left[R_n^{-1}\mathbf{Cov}\left[E_N(\rho_n)\right]\right] \quad (4)$$
$$\text{subject to} \quad \mathbf{Var}\left[r_n^2(t)\right] \leq \alpha.$$

The explicit solution of this problem is characterized by the following proposition.

Proposition 3. The optimal prefilter solving (4) satisfies the following relation:

$$|K_n^{opt}(e^{j\omega})|^4 = const \cdot |S(e^{j\omega}, \rho_n)H(e^{j\omega})|^2[1 + \lambda|C(e^{j\omega}, \rho_n)|^2]^2 \text{ Trace}\left\{R_n^{-1}\frac{\partial C}{\partial \rho}(e^{j\omega}, \rho_n) \times \frac{\partial C^*}{\partial \rho}(e^{j\omega}, \rho_n)\right\}, \quad (5)$$

where the constant is determined by the design restriction.

Proof. See (Hildebrand *et al.*, 2002). □

In order to compute the optimal prefilter in practice, one needs an estimate of the unknown spectral density $|S(e^{j\omega}, \rho_n)H(e^{j\omega})|^2$ of the signal $y(t, \rho_n)$ which is the output of the plant under normal operating conditions, i.e. with zero reference signal. The estimate can be obtained with standard techniques in the time or in the frequency domain (Ljung, 1999; Pintelon and Schoukens, 2001). Note that since the data needed to estimate this quantity do not stem from a special experiment they are available in large amounts. In fact periods of normal operating conditions can be interlaced with the IFT special experiments. By assuming these periods to be much longer than the length of the special experiment from which the gradient is estimated, the contribution of the variability in the estimate of $|S(e^{j\omega}, \rho_n)H(e^{j\omega})|^2$ to the variability of the gradient estimate can be considered as being negligible.
Having an estimate of $|S(e^{j\omega}, \rho_n)H(e^{j\omega})|^2$ one can construct the magnitude of the optimal prefilter by calculating the 4-th root of the right-hand side of (5). Then, there exist standard tools to approximate a given magnitude function by a stable minimum phase filter.

6. SIMULATION EXAMPLE

Consider the system described by

$$G(q) = \frac{q^{-1} - 0.5q^{-2}}{1 - 0.3q^{-1} - 0.28q^{-2}},$$
$$H(q) = \frac{1}{1 + 0.9q^{-1}}$$

with $\sigma^2 = 1$. Let the class of controllers be $C(q, \rho) = \rho^1 + \rho^2 q^{-1}$ and set $\lambda = 0.6$ in (2). The (local) minimizer $\bar{\rho} = [-0.69058\ 0.33105]$ has been found numerically. Let us assume that the constraint on the reference signal $r_n^2(t)$ during the IFT procedure is that this signal has to have one half the energy of the output of the first experiment. In the following we will quantify the performance improvement between the trivial constant filter satisfying the energy constraint and the optimal filter given by (5) when the criterion $\mathbf{E}\left[\Delta\bar{\rho}_n^T \mathbf{H}(\bar{\rho})\Delta\bar{\rho}_n\right]$ is used. We run the IFT procedure with experiment length $N = 512$, step sizes $\gamma_n = a/n$ with $a = 1$ and $R_n = \mathbf{H}(\bar{\rho})$.
Using the asymptotic approximations of the covariance of the gradient estimate given in Section 3 we can

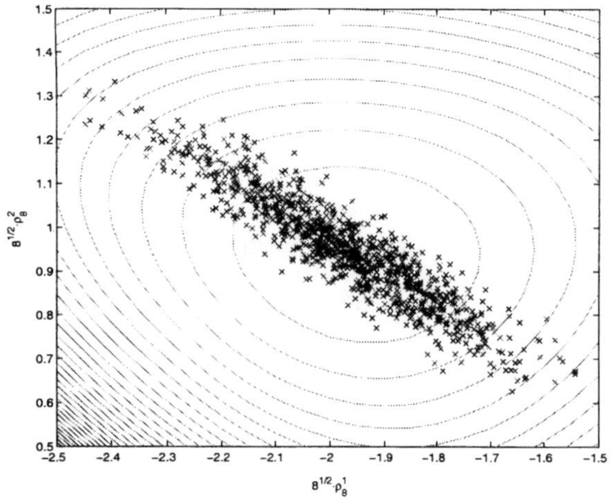

Fig. 2. The parameter $\bar{\rho}$ (•), the 1024 parameters ρ_8 obtained using the constant filter (×) and the contour lines of $J(\rho)$.

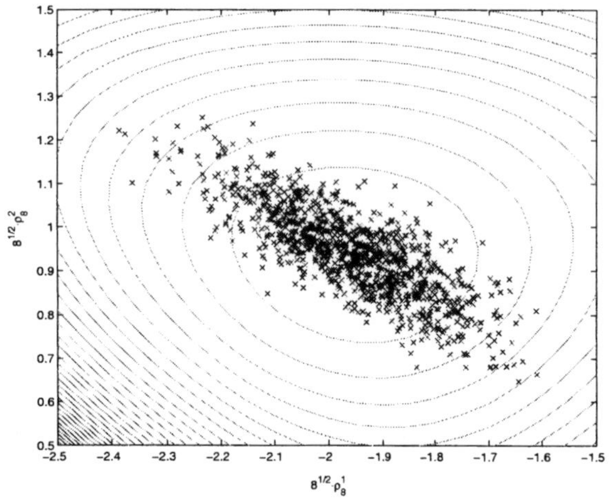

Fig. 3. The parameter $\bar{\rho}$ (•), the 1024 parameters ρ_8 obtained using the optimal filter (×) and the contour lines of $J(\rho)$.

find the asymptotic covariance of $\sqrt{n}\Delta\bar{\rho}_n$ according to (3). Then, for the constant prefilter we obtain that the asymptotic value of $n\mathbf{E}\left[\Delta\bar{\rho}_n^T\mathbf{H}(\bar{\rho})\Delta\bar{\rho}_n\right]$ is $8.39 \cdot 10^{-2}$. For the optimal prefilter the asymptotic value of $n\mathbf{E}\left[\Delta\bar{\rho}_n^T\mathbf{H}(\bar{\rho})\Delta\bar{\rho}_n\right]$ is $5.79 \cdot 10^{-2}$. The improvement in the value of $\mathbf{E}\left[\Delta\bar{\rho}_n^T\mathbf{H}(\bar{\rho})\Delta\bar{\rho}_n\right]$ between the two cases is 31%.

The above theoretical values can be illustrated by a Monte-Carlo simulation. The parameter vector ρ_8 has been extracted 1024 times. The extractions have been performed by starting the IFT procedure at $\rho_0 = \bar{\rho}$, in order to eliminate the transient effect of the initial condition, and then running 8 steps up to the parameter vector ρ_8. The 1024 parameter vectors obtained this way are shown in Figure 2 for the case of the constant prefilter. The corresponding sampled estimate of $8 \cdot \mathbf{E}\left[\Delta\bar{\rho}_8^T\mathbf{H}(\bar{\rho})\Delta\bar{\rho}_8\right]$ was $8.00 \cdot 10^{-2}$. The parameter vectors obtained for the case of the optimal prefilter are shown in Figure 3. In this case, the corresponding sampled estimate of $8 \cdot \mathbf{E}\left[\Delta\bar{\rho}_8^T\mathbf{H}(\bar{\rho})\Delta\bar{\rho}_8\right]$ was $5.48 \cdot 10^{-2}$. The estimated improvement between the two cases hence equals 31.5%, as predicted by the theoretical calculations made above.

7. CONCLUSIONS

In this contribution we have investigated the convergence properties of the IFT algorithm for disturbance rejection and shown how its performance can be improved by prefiltering the reference input for the special experiment.

The asymptotic convergence rate of the algorithm for a step size sequence proportional to $1/n$ was quantified as a function of the covariance of the gradient estimate at the optimum (Proposition 1). An expression for this covariance with an arbitrary controller in the loop was given (Proposition 2).

We investigated how to optimize the accuracy of the gradient estimate and the asymptotic convergence rate of the algorithm by a prefilter in the special experiment. An expression for the optimal prefilter was derived for constrained reference input energy (Proposition 3). It was shown how to construct the optimal prefilter from data collected during normal operating conditions of the process.

The effect of the prefilter amounts to decreasing the contribution of the process noise in the special experiment to the error in the gradient estimate at each step. Hence the prefilter will be the more effective the bigger the contribution of the process noise $v_n^2(t)$ is. This is the case if a low energy level of the reference signal for the special experiment is required.

REFERENCES

Blum, Julius R. (1954). Multidimensional stochastic approximation methods. *Annals of Mathematical Statistics* **25**, 737–744.

Hildebrand, R., A. Lecchini, G. Solari and M. Gevers (2002). Convergence analysis and optimal prefiltering in iterative feedback tuning. *Submitted.*

Hjalmarsson, H. (2002). Iterative Feedback Tuning - an overview. *International Journal of Adaptive Control and Signal Processing* **16**(5), 373–395.

Hjalmarsson, H., M. Gevers, S. Gunnarson and O. Lequin (1998). Iterative Feedback Tuning: theory and applications. *IEEE Control Systems* **18**(4), 26–41.

Ljung, L. (1999). *System Identification: Theory for the user*. Prentice Hall.

Nevelson, M. B. and R. Z. Khasminskii (1976). *Stochastic Approximation and Recursive Estimation*. Vol. 47 of *Translations of Mathematical Monographs*. third printing 2000 ed.. American Mathematical Society.

Pintelon, R. and J. Schoukens (2001). *System Identification: a frequency domain approach*. IEEE Press.

Wei, C.Z. (1985). Asymptotic properties of least-squares estimates in stochastic regression models. *The Annals of Statistics* **13**(4), 1498–1508.

Yin, G. (1988). A stopped stochastic approximation algorithm. *Systems and Control Letters* **11**, 107–115.

IFAC Publications
www.elsevier.com/locate/ifac

IDENTIFICATION OF PERFORMANCE LIMITATIONS IN CONTROL USING GENERAL SISO MODELS

Jonas Mårtensson and Håkan Hjalmarsson *

* *Department of Signals, Sensors and Systems, KTH*
S-100 44 Stockholm, Sweden
jonas.martensson@s3.kth.se

Abstract: Previously it has been shown for FIR and ARX models that the variance of identified non-minimum phase zeros depends very little on the model order. In this paper that result is extended to more general SISO model structures. An asymptotic, in the model order and the number of data, expression for the variance of non-minimum phase zeros is derived. The relevance of this expression for finite model orders and number of data is examined. *Copyright © 2003 IFAC*

Keywords: System identification, Performance limitations, Non-minimum phase systems, Transmission zeros, Variance, Asymptotic properties.

1. INTRODUCTION

Identification for control has received considerable interest in recent years. See (Gevers, 1993) for overviews of the activity in the early 1990's. Much of the attention has been on closed loop identification, see (Van den Hof and Schrama, 1995) and (Forssell and Ljung, 1999) and more recently on model validation/unfalsification (Bombois *et al.*, 1999), (Woodley *et al.*, 1998).

Experiment design in the context of control design has also received renewed interest, see e.g. (Hjalmarsson *et al.*, 1996), (Forssell and Ljung, 1998), (Lindqvist and Hjalmarsson, 2000), (Cooley *et al.*, 1998) and (Lindqvist, 2001).

An intrinsic problem in experiment design is that the optimal design depends on the system which is to be identified. Hence, even though these designs may be used to get intuition for how to design the identification experiment, they are in general infeasible.

Hence suboptimal methods must be developed. It is generally acknowledged that an accurate model is needed around the cross-over frequency *of the loop gain*. Since the loop gain depends on the yet to be designed controller, this frequency region is in generally unknown. However, for non-minimum phase systems, the non-minimum phase zeros restrict the achievable bandwidth, see e.g. (Freudenberg and Looze, 1998) or (Skogestad and Postlethwaite, 1996). A real single non-minimum phase zero at z restricts the bandwidth to approximately $z/2$. Hence, if the non-minimum phase zeros were known, the experiment design problem would be simplified considerably. Knowledge of the performance limitations would also ease the task of deciding on model structure, model order, noise model and pre-filters since one then knows the important frequency range.

Spurred by this observation it was in earlier work, (Lindqvist, 2001) and (Hjalmarsson and Lindqvist, 2002), established that the variance of estimated non-minimum phase zeros is not subject to the usual increase in variance when the model order is increased. For FIR and ARX models it was shown that as the model order is increased, the asymptotic variance (normalized with N) of an estimate $z_k(\widehat{\theta}_N^n)$ of a non-minimum phase zero at z_k^o converges to a finite limit given by the following expression:

$$\lim_{n\to\infty}\lim_{N\to\infty} N\mathbf{E}\left|z_k(\widehat{\theta}_N^n) - z_k^o\right|^2 = \frac{\sigma^2 |z_k^o|^2}{|b_0^o|^2(1-|z_k^o|^{-2})\left|Q(z_k^o)\right|^2 \prod_{i\neq k}^{n_o}\left|1-\frac{z_i^o}{z_k^o}\right|^2} \quad (1)$$

In this paper this result is extended to general single input/single output model structures.

In Section 2, the assumptions on which the following results rely on are stated. In Section 3, the settings of the parameter estimation are discussed. In Section 4 the system and model zeros are defined. The Asymptotic (in the number of data) variance of the estimated zeros is discussed in Section 5. Section 6 contains he main result, where an explicit expression for the asymptotic (in number of data and in model order) variance of non-minimum phase zeros is derived. The results are interpreted in Section 7. The results are verified, or at least exemplified, by some simulations in Section 8. Some conclusions are given in Section 9.

2. ASSUMPTIONS

The following assumptions on the system and identification procedure are used throughout this paper.

A1: The system is described by the Box-Jenkins structure, i.e.,

$$y(t) = \frac{B(q,\theta_o)}{F(q,\theta_o)}u(t) + \frac{C(q,\theta_o)}{D(q,\theta_o)}e_o(t) \quad (2)$$

where

$$B(q,\theta_o) = b_0^o + \sum_{k=1}^{n_o^b} b_k^o q^{-k},$$
$$F(q,\theta_o) = 1 + \sum_{k=1}^{n_o^f} f_k^o q^{-k},$$
$$C(q,\theta_o) = 1 + \sum_{k=1}^{n_o^c} c_k^o q^{-k},$$
$$D(q,\theta_o) = 1 + \sum_{k=1}^{n_o^d} d_k^o q^{-k}. \quad (3)$$

It is assumed that F, C and D are minimum phase, but that condition can be relaxed as long as the one-step ahead predictor is stable.
Note that the output-error structure is the special case where $C(q,\theta_o) = D(q,\theta_o) = 1$.

A2: The input is generated as $u(t) = Q(q)v(t)$ where $Q(q)$ is a minimum phase filter with no zeros on the unit circle and $v(t)$ is zero mean white noise with variance 1.

A3: The system noise $e_o(t)$ is zero mean white noise with variance σ^2.

A4: The model of the system is described by

$$y(t) = \frac{B(q,\theta)}{F(q,\theta)}u(t) + \frac{C(q,\theta)}{D(q,\theta)}e(t) \quad (4)$$

where B, F, C and D are defined analogously to (3), and $e(t)$ is zero mean white noise with variance σ^2. The true system belongs to the model set, i.e. $n_f \geq n_f^o$, $n_b \geq n_b^o$, $n_d \geq n_d^o$, $n_c \geq n_c^o$. Throughout this paper, n will denote the model order $\{n_b, n_f, n_c, n_d\}$ and n_o will denote the system order $\{n_b^o, n_f^o, n_c^o, n_d^o\}$. The parameter vector θ^n is defined as

$$\theta^n = \left[b_0 \cdots b_{n_b}\ f_1 \cdots f_{n_f}\ c_1 \cdots c_{n_c}\ d_1 \cdots d_{n_d}\right]^T.$$

When the model is over-parameterized, i.e. when $n > n_o$, the *true* parameter vector will be denoted

$$\theta_o^n = \left[b_0^o \cdots b_{n_b^o}^o\ 0 \cdots 0\ f_1^o \cdots f_{n_f^o}^o\ 0 \cdots 0\right.$$
$$\left. c_1^o \cdots c_{n_c^o}^o\ 0 \cdots 0\ d_1^o \cdots d_{n_d^o}^o\ 0 \cdots 0\right]^T.$$

Note that in the output-error case the model is described by

$$y(t) = \frac{B(q,\theta)}{F(q,\theta)}u(t) + e(t) \quad (5)$$

and the parameter vector is

$$\theta^n = \left[b_0 \cdots b_{n_b}\ f_1 \cdots f_{n_f}\right]^T.$$

3. PARAMETER ESTIMATION

The parameters are estimated with the prediction error method (PEM) using a least mean square (LMS) criterion to minimize the prediction error. The one-step-ahead predictor is used:

$$\widehat{y}(t|\theta) = \frac{D(q,\theta)B(q,\theta)}{C(q,\theta)F(q,\theta)}u(t) + \frac{C(q,\theta)-D(q,\theta)}{C(q,\theta)}y(t). \quad (6)$$

Let the prediction error be denoted $\varepsilon(t,\theta) = y - \widehat{y}(t|\theta)$. Then the least mean square-estimate of the parameters is

$$\widehat{\theta}_N = \arg\min_\theta \frac{1}{N}\sum_{t=1}^{N}\frac{1}{2}\varepsilon^2(t,\theta), \quad (7)$$

where N denotes the number of the data that is used for the estimation. Under assumptions A1 – A4 the parameter estimate has an asymptotic distribution, (asymptotic in N),

$$\sqrt{N}\left(\widehat{\theta}_N - \theta_o^n\right) \in \mathrm{AsN}(0,P)$$

where

$$P = \sigma^2\left(\mathbf{E}\{\psi(t,\theta_o^n)\psi^T(t,\theta_o^n)\}\right)^{-1}$$

and

$$\psi(t,\theta_o^n) = -\left.\frac{\partial}{\partial\theta}\varepsilon(t,\theta)\right|_{\theta=\theta_o^n} = \left.\frac{\partial}{\partial\theta}\widehat{y}(t|\theta)\right|_{\theta=\theta_o^n}.$$

Note that in the output-error case the one-step-ahead predictor is given by

$$\widehat{y}(t|\theta) = \frac{B(q,\theta)}{F(q,\theta)}u(t). \quad (8)$$

4. SYSTEM AND MODEL ZEROS

The main concern in this paper is not the parameter estimates themselves, but rather the estimation accuracy of the corresponding zeros. Introduce the polynomial

$$p(z, \theta^n) = b_0 z^{n_b} + b_1 z^{n_b - 1} \cdots + b_{n_b}. \quad (9)$$

The system zeros are then defined as the solutions z_i^o, $i = 1, \cdots, n_o$ to the equation

$$p(z, \theta_o^n) = 0.$$

All zeros are assumed to be unique. For a *model* represented by the parameter θ^n, the zeros $z_i(\theta^n)$, $i = 1, \cdots, n$ are the solutions to

$$p(z, \theta^n) = 0.$$

When the model is over-parameterized, i.e. when $n > n_o$, more parameters than necessary will be estimated, resulting in larger variances for the estimated parameters. The interest here is to see how the variances of the estimated system *zeros* are affected by the over-modelling. In particular, the distribution of estimates of non-minimum phase zeros is examined.

5. ASYMPTOTIC VARIANCE OF THE ESTIMATED MODEL ZEROS

The accuracy of the estimated zeros is evaluated by the variance of the estimates. Expressions for the variances are very complicated when data sets of finite length are used, but if the data length is allowed to grow unboundedly, simple expressions for the asymptotic variance can be derived. See (Ljung, 1999) for more background on the use of asymptotic expressions in the area of system identification. In (Lindqvist, 2001) it is shown that the asymptotic variance of an estimated zero can be expressed as

$$\lim_{N \to \infty} N \mathbf{E}|z_k(\widehat{\theta}_N^n) - z_k^o)|^2 = \frac{\sigma^2 |z_k^o|^2 \left[1 \cdots (z_k^o)^{-n}\right]}{|b_0^o|^2 \prod_{i \neq k}^{n_o} \left|1 - \frac{z_i^o}{z_k^o}\right|^2} \times \left[\mathbf{E}\{\psi_n(t)\psi_n^T(t)\}^{-1}\right]_{(1,1)} \begin{bmatrix} 1 \\ \vdots \\ (\overline{z_k^o})^{-n} \end{bmatrix}. \quad (10)$$

The subscript $(1,1)$ denotes the $(n+1) \times (n+1)$ upper left submatrix (corresponding to the gradient of the predictor w.r.t the b-parameters). Above, $z_k(\widehat{\theta}_N^n)$ is one of the estimated zeros and z_k^o is the corresponding system zero. The result is based on first order approximations of the estimated zero.

Note that whenever the term *asymptotic variance* is used in this text, the variance is normalized with N.

6. ORDER-ASYMPTOTIC VARIANCE OF THE ESTIMATED MODEL ZEROS

A simplified expression for the asymptotic variance of an estimated non-minimum phase zero can be formulated when the model order is allowed to grow unlimitedly. This means that many more poles and zeros than necessary will be estimated. This is often referred to as over-modelling. The following result was derived in (Lindqvist, 2001).

Theorem 6.1. Let $u(t) = Q(q)v(t)$ where $Q(q)$ is a stable minimum phase filter and $v(t)$ is zero mean white noise with variance 1. Further, let R_{uu}^n be the covariance matrix built up by the elements $r_u(k) = \mathbf{E}u(t)u(t-k)$, i.e., the (j,k)th element of R_{uu}^n is $r_u(j-k)$. Further let z be such that $|z| > 1$. Then it holds that

$$\lim_{n \to \infty} \left[1 \cdots (z)^{-n}\right] \left(R_{uu}^n\right)^{-1} \begin{bmatrix} 1 \\ \vdots \\ (\overline{z})^{-n} \end{bmatrix} = \frac{1}{(1 - |z|^{-2}) |Q(z)|^2}.$$

□

Theorem 6.2 is the main result presented in this paper. It shows that the asymptotic variance of an estimated non-minimum phase zero converges to a finite limit as the model order goes to infinity. This is interesting since the variance of the parameters θ_N^n grows unboundedly with n. This result was shown for FIR- and ARX-models in (Lindqvist, 2001) and (Hjalmarsson and Lindqvist, 2002) and here it is shown that those results can be extended to general model structures as well.

Theorem 6.2. Let the system be of Box-Jenkins structure, as defined in assumptions A1 – A4. Further, let $z_k(\widehat{\theta}_N^n)$ be one of the estimated zeros and let z_k^o be the corresponding system zero. If the system zero is non-minimum phase, i.e. $|z_k^o| > 1$, then it holds that

$$\lim_{n \to \infty} \lim_{N \to \infty} N \mathbf{E} \left| z_k(\widehat{\theta}_N^n) - z_k^o \right|^2 = \frac{\sigma^2 |F(z_k^o)|^2 |C(z_k^o)|^2 |z_k^o|^2}{|b_0^o|^2 |D(z_k^o)|^2 \left(1 - |z_k^o|^{-2}\right) |Q(z_k^o)|^2 \prod_{i \neq k}^{n_o} \left|1 - \frac{z_i^o}{z_k^o}\right|^2}.$$

The same result holds for the output-error model structure where $C(z_k^o) = D(z_k^o) = 1$.

Proof: see Appendix A □

More generally one can look at this result in the following way. Consider a SISO system

$$y = Gu + He,$$

where G and H are rational transfer functions and the input signal u is generated as Qv, where Q is a minimum phase filter. Let $\widetilde{G}$ denote the transfer function that has the same poles and zeros as G, except that the non-minimum phase zero in question is removed, i.e.

$$\widetilde{G} = \frac{G}{1 - \frac{z_k^o}{z}}. \tag{11}$$

Now, the expression in Theorem 6.2 can be written as

$$\lim_{n\to\infty} \lim_{N\to\infty} N\mathbf{E}\left|z_k(\widehat{\theta}_N^n) - z_k^o\right|^2 = \frac{|z_k^o|^2 \sigma^2 |H(z_k^o)|^2}{\left(1 - |z_k^o|^{-2}\right)\left|\widetilde{G}(z_k^o)\right|^2 |Q(z_k^o)|^2}. \tag{12}$$

This means that, for sufficiently large model order and data length, the variance of the zero estimate is approximately

$$\mathbf{E}\left|z_k(\widehat{\theta}_N^n) - z_k^o\right|^2 \approx \frac{1}{N}\frac{|z_k^o|^2 \sigma^2 |H(z_k^o)|^2}{\left(1 - |z_k^o|^{-2}\right)\left|\widetilde{G}(z_k^o)\right|^2 |Q(z_k^o)|^2}. \tag{13}$$

7. INTERPRETATIONS OF THE RESULTS OF THEOREM 6.2

On the basis of Equation (13) some conclusion can be drawn about the variance of estimated non-minimum phase zeros. Firstly, the location of the zero itself (or rather its distance from the origin) plays an important role. Zeros close to the unit circle give very large variances of the estimates. So do also zeros that are far from the origin.

Secondly, the locations of the poles and zeros of the transfer functions $\widetilde{G}$, H and Q have a quite complex impact on the variance. The signal $\widetilde{G}Qv$ can be considered as an input to the system $\left(1 - \frac{z_k^o}{z}\right)$, see Figure 1. The aim of the identification process is to identify the zero dynamics of the system, i.e. to get an estimate of z_k^o. An accurate estimate can be expected if the sequence of input and output data is able to capture the zero dynamics of the system. Zero dynamics means that for some kinds of input signals, the output signal will not be affected. A pole close to z_k^o would excite that kind of signals, whereas a zero close to z_k^o would cancel out those signals. This means that $\widetilde{G}Q$ should have poles close to z_k^o and zeros far from z_k^o for the variance to be low.

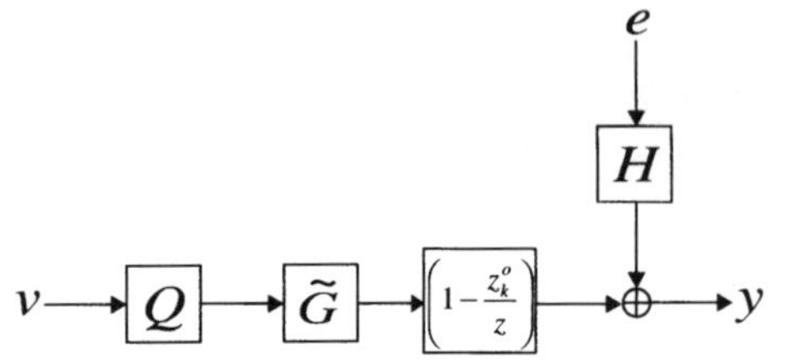

Fig. 1. The signal $\widetilde{G}Qv$ is considered as input to the non-minimum phase zero.

The opposite is desirable for the the noise term He. High accuracy is achieved when the noise term is small, or at least does not contain signals that would have been cancelled by the system zero. For instance, a pole close to z_k^o would create noise signals that contradict the system's true zero dynamics.

Worth noting here is that, in many cases, Q can be considered as a user's choice. Choosing a relevant input sequence is a part of the *experiment design* and the (somewhat heuristic) analysis above gives some insight in how to go about it.

8. NUMERICAL EXAMPLE

In this section, the results of Theorem 6.2 will be exemplified with some simulations. The system identification is performed using the System Identification Toolbox for MATLAB.

Consider the output-error system

$$y(t) = \frac{(1 - 1.2q^{-1})(1 + 0.8q^{-1})}{(1 - 0.5q^{-1})(1 + 0.6q^{-1})}u(t) + e(t), \tag{14}$$

which has a minimum phase zero in -0.8 and a non-minimum phase zero in 1.2. The input $u(t)$ for the identification experiment is generated as

$$u(t) = Q(q)v(t) = \left(1 - 0.8q^{-1}\right)v(t), \tag{15}$$

where $v(t)$ is white noise with variance 1. The noise $e(t)$ is white with variance 0.001. The system is identified using a model of order n (same order for both polynomials, B and F). The variance of the non-minimum phase zero estimate is examined for different values of n and compared with the theoretical asymptotic variance given by the expression in Theorem 6.2. Figure 2 shows the results of the simulations.

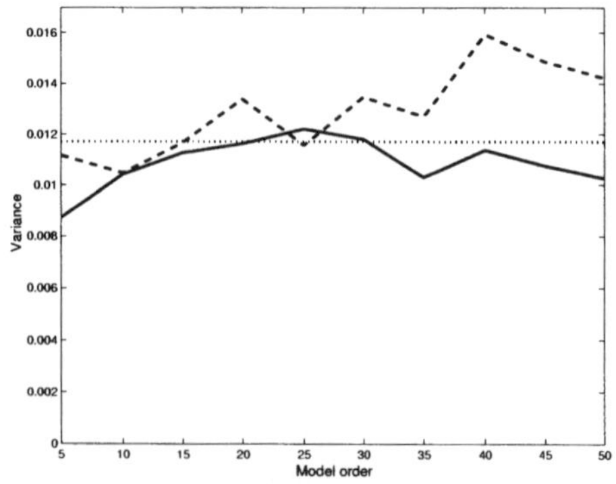

Fig. 2. Normalized variance of the estimated non-minimum phase zero plotted versus the model order. Monte Carlo simulated variances (500 runs) are used. Solid line: simulated variance, data length 10000. Dashed line: simulated variance, data length 1000. Dotted line: asymptotic variance (does not depend on N).

As Figure 2 indicates, the variance of the zero estimates does not only depend on the increasing model order, but also on the number of data that is used. Generally, the accuracy gets better with increasing data length and worse with increasing model order. This suggests that some relationship between n and N must hold for the asymptotic expression to be valid also for finite values of n and N. Figure 3 shows the variance as a function of data length for three different model orders. In this case the data length N should be at least a hundred times larger than the model order n for the asymptotic expression (13) to be valid. An interesting observation is that it seems to be a knee in the dependance on the number of data, regardless of the model order.

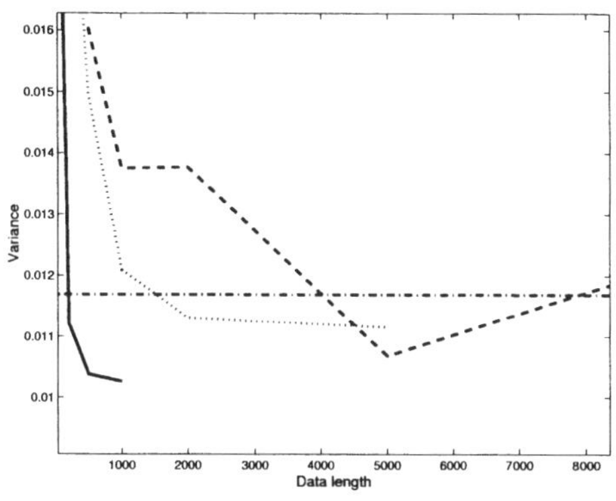

Fig. 3. Variance of the estimated non-minimum phase zero plotted versus the data length. Solid line: model order 5. Dotted line: model order 20. Dashed line: model order 40. Dot-dashed line: asymptotic variance.

Figure 4 clarifies the difference between the variance of minimum phase and non-minimum phase zeros as the model order increases. The system (14) is estimated using models of increasing order n. The estimated zeros from 500 identifications (for each value of n) are superimposed and one notices that the estimation of the non-minimum phase zero is much less affected by the increase in model order than the estimation of the minimum phase zero.

9. CONCLUSIONS

In this paper it is shown that the accuracy of estimated non-minimum phase zeros depends very little on the model order. This conclusion is based on the expression for the asymptotic variance of estimated zeros, given in Theorem 6.2. The result has been validated with simulations of an output-error system. An interpretation of Theorem 6.2 is given and the relevance of the asymptotic variance is examined for finite model orders and data lengths.

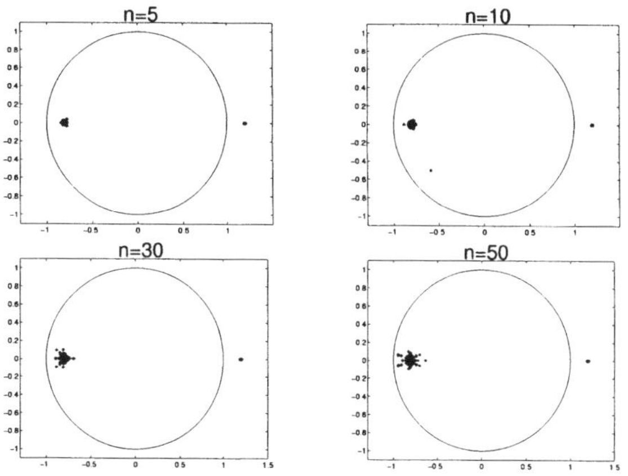

Fig. 4. Estimated zeros (at -0.8 and 1.2) from 500 identifications for model orders $n = 5, 10, 30$ and 50. Notice the increase in the variance of the minimum phase zero. The variance of the non-minimum phase zero is hardly affected at all.

REFERENCES

Bombois, X., M. Gevers and G. Scorletti (1999). Controller validation for a validated model set. In: *European Control Conference*.

Cooley, B.L., J.H. Lee and S.P. Boyd (1998). Control-relevant experiment design: A plant-friendly LMI-based approach. In: *Proceedings of the American Control Conference, Philadelphia, Pennsylvania*.

Forssell, U. and L. Ljung (1998). Identification for control: Some results on optimal experiment design. In: *Proceedings of the 37th IEEE Conference on Decision and Control*. Tampa, FL. pp. 3384–3389.

Forssell, U. and L. Ljung (1999). Closed-loop identification revisited. *Automatica*. To appear.

Freudenberg, J.S. and D.P. Looze (1998). *Frequency domain properties of scalar and multivariable feedback systems*. number 104 In: *Lecture notes in control and information sciences*. Springer.

Gevers, M. (1993). Towards a joint design of identification and control?. In: *Essays on Control: Perspectives in the Theory and its Applications* (H. L. Trentelman and J. C. Willems, Eds.). Birkhäuser.

Hjalmarsson, H. and K. Lindqvist (2002). Identification of performance limitations in control using arx-models. In: *Proceedings of The 15th IFAC World Congress*.

Hjalmarsson, H., M. Gevers and F. De Bruyne (1996). For model based control design criteria, closed loop identification gives better performance. *Automatica* **32**, 1659–1673.

Lindqvist, K. (2001). On Experiment Design in Identification of Smooth Linear Systems. Licentiate thesis. Royal Institute of Technology. Stockholm, Sweden.

Lindqvist, K. and H. Hjalmarsson (2000). Optimal input design using linear matrix inequalities. In: *Post print from the SYSID 2000 IFAC symposium on System Identification*. Santa Barbara, California, USA.

Ljung, L. (1999). *System Identification: Theory for the user*. second ed.. Prentice Hall.

Skogestad, S. and I. Postlethwaite (1996). *Multivariable Feedback Control—Analysis and Design*. John Wiley. chichester.

Van den Hof, P.M.J. and R.J.P. Schrama (1995). Identification and control – closed loop issues. *Automatica* **31**(12), 1751–1770.

Woodley, B.R., R.L. Kosut and J.P. How (1998). Uncertainty model unfalsification with simulation. In: *American Control Conference*. Vol. 5. Philadelfia. pp. 2754–2755.

APPENDIX A : PROOF OF THEOREM 6.2

Start by considering the output-error structure and then extend that result to the more general Box-Jenkins model structure. The gradient of the one-step ahead predictor is given by

$$\psi(t,\theta_o) = \frac{\partial}{\partial\theta}\left(\frac{B(q,\theta)}{F(q,\theta)}u(t)\right)\bigg|_{\theta=\theta_o} =$$

$$\left[1,\cdots,q^{-n_b},-\frac{B}{F}q^{-1},\cdots,-\frac{B}{F}q^{-n_f}\right]^T\frac{1}{F}u(t).$$

Introduce $\tilde{u}=\frac{1}{F}u$, $\widetilde{Q}=\frac{Q}{F}$ and $w=-\frac{B}{F}\tilde{u}$. Note that $\tilde{u}=\widetilde{Q}v$, where v is white noise with variance 1 and $\widetilde{Q}$ is a minimum phase filter. The covariance matrix can now be written as

$$\mathbf{E}\{\psi\psi^T\} = \mathbf{E}\left\{\begin{pmatrix}\tilde{u}(t)\\ \vdots\\ \tilde{u}(t-n_b)\\ w(t-1)\\ \vdots\\ w(t-n_f)\end{pmatrix}(\cdot)^T\right\} = \begin{bmatrix}R_{\tilde{u}\tilde{u}} & R_{\tilde{u}w}\\ R_{w\tilde{u}} & R_{ww}\end{bmatrix}.$$

Let the subindex $(1,1)$ denote the $(n_b+1)\times(n_b+1)$ upper left submatrix and invoke the well-known matrix inversion lemma to write

$$\begin{aligned}\left(\mathbf{E}\{\psi\psi^T\}^{-1}\right)_{(1,1)} &= R_{\tilde{u}\tilde{u}}^{-1} + R_{\tilde{u}\tilde{u}}^{-1}R_{\tilde{u}w}\times\\ &\left(R_{ww}-R_{w\tilde{u}}R_{\tilde{u}\tilde{u}}^{-1}R_{\tilde{u}w}\right)R_{w\tilde{u}}R_{\tilde{u}\tilde{u}}^{-1}.\end{aligned} \tag{16}$$

Since the system (3) is asymptotically stable it can be shown that

$$\left[R_{w\tilde{u}}R_{\tilde{u}\tilde{u}}^{-1}R_{\tilde{u}w}\right]_{j,k} = \left[R_{ww}\right]_{j,k} + o(\lambda^n) \tag{17}$$

for some $|\lambda|<1$. For details see (Hjalmarsson and Lindqvist, 2002). Now, as the model order is increased, i.e. $n\to\infty$, it follows that

$$R_{ww} - R_{w\tilde{u}}R_{\tilde{u}\tilde{u}}^{-1}R_{\tilde{u}w} \to 0.$$

This means that the second term of the right hand side in (16) is zero and that

$$\left(\mathbf{E}\{\psi\psi^T\}^{-1}\right)_{(1,1)} \to R_{\tilde{u}\tilde{u}}^{-1}. \tag{18}$$

Now turn to the Box-Jenkins structure. First introduce the notation $\Gamma_i^n(q)=\left(q^{-i}\ \cdots\ q^{-n}\right)^T, i=0,1$. The gradient of the predictor is

$$\psi = \frac{\partial}{\partial\theta}\left(\frac{DB}{CF}u(t)\right) + \frac{\partial}{\partial\theta}\left(\frac{C-D}{C}y(t)\right) = \begin{bmatrix}\frac{D}{CF}\Gamma_0^{n_b}(q)u(t)\\ -\frac{DB}{CF^2}\Gamma_1^{n_f}(q)u(t)\\ -\frac{DB}{C^2F}\Gamma_1^{n_c}(q)u(t)+\frac{D}{C^2}\Gamma_1^{n_c}(q)y(t)\\ \frac{B}{CF}\Gamma_1^{n_d}(q)u(t)-\frac{1}{C}\Gamma_1^{n_d}(q)y(t)\end{bmatrix}.$$

Introduce the variables $\tilde{u}=\frac{D}{CF}u$ and $\widetilde{Q}=\frac{DQ}{CF}$. Note that $\tilde{u}=\widetilde{Q}v$, where v is white noise with variance 1 and $\widetilde{Q}$ is a minimum phase filter. Also note that $y-\frac{B}{F}u=\frac{C}{D}e$. That gives that

$$\psi = \begin{bmatrix}\Gamma_0^{n_b}(q)\tilde{u}\\ -\Gamma_1^{n_f}(q)\frac{B}{F}\tilde{u}\\ \Gamma_1^{n_c}(q)\frac{1}{C}e\\ -\Gamma_1^{n_d}(q)\frac{1}{D}e\end{bmatrix} = \begin{bmatrix}\Gamma_0^{n_b}(q)\tilde{u}\\ \Gamma_1^{n_f}(q)w\\ \Gamma_1^{n_c}(q)e_1\\ \Gamma_1^{n_d}(q)e_2\end{bmatrix}, \tag{19}$$

where the variables $w=-\frac{B}{F}\tilde{u}$, $e_1=\frac{1}{C}e$ and $e_2=-\frac{1}{D}e$ are introduced. The driving noises v and e are uncorrelated which gives that $\tilde{u}$ and w are uncorrelated with e_1 and e_2. Hence,

$$\mathbf{E}\{\psi\psi^T\} = \begin{bmatrix}\begin{bmatrix}R_{\tilde{u}\tilde{u}} & R_{\tilde{u}w}\\ R_{w\tilde{u}} & R_{ww}\end{bmatrix} & 0\\ 0 & \begin{bmatrix}R_{e_1e_1} & R_{e_1e_2}\\ R_{e_2e_1} & R_{e_2e_2}\end{bmatrix}\end{bmatrix}. \tag{20}$$

The inverse of the block-diagonal matrix is compounded of the inverses of the blocks. Equivalently to the output-error case this gives that, see (16)-(18),

$$\left(\mathbf{E}\{\psi\psi^T\}^{-1}\right)_{(1,1)} = \left(\begin{bmatrix}R_{\tilde{u}\tilde{u}} & R_{\tilde{u}w}\\ R_{w\tilde{u}} & R_{ww}\end{bmatrix}^{-1}\right)_{(1,1)} \to R_{\tilde{u}\tilde{u}}^{-1} \tag{21}$$

as $n\to\infty$. Now, combining Theorem 6.1 and Equations (10), (18) and (21) proves Theorem 6.2. Note that $\tilde{u}$ is defined in different ways in (18) and (21).

IFAC Publications

CONTRO LOOP PERFORMANCE MONITORING BY CUSUM ALGORITHMS FOR LOCAL LINEAR HYPOTHESES

M. Kinnaert, R. Hanus and C. Parloir

Laboratoire d'Automatique et d'Analyse des Systèmes,
Université Libre de Bruxelles.
CP 165/55, 50 Av. F. D. Roosevelt, B-1050 Brussels, BELGIUM

Abstract: Monitoring the performance of a closed-loop by detecting whether the mean μ and the standard deviation σ of the control error (or the control signal) belong to a fixed triangular region in the (μ, σ)-plane is shown to be a sensible approach. This triangular region is approximated by a set of circles and for each of them, a CUSUM algorithm for local linear hypotheses is performed. *Copyright © 2003 IFAC*

Keywords: Change detection, performance monitoring, control loop, statistical local approach

1. INTRODUCTION

Monitoring closed-loop systems has attracted significant attention in the recent years. For the class of problems considered here, the starting point is a closed-loop system operating satisfactorily, and the aim is to detect deviations from the proper operating condition as time elapses. Such deviations may be due to sensor, actuator or process malfunction. Work on detection and diagnosis of oscillations in closed-loops systems belongs to this category (Hägglund, 1995), (Horch, 1999). The problem to be studied consists in detecting changes in the mean μ and/or the standard deviation σ of the control error (or the control signal) that bring the operating point outside of an acceptable domain in the (μ, σ)-plane. This detection problem results from the requirement that, for a given set point, the control error (control signal) should lie within a fixed interval with a certain probability. Classical tests for change detection such as CUSUM tests can be used to detect a change in the mean of a signal assuming no change in its variance or vice versa (Basseville and Nikiforov, 1993). Yet, practically simultaneous changes in mean and variance can occur. The detection problem can then be solved approximately by CUSUM algorithms for local linear hypotheses as explained in this paper. Our contribution is to clearly motivate the use of such an algorithm for the considered problem and to show how the tuning parameters can be set by solving an optimization problem.

The paper is organized as follows. The problem is stated in section 2. Section 3 presents a review of the CUSUM algorithm for local linear hypotheses. Section 4 particularizes the algorithm to our specific problem. The acceptable domain is approached by a set of circles. For each of them, a CUSUM algorithm for local linear hypotheses is performed. An approach for optimizing the tuning parameters of each of those algorithms is presented in section 5 and a method for determining its threshold is explained in section 6.

2. PROBLEM STATEMENT

The closed-loop system depicted in Fig. 1 is considered. f_a and f_s are two unknown deterministic signals modeling respectively actuator and sensor faults, while v is the measurement noise, a Gaus-

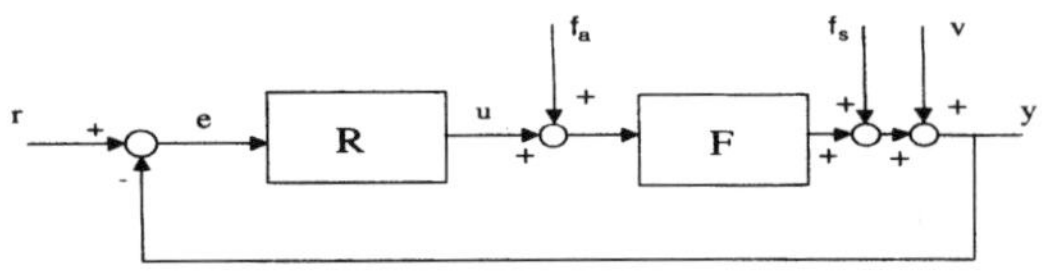

Fig. 1. Closed-loop system

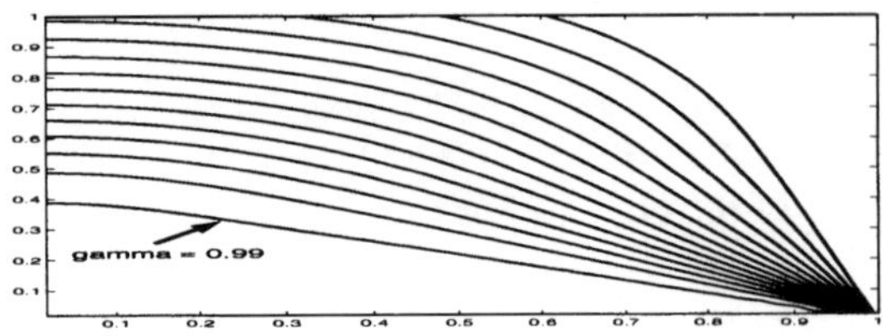

Fig. 2. Equi-probability curves corresponding to equation (2).Successive curves correspond to decreasing values of γ starting from $\gamma = 0.99$ next 0.96, 0.93, $\cdots$, up to 0.6

sian white noise process, of which the variance may increase upon occurrence of a fault (variance decrease, which may correspond to broken transmission wires, is normally detected easily without resorting to the kind of tools presented here). Our aim is to monitor the performance of the closed-loop system by supervising the control error e and the control signal u.

The control performance criterion on e amounts to requiring that, with a certain probability, say γ, the absolute value of the error lies below a given threshold h. The value of such a threshold can be provided by operators. Given the hypothesis on the measurement noise v and the inputs f_a and f_s , the error sequence $(e_k)_k$ is made of samples with normal probability law $\mathcal{L}(e) = \mathcal{N}(\mu, \sigma)$. The above requirement, namely $Pr(|e| < h) = \gamma$, can be written

$$\frac{1}{\sqrt{2\pi}\sigma} \int_{-h}^{h} \exp(-\frac{(e-\mu)^2}{2\sigma^2})de = \gamma. \quad (1)$$

After setting $x = \frac{e-\mu}{\sigma}$, (1) becomes

$$\frac{1}{\sqrt{2\pi}} \int_{\frac{-1-\mu/h}{\sigma/h}}^{\frac{1-\mu/h}{\sigma/h}} \exp(\frac{x^2}{2})dx = \gamma \quad (2)$$

Fig. 2 gives the equi-probability curves in the axis μ/h, σ/h which will be denoted μ_h and σ_h in the sequel. For a given Gaussian random variable with mean μ and variance σ, and for a given threshold h, it indicates the probability that a realization of this random variable be lower than h. In the range of probabilities typically considered, namely from 0.9 to 0.99, the equi-probability curves are close to straight lines. Thus, it makes sense to attempt to test whether the samples of the sequence $(e_k)_k$ correspond to a point below or above a given straight line in the plane (μ_h, σ_h). This straight line approximates the equi-probability curve associated to the desired value for the probability γ. Notice that, only positive values for the mean are considered here. But clearly a symmetric figure with respect to the σ_h axis holds if the sign of the mean is changed. Thus one should actually check whether the realization of the error signal corresponds to a point inside or outside of a triangle of the form depicted in Fig. 3.

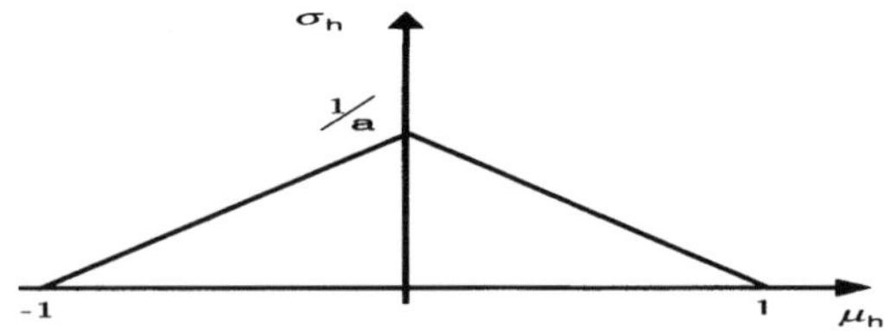

Fig. 3. Border of the region corresponding to the healthy working mode $\mu_h \pm a\sigma_h = \pm 1$

It may also be useful to check that the samples of the control sequence $(u_k)_k$ remain in a strip around the nominal value in the absence of set point change. Indeed, step-like faults on f_a or f_c do not have any effect on e at steady state when an integrating action is present in the controller, but they yield a change in the mean of u in the absence of set point change, when the process has no integrating action. As the processing of u is similar to that of e, we only deal with the latter one here.

The sequential change detection algorithm to be used requires the independence of the data samples. Hence the error signal should be processed by a whitening filter. Practically, simulation results have shown that the algorithm performs well even in the absence of whitening filter. Therefore, the design of such a filter is not further discussed here.

3. CUSUM ALGORITHM FOR LOCAL LINEAR HYPOTHESES

Let us start from the following simple hypothesis test. Given a sequence $(e_k)_k$ of independent random variables depending on the ℓ-dimensional vector θ with probability density function $p_\theta(e_k)$, choose between the following two hypotheses:

$$H_0 : \theta = \theta_0$$
$$H_1 : \theta = \theta_1$$

In this case, the decision function corresponding to the CUSUM algorithm can be written

$$t_a = \min\{k : g_k \geq q\} \quad (3)$$

$$g_k = \max(0, g_{k-1} + s_k) \quad (4)$$

where s_k is the log-likelihood ratio at time k,

$$s_k = ln\frac{p_{\theta_1}(e_k)}{p_{\theta_0}(e_k)},$$

g_k the decision function evaluated at time k and t_a the alarm time.

This test relies on the following key property of s_k:

$$E_{\theta_0}(s_k) < 0 \qquad E_{\theta_1}(s_k) > 0 \qquad (5)$$

where $E_{\theta_i}(s_k)$ is the expected value of s_k under the pdf $p_{\theta_i}(e_k)$. (5) implies that there is a separating surface in the parameter space such that

$$E_\theta(s_k) = 0 \qquad (6)$$

(Basseville and Nikiforov, 1993) (pages 257, 371).

Our aim is to develop a detection algorithm having ideally a straight line as separating curve in the first quadrant of the (μ_h, σ_h) plane. In a broader framework, this amounts to looking for an algorithm with an hyperplane as separating surface. To achieve this goal, a local approach is used: it is assumed that the changes in the θ space are small and they occur around a nominal parameter θ^*.

The local hypotheses amounts to assuming that θ_0 and θ_1 can be linked to a nominal parameter θ^* via

$$\theta_0 = \theta^* - \frac{1}{2}\nu\Upsilon$$
$$\theta_1 = \theta^* + \frac{1}{2}\nu\Upsilon$$

where $\nu > 0$, and Υ is the unit vector of known change direction. Under this hypothesis, the log-likelihood ratio for a sample of size N can be approximated by (Basseville and Nikiforov, 1993) (page 371)

$$\begin{aligned} &S_1^N(\theta^* - \frac{1}{2}\nu\Upsilon, \theta^* + \frac{1}{2}\nu\Upsilon) \\ &\approx \nu\Upsilon^T \frac{\partial \ln \prod_{i=1}^n p_\theta(e_i)}{\partial\theta}|_{\theta=\theta^*} \\ &\approx \nu\Upsilon^T \sum_{i=1}^N Z_i^* \end{aligned} \qquad (7)$$

where Z_i^* is the vector of efficient score defined as

$$Z_i^* = \frac{\partial \ln p_\theta(e_i)}{\partial\theta}|_{\theta=\theta^*}$$

One also has

$$S_1^N(\theta^* - \frac{1}{2}\nu\Upsilon, \theta^* + \frac{1}{2}\nu\Upsilon) = \sum_{i=1}^N s_i \qquad (8)$$

From (7) and (8), the following approximation of s_i results:

$$s_i \approx \nu\Upsilon^T Z_i^*$$

Without loss of generality, we consider the CUSUM algorithm associated with

$$s_i \approx \Upsilon^T Z_i^* \qquad (9)$$

which amounts to using a threshold q obtained by dividing the original threshold by the change magnitude ν. In the CUSUM algorithm for local linear hypotheses, the approximation (9) is substituted for s_k in (4). We shall now show that this approximation has the desired property that allows one to define two regions separated by one hyperplane in the θ-space.

Let $I(\theta)$ denote the Fisher information matrix defined as the $\ell \times \ell$ matrix with entries

$$\begin{aligned} I_{ij}(\theta) &= \int_{-\infty}^{\infty} [\frac{\partial}{\partial\theta_i}\ln p_\theta(e)][\frac{\partial}{\partial\theta_j}\ln p_\theta(e)]p_\theta(e)de \\ &\quad i = 1, \cdots, \ell, j = 1, \cdots, \ell \\ &= E_\theta(Z(i)Z(j)) \end{aligned} \qquad (10)$$

where $Z(i)$ is the ith component of the efficient score $Z = \frac{\partial \ln p_\theta(e)}{\partial\theta}$. As the expectation of the efficient score can be approximated by

$$E_\theta(Z_k^*) \approx I(\theta^*)(\theta - \theta^*) \qquad (11)$$

in the neighborhood of the point θ^* (Basseville and Nikiforov, 1993), (9) implies that the separating surface (6) can be approximated by the following hyperplane

$$\Upsilon^T I(\theta^*)(\theta - \theta^*) = 0 \qquad (12)$$

From (12), one deduces that, when the a priori information is not in terms of parameters θ_0 and θ_1, but in terms of a separating hyperplane, defined by θ^* and Υ, the CUSUM algorithm in which (9) is substituted for s_k in (4) appears to be an appropriate change detection algorithm.

The next section is devoted to the application of this approach to the particular case where the parameter vector is $\theta = [\mu_h \quad \sigma_h]^T$ and thus the hyperplane boils down to a straight line in the plane $\mu_h - \sigma_h$, namely $\mu_h + a\sigma_h = 1$ when $a > 0$.

4. DETECTION OF CHANGES IN MEAN AND/OR VARIANCE

The efficient score and the Fisher information matrix associated to the Gaussian pdf with mean μ and standard deviation σ must be computed.

A reparametrization in terms of the adimensional variables μ_h and σ_h is used, namely

$$p_{(\mu_h,\sigma_h)}(e) = \frac{1}{\sqrt{2\pi}\sigma_h h} \exp\left(-\frac{(e_h-\mu_h)^2}{2\sigma_h^2}\right) \quad (13)$$

Straightforward computations yield:

$$Z_{i,1}^* = \frac{\partial \ln p_{(\mu_h,\sigma_h)}(e_i)}{\partial \mu_h}|_{\theta=\theta_h^*}$$

$$= \frac{1}{\sigma_h^{*2}}\left(\frac{e_i}{h} - \mu_h^*\right) \quad (14)$$

$$Z_{i,2}^* = \frac{1}{\sigma_h^*}\left(\left(\frac{e_i}{h} - \mu_h^*\right)^2 \frac{1}{\sigma_h^{*2}} - 1\right) \quad (15)$$

$$I(\theta_h^*) = \begin{pmatrix} \frac{1}{\sigma_h^{*2}} & 0 \\ 0 & \frac{2}{\sigma_h^{*2}} \end{pmatrix} \quad (16)$$

For implementing the CUSUM algorithm (3),(4), (9), it remains to determine the vector Υ. To this end, the separating straight line

$$\mu_h + a\sigma_h = 1 \quad (17)$$

must be written in the form (12). (17) yields

$$[1 \;\; a] I(\theta_h^*)^{-1} I(\theta_h^*)\left(\begin{bmatrix}\mu_h \\ \sigma_h\end{bmatrix} - \begin{bmatrix}\mu_h^* \\ \sigma_h^*\end{bmatrix}\right) = 0 \quad (18)$$

where $\theta_h^* = [\mu_h^* \;\; \sigma_h^*]^T$ must fulfil equation (17), namely $\mu_h^* + a\sigma_h^* = 1$. Identifying (18) with (12), the following Υ vector is obtained

$$\Upsilon^T = [1 \;\; a] I(\theta_h^*)^{-1} = \begin{bmatrix}\sigma_h^{*2} & \frac{a\sigma_h^{*2}}{2}\end{bmatrix} \quad (19)$$

The final CUSUM algorithm is thus of the form (3),(4), with s_k given by (9) where (14),(15), and (19) are substituted for Z_i^* and Υ, namely

$$s_k \approx \frac{e_k}{h} - \mu_h^* + \frac{a}{2}\sigma_h^*\left(\left(\frac{e_k}{h} - \mu_h^*\right)^2 \frac{1}{\sigma_h^{*2}} - 1\right) \quad (20)$$

In order to gain insight in the influence of the choice of θ_h^*, we first investigate how well (11), and hence (12), holds in our particular case.

From (20), one deduces

$$E(s_k) = \mu_h - \mu_h^* + \frac{a}{2\sigma_h^*}(\mu_h - \mu_h^*)^2 + \frac{a}{2}\sigma_h^*\left(\frac{\sigma_h^2}{\sigma_h^{*2}} - 1\right) \quad (21)$$

The behaviour of the CUSUM algorithm is such that an alarm will be generated when $E(s_k) > 0$, in which case the decision function g_k exhibits a positive drift. Thus

$$E(s_k) = 0 \quad (22)$$

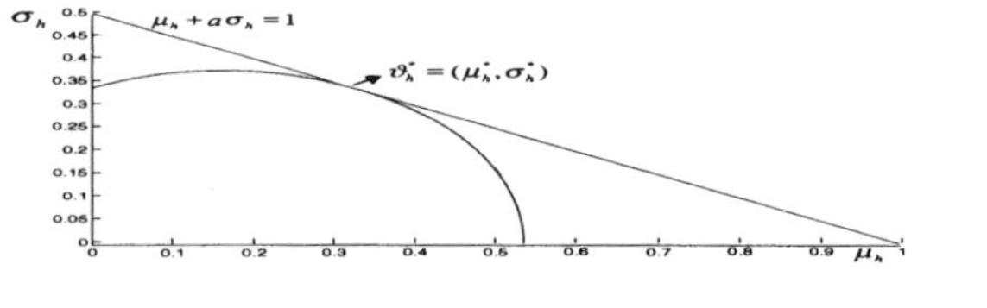

Fig. 4. Actual separating curve (21), (22) and desired straight line

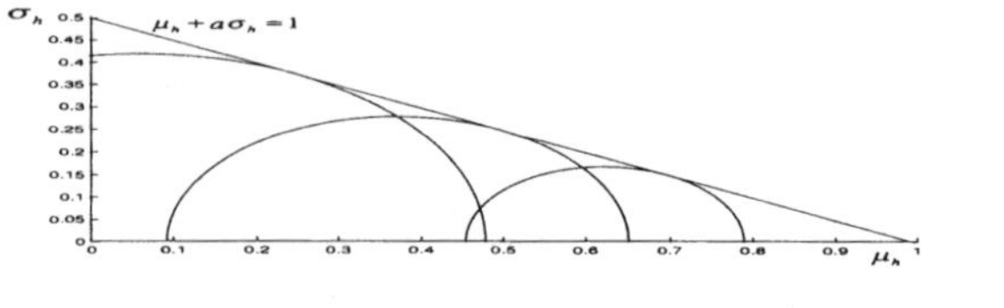

Fig. 5. Covering of the surface under the straight line by a set of circles

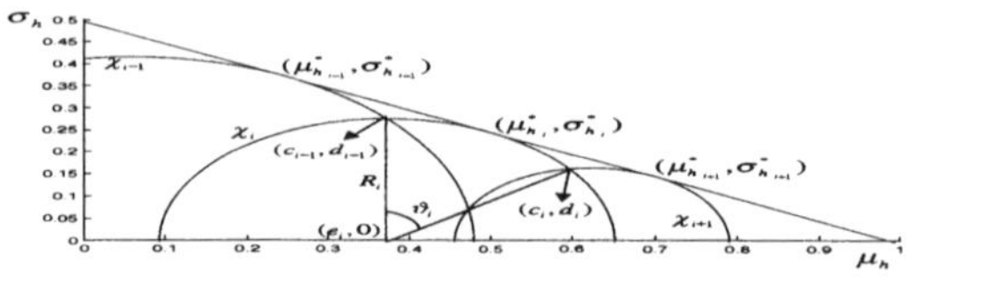

Fig. 6. Notations for the determination of the tangential points

is the separating curve. (21),(22) is actually the equation of a circle as it is clear from the following equivalent expression for the separating curve:

$$\left(\mu_h + \left(\frac{\sigma_h^*}{a} - \mu_h^*\right)\right)^2 + \sigma_h^2 = \left(\frac{\sigma_h^*}{a} - \mu_h^*\right)^2 + \sigma_h^{*2} + \frac{2}{a}\sigma_h^*\mu_h^* - \mu_h^{*2} \quad (23)$$

The arc of circle in the first quadrant is plotted together with the desired straight line (17) in Fig. 4 for $a = 2$, $\mu_h^* = 0.3290$ and $\sigma_h^* = 0.3355$.

There is a significant discrepancy between the straight line and the separating curve, which increases with the distance to θ_h^*. This means that the CUSUM algorithm (3),(4), (20) will generate an alarm as soon as the representative point of the processed signal is outside the circle. To avoid this problem, a set of CUSUM algorithms will be used, each one with a different circular separating curve, designed in such a way that the union of the set of circles approaches the triangular region satisfactorily (Fig. 5).

Each of those circles is tangential to the straight line, is centred on the μ_h axis, is characterized by its tangential point $\theta_h^* = (\mu_h^*, \sigma_h^*)$ and is associated to a CUSUM algorithm for local linear hypotheses.

The decision function g_k of each CUSUM algorithm increases on average when the point representative of the control error sequence is outside of the corresponding circle and an alarm is generated when the threshold q is exceeded. When all algorithms generate an alarm, the point representative of the control error lies outside the region made of the union of all circles. An adequate algorithm is needed to manage the non-simultaneity of the alarms generated by each algorithm.

Two parameters have to be tuned for each algorithm: the threshold q, and the nominal parameter θ_h^*. The choice of the threshold q is the object of section 6 while the determination of θ_h^* is explained in section 5.

5. DETERMINATION OF θ_H^*

For a given numbers of circles, the parameters of each circle could be determined by the minimization of the surface under the 2 straight lines of Fig. 3 which is not covered by those circles.

An odd number of circles is chosen to approximate this triangular surface. One of the circles is tangential to both straight lines, and there is an equal number, say N, of circles tangential to the straight line $\mu_h + a\sigma_h = 1$ only, and to the straight line $\mu_h - a\sigma_h = -1$ only.

For each circle, we have to find a value μ_h^*, the value of σ_h^* can then be deduced from the equations of the straight lines.

The parameter μ_{h0}^* of the circle which is tangential to the 2 straight lines can easily be determined: $\mu_{h0}^* = \frac{1}{1+a^2}$. This circle is denoted χ_0.

The symmetry of the problem allows one to consider only the case $\mu_h \geq 0$. So, it remains to determine N values of μ_h^* in order to minimize the surface under the straight line (17) which is not covered by the circles.

Fig. 6 represents three circles denoted χ_{i-1}, χ_i, χ_{i+1} tangential to the straight line (17). Let $\mu_{h_{i-1}}^*$, $\mu_{h_i}^*$, $\mu_{h_{i+1}}^*$ denote the abscissa of the point of tangency for the circles $\chi_{i-1}, \chi_i, \chi_{i+1}$.

The following notations have been used. $(e_i, 0)$ represents the centre of the circle χ_i. (c_{i-1}, d_{i-1}) (respectively (c_i, d_i)) the intersection between χ_{i-1} and χ_i (respectively between χ_i and χ_{i+1}), and R_i the radius of the circle χ_i. χ_{i-1}, χ_i, χ_{i+1} are sorted such that $R_{i-1} > R_i > R_{i+1}$ (and $e_{i-1} < e_i < e_{i+1}$). θ_i is the angle between the line which joins $(e_i, 0)$ and (c_{i-1}, d_{i-1}) and the line which joins $(e_i, 0)$ and (c_i, d_i).

$e_i, c_{i-1}, d_{i-1}, c_i, d_i, R_i, \theta_i$ can be given as a function of $\mu_{h_{i-1}}^*, \mu_{h_i}^*$ and $\mu_{h_{i+1}}^*$ only.

The following formula is based on the surface of simple geometrical figures and represents the surface situated above χ_i and under the straight line, for $c_{i-1} \leq \mu_h \leq c_i$.

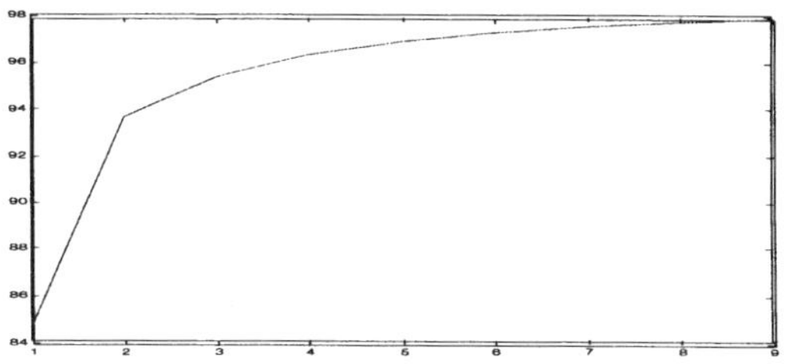

Fig. 7. Percentage of the surface under the straight lines which is covered by the circles versus N

$$S_i = \frac{c_i - c_{i-1}}{2a}(2 - c_i - c_{i-1}) - \frac{R_i^2 \theta_i}{2} - \frac{(e_i - c_{i-1})d_{i-1}}{2} - \frac{(c_i - e_i)d_i}{2} \qquad (24)$$

(24) is valid for $i = 1, 2, ..., N-1$ and can easily be adaptated for $i = 0$ or $i = N$.

The total surface (for $\mu_h \geq 0$) which is not covered by a circle is given by $S = \sum_{i=0}^{N} S_i$

In order to approximate at best the surface under the straight line with the circles, we have to minimize S with respect to $\mu_{h_1}^*, \mu_{h_2}^*, ..., \mu_{h_N}^*$ and to take into account the constraints (26) and (27) below.

$$\min_{\mu_{h_1}^*, \mu_{h_2}^*, ..., \mu_{h_N}^*} S \qquad (25)$$

$$0 \leq \mu_{h_i}^* \leq 1 \qquad (26)$$

$$(e_i - R_i) - (e_{i-1} + R_{i-1}) \leq 0 \qquad (27)$$

$$(i = 1, 2, ..., N)$$

The constraints (27) assures that two successive circles have an intersection (those conditions are not limitative because they are necessary to have a minimal surface S).

(25), (26) and (27) is a non linear optimization problem with non linear constraints which has been resolved numerically.

The percentage of the surface under the straight lines which is covered by the circles versus N is shown in Fig. 7.

6. DETERMINATION OF THE THRESHOLDS OF THE CUSUM ALGORITHMS

In this paragraph, a method to determine the thresholds q_i of the N CUSUM algorithms (3), (4) is explained (once again, only the case $\mu_h \geq 0$ is considered because of the symmetry of the problem). This method is based on the user require-

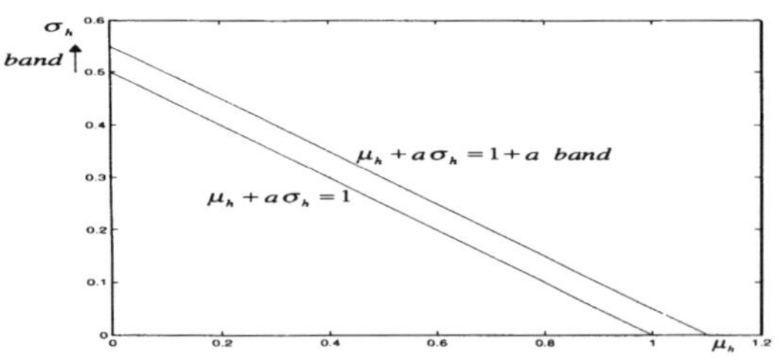

Fig. 8. Straight lines (17) and (28) for $a = 2$ and *band* $= 0.05$

ment of a chosen mean detection time in some particular faulty situations.

An approach in (Basseville and Nikiforov, 1993) (page 195) allows to obtain information about the mean detection time. It consists in computing a function, called average run length (ARL), from which the mean detection time can be deduced. The computation of the ARL function requires to solve some Fredholm integral equations. It must be performed numerically and it is quite heavy. Some approximations of the ARL function have been developed, as well as upper bounds on the mean detection time, but such results are not easily applicable practically. A more pragmatic approach is developed here. It is based on two informations which have to be provided by the user. We introduce the straight line

$$\mu_h + a\sigma_h = 1 + a\ band \qquad (28)$$

(with $band > 0$) which is parallel to (17) (Fig. 8). We will require that for all the points representative of the control error sequence on the straight line (28), the mean detection time is approximately equal to a chosen number of samples, say $t_{chosen} > 0$. Consequently, the mean detection time for a point representative of the control error sequence above the straight line (28) will be lower than t_{chosen} samples.

band and t_{chosen} are the two parameters which have to be given by the user in order to determine the thresholds q_i of the N CUSUM algorithms.

For each circle χ_i ($i = 1, 2, ..., N$), we consider the projection $\theta_{band_i} = (\mu_{h_{band_i}}, \sigma_{h_{band_i}})$ of the point of tangency $\theta^*_{h_i} = (\mu^*_{h_i}, \sigma^*_{h_i})$ on the straight line (28). θ_{band_i} is the point of the straight line (28) which is the nearest to χ_i and consequently the point for which the mean detection time of the CUSUM algorithm corresponding to the circle χ_i will be the largest. From (21), it can be deduced that the mean value of the decision function when the point representative of the control error is θ_{band_i} is given by the following.

$$E_{\theta_{band_i}}(s_k) = \mu_{h_{band_i}} - \mu^*_{h_i} + \frac{a(\mu_{h_{band_i}} - \mu^*_{h_i})^2}{2\sigma^*_{h_i}}$$

$$+ \frac{a}{2}\sigma^*_{h_i}\left(\frac{\sigma^2_{h_{band_i}}}{\sigma^{*2}_{h_i}} - 1\right) \qquad (29)$$

When the point representative of the control error is given by θ_{band_i}, it is desired to generate an alarm in t_{chosen} samples, on average. In this situation, the decision function will increase by $E_{\theta_{band_i}}(s_k)$, in mean value, at each sample. So, the following thresholds have to be chosen.

$$q_i = E_{\theta_{band_i}}(s_k)\ t_{chosen} \qquad (30)$$
$$(i = 1, 2, ..., N)$$

With this choice of the thresholds, for any point (μ_h, σ_h) representative of the control error sequence above the straight line (17), the mean detection time t_i for the CUSUM algorithm corresponding to the circle χ_i is approximated by

$$t_i(\mu_h, \sigma_h) \approx \frac{q_i}{E(s_k)_{|_{\mu^*_h = \mu^*_{h_i}\ \sigma^*_h = \frac{1-\mu^*_{h_i}}{a}}}} \qquad (31)$$

where $E(s_k)$ is given by (21).

7. CONCLUSION

A loop monitoring method aimed at testing that the operating point characterized by the mean μ and standard deviation σ of the control error remains in a triangular region in the (μ, σ)-plane has been studied. Such a test has been motivated from the information typically available via plant operators. To cover satisfactorily this triangular region, a set of circles is considered and for each of them, a CUSUM algorithm for local linear hypotheses is performed on-line. A method to determine the threshold of each of those CUSUM algorithms has been developed.

Acknowledgment: Partial support from the European project IFATIS (IST-2001-32122) is gratefully acknowledged. The first author is grateful to G.Moustakides for useful comments on this work.

8. REFERENCES

Basseville, M. and I.V. Nikiforov (1993). *Detection of Abrupt Changes: Theory and Applications.* Prentice Hall. Englewood Cliffs, N.J.

Hägglund, T. (1995). A control loop performance monitor. *Control Eng. Practice* **7**, 1505–1511.

Horch, A. (1999). A simple method for detection of stiction in control valve. *Control Eng. Practice* **7(10)**, 1221–1231.

IFAC Publications
www.elsevier.com/locate/ifac

MODEL APPROXIMATION OF PLANT AND NOISE DYNAMICS ON THE BASIS OF CLOSED-LOOP DATA

J. Zeng * R.A. de Callafon *

* *University of California, San Diego*
Dept. of Mechanical and Aerospace Engineering
9500 Gilman Drive
La Jolla, CA 92093-0411, U.S.A

Abstract: In this paper, we consider the problem of estimating low order and control relevant models of plant dynamics and additive noise dynamics on the basis of closed-loop experiments. Estimating low order models for both the plant and noise dynamics is important in control design applications that focus on disturbance rejection. Several methods for low order model estimation on the basis of closed-loop data exist in the literature, but fail to address the simultaneous estimation of low order noise models that are relevant in disturbance control problems. In this paper we evaluate and compare some of these methods and propose a new methodology that extends the results to low order noise model estimation. The new methodology is an extended two-stage method where the first stage is used to estimate high order models for filtering purposes. In the second stage, filtered signals are used for low order model approximation. The methodology is illustrated in a realistic simulation study based on the windage disturbance reduction of a flexible hard disk drive suspension. *Copyright © 2003 IFAC*

Keywords: identification; closed-loop; disturbance rejection; hard disk drive

1. INTRODUCTION

For the modeling purposes of a system with unknown or partially known dynamics, system identification techniques can be used to characterize the dynamic behavior of the system (Ljung 1992). Models obtained by system identification techniques can be used for simulation, prediction or control purposes. Models for simulation purposes focus mainly on system dynamics, whereas models for prediction purposes may require open-loop accurate models of both system and noise dynamics to provide reliable prediction of output signals (Ljung 1992). On the other hand, models intended for control purposes may require high quality system dynamic representations of critical closed-loop behavior to design reliable robust servo controllers (Van Den Hof and Schrama 1995).

The need for control oriented modeling has resulted in several methodologies that aim at iteratively improving closed-loop system behavior on the basis of closed-loop experiments (Gevers 2002). In most of the existing methods, the emphasis is placed on the control-relevant approximation of system dynamics only and ignore the approximate modeling of the disturbance dynamics that is relevant in disturbance control. For minimum variance and LQG control, successful modeling and control performance improvements have been shown in (Gevers and Ljung 1986, Hjalmarsson *et al.* 1994), but these results assume consistent estimation of system and disturbance dynamics.

In dealing with closed-loop data, one of the problems in approximate closed-loop identification of plant and noise dynamics is the correlation of the disturbance with any of the signals in the closed-loop. As a result, a so-called direct identification using input and output of the plant will lead to biased approximation results for the system and disturbance dynamics (Van

Den Hof and Schrama 1995, de Callafon 1998). Possible ways to overcome this problem is by assuming low noise correlation condition (Gevers 1993, Zang *et al.* 1995, Åström 1993) that might only be realistic in simulation studies.

A possible way to deal with closed-loop data is a reparametrization of the closed-loop identification problem. Reparametrization can be done by a direct parametrization of the closed-loop transfer function as done in (Donkelaar and Van Den Hof 1996) or in the recursive algorithms for closed-loop identification of (Landau and Karimi 1997, Landau and Karimi 1999). Although powerful for estimating control-relevant plant dynamics, bias approximation results similar to direct identification are obtained in case an approximate noise model is estimated (Karimi and Landau 1998).

The contribution of this paper is to propose an new estimation method that allows for a control-relevant estimation of low order models of both system and noise dynamics. Several other parameter identification methods are reviewed and the bias distribution of the low order plant and noise model estimates of the proposed extended two-stage method are presented for comparison. It is illustrated how control-relevant models for both the system and the disturbance dynamics can be obtained on a case study based on models of a flexible suspension and the windage disturbance found in a conventional hard disk drive.

2. PROBLEM FORMULATION

In order to discuss the problem of estimating low order and control relevant models of plant dynamics and additive noise dynamics on the basis of closed-loop experiments, a feedback connection $T(P_0, C)$ of an unknown plant P_0 and a feedback controller C will be considered here. The feedback connection $T(P_0, C)$ is described in Figure 1 where the output $y(t)$ of the plant is fed back to the input $u(t)$ of the plant. Additionally, an additive noise $v(t)$ acts on the output of the plant which is modeled as a monic stable and stably invertible noise filter H_0 having a white noise input $e(t)$.

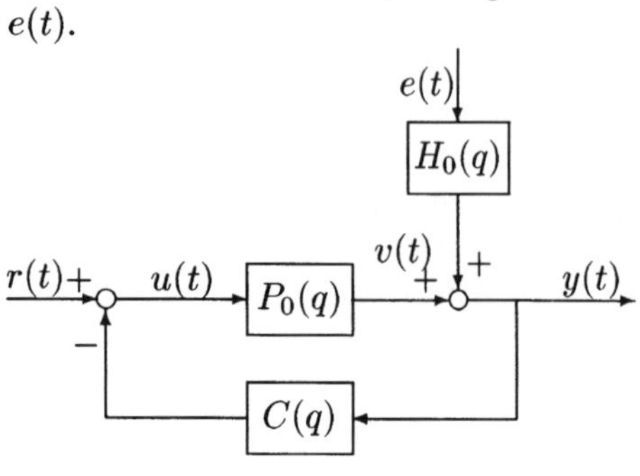

Fig. 1. Closed-loop system

Given Figure 1, the data coming from the plant $P_0(q)$ and subjected to external reference signal $r(t)$ and additive noise $H_0(q)e(t)$ operating under closed-loop condition can be described as follows:

$$y(t) = P_0(q)S_{in}(q)r(t) + S_{in}(q)H_0(q)e(t) \quad (1)$$
$$u(t) = S_{in}(q)r(t) - C(q)S_{in}H_0(q)e(t) \quad (2)$$

where q is the forward shift operator, and $S_{in}(q)$ is the input sensitivity function defined by

$$S_{in}(q) = \frac{1}{1 + C(q)P_0(q)}$$

Consider the feedback connection $T(P_0, C)$ in Figure 1 and the measurement of the reference $r(t)$ and the output $y(t)$, the objective is to estimate low order models of both plant P_0 and noise H_0 which are important in control design applications that focus on the disturbance rejection.

3. DIRECT METHOD

3.1 Method description

In this method, the input $u(t)$ and output $y(t)$ signals of the plant P_0 are used directly to identify the plant model P_θ and noise model H_θ. In this method the feedback is ignored, so the information of the controller C does not need to be known. Consider a general open loop input/output system with additive disturbances

$$y(t) = P_0 u(t) + H_0 e(t)$$

The one step predictor of the open-loop system is given by (Ljung 1992)

$$y(t|t-1, \theta) = H_\theta^{-1} P_\theta u(t) + (1 - H_\theta^{-1}) y(t).$$

To decouple the mutual influence between the plant model and noise model, estimation an independent parametrization of the plant model and noise model can be used. The parameter vector θ is split up in $\theta = [\xi^T \eta^T]^T$, and P_ξ and H_η are the models of P_0 and H_0, respectively. The prediction error is denoted by

$$\epsilon(t, \theta) = y(t) - y(t|t-1, \theta).$$

The resulting prediction error $\epsilon_\theta(t)$ is

$$\begin{aligned}\epsilon(t, \theta) = H_\theta^{-1}[(P_0 - P_\theta)u(t) + \\ +(H_0 - H_\theta)e(t)] + e(t).\end{aligned} \quad (3)$$

and the parameter estimate $\hat{\theta}$ is found by minimizing the 2-norm of prediction error:

$$\hat{\theta} = \arg\min_{\theta} ||\epsilon(t, \theta)||_2 \quad (4)$$

3.2 Biased distribution and conclusion

In the case of open loop identification, $u(t)$ and $e(t)$ are uncorrelated. The minimization (4) can be represented by an integral in the frequency domain, where minimizing argument of (4) is described by

$$\begin{aligned}\int_{-\pi}^{\pi} |H_\theta^{-1}(e^{jw})|^2 \Big[| P_0(e^{jw}) - P_\theta(e^{jw})|^2 \phi_u(w) \\ + |H_0(e^{jw}) - H_\theta(e^{jw})|^2 \phi_e(w) \Big] \, dw.\end{aligned} \quad (5)$$

For notational purposes, consider the model set

$$\mathcal{P} := \{P_\theta \mid \theta \in \Theta\} \quad (6)$$
$$\mathcal{H} := \{H_\theta \mid \theta \in \Theta\} \quad (7)$$

where Θ is the parameter space that guarantees stability of the prediction error (3). If the true system belongs to the model set, which is defined by $P_0 \in \mathcal{P}$ and $H_0 \in \mathcal{H}$, then a consistent identification of P_0 and H_0 is obtained (Ljung 1992). In case $P_0 \notin \mathcal{P}$ and $H_0 \in \mathcal{H}$, an expression for the approximate identification of the models P_θ and H_θ is derived as follows.

In the case the signals u and y are obtained under feedback, substitution (2) into (3) yields the following prediction error

$$\begin{aligned}\epsilon(t,\theta) &= H_\theta^{-1}[(P_0 - P_\theta)\, S_{in} r(t) \\ &+((P_\theta - P_0)CS_{in}H_0 + (H_0 - H_\theta))e(t)] + e(t)\end{aligned} \quad (8)$$

The last term $e(t)$ can be ignored because it does not contribute to the minimization. Because $r(t)$ and $e(t)$ are uncorrelated, we can get the bias expression of $\hat{\theta}$ as

$$\begin{aligned}\hat{\theta} &= \arg\min_\theta \int_{-\pi}^{\pi} |H_\theta^{-1}|^2[|P_0 - P_\theta|^2 |S_{in}|^2 \phi_r \\ &+|(P_\theta - P_0)CS_{in}H_0 + (H_0 - H_\theta)|^2 \phi_e]\, dw\end{aligned} \quad (9)$$

Compare (9) and (5), it shows that the estimation of P_0 and H_0 effect each other even if the plant model P_θ and noise model H_θ are parametrized independently. As a result, a biased estimation of the noise model will leads to a biased estimation of the plant model, and vice versa. This effect can be seen more clearly in case the noise model H_θ is not estimated and fixed to 1, as in an Output Error (OE) model, which yields a parameter estimate

$$\begin{aligned}\hat{\theta} &= \arg\min_\theta \int_{-\pi}^{\pi} [|P_0 - P_\theta|^2 |S_{in}|^2 \phi_r \\ &+ |(P_\theta C + 1)S_{in}H_0|^2 \phi_e]\, dw.\end{aligned} \quad (10)$$

When the reference signal $r(t) = 0$ and assuming that $-C^{-1} \in \mathcal{P}$, a biased estimation of $P_\theta = -C^{-1}$ is obtained, even if $P_0 \in \mathcal{P}$. As a result, the estimation of plant model P_θ will be biased, and it depends on the noise present on the closed-loop data.

4. TWO-STAGE IDENTIFICATION

4.1 Method description

In the two-stage method, identification of the plant model and noise model in closed loop is performed in two steps to eliminate the correlation between the input and the noise. The method can be summarized as follows (Van Den Hof and Schrama 1993). In the first step, one identifies a model S_{in}^* of the input sensitivity function S_{in} by considering the map from reference signal $r(t)$ to the plant input $u(t)$ in (2). The estimate S_{in}^* is then used to simulate a noise free input signal $u_r(t)$ via

$$u_r(t) = S_{in}^* r(t).$$

that will be uncorrelated with noise $e(t)$ on the closed-loop data. In case a consistent estimate $S_{in}^* = S_{in}$ is obtained in the first step, (1) rewrites into

$$y(t) = P_0 u_r(t) + S_{in} H_0 e(t)$$

Subsequently, in the second step of this method a plant model P_θ (and possibly a noise model H_θ) can be estimated by minimizing the two-norm of the prediction error

$$\begin{aligned}\epsilon(t,\theta) &= H_\theta^{-1}[P_0 u(t) - P_\theta u_r(t) \\ &+(H_0 - H_\theta)e(t)] + e(t)\end{aligned} \quad (11)$$

It should be noted that the model S_{in}^* is used only for filtering purposes. No specific restrictions on the order of this models is needed.

In general, the two-stage method is used only to estimate (low order) models P_θ of P_0 in the second step and the estimation of noise filters is omitted. For comparison and analysis purposes, we also consider the estimation of noise models in the standard two-stage method. Rewriting (11) in terms of the reference signal yields

$$\begin{aligned}\epsilon_\theta(t) &= H_\theta^{-1}\left[(P_0 S_{in} - P_\theta S_{in}^*)\, r(t)\right. \\ &+ \left.(H_0 S_{in} - H_\theta)\, e(t)\right] + e(t)\end{aligned} \quad (12)$$

and we will compare the results of noise model estimation with the direct method and the extended two-stage method proposed in this paper.

4.2 Biased distribution and conclusion

By minimizing the 2-norm of the prediction error (12) during the second step of the two-stage method, the parameter estimate $\hat{\theta}$ can be represented by the following integral expression (Van Den Hof and Schrama 1993)

$$\begin{aligned}\hat{\theta} &= \arg\min_\theta \int_{-\pi}^{\pi} |H_\theta^{-1}|^2 [|(P_0 - P_\theta)\, S_{in} \\ &+P_\theta\,(S_{in} - S_{in}^*)\,|^2 \phi_r + |H_0 S_{in} - H_\theta|^2\, \phi_e]\, dw\end{aligned} \quad (13)$$

From the above expression, the following remarks can be made with respect to the bias distribution of this method.

- In the case that the plant and noise model are identified independently, the bias of the noise model does not effect the estimation of the plant model.

- The estimation of the noise model H_θ is always biased, and it tends to $H_0 S_{in}$, which is the closed-loop noise model.

- The estimation of the plant model P_θ depends on the estimation of the input sensitivity function S_{in} obtained from the first step. In case $S_{in}^* \neq S_{in}$, it can be observed from equation (13) that the term $P_\theta\,(S_{in} - S_{in}^*)$ effects the fitting of $P_\theta \to P_0$. As a result, no explicit tunable expression for the misfit between P_0 and P_θ is obtained. However, this term can be made small by obtaining a consistent estimate of the sensitivity function in the first step of the method.

5. EXTENDED TWO-STAGE METHOD

5.1 Method description

The extended two-stage method, just as its name implies, is similar to the previously mentioned two-stage method. But the main difference lies in the use of a noise model estimate in the two steps of this method. To explain the extended two-stage method in more details, define $\bar{P} = P_0 S_{in}$ and $\bar{H} = H_0 S_{in}$ for notational convenience. $\bar{P}$ and $\bar{H}$ indicate the closed-loop transfer functions in (1). Using the knowledge of the controller C, (1) can be rewritten into the following two expressions:

$$y(t) = \bar{P}r(t) + \bar{H}e(t) \quad (14)$$
$$y(t) = P_0(1 - C\bar{P})r(t) + H_0(1 - C\bar{P})e(t) \quad (15)$$

From (15) it can be observed that with knowledge of $\bar{P}$, the controller C and a time realization of $e(t)$, the estimation of P_0 and H_0 becomes a standard open-loop identification problem. Furthermore, $\bar{P}$, $\bar{H}$ and a time realization of $e(t)$ are accessible from (14) by performing a consistent identification. From these observations, the extended two-stage method can be summarized as follows:

(1) Using the reference signal $r(t)$ and the output signal $y(t)$ according to (14) to perform an standard open-loop identification of $\bar{P}$ and $\bar{H}$. Using the estimated models $\bar{P}_*$ and $\bar{H}_*$, compute the closed-loop prediction error

$$\varepsilon_{cl}(t) = \bar{H}_*^{-1}\left(y(t) - \bar{P}_* r(t)\right). \quad (16)$$

(2) The estimated models $\bar{P}_*$ and $\bar{H}_*$ are used to create a filtered input $u_f(t)$ and a filtered prediction error $\varepsilon_f(t)$:

$$u_f(t) = (1 - C\bar{P}_*)r(t) \quad (17)$$
$$\varepsilon_f(t) = (1 - C\bar{P}_*)\varepsilon_{cl}(t). \quad (18)$$

Use the signals $u_f(t)$ and $\varepsilon_f(t)$ according to (15) to estimate low order models P_θ and H_θ by the output error minimization

$$\hat{\theta} = \arg\min_\theta \left\| y(t) - [P_\theta(q)\; H_\theta(q)] \begin{bmatrix} u_f(t) \\ \varepsilon_f(t) \end{bmatrix} \right\|_2 \quad (19)$$

During the open-loop identification of $\bar{P}$ and $\bar{H}$ in the first step of this method, a stable plant model $\bar{P}$ and a stable and stably-invertible noise model $\bar{H}$ are estimated. The reason for the construction of the closed loop residuals in (16) in the first step of the method is to allow control over the order of the estimated noise model in the second step. An alternative would be to compute a noise model from the estimate $\bar{H}_*$ using knowledge of C and a model P_θ, but this would lead to a higher order estimate of the noise model.

Compared to the two-stage method it can be observed that the extended two-stage method also uses the knowledge of $\bar{H}_*$ to estimate lower order approximations of H_0 in the second step. Only the signals $r(t)$ and $y(t)$ are used in this method, but it should be observed that the knowledge of the controller C can be replaced by an additional measurement of $u(t)$.

Similar to the two-stage method in (Karimi and Landau 1998), the models $\bar{P}_*$ and $\bar{H}_*$ are only used for filtering purposes. No specific restrictions on the order of these models is needed, as they are only used for filtering purposes. Moreover, the computation of the prediction error $\varepsilon_{cl}(t)$ can be used for model assessment purposes to validate the models $\bar{P}_*$ and $\bar{H}_*$ being estimated (Ljung 1992).

5.2 Biased distribution and conclusion

A result for the bias distribution of the estimation of plant model P_θ and noise model H_θ is given in the following theorem.

Theorem 1. Consider the first step in the extended two-stage method with estimates $\bar{P}_*$ and $\bar{H}_*$ with

$$\bar{P}_* \neq P_0 S_{in}, \quad \bar{H}_* \neq \bar{H} \quad (20)$$

then the minimization of (19) is equivalent to

$$\min_\theta \int_{-\pi}^{\pi} [|(P_0 - P_\theta)S_{in} + (\bar{P}_* - \bar{P})(P_\theta C + H_\theta(1 - C\bar{P}_*)\bar{H}_*^{-1})|^2 \phi_r(w) + |(H_0 - H_\theta)S_{in} + H_\theta(S_{in} - (1 - C\bar{P}_*)\bar{H}\bar{H}_*^{-1})|^2 \phi_e(w)]\, dw \quad (21)$$

where P_θ and H_θ denote the models estimated in the second step of the extended two-stage method.

Proof: With (20), (16) rewrites to

$$\varepsilon_{cl}(t) = \bar{H}_*^{-1}(\bar{P} - \bar{P}_*)r(t) + \bar{H}_*^{-1}\bar{H}e(t) \quad (22)$$

The prediction error can be computed as follows:

$$\begin{aligned}\epsilon_\theta(t) &= y(t) - P_\theta u_f - H_\theta \varepsilon_f \\ &= P_0 S_{in} r(t) + H_0 S_{in} e(t) - P_\theta u_f - H_\theta \varepsilon_f\end{aligned} \quad (23)$$

Using (17), (18) and (22), (23) can be written as

$$\begin{aligned}\epsilon_\theta(t) = {} & [(P_0 - P_\theta)S_{in} + (\bar{P}_* - \bar{P})(P_\theta C \\ & + H_\theta(1 - C\bar{P}_*)\bar{H}_*^{-1})]r(t) + [(H_0 - H_\theta)S_{in} \\ & + H_\theta(S_{in} - (1 - C\bar{P}_*)\bar{H}\bar{H}_*^{-1})]e(t)\end{aligned} \quad (24)$$

which leads to the bias distribution (21). □

This results gives the bias distribution for the general case. Useful insight in the bias distribution of P_θ and H_θ in the special cases are described in the following corollaries.

Corollary 1. Let $\bar{P}_* = P_0 S_{in}$, $\bar{H}_* = \bar{H}$ in the first step in the extended two-stage method, then minimization of (19) is equivalent to

$$\min_\theta \int_{-\pi}^{\pi} [\left| P_0(e^{jw}) - P_\theta(e^{jw}) \right|^2 |S_{in}|^2 \phi_r(w) + \left| H_0(e^{jw}) - H_\theta(e^{jw}) \right|^2 |S_{in}|^2 \phi_e(w)]\, dw \quad (25)$$

Proof: Substitute $\bar{P}_* = P_0 S_{in}$, $\bar{H}_* = \bar{H}$ into (21), then (25) is obtained. □

It is easily observed that in the case $\bar{P}_* = P_0 S_{in}$, $\bar{H}_* = \bar{H}$, the difference $| P_0 - P_\theta|^2$ is weighted by the reference spectrum ϕ_r and the difference $| H_0 - H_\theta|^2$ is weighted by noise spectrum ϕ_e. Both are weighted by the input sensitivity function S_{in}, which is advantageous for the closed-loop approximation of P_0 by P_θ and H_0 by H_θ.

Corollary 2. Let $\bar{P}_* \neq P_0 S_{in}$, $\bar{H}_* = \bar{H}$ in the first step in the extended two-stage method, then the minimization of (19) is equivalent to

$$\min_\theta \int_{-\pi}^{\pi} [|(P_0 - P_\theta)S_{in} + (\bar{P}_* - \bar{P})(P_\theta C + H_\theta(1 - C\bar{P}_*)\bar{H}_*^{-1})|^2 \phi_r(w) + |(H_0 - H_\theta)S_{in} + (\bar{P}_* - \bar{P})C H_\theta|^2 \phi_e(w)]\, dw \quad (26)$$

Proof: Substitute $\bar{P}_* \neq P_0 S_{in}$, $\bar{H}_* = \bar{H}$ into (21) and the result in (26) is obtained. □

In the case $\bar{P}_* \neq P_0 S_{in}$, $\bar{H}_* = \bar{H}$, two terms including $(\bar{P}_* - \bar{P})$ are introduced that will effect the fitting of P_θ to P_0, H_θ to H_0, respectively. The biased estimation of $\bar{P}$ will effect the estimation of the noise model. No explicit tunable expressions of the misfit between P_0 and P_θ, and between H_0 and H_θ are obtained. This situation is similar to the two-stage method.

Corollary 3. Let $\bar{P}_* = P_0 S_{in}$, $\bar{H}_* \neq \bar{H}$ in the first step in the extended two-stage method, then the minimization of (19) is equivalent to

$$\min_\theta \int_{-\pi}^{\pi} [|(P_0 - P_\theta)|^2 |S_{in}|^2 \phi_r(w) + |(H_0 - H_\theta)S_{in} + H_\theta(1 - \bar{H}\bar{H}_*^{-1})S_{in}|^2 \phi_e(w)]\, dw$$

Proof: Substitution of $\bar{P}_* = P_0 S_{in}$, $\bar{H}_* \neq \bar{H}$ into (21) yields the result. □

In the case $\bar{P}_* = P_0 S_{in}$, $\bar{H}_* \neq \bar{H}$, a term $H_\theta(1 - \bar{H}\bar{H}_*^{-1})$ is created to effect the fitting of H_θ to H_0, weighted by S_{in}. No explicit tunable expressions of the misfit between H_0 and H_θ is obtained.

Summarizing, the following result can be mentioned for the extended two-stage method.

- This method gives an unbiased estimation of the plant and noise model when they are in the model set.
- In case (high order) consistent estimates are obtained in the first step, a tunable bias expression for both the plant model P_θ and the noise model H_θ is obtained in the second step.
- Even though we use the higher order models to fit $\bar{P}$ and $\bar{H}$ in the first step, the orders of the models P_θ and H_θ can be reduced significantly in the second step.
- The bias of the noise model will effect the estimation of the plant model in some special cases, and vice versa.

6. APPLICATION TO CASE STUDY

The case study in this paper is a simulation study based on a model of a flexible mechanical suspension and the windage disturbance found in a conventional hard disk drive (HDD) (Crowder and de Callafon 2003). A schematic representation of the system under consideration is illustrated in Figure 2. Using the notation P_0 and H_0 to respectively represent the dynamics of the flexible suspension and the dynamics of the windage disturbances, a block diagram similar to Figure 1 is obtained.

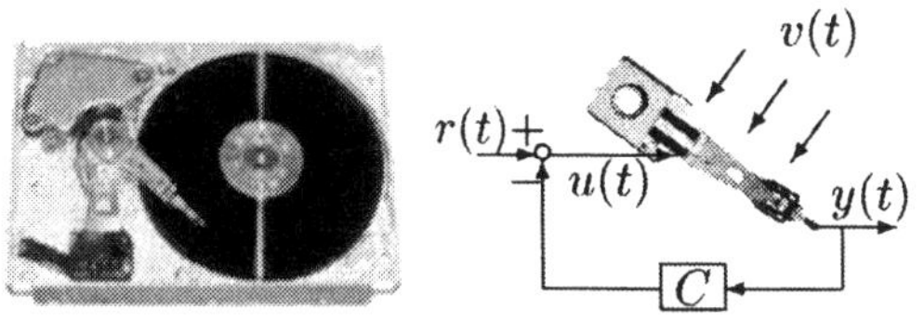

Fig. 2. Configuration of HDD (left) and schematic representation of flexible suspension, windage disturbance and servo controller C (right).

The consistent estimation of (relatively high 12th order) discrete time models for $P_0(q)$ and $H_0(q)$ on the basis of experimental data has been illustrated in (Crowder and de Callafon 2003). For illustrative purposes, an amplitude Bode plot of the 12th order models of P_0 and H_0 found in (Crowder and de Callafon 2003) is given in Figure 3.

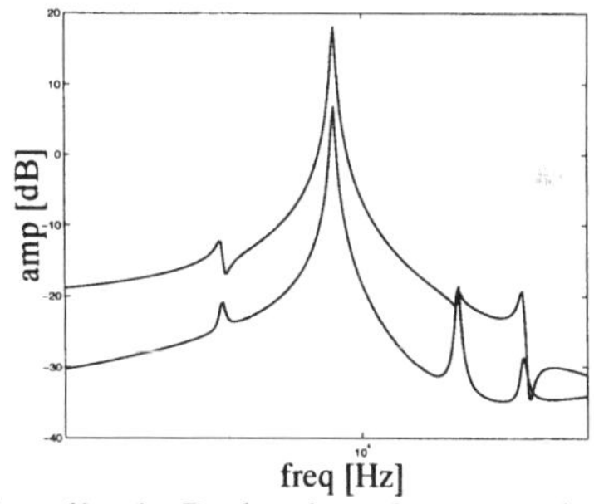

Fig. 3. Amplitude Bode plot of system dynamics P_0 (top line) and noise dynamics H_0 (bottom line)

The servo controller $C(q)$ used during the closed-loop experiments in our case study is a 3rd order lead/lag compensator given by

$$C(q) = \frac{5.33q^2 - 0.125q - 4.954}{q^3 - 0.2697q^2 - 0.5338q + 0.09558}$$

and closed-loop data of 4096 data points, sampled at 51.2kHz is obtained with r and e as independent white noise sequences with unit variance. The objective of the case study is to find low (4th) order models P_θ and H_θ of P_0 and H_0 depicted in Figure 3 on the basis of closed-loop experiments.

Figure 4, Figure 5, Figure 6 are the bode plots of the identified plant and noise models, compared with real system using direct method, two stage method and extended two-stage method respectively. By analyzing and comparing the characteristics of these methods, the following results can be summarized.

• Low order approximation with the direct method gives biased estimation results for both the plant and the disturbance dynamics.

• A low order approximation of plant dynamics is successful for the two-stage method. However, the method yields biased results for the noise filter.

• The extended two-stage method can be used to obtain low order approximations of both the plant model and the noise dynamics.

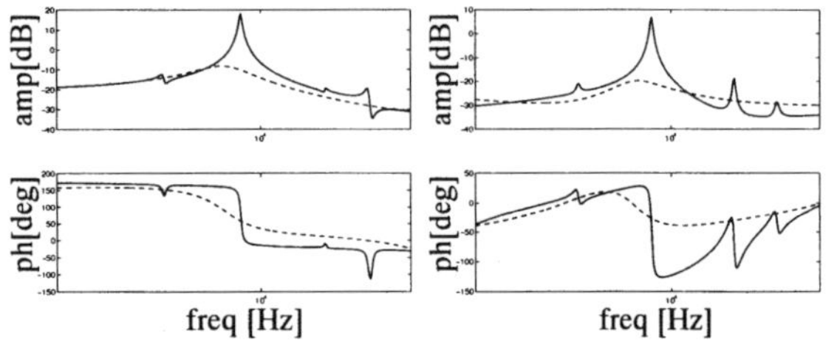

Fig. 4. Application of direct identification. Left: Bode plot of plant P_0 (solid) and the 4th order model P_θ (dashed). Right: Bode plot of H_0 (solid) and the 4th order noise model H_θ (dashed).

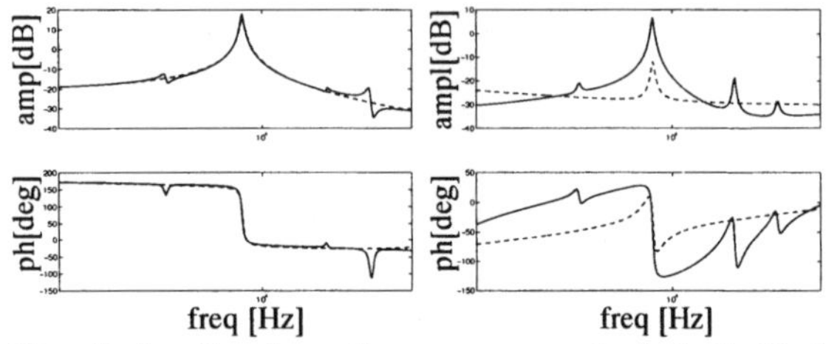

Fig. 5. Application of two-stage method. Left: Bode plot of plant P_0 (solid) and the 4th order model P_θ (dashed). Right: Bode plot of H_0 (solid) and the 4th order noise model H_θ (dashed).

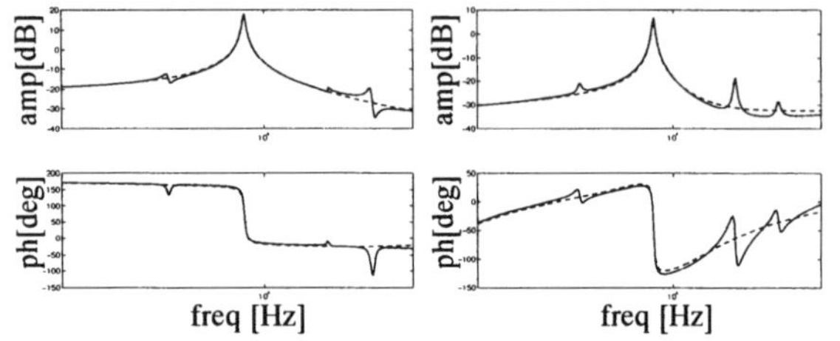

Fig. 6. Application of extended two-stage method. Left: Bode plot of plant P_0 (solid) and the 4th order model P_θ (dashed). Right: Bode plot of H_0 (solid) and the 4th order noise model H_θ (dashed).

7. CONCLUSIONS

In this paper, several methods for low order plant model and noise model identification are discussed and compared in terms of the bias distribution of the approximate estimation. A new extended two-stage estimation method is proposed to improve the approximate estimation of both plant and noise model dynamics. The method is evaluated on the basis of simulated closed-loop data from a hard disk drive experiment and shows improvements with respect to low order approximation of plant and noise models.

REFERENCES

Åström, K.J. (1993). Matching criteria for control and identification. In: *European Control Conference*. Groningen, the Netherlands. pp. 248–251.

Crowder, M. and R.A. de Callafon (2003). Estimation and prediction of windage induced suspension vibrations in a hard disk drive. *Journal of Information and Storage Processing Systems*. (submitted for publication).

de Callafon, R.A. (1998). *Feedback Oriented Identification for Enhanced and Robust Control*. Delft University Press. The Netherlands.

Donkelaar, E.T. and P.M.J. Van Den Hof (1996). Analysis of closed-loop indentification with a tailor-made parametrization. *Selected Topics in Identification, Modelling and Control* **9**, 17–24.

Gevers, M. (1993). Towards a joint design of identification and control? In: *Essays on Control, Perspectives in the Theory and its Application* (H.L Trentelman and J.C. Willems, Eds.). Birkhäuser. Boston, USA.

Gevers, M. (2002). A decade of progress in iterative process control design: From theory to practice. *Journal of Process Control* **12**(4), 519–531.

Gevers, M and L. Ljung (1986). Optimal experiment designs with respect to the intended model application. *Automatica* **22(5)**, 543–554.

Hjalmarsson, H., M. Gevers, F. De Bruyne and J. Leblond (1994). Identification for control: Closing the loop gives more accurate controllers. In: *Conference on Decision and Control*. Lake Buena Vista, FL, USA. pp. 4150–4155.

Karimi, A. and I. D. Landau (1998). Comparison of the closed-loop identification methods in terms of the bias distribution. *Systems and Control Letters* **34**, 159–167.

Landau, I.D. and A. Karimi (1997). Recursive algorithms for identification in closed loop: A unified approach and evaluation. *Automatica* **33**, 1499–1523.

Landau, I.D. and A. Karimi (1999). A recursive algoritnm for ARMAX model identification in closed loop. *IEEE Transactions on Automatic Control* **44**, 840–843.

Ljung, L. (1992). *System Identification, Theory for the User*. Prentice-Hall. Englewood Cliffs,NJ.

Van Den Hof, P.M.J. and R.J.P. Schrama (1993). An indirect method for transfer function estimation from closed loop data. *Automatica* **29**, 1523–1527.

Van Den Hof, P.M.J. and R.J.P. Schrama (1995). Identification and control–closed-loop issues. *Automatica* **31**, 1751–1770.

Zang, Z., R.R. Bitmead and M. Gevers (1995). Iterative weighted least-squares identification and weighted LQG control design. *Automatica* **31(11)**, 1577–1594.

IFAC Publications
www.elsevier.com/locate/ifac

IV METHODS FOR CLOSED-LOOP SYSTEM IDENTIFICATION

Marion Gilson * and **Paul Van den Hof** **

* *Centre de Recherche en Automatique de Nancy (CRAN), CNRS UMR7039, Université Henri Poincaré, Nancy 1, BP 239, F-54506 Vandoeuvre-lès-Nancy Cedex, France.*
Email: marion.gilson@cran.uhp-nancy.fr
** *Delft Center for Systems and Control, Delft University of Technology, Mekelweg 2, 2628 CD Delft, The Netherlands.*
Email: P.M.J.vandenHof@TN.TUDelft.NL

Abstract: In this paper, several instrumental variable (IV) and instrumental variable-related methods for closed-loop system identification are considered and set in an extended IV framework. Extended IV methods require the appropriate choice of particular design variables, as the number and type of instrumental signals, data prefiltering and the choice of an appropriate norm of the extended IV-criterion. The optimal IV estimator achieves minimum variance, but requires the exact knowledge of the noise model. For the closed-loop situation several IV methods, such as tailor-made IV, IV4 and BELS are put in an extended IV framework and characterized by different choices of design variables. Their variance properties are considered and illustrated with a simulation example. *Copyright © 2003 IFAC*

Keywords: Closed-loop system identification; linear estimators; instrumental variables.

1. INTRODUCTION

For many industrial production processes, safety and production restrictions are often strong reasons for not allowing identification experiments in open-loop. In such situations, experimental data can only be obtained under so-called closed-loop conditions. The main difficulty in closed-loop identification is due to the correlation between the disturbances and the control signal, induced by the loop. Several classical alternatives are available to cope with this problem, broadly classified into three main approaches: direct, indirect and joint input/output (Söderström and Stoica 1989, Ljung 1999). Some particular versions of these methods have been developed more recently in the area of control-relevant identification as e.g. the two-stage, the coprime factor, the dual-Youla methods. An overview of these recent developments can be found in Van den Hof (1998) and Forssell and Ljung (1999).

When looking at methods that can consistently identify plant models of systems operating in closed-loop while relying on simple linear (regression) algorithms, instrumental variable (IV) techniques seem to be rather attractive, but at the same time also not very often applied. On the other hand, when dealing with highly complex processes that are high dimensional in terms of inputs and outputs, it can be attractive to rely on methods that do not require non-convex optimization algorithms.

For closed-loop identification a basic IV estimator has been proposed (Söderström *et al.* 1987), and more recently a so-called tailor-made IV algorithm (Gilson and Van den Hof 2001), where the closed-loop plant is parametrized using (open-loop) plant parameters. The class of algorithms denoted by BELS (for Bias-Eliminated Least-Squares), e.g. Zheng (1996), is also directed towards the use of linear regression algorithms only. It has recently been shown that these algorithms are also particular forms of IV estimation schemes (Gilson and Van den Hof 2001). Then, when comparing the several available IV algorithms, the principal question to address should be: how to achieve the smallest variance of the estimate. For extended IV methods an optimality result has been developed in the open-loop case, showing consequences for the optimal choice of weights, filters, and instruments. This result can be extended to the closed-loop case.

In this paper the several IV and IV-related methods are set in an extended IV framework, and the consequences for the several design variables (related to optimal variance) are considered. Since for optimal variance, the noise model has to be known exactly, several bootstrap methods are proposed for approximating this required information from measurement data. The comparison between the different proposed methods is illustrated in a simulation example, showing that the optimal estimator can be accurately approximated by an appropriate choice of the design parameters.

2. PRELIMINARIES

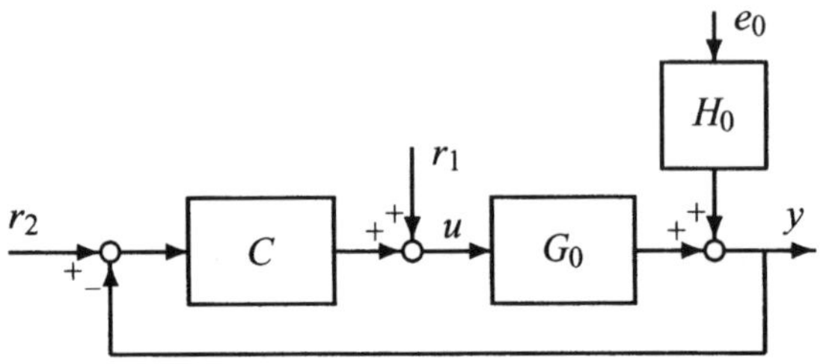

Fig. 1. Closed-loop configuration.

Consider a linear SISO closed-loop system shown in figure 1. The process is denoted by G_0 and the controller by C; $u(t)$ describes the process input signal, $y(t)$ the process output signal and $\{e_0(t)\}$ is a sequence of independent identically distributed random variables with variance λ_0. The external signals $r_1(t)$, $r_2(t)$ are assumed to be uncorrelated with $e_0(t)$. For ease of notation we also introduce the signal $r(t) = r_1(t) + C(q)r_2(t)$. With this notation, the data generating system becomes

$$\mathscr{S}: \begin{cases} y(t) = G_0(q)u(t) + H_0(q)e_0(t) \\ u(t) = r(t) - C(q)y(t) \end{cases} \tag{1}$$

The real plant G_0 is considered to satisfy $G_0(q) = B_0(q^{-1})/A_0(q^{-1})$, while in these expressions q^{-1} is the delay operator, and the numerator and denominator polynomials have degree n_0. The m-th order controller C is assumed to be known and specified by

$$C(q) = \frac{Q(q^{-1})}{P(q^{-1})} \tag{2}$$

with the pair of polynomials (P, Q) assumed to be coprime.

The closed-loop transfer function can be written as

$$y(t) = \frac{G_0}{1+CG_0} r(t) + \frac{H_0}{1+CG_0} e_0(t) \tag{3}$$

or in polynomial fraction form

$$y(t) = \frac{B^0_{cl}}{A^0_{cl}} r(t) + \frac{1}{A^0_{cl}} \xi(t) \tag{4}$$

with $\xi(t) = A_0 P H_0 e_0(t)$. The polynomials B^0_{cl} and A^0_{cl} will generically have orders $n_0 + m$. A parametrized process model is considered

$$\mathscr{G}: G(q,\theta) = \frac{B(q^{-1},\theta)}{A(q^{-1},\theta)} = \frac{b_1 q^{-1} + \cdots + b_n q^{-n}}{1 + a_1 q^{-1} + \cdots + a_n q^{-n}},$$

and the process model parameters are stacked columnwise in the parameter vector

$$\theta = \begin{bmatrix} a_1 & \cdots & a_n & b_1 & \cdots & b_n \end{bmatrix}^T \in \mathbb{R}^{2n}. \tag{5}$$

Furthermore, let us denote by $\varphi(t)$ and by $\psi(t)$ the closed-loop and open-loop regressors respectively, defined as

$$\varphi^T(t) = [-y(t-1) \cdots -y(t-n-m)\; r(t-1) \cdots r(t-r_B)]$$
$$\psi^T(t) = [-y(t-1) \cdots -y(t-n)\; u(t-1) \cdots u(t-n)]$$
$$\varphi_r^T(t) = [r(t-1)\; \cdots\; r(t-r_B)],$$

and r_B a user-specified integer. Additionally we use the following notation

$$\bar{\psi}(t) = P(q^{-1})\psi(t) \tag{6}$$
$$\bar{y}(t) = P(q^{-1})y(t). \tag{7}$$

3. IV METHODS IN AN EXTENDED IV FRAMEWORK

3.1 *Tailor-made IV identification (M1)*

The tailor-made IV method (referred to M1 in the following) as discussed in (Gilson and Van den Hof 2001) is designed to provide an unbiased estimate for the process model $G(q,\theta)$, while pertaining to simple linear regression type of estimates. Accurate noise modelling (i.e. estimating H_0) is not considered part of the problem.

Consider the (closed-loop) prediction error

$$\varepsilon(t,\theta) = \bar{A}_{cl}(q^{-1},\theta)y(t) - \bar{B}_{cl}(q^{-1},\theta)r(t) \tag{8}$$
$$\bar{B}_{cl}(q^{-1},\theta) = B(q^{-1},\theta)P(q^{-1})$$
$$\bar{A}_{cl}(q^{-1},\theta) = A(q^{-1},\theta)P(q^{-1}) + B(q^{-1},\theta)Q(q^{-1}),$$

parametrized in the plant parameter θ (tailor-made parametrization), and the prediction error alternatively written as

$$\varepsilon(t,\theta) = \bar{y}(t) - \bar{\psi}^T(t)\theta. \tag{9}$$

Then the tailor-made IV estimate of θ is determined as the solution to the set of equations

$$\frac{1}{N}\sum_{t=1}^{N} \varepsilon(t, \hat{\theta}_{tiv,F})\eta(t) = 0, \tag{10}$$

where $\eta(t) = F\varphi_r(t)$, and $F \in \mathbb{R}^{2n \times r_B}$ a user-chosen matrix with rank $2n$.

The choice $r_B = 2n$, $F = I_{2n}$ leads to a simple basic IV estimator, taking as instruments $2n$ delayed samples of the reference signal, that is supposed to bepersisntely exciting of sufficiently high order. For $r_B > 2n$, the matrix F constructs $2n$ instruments out of r_B delayed reference samples, by taking particular linear combinations.

3.2 *BELS method*

The so-called bias-eliminated least-squares method (BELS) as proposed by (Zheng and Feng 1995, Zheng 1996) has been shown to be a particular form of tailor-made IV estimator (Gilson and Van den Hof 2001). It has two different formats, dependent on the relation between n (model order) and m (controller order). For $m \leq n$ the BELS estimator is equivalent to the tailor-made estimate with $r_B = 2n$ and $F = I_{2n}$. For $m > n$ it is obtained by choosing $r_B = n + m$ and

$$F = M^T \hat{R}^T_{\varphi_r \varphi}(N)(\hat{R}_{\varphi_r \varphi}(N)\hat{R}^T_{\varphi_r \varphi}(N))^{-1}. \qquad (11)$$

with $\hat{R}_{\varphi_r \varphi}(N) = \frac{1}{N}\sum_{t=1}^{N} \varphi_r(t)\varphi^T(t)$, and $M \in \mathbb{R}^{(n+m+r_B)\times 2n}$ a full-column rank matrix dependent on controller dynamics. For a full description of the relation between BELS and tailor-made IV, see Gilson and Van den Hof (2001).

3.3 *Tailor-made and extended IV identification*

In order to analyse the variance properties of the estimators presented above, they are positioned in the framework of extended IV estimators. An extended IV estimate of θ_0 is obtained by generalizing the so-called basic IV estimates of θ by prefiltering the data and by using an augmented instrument $z(t) \in \mathbb{R}^{n_z}$ ($n_z \geq 2n$) so that an over-determined set of equations is obtained

$$\hat{\theta}_{eiv}(N) = \arg\min_{\theta} \left\| \left[\frac{1}{N}\sum_{t=1}^{N} z(t)L(q^{-1})\psi^T(t)\right]\theta - \left[\frac{1}{N}\sum_{t=1}^{N} z(t)L(q^{-1})y(t)\right] \right\|_Q^2,$$

where $L(q^{-1})$ is a stable prefilter and $\|x\|_Q^2 = x^T Q x$, with Q a positive definite weighting matrix.

Proposition 1. The tailor-made IV estimates presented in sections 3.1 and 3.2 with the particular choice of F given in (11) satisfies

$$\hat{\theta}_{iv}(N) = \arg\min_{\theta} \|\hat{R}_{\varphi_r \bar{\psi}}(N)\theta - \hat{R}_{\varphi_r \bar{y}}(N)\|_Q^2 \qquad (12)$$

with $\hat{R}_{\varphi_r \bar{\psi}}(N) = \frac{1}{N}\sum_{t=1}^{N} \varphi_r(t)\bar{\psi}^T(t)$, $\hat{R}_{\varphi_r \bar{y}}(N) = \frac{1}{N}\sum_{t=1}^{N} \varphi_r(t)\bar{y}(t)$, and

$$Q = \left(\hat{R}_{\varphi_r \varphi}\hat{R}^T_{\varphi_r \varphi}\right)^{-1} \in \mathbb{R}^{(n+m)\times(n+m)}.$$

It is equivalent to an extended IV estimator where

- the instruments are chosen such that $z(t) = \varphi_r(t)$ and $n_z = r_B = n + m$,
- the prefilter $L(q^{-1})$ is taken equal to the controller denominator $P(q^{-1})$,
- the regressor $\psi(t)$ depends on delayed input $u(t)$ and output $y(t)$ values.

Proof. A full proof is added in the appendix.

The notation (N) is omitted for ease of notation. The covariance matrix of this extended IV estimate is computed in the scalar case by using equation (8.30) in Söderström and Stoica (1989). Under the assumption $G_0 \in \mathscr{G}$

$$P_{eiv} = \lambda_0\left(R^T_{\varphi_r \bar{\psi}} Q R_{\varphi_r \bar{\psi}}\right)^{-1} R^T_{\varphi_r \bar{\psi}} Q R_{z_T z_T} Q R_{\varphi_r \bar{\psi}}\left(R^T_{\varphi_r \bar{\psi}} Q R_{\varphi_r \bar{\psi}}\right)^{-1} \qquad (13)$$

where [1]

$$R_{\varphi_r \bar{\psi}} = \bar{\mathbb{E}}\varphi_r(t)\bar{\psi}^T(t) = \bar{\mathbb{E}}\varphi_r(t)P(q^{-1})\psi^T(t) \qquad (14)$$

$$R_{z_T z_T} = \bar{\mathbb{E}} z_T(t) z_T^T(t) \qquad (15)$$

$$z_T(t) = \sum_{i=0}^{\infty} t_i \varphi_r(t-i) \qquad (16)$$

and $\{t_i\}$ is determined by the monic filter

$$T(q^{-1}) = P(q^{-1})A_0(q^{-1})H_0(q^{-1}) = \sum_{i=0}^{\infty} t_i q^{-i}. \qquad (17)$$

Remark. In the situation $r_B = 2n$ and $F = I_{2n}$, the tailor-made IV estimate is an extended IV estimate with $n_z = r_B = 2n$, $Q = I$ and $L(q^{-1}) = P(q^{-1})$. Thus, according to equation (13) and under the assumption $G_0 \in \mathscr{G}$, the expression for the covariance matrix of this estimate simplifies to

$$P_{tiv} = \lambda_0 R^{-1}_{\varphi_r \bar{\psi}} R_{z_T z_T} R^{-T}_{\varphi_r \bar{\psi}}. \qquad (18)$$

4. OPTIMAL CLOSED-LOOP IV

The choice of the instruments $z(t)$, of n_z, of the weighting matrix Q and of the prefilter $L(q^{-1})$ may have a considerable effect on the covariance matrix P_{eiv}. The lower bound of P_{eiv} for any unbiased identification method is given by the Cramer-Rao bound, which for the open-lop situation is specified in e.g. Ljung (1999) and Söderström and Stoica (1983). For the closed-loop case a lower bound of P_{eiv} has been provided in Forssell and Chou (1998), but restricted to the case of an ARMAX type of model. However, the more general case can also be analyzed in the closed-loop framework. Indeed, as explained in (Ljung 1999), the Cramer-Rao bound gives a lower bound for any unbiased estimation problem. Therefore, it applies also to the closed-loop IV technique and the lower bound of (13) is given by (normality is assumed here: $\sqrt{N}(\hat{\theta} - \theta^*) \in AsN(0, P_{eiv})$)

$$P_{eiv}^{opt} = \lambda_0\left[\bar{\mathbb{E}}\breve{\psi}(t)\breve{\psi}^T(t)\right]^{-1}, \qquad (19)$$

where

$$\breve{\psi}^T(t) = \left[\frac{d}{d\theta}\hat{y}(t|\theta)\right]^T\Bigg|_{\theta=\theta_0} = [A_0(q^{-1})H_0(q^{-1})]^{-1}\tilde{\psi}^T(t)$$

---

[1] The notation $\bar{\mathbb{E}}[.] = \lim_{N\to\infty}\frac{1}{N}\sum_{t=0}^{N-1}\mathbb{E}[.]$ is adopted from the prediction error framework of Ljung (1999)

is the gradient and $\tilde{\psi}(t)$ denotes the noise-free part of $\psi(t)$. P_{eiv}^{opt} is achieved by taking e.g.

$$z(t) = \lambda_0^{-1} \left[\left(A_0(q^{-1}) H_0(q^{-1}) \right)^{-1} \tilde{\psi}^T(t) \right]^T \quad (20)$$

$$n_z = 2n, \; Q = I \quad (21)$$

$$L(q^{-1}) = [A_0(q^{-1}) H_0(q^{-1})]^{-1}. \quad (22)$$

Then, the optimal IV estimator can only be obtained if the true noise model $A_0(q^{-1})H_0(q^{-1})$ is exactly known and therefore optimal accuracy cannot be achieved in practice.

5. APPROXIMATE IMPLEMENTATIONS OF THE OPTIMAL CLOSED-LOOP IV

In order to give some clues to the closed-loop identification method users, it would be interesting to compare the tailor-made IV method with the optimal IV one. However, as the latter cannot be achieved in practice, approximate implementaions of the optimal IV method will be considered. For this purpose one will need to take care that

- a noise model is available in order to construct the prefilter $L(q^{-1})$ and the instruments $z(t)$,
- a first model of $G_0(q)$ is needed to compute the noise free part of the regressor $\tilde{\psi}(t)$.

The choice of the instruments and prefilter in the IV method affects the asymptotic variance, while consistency properties are generically secured. This suggests that minor deviations from the optimal value (which is not available in practice) will only cause second-order effects in the resulting accuracy. Therefore it could be sufficient to use consistent, but not necessarily efficient estimates of G_0 and H_0 when constructing the instruments and the prefilter (Ljung 1999).
The following sections present two practical solutions to approximate the optimal closed-loop IV estimator. Only linear regressions are used for retaining the simplicity of the IV method.

5.1 *Extensions of the IV4 method (M2, M3)*

Several bootstrap IV methods have been proposed in the open-loop situation, in an attempt to approximate the optimal IV method, see e.g. (Young 1976, Söderström and Stoica 1983, Ljung 1999). A first solution consists thus in extending one of these algorithms to the closed-loop situation; here the IV4 method (Ljung 1999) will be considered. The major difference between open-loop and closed-loop cases is that in the latter, also the input is correlated with the noise. Therefore, the instruments have to be uncorrelated with the noise part of $u(t)$ but correlated with the noise-free part of $u(t)$.

Method M2.

Step 1. Write the model structure as a linear regression

$$\hat{y}(t, \theta) = \psi(t)^T \theta. \quad (23)$$

Estimate θ by a least-squares method and get $\hat{\theta}_1$ along with the corresponding transfer function $\hat{G}_1(q)$, of order n.

Step2. Generate the instruments $z_1(t)$ as

$$\tilde{y}_1(t) = \frac{C(q)\hat{G}_1(q)}{1 + C(q)\hat{G}_1(q)} r(t) \quad (24)$$

$$\tilde{u}_1(t) = \frac{1}{1 + C(q)\hat{G}_1(q)} r(t) \quad (25)$$

$$z_1(t) = [-\tilde{y}_1(t-1) \cdots -\tilde{y}_1(t-n) \tilde{u}_1(t-1) \cdots \tilde{u}_1(t-n)]^T$$

$z_1(t)$ can be seen as an estimation of the noise-free part of the regressor $\psi(t)$. Determine the IV estimate of θ in (23) as

$$\hat{\theta}_2 = \hat{R}_{z_1\psi}^{-1} \hat{R}_{z_1 y} \quad (26)$$

The corresponding estimated transfer function is given by $\hat{G}_2(q) = \frac{\hat{B}_2(q^{-1})}{\hat{A}_2(q^{-1})}$, of order n.

Step 3. Let $\hat{w}(t) = \hat{A}_2(q^{-1})y(t) - \hat{B}_2(q^{-1})u(t)$ and postulate an AR model of order $2n$ for $\hat{w}(t)$: $L(q^{-1})\hat{w}(t) = e(t)$. Estimate $L(q^{-1})$ using a least-squares method and denote the result by $\hat{L}(q^{-1})$.

Step 4. Generate the instruments $z_2(t)$ as

$$\tilde{y}_2(t) = \frac{C(q)\hat{G}_2(q)}{1 + C(q)\hat{G}_2(q)} r(t), \tilde{u}_2(t) = \frac{1}{1 + C(q)\hat{G}_2(q)} r(t)$$

$$z_2(t) = [-\tilde{y}_2(t-1) \cdots -\tilde{y}_2(t-n) \tilde{u}_2(t-1) \cdots \tilde{u}_2(t-n)]^T$$

Using these instruments $z_2(t)$ and the prefilter $\hat{L}(q^{-1})$, determine the IV estimate of θ in (23) as

$$\hat{\theta}_{M2} = \hat{R}_{z_2\psi_T}^{-1} \hat{R}_{z_2 y_T}, \quad (27)$$

where $\psi_T(t) = \hat{L}(q)\psi(t)$ and $y_T(t) = \hat{L}(q)y(t)$.

The asymptotic covariance matrix of the final estimates is the Cramer-Rao bound, provided the true noise model is an autoregression of order $2n$.

Method M3. This method can be improved by using a more sophisticated noise modeling procedure, e.g. by replacing the third step of the M2 algorithm by the `armasel` procedure developed in Broersen (2002), including an appropriate order selection step. This procedure consists in estimating several autoregressive models of different orders and in applying a nonasymptotic order selection criterion based on estimates of prediction error expectation.

5.2 *Another closed-loop "optimal" IV method (M4)*

Noise and process models have to be known in order to construct the instruments and the prefilter. Since, the second order statistical property is not of crucial importance, a simple solution consists in estimating these models by using a high-order least-squares estimator. The result will be obviously biased but a bias in the first step does not lead to a bias in the final model.

Method M4.

Step 1. Write the model structure as a linear regression (equation 23), and estimate θ by a high-order least-squares method.

Then, get $\hat{\theta}_1$ along with the process and noise models $\hat{G}_1(q) = \frac{\hat{B}_1(q^{-1})}{\hat{A}_1(q^{-1})}$, $\hat{H}_1(q) = \frac{1}{\hat{A}_1(q^{-1})}$ respectively.

Step 2. Compute the prefilter $\hat{L}(q^{-1}) = \hat{A}_1(q^{-1})\hat{H}_1(q) = 1$ in the case of an ARX model. Compute the noise-free part of the regressor

$$\tilde{\psi}(t) = [-\tilde{y}_1(t-1) \cdots -\tilde{y}_1(t-n)\tilde{u}_1(t-1) \cdots \tilde{u}_1(t-n)]^T$$

with $\tilde{y}_1(t)$ and $\tilde{u}_1(t)$ computed as in equations (24)-(25). Generate the instruments as

$$z(t) = \{[\hat{A}_1(q^{-1})\hat{H}_1(q^{-1})]^{-1}\tilde{\psi}^T(t)\}^T \tag{28}$$

Step 3. Using the instrument $z(t)$ and the prefilter $\hat{L}(q^{-1})$, determine the IV estimate in (23) as

$$\hat{\theta}_{M4} = \hat{R}_{z\psi}^{-1}\hat{R}_{zy}. \tag{29}$$

6. EXAMPLE

The following numerical example is used to compare the performance of the proposed approaches. The process to be identified is described by equation (1), where

$$G_0(q) = \frac{0.5q^{-1}}{1-0.8q^{-1}}, \quad n = 1 \tag{30}$$

$$C(q) = \frac{0.0012+0.0002q^{-1}-0.001q^{-2}}{0.5-0.9656q^{-1}+0.4656q^{-2}}, m=2 \tag{31}$$

$$H_0(q) = \frac{1-1.56q^{-1}+1.045q^{-2}-0.3338q^{-3}}{1-2.35q^{-1}+2.09q^{-2}-0.6675q^{-3}} \tag{32}$$

$r(t)$ is a deterministic sequence (realization of a random binary signal) and $e_0(t)$ is a white noise uncorrelated with $r(t)$. The process parameters are estimated by means of the M1 to M4 methods. Moreover, the results from the basic closed-loop IV method developed by Söderström *et al.* (1987) are also analyzed. This method referenced as M5, consists in using the delayed version of the reference signal as instruments; the estimate is thus given by

$$\hat{\theta}_{cliv} = \left[\sum_{t=1}^{N} \zeta(t)\psi^T(t)\right]^{-1} \left[\sum_{t=1}^{N} \zeta(t)y(t)\right] \tag{33}$$

$$\zeta(t) = \left[r(t)\ r(t-1)\ \cdots\ r(t-2n)\right]^T \tag{34}$$

For illustration purposes, all of these methods are compared to a benchmark which consists in applying the true noise and process models for generating the prefilter and the instruments.

The process parameters are estimated on the basis of closed-loop data sequences of length $N = 1000$. Monte Carlo simulations of 100 experiments have been performed for a signal to noise ratio

$$SNR = 10\log\left(\frac{P_{y_d}}{P_e}\right) = 15 \text{ dB}, \tag{35}$$

where P_x denotes the power of the signals x, and y_d is the noise-free output signal.

In figure 2, the Bode diagrams of the 100 models identified by the six methods are represented. Furthermore, the following function is computed and represented in figure 3 for each algorithm

$$g(\omega) = \frac{1}{MC}\sum_{k=1}^{MC} |G_0(e^{i\omega}) - \hat{G}_k(e^{i\omega})| \tag{36}$$

where MC denotes the number of Monte Carlo experimentations and $\hat{G}_k(e^{i\omega})$ the transfer function estimated during the kth Monte Carlo experimentation. Figures 2 and 3 show that M3 gives the best results (no bias, lower standard-deviation), really close to those of the benchmark. The two approximate versions of the optimal IV algorithm (M3, M4) and the closed-loop IV method (M5) give better results than the tailor-made IV one (M1) in that case. Moreover, the method based on the least-square high-order model (M4) seems to be more appropriate than the extension of the IV4 method to this closed-loop case (M2).

Furthermore, the 2-norm of the difference between the real and estimated transfer functions is also computed for each method

$$Norm = \frac{1}{MC}\sum_{k=1}^{MC} \int |G_0(e^{i\omega}) - \hat{G}_k(e^{i\omega})|^2 d\omega \tag{37}$$

The results are given in table 1 and confirm the previous graphic results: the bootstrap IV methods considered in the paper give better results than the tailor-made IV or the BELS techniques.

method	bench.	M1	M2	M3	M4	M5
Norm	1.921	4.766	2.893	2.223	2.591	3.685

Table 1. *Norm*

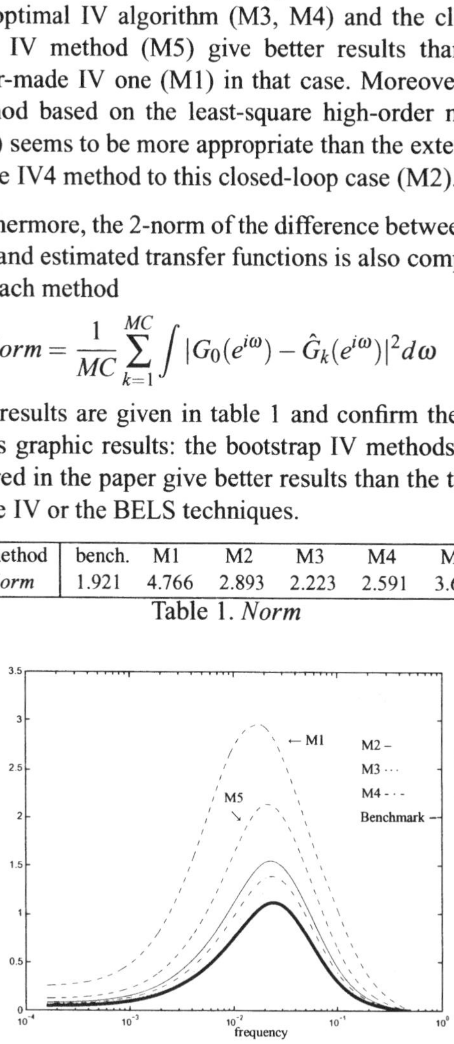

Fig. 3. $g(\omega)$

7. CONCLUSION

Several IV and IV-related estimators for closed-loop system identification have been studied and set in an extended-IV framework. Several methods have been developed to determine the design parameters which allow to approximate the optimal closed-loop IV estimator. In conclusion, the recently suggested Tailor-made IV methods and BELS methods lead to unbiased plant estimates in closed loop. However for arriving at estimates with attractive variance properties it is preferably to apply bootstrap IV methods as considered in this paper.

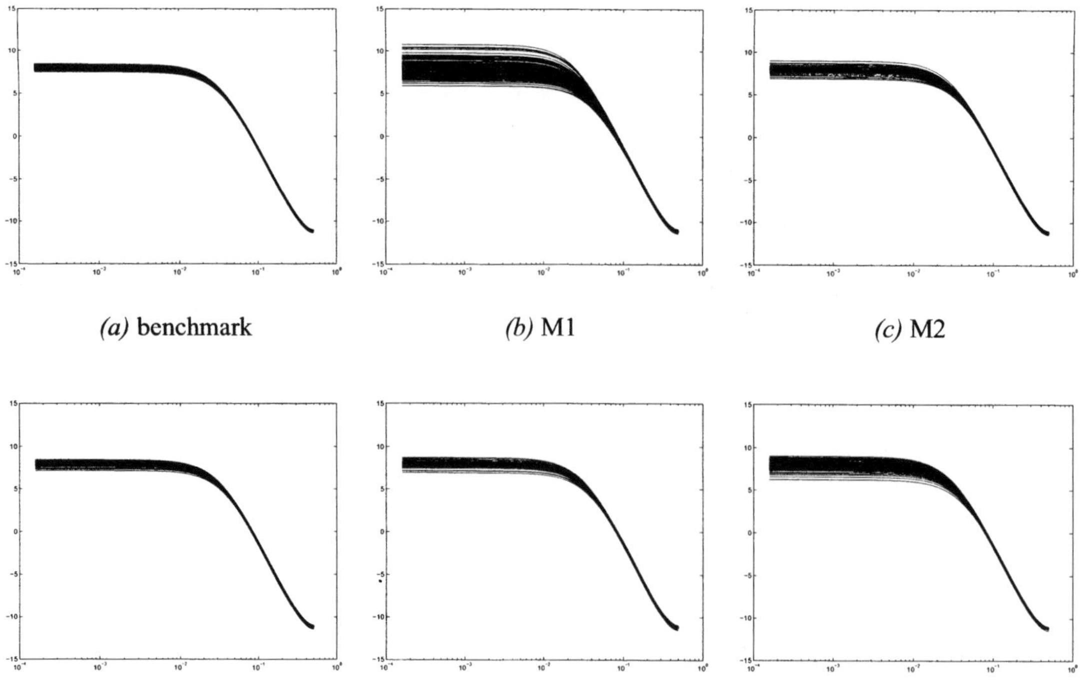

(a) benchmark (b) M1 (c) M2

(d) M3 (e) M4 (f) M5

Fig. 2. Bode amplitude plots of the process (black) and of the estimates (grey)

8. REFERENCES

Broersen, P.M.T. (2002). Automatic spectral analysis with time series models. *IEEE Transactions on Instrumental Measurement* **51**(2), 211–216.

Forssell, U. and L. Ljung (1999). Closed-loop identification revisited. *Automatica* **35**(7), 1215–1241.

Forssell, Urban and C. T. Chou (1998). Efficiency of prediction error and instrumental variable methods for closed-loop identification. Technical Report LiTH-ISY-R-2015. Dept of EE. Linköping University. S-581 83 Linköping, Sweden.

Gilson, M. and P. Van den Hof (2001). On the relation between a bias-eliminated least-squares BELS and an IV estimator in closed-loop identification. *Automatica* **37**(10), 1593–1600.

Ljung, L. (1999). *System Identification : Theory for the User - Second Edition*. Prentice-Hall.

Söderström, T. and P. Stoica (1983). *Instrumental Variable Methods for System Identification.* Springer-Verlag.

Söderström, T. and P. Stoica (1989). *System Identification.* Prentice-Hall.

Söderström, T., P. Stoica and E. Trulsson (1987). Instrumental variable methods for closed-loop systems. In: *10th IFAC World Congress.* Munich - Germany. pp. 363–368.

Van den Hof, P.M.J. (1998). Closed-loop issues in system identification. *Annual Reviews in Control* **22**, 173–186.

Young, P.C. (1976). Some observations on instrumental variable methods of time-series analysis. *Int. J. Control* **23**(5), 593–612.

Zheng, W.X. (1996). Identification of closed-loop systems with low-order controllers. *Automatica* **32**(12), 1753–1757.

Zheng, W.X. and C.B. Feng (1995). A bias-correction method for indirect identification of closed-loop systems. *Automatica* **31**(7), 1019–1024.

APPENDIX

Proof of Proposition 1. Using (9), the solution to (10) can be written as

$$\hat{\theta}_{tiv}(N) = \left[\sum_{t=1}^{N} \eta(t)\bar{\psi}^T(t)\right]^{-1} \left[\sum_{t=1}^{N} \eta(t)\bar{y}(t)\right].$$

By using equation (6), $\eta(t) = F\varphi_r(t)$, and then by replacing F by its expression (11), $\hat{\theta}_{tiv}$ can also be written as

$$\hat{\theta}_{tiv} = [\hat{R}^T_{\varphi_r\bar{\psi}}(\hat{R}_{\varphi_r\varphi}\hat{R}^T_{\varphi_r\varphi})^{-1}\hat{R}_{\varphi_r\bar{\psi}}]^{-1} \cdot \hat{R}^T_{\varphi_r\bar{\psi}}(\hat{R}_{\varphi_r\varphi}\hat{R}^T_{\varphi_r\varphi})^{-1}\hat{R}_{\varphi_r\bar{y}} \quad (38)$$

The structure of this expression is

$$\hat{\theta}_{tiv} = (A^T Q A)^{-1} A^T Q B \quad (39)$$

with

$$A = \hat{R}_{\varphi_r\bar{\psi}},\ Q = (\hat{R}_{\varphi_r\varphi}\hat{R}^T_{\varphi_r\varphi})^{-1},\ B = \hat{R}_{\varphi_r\bar{y}} \quad (40)$$

As a result, $\hat{\theta}_{tiv}$ is the solution to the extended IV problem

$$\hat{\theta}_{tiv} = \arg\min_{\theta} \left\|\hat{R}_{\varphi_r\bar{\psi}}\theta - \hat{R}_{\varphi_r\bar{y}}\right\|^2_Q \quad (41)$$

with weighting matrix Q given by equation (40).

IFAC
Publications
www.elsevier.com/locate/ifac

COPRIME FACTOR PERTURBATION MODELS FOR CLOSED-LOOP MODEL VALIDATION TECHNIQUES

M. Crowder * **R.A. de Callafon** *

* *University of California, San Diego*
Dept. of Mechanical and Aerospace Engineering
9500 Gilman Drive
La Jolla, CA 92093-0411, U.S.A

Abstract: This paper addresses the problem of checking the consistency of experimental closed-loop frequency-domain data with uncertainty models that are structured using coprime factorizations. The uncertainty models presented in this paper use the knowledge of a stabilizing feedback controller to structure and formulate the uncertainty on a model. Subsequently, the controller dependent coprime factor uncertainty model can be used to formulate model (in)validation tests on the basis of closed-loop data. Closed-loop model validation results are developed for the cases of noise-free and noise perturbed closed-loop data. The model validation tests involve the computation of a structured singular value over a finite frequency grid. It is also shown that the computation of the structured singular value simplifies considerably when the feedback controller used for the closed-loop experiments is the same as the controller used for formulating the controller dependent coprime factor uncertainty model. *Copyright © 2003 IFAC*

Keywords: identification; closed-loop; model validation; coprime factors

1. INTRODUCTION

Model validation is a critical procedure to establish whether or not a model can reliably predict the output of a system (Smith and Doyle 1992). In the last few years there has been much attention directed towards various techniques of performing uncertainty model validation. Specifically, the model validation of a general Linear Fractional Transformation (LFT) of discrete and continuous uncertain systems are studied in Poolla *et al.* (1994), Smith and Dullerud (1996) and Chen and Wang (1996). Model validation techniques using LFT's are applied to the frequency domain in Chen (1997) where the validation tests were illustrated to have a low level of computational complexity by formulating the model validation problem as a convex optimization.

In model (in)validation a distinction must be made between validating models on their open-loop and closed-loop behavior. A model suitable to predict and validate open-loop data may be different from a model that approximates the closed-loop behavior of a system. Closed-loop model validation techniques typically validates models on the basis of closed-loop data to verify the model for robust control applications. Control oriented or closed-loop model validation has been applied in Chen and Smith (1998) where it was observed that the convexity of the model validation problem can be preserved by using the knowledge of the controller.

In the line of development of closed-loop model validation techniques, a fractional representation approach is presented in this paper to address the control oriented identification and model validation problem. The fractional approach eliminates the effect of correlated noise on observed input and output signals that is unavoidable in feedback controlled systems. This approach allows enables a unified method to validate models for stable, marginally stable or unstable systems via the validation of stable coprime factorizations on the basis of closed-loop data. The work on fractional model identification was initiated by Hansen *et al.* (1989) and further developed in the work by Lee *et al.* (1993) de Callafon and Van den Hof (1997) and Lu *et al.* (1996). The fractional approach forms an excellent framework to address the identification of systems on the basis of closed-loop data (Anderson 1998) and control oriented model validation (de Callafon and Van den Hof 2000).

In this paper the problem of checking the consistency of experimental frequency-domain data is addressed with uncertainty models that are structured using coprime factorizations. Model validation techniques based on models formulated in a coprime factor framework have also been presented in Boulet and Francis (1998). The results of Boulet and Francis (1998) cover the noisy and noise-free conditions but are limited to open-loop frequency-response data that do not take into account the controller information. In this paper the coprime factorizations used in the uncertainty model depend on the knowledge of a stabilizing feedback controller to facilitate the closed-loop (in)validation of the uncertainty model.

The model validation tests presented in this paper involve the computation of a structured singular value $\mu(\cdot)$ over a finite frequency grid. Model validation techniques using (inverse) μ have also been studied in Lind and Brenner (1999) with applications towards aero-servoelastic systems. Model validation results using μ for SISO and MISO systems were also studied in Kumar and Balas (1994). Unfortunately, most of these results were applied to open-loop model validation and this paper extends these results to address the closed-loop model validation problem. It is also shown in this paper that the computation of the structured singular value for the model validation simplifies considerably when the knowledge of the feedback controller used for the closed-loop experiments is included in the coprime factor based uncertainty model.

2. PROBLEM FORMULATION

Given a nominal model $\hat{P}$ of a system P_0, an uncertainty structure Δ, and a set of input and output measurements (u, y) acting on the actual system P_0, the model (in)validation problem is to determine whether the measurements (u, y) could have been reproduced by the model with the uncertainty. The nominal model $\hat{P}$, along with the perturbation Δ will constitute an uncertainty model $\mathcal{P}$. Provided that $\mathcal{P}$ is constructed in such a way that $P_o \in \mathcal{P}$, such an uncertainty model $\mathcal{P}$ allows one to perform model (in)validation tests to verify the validity of $\mathcal{P}$. It is important to note that the model (in)validation test can be formulated as either an open-loop or closed-loop problem. The difference between open- and closed-loop data is not only determined by the data used for the model validation, but also depends on the way in which the uncertainty model $\mathcal{P}$ is structured. In this paper we consider a controller dependent coprime factor based uncertainty model that is presented in the following section.

2.1 *Use of Fractional Models*

An (upper) Linear Fractional Transformation (LFT)

$$\mathcal{F}_u(Q, \Delta) := Q_{22} + Q_{21}\Delta(I - Q_{11}\Delta)^{-1}Q_{12}$$

provides a general notation to represent all models $P \in \mathcal{P}$ as follows

$$\mathcal{P} = \{P \mid P = \mathcal{F}_u(Q, \Delta) \text{ with } \Delta \in \mathbb{R}H_\infty \text{ and } \|\Delta\|_\infty < 1\} \tag{1}$$

where Δ indicates an unknown (but bounded) uncertainty. The entries of the coefficient matrix Q in (1) dictate the way in which the uncertainty model $\mathcal{P}$ is being structured.

The uncertainty model $\mathcal{P}$ will be characterized by employing a fractional approach. A fractional based uncertainty model $\mathcal{P}$ is characterized by specifically using the knowledge of a controller C that stabilizes the nominal model $\hat{P}$. More specifically, the uncertainty model $\mathcal{P}$ proposed in this paper is structured as follows

$$\begin{aligned}\mathcal{P} = \{&P \mid P = ND^{-1} \text{ with} \\ &N = \hat{N} + D_c\bar{\Delta}\,, D = \hat{D} - N_c\bar{\Delta} \text{ and} \\ &\bar{\Delta} := V\Delta, \|\Delta\|_\infty < 1\}\end{aligned} \tag{2}$$

where (N_c, D_c) and $(\hat{N}, \hat{D})$ respectively denote a right coprime factorization (*rcf*) of the controller C and a nominal model $\hat{P}$. The weighting function V is used to normalize the unknown but bounded uncertainty.

The uncertainty model $\mathcal{P}$ in (2) is different from standard additive coprime factor perturbations as used in Boulet and Francis (1998). In the uncertainty model of (2), the perturbation $\bar{\Delta}$ is used to model a combined perturbation on the *rcf*(N, D) of the model P. It can be observed that $\hat{N}$ is perturbed by $\Delta_N = D_c\bar{\Delta}$ and $\hat{D}$ is perturbed by $\Delta_D = N_c\bar{\Delta}$ where the *rcf*(N_c, D_c) of the controller plays an important role in assigning the common perturbations in the *rcf*(N, D). From this representation, the coprime factors (N, D) can be expressed as

$$N = \hat{N} + \Delta_N \text{ and } D = \hat{D} - \Delta_D \tag{3}$$

where Δ_N and Δ_D are coupled and controller dependent additive perturbations on the coprime factorization $(\hat{N}, \hat{D})$ of the nominal model.

The reason to consider a combined perturbation Δ on *rcf*$(\hat{N}, \hat{D})$ of the nominal model $\hat{P}$ compared to independent perturbations (3) is two-fold. Firstly, independent additive perturbations of Δ_N and Δ_D in (3) would yield two components to the uncertainty, while it is the ratio of N and D that determines the model

P. One way to account for this effect is to choose a single weighting function V to bound both Δ_N and Δ_D in (3) as done in Boulet and Francis (1998). Such an approach might introduce conservatism in the uncertainty model, in case the perturbation Δ_N is significantly different from the perturbation Δ_D. The second reason to introduce a combined perturbation Δ in the uncertainty model (2) is to establish a link with the Youla-Kucera parameterization (Anderson 1998) that will facilitate closed-loop model validation of the uncertainty model.

A close relationship with the Youla-Kucera parameterization is obtained when the nominal model $\hat{P}$ is required to create a stable feedback connection with the controller C used in the uncertainty model (2). In this way, the coefficient matrix Q in (1) is formed by considering a model perturbation that is structured according to a Youla-Kucera parameterization as in Figure 1.

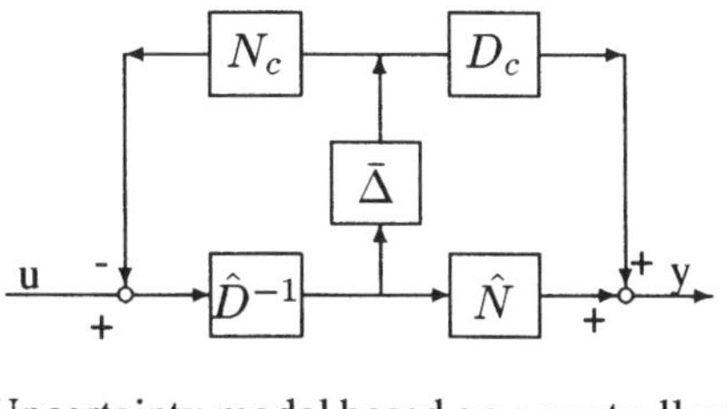

Fig. 1. Uncertainty model based on a controller related perturbations of coprime factorizations

With the uncertainty structure presented in Figure 1, the uncertainty model $\mathcal{P}$ can be written as follows

$$\begin{array}{c}\mathcal{P} = \{P \mid P = \mathcal{F}_u(Q,\Delta) \text{ with Q given by} \\ Q_{11} = V\hat{D}^{-1}N_c \qquad Q_{12} = V\hat{D}^{-1} \\ Q_{21} = D_c + \hat{N}\hat{D}^{-1}N_c \; Q_{22} = \hat{N}\hat{D}^{-1} \\ \text{and } \Delta \in \mathbb{R}H_\infty, \|\Delta\|_\infty < 1\}\end{array} \quad (4)$$

The coefficient matrix Q in (4) is formed by using the information of the *rcf* $(\hat{N},\hat{D})$ of the nominal model $\hat{P}$, the *rcf* (N_c, D_c) of the controller C and the perturbation given in(2). It can be observed that the uncertainty model is a generalization of an uncertainty description based on closed-loop transfer functions, in case either $\hat{P}$ or C is assumed to be stable and a trivial choice $(\hat{P}, I)$ or (C, I) is chosen for respectively the *rcf* of $\hat{P}$ or C.

2.2 *Closed-Loop Model Validation Problem*

Consider a feedback connection of the system P_o and a feedback controller $\bar{C}$, with $y = P_o u + v$ and $u = r - \bar{C}y$. It shoudl be noted that a distinction is made between the controller C and $\bar{C}$. The notation C is used to indicate the controller used in the construction of the uncertainty set (4), whereas the notation $\bar{C}$ is used to denote the controller used in the closed-loop experiments.

The input/output data $\{u, y\}$ of the system P_o controlled by the feedback $\bar{C}$ can be described by

$$\begin{bmatrix} y \\ u \end{bmatrix} = \begin{bmatrix} P_o \\ I \end{bmatrix}(I+\bar{C}P_o)^{-1}r + \begin{bmatrix} I \\ -\bar{C} \end{bmatrix}(I+P_o\bar{C})^{-1}v \quad (5)$$

where r denotes an external reference signal that provides sufficient excitation of the closed-loop system. The signal v denotes the additive noise on the output y that can be caused by sensor noise. For closed-loop model validation purposes, the reference signal r is considered as an input signal. The signal u and/or y can be considered as measurable closed-loop output signal. Without loss of generalization only the output signal y is considered here for closed-loop model validation to simplify the formulae.

Application of the feedback law $u = r - \bar{C}y$ to all models $P \in \mathcal{P}$ in (4) yields a set of closed-loop models $\mathcal{S}$ that is structured as follows.

$$\begin{array}{l}\mathcal{S} = \{S \mid S = \mathcal{F}_u(M,\Delta) \text{ with M given by} \\ \quad M_{11} = V(\hat{D}+\bar{C}\hat{N})^{-1}(\bar{C}-C)D_c \\ \quad M_{12} = V(\hat{D}+\bar{C}\hat{N})^{-1} \\ \quad M_{21} = (I+\hat{P}\bar{C})^{-1}(I+\hat{P}C)D_c \\ \quad M_{22} = (I+\hat{P}\bar{C})^{-1}\hat{P} \\ \quad\quad \text{and } \Delta \in \mathbb{R}H_\infty, \|\Delta\|_\infty < 1\}\end{array} \quad (6)$$

The entries of M in (6) are all known quantities and found by a *rcf* of a nominal model $\hat{P}$ and a bound for the unstructured uncertainty Δ in (4). As C is required to internally stabilize $\hat{P}$, it can be verified that all entries of M are stable *if and only if* the controller $\bar{C}$ also internally stabilizes the nominal model $\hat{P}$. The entries of the coefficient matrix M form the set of closed-loop models $\mathcal{S}$ that is structured due to the uncertainty model posed in (4) and application of the feedback law $u = r - \bar{C}y$. Note that in (6) the controller C is the controller used to create the uncertainty structure of (4), whereas $\bar{C}$ is the controller used in the closed-loop experiments. The availability of closed-loop data $\{r, y\}$ and the characterization of the closed-loop uncertainty set $\mathcal{S}$ form the basis of the closed-loop model (in)validation problem in this paper.

In case the feedback controller $\bar{C}$ used in the closed-loop experiments $u = r - \bar{C}y$ equals the controller C used in constructing the uncertainty set (4), the entries of M in (6) can be greatly reduced. In case $\bar{C} = C$, the set of closed-loop model $\mathcal{S}$ reduces to

$$\begin{array}{l}\mathcal{S} = \{S \mid S = \mathcal{F}_u(M,\Delta) \\ \text{with } \Delta \in \mathbb{R}H_\infty \text{ and } \|\Delta\|_\infty < 1\}\end{array} \quad (7)$$

where the entries of M are given by

$$\begin{array}{l} M_{11} = 0 \quad M_{12} = V(\hat{D}+\bar{C}\hat{N})^{-1} \\ M_{21} = D_c \; M_{22} = \hat{P}(I+\bar{C}\hat{P})^{-1}\end{array}$$

Note that for $\bar{C} = C$, $M_{11} = 0$ and stability robustness is trivially satisfied for the uncertainty set $\mathcal{S}$ in (7). However, when the controller C used to parameterize the models is redesigned to $\bar{C}$, $M_{11} = \mathcal{F}_l(Q, -\bar{C}) \neq 0$ and stability robustness is not trivially satisfied. The fact that $M_{11} = 0$ is an immediate consequence of the Youla-Kucera parameterization used in constructing the fractional model based uncertainty set (4). In case $\bar{C} = C$, the controller $\bar{C}$ used in the closed-loop experiments, is the same as the controller C used to create the uncertainty set.

Choosing $\bar{C} = C$ also carries the interpretation of performing closed-loop model validation for a set of models $\mathcal{P}$ that is know to be stabilized by $\bar{C} = C$. This information is used beneficially, as only models that are stabilized by $\bar{C}$ are considered. As a final remark it can be observed that $M_{11} = 0$ yields an LFT $\mathcal{F}_u(M, \Delta) = M_{22} + M_{21}\Delta M_{12}$ where the uncertainty Δ appears linearly in the closed-loop map from r to y. The affine representation of the uncertainty Δ can be exploited to formulate an affine model (in)validation problem on the basis of closed-loop data.

To account for noise during the observation of $y(t)$ we assume an unknown but bounded additive perturbation of the Discrete Fourier Transform $Y(\omega)$. In that case, the uncertainty model can be written as

$$Y(\omega) = (\mathcal{F}_u(M, \Delta) + W_{cl}(\omega)\delta(\omega))R(\omega) \quad (8)$$

where $\mathcal{F}_u(M, \Delta)$ is given in (6) and $W_{cl}(\omega)$ is a stable and stably invertible weighting function. In (8) the effect of the additive noise is bounded by $\delta(\omega)$ with $|\delta(\omega)| < 1 \ \forall \ \omega$ or $||\delta||_\infty < 1$, while $W_{cl}(\omega)$ is a frequency dependent weighting function to model the spectral content of the noise. Using the above assumptions with the knowledge of the uncertainty model represented in M and the knowledge of the noise contribution in $W_{cl}(\omega)$, the closed-loop model validation problem can be summarized as follows.

Closed-loop model validation problem: consider the closed-loop measurements $Y(\omega)$ and $R(\omega)$, $\omega \in \Omega$. The closed-loop uncertainty model is not invalidated by the data if there exists a Δ with $||\Delta||_\infty < 1$ and δ with $||\delta||_\infty < 1$ such that (8) holds.

For the closed-loop model validation problem the objective is to determine whether there exists a stable perturbation Δ with $||\Delta||_\infty < 1$ and an additive noise perturbation δ with $||\delta||_\infty < 1$, such that (8) holds. The model validation results are presented in the next section.

3. CLOSED-LOOP FREQUENCY RESPONSE MODEL INVALIDATION

3.1 Noise-Free Case

For the closed-loop model validation problem we consider a closed-loop system where a reference signal r is applied and a noise free system response y is measured. The frequency domain data of the closed-loop system can be described by

$$Y(\omega) = \mathcal{F}_u(\hat{M}, \Delta)R(\omega), \ \omega \in \Omega \quad (9)$$

where the entries of $\hat{M}$ are given by

$$\begin{aligned} \hat{M}_{11} &:= M_{11} \ \hat{M}_{12} := -M_{12}R(\omega) \\ \hat{M}_{21} &:= M_{21} \ \hat{M}_{22} := Y(\omega) - M_{22}(\omega)R(\omega) \end{aligned} \quad (10)$$

In (10) the entries $\hat{M}_{ij}$ are frequency dependent functions where $\omega \in \Omega$ and $Y(\omega)$ and $R(\omega)$ are the respective Discrete Fourier Transforms of the (noise free) signals $y(t)$ and $r(t)$.

The model validation problem is performed frequency point-wise and decomposed into consistency problems evaluated over the frequency grid Ω. The consistency problems check the existence of $\Delta(w)$ with $\overline{\sigma}(\Delta(w)) < 1$ for $w \in \Omega$. In order to guarantee the existence of a $\Delta \in \mathbb{R}H_\infty$ with $||\Delta||_\infty < 1$, a boundary interpolation result (Chen 1997, Boulet and Francis 1998) should be used and the result is summarized in the following lemma.

Lemma 1. Let $\overline{\sigma}(\Delta(w)) < 1 \, \forall \, w \in \Omega$, then $\exists \ \Delta \in \mathbb{R}H_\infty$ with $||\Delta||_\infty < 1$.

One is referred to Boulet and Francis (1998) for a proof of this result. The boundary interpolation result is used to formalize the model (in)validation result for the closed-loop data in (9) in the following theorem.

Theorem 1. Let $Y(\omega)$, $R(\omega)$ with $\omega \in \Omega$ denote the noise free frequency response measurement of the closed-loop system $T(P_0, C)$ and let $\hat{M}$ be defined as in (10). The closed-loop uncertainty model $\mathcal{S}$ in (6) is not invalidated iff $\mu_\Delta(\hat{M}_{11} - \hat{M}_{12}\hat{M}_{22}^{-1}\hat{M}_{21}) > 1$ where $\mu_\Delta(\cdot)$ is computed with respect to the uncertainty structure Δ given in (6).

Proof: With Lemma 1 it suffices to show the existence of a $\Delta(w)$ with $\overline{\sigma}(\Delta(w)) < 1$ for $w \in \Omega$. This is equivalent to $\mathcal{F}_u(\hat{M}, \Delta_s) = 0$ for $w \in \Omega$. The inverse of $[\mathcal{F}_u(\hat{M}, \Delta_s)]^{-1} = \mathcal{F}_u(N, \Delta_s)$ where the entries of N are given by

$$\begin{aligned} N_{11} &= \hat{M}_{11} - \hat{M}_{12}\hat{M}_{22}^{-1}\hat{M}_{21} \\ N_{12} &= -\hat{M}_{12}\hat{M}_{22}^{-1} \\ N_{21} &= \hat{M}_{22}^{-1}\hat{M}_{21} \\ N_{22} &= \hat{M}_{22}^{-1} \end{aligned}$$

With $\hat{M}_{22}$ invertible, it can be seen that $\mathcal{F}_u(N, \Delta_s)$ is ill-defined when $(I - \hat{M}_{11}\Delta + \hat{M}_{12}\hat{M}_{22}^{-1}\hat{M}_{21}\Delta)^{-1}$ is ill-defined. This condition can be replaced by the computation of the structured singular value $\mu_\Delta(\hat{M}_{11} - \hat{M}_{12}\hat{M}_{22}^{-1}\hat{M}_{21}) > 1$. □

The evaluation of $\mu_\Delta(\hat{M}_{11} - \hat{M}_{12}\hat{M}_{22}^{-1}\hat{M}_{21}) > 1$ is done frequency point-wise over $\omega \in \Omega$. In case Δ is unstructured, $\mu_\Delta(\cdot) > 1$ can be replaced with $\bar{\sigma}(\cdot) > 1$. The above model invalidation condition of $\mu_\Delta(\hat{M}_{11} - \hat{M}_{12}\hat{M}_{22}^{-1}\hat{M}_{21}) > 1$ is actually the inverse-μ result as derived in Lind and Brenner (1999) and is an adaptation of the result mentioned in Boulet and Francis (1998). Due to the noise free data, the value of $\Delta(\omega)$ can actually be computed over $\omega \in \Omega$. The result is summarized in the following corollary for an interpretation of the noise free closed-loop model validation problem.

Corollary 1. Let $\Phi_{cl}(\omega)$ denote the closed-loop noise free frequency response of the system and let Δ_V be defined as $\Delta_V = D_c{}^{-1}(I + \hat{P}C)^{-1}((I - \Phi_{cl}(\omega)\bar{C})^{-1}\Phi_{cl}(\omega) - \hat{P})\hat{D}$. Then the closed-loop uncertainty model $\mathcal{S}$ in (6) is not invalidated by $\Phi_{cl}(\omega)$ iff $|\Delta_V(\omega)| < |V(\omega)|$.

Proof: Let $P_0 = ND^{-1}$ with $N = \hat{N} + D_c\Delta_V$ and $D = \hat{D} - N_c\Delta_V$ where $(\hat{N}, \hat{D})$ and (N_c, D_c) are the respective right coprime factors of the model $\hat{P}$ and the controller C. Rearranging terms leads to the expression $(P_0\hat{D} - \hat{N}) = (\hat{P}N_c + D_c)\Delta$. Substituting $\hat{P} = \hat{N}\hat{D}^{-1}$ and $C = N_cD_c^{-1}$ and with further rearrangement leads to the uncertainty expression $\Delta_V = D_c{}^{-1}(I + PC)^{-1}(P_0 - \hat{P})\hat{D}$. In the case of noise free measurements, $(I - \Phi_{cl}(\omega)\bar{C})^{-1}\Phi_{cl}(\omega) = P_0$ and the result simplifies to $\Delta_V = D_c{}^{-1}(I + \hat{P}C)^{-1}((I - \Phi_{cl}(\omega)\bar{C})^{-1}\Phi_{cl}(\omega) - \hat{P})\hat{D}$. □

The result in Corollary 1 illustrates that an explicit and computable expression for the uncertainty Δ_V can be obtained. Note that the assumptions listed in Boulet and Francis (1998) for the noise-free open-loop model validation test are relaxed in our case. Since the validation test presented in this paper involves the measurements of $Y(\omega)$ and $U(\omega)$ explicitly, it is unnecessary to assume nonsingularity of $\Phi_{ol}(\omega) = Y(\omega)/U(\omega)$. The following corollary illustrates how the model validation results simplify when the controller $\bar{C}$ used to gather the closed-loop experiments is equal to the controller C used to characterize the uncertainty model $\mathcal{P}$.

Corollary 2. If the controller $\bar{C}$ used in the closed-loop experiments is equivalent to the controller C used in the uncertainty model $\mathcal{P}$ of (2) then the test on the structured singular value mentioned in Theorem 1 reduces to

$$\underline{\sigma}(V^{-1}D_c^{-1}\hat{D}\frac{P_0 - \hat{P}}{1 + P_0\bar{C}}) < 1 \tag{11}$$

Proof: Substituting $\hat{M}_{11}$, $\hat{M}_{12}$, $\hat{M}_{21}$, and $\hat{M}_{22}$ from (10) and letting $C = \bar{C}$ reduces the model invalidation test to $\bar{\sigma}(V(\hat{D} + \bar{C}\hat{N})^{-1}R(Y - (1 + \hat{P}\bar{C})^{-1}\hat{P}R)^{-1}D_c) > 1$. Note that $\bar{\sigma}(\cdot) = \mu_\Delta(\cdot)$ since we consider a full-block matrix Δ for the closed-loop noise-free case. Simplifying and noting that $\frac{Y}{R} = \frac{P_0}{1+P_0\bar{C}}$, the expression reduces to $\bar{\sigma}(VD_c\hat{D}^{-1}(1 + \bar{C}\hat{P})^{-1}(\frac{P_0}{1+P_0\bar{C}} - \frac{\hat{P}}{1+\hat{P}\bar{C}})^{-1} > 1$. Cancelling common terms and noting that $\bar{\sigma}(M) > 1 \Leftrightarrow \underline{\sigma}(M^{-1}) < 1$, the model invalidation test is written as $\underline{\sigma}(V^{-1}D_c^{-1}\hat{D}\frac{P_0 - \hat{P}}{1+P_0\bar{C}}) < 1$. □

The above result is expected as we note a dependence on the closed-loop sensitivity function and a weighted dependence on the difference between the system and nominal model. As mentioned earlier in the case $\bar{C} = C$, the term $M_{11} = 0$ and robust stability is trivially satisfied. This follows directly from (7) and is the result of the dual-Youla parameterization.

3.2 Noisy Case

The noise free case was used to illustrate the main results and ideas of the closed-loop model validation problem. For practical application of the model validation techniques, the noisy measurements of the closed-loop data in (8) need to be considered. In (8) the quantity $W_{cl}(\omega)\delta(\omega)$ models an additive perturbation of the closed-loop measurement $Y(\omega)$ and incorporates both the effect of an additive noise during a measurement and also the effect of initial conditions. The noise $\delta(\omega)$ is unknown, but bounded by $||\delta(\omega)|| < 1\ \forall\ \omega \in \Omega$.

In structuring the closed-loop uncertainty model as in (6), the closed-loop model validation problem for the noisy case can be determined. In order to take into account the effect of the unknown but bounded additive noise, we can recast the uncertainty model as an LFT $\mathcal{F}_u(K, \Delta_s)$ where

$$\Delta_s = \begin{bmatrix} \Delta & 0 \\ 0 & \delta \end{bmatrix} \text{with } ||\Delta||_\infty < 1 \text{ and } ||\delta||_\infty < 1 \tag{12}$$

and the entries of K are given by

$$K_{11} = \begin{bmatrix} M_{11} & 0 \\ 0 & 0 \end{bmatrix} \quad K_{12} = \begin{bmatrix} M_{12} \\ W_{cl} \end{bmatrix}$$
$$K_{21} = \begin{bmatrix} M_{21} & 1 \end{bmatrix} \quad K_{22} = \begin{bmatrix} M_{22} \end{bmatrix} \tag{13}$$

In structuring the closed-loop uncertainty model in this form, (8) can be written as

$$Y(\omega) = \mathcal{F}_u(\hat{K}, \Delta_s)R(\omega) \tag{14}$$

where the entries of $\hat{K}$ are given by

$$\begin{aligned} \hat{K}_{11} &:= K_{11}(\omega)\ \hat{K}_{12} := -K_{12}(\omega)R(\omega) \\ \hat{K}_{21} &:= K_{21}(\omega)\ \hat{K}_{22} := Y(\omega) - K_{22}(\omega)R(\omega). \end{aligned} \tag{15}$$

Similar as in the noise-free case, $\hat{K}_{ij}$ are frequency dependent functions of $\omega \in \Omega$. With this LFT formulation, a result similar to Theorem 1 can be formulated for the closed-loop model validation problem on the basis of noisy data.

Theorem 2. Let $Y(\omega)$ and $R(\omega)$ denote the closed-loop noisy frequency response measurement of the closed-loop system $T(P_0, C)$ and let $\hat{K}$ be defined as in (15). The closed-loop uncertainty model in (14) is not invalidated by the data iff $\mu_{\Delta_s}(\hat{K}_{11} - \hat{K}_{12}\hat{K}_{22}^{-1}\hat{K}_{21}) > 1$ where $\mu_{\Delta_s}(\cdot)$ is computed with respect to the uncertainty structure Δ_s given in (12).

The proof of Theorem 2 is similar to the proof of Theorem 1 and is omitted here for brevity. Consider again the case where the controller $\bar{C}$ used to gather the closed-loop experiments is equal to the controller C used to characterize the uncertainty model $\mathcal{P}$. With $C = \bar{C}$, a direct calculation of μ can be obtained by using the reduced rank μ problem (Fan and Tits 1985). Using the model invalidation result of Theorem 2, substitute the values of $\hat{K}_{12}$, $\hat{K}_{21}$ and $\hat{K}_{22}$ from (15) and note that $C = \bar{C} \Rightarrow M_{11} = 0 \Rightarrow \hat{K}_{11} = 0$. As a result, the model invalidation test is reduced to

$$\mu_{\Delta_s}(\begin{bmatrix} M_{12} \\ W_{cl} \end{bmatrix} R(Y - M_{22}R)^{-1}[M_{21}\ 1]\,) > 1 \quad (16)$$

where Δ_s is computed with respect to the uncertainty structure given in (12). Note that the dependency of ω is dropped for notational simplicity from the expression of $\mu_{\Delta_s}(\cdot)$ in (16). Since the argument of $\mu_{\Delta_s}(\cdot)$ in (16) is a diadic matrix (Fan and Tits 1985), it satisfies

$$\mu_{\Delta_s}(\begin{bmatrix} a \\ b \end{bmatrix} [c\ d]\,) = |ac| + |bd| \quad (17)$$

and the model invalidation test can be rewritten as

$$\begin{aligned} &|M_{12}R(Y - M_{22}R)^{-1}M_{21}| + \\ &|W_{cl}R(Y - M_{22}R)^{-1}|) > 1 \end{aligned} \quad (18)$$

It can be seen from (18) that for $C = \bar{C}$ the closed-loop model invalidation test involves checking whether the sum of two transfer functions is less than 1, which greatly simplifies the computation of the structured singular value for the model (in)validation problem.

4. CONCLUSIONS

Uncertainty models that are structured using coprime factorizations are used to address closed-loop relevant model (in)validation on the basis of closed-loop frequency domain data. Important in the formulation of the uncertainty model is the knowledge on the controller used to create a closed-loop oriented model (in)validation of the uncertainty model.

The controller dependent uncertainty model is used to formulate model (in)validation tests on the basis of closed-loop data. The model validation tests involve the computation of a structured singular value over a finite frequency grid. It is also shown that the computation of the structured singular value reduces to the sum of two transfer functions when the feedback controller used for the closed-loop experiments is the same as the controller used for formulating the controller dependent coprime factor uncertainty model.

REFERENCES

Anderson, B.D.O. (1998). From Youla–Kucera to identification, adaptive and non–linear control. *Automatica* **34**, 1485–1506.

Boulet, B. and B. Francis (1998). Consistency of open-loop experimental frequency-response data with coprime factor for plant models. *IEEE Trans. on Automatic Control* pp. 1680–1691.

Chen, J. (1997). Frequency-domain tests for validation of linear fractional uncertain models. *IEEE Trans. on Automatic Control* **42**, 748–760.

Chen, J. and S. Wang (1996). Validation of linear fractional uncertain models: Solutions via matrix inequalities. *IEEE Trans. on Automatic Control* **41**, 844–849.

Chen, L. and R.S. Smith (1998). Closed-loop model validation: Application to an unstable experimental system. In: *Proc. American Control Conference*. Philadelphia, PA, USA. pp. 618–622.

de Callafon, R.A. and P.M.J. Van den Hof (1997). Suboptimal feedback control by a scheme of iterative identification and control design. *Mathematical Modelling of Systems* **3**(1), 77–101.

de Callafon, R.A. and P.M.J. Van den Hof (2000). Closed-loop model validation using coprime factor uncertainty models. In: *Prepr. 12th IFAC Symposium on System Identification*. Santa Barbara, CA, USA.

Fan, M. and A. Tits (1985). Characterization and efficient computation of the structured singular value. Technical Report TR 85-2. The Institute for Systems Research.

Hansen, F.R., G.F. Franklin and R. Kosut (1989). Closed-loop identification via the fractional representation: Experiment design. In: *Proc. American Control Conference*. Pittsburgh, USA. pp. 1422–1427.

Kumar, A. and G. Balas (1994). An approach to model validation in the μ framework. In: *Proc. American Control Conference*. Baltimore, MY, USA. pp. 3021–3026.

Lee, W.S., B.D.O. Anderson, R.L. Kosut and I.M.Y. Mareels (1993). A new approach to adaptive robust control. *Int. Journal of Adaptive Control and Signal Processing* **7**(3), 183–211.

Lind, R. and M. Brenner (1999). *Robust Aeroservoelastic Stability Analysis*. Springer Verlag. London, UK.

Lu, W.M., K. Zhou and J. Doyle (1996). Stabilization of uncertain linear systems: An LFT approach. *IEEE Trans. on Automatic Control* pp. 50–65.

Poolla, K., P. P. Khargonekar, A. Tikku, J. Krause and K. Nagpal (1994). A time domain approach to model validation. *IEEE Trans. on Automatic Control* **39**, 951–959.

Smith, R.S. and G. Dullerud (1996). Validation of continuous time control models by finite experimental data. *IEEE Trans. on Automatic Control* **41**, 1094–1105.

Smith, R.S. and J.C. Doyle (1992). Model validation: A connection between robust control and identification. *IEEE Trans. on Automatic Control* **37**, 942–952.

IFAC
Publications
www.elsevier.com/locate/ifac

AN INTRODUCTION TO REPRODUCING KERNEL HILBERT SPACES AND WHY THEY ARE SO USEFUL

Grace Wahba *,[1]

* *Department of Statistics, University of Wisconsin, 1210 W. Dayton St., Madison WI, USA*

Abstract: We review some of the basic facts about reproducing kernel Hilbert spaces (RKHS), and the solution of optimization problems in RKHS. These facts provide some clues to how useful RKHS-based methods can be in curve fitting, function estimation, model description, model fitting and ill-posed inverse problems. A number of references are made to mostly older works of the author, colleagues and former students - not an attempt at a balanced review of the literature. The growth of the further development and application of RKHS-based methods is demonstrated in the other papers in the invited Session INV-2 of this 13th IFAC Symposium on System Identification (SYSID-2003), and elsewhere in SISID-2003. A recent advanced google search for "reproducing kernel" turned up over 4400 entries.

Keywords: RKHS, reproducing kernel Hilbert spaces, representers, bias-variance tradeoff, tuning parameters, ill-posed inverse problems, optimization

1. INTRODUCTION

Reproducing kernel Hilbert spaces (RKHS) are wonderful objects and can be used in a wide variety of curve fitting, function estimation, model description and model building applications. In this paper we briefly review a few important facts about RKHS, note some historically relevant works,and then note a selected list of contributions of the author and colleagues to the field. This is a self-referential list and in no way intended as a balanced survey of the field, something that would be impossible in the space available. (See the comments regarding the google search.) Mostly older contributions are noted, some more recent works of the author and colleagues are discussed in a companion article by the same author in this proceedings.

2. WHAT IS AN RKHS?

An RKHS is a Hilbert space (Akhiezer & Glazman (1963)) in which all the point evaluations are bounded linear functionals. Letting $\mathcal{H}$ be a Hilbert space of functions on some domain $\mathcal{T}$, this means, that for every $t \in \mathcal{T}$ there exists an element $\eta_t \in \mathcal{H}$, such that

$$f(t) = <\eta_t, f>, \quad \forall f \in \mathcal{H},$$

where $<,>$ is the inner product in $\mathcal{H}$. Let $<\eta_s, \eta_t> = K(s,t)$. Then $K(s,t)$ is positive definite on $\mathcal{T} \otimes \mathcal{T}$, that is, for $\forall t_1, \cdots, t_n \in \mathcal{T}$, $\sum_{i,j} a_i a_j K(t_i, t_j) \geq 0$. K is called the reproducing kernel (RK) for $\mathcal{H}$. Note that since $\eta_t \equiv K(t, \cdot)$, then $< K(t,\cdot), K(s,\cdot) > \equiv K(s,t)$, this being the origin of the term "reproducing kernel".

3. THE MOORE-ARONSZAJN THEOREM

The famous Moore-Aronszajn theorem (Aronszajn (1950)) theorem states that for every positive definite function $K(\cdot,\cdot)$ on $\mathcal{T} \otimes \mathcal{T}$, there exists a unique RKHS and vice versa. The Hilbert space associated with K can be constructed as containing all finite linear combinations of the form $\sum a_j K(t_j, \cdot)$, and their limits under the norm induced by the inner product

[1] Supported by NSF grant DMS-0072292, NIH grant EY09946, and NASA grant NAG51073

$< K(s,\cdot), K(t,\cdot >= K(s,t)$. Norm convergence implies pointwise convergence in a RKHS, as can be seen by observing that

$$|f_n(t) - f_m(t)| = | < K(t,\cdot), f_n - f_m > | \\ \leq K(t,t)\|f_n - f_m\|.$$

Thus, these limit functions are well defined pointwise. Note that absolutely nothing has been said about $\mathcal{T}$. The discussion above applies to any domain on which it is possible to define a positive definite function, a matrix being a special case when $\mathcal{T}$ has only a countable or finite number of points.

4. RELATION TO GAUSSIAN PROCESSES

Note that, for every positive definite $K(\cdot,\cdot)$ on $\mathcal{T} \otimes \mathcal{T}$ there exists a zero mean Gaussian process with K as its covariance, giving rise to the relation between Bayes estimates, Gaussian processes and optimization problems in RKHS. See Parzen (1970), Kimeldorf & Wahba (1971), Wahba (1990) and elsewhere.

5. MORE RKS

Tensor sums and products of RK's are RK's, which allow the building up of rather general spaces on rather general domains. Letting $s_1, t_1 \in \mathcal{T}^{(1)}$, $s_2, t_2 \in \mathcal{T}^{(2)}$, and letting $s = (s_1, s_2), t = (t_1, t_2)$, then

$$K(s,t) = K_1(s_1,t_1)K(s_2,t_2)$$

is an RK on $\mathcal{T} = \mathcal{T}^{(1)} \otimes \mathcal{T}^{(2)}$ whenever K_1 and K_2 are RK's on their respective domains. Subspaces of RKHS are also RKHS, and the RK's can be obtained by e. g. projecting the representers of evaluation in $\mathcal{H}$ onto the subspace. Thus, RK's can be built up on an endless variety of domains.

6. THE MERCER THEOREM, RADIAL BASIS FUNCTIONS

If $\int\int_{\mathcal{T}\otimes\mathcal{T}} K^2(s,t)dsdt \leq \infty$ then K will have a countable sequence of eigenvalues and eigenfunctions. See Riesz & Nagy (1955). This theorem is often stated in the Support Vector Machine literature without the above condition, but it doesn't always hold, for example for the Gaussian kernel $K(s,t) = e^{-\frac{1}{\sigma^2}\|s-t\|^2}$ on the infinite real line or the infinite Euclidean d-space. (The norm in the exponent here is the Euclidean norm.) See Halmos (1957). However,that doesn't stop the Gaussian kernel or other radial basis functions, from being useful. Radial basis functions (RBF's) are functions of s and $t \in E^d$ that depend only on $(s-t)$. Radial basis functions that are positive definite (as is the Gaussian kernel) were characterized by Micchelli (1986)). Other RBF's can be found in Wahba (1996) and in many places elsewhere.

7. THE SPECIAL REPRESENTER THEOREM

A special but important case of the representer theorem (Kimeldorf & Wahba (1971)) is the following: The solution to the problem: Find $f \in \mathcal{H}$ to minimize

$$\sum_{i=1}^{n} \mathcal{C}(y_i, f(t_i)) + \lambda\|f\|^2 \tag{1}$$

where $\mathcal{C}$ is convex in f, has a representation as

$$f_\lambda(\cdot) = \sum_{i=1}^{n} c_i K(t_i,\cdot). \tag{2}$$

Then (2) is substituted in (1) and the c_i's are found numerically. When $\mathcal{C}$ is quadratic, it is only necessary to solve a linear system, but otherwise a descent algorithm is used. Various tricks for getting good approximations to the solution when n is large are available, including using only a subset of the $K(t_i,\cdot)$ in the computations. See for example Lin, Wahba, Xiang, Gao, Klein & Klein (2000). The function $\eta_{t_i}(\cdot)$ defined by $\eta_{t_i}(t) = K(t_i,t)$ is known as *the representer of evaluation at* t_i. Convexity of $\mathcal{C}$ is not necessary for the theorem, but there will generally be uniqueness questions and numerical difficulties if it is not. Commonly used $\mathcal{C}(y,f)$ are as in penalized least squares: $(y-f)^2$, penalized likelihood for Bernoulli $(0,1)$ data, $\mathcal{C}(y,f) = -yf + log(1+e^f)$, ($f$ will be an estimate of the log odds ratio), the hinge function for ± 1 data) $\mathcal{C}(y,f) = (1-yf)_+$, where $(\tau)_+\tau$ for $\tau > 0$ and 0 otherwise, used in Support Vector Machines (see `http://kernel-machines.org`, Wahba (2002), where *sign* f is the classifier, $\mathcal{C}$ is a robust functional, and elsewhere.

8. THE REPRESENTER THEOREM, MORE GENERALLY

The first generalization of the representer theorem involves replacing $\|f\|^2$ by $\|Pf\|^2$, where P is the orthogonal projection onto a subspace of (small) codimension, (say m). This allows the elements in a subspace of dimension m (for example, low degree polynomials) to remain unpenalized. See Kimeldorf & Wahba (1971), Wahba (1990) for details. More generally, Let Lf be a bounded linear functional in $\mathcal{H}$ with representer η_L. Recall that a (real) bounded linear functional in a Hilbert space is a linear map from elements of the space to the real line satisfying $Lf \leq M\|f\|, \forall f$, for some M not depending on f. Every bounded linear functional in a Hilbert space has a unique representer in the Hilbert space (Riesz representation theorem, Akhiezer & Glazman (1963)), satisfying

$$Lf =< \eta_L, f > .$$

If we replace $f(t_i)$ by $L_i f$ in (1), then the minimizer is in (2) has a representation of the form

$$f_\lambda(\cdot) = \sum_{i=1}^{n} c_i \eta_i(\cdot) \qquad (3)$$

where η_i is the representer of L_i Furthermore, the representer η_L of a bounded linear functional L can be obtained by observing that

$$\eta_L(s) =< \eta_L, K(s,\cdot) >= LK(s,\cdot).$$

Here, L on the right hand side is applied to K as a function of "$\cdot$" while s is fixed. Furthermore

$$< \eta_i, \eta_j >= L_{i(s)} L_{j(t)} K(s,t) > .$$

where $L_{i(s)}$ means that L_i is applied to what follows considered as a function of s. This is useful in a variety of contexts, for example when the observations represent derivatives or averages. It is useful in the numerical solution of ill posed inverse problems in general. For example, let

$$y_i = \int H(t_i, u) f(u) du + \epsilon_i$$

where the ϵ_i are i.i.d Gaussian random variables. In this case $\mathcal{C}$ would correspond to least squares. Under appropriate regularity conditions,

$$L_i f = \int H(t_i, u) f(u) du,$$

$$\eta_i(s) = \int H(t_i, u) K(u, s) du.$$

and

$$< \eta_i, \eta_j >= \int\int H(t_i, u) H(t_j, v) K(u, v) du dv.$$

This setup is a generalized version of Tikhonov regularization (Tikhonov (1963) Wahba (1977a) O'Sullivan & Wahba (1985) Nychka, Wahba, Goldfarb & Pugh (1984)

9. THE BIAS-VARIANCE TRADEOFF

The parameter λ controls the tradeoff between the size of $\sum_{i=1}^{n} \mathcal{C}(y_i, f(t_i))$ and the size of $\|f\|^2$ in (1). More generally there may be other so-called tuning parameters (such as σ in the Gaussian RBF), or, different λ's penalizing components in different subspaces differently. Many approaches to choosing these tuning parameters in various contexts can be found in the literature. A selected list of the authors and students contributions include Craven & Wahba (1979) Golub, Heath & Wahba (1979) Wahba (1990) Gu & Wahba (1991) Wahba, Wang, Gu, Klein & Klein (1995) Xiang & Wahba (1996) Lin (1998) Wahba (1999) Gao, Wahba, Klein & Klein (2001) Lee, Lin & Wahba (2002). See also Wahba & Wang (1990).

10. BAYESIAN AND OTHER CONFIDENCE INTERVALS

Bayesian "confidence intervals" were proposed in Wahba (1983). The words "confidence intervals" are in quotes because these are not pointwise confidence intervals. They have the "across the function" property, which means, that on average, the n 95% confidence intervals about the fitted function at the n data points would contain, about 95% of the true values of the unknown f being estimated, see Nychka (1988). Simulation studies show that these confidence intervals do have this property, when λ is tuned optimally. with Gaussian observations as well as Bernoulli data, see Lin et al. (2000). Wang & Wahba (1994) examine parametric bootstrap and other confidence intervals for smoothing splines, which are a special case of the setup here.

11. A VARIETY OF SPLINES

As noted, splines are a special case of the setup considered here. Without going into any detail, polynomial smoothing splines on the unit interval involve penalty functionals (the term following λ in (1)) of the form $\int_0^1 (f^{(m)})^2(u)du$. See Schoenberg (1964a), Wahba (1990). Thin plate splines on the plane involve the penalty functional $J(f)$ given by

$$J(f) = \int_{-\infty}^{\infty}\int_{-\infty}^{\infty} f_{x_1x_1}^2 + 2f_{x_1x_2}^2 + f_{x_2x_2}^2 dx_1 dx_2$$

and its generalization to higher derivatives, and higher dimensions. Smoothing thin plate splines are discussed in Wahba & Wendelberger (1980). Splines on the sphere generalize $J(f)$ to the sphere (Wahba (1981) Wahba (1982). Vector splines on the sphere can be found in Wahba (1982b). Problems with correlated vector-valued Gaussian observations are discussed in Wahba (1992) and a general vector-valued Bernoulli case is discussed in Gao et al. (2001). Smoothing Spline ANOVA models are discussed in the companion article by the same author in this proceedings, and in references cited there.

12. CONCLUDING REMARKS

Solutions of optimization problems in RKHS have proved to be useful in a large class of curve fitting, function estimation, model building and system description problems. Interest in these wonderful objects have exploded in recent years, as increased computer capacity has made the solution of ever larger and more complex problems practical. New RK's, new approaches and new applications are being contributed at a rapid rate, as can be seen for example in the other papers in this Conference and elsewhere. Preprints of the author's' papers sincd mid 1993, and some earlier are available vi `http://www.stat.wisc.edu/~wahba` goto TRLIST. An April 2003 search of advanced google for the phrase "reproducing kernel" produced over 4400 entries!

13. REFERENCES

Akhiezer, N. & Glazman, I. (1963), *Theory of Linear Operators in Hilbert Space*, Ungar, New York.

Aronszajn, N. (1950), 'Theory of reproducing kernels', *Trans. Am. Math. Soc.* **68**, 337–404.

Craven, P. & Wahba, G. (1979), 'Smoothing noisy data with spline functions: estimating the correct degree of smoothing by the method of generalized cross-validation', *Numer. Math.* **31**, 377–403.

Gao, F., Wahba, G., Klein, R. & Klein, B. (2001), 'Smoothing spline ANOVA for multivariate Bernoulli observations, with applications to ophthalmology data, with discussion', *J. Amer. Statist. Assoc.* **96**, 127–160.

Golub, G., Heath, M. & Wahba, G. (1979), 'Generalized cross validation as a method for choosing a good ridge parameter', *Technometrics* **21**, 215–224.

Gu, C. & Wahba, G. (1991), 'Minimizing GCV/GML scores with multiple smoothing parameters via the Newton method', *SIAM J. Sci. Statist. Comput.* **12**, 383–398.

Halmos, P. (1957), *Introduction to Hilbert Space and the Theory of Spectral Multiplicity*, Chelsea, New York.

Kimeldorf, G. & Wahba, G. (1971), 'Some results on Tchebycheffian spline functions', *J. Math. Anal. Applic.* **33**, 82–95.

Lee, Y., Lin, Y. & Wahba, G. (2002), Multicategory support vector machines, theory, and application to the classification of microarray data and satellite radiance data, Technical Report 1064, Department of Statistics, University of Wisconsin, Madison WI.

Lin, X. (1998), Smoothing spline analysis of variance for polychotomous response data, Technical Report 1003, PhD thesis, Department of Statistics, University of Wisconsin, Madison WI. Available via G. Wahba's website.

Lin, X., Wahba, G., Xiang, D., Gao, F., Klein, R. & Klein, B. (2000), 'Smoothing spline ANOVA models for large data sets with Bernoulli observations and the randomized GACV', *Ann. Statist.* **28**, 1570–1600.

Micchelli, C. (1986), 'Interpolation of scattered data: distance matrices and conditionally positive definite functions', *Constructive Approximation* **2**, 11–22.

Nychka, D. (1988), 'Bayesian confidence intervals for smoothing splines', *J. Amer. Statist. Assoc.* **83**, 1134–1143.

Nychka, D., Wahba, G., Goldfarb, S. & Pugh, T. (1984), 'Cross-validated spline methods for the estimation of three dimensional tumor size distributions from observations on two dimensional cross sections', *J. Am. Stat. Assoc.* **79**, 832–846.

O'Sullivan, F. & Wahba, G. (1985), 'A cross validated Bayesian retrieval algorithm for non-linear remote sensing', *J. Comput. Physics* **59**, 441–455.

Parzen, E. (1970), Statistical inference on time series by rkhs methods, *in* R. Pyke, ed., 'Proceedings 12th Biennial Seminar', Canadian Mathematical Congress, Montreal. 1-37.

Riesz, F. & Nagy, B. S. (1955), *Functional Analysis*, Ungar, New York.

Schoenberg, I. (1964a), 'Spline functions and the problem of graduation', *Proc. Nat. Acad. Sci. U.S.A.* **52**, 947–950.

Tikhonov, A. (1963), 'Solution of incorrectly formulated problems and the regularization method', *Soviet Math. Dokl.* **4**, 1035–1038.

Wahba, G. (1977a), 'Practical approximate solutions to linear operator equations when the data are noisy', *SIAM J. Numer. Anal.* **14**, 651–667.

Wahba, G. (1981), 'Spline interpolation and smoothing on the sphere', *SIAM J. Sci. Stat. Comput.* **2**, 5–16.

Wahba, G. (1982), 'Erratum: Spline interpolation and smoothing on the sphere', *SIAM J. Sci. Stat. Comput.* **3**, 385–386.

Wahba, G. (1982b), Vector splines on the sphere, with application to the estimation of vorticity and divergence from discrete, noisy data, *in* W. Schempp & K. Zeller, eds, 'Multivariate Approximation Theory, Vol.2', Birkhauser Verlag, pp. 407–429.

Wahba, G. (1983), 'Bayesian "confidence intervals" for the cross-validated smoothing spline', *J. Roy. Stat. Soc. Ser. B* **45**, 133–150.

Wahba, G. (1990), *Spline Models for Observational Data*, SIAM. CBMS-NSF Regional Conference Series in Applied Mathematics, v. 59.

Wahba, G. (1992), Multivariate function and operator estimation, based on smoothing splines and reproducing kernels, *in* M. Casdagli & S. Eubank, eds, 'Nonlinear Modeling and Forecasting, SFI Studies in the Sciences of Complexity, Proc. Vol XII', Addison-Wesley, pp. 95–112.

Wahba, G. (1996), 'NIPS 1996 model complexity workshop notes', lecture overheads. Available via `http://www.stat.wisc.edu/ wahba` goto TALKKS goto 1996.

Wahba, G. (1999), Support vector machines, reproducing kernel Hilbert spaces and the randomized GACV, *in* B. Scholkopf, C. Burges & A. Smola, eds, 'Advances in Kernel Methods-Support Vector Learning', MIT Press, pp. 69–88.

Wahba, G. (2002), 'Soft and hard classification by reproducing kernel Hilbert space methods', *Proc.National Academy of Sciences* **99**, 16524–16530.

Wahba, G. & Wang, Y. (1990), 'When is the optimal regularization parameter insensitive to the choice of the loss function?', *Commun. Statist.-Theory Meth.* **19**, 1685–1700.

Wahba, G. & Wendelberger, J. (1980), 'Some new mathematical methods for variational objective analysis using splines and cross-validation', *Monthly Weather Review* **108**, 1122–1145.

Wahba, G., Wang, Y., Gu, C., Klein, R. & Klein, B. (1995), 'Smoothing spline ANOVA for exponential families, with application to the Wisconsin Epidemiological Study of Diabetic Retinopathy', *Ann. Statist.* **23**, 1865–1895. Neyman Lecture.

Wang, Y. & Wahba, G. (1994), Bootstrap confidence intervals for smoothing splines and their comparison to Bayesian 'confidence intervals', Technical Report 913, Dept. of Statistics, University of Wisconsin, Madison, WI, to appear, *J. Stat. Comp. Sim.*

Xiang, D. & Wahba, G. (1996), 'A generalized approximate cross validation for smoothing splines with non-Gaussian data', *Statistica Sinica* **6**, 675–692.

IFAC Publications
www.elsevier.com/locate/ifac

AN INTRODUCTION TO SMOOTHING SPLINE ANOVA MODELS IN RKHS, WITH EXAMPLES IN GEOGRAPHICAL DATA, MEDICINE, ATMOSPHERIC SCIENCES AND MACHINE LEARNING

Grace Wahba *,[1]

* *Department of Statistics, University of Wisconsin, 1210 W. Dayton St., Madison WI, USA*

Abstract: This paper is a brief introduction to smoothing spline ANOVA (SS-ANOVA) models in reproducing kernel Hilbert spaces (RKHS) and some of their applications. These models decompose a function of several variables as sums of functions of one variable plus sums of functions of two variables and so forth, analogous to the ordinary analysis of variance decomposition familiar to students in elementary Statistics classes. This is done in such a way that the individual terms are projections onto orthogonal subspaces in RKHS, and the relevant reproducing kernels may be found in many examples. Given the appropriate RKHS, various kinds of estimation and model fitting problems in several variables given observational data can be solved. *Copyright © 2003 IFAC*

Keywords: RKHS, functional decompositions, risk factor estimation, multiple correlated Bernoulli outcomes, splines on the sphere, multicategory support vector machines.

1. INTRODUCTION

Smoothing Spline ANOVA (SS-ANOVA) models in reproducing kernel Hilbert spaces (RKHS) provide a very general framework for data analysis, modeling and learning in a variety of fields. Discrete, noisy scattered, direct and indirect observations can be accommodated with multiple inputs and multiple possibly correlated outputs and a variety of meaningful structures. The purpose of this paper is to give a brief overview of the approach and describe and contrast a series of applications, while noting some recent results.

2. THE GENERAL SS-ANOVA MODEL

The SS-ANOVA model with Gaussian data has the form

$$y_i = f(t_1(i), \cdots, t_d(i)) + \epsilon_i, \quad i = 1, \cdots, n,$$

where $\epsilon = (\epsilon_1, \cdots, \epsilon_n)' \sim N(0, \sigma^2 I_{n \times n})$, $t_\alpha \in \mathcal{T}^{(\alpha)}$, where $\mathcal{T}^{(\alpha)}$ is a measurable space, $\alpha = 1, \cdots, d; (t_1, \cdots, t_d) = t \in \mathcal{T} = \mathcal{T}^{(1)} \otimes \cdots \otimes \mathcal{T}^{(d)}$, and σ^2 may be unknown. For f satisfying some measurability conditions a unique ANOVA decomposition of f of the form

$$f(t_1, \cdots, t_d) =$$
$$\mu + \sum_\alpha f_\alpha(t_\alpha) + \sum_{\alpha\beta} f_{\alpha\beta}(t_{\alpha\beta}) + \cdots \quad (1)$$

can always be defined as follows: Let $d\mu_\alpha$ be a probability measure on $\mathcal{T}^{(\alpha)}$ and define the averaging operator $\mathcal{E}_\alpha$ on $\mathcal{T}$ by

$$(\mathcal{E}_\alpha f)(t) = \int_{\mathcal{T}^{(\alpha)}} f(t_1, \cdots, t_d) d\mu_\alpha(t_\alpha).$$

Then the identity is decomposed as

$$I = \prod_\alpha (\mathcal{E}_\alpha + (I - \mathcal{E}_\alpha)) =$$

[1] Supported by NSF grant DMS-0072292, NIH grant EY09946, and NASA grant NAG51073

$$\prod_\alpha \mathcal{E}_\alpha + \sum_\alpha (I-\mathcal{E}_\alpha)\prod_{\beta\neq\alpha}\mathcal{E}_\beta$$

$$+\sum_{\alpha<\beta}(I-\mathcal{E}_\alpha)(I-\mathcal{E}_\beta)\prod_{\gamma\neq\alpha,\beta}\mathcal{E}_\gamma + \cdots + \prod_\alpha (I-\mathcal{E}_\alpha)$$

The components of this decomposition generate the ANOVA decomposition of f of the form (1) by $C = (\prod_\alpha \mathcal{E}_\alpha)f$, $f_\alpha = ((I-\mathcal{E}_\alpha)\prod_{\beta\neq\alpha}\mathcal{E}_\beta)f$, $f_{\alpha\beta} = ((I-\mathcal{E}_\alpha)(I-\mathcal{E}_\beta)\prod_{\gamma\neq\alpha,\beta}\mathcal{E}_\gamma)f$, and so forth. Further details in the RKHS context may be found in Wahba (1990)Gu & Wahba (1993)Wahba, Wang, Gu, Klein & Klein (1995)

The idea behind SS-ANOVA is to construct an RKHS $\mathcal{H}$ of functions on $\mathcal{T}$ so that the components of the SS-ANOVA decomposition represent an orthogonal decomposition of f in $\mathcal{H}$. Then RKHS methods can be used to explicitly impose smoothness penalties of the form $\sum_\alpha \lambda_\alpha J_\alpha(f_\alpha) + \sum_{\alpha\beta}\lambda_{\alpha\beta}J_{\alpha\beta}(f_{\alpha\beta}) + \cdots$, where, however, the series will be truncated at some point. This is done as follows: Let $\mathcal{H}^{(\alpha)}$ be an RKHS of functions on $\mathcal{T}^{(\alpha)}$ with $\int_{\mathcal{T}^{(\alpha)}} f_\alpha(t_\alpha)d\mu_\alpha = 0$ for $f_\alpha(t_\alpha)\in\mathcal{H}^{(\alpha)}$, and let $[1^{(\alpha)}]$ be the one dimensional space of constant functions on $\mathcal{T}^{(\alpha)}$. Construct $\mathcal{H}$ as

$$\mathcal{H} = \prod_{j=1}^{d}(\{[1^{(\alpha)}]\}\oplus\{\mathcal{H}^{(\alpha)}\})$$

$$= [1]\oplus\sum_j \mathcal{H}^{(\alpha)} \oplus \sum_{\alpha<\beta}[\mathcal{H}^{(\alpha)}\otimes\mathcal{H}^{(\beta)}]\oplus\cdots,$$

endequation where $[1]$ denotes the constant functions on $\mathcal{T}$. With some abuse of notation, factors of the form $[1^{(\alpha)}]$ are omitted whenever they multiply a term of a different form. Thus $\mathcal{H}^{(\alpha)}$ is a shorthand for $[1^{(1)}]\otimes\cdots\otimes[1^{(\alpha-1)}]\otimes\mathcal{H}^{(\alpha)}\otimes[1^{(\alpha+1)}]\otimes\cdots\otimes[1^{(d)}]$ (which is a subspace of $\mathcal{H}$). The components of the ANOVA decomposition are now in mutually orthogonal subspaces of $\mathcal{H}$. Note that the components will depend on the measures $d\mu_\alpha$ and these should be chosen in a specific application so that the fitted mean, main effects, two factor interactions, etc. have reasonable interpretations.

Next, $\mathcal{H}^{(\alpha)}$ is decomposed into a parametric part and a smooth part, by letting $\mathcal{H}^{(\alpha)} = \mathcal{H}_\pi^{(\alpha)}\oplus\mathcal{H}_s^{(\alpha)}$, where $\mathcal{H}_\pi^{(\alpha)}$ is finite dimensional (the "parametric" part) and $\mathcal{H}_s^{(\alpha)}$ (the "smooth" part) is the orthocomplement of $\mathcal{H}_\pi^{(\alpha)}$ in $\mathcal{H}^{(\alpha)}$. Elements of $\mathcal{H}_\pi^{(\alpha)}$ are not penalized through the device of letting $J_\alpha(f_\alpha) = \|P_s^{(\alpha)}f_\alpha\|^2$ where $P_s^{(\alpha)}$ is the orthogonal projector onto $\mathcal{H}_s^{(\alpha)}$. $[\mathcal{H}^{(\alpha)}\otimes\mathcal{H}^{(\beta)}]$ is now a direct sum of four orthogonal subspaces: $[\mathcal{H}^{(\alpha)}\otimes\mathcal{H}^{(\beta)}] = [\mathcal{H}_\pi^{(\alpha)}\otimes\mathcal{H}_\pi^{(\beta)}]\oplus[\mathcal{H}_\pi^{(\alpha)}\otimes\mathcal{H}_s^{(\beta)}]\oplus[\mathcal{H}_s^{(\alpha)}\otimes\mathcal{H}_\pi^{(\beta)}]\oplus[\mathcal{H}_s^{(\alpha)}\otimes\mathcal{H}_s^{(\beta)}]$. By convention the elements of the finite dimensional space $[\mathcal{H}_\pi^{(\alpha)}\otimes\mathcal{H}_\pi^{(\beta)}]$ will not be penalized. Continuing this way results in an orthogonal decomposition of $\mathcal{H}$ into sums of products of unpenalized finite dimensional subspaces, plus main effects 'smooth' subspaces, plus two factor interaction spaces of the form parametric $\otimes$ smooth $[\mathcal{H}_\pi^{(\alpha)}\otimes\mathcal{H}_s^{(\beta)}]$, smooth $\otimes$ parametric $[\mathcal{H}_s^{(\alpha)}\otimes\mathcal{H}_\pi^{(\beta)}]$ and smooth $\otimes$ smooth $[\mathcal{H}_s^{(\alpha)}\otimes\mathcal{H}_s^{(\beta)}]$ and similarly for the three and higher factor subspaces.

Now suppose that we have selected the model $\mathcal{M}$, that is, we have decided which subspaces will be included. Collect all of the included unpenalized subspaces into a subspace, call it $\mathcal{H}^0$, of dimension M, and relabel the other subspaces as $\mathcal{H}^\beta, \beta = 1,2,\cdots,p$. $\mathcal{H}^\beta$ may stand for a subspace $\mathcal{H}_s^{(\alpha)}$, or one of the three subspaces in the decomposition of $[\mathcal{H}^{(\alpha)}\otimes\mathcal{H}^{(\beta)}]$ which contains at least one 'smooth' component, or, a higher order subspace with at least one 'smooth' component. Collecting these subspaces as $\mathcal{M} = \mathcal{H}^0\oplus\sum_\beta\mathcal{H}^\beta$, the estimation problem in the Gaussian case becomes: Find f in $\mathcal{M} = \mathcal{H}^0\oplus\sum_\beta\mathcal{H}^\beta$ to minimize

$$\frac{1}{n}\sum_{i=1}^{n}(y_i - f(t(i)))^2 + \lambda\sum_{\beta=1}^{p}\theta_\beta^{-1}\|P^\beta f\|^2, \quad (2)$$

where P^β is the orthogonal projector in $\mathcal{M}$ onto $\mathcal{H}^\beta$, and choose the (overparameterized) tuning parameters λ, θ_β. Bayesian confidence intervals, with the so-called 'across the function' property, are available for these models.

The residual sum of squares (RSS) in (2) is replaced by the log likelihood

$$\mathcal{L}(y,f) = -\sum_{i=1}^{n}[y_i f(t(i)) - b(f(t)))]$$

for data from exponential families. Some of the examples below will involve Bernoulli $(0,1)$ data, in which case $b(f) = log(1+e^f)$. Software for computing and tuning SS-ANOVA models may be found in the codes GRKPACK, RKPACK and gss and elsewhere, links to these and other spline related codes can be found via `http://www.stat.wisc.edu/~wahba` go to "SOFTWARE". Tuning methods are discussed in the first talk in this session. RSS may be replaced by robust functionals, or any convex functionals satisfying some mild conditions insuring uniqueness, and, in recent work on classification by support vector machines, RSS is replaced by so-called hinge functions.

3. APPLICATIONS IN ENVIRONMENTAL DATA

Gu & Wahba (1993) considered data from the Eastern Lake Survey of 1984 which gave water acidity measurements and geographic locations, and other measurements of lakes in the Blue Ridge Mountains area. Of interest is the pH as it depends on the geographic location and calcium concentration in the lakes. Model diagnostics were proposed there, and the model

$$y_i = f_1(t_1(i)) + f_2(t_2(i)) + f_{1,2}(t_1(i),t_2(i)) + \epsilon_i$$

was chosen, where t_1 is calcium content and t_2 is the pair (latitude, longitude). The thin plate spline penalty was imposed on the spatial variable. The calcium content and geography main effects models were plotted, and it can be seen that geography is a near proxy for elevation along the Blue Ridge mountains.

4. RISK FACTOR ESTIMATION

Wahba et al. (1995) considered the risk of progression of diabetic retinopathy in a subpopulation of the Wisconsin Epidemiological Study of Diabetic Retinopathy, whose baseline retinopathy score was below (i. e. good) a prespecified level. The observations were $y_i = 1$ if the ith person's retinopathy progressed at the first followup, and 0 if it had not. Here f is the log odds ratio, $f = log[p/(1-p)]$. Three important variables were identified by informal means (see Section 9) and were t_1 = duration of diabetes, t_2 = glycosylated hemoglobin, and t_3 = body mass index, and was modeled as

$$f(t) = \mu + f_1(t_s) + a_2 t_2 + f_3(t_3) + f_{13}(t_1, t_3).$$

An interesting scientific result was found, that, persons in the study group with the longest duration of diabetes were at a lower risk, possibly because they had survived longest without exceeding the prespecified threshold.

5. TIME AND SPACE MODELS ON THE GLOBE

In Wahba & Luo (1997)Luo, Wahba & Johnson (1997) thirty years (1961-90) of Dec. Jan. Feb. average temperature measurements at 1000 stations around the globe (with missing data) was analyzed for spatial trends, as well as a global trend. Here $t = (t_1, t_2) = (x, P)$ where x is year, and P is (latitude, longitude). The RKHS of historical global temperature functions that was used is $\mathcal{H} = [[1^{(1)}] \oplus [\phi] \oplus \mathcal{H}_s^{(1)}] \otimes [[1^{(2)}] \oplus \mathcal{H}_s^{(2)}]$, a collection of functions $f(x, P)$, on $\{1, 2, ..., 30\} \otimes \mathcal{S}$, where $\mathcal{S}$ is the sphere. $\mathcal{H}$ and f have the corresponding (six term) decompositions given below:

$$\begin{array}{rcccccccc} \mathcal{H} = & [1] & \oplus & [\phi] & \oplus & [\mathcal{H}_s^{(1)}] & \oplus & [\mathcal{H}_s^{(2)}] \\ f(x,P) = & C & + & d\phi(x) & + & f_1(x) & + & f_2(P) \\ = & mean & + & global & + & time & + & space \\ & & & time & & main & & main \\ & & & trend & & effect & & effect \end{array}$$

$$\begin{array}{cccc} \oplus & [[\phi] \otimes \mathcal{H}_s^{(2)}] & \oplus & [\mathcal{H}_s^{(1)} \otimes \mathcal{H}_s^{(2)}] \\ + & \phi(x) f_{\phi,2}(P) & + & f_{12}(x, P) \\ + & trend & + & space- \\ & by\ space & & time \\ & effect & & interaction \end{array}$$

Here ϕ is a linear function which averages to 0. A sum of squares of second differences was applied to the time variable, and a spline on the sphere penalty (Wahba (1981)Wahba (1982)) was applied to the space variable. For a cross country skier in the Midwest, as this author is, the results were very disappointing, in that they clearly showed a warming trend stretching from the Midwest towards Alaska (trend by space term) which was stronger than the global mean trend.

6. MULTIPLE CORRELATED BERNOULLI OUTCOMES

Gao, Wahba, Klein & Klein (2001) were motivated by a demographic study involving a population with a variety of observed risk factors for several particular eye diseases, the outcomes were the incidence of one or more of several diseases or conditions in either or both of two eyes. Outcomes of the two eyes in a particular subject are presumed to be correlated, and incidences of the various outcomes may also be correlated. The amount of correlation may be of particular interest. The risk factors could be person specific or eye-specific. The "two-eye" methods are a special case of what might be called "k-eye" methods where one person (unit) has several component outcomes which might have correlated outcomes, depending on unit-specific and component specific risk factors.

The general log-linear model for multivariate Bernoulli data goes as follows: Assuming there are J different endpoints, and K_j repeated measurements for the jth endpoint, let Y_{jk} denote the kth measurement of the jth endpoint. For example, in ophthalmological studies, we have two repeated measurements for each disease: left eye and right eye. In a typical longitudinal study, we have repeated measurements over the time. $Y = (Y_{jk}, j = 1, ..., J, k = 1, ..., K_j)$ is a multivariate Bernoulli outcome variable. Let $X_{jk} = (X_{jk1}, X_{jk2}, ..., X_{jkD})$ be a vector of predictor variables ranging over the subset $\mathcal{X}$ of $\mathcal{R}^D$, where X_{jkd} denotes the dth predictor variable for the kth measurement of the jth endpoint. Some predictor variables may take different values for different measurements while others may be the same for all Y_{jk}'s. For example, in ophthalmology studies, there may be present both person-specific predictors and eye-specific predictors. The person-specific predictors are the same for each person. For the eye-specific predictors, the set of predictor variables is the same, but they may take different values for the left and right eyes. We can treat observations from both eyes as correlated repeated measurements in our model. Let $X = (X_{jk}, j = 1, ..., J, k = 1, ..., K_j)$. Then (X, Y) is a pair of random vectors. For a response vector $y = (y_{jk}, j = 1, ..., J, k = 1, ..., K_j)$, its joint probability distribution conditioning on the predictor variables X can be written as

$$P(Y = y|X) =$$

$$\exp\{\sum_{j=1}^{J}\sum_{k=1}^{K_j} f_{jk}y_{jk} + \sum_{j=1}^{J}\sum_{k_1<k_2}\alpha_{jk_1,jk_2}y_{jk_1}y_{jk_2}$$

$$+\sum_{j_1<j_2}\sum_{k_1,k_2}\alpha_{j_1k_1,j_2k_2}y_{j_1k_1}y_{j_2k_2}$$

$$+ ... + \alpha_{11,12,...,JK_J}y_{11}y_{12}....y_{JK_J} - b(f,\alpha)\},$$

where

$$b(f,\alpha) =$$

$$\log(1+\sum_{j,k}e^{f_{jk}} + \sum_{j_1,k_1}\sum_{j_2,k_2}e^{(f_{j_1k_1}+f_{j_2k_2}+\alpha_{j_1k_1,j_2k_2})} + ...$$

$$+e^{(\sum_{all\ f} f + \sum_{all\ \alpha}\alpha)}).$$

Let $M = \sum_{j=1}^{J} K_j$ be the length of the vector Y. There are in total $2^M - 1$ parameters: $(f,\alpha) = (f_{11}, f_{12}, ..., f_{JK_J}, \alpha_{11,12}, ..., \alpha_{11,12,...,JK_J})$, which may depend on X. The parameter space is unconstrained. They have straightforward interpretations in terms of conditional probabilities. For example,

$$f_{jk} = logit(P(Y_{jk} = 1|Y^{(-jk)} = 0, X))$$

is the conditional logit function;

$$\alpha_{j_1k_1,j_2k_2} =$$

$$\log OR(Y_{j_1k_1}, Y_{j_2k_2}|Y^{(-j_1k_1,-j_2k_2)} = 0, X)$$

is the conditional log odds ratio, which is a meaningful way to measure pairwise association; interpretations of other terms are given in the paper.

n independent observations $(x_i, y_i), i = 1, ..., n$, are given, where $y_i = (y_{i11}, y_{i12}, ..., y_{iJK_J})$ and $x_i = (x_{i11}, x_{i12}, ..., x_{iJK_J})$. Here y_{ijk} and $x_{ijk} = (x_{ijk1}, x_{ijk2}, ..., x_{ijkD})$ are the outcome variable and predictor vector for the kth measurement of the jth endpoint of the ith subject. Let $f_{jk}(i)$ be the conditional logit function for the kth measurement of the jth endpoint of the ith subject. There is little reason to believe the f_{jk} will take different functional forms for the same endpoint. Hence we can assume $f_{ijk} = f_j(x_{ijk})$. The same reasoning applies to the association terms. The f_{jk} were modeled via SS-ANOVA in the paper, and a leaving-out-one-person based generalized cross validation for the smoothing parameters was obtained.

7. MULTICHOTOMOUS RESPONSES

Lin (1998) considered multichotomous outcomes, the data is $(y_i, t(i))$ where y_i is coded to show that the i subject, with attribute vector $t(i)$ is in one of $k+1$ categories, $k > 1$. Let $p_j(t), j = 0, 1, \cdots, k$ be the probability that a subject with attribute vector t is in category k, $\sum_{j=0}^{k} p_j(t) = 1$. Let $f^j(t) = \log[p_j(t)/p_0(t)], j = 1, \cdots, k$. Then

$$p_j(t) = \frac{e^{f^j(t)}}{1+\sum_{j=1}^{k} e^{f^j(t)}}, j = 1, \cdots, k$$

$$p_0(t) = \frac{1}{1+\sum_{j=1}^{k} e^{f^j(t)}}.$$

The class label for the ith subject is coded as $y_i = (y_{i1}, \cdots, y_{ik})$ where $y_{ij} = 1$ if the ith subject is in class j and 0 otherwise. Letting $f = (f^1, \cdots, f^k)$ the negative log likelihood can be written as

$$\mathcal{L}(y,f) =$$

$$\sum_{i=1}^{n}\{-\sum_{j=1}^{k} y_{ij}f^j(t_i) + log(\sum_{j=1}^{k} 1 + e^{f^j(t_i)})\}. (3)$$

$f^j = \sum_{\nu_j=1}^{M}\phi_\nu + h^j$ where the h^j can have an ANOVA decomposition. Then the penalty functional in (2) is replaced by

$$\sum_{j=1}^{k}\sum_{\alpha}\lambda_{j\alpha}J_{j\alpha}(h_\alpha^j) +$$

$$\sum_{\alpha<\beta}\lambda_{j\alpha\beta}J_{j\alpha\beta}(h_{\alpha\beta}^j) + \cdots. \quad (4)$$

Ten year mortality data of a group of $n = 646$ subjects with the risk factors age (x_1), glycosylated hemoglobin (x_2) and systolic blood pressure (x_3) were (among other things) recorded at baseline and they were divided into four categories with respect to their status after ten years, as 0 =alive, 1 = died of diabetes, 2 =died of heart disease, and 3 =died of other causes. Each of the $f^j, j = 1, 2, 3$ was modeled as $f^j(x_1, x_2, x_3) = \mu^j + f_1^j(x_1) + f_2^j(x_2) + f_3^j(x_3) + f_{23}^j(x_2, x_3)$. The $p_j, j = 0, \cdots, 3$ were estimated by minimizing $\mathcal{I}(y, f) = (3) + (4)$ and the multiple smoothing parameters estimated by a generalized cross validation method for polychotomous data given in Lin (1998). The plots graphically convey the suggestion that the younger deaths are disproportionately diabetic, thus quickly raising further questions to confront the data base.

8. THE MULTICATEGORY SUPPORT VECTOR MACHINE

The multicategory support vector machine (MSVM) proposed in Lee, Lin & Wahba (2002),Lee, Lin & Wahba (2001) considers the case where each subject is in one of k categories labeled as $j = 1, \cdots, k$, as in the preceeding section, except for notational convenience there are k instead of $k+1$ categories. The support vector machine is an efficient method for classification - it is not estimating the probability of membership in a particular category as before, but its target is an indicator as to which category as subject is in (or most likely to be in)(see Lin (2002). The class label y_i is now coded as a k dimensional vector with 1 in the jth position if example i is in category j and $-\frac{1}{k-1}$ otherwise. For example $y_i = (1, -\frac{1}{k-1}, \cdots, -\frac{1}{k-1})$ indicates that the ith example is in category 1. We define a k-tuple of separating functions $f(t) = (f^1(t), \cdots f^k(t))$, with each $f^j = d^j + h^j$ with $h^j \in \mathcal{H}_K$, and which will be required to satisfy a sum-to-zero constraint, $\sum_{j=1}^k f^j(t) = 0$, for all t in $\mathcal{T}$. Note that, unlike the estimate of Section 7, all categories are treated symmetrically.

Let $L_{jr} = 1, r \neq j$, $L_{jj} = 0, j, r = 1, \cdots, k$. Let $cat(y_i) = j$ if y_i is from category j. Then, if y_i is from category j, $L_{cat(y_i)r} = 0$ if $r = j$ and 1 otherwise. Then the MSVM is defined as the vector of functions $f_\lambda = (f_\lambda^1, \cdots, f_\lambda^k)$, with each h^k in $\mathcal{H}_K$ satisfying the sum-to-zero constraint, which minimizes

$$\frac{1}{n}\sum_{i=1}^{n}\sum_{r=1}^{k} L_{cat(y_i)r}(f^r(t_i) - y_{ir})_+$$

$$+\lambda \sum_{j=1}^{k} \|h^j\|_{\mathcal{H}_K}^2.$$

Generalizations of the penalty term are possible, if necessary. It can be shown that the $k = 2$ case reduces to the usual 2-category SVM just discussed, and it is shown in Lee et al. (2001) that the target for the MSVM is $f(t) = (f^1(t), \cdots, f^k(t))$ with $f^j(t) = 1$ if $p_j(t)$ is bigger than the other $p_l(t)$ and $f^j(t) = -\frac{1}{k-1}$ otherwise. See also Wahba (2002).

9. SUMMARY

The SS-ANOVA models have proved to be useful in a variety of modeling situations, only a few described here. In each case a tuning method which governs the bias-variance tradeoff must be employed, and, for very large sample sizes, efficient approximate methods need to be devised. Model selection, that is, the determination of which variables and/or terms to include in the model is an important issue. Zhang, Wahba, Lin, Voelker, Ferris, Klein & Klein (2001)Zhang, Wahba, Lin, Voelker, Ferris, Klein & Klein (2002) have recently proposed likelihood basis pursuit, a nonparametric form of the LASSO, for the model selection problem associated with SS-ANOVA. Although a number of tuning methods for the various situations have been proposed, along with numerical methods for large data sets, a variety of problems remain to be investigated, including optimum nonlinear transformations of the variables, efficient computational methods, methods for covariates not missing at random, and public software for very large sample sizes and for some of the more complex structures.

10. REFERENCES

Gao, F., Wahba, G., Klein, R. & Klein, B. (2001), 'Smoothing spline ANOVA for multivariate Bernoulli observations, with applications to ophthalmology data, with discussion', *J. Amer. Statist. Assoc.* **96**, 127–160.

Gu, C. & Wahba, G. (1993), 'Smoothing spline ANOVA with component-wise Bayesian "confidence intervals"', *J. Computational and Graphical Statistics* **2**, 97–117.

Lee, Y., Lin, Y. & Wahba, G. (2001), Multicategory support vector machines, Technical Report 1043, Department of Statistics, University of Wisconsin, Madison WI. To appear, *Computing Science and Statistics*, 33.

Lee, Y., Lin, Y. & Wahba, G. (2002), Multicategory support vector machines, theory, and application to the classification of microarray data and satellite radiance data, Technical Report 1063, Department of Statistics, University of Wisconsin, Madison WI.

Lin, X. (1998), Smoothing spline analysis of variance for polychotomous response data, Technical Report 1003, PhD thesis, Department of Statistics, University of Wisconsin, Madison WI. Available via G. Wahba's website.

Lin, Y. (2002), 'Support vector machines and the Bayes rule in classification', *Data Mining and Knowledge Discovery* **6**, 259–275.

Luo, Z., Wahba, G. & Johnson, D. (1997), Spatial-temporal analysis of temperature using smoothing spline ANOVA, Technical Report 97-01, Pennsylvania State University Statistics Dept., State College PA.

Wahba, G. (1981), 'Spline interpolation and smoothing on the sphere', *SIAM J. Sci. Stat. Comput.* **2**, 5–16.

Wahba, G. (1982), 'Erratum: Spline interpolation and smoothing on the sphere', *SIAM J. Sci. Stat. Comput.* **3**, 385–386.

Wahba, G. (1990), *Spline Models for Observational Data*, SIAM. CBMS-NSF Regional Conference Series in Applied Mathematics, v. 59.

Wahba, G. (2002), Soft and hard classification by reproducing kernel hilbert space methods, Technical Report 1067, Department of Statistics, University of Wisconsin, Madison WI. to appear, Proceedings of the National Academy of Sciences.

Wahba, G. & Luo, Z. (1997), 'Smoothing spline ANOVA fits for very large, nearly regular data sets, with application to historical global climate data', *Ann. Numer. Math.* **4**, 579–597.

Wahba, G., Wang, Y., Gu, C., Klein, R. & Klein, B. (1995), 'Smoothing spline ANOVA for exponential families, with application to the Wisconsin Epidemiological Study of Diabetic Retinopathy', *Ann. Statist.* **23**, 1865–1895. Neyman Lecture.

Zhang, H., Wahba, G., Lin, Y., Voelker, M., Ferris, M., Klein, R. & Klein, B. (2001), Variable selection via basis pursuit for non-Gaussian data, Technical Report 1042, Statistics Department University of Wisconsin, Madison WI. In Proceedings of the ASA Joint Statistical Meetings 2001 (CDROM), available from the American Statistical Association.

Zhang, H., Wahba, G., Lin, Y., Voelker, M., Ferris, M., Klein, R. & Klein, B. (2002), Variable selection and model building via likelihood basis pursuit, Technical Report 1059, Statistics Department University of Wisconsin, Madison WI.

IFAC Publications
www.elsevier.com/locate/ifac

ROBUST DESIGN WITH NONPARAMETRIC MODELS: PREDICTION OF SECOND-ORDER CHARACTERISTICS OF PROCESS VARIABILITY BY KRIGING [1]

Luc Pronzato * Éric Thierry *

* *Laboratoire I3S, CNRS/Université de Nice–Sophia Antipolis,*
Les Algorithmes, 2000 route des Lucioles, BP 121,
06903 Sophia-Antipolis Cedex, France

Abstract: We use kriging to predict the mean and variance of a response $y(\mathbf{x})$ when the input factors $\mathbf{x}$ are subject to random variability. Uncertainty on these predictions is obtained by considering fluctuations along one trajectory y of the process due to fluctuations of $\mathbf{x}$, and then averaging over the possible trajectories, conditionally on input-output data. Possible applications include robust design engineering, where the data that are obtained from prototypes in laboratory experiments, or from simulation codes, are used to construct models for the responses of interest to the designer, but mass-production involves variability of input factors around the specifications the designer will indicate.

Keywords: Kriging, Robust Design, Prediction, Computer Experiments, Process Variability, Nonparametric Modelling, Propagation of errors, Taguchi Method

1. INTRODUCTION

Assume that, based on controlled, say laboratory, or prototype, experiments, we construct a model of the response $y(\mathbf{x})$ of a system to input factors $\mathbf{x} \in \mathbb{R}^d$. Several objectives may be considered, such as y should be maximized, y should be set equal to some target T, a constraint $y \leq c$ should be satisfied, etc., that is, y may correspond either to an objective or a constraint in a multi-objective optimization problem, see for instance (Bates *et al.* 1999).

We consider the situation where the prediction $\eta(\mathbf{x})$ of $y(\mathbf{x})$ at some unobserved $\mathbf{x}$ is obtained by kriging (Krige 1951, Matheron 1963). The approach is particularly attractive for computer experiments, see, *e.g.*, (Sacks *et al.* 1989*a*, Sacks *et al.* 1989*b*, Welch *et al.* 1992), due to its flexibility and the possibility to predict model accuracy (or rather, inaccuracy) from deterministic responses: roughly, the unknown response $y(\mathbf{x})$ is described as a mean value plus the trajectory of a zero-mean stationary process, and statistical inference is made from observations of the response at given (design) points $\mathbf{x}_1, \ldots, \mathbf{x}_N$. The second-order characteristics of the process are estimated from the data $\mathbf{Y} = [y(\mathbf{x}_1), \ldots, y(\mathbf{x}_N)]$, and maximum-likelihood can be used when the process is Gaussian and its covariance is suitably parameterized. The best linear unbiased predictor for the value of $y(\mathbf{x})$ at new inputs $\mathbf{x}$ is then constructed from $\mathbf{Y}$, together with the variance of $y(\mathbf{x})$ due to model uncertainty. Classical extensions concern the case where observation errors are present (which corresponds to physical, as opposed to computer, experiments) and the mean value of the response is parameterized, with possibly a Bayesian prior on these parameters. We refer to the papers mentioned above and to (Stein 1999) for a more accurate introduction. The choice of appropriate design points $(\mathbf{x}_1, \ldots, \mathbf{x}_n)$ has received little attention, see (Sacks and Schiller 1988, Shewry and Wynn 1988), but Latin Hypercube designs are generally adopted for their suitable space-filling property (it is a most attractive feature of kriging to be able to generate fairly accurate models from very

[1] This work is part of the European project 'TITOSIM' (TIme TO market reduction via Statistical Information Management), GRD1-2000-25724

few data, see (Costa *et al.* 2000) for an application in signal processing). In robust design problems, see below, this is a definite advantage over the so-called Taguchi method, see (Vuchkov and Boyadjieva 2001). When the design points are generated sequentially, the method to be used generally depends on the final objective of the model, which may for instance correspond to the optimization of the response, see (Schonlau *et al.* 1998, Bates and Pronzato 2001).

In this paper, we focuss our attention on the following robust design problem: in mass production, $\mathbf{x}$ cannot be chosen accurately and must be considered as a random variable [2]. We shall assume that its second-order characteristics are known (for instance, it may be normal with known mean and variance). This induces variability on $y(\mathbf{x})$, which we want to predict, again in terms of mean and variance. Such predictions can then be taken into account, for instance by choosing the mean value for $\mathbf{x}$ such that the mean value of $y(\mathbf{x})$ is maximized and the variability of $y(\mathbf{x})$ is minimized. More complex situations can of course be considered, leading to various multi-objective optimization problems. The crucial point here is that, starting from a single response (which may be an objective or a constraint in the original problem), we get two responses due to the variability of $\mathbf{x}$ in mass production: the mean and the variance of $y(\mathbf{x})$. Also, since the model is constructed from a finite data sample $\mathbf{Y}$, both responses are uncertain, and their variances are of interest.

Propagation of errors is standard for classical models, such as polynomials, see (Vuchkov and Boyadjieva 2001). It is the aim of the paper to show that *propagation of errors is also feasible when the model is obtained by kriging*. The predictions of the mean and variance of $y(\mathbf{x})$ when $\mathbf{x}$ varies are constructed in Section 2. Uncertainty due to estimation from a finite data sample is considered in Section 3. Illustrative examples are presented in Section 4. Throughout the paper we restrict our attention to the case of computer experiments, the extension to physical experiments where observation errors are present does not raise particular difficulties (only the covariance structure must be modified, see, *e.g.*, Costa *et al.* (2000)).

2. BAYESIAN KRIGING FOR PREDICTING MEAN AND VARIANCE

We remind the construction of the predictor $\eta(\mathbf{x})$, using a Bayesian approach (Bayesian kriging), and derive the joint posterior distribution of two responses $y(\mathbf{x}_a), y(\mathbf{x}_b)$. This is used later on to construct predictions that take the variability of $\mathbf{x}$ into account.

[2] In practise, it happens that some of the factors can still be controlled during mass production, however, this does not modify the methodology presented below.

2.1 *Bayesian kriging*

We model the observations by $Y_i = y(\mathbf{x}_i) = \beta^\top \mathbf{r}(\mathbf{x}_i) + Z(\mathbf{x}_i)$ with $Z_i = Z(\mathbf{x}_i)$ a zero mean second order stationary stochastic process and $\beta^\top \mathbf{r}(\mathbf{x})$ the deterministic part (which corresponds to *universal kriging*). The realizations Z_i, Z_j are correlated, and we define

$$V(Z_i, Z_j) = \mathrm{E}_Z\{Z(\mathbf{x}_i)Z(\mathbf{x}_j)\}\,. \qquad (1)$$

Since the process is assumed to be stationary, we write

$$V(Z_i, Z_j) = \sigma_z^2 C(\mathbf{x}_i - \mathbf{x}_j)\,.$$

An usual model for the covariance $C(\cdot)$ is

$$C(\mathbf{z}) = C(\theta, \mathbf{z}) = \exp\left(-\sum_{i=1}^{d} \theta_i z_i^2\right)\,, \qquad (2)$$

which gives a process $Z(\mathbf{x})$ infinitely mean square differentiable, see, *e.g.*, (Stein 1999). Other models for the covariance (Matérn class) yield exactly one, two or m times mean square differentiability [3].

We assume that the parameters β have a normal prior $\mathcal{N}(\mu, \sigma^2\Omega)$, and that the process Z is Gaussian and independent of β. We denote by $\mathbf{Y}$ the N observations $Y_1, Y_2, \ldots, Y_N$ and $\mathbf{V}$, $\mathbf{C}$ the matrices defined by $[\mathbf{V}]_{i,j} = V(Z_i, Z_j)$, $[\mathbf{C}]_{i,j} = C(\mathbf{x}_i - \mathbf{x}_j)$, $i,j = 1, \ldots, N$.

The posterior distribution of β (conditional to $\mathbf{Y}$) is normal $\mathcal{N}(\bar\beta, \mathbf{W})$, with

$$\mathbf{W} = [\mathbf{R}^\top \mathbf{V}^{-1}\mathbf{R} + (\sigma^2\Omega)^{-1}]^{-1}\,,$$
$$\bar\beta = \mathbf{W}[\mathbf{R}^\top \mathbf{V}^{-1}\mathbf{Y} + (\sigma^2\Omega)^{-1})\mu]\,,$$

where the i-th row of the matrix $\mathbf{R}$ equals $\mathbf{r}^\top(\mathbf{x}_i)$.

We want to predict two responses $y_a = y(\mathbf{x}_a), y_b = y(\mathbf{x}_b)$. The joint distribution $\pi(\mathbf{Y}, y_a, y_b|\beta)$ of $\mathbf{Y}, y_a$, and y_b conditional to β is normal $\mathcal{N}(\tilde\beta, \tilde{\mathbf{V}})$, with

$$\tilde\beta = \begin{pmatrix} \mathbf{R} \\ R_{ab} \end{pmatrix} \beta\,,\ \tilde{\mathbf{V}} = \sigma^2 \begin{pmatrix} \mathbf{C} & \Gamma \\ \Gamma^\top & \Sigma \end{pmatrix}$$

where

$$R_{ab} = \begin{pmatrix} \mathbf{r}^\top(\mathbf{x}_a) \\ \mathbf{r}^\top(\mathbf{x}_a) \end{pmatrix}$$

$\Sigma_{1,1} = \Sigma_{2,2} = 1$, $\Sigma_{1,2} = C(\mathbf{x}_a - \mathbf{x}_b)$, $\Gamma_{i,1} = C(\mathbf{x}_i - \mathbf{x}_a)$, and $\Gamma_{i,2} = C(\mathbf{x}_i - \mathbf{x}_b)$, $i = 1, \ldots, N$.

From this we can compute the conditional

$$\pi(y_a, y_b|\beta, \mathbf{Y}) = \frac{\pi(\mathbf{Y}, y_a, y_b|\beta)}{\pi(\mathbf{Y}|\beta)}$$

which is normal $\mathcal{N}(\tilde{\mathbf{y}}, \sigma_z^2\mathbf{H})$ with

[3] Note, however, that the analytic properties of the sample function, that is of individual trajectories, are not necessarily related to mean square properties. This is considered in (Cramér and Leadbetter 1967), Chapter 5.

$$\tilde{\mathbf{y}}(\beta) = \Gamma^\top \mathbf{C}^{-1}\mathbf{Y} + \mathbf{U}^\top \beta\,,$$
$$\mathbf{H} = \Sigma - \Gamma^\top \mathbf{C}^{-1}\Gamma\,,$$

where

$$\mathbf{U}^\top = [R_{ab} - \Gamma^\top \mathbf{C}^{-1}\mathbf{R}]\,.$$

The joint $\pi(y_a, y_b|\mathbf{Y})$ is finally obtained by

$$\pi(y_a, y_b|\mathbf{Y}) = \int \pi(y_a, y_b|\beta, \mathbf{Y})\pi(\beta|\mathbf{Y})d\beta\,,$$

which gives after some calculation the normal $\mathcal{N}(\eta_{ab}, \sigma_z^2\mathbf{H} + \mathbf{U}^\top\mathbf{W}\mathbf{U})$, with $\eta_{ab} = \tilde{\mathbf{y}}(\bar{\beta})$.

We return to non-Bayesian kriging by letting σ^2 tend to infinity (that is, using a non informative prior for β). This gives

$$(y_a, y_b) \sim \mathcal{N}(\eta_{ab}, \sigma_z^2\mathbf{P})$$

with

$$\eta_{ab} = \{\Gamma^\top\mathbf{C}^{-1} + \mathbf{U}^\top[\mathbf{R}^\top\mathbf{C}^{-1}\mathbf{R}]^{-1}[\mathbf{R}^\top\mathbf{C}^{-1}]\}\mathbf{Y}\,,$$
$$\mathbf{P} = \Sigma - \Gamma^\top\mathbf{C}^{-1}\Gamma + \mathbf{U}^\top[\mathbf{R}^\top\mathbf{C}^{-1}\mathbf{R}]^{-1}\mathbf{U}\,.$$

Note that the values of σ_z^2 and the parameters θ that appear in $C(\mathbf{z})$, see *e.g.* (2), can be estimated from the data, for instance by maximum likelihood when the process is assumed to be Gaussian.

In what follows we shall only consider the case of simple kriging, where $\beta = \beta_0 \in \mathbb{R}$, $\mathbf{R} = \mathbf{1}_N$ (the N dimensional vector with all components equal to 1) and $R_{ab} = \mathbf{1}_2$. Other situations could be treated similarly. This gives

$$\mathbf{P}_{1,1} = 1 - \mathbf{c}_a^\top\mathbf{C}^{-1}\mathbf{c}_a + \frac{(1-\mathbf{c}_a^\top\gamma)^2}{S}\,,$$
$$\mathbf{P}_{2,2} = 1 - \mathbf{c}_b^\top\mathbf{C}^{-1}\mathbf{c}_b + \frac{(1-\mathbf{c}_b^\top\gamma)^2}{S}\,,$$
$$\mathbf{P}_{1,2} = C(\mathbf{x}_a - \mathbf{x}_b) - \mathbf{c}_a^\top\mathbf{C}^{-1}\mathbf{c}_b + \frac{(1-\mathbf{c}_a^\top\gamma)(1-\mathbf{c}_b^\top\gamma)}{S}\,,$$

where $\mathbf{c}_a = (C(\mathbf{x}_a - \mathbf{x}_1), \ldots, C(\mathbf{x}_a - \mathbf{x}_N))^\top$, $\mathbf{c}_b = (C(\mathbf{x}_b - \mathbf{x}_1), \ldots, C(\mathbf{x}_b - \mathbf{x}_N))^\top$, $\gamma = \mathbf{C}^{-1}\mathbf{1}_N$ and $S = \sum_{i,j=1}^N [\mathbf{C}^{-1}]_{i,j}$. Note that $P_{1,1} = P_{1,2} = 0$ if $\mathbf{x}_a = \mathbf{x}_i$, the inputs used for the observation Y_i, $i = 1, \ldots, N$.

When one is only interested into prediction at some particular point $\mathbf{x}$, only the first component of η_{ab} has to be considered, which corresponds to $\eta(\mathbf{x})$, and the $(1,1)$ component of $\mathbf{P}$ gives the uncertainty about this prediction.

2.2 *Models for mean and variance*

We consider now $\mathbf{x}$ as a random variable, with $\mathrm{E}_x\{\cdot\}$ the expectation with respect to $\mathbf{x}$, and denote

$$\bar{\mathbf{x}} = \mathrm{E}_x\{\mathbf{x}\}\,,\ \Sigma_x = \mathrm{E}_x\{(\mathbf{x}-\bar{\mathbf{x}})(\mathbf{x}-\bar{\mathbf{x}})^\top\}\,.$$

In a robust design problem, both $\bar{\mathbf{x}}$ and Σ_x may depend on factors that have to be settled by the designer.

From the results above, $\eta(\mathbf{x})$ can be considered as the mean of $y(\mathbf{x})$ conditional on $\mathbf{Y}$, under a noninformative prior for β, that is,

$$\mathrm{E}_y\{y(\mathbf{x})|\mathbf{Y}\} = \eta(\mathbf{x})\,.$$

A naive approach for taking variability of $\mathbf{x}$ into account is to use a Taylor series development of $\eta(\mathbf{x})$ at $\bar{\mathbf{x}}$:

$$\eta(\mathbf{x}) = \eta(\bar{\mathbf{x}}) + (\mathbf{x}-\bar{\mathbf{x}})^\top \frac{\partial\eta(\mathbf{x})}{\partial\mathbf{x}}_{|\bar{\mathbf{x}}} + \frac{1}{2}(\mathbf{x}-\bar{\mathbf{x}})^\top \frac{\partial^2\eta(\mathbf{x})}{\partial\mathbf{x}\partial\mathbf{x}^\top}_{|\bar{\mathbf{x}}}(\mathbf{x}-\bar{\mathbf{x}}) + HOT$$

which gives

$$\mathrm{E}_x\{\eta(\mathbf{x})\} = \eta(\bar{\mathbf{x}}) + \frac{1}{2}\mathrm{trace}\left[\frac{\partial^2\eta(\mathbf{x})}{\partial\mathbf{x}\partial\mathbf{x}^\top}_{|\bar{\mathbf{x}}}\Sigma_x\right] + HOT$$

and, denoting Var_x the variance with respect to $\mathbf{x}$,

$$\mathrm{Var}_x\{\eta(\mathbf{x})\} = \frac{\partial\eta(\mathbf{x})}{\partial\mathbf{x}^\top}_{|\bar{\mathbf{x}}}\Sigma_x\frac{\partial\eta(\mathbf{x})}{\partial\mathbf{x}}_{|\bar{\mathbf{x}}} + HOT$$

However, a more careful analysis shows that this approach is inexact. Moreover, it does not permit to take uncertainty on the prediction into account, that is, to derive variance models for $\mathrm{E}_x\{\eta(\mathbf{x})\}$ and $\mathrm{Var}_x\{\eta(\mathbf{x})\}$, as it will be done in Section 3.

In fact, when $\mathbf{x}$ fluctuates, the true response, that is, the trajectory, remains the same for different values of $\mathbf{x}$. We assume that the fourth derivative of $C(\cdot)$ at zero exists and is finite and consider a process

$$y_2(\mathbf{x}) = y(\bar{\mathbf{x}}) + (\mathbf{x}-\bar{\mathbf{x}})^\top G(\bar{\mathbf{x}}) + \frac{1}{2}(\mathbf{x}-\bar{\mathbf{x}})^\top H(\bar{\mathbf{x}})(\mathbf{x}-\bar{\mathbf{x}})\,,$$

with suitable means and covariances for the processes $G(\cdot)$ and $H(\cdot)$, such that

$$y_{h,h',h'',h'''}(\mathbf{x}) = y(\bar{\mathbf{x}}) + \sum_i (\mathbf{x}-\bar{\mathbf{x}})_i \frac{y(\bar{\mathbf{x}}+h_i\mathbf{e}_i) - y(\bar{\mathbf{x}})}{h_i} + \frac{1}{2}\sum_{i,j}(\mathbf{x}-\bar{\mathbf{x}})_i(\mathbf{x}-\bar{\mathbf{x}})_j\frac{1}{h_j'''}\left[\frac{y(\bar{\mathbf{x}}+h_i'\mathbf{e}_i) - y(\bar{\mathbf{x}})}{h_i'} - \frac{y(\bar{\mathbf{x}}+h_i''\mathbf{e}_i - h_j'''\mathbf{e}_j) - y(\bar{\mathbf{x}}-h_j'''\mathbf{e}_j)}{h_i''}\right]$$

tends to $y_2(\mathbf{x})$ in quadratic mean when $h, h', h'', h''' \to 0$:

$$\lim_{h,h',h'',h'''\to 0} E_y\{[y_2(\mathbf{x}) - y_{h,h',h'',h'''}(\mathbf{x})]^2\} = 0\,,$$

see (Stein 1999), Chapter 2. This gives

$$\mathrm{E}_x\{y_2(\mathbf{x})\} = y(\bar{\mathbf{x}}) + \frac{1}{2}\,\mathrm{trace}\,[H(\bar{\mathbf{x}})\Sigma_x]$$

and

$$\text{Var}_x\{y_2(\mathbf{x})\} = G^\top(\bar{\mathbf{x}})\Sigma_x G(\bar{\mathbf{x}}) + HOT$$

where the higher order terms HOT will be neglected. Next step is to consider expectation with respect to the processes, that is, with respect to possible trajectories, conditional on the observations. We get

$$\begin{aligned}&\text{E}_y\{\text{E}_x\{y_2(\mathbf{x})\}|\mathbf{Y}\}\\ &= \eta(\bar{\mathbf{x}}) + \frac{1}{2}\,\text{trace}\,[\text{E}_y\{H(\bar{\mathbf{x}})|\mathbf{Y}\}\Sigma_x]\\ &= \eta(\bar{\mathbf{x}}) + \frac{1}{2}\,\text{trace}\left[\frac{\partial^2\eta(\mathbf{x})}{\partial\mathbf{x}\partial\mathbf{x}^\top}_{|\bar{\mathbf{x}}}\Sigma_x\right] \qquad (3)\end{aligned}$$

for the prediction of the mean response over $\mathbf{x}$. Note that it is larger or smaller than the prediction at the mean value $\bar{\mathbf{x}}$ depending on the sign of the second order derivative of the predictor $\eta(\cdot)$ at this point.

For the expected variance, neglecting terms of order in $\mathbf{x}$ higher than two, we consider

$$\begin{aligned}&\text{E}_y\{\text{Var}_x\{y_{h,h'}(\mathbf{x})\}|\mathbf{Y}\} =\\ &\sum_{i,j}[\Sigma_x]_{i,j}\,\text{E}_y\left\{\frac{[y(\bar{\mathbf{x}}+h_i\mathbf{e}_i)-y(\bar{\mathbf{x}})]}{h_i}\right.\\ &\times\left.\frac{[y(\bar{\mathbf{x}}+h'_j\mathbf{e}_j)-y(\bar{\mathbf{x}})]}{h'_j}\right\}.\end{aligned}$$

Using the results of Section 2.1 on the joint posterior distribution of $[y(\bar{\mathbf{x}}+h_i\mathbf{e}_i), y(\bar{\mathbf{x}})]$, $[y(\bar{\mathbf{x}}+h'_j\mathbf{e}_j), y(\bar{\mathbf{x}})]$ and $[y(\bar{\mathbf{x}}+h_i\mathbf{e}_i), y(\bar{\mathbf{x}}+h'_j\mathbf{e}_j)]$, we get

$$\begin{aligned}&\lim_{h\to 0,h'\to 0}\text{E}_y\left\{\frac{[y(\bar{\mathbf{x}}+h_i\mathbf{e}_i)-y(\bar{\mathbf{x}})]}{h_i}\right.\\ &\times\left.\frac{[y(\bar{\mathbf{x}}+h'_j\mathbf{e}_j)-y(\bar{\mathbf{x}})]}{h'_j}\right\} =\\ &\frac{\partial\eta(\mathbf{x})}{\partial x_i}_{|\bar{\mathbf{x}}}\frac{\partial\eta(\mathbf{x})}{\partial x_j}_{|\bar{\mathbf{x}}} - \sigma_z^2\frac{\partial C(\mathbf{z})}{\partial z_i\partial z_j}_{|\mathbf{0}}\\ &-\sigma_z^2\frac{\partial\mathbf{c}^\top(\mathbf{z})}{\partial z_i}_{|\bar{\mathbf{x}}}\left(\mathbf{C}^{-1}-\frac{\gamma\gamma^\top}{S}\right)\frac{\partial\mathbf{c}(\mathbf{z})}{\partial z_j}_{|\bar{\mathbf{x}}},\end{aligned}$$

where $\mathbf{c}(\mathbf{x}) = (C(\mathbf{x}-\mathbf{x}_1),\ldots,C(\mathbf{x}-\mathbf{x}_N))^\top$.

This gives the following prediction for the variance:

$$\begin{aligned}&\text{E}_y\{\text{Var}_x\{y_2(\mathbf{x})\}|\mathbf{Y}\} = \frac{\partial\eta(\mathbf{x})}{\partial\mathbf{x}^\top}_{|\bar{\mathbf{x}}}\Sigma_x\frac{\partial\eta(\mathbf{x})}{\partial\mathbf{x}}_{|\bar{\mathbf{x}}}\\ &-\sigma_z^2\,\text{trace}\left[\frac{\partial^2 C(\mathbf{z})}{\partial\mathbf{z}\partial\mathbf{z}^\top}_{|\mathbf{0}}\Sigma_x\right] \qquad (4)\\ &-\sigma_z^2\,\text{trace}\left[\frac{\partial\mathbf{c}^\top(\mathbf{z})}{\partial\mathbf{z}}_{|\bar{\mathbf{x}}}\left(\mathbf{C}^{-1}-\frac{\gamma\gamma^\top}{S}\right)\frac{\partial\mathbf{c}(\mathbf{z})}{\partial\mathbf{z}^\top}_{|\bar{\mathbf{x}}}\Sigma_x\right].\end{aligned}$$

It tends to the value $\text{Var}_x\{\eta(\mathbf{x})\} \simeq \partial\eta(\mathbf{x})/\partial\mathbf{x}^\top_{|\bar{\mathbf{x}}}\Sigma_x\,\partial\eta(\mathbf{x})/\partial\mathbf{x}_{|\bar{\mathbf{x}}}$ of the prediction of the variance for the deterministic model when σ_z tends to zero, and, of course, it tends to zero when Σ_x tends to zero. Note that the second term is usually positive, whereas the third one is negative, so that situations may exist where $\text{E}_y\{\text{Var}_x\{y_2(\mathbf{x})\}|\mathbf{Y}\}$ is smaller than $\text{Var}_x\{\eta(\mathbf{x})\}$.

3. MODEL UNCERTAINTY

Uncertainty on the true response due to the fact that the data sample is finite induces uncertainty in the predictions (3) and (4). Using the same approach as in Section 2, we obtain after some (lengthy) calculations

$$\begin{aligned}&\text{Var}_y\{\text{E}_x\{y_2(\mathbf{x})\}|\mathbf{Y}\} =\\ &\sigma_z^2\left[1-\mathbf{c}^\top(\bar{\mathbf{x}})\mathbf{C}^{-1}\mathbf{c}(\bar{\mathbf{x}})+\frac{(1-\mathbf{c}^\top(\bar{\mathbf{x}})\gamma)^2}{S}\right]\\ &+\frac{\sigma_z^2}{4}\sum_{i,j,k,l=1}^{d}[\Sigma_x]_{i,j}[\Sigma_x]_{k,l}\left[\frac{\partial^4 C(\mathbf{z})}{\partial z_i\partial z_j\partial z_k\partial z_l}_{|\mathbf{0}}\right.\\ &\left.-\frac{\partial^2\mathbf{c}^\top(\mathbf{z})}{\partial z_i\partial z_j}_{|\bar{\mathbf{x}}}\left(\mathbf{C}^{-1}-\frac{\gamma\gamma^\top}{S}\right)\frac{\partial^2\mathbf{c}(\mathbf{z})}{\partial z_k\partial z_k}_{|\bar{\mathbf{x}}}\right]\\ &+\sigma_z^2\sum_{i,j=1}^{d}[\Sigma_x]_{i,j}\left[\frac{\partial^2 C(\mathbf{z})}{\partial z_i\partial z_j}_{|\mathbf{0}}\right.\\ &\left.-\left(\mathbf{c}^\top(\bar{\mathbf{x}})\mathbf{C}^{-1}+\frac{1-\mathbf{c}^\top(\bar{\mathbf{x}})\gamma}{S}\gamma^\top\right)\frac{\partial^2\mathbf{c}(\mathbf{z})}{\partial z_i\partial z_j}_{|\bar{\mathbf{x}}}\right]\end{aligned}$$

for the variance of the mean response. Note that it tends to the variance of $y(\bar{\mathbf{x}})$ when Σ_x tends to zero. Also, the second correcting term is negligible with respect to the third when Σ_x is small.

Similarly, we get for the variance of the variance

$$\begin{aligned}&\text{Var}_y\{\text{Var}_x\{y(\mathbf{x})\}|\mathbf{Y}\} = 2\sigma_z^4\,\text{trace}\left(\left\{\Sigma_x\left[\frac{\partial^2 C(\mathbf{z})}{\partial\mathbf{z}\partial\mathbf{z}^\top}_{|\mathbf{0}}\right.\right.\right.\\ &\left.\left.\left.+\frac{\partial\mathbf{c}^\top(\mathbf{z})}{\partial\mathbf{z}}_{|\bar{\mathbf{x}}}\left(\mathbf{C}^{-1}-\frac{\gamma\gamma^\top}{S}\right)\frac{\partial\mathbf{c}(\mathbf{z})}{\partial\mathbf{z}^\top}_{|\bar{\mathbf{x}}}\right]\right\}^2\right)\\ &-4\sigma_z^2\frac{\partial\eta(\mathbf{x})}{\partial\mathbf{x}^\top}_{|\bar{\mathbf{x}}}\Sigma_x\left[\frac{\partial^2 C(\mathbf{z})}{\partial\mathbf{z}\partial\mathbf{z}^\top}_{|\mathbf{0}}+\frac{\partial\mathbf{c}^\top(\mathbf{z})}{\partial\mathbf{z}}_{|\bar{\mathbf{x}}}\right.\\ &\left.\times\left(\mathbf{C}^{-1}-\frac{\gamma\gamma^\top}{S}\right)\frac{\partial\mathbf{c}(\mathbf{z})}{\partial\mathbf{z}^\top}_{|\bar{\mathbf{x}}}\right]\Sigma_x\frac{\partial\eta(\mathbf{x})}{\partial\mathbf{x}}_{|\bar{\mathbf{x}}}.\end{aligned}$$

Notice that we ignored the fact that in practise the parameters θ and σ_z^2 in the covariance model, (2) for instance, are estimated values. Taking uncertainty on these estimates into account in the evaluation of predictions is a challenging but difficult task.

4. EXAMPLES

Consider a one-dimensional process ($x \in I\!R$), with zero mean and covariance $\exp(-0.2z^2)$ ($\sigma_z = 1$). We observe the responses at the design points $0, 2, 3, 4, 7$ and 10. Figure 1 presents a typical realization of the process (full line), with observed values $\mathbf{Y}$ indicated by stars, and the associated predicted response (dashed

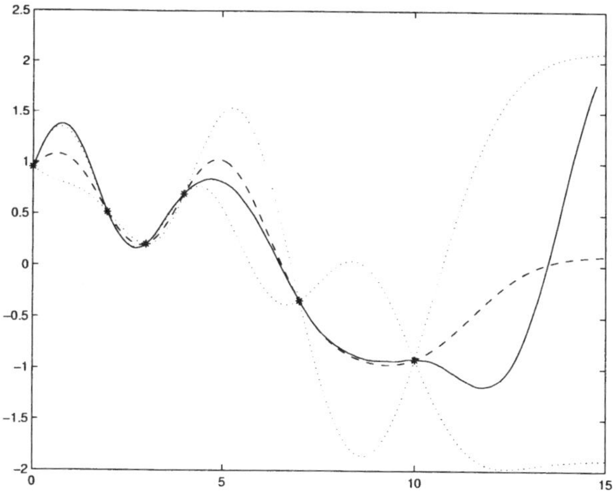

Fig. 1. Typical realization of the process (full line), observations (stars), prediction (dashed lines) and 2σ confidence bounds (dotted lines)

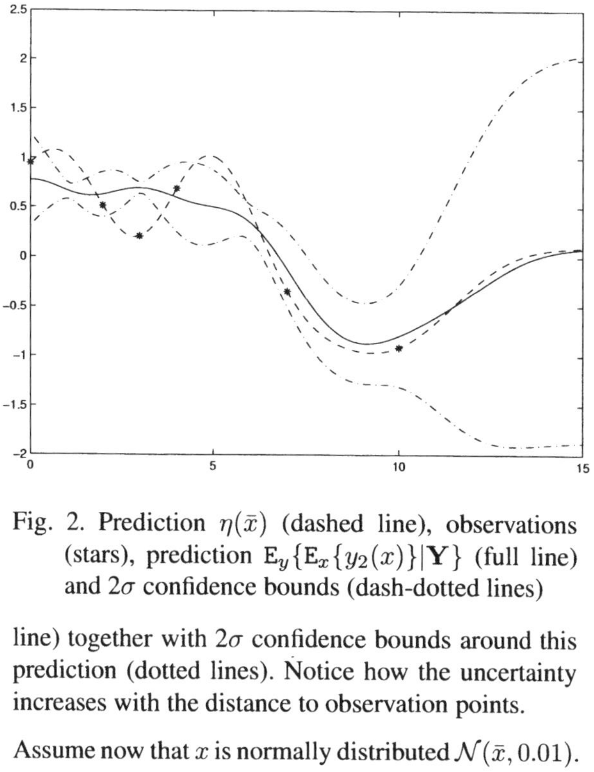

Fig. 2. Prediction $\eta(\bar{x})$ (dashed line), observations (stars), prediction $\mathrm{E}_y\{\mathrm{E}_x\{y_2(x)\}|\mathbf{Y}\}$ (full line) and 2σ confidence bounds (dash-dotted lines)

line) together with 2σ confidence bounds around this prediction (dotted lines). Notice how the uncertainty increases with the distance to observation points.

Assume now that x is normally distributed $\mathcal{N}(\bar{x}, 0.01)$.

Figure 2 presents the evolution of the prediction $\eta(\bar{x})$ at $\bar{x}$ (dashed line, as in Figure 1) and the prediction of the mean response $\mathrm{E}_y\{\mathrm{E}_x\{y_2(x)\}|\mathbf{Y}\}$ (full line), together with 2σ confidence bounds obtained from $\mathrm{Var}_y\{\mathrm{E}_x\{y_2(x)\}|\mathbf{Y}\}$ (dash-dotted lines), when $\bar{x}$ varies from 0 to 15.

A first observation is that averaging with respect to x smoothes the prediction, compare $\mathrm{E}_y\{\mathrm{E}_x\{y_2(x)\}|\mathbf{Y}\}$ to $\eta(\bar{x})$. Next, there is no clear relation between the distance to observation points and uncertainty of the prediction of the mean response: for instance, the uncertainty is close to a local maximum at the observed point $x = 2$, but close to local minimum at $x = 3$.

Figure 3 gives the evolution of the prediction of the variance $\mathrm{E}_y\{\mathrm{Var}_x\{y_2(x)\}|\mathbf{Y}\}$ (full line) together

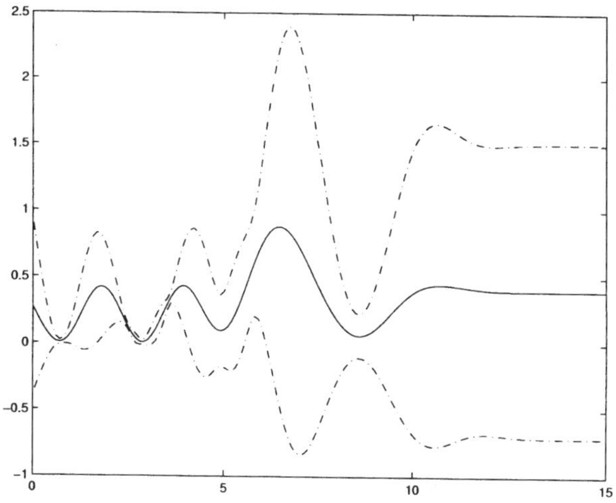

Fig. 3. Prediction $\mathrm{E}_y\{\mathrm{Var}_x\{y_2(x)\}|\mathbf{Y}\}$ (full line) and 2σ confidence bounds (dash-dotted lines)

with 2σ confidence bounds obtained from $\mathrm{Var}_y\{\mathrm{Var}_x\{ y_2(x)\}|\mathbf{Y}\}$ (dash-dotted lines)[4].

Observe that the prediction of the variance tends to be small when $\eta(x)$ is close to a stationary point, and to be large when the slope of $\eta(x)$ is large. The uncertainty on the variance tends to be small when the variance itself is small. If, for instance, the purpose were to maximize the expected response and minimize the variance, the point $x = 1$ would be a good compromise.

5. REFERENCES

Bates, R. and L. Pronzato (2001). Emulator-based global optimisation using lattices and Delaunay tesselation. In: *Proc. 3rd Int. Symp. on Sensitivity Analysis of Model Output* (P. Prado and R. Bolado, Eds.). Madrid. pp. 189–192.

Bates, R., R. Fontana, L. Pronzato and H.P. Wynn (1999). Computer experiments for concurrent engineering. In: *Proc. 6th Int. Conf. ATA on the New Role of Experimentation in the Modern Automotive Product Development Process*. Firenze. pp. 1585–1588.

Costa, J.-P., L. Pronzato and E. Thierry (2000). Nonlinear prediction by kriging, with application to noise cancellation. *Signal Processing* **80**, 553–566.

Cramér, H. and M.R. Leadbetter (1967). *Stationary and Related Stochastic Processes*. Wiley. New York.

Krige, D.G. (1951). A statistical approach to some mine valuation and allied problems on the Witwatersrand. Master Thesis, University of Witwatersrand.

Matheron, G. (1963). Principles of geostatistics. *Economic Geology* **58**, 1246–1266.

[4] Since the variance is a positive quantity, only the region above 0 gives an indication of the uncertainty on the prediction.

Sacks, J. and S. Schiller (1988). Spatial designs. In: *Statistical Decision Theory and Related Topics IV* (S.S. Gupta and J.O. Berger, Eds.). Vol. 2. pp. 385–399. Springer. Heidelberg.

Sacks, J., S.B. Schiller and W.J. Welch (1989*a*). Designs for computer experiments. *Technometrics* **31**, 41–47.

Sacks, J., W.J. Welch, T.J. Mitchell and H.P. Wynn (1989*b*). Design and analysis of computer experiments. *Statistical Science* **4**(4), 409–435.

Schonlau, M., W.J. Welch and D.R. Jones (1998). Global versus local search in constrained optimization of computer models. In: *New Developments and Applications in Experimental Design, Lecture Notes — Monograph Series, vol. 34.* pp. 11–25. IMS. Hayward.

Shewry, M.C. and H.P. Wynn (1988). Maximum entropy sampling with application to simulation codes. In: *Prep. 12th IMACS World Congress on Scientific Computation*. Paris. pp. 517–519.

Stein, M.L. (1999). *Interpolation of Spatial Data. Some Theory for Kriging*. Springer. Heidelberg.

Vuchkov, I.N. and L.N. Boyadjieva (2001). *Quality Improvement with Design of Experiments. A Response Surface Approach*. Kluwer. Dordrecht.

Welch, W.J., R.J. Buck, J. Sacks, H.P. Wynn, T.J. Mitchell and M.D. Morris (1992). Screening, predicting and computer experiments. *Technometrics* **34**(1), 15–25.

IFAC Publications
www.elsevier.com/locate/ifac

GEOSTATISTICAL MODELS AND KRIGING

Hans Wackernagel *

* *Centre de Géostatistique, Ecole des Mines de Paris*
Email: hans.wackernagel@ensmp.fr

Abstract: Geostatistics is an application of the theory of random functions to spatially distributed data. Geostatistical methods like kriging were initially proposed in mining and petroleum exploration and found their way back to mainstream statistics more than a decade ago. The geostatistical methodology can be subdivided into linear and multivariate geostatistics, non-stationary geostatistics, non-linear geostatistics, geostatistical simulation.

Keywords: Geostatistics, kriging, variogram, spatial data, simulation.

1. INTRODUCTION

Geostatistics is built around the concept of *regionalized variable*, a function $z(\mathbf{x})$ in a bounded spatial or temporal domain D, where $\mathbf{x}$ is a vector of spatial coordinates and z is the value of a quantity of interest. Geostatistics can be defined as the application of probabilistic methods to regionalized variables. A regionalized variable could be considered deterministic and randomness introduced by the sampling: this is the *transitive theory* of geostatistics that has gained new importance in fish abundance estimation (Petitgas, 2001; Bez and Rivoirard, 2001). However, the common model is to view a regionalized variable as a realization of a random function $Z(\mathbf{x})$, i.e. a family of random variables Z dependent on location $\mathbf{x} \in D$. The epistemological problem associated with the fact that usually only one realization is available has been answered by Matheron in his book *Estimating and Choosing* (Matheron, 1989) in distinguishing between objective quantities which are essentially integrals of the single realization and conventional parameters associated to the random function model. The former can be estimated while the latter have to be chosen.

Geostatistics originated in the field of mining and geology in the mid of the last century, when dealing with the difficult problem of assessing the ore reserves in a mineral deposit from samples taken at a few spatial locations and showing a skew distribution. First ideas of (Krige, 1951) and (de Wijs, 1951) were brought into a single setting by Matheron (Matheron, 1955; Matheron, 1965; Cressie, 1990*a*). Rapidly the methods spread into other domains of geosciences, and finally were brought back into mainstream statistics at the end of the century (Cressie, 1993).

Geostatistics is classically subdivided into linear and multivariate geostatistics, non-stationary geostatistics, non-linear geostatistics, simulation techniques. We review the four items in this order.

2. LINEAR GEOSTATISTICS

The spatial variation is analyzed in geostatistics using the *variogram* $\gamma(\mathbf{h})$, which is defined as the variance of increments $Z(\mathbf{x}+\mathbf{h})-Z(\mathbf{x})$ with $\mathbf{h}$ a vector between two points in D. Assuming that increments are stationary, i.e. that the variance of increments is translation invariant with respect to the vector $\mathbf{h}$ we have

$$\operatorname{var}(Z(\mathbf{x}+\mathbf{h})-Z(\mathbf{x})) = 2\gamma(\mathbf{h}). \qquad (1)$$

Usually the expectation of increments is assumed to be zero so that

$$\gamma(\mathbf{h}) = \frac{1}{2} E[(Z(\mathbf{x}+\mathbf{h})-Z(\mathbf{x}))^2)]. \qquad (2)$$

Random functions whose increments are second-order stationary and of zero mean are said to satisfy the *intrinsic hypothesis* (Chilès and Delfiner, 1999), which has been termed *intrinsic stationarity* in the statistical literature (Cressie, 1993). The variogram is not bounded, and is therefore preferred to the covariance function $C(\mathbf{h}) = E[(Z(\mathbf{x}+\mathbf{h}) - \mu)\cdot(Z(\mathbf{x}) - \mu)]$ of a second-order stationary random function with mean μ. For second-order stationary random functions the variogram is bounded and can be computed from the covariance function,

$$\gamma(\mathbf{h}) = C(0) - C(\mathbf{h}). \tag{3}$$

The variogram is a *conditionally negative definite* function, that is

$$\sum_{\alpha=0}^{n}\sum_{\beta=0}^{n} w_\alpha w_\beta \gamma(\mathbf{x}_\alpha - \mathbf{x}_\beta) \le 0 \tag{4}$$

for any set of $n+1$ coefficients summing up to zero,

$$\sum_{\alpha=0}^{n} w_\alpha = 0. \tag{5}$$

Linear combinations of random variables $Z(\mathbf{x}_\alpha)$ that satisfy the constraint (5) are qualified as *allowable* and the property (4) is obvious when computing the variance of an allowable linear combination.

So $-\gamma(\mathbf{h})$ is conditionally positive definite, while a covariance function is positive definite

$$\sum_{\alpha=0}^{n}\sum_{\beta=0}^{n} w_\alpha w_\beta C(\mathbf{x}_\alpha - \mathbf{x}_\beta) \ge 0$$

for any set of $n+1$ coefficients w_α; there is no condition on the sum of weights.

A variogram is *isotropic*, if $\gamma(\mathbf{h})$ depends only on the length $|\mathbf{h}|$ and not on the orientation of the vector $\mathbf{h}$. A simple way of obtaining an anisotropic variogram from an isotropic one is by linearly transforming the coordinates of the vector $\mathbf{h}$; this generates a *geometrical anisotropy*.

Variogram functions are defined for the isotropic case. A simple class of unbounded variograms is the class of power models

$$\gamma(\mathbf{h}) = b\,|\mathbf{h}|^\nu \qquad \text{with} \quad b>0,\ 0<\nu<2,$$

where ν is a shape parameter leading to a convex, linear or concave shape, respectively for values $\nu<1$, $\nu=1, \nu>1$.

Bounded variograms are constructed from covariance functions using formula (3). An example is the Matérn class constructed with Bessel functions $\mathcal{K}_\nu$ of the third kind

$$C(\mathbf{h}) = b\left(\frac{|\mathbf{h}|}{a}\right)^\nu \mathcal{K}_\nu\left(\frac{|\mathbf{h}|}{a}\right) \quad \text{with} \quad b>0,\ \nu>0.$$

where a is a scale parameter and ν the shape parameter, which measures the differentiability of the random function. For $\nu=1/2$ the covariance function is exponential and has a linear shape at the origin. In general the behavior near the origin is like $|h|^{2\nu}$ for non integer ν and like $|h|^{2\nu}\log|\mathbf{h}|$ for integer ν. See (Matheron, 1965; Yaglom, 1986; Stein, 1999; Chilès and Delfiner, 1999) for a detailed discussion of this and other models.

Having a set of n observations $z(\mathbf{x}_\alpha)$ it is advisable to perform an exploratory data analysis including an inspection of the *variogram cloud*, i.e. a plot of squared differences $(z(\mathbf{x}_\alpha)-z(\mathbf{x}_\beta))^2/2$ against corresponding vectors $\mathbf{h}_{\alpha\beta} = \mathbf{x}_\alpha - \mathbf{x}_\beta$, see (Haslett *et al.*, 1991). The sample (also: *experimental*) variogram can be obtained by sorting vectors $\mathbf{h}_{\alpha\beta}$ into distance (and angle) classes $\mathcal{H}_k$ and computing averages for each distance class,

$$\hat\gamma(\mathcal{H}_k) = \frac{1}{2|\mathcal{H}_k|}\sum_{\mathbf{h}_{\alpha\beta}\in\mathcal{H}_k}(z(\mathbf{x}_\alpha) - z(\mathbf{x}_\beta))^2,$$

where $|\mathcal{H}_k|$ is the cardinality of the distance class. A highly robust variogram estimator is suggested in (Genton, 1998) where further references on the topic are found.

The sequence of variogram estimates for different distance classes is not a conditionally negative definite function and has to be fitted with a parametric variogram function. Often the fit is done graphically by eye, but ways of fitting by least squares (Cressie, 1985), maximum likelihood (Mardia and Marshall, 1984; Zimmerman and Zimmerman, 1991) or in a Bayesian framework (Handcock and Stein, 1993; Diggle *et al.*, 1998) have been proposed.

At this stage we are in possession of a valid variogram model and can use it to predict a value $\hat z(\mathbf{x}_0)$ at an arbitrary location $\mathbf{x}_0$ of the domain D. We consider an intrinsic random function and the linear predictor

$$\hat Z(\mathbf{x}_0) = \sum_{\alpha=1}^{n} w_\alpha Z(\mathbf{x}_\alpha)$$

for which the weights are constrained to

$$\sum_{\alpha=1}^{n} w_\alpha = 1 \tag{6}$$

in order to obtain an allowable linear combination. The constraint (6) is equivalent to the condition (5) with $w_0=-1$, so that we may employ variograms when computing the variance of an allowable linear combination.

The prediction error is unbiased

$$\begin{aligned} E(\hat Z(\mathbf{x}_0) - Z(\mathbf{x}_0)) &= \sum_{\alpha=1}^{n} w_\alpha E(Z(\mathbf{x}_\alpha) - Z(\mathbf{x}_0)) \\ &= 0. \end{aligned}$$

The variance in our prediction problem is

$$\operatorname{var}(\widehat{Z}(\mathbf{x}_0) - Z(\mathbf{x}_0)) = -\sum_{\alpha=1}^{n}\sum_{\beta=1}^{n} w_\alpha w_\beta\, \gamma(\mathbf{x}_\alpha - \mathbf{x}_\beta) + 2\sum_{\alpha=1}^{n} w_\alpha \gamma(\mathbf{x}_\alpha - \mathbf{x}_0)$$

knowing that $\gamma(\mathbf{x}_0 - \mathbf{x}_0)$ is zero by definition (2).

Minimizing the prediction variance under the constraint (6) introducing a Lagrange parameter λ yields the *ordinary kriging* equations

$$\begin{cases} \sum_{\beta=1}^{n} w_\beta\, \gamma(\mathbf{x}_\alpha - \mathbf{x}_\beta) + \lambda = \gamma(\mathbf{x}_\alpha - \mathbf{x}_0) & \forall\alpha \\ \sum_{\beta=1}^{n} w_\beta = 1 \end{cases}$$

with the associated ordinary kriging variance

$$\sigma^2_{\mathrm{OK}} = \lambda + \sum_{\alpha=1}^{n} w_\alpha \gamma(\mathbf{x}_\alpha - \mathbf{x}_0).$$

For a second-order stationary random function we can use kriging to estimate the mean $\widehat{\mu}$ from the autocorrelated data and get the system

$$\begin{cases} \sum_{\beta=1}^{n} w_\beta C(\mathbf{x}_\alpha - \mathbf{x}_\beta) - \lambda = 0 & \forall\alpha \\ \sum_{\beta=1}^{n} w_\beta = 1 \end{cases}$$

with the corresponding minimal variance

$$\sigma^2_{\mathrm{KM}} = \lambda.$$

With second-order stationarity the prediction of values by an ordinary kriging may be called a *kriging with unknown mean*, because the underlying random function model is $Z(\mathbf{x}) = \mu + Y(\mathbf{x})$, with $Y(\mathbf{x})$ a random function of mean zero.

3. MULTIVARIATE GEOSTATISTICS

There are two ways of generalizing the variogram to model the cross-correlation between increments: the *cross variogram* (Matheron, 1965) and the *pseudo-cross variogram* (Myers, 1991). The cross variogram is defined as the cross covariance of the increments of two intrinsic random functions $Z_i(\mathbf{x})$ and $Z_j(\mathbf{x})$,

$$\begin{aligned} \gamma_{ij}(\mathbf{h}) &= \frac{1}{2}\operatorname{cov}((Z_i(\mathbf{x}+\mathbf{h}) - Z_i(\mathbf{x})) \times (Z_j(\mathbf{x}+\mathbf{h}) - Z_j(\mathbf{x}))) \\ &= \frac{1}{2}E((Z_i(\mathbf{x}+\mathbf{h}) - Z_i(\mathbf{x})) \times (Z_j(\mathbf{x}+\mathbf{h}) - Z_j(\mathbf{x}))) \end{aligned}$$

assuming joint stationarity of increments, while the pseudo-cross variogram $\pi_{ij}(\mathbf{h})$ is based on the variance of cross increments

$$\begin{aligned} \pi_{ij}(\mathbf{h}) &= \frac{1}{2}\operatorname{var}(Z_i(\mathbf{x}+\mathbf{h}) - Z_j(\mathbf{x})) \\ &= \frac{1}{2}E((Z_i(\mathbf{x}+\mathbf{h}) - Z_j(\mathbf{x}))^2) \end{aligned}$$

postulating stationary cross-increments with mean zero. Merits and disadvantages of both approaches have led to discussions of which (Cressie and Wikle, 1998) is one of the last episodes. For a pair of jointly second-order stationary random functions with cross-covariance function

$$C_{ij}(\mathbf{h}) = E((Z_i(\mathbf{x}+\mathbf{h}) - \mu_i) \times (Z_j(\mathbf{x}) - \mu_j)) \quad (7)$$

the two types of cross-variograms can be written

$$\begin{aligned} \gamma_{ij}(\mathbf{h}) &= C_{ij}(0) - \frac{1}{2}(C_{ij}(-\mathbf{h}) + C_{ij}(\mathbf{h})), \\ \pi_{ij}(\mathbf{h}) &= \frac{1}{2}(C_{ii}(0) + C_{jj}(0)) - C_{ij}(\mathbf{h}). \end{aligned}$$

Methods for fitting matrices of functions $C_{ij}(\mathbf{h})$, $\gamma_{ij}(\mathbf{h})$ or $\pi_{ij}(\mathbf{h})$ to corresponding sample functions are described in (Goulard and Voltz, 1992; Grzebyk and Wackernagel, 1994; Ver Hoef and Barry, 1998; Gelfand *et al.*, 2002).

In possession of a valid model for a set of random functions a value of a particular variable can be predicted by *cokriging* from available samples (Myers, 1982; Ver Hoef and Cressie, 1993). Many sample configurations are possible, depending on whether or not sample locations are shared by the different variables; in some of them the equations of the cokriging problem simplify, which is important both in view of the size and the numerical stability of the cokriging system (Rivoirard, 2001; Wackernagel *et al.*, 2002).

4. NON-STATIONARY GEOSTATISTICS

There are situations where a variable of primary interest, say $Z(\mathbf{x})$, is sampled at a few locations only, while a secondary variable $s(\mathbf{x})$ is available on a comparatively dense grid over the domain. Examples for $s(\mathbf{x})$ are seismic data in petroleum exploration (Chilès and Delfiner, 1999), altitude for mapping temperature (Hudson and Wackernagel, 1994), or, more interestingly, numerical output from a deterministic model attempting to predict the regionalized variable z on the basis of a physical process described by partial differential equations (Lajaunie *et al.*, 2001). Assuming a linear relation between the two variables, $E(Z(\mathbf{x})) = a + b s(\mathbf{x})$, we model $Z(\mathbf{x})$ with a non-stationary mean $s(\mathbf{x})$ termed an *external drift*. The corresponding kriging equations come as

$$\begin{cases} \sum_{\beta=1}^{n} w_\beta \gamma(\mathbf{x}_\alpha - \mathbf{x}_\beta) + \lambda_0 + \lambda_1 s(\mathbf{x}_\alpha) \\ \qquad\qquad = \gamma(\mathbf{x}_\alpha - \mathbf{x}_0) \qquad \forall\alpha \\ \sum_{\beta=1}^{n} w_\beta = 1 \\ \sum_{\beta=1}^{n} w_\beta s(x_\beta) = s(x_0) \end{cases}$$

Links with cokriging are explained in (Rivoirard, 2002), while in (Royle and Berliner, 1999) a generalization is proposed in the framework of a hierarchical model.

The variogram of the residual between primary variable and external drift in the kriging with external drift is problematic to infer as exposed in standard textbooks. Instead of external drift, it is also possible to use a polynomial drift (linear or quadratic) and this is termed *universal kriging*. For polynomial drift Matheron has set up a theory that solves the inference problems and in which the variogram turns out to be a particular form of *generalized covariance function* (Matheron, 1973; Chilès and Delfiner, 1999).

The equivalence between non-stationary kriging and splines is shown in (Matheron, 1981; Wahba, 1990*b*). See also the discussion in (Wahba, 1990*a*; Cressie, 1990*b*).

Space-time data has been modeled with geostatistical methods by using space-time covariance models (Kyriakidis and Journel, 1999). Links between kriging and Kalman filtering have soon been recognized (Mardia *et al.*, 1998; Wikle and Cressie, 1999). Geostatistical Kalman filtering and other methods of data assimilation are particularly interesting when a network of automatic stations and a deterministic numerical model describing the generally nonlinear time dynamics are available (Bertino *et al.*, 2002).

5. NON-LINEAR GEOSTATISTICS

The linear prediction methods described so far are designed for estimating values of Z at points or a mean function $\mu(\mathbf{x})$ over the domain. However, when it comes to estimate the proportion of values over a threshold for given spatial units, nonlinear geostatistical methods are preferable.

A threshold in an environmental regulation is defined with respect to a time interval, a surface or a volume called the *support*, which is usually of a size different from the sampling support. We are faced with the *change of support* problem consisting in having to anticipate the distribution of values for a selection volume knowing the distribution of sample values on pointwise support. The classification of predicted values above or below thresholds includes an *information effect*, due to the incomplete sampling of the spatial or temporal domain, whose strength will depend on the type of estimator used.

The nonlinear estimation is performed using various formulations of *disjunctive kriging*, which require only the knowledge of bivariate point-point and point-block distributions. Corresponding change of support models are given by so called *isofactorial models* of bivariate distributions that were initially derived by Matheron from analogous models in continuous correspondence analysis. The results are presented with different types of *selectivity curves* that resemble graphical presentations in econometrics or survival analysis and which are particularly convenient to compare the impact of change of support with respect to specific thresholds (Rivoirard, 1994; Chilès and Delfiner, 1999; Wackernagel, 2003).

6. GEOSTATISTICAL SIMULATION

Armed with a spatial stochastic model of random functions, sets, point processes or populations of objects there are numerous techniques and algorithms for generating realizations. A difficult problem is to condition the simulation to obtain realizations that coincide at data locations with the regionalized phenomenon under study.

Some algorithms generate directly simulations of a random function conditional upon the data. The alternative is to generate non-conditional simulations whose compatibility with the spatial stochastic model can be checked and to condition them in a second step: in the case of Gaussian random functions this conditioning can be performed by kriging using transformed data if necessary. Detailed reviews of geostatistical simulation are found in (Chilès and Delfiner, 1999; Lantuéjoul, 2001). The particular problem of simulating geological facies and lithotypes using truncated Gaussian and plurigaussian simulations is described in (Armstrong *et al.*, 2003).

7. REFERENCES

Armstrong, M, A G Galli, G Le Loc'h, F Geffroy and R Eschard (2003). *Plurigaussian Simulations in Geosciences*. Springer-Verlag. Berlin.

Bertino, L, G Evensen and H Wackernagel (2002). Combining geostatistics and Kalman filtering for data assimilation in an estuarine system. *Inverse problems* **18**, 1–23.

Bez, N and J Rivoirard (2001). Transitive geostatistics to characterise spatial aggregation with diffuse limits: an application on mackerel ichtyoplankton. *Fisheries Research* **50**, 41–58.

Chilès, J P and P Delfiner (1999). *Geostatistics: Modeling Spatial Uncertainty*. Wiley. New York.

Cressie, N (1985). Fitting variogram models by weighted least squares. *Mathematical Geology* **17**, 563–586.

Cressie, N (1990*a*). The origins of kriging. *Mathematical Geology* **22**, 239–252.
Cressie, N (1990*b*). Reply to letter by G. Wahba. *The American Statistician* **44**, 256–258.
Cressie, N (1993). *Statistics for Spatial Data*. revised ed.. Wiley. New York.
Cressie, N and C K Wikle (1998). The variance-based cross-variogram: you can add apples and oranges. *Mathematical Geology* **30**, 789–800.
de Wijs, H J (1951). Statistics of ore distribution, part I: frequency distribution of assay values. *Geologie en Mijnbouw* **13**, 365–375.
Diggle, P J, J A Tawn and R A Moyeed (1998). Model-based geostatistics (with discussion). *Applied Statistics* **47**, 299–350.
Gelfand, A E, A M Schmidt and C F Sirmans (2002). Multivariate spatial processes models: Conditional and unconditional bayesian approaches using coregionalization. Technical report. University of Connecticut.
Genton, M G (1998). Highly robust variogram estimation. *Mathematical Geology* **30**, 213–221.
Goulard, M and M Voltz (1992). Linear coregionalization model: tools for estimation and choice of multivariate variograms. *Mathematical Geology* **24**, 269–286.
Grzebyk, M and H Wackernagel (1994). Multivariate analysis and spatial/temporal scales: real and complex models. In: *Proceedings of XVIIth International Biometrics Conference*. Vol. 1. Hamilton, Ontario. pp. 19–33.
Handcock, M S and M L Stein (1993). A Bayesian analysis of kriging. *Technometrics* **35**, 403–410.
Haslett, J, R Bradley, P S Craig, G Wills and A R Unwin (1991). Dynamic graphics for exploring spatial data, with application to locating global and local anomalies. *The American Statistician* **45**, 234–242.
Hudson, G and H Wackernagel (1994). Mapping temperature using kriging with external drift: theory and an example from Scotland. *International J. Climatology* **14**, 77–91.
Krige, D G (1951). A statistical approach to some mine valuation and allied problems on the Witwatersrand. *J. Chem. Metal. Min. Soc. South Africa* **52**, 119–139.
Kyriakidis, P C and A G Journel (1999). Geostatistical space-time models: a review. *Mathematical Geology* **31**, 651–684.
Lajaunie, C, H Wackernagel and L Bertino (2001). Geostatistical normalization: Case studies. IMPACT Project Report N-31/01/G. Centre de Géostatistique, Ecole des Mines de Paris. http://www.mai.liu.se/impact.
Lantuéjoul, C (2001). *Geostatistical Simulation: Models and Algorithms*. Springer-Verlag. Berlin.
Mardia, K V and R J Marshall (1984). Maximum likelihood estimation of models for residual covariance in spatial regression. *Biometrika* **71**, 135–146.
Mardia, K V, C Goodall, E Redfern and F J Alonso (1998). The kriged Kalman filter (with discussion). *Test* **7**, 217–285.
Matheron, G (1955). Application des méthodes statistiques à l'évaluation des gisements. *Annales des Mines* **144 (12)**, 50–75.
Matheron, G (1965). *Les Variables Régionalisées et leur Estimation*. Masson. Paris.
Matheron, G (1973). The intrinsic random functions and their applications. *Advances in Applied Probability* **5**, 439–468.
Matheron, G (1981). Splines and kriging: their formal equivalence. In: *Down-to-Earth Statistics: Solutions Looking for Geological Problems* (D F Merriam, Ed.). Vol. 8 of *Syracuse University Geology Contribution*. New York. pp. 77–95.
Matheron, G (1989). *Estimating and Choosing*. Springer-Verlag. Berlin.
Myers, D E (1982). Matrix formulation of cokriging. *Mathematical Geology* **14**, 249–258.
Myers, D E (1991). Pseudo-cross variograms, positive-definiteness, and cokriging. *Mathematical Geology* **23**, 805–816.
Petitgas, P (2001). Geostatistics in fisheries survey design and stock assessment: models, variances and applications. *Fish and Fisheries* **2**, 231–249.
Rivoirard, J (1994). *Introduction to Disjunctive Kriging and Non-Linear Geostatistics*. Oxford University Press. Oxford.
Rivoirard, J (2001). Which models for collocated cokriging?. *Mathematical Geology* **33**, 117–131.
Rivoirard, J (2002). On the structural link between variables in kriging with external drift. *Mathematical Geology* **34**, 797–808.
Royle, J A and L M Berliner (1999). A hierarchical approach to multivariate spatial modeling and prediction. *J. Agricultural, Biological and Environmental Statistics* **4**, 29–56.
Stein, M L (1999). *Interpolation of Spatial Data: Some Theory for Kriging*. Springer-Verlag. New York.
Ver Hoef, J M and N Cressie (1993). Multivariable spatial prediction. *Mathematical Geology* **25**, 219–240.
Ver Hoef, J M and R P Barry (1998). Constructing and fitting models for cokriging and multivariable spatial prediction. *J. Statistical Planning and Inference* **69**, 275–294.
Wackernagel, H (2003). *Multivariate Geostatistics: an Introduction with Applications*. 3rd ed.. Springer-Verlag. Berlin.
Wackernagel, H, L Bertino, J P Sierra and J González del Río (2002). Multivariate kriging for interpolating data from different sources. In: *Quantitative Methods for Current Environmental Issues* (CW Anderson, V Barnett, P Chatwin and AH El Shaarawi, Eds.). Springer-Verlag.
Wahba, G (1990*a*). Letter to the editor. *The American Statistician* **44**, 255–256.

Wahba, G (1990*b*). *Spline Models for Observational Data*. Society for Industrial and Applied Mathematics. Philadelphia.

Wikle, C K and N Cressie (1999). A dimension-reduced approach to space-time Kalman filtering. *Biometrika* **86**, 815–829.

Yaglom, A M (1986). *Correlation Theory of Stationary and Related Random Functions*. Springer-Verlag. Berlin.

Zimmerman, D L and M B Zimmerman (1991). A comparison of spatial variogram estimators and corresponding ordinary kriging predictors. *Technometrics* **33**, 77–91.

IFAC Publications
www.elsevier.com/locate/ifac

HILBERT SPACE EMBEDDINGS IN DYNAMICAL SYSTEMS

Alexander J. Smola * and **S.V.N. Vishwanathan** **

* *RSISE, Australian National University, Canberra, Australia*
** *Computer Science and Automation, IISc, Bangalore, India*

Abstract: In this paper we study Hilbert space embeddings *of* dynamical systems and embeddings generated *via* dynamical systems. This is achieved by following the behavioural framework invented by Willems, namely by comparing trajectories of states. As important special cases we recover the diffusion kernels of Kondor and Lafferty, generalised versions of directed graph kernels of Gärtner, novel kernels on matrices and new similarity measures on Markov Models. *Copyright © 2003 IFAC*

Keywords: Dynamical Systems, Reproducing Kernel Hilbert Space, Kernel Methods, Behavioural Framework, Graphs, Markov Models

1. INTRODUCTION

When dealing with dynamical systems, one may compare their similarities by checking whether they satisfy similar functional dependencies. With suitable parameterizations this is useful in determining when systems are similar. However, it is not difficult to find rather different functional dependencies, which, nonetheless, behave almost identically, e.g., as long as the domain of initial conditions is sufficiently restricted. For instance, consider the maps

$$x \leftarrow a(x) = |x|^p \text{ and } x \leftarrow b(x) = \min(|x|^p, |x|) \quad (1)$$

for $p > 1$. While a and b clearly differ, the two systems behave identically for all initial conditions satisfying $|x| \leq 1$. This example may seem contrived and it is quite obvious from (1) that identical behaviour will occur in this case, yet for more complex maps and higher dimensional spaces such statements are not quite as easily formulated.

This leads to the approach taken in the present paper, namely to compare *trajectories* of dynamical systems and derive measures of similarity from them. On graphs, for instance, which are described by their adjacency matrix (or the so called graph Laplacian), this corresponds to comparing paths followed by a diffusion process (Kondor and Lafferty, 2002). The advantage of our approach is that it is independent of the parameterization of the system, its downside being that one needs efficient mechanisms to compute the trajectories, which may or may not always be available. This approach is in spirit similar to the behavioural framework of (Willems, 1986*a*; Willems, 1986*b*; Willems, 1987), which identifies systems by identifying trajectories.

We begin with some definitions. For the remainder of the paper we assume that the state space $\mathscr{X}$, with $x \in \mathscr{X}$, is a Hilbert space. This is not a major restriction, since e.g., any countable set S can be made into a Hilbert space by mapping it into ℓ_2^S. Similar choices can be made for non-countable sets. [1]

Moreover we denote by $\mathscr{A}$ the set of time evolution operators, such that

$$x_{\mathbf{A}}(t) := \mathbf{A}(t)x \text{ for } \mathbf{A} \in \mathscr{A}. \quad (2)$$

Here x is subjected to the dynamics introduced by $\mathbf{A}$ and $t \in \mathscr{T}$ is the time of the measurement. We will choose $\mathscr{T} = \mathbb{N}_0$ or $\mathscr{T} = \mathbb{R}_0^+$, depending on whether we wish to deal with discrete-time or continuous-time systems.

---

[1] Should further generality be required, one may use algebraic semi-ring methods proposed by (Cortes *et al.*, 2002), albeit at a significant technical complication. This leads to rational series and functions on them.

Note that $\mathbf{A}$ may be a *nonlinear* operator and that both x and $\mathbf{A}$ may be random variables rather than deterministic variables. In those cases, we assume that both x and $\mathbf{A}$ are endowed with suitable probability measures. For instance, we may want to consider initial conditions corrupted by additional noise or linear time invariant systems with additive noise.

Finally, we assume that there exists a *valid* probability measure μ on $\mathscr{T}$, such that $\mu(\mathscr{T}) = 1$ and no $t \in \mathscr{T}$ has nonzero measure (or density). For the sake of analytical tractability we will assume that for some $\lambda > 0$

$$\mu(t) = \lambda^{-1}e^{-\lambda t} \text{ for } \mathscr{T} = \mathbb{R}_0^+ \quad (3)$$

$$\mu(t) = \frac{e^{-\lambda t}}{1-e^{-\lambda}} \text{ for } \mathscr{T} = \mathbb{N}_0. \quad (4)$$

Such exponential "discounting" is a popular choice in reinforcement learning (Sutton and Barto, 1998; Baxter and Bartlett, 1999) and control theory. In our case it will ensure convergence of integrals.

2. DOT PRODUCTS ON TRAJECTORIES

We define a dot product on the trajectories by taking expectations over the dot product on the coordinates. This leads to a Hilbert space on $\mathscr{X}^{\mathscr{T}}$ via

$$\langle \theta, \theta' \rangle := \mathbf{E}_t[\langle \theta(t), \theta'(t) \rangle] \text{ for } \theta, \theta' \in \mathscr{X}^{\mathscr{T}}. \quad (5)$$

Here the expectation is taken with respect to the probability measure $\mu(t)$. If we now identify $(x, \mathbf{A})$ with the trajectory $x_{\mathbf{A}}(\mathscr{T}) \in \mathscr{X}^{\mathscr{T}}$, we have a dot product between $(x, \mathbf{A})$ pairs:

$$k((x,\mathbf{A}),(\tilde{x},\tilde{\mathbf{A}})) := \mathbf{E}_t\left[\langle \mathbf{A}(t)x, \tilde{\mathbf{A}}(t)\tilde{x} \rangle\right]. \quad (6)$$

Here k was used as a shorthand for a dot product. It is in general referred to as a Mercer kernel and it can be used to solve numerous problems in machine learning (Wahba, 1990; Schölkopf and Smola, 2002). Note that (6) or any other Mercer kernel k defines a dot product and hence allows us to define a metric on its arguments. This implies that, as per our model, two pairs of initial conditions and time propagation operators are identical if they produce the same trajectory.

One may derive further kernels from (6) simply by using any of the composition rules for kernels, such as taking integer powers of kernels, i.e., $k' = k^n$, exponentiation of kernels $k' = e^k$, convex combination, normalization in feature space, or any composition thereof. (Haussler, 1999) describes a large class of such transformations in great detail.

2.1 *Kernels on Dynamical Systems*

Rather than specifying a set of initial conditions explicitly we may wish to compare two dynamical systems, yielding a kernel $k(\mathbf{A}, \tilde{\mathbf{A}})$. This can be achieved by restricting $k((x,\mathbf{A})),(\tilde{x},\tilde{\mathbf{A}}))$ to $x = \tilde{x}$, i.e., we compare only trajectories for identical initial conditions, and the expectation over x, should we be interested in the behaviour of $\mathbf{A}, \tilde{\mathbf{A}}$ over a range of x. Consequently we obtain

$$k(\mathbf{A},\tilde{\mathbf{A}}) := \mathbf{E}_x\left[k((x,\mathbf{A}),(x,\tilde{\mathbf{A}}))\right]. \quad (7)$$

This is a convex combination of kernels, hence (7) is a kernel. From the fact that $\mathscr{X}$ is a Hilbert space itself and the requirement that no $t \in \mathscr{T}$ have nonzero measure (or density) it follows that the canonical metric

$$d(\mathbf{A},\tilde{\mathbf{A}})^2 := k(\mathbf{A},\mathbf{A}) + k(\tilde{\mathbf{A}},\tilde{\mathbf{A}}) - 2k(\mathbf{A},\tilde{\mathbf{A}}) \quad (8)$$

is zero only if the trajectories of $x_{\mathbf{A}}(t)$ and $x_{\tilde{\mathbf{A}}}(t)$ are identical for all $x \in \mathscr{X}$ with nonzero probability measure $p(x)$.

We are therefore able to determine the proximity between various dynamical systems with respect to their initial conditions. If $\mathbf{A}, \tilde{\mathbf{A}}$ should happen to be random variables, we can generalise the setting by taking expectations over each of the random variables independently to obtain

$$k(\mathscr{A},\tilde{\mathscr{A}}) := \mathbf{E}_{\mathbf{A}}\mathbf{E}_{\tilde{\mathbf{A}}}\mathbf{E}_x\left[k((x,\mathbf{A}),(x,\tilde{\mathbf{A}}))\right]. \quad (9)$$

Such situations may occur, e.g., when $\mathscr{A}, \tilde{\mathscr{A}}$ are so-called lattices. Nondeterministic finite state automata can also be described in this fashion. See also (Cortes *et al.*, 2002).

2.2 *Kernels via Dynamical Systems*

In complete symmetry to the reasoning in the previous section we might also want to compare initial conditions with respect to a fixed class of dynamical systems and therefore define

$$k(x,\tilde{x}) := \mathbf{E}_{\mathbf{A}}\left[k((x,\mathbf{A}),(\tilde{x},\mathbf{A}))\right]. \quad (10)$$

In this context, we will consider initial conditions as similar, if the trajectories they induce in a dynamical system are similar. For instance, source code can be considered similar, if the dynamical system (here the compiler or possibly a set of various compilers) treats the code similarly and produces similar parse trees and/or executables.

Again, if $(x, \tilde{x})$ themselves are random variables, we may extend (10) to

$$k(x,\tilde{x}) := \mathbf{E}_x\mathbf{E}_{\tilde{x}}\mathbf{E}_{\mathbf{A}}\left[k((x,\mathbf{A}),(\tilde{x},\mathbf{A}))\right] \quad (11)$$

by taking expectations over the random variables themselves. Again the canonical metric induced by k is zero only when two initial conditions lead to identical trajectories. This means that, e.g., for a dynamical system with various basins of attraction, rather than using the distance in the initial conditions, the overall distance of the trajectories is used.

3. LINEAR SYSTEMS

A special, yet important case to consider are linear, time-invariant systems, where time propagation occurs as a linear function of the current state. We begin with two technical lemmas:

Lemma 1. Denote by A, B linear operators on $\mathscr{X}$ with $\|A\|, \|B\| \leq \Lambda$ for some $\Lambda > 0$. Then for all λ with $e^{\lambda} > \Lambda^2$ and for all linear operators $W : \mathscr{X} \to \mathscr{X}$ the series

$$M := \sum_{t=0}^{\infty} e^{-\lambda t} A^t W B^t \tag{12}$$

converges and M can be computed by solving the Sylvester equation $e^{-\lambda} AMB + W = M$.

Note that Sylvester equations of type $AXB + CXD = E$ can be readily solved at $O(n^3)$ time (Gardiner *et al.*, 1992) with freely available code ($A, B, C, D \in \mathbb{R}^{n \times n}$).

Proof To show that M is well defined we use the triangle inequality, leading to

$$\|M\| = \left\| \sum_{t=0}^{\infty} e^{-\lambda t} A^t W B^t \right\| \leq \sum_{t=0}^{\infty} \left\| e^{-\lambda t} A^t W B^t \right\|$$
$$\leq \sum_{t=0}^{\infty} \left(e^{-\lambda} \Lambda^2 \right)^t \|W\| = \frac{\|W\|}{1 - e^{-\lambda} \Lambda^2}.$$

Next we decompose the sum in M to obtain

$$M = A^0 W B^0 + \sum_{t=1}^{\infty} e^{-\lambda t} A^t W B^t$$
$$= W + e^{-\lambda} A \left[\sum_{t=0}^{\infty} e^{-\lambda t} A^t W B^t \right] B = W + e^{-\lambda} AMB.$$

■

A similar result holds for continuous time systems:

Lemma 2. Denote by A, B, W linear operators $\mathscr{X} \to \mathscr{X}$ such that $\|A\|, \|B\| \leq \Lambda$. Then, for all $\lambda > 2\Lambda$ the integral

$$M := \int_0^{\infty} e^{-\lambda t} \exp(At)^\top W \exp(Bt)\, dt \tag{13}$$

converges and M is the solution of the Sylvester equation $(A^\top + \frac{\lambda}{2}\mathbf{1})M + M(B + \frac{\lambda}{2}\mathbf{1}) = -W$.

Proof (Sketch only) Convergence follows from the triangle inequality and M can be rewritten to satisfy the self-consistency condition by partial integration. ■

3.1 *Discrete-Time Systems*

Here we assume that time propagation occurs as

$$x_{\mathbf{A}}(t+1) = A x_{\mathbf{A}}(t) + a_t + \xi_t \tag{14}$$

where $A : \mathscr{X} \to \mathscr{X}$ is a linear operator (typically a matrix), $a_t \in \mathscr{X}$ is a linear offset, and the noise variables ξ_t are assumed independent with zero mean and covariance C. Repeated substitution of (14) allows us to write $x_{\mathbf{A}}(t)$ as

$$x_{\mathbf{A}}(t) = A^t x_0 + \sum_{i=0}^{t} A^{t-i} \xi_i + A^{t-i} a_t. \tag{15}$$

$\langle x_{\mathbf{A}}, \tilde{x}_{\tilde{\mathbf{A}}} \rangle$ can be computed in closed form by straightforward algebra (mainly rearranging sums and exploiting harmonic series expansions). To simplify our presentation we henceforth assume that $a_t = 0$ and obtain

$$\sum_{t=0}^{\infty} e^{-\lambda t} \langle A^t x_0, \tilde{A}^t \tilde{x}_0 \rangle$$
$$= x_0^\top \left[\sum_{t=0}^{\infty} e^{-\lambda t} (A^t)^\top \tilde{A}^t \right] \tilde{x}_0 = x_0^\top M \tilde{x}_0 = \operatorname{tr}[(\tilde{x}_0 x_0^\top) M]$$
$$\mathbf{E}_{\xi} \left[\sum_{t=0}^{\infty} \sum_{j,j'=0}^{t} e^{-\lambda t} \langle A^{t-j} \xi_j, \tilde{A}^{t-j'} \xi_{j'} \rangle \right]$$
$$= \operatorname{tr} \sum_{t=0}^{\infty} \sum_{i=0}^{\infty} \left[e^{-\lambda(t+i)} (A^{t+i})^\top \tilde{A}^{t+i} C \right]$$
$$= \operatorname{tr} \sum_{t=0}^{\infty} \left[e^{-\lambda t} (A^t)^\top M \tilde{A}^t C \right] = \operatorname{tr}(C\tilde{M}).$$

Here $M, \tilde{M}$ satisfy the conditions of Lemma 1, that is $e^{-\lambda} A^\top M \tilde{A} + \mathbf{1} = M$ and $e^{-\lambda} A^\top \tilde{M} \tilde{A} + M = \tilde{M}$. This leads to

$$k((A, x), (\tilde{A}, \tilde{x})) = \operatorname{tr}(C\tilde{M}) + x_0^\top M \tilde{x}_0. \tag{16}$$

Following the reasoning of Section 2.1 we obtain a kernel on the initial conditions $(x_0, \tilde{x}_0)$ by setting $A = \tilde{A}$. This simply results in a modified metric tensor M plus the offset $\operatorname{tr}(C\tilde{M})$ with respect to the normal scalar product. In other words

$$k(x_0, \tilde{x}_0) = \operatorname{tr}(C\tilde{M}) + x_0^\top M \tilde{x}_0. \tag{17}$$

Furthermore, following the reasoning of Section 2.2 we get a kernel on dynamical systems by taking the expectation over the initial conditions. Assuming a covariance $\tilde{C}$ for the initial conditions we have

$$k(A, \tilde{A}) = \operatorname{tr}(\tilde{C} M) + \operatorname{tr}(C\tilde{M}). \tag{18}$$

A nice side-effect is that this allows us to measure similarities between objects such as matrices based on their spectral properties. See (Vishwanathan, 2002, Chapter 6) for more details.

Two special cases are worth-while considering (we assume $C = 0$ for simplicity):

- If $A = \tilde{A}$ we can solve the Sylvester equation in closed form as $M = (1 - e^{-\lambda} A^\top A)^{-1}$. This shows that the RKHS norm of A, as given by $\sqrt{k(A,A)}$ depends on how close to e^{λ} the eigenvalues of $A^\top A$ are: for large singular values the RKHS norm is large.
- If $A^\top \tilde{A} = 0$ the dynamics induced by A and $\tilde{A}$ differ as much as possible. Here all terms in the infinite sum except for the one due to the initial conditions vanish and we obtain $k(A,\tilde{A}) = \mathrm{tr}\tilde{C}$.

3.2 *Continuous-Time Systems*

Next we study continuous-time systems, where $x_{\mathbf{A}}(t)$ satisfies the stochastic differential equation

$$\frac{d}{dt} x_{\mathbf{A}}(t) = A x_{\mathbf{A}}(t) + a(t) + \xi(t). \tag{19}$$

Here $\xi(t)$ is a stochastic process with $\mathbf{E}[\xi(t)] = 0$ (in many applications we set $\xi(t) = 0$). Standard algebra shows that (19) can be solved by

$$x_{\mathbf{A}}(t) = \exp(At)x_0 + \int_0^t \exp(A(t-\tau))(a(\tau) + \xi(\tau))d\tau.$$

As before, we need to compute $k((x,\mathbf{A}),(\tilde{x},\tilde{\mathbf{A}}))$ and take expectations over ξ to eliminate the dependency on the random variable $\xi(t)$. For certain types of $a(t)$ this can be done in closed form.

To keep equations simple and for the purpose of conveying the basic idea we will limit ourselves to the special case of $a(t) = \xi(t) = 0$. This leads to

$$k((x,A),(\tilde{x},\tilde{A})) = \lambda^{-1} \int_0^\infty e^{-\lambda t} \langle \exp(At)x_0, \exp(\tilde{A}t)x_0' \rangle dt.$$

The latter can be solved using Lemma 2 and we obtain

$$k((x,A),(\tilde{x},\tilde{A})) = \lambda^{-1} x^\top M x. \tag{20}$$

Here M satisfies $(A^\top + \frac{\lambda}{2}\mathbf{1})M + M^\top(\tilde{A} + \frac{\lambda}{2}\mathbf{1}) = \mathbf{1}$.

If we fix $A = \tilde{A}$ to obtain kernels on dynamical systems this leads to

$$k(x,\tilde{x}) = x^\top M x. \tag{21}$$

In this case we can solve the Sylvester equation to obtain $M = \frac{1}{2}(A + \frac{\lambda}{2}\mathbf{1})^{-1}$. Likewise, if we wish to study kernels on initial conditions we obtain

$$k(x,\tilde{x}) = \mathrm{tr} MC. \tag{22}$$

where C is the covariance over the initial conditions. It is easy to see that both discrete and continuous-time dynamical systems lead to kernels with very similar structure. This is not surprising, given the similar differential equations (in fact, the discrete case is often used as an approximation for the continuum).

In the following we study two special cases of dynamical systems:

Snapshots: If $p(t) = \delta_T(t)$, that is, if we only care about a snapshot of the dynamical system at time T, we obtain

$$k((x,\mathbf{A}),(\tilde{x},\mathbf{A})) = x_0 \exp(At)\exp(At)^\top x_0^\top. \tag{23}$$

For a suitable choice of A this leads to kernels on graphs (Kondor and Lafferty, 2002; Vishwanathan, 2002).

Differential Equations: It is well known that linear differential equations can be transformed into first order linear differential equation by including the state space (Hirsch and Smale, 1974). This means that differential equations thereby impose a kernel on the initial conditions. Similarity here corresponds to correlation of the state space trajectories.

Likewise we can introduce a metric on the space of linear differential equations based on the similarity properties in their time evolution behaviour.

The kernels described above, appear somewhat simple minded at first sight. After all, with regard to initial conditions we are only replacing the standard Euclidean metric by the covariance of two trajectories under a dynamic system. The next section will show several useful and relevant cases.

4. MARKOV PROCESSES

Markov processes have the property that their time evolution behaviour depends only on their current state and the state transition properties of the model. Denote by S the set of states, then for $x \in \mathbb{R}^S$ the dynamics are given by

$$x(t+1) = Ax(t) \text{ or } \frac{d}{dt}x(t) = Ax(t) \tag{24}$$

for discrete-time and continuous time processes respectively. In this particular case x is the vector of state probabilities and A is the state transition matrix, that is for discrete processes $A_{ij} = p(i|j)$. Here $p(i|j)$ is the probability of reaching state i from state j.

4.1 *General Properties*

In *Discrete-Time Markov Processes* $k(i,j)$ computes the average overlap between the states when originating from $x = e_i, \tilde{x} = e_j$. Here e_i are "pure" states, that is the system is guaranteed to be in state i rather than in a mixture of states.

Since A is a stochastic matrix (positive entries with row-sum 1), its eigenvalues are bounded by 1 and therefore, any discounting factor $\lambda > 0$ will lead to a well-defined kernel.

Note that the average overlap between state vectors originating from different initial states are used in the context of graph segmentation and clustering (Weiss, 1999; Harel and Koren, 2001). This means that $\mu(t)$ is nonzero only for some $t \leq t_0$, which is similar to heavy discounting.

Recall, however, if e^{λ} is much smaller than the mixing time, k will almost exclusively measure the overlap between the initial states $x, \tilde{x}$ and the transient distribution on the Markov process. The quantity of interest here will be the ratio between e^{λ} and the gap between 1 and the second largest eigenvalue of A (Graham, 1999).

An extension to *Continuous-Time Markov Chains* (CMTC) is straightforward. Again $x(t)$ corresponds to the state at time t and the matrix A (called the rate matrix in this context) denotes the differential change in the concentration of states.

When the CTMC reaches a state, it stays there for an exponentially distributed random time (called the state holding time) with a mean that depends only on the state. For instance, diffusion processes can be modelled in this fashion. Clearly we can define kernels using diffusion on CTMCs by plugging the rate matrix A into (20).

4.2 *Graphs*

An important special case of Markov processes are random walks on (directed) graphs. Here diffusion through each of the edges of the graph is constant (in the direction of the edge). This means that, given an adjacency matrix representation of a graph via D (here $D_{ij} = 1$ if an edge from j to i exists), we compute the Laplacian $L = D - \operatorname{diag}(D\mathbf{1})$ of the graph, and use the latter to define the diffusion process $\frac{d}{dt}x(t) = Lx(t)$.

- Using the snapshot weights where we measure the overlap at time t, we obtain
$$K = \exp(tL)^{\top} \exp(tL) \qquad (25)$$
as covariance matrix between the distributions over various states. K_{ij} therefore equals the probability that any other state l could have been reached jointly from i and j (Kondor and Lafferty, 2002).
- The time T chosen for the snapshot is typically user defined and consequently it tends to be debatable. On the other hand, we might as well average over a large range of T (as typically done in Section 3.2), leading to a kernel matrix
$$K = \frac{1}{2}\left(A + \frac{\lambda}{2}\mathbf{1}\right)^{-1}. \qquad (26)$$
This yields a kernel whose inverse differs by $\frac{\lambda}{2}\mathbf{1}$ from the normalized graph Laplacian (another important quantity used for graph segmentation). The attraction of (26) is that its inverse is easy to come by and sparse, translating into significant computational savings for estimation.
- The kernel proposed by (Gärtner, 2002) can be recovered by setting W to have entries $\{0,1\}$ according to whether vertices i and j bear the same label and considering a discrete time random walk rather than a continuous time diffusion processes (various measures $\mu(t)$ take care of the exponential and harmonic weights).
- Finally, discrete-time Markov processes can be treated in complete analogy, yielding either $K = (\mathbf{1} - A)^{-1}$ or similar variants, should snapshots be required.

(Kondor and Lafferty, 2002) suggested to study diffusion on *undirected* graphs. Their derivations are a special case of (23). Note that for undirected graphs the matrices D and L are symmetric. This has the advantage that (23) can be further simplified to
$$k(x, \tilde{x}) = x^{\top} \exp(LT)^{\top} \exp(LT)\tilde{x} = x^{\top} \exp(2LT)\tilde{x}.$$

5. DISCUSSION

5.1 *Nonlinear Systems*

In nonlinear systems, in general, a closed-form computation of (6) will not be possible. Instead, we will need to resort to approximate solutions, mainly via numerical quadrature. This means that we will need to *simulate* trajectories in order to compare them. The computational cost depends on the discounting factor λ, since we only need to compute summands of k until their contribution vanishes, which happens exponentially fast (with $e^{-\lambda}$ controlling the speed).

While this may not seem satisfactory at first, we are confident that further efficient methods for computing trajectories analytically will be found. Of particular interest in this context are techniques for dealing with piecewise linear systems, as described in (Johansson and Rantzer, 1998).

Moreover, the area of linear matrix inequalities and control is appealing for the design of kernels, given the similarities that our calculations exhibit with some standard techniques in this area.

5.2 *Pair-HMM Kernels*

(Watkins, 2000) introduced the concept of Pair-HMMs for the purpose of computing kernel functions on sequences. While the focus was mainly on biological sequence analysis, a kernel computing
$$k(x, \tilde{x}) := p(x, \tilde{x}) = \sum_{\sigma \in \Sigma} p(x|\sigma)p(\tilde{x}|\sigma)p(\sigma) \quad (27)$$
is clearly also useful for the comparison of temporal sequences. Here σ denotes a path through the Pair-HMM (a Hidden Markov Model with a conditional independence property for the emissions) and the average is taken over all paths. Efficient computation is

ensured via dynamic programming, thus bounding the (otherwise) exponential number of terms in (27).

It would be interesting to see whether, rather than using sequences as arguments, one may also use automata directly as inputs for such similarity measures, while being able to perform dynamic programming on the triple of states (for the two inputs plus the automaton). The advantage of (27) over our "naive" correlation approach is that it can deal better with temporally misaligned data, that is, if particular events occur in a qualitatively similar, yet slightly time-shifted version.

5.3 *Summary and Outlook*

The current paper sets the stage for kernels on dynamical systems as they occur frequently in linear and affine systems. By using correlations between trajectories we were able to compare various systems on a behavioural level rather than a mere functional description. This allowed us to define similarity measures in a natural way.

While the larger domain of kernels on dynamical systems is still untested, special instances of the theory have proven to be useful in areas as varied as classification with categorical data (Kondor and Lafferty, 2002; Gärtner, 2002) and speech processing (Cortes *et al.*, 2002). This gives reason to believe that further useful applications will be found shortly.

For instance, we could use kernels in combination with novelty detection to determine unusual initial conditions, or likewise, to find unusual dynamics. In addition, we can use the kernels to find a metric between objects such as HMMs, e.g., to compare various estimation methods.

Acknowledgements

Parts of this work were supported by the Australian Research Council. SVNV was supported by grants from Netscaler Inc. and Trivium India Software as well as an Infosys fellowship. We thank Laurent El Ghaoui, Patrick Haffner, Daniela Pucci de Farias, and Bob Williamson for helpful discussions.

6. REFERENCES

Baxter, J. and P.L. Bartlett (1999). Direct gradient-based reinforcement learning: Gradient estimation algorithms. Technical report. Research School of Information, ANU Canberra.

Cortes, C., P. Haffner and M. Mohri (2002). Rational kernels. In: *Proceedings of Neural Information Processing Systems 2002*. in press.

Gardiner, J. D., A. L. Laub, J. J. Amato and C. B. Moler (1992). Solution of the Sylvester matrix equation $AXB^\top + CXD^\top = E$. *ACM Transactions on Mathematical Software* **18**(2), 223–231.

Gärtner, T. (2002). Exponential and geometric kernels for graphs. In: *NIPS*02 workshop on unreal data*. Vol. Principles of modeling nonvectorial data. `http://mlg.anu.edu.au/unrealdata`.

Graham, F. C. (1999). Logarithmic Sobolev techniques for random walks on graphs. In: *Emerging Applications of Number Theory* (D. A. Hejhal, J. Friedman, M. C. Gutzwiller and A. M. Odlyzko, Eds.). number 109 In: *IMA Volumes in Mathematics and its Applications*. Springer. pp. 175–186. ISBN 0-387-98824-6.

Harel, D. and Y. Koren (2001). Clustering spatial data using random walks. In: *Knowledge Discovery and Data Mining (KDD01)*. pp. 281–286.

Haussler, D. (1999). Convolutional kernels on discrete structures. Technical Report UCSC-CRL-99-10. Computer Science Department, UC Santa Cruz.

Hirsch, M. W. and S. Smale (1974). *Differential equations, dynamical systems, and linear algebra*. Academic Press. New York.

Johansson, M. and A. Rantzer (1998). Computation of piecewise quadratic lyapunov functions for hybrid systems. *IEEE Trans. Automat. Control,* **43**, 555–559.

Kondor, R. S. and J. Lafferty (2002). Diffusion kernels on graphs and other discrete structures. In: *Proceedings of the ICML*. To appear.

Schölkopf, B. and A. J. Smola (2002). *Learning with Kernels*. MIT Press.

Sutton, R.S. and A.G. Barto (1998). *Reinforcement Learning: An Introduction*. MIT Press.

Vishwanathan, S. V. N. (2002). Kernel Methods: Fast Algorithms and Real Life Applications. PhD thesis. Indian Institute of Science. Bangalore, India.

Wahba, G. (1990). *Spline Models for Observational Data*. Vol. 59 of *CBMS-NSF Regional Conference Series in Applied Mathematics*. SIAM. Philadelphia.

Watkins, C. (2000). Dynamic alignment kernels. In: *Advances in Large Margin Classifiers* (A. J. Smola, P. L. Bartlett, B. Schölkopf and D. Schuurmans, Eds.). MIT Press. Cambridge, MA. pp. 39–50.

Weiss, Y. (1999). Segmentation using eigenvectors: A unifying view. In: *International Conference on Computer Vision ICCV*. pp. 975–982.

Willems, J. C. (1986*a*). From time series to linear system. I. Finite-dimensional linear time invariant systems. *Automatica J. IFAC* **22**(5), 561–580.

Willems, J. C. (1986*b*). From time series to linear system. II. Exact modelling. *Automatica J. IFAC* **22**(6), 675–694.

Willems, J. C. (1987). From time series to linear system. III. Approximate modelling. *Automatica J. IFAC* **23**(1), 87–115.

IFAC Publications
www.elsevier.com/locate/ifac

BAYESIAN INPUT SELECTION FOR NONLINEAR REGRESSION WITH LS-SVMS

T. Van Gestel *,[1] M. Espinoza *,[1] J.A.K. Suykens * C. Brasseur **
B. De Moor *

* *K.U. Leuven, ESAT-SCD-SISTA, Leuven, Belgium.*
** *Fortis Bank, Financial Markets, Brussels, Belgium.*

Abstract: Input selection for linear and nonlinear modelling is an important problem, related to the trade-off between model complexity and in sample model accuracy. For linear modelling, well-known complexity criteria like the Akaike and Bayesian Information Criteria have been developed. In this paper, we explain the Bayesian evidence framework for Least Squares Support Vector Machines (LS-SVMs) and explain its use for input selection.

Keywords: Bayesian Learning, Input Selection, Kernel Based Learning, Least Squares Support Vector Machines, Nonlinear Regression

1. INTRODUCTION

Input selection for linear and nonlinear modelling is an important problem. Given a set of candidate explanatory input variables, a the trade-off between model complexity and in sample model accuracy is made. For linear modelling well-known complexity criteria like the Akaike and Bayesian Information Criteria (AIC and BIC) have been developed consisting of an in sample error term and a complexity term that penalizes too complex models (Akaike, 1978; Schwarz, 1978). These complexity terms are typically estimated using asymptotical approximations that yield the optimal model with respect to some distance measure.

While powerful design techniques for multilayer perceptrons like the Bayesian evidence framework (MacKay, 1992) have been developed, the practical use of neural networks for nonlinear modelling suffers from drawbacks like the nonconvex optimization problem with multiple local minima and the choice of the number of hidden neurons. In Support Vector Machines (SVMs), Least Squares SVMs (LS-SVMs) and related kernel based techniques (Schölkopf and Smola, 2002; Suykens *et al.*, 2002; Vapnik, 1998; Wahba, 1990), the solution follows from a convex optimization problem. Basically these methods map the inputs in a nonlinear way first into a high kernel induced feature space, in which ridge regression is applied in the case of LS-SVMs [2]. The solution follows from a linear system in the dual space in terms of the positive definite kernel function by applying Mercer's Theorem.

In this paper, the Bayesian evidence framework (Van Gestel *et al.*, 2001*a*; Van Gestel *et al.*, 2002*a*) to perform input selection for nonlinear regression is explained. Bayes' formula is applied on different levels of inference to tune hyperparameters and to select the relevant inputs. The Bayesian framework embodies Occam's razor to find an optimal trade-off between in sample accuracy and model complexity in a similar way as the AIC and BIC.

This paper is organized as follows. The primal-dual LS-SVM formulations are reviewed in Section 2. The Bayesian framework is explained in Section 3 and applied in Section 4.

---

[1] Corresponding authors, email: {tony.vangestel,marcelo.espinoza}@esat.kuleuven.ac.be. T. Van Gestel is also a senior quantitative analyst with Dexia Group (Credit Risk Management): tony.vangestel@dexia.com.

[2] The LS-SVM formulation is related to regularization networks (Poggio and Girosi, 1990) and Gaussian Processes (Neal, 1996), but the explicit primal-dual formulations allow to make extensions to nonlinear control, recurrent predictions and kernel principal component analysis (Suykens *et al.*, 2002; Suykens *et al.*, 2003).

2. FROM LINEAR TO NONLINEAR KERNEL BASED MODELING AND PREDICTION

2.1 *Kernel Functions*

An important concept of nonlinear kernel based regression and prediction is the interpretation of applying linear regression techniques in the kernel induced feature space. The linear model formulation to predict the output y based on the n explanatory variables $x = [x_1; \ldots; x_n] = [x_1, \ldots, x_n]^T$ can be written as

$$y = w^T x + b + e, \qquad (1)$$

with output $y \in \mathbb{R}$, input vector $x \in \mathbb{R}^n$, coefficient vector $w \in \mathbb{R}^n$ and bias term $b \in \mathbb{R}$. A straightforward way to extend the linear model (1) to a nonlinear model is to preprocess the inputs x in an nonlinear way by the mapping

$$\varphi : \mathbb{R}^n \to \mathbb{R}^{n_f} : x \mapsto \varphi(x) \qquad (2)$$

where the feature vector $\varphi(x)$ is typically high (or even infinite) dimensional. In kernel based learning, the mapping is defined implicitly in terms of the positive definite kernel function

$$K(x_1, x_2) = \varphi(x_1)^T \varphi(x_2) \qquad (3)$$

from Mercer's theorem (Mercer, 1909) and the nonlinear mapping is tuned implicitly in terms of the parameters of the kernel function. Some commonly used kernel functions are the linear kernel $K(x_1,x_2) = x_1^T x_2$, the polynomial kernel $K(x_1,x_2) = (1 + x_1^T x_2/c)^d$ of degree d and the Radial Basis Function (RBF) kernel $K(x_1,x_2) = \exp(-\frac{\|x_1-x_2\|_2^2}{\sigma^2})$, where d and c, σ are constants. Notice that the Mercer condition holds for all $\sigma, c \in \mathbb{R}^+$ values in the RBF (Radial Basis Function) and the polynomial case.

Hence, one can first extend the results of linear modelling in a straightforward way to nonlinear modelling by first applying the (linear) reasoning in the kernel induced feature space. Practical expressions are obtained in the dual space by application of the Mercer condition (Mercer, 1909).

2.2 *Primal-Dual Formulations*

Given the training data $\mathscr{D} = \{(x_i, y_i)\}_{i=1}^{n_\mathscr{D}}$, one obtains the coefficient vector w and bias term b from minimization of the following ridge regression problem

$$\min_{w,b,e} \mathscr{J}(w,b) = \frac{\mu}{2} w^T w + \frac{\zeta}{2} \sum_{i=1}^{n_\mathscr{D}} e_i^2 \qquad (4)$$

$$\text{s.t.} \quad e_i = y_i - (w^T \varphi(x_i) + b), \quad i = 1, \ldots, n_\mathscr{D}. \qquad (5)$$

The hyperparameters $\mu, \zeta \in \mathbb{R}^+$ allow to find an appropriate trade-off between regularization $\frac{1}{2} w^T w$ and training set error minimization $\frac{1}{2} \sum_{i=1}^{n_\mathscr{D}} e_i^2$. This balance between (in sample) training set accuracy and model complexity is made on the second level of inference. The use of the regularization term in ridge regression (Hoerl and Kennard, 1970) has been used to increase the quality of the model parameter estimates in identification, e.g., in in (Golub *et al.*, 1979; Sjöberg *et al.*, 1995; Van Gestel *et al.*, 2001*b*). Observe that no regularization is applied on the bias term (Brown, 1977).

In order to solve the constrained optimization problem (4)-(5), one constructs the Lagrangian $\mathscr{L} = \frac{\mu}{2} w^T w + \frac{\zeta}{2} \sum_{i=1}^{n_\mathscr{D}} e_i^2 + \sum_{i=1}^{n_\mathscr{D}} \alpha_i (y_i - (w^T \varphi(x_i) + b) - e_i)$, where the support values $\alpha_i \in \mathbb{R}$ are the Lagrange multipliers associated with the equality constraints (5). Defining the matrix notation $\Phi = [\varphi(x_1), \ldots, \varphi(x_{n_\mathscr{D}})]^T \in \mathbb{R}^{n_\mathscr{D} \times n_f}$, $\alpha = [\alpha_1, \ldots, \alpha_{n_\mathscr{D}}]^T \in \mathbb{R}^{n_\mathscr{D}}$, $e = [e_1, \ldots, e_{n_\mathscr{D}}]^T \in \mathbb{R}^{n_\mathscr{D}}$, $y = [y_1, \ldots, y_{n_\mathscr{D}}]^T \in \mathbb{R}^{n_\mathscr{D}}$ and $1 = [1, \ldots, 1]^T \in \mathbb{R}^{n_\mathscr{D}}$, the conditions for optimality correspond to the symmetric linear system

$$\begin{bmatrix} -\mu I_{n_f} & \Phi^T & \mathbf{0} & \mathbf{0} \\ \Phi & \mathbf{0} & 1 & I_{n_\mathscr{D}} \\ \mathbf{0} & 1^T & \mathbf{0} & \mathbf{0} \\ \mathbf{0} & I_{n_\mathscr{D}} & \mathbf{0} & -\zeta I_{n_\mathscr{D}} \end{bmatrix} \cdot \begin{bmatrix} w \\ \alpha \\ b \\ e \end{bmatrix} = \begin{bmatrix} 0 \\ y \\ 0 \\ 0 \end{bmatrix}. \qquad (6)$$

For the linear case (e.g. a linear kernel) one typically has $n_f = n \ll n_\mathscr{D}$ and after elimination of e and α, one solves the $(n_f + 1) \times (n_f + 1)$ linear system in the primal space

$$\begin{bmatrix} \Phi^T \Phi + \frac{\mu}{\zeta} I_{n_f} & \Phi^T 1 \\ 1^T \Phi & n_\mathscr{D} \end{bmatrix} \begin{bmatrix} w \\ b \end{bmatrix} = \begin{bmatrix} \Phi^T y \\ 1^T y \end{bmatrix}. \qquad (7)$$

In nonlinear kernel based regression, one usually has $n_f \gg n_\mathscr{D}$ and, moreover, the feature vector $\varphi(x)$ is only implicitly defined in terms of the kernel function K from (3). Eliminating w and e from (6), one obtains the linear Karush-Kuhn-Tucker system of dimension $(n_\mathscr{D} + 1) \times (n_\mathscr{D} + 1)$ in the dual space (Suykens and Vandewalle, 1999; Suykens *et al.*, 2002)

$$\begin{bmatrix} \frac{1}{\mu} \Omega + \frac{1}{\zeta} I_{n_\mathscr{D}} & 1 \\ 1^T & 0 \end{bmatrix} \begin{bmatrix} \alpha \\ b \end{bmatrix} = \begin{bmatrix} y \\ 0 \end{bmatrix}, \qquad (8)$$

where the Mercer condition (3) is applied in the matrix $\Omega = \Phi\Phi^T \in \mathbb{R}^{n_\mathscr{D} \times n_\mathscr{D}}$ with elements $\Omega_{ij} = K(x_i, x_j)$, $i, j = 1, \ldots, n_\mathscr{D}$ and guarantees that $\Omega \geq 0$. Defining the idempotent centering matrix $\mathbf{N}_c = I_{n_\mathscr{D}} - \frac{1}{n_\mathscr{D}} 11^T \in \mathbb{R}^{nd \times n_\mathscr{D}}$, one obtains explicit formulas for α and b: $(\frac{1}{\mu} \mathbf{N}_c \Omega \mathbf{N}_c + \frac{1}{\zeta} I_{n_\mathscr{D}}) \alpha = \mathbf{N}_c y$ and $n_\mathscr{D} b = 1^T (\frac{1}{\mu} \Omega + \frac{1}{\zeta} I_{n_\mathscr{D}})^{-1} y$, resp. Up to the centering, these expressions correspond to the (only) dual expressions in Gaussian Processes (Neal, 1996). Applying no regularization on the bias term results into a zero mean training set error and is the preferred form of rigde regression as has been motivated in (Brown, 1977).

Given the support values α and bias term b, one obtains the predicted value $\hat{y}$ corresponding to a new input x as a weighted sum of the kernel functions evaluated in the new point and the training points:

$$\hat{y} = w^T \varphi(x) + b = \frac{1}{\mu} \sum_{t=1}^{n_\mathscr{D}} \alpha_t K(x, x_t) + b. \qquad (9)$$

3. BAYESIAN INFERENCE

Given the primal-dual formulations, it is clear how to estimate the model parameters w, b and point prediction y. However, the regularization and kernel function parameters still have to be tuned from the given training data. In this Section, the model parameters w, b, the hyperparameters μ, ζ and model structure $\mathscr{M}$ (corresponding, e.g., to the input set and/or tunable kernel parameters) are inferred from the data $\mathscr{D}$ by applying Bayes formula on three different levels (Van Gestel *et al.*, 2002*a*; Van Gestel *et al.*, 2001*a*).

3.1 *Inference of the Model Parameters (Level 1)*

3.1.1. *Probabilistic Interpretation of the LS-SVM Formulation* Given the data points $\mathscr{D} = \{(x_t, y_t)\}_{t=1}^{n_\mathscr{D}}$ and the hyperparameters μ and ζ of the model $\mathscr{M}$ (LS-SVM with given input set and kernel function K), we obtain the model parameters w and b by maximizing the posterior $p(w, b|\mathscr{D}, \log\mu, \log\zeta, \mathscr{M})$. Application of Bayes' rule at the first level of inference (MacKay, 1992) gives

$$p(w,b|\mathscr{D},\log\mu,\log\zeta,\mathscr{M}) = \frac{p(\mathscr{D}|w,b,\log\mu,\log\zeta,\mathscr{M})p(w,b|\log\mu,\log\zeta,\mathscr{M})}{p(\mathscr{D}|\log\mu,\log\zeta,\mathscr{M})}, \quad (10)$$

where the evidence $p(\mathscr{D}|\log\mu,\log\zeta,\mathscr{M})$ is a normalizing constant that will be related to model complexity.

The regularization parameters μ and ζ are related to the prior $p(w,b|\log\mu,\log\zeta,\mathscr{M})$ and likelihood $p(\mathscr{D}|w,b,\log\mu,\log\zeta,\mathscr{M})$, respectively. We take the prior $p(w,b|\log\mu,\log\zeta, \mathscr{M})$ independent of the hyperparameter ζ, i.e., $p(w,b|\log\mu,\log\zeta,\mathscr{M}) = p(w,b|\log\mu,\mathscr{M})$. Both w and b are assumed to be independent. The coefficients w are assumed to be multivariate Gaussian distributed with mean zero and covariance matrix $\frac{1}{\mu}I_{n_f}$:

$$p(w|\log\mu,\mathscr{M}) = \left(\tfrac{\mu}{2\pi}\right)^{\frac{n_f}{2}}\exp\left(-\tfrac{\mu}{2}w^Tw\right). \quad (11)$$

This means that a priori we do not expect a functional relation between the feature vector φ and the observation y. Before the data are available, the most likely model has zero weights $w_k = 0$ ($k = 1,\ldots,n_f$). A uniform distribution for the prior on b is taken, which can also be approximated as a Gaussian distribution $p(b|\log\sigma_b,\mathscr{M}) = (2\pi\sigma_b)^{-1/2}\exp(-b^2/(2\sigma_b^2))$, with $\sigma_b \to \infty$. This is called a uniform prior and explains why there is no regularization term on b in (4)-(5). The negative logarithm of the prior corresponds to the regularization term $\frac{\mu}{2}w^Tw$ in (4)-(5).

We take the likelihood of the observed data $\mathscr{D}$ independent of the hyperparameter μ and assume that all data points (x_t, y_t) are independent:

$$\begin{aligned} &p(\mathscr{D}|w,b,\log\zeta,\mathscr{M}) \\ &= \textstyle\prod_{t=1}^{n_\mathscr{D}} p(y_t|x_t,w,b,\log\zeta,\mathscr{M})p(x_t|w,b,\log\zeta,\mathscr{M}) \\ &\propto \textstyle\prod_{t=1}^{n_\mathscr{D}} p(y_t|x_t,w,b,\log\zeta,\mathscr{M}), \end{aligned} \quad (12)$$

with $p(x_t|w,b,\log\zeta,\mathscr{M}) = p(x_t)$. Assuming that the additive noise e_t is drawn from a Gaussian distribution with variance $1/\zeta$, we have $p(y_t|x_t,w,b,\log\zeta,\mathscr{M}) = (\frac{\zeta}{2\pi})^{1/2}\exp(-\frac{\zeta}{2}(y_t - w^T\varphi(x_t) - b)^2)$.. The negative logarithm of the likelihood (12) corresponds to the sum squared error term $\frac{\zeta}{2}\sum_{t=1}^{n_\mathscr{D}} e_t^2$.

Substituting (11) and (12) into (10), neglecting all constants and taking the negative logarithm, Bayes' rule at the first level of inference corresponds to the constrainted minimization problem (4)-(5) that can be solved for w and b in the primal space from (7) in the linear case when $n \leq n_\mathscr{D}$ or α and b in the dual space from (8) in the nonlinear kernel-based regression case and in the linear case when $n \geq n_\mathscr{D}$. In the remainder of this paper, the maximum a posteriori parameter estimates are denoted by the subscript 'mp', e.g., w_{mp} and b_{mp}.

We discuss now an alternative representation of the LS-SVM cost function (4)-(5) and the posterior probability (10), which will be used in the next Subsection. By observing that the LS-SVM cost function is a quadratic cost function, the posterior $p(w,b|\mathscr{D},\log\mu, \log\zeta, \mathscr{M})$ can also be written as the Gaussian distribution with mean $[w_{mp}; b_{mp}]$ and covariance matrix Q. $Q = \text{covar}([w;b]) = \mathscr{E}([w - w_{mp}; b - b_{mp}][w - w_{mp}; b - b_{mp}]^T)$, The covariance matrix Q is related to the Hessian H of the LS-SVM cost function (4)-(5):

$$Q^{-1} = H = \begin{bmatrix} \mu I + \zeta\Phi^T\Phi & \zeta\Phi^T 1 \\ \zeta 1^T\Phi & \zeta n_\mathscr{D} \end{bmatrix}, \quad (13)$$

where the inverse is calculated using, e.g., a Schur complement type of argument.

3.1.1.1. *Model Evidence and Complexity Terms*
The evidence $p(D|\log\mu,\log\zeta,\mathscr{M})$ is obtained by using the expressions for the prior, likelihood and posterior from Bayes formula (10)

$$p(\mathscr{D}|\log\mu,\log\zeta,\mathscr{M}) = \frac{p(\mathscr{D}|w,b,\log\mu,\log\zeta,\mathscr{M})p(w,b|\log\mu,\log\zeta,\mathscr{M})}{p(w,b|D,\log\mu,\log\zeta,\mathscr{M})}. \quad (14)$$

Substituting (11) and (12) into (14) yields (Van Gestel *et al.*, 2002*a*; Van Gestel *et al.*, 2001*a*)

$$p(\mathscr{D}|\log\mu,\log\zeta,\mathscr{M}) \propto \underbrace{p(\mathscr{D}|w_{mp},b_{mp},\log\zeta,\mathscr{M})}_{\text{model fit}}\underbrace{p(w_{mp}|\log\mu,\mathscr{M})(\det H)^{-\frac{1}{2}}}_{\text{Occam factor}}. \quad (15)$$

Within the Bayesian framework, the evidence of the model is equal to the best fit likelihood of the posterior model parameters times the Occam factor that penalizes models that are too complex (MacKay, 1992) and have too many degrees of freedom. As the determinant of the covariance matrix $(\det Q)^{1/2} = (\det H)^{-1/2}$ is a measure for the volume of the posterior distribution, a large contraction in the volume of the prior distribution towards the posterior may tend to cause overfitting because there are too many free parameters, the Occam factor penalizes complex models with too many degrees of freedom. Taking the negative logarithm of (15), the model evidence consists of the training set error plus a complexity term. This is related to well known complexity criteria like the Akaike and Bayesian Information Criterion (Akaike, 1978; Schwarz, 1978) that are obtained using appropriate (asymptotical) simplifications.

3.2 *Inference of the Hyperparameters (Level 2)*

The dimension of the kernel induced feature space can be larger than the number of training data points and for some nonlinear kenrel functions, like the RBF kernel, it can even become infinite dimensional. Therefore, it is necessary to find an appropriate trade-off between regularization and training set error by tuning the hyperparameters μ and ζ.

3.2.1. *Optimization Problem for μ and ζ* The regularization hyperparameters μ and ζ (MacKay, 1992) are inferred from the data $\mathscr{D}$ by applying Bayes' rule on the second level:

$$\begin{aligned} p(\log\mu, \log\zeta|\mathscr{D}, \mathscr{M}) \quad (16) \\ = \frac{p(\mathscr{D}|\log\mu, \log\zeta, \mathscr{M})p(\log\mu, \log\zeta|\mathscr{M})}{p(\mathscr{D}|\mathscr{M})} \\ \propto p(\mathscr{D}|\log\mu, \log\zeta, \mathscr{M}), \end{aligned}$$

where a flat, non-informative prior is assumed on the hyperparameters μ and ζ. The probability $p(\mathscr{D}|\log\mu, \log\zeta, \mathscr{M})$ is equal to the evidence of the previous level. Taking the negative logarithm of (16) and using (10), one obtains the following cost function

$$\begin{aligned} \mathscr{J}(\mu, \zeta) = \mu \mathscr{J}_w(w_{mp}) + \zeta \mathscr{J}_e([w_{mp}; b_{mp}]) \\ + \frac{1}{2}\log\det H - \frac{n_f}{2}\log\mu - \frac{n_{\mathscr{D}}}{2}\log\zeta, \quad (17) \end{aligned}$$

where we used the shorthand notation $\mathscr{J}_w(w) = \frac{1}{2}w^T w$ and $\mathscr{J}_e([w;b]) = \frac{1}{2}e^T e$ with $e = y - [\Phi; 1^T][w;b]$. The expression for $\det H$ is equal to $\det H = n_{\mathscr{D}}\zeta\mu^{n_f - n_f^e} \prod_{i=1}^{n_f^e}(\mu + \zeta\lambda_{\Omega_c,i})$, where $\lambda_{\Omega_c,i}$, $i = 1, \ldots, n_f^e$ are the n_f^e non-zero eigenvalues of $\mathbf{N}_c\Phi\Phi^T\mathbf{N}_c = \mathbf{N}_c\Omega\mathbf{N}_c = \Omega_c$.

3.2.2. *Scalar Optimization Problem for γ* In the dual space, the regularization parameter $\gamma = \zeta/\mu$ is obtained by minimizing (Van Gestel *et al.*, 2002*a*; Van Gestel *et al.*, 2001*a*)

$$\begin{aligned} \min_{\gamma} \mathscr{J}_2(\gamma) = \Sigma_{i=1}^{n_{\mathscr{D}}-1}\log(\lambda_{\Omega,i} + \gamma^{-1}) \\ + (n_{\mathscr{D}} - 1)\log[\mathscr{J}_w(w) + \gamma\mathscr{J}_e(w,b)], \quad (18) \end{aligned}$$

with $\mathscr{J}_w = \frac{1}{2}w^T w$, $\mathscr{J}_e = \frac{1}{2}e^T e$ and where $\lambda_{\Omega,i}$ ($i = 1, \ldots, n_{\mathscr{D}} - 1$) are the $n_{\mathscr{D}} - 1$ eigenvalues of the centered kernel matrix $M_c\Omega M_c$, with idempotent centering matrix $M_c = I - 1/n_{\mathscr{D}}11^T$. The level 1 cost function can be expressed as $\mathscr{J}_w(w) + \gamma\mathscr{J}_e(w,b) = \frac{1}{2}y^T M_c(M_c\Omega M_c + \gamma^{-1}I_{n_{\mathscr{D}}})^{-1}M_c y$, which avoids the explicit solution of the linear system (8) for all possible γ values. As these expressions need to be evaluated for different values of γ one can compute first an eigenvalue decomposition of $M_c\Omega M_c$. This allows then to compute $\mathscr{J}_w$, $\mathscr{J}_e$ and $\mathscr{J}_w + \gamma\mathscr{J}_e$ from matrix vector multiplications only, avoiding the (computationally intensive) calculation of the inverse $(M_c\Omega M_c + \gamma^{-1}I_{n_{\mathscr{D}}})^{-1}$ (Van Gestel *et al.*, 2002*a*; Van Gestel *et al.*, 2001*a*) for different values of γ.

Given the optimal γ from (18) one finds the effective number of parameters[3] d_{eff} from $d_{\text{eff}} = (n_{\mathscr{D}} + \gamma\mathscr{J}_e/\mathscr{J}_w)/(1 + \gamma\mathscr{J}_e/\mathscr{J}_w)$. The optimal μ and ζ are obtained from $\mu = (d_{\text{eff}} - 1)/(2\mathscr{J}_w(w))$ and $\zeta = (n_{\mathscr{D}} - d_{\text{eff}})/(2\mathscr{J}_e(w,b))$, where the latter is related to the unbiased sample estimate of the noise variance.

3.3 *Model Comparison (Level 3)*

The model structure $\mathscr{M}$ with RBF-kernel parameter σ is inferred on Level 3:

$$p(\mathscr{M}|\mathscr{D}) \propto p(\mathscr{D}|\mathscr{M})p(\mathscr{M}), \quad (19)$$

where typically equal prior probabilities are assumed, meaning that $p(\mathscr{M}|\mathscr{D}) \propto p(\mathscr{D}|\mathscr{M})$. The posterior level 3 probability is obtained as the level 2 probability with an additional Occam factor due to the posterior uncertainty on the inferred hyperparameters μ and ζ

$$p(\mathscr{D}|\mathscr{M}) \simeq p(\mathscr{D}|\log\mu, \log\zeta, \mathscr{M})\frac{\sigma_{\log\mu|\mathscr{D}}\sigma_{\log\zeta|\mathscr{D}}}{\sigma_{\log\mu}\sigma_{\log\zeta}},$$

with $\sigma_{\log\mu|\mathscr{D}} \simeq 2/(d_{\text{eff}} - 1)$ and $\sigma_{\log\zeta|\mathscr{D}} \simeq 2/(n_{\mathscr{D}} - d_{\text{eff}})$. When flat, uninformative priors are used on level 2, $\sigma_{\log\mu}$ and $\sigma_{\log\zeta}$ are omitted in the comparisons of different models. In the case of equal prior model probabilities $p(\mathscr{M}_i) = p(\mathscr{M}_j)$ ($\forall i, j$) the models $\mathscr{M}_i$ and $\mathscr{M}_j$ are compared according to their Bayes factor $\mathscr{B}_{ij} = p(\mathscr{D}|\mathscr{M}_i)/p(\mathscr{D}|\mathscr{M}_j)$:

$$\mathscr{B}_{ij} = \frac{p(\mathscr{D}|\log\mu_i, \log\zeta_i, \mathscr{M}_i)}{p(\mathscr{D}|\log\mu_j, \log\zeta_j, \mathscr{M}_j)}\frac{\sigma_{\log\mu_i|\mathscr{D}}\sigma_{\log\zeta_i|\mathscr{D}}}{\sigma_{\log\mu_j|\mathscr{D}}\sigma_{\log\zeta_j|\mathscr{D}}}. \quad (20)$$

3.4 *Bayesian Input Selection*

Model comparison is also used to infer the set of most relevant inputs (Van Gestel *et al.*, 2001*a*) out of the given set of candidate explanatory variables by making pairwise comparisons of models with different input sets. In a backward input selection procedure, one starts from the full candidate input set and removes in each input pruning step that input that yields the best model improvement (or smallest decrease) in terms of the model probability (19). The procedure is stopped when no significance decrease of the model probability is observed.

---

[3] The effective number of parameters can be defined in terms of the trace of the smoother matrix or in terms of the ratios of the eigenvalues of the Hessian of the unregulated and regulated cost function: $d_{\text{eff}} = 1 + \Sigma_{i=1}^{n_f^e}(\zeta\lambda_{\Omega_c,i})/(\mu + \zeta\lambda_{\Omega_c,i})$.

4. DESIGN & IMPLEMENTATION

A practical scheme for the design of the LS-SVM within the evidence framework consists of the following steps:

(1) The inputs are standardized to zero mean and unit variance. The normalized training data are denoted by $D = \{(x_t, y_t)\}_{t=1}^{n_{\mathscr{D}}}$.
(2) Select the model $\mathscr{M}_j$ by choosing a kernel type K_j (possibly with a kernel parameter, e.g., σ_j for an RBF-kernel). For this model $\mathscr{M}_j$, the optimal hyperparameters μ_{mp} and ζ_{mp} are estimated on the second level of inference. This is done as follows: a) Solve the scalar optimization problem (18) in $\gamma = \zeta/\mu$ using, e.g., a quasi-Newton method. b) Calculate the effective number of parameters d_{eff} and the regularization parameters μ_{mp} and $\zeta_{mp} = \mu_{mp}\gamma_{mp}$.
(3) Calculate the model evidence $p(\mathscr{M}_j|\mathscr{D})$ from (19).
(4) For a kernel K_j with tuning parameters, refine the tuning parameters. E.g., for the RBF-kernel with tuning parameter σ_j, refine σ_j such that a higher model evidence $p(\mathscr{D}|\mathscr{M}_j)$ is obtained. This can be done by maximizing the model evidence with respect to σ_j by evaluating the model evidence for the refined kernel parameter starting from step 2(a).
(5) Select the model $\mathscr{M}_j$ with maximal model evidence $p(\mathscr{D}|\mathscr{M}_j)$. Go to step 2, unless the best model has been selected.

Input selection is done by starting from an initial candidate input set and calculating the model evidence $p(\mathscr{M}_j|\mathscr{D})$ by taking steps 2-4. In a backward input selection procedure, one step consists of comparing the model probabilities of the model with full candidate input sets and all the models where in turn one of the candidate inputs has been removed. The backward input steps are repeated until $p(\mathscr{M}_j|D)$ stops increasing.

5. EXAMPLES

We apply Bayesian input selection on the estimation of a noisy sinc function $y = \text{sinc}(x_3) + e$, (with $\mathscr{E}(e) = 0$ and $\sigma_e = 0.1$), where the correct input x_3 has to be determined from 5 candidate inputs $x = [x_1; x_2; x_3; x_4; x_5]$. All inputs were drawn from a uniform distribution in the interval $[-3, 3]$. In each backward input selection step the optimal hyper and kernel parameters of the LS-SVM with RBF-kernel are determined. These are used as a starting point to calculate the level 3 cost function when removing in turn the candidate inputs (Figure 1.a). The backward input selection is stopped when the level 3 cost function starts increasing (Figure 1.b), in this case after 5 input pruning steps, selecting the 'true' input x_3. The estimated functions using all 5 candidate inputs and only the 'true' input are depiced in Figures 1.c and 1.d, respectively.

The real-life problem we are concerned with is to weekly predict an aggregated index for the European chemical sector based on a set of 27 candidate explanatory input variables selected by a financial analyst and resulting from econometric time series analysis (Van Gestel *et al.*, 2002*b*). Performing backward

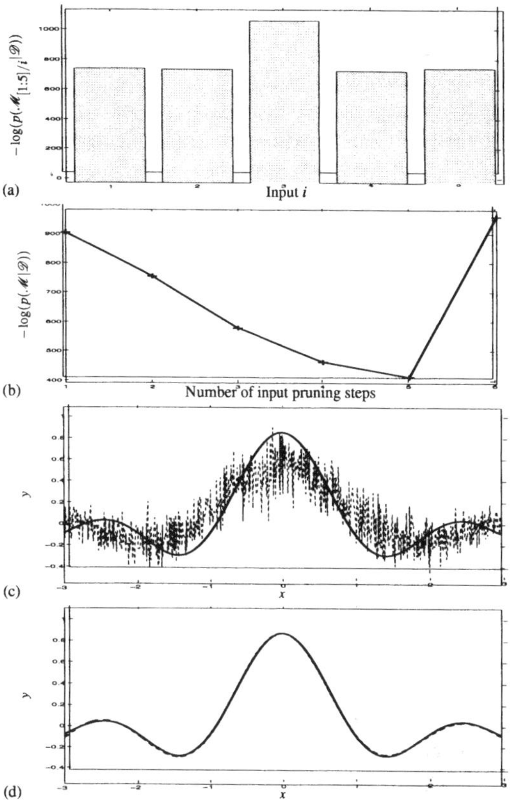

Fig. 1. Bayesian input selection. (a) Level 3 cost function when removing different inputs at step 1. (b) Evolution of level 3 cost function as a function of the number of input pruning steps. (c-d) True and estimated sinc function (full and dashed line, resp.) using 5 and 1 inputs.

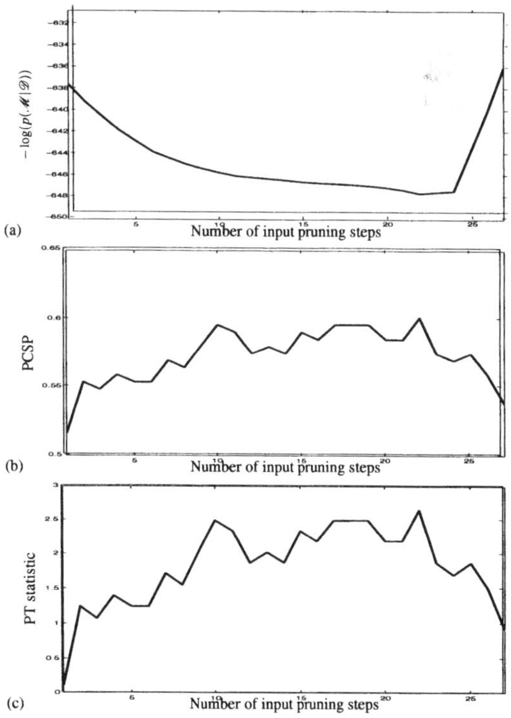

Fig. 2. Evolution of level 3 cost function, percentage of correct sign prediction (PCSP), Pesaran-Timmerman test statistic (PTstat) as a function of the number of input pruning steps.

input selection, the evolution of the level 3 cost function $-\log(p(\mathscr{M}|\mathscr{D}))$, the percentage of out of sample sign predictions and the Pesaran-Timmerman test statistic[4] are depicted in Figure 2 as a function of the number of input pruning steps. Observe that the PT-stat value becomes maximal when the level 3 cost function becomes minimal.

6. CONCLUSIONS

The primal-dual LS-SVM formulations allow to extend the Bayesian framework for linear ridge regression to nonlinear kernel based regression. In this paper, input selection is performed for nonlinear regression within this Bayesian framework. Basically, a trade-off is made between in sample model accuracy and model complexity in a similar way as in the Akaike and Bayesian Information Criteria.

7. ACKNOWLEDGEMENTS

T. Van Gestel is a senior quantitative analyst with Dexia Group (Risk Management Methodology) and holds a honorary position as a Postdoctoral Researchers with the Fund for Scientific Research-Flanders (FWO-Vlaanderen). J. Suykens is a Postdoctoral Researcher with the FWO-Vlaanderen. This work was partially supported by grants and projects from the Research Council K.U.Leuven (GOA-Mefisto 666, IDO, PhD/postdoc & fellow grants), the Flemish Government (FWO-Vlaanderen: PhD/postdoc grants, projects G.0256.97, G.0115.01, G.0240.99, G.0197.02, G.0407.02, ICCoS, ANMMM; AWI, IWT: Soft4s, STWW-Genprom, GBOU-McKnow, Eureka-Impact/FLiTE, PhD grants), the Belgian Federal Government (DWTC: IUAP IV-02, IUAP V-22; PODO-II CP/40). The scientific responsibility is assumed by its authors.

8. REFERENCES

Akaike, H. (1978). A Bayesian analysis of the minimum AIC procedure. *Annals of the Institute of Statistical Mathematics* **30**, 9–14.

Brown, P.J. (1977). Centering and scaling in ridge regression. *Technometrics* **19**, 35–36.

Golub, G.H., M. Heath and G. Wahba (1979). Generalized cross-validation: a method for choosing a good ridge regression parameter. *Technometrics* **21**, 215–223.

Hoerl, A.E. and R.W. Kennard (1970). Ridge regression: Biased estimation for nonorthogonal problems. *Technometrics* **12**, 55–67.

MacKay, D.J.C. (1992). Bayesian interpolation. *Neural Computation* **4**, 415–447.

Mercer, J. (1909). Functions of positive and negative type and their connection with the theory of integral equations. *Philos. Trans. Roy. Soc. London* **209**, 415–446.

Neal, R.M. (1996). *Bayesian Learning for Neural Networks*. Vol. 118 of *Lecture Notes in Statistics*. Springer. New York.

Pesaran, M.H. and A. Timmerman (1992). A simple nonparametric test of predictive performance. *Journal of Business and Economic Statistics* **10**, 461–465.

Poggio, T. and F. Girosi (1990). Networks for approximation and learning. *Proceedings of the IEEE* **78**, 1481–1497.

Schölkopf, B. and A. Smola (2002). *Learning with Kernels*. MIT Press. Cambridge, MA.

Schwarz, G. (1978). Estimating the dimension of a model. *Annals of Statistics* **6**, 461–464.

Sjöberg, J., Q. Zhang, L. Ljung, A. Benveniste, B. Deylon, P. Glorennec, H. Hjalmarsson and A. Juditsky (1995). Nonlinear black-box modelling in system identification: a unified overview. *Automatica* **31**, 1691–1724.

Suykens, J.A.K. and J. Vandewalle (1999). Least squares support vector machine classifiers. *Neural Processing Letters* **9**, 293–300.

Suykens, J.A.K., T. Van Gestel, J. De Brabanter, B. De Moor and J. Vandewalle (2002). *Least Squares Support Vector Machines*. World Scientific. Singapore.

Suykens, J.A.K., T. Van Gestel, J. Vandewalle and B. De Moor (2003). A support vector machine formulation to pca analysis and its kernel version. *IEEE Transactions on Neural Networks*. Accepted for publication.

Van Gestel, T., J.A.K. Suykens, D.-E. Baestaens, A. Lambrechts, G. Lanckriet, B. Vandaele, B. De Moor and J. Vandewalle (2001*a*). Predicting financial time series using least squares support vector machines within the evidence framework. *IEEE Transactions on Neural Networks (Special Issue on Financial Engineering)* **12**, 809–821.

Van Gestel, T., J.A.K. Suykens, G. Lanckriet, A. Lambrechts, B. De Moor and J. Vandewalle (2002*a*). A Bayesian framework for least squares support vector machine classifiers, Gaussian processes and kernel Fisher discriminant analysis. *Neural Computation* **14**, 1115–1147.

Van Gestel, T., J.A.K. Suykens, P. Van Dooren and B. De Moor (2001*b*). Identification of stable models in subspace identification by using regularization. *IEEE Transactions on Automatic Control* **46**, 1416–1420.

Van Gestel, T., M. Espinoza, J. Suykens, B. Carine and B. De Moor (2002*b*). Bayesian nonlinear kernel based stock market prediction. Technical Report 02-151. K.U.Leuven, ESAT/SCD/SISTA. Submitted for publication.

Vapnik, V. (1998). *Statistical Learning Theory*. Wiley. New-York.

Wahba, G. (1990). *Spline Models for Observational Data*. SIAM. Philadelphia.

[4] This test statistic (PTstat) is standard normal distributed under the null hypotesis of random sign predictions (Pesaran and Timmerman, 1992), values above 1.96 and 2.58 are significant at 5% and 1%.